龚昇论文选集

龚 昇 / 著

中国科学技术大学出版社

图书在版编目（CIP）数据

龚昇论文选集/龚昇著.—合肥：中国科学技术大学出版社，2008.7
ISBN 978-7-312-02296-8

Ⅰ.龚…　Ⅱ.龚…　Ⅲ.数学－文集　Ⅳ.O1-53

中国版本图书馆CIP数据核字（2008）第100209号

出版发行　中国科学技术大学出版社
　　地址　安徽省合肥市金寨路96号，邮编　230026
　　网址　http://press.ustc.edu.cn
印　刷　中国科学技术大学印刷厂
经　销　全国新华书店
开　本　880mm×1230mm　1/16
印　张　43.5　插页　2
字　数　1057千
版　次　2008年7月第1版
印　次　2008年7月第1次印刷
印　数　1－1200册
定　价　158.00元

⊙ 本书作者年轻时照片

⊙ 1984年4月，华罗庚（右）与本书作者于合肥稻香楼宾馆前合影

前　　言

我于 1930 年 1 月 16 日在上海川沙县北蔡镇出生. 1950 年夏毕业于上海交通大学数学系，之后在中国科学院数学研究所工作至 1958 年. 从 1958 年至今一直在中国科学技术大学工作. 先后师从陈建功老师与华罗庚老师.

上世纪六十年代罗庚师到科大来任专职副校长，使我聆听他的教诲的机会增多，有一次他对我讲，一个人最后余下的就是一本选集. 我当时年轻，以为这是遥远的事，他这样说是为了教育我淡泊名利，努力学习与工作. 时间匆匆，到 2002 年我得病之后，就经常想起罗庚师四十多年前对我讲的这句话.

我一生坎坷，平庸，受到种种欺凌与屈辱，在此盘点人生的时候，尽管自认为一直很努力，但与那些大数学家比，我是如此微不足道，有愧师恩.

我无学位，非院士，不过是一个普普通通的老教书匠，在后人评说的时候，不会对这样一个普普通通的教书匠的选集，过于苛求吧！

选文五十篇，为了便于查阅，按课题分类列出.

杨乐院士、叶向东教授与中国科学技术大学出版社对出版本书大力支持，戚素芬同志为本书付出了大量辛勤劳动，对他们致以深切的感谢！

2008. 6. 8

目　　次

I. 典型群上的调和分析

II. 多复变数的奇异积分与积分方程

[1]这里及以后号码都是第 676 页到第 687 页中论文目录的编号.

III. 多复变数的 Bloch 常数

IV. 多复变数的 Schwarz 导数

V. 一类 Reinhardt 域的 Bergman 核函数

VI. 多复变数的凸映照与星形映照

VII. 其　他

VIII. 附　录

I. 典型群上的调和分析

华罗庚创立了多复变数函数论中的典型域上的调和分析的理论 (1)，这是对多复变数的巨大贡献，作为这个理论的发展之一，是典型群上的调和分析．这起源于他的一篇文章 (2)，在此文中证明了酉群上的连续函数的 Fourier 级数可以 Abel 求和于它自己．以此为出发点，文献 [73.1][1], [72.1], [71.1], [70.1], [69.1] 建立了酉群上的调和分析．其后，王世坤、陈广晓、董道珍、贺祖琪等人按照上述的思想与方法，在旋转群上以及酉辛群上建立了对应的调和分析．上述这些工作总结在 (3) (即 [M8], 英文版 [M4]) 中．之后，李世雄、郑学安等在这些工作的基础上，对紧致李群上的调和分析进行研究，他们的工作，总结在 (4) 中．这项研究获国家自然科学三等奖，得奖人：龚昇, 郑学安.

参　考　文　献

(1) 华罗庚．多复变数函数论中的典型域的调和分析．北京：科学出版社，1958.

(2) 华罗庚. 紧致群上的连续函数所成的空间中的一条收敛定理. 科学记录, 1958, 2: 341-344.

(3) 龚昇. 典型群上的调和分析．北京：科学出版社，1983.

(4) 郑学安．紧致齐性空间上的调和分析．上海：上海科学技术出版社，2000.

[1]这里及以后的号码都是第 673 页到第 687 页中著作论文目录的编号.

《典型群上的调和分析》序

H. Weyl 研究了有限维的紧致群上的调和分析，证明了有一组正交有则的系存在，以及任何一个连续函数都可以用这个系的有限个函数的线性式来逼近. 由于他研究的对象太抽象了，没有方法表达出这一正交有则系，当然谈不上其逼近表达式是什么，更谈不上深入的收敛性了.

紧致群的第一个例子是 $\{e^{i\theta}\}$. 这上面有一个正交正则系

$$\{e^{in\theta}\}, \qquad n=0,\pm1,\pm2,\cdots,$$

也就是

$$\frac{1}{2\pi}\int_0^{2\pi} e^{in\theta}\cdot e^{-im\theta}d\theta=\begin{cases}1, & 当 \quad n=m\\ 0, & 当 \quad n\neq m\end{cases}$$

这一紧致群的调和分析就是整套的 Fourier 分析，已有大量的文献，由 A. Zygmund 于 1959 年写的两卷七百多页的巨著就可以看到其发展的情况.

$\{e^{i\theta}\}$ 的直接推广就是 U_n 群，也就是 n 行 n 列的方阵 U, 它适合于

$$U\overline{U}'=I,$$

$\overline{U}'$ 是 U 的共轭及转置，同时 U_2 与旋转群 SO_3 有同构处，在物理上 U_n 群的应用越来越重要.

除掉抽象的紧致群及最简单的 $U_1=\{e^{i\theta}\}$ 之外，无人深入研究. 笔者在研究多复变数典型域的调和分析时，定出了 U_n 上的明确的正交正则完全系而且证明了 Abel 求和法，由此当然可以推出逼近的线性式.

同时任何紧致群可以“嵌入”到 U_n 之中，因此这一方面的研究既具体化了又不失其抽象普遍性.

但 Abel 求和法是我们处理调和分析的最初一步的方法，在 Fourier 分析中存在着大量基本的问题，例如在求和法中， $(c,1)$ 求和是最合适的方法，这是 Fejér Lebesgue 的重要贡献.

Casàro 求和怎样推广到 U_n 上的调和分析中来，龚昇同志用了创造性的技巧得到了相当于 Lebesgue 等人的结果．他还对 U_n 上的调和分析得到了一系列的结果．也许对 U_n 上的调和分析运用到近代物理会有所帮助．因此，人们都会问： Fourier 分析的经典工作对 U_n 上的调和分析有没有对应的定理？

这是一个有丰富前途的方向，乐之为序．

华罗庚

1979.2.26

第10卷 第2期　　数学学报　　Vol. 10, No. 2
1960年6月　　ACTA MATHEMATICA SINICA　　June, 1960

酉羣上的富理埃分析*

I. 富理埃級数的Abel求和及Dirichlet核

龔　昇
（中国科学技术大学）

目　　次

§1.1. 引言

一个变数的Fourier分析，現在已經有了丰富的成果，不少問題得到了圓滿而完整的解决，在很多人的研究中达到了很深刻的地步．当然，一个变数的Fourier分析在数学的不少其他領域起着重要的作用．可是多个变数的Fourier分析情况就不完全是如此，在有些內容上是有些不完整之处的．至于更一般的在任意紧致羣上的Fourier分析，那末，众所周知的，重要的結果只是Peter-Weyl[6]定理，它告訴我們說：紧致羣上的連續函数可以用有限綫性式来逼近之．最近，华罗庚[2]証明了酉羣上的連續函数的Fourier級数可以Abel求和于它自己．这是有关有限維(維数大于1)的紧致羣上的函数的Fourier級数的第一条收斂定理．当然一条收斂定理是优于逼近定理的．

我們将以此为依据，对酉羣上的Fourier分析进行系統的研究．一个变数的Fourier分析可以看作一維酉羣上的Fourier分析，所以我們所研究的一般的酉羣上的Fourier分析，将一个变数的Fourier分析作为一个很特殊的例子，而一个变数的Fourier分析上所用的那些技巧，在此将不再起有影响的作用．同时，还須指出的是，任意紧致子羣上的連續函数可以推展成酉羣上的連續函数，以至对于酉羣上的Fourier分析进行研究，对于一般的紧致羣的情形也是有意义的．这里及以后所用的主要思想是羣表示，这首先是由华罗庚所成功处理的．

本文是对于酉羣上可积函数的Fourier級数的研究的一部分成果．主要是对华罗庚所定义的Abel求和进行进一步的研究，給出了Abel求和的具体表达式，此外，还給出了Fourier級数的Dirichlet核．

具体的說，在§1.3—§1.7的这几节中，我們給出了Abel求和的最終具体表达式，这是經过很复杂的計算而得来的．§1.8是用分析的方法求出了Fourier級数的Dirichlet核，

* 1960年5月6日收到．

这是借助于 Abel 求和中的 Poisson 核而求得的. §1.9 給于上述結果以另一个証明(代数的証明),这个証明是属于华罗庚的. 应用以前同样的方法,我們在§1.10中定义了一种求和法,并定出了它們相应的核.

为了不使本文篇幅太长,关于以这些核为依据的收斂定理及求和定理,都将另文发表.

本文中所考虑的部分和及求和都是"方块"的,至于别种形式的求和(例如,球求和)将在另文重新加以討論. 文中所用符号与[1]中相同,处理問題的方法及技巧也是与[1]深切相关的,文中的結果是在科学記录上发表过的結果的一部分[4,5].

作者愿向导师华罗庚教授致以深切的感謝,本文是在他的指导下完成的,也愿向中国科学院数学研究所多复变数函数論討論班的同事們致謝,他們的宝貴意見,使我得益匪浅.

§1.2. 酉羣上的 Fourier 級数

設 U_n 是 n 阶酉羣. 若 $U\in U_n$, 我們以 $A_{f_1,\cdots,f_n}(U)$ 表示 U 的标記为 $(f_1, f_2, \cdots, f_n)$ 的酉表示,这里 $f_1, f_2, \cdots, f_n$ 是滿足 $f_1\geqslant f_2\geqslant\cdots\geqslant f_n$ 的整数. 記 $N(f_1, f_2, \cdots, f_n)$ 为 $A_{f_1\cdots f_n}(U)$ 的阶. 我們常用 f 来代表 $(f_1, f_2, \cdots, f_n)$. 若

$$A_f(U)=(a_{ij}(U))_{1\leqslant i,\, j\leqslant N(f)},$$

那末我們知道 $\{\varphi^f_{ij}(U)\}$ 是 U_n 上的就范直交系,这里

$$\varphi^f_{ij}(U)=\sqrt{\frac{N(f)}{C}}\, a^f_{ij}(U),$$

而 C 是 U_n 的体积,即

$$C=(2\pi)^{\frac{1}{2}n(n+1)}/((n-1)!(n-2)!\cdots 2!1!)$$

$\{\varphi^f_{ij}\}$ 的全体对可积函数来講是完整的. 記

$$\Phi_f(U)=(\varphi^f_{ij}(U))_{1\leqslant i,j\leqslant N(f)}=\sqrt{\frac{N(f)}{C}}\,A_f(U).$$

若 $u(U)$ 是可积函数,展开成 Fourier 級数

$$\sum_{f_1\geqslant f_2\geqslant\cdots\geqslant f_n}\operatorname{tr}\,(C_{f_1\cdots f_n}\Phi'_{f_1\cdots f_n}(U)),$$

这里

$$C_{f_1\cdots f_n}=\int_{U_n}u(V)\Phi_{f_1\cdots f_n}(\overline{V})\dot{V},$$

$\operatorname{tr} B$ 表示 B 的迹, B' 表示 B 的轉置.

更明白些,

$$\begin{aligned}u(U)&\sim\sum_{f_1\geqslant\cdots\geqslant f_n}\operatorname{tr}\left(\frac{N(f)}{C}\int_{U_n}u(V)A_{f_1\cdots f_n}(\overline{V}U')\dot{V}\right)\\&=\frac{1}{C}\int_{U_n}u(V)\sum_{f_1\geqslant\cdots\geqslant f_n}N(f)\chi_{f_1\cdots f_n}(\overline{V}U')\dot{V},\end{aligned}$$

这里 $\chi_{f_1\cdots f_n}(U)$ 为表示 $A_{f_1\cdots f_n}(U)$ 的特征.

首先我們很容易的可以証明 $u(U)$ 的 Fourier 級数等于它的实部,即

$$\sum_{f_1\geqslant\cdots\geqslant f_n}\operatorname{tr}\,(C_{f_1\cdots f_n}\Phi'_{f_1\cdots f_n}(U))=\sum_{f_1\geqslant\cdots\geqslant f_n}\operatorname{Re}\operatorname{tr}\,(C_{f_1\cdots f_n}\Phi'_{f_1\cdots f_n}(U)).$$

若 $U \in U_n$, $V \in U_n$, 那末

$$u(VU) \sim \frac{1}{C}\int_{U_n} u(W) \sum_{f_1 \geqslant \cdots \geqslant f_n} N(f)\chi_{f_1\cdots f_n}(\overline{W}U'V')\dot{W}$$

$$= \sum_{f_1 \geqslant \cdots \geqslant f_n} \frac{N(f)}{C} \operatorname{tr}\left(\int_{U_n} u(W) A_{f_1\cdots f_n}(\overline{W}) A_{f_1\cdots f_n}(U') A_{f_1\cdots f_n}(V')\dot{W}\right)$$

$$= \sum_{f_1 \geqslant \cdots \geqslant f_n} \sqrt{\frac{N(f)}{C}} \operatorname{tr}(C_{f_1\cdots f_n} A_{f_1\cdots f_n}(U') A_{f_1\cdots f_n}(V')).$$

但是

$$u(\overline{V}U) \sim \sum_{f_1 \geqslant \cdots \geqslant f_n} \frac{N(f)}{C} \operatorname{tr}\left(\int_{U_n} u(W) A_{f_1\cdots f_n}(\overline{W}U'\overline{V}')\dot{W}\right)$$

$$= \sum_{f_1 \geqslant \cdots \geqslant f_n} \frac{N(f)}{C} \operatorname{tr}\left(\int_{U_n} u(W) A_{-f_n,\cdots,-f_1}(W\overline{U}'V')\dot{W}\right)$$

$$= \sum_{f_1 \geqslant \cdots \geqslant f_n} \frac{N(f)}{C} \operatorname{tr}\left(\int_{U_n} u(W) A_{f_1\cdots f_n}(W) A'_{f_1\cdots f_n}(\overline{U}) A'_{f_1\cdots f_n}(V)\dot{W}\right)$$

$$= \sum_{f_1 \geqslant \cdots \geqslant f_n} \sqrt{\frac{N(f)}{C}} \operatorname{tr}(\overline{C}_{f_1\cdots f_n} A'_{f_1\cdots f_n}(\overline{U}) A'_{f_1\cdots f_n}(V)),$$

这是因为

$$A_{f_1\cdots f_n}(U) = A_{-f_n\cdots -f_1}(\overline{U})$$

及

$$N(f_1\cdots f_n) = N(-f_n, \cdots, -f_1).$$

所以

$$\frac{u(VU) + u(\overline{V}U)}{2} \sim \sum_{f_1 \geqslant \cdots \geqslant f_n} \sqrt{\frac{N(f)}{C}} (\operatorname{Re}[\operatorname{tr}(C_{f_1\cdots f_n} A'_{f_1\cdots f_n}(U))] A'_{f_1\cdots f_n}(V)).$$

命 $V = I$, 即得

$$u(U) \sim \sum_{f_1 \geqslant \cdots \geqslant f_n} \operatorname{Re}\operatorname{tr}(C_{f_1\cdots f_n} \Phi'_{f_1\cdots f_n}(U)).$$

§1.3. Abel 求和

华罗庚[2]定义了在 n 阶酉群 U_n 上可积函数 $u(U)$ $(U \in U_n)$ 的 Fourier 級数

$$\sum_{f_1 \geqslant \cdots \geqslant f_n} \operatorname{tr}(C_{f_1\cdots f_n} \Phi'_{f_1\cdots f_n}(U)) \tag{1.3.1}$$

的 Abel 和为

$$\sum_{f_1 \geqslant \cdots \geqslant f_n} \rho^f(r) \operatorname{tr}(C_{f_1\cdots f_n} \Phi'_{f_1\cdots f_n}(U)), \tag{1.3.2}$$

这里

$$\rho^f(r) = \frac{1}{N(f)} \cdot \frac{1}{C}\int_{U_n} \frac{(1-r^2)^{n^2}}{|\det(I - rU)|^{2n}} \chi_f(U)\dot{U}. \tag{1.3.3}$$

并且証明了当 $r \to 1$ 时, (1.3.2)的值趋于 $u(U)$, 即(1.3.1)是可以 Abel 求和, 其和为 $u(U)$.

至于 $\rho^f(r)$ 的值，他指出，

(i) $$\rho^f(r)\to 1,\quad 当\quad r\to 1;$$

(ii) $$\rho^f(r)=\begin{cases} r^{f_1+\cdots+f_n} & 当\ f_1\geqslant f_2\geqslant\cdots\geqslant f_n\geqslant 0,\\ r^{-f_1-\cdots-f_n} & \ 0\geqslant f_1\geqslant\cdots\geqslant f_{n-1}\geqslant f_n.\end{cases}$$

我們要証明

定理 1.3.1. 当 $l_1>l_2>\cdots>l_s\geqslant 0>l_{s+1}>\cdots>l_n(n\geqslant s\geqslant 0)$ 时，那末

$$\rho^f(r)=r^{f_1+\cdots+f_s-f_{s+1}-\cdots-f_n}\sum_{s\geqslant g_{s+1}>\cdots>g_n\geqslant 0}\frac{N_s(f,g)N_s(g,f)}{N(f)N(g)}r^{2(g_{s+1}+\cdots+g_n)},\tag{1.3.4}$$

这里 $N_s(a,b)$ 表示 $N(a_1,\cdots,a_s,b_{s+1},\cdots,b_n)$，$g=(g_1,\cdots,g_n)$ 而 $(g_1+n-1,g_2+$ $+n-2,\cdots,g_n)$ 是 $(0,1,\cdots,n-1)$ 的一个排列。$0\geqslant g_1\geqslant g_2\geqslant\cdots\geqslant g_s\geqslant s-n$；$l_1=f_1+n-1,\cdots,l_{n-1}=f_{n-1}+1,\ l_n=f_n$.

$\rho^f(r)$ 还可以表成以下这些形式.

定理 1.3.2. 若 $(v_1,v_2,\cdots,v_n)$ 是 $(0,1,\cdots,n-1)$ 的一个排列，而 $v_1>\cdots>v_s$，那末，

$$\rho^f(r)=r^{|l_1|+\cdots+|l_n|-\frac{n(n-1)}{2}}\sum_{n-1\geqslant v_{s+1}>\cdots>v_n\geqslant 0}\prod_{j=1}^{s}\prod_{k=s+1}^{n}\frac{(l_j-v_k)(v_j-l_k)}{(v_j-v_k)(l_j-l_k)}r^{2(v_{s+1}+\cdots+v_n)}.\tag{1.3.5}$$

定理 1.3.3. 如同定理 1.3.1 中的假設，那末

$$\rho^f(r)=r^{f_1+\cdots+f_s-f_{s+1}-\cdots-f_n}\sum_{\nu=0}^{s(n-s)}b_\nu r^{2\nu},\tag{1.3.6}$$

这里

$$\begin{aligned}b_\nu&=\sum_{\substack{n-1\geqslant v_{s+1}>\cdots>v_n\geqslant 0\\ v_{s+1}+\cdots+v_n=\nu+\frac{(n-s)(n-s-1)}{2}}}\prod_{j=1}^{s}\prod_{k=s+1}^{n}\frac{(l_j-v_k)(v_j-l_k)}{(v_j-v_k)(l_j-l_k)}\\ &=\sum_{\substack{s\geqslant g_{s+1}>\cdots>g_n\geqslant 0\\ g_{s+1}+\cdots+g_n=\nu}}\frac{N_s(f,g)N_s(g,f)}{N(f)N(g)}.\end{aligned}$$

§ 1.4. $\rho^f(r)$ 的計算

在这几节中，我們将来証明上节中的这些定理.

首先我們看到(参閱[2])

$$\rho^f(r)N(f)=\frac{1}{(2\pi)^n}\int\cdots\int_{\pi\geqslant\theta_1>\cdots>\theta_n>-\pi}\frac{(1-r^2)^{n^2}}{\prod_{\nu=1}^{n}|1-re^{i\theta_\nu}|^{2n}}\cdot$$

$$\cdot\begin{vmatrix}e^{il_1\theta_1},&\cdots,&e^{il_1\theta_n}\\ e^{il_2\theta_1},&\cdots,&e^{il_2\theta_n}\\ \cdots&\cdots&\cdots\\ e^{il_n\theta_1},&\cdots,&e^{il_n\theta_n}\end{vmatrix}\prod_{\nu<\mu}(e^{-i\theta_\nu}-e^{-i\theta_\mu})d\theta_1\cdots d\theta_n.\tag{1.4.1}$$

将(1.4.1)中被积函数的行列式展开，于是得到

$$\rho^{l}(r)N(f)=\sum \delta^{1,2,\cdots,n}_{a_1,a_2,\cdots,a_n}\frac{1}{(2\pi)^n}\int\cdots\int_{\pi\geqslant\theta_1\geqslant\cdots\geqslant\theta_n\geqslant-\pi}\frac{(1-r^2)^{n^2}}{\prod_{\nu=1}^{n}|1-re^{i\theta_\nu}|^{2n}}\cdot e^{i(l_{a_1}\theta_1+\cdots+l_{a_n}\theta_n)}D(e^{-i\theta_1},\cdots,e^{-i\theta_n})d\theta_1\cdots d\theta_n. \tag{1.4.2}$$

上述积分进行变数变换，使 $l_{a_1}\theta_1+\cdots+l_{a_n}\theta_n$ 变成 $l_1\psi_1+\cdots+l_n\psi_n$，于是可以看出(为了簡便起見，我們常写 $\rho^{l}(r)$ 为 $\rho(r)$)．

$$\rho(r)N(f)=\frac{1}{(2\pi)^n}\int_0^{2\pi}\cdots\int_0^{2\pi}\frac{(1-r^2)^{n^2}}{\prod_{\nu=1}^{n}|1-re^{i\theta_\nu}|^{2n}}\cdot e^{i(l_1\theta_1+\cdots+l_n\theta_n)}D(e^{-i\theta_1},\cdots,e^{-i\theta_n})d\theta_1\cdots d\theta_n. \tag{1.4.3}$$

在(1.4.3)中被积函数的行列式进行展开，于是

$$\rho(r)N(f)=(-1)^{\frac{1}{2}n(n-1)}\sum \delta^{0,1,\cdots,n-1}_{k_0,k_1,\cdots,k_{n-1}}\frac{1}{(2\pi)^n}\int_0^{2\pi}\cdots\int_0^{2\pi}\cdot\frac{(1-r^2)^{n^2}}{\prod_{\nu=1}^{n}|1-re^{i\theta_\nu}|^{2n}}e^{i[(l_1-k_0)\theta_1+\cdots+(l_n-k_{n-1})\theta_n]}d\theta_1\cdots d\theta_n,$$

而这就是

$$\rho(r)N(f)=(-1)^{\frac{1}{2}n(n-1)}(1-r^2)^{n^2}\cdot\begin{vmatrix}\frac{1}{2\pi}\int_0^{2\pi}\frac{e^{i(l_1-0)\theta}d\theta}{|1-re^{i\theta}|^{2n}}, & \frac{1}{2\pi}\int_0^{2\pi}\frac{e^{i(l_2-0)\theta}d\theta}{|1-re^{i\theta}|^{2n}}, & \cdots, & \frac{1}{2\pi}\int_0^{2\pi}\frac{e^{i(l_n-0)\theta}d\theta}{|1-re^{i\theta}|^{2n}}\\ \frac{1}{2\pi}\int_0^{2\pi}\frac{e^{i(l_1-1)\theta}d\theta}{|1-re^{i\theta}|^{2n}}, & \frac{1}{2\pi}\int_0^{2\pi}\frac{e^{i(l_2-1)\theta}d\theta}{|1-re^{i\theta}|^{2n}}, & \cdots, & \frac{1}{2\pi}\int_0^{2\pi}\frac{e^{i(l_n-1)\theta}d\theta}{|1-re^{i\theta}|^{2n}}\\ \cdots\cdots & \cdots\cdots & \cdots\cdots & \cdots\cdots\\ \frac{1}{2\pi}\int_0^{2\pi}\frac{e^{i(l_1-(n-1))\theta}d\theta}{|1-re^{i\theta}|^{2n}}, & \frac{1}{2\pi}\int_0^{2\pi}\frac{e^{i(l_2-(n-1))\theta}d\theta}{|1-re^{i\theta}|^{2n}}, & \cdots, & \frac{1}{2\pi}\int_0^{2\pi}\frac{e^{i(l_n-(n-1))\theta}d\theta}{|1-re^{i\theta}|^{2n}}\end{vmatrix}.$$

在上面这个行列式中，将第一列乘以 $C_0^{n-1}r^0(-1)^0$，第二列乘以 $C_1^{n-1}r^1(-1)^1$，$\cdots$，第 n 列乘以 $C_{n-1}^{n-1}r^{n-1}(-1)^{n-1}$，然后开始二列相加到第二列上，开始三列相加到第三列上，$\cdots$，开始 n 列相加到第 n 列上，于是得到

$$\rho(r)N(f)=(1-r^2)^{n^2}r^{-\frac{n(n-1)}{2}}\cdot\begin{vmatrix}\frac{1}{2\pi}\int_0^{2\pi}\frac{e^{il_1\theta}d\theta}{(1-re^{i\theta})^n(1-re^{-i\theta})^n}, & \frac{1}{2\pi}\int_0^{2\pi}\frac{e^{il_2\theta}d\theta}{(1-re^{i\theta})^n(1-re^{-i\theta})^n}, & \cdots, & \frac{1}{2\pi}\int_0^{2\pi}\frac{e^{il_n\theta}d\theta}{(1-re^{i\theta})^n(1-re^{-i\theta})^n}\\ \frac{1}{2\pi}\int_0^{2\pi}\frac{e^{il_1\theta}d\theta}{(1-re^{i\theta})^n(1-re^{-i\theta})^{n-1}}, & \frac{1}{2\pi}\int_0^{2\pi}\frac{e^{il_2\theta}d\theta}{(1-re^{i\theta})^n(1-re^{-i\theta})^{n-1}}, & \cdots, & \frac{1}{2\pi}\int_0^{2\pi}\frac{e^{il_n\theta}d\theta}{(1-re^{i\theta})^n(1-re^{-i\theta})^{n-1}}\\ \cdots\cdots & \cdots\cdots & \cdots\cdots & \cdots\cdots\\ \frac{1}{2\pi}\int_0^{2\pi}\frac{e^{il_1\theta}d\theta}{(1-re^{i\theta})^n(1-re^{-i\theta})}, & \frac{1}{2\pi}\int_0^{2\pi}\frac{e^{il_2\theta}d\theta}{(1-re^{i\theta})^n(1-re^{-i\theta})}, & \cdots, & \frac{1}{2\pi}\int_0^{2\pi}\frac{e^{il_n\theta}d\theta}{(1-re^{i\theta})^n(1-re^{-i\theta})}\end{vmatrix}. \tag{1.4.4}$$

事实上,我們还可以从(1.4.3)更直接的来証明(1.4.4),从 $D(x_1, x_2, \cdots, x_n)$ 的性质,我們知道

$$D(e^{-i\theta_1}, \cdots, e^{-i\theta_n}) = r^{-\frac{1}{2}n(n-1)}D(re^{-i\theta_1}, \cdots, re^{-i\theta_n}) = r^{-\frac{1}{2}n(n-1)}D(re^{-i\theta_1}-1, \cdots, re^{-i\theta_n}-1) =$$

$$= (-r)^{-\frac{1}{2}n(n-1)}D(1-re^{-i\theta_1}, \cdots, 1-re^{-i\theta_n}) = (-r)^{\frac{1}{2}n(n-1)}\prod_{j=1}^{n}(1-re^{-i\theta_j})^{n-1}D\left(\frac{1}{1-re^{-i\theta_1}}, \cdots, \frac{1}{1-re^{-i\theta_n}}\right).$$

将此式代入(1.4.3),得到

$$\rho(r)V(f) = (-1)^{\frac{1}{2}n(n-1)}\frac{1}{(2\pi)^n}\int_0^{2\pi}\cdots\int_0^{2\pi}\frac{(1-r^2)^{n^2}e^{i(l_1\theta_1+\cdots+l_n\theta_n)}}{\prod_{\nu=1}^{n}(1-re^{i\theta_\nu})^n\prod_{\nu=1}^{n}(1-re^{-i\theta_\nu})}\cdot D\left(\frac{1}{1-re^{-i\theta_1}}, \cdots, \frac{1}{1-re^{-i\theta_n}}\right)d\theta_1\cdots d\theta_n$$

对被积函数中的行列式展开,即得(1.4.4).

引入符号

$$\Lambda_{q_1,\cdots,q_n}^{l_1,\cdots,l_n} = \begin{vmatrix} \frac{1}{2\pi}\int_0^{2\pi}\frac{e^{ip_1\theta}d\theta}{(1-re^{i\theta})^{q_1}(1-re^{-i\theta})^n}, & \frac{1}{2\pi}\int_0^{2\pi}\frac{e^{ip_2\theta}d\theta}{(1-re^{i\theta})^{q_2}(1-re^{-i\theta})^n}, & \cdots, & \frac{1}{2\pi}\int_0^{2\pi}\frac{e^{ip_n\theta}d\theta}{(1-re^{i\theta})^{q_n}(1-re^{-i\theta})^n} \\ \frac{1}{2\pi}\int_0^{2\pi}\frac{e^{ip_1\theta}d\theta}{(1-re^{i\theta})^{q_1}(1-re^{-i\theta})^{n-1}}, & \frac{1}{2\pi}\int_0^{3\pi}\frac{e^{ip_2\theta}d\theta}{(1-re^{i\theta})^{q_2}(1-re^{-i\theta})^{n-1}}, & \cdots, & \frac{1}{2\pi}\int_0^{2\pi}\frac{e^{ip_n\theta}d\theta}{(1-re^{i\theta})^{q_n}(1-re^{-i\theta})^{n-1}} \\ \cdots\cdots & \cdots\cdots & \cdots\cdots & \cdots\cdots \\ \frac{1}{2\pi}\int_0^{2\pi}\frac{e^{ip_1\theta}d\theta}{(1-re^{i\theta})^{q_1}(1-re^{-i\theta})}, & \frac{1}{2\pi}\int_0^{2\pi}\frac{e^{ip_2\theta}d\theta}{(1-re^{i\theta})^{q_2}(1-re^{-i\theta})}, & \cdots, & \frac{1}{2\pi}\int_0^{2\pi}\frac{e^{ip_n\theta}d\theta}{(1-re^{i\theta})^{-q_n}(1-re^{-i\theta})} \end{vmatrix}. \qquad (1.4.5)$$

于是(1.4.4)就是

$$\rho(r)V(f) = (1-r^2)^{n^2}r^{-\frac{n(n-1)}{2}}\Lambda_{n,\cdots,n}^{l_1,\cdots,l_n}. \qquad (1.4.6)$$

§1.5. 几个代数恒等式

在这一节中,我們已經假定了

$$l_1 > l_2 > \cdots > l_s \geqslant 0 > l_{s+1} > \cdots > l_n \quad (n \geqslant s \geqslant 0),$$

由(1.4.5)的 Λ 的定义,我們立刻知道

$$\Lambda_{n,\cdots,n}^{l_1,\cdots,l_n} = \Lambda_{n,\cdots,n,n-1}^{l_1,\cdots,l_{n-1},l_n} + r\Lambda_{n,\cdots,n,n}^{l_1,\cdots,l_{n-1},l_n+1}.$$

但是

$$A_{n,\cdots,n,n-1}^{l_1,\cdots,l_{n-1},l_n} = A_{n,\cdots,n,n-2}^{l_1,\cdots,l_{n-1},l_n} + rA_{n,\cdots,n,n-1}^{l_1,\cdots,l_{n-1},l_n+1},$$

$$\cdots\cdots\cdots\cdots\cdots\cdots$$

$$\cdots\cdots\cdots\cdots\cdots\cdots$$

$$A_{n,\cdots,n,1}^{l_1,\cdots,l_{n-1},l_n} = A_{n,\cdots,n,0}^{l_1,\cdots,l_{n-1},l_n} + rA_{n,\cdots,n,1}^{l_1,\cdots,l_{n-1},l_n+1}.$$

将这些恆等式一起相加，考虑到当 $l_n < 0$ 时，

$$A_{n,\cdots,n,0}^{l_1,\cdots,l_{n-1},l_n} = 0,$$

于是我們得到恆等式

$$A_{n,\cdots,n}^{l_1,\cdots,l_n} = r\sum_{j=0}^{n-1} A_{n,\cdots,n,n-j}^{l_1,\cdots,l_{n-1},l_n+1}. \tag{1.5.1}$$

但是

$$A_{n,\cdots,n,n-j}^{l_1,\cdots,l_{n-1},l_n+1} = A_{n,\cdots,n,n-j-1}^{l_1,\cdots,l_{n-1},l_n+1} + rA_{n,\cdots,n,n-j}^{l_1,\cdots,l_{n-1},l_n+2},$$

$$A_{n,\cdots,n,n-j-1}^{l_1,\cdots,l_{n-1},l_n+1} = A_{n,\cdots,n,n-j-2}^{l_1,\cdots,l_{n-1},l_n+1} + rA_{n,\cdots,n,n-j-1}^{l_1,\cdots,l_{n-1},l_n+2},$$

$$\cdots\cdots\cdots\cdots\cdots\cdots$$

$$\cdots\cdots\cdots\cdots\cdots\cdots$$

$$A_{n,\cdots,n,1}^{l_1,\cdots,l_{n-1},l_n+1} = A_{n,\cdots,n,0}^{l_1,\cdots,l_{n-1},l_n+1} + rA_{n,\cdots,n,1}^{l_1,\cdots,l_{n-1},l_n+2}.$$

将这些恆等式相加，由于考虑到

$$A_{n,\cdots,n,0}^{l_1,\cdots,l_{n-1},l_n+1} = 0,$$

所以

$$A_{n,\cdots,n,n-j}^{l_1,\cdots,l_{n-1},l_n+1} = r\sum_{k=j}^{n-1} A_{n,\cdots,n,n-k}^{l_1,\cdots,l_{n-1},l_n+2},$$

以此式代入(1.5.1)即得

$$A_{n,\cdots,n}^{l_1,\cdots,l_n} = r^2\sum_{j=0}^{n-1}\sum_{k=j}^{n-1} A_{n,\cdots,n,n-k}^{l_1,\cdots,l_{n-1},l_n+2} = r^2\sum_{k=0}^{n-1}(k+1)A_{n,\cdots,n,n-k}^{l_1,\cdots,l_{n-1},l_n+2}.$$

継續应用上述的步驟．最后我們可以得到

$$\begin{aligned} A_{n,\cdots,n}^{l_1,\cdots,l_n} &= r^{-l_n}\sum_{k=0}^{n-1}\frac{(k+1)\cdots(k-l_n-1)}{(-l_n-1)!}A_{n,\cdots,n,n-k}^{l_1,\cdots,l_{n-1},0} \\ &= r^{-l_n}\sum_{k=0}^{n-1}\frac{(k-l_n-1)!}{k!(-l_n-1)!}A_{n,\cdots,n,n-k}^{l_1,\cdots,l_{n-1},0}. \end{aligned} \tag{1.5.2}$$

注意到 $l_1 > \cdots > l_s \geqslant 0 > l_{s+1} > \cdots > l_n$，我們在 (1.5.2) 的右边，再对 l_{n-1} 进行上述步驟，使之递加到 0，…，一直到对 l_{s+1}，进行上述步驟，使之递加到 0，最后，我們得到

$$A_{n,\cdots,n}^{l_1,\cdots,l_n} = r^{-l_{s+1}-\cdots-l_n}\sum_{k_{s+1}=0}^{n-1}\cdots\sum_{k_n=0}^{n-1}\frac{(k_{s+1}-l_{s+1}-1)!}{k_{s+1}!(-l_{s+1}-1)!}\cdots$$

$$\cdots\frac{(k_n-l_n-1)!}{k!(-l_n-1)!}A_{n,\cdots,n,n-k_{s+1},\cdots,n-k_n}^{l_1,\cdots,l_s,0,\cdots,0}. \tag{1.5.3}$$

§ 1.6. Λ 的值

为了計算 $\rho^l(r)$ 的值,我們根据(1.4.6),只要对 $\Lambda_{n,\cdots,n}^{l_1,\cdots,l_n}$ 进行計算即可，而 $\Lambda_{n,\cdots,n}^{l_1,\cdots,l_n}$ 根据(1.5.3)知道，只要求出 $\Lambda_{n,\cdots,n,n-k_{s+1},\cdots,n-k_n}^{l_1,\cdots,l_s,0,\cdots,0}$ 就可以了，但是 $\Lambda_{n,\cdots,n,n-k_{s+1},\cdots,n-k_n}^{l_1,\cdots,l_s,0,\cdots,0}$ 等于

$$\begin{vmatrix} \frac{1}{2\pi}\int_0^{2\pi}\frac{e^{il_1\theta}d\theta}{(1-re^{i\theta})^n(1-re^{-i\theta})^n}, & \cdots, & \frac{1}{2\pi}\int_0^{2\pi}\frac{e^{il_s\theta}d\theta}{(1-re^{i\theta})^n(1-re^{-i\theta})^n}, & \frac{1}{2\pi}\int_0^{2\pi}\frac{d\theta}{(1-re^{i\theta})^{n-k_{s+1}}(1-re^{-i\theta})^n}, & \cdots, & \frac{1}{2\pi}\int_0^{2\pi}\frac{d\theta}{(1-re^{i\theta})^{n-k_n}(1-re^{-i\theta})^n} \\ \frac{1}{2\pi}\int_0^{2\pi}\frac{e^{il_1\theta}d\theta}{(1-re^{i\theta})^n(1-re^{-i\theta})^{n-1}}, & \cdots, & \frac{1}{2\pi}\int_0^{2\pi}\frac{e^{il_s\theta}d\theta}{(1-re^{i\theta})^n(1-re^{-i\theta})^{n-1}}, & \frac{1}{2\pi}\int_0^{2\pi}\frac{d\theta}{(1-re^{i\theta})^{n-k_{s+1}}(1-re^{-i\theta})^{n-1}}, & \cdots, & \frac{1}{2\pi}\int_0^{2\pi}\frac{d\theta}{(1-re^{i\theta})^{n-k_n}(1-re^{-i\theta})^{n-1}} \\ \cdots\cdots & & & & & \\ \cdots\cdots & & & & & \\ \frac{1}{2\pi}\int_0^{2\pi}\frac{e^{il_1\theta}d\theta}{(1-re^{i\theta})^n(1-re^{-i\theta})}, & \cdots, & \frac{1}{2\pi}\int_0^{2\pi}\frac{e^{il_s\theta}d\theta}{(1-re^{i\theta})^n(1-re^{-i\theta})}, & \frac{1}{2\pi}\int_0^{2\pi}\frac{d\theta}{(1-re^{i\theta})^{n-k_{s+1}}(1-re^{-i\theta})}, & \cdots, & \frac{1}{2\pi}\int_0^{2\pi}\frac{d\theta}{(1-re^{i\theta})^{n-k_n}(1-re^{-i\theta})}. \end{vmatrix} \tag{1.6.1}$$

为了对上述行列式进行計算,我們先証明下面的

引理 1.6.1. 若 $p \geqslant 0, q > 0, m > 0, 1 > r > 0$, 那末

$$\frac{1}{2\pi}\int_0^{2\pi}\frac{e^{ip\theta}d\theta}{(1-re^{i\theta})^q(1-re^{-i\theta})^m} = \frac{r^{-p}}{(m-1)!}\left[\frac{d^{m-1}}{dt^{m-1}}(t^{m+p-1}(1-t)^{-q})\right]_{t=r^2} \tag{1.6.2}$$

成立.

証. 因为 $1 > r > 0$, 所以

$$(1-re^{i\theta})^{-q} = \sum_{k=0}^{\infty}\frac{(q+k-1)!\,r^k e^{ik\theta}}{(q-1)!\,k!},$$

$$(1-re^{-i\theta})^{-m} = \sum_{k=0}^{\infty}\frac{(m+l-1)!\,r^l e^{-il\theta}}{(m-1)!\,l!}.$$

于是

$$\frac{1}{2\pi}\int_0^{2\pi}\frac{e^{ip\theta}d\theta}{(1-re^{i\theta})^q(1-re^{-i\theta})^m} = \sum_{\substack{l-k=p \\ l\geqslant 0,\ k\geqslant 0}}\frac{(q+k-1)!\,(m+l-1)!\,r^{k+l}}{(m-1)!\,l!\,(q-1)!\,k!} = r^p\sum_{k=0}^{\infty}\frac{(q+k-1)!\,(m+k+p-1)!}{(m-1)!\,(k+p)!\,(q-1)!\,k!}\,r^{2k}.$$

考虑 $(1-t)^{-q}$ 的展开式，容易証明

$$\sum_{k=0}^{\infty}\frac{(q+k-1)!(m+k+p-1)!}{(m-1)!(k+p)!(q-1)!k!}t^k=\frac{t^{-p}}{(m-1)!}\frac{d^{m-1}}{dt^{m-1}}(t^{m+p-1}(1-t)^{-q}).$$

而此即(1.6.2).

将(1.6.2)的值代入(1.6.1)，我們就得到 $A_{n,\cdots,n,n-k_{s+1},\cdots,n-k_n}^{l_1,\cdots,l_s,0,\cdots,0}$ 等于

$$\frac{r^{-l_1-l_2-\cdots-l_s}}{(n-1)!\cdots 2!\,1!}\begin{vmatrix}\frac{d^{n-1}}{dt^{n-1}}(t^{n+l_1-1}(1-t)^{-n}),\cdots,\frac{d^{n-1}}{dt^{n-1}}(t^{n+l_s-1}(1-t)^{-n}),\frac{d^{n-1}}{dt^{n-1}}(t^{n-1}(1-t)^{-n+k_{s+1}}),\cdots,\frac{d^{n-1}}{dt^{n-1}}(t^{n-1}(1-t)^{-n+k_n})\\ \frac{d^{n-2}}{dt^{n-2}}(t^{n+l_1-2}(1-t)^{-n}),\cdots,\frac{d^{n-3}}{dt^{n-3}}(t^{n+l_s-2}(1-t)^{-n}),\frac{d^{n-2}}{dt^{n-2}}(t^{n-2}(1-t)^{-n+k_{s+1}}),\cdots,\frac{d^{n-2}}{dt^{n-2}}(t^{n-2}(1-t)^{-n+k_n})\\ \cdots\cdots\cdots\cdots\cdots\cdots\cdots\cdots\cdots\cdots\cdots\cdots\\ \cdots\cdots\cdots\cdots\cdots\cdots\cdots\cdots\cdots\cdots\cdots\cdots\\ \frac{d^{0}}{dt^{0}}(t^{n+l_1-n}(1-t)^{-n}),\cdots,\frac{d^{0}}{dt^{0}}(t^{n+l_s-n}(1-t)^{-n}),\frac{d^{0}}{dt^{0}}(t^{n-n}(1-t)^{-n+k_{s+1}}),\cdots,\frac{d^{0}}{dt^{0}}(t^{n-n}(1-t)^{-n+k_n}).\end{vmatrix}\tag{1.6.3}$$

但是

$$\frac{d^{n-j}}{dt^{n-j}}(t^{n+l-j}(1-t)^{-n})=\sum_{k=0}^{n-j}C_k^{n-j}(t^{n-j})^{(k)}(t^l(1-t)^{-n})^{(n-j-k)}$$

及

$$\frac{d^{n-j}}{dt^{n-j}}(t^{n-j}(1-t)^{-n+v})=\sum_{k=0}^{n-j}C_k^{n-j}(t^{n-j})^{(k)}((1-t)^{-n+v})^{(n-j-k)}.$$

代入(1.6.3)，应用行列式的性质，立刻証得 $(t=r^2)$.

$$A_{n,\cdots,n,n-k_{s+1},\cdots,n-k_n}^{l_1,\cdots,l_s,0,\cdots,0}=$$

$$=\frac{r^{-l_1-\cdots-l_s+n(n-1)}}{(n-1)!\cdots 2!\,1!}\cdot\begin{vmatrix}\frac{d^{n-1}}{dt^{n-1}}(t^{l_1}(1-t)^{-n}),\cdots,\frac{d^{n-1}}{dt^{n-1}}(t^{l_s}(1-t)^{-n}),\frac{d^{n-1}}{dt^{n-1}}((1-t)^{-n+k_{s+1}}),\cdots,\frac{d^{n-1}}{dt^{n-1}}((1-t)^{-n+k_n})\\ \frac{d^{n-2}}{dt^{n-2}}(t^{l_1}(1-t)^{-n}),\cdots,\frac{d^{n-2}}{dt^{n-2}}(t^{l_s}(1-t)^{-n}),\frac{d^{n-2}}{dt^{n-2}}((1-t)^{-n+k_{s+1}}),\cdots,\frac{d^{n-2}}{dt^{n-2}}((1-t)^{-n+k_n})\\ \cdots\cdots\cdots\cdots\cdots\cdots\cdots\cdots\cdots\cdots\cdots\cdots\\ \cdots\cdots\cdots\cdots\cdots\cdots\cdots\cdots\cdots\cdots\cdots\cdots\\ \frac{d^{0}}{dt^{0}}(t^{l_1}(1-t)^{-n}),\cdots,\frac{d^{0}}{dt^{0}}(t^{l_s}(1-t)^{-n}),\frac{d^{0}}{dt^{0}}((1-t)^{-n+k_{s+1}}),\cdots,\frac{d^{0}}{dt^{0}}((1-t)^{-n+k_n}).\end{vmatrix}.$$

应用华罗庚的一条定理([1],定理 1.2.4),可以知道上式右边等于

$$\frac{r^{-l_1-\cdots-l_s+n(n-1)}}{(n-1)!\cdots 2!\,1!}\cdot\lim_{\substack{t_1\to t\\ \cdots\\ t_n\to t}}\left|\begin{matrix} t_1^{l_1}(1-t_1)^{-n}, \cdots, t_1^{l_s}(1-t_1)^{-n}, (1-t_1)^{-n+k_{s+1}}, \cdots, (1-t_1)^{-n+k_n} \\ t_2^{l_1}(1-t_2)^{-n}, \cdots, t_2^{l_s}(1-t_2)^{-n}, (1-t_2)^{-n+k_{s+1}}, \cdots, (1-t_2)^{-n+k_n} \\ \cdots\cdots\cdots\cdots\cdots\cdots \\ \cdots\cdots\cdots\cdots\cdots\cdots \\ t_n^{l_1}(1-t_n)^{-n}, \cdots, t_n^{l_s}(1-t_n)^{-n}, (1-t_n)^{-n+k_{s+1}}, \cdots, (1-t_n)^{-n+k_n} \end{matrix}\right| \Big/ D(t_1, \cdots, t_n) =$$

$$= r^{-l_1-\cdots-l_s+n(n-1)}(1-t)^{n^2}\cdot\lim_{\substack{t_1\to t\\ \cdots\\ t_n\to t}}\left|\begin{matrix} t_1^{l_1}, \cdots, t_1^{l_s}, (1-t_1)^{k_{s+1}}, \cdots, (1-t_1)^{k_n} \\ t_2^{l_1}, \cdots, t_2^{l_s}, (1-t_2)^{k_{s+1}}, \cdots, (1-t_2)^{k_n} \\ \cdots\cdots\cdots\cdots\cdots\cdots \\ \cdots\cdots\cdots\cdots\cdots\cdots \\ t_n^{l_1}, \cdots, t_n^{l_s}, (1-t_n)^{k_{s+1}}, \cdots, (1-t_n)^{k_n} \end{matrix}\right| \Big/ D(t_1, \cdots, t_n). \tag{1.6.4}$$

以(1.6.4)代入(1.5.3),注意到

$$g_p(t) = \frac{1}{(-p-1)!}\sum_{k=0}^{n-1}\frac{(k-p-1)!}{k!}(1-t)^k = \sum_{\nu=0}^{n-1} A_\nu^p t^\nu,$$

而

$$p < 0,\ A_\nu^p = \frac{(-1)^\nu (n-p-1)!}{(\nu-p)\cdot \nu!\,(-p-1)!\,(n-\nu-1)!}. \tag{1.6.5}$$

我們有

$$A_{n,\cdots,n}^{l_1,\cdots,l_n} = r^{-l_1-\cdots-l_n+n(n-1)}(1-r^2)^{-n^2}\cdot\lim_{\substack{t_1\to t\\ \cdots\\ t_n\to t}}\left|\begin{matrix} t_1^{l_1}, \cdots, t_1^{l_s}, g_{l_{s+1}}(t_1), \cdots, g_{l_n}(t_1) \\ t_2^{l_1}, \cdots, t_2^{l_s}, g_{l_{s+1}}(t_2), \cdots, g_{l_n}(t_2) \\ \cdots\cdots\cdots\cdots\cdots\cdots \\ \cdots\cdots\cdots\cdots\cdots\cdots \\ t_n^{l_1}, \cdots, t_n^{l_s}, g_{l_{s+1}}(t_n), \cdots, g_{l_n}(t_n) \end{matrix}\right| \Big/ D(t_1, \cdots, t_n). \tag{1.6.6}$$

我們来証明(1.6.5),首先我們注意到一个代数恆等式

$$\sum_{r=0}^{s}\frac{(r+l)!}{r!} = \frac{(s+l-1)}{(l+1)\cdot s!}, \tag{1.6.7}$$

这里 s, l 都是正整数。

証. 显然

$$\sum_{r=0}^{s}\frac{(r+l+1)!}{r!} = \sum_{r=1}^{s}\frac{(r+l)!}{(r-1)!} + (l+1)\sum_{r=0}^{s}\frac{(r+l)!}{r!} =$$

$$= \sum_{r=0}^{s-1}\frac{(r+l+1)!}{r!} + (l+1)\sum_{r=0}^{s}\frac{(r+l)!}{r!},$$

立得(1.6.7).

$$
\begin{aligned}
g_p(t) &= \frac{1}{(-p-1)!}\sum_{k=0}^{n-1}\frac{(k-p-1)!}{k!}(1-t)^k = \\
&= \sum_{k=0}^{n-1}\frac{(k-p-1)!}{k!(-p-1)!}\sum_{\nu=0}^{k}\frac{k!(-t)^\nu}{(k-\nu)!\nu!} = \\
&= \sum_{k=0}^{n-1}\sum_{\nu=0}^{k}\frac{(k-p-1)!(-t)^\nu}{(-p-1)!(k-\nu)!\nu!} \\
&= \sum_{\nu=0}^{n-1}\sum_{k=\nu}^{n-1}\frac{(k-p-1)!(-t)^\nu}{(-p-1)!(k-\nu)!\nu!}.
\end{aligned}
$$

所以

$$
A_\nu^p = \sum_{k=\nu}^{n-1}\frac{(k-p-1)!(-1)^\nu}{(-p-1)!(k-\nu)!\nu!} = \frac{(-1)^\nu}{\nu!(-p-1)!}\sum_{r=0}^{n-\nu-1}\frac{(r+\nu-p-1)!}{r!}
$$

$$
(\text{取 } k-\nu=r).
$$

在(1.6.7)中取 $s=n-\nu-1$, $l=\nu-p-1$, 于是得到

$$
A_\nu^p = \frac{(-1)^\nu(n-p-1)!}{\nu!(-p-1)!(n-\nu-1)!(\nu-p)}
$$

而这就是(1.6.5). 由[1]中的定理,我們知道(1.6.6)就是

$$
A_{n,\cdots,n}^{l_1,\cdots,l_n} = r^{-l_1-\cdots-l_n+n(n-1)}(1-r^2)^{n^2}\cdot
$$

$$
\cdot\begin{vmatrix}
\frac{d^{n-1}t^{l_1}}{dt^{n-1}}, & \cdots, & \frac{d^{n-1}t^{l_s}}{dt^{n-1}}, & \frac{d^{n-1}g_{l_{s+1}}(t)}{dt^{n-1}}, & \cdots, & \frac{d^{n-1}g_{l_n}(t)}{dt^{n-1}} \\
\frac{d^{n-2}t^{l_1}}{dt^{n-2}}, & \cdots, & \frac{d^{n-2}t^{l_s}}{dt^{n-2}}, & \frac{d^{n-2}g_{l_{s+1}}(t)}{dt^{n-2}}, & \cdots, & \frac{d^{n-2}g_{l_n}(t)}{dt^{n-2}} \\
\cdots & \cdots & \cdots & \cdots & \cdots & \cdots \\
t^{l_1}, & \cdots, & t^{l_s}, & g_{l_{s+1}}(t), & \cdots, & g_{l_n}(t)
\end{vmatrix}. \tag{1.6.8}
$$

§1.7. §1.3 中的定理的証明

将(1.6.8)的值代入(1.4.6),于是就得到

$$
\rho(r)N(f) = \frac{r^{-l_1-\cdots-l_n+\frac{n(n-1)}{2}}}{(n-1)!\cdots2!\,1!}\sum_{\nu_{s+1}=0}^{n-1}\cdots\sum_{\nu_n=0}^{n-1}A_{\nu_{s+1}}^{l_{s+1}}\cdots A_{\nu_n}^{l_n}\cdot
$$

$$
\cdot\begin{vmatrix}
\frac{d^{n-1}t^{l_1}}{dt^{n-1}}, & \cdots, & \frac{d^{n-1}t^{l_s}}{dt^{n-1}}, & \frac{d^{n-1}t^{\nu_{s+1}}}{dt^{n-1}}, & \cdots, & \frac{d^{n-1}t^{\nu_n}}{dt^{n-1}} \\
\frac{d^{n-2}t^{l_1}}{dt^{n-2}}, & \cdots, & \frac{d^{n-2}t^{l_s}}{dt^{n-2}}, & \frac{d^{n-2}t^{\nu_{s+1}}}{dt^{n-2}}, & \cdots, & \frac{d^{n-2}t^{\nu_n}}{dt^{n-2}} \\
\cdots & \cdots & \cdots & \cdots & \cdots & \cdots \\
\cdots & \cdots & \cdots & \cdots & \cdots & \cdots \\
t^{l_1}, & \cdots, & t^{l_s}, & t^{\nu_{s+1}}, & \cdots, & t^{\nu_n}
\end{vmatrix}. \tag{1.7.1}
$$

从(1.7.1)可以看出

$$\rho(r)N(f)=\frac{r^{-l_1-\cdots-l_n+\frac{n(n-1)}{2}}}{(n-1)!\cdots 2!\,1!}\sum_{n-1>\nu_{s+1}>\cdots>\nu_n>0}B^{l_{s+1},\cdots,l_n}_{\nu_{s+1},\cdots,\nu_n}\cdot$$

$$\cdot\begin{vmatrix}\frac{d^{n-1}t^{l_1}}{dt^{n-1}}, & \cdots, & \frac{d^{n-1}t^{l_s}}{dt^{n-1}}, & \frac{d^{n-1}t^{\nu_{s+1}}}{dt^{n-1}}, & \cdots, & \frac{d^{n-1}t^{\nu_n}}{dt^{n-1}}\\ \frac{d^{n-2}t^{l_1}}{dt^{n-2}}, & \cdots, & \frac{d^{n-2}t^{l_s}}{dt^{n-2}}, & \frac{d^{n-2}t^{\nu_{s+1}}}{dt^{n-2}}, & \cdots, & \frac{d^{n-2}t^{\nu_n}}{dt^{n-2}}\\ \cdots & \cdots & \cdots & \cdots & \cdots & \cdots\\ \cdots & \cdots & \cdots & \cdots & \cdots & \cdots\\ t^{l_1}, & \cdots, & t^{l_s}, & t^{\nu_{s+1}}, & \cdots, & t^{\nu}\end{vmatrix}, \tag{1.7.2}$$

这里

$$B^{l_{s+1},\cdots,l_n}_{\nu_{s+1},\cdots,\nu_n}=\begin{vmatrix}A^{l_{s+1}}_{\nu_{s+1}}, & A^{l_{s+1}}_{\nu_{s+2}}, & \cdots, & A^{l_{s+1}}_{\nu_n}\\ A^{l_{s+2}}_{\nu_{s+1}}, & A^{l_{s+2}}_{\nu_{s+2}}, & \cdots, & A^{l_{s+2}}_{\nu_n}\\ \cdots & \cdots & \cdots & \cdots\\ \cdots & \cdots & \cdots & \cdots\\ A^{l_n}_{\nu_{s+1}}, & A^{l_n}_{\nu_{s+2}}, & \cdots, & A^{l_n}_{\nu_n}\end{vmatrix}=$$

$$=\frac{(-1)^{\nu_{s+1}+\cdots+\nu_n}(n-l_{s+1}-1)!\cdots(n-l_n-1)!}{\nu_{s+1}!\cdots\nu_n!\,(-l_{s+1}-1)!\cdots(-l_n-1)!\,(n-\nu_{s+1}-1)!\cdots(n-\nu_n-1)!}\cdot$$

$$\cdot\begin{vmatrix}\frac{1}{\nu_{s+1}-l_{s+1}}, & \cdots, & \frac{1}{\nu_n-l_{s+1}}\\ \frac{1}{\nu_{s+1}-l_{s+2}}, & \cdots, & \frac{1}{\nu_n-l_{s+2}}\\ \cdots & \cdots & \cdots\\ \cdots & \cdots & \cdots\\ \frac{1}{\nu_{s+1}-l_n}, & \cdots, & \frac{1}{\nu_n-l_n}\end{vmatrix}.$$

应用[1]中的定理 1.1.3. 我們得到

$$B^{l_{s+1},\cdots,l_n}_{\nu_{s+1},\cdots,\nu_n}=\frac{(-1)^{\nu_{s+1}+\cdots+\nu_n}(n-l_{s+1}-1)!\cdots(n-l_n-1)!}{\nu_{s+1}!\cdots\nu_n!\,(-l_{s+1}-1)!\cdots(-l_n-1!)}\times$$

$$\times\frac{D(\nu_{s+1},\cdots,\nu_n)D(-l_{s+1},\cdots,l_n)}{(n-\nu_{s+1}-1)!\cdots(n-\nu_n-1)!\prod_{i=1}^{n-s}\prod_{j=1}^{n-s}(\nu_{s+i}-l_{s+j})}=$$

$$=(-1)^{\nu_{s+1}+\cdots+\nu_n}\frac{(n-l_{s+1}-1)!}{(-l_{s+1}-1)!\,n!}n!\,\frac{(n-1)!}{\nu_{s+1}!(n-\nu_{s+1}-1)!}\cdot$$

$$\cdot\frac{1}{(n-1)!}\cdots\frac{(n-l_n-1)!}{(-l_n-1)!\,n!}n!\,\frac{(n-1)!}{\nu_n!\,(n-\nu_n-1)!}\cdot$$

$$\cdot\frac{1}{(n-1)!}\frac{D(\nu_{s+1},\cdots,\nu_n)D(-l_{s+1},\cdots,-l_n)}{\prod_{j=s+1}^{n}\prod_{k=s+1}^{n}(\nu_j-l_k)}=$$

$$= (-1)^{\nu_{s+1}+\cdots+\nu_n} n^{n-s} \cdot \prod_{j=s+1}^{n} C_n^{n-l_j-1} C_{\nu_j}^{n-1} \cdot$$

$$\cdot \frac{D(\nu_{s+1}, \cdots, \nu_n) D(-l_{s+1}, \cdots, -l_n)}{\prod_{j=s+1}^{n} \prod_{k=s+1}^{n} (\nu_j - l_k)}. \tag{1.7.3}$$

另一方面，由于

$$N(\mu_1, \cdots, \mu_n) = \lim_{\substack{x_1 \to 1 \\ \cdots\cdots \\ x_n \to 1}} \frac{M_{\mu_1 \cdots \mu_n}(x_1, \cdots, x_n)}{D(x_1, \cdots, x_n)}.$$

我們得到

$$\begin{vmatrix} \frac{d^{n-1} t^{l_1}}{dt^{n-1}}, & \cdots, & \frac{d^{n-1} t^{l_s}}{dt^{n-1}}, & \frac{d^{n-1} t^{\nu_{s+1}}}{dt^{n-1}}, & \cdots, & \frac{d^{n-1} t^{\nu_n}}{dt^{n-1}} \\ \frac{d^{n-2} t^{l_1}}{dt^{n-2}}, & \cdots, & \frac{d^{n-2} t^{l_s}}{dt^{n-2}}, & \frac{d^{n-2} t^{\nu_{s+1}}}{dt^{n-2}}, & \cdots, & \frac{d^{n-2} t^{\nu_n}}{dt^{n-2}} \\ \cdots & \cdots & \cdots & \cdots & \cdots & \cdots \\ \cdots & \cdots & \cdots & \cdots & \cdots & \cdots \\ t^{l_1}, & \cdots, & t^{l_s}, & t^{\nu_{s+1}}, & \cdots, & t^{\nu_n} \end{vmatrix} =$$

$$= \lim_{\substack{t_1 \to t \\ \cdots\cdots \\ t_n \to t}} \begin{vmatrix} t_1^{l_1}, & \cdots, & t_n^{l_1} \\ \cdots & \cdots & \cdots \\ t_1^{l_s}, & \cdots, & t_n^{l_s} \\ t_1^{\nu_{s+1}}, & \cdots, & t_n^{\nu_{s+1}} \\ \cdots & \cdots & \cdots \\ t_1^{\nu_n}, & \cdots, & t_n^{\nu_n} \end{vmatrix} \Big/ D(t_1, \cdots, t_n) =$$

$$= t^{l_1+\cdots+l_s+\nu_{s+1}+\cdots+\nu_n-\frac{1}{2}n(n-1)} N(f_1, \cdots, f_s, \nu_{s+1}+s+1-n, \cdots, \nu_n). \tag{1.7.4}$$

将(1.7.4)代入(1.7.2)，即得

$$\rho(r) N(f) = r^{|l_1|+\cdots+|l_n|-\frac{n(n-1)}{2}} \sum_{n-1 \geqslant \nu_{s+1} > \cdots > \nu_n \geqslant 0} B_{\nu_{s+1}, \cdots, \nu_n}^{l_{s+1}, \cdots, l_n} N(f_1, \cdots, f_s, \nu_{s+1}+s+$$

$$+1-n, \cdots, \nu_n) r^{2(\nu_{s+1}+\cdots+\nu_n)}. \tag{1.7.5}$$

我們注意到二个重要而显然的代数恆等式

$$C_n^{n-l_k-1} = \frac{1}{n!} \prod_{j=1}^{n} (\nu_j - l_k)$$

及

$$\nu_k! \, (n - \nu_k - 1)! = \prod_{\substack{j=1 \\ j \neq k}}^{n} (\nu_j - \nu_k)(-1)^{\nu_k},$$

以此代入(1.7.3)，得到

$$B_{\nu_{s+1},\cdots,\nu_n}^{l_{s+1},\cdots,l_n}=(-1)^{\nu_{s+1}+\cdots+\nu_n}\prod_{k=s+1}^{n}\left[\frac{1}{n!}\prod_{j=1}^{n}(\nu_j-l_k)(n-1)!\Big/\prod_{\substack{j=1\\j\neq k}}^{n}(\nu_i-\nu_k)(-1)^{\nu_k}\right]\times$$

$$\times\frac{n^{n-s}D(-l_{s+1},\cdots,l_n)D(\nu_{s+1},\cdots,\nu_n)}{\prod_{j=s+1}^{n}\prod_{k=s+1}^{n}(\nu_j-l_k)}=$$

$$=\frac{D(-l_{s+1},\cdots,-l_n)D(\nu_{s+1},\cdots,\nu_n)\prod_{k=s+1}^{n}\prod_{j=1}^{n}(\nu_j-l_k)}{\prod_{k=s+1}^{n}\prod_{\substack{j=1\\j\neq k}}^{n}(\nu_j-\nu_k)\prod_{j=s+1}^{n}\prod_{k=s+1}^{n}(\nu_j-l_k)}=$$

$$=\frac{D(-l_{s+1},\cdots,-l_n)\prod_{k=s+1}^{n}\prod_{j=1}^{s}(\nu_j-l_k)}{\prod_{k=s+1}^{n}\prod_{j=1}^{s}(\nu_j-\nu_k)D(\nu_{s+1},\cdots,\nu_n)}=$$

$$=\frac{D(\nu_1,\cdots,\nu_s,l_{s+1},\cdots,l_n)}{D(\nu_1,\cdots,\nu_s,\nu_{s+1},\cdots,\nu_n)},\tag{1.7.6}$$

这里 $\nu_1>\nu_2>\cdots>\nu_s$，而 $(\nu_1,\cdots,\nu_n)$ 是 $(0,1,\cdots,n-1)$ 的一个排列. 从 (1.7.6) 及注意到

$$N(\mu_1,\cdots,\mu_n)=\frac{D(\mu_1+n-1,\mu_2+n-2,\cdots,\mu_{n-1}+1,\mu_n)}{D(n-1,n-2,\cdots,1,0)},$$

我們立刻得到定理 1.3.1. 按照 $N(\mu_1,\cdots,\mu_n)$ 的定义及

$$D(x_1,\cdots,x_n)=\prod_{j<k}(x_j-x_k)$$

的定义，我們經过化簡手續后，就可以得到定理 1.3.2. 至于定理 1.3.3 的証明，可以直接从定理 1.3.1 及 1.3.2 中将指数相同的項相归并而得之.

作为定理 1.3.1 及 1.3.2 的推論，我們有如下的代数恆等式:

設 $n\geqslant s\geqslant 0$, $f_1\geqslant f_2\geqslant\cdots\geqslant f_n$, $l_j=f_j+n-j$, $j=1,\cdots,n$ 那末

$$\sum_{s\geqslant g_{s+1}\geqslant\cdots\geqslant g_n\geqslant 0}\frac{N_s(f,g)N_s(g,f)}{N(f)N(g)}=1,\tag{1}$$

这里 $g=(g_1,g_2,\cdots,g_n)$, $(g_1+n-1,g_2+n-2,\cdots,g_n)$ 是 $(0,1,\cdots,n-1)$ 的一个排列, $0\geqslant g_1\geqslant g_2\geqslant\cdots\geqslant g_s\geqslant s-n$.

$$\sum_{n-1\geqslant\nu_{s+1}>\cdots>\nu_n\geqslant 0}\prod_{j=1}^{s}\prod_{k=s+1}^{n}\frac{(l_j-\nu_k)(\nu_j-l_k)}{(\nu_j-\nu_k)(l_j-l_k)}=1,\tag{2}$$

这里 $(\nu_1,\cdots,\nu_n)$ 是 $(0,1,\cdots,n-1)$ 的一个排列，而 $\nu_1>\cdots>\nu_s$.

当然(1)与(2)是二个有趣的恆等式.

可以驗証 § 1.3 中的 $\rho^f(r)$ 的性質(ii)是滿足的.

从定理 1.3.1 等，我們自然可以得到 Tauberian 型的收斂定理，为了不使篇幅太长，将在另文中叙述并証明之.

§ 1.8. Dirichlet 核

現在考虑可积函数 $u(U)$ 的 Fourier 級数

$$\sum_{f_1\geq f_2\geq\cdots\geq f_n}\mathrm{tr}(C_{f_1\cdots f_n}\Phi'_{f_1\cdots f_n}(U)). \tag{1.8.1}$$

的部分和

$$S_N^{(u)}(U)=\sum_{N\geq l_1>l_2>\cdots>l_n\geq -N}\mathrm{tr}(C_{f_1\cdots f_n}\Phi'_{f_1\cdots f_n}(U)), \tag{1.8.2}$$

这里 $l_1=f_1+n-1,\ l_2=f_2+n-2,\ \cdots,\ l_n=f_n$.

我們将要証明

定理 1.8.1. 可积函数 $u(U)$ 的 Fourier 級数(1.8.1)的部分和(1.8.2)可以表为

$$\frac{1}{C}\int_{U_n}u(VU)\mathscr{D}_N(V)\dot{V},$$

这里 $\mathscr{D}_N(V)$ 是 Dirichlet 核，它等于

$$\frac{(-i)^{\frac{1}{2}n(n-1)}}{(n-1)!\cdots 2!\,1!\,D(e^{i\theta_1},\cdots,e^{i\theta_n})}\begin{vmatrix} d_N(\theta_1), & \cdots, & d_N(\theta_n)\\ d'_N(\theta_1), & \cdots, & d'_N(\theta_n)\\ \cdots & \cdots & \cdots\\ \cdots & \cdots & \cdots\\ d_N^{(n-1)}(\theta_1), & \cdots, & d_N^{(n-1)}(\theta_n)\end{vmatrix},$$

这里 $e^{i\theta_1},\cdots,e^{i\theta_n}$ 是 $\overline{V}$ 的特征根，$d_N(\theta)=\dfrac{\sin\left(N+\frac{1}{2}\right)\theta}{\sin\frac{1}{2}\theta}$ 就是一个变数的 Dirichlet 核.

証. 首先从 § 1.2，我們看到 (1.8.1) 就是

$$\frac{1}{C}\int_{U_n}u(VU)\sum_{f_1\geq\cdots\geq f_n}N(f)\chi_{f_1\cdots f_n}(\overline{V})\dot{V}, \tag{1.8.3}$$

所以(1.8.2)也就是

$$\frac{1}{C}\int_{U_n}u(VU)\sum_{N\geq l_1>l_2>\cdots>l_n\geq -N}N(f)\chi_{f_1\cdots f_n}(\overline{V})\dot{V}.$$

我們来計算

$$\sum_{N\geq l_1>l_2>\cdots>l_n\geq -N}N(f)\chi_{f_1\cdots f_n}(\overline{V}).$$

从(1.3.2)及(1.8.3)知道，所謂級数(1.8.1)的 Abel 求和就是

$$\frac{1}{C}\int_{U_n}u(VU)\sum_{f_1\geq\cdots\geq f_n}\rho^f(r)N(f)\chi_{f_1\cdots f_n}(\overline{V})\dot{V}. \tag{1.8.4}$$

由 [2] 及(1.8.4)，知道

$$\sum_{f_1\geq f_2\geq\cdots\geq f_n}\rho^f(r)N(f)\chi_{f_1\cdots f_n}(\overline{V})=\frac{(1-r^2)^{n^2}}{|\det(I-r\overline{V}')|^{2n}}.$$

于是，从 $\chi_f(\overline{V})$ 的定义，我們知道，若 $e^{i\theta_1},\cdots,e^{i\theta_n}$ $(\theta_1\geq\theta_2\geq\cdots\geq\theta_n)$ 是 $\overline{V}$ 的特征根，

那末

$$\frac{1}{D(e^{i\theta_1}, \cdots, e^{i\theta_n})} \sum_{l_1>l_2>\cdots>l_n} \rho^f(r)N(f)M_{l_1\cdots l_n}(e^{i\theta_1}, \cdots, e^{i\theta_n}) =$$

$$= \frac{(1-r^2)^{n^2}}{|1-re^{i\theta_1}|^{2n}\cdots|1-re^{i\theta_n}|^{2n}},$$

这里

$$M_{l_1\cdots l_n}(x_1, \cdots, x_n) = \begin{vmatrix} x_1^{l_1}, & \cdots, & x_n^{l_1} \\ x_1^{l_2}, & \cdots, & x_n^{l_2} \\ \cdots & \cdots & \cdots \\ \cdots & \cdots & \cdots \\ x_1^{l_n}, & \cdots, & x_n^{l_n} \end{vmatrix}.$$

考虑 $\theta_1, \theta_2, \cdots, \theta_n$ 的函数

$$g(\theta_1, \cdots, \theta_n) = \frac{D(e^{i\theta_1}, \cdots, e^{i\theta_n})(1-r^2)^{n^2}}{|1-re^{i\theta_1}|^{2n}\cdots|1-re^{i\theta_n}|^{2n}}, \tag{1.8.5}$$

展开成多重 Fourier 級数,

$$\sum_{\infty>\nu_1,\cdots,\nu_n>-\infty} a_{\nu_1,\nu_2,\cdots,\nu_n} e^{i(\nu_1\theta_1+\cdots+\nu_n\theta_n)}. \tag{1.8.6}$$

研究它的部分和

$$\sum_{\nu_1=-N}^{N} \cdots \sum_{\nu_n=-N}^{N} a_{\nu_1\cdots\nu_n} e^{i(\nu_1\theta_1+\cdots+\nu_n\theta_n)} =$$

$$= \frac{1}{(2\pi)^n}\int_0^{2\pi}\cdots\int_0^{2\pi} g(\psi_1, \cdots, \psi_n) \prod_{j=1}^{n} d_N(\psi_j-\theta_j)\, d\psi_1\cdots d\psi_n, \tag{1.8.7}$$

这里

$$d_N(\theta) = \frac{\sin\left(N+\frac{1}{2}\right)\theta}{\sin\frac{1}{2}\theta},$$

即是一个变数的 Dirichlet 核.

由于

$$\frac{D(e^{i\theta_1}, \cdots, e^{i\theta_n})(1-r^2)^{n^2}}{|1-re^{i\theta_1}|^{2n}\cdots|1-re^{i\theta_n}|^{2n}} =$$

$$= \sum_{l_1>\cdots>l_n} \rho^f(r)N(f)\left(\sum_{\mu} \delta^{1,2,\cdots,n}_{\mu_1,\mu_2,\cdots,\mu_n} e^{i(l_{\mu_1}\theta_1+l_{\mu_2}\theta_2+\cdots+l_{\mu_n}\theta_n)}\right) =$$

$$= \sum_{\infty>\nu_1,\cdots,\nu_n>-\infty} a_{\nu_1,\cdots,\nu_n} e^{i(\nu_1\theta_1+\cdots+\nu_n\theta_n)},$$

可以看出(1.8.7)的左边就是

$$\sum_{N>l_1>l_2>\cdots>l_n>-N} \rho^f(r)N(f)M_{l_1\cdots l_n}(e^{i\theta_1}, \cdots, e^{i\theta_n}). \tag{1.8.8}$$

于是

$$\sum_{N\geqslant l_1>l_2>\cdots>l_n\geqslant -N}\rho^{l}(r)N(f)\chi_{l_1\cdots l_n}([e^{i\theta_1},\cdots,e^{i\theta_n}])=$$

$$=\frac{1}{(2\pi)^n}\int_0^{2\pi}\cdots\int_0^{2\pi}\frac{D(e^{i\psi_1},\cdots,e^{i\psi_n})}{D(e^{i\theta_1},\cdots,e^{i\theta_n})}\cdot$$

$$\cdot\frac{(1-r^2)^{n^2}}{\prod\limits_{\nu=1}^{n}|1-re^{i\psi_\nu}|^{2n}}\prod_{j=1}^{n}d_N(\psi_j-\theta_j)\,d\psi_1\cdots d\psi_n \tag{1.8.9}$$

应用以前我們一再使用的积分技巧，可以証明(1.8.9)的右边就是

$$\frac{(1-r^2)^{n^2}}{D(e^{i\theta_1},\cdots,e^{i\theta_n})}\cdot\frac{1}{(2\pi)^n}\int\limits_{2\pi>\psi_1>\psi_2>\cdots>\psi_n\geqslant 0}\cdots\int\frac{D(e^{i\psi_1},\cdots,e^{i\psi_n})}{\prod\limits_{j=1}^{n}|1-re^{i\psi_j}|^{2n}}$$

$$P(\psi_1,\cdots,\psi_n,\theta_1,\cdots,\theta_n)\,d\psi_1\,d\psi_2,\cdots,d\psi_n, \tag{1.8.10}$$

而 $P(\psi_1,\cdots,\psi_n,\theta_1,\cdots,\theta_n)$ 是

$$\begin{vmatrix} d_N(\psi_1-\theta_1), & \cdots, & d_N(\psi_1-\theta_n)\\ d_N(\psi_2-\theta_1), & \cdots, & d_N(\psi_2-\theta_n)\\ \cdots & \cdots & \cdots\\ \cdots & \cdots & \cdots\\ d_N(\psi_n-\theta_1), & \cdots, & d_N(\psi_n-\theta_n)\end{vmatrix}.$$

将(1.8.10)的右边的积分对酉羣的旁系积分，于是得到

$$\frac{1}{D(e^{i\theta_1},\cdots,e^{i\theta_n})}\cdot\frac{1}{C}\int_{U_n}\frac{(1-r^2)^{n^2}}{|\det(I-r\overline{W}')|^{2n}}\cdot\frac{P(\psi_1,\cdots\psi_n,\theta_1,\cdots\theta_n)}{D(e^{-i\psi_1},\cdots,e^{-i\psi_n})}\dot{W}, \tag{1.8.11}$$

让 $r\to 1$，于是 $\rho^{l}(r)\to 1$，这样(1.8.8)就是 $S_N^{(n)}(U)$，而(1.8.11)根据调和函数理論[3]，当 $r\to 1$ 时，这就是

$$\frac{1}{D(e^{i\theta_1},\cdots,e^{i\theta_n})}\lim_{\substack{\psi_1\to\theta_1\\ \cdots\cdots\\ \psi_n\to\theta_n}}\frac{P(\psi_1,\cdots,\psi_n,\theta_1,\cdots,\theta_n)}{D(e^{-i\psi_1},\cdots,e^{-i\psi_n})},$$

而这就是

$$\frac{1}{D(e^{i\theta_1},\cdots,e^{i\theta_n})}\lim_{\substack{\psi_1\to\theta_1\\ \cdots\cdots\\ \psi_n\to\theta_n}}\frac{P(\psi_1,\cdots,\psi_n,\theta_1,\cdots,\theta_n)}{D(\psi_1,\cdots,\psi_n)}\cdot\frac{D(\psi_1,\cdots,\psi_n)}{D(e^{-i\psi_1},\cdots,e^{-i\psi_n})},$$

应用[1]中的定理 1.2.4，即得 $\mathscr{D}_N(v)$ 等于

$$\frac{(-i)^{\frac{1}{2}n(n-1)}}{(n-1)!\cdots 2!\,1!\,D(e^{i\theta_1},\cdots,e^{i\theta_n})}\begin{vmatrix} d_N(\theta_1), & \cdots, & d_N(\theta_n)\\ d'_N(\theta_1), & \cdots, & d'_N(\theta_n)\\ \cdots & \cdots & \cdots\\ \cdots & \cdots & \cdots\\ d_N^{(n-1)}(\theta_1), & \cdots, & d_N^{(n-1)}(\theta_n)\end{vmatrix}.$$

§ 1.9. Dirichlet 核的代数証明

这一节的内容是属于华罗庚的。作者証明了定理 1.8.1 之后，华罗庚給出了一个如下的简单的代数証明，而且他的証明并不依赖于 Poisson 核。

由于定义，我們知道

$$\sum_{N\geqslant l_1>\cdots>l_n\geqslant -N} N(f)M_{f_1\cdots f_n}(e^{i\theta_1},\cdots,e^{i\theta_n})=$$

$$=\sum_{N\geqslant l_1>l_2>\cdots>l_n\geqslant -N}\begin{vmatrix}1, & 1, & \cdots, & 1\\ l_1, & l_2, & \cdots, & l_n\\ \cdots & \cdots & \cdots & \cdots\\ \cdots & \cdots & \cdots & \cdots\\ l_1^{n-1}, & l_2^{n-1}, & \cdots, & l_n^{n-1}\end{vmatrix}\cdot\begin{vmatrix}\lambda_1^{l_1}, & \cdots, & \lambda_n^{l_1}\\ \lambda_1^{l_2}, & \cdots, & \lambda_n^{l_2}\\ \cdots & \cdots & \cdots\\ \cdots & \cdots & \cdots\\ \lambda_1^{l_n}, & \cdots, & \lambda_n^{l_n}\end{vmatrix}\Bigg/ D(n-1,\cdots,1,0), \tag{1.9.1}$$

这里 $\lambda_1=e^{i\theta_1},\ \lambda_2=e^{i\theta_2},\ \cdots,\ \lambda_n=e^{i\theta_n}$.

由于行列式展开的对称性，将 $(l_1, l_2, \cdots, l_n)$ 进行另外一种排列，(1.9.1)的右边的值不变，所以

$$\sum_{N\geqslant l_1>\cdots>l_n\geqslant -N} N(f)M_{f_1\cdots f_n}(e^{i\theta_1},\cdots,e^{i\theta_n})=$$

$$=\frac{1}{n!}\sum_{l_1=-N}^{N}\cdots\sum_{l_n=-N}^{N}\begin{vmatrix}1, & 1, & \cdots, & 1\\ l_1, & l_3, & \cdots, & l_n\\ \cdots & \cdots & \cdots & \cdots\\ \cdots & \cdots & \cdots & \cdots\\ l_1^{n-1}, & l_2^{n-1}, & \cdots, & l_n^{n-1}\end{vmatrix}\cdot\begin{vmatrix}\lambda_1^{l_1}, & \cdots, & \lambda_n^{l_1}\\ \lambda_1^{l_3}, & \cdots, & \lambda_n^{l_3}\\ \cdots & \cdots & \cdots\\ \cdots & \cdots & \cdots\\ \lambda_1^{l_n}, & \cdots, & \lambda_n^{l_n}\end{vmatrix}\Bigg/ D(n-1,\cdots,1,0)=$$

$$=\frac{1}{n!}\sum_{l_1=-N}^{N}\cdots\sum_{l_n=-N}^{N}\begin{vmatrix}\lambda_1^{l_1}+\lambda_1^{l_2}+\cdots+\lambda_1^{l_n}, & l_1\lambda_1^{l_1}+l_2\lambda_1^{l_2}+\cdots+l_n\lambda_1^{l_n}, & \cdots, & l_1^{n-1}\lambda_1^{l_1}+l_2^{n-1}\lambda_1^{l_2}+\cdots+l_n^{n-1}\lambda_1^{l_n}\\ \lambda_2^{l_1}+\lambda_2^{l_2}+\cdots+\lambda_2^{l_n}, & l_1\lambda_2^{l_1}+l_2\lambda_2^{l_2}+\cdots+l_n\lambda_2^{l_n}, & \cdots, & l_1^{n-1}\lambda_2^{l_1}+l_2^{n-1}\lambda_2^{l_2}+\cdots+l_n^{n-1}\lambda_2^{l_n}\\ \cdots & \cdots & \cdots & \cdots\\ \lambda_n^{l_1}+\lambda_n^{l_2}+\cdots+\lambda_n^{l_n}, & l_1\lambda_n^{l_1}+l_2\lambda_n^{l_2}+\cdots+l_n\lambda_n^{l_n}, & \cdots & l_1^{n-1}\lambda_n^{l_1}+l_2^{n-1}\lambda_n^{l_2}+\cdots+l_n^{n-1}\lambda_n^{l_n}\end{vmatrix}\Bigg/ D(n-1,\cdots,1,0). \tag{1.9.2}$$

将(1.9.2)的右边的分子析成 n^2 个行列式,那末,这些行列式都成为如下的形式:

$$\sum_{l_1=-N}^{N}\cdots\sum_{l_n=-N}^{N}\begin{vmatrix} \lambda_1^{l_{\nu_1}}, & \lambda_2^{l_{\nu_1}}, & \cdots, & \lambda_n^{l_{\nu_1}} \\ l_{\nu_2}\lambda_1^{l_{\nu_2}}, & l_{\nu_2}\lambda_2^{l_{\nu_2}}, & \cdots, & l_{\nu_2}\lambda_n^{l_{\nu_2}} \\ \cdots & \cdots & \cdots & \cdots \\ \cdots & \cdots & \cdots & \cdots \\ l_{\nu_n}^{n-1}\lambda_1^{l_{\nu_n}}, & l_{\nu_n}^{n-1}\lambda_2^{l_{\nu_n}}, & \cdots, & l_{\nu_n}^{n-1}\lambda_n^{l_{\nu_n}} \end{vmatrix}, \tag{1.9.3}$$

这里 $\nu_1, \nu_2, \cdots, \nu_n$ 是 $1, 2, \cdots, n$ 中取值,显然(1.9.3)中所有的行列式都等于零,除了 $(\nu_1, \nu_2, \cdots, \nu_n)$ 是 $(1, 2, \cdots, n)$ 的一个排列,于是

$$\sum_{N\geqslant l_1>\cdots>l_n\geqslant -N} N(f)M_{l_1\cdots l_n}(e^{i\theta_1}, \cdots, e^{i\theta_n}) =$$

$$= \sum_{l_1=-N}^{N}\cdots\sum_{l_n=-N}^{N}\begin{vmatrix} \lambda_1^{l_1}, & \lambda_2^{l_1}, & \cdots, & \lambda_n^{l_1} \\ l_2\lambda_1^{l_2}, & l_2\lambda_2^{l_2}, & \cdots, & l_2\lambda_n^{l_2} \\ \cdots & \cdots & \cdots & \cdots \\ \cdots & \cdots & \cdots & \cdots \\ l_n^{n-1}\lambda_1^{l_n}, & l_n^{n-1}\lambda_2^{l_n}, & \cdots, & l_n^{n-1}\lambda_n^{l_n} \end{vmatrix} \Big/ D(n-1, \cdots, 1, 0) =$$

$$= \begin{vmatrix} \sum_{-N}^{N}\lambda_1^l, & \sum_{-N}^{N}\lambda_2^l, & \cdots, & \sum_{-N}^{N}\lambda_n^l \\ \sum_{-N}^{N} l\lambda_1^l, & \sum_{-N}^{N} l\lambda_2^l, & \cdots, & \sum_{-N}^{N} l\lambda_n^l \\ \cdots & \cdots & \cdots & \cdots \\ \cdots & \cdots & \cdots & \cdots \\ \sum_{-N}^{N} l^{n-1}\lambda_1^l, & \sum_{-N}^{N} l^{n-1}\lambda_2^l, & \cdots, & \sum_{-N}^{N} l^{n-1}\lambda_n^l \end{vmatrix} \Big/ D(n-1, \cdots, 1, 0) =$$

$$= \frac{1}{i^{\frac{1}{2}n(n-1)}D(n-1, \cdots, 1, 0)}\begin{vmatrix} d_N(\theta_1), & d_N(\theta_2), & \cdots, & d_N(\theta_n) \\ d'_N(\theta_1), & d'_N(\theta_2), & \cdots, & d'_N(\theta_n) \\ \cdots & \cdots & \cdots & \cdots \\ \cdots & \cdots & \cdots & \cdots \\ d_N^{(n-1)}(\theta_1), & d_N^{(n-1)}(\theta_2), & \cdots, & d_N^{(n-1)}(\theta_n) \end{vmatrix}.$$

这就証明了定理 1.8.1,显然,这个証明比起 §1.8 中的証明来要簡洁而直接,但 §1.8 中的証明是一种处理矩陣积分有用的方法,因此途径虽为曲折一些,还是写了出来。

§1.10. 求和法的一种定义及它的核

前面二节所用来証明 Dirichlet 核的方法,立刻建議我們对于酉羣 U_n 上可积函数的 Fourier 級数,可以作出一个一般的求和法来,并且可以应用前二节的方法容易地求出它們的核来。

設 $u(\theta)$ 是在 $0 \leqslant \theta \leqslant 2\pi$ 上可积的函数,它的 Fourier 級数是

$$\sum_{-\infty}^{\infty} a_p e^{ip\theta}, \tag{1.10.1}$$

T 是一种求和法,求和

$$\tau_m = \sum_{-\infty}^{\infty} \mu_{mp} a_p e^{ip\theta},$$

它的核是

$$k_m(\theta) = \sum_{-\infty}^{\infty} \mu_{mp} e^{ip\theta} \tag{1.10.2}$$

(当然,我們假設核是存在的, 即 (1.10.2) 是收斂的) 假使当 m 趋于极限时, $\tau_m \to s$, 那末 (1.10.1) 称为可以 T 求和,其和为 s.

根据这些,我們来对 U_n 上的可积函数的 Fourier 級数,給出一种求和法的

定义 1.10.1. 設 $u(U)$ 为在 U_n 上可积的函数,它的 Fourier 級数为

$$\sum_{l_1 \geqslant \cdots \geqslant l_n} \operatorname{tr}\left(C_{l_1\cdots l_n} \Phi'_{l_1\cdots l_n}(U)\right), \tag{1.10.3}$$

作和

$$\tau_m = \sum_{l_1 \geqslant \cdots \geqslant l_n} \mu_{m,l_1}\mu_{m,l_2}\cdots\mu_{m,l_n} \operatorname{tr}\left(C_{l_1\cdots l_n} \Phi'_{l_1\cdots l_n}(U)\right),$$

若当 m 趋于极限时, $\tau_m \to s$, 那末我們称(1.10.3)可以 T 求和, 其和为 s.

例 1.10.1. (α 次 Cesèro 求和, $\alpha > -1$)

$$\tau_N = \sum_{N \geqslant l_1 \geqslant \cdots \geqslant l_n \geqslant -N} A^{\alpha}_{l_1} \cdots A^{\alpha}_{l_n} \operatorname{tr}\left(C_{l_1\cdots l_n} \Phi'_{l_1\cdots l_n}(U)\right),$$

这里

$$A^{\alpha}_{l_j} = \frac{\Gamma(\alpha + N - |l_j| + 1)\Gamma(N+1)}{\Gamma(\alpha + N + 1)\Gamma(N - |l_j| + 1)}.$$

例 1.10.2. (Abel-Poisson 求和)[1)]

$$\tau_r = \sum_{l_1 \geqslant l_2 \geqslant \cdots \geqslant l_n} r^{|l_1| + \cdots + |l_n|} \operatorname{tr}\left(C_{l_1\cdots l_n} \Phi'_{l_1\cdots l_n}(U)\right).$$

应用証明定理 1.8.1 的方法可以証明如下的

定理 1.10.1. 由定义 1.10.1 所給出的 T 求和法的核 $K_m(V)$ 为

$$\frac{(-i)^{\frac{1}{2}n(n-1)}}{1!\,2!\cdots(n-1)!\,D(e^{i\theta_1}, \cdots, e^{i\theta_n})} \begin{vmatrix} k_m(\theta_1), & \cdots, & k_m(\theta_n) \\ k'_m(\theta_1), & \cdots, & k'_m(\theta_n) \\ \cdots\cdots & \cdots\cdots & \cdots\cdots \\ \cdots\cdots & \cdots\cdots & \cdots\cdots \\ k_m^{(n-1)}(\theta_1), & \cdots, & k_m^{(n-1)}(\theta_n) \end{vmatrix}, \tag{1.10.4}$$

这里 $e^{i\theta_1}, \cdots, e^{i\theta_n}$ 是 $\overline{V}$ 的特征根. $k_m(\theta) = \sum_{-\infty}^{\infty} \mu_{mp} e^{ip\theta}$ 是一个变数的 Fourier 級数的 T 求和法的核.

証. 我們知道(1.8.5)的多重 Fourier 級数为(1.8.6), 求(1.8.6)的和

$$\sum_{\infty \geqslant \nu_1, \cdots, \nu_n \geqslant -\infty} a_{\nu_1, \cdots, \nu_n} \mu_{m,\nu_1} \mu_{m,\nu_2} \cdots \mu_{m,\nu_n} e^{i(\nu_1\theta_1 + \cdots + \nu_n\theta_n)}, \tag{1.10.5}$$

1) 注意这与 § 1.3 中华罗庚所定义的 Abel 求和是不同的.

于是，这就是

$$\frac{1}{(2\pi)^n}\int_0^{2\pi}\cdots\int_0^{2\pi} g(\psi_1,\cdots,\psi_n)\prod_{j=1}^{n} k_m(\psi_j-\theta_j)\,d\psi_1\cdots d\psi_n,$$

这里 $k_m(\theta)=\sum_{-\infty}^{\infty}\mu_{mp}e^{ip\theta}$ 为一个变数的 Fourier 级数的 T 求和法的核。

可以看出(1.10.5)就是

$$\sum_{l_1>\cdots>l_n}\mu_{m,l_1}\cdots\mu_{m,l_n}\rho^f(r)N(f)M_{f_1\cdots f_n}(e^{i\theta_1},\cdots,e^{i\theta_n}),$$

因之，

$$\sum_{l_1>\cdots>l_n}\mu_{m,l_1}\cdots\mu_{m,l_n}\rho^f(r)N(f)\chi_{f_1\cdots f_n}([e^{i\theta_1},\cdots,e^{i\theta_n}])=$$

$$=\frac{1}{(2\pi)^n}\int_0^{2\pi}\cdots\int_0^{2\pi}\frac{D(e^{i\psi_1},\cdots,e^{i\psi_n})}{D(e^{i\theta_1},\cdots,e^{i\theta_n})}\frac{(1-r^2)^{n^2}}{\prod_{\nu=1}^{n}|1-e^{i\psi_\nu}|^{2n}}\prod_{j=1}^{n}k_m(\psi_j-\theta_j)\,d\psi_1\cdots d\psi_n. \tag{1.10.6}$$

应用以往我們慣用的技巧(1.10.6)的右边就是

$$\frac{(1-r^2)^{n^2}}{D(e^{i\theta_1},\cdots,e^{i\theta_n})}\frac{1}{(2\pi)^n}\int\cdots\int_{\psi_1\geq\cdots\geq\psi_n}\frac{D(e^{i\psi_1},\cdots,e^{i\psi_n})}{\prod_{j=1}^{n}|1-re^{i\psi_j}|^{2n}}Q(\psi_1,\cdots,\psi_n,\theta_1,\cdots,\theta_n)\times$$

$$\times\,d\psi_1\cdots d\psi_n, \tag{1.10.7}$$

而 $Q(\psi_1,\cdots,\psi_n,\theta_1,\cdots,\theta_n)$ 是

$$\begin{vmatrix} k_m(\psi_1-\theta_1), & \cdots, & k_m(\psi_1-\theta_n)\\ k_m(\psi_2-\theta_1), & \cdots, & k_m(\psi_2-\theta_n)\\ \cdots\cdots & \cdots\cdots & \cdots\cdots\\ k_m(\psi_n-\theta_1), & \cdots, & k_m(\psi_n-\theta_n)\end{vmatrix}.$$

将(1.10.7)对酉群的旁系积分，于是得到

$$\frac{1}{D(e^{i\theta_1},\cdots,e^{i\theta_n})}\cdot\frac{1}{C}\int_{U_n}\frac{(1-r^2)^{n^2}}{|\det(I-r\overline{W'})|^{2n}}\cdot\frac{Q(\psi_1\cdots\psi_n,\theta_1\cdots\theta_n)}{D(e^{-i\psi_1},\cdots,e^{-i\psi_n})}\dot{W}.$$

註 $r\to1$，由于 $\rho^f(r)\to1$，应用与 § 1.8 中一样的方法，即得定理 1.10.1.

当然定理 1.10.1 还可以用 § 1.9 中的方法給与一个代数的証明，这里从略。

例 1.10.1. 在 (1.10.4) 中以

$$k_N(\theta)=\frac{1}{N+1}\left\{\frac{\sin\frac{1}{2}(N+1)\theta}{\sin\frac{1}{2}\theta}\right\}^2$$

代入，即得到$(C,1)$求和的 Fejer 核 $F_N(V)$，(事实上，我們在下一篇文章中将給出一个重要的 Fejer 核，与此不相同)。

例 1.10.2. 在(1.10.4)中以

$$k_r(\theta)=\frac{1-r^2}{1-2r\cos\theta+r^2}$$

代入，即得 Abel-Poisson 求和的 Poisson 核 $P(r, V)$（事实上，这与 §1.3 中所給出的 Poisson 核也是不同的）.

参 考 文 献

[1] 华罗庚：多复变数函数論中的典型域的調和分析，科学出版社，1958 年.
[2] 华罗庚：紧致羣上的連續函数所成的空間中的一条收斂定理，科学記录，2 (1958)，341—344.
[3] 华罗庚：一組偏微分方程，科学記录，1 (1957)，339—340.
[4] 龔昇：酉羣上的富理埃分析，I. Fourier 級数的部分和及求和.科学記录，4 (1960)，
[5] 龔昇：酉羣上的富理埃分析，II. 华— Abel 求和.科学記录，4(1960).
[6] Peter, E. and Weyl, H., *Math. Anns.*, 97 (1927), 735—755.

FOURIER ANALYSIS ON UNITARY GROUP

I. ABEL SUMMABILITY AND DIRICHLET KERNEL OF FOURIER SERIES

Kung Sun

(The Chinese University of Science and Technology)

Abstract

It is known[6] that every continuous function on a compact group can be approximated by finite polynomials. Prof. Hua[2] has established that the Fourier series of continuous functions on unitary group can be summable to itself by means of the Abel summability. Hua's theorem is the first one for the convergence of Fourier series of functions on compact groups of a finite dimension. Certainly, a convergence theorem is better than an approximation theorem.

Let $u(U)$ be an integrable function on n-dimensional unitary group U_n, Its Fourier series is

$$\sum_{f_1\geqslant f_2\geqslant\cdots\geqslant f_n} \operatorname{tr}\left(C_{f_1\cdots f_n}\Phi'_{f_1\cdots f_n}(U)\right), \tag{1}$$

where

$$\Phi_{f_1\cdots f_n}(U)=\sqrt{\frac{N(f)}{C}}\,A_{f_1\cdots f_n}(U),$$

$$C_{f_1\cdots f_n}=\int_{U_n} u(V)\Phi_{f_1\cdots f_n}(\overline{V})\dot{V}.$$

$A_{f_1\cdots f_n}(U)$ is the unitary representation with signature $f=(f_1, f_2, \cdots, f_n)$ where $f_1, f_2, \cdots, f_n$ are integers satisfying $f_1\geqslant f_2\geqslant\cdots\geqslant f_n$; $N(f)=N(f_1, \cdots, f_n)$ is the order of the matrix $A_{f_1\cdots f_n}(U)$ and C_c is the volume of the unitary group U_n, i.e.

$$C=(2\pi)^{\frac{1}{2}n(n+1)}/((n-1)!\,(n-2)!\cdots 2!\,1!).$$

Hua[2] defined the Abel summability of Fourier series (1) as

$$\sum_{f_1\geqslant\cdots\geqslant f_n}\rho^f(r)\operatorname{tr}\left(C_{f_1\cdots f_n}\Phi'_{f_1\cdots f_n}(U)\right)=\frac{1}{C}\int_{U_n}u(VU)\frac{(1-r^2)^{n^2}\dot{V}}{|\det(1-r\overline{V}')|^{2n}},$$

where

$$\rho^f(r) = \frac{1}{N(f)} \cdot \frac{1}{C}\int_{U_n} \frac{(1-r^2)^{n^2}\chi_f(V)\dot{V}}{|\det(I - r\bar{V}')|^{2n}}, \tag{2}$$

Moreover, he pointed out that

(i) $\rho^f(r) \to 1$ as $r \to 1$,

(ii) $\rho^f(r) = \begin{cases} r^{f_1+\cdots+f_n} & f_1 \geqslant f_2 \geqslant \cdots \geqslant f_n \geqslant 0; \\ \quad\text{as} & \\ r^{-f_1-\cdots-f_n} & 0 \geqslant f_1 \geqslant \cdots \geqslant f_{n-1} \geqslant f_n. \end{cases}$

In the present paper, by a complicate calculation, we proved

Theorem I.

$$\rho^f(r) = r^{f_1+\cdots+f_s-f_{s+1}-\cdots-f_n} \sum_{s\geqslant g_{s+1}\geqslant\cdots\geqslant g_n\geqslant 0} \frac{N_s(f, g)N_s(g, f)}{N(f)N(g)} r^{2(g_{s+1}+\cdots+g_n)}$$

provided $l_1 > l_2 > \cdots > l_s \geqslant 0 > l_{s+1} > \cdots > l_n$ $(n \geqslant s \geqslant 0)$, where $N_s(a, b)$ denote $N(a_1, \cdots, a_s, b_{s+1}, \cdots, b_n)$, $g = (g_1, g_2, \cdots, g_n)$, $(g_1 + n - 1, g_2 + n - 2, \cdots, g_n)$ is a permutation of $(0, 1, \cdots, n-1)$, $0 \geqslant g_1 \geqslant g_2 \geqslant \cdots \geqslant g_s \geqslant s - n$.

Many various forms of $\rho^f(r)$ are given also.

Moreover we establish the following

Theorem II. The partial sum

$$S_N^{(n)}(U) = \sum_{N\geqslant l_1>\cdots>l_n\geqslant -N} \operatorname{tr}(C_{f_1\cdots f_n}\Phi'_{f_1\cdots f_n}(U))$$

of the Fourier series (1) is

$$\frac{1}{C}\int_{U_n} u(VU)\mathscr{D}_N(V)\dot{V},$$

where $\mathscr{D}_N(V)$ is the Dirichlet kernel

$$\frac{(-i)^{\frac{1}{2}n(n-1)}}{1!\,2!\cdots(n-1)!\,D(e^{i\theta_1}, \cdots, e^{i\theta_n})} \begin{vmatrix} d_N(\theta_1), & \cdots, & d_N(\theta_n) \\ d'_N(\theta_1), & \cdots, & d'_N(\theta_n) \\ \cdots\cdots & \cdots\cdots & \cdots\cdots \\ \cdots\cdots & \cdots\cdots & \cdots\cdots \\ d_N^{(n-1)}(\theta_1), & \cdots, & d_N^{(n-1)}(\theta_n) \end{vmatrix},$$

$e^{i\theta_1}, e^{i\theta_2}, \cdots, e^{i\theta_n}$ are the characteristic roots of $\bar{V}$, and $d_N(\theta) = \sin\left(N+\frac{1}{2}\right)\Big/\sin\frac{1}{2}\theta$, the Dirichlet kernel of one variable.

Two proofs of theorem II are given, one is analytic, and the other is algebraic which is given by L. K. Hua.

第 12 卷 第 1 期　　数 学 学 报　　Vol. 12, No. 1
1962 年 3 月　　ACTA MATHEMATICA SINICA　　March, 1962

酉羣上的富里埃分析(II)
富里埃級数的Cesàro求和*

龚　升
(中国科学技术大学)

§2.1. 引　言

若 $u(U)$ 是 n 阶酉羣 U_n 上的可积函数,它的 Fourier 級数为

$$\sum_{f_1 \geqslant f_2 \geqslant \cdots \geqslant f_n} \operatorname{tr}\left(C_{f_1\cdots f_n} \Phi'_{f_1\cdots f_n}(U)\right) \tag{2.1.1}$$

而

$$C_{f_1\cdots f_n} = \int_{U_n} u(V)\, \overline{\Phi_{f_1\cdots f_n}(V)}\, \dot{V}.$$

华罗庚[2]定义了 (2.1.1) 的 Abel 平均为

$$\sum_{f_1 \geqslant f_2 \geqslant \cdots \geqslant f_n} \rho^f(r) \operatorname{tr}\left(C_{f_1\cdots f_n} \Phi'_{f_1\cdots f_n}(U)\right). \tag{2.1.2}$$

并且証明了:如果 $u(U)$ 在 U_n 上連續,那末,当 $r \to 1$ 时,(2.1.2)的极限是存在的,且等于 $u(U)$. 即,$u(U)$ 的 Fourier 級数 (2.1.1) 是可以 Abel 求和于它自己.

我們在前文[3]中通过了复杂的矩陣积分的技巧,給出了 $\rho^f(r)$ 的具体表达式. 并且給出了 (2.1.1) 的 Dirichlet 核.

在本文中,我們在 Abel 求和的启发下,研究了 (2.1.1) 的 Cesàro 求和.

在 § 2.2 中,我們給出了 Cesàro 求和的定义和对应的核. § 2.3—2.4 証明了如下的 Riesz 型定理:若 $u(U)$ 在 U_n 上連續,那末,$u(U)$ 的 Fourier 級数 (2.1.1) 可以 (c, α) 求和于它自己,但 $\alpha > \frac{n-1}{n}$. 这是本文的中心之一.

在 Cesàro 求和的定义中,不仅核可以用矩陣形式明白的表示出来,其求和系数及一些有关积分常数也都可以具体的算出来. 为了对一般的 Cesàro 求和有直观的了解,我們比較仔細的研究了 Cesàro 求和中最典型而最重要的情形,即 Fejér 求和的情形作为例子. 通过这个例子,其他一般的 Cesàro 求和中的一些系数及常数也可以用相类似的方法定出来. 这是 § 2.5—2.7 的內容,所用的方法是上文[1]中所用过的矩陣积分的进一步应用,这是本文中心之二.

为了不使篇幅过长,沿着这个思想发展下去的一些工作,例如:一般紧致羣上的逼近理論等将在另文中叙述之.

本文中所用的符号与前文[1]相同,不再予以一一交代.

---

* 1960 年 12 月 19 日收到,1961 年 11 月 6 日收到修改稿.

§2.2. Cesàro 求和的定义和核

我們定义在 U_n 上可积函数 $u(U)$ 的 Fourier 級数 (2.1.1) 的 Cesàro (c, α) 和为

$$\sum_{nN\geqslant f_1\geqslant f_2\geqslant\cdots\geqslant f_n\geqslant -nN} B^{\alpha}_{f_1\cdots f_n}\operatorname{tr}(C_{f_1\cdots f_n}\Phi'_{f_1\cdots f_n}(U)) \tag{2.2.1}$$

这里

$$B^{\alpha}_{f_1\cdots f_n} = \frac{1}{CN(f)}\int_{U_n}\chi_{f_1\cdots f_n}(\bar{V})\mathbf{K}^{\alpha}_{N}(V)\dot{V} \tag{2.2.2}$$

(2.2.2) 式中的 $\mathbf{K}^{\alpha}_{N}(V)$ 为 Cesàro (c, α) 核，它等于

$$\frac{\det^n\left[\sum_{k=0}^{N} A^{\alpha-1}_{N-k}V^k(I-\bar{V}'^{2k+1})\right]}{B^{\alpha}_{N}(2A^{\alpha}_{N})^{n^2}\det^n(I-\bar{V}')} \tag{2.2.3}$$

这里

$$B^{\alpha}_{N} = \frac{1}{C}\int_{U_n}\frac{\det^n\left[\sum_{k=0}^{N} A^{\alpha-1}_{N-k}V^k(I-\bar{V}'^{2k+1})\right]}{(2A^{\alpha}_{N})^{n^2}\det^n(I-\bar{V}')}\dot{V} \tag{2.2.4}$$

而

$$A^{\alpha}_{N} = C^{N+\alpha}_{N} = \frac{(\alpha+N)\cdots(\alpha+1)}{N!}.$$

当然在以上的定义中我們要求 $\alpha > -1$.

首先我們要証明

定理 2.2.1. 在 U_n 上可积的函数 $u(U)$ 的 Fourier 級数 (2.1.1) 的 Cesàro (c, α) 和 (2.2.1) 可以表达成为

$$\frac{1}{C}\int_{U_n}u(VU)\mathbf{K}^{\alpha}_{N}(V)\dot{V}. \tag{2.2.5}$$

这里 $\mathbf{K}^{\alpha}_{N}(V)$ 由 (2.2.3) 所定义.

証. 若 $e^{i\theta_1}, \cdots, e^{i\theta_n}$ 为 $\bar{V}$ 的特征根，那末

$$\sum_{nN\geqslant f_1\geqslant\cdots\geqslant f_n\geqslant -nN} B^{\alpha}_{f_1\cdots f_n}N(f)M_{f_1\cdots f_n}(\bar{V}) = \mathbf{K}^{\alpha}_{N}(V)D(e^{i\theta_1}, \cdots, e^{i\theta_n}) \tag{2.2.6}$$

这里 $M_{f_1\cdots f_n}(\bar{V})$ 的定义見[1].

我們现在来証明 (2.2.6). 把 (2.2.6) 的二边看成 $e^{i\theta_1}, \cdots, e^{i\theta_n}$ 的函数，由于 (2.2.6) 左右二边都是 $e^{i\theta_1}, \cdots, e^{i\theta_n}$ 的多項式，比較系数立刻得到

$$B^{\alpha}_{f_1\cdots f_n}N(f) = \frac{1}{(2\pi)^n}\int_0^{2\pi}\cdots\int_0^{2\pi} e^{i(f_1\theta_1+\cdots+f_n\theta_n)}\mathbf{K}^{\alpha}_{N}(V)D(e^{-i\theta_1}, \cdots, e^{-i\theta_n})\,d\theta_1\cdots d\theta_n$$

应用在[3]中慣用的矩陣积分的技巧，上式即为 (2.2.2)，所以 (2.2.6) 是成立的.

从 (2.2.6) 出发，左右除以 $D(e^{i\theta_1}, \cdots, e^{i\theta_n})$，即得

$$\sum_{nN\geqslant f_1\geqslant\cdots\geqslant f_n\geqslant -nN} B^{\alpha}_{f_1\cdots f_n}N(f)\chi_{f_1\cdots f_n}(\bar{V}) = \mathbf{K}^{\alpha}_{N}(V). \tag{2.2.7}$$

从[3]中 §1.2 中的公式，可以知道 (2.2.1) 就是

$$\frac{1}{C}\int_{U_n} u(VU) \sum_{nN \geq l_1 \geq \cdots \geq l_n \geq -nN} B^{\alpha}_{l_1 \cdots l_n} N(f) \chi_{l_1 \cdots l_n}(\overline{V}) \dot{V}$$

而从 (2.2.7)，此即 (2.2.5).

立刻可以看到，我們这样所定义的 Cesàro 和 (c, α)，当 α 趋于无穷时，即为华罗庚所定义的 Abel 和，且 (c, α) 核也成为 Poisson 核.

如果当 N 趋于无穷时，(2.2.1) 的极限是存在的，那末我們称 $u(U)$ 的 Fourier 級数 (2.1.1) 是可以 (c, α) 求和的.

§ 2.3. Cesàro 核的半定正性

我們将要着手証明如下的 Riesz 型定理.

定理 2.3.1. 設 $u(U)$ 在 U_n 上連續，那末它的 Fourier 級数 (2.1.1) 可以 (c, α) 求和于它自己，但 $\alpha > \frac{n-1}{n}$.

为了証明上述定理，我們先来証明：当 $\alpha > \frac{n-1}{n}$ 时，Cesàro 核是"半定正"的(Quasi-Positive)，即要証明如下的

定理 2.3.2. (c, α) 核 $\mathbf{K}^{\alpha}_N(V)$ 是半定正的，即

$$\frac{1}{C}\int_{U_n} |\mathbf{K}^{\alpha}_N(V)| \dot{V} \leqslant M, \tag{2.3.1}$$

这里 M 只与 n, α 有关，而与 N 无关. 但 $\alpha > \frac{n-1}{n}$.

証. 当 n 是偶数时，(2.3.1) 是显然的，因为在这种情形时，$\mathbf{K}^{\alpha}_N(V)$ 本身是非負的，以至取 $M = 1$ 即可.

当 n 是奇数时，我們应用数学归納法. 当 $n = 1$ 时，(2.3.1) 显然成立，现在我們假設 (2.3.1) 当酉群的阶数小于 n 时都成立，从而导出阶数等于 n 时也成立.

容易看出，$|B^{\alpha}_N|$ 当 N 趋于无穷时，不趋于零，因之，我們只要証明

$$I = \frac{1}{(2\pi)^n} \int \cdots \int_{\pi \geq \theta_1 \geq \cdots \geq \theta_n \geq -\pi} |\sigma^{\alpha}_N(\theta_1) \cdots \sigma^{\alpha}_N(\theta_n)|^n |D(e^{i\theta_1}, \cdots, e^{i\theta_n})|^2 d\theta_1 \cdots d\theta_n \tag{2.3.2}$$

有界即可. 这里

$$\sigma^{\alpha}_N(\theta) = \frac{1}{2A^{\alpha}_N \sin\frac{1}{2}\theta} \vartheta \sum_{k=0}^{N} A^{\alpha-1}_{N-k} e^{i(k+\frac{1}{2})\theta}.$$

先取 $\delta \geqslant \frac{1}{N}$，将积分区域分解为

R_1:　$\delta \geqslant \theta_1 \geqslant \theta_2 \geqslant \cdots \geqslant \theta_n \geqslant -\pi$.

R_2:　$\pi \geqslant \theta_1 \geqslant \delta \geqslant \theta_2 \geqslant \cdots \geqslant \theta_n \geqslant -\pi$.

....

R_n:　$\pi \geqslant \theta_1 \geqslant \theta_2 \geqslant \cdots \geqslant \delta \geqslant \theta_n \geqslant -\pi$.

R_{n+1}:　$\pi \geqslant \theta_1 \geqslant \theta_2 \geqslant \cdots \geqslant \theta_n \geqslant \delta$.

記 (2.3.2) 的积分,其积分区域在 R_j 上者为 I_j, 于是

$$I = I_1 + I_2 + \cdots + I_{n+1}.$$

我們先考虑 I_2, 将 R_2 再分解为

R_{2n}: $\pi \geqslant \theta_1 \geqslant \delta \geqslant \theta_2 \geqslant \cdots \geqslant \theta_n \geqslant -\delta$,

$R_{2\overline{n-1}}$: $\pi \geqslant \theta_1 \geqslant \delta \geqslant \theta_2 \geqslant \cdots \geqslant -\delta \geqslant \theta_n \geqslant -\pi$,

……………………………………

R_{22}: $\pi \geqslant \theta_1 \geqslant \delta \geqslant \theta_2 \geqslant -\delta \geqslant \cdots \geqslant \theta_n \geqslant -\pi$,

R_{21}: $\pi \geqslant \theta_1 \geqslant \delta > -\delta \geqslant \theta_2 \geqslant \cdots \geqslant \theta_n \geqslant -\pi$.

記 (2.3.2) 的积分,其积分区域在 R_{ij} 上者为 I_{ij}. 我們来研究 I_{2n}.

$$I_{2n} = \frac{1}{(2\pi)^n}\int\cdots\int_{R_{2n}} |\sigma_N^\alpha(\theta_1)\cdots\sigma_N^\alpha(\theta_n)|^n |D(e^{i\theta_1}, \cdots, e^{i\theta_n})|^2 d\theta_1\cdots d\theta_n =$$

$$= \frac{1}{(2\pi)^n}\int\cdots\int_{\delta\geqslant\theta_2\geqslant\cdots\geqslant\theta_n\geqslant-\delta} |\sigma_N^\alpha(\theta_2)\cdots\sigma_N^\alpha(\theta_n)|^n |D(e^{i\theta_2}, \cdots, e^{i\theta_n})|^2 d\theta_2\cdots d\theta_n \times$$

$$\times\int_\delta^\pi |\sigma_N^\alpha(\theta_1)|^n \prod_{j=2}^n |e^{i\theta_1} - e^{i\theta_j}|^2 d\theta_1.$$

但是由于

$$\prod_{j=2}^n |e^{i\theta_1} - e^{i\theta_j}|^2 = \prod_{j=2}^n |(1 - e^{i\theta_1}) - (1 - e^{i\theta_j})|^2 =$$

$$= \sum_{m_1=0}^{n-1}\sum_{l_1=0}^{n-1}\sum_{m_2=0}^{1}\sum_{l_2=0}^{1}\cdots\sum_{\substack{m_n=0\\ m_1+l_1+\cdots+m_n+l_n=2(n-1)}}^{1}\sum_{l_n=0}^{1} (-1)^{2(n-1)-m_1-l_1}(1 - e^{i\theta_1})^{m_1}(1 - e^{-i\theta_1})^{l_1}(1 - e^{i\theta_2})^{m_2}\times$$

$$\times (1 - e^{-i\theta_2})^{l_2}\cdots(1 - e^{i\theta_n})^{m_n}(1 - e^{-i\theta_n})^{l_n}.$$

因之

$$I_{2n} \leqslant \sum_{s_1=0}^{2(n-1)}\sum_{s_2=0}^{2}\cdots\sum_{\substack{s_n=0\\ s_1+s_2+\cdots+s_n=2(n-1)}}^{2} a_{s_1\cdots s_n}\int\cdots\int_{\delta\geqslant\theta_2\geqslant\cdots\geqslant\theta_n\geqslant-\delta} |1 - e^{i\theta_2}|^{s_2}\cdots|1 - e^{i\theta_n}|^{s_n}|\sigma_N^\alpha(\theta_2)|^n\cdots$$

$$\cdots|\sigma_N^\alpha(\theta_n)|^n|D(e^{i\theta_2}, \cdots, e^{i\theta_n})|^2 d\theta_2\cdots d\theta_n \int_\delta^\pi |1 - e^{i\theta_1}|^{s_1}|\sigma_N^\alpha(\theta_1)|^n d\theta_1,$$

这里 $a_{s_1\cdots s_n}$ 为只与 $s_1, s_2, \cdots, s_n$ 有关的絕对常数.

为了估計上述积分,我們取其中的任意一个积分

$$I_{s_1\cdots s_n} = \int_\delta^\pi |1 - e^{i\theta_1}|^{s_1}|\sigma_N^\alpha(\theta_1)|^n d\theta_1 \int\cdots\int_{\delta\geqslant\theta_2\geqslant\cdots\geqslant\theta_n\geqslant-\delta} |1 - e^{i\theta_2}|^{s_2}\cdots|1 - e^{i\theta_n}|^{s_n}\times$$

$$\times|\sigma_N^\alpha(\theta_2)|^n\cdots|\sigma_N^\alpha(\theta_n)|^n|D(e^{i\theta_2}, \cdots, e^{i\theta_n})|^2 d\theta_2\cdots d\theta_n. \tag{2.3.3}$$

由于 $\delta \geqslant \frac{1}{N}$, 所以

$$|\sigma_N^\alpha(\theta)| \leqslant aN^{-\alpha}|\theta|^{-\alpha-1} \tag{2.3.4}$$

这里 a 表示絕对常数,我們以后常用 a 表示不同的絕对常数. 因之

$$I_{s_1\cdots s_n}\leqslant a\int_\delta^\pi N^{-\alpha n}\theta^{-\alpha n-n+s_1}d\theta\cdot N^{n-1}\delta^{s_2+\cdots+s_n},$$

这里我用到了归纳假定

$$\int_{\delta\geqslant\theta_2\geqslant\cdots\geqslant\theta_n\geqslant-\delta}\cdots\int|\sigma_N^\alpha(\theta_2)\cdots\sigma_N^\alpha(\theta_n)|^{n-1}|D(e^{i\theta_2},\cdots,e^{i\theta_n})|^2d\theta_2\cdots d\theta_n$$

是有界的．由于 $\alpha>\frac{n-1}{n}$, $0\leqslant s_1\leqslant 2(n-1)$, 所以

$$-\alpha n-n+s_1<-(n-1)-n+s_1<-1,$$

因之，

$$I_{s_1\cdots s_n}\leqslant aN^{-\alpha n+n-1}\delta^{-\alpha n-n+s_1+1+s_2+\cdots+s_n}=O((N\delta)^{-\alpha n+n-1}),$$

这是因为 $s_1+s_2+\cdots+s_n=2(n-1)$.

于是我們得到

$$I_{2n}=O((N\delta)^{-\alpha n+n-1}).$$

我們再来考慮 $I_{2\overline{n-1}}$，这时候

$$I_{2\overline{n-1}}=\frac{1}{(2\pi)^n}\int_{\delta\geqslant\theta_2\geqslant\cdots\geqslant\theta_n\geqslant-\delta}\cdots\int|\sigma_N^\alpha(\theta_2)\cdots\sigma_N^\alpha(\theta_{n-1})|^n|D(e^{i\theta_2},\cdots,e^{i\theta_{n-1}})|^2d\theta_2\cdots d\theta_{n-1}$$

$$\int_\delta^\pi|\sigma_N(\theta_n)|^n\prod_{j=2}^{n-1}|e^{i\theta_j}-e^{i\theta_n}|^2d\theta_n\int_\delta^\pi|\sigma_N(\theta_1)|^n\prod_{j=2}^{n}|e^{i\theta_1}-e^{i\theta_j}|^2d\theta_1.$$

与考慮 I_{2n} 时一样

$$\prod_{j=2}^{n}|e^{i\theta_1}-e^{i\theta_j}|^2\prod_{k=2}^{n-1}|e^{i\theta_k}-e^{i\theta_n}|^2=$$

$$=\prod_{j=2}^{n}|(1-e^{i\theta_1})-(1-e^{i\theta_j})|^2\prod_{k=2}^{n-1}|(1-e^{i\theta_k})-(1-e^{i\theta_n})|^2\leqslant$$

$$\leqslant\sum_{\substack{m_1=0\\ m_1+\cdots+m_n=4n-6}}^{2n-2}\sum_{m_2=0}^{4}\cdots\sum_{m_{n-1}=0}^{4}\sum_{m_n=0}^{2n-2}b_{m_1\cdots m_n}|1-e^{i\theta_1}|^{m_1}|1-e^{i\theta_2}|^{m_2}\cdots|1-e^{i\theta_n}|^{m_n}$$

这里 $b_{m_1\cdots m_n}$ 为只与 $m_1,\cdots,m_n$ 有关的絕对常数．

但是

$$\int_\delta^\pi|1-e^{i\theta_1}|^{m_1}|\sigma_N(\theta_1)|^nd\theta_1\int_{-\pi}^{-\delta}|1-e^{i\theta_n}|^{m_n}|\sigma_N(\theta_n)|^nd\theta_n\times$$

$$\times\int_{\delta\geqslant\theta_2\geqslant\cdots\geqslant\theta_{n-1}\geqslant-\delta}\cdots\int|1-e^{i\theta_2}|^{m_2}\cdots|1-e^{i\theta_{n-1}}|^{m_{n-1}}|\sigma_N(\theta_2)|^n\cdots|\sigma_N(\theta_{n-1})|^n\times$$

$$\times|D(e^{i\theta_2},\cdots,e^{i\theta_{n-1}})|^2d\theta_2\cdots d\theta_{n-1}\leqslant$$

$$\leqslant\int_\delta^\pi N^{-\alpha n}\theta_1^{-\alpha n-n+m_1}d\theta_1\int_{-\pi}^{-\delta}N^{-\alpha n}\theta_n^{-\alpha n-n+m_n}d\theta_n\cdot\delta^{4n-6-m_1-m_n}N^{2(n-2)}=$$

$$=O(N^{-\alpha n}\delta^{-\alpha n-n+m_1+1}N^{-\alpha n}\delta^{-\alpha n-n+m_n+1}\delta^{4n-6-m_1-m_n}N^{2(n-2)})=O((N\delta)^{2(-\alpha n+n-2)})$$

这里我們用到了归納假設

$$\int_{\delta\geqslant\theta_2\geqslant\cdots\geqslant\theta_{n-1}\geqslant-\delta}\cdots\int|\sigma_N^\alpha(\theta_2)|^{n-2}\cdots|\sigma_N^\alpha(\theta_{n-1})|^{n-2}|D(e^{i\theta_2},\cdots,e^{i\theta_{n-1}})|^2d\theta_2\cdots d\theta_{n-1}$$

是有界的，所以得到

$$I_{2\overline{n-1}}=O((N\delta)^{2(-an+n-2)}).$$

同样可以証明

$$I_{2\overline{n-j}}=O((N\delta)^{(j+1)(-an+n-j-1)}).$$

以至最后得到

$$I_2=O((N\delta)^{-an+n-1})+O((N\delta)^{2(-an+n-2)})+\cdots$$
$$+\cdots+O((N\delta)^{(j+1)(-an+n-j-1)})+\cdots+O((N\delta)^{-an^2}).$$

如同处理 I_2 一样，我們同样可以証明：

$$I_3=O((N\delta)^{2(-an+n-2)})+O((N\delta)^{3(-an+n-3)})+\cdots+O((N\delta)^{-an^2}),$$

$$\cdots\cdots\cdots\cdots\cdots\cdots\cdots\cdots\cdots\cdots\cdots\cdots$$

$$I_{n+1}=O((N\delta)^{-an^2}).$$

我們現在来处理 R_1 上的积分 I_1，将 R_1 再划分成

$$s_1:\quad \delta\geqslant\theta_1\geqslant\cdots\geqslant\theta_{n-1}\geqslant\theta_n\geqslant-\delta,$$
$$s_2:\quad \delta\geqslant\theta_1\geqslant\cdots\geqslant\theta_{n-1}\geqslant-\delta\geqslant\theta_n\geqslant-\pi,$$
$$\cdots\cdots\cdots\cdots\cdots\cdots\cdots\cdots\cdots\cdots$$
$$s_{n+1}:\ -\delta\geqslant\theta_1\geqslant\cdots\geqslant\theta_n\geqslant-\pi.$$

記 (2.3.2) 的积分，其积分区域在 s_k 上者为 J_k，于是我們如同处理 I_{jk} 的情况一样，可以証明

$$J_2=O((N\delta)^{-an+n+1})+O((N\delta)^{2(-an+n-2)})+\cdots+O((N\delta)^{-an^2}),$$
$$J_3=O((N\delta)^{2(-an+n-2)})+\cdots+O((N\delta)^{-an^2}),$$
$$\cdots\cdots\cdots\cdots\cdots\cdots\cdots\cdots\cdots\cdots\cdots\cdots$$
$$J_{n+1}=O((N\delta)^{-an^2}).$$

最后我們来考察 J_1，

$$J_1=\frac{1}{(2\pi)^n}\int\cdots\int_{\delta\geqslant\theta_1\geqslant\cdots\geqslant\theta_n\geqslant-\delta}|\sigma_N^a(\theta_1)\cdots\sigma_N^a(\theta_n)|^n|D(e^{i\theta_1},\cdots,e^{i\theta_n})|^2d\theta_1\cdots d\theta_n\leqslant$$
$$\leqslant\int\cdots\int_{\delta\geqslant\theta_2\geqslant\cdots\geqslant\theta_n\geqslant-\delta}|\sigma_N^a(\theta_2)\cdots\sigma_N^a(\theta_n)|^n\ D(e^{i\theta_2},\cdots,e^{i\theta_n})|^2d\theta_2\cdots d\theta_n\times$$
$$\times\int_{\theta_2}^{\delta}|\sigma_N^a(\theta_1)|^n\prod_{j=2}^{n}|e^{i\theta_1}-e^{i\theta_j}|^2d\theta_1\leqslant$$
$$\leqslant a(N\delta)^{2(n-1)}\int\cdots\int_{\delta\geqslant\theta_2\geqslant\cdots\geqslant\theta_n\geqslant-\delta}|\sigma_N^a(\theta_2)\cdots\sigma_N^a(\theta_n)|^{n-1}|D(e^{i\theta_2},\cdots,e^{i\theta_n})|^2d\theta_2\cdots d\theta_n$$

由归納假設，上式为：

$$O((N\delta)^{2(n-1)}).$$

总結起来

$$I=O((N\delta)^{2(n-1)})+O((N\delta)^{-an+n-1})+O((N\delta)^{2(-an+n-2)})+\cdots+O((N\delta)^{-an^2}).$$

現在我們取 $\delta=\dfrac{1}{N}$，即得

$$I=O(1),$$

此即証明了 (2.3.1).

§2.4. Riesz 型定理的証明

现在我們来証明定理 2.3.1.

由 §2.2 知道

$$\frac{1}{C}\int_{U_n} \mathrm{K}_N^\alpha(V)\dot{V} = 1. \tag{2.4.1}$$

并且在 §2.2 中已經証明了 $u(U)$ 的 Fourier 級数 (2.1.1) 的 Cesàro (c, α) 和 (2.2.1) $\Sigma_N^\alpha(U)$ 可以写成

$$\frac{1}{C}\int_{U_n} u(VU)\mathrm{K}_N^\alpha(V)\dot{V}.$$

因之,从 (2.4.1) 得到

$$\Sigma_N^\alpha(U) - u(U) = \frac{1}{C}\int_{U_n} (u(VU) - u(U))\mathrm{K}_N^\alpha(V)\dot{V} \tag{2.4.2}$$

我們要証:当 N 趋于无穷时,

$$\Sigma_N^\alpha(U) \to u(U);$$

只要証明:当 N 趋于无穷时

$$\int_{U_n} (u(VU) - u(U))\mathrm{K}_N^\alpha(V)\dot{V} \tag{2.4.3}$$

趋于零即可. 而 (2.4.3) 即是

$$I = \frac{1}{(2\pi)^n B_N^\alpha} \int\cdots\int_{\pi \geqslant \theta_1 \geqslant \cdots \geqslant \theta_n \geqslant -\pi} \varphi(e^{i\theta_1}, \cdots, e^{i\theta_n})(\sigma_N^\alpha(\theta_1)\cdots\sigma_N^\alpha(\theta_n))^n \times$$
$$\times |D(e^{i\theta_1}, \cdots, e^{i\theta_n})|^2 d\theta_1 \cdots d\theta_n \tag{2.4.4}$$

这里

$$\varphi(e^{i\theta_1}, \cdots, e^{i\theta_n}) = \frac{1}{\omega_n'}\int_{[U_n]} (u(VU) - u(U))[\dot{V}]. \tag{2.4.5}$$

如同証明定理 2.3.2 一样,将积分区域

$$\pi \geqslant \theta_1 \geqslant \cdots \geqslant \theta_n \geqslant -\pi$$

分解为 $R_1, R_2, \cdots, R_{n+1}$.

記 (2.4.4) 的积分,其积分区域在 R_i 上者为 I_i'. 由于 $u(U)$ 在 U_n 上連續, 以至在 U_n 上,我們有

$$|\varphi(e^{i\theta_1}, \cdots, e^{i\theta_n})| \leqslant L$$

这里 L 是一个絕对常数. 因之,当我們选取 $\delta \geqslant \frac{1}{N}$ 时,如同在 §2.3 中处理 $I_2, \cdots, I_{n+1}$ 时一样,我們来处理 $I_2', \cdots, I_{n+1}'$.

我們来考慮 I_2'. 如同上节中处理 I_2 一样,将 R_2 分解为 $R_{2n}, R_{2\,\overline{n-1}}, \cdots, R_{22}, R_{21}$. 記 (2.4.4) 的积分,其积分区域在 R_{ij} 上者为 I_{ij}'. 我們来研究 I_{2n}'.

$$I_{2n}' = \frac{1}{B_N^\alpha(2\pi)^n}\int\cdots\int_{R_{2n}} \varphi(e^{i\theta_1}, \cdots, e^{i\theta_n})(\sigma_N^\alpha(\theta_1)\cdots\sigma_N^\alpha(\theta_n))^n \times$$
$$\times |D(e^{i\theta_1}, \cdots, e^{i\theta_n})|^2 d\theta_1 \cdots d\theta_n. \tag{2.4.6}$$

从 (2.4.6) 可得

$$|I'_{2n}| \leqslant \frac{L}{|B_N^\alpha|(2\pi)^n} \int\cdots\int_{\delta\geqslant\theta_2\geqslant\cdots\geqslant\theta_n\geqslant-\delta} |\sigma_N^\alpha(\theta_2)|^n\cdots|\sigma_N^\alpha(\theta_i)|^n\times$$

$$\times|D(e^{i\theta_2},\cdots,e^{i\theta_n})|^2d\theta_2\cdots d\theta_n\int_\delta^\pi|\sigma_N^\alpha(\theta_1)|^n\prod_{j=2}^n|e^{i\theta_1}-e^{i\theta_j}|^2d\theta_1.$$

如同上节一样,我們考虑 $I_{s_1,\cdots,s_n}$, 这里 $I_{s_1\cdots s_n}$ 由 (2.3.3) 所定义,利用 (2.3.4) 得到

$$I_{s_1\cdots s_n}\leqslant a\int_\delta^\pi N^{-\alpha n}\theta_1^{-\alpha n-n+s_1}d\theta_1 N^{n-1}\delta^{s_2+\cdots+s_n}\times$$

$$\times\int\cdots\int_{\delta\geqslant\theta_2\geqslant\cdots\geqslant\theta_n\geqslant-\delta}|\sigma_N^\alpha(\theta_2)\cdots\sigma_N^\alpha(\theta_n)|^{n-1}|D(e^{i\theta_2},\cdots,e^{i\theta_n})|^2d\theta_2\cdots d\theta_n.$$

由定理 2.3.2, 我們知道

$$I_{s_1\cdots s_n}=O((N\delta)^{-\alpha n+n-1}).$$

因之,我們得到

$$|I'_{2n}|=O((N\delta)^{-\alpha n+n-1})+O((N\delta)^{2(-\alpha n+n-2)})+\cdots+O((N\delta)^{-\alpha n^2}).$$

同样可以証明

$$|I'_{2\,\overline{n-1}}|=O((N\delta)^{2(-\alpha n+n-2)})+\cdots+O((N\delta)^{-\alpha n^2}).$$

因之

$$|I'_2|=O((N\delta)^{-\alpha n+n-1})+\cdots+O((N\delta)^{-\alpha n^2}).$$

与以上一样可以証明

$$|I'_3|=O((N\delta)^{2(-\alpha n+n-1)})+\cdots+O((N\delta)^{-\alpha n^2}),$$

$$\cdots\cdots\cdots\cdots\cdots\cdots\cdots\cdots\cdots\cdots$$

$$|I'_{n+1}|=O((N\delta)^{-\alpha n^2}).$$

因之

$$|I'_2|=O((N\delta)^{-\alpha n+n-1})+\cdots+O((N\delta)^{-\alpha n^2}).$$

同样的方法,可以証明

$$|I'_3|=O((N\delta)^{2(-\alpha n+n-2)})+\cdots+O((N\delta)^{-\alpha n^2}),$$

$$\cdots\cdots\cdots\cdots\cdots\cdots\cdots\cdots\cdots\cdots$$

$$|I'_{n+1}|=O((N\delta)^{-\alpha n^2}).$$

因之,

$$|I'_2|+\cdots+|I'_{n+1}|=O((N\delta)^{-\alpha n+n-1})+\cdots$$

$$\cdots+O((N\delta)^{i(-\alpha n+n-i)})+\cdots+O((N\delta)^{-\alpha n^2}).$$

现在来考虑 I'_1. 如同在 § 2.3 中处理 I_1 一样先将 R_1 分解为 $s_1,\cdots,s_{n+1}$. 并且記 (2.4.4) 的积分,其积分区域在 s_k 上者为 J'_k. 如同上述的証明一样,可以得到

$$|J'_2|+\cdots+|J'_{n+1}|=O((N\delta)^{-\alpha n+n-1})+O((N\delta)^{2(-\alpha n+n-2)})+\cdots+O((N\delta)^{-\alpha n^2}).$$

最后来考虑 J'_1,

$$J'_1=\frac{1}{(2\pi)^nB_N^\alpha}\int\cdots\int_{\delta\geqslant\theta_1\geqslant\cdots\geqslant\theta_n\geqslant-\delta}\varphi(e^{i\theta_1},\cdots,e^{i\theta_n})(\sigma_N^\alpha(\theta_1)\cdots\sigma_N^\alpha(\theta_n))^n\times$$

$$\times|D(e^{i\theta_1},\cdots,e^{i\theta_n})|^2d\theta_1\cdots d\theta_n.$$

由于 $u(U)$ 在 U_n 上連續，以至对于任意給出的 $\eta > 0$，我們一定可以选取 δ 充分小，使得当

$$\delta \geqslant \theta_1 \geqslant \cdots \geqslant \theta_n \geqslant -\delta$$

时

$$|\varphi(e^{i\theta_1}, \cdots, e^{i\theta_n})| < \eta.$$

因之，由定理 2.3.2，当 $\alpha > \dfrac{n-1}{n}$ 时，

$$\frac{1}{C}\int_{U_n} |K_N^\alpha(V)|\dot{V} < M$$

以至对于任意 $\varepsilon > 0$，我們可以选取 δ，使得

$$|J_1'| < \varepsilon/2.$$

因而对于任給的 $\varepsilon > 0$，选定了 $\delta = \delta(\varepsilon)$ 之后，由于 $\alpha n - n + 1 > 0$，因之，我們总可以选取 N 充分大，使

$$|I_2| + \cdots + |I_{n+1}| + |J_2| + \cdots + |J_{n+1}| < \varepsilon/2.$$

总結起来，于是得到，对于任意已給的 $\varepsilon > 0$，我們有

$$|I| < \varepsilon,$$

这就証明了定理 2.3.1.

§ 2.5. Fejér 求和

为了对 (c, α) 求和有更直观的了解，我們对其中最簡单而最重要的 Fejér 求和进行仔細的研究．特別是 (c, α) 求和中的系数及积分常数都可以参照以下 Fejér 求和中的系数及积分常数把它定出来．

从 § 2.2 中的定义容易看出：Fejér 核为

$$\frac{1}{B_N(N+1)^{n^2}}\left|\frac{\det(I - V^{N+1})}{\det(I - V)}\right|^{2n}, \tag{2.5.1}$$

这里

$$B_N = B_N' = \frac{1}{C(N+1)^{n^2}}\int_{U_n}\left|\frac{\det(I - V^{N+1})}{\det(I - V)}\right|^{2n}\dot{V}. \tag{2.5.2}$$

特別有兴趣的是：Fejér 核也是定正的．而且具有如此簡洁的形式．并且从此推出：当 $\alpha \geqslant 1$ 时，所有的 (c, α) 核都是定正的．

从定义知道，Fejér 求和的系数为

$$B_{f_1\cdots f_n} = B_{f_1\cdots f_n}^1 = \frac{1}{B_N N(f)(N+1)^{n^2}C}\int_{U_n} x_{f_1\cdots f_n}(\bar{V})\left|\frac{\det(I - V^{N+1})}{\det(I - V)}\right|^{2n}\dot{V}, \tag{2.5.2}$$

而 $u(U)$ 的 Fourier 級数 (2.1.1) 的 Fejér 和成为

$$\sum_{nN\geqslant f_1\geqslant\cdots\geqslant f_n\geqslant -nN} B_{f_1\cdots f_n}\operatorname{tr}(C_{f_1\cdots f_n}\Phi_{f_1\cdots f_n}'(U)) =$$

$$= \frac{1}{B_N(N+1)^{n^2}C}\int_{U_n} u(VU)\left|\frac{\det(I - V^{N+1})}{\det(I - V)}\right|^{2n}\dot{V} \tag{2.5.3}$$

Riesz 型定理（定理 2.3.1）成为下述的 Fejér 型定理：若 $u(U)$ 在 U_n 上連續，那末 $u(U)$ 的

Fourier 級数 (2.1.1) 可以 Fejér 求和于它自己.

我們利用 Fejér 核的定正性可以比証明定理 2.3.1 簡便得多的直接証明上述的 Fejér 型定理,而不必事先証明定理 2.3.2. 这里从略.

§2.6. 系数的具体表达式

从这一节开始,我們将应用复杂而精巧的矩陣积分的技巧来給出 Fejér 求和中的系数及积分常数的具体表达式. 通过这个計算过程,我們可以看出如何来写出一般的 $B^{\alpha}_{l_1\cdots l_n}$ 的具体表达式及 B^{α}_N 的值.

由 (2.5.2) 及

$$D(e^{-i\theta_1}, \cdots, e^{-i\theta_n}) = (-1)^{\frac{1}{2}n(n-1)}D(1-e^{-i\theta_1}, \cdots, 1-e^{-i\theta_n})$$

出发,应用我們在 [3] 中常用的技巧,得到(参閱 §1.4—§1.6)

$$B_{l_1\cdots l_n}N(f) = (N+1)^{-n^2}(-1)^{\frac{1}{2}n(n-1)}B_N^{-1}\times$$

$$\times\begin{vmatrix} \frac{1}{2\pi}\int_0^{2\pi}\frac{e^{il_1\theta}e^{inN\theta}(1-e^{-i(N+1)\theta})^{2n}d\theta}{(1-e^{-i\theta})^{2n}}, & \cdots, & \frac{1}{2\pi}\int_0^{2\pi}\frac{e^{il_n\theta}e^{inN\theta}(1-e^{-i(N+1)\theta})^{2n}d\theta}{(1-e^{-i\theta})^{2n}} \\ \frac{1}{2\pi}\int_0^{2\pi}\frac{e^{il_1\theta}e^{inN\theta}(1-e^{-i(N+1)\theta})^{2n}d\theta}{(1-e^{-i\theta})^{2n-1}}, & \cdots, & \frac{1}{2\pi}\int_0^{2\pi}\frac{e^{il_n\theta}e^{inN\theta}(1-e^{-i(N+1)\theta})^{2n}d\theta}{(1-e^{-i\theta})^{2n-1}} \\ \cdots & \cdots & \cdots \\ \frac{1}{2\pi}\int_0^{2\pi}\frac{e^{il_1\theta}e^{inN\theta}(1-e^{-i(N+1)\theta})^{2n}d\theta}{(1-e^{-i\theta})^{n+1}}, & \cdots, & \frac{1}{2\pi}\int_0^{2\pi}\frac{e^{il_n\theta}e^{inN\theta}(1-e^{-i(N+1)\theta})^{2n}d\theta}{(1-e^{-i\theta})^{n+1}} \end{vmatrix}. \tag{2.6.1}$$

記

$$a_p^q = \frac{1}{2\pi}\int_0^{2\pi}\frac{e^{iq\theta}(1-e^{-i(N+1)\theta})^{2n}}{(1-e^{-i\theta})^p}, \tag{2.6.2}$$

那末 (2.6.1) 可以写成

$$B_{l_1\cdots l_n}N(f) = (N+1)^{-n^2}(-1)^{\frac{1}{2}n(n-1)}B_N^{-1}\times$$

$$\times\begin{vmatrix} a_{2n}^{l_1+nN}, & a_{2n}^{l_2+nN}, & \cdots, & a_{2n}^{l_n+nN} \\ a_{2n-1}^{l_1+nN}, & a_{2n-1}^{l_2+nN}, & \cdots, & a_{2n-1}^{l_n+nN} \\ \cdots & \cdots & \cdots & \cdots \\ a_{n+1}^{l_1+nN}, & a_{n+1}^{l_2+nN}, & \cdots, & a_{n+1}^{l_n+nN} \end{vmatrix}. \tag{2.6.3}$$

引理 2.6.1. 当 $p \leqslant 2n$ 时

$$a_p^q = \frac{(2n)!}{(p-1)!}\sum_{k=0}^{\infty}\sum_{\substack{s=0 \\ k+(N+1)s=q}}^{2n}\frac{(p+k-1)!}{k!\,s!\,(2n-s)!}. \tag{2.6.4}$$

証. 取 $z = re^{-i\theta}$, $0<r<1$, 那末

$$a_p^q = \frac{1}{2\pi}\int_0^{2\pi}\frac{e^{iq\theta}(1-e^{-i(N+1)\theta})^{2n}}{(1-e^{-i\theta})^p}d\theta,$$

由于 $p \leqslant 2n$, 应用 Lebesgue 从属定理,上式等于

$$\lim_{r\to 1}\frac{1}{2\pi}\int_0^{2\pi}\frac{e^{iq\theta}(1-z^{N+1})^{2n}}{(1-z)^p}d\theta, \tag{2.6.5}$$

但

$$(1-z)^{-p}=\sum_{k=0}^{\infty}\frac{(p+k-1)!}{k!(p-1)!}z^{k}$$

及

$$(1-z^{N+1})^{2n}=\sum_{s=0}^{2n}\frac{(2n)!}{s!(2n-s)!}z^{(N+1)s};$$

所以

$$\frac{(1-z^{N+1})^{2n}}{(1-z)^{p}}=\sum_{k=0}^{\infty}\sum_{s=0}^{2n}\frac{(p+k-1)!(2n)!}{k!(p-1)!s!(2n-s)!}z^{k+(N+1)s}.$$

我們从 (2.6.2) 及 (2.6.5) 立刻得到 (2.3.4).

a_p^q 还可以写成

$$\frac{(2n)!}{(p-1)!}\sum_{\substack{s=0\\q-(N+1)s\geq 0}}^{2n}\frac{(p+q-(N+1)s-1)!}{(q-(N+1)s)!\,s!(2n-s)!}.$$

根据引理 2.6.1, 我們得到 (2.6.3) 就是

$$B_{l_1\cdots l_n}N(f)=\frac{((2n)!)^{n}(-1)^{\frac{1}{2}n(n-1)}}{(2n-1)!(2n-2)!\cdots n!(N+1)^{n^2}B_N}\times$$

$$\times\begin{vmatrix}\sum\limits_{\substack{s_1=0\\k_1\geq0}}^{2n}\frac{(2n+k_1-1)!}{k_1!s_1!(2n-s_1)!}, & \cdots, & \sum\limits_{\substack{s_n=0\\k_n\geq0}}^{2n}\frac{(2n+k_n-1)!}{k_n!s_n!(2n-s_n)!}\\ \sum\limits_{\substack{s_1=0\\k_1\geq0}}^{2n}\frac{(2n-1+k_1-1)!}{k_1!s_1!(2n-s_1)!}, & \cdots, & \sum\limits_{\substack{s_n=0\\k_n\geq0}}^{2n}\frac{(2n-1+k_n-1)!}{k_n!s_n!(2n-s_n)!}\\ \cdots\cdots & \cdots\cdots & \cdots\cdots\\ \sum\limits_{\substack{s_1=0\\k_1\geq0}}^{2n}\frac{(n+1+k_1-1)!}{k_1!s_1!(2n-s_1)!}, & \cdots, & \sum\limits_{\substack{s_n=0\\k_n\geq0}}^{2n}\frac{(n+1+k_n-1)!}{k_n!s_n!(2n-s_n)!}\end{vmatrix}. \qquad (2.6.6)$$

这里

$$k_1=l_1+nN-(N+1)s_1,\ \cdots,\ k_n=l_n+nN-(N+1)s_n.$$

簡化 (2.6.6), 它的右边等于

$$\frac{((2n)!)^{n}(-1)^{\frac{1}{2}n(n+1)}}{(2n-1)!(2n-2)!\cdots n!(N+1)^{n^2}B_N}\sum_{\substack{s_1=0\\k_1\geq0}}^{2n}\cdots\sum_{\substack{s_n=0\\k_n\geq0}}^{2n}\frac{1}{k_1!\,s_1!(2n-s_1)!\cdots k_n!\,s_n!(2n-s_n)!}$$

$$\begin{vmatrix}(2n-1+k_1)!\cdots(2n-1+k_n)!\\ (2n-2+k_1)!\cdots(2n-2+k_n)!\\ \cdots\cdots\cdots\\ (n+k_1)!\quad\cdots\quad(n+k_n)!\end{vmatrix}=$$

$$=\frac{(n!)^{n}(-1)^{\frac{1}{2}n(n-1)}}{(2n-1)!\cdots n!\,(N+1)^{n^2}B_N}\sum_{\substack{s_1=0\\k_1\geq0}}^{2n}\cdots\sum_{\substack{s_n=0\\k_n\geq0}}^{2n}C_{s_1}^{2n}\,C_{n}^{n+k_1}\cdots C_{s_n}^{2n}C_{n}^{n+k_n}\times$$

$$\times\begin{vmatrix}(2n-1+k_1)\cdots(n+1+k_1), & \cdots, & (2n-1+k_n)\cdots(n+1+k_n)\\ \cdots\cdots & \cdots & \cdots\cdots\\ (n+2+k_1)(n+1+k_1), & \cdots, & (n+2+k_n)(n+1+k_n)\\ n+1+k_1 & ,\cdots, & n+1+k_n\\ 1 & ,\cdots, & 1\end{vmatrix}$$

上式右边的行列式等于

$$\lim_{x\to 1}(-1)^{\frac{1}{2}n(n-1)}\begin{vmatrix}(x^{-(n+1+k_1)})^{(n-1)}, & \cdots, & (x^{-(n+1+k_n)})^{n-1}\\ \cdots\cdots & \cdots & \cdots\cdots\\ (x^{-(n+1+k_1)})', & \cdots, & (x^{-(n+1+k_n)})'\\ x^{-(n+1+k_1)}, & \cdots, & x^{-(n+1+k_n)}\end{vmatrix}$$

根据 [1] 中的一条定理，它等于

$$\lim_{\substack{x_1\to 1\\ \cdots\\ x_n\to 1}}\frac{(n-1)!\cdots 2!1!}{D(x_1,\cdots,x_n)}\begin{vmatrix}x_1^{-(n+1+k_1)}, & \cdots, & x_n^{-(n+1+k_1)}\\ x_1^{-(n+1+k_2)}, & \cdots, & x_n^{-(n+1+k_2)}\\ \cdots\cdots & \cdots & \cdots\cdots\\ x_1^{-(n+1+k_n)}, & \cdots, & x_n^{-(n+1+k_n)}\end{vmatrix}=$$

$$=\lim_{\substack{x_1\to 1\\ \cdots\\ x_n\to 1}}\frac{(n-1)!\cdots 2!1!}{D(x_1,\cdots,x_n)}\begin{vmatrix}x_1^{(N+1)s_1-l_1}, & \cdots, & x_n^{(N+1)s_1-l_1}\\ x_1^{(N+1)s_2-l_2}, & \cdots, & x_n^{(N+1)s_2-l_2}\\ \cdots\cdots & \cdots & \cdots\cdots\\ x_1^{(N+1)s_n-l_n}, & \cdots, & x_n^{(N+1)s_n-l_n}\end{vmatrix}.$$

根据 $N(g_1,\cdots,g_n)$ 的定义，上式右边就是

$$(-1)^{\frac{1}{2}n(n-1)}(n-1)!\cdots 2!1!N((N+1)s_1-l_1,(N+1)s_2-$$
$$-l_2-1,\cdots,(N+1)s_n-l_n-n+1)=(-1)^{\frac{1}{2}n(n-1)}(n-1)!\cdots 2!1!N((N+1)s_1-$$
$$-f_1-n+1,(N+1)s_2-f_2-n+1,\cdots,(N+1)s_n-f_n-n+1)=$$
$$=(-1)^{\frac{1}{2}n(n-1)}(n-1)!\cdots 2!1!N((N+1)s_1-f_1,(N+1)s_2-f_2,\cdots,(N+1)s_n-f_n).$$

最后，我們得到

$$B_{f_1\cdots f_n}N(f)=\frac{(n!)^n(n-1)!\cdots 2!1!}{(N+1)^{n^2}(2n-1)!\cdots n!A}\sum_{\substack{s_1=0\\ k_1\geqslant 0}}^{2n}\cdots\sum_{\substack{s_n=0\\ k_n\geqslant 0}}^{2n}C_{s_1}^{2n}\cdots C_n^{n+k_1}\cdots$$
$$\cdots C_{s_n}^{2n}C_n^{n+k_n}N((N+1)s_1-f_1,\cdots,(N+1)s_n-f_n)$$

这里

$$k_j=l_j+nN-(N+1)s_j\quad j=1,2,\cdots,n.$$

§2.7. 积分常数的計算

从 $B_{f_1\cdots f_n}$ 的定义知道，$B_{0,\cdots,0}=1$，所以

$$B_N=\frac{(n!)^n(n-1)!\cdots 2!1!}{(N+1)^{\frac{1}{2}n(n+1)}(2n-1)\cdots n!}\sum_{\substack{s_1=0\\ k_1\geqslant 0}}^{2n}\cdots\sum_{\substack{s_n=0\\ k_n\geqslant 0}}^{2n}C_{s_1}^{2n}C_n^{n+k_1}\cdots$$
$$\cdots C_{s_n}^{2n}C_n^{n+k_n}N(s_1,\cdots,s_n).\tag{2.7.1}$$

这里

$$k_j = n - j + nN - (N+1)s_j = (N+1)(n - s_j) - j$$

而 $j = 1, 2, \cdots, n$.

从 k_j 的定义知道，当 $N \geqslant n-1$ 时，要 $k_j \geqslant 0$，必须 $s_j = 0, 1, \cdots, n-1$；再应用我們以往慣用的方法 (2.7.1) 成为

$$B_N = \frac{(n!)^n(n-1)!\cdots 2!1!}{(N+1)^{\frac{1}{2}n(n+1)}(2n-1)!\cdots n!} \times$$

$$\times \sum_{n-1 \geqslant s_1 \geqslant \cdots \geqslant s_n \geqslant 0} \begin{vmatrix} A_{s_1}^1 A_{s_1}^2 \cdots A_{s_1}^n \\ A_{s_2}^1 A_{s_2}^2 \cdots A_{s_2}^n \\ \cdots\cdots\cdots\cdots \\ A_{s_n}^1 A_{s_n}^2 \cdots A_{s_n}^n \end{vmatrix} N(s_1, \cdots, s_n), \qquad (2.7.2)$$

这里

$$A_{s_j}^k = C_{s_j}^{2n} C_n^{n+(N+1)(n-s_j)-k}.$$

在 (2.7.2) 的行列式中，有二个 s_l 与 $s_{l'}$ 是相同的时候，行列式为零，但 $n-1 \geqslant s_1 \geqslant s_2 \geqslant \geqslant \cdots \geqslant s_n \geqslant 0$，所以只有 $s_1 = n-1$, $s_2 = n-2$, $\cdots$, $s_n = 0$ 时，行列式不为零. 于是 (2.7.2) 成为

$$B_N = \frac{(n!)^n(n-1)!\cdots 2!1!}{(N+1)^{\frac{1}{2}n(n+1)}(2n-1)!\cdots n!} N((n-1), \cdots, 1, 0) \begin{vmatrix} A_{n-1}^1, & A_{n-1}^2, & \cdots, & A_{n-1}^n \\ A_{n-2}^1, & A_{n-2}^2, & \cdots, & A_{n-2}^n \\ \cdots & \cdots & \cdots & \cdots \\ A_0^1, & A_0^2, & \cdots, & A_0^n \end{vmatrix},$$

而 $N(n-1, \cdots, 1, 0) = 2^{\frac{1}{2}n(n-1)}$，所以

$$B_N = \frac{(n!)^n(n-1)!\cdots 2!1!((2n)!)^n 2^{\frac{1}{2}n(n-1)}}{(N+1)^{\frac{1}{2}n(n+1)}(2n-1)!\cdots n!(2n)!\cdots(n+1)!(n-1)!\cdots 1!} \times$$

$$\times \begin{vmatrix} C_n^{n+(N+1)-1}, & C_n^{n+(N+1)-2}, & \cdots, & C_n^{n+(N+1)-n} \\ \cdots & \cdots & \cdots & \cdots \\ C_n^{n+(N+1)(n-1)-1}, & C_n^{n+(N+1)(n-1)-2}, & \cdots, & C_n^{n+(N+1)(n-1)-n} \\ C_n^{n+(N+1)n-1}, & C_n^{n+(N+1)n-2}, & \cdots, & C_n^{n+(N+1)n-n} \end{vmatrix}. \qquad (2.7.3)$$

記 (2.7.3) 的右边的行列式为 Δ，那末 (2.7.3) 就是

$$B_N = \frac{(n!)^{n-1}((2n)!)^{n-1}2^{\frac{1}{2}n(n-1)}\Delta}{(N+1)^{\frac{1}{2}n(n+1)}((2n-1)!\cdots(n+1)!)^2}. \qquad (2.7.4)$$

我們来对 Δ 进行計算，由于

$$\sum_{s=0}^{p} C_t^{t+s} = C_{t+1}^{t+p+1},$$

所以 (2.7.4) 成为

$$\begin{vmatrix} \sum_{k=0}^{(N+1)-1} C_{n-1}^{n-1+k}, & \sum_{k=0}^{(N+1)-2} C_{n-1}^{n-1+k}, & \cdots, & \sum_{k=0}^{(N+1)-n} C_{n-1}^{n-1+k} \\ \cdots & \cdots & \cdots & \cdots \\ \sum_{k=0}^{n(N+1)-1} C_{n-1}^{n-1+k}, & \sum_{k=0}^{n(N+1)-2} C_{n-1}^{n-1+k}, & \cdots, & \sum_{k=0}^{n(N+1)-n} C_{n-1}^{n-1+k} \end{vmatrix}.$$

相邻行相减，得到

$$\begin{vmatrix} C_{n-1}^{n-1+(N+1)-1}, & C_{n-1}^{n-1+(N+1)-2}, & \cdots, & C_{n-1}^{n-1+(N+1)-n+1}, & C_n^{N+1} \\ \cdots & \cdots & \cdots & \cdots & \cdots \\ C_{n-1}^{n-1+n(N+1)-1}, & C_{n-1}^{n-1+n(N+1)-2}, & \cdots, & C_{n-1}^{n-1+n(N+1)-n+1}, & C_n^{(N+1)n} \end{vmatrix}$$

继續进行这个步驟，最后得到

$$\Delta = \begin{vmatrix} C_1^{N+1}, & \cdots, & C_{n-1}^{N+1}, & C_n^{N+1} \\ \cdots & \cdots & \cdots & \cdots \\ C_1^{(n-1)(N+1)}, & \cdots, & C_{n-1}^{(n-1)(N+1)}, & C_n^{(n-1)(N+1)} \\ C_1^{n(N+1)}, & \cdots, & C_{n-1}^{n(N+1)}, & C_n^{n(N+1)} \end{vmatrix}.$$

应用以往慣用的技巧，Δ 等于

$$\frac{n!(N+1)^n}{n!(n-1)!\cdots 1!}\lim_{x\to 1}\begin{vmatrix} x^{(N+1)-1}, & \cdots, & (x^{(N+1)-1})^{(n-1)} \\ \cdots & \cdots & \cdots \\ x^{(n-1)(N+1)-1}, & \cdots, & (x^{(n-1)(N+1)-1})^{(n-1)} \\ x^{n(N+1)-1}, & \cdots, & (x^{n(N+1)-1})^{(n-1)} \end{vmatrix} =$$

$$= \frac{n!(N+1)^n(n-1)!\cdots 1!}{n!(n-1)!\cdots 1!}\lim_{\substack{x_1\to 1\\ \cdots \\ x_n\to 1}}\begin{vmatrix} x_1^{(N+1)-1}, & \cdots, & (x_1^{(N+1)-1})^{(n-1)} \\ \cdots & \cdots & \cdots \\ x_1^{(n-1)(N+1)-1}, & \cdots, & (x_1^{(n-1)(N+1)-1})^{(n-1)} \\ x_1^{n(N+1)-1}, & \cdots, & (x_1^{n(N+1)-1})^{(n-1)} \end{vmatrix} \Big/ D(x_1,\cdots,x_n) =$$

$$= (N+1)^n N(nN, (n-1)N, \cdots, N) = (N+1)^n N^{\frac{1}{2}n(n-1)} 2^{\frac{1}{2}n(n-1)}. \tag{2.7.5}$$

于是最后从 (2.7.4) 得到

定理 2.7.1. 当 $N \geqslant n-1$ 时，我們有

$$B_N = \frac{(n!)^{n-1}((2n)!)^{n-1}2^{n(n-1)}}{((2n-1)!\cdots(n+1)!)^2}\left(1-\frac{1}{N+1}\right)^{\frac{1}{2}n(n-1)}. \tag{2.7.6}$$

由上述定理及 (2.6.7)，我們有

定理 2.7.2. 当 $N \geqslant n-1$ 时，我們有

$$B_{f_1\cdots f_n} = \frac{(2n)!\cdots 2!1!}{N(f)((2n)!)^n n! 2^{n(n-1)}(N+1)^{\frac{3}{2}n(n+1)}N^{\frac{1}{2}n(n-1)}}\sum_{\substack{s_1=0\\ k_1\geqslant 0}}^{2n}\cdots$$

$$\cdots\sum_{\substack{s_n=0\\ k_n\geqslant 0}}^{2n} C_{s_1}^{2n}C_n^{n+k_1}\cdots C_{s_n}^{2n}C_n^{n+k_n}N((N+1)s_1-f_1,\cdots,(N+1)s_n-f_n) \tag{2.7.7}$$

从 $N(f)$ 的定义及定理 2.7.2 还可以看出：$B_{f_1\cdots f_n} = 1 + O\left(\frac{1}{N}\right)$.

§2.8. 几点注記

(1) 本文中的 Rièsz 型定理(定理 2.3.1)的条件："$u(U)$ 在酉羣 U_n 上連續" 可以減輕为 "$u(U)$ 在 U_n 上可积且有界"，那末 $\sum_N^n(U)$ 在 $u(U)$ 連續的那些点上当 N 趋于无穷时，还是趋于 $u(U)$，即，Rièsz 求和是局部性质.

(2) 华罗庚在[2]中定义了 Abel 求和，且証明了若 $u(U)$ 在 U_n 上連續，则可以 Abel

求和于它自己，从而得到了一条收斂定理．当 r 充分靠近 1 时，我們可以将 Abel 和

$$\sum_{N \geq l_1 \geq \cdots \geq l_n \geq -N} \rho^l(r) \operatorname{tr}\left(C_{l_1 \cdots l_n} \Phi'_{l_1 \cdots l_n}(U)\right)$$

作为 $u(U)$ 的近似值．但在这近似值中含参数 r．本文中的 Rièsz 定理是又一次的得到了一条收斂定理且作为 $u(U)$ 的近似值的 (c, α) 和 (2.2.1) 中是不含有 r 的．

(3) Rièsz 型定理(定理 2.3.1)中的条件 $\alpha > \frac{n-1}{n}$ 是否可以再往前推进？当 $n = 1$ 时，这个結果众所周知的已不能推进，当 $n > 1$ 时能否前进，这是一个值得研究的問題．

(4) 从 Rièsz 型定理（定理 2.3.1）的証明过程中，我們可以得出一条更一般的收斂定理，为了本文篇幅不至于过长，拟在以后的文章中証述之．并将它用之于紧致羣上的逼近理論．

参　考　文　献

[1] 华罗庚，多复变数函数論中的典型域的調和分析，科学出版社，北京，1958．
[2] 华罗庚，紧致羣上的連續函数所成的空間中的一条收斂定理，科学記录，**2** (1958)，341—344．
[3] 龔升，酉羣上的富里埃分析 I. 富里埃級数的 Abel 求和及 Dirichlet 核，数学学报 **10** (1960)，239—261．

第13卷 第2期　　数学学报　　Vol. 13, No. 2
1963 年 5 月　　ACTA MATHEMATICA SINICA　　May, 1963

酉羣上的富里埃分析
(III) 富里埃級数的收斂判别法*

龔　昇
(中国科学技术大学)

§ 3.1. 引　言

在"酉羣上的富里埃分析 I, II"二文中，我們已經对酉羣上的富里埃級数的 Abel 求和及 Cesàro 求和，作了比較仔細的研究，并且給出了富里埃級数的 Dirichlet 核．本文的主要目的是依靠已知的 Dirichlet 核，給出一个比較簡单的收斂判别法．記在 n 阶酉羣 U_n 上定义的具有 k 阶連續微商的函数 $u(U)$ 的全体为 C^k，那末，这个判别法可以叙述为：若 $u(U)\in C^{\frac{n(n-1)}{2}}$，則其 Fourier 級数的部分和收斂于它自己，但 $n>1$．显然当 $n=1$ 时，上述結果是不对的．这是本文的主要目的．

設在酉羣 U_n 上可积函数 $u(U)$ 的 Fourier 級数为

$$\sum_{f_1\geqslant f_2\geqslant\cdots\geqslant f_n}\operatorname{tr}\left(C_{f_1\cdots f_n}\Phi'_{f_1\cdots f_n}(U)\right), \tag{3.1.1}$$

这里

$$C_{f_1\cdots f_n}=\left(C_f^{(jk)}\right)_{1\leqslant j,\,k\leqslant N(f)},\quad \Phi_{f_1\cdots f_n}(U)=\left(\varphi_f^{(jk)}\right)_{1\leqslant j,\,k\leqslant N(f)}.$$

我們称 $u(U)$ 的 Fourier 級数(3.1.1)是絕对收斂的，若級数

$$\sum_{f_1\geqslant f_2\geqslant\cdots\geqslant f_n}\sum_{j,k=1}^{N(f)}\left|C_f^{(jk)}\right|\cdot\left|\varphi_f^{(jk)}(U)\right| \tag{3.1.2}$$

是收斂的．

本文的后一部分，是給出了一个关于 $u(U)$ 的 Fourier 級数是絕对收斂的判别法．这个判别法，可以大致上陈述如下：若 $u(U)\in C^{2n(n-1)}$，且对于某一函数滿足适当的 Lipschitz 条件．这可以看作 Бернштейн 定理的在羣上推广的嘗試（參閱[4]）．在証明这个定理的过程中，我們还可以看到，可以得到一些别的关于絕对收斂的判别法，但是我們并不在此多加叙述了（例如，可以參閱 J. Musielak[5] 的工作，我們可以得到关于 Zygmund 定理，Szasz 定理在酉羣上的推广）．

当然前述的种种結果中的很多条件，特别是对于可微次数的要求，都是有減輕的可能的．

本文中所应用的符号与前二文相同，不再一一申明．

§ 3.2. Fourier 級数的部分和

我們已經知道[2]，在酉羣 U_n 上可积函数 $u(U)$ 的 Fourier 級数 (3.1.1) 的部分和

* 1961 年 9 月 29 日收到，1962 年 10 月 31 日收到修改稿．

为

$$S_N^{(n)}(U)=\sum_{N\geqslant l_1>\cdots>l_n>-N}\operatorname{tr}\left(C_{l_1\cdots l_n}\Phi'_{l_1\cdots l_n}(U)\right). \tag{3.2.1}$$

巳經証明[2] $S_N^{(n)}(U)$ 可以表为

$$\frac{1}{C}\int_{U_n}u(VU)\mathscr{D}_N(V)\dot{V}, \tag{3.2.2}$$

这里 $\mathscr{D}_N(V)$ 为 Dirichlet 核，它等于1)

$$\frac{(-i)^{\frac{n(n-1)}{2}}D\left(\frac{\partial}{\partial\theta_1},\cdots,\frac{\partial}{\partial\theta_n}\right)(d_N(\theta_1)\cdots d_N(\theta_N))}{(n-1)!\cdots 2!1!\,D(e^{i\theta_1},\cdots,e^{i\theta_n})},$$

这里 $D\left(\frac{\partial}{\partial\theta_1},\cdots,\frac{\partial}{\partial\theta_n}\right)$ 表示算子

$$(-1)^{\frac{n(n-1)}{2}}\begin{vmatrix}1, & 1, & \cdots, & 1\\ \frac{\partial}{\partial\theta_1}, & \frac{\partial}{\partial\theta_2}, & \cdots, & \frac{\partial}{\partial\theta_n}\\ \cdots & \cdots & \cdots & \cdots\\ \cdots & \cdots & \cdots & \cdots\\ \frac{\partial^{n-1}}{\partial\theta_1^{n-1}}, & \frac{\partial^{n-1}}{\partial\theta_2^{n-1}}, & \cdots, & \frac{\partial^{n-1}}{\partial\theta_n^{n-1}}\end{vmatrix}=\prod_{j<k}\left(\frac{\partial}{\partial\theta_j}-\frac{\partial}{\partial\theta_k}\right),$$

而 $e^{i\theta_1},\cdots,e^{i\theta_n}$ 为 $\bar{V}$ 的特征根，$\theta_1\geqslant\theta_2\geqslant\cdots\geqslant\theta_n$，即

$$\bar{V}=W\begin{pmatrix}e^{i\theta_1} & & \\ & \ddots & \\ & & e^{i\theta_n}\end{pmatrix}\bar{W}',$$

这里 $W\in U_n$. 作 $\Gamma(\in U_n)$，使

$$\Gamma\begin{pmatrix}e^{i\theta_1} & & \\ & \ddots & \\ & & e^{i\theta_n}\end{pmatrix}\bar{\Gamma}'=\begin{pmatrix}e^{i\theta_{\nu_1}} & & \\ & \ddots & \\ & & e^{i\theta_{\nu_n}}\end{pmatrix},$$

这里 $(\nu_1,\cdots,\nu_n)$ 为 $(1,\cdots,n)$ 的一个排列，記

$$\bar{V}^*=W\Gamma\begin{pmatrix}e^{i\theta_1} & & \\ & \ddots & \\ & & e^{i\theta_n}\end{pmatrix}\bar{\Gamma}'\bar{W}'$$

由羣上积分的定义[1]，知道

$$\frac{1}{C}\int_{U_n}u(V^*U)\mathscr{D}_N(V)\dot{V}=\frac{1}{C}\int_{U_n}u(VU)\mathscr{D}_N(V)\dot{V}.$$

1) 在[2]中，誤为

$$\frac{i^{\frac{n(n-1)}{2}}D\left(\frac{\partial}{\partial\theta_1},\cdots,\frac{\partial}{\partial\theta_n}\right)(d_N(\theta_1)\cdots d_N(\theta_n))}{(n-1)!\cdots 2!1!\,D(e^{i\theta_1},\cdots,e^{i\theta_n})}$$

作所有可能的 V^*，将所有的 $u(V^*U)$ 加起来，除以 $n!$，記之为 $u^*(VU)$，于是

$$S_N^{(n)}(U)=\frac{1}{C}\int_{U_n}u^*(VU)\mathscr{D}_N(V)\dot{V}. \tag{3.2.3}$$

記

$$g(\theta_1,\cdots,\theta_n)=\frac{1}{\omega_n'}\int_{[\dot V]}u^*(VU)[\dot V], \tag{3.2.4}$$

于是得到

$$S_N^{(n)}(U)=\frac{1}{(2\pi)^n}\underset{\pi\geqslant\theta_1\geqslant\cdots\geqslant\theta_n\geqslant-\pi}{\int\cdots\int}g(\theta_1,\cdots,\theta_n)\mathscr{D}_N(V)|D(e^{i\theta_1},\cdots,e^{i\theta_n})|^2d\theta_1\cdots d\theta_n.$$

注意到 u^* 的定义，可以看出 $g(\theta_1,\cdots,\theta_n)$ 是 $\theta_1,\cdots,\theta_n$ 的对称函数，于是应用我們慣用的技巧，得到

$$S_N^{(n)}(U)=\frac{(i)^{\frac{1}{2}n(n-1)}}{(2\pi)^n(n-1)!\cdots2!1!}\int_0^{2\pi}\cdots\int_0^{2\pi}g(\theta_1,\cdots,\theta_n)D(e^{-i\theta_1},\cdots,e^{-i\theta_n})\times$$
$$\times d_N(\theta_1)d_N'(\theta_2)\cdots d_N^{(n-1)}(\theta_n)d\theta_1\cdots d\theta_n. \tag{3.2.5}$$

所以我們有

$$S_N^{(n)}(U)-u(U)=\frac{1}{C}\int_{U_n}(u(VU)-u(U))\mathscr{D}_N(V)\dot{V}=$$
$$=\frac{(i)^{\frac{1}{2}n(n-1)}}{(2\pi)^n(n-1)!\cdots2!1!}\int_0^{2\pi}\cdots\int_0^{2\pi}(g(\theta_1,\cdots,\theta_n)-$$
$$-u(U))D(e^{-i\theta_1},\cdots,e^{-i\theta_n})d_N(\theta_1)d_N'(\theta_2)\cdots d_N^{(n-1)}(\theta_n)d\theta_1\cdots d\theta_n. \tag{3.2.6}$$

§3.3. Fourier 級数的收斂判别法

記在 U_n 上定义的具有 $\frac{n(n-1)}{2}$ 阶連續微商的函数的全体为 $C^{\frac{n(n-1)}{2}}$，若 $u(U)\in C^{\frac{n(n-1)}{2}}$，那末，应用 $\frac{1}{2}n(n-1)$ 次分部积分法及考虑到 $g(\theta_1,\cdots,\theta_n)$ 的周期性，我們有

$$S_N^{(n)}(U)-u(U)=\frac{(-i)^{\frac{1}{2}n(n-1)}}{(2\pi)^n(n-1)!\cdots2!1!}\int_0^{2\pi}\cdots\int_0^{2\pi}\frac{\partial^0}{\partial\theta_1^0}\frac{\partial^1}{\partial\theta_2^1}\cdots\frac{\partial^{n-1}}{\partial\theta_n^{n-1}}\times$$
$$\times[(g(\theta_1,\cdots,\theta_n)-u(U))D(e^{-i\theta_1},\cdots,e^{-i\theta_n})]\times$$
$$\times d_N(\theta_1)\cdots d_N(\theta_n)d\theta_1\cdots d\theta_n. \tag{3.3.1}$$

記 $I_{\mu_1\cdots\mu_n}$ 为

$$\frac{(-i)^{\frac{1}{2}n(n-1)}}{(2\pi)^n(n-1)!\cdots2!1!}\int_0^{2\pi}\cdots\int_0^{2\pi}\frac{\partial^{\mu_1}}{\partial\theta_1^{\mu_1}}\frac{\partial^{\mu_2}}{\partial\theta_2^{\mu_2}}\cdots\frac{\partial^{\mu_n}}{\partial\theta_n^{\mu_n}}(g(\theta_1,\cdots,\theta_n)-u(U))\times$$
$$\times\frac{\partial^{\nu_1}}{\partial\theta_1^{\nu_1}}\frac{\partial^{\nu_2}}{\partial\theta_2^{\nu_2}}\cdots\frac{\partial^{\nu_n}}{\partial\theta_n^{\nu_n}}D(e^{-i\theta_1},\cdots,e^{-i\theta_n})d_N(\theta_1)\cdots d_N(\theta_n)d\theta_1\cdots d\theta_n. \tag{3.3.2}$$

这里 μ_k,ν_k 为非負整数，且滿足

$$\mu_k + \nu_k = k - 1,$$

$$\mu_k \geqslant 0, \quad \nu_k \geqslant 0, \quad k = 1, 2, \cdots, n.$$

于是

$$S_N^{(n)}(U) - u(U) = \sum_{\mu_1, \cdots, \mu_n} I_{\mu_1, \cdots, \mu_n}.$$

我們来分别研究 $I_{\mu_1, \cdots, \mu_n}$. 先考慮 $I_{0, \cdots, 0}$, $I_{0, \cdots, 0}$ 等于

$$\frac{(-i)^{\frac{1}{2}n(n-1)}}{(2\pi)^n (n-1)! \cdots 2!1!} \int_0^{2\pi} \cdots \int_0^{2\pi} (g(\theta_1, \cdots, \theta_n) - u(U)) \frac{\partial^0}{\partial \theta_1^0} \frac{\partial^1}{\partial \theta_2^1} \cdots \frac{\partial^{n-1}}{\partial \theta_n^{n-1}} \times$$
$$\times D(e^{-i\theta_1}, \cdots, e^{-i\theta_n}) d_N(\theta_1) d_N(\theta_2) \cdots d_N(\theta_n) d\theta_1 \cdots d\theta_n. \tag{3.3.3}$$

将上述积分区域分解为以下一些区域:

$$R_1: \quad \delta \geqslant \theta_1 \geqslant \theta_2 \geqslant \cdots \geqslant \theta_n \geqslant -\pi;$$
$$R_2: \quad \pi \geqslant \theta_1 \geqslant \delta \geqslant \theta_2 \geqslant \cdots \geqslant \theta_n \geqslant -\pi;$$
$$R_3: \quad \pi \geqslant \theta_1 \geqslant \theta_2 \geqslant \delta \geqslant \cdots \geqslant \theta_n \geqslant -\pi;$$
$$\cdots\cdots\cdots\cdots\cdots\cdots$$
$$R_n: \quad \pi \geqslant \theta_1 \geqslant \theta_2 \geqslant \cdots \geqslant \delta \geqslant \theta_n \geqslant -\pi;$$
$$R_{n+1}: \quad \pi \geqslant \theta_1 \geqslant \theta_2 \geqslant \cdots \geqslant \theta_n \geqslant \delta.$$

記 I_j 为

$$\frac{(-i)^{\frac{1}{2}n(n-1)}}{(2\pi)^n (n-1)! \cdots 2!1!} \int \cdots \int_{R_j} (g(\theta_1, \cdots, \theta_n) - u(U)) \frac{\partial^0}{\partial \theta_1^0} \frac{\partial^1}{\partial \theta_2^1} \cdots \frac{\partial^{n-1}}{\partial \theta_n^{n-1}} \times$$
$$\times D(e^{-i\theta_1}, \cdots, e^{-i\theta_n}) d_N(\theta_1) d_N(\theta_2) \cdots d_N(\theta_n) d\theta_1 \cdots d\theta_n. \tag{3.3.4}$$

于是

$$I_{0, 0, \cdots, 0} = I_1 + I_2 + \cdots + I_{n+1}.$$

我們来考察 I_2, 由于(参閱[4])

$$\int_\delta^\pi f(t) d_N(t) dt = O\left(\frac{1}{\delta} \omega\left(\frac{1}{N}\right)\right),$$

这里 ω 是 f 的連續模;及

$$\int_{-\pi}^{\pi} |d_N(\theta)| d\theta = O(\ln N),$$

所以

$$I_2 = O\left(\frac{\omega\left(\frac{1}{N}\right) \ln^{n-1} N}{\delta}\right).$$

由于 $u(U) \in C^{n-1}$, 所以当 $n > 1$ 时, $\omega\left(\frac{1}{N}\right) = O\left(\frac{1}{N}\right)$. 因之

$$I_2 = O\left(\frac{\ln^{n-1} N}{\delta N}\right),$$

同样可以証明

$$I_3 = O\left(\frac{\ln^{n-2} N}{\delta^2 N}\right),$$

$$\cdots\cdots\cdots\cdots$$

$$I_p = O\left(\frac{\ln^{n-p+1}N}{\delta^{p-1}N}\right),$$

$$\cdots\cdots\cdots\cdots\cdots$$

$$I_{n+1} = O\left(\frac{1}{\delta^n N}\right).$$

将 R_1 分解为:

$$S_1:\quad \delta \geqslant \theta_1 \geqslant \theta_2 \geqslant \cdots \geqslant \theta_n \geqslant -\delta;$$

$$S_2:\quad \delta \geqslant \theta_1 \geqslant \cdots \geqslant \theta_{n-1} \geqslant -\delta \geqslant \theta_n \geqslant -\pi;$$

$$\cdots\cdots\cdots\cdots\cdots\cdots\cdots\cdots\cdots\cdots$$

$$S_{n+1}:\quad \delta \geqslant \theta_1 \geqslant -\delta \geqslant \theta_2 \geqslant \cdots \geqslant \theta_n \geqslant -\pi.$$

于是 I_1 分解为

$$J_p = \frac{(-i)^{\frac{1}{2}n(n-1)}}{(2\pi)^n(n-1)!\cdots 2!1!}\int\cdots\int_{S_p}(g(\theta_1,\cdots,\theta_n)-u(U))\frac{\partial^0}{\partial\theta_1^0}\frac{\partial^1}{\partial\theta_2^1}\cdots\frac{\partial^{n-1}}{\partial\theta_n^{n-1}}\times$$
$$\times D(e^{-i\theta_1},\cdots,e^{-i\theta_n})d_N(\theta_1)d_N(\theta_2)\cdots d_N(\theta_n)d\theta_1 d\theta_2\cdots d\theta_n. \tag{3.3.5}$$
$$p = 1, 2, \cdots, n+1.$$

应用与上述一样的方法,可以証明

$$J_p = O\left(\frac{\ln^{n-p+1}N}{\delta^{p-1}N}\right)$$
$$p = 2, 3, \cdots, n+1.$$

最后考虑 J_1,由于 $\delta \geqslant \theta_1 \geqslant \cdots \geqslant \theta_n \geqslant -\delta$ 时,

$$|g(\theta_1,\theta_2,\cdots,\theta_n) - u(U)| = O(\delta),$$

这是因为 $u(U) \in C^{n-1}$ 的緣故. 因之

$$J_1 = O(\delta \ln^n N).$$

最后得到

$$I_{0,\cdots,0} = O(\delta\ln^n N) + O\left(\frac{\ln^{n-1}N}{\delta N}\right) + \cdots + O\left(\frac{\ln^{n-p+1}N}{\delta^{p-1}N}\right) + \cdots + O\left(\frac{1}{\delta^n N}\right).$$

取 $\delta = (N\ln^n N)^{\frac{-1}{n+1}}$, 于是

$$I_{0,0,\cdots,0} = O\left(\left(\frac{\ln^{n^2}N}{N}\right)^{\frac{1}{n+1}}\right).$$

我們再来研究 $I_{\mu_1,\cdots,\mu_n}$, 当 $\mu_1,\cdots,\mu_n$ 不全为零的时候, 由于 $0 \leqslant \mu_k \leqslant k-1$, 且不能全为零, 所以至少有二个 v_j $(j = 1, 2, \cdots, n)$ 是相等的. 例如 $v_j = v_k$, 这样我們从 $D(e^{-i\theta_1},\cdots,e^{-i\theta_n})$ 的定义知道:

$$\xi = \frac{\partial^{v_1}}{\partial\theta_1^{v_1}}\frac{\partial^{v_2}}{\partial\theta_2^{v_2}}\cdots\frac{\partial^{v_n}}{\partial\theta_n^{v_n}}D(e^{-i\theta_1},\cdots,e^{-i\theta_n}).$$

有因子 $(e^{-i\theta_j} - e^{-i\theta_k})$, 因之,

$$\xi = (e^{-i\theta_j} - e^{-i\theta_k})F(e^{-i\theta_1},\cdots,e^{-i\theta_n}),$$

而

$$e^{-i\theta_j} - e^{-i\theta_k} = (1 - e^{-i\theta_k}) - (1 - e^{-i\theta_j}),$$

因之

$$I_{\mu_1,\cdots,\mu_n}=\frac{(-i)^{\frac{1}{2}n(n-1)}}{(2\pi)^n(n-1)!\cdots2!1!}\int_0^{2\pi}\cdots\int_0^{2\pi}\frac{\partial^{\mu_1}}{\partial\theta_1^{\mu_1}}\frac{\partial^{\mu_2}}{\partial\theta_2^{\mu_2}}\cdots\frac{\partial^{\mu_n}}{\partial\theta_n^{\mu_n}}(g(\theta_1,\cdots,\theta_n)-$$

$$-u(U))[(1-e^{-i\theta_k})-(1-e^{-i\theta_j})]F(e^{-i\theta_1},\cdots,e^{-i\theta_n})\times$$

$$\times d_N(\theta_1)d_N(\theta_2)\cdots d_N(\theta_n)d\theta_1d\theta_2\cdots d\theta_n.$$

因为 $u(V)\in C^{n-1}$, $F(e^{-i\theta_1},\cdots,e^{-i\theta_n})$ 显然是可以微商的，因之

$$I_{\mu_1,\cdots,\mu_n}=O\left(\frac{\ln^{n-1}N}{N}\right).$$

于是我們得到 $S_N^{(n)}(U)$ 是收斂的，且收斂于 $u(U)$. 并且 $S_N^{(n)}(U)$ 与 $u(U)$ 的誤差程度为 $O\left(\left(\frac{\ln^{n^2}N}{N}\right)^{\frac{1}{n+1}}\right)$.

定理 3.3.1. 若 $u(U)\in C^{\frac{n(n-1)}{2}}$，$(n\geqslant 2)$ 那末它的 Fourier 級数的部分和 $S_N^{(n)}(U)$ 是收斂于 $u(U)$ 的，并且有

$$\left|S_N^{(n)}(U)-u(U)\right|\leqslant A\left(\frac{\ln^{n^2}N}{N}\right)^{\frac{1}{n+1}},$$

这里 A 为絕对常数.

§3.4. Fourier 級数的絕对收斂

設 $u(U)$ 在 n 阶酉羣 U_n 上可积，那末它有 Fourier 級数(3.1.1)，这里

$$C_{f_1\cdots f_n}=\int_{U_n}u(V)\overline{\Phi_{f_1\cdots f_n}(V)}\dot V,\quad \Phi_{f_1\cdots f_n}(U)=\sqrt{\frac{N(f)}{C}}A_{f_1\cdots f_n}(U),$$

而 $A_{f_1\cdots f_n}(U)$ 为 U 的标記为 $(f_1,\cdots,f_n)$ 的酉表示，于是我們有

$$C_f\bar C_f'=\iint u(V)u(U)\overline{\Phi_f(V)}\Phi_f'(U)\dot V\dot U=$$

$$=\frac{N(f)}{C}\iint u(V)u(U)\overline{A_f(V)}A_f'(U)\dot V\dot U.$$

記 $U\bar V'=W$，于是得到

$$\operatorname{tr}(C_f\bar C_f')=\frac{N(f)}{C}\iint u(V)u(WV)\chi_f(W)\dot W\dot V$$

記

$$g(W)=\int u(V)u(WV)\dot V.$$

于是

$$\operatorname{tr}(C_f\bar C_f')=\frac{N(f)}{C}\int g(W)\chi_f(W)\dot W.$$

設 W 的特征根为 $e^{i\theta_1},\cdots,e^{i\theta_n}$，記

$$h(\theta_1,\cdots,\theta_n)=\frac{1}{\omega_n'}\int_{[W]}g(W)[\dot W],$$

于是

$$\operatorname{tr}(C_f\bar{C}_f') = \\ = \frac{N(f)}{(2\pi)^n}\int\cdots\int_{\theta_1\geqslant\cdots\geqslant\theta_n} h(\theta_1,\cdots,\theta_n)M_f(e^{i\theta_1},\cdots,e^{i\theta_n})D(e^{-i\theta_1},\cdots,e^{-i\theta_n})d\theta_1\cdots d\theta_n.$$

如同以前一样，我們用 h^* 来替代 h，并应用我們以往慣用的积分技巧，即得 $\operatorname{tr}(C_f\bar{C}_f')$ 等于

$$\frac{N(f)}{(2\pi)^n}\int_{-\pi}^{\pi}\cdots\int h^*(\theta_1,\cdots,\theta_n)e^{i(l_1\theta_1+\cdots+l_n\theta_n)}D(e^{-i\theta_1},\cdots,e^{-i\theta_n})d\theta_1\cdots d\theta_n.$$

記

$$H(\theta_1,\cdots,\theta_n)=h^*(\theta_1,\cdots,\theta_n)D(e^{-i\theta_1},\cdots,e^{-i\theta_n})= \\ =D(e^{-i\theta_1},\cdots,e^{-i\theta_n})\frac{1}{\omega_n'}\int_{[\dot{V}_n]}\int_{V_n}u(V)u^*(WV)\dot{V}[\dot{W}] \tag{3.4.1}$$

于是我們有

$$\frac{\operatorname{tr}(C_f\bar{C}_f')}{N(f)}=\frac{1}{(2\pi)^n}\int_{-\pi}^{\pi}\cdots\int H(\theta_1,\cdots,\theta_n)e^{i(l_1\theta_1+\cdots+l_n\theta_n)}d\theta_1\cdots d\theta_n \tag{3.4.2}$$

但 $l_1>\cdots>l_n$。如果 $(i_1,\cdots,i_n)$ 是 $(1,\cdots,n)$ 的一个排列，那末容易看出

$$\frac{1}{(2\pi)^n}\int_{-\pi}^{\pi}\cdots\int H(\theta_1,\cdots,\theta_n)e^{i(l_{i_1}\theta_1+\cdots+l_{i_n}\theta_n)}d\theta_1\cdots d\theta_n=\delta_{i_1,\cdots,i_n}^{1,\cdots,n}\frac{\operatorname{tr}(C_f\bar{C}_f')}{N(f)}.$$

这样我們就得到：作为多变数可积函数 $H(\theta_1,\cdots,\theta_n)$，它的多重 Fourier 級数为

$$\sum_{l_1,\cdots,l_n} d_{l_1\cdots l_n}e^{-i(l_1\theta_1+\cdots+l_n\theta_n)}, \tag{3.4.3}$$

这里

$$d_{l_1\cdots l_n}=\frac{\operatorname{tr}(C_f\bar{C}_f')}{N(f)},\quad \text{若 } l_1>l_2>\cdots>l_n$$

$$=0,\qquad \text{若 } l_1,\cdots,l_n \text{ 中至少有二个相同.}$$

$$d_{l_{i_1},\cdots,l_{i_n}}=\delta_{i_1,\cdots,i_n}^{1,\cdots,n}\frac{\operatorname{tr}(C_f\bar{C}_f')}{N(f)},\quad \text{若 }(l_{i_1},\cdots,l_{i_n})\text{ 为 }(1,\cdots,n)\text{ 的一个排列，且 } l_1>l_2>\cdots>l_n.$$

这样就可以得到：(3.4.3)可以写成

$$\sum_{l_1>\cdots>l_n}\frac{\operatorname{tr}(C_f\bar{C}_f')}{N(f)}M_{l_1\cdots l_n}(e^{-i\theta_1},\cdots,e^{-i\theta_n}). \tag{3.4.4}$$

同样可以証明：函数 $D\left(\frac{\partial}{\partial\theta_1},\cdots,\frac{\partial}{\partial\theta_n}\right)H(\theta_1,\cdots,\theta_n)$ 的多重 Fourier 級数为

$$D(n-1,\cdots,1,0)\sum_{l_1>\cdots>l_n}(-i)^{\frac{n(n-1)}{2}}\operatorname{tr}(C_f\bar{C}_f')M_{l_1\cdots l_n}(e^{-i\theta_1},\cdots,e^{-i\theta_n}). \tag{3.4.5}$$

設 $f(\theta_1,\cdots,\theta_n)$ 为在 $-\pi\leqslant\theta_1,\cdots,\theta_n\leqslant\pi$ 中定义的連續函数，命

$$A^{(k)}(f;\theta_1,\cdots,\theta_n;h_1,\cdots,h_n)= \\ =f(\theta_1,\cdots,\theta_{k-1},\theta_k+h_k,\theta_{k+1},\cdots,\theta_n)-f(\theta_1,\cdots,\theta_n);$$

$$A^{(j,k)}(f;\theta_1,\cdots,\theta_n;h_1,\cdots,h_n)=A^{(j)}(A^{(k)}(f;\theta_1,\cdots,\theta_n;h_1,\cdots,h_n));\ (j<k)$$

同样定义 $A^{(j,k,l)}(f;\theta_1,\cdots,\theta_n;h_1,\cdots,h_n)$ $(j<k<l)$ 等等．記

$$A(f;\theta_1,\cdots,\theta_n;h_1,\cdots,h_n)=A^{(1,2,\cdots,n)}(f;\theta_1,\cdots,\theta_n;h_1,\cdots,h_n).$$

我們分別称

$$\omega(f;\theta_1,\cdots,\theta_n;h_1,\cdots,h_n)=\sup_{\substack{|\delta_1|\leqslant h_1,\cdots,|\delta_n|\leqslant h_n\\ |\theta_1|\leqslant\pi,\cdots,|\theta_n|\leqslant\pi}}|A(f;\theta_1,\cdots,\theta_n;h_1,\cdots,h_n)|$$

及

$$\omega_p(f;h_1,\cdots,h_n)=\left[\sup_{|\delta_1|\leqslant h_1,\cdots,|\delta_n|\leqslant h_n}\int_{-\pi}^{\pi}\cdots\int_{-\pi}^{\pi}|A(f;\theta_1,\cdots,\theta_n;\delta_1,\cdots,\delta_n)|^p d\theta_1,\cdots,d\theta_n\right]^{1/p}\quad(p\geqslant1)$$

为在 $-\pi\leqslant\theta_1,\cdots,\theta_n\leqslant\pi$ 中定义的連續函数 $f(\theta_1,\cdots,\theta_n)$ 的連續模及 p 次积分模．

若对于任意 $h_1>0,\cdots,h_n>0$，存在 α，使

$$\omega_p(f;h_1,\cdots,h_n)\leqslant K(h_1,\cdots,h_n)^{\alpha}\quad(K\text{ 为絕对常数})$$

恆成立，那末称 $f(\theta_1,\cdots,\theta_n)$ 是滿足 Lipschtz 条件 (p,α)．記其全部的函数成类为 Lip (p,α)．

有了这些作准备之后，我們可以証明以下的三条定理．

定理 3.4.1. 若 $u(U)\in C^{\frac{n(n-1)}{2}}$，且

$$D\left(\frac{\partial}{\partial\theta_1},\cdots,\frac{\partial}{\partial\theta_n}\right)H(\theta_1,\cdots,\theta_n)\in\text{Lip}\ (2,\alpha),$$

这里H由(3.4.1)所定义．那末当

$$\alpha>\frac{1}{r}-\frac{1}{2}$$

时，則級数

$$\sum_{f_1\geqslant\cdots\geqslant f_n}|\text{tr}(C_{f_1\cdots f_n}\overline{C}'_{f_1\cdots f_n})|^r\tag{3.4.6}$$

收斂，但

$$1>r>\frac{2}{3}.$$

定理 3.4.2. 若 $u(U)\in C^{\frac{3n(n-1)}{2}}$，且

$$\left(D\left(\frac{\partial}{\partial\theta_1},\cdots,\frac{\partial}{\partial\theta_n}\right)\right)^3H(\theta_1,\cdots,\theta_n)\in\text{Lip}\ (2,\alpha),$$

这里H由(3.4.1)所定义．那末当

$$\alpha>\frac{1}{r}-\frac{1}{2}$$

时，則級数

$$\sum_{f_1\geqslant\cdots\geqslant f_n}\left(\sum_{j,k=1}^{N(f)}|C_f^{(jk)}|\cdot|\varphi_f^{(jk)}|\right)^{2r}\tag{3.4.7}$$

收敛,但

$$1 > r > \frac{2}{3}.$$

定理 3.4.3. 若 $u(U) \in C^{2n(n-1)}$,且

$$\left(D\left(\frac{\partial}{\partial\theta_1}, \cdots, \frac{\partial}{\partial\theta_n}\right)\right)^4 H(\theta_1, \cdots, \theta_n) \in \mathrm{Lip}\ (2, \alpha)$$

則 $u(U)$ 的 Fourier 級数絕对收斂,但

$$\alpha > \frac{1}{2}.$$

§ 3.5. 上节定理的証明

由于函数 $D\left(\frac{\partial}{\partial\theta_1}, \cdots, \frac{\partial}{\partial\theta_n}\right) H(\theta_1, \cdots, \theta_n)$ 的多重 Fourier 級数为 (3.4.5) 的形式,且 $D\left(\frac{\partial}{\partial\theta_1}, \cdots, \frac{\partial}{\partial\theta_n}\right) H(\theta_1, \cdots, \theta_n) \in \mathrm{Lip}\ (2, \alpha)$, 所以由于多重 Fourier 級数的絕对收斂的定理(例如,参閱[5], 定理 2),我們知道,当 $\alpha > \frac{1}{r} - \frac{1}{2}$ 时, 級数(3.4.6)是收斂的,故得定理 3.4.1.

应用 Буняковский-Schwarz 不等式,我們知道

$$|\mathrm{tr}(C_{f_1\cdots f_n}\Phi'_{f_1\cdots f_n}(U))|^2 \leqslant \frac{(N(f))^2}{C}\mathrm{tr}\,(C_{f_1\cdots f_n}\bar{C}'_{f_1\cdots f_n}), \tag{3.5.1}$$

而如同以往一样可以証明: 函数

$$\left(D\left(\frac{\partial}{\partial\theta_1}, \cdots, \frac{\partial}{\partial\theta_n}\right)\right)^3 H(\theta_1, \cdots, \theta_n)$$

的多重 Fourier 級数为

$$(D(n-1, \cdots, 1, 0))^3 \sum_{f_1\geqslant\cdots\geqslant f_n} (-i)^{\frac{3n(n-1)}{2}} (N(f))^2\,\mathrm{tr}\,(C_f\bar{C}'_f)M_f(e^{-i\theta_1}, \cdots, e^{-i\theta_n}),$$

由于

$$\left(D\left(\frac{\partial}{\partial\theta_1}, \cdots, \frac{\partial}{\partial\theta_n}\right)\right)^3 H(\theta_1, \cdots, \theta_n) \in \mathrm{Lip}\ (2, \alpha),$$

所以由于多重 Fourier 級数的絕对收斂的定理,立刻得到,当 $\alpha > \frac{1}{r} - \frac{1}{2}$ 时,級数

$$\sum_{f_1\geqslant\cdots\geqslant f_n} (N(f))^2\,\mathrm{tr}\,(C_{f_1\cdots f_n}\bar{C}'_{f_1\cdots f_n})$$

收斂,但这也就保証了級数(3.4.7)的收斂性,因为由 Буняковский-Schwarz 不等式,我們知道

$$\left|\sum_{j,k=1}^{N(f)} |C_f^{(jk)}|\cdot|\varphi_f^{(jk)}|\right|^2 \leqslant \frac{(N(f))^2}{C}\mathrm{tr}\,(C_{f_1\cdots f_n}\bar{C}'_{f_1\cdots f_n}) \tag{3.5.2}$$

成立. 这就証明了定理 3.4.2.

同样由于 Буняковский-Schwarz 不等式

$$\left|\sum_{l_1\geqslant\cdots\geqslant l_n} N(f)(\mathrm{tr}\,(C_f\bar{C}_f'))^{\frac{1}{2}}\right| \leqslant \left(\sum N^3(f)\,\mathrm{tr}\,(C_f\bar{C}_f')\right)^{\frac{1}{2}}\left(\sum\frac{1}{(N(f))^2}\right)^{\frac{1}{2}}, \qquad (3.5.3)$$

而函数

$$\left(D\left(\frac{\partial}{\partial\theta_1},\cdots,\frac{\partial}{\partial\theta_n}\right)\right)^4 H(\theta_1,\cdots,\theta_n)$$

的多重 Fourier 級数为

$$(D(n-1,\cdots,1,0))^4\sum_{l_1>\cdots>l_n}(N(f))^3\,\mathrm{tr}\,(C_f\bar{C}_f')M_{l_1\cdots l_n}(e^{-i\theta_1},\cdots,e^{-i\theta_n}).$$

同样应用多重 Fourier 級数的絕对收斂的定理,就得到,当

$$\left(D\left(\frac{\partial}{\partial\theta_1},\cdots,\frac{\partial}{\partial\theta_n}\right)\right)^4 H(\theta_1,\cdots,\theta_n)\in \mathrm{Lip}\ (2,\alpha)$$

时,級数

$$\sum_{l_1>\cdots>l_n}(N(f))^3\ \mathrm{tr}\ (C_f\bar{C}_f')$$

收斂,但 $\alpha>\frac{1}{2}$. 从不等式(3.5.2)及(3.5.3),立得定理 3.4.3.

§3.6. 几点注記

(1) 在本文中,所有的条件,特别是关于可微次数的条件都可能削弱.

(2) 在关于絕对收斂的討論中,如果我們另外一些关于多重 Fourier 級数的絕对收斂定理,例如 Zygmund 定理, Szasz 定理等的推广形式,可以获得不少关于酉羣上 Fourier 級数絕对收斂的充分条件. 但有了本文的討論,来进行这些工作已是輕而易举的事了,故不此多加叙述.

(3) 这里討論收斂或絕对收斂,都是在"方体"求和意义下进行的. 至于"球求和"就完全不同,这将在第 V 文中进行討論. 利用本文的想法,参閱将发表的第 V 文及[6]中第 9 章,可以容易的得出关于球求和的絕对收斂的結果来.

参 考 文 献

[1] 华罗庚,多复变数函数論中的典型域的調和分析,科学出版社,1958 年.

[2] 龔 昇,酉羣上的富里埃分析 (I),数学学报,**10** (1960), 239—261.

[3] 龔 昇,酉羣上的富里埃分析 (II),数学学报,**12** (1962), 17—31.

[4] Zygmund, A., Trigonometric series, 1959.

[5] Musielak, J., On absolute convergence of multiple Fourier series, *Annales Polonici Math.*, **3** (1958—59), 107—120.

[6] Chandrasekharan, K. and Minakshisundaren, S., Typical means, Oxford University Press, 1952.

第13卷 第3期　　数 学 学 报　　Vol. 13, No. 3
1963年8月　　ACTA MATHEMATICA SINICA　　August, 1963

酉羣上的富理埃分析

IV. 关于 Peter-Weyl 定理*

龔　昇
(中国科学技术大学)

§4.1. Peter-Weyl 定理

若G是紧致李羣，f是G上的連續函数. 对于任意已給的$\varepsilon > 0$，一定存在在G的表示环（representative ring）上定义的一个函数g，使得$|f(\sigma) - g(\sigma)| < \varepsilon$对于任意的$\sigma \in G$都成立. 这是众所周知的 H. Peter 与 H. Weyl 的定理(参閱[7]第VI章).

事实上，上述結果对于一般的紧致拓扑羣也是对的（参閱[5]第五章或[8]第VIII章).

当然这是一条极为重要的紧致拓扑羣上逼近理論的存在定理.

在[2]中，华罗庚証明了在酉羣上連續函数的一条收斂定理，并指出了其在一般紧致拓扑羣上的意义.

本文是作者試图将酉羣上的調和分析应用于紧致拓扑羣的逼近理論上的尝試.

在本文中，我們不仅要具体作出在G的表示环上定义的函数$g(\sigma)$，而且还要指出f与g之間的誤差程度有多大.

在§4.2中，我們定义了在紧致拓扑羣G上連續的函数的連續模. 在§4.3中，我們研究了在紧致拓扑羣G上定义的属于 Lip p 的連續函数用 Cesàro 平均而得到的逼近式及其逼近程度. 这是本文的中心. 例如，我們可以借助于[3]中的結果，証明了：如果$f(\sigma) \in \mathrm{Lip}\, p$（定义見§4.2)，那末可以找到一个在$G$的表示环上定义的有限綫性式$g_N(\sigma)$，使得

$$|f(\sigma) - g_N(\sigma)| \leqslant AN^{-p}$$

成立，这里A为絕对常数. 而$g_N(\sigma)$是完全可以具体写出来的(見(4.3.8)).

事实上，这里还指出了如何去构造其它的逼近式，以及如何来求出逼近誤差的程度，在§4.4中我們作了一些說明，而为了不使篇幅过长，略去了这些繁复的計算.

§4.5是我們举了一个如何处理酉羣上的插值問題的一例，最后把一部分結果应用于多复变数矩陣双曲空間中的解析函数的逼近理論中去.

文中所用符号与前几篇相同，不再一一說明.

§4.2. 紧致拓扑羣上的連續函数

設G是紧致拓扑羣，$\sigma \in G$. 若$f(\sigma)$是G上的連續函数，那末$f(\sigma)$一定是在G上均

* 1962年2月3日收到，1962年10月30日收到修改稿.

匀連續，也就是对于每一个 $\varepsilon>0$，存在 G 的单位 e 的邻域 V，使得 $xy^{-1}\in V$ 及 $x\in G$，$y\in G$ 时，恆有

$$|f(x)-f(y)|<\varepsilon. \tag{4.2.1}$$

与这等价的説法是：对于每一个 $\varepsilon>0$，存在 G 的单位 e 的邻域 V'，使得当 $x^{-1}y\in V'$ 及 $x\in G$，$y\in G$ 时，恆有(4.2.1)(参閱[5]).

我們知道对于任一紧致拓扑羣 G，至少有一个实在的表示 (faithful representation)，当然这个表示可以是酉表示，且与 G 同构. 因之，在同构的意义下，我們可以把紧致拓扑羣"嵌入"酉羣内作为一个子羣. 即，若 $\sigma\in G$，則 σ 的酉表示 $U(\sigma)$ 使 G 与 $\{U(\sigma)|\sigma\in G\}$ 同构.

記 $W=\{U\in U_n|U=U(\sigma),\sigma\in G\}$. 于是 $W\subset U_n$，我們在 W 上定义連續函数 $u(U)=f(\sigma)$. 当然 $u(U)$ 不仅在 W 上可以定义，且可以連續的拓展到整个 U_n 上 (参閱 [5]，[7]). 必須注意的是：由于实在的表示可以不止一个，以至这样的 W 也可以不止一个.

設 U，V 为 W 中的二点，$U=(u_{ij})_{1\leqslant i,j\leqslant n}$，$V=(v_{ij})_{1\leqslant i,j\leqslant n}$. 于是 U，V 二点之間的欧氏距离 $d(U,V)$ 的平方为

$$\sum_{i,j=1}^{n}|u_{ij}-v_{ij}|^2,$$

而此即

$$\begin{aligned}&\operatorname{tr}((U-V)\overline{(U-V)}')\\&=\operatorname{tr}(2I-U\bar{V}'-V\bar{U}').\end{aligned}$$

若 $U\bar{V}'$ 的特征根为 $e^{i\theta_1},\cdots,e^{i\theta_n}$，那末 $d(U,V)$ 的平方就是

$$2\sum_{j=1}^{n}(1-\cos\theta_j). \tag{4.2.2}$$

設 $u(U)$ 是 U_n 上的連續函数，我們称

$$\omega(u,\delta)=\max_{d(U,V)\leqslant\delta}|u(U)-u(V)|$$

为函数 $u(U)$ 的連續模.

設 $f(\sigma)$ 在紧致拓扑羣 G 上連續，$(\sigma\in G)$ 則 G 至少有一个实在的酉表示，而且可以在同构的意义下，将 G"嵌入"某一酉羣(例如 m 阶酉羣)作为子羣. 如前那样記

$$W=\{U\in U_m|U=U(\sigma),\sigma\in G\},$$

于是对应于一个实在的 m 阶的酉表示，可以有

$$\omega_m'(f,\delta)=\max_{\substack{x\in G,y\in G\\ d(U(x),U(y))\leqslant\delta}}|f(x)-f(y)|.$$

更进一步，若 $f(\sigma)$ 在 G 上連續，且

$$|f(x)-f(y)|\leqslant A(d(U(x),U(y)))^{p'},$$

这里 A 为絕对常数. 在各种可能的实在的酉表示中，选择一个使 p' 最大的酉表示，若此酉表示所对应的 p' 的值記之为 p，則称函数 $f(\sigma)$ 滿足 Lipschtz 条件，記之为 $f(\sigma)\in\mathrm{Lip}\,p$.

§4.3. 用 Cesàro 平均得到的逼近

我們在[3]中，仔細的研究了酉羣 U_n 上的 Fourier 級数的 Cesàro (c,α) 平均，并且証

明了：如果 $u(U)$ 在 U_n 上連續，則 $u(U)$ 的 Fourier 級数可以 (c, α) 求和于它自己，但 $\alpha > \frac{n-1}{n}$.

在証明定理 2.3.1 的过程中(見[3]§2.4)，我們实际上証明了 $u(U)$ 与它的 Fourier 級数的 Cesàro (c, α) 平均的第 N 项 $\sum_N^\alpha(U)$ 之差的絕对值不大于

$$O(\omega(u, \delta)) + O\left(\frac{1}{(N\delta)^{\alpha n-n+1}}\right) + O\left(\frac{1}{(N\delta)^{2(\alpha n-n+2)}}\right) + \cdots + O\left(\frac{1}{(N\delta)^{\alpha n^2}}\right).$$

因之，若 $u(U) \in \mathrm{Lip}\, p$，則取

$$\delta = N^{\frac{n-\alpha n-1}{p+\alpha n-n+1}},$$

于是得到

$$\left|u(U) - \sum_N^\alpha(U)\right| \leqslant AN^{-\frac{p(\alpha n-n+1)}{p+\alpha n-n+1}}.$$

这里 A 为絕对常数及 $\alpha > \frac{n-1}{n}$.

这是比較粗糙的結果，我們回顾定理 2.3.1 的証明过程中，可以証明得到以下更仔細的結果。

定理 4.3.1. 若 $u(U)$ 是酉羣 U_n 上的連續函数，且屬于 $\mathrm{Lip}\, p$，$(0 < p \leqslant 1)$，則它的 Fourier 級数的 (c, α) 平均的第 N 项 $\sum_N^\alpha(U)$ 滿足

(1) 若 $\alpha n - n + 1 > p$，則

$$\left|u(U) - \sum_N^\alpha(U)\right| \leqslant A_1 N^{-p}; \tag{4.3.1}$$

(2) 若 $\alpha n - n + 1 = p$，則

$$\left|u(U) - \sum_N^\alpha(U)\right| \leqslant A_2 N^{-p} \ln N; \tag{4.3.2}$$

(3) 若 $\alpha n - n + 1 < p$，則

$$\left|u(U) - \sum_N^\alpha(U)\right| \leqslant A_3 N^{-\alpha n+n-1}; \tag{4.3.3}$$

这里 A_1, A_2, A_3 为絕对常数.

証. 記(參閱[3](2.3.3))

$$I'_{s_1,\cdots,s_n} = \int_\delta^\pi |\varphi(e^{i\theta_1}, \cdots, e^{i\theta_n})|\,|1 - e^{i\theta_1}|^{s_1}|\sigma_N^\alpha(\theta_1)|^n d\theta_1 \int_{\delta \geqslant \theta_2 \geqslant \cdots \geqslant \theta_n \geqslant -\delta}\cdots\int \times$$

$$\times |1 - e^{i\theta_2}|^{s_2}\cdots|1 - e^{i\theta_n}|^{s_n}|\sigma_N^\alpha(\theta_2)|^n\cdots|\sigma_N^\alpha(\theta_n)|^n \cdot |D(e^{i\theta_2}, \cdots, e^{i\theta_n})|^2 d\theta_2\cdots d\theta_n, \tag{4.3.4}$$

这里 φ, σ_N^α, D 的意义請参閱[3].

由于 $u(U) \in \mathrm{Lip}\, p$，及当 $\delta \geqslant \frac{1}{N}$ 时

$$|\sigma_N^\alpha(\theta)| \leqslant a_1 N^{-\alpha}\theta^{-\alpha-1},$$

所以得到 $I'_{s_1,\cdots,s_n}$ 不大于

$$a_2\int_\delta^\pi N^{-\alpha n}\,\theta_1^{-\alpha n-n+s_1+p}\,d\theta_1\,N^{n-1}\,\delta^{s_2+\cdots+s_n}\underset{\delta\geq\theta_2\geq\cdots\geq\theta_n\geq-\delta}{\int\cdots\int}|\sigma_N^\alpha(\theta_2)\cdots\sigma_N^\alpha(\theta_n)|^{n-1}\times$$

$$\times\,|D(e^{i\theta_2},\cdots,e^{i\theta_n})|^2\,d\theta_2\cdots d\theta_n+a_3\delta^p\int_\delta^\pi N^{-\alpha n}\,\theta_1^{-\alpha n-n+s_1}\,d\theta_1\,N^{n-1}\times$$

$$\times\,\delta^{s_2+\cdots+s_n}\underset{\delta\geq\theta_2\geq\cdots\geq\theta_n\geq-\delta}{\int\cdots\int}|\sigma_N^\alpha(\theta_2)\cdots\sigma_N^\alpha(\theta_n)|^{n-1}|D(e^{i\theta_2},\cdots,e^{i\theta_n})|^2\,d\theta_2\cdots d\theta_n,\quad(4.3.5)$$

这里 a_1, a_2, a_3 表示三个不同的絕对常数.

(1) 当 $\alpha n-n+1>p$, 由于 $s_1\leqslant 2n-2$, 所以

$$-\alpha n-n+s_1+p<-\alpha n-n+s_1+\alpha n-n+1=-2n+1+s_1\leqslant-1.$$

因之,从(4.3.5)得到

$$I'_{s_1,\cdots,s_n}=O((\delta N)^{-\alpha n+n-1}\delta^p).$$

取 $\delta=\dfrac{1}{N}$, 卽得(4.3.1).

(2) 当 $\alpha n-n+1=p$, 则

$$-\alpha n-n+s_1+p=-\alpha n-n+s_1+\alpha n-n+1=-2n+1+s_1,$$

所以从(4.3.5)知道

$$I'_{2n-2,s_2,\cdots,s_n}=O(N^{-\alpha n+n-1}\ln\delta)+O((\delta N)^{-\alpha n+n-1}\delta^p),$$

而当 $s_1\neq 2n-2$ 时,

$$I'_{s_1,\cdots,s_n}=O((\delta N)^{-\alpha n+n-1}\delta^p),$$

卽

$$\left|u(U)-\sum_N^\alpha(U)\right|\leqslant a_4N^{-p}\ln\delta+a_5N^{-p},$$

这里 a_4, a_5 为絕对常数. 取 $\delta=\dfrac{1}{N}\ln N$, 则得(4.3.2).

(3) 当 $\alpha n-n+1<p$ 时,

$$-\alpha n-n+s_1+p>-\alpha n-n+s_1+\alpha n-n+1=-2n+1+s_1,$$

所以

$$I'_{2n-2,s_2,\cdots,s_n}=O(N^{-\alpha n+n-1})+O((N\delta)^{-\alpha n+n-1}\delta^p).$$

于是

$$\left|u(U)-\sum_N^\alpha(U)\right|\leqslant a_6N^{-\alpha n+n-1}+a_7(N\delta)^{-\alpha n+n-1}\delta^p\ln\delta+$$
$$+a_8(N\delta)^{-\alpha n+n-1}\delta^p.$$

考虑到 $p>\alpha n-n+1$ 而且 $\delta\geqslant\dfrac{1}{N}$, 所以得到(4.3.3).

回忆一下酉羣 U_n 上可积函数 $u(U)$ 的 Fourier 級数的 Cesàro (c,α) 平均的第 N 項为

$$\sum_{nN\geq f_1\geq\cdots\geq f_n\geq-nN}B^\alpha_{f_1\cdots f_n}\operatorname{tr}(C_{f_1\cdots f_n}\Phi'_{f_1\cdots f_n}(U)),\qquad(4.3.6)$$

这里

$$B^{\alpha}_{f_1\cdots f_n}=\frac{1}{CN(f)}\int_{U_n}\chi_{f_1\cdots f_n}(\bar{V})\mathbf{K}^{\alpha}_N(V)\dot{V}, \tag{4.3.7}$$

而 $\mathbf{K}^{\alpha}_N(V)$ 为 Cesàro (c,α) 核由[3](2.2.3)所定义。

若 G 为紧致拓扑群，$f(\sigma)$ 为 G 上的連續函数，且 $f(\sigma)\in \mathrm{Lip}\,p$，則可以作在 G 的表示环上定义的函数

$$g_N(\sigma)=\sum_{nN\geqslant f_1\geqslant\cdots\geqslant f_n\geqslant -nN}B^{\alpha}_{f_1\cdots f_n}\operatorname{tr}\left(C_{f_1\cdots f_n}\Phi'_{f_1\cdots f_n}(U(\sigma))\right), \tag{4.3.8}$$

这里 $B^{\alpha}_{f_1\cdots f_n}$ 由(4.3.7)所定义。于是我們可以把定理 4.3.1 写成为紧致拓扑羣的形式为

定理 4.3.2. 設 G 为紧致拓扑羣，$f(\sigma)$ 为 G 上的連續函数，且 $f(\sigma)\in \mathrm{Lip}\,p\,(\sigma\in G)$，在 G 的表示环上作由(4.3.8)所定义的函数 $g_N(\sigma)$，于是得到：

(1) 若　$p\neq \alpha n-n+1$，則

$$|f(\sigma)-g_N(\sigma)|\leqslant A_4N^{-q},$$

这里 $q=\min(p,\alpha n-n+1)$，A_4 为絕对常数。

(2) 若　$p=\alpha n-n+1$，則

$$|f(\sigma)-g_N(\sigma)|\leqslant A_2N^{-p}\ln N.$$

作为上述定理的推論，已知 p 之后，只要取 α 充分大，使 $\alpha n-n+1>p$，就可以得到

定理 4.3.3. 設 G 为紧致拓扑羣，$f(\sigma)$ 为 G 上的連續函数，且 $f(\sigma)\in \mathrm{Lip}\,p(\sigma\in G)$，則一定可以在 G 的表示环上作函数 $g_N(\sigma)$，使

$$|f(\sigma)-g_N(\sigma)|\leqslant A_5N^{-p},$$

这里 A_5 为絕对常数。

作为定理 4.3.2 的另一个推論，特别取 $\alpha=1$，即我們取 Fejér 平均，就能够得到

定理 4.3.4. 設 G 为局部連通的紧致拓扑羣，$f(\sigma)$ 为 G 上的連續函数，且 $f(\sigma)\in \mathrm{Lip}\,p$ $(\sigma\in G)$，在 G 的表示环上作

$$g_N(\sigma)=\sum_{nN\geqslant f_1\geqslant\cdots\geqslant f_n\geqslant -nN}B_{f_1\cdots f_n}\operatorname{tr}\left(C_{f_1\cdots f_n}\Phi'_{f_1\cdots f_n}(U(\sigma))\right), \tag{4.3.9}$$

这里 $B_{f_1\cdots f_n}$ 由[3]中(2.5.2)所定义(当 $N\geqslant n-1$ 时，由[3]中(2.7.7)具体表示出来)。

(1) 若　$p<1$ 时，則

$$|f(\sigma)-g_N(\sigma)|<A_6N^{-p}. \tag{4.3.10}$$

(2) 若　$p=1$ 时，則

$$|f(\sigma)-g_N(\sigma)|<A_7\ln N/N, \tag{4.3.11}$$

这里 A_6，A_7 为絕对常数。

附带一提的是定理 4.3.4 把一个变数的函数的 Fourier 級数的 C. H. Бернштейн 的定理作为最特殊的情形。

§ 4.4.　一些一般的推論

依賴于酉羣上定义的函数 $u(U)$ 的 Fourier 級数的 Cesàro 求和，可以得到紧致拓扑

羣上的逼近定理，这样不但具体的作出了在 G 的表示环上定义的函数，而且算出了誤差程度．这样所作的函数，不一定就是最优逼近．但是其它的求和所得的逼近式，也可以依据这条路綫作出，而且同样的可以給出它的誤差程度．为此，我們在这一节中举出一些一般的結果，作为前面所得的結果所用方法的推論．例如，我們可以証明如下的一条一般收斂定理．

設 $K_N(\theta)$ 是一个变数的 Fourier 級数的 T 求和法的核．設 $V \in U_n$，$\bar{V}$ 的特征根为 $e^{i\theta_1}, \cdots, e^{i\theta_n}$，記

$$D = \frac{1}{C}\int_{U_n} (K_N(\theta_1)\cdots K_N(\theta_n))^n \dot{V} \tag{4.4.1}$$

定义

$$\sum_{nN\geqslant f_1\geqslant\cdots\geqslant f_n\geqslant -nN} D_{f_1\cdots f_n}\,\mathrm{tr}\,(C_{f_1\cdots f_n}\Phi'_{f_1\cdots f_n}(U)) \tag{4.4.2}$$

为 $u(U)$ 的 Fourier 級数的 T 平均．这里

$$D_{f_1\cdots f_n} = \frac{1}{DN(f)C}\int_{U_n} \chi_{f_1\cdots f_n}(\bar{V})(K_N(\theta_1)\cdots K_N(\theta_n))^n \dot{V}.$$

(4.4.2)可以写成

$$\int_{U_n} u(VU)T_N(\bar{V})\dot{V},$$

这里

$$T_N(\bar{V}) = \frac{1}{DC}(K_N(\theta_1)\cdots K_N(\theta_n))^n. \tag{4.4.3}$$

如果我們再假設

(i) 当 $\pi \geqslant \theta \geqslant \delta$ 时，对于任意固定的 $\delta > 0$，我們有

$$|K_N(\theta)| = O(N^{-\eta}) \tag{4.4.4}$$

成立，而 $\eta > 0$．

(ii) 对于任意的 θ，

$$|K_N(\theta)| = O(N^{\xi}) \tag{4.4.5}$$

成立，而 $\xi > 0$．

于是我們有

定理 4.4.1. 若 $u(U)$ 在 U_n 上連續，它的 Fourier 級数的 T 平均(4.4.2)收斂于 $u(U)$，假使由(4.4.3)所定义的 $T_N(\bar{V})$ 満足

$$\int_{U_m} |T_N(\bar{V})|\dot{V} \leqslant H_m \quad (m = 1, 2, \cdots, n), \tag{4.4.6}$$

这里 H_m 是只与 m 有关的絕对常数，而且 $K_N(\theta)$ 满足(4.4.4)及(4.4.5)且 $\eta \geqslant \xi$．

这个定理的証明从略，可以参閱[3]中定理 2.3.1 的証明．

作为上述定理的举例，我們来定义在酉羣 U_n 上定义的函数 $u(U)$ 的 Fourier 級数的 Jackson 平均为

$$\begin{aligned} J_N(U) &= \sum_{2nN\geqslant f_1\geqslant\cdots\geqslant f_n\geqslant -2nN} J_{f_1\cdots f_n}\,\mathrm{tr}\,(C_{f_1\cdots f_n}\Phi'_{f_1\cdots f_n}(U)) \\ &= \frac{J}{C}\int_{U_n} u(VU)\left|\frac{\det(I - V^{N+1})}{\det(I - V)}\right|^{4n}\dot{V} \end{aligned} \tag{4.4.7}$$

而

$$J=\left(\frac{3}{N(2N^2+1)}\right)^{n^2},$$

当然 $J_{f_1\cdots f_n}$ 的具体数值可以用积分表达出来. 首先我們看到(4.4.6)是滿足的, 因为此时 $H_m=1$, $m=1,2,\cdots,n$. 其次我們看到当 $\pi\geqslant\theta\geqslant\delta$ 时,

$$K_N(\theta)=O(N^{-3}\delta^{-4}),$$

而对于任意的 θ

$$K_N(\theta)=O(N).$$

所以定理 4.4.1 中的条件全滿足,以至(4.4.7)是收斂的,且收斂于 $u(U)$.

当然我們可以指出 T 平均与 $u(U)$ 的誤差有多大,例如,我們可以不大費力粗糙的算出

$$|J_N(U)-u(U)|\leqslant AN^{-\frac{p(n+2)}{4n+p-1}},$$

若 $u(U)\in \mathrm{Lip}\,p$.

至于滿足定理 4.1.1 的条件的 T 求和与 $u(U)$的誤差估計等等为了本文不至于篇幅过长,都略而不去进行了. 当然这些結果都可以推到緊致拓扑羣上.

§4.5. 酉羣上插值理論一例

对于酉羣上插值問題应該如何提? 这是个値得研究的問題,这里举一个最簡单的例.

如果已知在酉羣 U_n 上定义的函数 $u(U)$, 要求一个有限綫性式,使之与 $u(U)$ 在以下 $n(2N+1)$个点上有相同的数值. 这 $n(2N+1)$ 个点是

$$U_{p_1\cdots p_n}=V_0\begin{pmatrix}e^{\frac{2p_1\pi i}{2N+1}} & & \\ & \ddots & \\ & & e^{\frac{2p_n\pi i}{2N+1}}\end{pmatrix}\bar{V}_0'$$

这里 p_j 是滿足 $1\leqslant p_j\leqslant 2N+1$ 的正整数($j=1,2,\cdots,n$), V_0 是固定的一个酉矩陣. 换句話說:我們考慮的是一种特殊的"等距"結点的插値問題.

回答这个問題是不困难的,因为这 $n(2N+1)$ 个点,不論 N 取多么大, 都是在一个 n 維流形上的,而 U_n 本身是 n^2 維的.

考慮

$$\frac{\det^n(I-U^{2N+1})}{A^0C\det^{nN}U\det^n(I-U)}=\frac{1}{A^0C}\prod_{j=1}^n\frac{\sin^n\left(N+\frac{1}{2}\right)\theta_j}{\sin^n\theta_j/2},$$

这里

$$A^0=\frac{1}{C}\int_{U_n}\det^{nN}V\frac{\det^n(I-\bar{V}'^{2N+1})}{\det^n(I-\bar{V}')}\dot{V}$$

而 $e^{i\theta_1},\cdots,e^{i\theta_n}$ 为 U 的特征根.

可以証明

$$\sum_{nN\geqslant f_1\geqslant\cdots\geqslant f_n\geqslant -nN}A^0_{f_1\cdots f_n}N(f)\chi_{f_1\cdots f_n}(U)=\frac{\det^n(I-U^{2N+1})}{A^0\det^{nN}U\det^n(I-U)},$$

而

$$A^0_{f_1\cdots f_n}=\frac{1}{A^0N(f)C}\int_{U_n}\chi_{f_1\cdots f_n}(\bar V)\frac{\det^{nN}V\det^n(I-\bar V'^{2N+1})}{\det^n(I-\bar V')}\dot V.$$

作

$$\varphi_{p_1\cdots p_n}(U)=\frac{\det^n(I-U^{2N+1}U^{2N+1}_{p_1\cdots p_n})}{(2N+1)^{n^2}\det^{nN}U\cdot U_{p_1\cdots p_n}\det^n(I-U\cdot U_{p_1\cdots p_n})}.$$

从这个定义可以看出

(i)

$$\varphi_{p_1\cdots p_n}(U)=\frac{A^0C}{(2N+1)^{n^2}}\sum_{nN\geqslant f_1\geqslant\cdots\geqslant f_n\geqslant -nN}A^0_{f_1\cdots f_n}\operatorname{tr}(\Phi'_{f_1\cdots f_n}(U_{p_1\cdots p_n})\Phi'_{f_1\cdots f_n}(U)).$$

即 $\varphi_{p_1\cdots p_n}(U)$ 是一个有限綫性式，它可以写成由指标在 $nN\geqslant f_1\geqslant\cdots\geqslant f_n\geqslant -nN$ 之中的酉表示的有限綫性式.

(ii)

$$\varphi_{p_1\cdots p_n}(U_{q_1\cdots q_n})=\begin{cases}1 & \text{当 } (p_1\cdots p_n)=(q_1\cdots q_n),\\ 0 & (p_1\cdots p_n)\neq(q_1\cdots q_n),\end{cases}$$

因之作函数

$$P_N(U)=\sum_{p_1,p_2,\cdots,p_n}u(U_{p_1\cdots p_n})\varphi_{p_1\cdots p_n}(U),$$

即是我們所要求的函数. 写得更清楚一些,此即

$$P_N(U)=\frac{A^0C}{(2N+1)^{n^2}}\sum_{nN\geqslant f_1\geqslant\cdots\geqslant f_n\geqslant -nN}A^0_{f_1\cdots f_n}\operatorname{tr}[E_{f_1\cdots f_n}\Phi'_{f_1\cdots f_n}(U)],$$

此处

$$E_{f_1\cdots f_n}=\sum_{p_1,p_2,\cdots,p_n}u(U_{p_1\cdots p_n})\Phi'_{f_1\cdots f_n}(U_{p_1\cdots p_n}).$$

最后,还須指出的是:我們还可以用別的方法来构造 $P_N(U)$ 来, 例如: 我們可以采用 C. H. Бернштейн 的第一方法, 应用 [3] 中的 Fejér 核来造出另一个 $P_N(U)$ 来 (参閱 [6]第三卷,第四章).

§4.6. 多复变数矩陣双曲空間上的逼近

作为以前这些結果的直接推論, 可以得到多复变数矩陣双曲空間上的逼近. 多复变数典型域的第一类 R_I 被定义为

$$I-z\bar z'>0,\quad z=z^{(m,n)},$$

于是 n 阶的酉羣就是 R_I 的特征流形.

例如,从[4]就可以得到

定理 4.6.1. 設 $u(z)$ 在 R_I 及其特征流形上都是解析的,那末 $u(z)$ 的开始 N 項 Taylor 展开 $s_N^{(u)}(z)$ 滿足

$$|u(z)-s_N^{(u)}(z)|<A_8\left(\frac{\ln^{n^2}N}{N}\right)^{\frac{1}{n+1}},$$

这里 A_8 为絕对常数，所謂开始 N 項 Taylor 展开是指 z 的表示 $\Phi_{f_1\cdots f_n}(z)$，而 $N \geqslant l_1 > \cdots > l_n \geqslant 0$.

再例如从定理 4.3.4 立刻得到

定理 4.6.2. 設 $u(z)$ 在 R_I 中解析，在 U_n 上屬于 Lip p，則 $u(z)$ 的 Taylor 展开的 Fejér 平均 $\sum\limits_N^{(r)}(z)$ 满足

$$\left|u(z)-\sum_N^{(r)}(z)\right|<A_9N^{-p}.$$

又若 $u(z)$ 在 R_I 及 U_n 上解析，那末

$$\left|u(z)-\sum_N^{(r)}(z)\right|<A_{10}\ln N/N,$$

这里 A_9, A_{10} 是絕对常数. 这里

$$\sum_N^{(r)}(z)=\sum_{nN\geqslant f_1\geqslant\cdots\geqslant f_n\geqslant 0}B_{f_1\cdots f_n}\operatorname{tr}(C_{f_1\cdots f_n}\Phi'_{f_1\cdots f_n}(z)),$$

而 $B_{f_1\cdots f_n}$ 由[3]中的(2.5.2)所定义.

当然其它的結果也可以推演到 R_I 上，不再一一叙述了.

参　考　文　献

[1] 华罗庚，多复变数函数論中典型域的調和分析，科学出版社，1958.
[2] 华罗庚，紧致羣上的連續函数所成的空間中的一条收斂定理，科学記录，**2** (1958), 341—344 頁.
[3] 龔昇，酉羣上的富理埃分析 II, 数学学报，**12** (1962), 17—31.
[4] 龔昇，酉羣上的富理埃分析 III, 数学学报，**13** (1963), 152—161.
[5] Понтрягин Л. С., Непрерывные группы, Москва, 1954.
[6] Натансон И. Н., Конструктивная теория функций, 1949.
[7] Chevalley C., Theory of Lie groups I, Princeton, 1946.
[8] Loomis L. H., An introduction to abstract harmonic analysis, 1953.

第15卷 第3期 数学学报 Vol. 15, No. 3
1965年5月 ACTA MATHEMATICA SINICA May, 1965

酉羣上的富理埃分析(V)

Fourier 級数的球求和及 Fourier 积分*

龔 昇
(中国科学技术大学)

§5.1. 引 言

設 $u(U)$ 在 n 阶酉羣 U_n 上可积,那末它有 Fourier 級数

$$\sum_{f_1\geqslant\cdots\geqslant f_n} \operatorname{tr}\left(C_{f_1\cdots f_n}\Phi'_{f_1\cdots f_n}(U)\right). \tag{5.1.1}$$

在[2, 3, 4]中,我們考虑的都是"方体"求和,卽从

$$\sum_{N\geqslant f_1>\cdots>f_n\geqslant -N} \operatorname{tr}\left(C_{f_1\cdots f_n}\Phi'_{f_1\cdots f_n}(U)\right) \tag{5.1.2}$$

出发加以考虑.在[2, 3, 4]中,我們已經得到一系列結果,这些結果以具有整洁的核作为特色.本文考虑"球"求和.所謂球求和,乃是把 Fourier 級数考虑为

$$\sum_{m=0}^{\infty}\sum_{\substack{f_1\geqslant\cdots\geqslant f_n\\ l_1^2+\cdots+l_n^2=m}} \operatorname{tr}\left(C_{f_1\cdots f_n}\Phi'_{f_1\cdots f_n}(U)\right). \tag{5.1.3}$$

以此出发,加以研究.如果考虑的区域不是酉羣,而是 n 个单位圆周的拓扑积,那末前一种考虑相当于以 A. Zygmund[10] 为代表的对于多重 Fourier 級数的研究,而后一种考虑相当于以 S. Bochner[6,7] 为代表的对于多重 Fourier 級数的研究.

本文首先給出在 n 阶酉羣 U_n 上可积函数 $u(U)$ 的 Fourier 級数的一般球平均的积分表达式,当然对于球平均的定义本身需要有一些要求.从这点出发,我們証明了:如果 $u(U)$ 在 U_n 上連續,那末它的 Fourier 級数可以 δ 次 Riesz 求和于它自己,但 $\delta>\frac{n-1}{2}$.

同时还証明了:(1) 如果 $u(U)$ 在 U_n 上連續,那末它的 Fourier 級数可以 Gauss-Sommerfeld 求和于它自己.

(2) 如果 $u(U)$ 在 U_n 上連續,那末它的 Fourier 級数可以 Abel 求和于它自己.

这是本文的第一部分,必須注意的是:这里所說的求和,全是指"球"求和,卽从(5.1.3)出发加以考虑.与以往几篇文章所說的求和的意义是完全不同的.在以后,除非特别声明,所說的求和,全是指"球"求和.在第一部分的最后,还給出了一般的球求和的收斂定理.

以往考虑"方体"求和时,核是非常簡洁的,而求和的系数比較复杂.在"球"求和的情形下,这点是可以避免的.

* 1962年4月23日收到.

本文的第二部分定义了酉群上的 Fourier 积分，給出了 Dirichlet 核；定义了 Fourier 积分的“方体”求和及“球”求和，給出了相应的收斂定理；并指出了与 Fourier 級数的关系。

本文継續应用矩陣积分的技巧，此外还結合了 S. Bochner 处理多重 Fourier 級数与积分的技巧。

本文所用的符号与 [2, 3, 4] 中的符号相同，此外还采用了多重 Fourier 級数与积分中一些常用的符号。

本文是 1962 年 1 月至 2 月間作者在廈門大学完成的，作者謹向廈門大学，特别是数学系，表示衷心的感謝。

§ 5.2. Fourier 級数的球求和

为了考虑 Fourier 級数的球求和，我們限制应用 Toeplitz 矩陣 (a_{pq})，这里 $a_{pq}=\phi(p/q)$。当然 $\phi(t)$ 是定义在 $0\leqslant t<\infty$ 上的一个固定的函数，且 $\phi(0)=1$。所謂考虑某一种“球”求和，乃是考虑某一种平均

$$\sum_{m=0}^{\infty}\phi\left(\frac{m}{R}\right)\sum_{\substack{f_1\geqslant\cdots\geqslant f_n\\ l_1^2+\cdots+l_n^2=m}}\operatorname{tr}\left(C_{f_1\cdots f_n}\Phi'_{f_1\cdots f_n}(U)\right)\tag{5.2.1}$$

当 R 趋于无穷时的情形。如果当 $R\to\infty$ 时 (5.2.1)的极限存在，那末称 $u(U)$ 的 Fourier 級数可以按照某一种球求和于它的极限值。

最有兴趣而具体的几个 $\phi(t)$ 是以下的几个函数：

(1) $\phi(t)=e^{-t}$，这是 Abel 求和；

(2) $\phi(t)=e^{-t^2}$，这是 Gauss-Sommerfeld 求和；

(3) $\phi(t)=\begin{cases}(1-t^2)^\delta, & \text{当 } 0\leqslant t<1,\\ 0, & \text{当 } 1\leqslant t,\end{cases}$ 这是 δ 次 Riesz 求和。

說得更清楚些，所謂考虑 $u(U)$ 的 Fourier 級数的 Abel 求和，乃是考虑 Abel 平均

$$\sum_{\substack{f_1\geqslant\cdots\geqslant f_n\\ m=l_1^2+\cdots+l_n^2\leqslant R^2}}e^{-\frac{\sqrt{m}}{R}}\operatorname{tr}\left(C_{f_1\cdots f_n}\Phi'_{f_1\cdots f_n}(U)\right)\tag{5.2.2}$$

当 $R\to\infty$ 时的情形；Gauss-Sommerfeld 求和乃是考虑 Gauss-Sommerfeld 平均

$$\sum_{\substack{f_1\geqslant\cdots\geqslant f_n\\ m=l_1^2+\cdots+l_n^2\leqslant R^2}}e^{-\frac{m}{R^2}}\operatorname{tr}\left(C_{f_1\cdots f_n}\Phi'_{f_1\cdots f_n}(U)\right)\tag{5.2.3}$$

当 $R\to\infty$ 时的情形；δ 次 Riesz 求和，乃是考虑 δ 次 Riesz 平均

$$\sum_{\substack{f_1\geqslant\cdots\geqslant f_n\\ m=l_1^2+\cdots+l_n^2\leqslant R^2}}\left(1-\frac{m}{R^2}\right)^\delta\operatorname{tr}\left(C_{f_1\cdots f_n}\Phi'_{f_1\cdots f_n}(U)\right)\tag{5.2.4}$$

当 $R\to\infty$ 时的情形。

我們将要証明

定理 5.2.1. 設 $u(U)$ 在 U_n 上連續，那末它的 Fourier 級数可以 Abel 求和于它自

己，即(5.2.2)当 $R\to\infty$ 时趋于 $u(U)$ 自己.

定理 5.2.2. 設 $u(U)$ 在 U_n 上連續，那末它的 Fourier 級数可以 Gauss-Sommerfeld 求和于它自己，即(5.2.3)当 $R\to\infty$ 时趋于 $u(U)$ 自己.

定理 5.2.3. 設 $u(U)$ 在 U_n 上連續，那末它的 Fourier 級数可以 δ 次 Riesz 求和于它自己，但 $\delta>\frac{n-1}{2}$. 即当 $\delta>\frac{n-1}{2}$ 时，(5.2.4)当 $R\to\infty$ 时趋于 $u(U)$ 自己.

在定理証明过程中，我們可以发现，不論是 Abel 求和，Gauss-Sommerfeld 求和或 δ 次 Riesz 求和 $\left(\delta>\frac{n-1}{2}\right)$，都是"局部"性质.

§5.3. 积分表达式

考虑(5.2.1)，記之为 $S_R^{\phi}(U)$，这可以写成为

$$\frac{1}{c}\int_{U_n}u(WU)\sum_{\substack{f_1\geqslant\cdots\geqslant f_n\\ m=f_1^2+\cdots+f_n^2\leqslant R^2}}\phi\left(\frac{m}{R}\right)N(f)\chi_{f_1\cdots f_n}(\overline{W})\dot{W}. \tag{5.3.1}$$

記 $\overline{W}$ 的特征根为 $e^{i\theta_1},\cdots,e^{i\theta_n}$，那末由于[2]，我們知道

$$\sum_{f_1\geqslant\cdots\geqslant f_n}r^{|l_1|+\cdots+|l_n|}N(f)\chi_{f_1\cdots f_n}(\overline{W})$$

$$=\frac{i^{\frac{n(n-1)}{2}}}{(n-1)!\cdots2!1!D(e^{i\theta_1},\cdots,e^{i\theta_n})}\begin{vmatrix}p_r(\theta_1)&\cdots&p_r(\theta_n)\\ p_r'(\theta_1)&\cdots&p_r'(\theta_n)\\ \cdots&\cdots&\cdots\\ p_r^{(n-1)}(\theta_1)&\cdots&p_r^{(n-1)}(\theta_n)\end{vmatrix},$$

这里 $p_r(\theta)=\frac{1-r^2}{1-2r\cos\theta+r^2}$.

記 $g(\theta_1,\cdots,\theta_n)$ 为

$$\frac{i^{\frac{n(n-1)}{2}}}{(n-1)!\cdots2!1!}\begin{vmatrix}p_r(\theta_1)&\cdots&p_r(\theta_n)\\ p_r'(\theta_1)&\cdots&p_r'(\theta_n)\\ \cdots&\cdots&\cdots\\ p_r^{(n-1)}(\theta_1)&\cdots&p_r^{(n-1)}(\theta_n)\end{vmatrix}$$

$$=\sum_{f_1\geqslant\cdots\geqslant f_n}r^{|l_1|+\cdots+|l_n|}N(f)M_{f_1\cdots f_n}([e^{i\theta_1},\cdots,e^{i\theta_n}]), \tag{5.3.2}$$

設 $g(\theta_1,\cdots,\theta_n)$ 的 Fourier 級数为

$$\sum_{\infty>\nu_1,\cdots,\nu_n>-\infty}a_{\nu_1,\cdots,\nu_n}e^{i(\nu_1\theta_1+\cdots+\nu_n\theta_n)}.$$

由于[6]，我們知道

$$\sum_{\substack{f_1\geqslant\cdots\geqslant f_n\\ \nu_1^2+\cdots+\nu_n^2\leqslant R^2}}\phi\left(\frac{m}{R}\right)a_{\nu_1,\cdots,\nu_n}e^{i(\nu_1\theta_1+\cdots+\nu_n\theta_n)}=R\int_0^{\infty}g_0(t)H_{\phi}(tR)dt. \tag{5.3.3}$$

如果 $\phi(t)$ 在每个有限区間都絕对連續，且

$$\int_0^{\infty}|\phi(t)|t^{\frac{n-1}{2}}dt<\infty, \tag{5.3.4}$$

这里

$$g_\theta(t)=\frac{\Gamma\left(\frac{n}{2}\right)}{2(\pi)^{\frac{n}{2}}}\int_\sigma g(\theta_1+t\eta_1,\cdots,\theta_n+t\eta_n)d\sigma_\eta, \tag{5.3.5}$$

$$H_\phi(tR)=\frac{(tR)^{\frac{n}{2}}}{2^{\frac{n}{2}-1}\Gamma\left(\frac{n}{2}\right)}\int_0^\infty\phi(u)u^{\frac{n}{2}}J_{\frac{n-2}{2}}(utR)du, \tag{5.3.6}$$

而 $J_\mu(s)$ 是第一类 μ 阶 Bessel 函数，σ 表球面：$\eta_1^2+\cdots+\eta_n^2=1$，$d\sigma_\eta$ 表球面上的体积元素；另一方面，(5.3.3)的左边就是

$$\sum_{m=l_1^2+\cdots+l_n^2\leqslant R^2}\phi\left(\frac{m}{R}\right)r^{|l_1|+\cdots+|l_n|}N(f)M_{f_1\cdots f_n}([e^{i\theta_1},\cdots,e^{i\theta_n}]).$$

因而得到

$$\sum_{\substack{f_1\geqslant\cdots\geqslant f_n\\ m=l_1^2+\cdots+l_n^2\leqslant R^2}}\phi\left(\frac{m}{R}\right)r^{|l_1|+\cdots+|l_n|}N(f)\chi_{f_1\cdots f_n}(\overline{W})$$

$$=\frac{R}{D(e^{i\theta_1},\cdots,e^{i\theta_n})}\int_0^\infty g_\theta(t)H_\phi(tR)dt. \tag{5.3.7}$$

記上面的值为 $F(r,R,\overline{W})$，于是 $F(r,R,\overline{W})$ 又可以写成([7]第 IV 章)

$$\frac{R\Gamma\left(\frac{n}{2}\right)}{2(\pi)^{\frac{n}{2}}D(e^{i\theta_1},\cdots,e^{i\theta_n})}\int_0^\infty\int_\sigma g(\theta+t\eta)H_\phi(tR)d\sigma_\eta dt, \tag{5.3.8}$$

这里 $\theta=(\theta_1,\theta_2,\cdots,\theta_n)$，$\eta=(\eta_1,\cdots,\eta_n)$，而(5.3.8)又可以写成

$$\frac{R\Gamma\left(\frac{n}{2}\right)}{2\pi^{\frac{n}{2}}D(e^{i\theta_1},\cdots,e^{i\theta_n})}\int_{-\infty}^{\infty}\cdots\int g(\theta+\xi)\frac{H_\phi(|\xi|R)}{|\xi|^{n-1}}d\xi_1\cdots d\xi_n. \tag{5.3.9}$$

这里 $\xi=(\xi_1,\xi_2,\cdots,\xi_n)$，$|\xi|=(\xi_1^2+\cdots+\xi_n^2)^{\frac{1}{2}}$.

(5.3.9)又可以写成

$$\frac{\Gamma\left(\frac{n}{2}\right)R\cdot i^{\frac{n(n-1)}{2}}}{2\pi^{\frac{n}{2}}D(e^{i\theta_1},\cdots,e^{i\theta_n})(n-1)!\cdots2!1!}\sum_{i_0,i_1,\cdots,i_n}\int_{-\infty}^{\infty}\cdots\int\delta_{i_0,i_1,\cdots,i_{n-1}}^{0,1,\cdots,n-1}p_r^{(i_0)}(\theta_1+\xi_1)\cdots$$

$$p_r^{(i_n)}(\theta_n+\xi_n)\frac{H_\phi(|\xi|R)}{|\xi|^{n-1}}d\xi_1\cdots d\xi_n. \tag{5.3.10}$$

在上述积分中，对 p_r 的微商是对 θ 进行的，然后以 $\theta+\xi$ 代入．事实上，如果对 p_r 的微商是对 ξ 进行，然后以 $\theta+\xi$ 代入，結果是一样的．对上述每一个积分进行分部积分，如果 H_ϕ 滿足以下的性质：

$$\frac{\partial^{j_1}}{\partial\xi_1^{j_1}}\cdots\frac{\partial^{j_n}}{\partial\xi_n^{j_n}}\frac{H_\phi(|\xi|R)}{|\xi|^{n-1}}\bigg|_{|\xi|=\infty}=0, \tag{5.3.11}$$

这里 $0\leqslant j_1,\cdots,j_n\leqslant n-1$.

于是，(5.3.10)就成为

$$\frac{\Gamma\left(\frac{n}{2}\right) i^{\frac{n(n-1)}{2}} R}{2\pi^{\frac{n}{2}} D(e^{i\theta_1}, \cdots, e^{i\theta_n})(n-1)!\cdots 2!1!} \int_{-\infty}^{\infty} \cdots \int p_r(\theta_1+\xi_1)\cdots p_r(\theta_n+\xi_n)$$

$$\times D\left(\frac{\partial}{\partial \xi_1}, \cdots, \frac{\partial}{\partial \xi_n}\right) \frac{H_\phi(|\xi| R)}{|\xi|^{n-1}} d\xi_1 \cdots d\xi_n, \tag{5.3.12}$$

这里

$$D\left(\frac{\partial}{\partial \xi_1}, \cdots, \frac{\partial}{\partial \xi_n}\right) = (-1)^{\frac{n(n-1)}{2}} \begin{vmatrix} 1 & 1 & \cdots & 1 \\ \frac{\partial}{\partial \xi_1} & \frac{\partial}{\partial \xi_2} & \cdots & \frac{\partial}{\partial \xi_n} \\ \cdots & \cdots & \cdots & \cdots \\ \frac{\partial^{n-1}}{\partial \xi_1^{n-1}} & \frac{\partial^{n-1}}{\partial \xi_2^{n-1}} & \cdots & \frac{\partial^{n-1}}{\partial \xi_n^{n-1}} \end{vmatrix}.$$

記

$$S_R^\phi(r, U) = \sum_{\substack{f_1 \geqslant \cdots \geqslant f_n \\ m = l_1^2 + \cdots + l_n^2 \leqslant R^2}} \phi\left(\frac{m}{R}\right) r^{|l_1| + \cdots + |l_n|} \operatorname{tr}\left(C_{f_1 \cdots f_n} \Phi'_{f_1 \cdots f_n}(U)\right),$$

那末 $S_R^\phi(r, U)$ 等于

$$\frac{1}{c} \int_{U_n} u(WU) F(r, R, \overline{W}) \dot{W}, \tag{5.3.13}$$

而 $F(r, R, \overline{W})$ 为(5.3.8)所定义。

由于 $F(r, R, \overline{W})$ 只是 $\overline{W}$ 的特征根的对称函数，所以我們可以将 u 对于 $\overline{W}$ 的特征根进行对称化，如同在[4]中进行的那样，得到另一个函数 $u^*(WU)$. 可以証明(5.3.13)也等于

$$\frac{1}{c} \int_{U_n} u^*(WU) F(r, R, \overline{W}) \dot{W}.$$

記

$$\psi_U(e^{-i\theta_1}, \cdots, e^{-i\theta_n}) = \frac{1}{\varpi_n'} \int_{[\dot{W}]} u^*(WU)[\dot{W}],$$

显然 ψ_U 为 $\theta_1, \cdots, \theta_n$ 的对称函数，于是得到

$$S_R^\phi(r, U) = \frac{1}{(2\pi)^n} \int \cdots \int_{\pi \geqslant \theta_1 \geqslant \cdots \geqslant \theta_n \geqslant -\pi} \psi_U(e^{-i\theta_1}, \cdots, e^{-i\theta_n}) F(r, R, \overline{W}) |D(e^{i\theta_1}, \cdots, e^{i\theta_n})|^2$$

$$\times d\theta_1 \cdots d\theta_n. \tag{5.3.14}$$

将(5.3.12)的値代入(5.3.14)，应用 Fubini 定理，得到 $S_R^\phi(r, U)$ 等于

$$\frac{i^{\frac{n(n-1)}{2}} R \Gamma\left(\frac{n}{2}\right)}{2\pi^{\frac{n}{2}} (2\pi)^n (n-1)!\cdots 2!1!} \int_{-\infty}^{\infty} \cdots \int \int \cdots \int_{\pi \geqslant \theta_1 \geqslant \cdots \geqslant \theta_n \geqslant -\pi} \psi_U(e^{-i\theta_1}, \cdots, e^{-i\theta_n}) D(e^{-i\theta_1}, \cdots, e^{-i\theta_n})$$

$$\times p_r(\theta_1+\xi_1)\cdots p_r(\theta_n+\xi_n) D\left(\frac{\partial}{\partial \xi_1}, \cdots, \frac{\partial}{\partial \xi_n}\right) \frac{H_\phi(|\xi| R)}{|\xi|^{n-1}} d\theta_1 \cdots d\theta_n d\xi_1 \cdots d\xi_n. \tag{5.3.15}$$

在上述积分中，先将 θ 的次序进行各种不同的排列之后对 ξ 进行相应的排列，得到 $S_R^\phi(r, U)$ 等于

$$\frac{i^{\frac{n(n-1)}{2}}R\Gamma\left(\frac{n}{2}\right)}{2\pi^{\frac{n}{2}}n!(n-1)!\cdots 2!1!(2\pi)^n}\int_{-\infty}^{\infty}\cdots\int\int_{-\pi}^{\pi}\cdots\int\psi_U(e^{-i\theta_1},\cdots,e^{-i\theta_n})D(e^{-i\theta_1},\cdots,e^{-i\theta_n})$$
$$\times p_r(\theta_1+\xi_1)\cdots p_r(\theta_n+\xi_n)D\left(\frac{\partial}{\partial\xi_1},\cdots,\frac{\partial}{\partial\xi_n}\right)$$
$$\times\frac{H_\phi(|\xi|R)}{|\xi|^{n-1}}d\theta_1\cdots d\theta_n d\xi_1\cdots d\xi_n. \tag{5.3.16}$$

取 $\lim\limits_{r\to1}S_R^\phi(r,U)$，由 Lebesgue 从属定理及 Zygmund 关于多重 Fourier 級数的 Abel 求和定理[10]，知道(5.3.16)当 $r\to1$ 时为

$$\frac{\Gamma\left(\frac{n}{2}\right)i^{\frac{n(n-1)}{2}}R}{n!(n-1)!\cdots 2!1!2\pi^{\frac{n}{2}}}\int_{-\infty}^{\infty}\cdots\int D(e^{i\xi_1},\cdots,e^{i\xi_n})\psi_U(e^{i\xi_1},\cdots,e^{i\xi_n})$$
$$\times D\left(\frac{\partial}{\partial\xi_1},\cdots,\frac{\partial}{\partial\xi_n}\right)\frac{H_\phi(|\xi|R)}{|\xi|^{n-1}}d\xi_1\cdots d\xi_n. \tag{5.3.17}$$

于是得到(5.3.17)还可以表为

$$\frac{\Gamma\left(\frac{n}{2}\right)i^{\frac{n(n-1)}{2}}R}{2\pi^{\frac{n}{2}}(n-1)!\cdots 2!1!}\int\cdots\int_{\infty>\xi_1>\cdots>\xi_n>-\infty}D\left(\frac{\partial}{\partial\xi_1},\cdots,\frac{\partial}{\partial\xi_n}\right)\frac{H_\phi(|\xi|R)}{|\xi|^{n-1}}$$
$$\times D(e^{i\xi_1},\cdots,e^{i\xi_n})\psi_U(e^{i\xi_1},\cdots,e^{i\xi_n})d\xi_1\cdots d\xi_n. \tag{5.3.18}$$

最后得到

定理 5.3.1. 設 $u(U)$ 在 U_n 上可积，那末它的 Fourier 級数的某一种球平均(5.2.1)可以表为

$$\frac{\Gamma\left(\frac{n}{2}\right)R}{2\pi^{\frac{n}{2}}(n-1)!\cdots 2!1!}\int\cdots\int_{\infty>\xi_1>\cdots>\xi_n>-\infty}D(\lambda_1,\cdots,\lambda_n)\psi_U(\lambda_1,\cdots,\lambda_n)$$
$$\times D\left(\frac{\partial}{\partial\lambda_1}\cdots\frac{\partial}{\partial\lambda_n}\right)\frac{H_\phi(|\xi|R)}{|\xi|^{n-1}}d\xi_1\cdots d\xi_n, \tag{5.3.19}$$

这里 $\psi_U(\lambda_1,\cdots,\lambda_n)=\frac{1}{\varpi_n'}\int_{[\dot{W}]}u(WU)[\dot{W}]$，而 W 的特征根为 $\lambda_1,\cdots,\lambda_n$，且 $\lambda_j=e^{i\xi_j}$. 但是 $\phi(t)$ 必須滿足(5.3.4)及(5.3.11)，且在每一有限区間絕对連續.

(5.3.19)还可以表成(参閱[7]第 IV 章)

$$\frac{i^{\frac{n(n-1)}{2}}R}{n!(n-1)!\cdots 2!1!}\int_0^\infty h_U(t)t^{n-1}dt, \tag{5.3.20}$$

这里

$$h_U(t)=\frac{\Gamma\left(\frac{n}{2}\right)}{2(\pi)^{\frac{n}{2}}}\int_\sigma D(e^{it\eta_1},\cdots,e^{it\eta_n})\psi_U(e^{it\eta_1},\cdots,e^{it\eta_n})D\left(\frac{\partial}{\partial\xi_1}\cdots\frac{\partial}{\partial\xi_n}\right)$$
$$\times\frac{H_\phi(|\xi|R)}{|\xi|^{n-1}}\bigg|_{\xi=t\eta}d\sigma_\eta. \tag{5.3.21}$$

但 σ 表示球面 $\eta_1^2+\cdots+\eta_n^2=1$.

§ 5.4. Riesz 平均的表达式

从上一节的結果，我們立刻得到[6,7]

(1) 当 $\phi(t)=e^{-t}$，卽 Abel 求和时，

$$H_\phi=H_R^\phi(v)=\frac{(n-1)!}{2^{n-2}\left(\Gamma\left(\frac{n}{2}\right)\right)^2}\cdot\frac{v^{n-1}}{(1+v^2)^{\frac{n+1}{2}}};\tag{5.4.1}$$

(2) 当 $\phi(t)=e^{-t^2}$，卽 Gauss-Sommerfeld 求和时，

$$H_\phi=H_R^\phi(v)=\frac{v^{n-1}e^{-\frac{v^2}{4}}}{2^{n-1}\Gamma\left(\frac{n}{2}\right)};\tag{5.4.2}$$

(3) 当 $\phi(t)=\begin{cases}(1-t^2)^\delta, & \text{当 } 0\leqslant t<1,\\ 0, & \text{当 } 1\leqslant t,\end{cases}$

卽 δ 次 Riesz 求和时，

$$H_\phi=H_R^\phi(v)=\frac{2^{\delta-\frac{n}{2}+1}\Gamma(\delta+1)}{\Gamma\left(\frac{n}{2}\right)}v^{n-1}V_{\delta+\frac{n}{2}}(v),\tag{5.4.3}$$

这里

$$V_s(v)=\frac{J_s(v)}{v^s},$$

而 J_s 为第一类 s 阶 Bessel 函数.

显然，这三个 $\phi(t)$ 都在有限区間絕对連續，且均满足条件(5.3.4)，卽

$$\int_0^\infty|\phi(t)|t^{\frac{n-1}{2}}dt<\infty$$

(对于 Riesz 求和，δ 必須大于 $\frac{n-1}{2}$，$\phi(t)$ 才满足(5.3.4))．至于条件(5.3.11)，在这三种情形，分别为

(1) $\dfrac{\partial^{j_1}}{\partial\xi_1^{j_1}}\cdots\dfrac{\partial^{j_n}}{\partial\xi_n^{j_n}}\cdot\dfrac{1}{(1+|\xi|^2R^2)^{\frac{n+1}{2}}}$

当 $|\xi|=\infty$ 时均为零，但 $0\leqslant j_1,\cdots,j_n\leqslant n-1$;

(2) $\dfrac{\partial^{j_1}}{\partial\xi_1^{j_1}}\cdots\dfrac{\partial^{j_n}}{\partial\xi_n^{j_n}}e^{-\frac{|\xi|^2R^2}{4}}$

当 $|\xi|=\infty$ 时均为零，但 $0\leqslant j_1,\cdots,j_n\leqslant n-1$;

(3) $\dfrac{\partial^{j_1}}{\partial\xi_1^{j_1}}\cdots\dfrac{\partial^{j_n}}{\partial\xi_n^{j_n}}V_{\delta+\frac{n}{2}}(|\xi|R)$

当 $|\xi|=\infty$ 时均为零，但 $0\leqslant j_1,\cdots,j_n\leqslant n-1$.

(1)，(2)二种情形显然成立．至于(3)，我們注意到

$$\frac{d}{xdx}V_\mu(x)=-V_{\mu+1}(x)$$

及当 $t\to\infty$ 时

$$J_\mu(t)=O(t^{-\frac{1}{2}}),$$

就能得到(5.3.11)成立.

于是得到: $u(U)$ 的 Fourier 级数的 Abel 平均(5.2.2)可以表为

$$\frac{i^{\frac{n(n-1)}{2}}R^n}{2\pi^{\frac{n}{2}}(n-2)!\cdots2!1!2^{n-2}\Gamma\left(\frac{n}{2}\right)}\int\cdots\int_{\infty>\xi_1>\cdots>\xi_n>-\infty}D\left(\frac{\partial}{\partial\xi_1},\cdots,\frac{\partial}{\partial\xi_n}\right)$$

$$\times\frac{1}{(1+|\xi|^2R^2)^{\frac{n+1}{2}}}D(e^{i\xi_1},\cdots,e^{i\xi_n})\psi_U(e^{i\xi_1},\cdots,e^{i\xi_n})d\xi_1\cdots d\xi_n,\qquad(5.4.4)$$

这还可以表成(5.3.20),但 $h_U(t)$ 为

$$\frac{(n-1)!R^{n-1}}{2^{n-1}\pi^{\frac{n}{2}}\Gamma\left(\frac{n}{2}\right)}\int_\sigma D(e^{it\eta_1},\cdots,e^{it\eta_n})\psi_U(e^{it\eta_1},\cdots,e^{it\eta_n})$$

$$\times D\left(\frac{\partial}{\partial\xi_1},\cdots,\frac{\partial}{\partial\xi_n}\right)\frac{1}{(1+R^2|\xi|^2)^{\frac{n+1}{2}}}\bigg|_{\xi=t\eta}d\sigma_\eta.\qquad(5.4.5)$$

$u(U)$ 的 Fourier 级数的 Gauss-Sómmerfeld 平均(5.2.3)可以表为

$$\frac{i^{\frac{n(n-1)}{2}}R^n}{\pi^{\frac{n}{2}}(n-1)!\cdots2!1!2^n}\int\cdots\int_{\infty>\xi_1\geqslant\cdots\geqslant\xi_n>-\infty}D\left(\frac{\partial}{\partial\xi_1},\cdots,\frac{\partial}{\partial\xi_n}\right)e^{-\frac{|\xi|^2R^2}{4}}$$

$$\times D(e^{i\xi_1},\cdots,e^{i\xi_n})\psi_U(e^{i\xi_1},\cdots,e^{i\xi_n})d\xi_1\cdots d\xi_n.\qquad(5.4.6)$$

由于

$$\frac{\partial}{\partial\xi_j}e^{-\frac{R^2|\xi|^2}{4}}=-\frac{R^2\xi_j}{2}e^{-\frac{R^2|\xi|^2}{4}},$$

$$\frac{\partial^2}{\partial\xi_j^2}e^{-\frac{R^2|\xi|^2}{4}}=\left(-\frac{R^2}{2}+\frac{R^4}{4}\xi_j^2\right)e^{-\frac{R^2|\xi|^2}{4}},$$

$$\cdots\cdots\cdots\cdots\cdots\cdots\cdots\cdots.$$

应用行列式性质,得到

$$D\left(\frac{\partial}{\partial\xi_1},\cdots,\frac{\partial}{\partial\xi_n}\right)e^{-\frac{R^2|\xi|^2}{4}}=\left(\frac{-R^2}{2}\right)^{\frac{n(n-1)}{2}}D(\xi_1,\cdots,\xi_n)e^{-\frac{R^2|\xi|^2}{4}}.$$

所以(5.2.3)可以表为

$$\frac{(-i)^{\frac{n(n-1)}{2}}R^{n^2}}{\pi^{\frac{n}{2}}(n-1)!\cdots2!1!2^{\frac{n(n+1)}{2}}}\int\cdots\int_{\infty>\xi_1>\cdots>\xi_n>-\infty}e^{-\frac{R^2|\xi|^2}{4}}D(\xi_1,\cdots,\xi_n)$$

$$\times D(e^{i\xi_1},\cdots,e^{i\xi_n})\psi_U(e^{i\xi_1},\cdots,e^{i\xi_n})d\xi_1\cdots d\xi_n,\qquad(5.4.7)$$

这还可以表成(5.3.20),但 $h_U(t)$ 为

$$\frac{(-t)^{\frac{n(n-1)}{2}}R^{n^2-1}}{2^{\frac{n(n+1)}{2}}\pi^{\frac{n}{2}}e^{\frac{R^2t^2}{4}}}\int_\sigma D(e^{it\eta_1},\cdots,e^{it\eta_n})D(\eta_1,\cdots,\eta_n)\psi_U(e^{it\eta_1},\cdots,e^{it\eta_n})d\sigma_\eta.\qquad(5.4.8)$$

至于 $u(U)$ 的 Fourier 級数的 δ 次 Riesz 平均，当 $\delta > \frac{n-1}{2}$ 时，可以表为

$$\frac{i^{\frac{n(n-1)}{2}} R^n 2^{\delta-\frac{n}{2}} \Gamma(\delta+1)}{\pi^{n/2}(n-1)!\cdots 2!1!} \int\cdots\int_{\infty>\xi_1\geqslant\cdots\geqslant\xi_n>-\infty} D\left(\frac{\partial}{\partial\xi_1}, \cdots, \frac{\partial}{\partial\xi_n}\right) V_{\delta+\frac{n}{2}}(|\xi|R)$$

$$\times D(e^{i\xi_1}, \cdots, e^{i\xi_n})\psi_U(e^{i\xi_1}, \cdots, e^{i\xi_n}) d\xi_1\cdots d\xi_n, \tag{5.4.9}$$

这还可以表为(5.3.20)，但 $h_U(t)$ 为

$$\frac{2^{\delta-\frac{n}{2}}\Gamma(\delta+1)R^{n-1}}{\pi^{\frac{n}{2}}}\int_\sigma D(e^{it\eta_1}, \cdots, e^{it\eta_n})\psi_U(e^{it\eta_1}, \cdots, e^{it\eta_n})$$

$$\times D\left(\frac{\partial}{\partial\xi_1}, \cdots, \frac{\partial}{\partial\xi_n}\right) V_{\delta+\frac{n}{2}}(|\xi|R)\Big|_{\xi=t\eta} d\sigma_\eta. \tag{5.4.10}$$

§ 5.5. 定理 5.2.2 的証明

我們首先証明以下的

定理 5.5.1. 設 $\phi(t)$ 在有限区間絕对連續且滿足 (5.3.4) 及 (5.3.11)，那末当 $R^2 \geqslant \frac{(2n-1)n(n-1)}{6}$ 时，

$$\frac{\Gamma\left(\frac{n}{2}\right) i^{\frac{n(n-1)}{2}} R}{2\pi^{\frac{n}{2}}(n-1)!\cdots 2!1!} \int\cdots\int_{\infty>\xi_1\geqslant\cdots\geqslant\xi_n>-\infty} D\left(\frac{\partial}{\partial\xi_1}, \cdots, \frac{\partial}{\partial\xi_n}\right)\frac{H_\phi(|\xi|R)}{|\xi|^{n-1}}$$

$$\times D(e^{i\xi_1}, \cdots, e^{i\xi_n}) d\xi_1\cdots d\xi_n = 1. \tag{5.5.1}$$

証. (5.5.1)的左边就是

$$\frac{\Gamma\left(\frac{n}{2}\right)(-i)^{\frac{n(n-1)}{2}} R}{2\pi^{\frac{n}{2}}(n-1)!\cdots 2!1!}\int_{-\infty}^{\infty}\cdots\int \frac{\partial^0}{\partial\xi_1^0}\frac{\partial^1}{\partial\xi_2^1}\cdots\frac{\partial^{n-1}}{\partial\xi_n^{n-1}}\frac{H_\phi(|\xi|R)}{|\xi|^{n-1}}$$

$$\times D(e^{i\xi_1}, \cdots, e^{i\xi_n}) d\xi_1\cdots d\xi_n.$$

由于 H_ϕ 滿足(5.3.11)，所以上式就是

$$\frac{\Gamma\left(\frac{n}{2}\right) i^{\frac{n(n-1)}{2}} R}{2\pi^{\frac{n}{2}}(n-1)!\cdots 2!1!}\int_{-\infty}^{\infty}\cdots\int \frac{H_\phi(|\xi|R)}{|\xi|^{n-1}}\frac{\partial^0}{\partial\xi_1^0}\cdots\frac{\partial^{n-1}}{\partial\xi_n^{n-1}}$$

$$\times D(e^{i\xi_1}, \cdots, e^{i\xi_n}) d\xi_1\cdots d\xi_n. \tag{5.5.2}$$

(5.5.2)还可以写成

$$R\int_0^\infty f_0(t) H_\phi(tR)\, dt, \tag{5.5.3}$$

这里

$$f_0(t) = \frac{i^{\frac{n(n-1)}{2}}}{(n-1)!\cdots 2!1!}\cdot\frac{\Gamma\left(\frac{n}{2}\right)}{2(\pi)^{\frac{n}{2}}}\int_\sigma \frac{\partial^0}{\partial\xi_1^0}\frac{\partial^1}{\partial\xi_2^1}\cdots\frac{\partial^{n-1}}{\partial\xi_n^{n-1}} D(e^{i\xi_1}, \cdots, e^{i\xi_n})\Big|_{\xi=t\eta} d\sigma_\eta.$$

記 $f(\theta_1, \cdots, \theta_n)$ 为

$$\frac{i^{\frac{n(n-1)}{2}}}{(n-1)!\cdots 2!1!} \frac{\partial^0}{\partial\theta_1^0} \frac{\partial^1}{\partial\theta_2^1} \cdots \frac{\partial^{n-1}}{\partial\theta_n^{n-1}} D(e^{i\theta_1}, \cdots, e^{i\theta_n}),$$

它的多重 Fourier 級数为

$$\sum_{\nu_1,\cdots,\nu_n} b_{\nu_1,\cdots,\nu_n} e^{i(\nu_1\theta_1+\cdots+\nu_n\theta_n)}. \tag{5.5.4}$$

显然，如果 $\nu^2 \neq \frac{(2n-1)n(n-1)}{6}$，那末 $b_{\nu_1,\cdots,\nu_n} = 0$，由于 $\phi(t)$ 滿足(5.3.4)，所以(5.5.3)实际上就是 $f(\theta_1, \cdots, \theta_n)$ 的多重 Fourier 級数(5.5.4)在 ϕ 意义下的球平均在 $\theta = 0$ 处的値(参閱[6])，卽(5.5.3)等于

$$\sum_{\substack{\nu_1,\cdots,\nu_n \\ \nu_1^2+\cdots+\nu_n^2 \leqslant R^2}} \phi\left(\frac{\nu}{R}\right) b_{\nu_1,\cdots,\nu_n} e^{i(\nu_1\theta_1+\cdots+\nu_n\theta_n)} \tag{5.5.5}$$

在 $\theta = 0$ 处的値. 有鉴于 $\phi(0) = 1$ 及 D 的定义[4]，所以得到当 $R^2 \geqslant \frac{(2n-1)n(n-1)}{6}$ 时(5.5.1)成立.

现在証明定理 5.2.2.

記

$$\varphi_U(e^{i\theta_1}, \cdots, e^{i\theta_n}) = \frac{1}{\mathfrak{V}_n'} \int_{[\dot W]} (u^*(WU) - u(U))[\dot W], \tag{5.5.6}$$

于是根据定理 5.5.1，得到 $u(U)$ 的 Fourier 級数的 Gauss-Sommerfeld 平均(5.2.3) $G_R(U)$，与 $u(U)$ 之差，卽 $G_R(U) - u(U)$ 等于

$$\frac{i^{\frac{n(n-1)}{2}} R}{n!(n-1)!\cdots 2!1!} \int_0^\infty h_U^{(2)}(t) t^{n-1} dt, \tag{5.5.7}$$

这里 $h_U^{(2)}(t)$ 等于

$$\frac{(-t)^{\frac{n(n-1)}{2}} R^{n^2-1}}{2^{\frac{n(n+1)}{2}} \pi^{\frac{n}{2}} e^{\frac{R^2t^2}{4}}} \int_\sigma D(e^{it\eta_1}, \cdots, e^{it\eta_n}) D(\eta_1, \cdots, \eta_n) \varphi_U(e^{it\eta_1}, \cdots, e^{it\eta_n}) d\sigma_\eta. \tag{5.5.8}$$

将(5.5.7)的积分分成三部分，卽 $G_R(U) - u(U)$ 等于

$$\frac{i^{\frac{n(n-1)}{2}} R}{n!(n-1)!\cdots 2!1!} \left(\int_0^{\frac{1}{R}} + \int_{\frac{1}{R}}^\eta + \int_\eta^\infty\right) = I_1 + I_2 + I_3,$$

这里 η 为大于 $\frac{1}{R}$ 的一个固定常数.

先来考虑 I_1，記 $Rt = u$，于是 $h_U^{(2)}(t)$ 等于

$$c_1 u^{\frac{n(n-1)}{2}} R^{\frac{(n-1)(n+2)}{2}} e^{-\frac{u^2}{4}} \int_\sigma D(e^{\frac{i\eta_1 u}{R}}, \cdots, e^{\frac{i\eta_n u}{R}}) D(\eta_1, \cdots, \eta_n) \varphi_U(e^{\frac{i\eta_1 u}{R}}, \cdots, e^{\frac{i\eta_n u}{R}}) d\sigma_\eta,$$

这里 c_1 为絕对常数，根据 D 的定义，并注意到在 I_1 中 $0 \leqslant u \leqslant 1$，所以

$$h_U^{(2)}(t) = o(u^{\frac{n(n-1)}{2}} R^{n-1} e^{-\frac{u^2}{4}}).$$

这里用到了 $u(U)$ 的連續性，于是 I_1 等于

$$o\left(\int_0^1 e^{-\frac{u^2}{4}} u^{\frac{(n+2)(n-1)}{2}} du\right) = o(1).$$

至于 I_2，記

$$F(t) = \int_0^t |h_U^{(2)}(t)| t^{n-1} dt,$$

于是

$$F(t) \leqslant \frac{c_2 R^{n^2-1}}{e^{\frac{R^2t^2}{4}}} \int\cdots\int_{\eta_1^2+\cdots+\eta_n^2 \leqslant t^2} |D(e^{i\eta_1}, \cdots, e^{i\eta_n}) D(\eta_1, \cdots, \eta_n) \varphi_U(e^{i\eta_1}, \cdots, e^{i\eta_n})| d\eta_1 \cdots d\eta_n,$$

这里 c_2 为絕对常数. 由此得到当 $t \to 0$ 时，

$$F(t) = o(R^{n^2-1} e^{-\frac{R^2t^2}{4}} t^{n^2}).$$

因之

$$|I_2| \leqslant c_3 R \int_{\frac{1}{R}}^{\eta} dF(t) = c_3 R F(t) \Big|_{\frac{1}{R}}^{\eta} - c_3 R \int_{\frac{1}{R}}^{\eta} F(t) dt.$$

上式右边第一項为

$$O(R^{n^2} e^{-\frac{R^2\eta^2}{4}} \eta^{n^2}) + o(1),$$

第二項为

$$o\left(\int_{\frac{1}{R}}^{\eta} R^{n^2} t^{n^2} e^{-\frac{R^2t^2}{4}} dt\right) = o(1),$$

于是得到 $I_2 = o(1)$.

最后，对于 I_3，則有

$$|I_3| \leqslant c_4 \int_{\eta}^{\infty} R^{n^2} t^{\frac{n^2+n-2}{2}} e^{-\frac{R^2t^2}{4}} dt = o(1),$$

这里 c_3, c_4 均为絕对常数. 这样就証明了定理 5.2.2. 由于 $I_3 = o(1)$ 的証明并未用到 $u(U)$ 的性质，以致事实上还証明了：$u(U)$ 的 Fourier 級数的 Gauss-Sommerfeld 求和是局部性质.

§ 5.6. 定理 5.2.3 的証明

定理 5.2.3 的証明比定理 5.2.2 的証明略为困难一些，因为这时

$$D\left(\frac{\partial}{\partial \xi_1}, \cdots, \frac{\partial}{\partial \xi_n}\right) \frac{H_\phi(|\xi| R)}{|\xi|^{n-1}}$$

不能簡单地写出来. 但是有了定理 5.2.2 的証明，再来証明定理 5.2.3 就清楚得多了.

从 H_ϕ 的定义及(5.4.9)出发，根据 Bessel 函数的性质[7,8]，我們知道

$$\frac{\partial}{\partial \xi_j} V_{\delta+\frac{n}{2}}(|\xi| R) = -R^2 \xi_j V_{\delta+\frac{n}{2}+1}(|\xi| R),$$

$$\frac{\partial^2}{\partial \xi_j^2} V_{\delta+\frac{n}{2}}(|\xi| R) = -R^2 V_{\delta+\frac{n}{2}+1}(|\xi| R) + R^4 \xi_j^2 V_{\delta+\frac{n}{2}+2}(|\xi| R),$$

$$\left.\begin{aligned}&\frac{\partial^3}{\partial\xi_j^3}V_{\delta+\frac{n}{2}}(|\xi|R)=3R^4\xi_j V_{\delta+\frac{n}{2}+2}(|\xi|R)-R^6\xi_j^3 V_{\delta+\frac{n}{2}+3}(|\xi|R),\\&\frac{\partial^4}{\partial\xi_j^4}V_{\delta+\frac{n}{2}}(|\xi|R)=3R^4 V_{\delta+\frac{n}{2}+2}(|\xi|R)-6R^6\xi_j^2 V_{\delta+\frac{n}{2}+3}(|\xi|R)\\&\qquad+R^8\xi_j^4 V_{\delta+\frac{n}{2}+4}(|\xi|R),\\&\cdots\cdots\cdots\cdots\cdots\cdots\cdots\cdots.\end{aligned}\right\}\tag{5.6.1}$$

根据定理 5.5.1，得到 $u(U)$ 的 Fourier 级数的 δ 次 Riesz 平均 (5.2.4)．$S_R^\delta(U)$ 在 $\delta>\frac{n-1}{2}$ 时与 $u(U)$ 之差，即 $S_R^\delta(U)-u(U)$ 等于

$$\frac{i^{\frac{n(n-1)}{2}}R}{n!(n-1)!\cdots2!1!}\int_0^\infty h_U^{(3)}(t)t^{n-1}dt,\tag{5.6.2}$$

这里 $h_U^{(3)}(t)$ 等于

$$\frac{2^{\delta-\frac{n}{2}}\Gamma(\delta+1)R^{n-1}}{\pi^{n/2}}\int_\sigma D(e^{it\eta_1},\cdots,e^{it\eta_n})\varphi_U(e^{it\eta_1},\cdots,e^{it\eta_n})$$
$$\times D\left(\frac{\partial}{\partial\xi_1},\cdots,\frac{\partial}{\partial\xi_n}\right)V_{\delta+\frac{n}{2}}(|\xi|R)\Big|_{\xi=t\eta}d\sigma_\eta.\tag{5.6.3}$$

将(5.6.2)的积分分成三部分，即 $S_R^\delta(U)-u(U)$ 等于

$$\frac{i^{\frac{n(n-1)}{2}}R}{n!(n-1)!\cdots2!1!}\left(\int_0^{\frac{1}{R}}+\int_{\frac{1}{R}}^{\eta}+\int_\eta^\infty\right)=I_1+I_2+I_3,$$

这里 η 为大于 $\frac{1}{R}$ 的一个固定常数.

先研究 I_1，作变换 $Rt=u$，于是

$$I_1=\frac{i^{\frac{n(n-1)}{2}}}{n!(n-1)!\cdots2!1!}\int_0^1 h_U^{(3)}\left(\frac{u}{R}\right)\frac{u^{n-1}du}{R^{n-1}}.$$

根据 Bessel 函数的性质[7,8]，我們有

$$h_U^{(3)}\left(\frac{u}{R}\right)=o(R^{-\frac{n(n-1)}{2}}R^{n-1}R^{\frac{n(n-1)}{2}}u^{\frac{n(n-1)}{2}})+o(R^{n-3}u^{\frac{n(n-1)}{2}-2})$$
$$+\cdots+o(R^{-\frac{n(n-1)}{2}+n}u),$$

所以得到 $I_1=o(1)$.

再来考虑 I_2，这时 $\frac{1}{R}\leqslant t\leqslant\eta$. 根据 Bessel 函数的性质[7,8]，我們有

$$\left|D\left(\frac{\partial}{\partial\xi_1},\cdots,\frac{\partial}{\partial\xi_n}\right)\frac{H_\phi(|\xi|R)}{(R|\xi|)^{n-1}}\right|_{\xi=t\eta}=O\left(\frac{R^{n(n-1)}t^{\frac{n(n-1)}{2}}}{(Rt)^{n(\delta+\frac{n}{2})+\frac{n(n-1)}{2}+\frac{n}{2}}}\right)$$
$$+O\left(\frac{R^{n(n-1)}t^{\frac{n(n-1)}{2}}R^{-2}t^{-2}}{(Rt)^{n\delta+n^2}}\right)+\cdots+O\left(\frac{R^{n(n-1)}t^{\frac{n(n-1)}{2}}(Rt)^{\frac{-n(n-1)}{2}+1}}{(Rt)^{n\delta+n^2}}\right)$$

$$= O(R^{-n\delta-n}t^{-n\delta-\frac{n^2}{2}-\frac{n}{2}}) + O(R^{-n\delta-n-2}t^{-n\delta-\frac{n^2}{2}-\frac{n}{2}-2}) + \cdots$$
$$+ O(R^{-n\delta-\frac{n^2}{2}-\frac{n}{2}+1}t^{-n\delta-n^2+1}). \tag{5.6.4}$$

因之

$$R^{-n+1}|h_U^{(3)}(t)| \leqslant \frac{\Gamma\left(\frac{n}{2}\right)}{2\pi^{\frac{n}{2}}}\int_\sigma |D(e^{it\eta_1},\cdots,e^{it\eta_n})\varphi_U(e^{it\eta_1},\cdots,e^{it\eta_n})|d\sigma_\eta$$
$$\times[O(R^{-n\delta-n}t^{-n\delta-\frac{n^2}{2}-\frac{n}{2}}) + O(R^{-n\delta-n-2}t^{-n\delta-\frac{n^2}{2}-\frac{n}{2}-2}) + \cdots$$
$$+ O(R^{-n\delta-\frac{n^2}{2}-\frac{n}{2}+1}t^{-n\delta-n^2+1})]. \tag{5.6.5}$$

記

$$h_U^{(4)}(t) = \frac{\Gamma\left(\frac{n}{2}\right)}{2\pi^{\frac{n}{2}}}\int_\sigma |D(e^{it\eta_1},\cdots,e^{it\eta_n})\varphi_U(e^{it\eta_1},\cdots,e^{it\eta_n})|d\sigma_\eta \tag{5.6.6}$$

及

$$F(t) = \int_0^t h_U^{(4)}(t)t^{n-1}dt, \tag{5.6.7}$$

于是当 $t\to 0$ 时，我們有

$$F(t) = o(t^n t^{\frac{n(n-1)}{2}}) = o(t^{\frac{n(n+1)}{2}}). \tag{5.6.8}$$

从(5.6.2)，(5.6.4)，(5.6.5)，(5.6.6)及(5.6.7)，我們得到

$$|I_2| = O\left(\int_{\frac{1}{R}}^{\eta} R^{-n\delta}t^{-n\delta-\frac{n^2}{2}-\frac{n}{2}}dF(t)\right) + O\left(\int_{\frac{1}{R}}^{\eta} R^{-n\delta-2}t^{-n\delta-\frac{n^2}{2}-\frac{n}{2}-2}dF(t)\right)$$
$$+ \cdots + O\left(\int_{\frac{1}{R}}^{\eta} R^{-n\delta-\frac{n^2}{2}+\frac{n}{2}+1}t^{-n\delta-n^2+1}dF(t)\right).$$

上述各項分开来考慮，例如，我們考慮第一項

$$\int_{\frac{1}{R}}^{\eta} R^{-n\delta}t^{-n\delta-\frac{n^2}{2}-\frac{n}{2}}dF(t).$$

分部积分之，这等于

$$R^{-n\delta}t^{-n\delta-\frac{n^2}{2}-\frac{n}{2}}F(t)\Big|_{\frac{1}{R}}^{\eta} + \left(n\delta+\frac{n^2}{2}+\frac{n}{2}\right)\int_{\frac{1}{R}}^{\eta} R^{-n\delta}t^{-n\delta-\frac{n^2}{2}-\frac{n}{2}-1}F(t)dt = o(1).$$

同样可以处理其他各項，所以得到

$$I_2 = o(1).$$

至于 I_3，可以象处理 I_2 一样，只是在这时

$$F(t) = O(t^n). \tag{5.6.9}$$

于是

$$I_3 = O(R^{-\frac{n(n-1)}{2}}).$$

这样就証明了定理 5.2.3. 我們注意到証明 $I_3 = o(1)$ 时是用(5.6.9)代替(5.6.8)，所以，事实上我們还証明了：$u(U)$ 的 Fourier 級数的 δ 次 Riesz 求和是"局部"性质，但

$\delta > \frac{n-1}{2}$.

§5.7. 一条一般的收斂定理

如果 $\phi(t)$ 在有限区間絕对連續，且滿足(5.3.4)及(5.3.11)，那末 $u(U)$ 的 Fourier 級数在 ϕ 意义下的球平均可以写为(5.3.18)或(5.3.19)。

事实上，这还可以写成 (5.3.20)．如果記 (5.2.1) 为 $S_R^\phi(U)$，那末根据定理 5.5.1，$S_R^\phi(U)-u(U)$ 可以写成

$$\frac{i^{\frac{n(n-1)}{2}}R}{n!(n-1)!\cdots 2!1!}\int_0^\infty h_U^{(0)}(t)t^{n-1}dt, \tag{5.7.1}$$

而

$$h_U^{(0)}(t)=\frac{\Gamma\left(\frac{n}{2}\right)}{2\pi^{\frac{n}{2}}}\int_\sigma D(e^{it\eta_1},\cdots,e^{it\eta_n})\varphi_U(e^{it\eta_1},\cdots,e^{it\eta_n})$$
$$\times D\left(\frac{\partial}{\partial\xi_1},\cdots,\frac{\partial}{\partial\xi_n}\right)\frac{H_\phi(|\xi|R)}{|\xi|^{n-1}}\bigg|_{\xi=t\eta}d\sigma_\eta,$$

这里 φ_U 由(5.5.6)所定义。

将(5.7.1)的积分分成三部分：

$$\frac{i^{\frac{n(n-1)}{2}}R}{n!(n-1)!\cdots 2!1!}\left(\int_0^{\frac{1}{R}}+\int_{\frac{1}{R}}^\eta+\int_\eta^\infty\right)=I_1+I_2+I_3,$$

这里 η 为大于 $\frac{1}{R}$ 的一个固定的数。

先来研究 I_1．作变换 $Rt=u$，于是

$$I_1=\frac{i^{\frac{n(n-1)}{2}}}{n!(n-1)!\cdots 2!1!}\int_0^1 h_U^{(0)}\left(\frac{u}{R}\right)\frac{u^{n-1}du}{R^{n-1}}.$$

如果

$$h_U^{(0)}\left(\frac{u}{R}\right)=o(R^{n-1}), \tag{5.7.2}$$

那末就得到 $I_1=o(1)$．但是

$$h_U^{(0)}\left(\frac{u}{R}\right)=o\left(R^{\frac{-n(n-1)}{2}}D\left(\frac{\partial}{\partial\xi_1},\cdots,\frac{\partial}{\partial\xi_n}\right)\frac{H_\phi(|\xi|R)}{|\xi|^{n-1}}\bigg|_{\xi=\frac{u\eta}{R}}\right),$$

因而，若

$$D\left(\frac{\partial}{\partial\xi_1},\cdots,\frac{\partial}{\partial\xi_n}\right)\frac{H_\phi(|\xi|R)}{|\xi|^{n-1}}\bigg|_{\xi=\frac{u\eta}{R}}=O(R^{\frac{(n-1)(n+2)}{2}}) \tag{5.7.3}$$

成立，那末(5.7.2)成立。

其次我們来考虑 I_2．如果此时

$$\left|D\left(\frac{\partial}{\partial\xi_1},\cdots,\frac{\partial}{\partial\xi_n}\right)\frac{H_\phi(|\xi|R)}{(R|\xi|)^{n-1}}\right|_{\xi=t\eta}=O(R^{-p-n}t^{-p-\frac{n(n+1)}{2}}), \tag{5.7.4}$$

那末

$$|I_2| = O\left(\int_{\frac{1}{R}}^{\eta} R^{-p} t^{-p-\frac{n(n+1)}{2}} dF(t)\right).$$

分部积分之，即得

$$|I_2| = O\left(R^{-p} t^{-p-\frac{n(n+1)}{2}} F(t)\Big|_{\frac{1}{R}}^{\eta} + \left(p + \frac{n(n+1)}{2}\right)\int_{\frac{1}{R}}^{\eta} R^{-p} t^{-p-\frac{n(n+1)}{2}-1} F(t)dt\right) = o(1),$$

如果 $p > 0$。

至于 I_3，象处理 I_2 一样，只是这时

$$F(t) = O(t^n).$$

于是若(5.7.4)当 $\eta \leqslant t < \infty$ 时依然成立，且 $p > 0$，那末 $I_3 = o(1)$.

綜合起来，得到

定理 5.7.1. 設 $u(U)$ 在 U_n 上連續，它的 Fourier 級数在 ϕ 意义下的球平均为(5.2.1)，記作 $S_R^\phi(U)$，如果 $\phi(t)$ 在任一有限区間絕对連續且滿足(5.3.4)，(5.3.11). 此外，由 $\phi(t)$ 决定的 H_ϕ（見(5.3.6)）滿足以下二个条件：

(1) 当 $0 \leqslant t \leqslant \dfrac{1}{R}$ 时，

$$D\left(\frac{\partial}{\partial \xi_1}, \cdots, \frac{\partial}{\partial \xi_n}\right)\frac{H_\phi(|\xi|R)}{|\xi|^{n-1}}\Big|_{\xi=t\eta} = O(R^{\frac{(n-1)(n+2)}{2}}); \tag{5.7.5}$$

(2) 当 $\dfrac{1}{R} \leqslant t < \infty$ 时，

$$D\left(\frac{\partial}{\partial \xi_1}, \cdots, \frac{\partial}{\partial \xi_n}\right)\frac{H_\phi(|\xi|R)}{|\xi|^{n-1}}\Big|_{\xi=t\eta} = O(R^{-p-1} t^{-p-\frac{n(n+1)}{2}}), \tag{5.7.6}$$

而 $p > 0$. 那末当 $R \to \infty$ 时 $S_R^\phi(U)$ 收斂于 $u(U)$. 从此还可以看出，如果 $\phi(t)$ 滿足(5.3.4)，(5.3.11)，(5.7.5)及(5.7.6)，那末 Fourier 級数按照 ϕ 的意义球求和是"局部"性质.

从定理 5.7.1，立刻导出定理 5.2.1，因为我們注意到

$$\frac{\partial}{\partial \eta_i}\frac{1}{(1 + R^2|\eta|^2)^{\frac{n+1}{2}}} = \frac{-(n+1)R^2\eta_i}{(1 + R^2|\eta|^2)^{\frac{n+3}{2}}},$$

$$\frac{\partial^2}{\partial \eta_i^2}\frac{1}{(1 + R^2|\eta|^2)^{\frac{n+1}{2}}} = \frac{-(n+1)R^2}{(1 + R^2|\eta|^2)^{\frac{n+3}{2}}} + \frac{(n+1)(n+3)R^4\eta_i^2}{(1 + R^2|\eta|^2)^{\frac{n+5}{2}}},$$

………………………………

于是得到定理 5.7.1 中的条件全部滿足，此时 $p = -\dfrac{n(n+1)}{2}$.

§5.8. 一条 Tauber 型的收斂定理

在[5]中，我們用"方体"求和得到了一些逼近定理，对于球求和，显然也有相应的結果. 例如，我們可以有(参閱[5])

定理 5.8.1. 若 $u(U) \in \mathrm{Lip}\, p$，那末当 $\delta > \dfrac{n-1}{2}$ 时，有

$$S_R^\delta(U) - u(U) = O(R^{-p}). \tag{5.8.1}$$

从定理 5.2.3 的証明过程容易証明上述結果.

以此出发，可以証明以下一条收斂定理.

定理 5.8.2. 若 $u(U)\in \mathrm{Lip}\,p$，且

$$N(f)\,\mathrm{tr}\,(C_{f_1\cdots f_n}\overline{C}'_{f_1\cdots f_n})=o((l_1^2+\cdots+l_n^2)^{-\alpha}),\tag{5.8.2}$$

这里

$$\alpha=n-\frac{2p(1+\beta)}{n-1-2\beta+\varepsilon},\quad \beta\geqslant 0,\quad \varepsilon>0,$$

那末

$$S_R^{\beta}(U)-u(U)=o(1).\tag{5.8.3}$$

証. 由(5.8.1)得到：当 $\delta>\dfrac{n-1}{2}$ 时，

$$S_R^{\delta}(U)-u(U)=O(R^{-p}).$$

当 $t=O(R)$ 时，

$$|S((R+t)^{\frac{1}{2}})-S(R^{\frac{1}{2}})|\leqslant\sum_{R<m\leqslant R+t}|A_m(U)|,$$

这里

$$S(R)\equiv S_R^0(U)=S_R(U),$$

$$A_m(U)=\sum_{\substack{f_1\geqslant\cdots\geqslant f_n\\ m=l_1^2+\cdots+l_n^2}}\mathrm{tr}\,(C_{f_1\cdots f_n}\Phi'_{f_1\cdots f_n}(U)).$$

由 Буняковский 不等式

$$|\mathrm{tr}\,(C_{f_1\cdots f_n}\Phi'_{f_1\cdots f_n}(U))|^2\leqslant\mathrm{tr}\,(C_{f_1\cdots f_n}\overline{C}'_{f_1\cdots f_n})\,\mathrm{tr}\,(\Phi_{f_1\cdots f_n}(U)\overline{\Phi_{f_1\cdots f_n}(U)}')$$

$$=\frac{N(f)}{c}\,\mathrm{tr}\,(C_{f_1\cdots f_n}\overline{C}'_{f_1\cdots f_n}),$$

根据(5.8.2)，我們有(参閱[7]第 IV 章)

$$|S((R+t)^{\frac{1}{2}})-S(R^{\frac{1}{2}})|=o(\sum(l_1^2+\cdots+l_n^2)^{-\frac{\alpha}{2}})$$

$$=o\left(\sum_{R<m\leqslant R+t}r_n(m)m^{\frac{-\alpha}{2}}\right)=o\left(\int_R^{R+t}x^{-\frac{\alpha}{2}}dR_n(x)\right)=o(tR^{\frac{n}{2}-\frac{\alpha}{2}-1}),$$

这里 $r_n(m)$ 表示球面 $x_1^2+\cdots+x_n^2=m$ 上的整点个数. 我們知道(参閱[7])

$$r_n(m)=O(m^{\frac{n}{2}-1+\eta}),\quad \eta>0$$

及

$$R_n(x)=\sum_{m\leqslant x}r_n(m)=\pi^{\frac{n}{2}}\left\{\Gamma\left(\frac{n}{2}+1\right)\right\}^{-1}x^{\frac{n}{2}}+O(x^{\frac{n}{2}-\frac{n}{m+1}}).$$

应用[7]中定理 1.81，于是当 $0\leqslant\beta<\delta$ 时，

$$S_R^{\beta}(U)-u(U)=o(R^{\frac{2}{\delta+1}\left\{\delta\beta+(\delta-\beta)\left(\frac{n}{2}-\frac{\alpha}{2}\right)+\beta\left(1-\frac{p}{2}\right)-\frac{p}{2}\right\}-2\beta}).$$

取

$$\alpha=n-\frac{p(1+\beta)}{\delta-\beta},$$

即得(5.8.3)，但 $\delta > \frac{1}{2}(n-1)$，现在特别取

$$\delta = \frac{1}{2}(n-1) + \frac{\varepsilon}{2},$$

这里 $\varepsilon > 0$，即得定理．

从上面的定理，立刻得到以下的定理．这个定理作为定理 5.2.3 的补充．

定理 5.8.3. 若 $u(U) \in \mathrm{Lip}\, p$，且

$$N(f)\,\mathrm{tr}\,(C_{f_1\cdots f_n}\bar{C}'_{f_1\cdots f_n}) = o(1), \tag{5.8.4}$$

那末当 $\delta \leqslant \frac{n-1}{2}$ 而 $R \to \infty$ 时，$S_R^\delta(U)$ 以 $u(U)$ 为极限．

在定理 5.8.2 中取 $\beta = 0$，則得一条收斂定理．

定理 5.8.4. 若 $u(U) \in \mathrm{Lip}\, p$，且

$$N(f)\,\mathrm{tr}\,(C_{f_1\cdots f_n}\bar{C}'_{f_1\cdots f_n}) = o((l_1^2 + \cdots + l_n^2)^{-\alpha}), \tag{5.8.5}$$

那末 $u(U)$ 的 Fourier 級数收斂于 $u(U)$ 自己，但

$$\alpha = n - \frac{2p}{n-1+\varepsilon}, \quad \varepsilon > 0.$$

§ 5.9. Fourier 积分

我們来定义酉羣上的 Fourier 积分．

我們知道在 U_n 上可积的函数 $u(U)$ 的 Fourier 系数为

$$C_{f_1\cdots f_n} = \int_{U_n} u(V)\overline{\Phi_{f_1\cdots f_n}(V)}\dot{V}.$$

把 Fourier 級数中 $(N(f))^2$ 項

$$\mathrm{tr}\,(C_{f_1\cdots f_n}\Phi'_{f_1\cdots f_n}(U)) = \frac{1}{c}\int_{U_n} u(V)N(f)\chi_{f_1\cdots f_n}(\bar{V}U')\dot{V}$$

看作一項．

設 $V = WU$，記

$$\frac{1}{\varpi'_n}\int_{[\dot{W}]} u(WU)[\dot{W}] = \psi_U(e^{-i\theta_1}, \cdots, e^{-i\theta_n}), \tag{5.9.1}$$

$$\underset{N \geqslant l_1 > \cdots > l_n > -N}{\int\cdots\int} N(f)\chi_{f_1\cdots f_n}(\overline{W})df_1\cdots df_n = \vartheta_N(\theta_1, \cdots, \theta_n), \tag{5.9.2}$$

这里 $e^{i\theta_1}, \cdots, e^{i\theta_n}$ 为 $\overline{W}$ 的特征根，而 $f_1, \cdots, f_n$ 为滿足 $f_1 \geqslant \cdots \geqslant f_n$ 的实数，$\chi_{f_1\cdots f_n}(\overline{W})$ 被定义为

$$\begin{vmatrix} e^{il_1\theta_1} & \cdots & e^{il_1\theta_n} \\ e^{il_2\theta_1} & \cdots & e^{il_2\theta_n} \\ \cdots & \cdots & \cdots \\ e^{il_n\theta_1} & \cdots & e^{il_n\theta_n} \end{vmatrix} \Bigg/ \begin{vmatrix} e^{i(n-1)\theta_1} & \cdots & e^{i(n-1)\theta_n} \\ \cdots & \cdots & \cdots \\ e^{i\theta_1} & \cdots & e^{i\theta_n} \\ 1 & \cdots & 1 \end{vmatrix}. \tag{5.9.3}$$

于是我們定义 $u(U)$ 在 U_n 上的 Fourier 积分为

$$\lim_{N\to\infty}\frac{1}{(2\pi)^n}\underset{\infty>\theta_1\geqslant\cdots>\theta_n>-\infty}{\int\cdots\int}\psi_U(e^{-i\theta_1},\cdots,e^{-i\theta_n})\mathscr{D}_N(\theta_1,\cdots,\theta_n)$$
$$\times|D(e^{i\theta_1},\cdots,e^{i\theta_n})|^2d\theta_1\cdots d\theta_n, \tag{5.9.4}$$

而 $\mathscr{D}_N(\theta_1,\cdots,\theta_n)$ 称为 Fourier 积分的 Dirichlet 核.

我們也可以如同 Fourier 級数那样定义 $u(U)$ 的 Fourier 积分的"方体"求和及"球"求和. 例如, 我們可以定义 $u(U)$ 的 Fourier 积分的"方体"的 Fejer 平均为(参閱[3]及[9])

$$\frac{1}{(\pi)^n}\underset{\infty>\theta_1\geqslant\cdots\geqslant\theta_n>-\infty}{\int\cdots\int}\psi_U(e^{-i\theta_1},\cdots,e^{-i\theta_n})\frac{(\sin N\theta_1\cdots\sin N\theta_n)^{2n}}{N^n(\theta_1\cdots\theta_n)^{2n}}$$
$$\times|D(e^{i\theta_1},\cdots,e^{i\theta_n})|^2d\theta_1\cdots d\theta_n. \tag{5.9.5}$$

給定 $\phi(t)$,如果滿足

$$\int_0^\infty|\phi(t)|t^{n-1}dt<\infty, \tag{5.9.6}$$

那末我們定义 $u(U)$ 的 Fourier 积分在 ϕ 意义下的球平均为

$$\frac{1}{(2\pi)^n}\underset{\infty>\theta_1\geqslant\cdots\geqslant\theta_n>-\infty}{\int\cdots\int}\psi_U(\theta_1,\cdots,\theta_n)K_R^\phi(\theta_1,\cdots,\theta_n)$$
$$\times|D(e^{i\theta_1},\cdots,e^{i\theta_n})|^2d\theta_1\cdots d\theta_n, \tag{5.9.7}$$

这里 $K_R^\phi(\theta_1,\cdots,\theta_n)$ 被定义为

$$\underset{\infty>l_1>\cdots>l_n>-\infty}{\int\cdots\int}\phi\left(\frac{v}{R}\right)N(f)\chi_{l_1\cdots l_n}(\overline{W})df_1\cdots df_n, \tag{5.9.8}$$

而 $v^2=l_1^2+\cdots+l_n^2$.

§5.10. Fourier 积分的 Dirichlet 核

我們計算由(5.9.2)定义的 Dirichlet 核. 首先看到 $\mathscr{D}_N(\theta_1,\cdots,\theta_n)$ 等于 $P(\theta_1,\cdots,\theta_n)/D(n-1,\cdots,1,0)D(e^{i\theta_1},\cdots,e^{i\theta_n})$ 而

$$P(\theta_1,\cdots,\theta_n)=\underset{N>l_1>\cdots>l_n\geqslant-N}{\int\cdots\int}\begin{vmatrix}l_1^{n-1}&\cdots&l_n^{n-1}\\ \cdots&\cdots&\cdots\\ l_1&\cdots&l_n\\ 1&\cdots&1\end{vmatrix}\cdot\begin{vmatrix}e^{il_1\theta_1}&\cdots&e^{il_1\theta_n}\\ \cdots&\cdots&\cdots\\ e^{il_{n-1}\theta_1}&\cdots&e^{il_{n-1}\theta_n}\\ e^{il_n\theta_1}&\cdots&e^{il_n\theta_n}\end{vmatrix}dl_1\cdots dl_n$$

$$=\int_{-N}^N\cdots\int_{-N}^N l_1^{n-1}l_2^{n-2}\cdots l_{n-1}^1l_n^0\begin{vmatrix}e^{il_1\theta_1}&\cdots&e^{il_1\theta_n}\\ \cdots&\cdots&\cdots\\ e^{il_n\theta_1}&\cdots&e^{il_n\theta_n}\end{vmatrix}dl_1\cdots dl_n$$

$$=\sum\delta_{i_1,i_2,\cdots,i_n}^{1,2,\cdots,n}\int_{-N}^N\cdots\int_{-N}^N l_1^{n-1}\cdots l_{n-1}^1l_n^0e^{il_1\theta_{i_1}}\cdots e^{il_n\theta_{i_n}}dl_1\cdots dl_n.$$

而这又等于

$$(-1)^{\frac{n(n-1)}{2}}\sum \delta^{1,2,\cdots,n}_{j_1,j_2,\cdots,j_n}\int_{-N}^{N}\cdots\int_{-N}^{N} l_n^{n-1}\cdots l_2^1 l_1^0 e^{il_{j_1}\theta_1}\cdots e^{il_{j_n}\theta_n}dl_1\cdots dl_n$$

$$=(-1)^{\frac{n(n-1)}{2}}\sum \delta^{1,\cdots,n}_{j_1,\cdots,j_n}\prod_{k=1}^{n}\int_{-N}^{N} l_{j_k}^{j_k-1}e^{il_{j_k}\theta_k}dl_{j_k}$$

$$=(-1)^{\frac{n(n-1)}{2}}\sum \delta^{1,\cdots,n}_{j_1,\cdots,j_n}\frac{1}{i^{j_1-1}}\left(\frac{2\sin N\theta_1}{\theta_1}\right)^{(j_1-1)}\cdots\frac{1}{i^{j_n-1}}\left(\frac{2\sin N\theta_n}{\theta_n}\right)^{(j_n-1)}.$$

把它写成行列式，这就是

$$i^{\frac{n(n-1)}{2}}\begin{vmatrix} k_N(\theta_1) & k_N(\theta_2) & \cdots & k_N(\theta_n)\\ k_N'(\theta_1) & k_N'(\theta_2) & \cdots & k_N'(\theta_n)\\ \cdots & \cdots & \cdots & \cdots\\ k_N^{(n-1)}(\theta_1) & k_N^{(n-1)}(\theta_2) & \cdots & k_N^{(n-1)}(\theta_n)\end{vmatrix},$$

这里

$$k_N(\theta)=2\sin N\theta/\theta.$$

于是得到 Fourier 积分的 Dirichlet 核为

$$\frac{(-i)^{\frac{n(n-1)}{2}}}{(n-1)!\cdots 2!1!D(e^{i\theta_1},\cdots,e^{i\theta_n})}D\left(\frac{\partial}{\partial\theta_1},\cdots,\frac{\partial}{\partial\theta_n}\right)(k_N(\theta_1)\cdots k_N(\theta_n)) \quad (5.10.1)$$

或写成

$$\frac{D\left(\frac{\partial}{\partial\lambda_1},\cdots,\frac{\partial}{\partial\lambda_n}\right)(k_N(\theta_1)\cdots k_N(\theta_n))}{(n-1)!\cdots 2!1!D(\bar\lambda_1,\cdots,\bar\lambda_n)}, \quad (5.10.2)$$

这里 $\lambda_1,\cdots,\lambda_n$ 为 W 的特征根。

值得注意的是 Fourier 积分的 Dirichlet 核与 Fourier 級数的 Dirichlet 核(参閱 § 1.8)具有同一形式，只是将 $d_N(\theta)$ 换成了 $k_N(\theta)$。

如同[4]中一样，从(5.10.1)出发并应用多重 Fourier 积分理論(例如参閱[9]，[10])可以証明如下的收斂定理。

定理 5.10.1. 設 $u(U)\in C^{n-1}$，那末 $u(U)$ 的 Fourier 积分收斂于 $u(U)$。

如同[3]中一样，从(5.9.5)出发并应用多重 Fourier 积分理論，可以証明

定理 5.10.2. 設 $u(U)$ 在 U_n 上連續，那末它的 Fourier 积分的 Fejer 平均(5.9.5)(在"方体"的意义下)收斂于 $u(U)$。

§ 5.11. Fourier 积分的球平均

我們計算由(5.9.8)定义的 $K_R^{\phi}(\theta_1,\cdots,\theta_n)$，这时 $K_R^{\phi}(\theta_1,\cdots,\theta_n)=Q(\theta_1,\cdots,\theta_n)/D(n-1,\cdots,1,0)D(e^{i\theta_1},\cdots,e^{i\theta_n})$，而 $Q(\theta_1,\cdots,\theta_n)$ 等于

$$\int\cdots\int_{l_1>\cdots>l_n}\phi\left(\frac{v}{R}\right)\begin{vmatrix} l_1^{n-1} & \cdots & l_n^{n-1}\\ \cdots & \cdots & \cdots\\ \cdots & \cdots & \cdots\\ l_1 & \cdots & l_n\\ 1 & \cdots & 1\end{vmatrix}\cdot\begin{vmatrix} e^{il_1\theta_1} & \cdots & e^{il_1\theta_n}\\ \cdots & \cdots & \cdots\\ e^{il_{n-1}\theta_1} & \cdots & e^{il_{n-1}\theta_n}\\ e^{il_n\theta_1} & \cdots & e^{il_n\theta_n}\end{vmatrix}dl_1\cdots dl_n.$$

上式又等于

$$\int_{-\infty}^{\infty}\cdots\int\phi\left(\frac{v}{R}\right)l_1^{n-1}l_2^{n-2}\cdots l_n^0\begin{vmatrix}e^{il_1\theta_1}&\cdots&e^{il_1\theta_n}\\ \cdots&\cdots&\cdots\\ e^{il_n\theta_1}&\cdots&e^{il_n\theta_n}\end{vmatrix}dl_1\cdots dl_n$$

$$=\sum\delta_{j_1,\cdots,j_n}^{1,\cdots,n}\int_{-\infty}^{\infty}\cdots\int\phi\left(\frac{v}{R}\right)l_1^{n-1}\cdots l_n^0\,e^{il_{j_1}\theta_1}\cdots e^{il_{j_n}\theta_n}dl_1\cdots dl_n$$

$$=(-1)^{\frac{n(n-1)}{2}}\sum\delta_{j_1,\cdots,j_n}^{1,\cdots,n}\int_{-\infty}^{\infty}\cdots\int\phi\left(\frac{v}{R}\right)l_n^{n-1}\cdots l_1^0\,e^{il_{j_1}\theta_1}\cdots e^{il_{j_n}\theta_n}dl_1\cdots dl_n. \tag{5.11.1}$$

但是

$$\int_{-\infty}^{\infty}\cdots\int\phi\left(\frac{v}{R}\right)l_n^{n-1}\cdots l_1^0e^{il_{j_1}\theta_1}\cdots e^{il_{j_n}\theta_n}dl_1\cdots dl_n$$

$$=i^{\frac{-n(n-1)}{2}}\left(\frac{\partial}{\partial\theta_1}\right)^{j_1-1}\cdots\left(\frac{\partial}{\partial\theta_n}\right)^{j_n-1}\int_{-\infty}^{\infty}\cdots\int\phi\left(\frac{v}{R}\right)e^{il_{j_1}\theta_1}\cdots e^{il_{j_n}\theta_n}dl_1\cdots dl_n. \tag{5.11.2}$$

根据[6]引理 1，当 $\phi(t)$ 满足(5.9.6)时，上式即为

$$\frac{\Gamma\left(\frac{n}{2}\right)i^{\frac{-n(n-1)}{2}}R(2\pi)^{\frac{n}{2}}}{2^{1-\frac{n}{2}}}\left(\frac{\partial}{\partial\theta_1}\right)^{j_1-1}\cdots\left(\frac{\partial}{\partial\theta_n}\right)^{j_n-1}|\theta|^{-n+1}H_\phi(|\theta|R), \tag{5.11.3}$$

这里 H_ϕ 由(5.3.6)所定义. $|\theta|^2=\theta_1^2+\cdots+\theta_n^2$.

于是由(5.11.3)代入(5.11.1)，得到 $Q(\theta_1,\cdots,\theta_n)$ 等于

$$\Gamma\left(\frac{n}{2}\right)2^{\frac{n}{2}-1}(-i)^{\frac{n(n-1)}{2}}R(2\pi)^{\frac{n}{2}}D\left(\frac{\partial}{\partial\theta_1},\cdots,\frac{\partial}{\partial\theta_n}\right)\frac{H_\phi(|\theta|R)}{|\theta|^{n-1}}.$$

于是得到 $K_R^\phi(\theta_1,\cdots,\theta_n)$ 等于

$$\frac{\Gamma\left(\frac{n}{2}\right)2^{\frac{n}{2}-1}(-i)^{\frac{n(n-1)}{2}}R(2\pi)^{\frac{n}{2}}}{D(n-1,\cdots,2,1)D(e^{i\theta_1},\cdots,e^{i\theta_n})}D\left(\frac{\partial}{\partial\theta_1},\cdots,\frac{\partial}{\partial\theta_n}\right)\frac{H_\phi(|\theta|R)}{|\theta|^{n-1}}. \tag{5.11.4}$$

以(5.11.4)代入(5.9.7)，得到 $u(U)$ 的 Fourier 积分在 ϕ 意义下的球平均为

$$\frac{\Gamma\left(\frac{n}{2}\right)(-i)^{\frac{n(n-1)}{2}}R}{2\pi^{\frac{n}{2}}(n-1)!\cdots2!1!}\int_{\infty>\theta_1\geqslant\cdots\geqslant\theta_n>-\infty}\cdots\int\psi_U(e^{-i\theta_1},\cdots,e^{-i\theta_n})D\left(\frac{\partial}{\partial\theta_1},\cdots,\frac{\partial}{\partial\theta_n}\right)\frac{H_\phi(|\theta|R)}{|\theta|^{n-1}}$$
$$\times D(e^{-i\theta_1},\cdots,e^{-i\theta_n})d\theta_1\cdots d\theta_n. \tag{5.11.5}$$

而这又可以写成

$$\frac{\Gamma\left(\frac{n}{2}\right)i^{\frac{n(n-1)}{2}}R}{2\pi^{\frac{n}{2}}(n-1)!\cdots2!1!}\int_{\infty>\xi_1\geqslant\cdots\geqslant\xi_n>-\infty}\cdots\int\psi_U(e^{i\xi_1},\cdots,e^{i\xi_n})D\left(\frac{\partial}{\partial\xi_1},\cdots,\frac{\partial}{\partial\xi_n}\right)$$
$$\times\frac{H_\phi(|\xi|R)}{|\xi|^{n-1}}D(e^{i\xi_1},\cdots,e^{i\xi_n})d\xi_1\cdots d\xi_n,$$

而 $|\xi|^2=\xi_1^2+\cdots+\xi_n^2$，此即(5.3.18)．当然，由此还可以得到(5.9.7)，还可以写成

(5.3.19)或(5.3.20).

这里特别值得注意的是：当 $\phi(t)$ 满足 (5.9.6) 时，$u(U)$ 的 Fourier 积分按 ϕ 意义的球平均与 $u(U)$ 的 Fourier 級数按 ϕ 意义的球平均是完全一样的，但后者的 ϕ 必須满足(5.3.4)及在任一有限区域是絕对連續的.

在多重 Fourier 級数与多重 Fourier 积分之間，在球求和的意义下，也有相应的結果.

这也可以从另一个角度来說明，我們为什么定义(5.9.4)作为 $u(U)$ 的 Fourier 积分.

§5.12. 附　記

(1) 所有定理中的条件：$u(U)$ 在 U_n 上連續，显然是可以減輕的.

(2) Fourier 級数的球求和理論，可以沿着本文的方法継續发展. 例如，建立絕对求和概念，现在已无太大困难，求和的充要条件也是一样. 所有这些，可以参照[6]或[7].

(3) 关于 Fourier 积分理論，本文只是开了一个头，可以进一步发展下去.

参　考　文　献

[1] 华罗庚，多复变数函数论中的典型域的调和分析，科学出版社，北京，1958.
[2] 龚　昇，酉群上的富理埃分析(I)，数学学报，**10** (1960)，239—261.
[3] 龚　昇，酉群上的富理埃分析(II)，数学学报，**12** (1962)，17—31.
[4] 龚　昇，酉群上的富理埃分析(III)，数学学报，**13** (1963)，152—161.
[5] 龚　昇，酉群上的富理埃分析(IV)，数学学报，**13** (1963)，323—331.
[6] Bochner S., Summation of multiple Fourier series by spherical means, *Tran. Amer. Math. Soc.*, **40** (1936), 175—267.
[7] Chandrasekharan K. and Minakshisundaran S., Typical means, Oxford University Press, Oxford, 1952.
[8] Erdélyi A., Higher Transcendental Functions, Vol. II, McGraw-Hill, New York, 1953.
[9] Titchmarch E. C., An introduction to the theory of Fourier integrale, Oxford University Press, Oxford, 1948.
[10] Zygmund A., Trigonometric Series, Vol. I and II, Cambridge University Press, Cambridge, 1959.

旋转群上的富理埃级数的部分和

龚　　昇

（中国科技大学，中国科学院应用数学研究推广办公室）

1958年，华罗庚教授[1]証明了如下的重要定理：酉群上的連續函数的Fourier級数可以Abel求和于它自己，这是比著名的Peter—Weyl定理更为深刻的定理、作者在此基础上研究了酉群上的富理埃分析[4,5,6,7,8]由于任意紧致子群上的連續函数可以拓展成酉群上的連續函数，所以对酉群上的Fourier分析的研究，对于一般的紧致群也是有意义的。

对于其它的典型群，也可以参照酉群的情形相应地建立起它們的調和分析，但具体討論时，各有其特殊的困难。

陆启鏗教授[2]給出了n維旋轉群$SO(n)$的Poisson核，这就有了$SO(n)$上的Fourier級数的Abel求和，1965年，鐘家庆同志[3]用母函数的办法定出了Abel求和的系数，这是由于$SO(n)$的表示的特征比較复杂，所以采用簡接的办法来定出Abel求和的系数。

本文的目的是应用华罗庚教授[4]定出酉群上的Fourier級数的部分和的Dirichlet核的办法定出了$SO(n)$的相应的Dirichlet核。

所謂旋轉群是指$SO(n)=O^+(n)=\{\Gamma \mid \Gamma\Gamma'=I,\ \det\Gamma=1\}$．$SO(n)$中元素$\Gamma$的共軛类代表視$n$为奇、偶而分为两种情形：

1°，$n=2k$

$$\Gamma\sim\begin{pmatrix} C(\theta_1) & & \\ & \ddots & \\ & & C(\theta_k)\end{pmatrix},\ C(\theta_i)=\begin{pmatrix}\cos\theta i, & -\sin\theta_i\\ \sin\theta i, & \cos\theta i\end{pmatrix}.$$

2°，$n=2k+1$

$$\Gamma\sim\begin{pmatrix} C(\theta_1) & & \\ & \ddots & \\ & & C(\theta_k)\end{pmatrix}.$$

$SO(n)$的单值既約表示由k个非負整数$m=(m_1,\cdots,m_k)m_1\geqslant\cdots\geqslant m_k\geqslant 0$表征，$m$称为标签$\Gamma\to(\varphi^m_{ij}(\Gamma))$为$SO(n)$的以$m=(m_1,\ \cdots,\ m_k)$为标签的既約表示，$N(m)$是該表示的維数，表示的特征記作$\sigma_m(\Gamma)$．若$u(\Gamma)$为$SO(n)$上的可积函数，$C$为$SO(n)$的体积，令

$$a^m_{ij}=\frac{N(m)}{C}\int_{SO(n)}\varphi^m_{ij}(\Gamma)u(\Gamma)\dot{\Gamma},$$

則

$$\sum_{m\geqslant 0}\sum_{i,j}^{N(m)}a^m_{ij}\varphi^m_{ij}(\Gamma)$$

称为$u(\Gamma)$的Fourier級数、a^m_{ij}为$u(\Gamma)$的Fourier系数、記$l_1=m_1+k-1,\cdots,l_k=m_k$,

则

$$S_N(\Gamma)=\sum_{N\geqslant l_1>\cdots>l_k\geqslant 0}\sum_{i,j}^{N(m)}a_{ij}^m\varphi_{ij}^m(\Gamma)$$

为 $u(\Gamma)$ 的 Fourier 級数的部分和。

显然

$$S_N(\Gamma_0)=\frac{1}{C}\int_{SO(n)}u(\Gamma\Gamma_0)\sum_{N\geqslant l_1>\cdots>l_k\geqslant 0}N(m)\sigma_m(\Gamma)\dot{\Gamma}.$$

因之，

$$\mathscr{D}_N(\Gamma)=\sum_{N\geqslant l_1>\cdots>l_k\geqslant 0}N(m)\sigma_m(\Gamma)$$

就是 $SO(n)$ 的 Fourier 級数的 $\mathscr{D}$irichlet 核，我們要証明如下的

定理 若 $n=2k$，则

$$\mathscr{D}_N(\Gamma)=\frac{\det(d_N^{(2i-2)}(\theta_j))1\leqslant i,j\leqslant k}{2\det(d_{k-1}^{(2i-2)}(\theta_j))_{1\leqslant i,j\leqslant k}};$$

若 $n=2k+1$，则

$$\mathscr{D}_N(\Gamma)=\frac{\det(e_N^{(2i-1)}(\theta_j))1\leqslant i,j\leqslant k}{\det(e^{(2i-1)}(\theta_j))_{1\leqslant i,j,k}}$$

这里 $d_v(\theta)$ 为一个变数的 Dirichlet 核 $\sin\left(v+\frac{1}{2}\right)\theta\Big/\sin\frac{\theta}{2}$，而

$$e_v(\theta)=\frac{\sin(v+1)\theta}{\sin\frac{\theta}{2}}$$

先来証明一个恆等式

引理 若 $l_1,\cdots,l_k$ 为满足 $l_1>l_2>\cdots l_k\geqslant 0$ 的整数，$P_j(l_i)$ 为只与 l_i 有关的函数 $(j=1,\cdots,k)$ N 为正整数，a,b 为任意实数则

$$\sum_{N\geqslant l_1>l_2>\cdots>l_k\geqslant 0}\begin{vmatrix} l_1^{a-b}, & l_2^{a-b}, & \cdots, & l_k^{a-b},\\ l_1^{2a-b}, & l_2^{2a-b}, & \cdots, & l_k^{2a-b},\\ \cdots & \cdots & \cdots & \cdots\\ \cdots & \cdots & \cdots & \cdots\\ l_1^{ka-b}, & l_k^{ka-b}, & \cdots, & l_k^{ka-b},\end{vmatrix}\begin{vmatrix} P_1(l_1), & P_2(l_1), & \cdots, & P_k(l_1)\\ P_1(l_2), & P_2(l_2), & \cdots, & P_k(l_2)\\ \cdots & \cdots & \cdots, & \cdots\\ \cdots & \cdots & \cdots, & \cdots\\ P_1(l_k), & P_2(l_k), & \cdots, & P_k(l_k)\end{vmatrix}$$

$$=\det\left(\sum_{l=0}^{N}l^{ai-b}P_j(l)\right)1\leqslant i,\ j\leqslant k.$$

証： 上式左边等于

$$\frac{1}{k!}\sum_{l_1=0}^{N}\cdots\sum_{l_k=0}^{N}\det\left(\sum_{\mu=1}^{k}l_\mu^{ai-b}P_j(l_\mu)\right)_{1\leqslant i,j\leqslant k}$$

而
$$\det\left(\sum_{\mu=1}^{k}l_\mu^{ai-b}P_j(l_\mu)\right)_{1\leqslant i,j\leqslant k}$$

可以拆成 k^k 个行列式，每个为

$$\begin{vmatrix} l_{i_1}^{a-b}P_1(l_{i_1}), & l_{i_1}^{a-b}P_2(l_{i_1}), & \cdots, & l_{i_1}^{a-b}P_k(l_{i_1}) \\ l_{i_2}^{2a-b}P_1(l_{i_2}), & l_{i_2}^{2a-b}P_2(l_{i_2}), & \cdots, & l_{i_2}^{2a-b}P_k(l_{i_2}) \\ \cdots & \cdots & \cdots, & \cdots \\ l_{i_k}^{ka-b}P_1(l_{i_k}), & l_{i_k}^{ka-b}P_2(l_{i_k}), & \cdots, & l_{i_k}^{ka-b}P_k(l_{i_k}) \end{vmatrix}$$

的形式，显然除了 $(i_1, i_2, \cdots, i_k)$ 为 $(1, 2, \cdots, k)$ 的一个排列外，其余全为零，所以

$$\frac{1}{k!}\sum_{l_1=0}^{N}\cdots\sum_{l_k=0}^{N}\det\left(\sum_{\mu=1}^{k} l_\mu^{ai-b}P_j(l_\mu)\right)$$

等于

$$\sum_{l_1=0}^{N}\cdots\sum_{l_k=0}^{N}\det\left(l_i^{ai-b}P_j(l_i)\right)_{1\leqslant i,j\leqslant k}$$

而这就是

$$\det\left(\sum_{l=0}^{N} l^{ai-1}P_j(l)\right)_{1\leqslant i,j\leqslant k}$$

現在来証明这条定理.

先来考虑 $n=2k+1$ 的情形。以 $m=(m_1,\cdots,m_k)$ 为标签的旣約表示的特征标为[5]

$$[m]=\frac{s\left(l+\frac{1}{2}\right)}{s\left(K-\frac{1}{2}, \cdots, \frac{3}{2}, \frac{1}{2}\right)}$$

这里 $l=(l_1,\cdots,l_k)$，$l_1=m_1+k-1,\cdots,l_{k-1}=m_{k-1}+1$，$l_k=m_k$，

$$s[l]=\begin{vmatrix} s_{l_1}(\theta_1), & \cdots, & s_{l_1}(\theta_k) \\ \cdots & \cdots, & \cdots \\ \cdots & \cdots, & \cdots \\ s_{l_k}(\theta_1), & \cdots & s_{l_k}(\theta_k) \end{vmatrix},$$

$s_q(\alpha)=2i\sin q\alpha$. 其維数为 $N(m)=\lim\limits_{\substack{\theta_1\to 0\\ \cdots\cdots\\ \theta_k\to 0}}[m]$，此为

$$l'_1\cdots l'_k\begin{vmatrix} 1, & l_1'^2, & \cdots, & l_1'^{2k-2} \\ 1, & l_2'^2, & \cdots, & l_2'^{2k-2} \\ \cdots & \cdots & \cdots, & \cdots \\ 1, & l_k'^2, & \cdots, & l_k'^{2k-2} \end{vmatrix} \Bigg/ \left(k-\frac{1}{2}\right)\cdots\frac{1}{2}\begin{vmatrix} 1, & \left(k-\frac{1}{2}\right)^2, & \cdots, & \left(k-\frac{1}{2}\right)^{2k-2} \\ 1, & \left(k-\frac{3}{2}\right)^2, & \cdots, & \left(k-\frac{3}{2}\right)^{2k-2} \\ \cdots & \cdots & \cdots, & \cdots \\ \cdots, & \left(\frac{1}{2}\right)^2, & \cdots, & \left(\frac{1}{2}\right)^{2k-2} \end{vmatrix}$$

这里 $l'_j=l_j+\frac{1}{2}$. 于是 Dirichlet 核

$$\mathscr{D}_N(\Gamma)=\sum_{N\geqslant l_1>\cdots>l_k\geqslant 0}N(m)[m]=\frac{P}{Q},$$

而

$$P=\sum_{N\geqslant l_1>\cdots>l_k>0}\begin{vmatrix} l_1', & l_1'^3, & \cdots, & l_1'^{2k-1} \\ l_2', & l_2'^3, & \cdots, & l_2'^{2k-1} \\ \cdots & \cdots & \cdots & \cdots \\ l_k', & k_k'^3, & \cdots, & l_k'^{2k-1}\end{vmatrix}\begin{vmatrix} \sin l_1'\theta_1, & \cdots, & \sin l_1'\theta_k \\ \sin l_2'\theta_1, & \cdots, & \sin l_2'\theta^k \\ \cdots & \cdots & \cdots \\ \sin l_k'\theta_1, & \cdots, & \sin l_k'\theta_k\end{vmatrix},$$

$$Q=\begin{vmatrix} \left(k-\frac{1}{2}\right), & \left(k-\frac{1}{2}\right)^3, & \cdots, & \left(k-\frac{1}{2}\right)^{2k-1} \\ \left(k-\frac{3}{2}\right), & \left(k-\frac{3}{2}\right)^3, & \cdots, & \left(k-\frac{3}{2}\right)^{2k-1} \\ \cdots & \cdots & \cdots & \cdots \\ \frac{1}{2}, & \left(\frac{1}{2}\right)^3, & \cdots, & \left(\frac{1}{2}\right)^{2k-1}\end{vmatrix}\begin{vmatrix} \sin\left(k-\frac{1}{2}\right)\theta_1, & \cdots, & \sin\left(k-\frac{1}{2}\right)\theta_k \\ \sin\left(k-\frac{3}{2}\right)\theta_1, & \cdots, & \sin\left(k-\frac{3}{2}\right)\theta_k \\ \cdots & \cdots & \cdots \\ \sin\frac{1}{2}\theta_1, & \cdots, & \sin\frac{1}{2}\theta_k\end{vmatrix}$$

对 P 应用引理，取 $a=2$，$b=1$，$P_j(l)=\sin l\theta_j$，就得到

$$P=\det\left(\sum_{l=0}^{N} l'^{2i-1}\sin l'\theta_j\right)_{1\leqslant i,\ j\leqslant k.}$$

显然 $$\sum_{l=0}^{N} l'^{2i-1}\sin l'\theta=(-1)^i\frac{d^{2i-1}}{d\theta^{2i-1}}\left(\frac{\sin(N+1)\theta}{2\sin\theta/2}\right).$$

因之， $$P=(-1)^{\frac{k(k+1)}{2}}\det\left(\left(\frac{\sin(N+1)\theta_j}{2\sin\theta_j/2}\right)^{(2i-1)}\right)_{1\leqslant i,\ j\leqslant k.}$$

同样可得

$$Q=(-1)^{\frac{k(k+1)}{2}}\det\left(\left(\frac{\sin k\theta_j}{2\sin\theta_j/2}\right)^{(2i-1)}\right)_{1\leqslant i,\ j\leqslant k.}$$

于是証明了当 $n=2k+1$ 的情形。

再来考虑 $n=2k$ 的情形。

当 $l_k=0$ 时，以 $m=(m_1,\cdots,m_{k-1},0)$ 为标签的既約表示的特征标为[9]

$$[m]=\frac{c(l_1,\cdots,l_k)}{c(k-1,\cdots,1,0)}$$

这里 $l=(l_1,\cdots,l_k)$，$l_1=m_1+k-1$，$\cdots$，$l_{k-1}=m_{k-1}+1$，$l_k=m_k$.

$$c(l_1,\cdots,l_k)=\begin{vmatrix} c_{l_1}(\theta_1), & \cdots, & c_{l_1}(\theta_k) \\ \cdots & \cdots & \cdots \\ c_{l_k}(\theta_1), & \cdots, & c_{l_k}(\theta_k)\end{vmatrix}$$

当 $q>1$ 时，$c_q(\alpha)=2\cos q\alpha$，$c_0(\alpha)=1$. 表示的維数为 $N(m)=\lim\limits_{\substack{\theta_1\to 0\\ \cdots\\ \theta_k\to 0}}[m]$，此为

$$\begin{vmatrix} 1, & l_1^2, & \cdots, & l_1^{2k-2} \\ 1, & l_2^2, & \cdots, & l_2^{2k-2} \\ \cdots & \cdots & \cdots & \cdots \\ 1, & l_k^2, & \cdots, & l_k^{2k-2}\end{vmatrix}\Bigg/ 2\begin{vmatrix} 1, & (k-1)^2, & \cdots, & (k-1)^{2k-2} \\ 1, & (k-2)^2, & \cdots, & (k-2)^{2k-2} \\ \cdots & \cdots & \cdots & \cdots \\ 1, & 1, & \cdots, & 1 \\ 1, & 0, & \cdots, & 0\end{vmatrix} \qquad (*)$$

当 $l_k>0$ 时，以 $m=(m_1,\cdots,m_k)$ 为标签的既約表示的特征标为[9]

$$[m]_+=\frac{\frac{1}{2}\left\{c(l)+s(l)\right\}}{c(k-1,\cdots,1,0)}$$

及

$$[m]_-=\frac{\frac{1}{2}\left\{c(l)-s(l)\right\}}{c(k-1,\cdots,1,0)}$$

它們的維数均为 $N(m)$．由（$*$）所定义。

于是 Dirichlet 核

$$\mathscr{D}_N(\Gamma)=\sum_{N\geqslant l_1>\cdots>l_k\geqslant 0}N(m)\,[m]=\frac{P_1}{Q_1},$$

而

$$P_1=\sum_{N\geqslant l_1>\cdots>l_k\geqslant 0}\begin{vmatrix}1, & l_1^2, & \cdots, & l_1^{2k-2}\\ 1, & l_2^2, & \cdots, & l_2^{2k-2}\\ \cdots & \cdots & \cdots & \cdots\\ 1, & l_k^2, & \cdots, & l_k^{2k-2}\end{vmatrix}\begin{vmatrix}c_{l_1}(\theta_1), & \cdots, & c_{l_1}(\theta_k)\\ c_{l_2}(\theta_1), & \cdots, & c_{l_2}(\theta_k)\\ \cdots & \cdots & \cdots\\ c_{l_k}(\theta_1), & \cdots, & c_{l_k}(\theta_k)\end{vmatrix},$$

及

$$Q_1=2\begin{vmatrix}1, & (k-1)^2, & \cdots, & (k-1)^{2k-2}\\ 1, & (k-2)^2, & \cdots, & (k-2)^{2k-2}\\ \cdots & \cdots & \cdots & \cdots\\ 1, & 1, & \cdots, & 1\\ 1, & 0, & \cdots, & 0\end{vmatrix}\begin{vmatrix}c_{k-1}(\theta_1), & \cdots, & c_{k-1}(\theta_k)\\ \cdots & \cdots & \cdots\\ c_1(\theta_1), & \cdots, & c_1(\theta_k)\\ c_0(\theta_1), & \cdots, & c_0(\theta_k)\end{vmatrix}$$

对 P_1 应用引理，取 $a=2$，$b=2$，$p_j(l)=c_l(\theta_j)$，就得到

$$P_1=\det\left(\sum_{l=0}^{N}l^{2s-2}c_l(\theta_t)\right)_{1\leqslant s,t\leqslant k}.$$

显然，当 $s>1$ 时，

$$\sum_{l=0}^{N}l^{2s-2}c_l(\theta)=2\sum_{l=1}^{N}\cos l\theta=(-1)^{s-1}\left(\frac{\sin\left(N+\frac{1}{2}\right)\theta}{\sin\frac{1}{2}\theta}\right)^{(2s-2)},$$

当 $s=1$ 时，

$$\sum_{l=1}^{N}l^{2s-2}c_l(\theta)=\sum_{l=0}^{N}c_l(\theta)=\frac{\sin\left(N+\frac{1}{2}\right)\theta}{\sin\frac{\theta}{2}}=d_N(\theta).$$

因之，

$$P_1=(-1)^{\frac{k(k-1)}{2}}\det\left(\left(\frac{\sin\left(N+\frac{1}{2}\right)\theta_j}{\sin\theta_j/2}\right)^{(2i-2)}\right)_{1\leqslant i,\,j\leqslant k}.$$

同样可得

$$Q_1=(-1)^{\frac{k(k-1)}{2}}\cdot 2\det\left(\left(\frac{\sin\left(k-\frac{1}{2}\right)\theta_i}{\sin\theta_i/2}\right)^{(2i-2)}\right)_{1\leqslant i,\ j\leqslant k.}$$

于是証明了当 $n=2k$ 的情形。

由这条定理，以及注意以下的事实，可以如同在[6]中所进行的那样，得出在旋轉群上富理埃級数的收斂判别法，这将另文討論之。

当 $n=2k$ 时，
$$\Gamma=\Gamma_1\begin{pmatrix}c_1 & & \\ & \ddots & \\ & & c_k\end{pmatrix}\Gamma'_1,\quad c_i=\begin{pmatrix}\cos\theta_i, & -\sin\theta_i\\ \sin\theta_i, & \cos\theta_i\end{pmatrix};$$

当 $n=2k+1$ 时
$$\Gamma=\Gamma_1\begin{pmatrix}c_1 & & \\ & \ddots & \\ & & c_k\end{pmatrix}\Gamma'_1,$$

而 $\Gamma_1\in SO(n)$、記对傍集的积分为

$$\frac{1}{C}\int_{[\Gamma_1]}u\left(\Gamma_1\begin{pmatrix}c_1 & & \\ & \ddots & \\ & & c_k\end{pmatrix}\Gamma_1\Gamma_0\right)\dot{\Gamma}_1=g_{\Gamma_0}(\theta_1,\cdots,\theta_k)$$

若 $n=2k$；

$$\frac{1}{C}\int_{[\Gamma_1]}u\left(\Gamma_1\begin{pmatrix}c_1 & & \\ & \ddots & \\ & & ck_1\end{pmatrix}\Gamma'_1\Gamma_0\right)\dot{\Gamma}_1=g_{\Gamma_0}(\theta_1,\cdots,\theta_k)$$

若 $n=2k+1$；則

$$S_N(\Gamma_0)=\int\cdots\int g_{\Gamma_0}(\theta_1,\cdots,\theta_k)\mathscr{D}_N(\Gamma)d\tau$$

这里 $d\tau$ 当 $n=2k$ 时为

$$\frac{2^{(k-1)^2}}{k!}\prod_{p<q}^{k}(\cos\theta_p-\cos\theta_q)^2d\theta_1\cdots d\theta_k,$$

当 $n=2k+1$ 时为

$$\frac{2^{k(k-1)}}{k!}\prod_{p=1}^{k}(1-\cos\theta_p)\prod_{p<q}^{k}(\cos\theta_p-\cos\theta_q)^2d\theta_1\cdots d\theta_k.$$

参 考 文 献

[1] 华罗庚，紧致群上的連續函数所成的空間中的一条收斂定理，科学記录，2(1958) 341—344.

[2] 陆启鏗，典型域与典型流形，上海科技出版社出版.

[3] 鐘家庆，旋轉群的調和分析——Abel 求和，中国科技大学学报 1979 32—44.

[4] 龔 昇，酉群上的富里埃分析Ⅰ，数学学报 10 (1960) 239—261.

[5] 龔 昇，酉群上的富里埃分析Ⅱ，数学学报 12 (1962) 17—31.

[6] 龔 昇，酉群上的富里埃分析Ⅲ，数学学报 13 (1963) 152—161.

[7] 龔 昇，酉群上的富里埃分析Ⅳ，数学学报 13 (1963) 323—331.

[8] 龔 昇，酉群上的富里埃分析Ⅴ，数学学报15 (1965) 305—325.

[9] F. D. Murneghan, The theory of group nepsesentations.

II. 多复变数的奇异积分与积分方程

与单复变数不一样，多复变数有众多个 Cauchy 核与 Cauchy 积分．因之，有众多个 Cauchy 型积分．华罗庚 (1) 给出了典型域上的 Cauchy-Szegö 核，这是他对多复变数函数论的重大贡献．有了 Cauchy 核， Cauchy 积分顺理成章地要讨论其 Cauchy 型积分．所以讨论典型域上的 Cauchy 型积分也是华罗庚的典型域上调和分析的另一个发展． [67.1], [66.1], [65.1], [63.1], [54.1], [52], [51], [50] 就是对典型域上的 Cauchy 型积分以及强拟凸域上的 Henkin 型积分的系统研究．要指出两点： 1. 这是第一次用 Cauchy 型积分来解边界上奇异积分方程， 2. 不同于单复变数，这里所讨论的多复变数奇异积分有众多方法定义其主值，因此可以得到众多的 Plemelj 公式．

这些工作总结在 (2) ([M9]) 英文版 [M3]) 中．这项工作获 1996 年中国科学院自然科学一等奖，得奖人：龚昇, 孙继广，史济怀．

参 考 文 献

(1) 华罗庚．多复变数函数论中的典型域的调和分析．北京：科学出版社，1958.

(2) 龚昇. 多复变数的奇异积分．上海：上海科学技术出版社，1982.

Preface of *Integrals of Cauchy Type on the Ball*

This book appeared in Chinese about ten years ago, and most of the results in it were found by the author and his collaborators in the sixties and early eighties, the intervening decade having been lost to the Cultural Revolution. The reader who knows about the enormous amount of work on singular integrals done the world over in the past twenty years will inevitably ask: Can this book be at the front line of present-day research? Must it not be more or less obsolete ?

In actual fact, the book is full of highly original results and has only minimal overlap with the work of other researchers. The ultimate reason for this is perhaps that the author lets himself be guided throughout by the Cauchy-Szegö integral; a "singular integral" for him is the Cauchy integral pushed to the boundary. The domains he considered are the bounded symmetric domains in the sense of Cartan and, in a chapter at the end, smooth strongly pseudoconvex domains. It is this restriction of generality which lies at the source of the originality and strength of the book. For these very special, and very interesting, singular integrals the author obtains results far stronger than those in the rest of the literature where, in the tradition of the Calderón-Zygmund theory, the investigations are usually about large classes of kernels.

On the simplest bounded symmetric domain, the unit ball in C^n, the author was the first to investigate singular integrals associated to non-isotropic metrics. These have been greatly generalized to the context of the Heisenberg group and of spaces of homogeneous type in studies vigorously pursued worldwide since the early seventies. But these questions occupy only a very small part of the present book—about half of Chapter I and VI. In these chapters already there are also results involving modified metric that no one else has studied. Chapter II and III, which are still devoted entirely to the unit ball, are completely original. They contain a study of certain systems of singular integral equations and results about a generalized version of Hadamard's "parties finies".

Most interesting, however, are Chapters IV and V, where the author considers the singular integrals obtained from the Cauchy-Szegö integral on certain Cartan domains of rank higher than one. This is more difficult by an order of magnitude than the case of the ball; the integrals have singularities on entire surfaces instead of only at isolated points. In the last few years a number of important results about integrals with singularities on surfaces have appeared in the literature, but they are not strong enough to be applicable to these Cauchy-type integrals. Yet the author succeeds here in proving very satisfying results on convergence and on preservation of Lipschitz classes.

Besides being interesting in themselves, these results raise a host of new questions. To mention only some of the most obvious ones: What is the effect of these operators on Sobolev spaces? On non-isotropic Lipschitz spaces? Can the proofs be made more conceptual, to include all the Cartan domains at once? A whole new line of investigation seems to be opening up here, and the reader will be forced to conclude that this is indeed an inspiring book.

Adam Korányi July 1993

第15卷 第3期 1965年5月
数学学报 ACTA MATHEMATICA SINICA
Vol. 15, No. 3 May, 1965

多复变数的 Cauchy 型积分 (I) 超球的 Cauchy 型积分*

龔 昇　　孙继广
(中国科学技术大学)(中国科学院数学研究所)

§1.1. 引　　言

Cauchy 型积分在复变函数論中的重要性不必多説了，它不但有着函数論本身的重要意义，而且是奇异积分方程、边界值問題等学科中不可缺少的工具．但是对于多复变数 Cauchy 型积分的研究是不多的．陆启鏗与鍾同德在［3］中研究了由 Bochner 定义的 Cauchy 核所生成的 Cauchy 型积分，并得出了相应的 Привалов 定理．这时候 Cauchy 型积分所定义的函数一般来説并不是解析函数，积分是在区域的整个边界上进行．Какичев[4] 曾經对多圆柱的 Cauchy 型积分进行研究．

华罗庚[1]作出了具有相当一般性的圆型域的 Cauchy 核(参閱[2])，从而使这些域的 Cauchy 型积分的定义有了可能．值得指出的是，这时的 Cauchy 型积分是通过特征流形来定义的，并且所定义的确是解析函数．

本文对超球的 Cauchy 型积分

$$\frac{1}{\omega_{2n-1}}\int_{u\bar{u}'=1}\frac{f(u)\dot{u}}{(1-z\bar{u}')^n}\qquad(n\geqslant 2)\tag{1.1.1}$$

进行研究，其中 $u=(u_1,\cdots,u_n)$, $z=(z_1,\cdots,z_n)$, $z\bar{z}'<1$. $\dot{u}$ 表示 $u\bar{u}'=1$ 的体积元素，$\omega_{2n-1}=2\pi^n/\Gamma(n)$ 表示 $u\bar{u}'=1$ 的体积．在本文中首先定义了 Cauchy 主値，并且証明了：滿足 Lipschitz 条件的函数 $f(u)$，其 Cauchy 主値是存在的，并且給出了从球内部趋于边界时 Cauchy 型积分的极限値．

当 $u\bar{u}'=1$ 而 $z\bar{z}'>1$ 时，Cauchy 核 $(1-z\bar{u}')^{-n}$ 就失去意义．事实上，任取一点 $z=(\lambda,0,\cdots,0)U$, $\lambda>1$, $U\bar{U}'=I$, I 表示单位方陣，则当 $u=\left(\frac{1}{\lambda},\sqrt{1-\frac{1}{\lambda^2}},0,\cdots,0\right)U$ 时，$1-z\bar{u}'=0$．所以，此时由(1.1.1)式所定义的 Cauchy 型积分是不存在的．

至于 Cauchy 型积分的应用，我們将陆續另文发表．

我們愿向我們的导师华罗庚先生表示衷心的感謝，因为他对这項工作給予了十分宝貴的指示．本文内容曾在中国科学技术大学数学系綜合討論班上报告过，得到了大家不少帮助，特别是曾肯成同志与陆启鏗同志，我們愿向他們致謝．

§1.2. 一 条 引 理

根据 Cauchy 积分公式[1]，我們有

* 1964年3月8日收到，1964年5月20日收到修改稿．

$$\frac{1}{\omega_{2n-1}}\int_{u\bar{u}'=1}\frac{\dot{u}}{(1-z\bar{u}')^n}=1.$$

特别取 $z=(\rho, 0, \cdots, 0)$, $0\leqslant\rho<1$, 于是

$$\frac{1}{\omega_{2n-1}}\int_{u\bar{u}'=1}\frac{\dot{u}}{(1-\rho\bar{u}_1)^n}=1.$$

我們証明如下的

引理 1.1.

$$\lim_{\varepsilon\to 0}\frac{1}{\omega_{2n-1}}\int_{\substack{u\bar{u}'=1\\|1-\bar{u}_1|\geqslant\varepsilon}}\frac{\dot{u}}{(1-\bar{u}_1)^n}$$

是存在的,且等于 1/2.

証.

已知有

$$1=\frac{1}{\omega_{2n-1}}\left(\int_{\Sigma}+\int_{\sigma}\right)\frac{\dot{u}}{(1-\rho\bar{u}_1)^n}=I_1+I_2, \tag{1.2.1}$$

这里

$$\Sigma:\begin{cases}u\bar{u}'=1,\\|1-\bar{u}_1|\geqslant\varepsilon,\end{cases}\quad \sigma:\begin{cases}u\bar{u}'=1,\\|1-\bar{u}_1|\leqslant\varepsilon.\end{cases}$$

先来看 I_2:

$$I_2=\frac{1}{\omega_{2n-1}}\int_{\sigma}\frac{\dot{u}}{(1-\rho\bar{u}_1)^n}. \tag{1.2.2}$$

令 $\bar{u}_1=re^{i\theta}$, $u_2=v_2$, $\cdots$, $u_n=v_n$, 記 $v=(v_2, \cdots, v_n)$, 于是 σ 为

$$\begin{cases}v\bar{v}'=1-r^2,\\ \cos\theta\geqslant\dfrac{1+r^2-\varepsilon^2}{2r}.\end{cases}$$

这样,

$$I_2=\frac{1}{\omega_{2n-1}}\int_{v\bar{v}'\leqslant 2\varepsilon-\varepsilon^2}\dot{v}\int_{-a}^{a}\frac{d\theta}{(1-\rho re^{i\theta})^n}\left(a=\cos^{-1}\frac{1+r^2-\varepsilon^2}{2r}\right).$$

記

$$\begin{aligned}J&=\int_{-a}^{a}\frac{d\theta}{(1-\rho re^{i\theta})^n}=\sum_{p=0}^{\infty}\frac{\Gamma(n+p)}{\Gamma(p+1)\Gamma(n)}\int_{-a}^{a}(\rho r)^pe^{ip\theta}d\theta\\&=2\sum_{p=1}^{\infty}\frac{\Gamma(n+p)}{\Gamma(p+1)\Gamma(n)}(\rho r)^p\frac{\sin pa}{p}+2a\\&=2\,\mathrm{Im}\left\{\sum_{p=1}^{\infty}\frac{\Gamma(n+p)}{\Gamma(p+1)\Gamma(n)}\frac{(\rho re^{ia})^p}{p}\right\}+2a.\end{aligned}$$

由于

$$\sum_{p=0}^{\infty}\frac{\Gamma(n+p)x^p}{\Gamma(p+1)\Gamma(n)}=\frac{1}{(1-x)^n},$$

$$\begin{aligned}\sum_{p=1}^{\infty}\frac{\Gamma(n+p)x^{p-1}}{\Gamma(p+1)\Gamma(n)}&=\frac{1}{x(1-x)^n}-\frac{1}{x}=\frac{1-(1-x)^n}{x(1-x)^n}=\frac{1-y^n}{(1-y)y^n}\\&=\frac{1+y+\cdots+y^{n-1}}{y^n}=\frac{1}{y}+\cdots+\frac{1}{y^n}\quad(y=1-x)\end{aligned}$$

$$= \frac{1}{(1-x)^n} + \cdots + \frac{1}{1-x},$$

$$\sum_{p=1}^{\infty} \frac{\Gamma(n+p)x^p}{\Gamma(p+1)\Gamma(n)p} = \int_0^x \frac{1-(1-x)^n}{x(1-x)^n} dx = \frac{1}{(n-1)(1-x)^{n-1}}$$

$$+ \cdots + \frac{1}{1-x} + \log \frac{1}{1-x} - \sum_{\nu=1}^{n-1} \frac{1}{\nu},$$

所以

$$J = 2\,\mathrm{Im}\left\{\frac{1}{(n-1)(1-\rho re^{ia})^{n-1}} + \cdots + \frac{1}{1-\rho re^{ia}} + \log \frac{1}{1-\rho re^{ia}}\right\} + 2a,$$

$$I_2 = \frac{2}{\omega_{2n-1}} \int_{v\bar{v}' \leqslant 2\varepsilon-\varepsilon^2} \mathrm{Im}\left\{\sum_{\nu=1}^{n-1} \frac{1}{\nu(1-\rho re^{ia})^\nu} + \log \frac{1}{1-\rho re^{ia}}\right\} \dot{v} + \frac{2a}{\omega_{2n-1}} \int_{v\bar{v}' \leqslant 2\varepsilon-\varepsilon^2} \dot{v}$$

$$= \frac{2}{\omega_{2n-1}} \mathrm{Im}\left\{\sum_{\nu=1}^{n-1} \int_{v\bar{v}' \leqslant 2\varepsilon-\varepsilon^2} \frac{\dot{v}}{\nu(1-\rho re^{ia})^\nu} + \int_{v\bar{v}' \leqslant 2\varepsilon-\varepsilon^2} \log \frac{1}{1-\rho re^{ia}} \dot{v}\right\}$$

$$+ \frac{2a}{\omega_{2n-1}} \int_{v\bar{v}' \leqslant 2\varepsilon-\varepsilon^2} \dot{v}$$

$$= \frac{2}{\omega_{2n-1}} \mathrm{Im}\left\{\sum_{p=1}^{n-1} K_p + K_0\right\} + \frac{2a}{\omega_{2n-1}} \int_{v\bar{v}' \leqslant 2\varepsilon-\varepsilon^2} \dot{v}, \tag{1.2.3}$$

此处

$$K_p = \int_{v\bar{v}' \leqslant 2\varepsilon-\varepsilon^2} \frac{\dot{v}}{p(1-\rho re^{ia})^p}, \quad K_0 = \int_{v\bar{v}' \leqslant 2\varepsilon-\varepsilon^2} \log \frac{1}{1-\rho re^{ia}} \dot{v}.$$

先来看 I_2 表达式中的最后一項

$$\frac{2a}{\omega_{2n-1}} \int_{v\bar{v}' \leqslant 2\varepsilon-\varepsilon^2} \dot{v},$$

注意到 $\int_{v\bar{v}' \leqslant 2\varepsilon-\varepsilon^2} \dot{v}$ 是以 $\sqrt{2\varepsilon-\varepsilon^2}$ 为半径，以原点为中心的 $2n-2$ 維球的体积，由定义知

$$a = \cos^{-1}\left(\frac{1+r^2-\varepsilon^2}{2r}\right).$$

由于 $\frac{1+r^2-\varepsilon^2}{2r} \leqslant 1$，所以 $(1-r)^2 \leqslant \varepsilon^2$，以至当 $\varepsilon \to 0$ 时，$a = O(\varepsilon)$，于是

$$\frac{2a}{\omega_{2n-1}} \int_{v\bar{v}' \leqslant 2\varepsilon-\varepsilon^2} \dot{v} = O(\varepsilon^{\frac{2n-2}{2}+1}) = O(\varepsilon^n).$$

再来考慮其他項．由于 a 的定义，

$$re^{ia} = r\cos a + ir\sqrt{1-\cos^2 a} = \frac{1+r^2-\varepsilon^2}{2} + ir\sqrt{1-\left(\frac{1+r^2-\varepsilon^2}{2r}\right)^2}$$

$$= \frac{1+r^2-\varepsilon^2}{2} + \frac{i}{2}\sqrt{4r^2-(1+r^2)^2+2\varepsilon^2(1+r^2)-\varepsilon^4}$$

$$= \frac{1+r^2-\varepsilon^2}{2} + \frac{i}{2}[2\varepsilon^2(1+r^2)-(1-r^2)^2-\varepsilon^4]^{\frac{1}{2}}$$

$$= \frac{2-v\bar{v}'-\varepsilon^2}{2} + \frac{i}{2}[2\varepsilon^2(2-v\bar{v}')-(v\bar{v}')^2-\varepsilon^4]^{\frac{1}{2}}.$$

所以,当 $p \geqslant 1$ 时

$$K_p = \frac{1}{p}\int_{v\bar{v}'\leqslant 2\varepsilon-\varepsilon^2} \frac{\dot{v}}{\left[1-\rho\frac{2-v\bar{v}'-\varepsilon^2}{2}-\frac{i\rho}{2}\left(2\varepsilon^2(2-v\bar{v}')-(v\bar{v}')^2-\varepsilon^4\right)^{\frac{1}{2}}\right]^p}$$

$$= \frac{1}{p}\int_{v\bar{v}'\leqslant 2\varepsilon-\varepsilon^2} \frac{\dot{v}}{\left[1-\rho+\rho\frac{v\bar{v}'+\varepsilon^2}{2}-\frac{i\rho}{2}\left(2\varepsilon^2(2-v\bar{v}')-(v\bar{v}')^2-\varepsilon^4\right)^{\frac{1}{2}}\right]^p}.$$

用球坐标,则

$$K_p = \frac{1}{p}\int_0^{\sqrt{2\varepsilon-\varepsilon^2}} \frac{S^{2n-3}dS}{\left(1-\rho+\rho\frac{\varepsilon^2+S^2}{2}-\frac{i\rho}{2}\sqrt{2\varepsilon^2(2-\varepsilon^2)-S^4-\varepsilon^4}\right)^p}$$

$$\times\int_0^\pi \sin^{2n-4}\varphi_1 d\varphi_1\cdots\int_0^\pi \sin^2\varphi_{2n-5}d\varphi_{2n-5}\int_0^\pi \sin\varphi_{2n-4}d\varphi_{2n-4}\int_0^{2\pi} d\varphi_{2n-3}$$

$$= \frac{2\pi^{n-1}}{p(n-2)!}\int_0^{\sqrt{2\varepsilon-\varepsilon^2}} \frac{S^{2n-3}dS}{\left[(1-\rho)+\rho\frac{\varepsilon^2+S^2}{2}-\frac{i\rho}{2}\sqrt{4\varepsilon^2-(S^2+\varepsilon^2)^2}\right]^p}.$$

当 $\varepsilon>0$ 固定时,由 Lebesgue 定理

$$\lim_{\rho\to 1}\int = \int\lim_{\rho\to 1},$$

于是

$$J_p = \lim_{\rho\to 1}K_p = \lim_{\rho\to 1}\int_{v\bar{v}'\leqslant 2\varepsilon-\varepsilon^2}\frac{\dot{v}}{p(1-\rho re^{i\alpha})^p}$$

$$= \frac{2\pi^{n-1}}{(n-2)!p}\int_0^{\sqrt{2\varepsilon-\varepsilon^2}} \frac{S^{2n-3}dS}{\left[\frac{\varepsilon^2+S^2}{2}-\frac{i}{2}\sqrt{4\varepsilon^2-(S^2+\varepsilon^2)^2}\right]^p}.$$

可以用 Lebesgue 定理的原因,是这时被积函数的絶对值为

$$\frac{S^{2n-3}}{\left[(1-\rho)^2+2\rho(1-\rho)\frac{\varepsilon^2+S^2}{2}+\rho^2\varepsilon^2\right]^{p/2}}$$

$$= \frac{S^{2n-3}}{((1-\rho)^2+\rho(1-\rho)S^2+\rho\varepsilon^2)^{p/2}} \leqslant \frac{2S^{2n-3}}{\varepsilon^p}\quad(\text{只要 } 1\geqslant\rho\geqslant 2^{-\frac{2}{n-1}}).$$

由于

$$|J_p| \leqslant \frac{2\pi^{n-1}}{(n-2)!p}\int_0^{\sqrt{2\varepsilon-\varepsilon^2}}\frac{S^{2n-3}dS}{\varepsilon^p} = \frac{\pi^{n-1}(2-\varepsilon)^{n-1}\varepsilon^{n-p-1}}{(n-1)!p},$$

所以当 $p<n-1$ 时, $J_p = O(\varepsilon^{n-p-1})$; 而当 $p=n-1$ 时, 記 $\eta=\sqrt{2\varepsilon-\varepsilon^2}$, $S=\eta R$, 于是

$$J_{n-1} = \frac{2\pi^{n-1}}{(n-1)!}\int_0^1 \frac{R^{2n-3}dR}{\left[\frac{R^2+\varepsilon^2\eta^{-2}}{2}-\frac{i}{2}\sqrt{4\eta^{-4}\varepsilon^2-(R^2+\varepsilon^2\eta^{-2})^2}\right]^{n-1}}.$$

再应用 Lebesgue 定理,得到

$$\lim_{\varepsilon\to 0}J_{n-1} = \frac{2^n\pi^{n-1}}{(n-1)!}\int_0^1\frac{R^{2n-3}dR}{(R^2-i\sqrt{1-R^4})^{n-1}}.$$

令 $R^2 = t$，上式即为

$$\frac{(2\pi)^{n-1}}{(n-1)!}\int_0^1 \frac{t^{n-2}dt}{(t-i\sqrt{1-t^2})^{n-1}} = \frac{(2\pi)^{n-1}}{(n-1)!}\int_0^{\pi/2} e^{i(n-1)\theta}\cos^{n-2}\theta\sin\theta d\theta.$$

注意到

$$\int_0^{\pi/2}\cos^{n-1}\theta\sin\theta e^{i(n-1)\theta}d\theta = \int_0^{\pi/2}e^{i(n-1)\theta}d\left(\frac{-\cos^{n-1}\theta}{n-1}\right)$$

$$= \frac{-e^{i(n-1)\theta}\cos^{n-1}\theta}{n-1}\bigg|_0^{\pi/2} + i\int_0^{\pi/2}\cos^{n-1}\theta\cdot e^{i(n-1)\theta}d\theta$$

$$= \frac{1}{n-1} + i\int_0^{\pi/2}\cos^{n-1}\theta\cdot e^{i(n-1)\theta}d\theta,$$

所以

$$\operatorname{Im}(\lim_{\varepsilon\to 0}J_{n-1}) = \frac{(2\pi)^{n-1}}{(n-1)!}\int_0^{\pi/2}\cos^{n-1}\theta\cdot\cos(n-1)\theta\, d\theta$$

$$= \frac{(2\pi)^{n-1}}{(n-1)!}\cdot\frac{1}{2^{n-1}}\int_0^{\pi/2}d\theta = \frac{\pi^n}{(n-1)!2},$$

即

$$\lim_{\varepsilon\to 0}\operatorname{Im}J_{n-1} = \frac{\pi^n}{(n-1)!2}. \tag{1.2.4}$$

我們还看到

$$K_0 = \int_{v\bar v'\leqslant 2\varepsilon-\varepsilon^2}\ln\frac{1}{1-\rho re^{i\alpha}}\dot v$$

$$= \int_{v\bar v'\leqslant 2\varepsilon-\varepsilon^2}\ln\frac{1}{1-\rho+\rho\dfrac{v\bar v'+\varepsilon^2}{2}-\dfrac{i\rho}{2}\sqrt{2\varepsilon^2(2-v\bar v')-(v\bar v')^2-\varepsilon^4}}\dot v,$$

用球坐标：

$$J_0 = \int_0^{\sqrt{2\varepsilon-\varepsilon^2}} S^{2n-3}\ln\frac{1}{\dfrac{\varepsilon^2+S^2}{2}-\dfrac{i}{2}\sqrt{2\varepsilon^2(2-S^2)-S^4-\varepsilon^4}}dS\int_0^{\pi}\sin^{2n-4}\varphi_1 d\varphi_1\cdots$$

$$\cdots\int_0^{2\pi}d\varphi_{2n-3} = \frac{2\pi^{n-1}}{(n-2)!}\int_0^{\sqrt{2\varepsilon-\varepsilon^2}}S^{2n-3}\ln\frac{1}{\dfrac{\varepsilon^2+S^2}{2}-\dfrac{i}{2}\sqrt{2\varepsilon^2(2-S^2)-S^4-\varepsilon^4}}dS,$$

記 $S=\eta R$，

$$J_0 = \frac{2\pi^{n-1}}{(n-2)!}\eta^{2n-2}\int_0^1 R^{2n-3}\ln\frac{1}{\eta^2\left(\dfrac{R^2+\varepsilon^2\eta^{-2}}{2}-\dfrac{i}{2}\sqrt{4\eta^{-4}\varepsilon^2-(R^2+\varepsilon^2\eta^{-2})^2}\right)}dR$$

$$= \frac{2\pi^{n-1}}{(n-2)!}\frac{2\eta^{2n-2}\ln\dfrac{\sqrt2}{\eta}}{2n-2} + \frac{2\pi^{n-1}}{(n-2)!}\eta^{2n-2}$$

$$\times\int_0^1 R^{2n-3}\ln\frac{1}{R^2+\varepsilon^2\eta^{-2}-i\sqrt{4\eta^{-4}\varepsilon^2-(R^2+\varepsilon^2\eta^{-2})^2}}dR = o(1),$$

最后得到

$$\lim_{\varepsilon\to 0}\lim_{\rho\to 1}I_2 = \frac{2}{\omega_{2n-1}}\cdot\frac{\pi^n}{(n-1)!2} = \frac{1}{2}. \tag{1.2.5}$$

这样，从

$$1=\frac{1}{\omega_{2n-1}}\int_{\Sigma}\frac{\dot{u}}{(1-\rho\bar{u}_1)^n}+\frac{1}{\omega_{2n-1}}\int_{\sigma}\frac{\dot{u}}{(1-\rho\bar{u}_1)^n}$$

得到

$$\lim_{\varepsilon\to 0}\lim_{\rho\to 1}\frac{1}{\omega_{2n-1}}\int_{\substack{u\bar{u}'=1\\|1-\bar{u}_1|>\varepsilon}}\frac{\dot{u}}{(1-\rho\bar{u}_1)^n}=\frac{1}{2}. \tag{1.2.6}$$

今定义 Cauchy 主值

$$\text{V. P.}\frac{1}{\omega_{2n-1}}\int_{u\bar{u}'=1}\frac{\dot{u}}{(1-\bar{u}_1)^n}=\lim_{\varepsilon\to 0}\frac{1}{\omega_{2n-1}}\int_{\substack{u\bar{u}'=1\\|1-\bar{u}_1|>\varepsilon}}\frac{\dot{u}}{(1-\bar{u}_1)^n}, \tag{1.2.7}$$

于是有

$$\text{V. P.}\frac{1}{\omega_{2n-1}}\int_{u\bar{u}'=1}\frac{\dot{u}}{(1-\bar{u}_1)^n}=\frac{1}{2}. \tag{1.2.8}$$

§1.3. Cauchy 主 値

設函数 $f(u)$ 定义在 $u\bar{u}'=1$ 上，如果对于任意二点 u 及 u^*，$u\bar{u}'=1$，$u^*\bar{u}'^*=1$，恆有

$$|f(u)-f(u^*)|=O([(u-u^*)\overline{(u-u^*)'}]^{\alpha/2})=O((2-u\bar{u}'^*-u^*\bar{u}')^{\alpha/2}),$$

則称 $f(u)$ 在 $u\bar{u}'=1$ 上滿足 Lip α，此处 $0<\alpha\leqslant 1$.

利用 §1.2 的結果，可以建立如下的

定理 1.1. 若 $f(u)$ 在 $u\bar{u}'=1$ 上滿足 Lip α，則 Cauchy 主値

$$\text{V. P.}\frac{1}{\omega_{2n-1}}\int_{u\bar{u}'=1}\frac{f(u)}{(1-v\bar{u}')^n}\dot{u}\qquad(v\bar{v}'=1)$$

是存在的，并且等于

$$\frac{1}{\omega_{2n-1}}\int_{u\bar{u}'=1}\frac{f(u)-f(v)}{(1-v\bar{u}')^n}\dot{u}+\frac{1}{2}f(v), \tag{1.3.1}$$

这里 Cauchy 主値的意义为

$$\lim_{\varepsilon\to 0}\frac{1}{\omega_{2n-1}}\int_{\substack{u\bar{u}'=1\\|1-v\bar{u}'|\geqslant\varepsilon}}\frac{f(u)}{(1-v\bar{u}')^n}\dot{u}.$$

証.

先取 $v=(1,0,\cdots,0)$，于是

$$\frac{1}{\omega_{2n-1}}\int_{\substack{u\bar{u}'=1\\|1-\bar{u}_1|\geqslant\varepsilon}}\frac{f(u)}{(1-\bar{u}_1)^n}\dot{u}=\frac{1}{\omega_{2n-1}}\int_{\substack{u\bar{u}'=1\\|1-\bar{u}_1|\geqslant\varepsilon}}\frac{f(u)-f(1,0,\cdots,0)}{(1-\bar{u}_1)^n}\dot{u}$$
$$+\frac{f(1,0,\cdots,0)}{\omega_{2n-1}}\int_{\substack{u\bar{u}'=1\\|1-\bar{u}_1|\geqslant\varepsilon}}\frac{\dot{u}}{(1-\bar{u}_1)^n}=I_1+I_2. \tag{1.3.2}$$

由于

$$|f(u)-f(1,0,\cdots,0)|=O[(2-u_1-\bar{u}_1)^{\alpha/2}]=O(|1-u_1|^{\alpha/2}),$$

因之

$$\left|\frac{1}{\omega_{2n-1}}\int_{\substack{u\bar{u}'=1\\|1-\bar{u}_1|\geqslant\varepsilon}}\frac{f(u)-f(1,0,\cdots,0)}{(1-\bar{u}_1)^n}\dot{u}\right|\leqslant\frac{K}{\omega_{2n-1}}\int_{\substack{u\bar{u}'=1\\|1-\bar{u}_1|\geqslant\varepsilon}}|1-\bar{u}_1|^{\frac{\alpha}{2}-n}\dot{u}$$
$$=\frac{K}{\omega_{2n-1}}\int_{v\bar{v}'<1}\dot{v}\int_{|\theta|\geqslant a}\frac{d\theta}{|1-re^{i\theta}|^{n-\frac{\alpha}{2}}}, \tag{1.3.3}$$

此处令 $\bar{u}_1 = re^{i\theta}$, $u_2 = v_2, \cdots, u_n = v_n$, $v = (v_2, \cdots, v_n)$, K 为一正的常数.

$$a = \cos^{-1}\left(\frac{1 + r^2 - \varepsilon^2}{2r}\right).$$

我們来考虑

$$\int_{|\theta| \geqslant a} \frac{d\theta}{|1 - re^{i\theta}|^{n - \frac{\alpha}{2}}}. \tag{1.3.4}$$

先設 $n = 2$, 于是由 Hölder 不等式

$$\int_{|\theta| \geqslant a} \frac{d\theta}{|1 - re^{i\theta}|^{2 - \frac{\alpha}{2}}} \leqslant \left(\int_{|\theta| \geqslant a} \frac{d\theta}{|1 - re^{i\theta}|^2}\right)^{\frac{2 - \frac{\alpha}{2}}{2}} \leqslant \left(\frac{2\pi}{1 - r^2}\right)^{1 - \frac{\alpha}{4}}.$$

因而当 $n = 2$ 时

$$\frac{1}{\omega_{2n-1}} \int_{\substack{u\bar{u}'=1 \\ |1-\bar{u}_1| \geqslant \varepsilon}} |1 - \bar{u}_1|^{\frac{\alpha}{2} - n} \dot{u} \leqslant \frac{2\pi}{\omega_3} \int_{v\bar{v}' < 1} \frac{\dot{v}}{(1 - r^2)^{1 - \frac{\alpha}{4}}} = \frac{2\pi}{\omega_3} \int_{v\bar{v}' < 1} \frac{\dot{v}}{(v\bar{v}')^{1 - \frac{\alpha}{4}}}.$$

用球坐标, 此为

$$O\left(\int_0^1 \frac{s\,ds}{s^{2 - \frac{\alpha}{2}}}\right) = O(1).$$

所以我們得到, 当 $n = 2$ 时

$$\frac{1}{\omega_3} \int_{\substack{u\bar{u}'=1 \\ |1-\bar{u}_1| \geqslant \varepsilon}} |1 - \bar{u}_1|^{\frac{\alpha}{2} - 2} \dot{u} = O(1). \tag{1.3.5}$$

当 $n > 2$ 时,

$$\int_{|\theta| \geqslant a} \frac{d\theta}{|1 - re^{i\theta}|^{n - \frac{\alpha}{2}}} \leqslant \int_0^{2\pi} \frac{d\theta}{|1 - re^{i\theta}|^{n - \frac{\alpha}{2}}} \leqslant \frac{1}{(1 - r)^{n - 2 - \frac{\alpha}{2}}} \int_0^{2\pi} \frac{d\theta}{|1 - re^{i\theta}|^2}$$

$$= \frac{2\pi}{(1 - r)^{n - \frac{\alpha}{2} - 1}(1 + r)} = O\left(\left(\frac{1}{1 - r^2}\right)^{n - \frac{\alpha}{2} - 1}\right).$$

所以

$$\frac{1}{\omega_{2n-1}} \int_{\substack{u\bar{u}'=1 \\ |1-\bar{u}_1| \geqslant \varepsilon}} |1 - \bar{u}_1|^{\frac{\alpha}{2} - n} \dot{u} = O\left(\int_{v\bar{v}' \leqslant 1} (1 - r^2)^{\frac{\alpha}{2} - n + 1} \dot{v}\right)$$

$$= O\left(\int_0^1 s^{2n-3} \cdot s^{\alpha - 2n + 2} ds\right) \quad \text{(用球坐标)}$$

$$= O\left(\int_0^1 s^{\alpha - 1} ds\right) = O(1). \tag{1.3.6}$$

因而 $\lim\limits_{\varepsilon \to 0} I_1$ 是一个通常积分. 根据 § 1.2 之結果, $\lim\limits_{\varepsilon \to 0} \int_{\substack{u\bar{u}'=1 \\ |1-\bar{u}_1| \geqslant \varepsilon}} \frac{\dot{u}}{(1 - \bar{u}_1)^n} = \frac{1}{2}$. 所以在(1.3.2)中令 $\varepsilon \to 0$, 則当 $v = (1, 0, \cdots, 0)$ 时証明了(1.3.1).

对于一般的 $v: v\bar{v}' = 1$, 一定存在酉方陣 U, 使 $vU = (1, 0, \cdots, 0)$, 故得定理 1.1.

§ 1.4. Cauchy 型积分的极限值

記 $\rho(z, v) = ((z - v)\overline{(z - v)'})^{\frac{1}{2}}$, $z\bar{z}' < 1$, $v\bar{v}' = 1$. 并定义 $d(z, \mathfrak{L}) = \min\limits_{u \in \mathfrak{L}} |1 - z\bar{u}'|$, $\mathfrak{L}$ 表示 $u\bar{u}' = 1$, $z\bar{z}' < 1$. 先来証明如下的

定理 1.2. 設 $f(u)\in \mathrm{Lip}\,\alpha,\ 0<\alpha\leqslant 1$ (在 $u\bar{u}'=1$ 上). 令 z 从 $z\bar{z}'<1$ 內趋于点 $v(v\bar{v}'=1)$, 只要

$$\frac{\rho(z,v)}{d(z,\mathfrak{L})}<M, \tag{1.4.1}$$

M 为一常数,則必有

$$\lim_{z\to v}\int_{u\bar{u}'=1}\frac{f(u)-f(v)}{(1-z\bar{u}')^n}\dot{u}=\int_{u\bar{u}'=1}\frac{f(u)-f(v)}{(1-v\bar{u}')^n}\dot{u}. \tag{1.4.2}$$

証.

首先有

$$\begin{aligned}I&=\int_{u\bar{u}'=1}\left(\frac{f(u)-f(v)}{(1-z\bar{u}')^n}-\frac{f(u)-f(v)}{(1-v\bar{u}')^n}\right)\dot{u}\\&=\int_{u\bar{u}'=1}\frac{[(1-v\bar{u}')^n-(1-z\bar{u}')^n](f(u)-f(v))}{(1-z\bar{u}')^n(1-v\bar{u}')^n}\dot{u}\\&=\left(\int_{\sigma_\delta}+\int_{\Sigma_\delta}\right)\frac{[(1-v\bar{u}')^n-(1-z\bar{u}')^n](f(u)-f(v))}{(1-z\bar{u}')^n(1-v\bar{u}')^n}\dot{u}=I_1+I_2,\end{aligned} \tag{1.4.3}$$

此处

$$\sigma_\delta:\begin{cases}u\bar{u}'=1,\\|1-v\bar{u}'|<\delta,\end{cases}\qquad \Sigma_\delta:\begin{cases}u\bar{u}'=1,\\|1-v\bar{u}'|>\delta.\end{cases}$$

但

$$|I_1|\leqslant\int_{\sigma_\delta}\frac{|(v-z)\bar{u}'|(|1-v\bar{u}'|^{n-1}+|1-v\bar{u}'|^{n-2}\cdot|1-z\bar{u}'|+\cdots+|1-z\bar{u}'|^{n-1})\dot{u}}{|1-z\bar{u}'|^n|1-v\bar{u}'|^{n-\frac{\alpha}{2}}}$$

$$\leqslant\sum_{k=1}^n\int_{\sigma_\delta}\frac{\rho(z,v)\dot{u}}{|1-z\bar{u}'|^k|1-v\bar{u}'|^{n-(k-1)-\frac{\alpha}{2}}}.$$

注意到 $|1-v\bar{u}'|\leqslant|1-z\bar{u}'|+|(z-v)\bar{u}'|$, 而

$$|(z-v)\bar{u}'|\leqslant[(z-v)\overline{(z-v)}']^{\frac{1}{2}}<Md(z,\mathfrak{L})\leqslant M|1-z\bar{u}'|,$$

所以 $|1-v\bar{u}'|\leqslant(M+1)|1-z\bar{u}'|$, 即 $|1-z\bar{u}'|\geqslant\dfrac{1}{M+1}|1-v\bar{u}'|$. 再由于

$$\frac{\rho(z,v)}{|1-z\bar{u}'|}\leqslant\frac{\rho(z,v)}{d(z,\mathfrak{L})}<M,$$

于是

$$|I_1|\leqslant\sum_{k=1}^n M(M+1)^k\int_{\sigma_\delta}\frac{\dot{u}}{|1-v\bar{u}'|^{n-\frac{\alpha}{2}}}=M_1\int_{\sigma_\delta}\frac{\dot{u}}{|1-v\bar{u}'|^{n-\frac{\alpha}{2}}}, \tag{1.4.4}$$

其中 $M_1=\sum\limits_{k=1}^n M(M+1)^k$.

由下面的引理 1.2 知,

$$\int_{\sigma_\delta}\frac{\dot{u}}{|1-v\bar{u}'|^{n-\frac{\alpha}{2}}}=O(\delta^{\alpha/2})\qquad(\delta\to 0), \tag{1.4.5}$$

所以, 对于任与的 $\varepsilon>0$, 可取一 $\delta>0$, 使在 σ_δ 上的积分 I_1 滿足 $|I_1|<\dfrac{\varepsilon}{2}$. 在 δ 取定

之后，由于

$$\int_{\Sigma_\delta} \frac{f(u) - f(v)}{(1 - z\bar{u}')^n} \dot{u}$$

作为 z 的函数，在 $z = v$ 連續，則当 $\rho(z, v)$ 充分小时，必有 $|I_2| < \frac{\varepsilon}{2}$.

綜上所述，当 $\rho(z, v)$ 充分小时，有 $|I| < \varepsilon$. 定理得証.

現在証明

引理 1.2.

$$\varphi(\delta) = \int_{\sigma_\delta} \frac{\dot{u}}{|1 - v\bar{u}'|^{n-\frac{\alpha}{2}}} = O(\delta^{\alpha/2}) \qquad (\delta \to 0), \tag{1.4.5}$$

其中

$$\sigma_\delta: \begin{cases} u\bar{u}' = 1, \\ |1 - v\bar{u}'| \leqslant \delta, \end{cases} \quad v\bar{v}' = 1,\ u\bar{u}' = 1,\ n \geqslant 2,\ 0 < \alpha \leqslant 1.$$

証.

因为存在酉方陣 U，使得 $v = (1, 0, \cdots, 0)U$，所以不妨假設

$$\varphi(\delta) = \int_{\substack{u\bar{u}'=1 \\ |1-\bar{u}_1| \leqslant \delta}} \frac{\dot{u}}{|1 - \bar{u}_1|^{n-\frac{\alpha}{2}}}.$$

記 $u_j = x_j + iy_j$, $j = 1, 2, \cdots, n$，并令 $x_1 = \cos\theta_1$, $y_1 = \sin\theta_1\cos\theta_2$, $x_2 = \sin\theta_1\sin\theta_2\cos\theta_3$, $\cdots$, $y_n = \sin\theta_1\sin\theta_2\cdots\sin\theta_{2n-1}$，其中 $0 \leqslant \theta_1, \cdots, \theta_{2n-2} < \pi$, $0 \leqslant \theta_{2n-1} < 2\pi$，則有

$$\dot{u} = \sin^{2n-2}\theta_1 \sin^{2n-3}\theta_2 \cdots \sin\theta_{2n-2}\, d\theta_1\, d\theta_2 \cdots d\theta_{2n-1},$$

$$|1 - u_1|^2 = (1 - \cos\theta_1)^2 + \sin^2\theta_1\cos^2\theta_2.$$

于是当 $\delta \to 0$ 时，

$$\varphi(\delta) = O\left(\iint\limits_{\substack{0 \leqslant \theta_1,\ \theta_2 < \pi \\ (1-\cos\theta_1)^2 + \sin^2\theta_1\cos^2\theta_2 \leqslant \delta^2}} \frac{\sin^{2n-2}\theta_1 \sin^{2n-3}\theta_2\, d\theta_1\, d\theta_2}{[(1 - \cos\theta_1)^2 + \sin^2\theta_1\cos^2\theta_2]^{\frac{2n-\alpha}{4}}} \right). \tag{1.4.6}$$

注意到

$$\delta^2 \geqslant (1 - \cos\theta_1)^2 + \sin^2\theta_1\cos^2\theta_2 \geqslant (1 - \cos\theta_1)^2 = 4\sin\frac{4\theta_1}{2},$$

习知当 $0 \leqslant x \leqslant \frac{\pi}{2}$ 时，$\sin x \geqslant \frac{2}{\pi}x$，所以 $\delta^2 \geqslant \frac{4}{\pi^4}\theta_1^4$，即 $\theta_1 \leqslant \frac{\pi}{\sqrt{2}}\delta^{\frac{1}{2}} < \pi\delta^{\frac{1}{2}}$.

記 $(1 - \cos\theta_1)^2 = a$, $\sin^2\theta_1 = b$，并令 $\cos\theta_2 = t$，則

$$\varphi(\delta) = O\left(\int_0^{\pi\delta^{\frac{1}{2}}} d\theta_1 \int_0^1 \frac{b^{n-1}\,dt}{(a + bt^2)^{\frac{2n-\alpha}{4}}} \right)$$

$$= O\left(\int_0^{\pi\delta^{\frac{1}{2}}} \frac{b^{n-1}}{a^{\frac{2n-\alpha}{4}}}\, d\theta_1 \int_0^1 \frac{dt}{\left(1 + \left(\sqrt{\frac{b}{a}}\, t\right)^2\right)^{\frac{2n-\alpha}{4}}} \right).$$

令 $\sqrt{\frac{b}{a}}\, t = \operatorname{tg}\psi$，便有

$$\varphi(\delta) = O\left(\int_0^{\pi\delta^{\frac{1}{2}}} \frac{b^{n-1-\frac{1}{2}}}{a^{\frac{2n-a-2}{4}}} d\theta_1 \int_0^{\operatorname{tg}^{-1}\sqrt{\frac{b}{a}}} \cos^{\frac{2n-a}{2}-2}\psi\, d\psi\right)$$

$$= O\left(\int_0^{\pi\delta^{\frac{1}{2}}} \frac{\sin^{2n-3}\theta_1\, d\theta_1}{(1-\cos\theta_1)^{\frac{2n-a-2}{2}}}\right) = O\left(\int_0^{\pi\delta^{\frac{1}{2}}} \theta_1^{2n-3-(2n-a-2)} d\theta_1\right) = O(\delta^{a/2}).$$

証毕.

进而我們可以得到本节之主要結果:

定理 1.3. 若 $f(u)$ 在 $u\bar{u}'=1$ 上滿足 $\operatorname{Lip}\alpha$, 則当 z 从 $z\bar{z}'<1$ 內部趋于 $v(v\bar{v}'=1)$ 时,只要保持

$$\frac{\rho(z, v)}{d(z, \mathfrak{L})} < M \qquad (M\ 为一常数),$$

則有

$$\lim_{z\to v} \frac{1}{\omega_{2n-1}} \int_{u\bar{u}'=1} \frac{f(u)}{(1-z\bar{u}')^n}\dot{u} = \text{V. P.} \frac{1}{\omega_{2n-1}} \int_{u\bar{u}'=1} \frac{f(u)\dot{u}}{(1-v\bar{u}')^n} + \frac{1}{2}f(v). \tag{1.4.7}$$

証.

首先可以写成

$$\frac{1}{\omega_{2n-1}} \int_{u\bar{u}'=1} \frac{f(u)\dot{u}}{(1-z\bar{u}')^n} = \frac{1}{\omega_{2n-1}} \int_{u\bar{u}'=1} \frac{f(u)-f(v)}{(1-z\bar{u}')^n}\dot{u}$$

$$+ \frac{f(v)}{\omega_{2n-1}} \int_{u\bar{u}'=1} \frac{\dot{u}}{(1-z\bar{u}')^n} = I_1 + I_2. \tag{1.4.8}$$

由定理 1.2 及定理 1.1, 有

$$\lim_{z\to v} I_1 = \frac{1}{\omega_{2n-1}} \int_{u\bar{u}'=1} \frac{f(u)-f(v)}{(1-v\bar{u}')^n}\dot{u} = \text{V. P.} \frac{1}{\omega_{2n-1}} \int_{u\bar{u}'=1} \frac{f(u)\dot{u}}{(1-v\bar{u}')^n} - \frac{1}{2}f(v). \tag{1.4.9}$$

再注意到

$$\frac{1}{\omega_{2n-1}} \int_{u\bar{u}'=1} \frac{\dot{u}}{(1-z\bar{u}')^n} = 1,$$

便可立得公式(1.4.7).

§ 1.5. Cauchy 型积分的极限函数的連續性質

更深入一步,我們有如下的

定理 1.4[1]. 若在 $u\bar{u}'=1$ 上 $f(u)\in\operatorname{Lip}\alpha$, $0<\alpha\leqslant 1$, 則由(1.4.7)式所示的极限函数

$$F(v) = \lim_{z\to v} \frac{1}{\omega_{2n-1}} \int_{u\bar{u}'=1} \frac{f(u)\dot{u}}{(1-z\bar{u}')^n}$$

必属于 $\operatorname{Lip}(\alpha-\delta)$, 这里 $v\bar{v}'=1$, δ 是小于 α 的任一正数.

为了証明这一定理,我們首先証明

引理 1.3. 設 $u=(u_1, \cdots, u_n)$, 則

1) 在一个变数时,这称之为 J. Plemelj-Привалов 定理.

$$\Psi(\varepsilon)=\int_{\substack{u\bar{u}'=1\\|1-\bar{u}_1|\geqslant\varepsilon}}\frac{\dot{u}}{|1-\bar{u}_1|^n}=O(\log\varepsilon)\qquad\varepsilon\to 0.\tag{1.5.1}$$

如同引理 1.2 的証法, 对 $u\bar{u}'=1$ 的点采用球坐标表示, 便有

$$\Psi(\varepsilon)=O\left(\iint_{\substack{0\leqslant\theta_1,\theta_2<\pi\\(1-\cos\theta_1)^2+\sin^2\theta_1\cos^2\theta_2\geqslant\varepsilon^2}}\frac{\sin^{2n-2}\theta_1\sin^{2n-3}\theta_2\,d\theta_1\,d\theta_2}{[(1-\cos\theta_1)^2+\sin^2\theta_1\cos^2\theta_2]^{n/2}}\right).\tag{1.5.2}$$

因为当 $\frac{\pi}{2}\leqslant\theta_1<\pi$ 时, $(1-\cos\theta_1)^2+\sin^2\theta_1\cos^2\theta_2\geqslant 1$, 同时由于

$$\begin{aligned}\varepsilon^2&\leqslant(1-\cos\theta_1)^2+\sin^2\theta_1\cos^2\theta_2\leqslant(1-\cos\theta_1)^2+\sin^2\theta_1\\&=2(1-\cos\theta_1)=4\sin^2\frac{\theta_1}{2}\leqslant\theta_1^2,\end{aligned}$$

所以, 若記 $(1-\cos\theta_1)^2=a$, $\sin^2\theta_1=b$, 并令 $\cos\theta_2=t$, 則

$$\Psi(\varepsilon)=O(1)+O\left(\int_\varepsilon^{\pi/2}b^{n-1}d\theta_1\int_0^1\frac{dt}{(a+bt^2)^{n/2}}\right).$$

令 $\sqrt{\frac{b}{a}}\,t=\operatorname{tg}\psi$, 便有

$$\begin{aligned}\Psi(\varepsilon)&=O(1)+O\left(\int_\varepsilon^{\pi/2}\frac{b^{n-1-\frac{1}{2}}}{a^{\frac{n-1}{2}}}d\theta_1\int_0^{\operatorname{tg}^{-1}\sqrt{\frac{b}{a}}}\cos^{n-2}\psi\,d\psi\right)\\&=O(1)+O\left(\int_\varepsilon^{\pi/2}\frac{\sin^{2n-3}\theta_1}{(1-\cos\theta_1)^{n-1}}d\theta_1\right)\\&=O(1)+O\left(\int_\varepsilon^{\pi/2}\theta_1^{-1}\,d\theta_1\right)=O(\log\varepsilon)\qquad\varepsilon\to 0.\end{aligned}$$

引理得証.

现在来証明定理 1.4.

令

$$F_1(v)=\frac{1}{\omega_{2n-1}}\int_{u\bar{u}'=1}\frac{f(u)-f(v)}{(1-v\bar{u}')^n}\dot{u},\tag{1.5.3}$$

显然, 只需証明 $F_1(v)$ 在点 $e_1=(1,0,\cdots,0)$ 属于 $\mathrm{Lip}(\alpha-\delta)$, 其中 δ 为小于 α 的任一正数. 也就是对于任一点 v, $v\bar{v}'=1$, 必有

$$|F_1(v)-F_1(e_1)|\leqslant M\rho\log\rho,\tag{1.5.4}$$

其中 M 为一絕对常数, $\rho=\rho(v,e_1)=((v-e_1)\overline{(v-e_1)}')^{1/2}$.

必須指出, 不妨設 $v=(v_1,v_*,0,\cdots,0)=e_1V$, 其中 $V=\begin{pmatrix}v_1&v_*\\-\bar{v}_*&\bar{v}_1\end{pmatrix}\dotplus I^{(n-2)}$ 为一酉方陣. 这是因为对于任一定点 $v=(v_1,\cdots,v_n)$, $v\bar{v}'=1$, 必有酉方陣 $U=\begin{pmatrix}1&0\\0&U_1\end{pmatrix}$, 使 $v=(v_1,v_*,0,\cdots,0)U$, 所以可先作一变换 $u=\tilde{u}U$, 相应地, 有 $v=\tilde{v}U$, $e_1=e_1U$, 其中 $\tilde{v}=(v_1,v_*,0,\cdots,0)=e_1V$, V 即如上所述.

由于

$$F_1(v) - F_1(e_1) = \frac{1}{\omega_{2n-1}} \int_{w\bar{w}'=1} \frac{f(w) - f(v)}{(1 - v\bar{w}')^n} \dot{w} - \frac{1}{\omega_{2n-1}} \int_{u\bar{u}'=1} \frac{f(u) - f(e_1)}{(1 - e_1\bar{u}')^n} \dot{u}$$

$$= \frac{1}{\omega_{2n-1}} \int_{u\bar{u}'=1} \frac{f(uV) - f(e_1V)}{(1 - e_1\bar{u}')^n} \dot{u} - \frac{1}{\omega_{2n-1}} \int_{u\bar{u}'=1} \frac{f(u) - f(e_1)}{(1 - e_1\bar{u}')^n} \dot{u},$$

于是

$$|F_1(v) - F_1(e_1)| \leqslant \frac{1}{\omega_{2n-1}} \left| \int_\sigma \frac{f(uV) - f(e_1V)}{(1 - e_1\bar{u}')^n} \dot{u} \right| + \frac{1}{\omega_{2n-1}} \left| \int_\sigma \frac{f(u) - f(e_1)}{(1 - e_1\bar{u}')^n} \dot{u} \right|$$

$$+ \frac{1}{\omega_{2n-1}} \left| \int_\Sigma \frac{f(uV) - f(u)}{(1 - e_1\bar{u}')^n} \dot{u} \right| + \frac{1}{\omega_{2n-1}} \left| \int_\Sigma \frac{f(e_1V) - f(e_1)}{(1 - e_1\bar{u}')^n} \dot{u} \right|$$

$$= J_1 + J_2 + J_3 + J_4, \tag{1.5.5}$$

此处

$$\sigma: \begin{cases} u\bar{u}' = 1, \\ |1 - e_1\bar{u}'| \leqslant \varepsilon, \end{cases} \quad \Sigma: \begin{cases} u\bar{u}' = 1, \\ |1 - e_1\bar{u}'| \geqslant \varepsilon, \end{cases} \quad \varepsilon = \rho^2. \tag{1.5.6}$$

由引理 1.2 可知

$$\left.\begin{aligned} J_1 &\leqslant M_1\varepsilon^{\alpha/2} = M_1\rho^\alpha, \\ J_2 &\leqslant M_1\varepsilon^{\alpha/2} = M_1\rho^\alpha. \end{aligned}\right\} \tag{1.5.7}$$

由引理 1.3 可知

$$J_4 \leqslant \frac{1}{\omega_{2n-1}} |f(e_1V) - f(e_1)| \int_\Sigma \frac{\dot{u}}{|1 - e_1\bar{u}'|^n} \leqslant M_2\rho^\alpha \log \rho. \tag{1.5.8}$$

上面的 M_1, M_2 皆为絕对常数.

注意到

$$J_3 \leqslant \frac{1}{\omega_{2n-1}} \int_\Sigma \frac{|f(uV) - f(u)|}{|1 - e_1\bar{u}'|^n} \dot{u} \leqslant \frac{M_3}{\omega_{2n-1}} \int_\Sigma \frac{((uV - u)\overline{(uV - u)}')^{\alpha/2}}{|1 - e_1\bar{u}'|^n} \dot{u},$$

由于

$$(uV - u)\overline{(uV - u)}' = u(2I - V - \bar{V}')\bar{u}' \leqslant 2\mathrm{Re}(1 - v_1)$$

而

$$\rho = \rho(v, e_1) = (e_1(V - I)(\bar{V}' - I)e_1')^{\frac{1}{2}} = (2\mathrm{Re}(1 - v_1))^{\frac{1}{2}},$$

所以

$$J_3 \leqslant \frac{M_3\rho^\alpha}{\omega_{2n-1}} \int_\Sigma \frac{\dot{u}}{|1 - e_1\bar{u}'|^n} \leqslant M_4\rho^\alpha \log \rho. \tag{1.5.9}$$

上面的 M_3, M_4 亦皆为絕对常数.

取 $M = \max\{M_1, M_2, M_4\}$, 便有(1.5.4). 定理証毕.

此外还要証明一个以后用得到的

定理 1.5. 若 $\varphi(w, u)$ 为定义在 $w\bar{w}' = 1$, $u\bar{u}' = 1$ 上的連續函数,作为 w 的函数及作为 u 的函数都滿足 Lipschitz 条件,那末

$$\Phi(v, u) = \frac{1}{\omega_{2n-1}} \int_{w\bar{w}'=1} \frac{\varphi(w, u)}{(1 - v\bar{w}')^n} \dot{w}$$

作为 v 的函数及 u 的函数也都滿足 Lipschitz 条件.

証. 根据定理 1.4, 我們只需証明作为 u 的函数, $\Phi(v, u) = \psi(u)$ 是滿足 Lipschitz

条件的. 若 $u_0\bar{u}_0' = 1$, 则

$$\psi(u) - \psi(u_0) = \frac{1}{\omega_{2n-1}}\int_{w\bar{w}'=1}\frac{\varphi(w,u) - \varphi(w,u_0)}{(1-v\bar{w}')^n}\dot{w}$$

$$= \frac{1}{\omega_{2n-1}}\int_{w\bar{w}'=1}\frac{[\varphi(w,u) - \varphi(v,u)] - [\varphi(w,u_0) - \varphi(v,u_0)]}{(1-v\bar{w}')^n}\dot{w}$$

$$+ [\varphi(v,u) - \varphi(v,u_0)]\frac{1}{\omega_{2n-1}}\int_{w\bar{w}'=1}\frac{\dot{w}}{(1-v\bar{w}')^n} = I_1 + I_2.$$

令

$$\rho = \rho(u, u_0) = ((u-u_0)\overline{(u-u_0)}')^{\frac{1}{2}},$$

据假設, $\varphi(w, u)$ 作为 u 的函数属于 $\mathrm{Lip}\,\alpha\ (0 < \alpha \leqslant 1)$, 那末显然有

$$I_2 = O(\rho^\alpha) \qquad (\rho \to 0).$$

命

$$\Sigma: \begin{cases} w\bar{w}' = 1, \\ |1 - v\bar{w}'| \geqslant \varepsilon, \end{cases} \quad \sigma: \begin{cases} w\bar{w}' = 1, \\ |1 - v\bar{w}'| \leqslant \varepsilon, \end{cases}$$

则

$$I_1 = \frac{1}{\omega_{2n-1}}\left(\int_\sigma + \int_\Sigma\right) = I_3 + I_4,$$

显然我們有

$$I_3 = O\left(\int_\sigma \frac{\dot{w}}{|1 - v\bar{w}'|^{n-\frac{\alpha}{2}}}\right) = O(\varepsilon^{\alpha/2}) \qquad \varepsilon \to 0,$$

及

$$I_4 = \frac{1}{\omega_{2n-1}}\int_\Sigma \frac{\varphi(w,u) - \varphi(w,u_0)}{(1-v\bar{w}')^n}\dot{w} - (\varphi(v,u) - \varphi(v,u_0))\frac{1}{\omega_{2n-1}}\int_\Sigma\frac{\dot{w}}{(1-v\bar{w}')^n}$$

$$= I_5 + I_6.$$

容易看出

$$I_5 = O(\rho^\alpha \log \varepsilon)$$

及

$$I_6 = O(\rho^\alpha).$$

取 $\varepsilon = \rho^2$, 则最后得到

$$\psi(u) - \psi(u_0) = O(((u-u_0)\overline{(u-u_0)}')^{\alpha/2}\log(u-u_0)\overline{(u-u_0)}'),$$

故对于任意小于 α 的正数 δ, 有

$$\psi(u) \in \mathrm{Lip}(\alpha - \delta).$$

参　考　文　献

[1] 华罗庚, 多复变数函数论中的典型域的调和分析, 科学出版社, 1958.

[2] 龚　昇、孙继广, 圆型域中解析函数的积分表示, 中国科学技术大学学报.

[3] 陆启鏗、鍾同德, Привалов 定理的拓广, 数学学报, 7 (1957), 143—165.

[4] Какичев В. А., Граничные свойства интеграла типа Коши многих переменых, Уч. зап. шахтинск. педин-та, 2: 6 (1959), 29—90.

第15卷 第6期 数 学 学 报 Vol. 15, No. 6
1965年11月 ACTA MATHEMATICA SINICA Nov., 1965

多复变数的 Cauchy 型积分(II)
Lie 球双曲空間的 Cauchy 型积分*

龚 昇 孙继广
(中国科学技术大学) (中国科学院数学研究所)

§2.1. 引 言

Lie 球双曲空间 $\mathfrak{R}_{\mathrm{IV}}(N)$(以下记为 $\mathfrak{R}_{\mathrm{IV}}$),由满足

$$\begin{cases}1+|zA_0z'|^2-z\bar{z}'>0,\\ 1-|zA_0z'|>0\end{cases}\quad 而\ A_0=\frac{1}{2}\begin{pmatrix}0&1\\-1&0\end{pmatrix}+\frac{1}{2}I^{(N-2)},$$

的 $N(\geqslant 2)$ 个复元素矢量 $z=(z_1, z_2, \cdots, z_N)$ 所构成. $\mathfrak{R}_{\mathrm{IV}}$ 的边界,记为 $\mathfrak{B}_{\mathrm{IV}}$,特别地,以 $\mathfrak{L}_{\mathrm{IV}}$ 表示 $\mathfrak{R}_{\mathrm{IV}}$ 的特征流形. 此外,$N(\geqslant 2)$ 个复变数 $z_1, \cdots, z_N$ 空间中的域 $\mathfrak{R}^*_{\mathrm{IV}}$, $\mathfrak{D}_1$ 及 $\mathfrak{D}_2$ 分别指

$$\mathfrak{R}^*_{\mathrm{IV}}:\quad \begin{cases}1+|zA_0z'|^2-z\bar{z}'>0\\ 1-|zA_0z'|<0\end{cases},$$

$$\mathfrak{D}_1:\quad \begin{cases}1+|zA_0z'|^2-z\bar{z}'<0\\ 1-|zA_0z'|>0\end{cases}$$

和

$$\mathfrak{D}_2:\quad \begin{cases}1+|zA_0z'|^2-z\bar{z}'<0\\ 1-|zA_0z'|<0.\end{cases}$$

$\mathfrak{R}^*_{\mathrm{IV}}$ 的边界,记为 $\mathfrak{B}^*_{\mathrm{IV}}$,而 $\mathfrak{L}_{\mathrm{IV}}$ 同时也是 $\mathfrak{R}^*_{\mathrm{IV}}$ 的特征流形.

习知,如果 $\varphi(\xi)$ 是 $\mathfrak{L}_{\mathrm{IV}}$ 上的连续函数,则当 $z\in\mathfrak{R}_{\mathrm{IV}}$ 或 $z\in\mathfrak{R}^*_{\mathrm{IV}}$ 时, Cauchy 型积分

$$V(\mathfrak{L}_{\mathrm{IV}})^{-1}\int_{\mathfrak{L}_{\mathrm{IV}}}H_{\mathrm{IV}}(z,\xi)\varphi(\xi)\dot{\xi}\tag{2.1.1}$$

是存在的[1],此处 $H_{\mathrm{IV}}(z,\xi)$ 表示 Cauchy 核$(1+zA_0z'\overline{\xi A_0\xi'}-z\bar{\xi}')^{-\frac{N}{2}}$, $V(\mathfrak{L}_{\mathrm{IV}})$是 $\mathfrak{L}_{\mathrm{IV}}$ 的体积,$\dot{\xi}$ 表示 $\mathfrak{L}_{\mathrm{IV}}$ 的体积元素.

但当 $z\in\mathfrak{D}_1$ 或 $\mathfrak{D}_2$,而 $N>2$ 时, Cauchy 核 $H_{\mathrm{IV}}(z,\xi)$ 失去意义,因而 Cauchy 型积分(2.1.1)不复存在. 事实上,任与一点 $z\in\mathfrak{D}_1$(或 $z\in\mathfrak{D}_2$),必有分解式 $z=e^{i\theta}(\lambda_1, \lambda_2, 0, \cdots, 0)P_0^{-1}\Gamma P_0$,其中 $0\leqslant\lambda_1<1<\lambda_2$(当 $z\in\mathfrak{D}_2$ 时,$0<\lambda_1$), $P_0=\begin{pmatrix}1&1\\i&-i\end{pmatrix}\dot{+}\sqrt{2}\,I^{(N-2)}$, Γ 为一实正交方阵. 对所与之点 z,我们取 $\mathfrak{L}_{\mathrm{IV}}$ 上的点

$$\xi=e^{i\theta}\left(\frac{1+\lambda_1\lambda_2}{\lambda_1+\lambda_2},\ 0,\ \sqrt{1-\left(\frac{1+\lambda_1\lambda_2}{\lambda_1+\lambda_2}\right)^2},\ 0,\ \cdots 0\right)P_0(P_0^{-1}\Gamma P_0),$$

* 1964年3月8日收到,1964年11月21日收到修改稿.

则 $1+zA_0z'\overline{\xi A_0\xi'}-z\bar{\xi}'=0$.（当 $N=2$ 时，$\mathfrak{R}_{IV}$ 即双圆柱：$|z_1|<1$，$|z_2|<1$；$\mathfrak{R}_{IV}^*$，$\mathfrak{D}_1$ 及 $\mathfrak{D}_2$ 分别为 $|z_1|>1$，$|z_2|>1$；$(1-|z_1|^2)(1-|z_2|^2)<0$，$|z_1z_2|<1$；$(1-|z_1|^2)(1-|z_2|^2)<0$，$|z_1z_2|>1$. 与之相应的 Cauchy 型积分的问题，实即双圆柱的 Cauchy 型积分的问题，Какичев[5]已作过研究.）

本文主要是对 $\mathfrak{R}_{IV}$ 上的 Cauchy 型积分(2.1.1)进行研究. 在定义了 $\mathfrak{B}_{IV}-\mathfrak{L}_{IV}$ 上的 Cauchy 主值之后，我们证明了：满足 Lipschitz 条件的函数，在 $\mathfrak{B}_{IV}-\mathfrak{L}_{IV}$ 上的 Cauchy 主值是存在的，而且给出了从 $\mathfrak{R}_{IV}$ 的内部趋于 $\mathfrak{B}_{IV}-\mathfrak{L}_{IV}$ 时，Cauchy 型积分的极限值. 并由此推出：$\mathfrak{R}_{IV}$ 上的任一 B-调和函数在 $\mathfrak{B}_{IV}-\mathfrak{L}_{IV}$ 上的值，必可用它的 B-调和共轭函数在 $\mathfrak{L}_{IV}$ 上的值通过积分表出. 我们还证明了：当 $N=2n(n\geqslant 1)$ 时，如果 $\varphi(\xi)=\varphi(e^{i\theta}xp_0)=\varphi_0(e^{i\theta})$ 对 $e^{i\theta}$ 具有 $n-1$ 阶连续微商，并且它对 $e^{i\theta}$ 的 k 阶微商 $(0\leqslant k\leqslant n-1)$ 在 $\mathfrak{L}_{IV}$ 上属于 Lip $p(0<p\leqslant 1)$；当 $N=2n+1$ $(n\geqslant 1)$ 时，如果 $\varphi(\xi)=\varphi(e^{i\theta}xp_0)=\varphi_0(e^{i\theta})$ 对 $e^{i\theta}$ 具有 $n+1$ 阶连续微商，并且它对 $e^{i\theta}$ 的 k 阶微商 $(0\leqslant k\leqslant n+1)$ 在 $\mathfrak{L}_{IV}$ 上连续，则 Cauchy 型积分(2.1.1)在 $\mathfrak{L}_{IV}$ 上的极限值是存在的. 利用解析变换 $w=(zA_0z')^{-1}z(z\in\mathfrak{R}_{IV}$, $w\in\mathfrak{R}_{IV}^*)$，不难得到 $\mathfrak{R}_{IV}^*$ 上的 Cauchy 型积分的相应结果.

在讨论中所用到的有关 $\mathfrak{R}_{IV}$ 的已知事实（例如 $\mathfrak{R}_{IV}$ 的运动群 Γ^{IV}、原点的固定分群 Γ_0^{IV} 等等），可参阅[1]，[2].

本文曾在中国科学技术大学数学系综合讨论班上报告过，得到了不少帮助. 我们愿向我们的导师华罗庚先生表示衷心的感谢，他自始至终关怀着这项工作，并给与了宝贵的指示. 陆启铿先生曾对本文提供了有益的意见，我们向他致谢.

§2.2. 一 条 引 理

根据 Cauchy 积分公式[1]，我们有

$$V(\mathfrak{L}_{IV})^{-1}\int_{\mathfrak{L}_{IV}}H_{IV}(z,\xi)\dot{\xi}=1, \tag{2.2.1}$$

其中 $z=(z_1,z_2,\cdots,z_N)\in\mathfrak{R}_{IV}$，$\xi=(\xi_1,\cdots,\xi_N)=e^{i\theta}xP_0\in\mathfrak{L}_{IV}$，$xx'=1$，$\dot{\xi}=2^{\frac{N}{2}}d\theta\cdot\dot{x}$.

特别地，取 $z=(0,\rho,0,\cdots,0)$，$0\leqslant\rho<1$，有 $zA_0z'=0$，于是

$$1=V(\mathfrak{L}_{IV})^{-1}\int_{\mathfrak{L}_{IV}}(1-\rho\bar{\xi}_2)^{-\frac{N}{2}}\dot{\xi}=2^{\frac{N}{2}}V(\mathfrak{L}_{IV})^{-1}\int_{xx'=1}\dot{x}\int_0^{\pi}(1-\rho e^{-i\theta}(x_1+ix_2))^{-\frac{N}{2}}d\theta. \tag{2.2.2}$$

作变换 $y_1+iy_2=e^{-i\theta}(x_1+ix_2)$，$y_3=x_3,\cdots,y_N=x_N$，$\theta=\theta$，则有

$$\begin{aligned}1&=2^{\frac{N}{2}}V(\mathfrak{L}_{IV})^{-1}\int_{yy'=1}(1-\rho(y_1+iy_2))^{-\frac{N}{2}}\dot{y}\int_0^{\pi}d\theta\\&=\varpi_{N-1}^{-1}\int_{yy'=1}(1-\rho(y_1+iy_2))^{-\frac{N}{2}}\dot{y}.\end{aligned} \tag{2.2.3}$$

此处 $\varpi_{N-1}=2^{-\frac{N}{2}}\pi^{-1}V(\mathfrak{L}_{IV})=2\pi^{\frac{N}{2}}\Big/\Gamma\left(\frac{N}{2}\right)$ 是 $yy'=1$ 的体积.

我们来证明如下的

引理 2.1. $\lim\limits_{\varepsilon\to 0}\varpi_{N-1}^{-1}\int\limits_{\substack{yy'=1\\|1-y_1-iy_2|>\varepsilon}}(1-(y_1+iy_2))^{-\frac{N}{2}}\dot{y}$ 是存在的，且等于 1/2.

證．当 $N=2$ 时，引理之结论是复变数函数论中熟知的事实．当 $N=2n$ 时，实即引理 1.1[3]．因此，只需讨论 $N=2n+1$，$n\geqslant 1$ 的情形．

已知有

$$\begin{aligned} 1 &= \varpi_{N-1}^{-1}\int_{yy'=1}(1-\rho(y_1+iy_2))^{-\frac{N}{2}}\dot{y} \\ &= \varpi_{N-1}^{-1}\left(\int_{\Sigma}+\int_{\sigma}\right)(1-\rho(y_1+iy_2))^{-\frac{N}{2}}\dot{y}=I_1+I_2. \end{aligned} \tag{2.2.4}$$

其中

$$\Sigma:\ \begin{cases} yy'=1, \\ |1-y_1-iy_2|\geqslant\varepsilon, \end{cases}\qquad \sigma:\ \begin{cases} yy'=1, \\ |1-y_1-iy_2|\leqslant\varepsilon. \end{cases} \tag{2.2.5}$$

先看 I_2：

$$\begin{aligned} I_2 &= \varpi_{N-1}^{-1}\int_{\sigma}(1-\rho(y_1+iy_2))^{-\frac{N}{2}}\dot{y} \\ &= \varpi_{N-1}^{-1}\int_{y^{(0)}y^{(0)'}\leqslant 2\varepsilon-\varepsilon^2}\dot{y}^{(0)}\int_{-a}^{a}(1-\rho re^{i\varphi})^{-\frac{N}{2}}d\varphi, \end{aligned} \tag{2.2.6}$$

此处 $y^{(0)}=(y_3,\cdots,y_N)$，$y_1+iy_2=re^{i\varphi}$，$a=\cos^{-1}\dfrac{1+r^2-\varepsilon^2}{2r}$．记

$$J=\int_{-a}^{a}(1-\rho re^{i\varphi})^{-\frac{N}{2}}d\varphi=\sum_{p=0}^{\infty}\frac{\Gamma\left(\frac{N}{2}+p\right)}{\Gamma(p+1)\Gamma\left(\frac{N}{2}\right)}\int_{-a}^{a}(\rho r)^p e^{ip\varphi}d\varphi$$

$$=2\sum_{p=1}^{\infty}\frac{\Gamma\left(\frac{N}{2}+p\right)}{\Gamma(p+1)\Gamma\left(\frac{N}{2}\right)}(\rho r)^p\frac{\sin pa}{p}+2a$$

$$=2\,\mathrm{Im}\left\{\sum_{p=1}^{\infty}\frac{\Gamma\left(\frac{N}{2}+p\right)}{\Gamma(p+1)\Gamma\left(\frac{N}{2}\right)}\frac{(\rho re^{ia})^p}{p}\right\}+2a. \tag{2.2.7}$$

令 $1-x=y^2$，利用[3]引理 1.1 证明中已得的事实，易知

$$\begin{aligned} \sum_{p=1}^{\infty}\frac{\Gamma\left(\frac{N}{2}+p\right)x^{p-1}}{\Gamma(p+1)\Gamma\left(\frac{N}{2}\right)} &= \frac{1-y^N}{(1-y^2)y^N}=\frac{1+y+y^2+\cdots+y^{N-1}}{(1+y)y^N} \\ &= \frac{1}{(1+y)y}+\frac{1}{y^3}+\frac{1}{y^5}+\cdots+\frac{1}{y^{2n+1}} \\ &= \sum_{k=0}^{n}\frac{1}{(1-x)^{k+\frac{1}{2}}}-\frac{1}{1+\sqrt{1-x}}, \end{aligned} \tag{2.2.8}$$

因而

$$\sum_{p=1}^{\infty}\frac{\Gamma\left(\frac{N}{2}+p\right)x^p}{\Gamma(p+1)\Gamma\left(\frac{N}{2}\right)p}=\int_0^x\sum_{p=1}^{\infty}\frac{\Gamma\left(\frac{N}{2}+p\right)x^{p-1}}{\Gamma(p+1)\Gamma\left(\frac{N}{2}\right)}dx$$

$$= \sum_{k=1}^{n} \frac{1}{\left(k-\frac{1}{2}\right)(1-x)^{k-\frac{1}{2}}} + 2\log\frac{2}{1+\sqrt{1-x}} - 2, \tag{2.2.9}$$

所以

$$J = 2\,\mathrm{Im}\left\{\sum_{k=1}^{n} \frac{1}{\left(k-\frac{1}{2}\right)(1-\rho re^{ia})^{k-\frac{1}{2}}} + 2\log\frac{2}{1+\sqrt{1-\rho re^{ia}}}\right\} + 2a, \tag{2.2.10}$$

于是

$$\begin{aligned} I_2 = {} & \frac{2}{\varpi_{N-1}}\int_{y^{(0)}y^{(0)\prime}<2\varepsilon-\varepsilon^2} \mathrm{Im}\left\{\sum_{k=1}^{n} \frac{1}{\left(k-\frac{1}{2}\right)(1-\rho re^{ia})^{k-\frac{1}{2}}}\right. \\ & \left. + 2\log\frac{2}{1+\sqrt{1-\rho re^{ia}}}\right\}\dot{y}^{(0)} + \frac{2a}{\varpi_{N-1}}\int_{y^{(0)}y^{(0)\prime}<2\varepsilon-\varepsilon^2}\dot{y}^{(0)} \\ = {} & \frac{2}{\varpi_{N-1}}\mathrm{Im}\left\{\sum_{k=1}^{n} L_k + L_0\right\} + L_*, \end{aligned} \tag{2.2.11}$$

其中

$$L_* = \frac{2a}{\varpi_{N-1}}\int_{y^{(0)}y^{(0)\prime}<2\varepsilon-\varepsilon^2}\dot{y}^{(0)} = O(\varepsilon^{\frac{N}{2}}), \tag{2.2.12}$$

$$L_k = \int_{y^{(0)}y^{(0)\prime}<2\varepsilon-\varepsilon^2} \frac{\dot{y}^{(0)}}{\left(k-\frac{1}{2}\right)(1-\rho re^{ia})^{k-\frac{1}{2}}}, \quad 1 \leqslant k \leqslant n, \tag{2.2.13}$$

$$L_0 = 2\int_{y^{(0)}y^{(0)\prime}<2\varepsilon-\varepsilon^2} \log\frac{2}{1+\sqrt{1-\rho re^{ia}}}\dot{y}^{(0)}. \tag{2.2.14}$$

应用[3]引理 1.1 的证明方法,易得

$$\lim_{\rho\to 1} L_k = O(\varepsilon^{n-k}), \quad 1 \leqslant k \leqslant n-1. \tag{2.2.15}$$

$$\mathrm{Im}\, L_0 = o(1), \quad (\varepsilon \to 0). \tag{2.2.16}$$

并由 a 之定义知

$$re^{ia} = \frac{2-y^{(0)}y^{(0)\prime}-\varepsilon^2}{2} + \frac{i}{2}(2\varepsilon^2(2-y^{(0)}y^{(0)\prime}) - (y^{(0)}y^{(0)\prime})^2 - \varepsilon^4)^{\frac{1}{2}},$$

于是

$$L_n = \frac{1}{n-\frac{1}{2}}\int_{y^{(0)}y^{(0)\prime}<2\varepsilon-\varepsilon^2} \frac{\dot{y}^{(0)}}{\left[1-\rho+\rho\frac{y^{(0)}y^{(0)\prime}+\varepsilon^2}{2}-\frac{i\rho}{2}(2\varepsilon^2(2-y^{(0)}y^{(0)\prime})-(y^{(0)}y^{(0)\prime})^2-\varepsilon^4)^{\frac{1}{2}}\right]^{n-\frac{1}{2}}},$$

用球坐标 $y_3 = s\cos\varphi_1$, $y_4 = s\sin\varphi_1\cos\varphi_2$, $\cdots$, $y_{2n+1} = s\sin\varphi_1\cdots\sin\varphi_{2n-2}$, 其中 $s^2 = 1-r^2 = y^{(0)}y^{(0)\prime}$,则

$$L_n = \frac{2\pi^{\frac{2n-1}{2}}}{\left(n-\frac{1}{2}\right)\Gamma\left(\frac{2n-1}{2}\right)}\int_0^{\sqrt{2\varepsilon-\varepsilon^2}} \frac{s^{2n-2}ds}{\left(1-\rho+\rho\frac{\varepsilon^2+s^2}{2}-\frac{i\rho}{2}\sqrt{2\varepsilon^2(2-s^2)-s^4-\varepsilon^4}\right)^{n-\frac{1}{2}}},$$

当 $\varepsilon > 0$ 固定时,同引理 1.1 之证明,可应用 Lebesgue 定理,得

$$J_n = \lim_{\rho\to 1} L_n = \frac{2\pi^{\frac{2n-1}{2}}}{\left(n-\frac{1}{2}\right)\Gamma\left(\frac{2n-1}{2}\right)}\int_0^{\sqrt{2\varepsilon-\varepsilon^2}} \frac{s^{2n-2}ds}{\left(\frac{\varepsilon^2+s^2}{2}-\frac{i}{2}\sqrt{4\varepsilon^2-(s^2+\varepsilon^2)^2}\right)^{n-\frac{1}{2}}}.$$

记 $\eta=\sqrt{2\varepsilon-\varepsilon^2}$, $s=\eta R$, 于是

$$J_n = \frac{2\pi^{\frac{2n-1}{2}}}{\left(n-\frac{1}{2}\right)\Gamma\left(\frac{2n-1}{2}\right)}\int_0^1 \frac{R^{2n-2}dR}{\left(\frac{R^2+\varepsilon^2\eta^{-2}}{2}-\frac{i}{2}\sqrt{4\eta^{-4}\varepsilon^2-(R^2+\varepsilon^2\eta^{-2})^2}\right)^{n-\frac{1}{2}}}.$$

再应用 Lebesgue 定理，可得

$$\lim_{\varepsilon\to 0} J_n = \frac{2\pi^{\frac{2n-1}{2}}}{\left(n-\frac{1}{2}\right)\Gamma\left(\frac{2n-1}{2}\right)}\int_0^1 \frac{R^{2n-2}dR}{\left(\frac{R^2}{2}-\frac{i}{2}\sqrt{1-R^4}\right)^{n-\frac{1}{2}}}$$

$$= \frac{(2\pi)^{\frac{2n-1}{2}}}{\Gamma\left(n+\frac{1}{2}\right)}\int_0^1 \frac{t^{n-\frac{3}{2}}dt}{(t-i\sqrt{1-t^2})^{n-\frac{1}{2}}}$$

$$= \frac{(2\pi)^{\frac{n-1}{2}}}{\Gamma\left(n+\frac{1}{2}\right)}\int_0^{\frac{\pi}{2}} e^{i(n-\frac{1}{2})\theta}\cos^{n-\frac{3}{2}}\theta\sin\theta d\theta.$$

注意到

$$\int_0^{\frac{\pi}{2}} \cos^{n-\frac{3}{2}}\theta\sin\theta e^{i(n-\frac{1}{2})\theta}d\theta$$

$$= \left.\frac{-e^{i(n-\frac{1}{2})\theta}\cos^{n-\frac{1}{2}}\theta}{n-\frac{1}{2}}\right|_0^{\frac{\pi}{2}} + i\int_0^{\frac{\pi}{2}}\cos^{n-\frac{1}{2}}\theta e^{i(n-\frac{1}{2})\theta}d\theta$$

$$= \frac{1}{n-\frac{1}{2}} + i\int_0^{\frac{\pi}{2}}\cos^{n-\frac{1}{2}}\theta e^{i(n-\frac{1}{2})\theta}d\theta,$$

所以

$$\operatorname{Im}\left(\lim_{\varepsilon\to 0} J_n\right) = \frac{(2\pi)^{n-\frac{1}{2}}}{\Gamma\left(n+\frac{1}{2}\right)}\int_0^{\frac{\pi}{2}}\cos^{n-\frac{1}{2}}\theta\cos\left(n-\frac{1}{2}\right)\theta\, d\theta$$

$$= \frac{(2\pi)^{n-\frac{1}{2}}}{\Gamma\left(n+\frac{1}{2}\right)}\cdot\frac{\pi}{2^{n+\frac{1}{2}}} = \pi^{n+\frac{1}{2}}\Big/2\Gamma\left(n+\frac{1}{2}\right). \tag{2.2.17}$$

由上可知

$$\lim_{\varepsilon\to 0}\lim_{\rho\to 1} I_2 = \frac{2}{\omega_{N-1}}\cdot\frac{\pi^{n+\frac{1}{2}}}{2\Gamma\left(n+\frac{1}{2}\right)} = \frac{2\Gamma\left(\frac{N}{2}\right)}{2\pi^{\frac{N}{2}}}\cdot\frac{\pi^{\frac{N}{2}}}{2\Gamma\left(\frac{N}{2}\right)} = \frac{1}{2}. \tag{2.2.18}$$

因此

$$\lim_{\varepsilon\to 0}\lim_{\rho\to 1} I_1 = \lim_{\varepsilon\to 0}\varpi_{N-1}^{-1}\int_{\substack{yy'=1\\ |1-y_1-iy_2|>\varepsilon}}(1-(y_1+iy_2))^{-\frac{N}{2}}\dot{y} = \frac{1}{2}. \tag{2.2.19}$$

引理证毕.

今定义 Cauchy 主值

$$\text{V.P.}V(\mathfrak{L}_{\text{IV}})^{-1}\int_{\mathfrak{L}_{\text{IV}}}(1-\xi_2)^{-\frac{N}{2}}\dot{\xi} = \lim_{\varepsilon\to 0}V(\mathfrak{L}_{\text{IV}})^{-1}\int_{\substack{\mathfrak{L}_{\text{IV}}\\ |1-\xi_2|\geqslant\varepsilon}}(1-\xi_2)^{-\frac{N}{2}}\dot{\xi}, \tag{2.2.20}$$

引理 2.1 便是

$$\text{V.P.}V(\mathfrak{L}_{\text{IV}})^{-1}\int_{\mathfrak{L}_{\text{IV}}}(1-\xi_2)^{-\frac{N}{2}}\dot{\xi} = \frac{1}{2}. \tag{2.2.21}$$

§2.3. 在 $\mathfrak{B}_{\text{IV}}-\mathfrak{L}_{\text{IV}}$ 上的 Cauchy 主值

在 $\mathfrak{L}_{\text{IV}}$ 上任与一函数 $\varphi(\xi)$，如果对于 $\mathfrak{L}_{\text{IV}}$ 上任意两点 ξ 与 ξ^*，恆有

$$|\varphi(\xi)-\varphi(\xi^*)| = O(((\xi-\xi^*)(\overline{\xi-\xi^*})')^{\frac{p}{2}}), \tag{2.3.1}$$

$0<p\leqslant 1$，则称 $\varphi(\xi)$ 在 $\mathfrak{L}_{\text{IV}}$ 上属于 Lip p.

定理 2.1. 若 $\varphi(\xi)$ 在 $\mathfrak{L}_{\text{IV}}$ 上属于 Lip p $(0<p\leqslant 1)$，则当 $\eta_0 = e^{i\theta_0}(t, 1, 0, \cdots, 0)P_0^{-1}TP_0\in\mathfrak{B}_{\text{IV}}-\mathfrak{L}_{\text{IV}}$ 时，Cauchy 主值

$$\text{V.P.}V(\mathfrak{L}_{\text{IV}})^{-1}\int_{\mathfrak{L}_{\text{IV}}}H_{\text{IV}}(\eta_0,\xi)\varphi(\xi)\dot{\xi}$$

$$= \lim_{\varepsilon\to 0}V(\mathfrak{L}_{\text{IV}})^{-1}\int_{\substack{\mathfrak{L}_{\text{IV}}\\ |1+\eta_0A_0\eta_0'\overline{\xi A_0\xi'}-\eta_0\bar{\xi}'|\geqslant\varepsilon|1-z_0\bar{\xi}'|}}H_{\text{IV}}(\eta_0,\xi)\varphi(\xi)\dot{\xi} \tag{2.3.2}$$

是存在的，其中 $z_0 = e^{i\theta_0}(t, 0, \cdots, 0)P_0^{-1}TP_0$. 特别地，当 $\eta_0 = (0, 1, 0, \cdots, 0)$ 时，有

$$\text{V.P.}V(\mathfrak{L}_{\text{IV}})^{-1}\int_{\mathfrak{L}_{\text{IV}}}(1-\xi_2)^{-\frac{N}{2}}\varphi(\xi)\dot{\xi} = \lim_{\varepsilon\to 0}V(\mathfrak{L}_{\text{IV}})^{-1}\int_{\substack{\mathfrak{L}_{\text{IV}}\\ |1-\xi_2|\geqslant\varepsilon}}(1-\xi_2)^{-\frac{N}{2}}\varphi(\xi)\dot{\xi}$$

$$= V(\mathfrak{L}_{\text{IV}})^{-1}\int_{\mathfrak{L}_{\text{IV}}}(1-\xi_2)^{-\frac{N}{2}}(\varphi(\xi)-\varphi(\xi^*))\dot{\xi} + \frac{1}{2\pi}\int_0^{\pi}\varphi(\xi^*)d\theta, \tag{2.3.3}$$

这里 $\xi^* = (e^{2i\theta}, 1, 0, \cdots, 0)$.

證. (i) 首先取 $\eta_0 = (0, 1, 0, \cdots, 0)\in\mathfrak{B}_{\text{IV}}-\mathfrak{L}_{\text{IV}}$.

书 $\xi = e^{i\theta}xP_0$, $0\leqslant\theta<\pi$, $xx'=1$. 取 $\xi^* = (e^{2i\theta}, 1, 0, \cdots, 0)\in\mathfrak{L}_{\text{IV}}$，于是

$$V(\mathfrak{L}_{\text{IV}})^{-1}\int_{\substack{\mathfrak{L}_{\text{IV}}\\ |1-\xi_2|\geqslant\varepsilon}}(1-\xi_2)^{-\frac{N}{2}}\varphi(\xi)\dot{\xi} = V(\mathfrak{L}_{\text{IV}})^{-1}\int_{\substack{\mathfrak{L}_{\text{IV}}\\ |1-\xi_2|\geqslant\varepsilon}}(1-\xi_2)^{-\frac{N}{2}}(\varphi(\xi)-\varphi(\xi^*))\dot{\xi}$$

$$+ V(\mathfrak{L}_{\text{IV}})^{-1}\int_{\substack{\mathfrak{L}_{\text{IV}}\\ |1-\xi_2|\geqslant\varepsilon}}(1-\xi_2)^{-\frac{N}{2}}\varphi(\xi^*)\dot{\xi} = I_1+I_2. \tag{2.3.4}$$

令 $y_1+iy_2 = e^{-i\theta}(x_1+ix_2)$, $y_3 = x_3$, $\cdots$, $y_N = x_N$，则由引理 2.1，有

$$\lim_{\varepsilon\to 0} I_2=\lim_{\varepsilon\to 0}\frac{2^{\frac{N}{2}}\pi}{V(\mathfrak{L}_{\mathrm{IV}})}\int_{\substack{yy'=1\\|1-y_1-iy_2|>\varepsilon}}(1-y_1-iy_2)^{-\frac{N}{2}}\dot{y}\cdot\frac{1}{\pi}\int_0^{\pi}\varphi(\xi^*)d\theta$$

$$=\frac{1}{2\pi}\int_0^{\pi}\varphi((e^{2i\theta},1,0,\cdots,0))d\theta.\tag{2.3.5}$$

注意到

$$\begin{aligned}(\xi-\xi^*)\overline{(\xi-\xi^*)'}&=\xi\bar{\xi}'+\xi^*\bar{\xi}'^*-\xi\bar{\xi}'^*-\xi^*\bar{\xi}'=4-e^{2i\theta}\bar{\xi}_1-\bar{\xi}_2-e^{-2i\theta}\xi_1-\xi_2\\&=4-e^{i\theta}(x_1-ix_2)-e^{-i\theta}(x_1+ix_2)-e^{-i\theta}(x_1+ix_2)-e^{i\theta}(x_1-ix_2)\\&=2[2-(y_1-iy_2)-(y_1+iy_2)]=4(1-y_1),\end{aligned}$$

所以有

$$|\varphi(\xi)-\varphi(\xi^*)|=O(((\xi-\xi^*)\overline{(\xi-\xi^*)'})^{\frac{p}{2}})=O((1-y_1)^{\frac{p}{2}}).$$

于是

$$|I_1|\leqslant K\int_{\substack{\mathfrak{L}_{\mathrm{IV}}\\|1-\xi_n|\geqslant\varepsilon}}\frac{(1-y_1)^{\frac{p}{2}}}{|1-y_1-iy_2|^{\frac{N}{2}}}\dot{\xi}\leqslant K\int_{\substack{\mathfrak{L}_{\mathrm{IV}}\\|1-y_1-iy_2|\geqslant\varepsilon}}\frac{\dot{\xi}}{|1-y_1-iy_2|^{\frac{N-p}{2}}}$$

$$=K\int_{y^{(0)}y^{(0)'}<1}\dot{y}^{(0)}\int_{|\alpha|\geqslant a}\frac{d\alpha}{|1-re^{i\alpha}|^{\frac{N-p}{2}}},\tag{2.3.6}$$

此处 $y_1+iy_2=re^{i\alpha}$，$y^{(0)}=(y_3,\cdots,y_N)$，$a=\cos^{-1}\dfrac{1+r^2-\varepsilon^2}{2r}$.

当 $N\geqslant 5$ 时，有

$$\int_{|\alpha|\geqslant a}\frac{d\alpha}{|1-re^{i\alpha}|^{\frac{N-p}{2}}}\leqslant\int_0^{2\pi}\frac{d\alpha}{|1-re^{i\alpha}|^{\frac{N-p}{2}}}$$

$$\leqslant\frac{1}{(1-r)^{\frac{N-p}{2}-2}}\int_0^{2\pi}\frac{d\alpha}{|1-re^{i\alpha}|^2}=O((1-r^2)^{\frac{p}{2}+1-\frac{N}{2}}),\tag{2.3.7}$$

利用球坐标，令 $s=\sqrt{1-r^2}$，可得

$$\int_{y^{(0)}y^{(0)'}<1}(1-r^2)^{\frac{p-N}{2}+1}\dot{y}^{(0)}=O\left(\int_0^1 s^{N-3}\cdot s^{p+2-N}ds\right)$$

$$=O\left(\int_0^1 s^{p-1}ds\right)=O(1)\quad(\varepsilon\to 0).\tag{2.3.8}$$

当 $N=4$ 时，可按[3] § 1.3 中的方法处理，同样得到 $|I_1|=O(1)$ $(\varepsilon\to 0)$.

当 $N=3$ 时，取球坐标 $y_1=\cos\theta_1$，$y_2=\sin\theta_1\cos\theta_2$，$y_3=\sin\theta_1\sin\theta_2$ $(0\leqslant\theta_1<\pi,$ $0\leqslant\theta_2<2\pi)$，于是有

$$\int_{\substack{yy'=1\\|1-y_1-iy_2|\geqslant\varepsilon}}\frac{\dot{y}}{|1-y_1-iy_2|^{\frac{3-p}{2}}}=\iint_{\substack{0\leqslant\theta_1<\pi\\0\leqslant\theta_2<2\pi\\(1-\cos\theta_1)^2+\sin^2\theta_1\cos^2\theta_2\geqslant\varepsilon^2}}\frac{\sin\theta_1\,d\theta_1\,d\theta_2}{[(1-\cos\theta_1)^2+\sin^2\theta_1\cos^2\theta_2]^{\frac{3-p}{4}}}$$

$$=\iint_{\substack{\frac{\pi}{2}\leqslant\theta_1<\pi\\0\leqslant\theta_2<2\pi\\(1-\cos\theta_1)^2+\sin^2\theta_1\cos^2\theta_2\geqslant\varepsilon^2}}+\iint_{\substack{0\leqslant\theta_1<\frac{\pi}{2}\\0\leqslant\theta_2<2\pi\\(1-\cos\theta_1)^2+\sin^2\theta_1\cos^2\theta_2\geqslant\varepsilon^2}}=J_1+J_2,\tag{2.3.9}$$

J_1 显然是一有界量，而且对于 J_2，只需估计

$$\phi(\varepsilon)=\int_{\varepsilon}^{\frac{\pi}{2}}\sin\theta_1\,d\theta_1\int_0^{\frac{\pi}{2}}\frac{d\theta_2}{((1-\cos\theta_1)^2+\sin^2\theta_1\cos^2\theta_2)^{\alpha}},\tag{2.3.10}$$

此处 $\alpha=\frac{3-p}{4}$，$\frac{1}{2}\leqslant\alpha<\frac{3}{4}$. 令 $\cos\theta_1=x_1$，$\cos 2\theta_2=x_2$，则

$$\phi(\varepsilon)=\frac{1}{2}\int_0^{\cos\varepsilon}\frac{dx_1}{(1-x_1)^{\alpha}}\int_{-1}^{1}\frac{dx_2}{\left[(1-x_1)+(1+x_1)\frac{1+x_2}{2}\right]^{\alpha}\sqrt{1-x_2^2}},\tag{2.3.11}$$

再令 $\frac{1+x_2}{2}=y$，则

$$\phi(\varepsilon)=\frac{1}{2}\int_0^{\cos\varepsilon}\frac{dx_1}{(1-x_1)^{\alpha}}\int_0^1\frac{dy}{\sqrt{y(1-y^2)}((1-x_1)+(1+x_1)y)^{\alpha}}.\tag{2.3.12}$$

记

$$J_3=\int_0^1\frac{dy}{\sqrt{y(1-y^2)}((1-x_1)+(1+x_1)y)^{\alpha}},\tag{2.3.13}$$

并取 $\lambda=\frac{4-p}{2}$，$\mu=\frac{4-p}{2-p}$，显然有 $\lambda>1$，$\mu>1$，$\frac{1}{\lambda}+\frac{1}{\mu}=1$ 以及 $\frac{\lambda}{2}<1$，$\alpha\mu>1$，$2\alpha-\frac{1}{\mu}<1$. 于是应用 Hölder 不等式，有

$$\begin{aligned}J_3&\leqslant\left(\int_0^1\frac{dy}{(y(1-y^2))^{\frac{\lambda}{2}}}\right)^{\frac{1}{\lambda}}\left(\int_0^1\frac{dy}{((1-x_1)+(1+x_1)y)^{\alpha\mu}}\right)^{\frac{1}{\mu}}\\&=K_1\left(\frac{1}{\alpha\mu-1}((1-x_1)+(1+x_1)y)^{1-\alpha\mu}\Big|_1^0\right)^{\frac{1}{\mu}}\\&=K_2\left(\frac{1}{(1-x_1)^{\alpha\mu-1}}-\frac{1}{2^{\alpha\mu-1}}\right)^{\frac{1}{\mu}}\\&\leqslant K_3\frac{1}{(1-x_1)^{\alpha-\frac{1}{\mu}}},\end{aligned}\tag{2.3.14}$$

其中 K_1, K_2, K_3 皆为绝对常数. 所以有

$$\phi(\varepsilon)=O\left(\int_0^{\cos\varepsilon}\frac{dx_1}{(1-x_1)^{2\alpha-\frac{1}{\mu}}}\right)=O(1),\tag{2.3.15}$$

因而 $|I_1|=O(1)$ $(\varepsilon\to 0)$ 对于 $N=3$ 时亦成立.

当 $N=2$ 时，令 $y_1=\cos\theta$，$y_2=\sin\theta$，则

$$\begin{aligned}\int_{\substack{yy'=1\\|1-y_1-iy_2|>\varepsilon}}\frac{\dot{y}}{|1-y_1-iy_2|^{\frac{2-p}{2}}}&=\int_{\substack{0\leqslant\theta<2\pi\\4\sin^2\frac{\theta}{2}>\varepsilon^2}}\frac{d\theta}{\left(4\sin^2\frac{\theta}{2}\right)^{\frac{2-p}{4}}}\\&=O\left(\int_{\varepsilon}^{\frac{\pi}{2}}\theta^{-1+\frac{p}{2}}d\theta\right)=O(1)\ (\varepsilon\to 0).\end{aligned}\tag{2.3.16}$$

于是，当 $N \geqslant 2$ 时，$|I_1| = O(1)$ $(\varepsilon \to 0)$ 恒成立. 所以我们得到

$$\mathrm{V.P.}V(\mathfrak{L}_{\mathrm{IV}})^{-1}\int_{\mathfrak{L}_{\mathrm{IV}}}(1-\bar{\xi}_2)^{-\frac{N}{2}}\varphi(\xi)\dot{\xi}$$

$$=\lim_{\varepsilon\to 0}V(\mathfrak{L}_{\mathrm{IV}})^{-1}\int_{\substack{\mathfrak{L}_{\mathrm{IV}}\\ |1-\bar{\xi}_2|>\varepsilon}}(1-\bar{\xi}_2)^{-\frac{N}{2}}(\varphi(\xi)-\varphi(\xi^*))\dot{\xi}+\frac{1}{2\pi}\int_0^{\pi}\varphi(\xi^*)d\theta. \qquad (2.3.17)$$

必须指出，上式右端第一项，乃是通常积分

$$I=V(\mathfrak{L}_{\mathrm{IV}})^{-1}\int_{\mathfrak{L}_{\mathrm{IV}}}(1-\bar{\xi}_2)^{-\frac{N}{2}}(\varphi(\xi)-\varphi(\xi^*))\dot{\xi}. \qquad (2.3.18)$$

事实上，可书

$$I=V(\mathfrak{L}_{\mathrm{IV}})^{-1}\int_{\substack{\mathfrak{L}_{\mathrm{IV}}\\ |1-\bar{\xi}_2|>\varepsilon}}(1-\bar{\xi}_2)^{-\frac{N}{2}}(\varphi(\xi)-\varphi(\xi^*))\dot{\xi}$$

$$+V(\mathfrak{L}_{\mathrm{IV}})^{-1}\int_{\substack{\mathfrak{L}_{\mathrm{IV}}\\ |1-\bar{\xi}_2|\leqslant\varepsilon}}(1-\bar{\xi}_2)^{-\frac{N}{2}}(\varphi(\xi)-\varphi(\xi^*))\dot{\xi}=I_1+I_0, \qquad (2.3.19)$$

而

$$|I_0|\leqslant K\int_{\substack{yy'=1\\ |1-y_1-iy_2|\leqslant\varepsilon}}\frac{\dot{y}}{|1-y_1-iy_2|^{\frac{N-p}{2}}}\quad (K\text{ 为一常数}), \qquad (2.3.20)$$

取球坐标 $y_1=\cos\theta_1$, $y_2=\sin\theta_1\cos\theta_2$, $\cdots$, $y_N=\sin\theta_1\cdots\sin\theta_{N-1}$ $(0\leqslant\theta_1,\cdots,\theta_{N-2}<\pi,\ 0\leqslant\theta_{N-1}<2\pi)$，则当 $N\geqslant 4$ 时，有

$$|I_0|=O\Bigg(\iint\limits_{\substack{0\leqslant\theta_1,\theta_2<\pi\\ (1-\cos\theta_1)^2+\sin^2\theta_1\cos^2\theta_2\leqslant\varepsilon^2}}\frac{\sin^{N-2}\theta_1\sin^{N-3}\theta_2\,d\theta_1\,d\theta_2}{[(1-\cos\theta_1)^2+\sin^2\theta_1\cos^2\theta_2]^{\frac{N-p}{4}}}\Bigg), \qquad (2.3.21)$$

利用与[3] § 1.4 中完全相同的计算，可得

$$|I_0|=O\Big(\int_0^{\sqrt{\varepsilon}}\theta_1^{-1+p}d\theta_1\Big)=O(\varepsilon^{\frac{p}{2}}). \qquad (2.3.22)$$

当 $N=3$ 时，应用本节估计 $|I_1|$ 的方法，采取同样的符号，易知有

$$|I_0|=O\Big(\int_0^{\sqrt{\varepsilon}}\sin\theta_1 d\theta_1\int_0^{\frac{\pi}{2}}\frac{d\theta_2}{((1-\cos\theta_1)^2+\sin^2\theta_1\cos^2\theta_2)^{\alpha}}\Big)$$

$$=O\Big(\int_{\cos\sqrt{\varepsilon}}^1\frac{dx_1}{(1-x_1)^{\alpha}}\int_0^1\frac{dy}{\sqrt{y(1-y^2)}((1-x_1)+(1+x_1)y)^{\alpha}}\Big)$$

$$=O\Big(\int_{\cos\sqrt{\varepsilon}}^1\frac{dx_1}{(1-x_1)^{2\alpha-\frac{1}{\mu}}}\Big)$$

$$=O((1-\cos\sqrt{\varepsilon})^{1+\frac{1}{\mu}-2\alpha})=O(\varepsilon^{1+\frac{1}{\mu}-2\alpha}), \qquad (2.3.23)$$

其中 $1+\frac{1}{\mu}-2\alpha>0$.

当 $N=2$ 时，令 $y_1=\cos\theta$, $y_2=\sin\theta$，则

$$|I_0| = O\left(\int_{\substack{0\leqslant\theta<2\pi \\ 4\sin^2\frac{\theta}{2}\leqslant\varepsilon^2}} \frac{d\theta}{\left(4\sin^2\frac{\theta}{2}\right)^{\frac{2-p}{4}}}\right) = O\left(\int_{-\varepsilon}^{\varepsilon} \theta_1^{-1+\frac{p}{2}}\, d\theta_1\right) = O\left(\varepsilon^{\frac{p}{2}}\right). \tag{2.3.24}$$

总之，我们有 $\lim\limits_{\varepsilon\to 0} I_0 = 0$，因而 $\lim\limits_{\varepsilon\to 0} I_1 = I$，即(2.3.17)就是(2.3.3).

(ii) 在 $\mathfrak{B}_{\rm IV} - \mathfrak{L}_{\rm IV}$ 上任取一点 $\eta_0 = e^{i\theta_0}(t,1,0,\cdots 0)P_0^{-1}\Gamma P_0$，首先利用 $\Gamma_0^{\rm IV}$ 中的变换 $z = e^{i\theta_0}\tilde{z}P_0^{-1}\Gamma P_0$，使 $\mathfrak{R}_{\rm IV}$ 内的点 $z = e^{i\theta_0}(t,\rho,0,\cdots 0)P_0^{-1}\Gamma P_0$ 变为 $\tilde{z} = (t,\rho,0,\cdots 0)$，使 η_0 变为 $\tilde{\eta}_0 = (t,1,0,\cdots 0)$. 再利用 $\Gamma^{\rm IV}$ 中将 $\tilde{z}_0 = (t,0,\cdots 0)$ 变为原点的变换 $w = \Phi(\tilde{z})$，将 $\tilde{z}$ 变为 $w = (0,\rho,0\cdots 0)$，将 $\tilde{\eta}_0$ 变为 $\zeta_0 = (0,1,0,\cdots 0)$. 在 $\mathfrak{L}_{\rm IV}$ 上，上述变换为 $\xi = e^{i\theta_0}\tilde{\xi}P_0^{-1}\Gamma P_0$ 和 $\zeta = \Phi(\tilde{\xi})$.

由于在变换 $w = \Phi(\tilde{z})$ 下，有

$$1 + wA_0w'\overline{\zeta A_0\zeta'} - w\bar{\zeta}' = \frac{(1+\tilde{z}A_0\tilde{z}'\overline{\tilde{\xi}A_0\tilde{\xi}'} - \tilde{z}\bar{\tilde{\xi}}')(1+|\tilde{z}_0A_0\tilde{z}_0'|^2 - \tilde{z}_0\bar{\tilde{z}}_0')}{(1+\tilde{z}A_0\tilde{z}'\overline{\tilde{z}_0A_0\tilde{z}_0'} - \tilde{z}\bar{\tilde{z}}_0')(1+\tilde{z}_0A_0\tilde{z}_0'\overline{\tilde{\xi}A_0\tilde{\xi}'} - \tilde{z}_0\bar{\tilde{\xi}}')}, \tag{2.3.25}$$

(见本节末尾附注)并注意到 $wA_0w' = 0$, $\tilde{z}_0A_0\tilde{z}_0' = 0$, $\tilde{z}_0A_0\tilde{\xi}_0' = 0$, $\tilde{z}_0\bar{\tilde{\xi}}_0' = \tilde{z}_0\bar{\tilde{z}}_0' = t^2$, 于是有

$$1 - w\bar{\zeta}' = (1+\tilde{z}A_0\tilde{z}'\overline{\tilde{\xi}A_0\tilde{\xi}'} - \tilde{z}\bar{\tilde{\xi}}')(1-\tilde{z}_0\bar{\tilde{\xi}}')^{-1}. \tag{2.3.26}$$

令 $w\to\zeta_0$，相应地 $\tilde{z}\to\tilde{\eta}_0$，便得到

$$|1-\zeta_0\bar{\zeta}'| = |1+\tilde{\eta}_0A_0\tilde{\eta}_0'\overline{\tilde{\xi}A_0\tilde{\xi}'} - \tilde{\eta}_0\bar{\tilde{\xi}}'|/|1-\tilde{z}_0\bar{\tilde{\xi}}'|. \tag{2.3.27}$$

因此，

$$\begin{aligned} &V(\mathfrak{L}_{\rm IV})^{-1}\int_{\substack{\mathfrak{L}_{\rm IV} \\ |1+\eta_0A_0\eta_0'\overline{\xi A_0\xi'}-\eta_0\bar{\xi}'|\geqslant\varepsilon|1-z_0\bar{\xi}'|}} H_{\rm IV}(z,\xi)\varphi(\xi)\dot{\xi} \\ &= V(\mathfrak{L}_{\rm IV})^{-1}\int_{\substack{\mathfrak{L}_{\rm IV} \\ |1+\tilde{\eta}_0A_0\tilde{\eta}_0'\overline{\tilde{\xi}A_0\tilde{\xi}'}-\tilde{\eta}_0\bar{\tilde{\xi}}'|\geqslant\varepsilon|1-\tilde{z}_0\bar{\tilde{\xi}}'|}} H_{\rm IV}(\tilde{z},\tilde{\xi})\varphi(e^{i\theta_0}\tilde{\xi}P_0^{-1}\Gamma P_0)\dot{\tilde{\xi}} \\ &= V(\mathfrak{L}_{\rm IV})^{-1}\int_{\substack{\mathfrak{L}_{\rm IV} \\ |1-\zeta_0\bar{\zeta}'|\geqslant\varepsilon}} \varphi(e^{i\theta_0}\Phi^{-1}(\zeta)P_0^{-1}\Gamma P_0)\cdot(1-\tilde{\xi}\bar{\tilde{z}}_0')^{\frac{N}{2}} \\ &\quad\times(1-\tilde{z}\bar{\tilde{z}}_0')^{-\frac{N}{2}}\cdot(1-w\bar{\zeta}')^{-\frac{N}{2}}\dot{\zeta}, \end{aligned} \tag{2.3.28}$$

再在上式中令 $z\to\eta_0$，相应地，$\tilde{z}\to\tilde{\eta}_0$, $w\to\zeta_0$，并记

$$\psi(\zeta) = \varphi(e^{i\theta_0}\Phi^{-1}(\zeta)P_0^{-1}\Gamma P_0)\cdot(1-\Phi^{-1}(\zeta)\bar{\tilde{z}}_0')^{\frac{N}{2}}(1-\tilde{\eta}_0\bar{\tilde{z}}_0')^{-\frac{N}{2}}, \tag{2.3.29}$$

则有

$$\begin{aligned} &V(\mathfrak{L}_{\rm IV})^{-1}\int_{\substack{\mathfrak{L}_{\rm IV} \\ |1+\eta_0A_0\eta_0'\overline{\xi A_0\xi'}-\eta_0\bar{\xi}'|\geqslant\varepsilon|1-z_0\bar{\xi}'|}} H_{\rm IV}(\eta_0,\xi)\varphi(\xi)\dot{\xi} \\ &= V(\mathfrak{L}_{\rm IV})^{-1}\int_{\substack{\mathfrak{L}_{\rm IV} \\ |1-\zeta_0\bar{\zeta}'|\geqslant\varepsilon}} (1-\zeta_0\bar{\zeta}')^{-\frac{N}{2}}\psi(\zeta)\dot{\zeta}. \end{aligned} \tag{2.3.30}$$

令 $\varepsilon\to 0$，由于 $\psi(\zeta)$ 显然在 $\mathfrak{L}_{\rm IV}$ 上满足 Lip p，所以由(i)可知

$$\mathrm{V.P.}V(\mathfrak{L}_{\mathrm{IV}})^{-1}\int_{\mathfrak{L}_{\mathrm{IV}}} H_{\mathrm{IV}}(\eta_0,\xi)\varphi(\xi)\dot{\xi}$$

是存在的，而且不难求出与(2.3.3)式相似的表达式，此处从略.

（附注）. 我们可以在 Γ^{IV} 中取出一个将 $z_0\in\mathfrak{R}_{\mathrm{IV}}$ 变至原点的变换 $w=\Phi(z)$, w, $z\in\mathfrak{R}_{\mathrm{IV}}$，相应地，$\zeta=\Phi(\xi)$, ζ, $\xi\in\overline{\mathfrak{R}_{\mathrm{IV}}}$，使得经过这个变换，有

$$1+wA_0w'\overline{\zeta A_0\zeta'}-w\bar{\zeta}'=\frac{(1+zA_0z'\overline{\xi A_0\xi'}-z\bar{\xi}')(1+|z_0A_0z_0'|^2-z_0\bar{z}_0')}{(1+zA_0z'\overline{z_0A_0z_0'}-z\bar{z}_0')(1+z_0A_0z_0'\overline{\xi A_0\xi'}-z_0\bar{\xi}')}. \tag{2.3.31}$$

事实上，Γ^{IV} 中将 $z_0\in\mathfrak{R}_{\mathrm{IV}}$ 变至原点的变换可书为[2]

$$\left(\frac{1+wA_0w'}{2},\ \frac{1-wA_0w'}{2i},\ wP_0^{-1}\right)$$
$$=\rho_z\left(\frac{1+zA_0z'}{2},\ \frac{1-zA_0z'}{2i},\ zP_0^{-1}\right)\begin{pmatrix}A & C\\ B & D\end{pmatrix}, \tag{2.3.32}$$

此处实方阵

$$M=\begin{pmatrix}A^{(2)} & C^{(2,N)}\\ B^{(N,2)} & D^{(N)}\end{pmatrix} \tag{2.3.33}$$

满足

$$M\begin{pmatrix}I^{(2)} & \\ & -I^{(N)}\end{pmatrix}M'=\begin{pmatrix}I^{(2)} & \\ & -I^{(N)}\end{pmatrix}, \tag{2.3.34}$$

$$\rho_z=\left\{\left[\left(\frac{1+zA_0z'}{2},\ \frac{1-zA_0z'}{2i}\right)-zP_0^{-1}X_0'\right]A\begin{pmatrix}1\\ i\end{pmatrix}\right\}^{-1}, \tag{2.3.35}$$

$$X_0=\frac{-1}{1-|z_0A_0z_0'|^2}\begin{pmatrix}(\overline{z_0A_0z_0'}-1)z_0P_0^{-1}+(z_0A_0z_0'-1)\bar{z}_0\bar{P}_0^{-1}\\ i(z_0A_0z_0'+1)\bar{z}_0\bar{P}_0^{-1}-i(\overline{z_0A_0z_0'}+1)z_0P_0^{-1}\end{pmatrix}. \tag{2.3.36}$$

我们特别取

$$A=\frac{1}{\sqrt{1+|z_0A_0z_0'|^2-z_0\bar{z}_0'}}\times\begin{pmatrix}\sqrt{1+z_0A_0z_0'+\overline{z_0A_0z_0'}+|z_0A_0z_0'|^2} & 0\\ \dfrac{i(z_0A_0z_0'-\overline{z_0A_0z_0'})}{\sqrt{1+z_0A_0z_0'+\overline{z_0A_0z_0'}+|z_0A_0z_0'|^2}} & \dfrac{1-|z_0A_0z_0'|^2}{\sqrt{1+z_0A_0z_0'+\overline{z_0A_0z_0'}+|z_0A_0z_0'|^2}}\end{pmatrix}, \tag{2.3.37}$$

它满足 $AA'=(I-X_0X_0')^{-1}$. 于是由(2.3.32)，(2.3.34)立知

$$1+wA_0w'\overline{\zeta A_0\zeta'}-w\bar{\zeta}'$$
$$=2\left(\frac{1+wA_0w'}{2},\ \frac{1-wA_0w'}{2i},\ wP_0^{-1}\right)\begin{pmatrix}I^{(2)} & \\ & -I^{(N)}\end{pmatrix}\overline{\left(\frac{1+\zeta A_0\zeta'}{2},\ \frac{1-\zeta A_0\zeta'}{2i},\ \zeta P_0^{-1}\right)}'$$
$$=\rho_z\bar{\rho}_\xi(1+zA_0z'\overline{\xi A_0\xi'}-z\bar{\xi}'). \tag{2.3.38}$$

将(2.3.37)代入(2.3.35)中，可算出

$$\rho_z=\frac{\sqrt{1+|z_0A_0z_0'|^2-z_0\bar{z}_0'}}{1+zA_0z'\overline{z_0A_0z_0'}-z\bar{z}_0'}\sqrt{\frac{1+\overline{z_0A_0z_0'}}{1+z_0A_0z_0'}}, \tag{2.3.39}$$

相应地，有

$$\rho_\xi=\frac{\sqrt{1+|z_0A_0z_0'|^2-z_0\bar{z}_0'}}{1+\xi A_0\xi'\overline{z_0A_0z_0'}-\xi\bar{z}_0'}\sqrt{\frac{1+\overline{z_0A_0z_0'}}{1+z_0A_0z_0'}},\tag{2.3.40}$$

代入(2.3.38),即可得到(2.3.31).

附带指出,在所取的变换之下,由于[2]

$$\dot{\zeta}=(1+|z_0A_0z_0'|^2-z_0\bar{z}')^{\frac{N}{2}}\cdot|1+\xi A_0\xi'\overline{z_0A_0z_0'}-\xi\bar{z}_0'|^{-N}\dot{\xi},\quad(\zeta,\xi\in\mathfrak{L}_{\rm IV})$$

再利用(2.3.31)即可得到

$$(1+wA_0w'\overline{\zeta A_0\zeta'}-w\bar{\zeta}')^{-\frac{N}{2}}\dot{\zeta}=(1+zA_0z'\overline{z_0A_0z_0'}-z\bar{z}_0')^{\frac{N}{2}}$$
$$\times(1+zA_0z'\overline{\xi A_0\xi'}-z\bar{\xi}')^{-\frac{N}{2}}(1+\xi A_0\xi'\overline{z_0A_0z_0'}-\xi\bar{z}_0')^{-\frac{N}{2}}\dot{\xi}.\tag{2.3.41}$$

§ 2.4. Cauchy 型积分在 $\mathfrak{B}_{\rm IV}-\mathfrak{L}_{\rm IV}$ 上的极限值

定理 2.2. 设 $\varphi(\xi)$ 在 $\mathfrak{L}_{\rm IV}$ 上满足 Lip $p(0<p\leqslant1)$,并令

$$F(z)=V(\mathfrak{L}_{\rm IV})^{-1}\int_{\mathfrak{L}_{\rm IV}}H_{\rm IV}(z,\xi)\varphi(\xi)\dot{\xi},\quad z\in\mathfrak{R}_{\rm IV}\tag{2.4.1}$$

则当 z 从 $\mathfrak{R}_{\rm IV}$ 的内部趋于点 $\eta_0=e^{i\theta_0}(t,1,0,\cdots,0)P_0^{-1}TP_0\in\mathfrak{B}_{\rm IV}-\mathfrak{L}_{\rm IV}$ 时,只要

$$|1+\eta_0A_0\eta_0'\overline{\xi A_0\xi'}-\eta_0\bar{\xi}'|\cdot|1+zA_0z'\overline{\xi A_0\xi'}-z\bar{\xi}'|^{-1}\leqslant M\ (\text{常数}),\tag{2.4.2}$$

则必有

$$\lim_{z\to\eta_0}F(z)=\text{V.P.}V(\mathfrak{L}_{\rm IV})^{-1}\int_{\mathfrak{L}_{\rm IV}}H_{\rm IV}(\eta_0,\xi)\varphi(\xi)\dot{\xi}$$
$$+\frac{1}{2}\cdot\frac{1}{2\pi}\int_0^{2\pi}\varphi\left(e^{i\theta_0}\left(e^{i\theta}\frac{1+te^{-i\theta}}{1+te^{i\theta}},1,0,\cdots,0\right)P_0^{-1}TP_0\right)\frac{d\theta}{(1+te^{i\theta})^{\frac{N}{2}}},\tag{2.4.3}$$

其中主值是按 § 2.3 中的定义取的.

证. (i) 先取 $\eta_0=(0,1,0,\cdots,0)\in\mathfrak{B}_{\rm IV}-\mathfrak{L}_{\rm IV}$,并令 $\xi^*=(e^{2i\theta},1,0,\cdots,0)\in\mathfrak{L}_{\rm IV}$,则

$$F(z)=V(\mathfrak{L}_{\rm IV})^{-1}\int_{\mathfrak{L}_{\rm IV}}H_{\rm IV}(z,\xi)(\varphi(\xi)-\varphi(\xi^*))\dot{\xi}$$
$$+V(\mathfrak{L}_{\rm IV})^{-1}\int_{\mathfrak{L}_{\rm IV}}H_{\rm IV}(z,\xi)\varphi(\xi^*)\dot{\xi}=F_1(z)+F_2(z).\tag{2.4.4}$$

我们首先证明:

$$\lim_{z\to\eta_0}F_1(z)=\frac{1}{V(\mathfrak{L}_{\rm IV})}\int_{\mathfrak{L}_{\rm IV}}\frac{\varphi(\xi)-\varphi(\xi^*)}{(1-\eta_0\bar{\xi}')^{\frac{N}{2}}}\dot{\xi},\tag{2.4.5}$$

此处 z 所沿之途径,满足已设条件(2.4.2). 考察

$$\int_{\mathfrak{L}_{\rm IV}}(\varphi(\xi)-\varphi(\xi^*))H_{\rm IV}(z,\xi)\dot{\xi}-\int_{\mathfrak{L}_{\rm IV}}\frac{(\varphi(\xi)-\varphi(\xi^*))\dot{\xi}}{(1-\eta_0\bar{\xi}')^{\frac{N}{2}}}$$
$$=\left(\int_{\sigma_\delta}+\int_{\Sigma_\delta}\right)(\varphi(\xi)-\varphi(\xi^*))[(1-\eta_0\bar{\xi}')^{\frac{N}{2}}-(1+zA_0z'\overline{\xi A_0\xi'}-z\bar{\xi}')^{\frac{N}{2}}]$$
$$\times H_{\rm IV}(z,\xi)(1-\eta_0\bar{\xi}')^{-\frac{N}{2}}\dot{\xi}=I_1+I_2.\tag{2.4.6}$$

其中

$$\sigma_\delta:\ \begin{cases}\xi\in\mathfrak{L}_{\rm IV},\\ |1-\eta_0\bar{\xi}'|\leqslant\delta,\end{cases}\qquad \Sigma_\delta:\ \begin{cases}\xi\in\mathfrak{L}_{\rm IV},\\ |1-\eta_0\bar{\xi}'|\geqslant\delta.\end{cases}\tag{2.4.7}$$

有

$$\begin{aligned}|I_1|=&\Big|\int_{\sigma_\delta}(\varphi(\xi)-\varphi(\xi^*))(\sqrt{1-\eta_0\bar{\xi}'}-\sqrt{1+zA_0z'\overline{\xi A_0\xi'}-z\bar{\xi}'})\\&\times\sum_{l=0}^{N-1}\sqrt{1-\eta_0\bar{\xi}'}^{N-1-l}\sqrt{1+zA_0z'\overline{\xi A_0\xi'}-z\bar{\xi}'}^{\,l}H_{\rm IV}(z,\xi)(1-\eta_0\bar{\xi}')^{-\frac{N}{2}}\dot{\xi}\Big|\\ \leqslant&\sum_{k=1}^{N}\int_{\sigma_\delta}|\varphi(\xi)-\varphi(\xi^*)|\cdot|\sqrt{1-\eta_0\bar{\xi}'}-\sqrt{1+zA_0z'\overline{\xi A_0\xi'}-z\bar{\xi}'}|\\&\times|1-zA_0z'\overline{\xi A_0\xi'}-z\bar{\xi}'|^{-\frac{k}{2}}|1-\eta_0\bar{\xi}'|^{\frac{k-1-N}{2}}\dot{\xi},\end{aligned}$$

因为 $\varphi(\xi)$ 满足 Lip p,并且 z 所沿之途径满足

$$|1-\eta_0\bar{\xi}'|\cdot|1+zA_0z'\overline{\xi A_0\xi'}-z\bar{\xi}'|\leqslant M,$$

所以

$$\begin{aligned}|I_1|\leqslant&K\int_{\sigma_\delta}|\sqrt{1-\eta_0\bar{\xi}'}-\sqrt{1+zA_0z'\overline{\xi A_0\xi'}-z\bar{\xi}'}|\\&\times|1+zA_0z'\overline{\xi A_0\xi'}-z\bar{\xi}'|^{-\frac{1}{2}}\cdot|1-\eta_0\bar{\xi}'|^{\frac{p-N}{2}}\dot{\xi}\\ \leqslant&K_1\int_{\sigma_\delta}|1-\eta_0\bar{\xi}'|^{\frac{p-N}{2}}\dot{\xi}\\ \leqslant&K_2\int_{\substack{yy'=1\\|1-y_1-iy_2|\leqslant\delta}}|1-y_1-iy_2|^{\frac{p-N}{2}}\dot{y}=O(\delta^{\frac{p}{2}}),\end{aligned}\tag{2.4.8}$$

这是在 § 2.2 中已知的事实(K, K_1, K_2 皆为常数).

任与 $\varepsilon>0$, 可取一 $\delta>0$, 使在 σ_δ 上的积分 I_1 满足 $|I_1|<\frac{\varepsilon}{2}$. 在 δ 取定之后，由于

$$\int_{\Sigma_\delta}H_{\rm IV}(z,\xi)(\varphi(\xi)-\varphi(\xi^*))\dot{\xi}$$

作为 z 的函数,在 $z=\eta_0$ 处连续,所以当 z 充分接近 η_0 时,必有 $|I_2|<\frac{\varepsilon}{2}$. 因此(2.4.5)成立.

再来证明 $F_2(z)$ 与 z 无关,而且有

$$F_2(z)=\frac{1}{\pi}\int_0^\pi\varphi(\xi^*)d\theta.\tag{2.4.9}$$

设 $z=e^{i\theta_0}(t,\rho,0,\cdots,0)P_0^{-1}\Gamma P_0\in\mathfrak{R}_{\rm IV}$. 先令 $z=e^{i\theta_0}\tilde{z}P_0^{-1}\Gamma P_0$, 相应地, $\xi=e^{i\theta_0}\tilde{\xi}P_0^{-1}\Gamma P_0$, $\xi^*=e^{-i\theta_0}\xi^*P_0^{-1}\Gamma' P_0$. 记 $\varphi(\xi^*)=\tilde{\varphi}(\tilde{\xi}^*)$,于是有

$$F_2(z)=V(\mathfrak{L}_{\rm IV})^{-1}\int_{\mathfrak{L}_{\rm IV}}H_{\rm IV}(\tilde{z},\tilde{\xi})\tilde{\varphi}(\tilde{\xi}^*)\dot{\tilde{\xi}}.\tag{2.4.10}$$

再作变换 $w=\Phi(\tilde{z})$, 使 $\tilde{z}_0=(t,0,\cdots,0)$ 变为原点, $\tilde{z}=(t,\rho,0,\cdots,0)$ 变为 $w=(0,\rho,0,\cdots,0)$, 相应地, $\zeta=\Phi(\tilde{\xi})$, $\tilde{\xi}^*$ 变为 ζ^*, $\zeta^*=\zeta^*(e^{2i\theta},t)$ 是 $e^{2i\theta}$ 及 t 的

有理函数。记 $\tilde{\varphi}(\xi^*)=\psi(\zeta^*)$，并注意到逆变换 $\tilde{z}=\Phi^{-1}(w)$ 将 $w_0=(-t, 0, \cdots, 0)$ 变为原点，使得(参见(2.3.41)式)

$$\begin{aligned} H_{\mathrm{IV}}(\tilde{z}, \xi)\dot{\xi} &= H_{\mathrm{IV}}(w, w_0)^{-1}H_{\mathrm{IV}}(w, \zeta)H_{\mathrm{IV}}(\zeta, w_0)\dot{\zeta} \\ &= (1-w\xi')^{-\frac{N}{2}}(1-\zeta\overline{w}_0')^{-\frac{N}{2}}\dot{\zeta}=(1-\rho\xi_2)^{-\frac{N}{2}}(1+t\zeta_1)^{-\frac{N}{2}}\dot{\zeta}, \end{aligned} \tag{2.4.11}$$

这里用到 $wA_0w'=0$, $w_0A_0w_0'=0$, $w\overline{w}_0'=0$. 于是

$$F_2(z)=V(\mathfrak{L}_{\mathrm{IV}})^{-1}\int_{\mathfrak{L}_{\mathrm{IV}}}(1-\rho\xi_2)^{-\frac{N}{2}}(1+t\zeta_1)^{-\frac{N}{2}}\psi(\zeta^*)\dot{\zeta}. \tag{2.4.12}$$

记 $\zeta=e^{i\theta}xP_0$, $0\leqslant\theta<\pi$, $xx'=1$, 令 $y_1+iy_2=e^{-i\theta}(x_1+ix_2)$, $y_3=x_3, \cdots, y_N=x_N$, 则

$$F_2(z)=2^{\frac{N}{2}}V(\mathfrak{L}_{\mathrm{IV}})^{-1}\int_{\mathfrak{L}_{\mathrm{IV}}}(1-\rho(y_1+iy_2))^{-\frac{N}{2}}(1+te^{2i\theta}(y_1+iy_2))^{-\frac{N}{2}}\psi(\zeta^*)d\theta\cdot\dot{y}, \tag{2.4.13}$$

再令 $y_1+iy_2=re^{i\varphi}$, $0\leqslant\varphi<2\pi$, $r=\sqrt{1-y_3^2-\cdots-y_N^2}$, 则有 $\dot{y}=d\varphi dy_3\cdots dy_N$, 以致

$$\begin{aligned} F_2(z) &= 2^{\frac{N}{2}}V(\mathfrak{L}_{\mathrm{IV}})^{-1}\int_0^{\pi}d\theta\int_{y_3^2+\cdots+y_N^2<1}dy_3\cdots dy_N \\ &\quad\times\int_0^{2\pi}(1-\rho re^{i\varphi})^{-\frac{N}{2}}(1+tre^{i\varphi}\cdot e^{2i\theta})^{-\frac{N}{2}}\psi(\zeta^*)d\varphi \\ &= \frac{2^{\frac{N}{2}}}{V(\mathfrak{L}_{\mathrm{IV}})}\sum_{k=0}^{\infty}\frac{\Gamma\left(\frac{N}{2}+k\right)}{\Gamma\left(\frac{N}{2}\right)\Gamma(k+1)}t^k\int_0^{\pi}\psi(\zeta^*)e^{2ik\theta}d\theta \\ &\quad\times\int_{y_3^2+\cdots+y_N^2<1}r^kdy_3\cdots dy_N\int_0^{2\pi}(1-\rho re^{i\varphi})^{-\frac{N}{2}}e^{ik\varphi}d\varphi, \end{aligned} \tag{2.4.14}$$

因为

$$\begin{aligned} \int_0^{2\pi}(1-\rho re^{i\varphi})^{-\frac{N}{2}}e^{ik\varphi}d\varphi &= \sum_{l=0}^{\infty}\frac{\Gamma\left(\frac{N}{2}+l\right)}{\Gamma\left(\frac{N}{2}\right)\Gamma(l+1)}\rho^l r^l\int_0^{2\pi}e^{i(k+l)\varphi}d\varphi \\ &= \begin{cases} 2\pi & \text{当 } k=0 \text{ 时}, \\ 0 & \text{当 } k>0 \text{ 时}, \end{cases} \end{aligned}$$

所以

$$\begin{aligned} F_2(z) &= 2^{\frac{N}{2}}V(\mathfrak{L}_{\mathrm{IV}})^{-1}\int_0^{\pi}\psi(\zeta^*)d\theta\cdot 2\pi\int_{y_3^2+\cdots+y_N^2<1}dy_3\cdots dy_N \\ &= \frac{2^{\frac{N}{2}}\cdot\Gamma\left(\frac{N}{2}\right)\cdot 2\pi}{2^{\frac{N}{2}+1}\pi^{\frac{N}{2}+1}}\cdot\frac{\pi^{\frac{N}{2}-1}}{\Gamma\left(\frac{N}{2}\right)}\int_0^{\pi}\varphi(\zeta^*)d\theta=\frac{1}{\pi}\int_0^{\pi}\varphi(\xi^*)d\theta. \end{aligned} \tag{2.4.15}$$

根据定理 2.1，有

$$V(\mathfrak{L}_{\rm IV})^{-1}\int_{\mathfrak{L}_{\rm IV}}(1-\eta_0\bar{\xi}')^{-\frac{N}{2}}(\varphi(\xi)-\varphi(\xi^*))\dot{\xi}$$

$$=\text{V.P.}\,V(\mathfrak{L}_{\rm IV})^{-1}\int_{\mathfrak{L}_{\rm IV}}(1-\eta_0\bar{\xi}')^{-\frac{N}{2}}\varphi(\xi)\dot{\xi}-\frac{1}{2\pi}\int_0^{\pi}\varphi(\xi^*)d\theta, \tag{2.4.16}$$

所以最后得到

$$\lim_{z\to\eta_0}F(z)=\text{V.P.}\,V(\mathfrak{L}_{\rm IV})^{-1}\int_{\mathfrak{L}_{\rm IV}}(1-\eta_0\bar{\xi}')^{-\frac{N}{2}}\varphi(\xi)\dot{\xi}+\frac{1}{2\pi}\int_0^{\pi}\varphi(\xi^*)d\theta, \tag{2.4.17}$$

于是当 $\eta_0=(0,1,0,\cdots,0)$ 时，定理得证.

(ii) 考虑 $\mathfrak{B}_{\rm IV}-\mathfrak{L}_{\rm IV}$ 上任一点 $\eta_0=e^{i\theta_0}(t,1,0,\cdots,0)P_0^{-1}\Gamma P_0$，因可利用 $\Gamma_0^{\rm IV}$ 中的变换 $z=e^{i\theta_0}\tilde{z}P_0^{-1}\Gamma P_0$，故不妨设 $\eta_0=(t,1,0,\cdots,0)$．作 $\Gamma^{\rm IV}$ 中的变换 $w=\Phi(z)$，使 $z_0=(t,0,\cdots,0)$ 变为原点，使 η_0 变为 $\zeta_0=(0,1,0,\cdots,0)$．于是

$$F(z)=V(\mathfrak{L}_{\rm IV})^{-1}H_{\rm IV}(z,\xi)\int_{\mathfrak{L}_{\rm IV}}H_{\rm IV}(w,\zeta)\psi(\zeta)\dot{\zeta}, \tag{2.4.18}$$

其中

$$\psi(\zeta)=\varphi(\Phi^{-1}(\zeta))\cdot(1+\Phi^{-1}(\zeta)A_0\Phi^{-1}(\zeta)'\overline{z_0A_0z_0'}-\Phi^{-1}(\zeta)\bar{z}_0')^{\frac{N}{2}}, \tag{2.4.19}$$

它显然在 $\mathfrak{L}_{\rm IV}$ 上属于 Lip p.

注意到

$$\frac{|1-\zeta_0\bar{\zeta}'|}{|1+wA_0w'\overline{\zeta A_0\zeta'}-w\bar{\zeta}'|}=\frac{|1+zA_0z'\overline{z_0A_0z_0'}-z\bar{z}_0'|\cdot|1+\eta_0A_0\eta_0'\overline{\xi A_0\xi'}-\eta_0\bar{\xi}'|}{(1+|z_0A_0z_0'|^2-z_0\bar{z}_0')\cdot|1+zA_0z'\overline{\xi A_0\xi'}-z\bar{\xi}'|}, \tag{2.4.20}$$

其中 $|1+zA_0z'\overline{z_0A_0z_0'}-z\bar{z}_0'|$ 显然有界，而 $1+|z_0A_0z_0'|^2-z_0\bar{z}_0'$ 为一正数，所以根据已设可知，在 w 趋于 ζ_0 时，恒有

$$|1-\zeta_0\bar{\zeta}'|\cdot|1+wA_0w'\overline{\zeta A_0\zeta'}-w\bar{\zeta}'|^{-1}\leqslant M_1\quad(\text{常数}),$$

因此可利用(i)的结果，得到

$$\lim_{z\to\eta_0}F(z)=(1+\eta_0A_0\eta_0'\overline{z_0A_0z_0'}-\eta_0\bar{z}_0')^{-\frac{N}{2}}$$

$$\times\left\{\text{V.P.}\,V(\mathfrak{L}_{\rm IV})^{-1}\int_{\mathfrak{L}_{\rm IV}}(1-\zeta_0\bar{\zeta}')^{-\frac{N}{2}}\psi(\zeta)\dot{\zeta}+\frac{1}{2\pi}\int_0^{\pi}\psi(\zeta^*)d\theta\right\}$$

$$=\lim_{\varepsilon\to 0}V(\mathfrak{L}_{\rm IV})^{-1}\int_{\substack{\mathfrak{L}_{\rm IV}\\|1-\zeta_0\bar{\zeta}'|\geqslant\varepsilon}}\varphi(\Phi^{-1}(\zeta))\cdot(1-\Phi^{-1}(\zeta)\bar{z}_0')^{\frac{N}{2}}\cdot(1-\eta_0\bar{z}_0')^{-\frac{N}{2}}$$

$$\times(1-\zeta_0\bar{\zeta}')^{-\frac{N}{2}}\dot{\zeta}+\frac{1}{2\pi}\int_0^{\pi}\varphi(\Phi^{-1}(\zeta^*))\cdot(1-\Phi^{-1}(\zeta^*)\bar{z}_0')^{\frac{N}{2}}(1-\eta_0\bar{z}_0')^{-\frac{N}{2}}d\theta, \tag{2.4.21}$$

此处 $\zeta^*=(e^{2i\theta},1,0,\cdots,0)$．再利用变换 $\xi=\Phi^{-1}(\zeta)$，并注意到 $\Phi^{-1}(\zeta^*)=\left(\dfrac{t+e^{2i\theta}}{1+te^{2i\theta}},1,0,\cdots,0\right)$，$1-\Phi^{-1}(\zeta^*)\bar{z}_0'=\dfrac{1-t^2}{1+te^{2i\theta}}$ 和 $1-\eta_0\bar{z}_0'=1-t^2$，再以 θ 代 2θ，便可得出公式(2.4.3).

§2.5. $\mathfrak{B}$-调和函数的边界值

设 $f(z)=u(z)+iv(z)$ 在 $\mathfrak{R}_{\rm IV}$ 内解析，在 $\mathfrak{R}_{\rm IV}+\mathfrak{L}_{\rm IV}$ 上连续，且 $u(\xi)$ 在 $\mathfrak{L}_{\rm IV}$ 上属于

Lip p $(0 < p \leqslant 1)$，我们可利用 § 2.4 中的结果，导出 $v(z)$ 在 $\mathfrak{B}_{IV} - \mathfrak{L}_{IV}$ 上的值与 $u(\xi)$ 间的一个有趣的关系式（如(2.5.4)所示）.

利用 $\mathfrak{R}_{IV}$ 上的解析函数的 Schwarz 公式[4]

$$f(z) = V(\mathfrak{L}_{IV})^{-1}\int_{\mathfrak{L}_{IV}} u(\xi)S(z, \xi)\dot{\xi} + iv(0), \tag{2.5.1}$$

其中 Schwarz 核 $S(z, \xi) = 2(1 + zA_0z'\overline{\xi A_0\xi'} - z\bar{\xi}')^{-1} - 1$. 于是

$$f(z) = 2V(\mathfrak{L}_{IV})^{-1}\int_{\mathfrak{L}_{IV}} H_{IV}(z,\xi)u(\xi)\dot{\xi} - V(\mathfrak{L}_{IV})^{-1}\int_{\mathfrak{L}_{IV}} u(\xi)\dot{\xi} + iv(0). \tag{2.5.2}$$

在上式中取 $z = e^{i\theta_0}(t, \rho, 0, \cdots, 0)P_0^{-1}TP_0 \in \mathfrak{R}_{IV}$，使之趋于 $\mathfrak{B}_{IV} - \mathfrak{L}_{IV}$ 的点 $\eta = e^{i\theta_0}(t, 1, 0, \cdots, 0)P_0^{-1}TP_0$（易知此一路径是满足(2.4.2)的），根据定理 2.2，便有

$$\begin{aligned} f(\eta) = {} & \text{V.P. } 2V(\mathfrak{L}_{IV})^{-1}\int_{\mathfrak{L}_{IV}} H_{IV}(\eta, \xi)u(\xi)\dot{\xi} \\ & + \frac{1}{2\pi}\int_0^{2\pi} u\left(e^{i\theta_0}\left(e^{i\theta}\frac{1+te^{-i\theta}}{1+te^{i\theta}}, 1, 0, \cdots, 0\right)P_0^{-1}TP_0\right) \\ & \times (1+te^{i\theta})^{-\frac{N}{2}}d\theta - V(\mathfrak{L}_{IV})^{-1}\int_{\mathfrak{L}_{IV}} u(\xi)\dot{\xi} + iv(0). \end{aligned} \tag{2.5.3}$$

在上式两端取虚部，得到

$$\begin{aligned} v(\eta) = {} & \text{V.P.} V(\mathfrak{L}_{IV})^{-1}\int_{\mathfrak{L}_{IV}} u(\xi)\frac{H_{IV}(\eta, \xi) - H_{IV}(\xi, \eta)}{i}\dot{\xi} \\ & + \frac{1}{2}\cdot\frac{1}{2\pi}\int_0^{2\pi} u\left(e^{i\theta_0}\left(e^{i\theta}\frac{1+te^{-i\theta}}{1+te^{i\theta}}, 1, 0, \cdots, 0\right)P_0^{-1}TP_0\right) \\ & \times \frac{T(t, \theta) - T(t, -\theta)}{i}d\theta + v(0). \end{aligned} \tag{2.5.4}$$

其中 $T(t, \theta) = (1 + te^{i\theta})^{-\frac{N}{2}}$.

§ 2.6. Cauchy 型积分在 $\mathfrak{L}_{IV}$ 上的极限值

首先证明两个引理.

引理 2.2. 设 ζ, z 为二复数，并且 $(1 - \zeta z)\left(1 - \frac{1}{\zeta}z\right) \neq 0$，则有

$$\frac{1}{(1-\zeta z)^n\left(1-\frac{1}{\zeta}z\right)^n} = (\alpha\beta)^n\sum_{\mu=1}^{n} C_{n-1}^{2n-\mu-1}\left(\frac{1}{\alpha^\mu(1-\zeta z)^\mu} + \frac{1}{\beta^\mu\left(1-\frac{1}{\zeta}z\right)^\mu}\right), \tag{2.6.1}$$

其中 n 为任一正整数，α, β 为

$$\alpha = \frac{1}{1-\zeta^2}, \quad \beta = \frac{\zeta^2}{\zeta^2 - 1}. \tag{2.6.2}$$

证. 利用归纳法，首先可证

$$\frac{1}{(1-\zeta z)^n\left(1-\frac{1}{\zeta}z\right)} = \sum_{\nu=1}^{n}\frac{\alpha^{n-\nu}\beta}{(1-\zeta z)^\nu} + \frac{\alpha^n}{1-\frac{1}{\zeta}z} \tag{2.6.3}$$

和

$$\frac{1}{\left(1-\frac{1}{\zeta}z\right)^n(1-\zeta z)}=\sum_{\nu=1}^{n}\frac{\beta^{n-\nu}\alpha}{\left(1-\frac{1}{\zeta}z\right)^\nu}+\frac{\beta^n}{1-\zeta z}. \tag{2.6.4}$$

事实上，当 $n=1$ 时，

$$\frac{1}{\left(1-\frac{1}{\zeta}z\right)(1-\zeta z)}=\frac{\alpha}{1-\frac{1}{\zeta}z}+\frac{\beta}{1-\zeta z} \tag{2.6.5}$$

显然成立．假定对于某一自然数 n，(2.6.3)式成立，则因

$$\begin{aligned}\frac{1}{(1-\zeta z)^{n+1}\left(1-\frac{1}{\zeta}z\right)}&=\frac{1}{1-\zeta z}\left(\sum_{\nu=1}^{n}\frac{\alpha^{n-\nu}\beta}{(1-\zeta z)^\nu}+\frac{\alpha^n}{1-\frac{1}{\zeta}z}\right)\\&=\sum_{\nu=2}^{n+1}\frac{\alpha^{n+1-\nu}\beta}{(1-\zeta z)^\nu}+\frac{\alpha^{n+1}}{1-\frac{1}{\zeta}z}+\frac{\alpha^n\beta}{1-\zeta z}\\&=\sum_{\nu=1}^{n+1}\frac{\alpha^{n+1-\nu}\beta}{(1-\zeta z)^\nu}+\frac{\alpha^{n+1}}{1-\frac{1}{\zeta}z},\end{aligned} \tag{2.6.6}$$

所以(2.6.3)式对于任一自然数 n 皆成立．同理可证(2.6.4)．

用归纳法证明(2.6.1)如下：当 $n=1$ 时，(2.6.1)即(2.6.5)．假设对于某一自然数 n，(2.6.1)成立，便有

$$\begin{aligned}\frac{1}{(1-\zeta z)^{n+1}\left(1-\frac{1}{\zeta}z\right)^{n+1}}&=(\alpha\beta)^n\sum_{\mu=1}^{n}C_{n-1}^{2n-\mu-1}\left(\frac{\beta}{\alpha^\mu(1-\zeta z)^{\mu+1}}\right.\\&\quad\left.+\frac{\alpha}{\beta^\mu\left(1-\frac{1}{\zeta}z\right)^{\mu+1}}+\frac{1}{\alpha^{\mu-1}(1-\zeta z)^\mu\left(1-\frac{1}{\zeta}z\right)}+\frac{1}{\beta^{\mu-1}\left(1-\frac{1}{\zeta}z\right)^\mu(1-\zeta z)}\right)\\&=(\alpha\beta)^{n+1}\sum_{\mu=1}^{n}C_{n-1}^{2n-\mu-1}\left[\frac{1}{\alpha^{\mu+1}(1-\zeta z)^{\mu+1}}+\frac{1}{\beta^{\mu+1}\left(1-\frac{1}{\zeta}z\right)^{\mu+1}}\right.\\&\quad\left.+\frac{1}{\alpha^\mu\beta}\left(\sum_{\nu=1}^{\mu}\frac{\alpha^{\mu-\nu}\beta}{(1-\zeta z)^\nu}+\frac{\alpha^\mu}{1-\frac{1}{\zeta}z}\right)+\frac{1}{\alpha\beta^\mu}\left(\sum_{\nu=1}^{\mu}\frac{\beta^{\mu-\nu}\alpha}{\left(1-\frac{1}{\zeta}z\right)^\nu}+\frac{\beta^\mu}{1-\zeta z}\right)\right]\\&=(\alpha\beta)^{n+1}\left(\sum_{\mu=2}^{n+1}\sum_{\nu=1}^{\mu}C_{n-1}^{2n-\mu}\frac{1}{\alpha^\nu(1-\zeta z)^\nu}+\sum_{\mu=1}^{n}C_{n-1}^{2n-\mu-1}\frac{1}{\alpha(1-\zeta z)}\right.\\&\quad\left.+\sum_{\mu=2}^{n+1}\sum_{\nu=1}^{\mu}C_{n-1}^{2n-\mu}\frac{1}{\beta^\nu\left(1-\frac{1}{\zeta}z\right)^\nu}+\sum_{\mu=1}^{n}C_{n-1}^{2n-\mu-1}\frac{1}{\beta\left(1-\frac{1}{\zeta}z\right)}\right)\\&=(\alpha\beta)^{n+1}\left[\sum_{\nu=1}^{n+1}\sum_{\mu=\nu}^{n+1}C_{n-1}^{2n-\mu}\frac{1}{\alpha^\nu(1-\zeta z)^\nu}+\left(C_n^{2n-1}-C_{n-1}^{2n-1}\right)\frac{1}{\alpha(1-\zeta z)}\right.\end{aligned}$$

$$+\sum_{\nu=1}^{n+1}\sum_{\mu=\nu}^{n+1}C_{n-1}^{2n-\mu}\frac{1}{\beta^{\nu}\left(1-\frac{1}{\zeta}z\right)^{\nu}}+(C_{n}^{2n-1}-C_{n-1}^{2n-1})\frac{1}{\beta\left(1-\frac{1}{\zeta}z\right)}\Bigg]$$

$$=(\alpha\beta)^{n+1}\sum_{\nu=1}^{n+1}C_{n}^{2n+1-\nu}\left(\frac{1}{\alpha^{\nu}(1-\zeta z)^{\nu}}+\frac{1}{\beta^{\nu}\left(1-\frac{1}{\zeta}z\right)^{\nu}}\right),\tag{2.6.7}$$

此处用到

$$C_r^r+C_r^{r+1}+\cdots+C_r^{n-1}=C_{r+1}^{n}.\tag{2.6.8}$$

于是，(2.6.1)对所有的自然数 n 皆成立．引理得证．

特别地，若 $|\zeta|=1$，则 $\alpha=\dfrac{\bar\zeta}{\bar\zeta-\zeta}$，$\beta=\dfrac{\zeta}{\zeta-\bar\zeta}$．应用公式(2.6.1)可得

$$\frac{1}{(1-\bar\zeta z)^n(1-\zeta z)^n}=\sum_{k=0}^{n-1}\frac{(-1)^kC_{n-1}^{n+k-1}}{(\zeta-\bar\zeta)^{n+k}(z-\zeta)^{n-k}}$$

$$+(-1)^n\sum_{k=0}^{n-1}\frac{C_{n-1}^{n+k-1}}{(\zeta-\bar\zeta)^{n+k}(z-\bar\zeta)^{n-k}}.\tag{2.6.9}$$

当 $N=2$ 时，$\mathfrak{R}_{\mathrm{IV}}$ 即为双圆柱：$|z_1|<1$，$|z_2|<1$．设 $\varphi(\xi)$ 在 $\mathfrak{L}_{\mathrm{IV}}(2)$ 上连续，作 Cauchy 型积分

$$F(z_1,z_2)=V(\mathfrak{L}_{\mathrm{IV}}(2))^{-1}\int_{\mathfrak{L}_{\mathrm{IV}}(2)}(1+zA_0z'\overline{\xi A_0\xi'}-z\bar\xi')^{-1}\varphi(\xi)\dot\xi.\tag{2.6.10}$$

其中 $\xi=e^{i\theta}(x_1+ix_2,\ x_1-ix_2)$，$x_1^2+x_2^2=1$，$x_2\geqslant 0$，$0\leqslant\theta<\pi$．记 $x_1+ix_2=e^{i\varphi}$，并取 $z_1=z_2=\rho$，$0<\rho<1$，代入(2.6.10)，便有

$$F(\rho,\rho)=\frac{1}{2\pi^2}\int_0^{\pi}d\varphi\int_0^{2\pi}\frac{\varphi(e^{i(\theta+\varphi)},e^{i(\theta-\varphi)})d\theta}{(1-\rho e^{-i(\theta+\varphi)})(1-\rho e^{-i(\theta-\varphi)})}.\tag{2.6.11}$$

作变换 $\theta+\varphi=\theta_1$，$\theta-\varphi=\theta_2$，(2.6.11)式即为

$$F(\rho,\rho)=\frac{1}{4\pi^2}\int_0^{2\pi}\frac{d\theta_2}{1-\rho e^{-i\theta_2}}\int_0^{2\pi}\frac{\varphi(e^{i\theta_1},e^{i\theta_2})}{1-\rho e^{-i\theta_2}}d\theta_1.\tag{2.6.12}$$

根据双圆柱的 Cauchy 型积分的熟知结果[5]，有

引理 2.3. 设 $\varphi(\xi)$ 在 $\mathfrak{L}_{\mathrm{IV}}(2)$ 上满足 Lipschitz 条件，则 Cauchy 型积分(2.6.11)的极限值

$$\lim_{\rho\to1-0}F(\rho,\rho)=\frac{1}{4\pi^2}\int_0^{2\pi}\frac{d\theta_2}{1-e^{-i\theta_2}}\int_0^{2\pi}\frac{\varphi(e^{i\theta_1},e^{i\theta_2})}{1-e^{-i\theta_1}}d\theta_1+\frac{1}{2}\cdot\frac{1}{2\pi}\int_0^{2\pi}\frac{\varphi(e^{i\theta_1},1)}{1-e^{-i\theta_1}}d\theta_1$$

$$+\frac{1}{2}\cdot\frac{1}{2\pi}\int_0^{2\pi}\frac{\varphi(1,e^{i\theta_2})}{1-e^{-i\theta_2}}d\theta_2+\frac{1}{4}\varphi(1,1).\tag{2.6.13}$$

本段主要结果是下述二定理．

定理 2.3. $N=2n$ $(n\geqslant1)$ 时，在 $\mathfrak{L}_{\mathrm{IV}}(N)$ 上任与一函数 $\varphi(\xi)$，如果 $\varphi(\xi)=\varphi(e^{i\theta}xP_0)=\varphi_0(e^{i\theta})$ 作为 $e^{i\theta}$ 的函数具有 $n-1$ 阶连续微商，并且它对 $e^{i\theta}$ 的 k 阶微商 $(0\leqslant k\leqslant n-1)$ 在 $\mathfrak{L}_{\mathrm{IV}}$ 上属于 $\mathrm{Lip}\,p(0<p\leqslant1)$，则当 $z=\rho\xi_0$ 从 $\mathfrak{R}_{\mathrm{IV}}(N)$ 的内部或从 $\mathfrak{R}_{\mathrm{IV}}^*(N)$ 的内部趋于 $\mathfrak{L}_{\mathrm{IV}}$ 上的点 ξ_0 时，Cauchy 型积分

$$F(z) = V(\mathfrak{L}_{\mathrm{IV}})^{-1}\int_{\mathfrak{L}_{\mathrm{IV}}} H_{\mathrm{IV}}(z,\xi)\varphi(\xi)\dot{\xi} \tag{2.6.14}$$

的极限值 $\lim\limits_{z\to\xi_0}F(z)$ 是存在的．特别当 $\xi_0=(1,1,0,\cdots,0)$ 时，$\lim\limits_{\rho\to1-0}F(\rho,\rho,0,\cdots,0)$ 和 $\lim\limits_{\rho\to1+0}F(\rho,\rho,0,\cdots,0)$ 的表达式如(2.6.28)所示．

注．(1) $\varphi(\xi)$ 在 $\mathfrak{L}_{\mathrm{IV}}$ 上属于 Lip p，指按 § 2.3 中的定义．记 $\xi=e^{i\theta}xP_0$，$\xi^*=e^{i\tau}yP_0$，由于

$$\begin{aligned}(\xi-\xi^*)\overline{(\xi-\xi^*)}' &= 2((x-y)(x-y)'+2\mathrm{Re}(1-e^{i(\theta-\tau)})xy')\\ &\leqslant 4((x-y)(x-y)'+|e^{i\theta}-e^{i\tau}|),\end{aligned}$$

以及当 $a,b\geqslant0$，$0<\alpha\leqslant1$ 时，有

$$(a+b)^\alpha\leqslant a^\alpha+b^\alpha,$$

所以，若 $\varphi(\xi)$ 属于 Lip p，则必有

$$|\varphi(\xi)-\varphi(\xi^*)| = O(((x-y)(x-y)')^{\frac{p}{4}}+|e^{i\theta}-e^{i\tau}|^{\frac{p}{2}}).$$

(2) 所述 $\varphi_0(e^{i\theta})$ 对 $e^{i\theta}$ 可以微商，指存在

$$\varphi_0'(e^{i\theta_0}) = \left.\frac{d\varphi_0(e^{i\theta})}{de^{i\theta}}\right|_{\theta=\theta_0} = \lim_{\theta\to\theta_0}\frac{\varphi_0(e^{i\theta})-\varphi_0(e^{i\theta_0})}{e^{i\theta}-e^{i\theta_0}},$$

这与 $\varphi_0(e^{i\theta})=\phi(\theta)$ 作为 θ 的函数的可微性是一致的．

定理证明如下．

无妨考虑 $\xi_0=(1,1,0,\cdots0)\in\mathfrak{L}_{\mathrm{IV}}$，以 $z=(\rho,\rho,0,\cdots0)\in\mathfrak{R}_{\mathrm{IV}}$ 代入(2.6.14)，得到

$$\begin{aligned}F_\rho = F((\rho,\rho,0,\cdots0)) &= \frac{2}{\varpi_{N-1}}\int_{\substack{xx'=1\\x_N\geqslant0}}\dot{x}\frac{1}{2\pi}\int_0^{2\pi}\frac{\varphi(e^{i\theta}xP_0)d\theta}{(\rho^2e^{-2i\theta}-2\rho x_1e^{-i\theta}+1)^{\frac{N}{2}}}\\ &= \frac{2}{\varpi_{N-1}}\int_{\substack{xx'=1\\x_N\geqslant0}}\dot{x}\frac{1}{2\pi}\int_0^{2\pi}\frac{\varphi(e^{i\theta}xP_0)d\theta}{(1-\rho e^{-i(\theta+\varphi_1)})^{\frac{N}{2}}(1-\rho e^{-i(\theta-\varphi_1)})^{\frac{N}{2}}},\end{aligned} \tag{2.6.15}$$

其中

$$e^{i\varphi_1} = x_1+i\sqrt{1-x_1^2}. \tag{2.6.16}$$

应用公式(2.6.9)，有

$$\begin{aligned}F_\rho = &\sum_{k=0}^{n-2}\frac{(-1)^kC_{n-1}^{n+k-1}}{(2i)^{n+k}}\cdot\frac{2}{\varpi_{N-1}}\int_{\substack{xx'=1\\x_N\geqslant0}}\frac{\dot{x}}{(\sqrt{1-x_1^2})^{n+k}}\cdot\frac{1}{2\pi}\int_0^{2\pi}\frac{\varphi_0(e^{i\theta})d\theta}{(\rho e^{-i\theta}-e^{i\varphi_1})^{n-k}}\\ &+(-1)^n\sum_{k=0}^{n-2}\frac{C_{n-1}^{n+k-1}}{(2i)^{n+k}}\frac{2}{\varpi_{N-1}}\int_{\substack{xx'=1\\x_N\geqslant0}}\frac{\dot{x}}{(\sqrt{1-x_1^2})^{n+k}}\frac{1}{2\pi}\int_0^{2\pi}\frac{\varphi_0(e^{i\theta})d\theta}{(\rho e^{-i\theta}-e^{-i\varphi_1})^{n-k}}\\ &+\frac{(-1)^{n-1}C_{n-1}^{2(n-1)}}{(2i)^{2(n-1)}}\frac{2}{\varpi_{N-1}}\int_{\substack{xx'=1\\x_N\geqslant0}}\frac{\dot{x}}{(\sqrt{1-x_1^2})^{2(n-1)}}\\ &\times\frac{1}{2\pi}\int_0^{2\pi}\frac{\varphi(e^{i\theta}xP_0)d\theta}{(1-\rho e^{-i(\theta+\varphi_1)})(1-\rho e^{-i(\theta-\varphi_1)})} = I_1+I_2+I_3.\end{aligned} \tag{2.6.17}$$

注意到

$$\frac{1}{2\pi}\int_0^{2\pi}\frac{\varphi_0(e^{i\theta})d\theta}{(\rho e^{-i\theta}-e^{i\varphi_1})^{n-k}}\quad(0\leqslant k\leqslant n-2)$$

$$=\frac{(-1)^{n-k}}{e^{i(n-k)\varphi_1}}\cdot\frac{1}{2\pi i}\int_{|\zeta|=1}\frac{\Phi(\zeta)d\zeta}{(\zeta-z)^{n-k}},\tag{2.6.18}$$

此处 $\zeta=e^{i\theta}$, $z=\rho e^{-i\varphi_1}$, $\Phi(\zeta)=\varphi_0(\zeta)\zeta^{n-k-1}$. 利用分部积分可得

$$\int_{|\zeta|=1}\frac{\Phi(\zeta)d\zeta}{(\zeta-z)^{n-k}}=\frac{1}{(n-k-1)!}\int_{|\zeta|=1}\frac{\Phi^{(n-k-1)}(\zeta)d\zeta}{\zeta-z},\tag{2.6.19}$$

其中

$$\Phi^{(n-k-1)}(\zeta)=\frac{d^{n-k-1}}{d\zeta^{n-k-1}}(\varphi_0(\zeta)\zeta^{n-k-1})=\sum_{l=0}^{n-k-1}C_l^{n-k-1}\frac{(n-k-1)!}{l!}\zeta^l\varphi_0^{(l)}(\zeta).\tag{2.6.20}$$

由于

$$J_\rho=\int\limits_{\substack{xx'=1\\x_N\geqslant0}}\frac{\dot{x}}{(\sqrt{1-x_1^2})^{n+k}}\frac{1}{2\pi i}\int_{|\zeta|=1}\frac{\zeta^l\varphi_0^{(l)}(\zeta)d\zeta}{\zeta-\rho e^{-i\varphi_1}}\quad(0\leqslant k\leqslant n-2,0\leqslant l\leqslant n-k-1)$$

$$=\int\limits_{\substack{xx'=1\\x_N\geqslant0}}\frac{\dot{x}}{(\sqrt{1-x_1^2})^{n+k}}\frac{1}{2\pi i}\int_{|\zeta|=1}\frac{\zeta^l\varphi_0^{(l)}(\zeta)-e^{-il\varphi_1}\varphi_0^{(l)}(e^{-i\varphi_1})}{\zeta-\rho e^{-i\varphi_1}}d\zeta$$

$$+e^{-il\varphi_1}\int\limits_{\substack{xx'=1\\x_N\geqslant0}}\frac{\varphi_0^{(l)}(e^{-i\varphi_1})}{(\sqrt{1-x_1^2})^{n+k}}\dot{x},$$

根据假设，上式第一项的内层积分为一通常积分，并且当 $0\leqslant\rho\leqslant1$ 时，它对 x 属于 $\mathrm{Lip}\,\frac{p}{2}(0<p\leqslant1)$, 而第二项积分是一个通常积分. 因而

$$\lim_{\rho\to1-0}J_\rho=\int\limits_{\substack{xx'=1\\x_N\geqslant0}}\frac{\dot{x}}{(\sqrt{1-x_1^2})^{n+k}}\frac{1}{2\pi i}\int_{|\zeta|=1}\frac{\zeta^l\varphi_0^{(l)}(\zeta)-e^{-il\varphi_1}\varphi_0^{(l)}(e^{-i\varphi_1})}{\zeta-e^{-i\varphi_1}}d\zeta$$

$$+e^{-il\varphi_1}\int\limits_{\substack{xx'=1\\x_N>0}}\frac{\varphi_0^{(l)}(e^{-i\varphi_1})\dot{x}}{(\sqrt{1-x_1^2})^{n+k}},\tag{2.6.21}$$

或书为

$$\lim_{\rho\to1-0}J_\rho=\int\limits_{\substack{xx'=1\\x_N>0}}\frac{\dot{x}}{(\sqrt{1-x_1^2})^{n+k}}\mathrm{V.P.}\frac{1}{2\pi}\int_0^{2\pi}\frac{e^{i(l+1)\theta}\varphi_0^{(l)}(e^{i\theta})d\theta}{e^{i\theta}-e^{-i\varphi_1}}$$

$$+\frac{1}{2}e^{-il\varphi_1}\int\limits_{\substack{xx'=1\\x_N>0}}\frac{\varphi_0^{(l)}(e^{-i\varphi_1})\dot{x}}{(\sqrt{1-x_1^2})^{n+k}}.\tag{2.6.22}$$

上式第一项的内层积分是取习知的 Cauchy 主值，根据假设可知，它在 $xx'=1$ 上必属于 $\mathrm{Lip}\,\frac{p}{2}$ (参见[6] p. 74—75), 因而第一项积分是有意义的. 于是

$$\lim_{\rho\to 1-0} I_1 = (-1)^n \sum_{k=0}^{n-2}\sum_{l=0}^{n-k-1} \frac{C_{n-1}^{n+k-1}C_l^{n-k-1}}{(2i)^{n+k}l!\,e^{i(n-k)\varphi_1}}$$

$$\times\left(\frac{2}{\varpi_{N-1}}\int_{\substack{xx'=1\\x_N>0}}\frac{\dot{x}}{(\sqrt{1-x_1^2})^{n+k}}\text{V.P.}\frac{1}{2\pi}\int_0^{2\pi}\frac{e^{i(l+1)\theta}\varphi_0^{(l)}(e^{i\theta})d\theta}{e^{i\theta}-e^{-i\varphi_1}}\right.$$

$$\left.+\frac{1}{e^{il\varphi_1}\varpi_{N-1}}\int_{\substack{xx'=1\\x_N>0}}\frac{\varphi_0^{(l)}(e^{-i\varphi_1})\dot{x}}{(\sqrt{1-x_1^2})^{n+k}}\right). \qquad (2.6.23)$$

同理可得

$$\lim_{\rho\to 1-0} I_2 = \sum_{k=0}^{n-2}\sum_{l=0}^{n-k-1}\frac{(-1)^k C_{n-1}^{n+k-1}C_l^{n-k-1}e^{i(n-k)\varphi_1}}{(2i)^{n+k}l!}$$

$$\times\left(\frac{2}{\varpi_{N-1}}\int_{\substack{xx'=1\\x_N>0}}\frac{\dot{x}}{(\sqrt{1-x_1^2})^{n+k}}\text{V.P.}\frac{1}{2\pi}\int_0^{2\pi}\frac{e^{i(l+1)\theta}\varphi_0^{(l)}(e^{i\theta})d\theta}{e^{i\theta}-e^{i\varphi_1}}\right.$$

$$\left.+\frac{e^{il\varphi_1}}{\varpi_{N-1}}\int_{\substack{xx'=1\\x_N>0}}\frac{\varphi_0^{(l)}(e^{i\varphi_1})\dot{x}}{(\sqrt{1-x_1^2})^{n+k}}\right). \qquad (2.6.24)$$

令

$$I_3^* = \frac{2}{\varpi_{N-1}}\int_{\substack{xx'=1\\x_N>0}}\frac{\dot{x}}{(1-x_1^2)^{n-1}}\cdot\frac{1}{2\pi}\int_0^{2\pi}\frac{\varphi_0(e^{i\theta})d\theta}{(1-\rho e^{-i(\theta+\varphi_1)})(1-\rho e^{-i(\theta-\varphi_1)})},$$

取球坐标 $x_1=\cos\varphi_1$, $x_2=\sin\varphi_1\cos\varphi_2$, $\cdots$, $x_N=\sin\varphi_1\cdots\sin\varphi_{N-1}$, $0\leqslant\varphi_1,\varphi_2,\cdots,\varphi_{N-1}<\pi$. 于是

$$I_3^* = \frac{2\pi}{\varpi_{N-1}}\int_0^\pi\sin^{N-3}\varphi_2\,d\varphi_2\cdots\int_0^\pi\sin\varphi_{N-2}d\varphi_{N-2}\int_0^\pi d\varphi_{N-1}$$

$$\times\frac{1}{\pi}\int_0^\pi d\varphi_1\frac{1}{2\pi}\int_0^{2\pi}\frac{\varphi_0(e^{i\theta})d\theta}{(1-\rho e^{-i(\theta+\varphi_1)})(1-\rho e^{-i(\theta-\varphi_1)})}.$$

作变换 $\theta+\varphi_1=\theta_1$, $\theta-\varphi_1=\theta_2$, 易知, 经变换 $\theta=\frac{\theta_1+\theta_2}{2}$, $x_1=\cos\frac{\theta_1-\theta_2}{2}$, $x_2=\sin\frac{\theta_1-\theta_2}{2}\cos\varphi_2$, $\cdots$, $x_N=\sin\frac{\theta_1-\theta_2}{2}\sin\varphi_2\cdots\sin\varphi_{N-1}$ 后, 函数

$$\varphi(e^{i\theta}xP_0)=\varphi\left(e^{i\frac{\theta_1+\theta_2}{2}}\left(\cos\frac{\theta_1-\theta_2}{2}+i\sin\frac{\theta_1-\theta_2}{2}\cos\varphi_2\right),\cdots,\right.$$

$$\left.\sqrt{2}\,e^{i\frac{\theta_1+\theta_2}{2}}\sin\frac{\theta_1-\theta_2}{2}\sin\varphi_2\cdots\sin\varphi_{N-1}\right)=\varphi^*(\theta_1,\theta_2,\varphi_2,\cdots,\varphi_{N-1})\quad(2.6.25)$$

在 $\zeta_1=e^{i\theta_1}$, $\zeta_2=e^{i\theta_2}(0\leqslant\theta_1,\theta_2<2\pi)$ 上属于 Lip $\frac{p}{2}$. 这可如下看出: 设 $\tau=\frac{\tau_1+\tau_2}{2}$, $y_1=\cos\frac{\tau_1-\tau_2}{2}$, $y_2=\sin\frac{\tau_1-\tau_2}{2}\cos\varphi_2$, $\cdots$, $y_N=\sin\frac{\tau_1-\tau_2}{2}\sin\varphi_2\cdots\sin\varphi_{N-1}$, 则因 $\varphi(\xi)$ 属于 Lip p, 便有

$$
\begin{aligned}
&|\varphi^*(\theta_1, \theta_2, \varphi_2, \cdots, \varphi_{N-1}) - \varphi^*(\tau_1, \tau_2, \varphi_2, \cdots, \varphi_{N-1})| \\
&\quad = O\left(|\theta - \tau|^{\frac{p}{2}} + ((x-y)(x-y)')^{\frac{p}{4}}\right) \\
&\quad = O\left((|\theta_1 - \tau_1| + |\theta_2 - \tau_2|)^{\frac{p}{2}}\right) \\
&\quad = O\left(|\theta_1 - \tau_1|^{\frac{p}{2}} + |\theta_2 - \tau_2|^{\frac{p}{2}}\right).
\end{aligned}
$$

应用引理 2.3,立得

$$
\begin{aligned}
&\lim_{\rho\to 1-0} \frac{1}{\pi}\int_0^{\pi} d\varphi_1 \frac{1}{2\pi}\int_0^{2\pi} \frac{\varphi_0(e^{i\theta})d\theta}{(1-\rho e^{-i(\theta+\varphi_1)})(1-\rho e^{-i(\theta-\varphi_1)})} \\
&= \text{V.P.}\frac{1}{4\pi^2}\int_0^{2\pi}\frac{d\theta_2}{1-e^{-i\theta_2}}\int_0^{2\pi}\frac{\varphi^*(\theta_1, \theta_2, \varphi_2, \cdots, \varphi_{N-1})d\theta_1}{1-e^{-i\theta_1}} \\
&\quad + \frac{1}{2}\text{V.P.}\frac{1}{2\pi}\int_0^{2\pi}\frac{\varphi^*(\theta_1, 0, \varphi_2, \cdots, \varphi_{N-1})d\theta_1}{1-e^{-i\theta_1}} \\
&\quad + \frac{1}{2}\text{V.P.}\frac{1}{2\pi}\int_0^{2\pi}\frac{\varphi^*(0, \theta_2, \varphi_2, \cdots, \varphi_{N-1})d\theta_2}{1-e^{-i\theta_2}} \\
&\quad + \frac{1}{4}\varphi^*(0, 0, \varphi_2, \cdots, \varphi_{N-1}). \qquad (2.6.26)
\end{aligned}
$$

由于 $\varphi^*(0, 0, \varphi_2, \cdots, \varphi_{N-1}) = \varphi((1, 1, 0, \cdots 0))$,并且

$$
\begin{aligned}
&\frac{(-1)^{n-1}C_{n-1}^{2(n-1)}}{(2i)^{2(n-1)}}\frac{2\pi}{\varpi_{2n-1}}\int_0^{\pi}\sin^{2n-3}\varphi_2 d\varphi_2\cdots\int_0^{\pi}\sin\varphi_{2n-2}d\varphi_{2n-2}\int_0^{\pi}d\varphi_{2n-1} \\
&\quad = \frac{\Gamma(2n-1)\sqrt{\pi}}{2^{2(n-1)}\Gamma(n)\,\Gamma\left(\dfrac{2n-1}{2}\right)} = 1,
\end{aligned}
$$

所以

$$
\begin{aligned}
\lim_{\rho\to 1-0} I_3 &= \frac{\Gamma(2n-1)}{2^{2(n-1)}\pi^{n-1}\Gamma(n)}\int_0^{\pi}\sin^{N-3}\varphi_2 d\varphi_2\cdots\int_0^{\pi}\sin\varphi_{N-2}d\varphi_{N-2} \\
&\quad\times\int_0^{\pi}\left\{\text{V.P.}\frac{1}{4\pi^2}\int_0^{2\pi}\frac{d\theta_2}{1-e^{-i\theta_2}}\int_0^{2\pi}\frac{\varphi^*(\theta_1, \theta_2, \varphi_2, \cdots, \varphi_{N-1})d\theta_1}{1-e^{-i\theta_1}}\right. \\
&\quad + \frac{1}{2}\text{V.P.}\frac{1}{2\pi}\int_0^{2\pi}\frac{\varphi^*(\theta_1, 0, \varphi_2, \cdots, \varphi_{N-1})}{1-e^{-i\theta_1}}d\theta_1 \\
&\quad \left. + \frac{1}{2}\text{V.P.}\frac{1}{2\pi}\int_0^{2\pi}\frac{\varphi^*(0, \theta_2, \varphi_2, \cdots, \varphi_{N-1})}{1-e^{-i\theta_2}}d\theta_2\right\}d\varphi_{N-1} \\
&\quad + \frac{1}{4}\varphi((1, 1, 0, \cdots 0)). \qquad (2.6.27)
\end{aligned}
$$

同样可考虑 $\rho\to 1+0$ 时的极限值. 最后得到

$$
\begin{aligned}
\lim_{\rho\to 1\mp 0} F_\rho &= \sum_{k=0}^{n-2}\sum_{l=0}^{n-k-1}\frac{C_{n-1}^{n+k-1}C_l^{n-k-1}}{(2i)^{n+k}l!}\frac{2}{\varpi_{N-1}}\int_{\substack{xx'=1\\ x_N>0}}\frac{1}{(\sqrt{1-x_1^2})^{n+k}} \\
&\quad\times\left[(-1)^n e^{-i(n-k)\varphi_1}\text{V.P.}\frac{1}{2\pi}\int_0^{2\pi}\frac{e^{i(l+1)\theta}\varphi_0^{(l)}(e^{i\theta})d\theta}{e^{i\theta}-e^{-i\varphi_1}}\right. \\
&\quad\left. + (-1)^k e^{i(n-k)\varphi_1}\text{V.P.}\frac{1}{2\pi}\int_0^{2\pi}\frac{e^{i(l+1)\theta}\varphi_0^{(l)}(e^{i\theta})d\theta}{e^{i\theta}-e^{i\varphi_1}}\right]\dot{x}
\end{aligned}
$$

$$\pm \sum_{k=0}^{n-2}\sum_{l=0}^{n-k-1}\frac{C_{n-1}^{n+k-1}C_l^{n-k-1}}{(2i)^{n+k}l!}\frac{1}{\varpi_{N-1}}\int_{\substack{xx'=1\\x_N>0}}\frac{1}{(\sqrt{1-x_1^2})^{n+k}}$$

$$\times[(-1)^n e^{-i(n-k+l)\varphi_1}\varphi_0^{(l)}(e^{-i\varphi_1})+(-1)^k e^{i(n-k+l)\varphi_1}\varphi_0^{(l)}(e^{i\varphi_1})]\dot{x}$$

$$+\frac{(2n-3)!!}{(2\pi)^{n-1}}\int_0^{\pi}\sin^{N-3}\varphi_2 d\varphi_2\cdots\int_0^{\pi}\sin\varphi_{N-2}d\varphi_{N-2}$$

$$\times\int_0^{\pi}\Big\{\text{V.P.}\frac{1}{4\pi^2}\int_0^{2\pi}\frac{d\theta_2}{1-e^{-i\theta_2}}\int_0^{2\pi}\frac{\varphi^*(\theta_1,\theta_2,\varphi_2,\cdots,\varphi_{N-1})d\theta_1}{1-e^{-i\theta_1}}$$

$$\pm\frac{1}{2}\text{V.P.}\frac{1}{2\pi}\int_0^{2\pi}\frac{\varphi^*(\theta_1,0,\varphi_2,\cdots,\varphi_{N-1})d\theta_1}{1-e^{-i\theta_1}}$$

$$\pm\frac{1}{2}\text{V.P.}\frac{1}{2\pi}\int_0^{2\pi}\frac{\varphi^*(0,\theta_2,\varphi_2,\cdots,\varphi_{N-1})d\theta_2}{1-e^{-i\theta_2}}\Big\}d\varphi_{N-1}$$

$$+\frac{1}{4}\varphi((1,1,0,\cdots,0)). \tag{2.6.28}$$

其中 $\varphi_0^{(l)}(e^{ia})=\left.\dfrac{d^l\varphi_0(e^{i\theta})}{d(e^{i\theta})^l}\right|_{\theta=a}$，$e^{i\varphi_1}$ 和 $\varphi^*(\theta_1,\theta_2,\varphi_2,\cdots,\varphi_{N-1})$ 分别如(2.6.16)和(2.6.25)所示．

定理 2.4. $N=2n+1\ (n\geqslant 1)$ 时，在 $\mathfrak{L}_{\text{IV}}(N)$ 上任与一函数 $\varphi(\xi)$，如果 $\varphi(\xi)=\varphi(e^{i\theta}xP_0)=\varphi_0(e^{i\theta})$ 对 $e^{i\theta}$ 具有 $n+1$ 阶连续微商，且 $\varphi_0^{(k)}(e^{i\theta})=\dfrac{d^k\varphi_0(e^{i\theta})}{d(e^{i\theta})^k}(0\leqslant k\leqslant n+1)$ 在 $\mathfrak{L}_{\text{IV}}$ 上连续，则当 $z=\rho\xi_0$ 从 $\mathfrak{R}_{\text{IV}}(N)$ 的内部或从 $\mathfrak{R}_{\text{IV}}^*(N)$ 的内部趋于 $\mathfrak{L}_{\text{IV}}$ 上的点 ξ_0 时，Cauchy 型积分(2.6.14)的极限值 $\lim\limits_{z\to\xi_0}F(z)$ 是存在的．特别当 $\xi_0=(1,1,0,\cdots0)$ 时，$\lim\limits_{\rho\to1-0}F((\rho,\rho,0,\cdots,0))$ 和 $\lim\limits_{\rho\to1+0}F((\rho,\rho,0,\cdots0))$ 的表达式分别如(2.6.38)和(2.6.40)所示．

證．应用公式(2.6.9)，有

$$F_\rho=F((\rho,\rho,0,\cdots0))=\sum_{k=0}^{n-1}\frac{(-1)^kC_{n-1}^{n+k-1}}{(2i)^{n+k}}\frac{2}{\varpi_{N-1}}\int_{\substack{xx'=1\\x_N>0}}\frac{\dot{x}}{(\sqrt{1-x_1^2})^{n+k}}$$

$$\times\frac{1}{2\pi}\int_0^{2\pi}\frac{\varphi_0(e^{i\theta})d\theta}{(\rho e^{-i\theta}-e^{i\varphi_1})^{n-k+\frac{1}{2}}(\rho e^{-i\theta}-e^{-i\varphi_1})^{\frac{1}{2}}}+(-1)^n\sum_{k=0}^{n-1}\frac{C_{n-1}^{n+k-1}}{(2i)^{n+k}}$$

$$\times\frac{2}{\varpi_{N-1}}\int_{\substack{xx'=1\\x_N>0}}\frac{\dot{x}}{(\sqrt{1-x_1^2})^{n+k}}\frac{1}{2\pi}\int_0^{2\pi}\frac{\varphi_0(e^{i\theta})d\theta}{(\rho e^{-i\theta}-e^{i\varphi_1})^{\frac{1}{2}}(\rho e^{-i\theta}-e^{-i\varphi_1})^{n-k+\frac{1}{2}}}$$

$$=J_1+J_2. \tag{2.6.29}$$

据假设，可书

$$\varphi_0(e^{i\theta})=\sum_{j=0}^{n+1}\frac{\varphi_0^{(j)}(e^{i\varphi_1})}{j!}(e^{i\theta}-e^{i\varphi_1})^j+\psi(e^{i\theta},e^{i\varphi_1}), \tag{2.6.30}$$

和

$$\varphi_0(e^{i\theta})=\sum_{j=0}^{n+1}\frac{\varphi_0^{(j)}(e^{-i\varphi_1})}{j!}(e^{i\theta}-e^{-i\varphi_1})^j+\psi(e^{i\theta},e^{-i\varphi_1}). \tag{2.6.31}$$

根据假设，$\phi(e^{i\theta}, e^{i\varphi_1})$ 和 $\phi(e^{i\theta}, e^{-i\varphi_1})$ 在 $\mathfrak{L}_{IV}$ 上连续，并且易知

$$\phi(e^{i\theta}, e^{i\varphi_1}) = o\left(|e^{i\theta} - e^{i\varphi_1}|^{n+1}\right), \tag{2.6.32}$$

$$\phi(e^{i\theta}, e^{-i\varphi_1}) = o\left(|e^{i\theta} - e^{-i\varphi_1}|^{n+1}\right). \tag{2.6.33}$$

于是

$$J_{1,k} = \frac{2}{\varpi_{N-1}} \int_{\substack{xx'=1 \\ x_N \geq 0}} \frac{\dot{x}}{(\sqrt{1-x_1^2})^{n+k}} \cdot \frac{1}{2\pi}\int_0^{2\pi} \frac{\varphi_0(e^{i\theta})d\theta}{(\rho e^{-i\theta}-e^{i\varphi_1})^{n-k+\frac{1}{2}}(\rho e^{-i\theta}-e^{-i\varphi_1})^{\frac{1}{2}}}$$

$$= \sum_{j=0}^{n+1} \frac{1}{j!}\frac{2}{\varpi_{N-1}} \int_{\substack{xx'=1 \\ x_N \geq 0}} \frac{\varphi_0^{(j)}(e^{-i\varphi_1})}{(\sqrt{1-x_1^2})^{n+k}}\dot{x}\, \frac{1}{2\pi}\int_0^{2\pi} \frac{(e^{i\theta}-e^{-i\varphi_1})^j d\theta}{(\rho e^{-i\theta}-e^{i\varphi_1})^{n-k+\frac{1}{2}}(\rho e^{-i\theta}-e^{-i\varphi_1})^{\frac{1}{2}}}$$

$$+ \frac{2}{\varpi_{N-1}} \int_{\substack{xx'=1 \\ x_N \geq 0}} \frac{\dot{x}}{(\sqrt{1-x_1^2})^{n+k}} \cdot \frac{1}{2\pi}\int_0^{2\pi} \frac{\phi(e^{i\theta}, e^{-i\varphi_1})d\theta}{(\rho e^{-i\theta}-e^{i\varphi_1})^{n-k+\frac{1}{2}}(\rho e^{-i\theta}-e^{-i\varphi_1})^{\frac{1}{2}}}. \tag{2.6.34}$$

由于当 $0 \leqslant \rho < 1$ 时，

$$\frac{1}{2\pi}\int_0^{2\pi} \frac{(e^{i\theta}-e^{-i\varphi_1})^j d\theta}{(\rho e^{-i\theta}-e^{i\varphi_1})^{n-k+\frac{1}{2}}(\rho e^{-i\theta}-e^{-i\varphi_1})^{\frac{1}{2}}} \qquad (0 \leqslant j \leqslant n+1,\ 0 \leqslant k \leqslant n-1)$$

$$= \frac{(-1)^{n-k+1}}{e^{i(n-k)\varphi_1}} \sum_{l=0}^{j}\sum_{\mu=0}^{l} \frac{(-1)^{j-l}\Gamma(j+1)\Gamma(n-k+\mu+\frac{1}{2})\Gamma(l-\mu+\frac{1}{2})}{\Gamma(l+1)\Gamma(j-l+1)\Gamma(n-k+\frac{1}{2})\Gamma(\mu+1)\Gamma(l-\mu+1)\Gamma(\frac{1}{2})}$$

$$\times \rho^l e^{i(2l-2\mu-j)\varphi_1}, \tag{2.6.35}$$

代入(2.6.34)，(2.6.29)中，令 $\rho \to 1-0$，立得

$$\lim_{\rho\to 1-0} J_1 = \sum_{k=0}^{n-1}\sum_{j=0}^{n+1} \frac{(-1)^{n+1}\Gamma(n+k)}{\Gamma(n-k+\frac{1}{2})\Gamma(n)\Gamma(k+1)(2i)^{n+k}} \frac{2}{\varpi_{N-1}} \int_{\substack{xx'=1 \\ x_N \geq 0}} \frac{\varphi_0^{(j)}(e^{-i\varphi_1})}{(\sqrt{1-x_1^2})^{n+k}}$$

$$\times \sum_{l=0}^{j}\sum_{\mu=0}^{l} \frac{(-1)^{j-l}\Gamma(n-k+\mu+\frac{1}{2})\Gamma(l-\mu+\frac{1}{2})e^{i(2l+k-2\mu-j-n)\varphi_1}}{\Gamma(l+1)\Gamma(j-l+1)\Gamma(\mu+1)\Gamma(l-\mu+1)\Gamma(\frac{1}{2})}\dot{x}$$

$$+ \sum_{k=0}^{n-1} \frac{(-1)^k\Gamma(n+k)}{\Gamma(n)\Gamma(k+1)(2i)^{n+k}} \frac{2}{\varpi_{N-1}} \int_{\substack{xx'=1 \\ x_N \geq 0}} \frac{\dot{x}}{(\sqrt{1-x_1^2})^{n+k}}$$

$$\times \frac{1}{2\pi}\int_0^{2\pi} \frac{\phi(e^{i\theta}, e^{-i\varphi_1})d\theta}{(e^{-i\theta}-e^{i\varphi_1})^{n-k+\frac{1}{2}}(e^{-i\theta}-e^{-i\varphi_1})^{\frac{1}{2}}}. \tag{2.6.36}$$

同理可得

$$\lim_{\rho\to 1-0} J_2 = \sum_{k=0}^{n-1}\sum_{j=0}^{n+1} \frac{(-1)^{k+1}\Gamma(n+k)}{\Gamma(n-k+\frac{1}{2})\Gamma(n)\Gamma(k+1)(2i)^{n+k}} \frac{2}{\varpi_{N-1}} \int_{\substack{xx'=1 \\ x_N \geq 0}} \frac{\varphi_0^{(j)}(e^{i\varphi_1})}{(\sqrt{1-x_1^2})^{n+k}}$$

$$\times \sum_{l=0}^{j}\sum_{\mu=0}^{l} \frac{(-1)^{j-l}\Gamma(n-k+\mu+\frac{1}{2})\Gamma(l-\mu+\frac{1}{2})e^{-i(2l+k-2\mu-j-n)\varphi_1}}{\Gamma(l+1)\Gamma(j-l+1)\Gamma(\mu+1)\Gamma(l-\mu+1)\Gamma(\frac{1}{2})}\dot{x}$$

$$+ (-1)^n \sum_{k=0}^{n-1} \frac{\Gamma(n+k)}{\Gamma(n)\Gamma(k+1)(2i)^{n+k}} \frac{2}{\varpi_{N-1}} \int_{\substack{xx'=1 \\ x_N > 0}} \frac{\dot{x}}{(\sqrt{1-x_1^2})^{n+k}}$$

$$\times\frac{1}{2\pi}\int_0^{2\pi}\frac{\psi(e^{i\theta},e^{i\varphi_1})d\theta}{(e^{-i\theta}-e^{i\varphi_1})^{\frac{1}{2}}(e^{-i\theta}-e^{-i\varphi_1})^{n-k+\frac{1}{2}}}. \tag{2.6.37}$$

于是

$$\lim_{\rho\to 1-0}F_\rho=\sum_{k=0}^{n-1}\sum_{j=0}^{n+1}\sum_{l=0}^{j}\sum_{\mu=0}^{l}\frac{(-1)^{j-l}C_k^{n+k-1}C_l^jC_\mu^l\Gamma(n-k+\mu+\frac{1}{2})\Gamma(l-\mu+\frac{1}{2})}{\sqrt{\pi}(2i)^{n+k}\Gamma(n-k+\frac{1}{2})\Gamma(j+1)\Gamma(l+1)}$$

$$\times\frac{2}{\varpi_{N-1}}\int_{\substack{xx'=1\\x_N\geqslant 0}}\frac{(-1)^{n+1}\varphi_0^{(j)}(e^{-i\varphi_1})e^{i(2l+k-2\mu-j-n)\varphi_1}+(-1)^{k+1}\varphi_0^{(j)}(e^{i\varphi_1})e^{-i(2l+k-2\mu-j-n)\varphi_1}}{(\sqrt{1-x_1^2})^{n+k}}\dot{x}$$

$$+\sum_{k=0}^{n-1}\frac{C^{n+k-1}}{(2i)^{n+k}}\frac{2}{\varpi_{N-1}}\int_{\substack{xx'=1\\x_N\geqslant 0}}\frac{\dot{x}}{(\sqrt{1-x_1^2})^{n+k}}$$

$$\times\frac{1}{2\pi}\int_0^{2\pi}\frac{(-1)^k(e^{-i\theta}-e^{i\varphi_1})^{k-n}\psi(e^{i\theta},e^{-i\varphi_1})+(-1)^n(e^{-i\theta}-e^{-i\varphi_1})^{k-n}\psi(e^{i\theta},e^{i\varphi_1})}{(e^{-i\theta}-e^{i\varphi_1})^{\frac{1}{2}}(e^{-i\theta}-e^{-i\varphi_1})^{\frac{1}{2}}}d\theta. \tag{2.6.38}$$

由于当 $\rho>1$ 时

$$\frac{1}{2\pi}\int_0^{2\pi}\frac{(e^{i\theta}-e^{-i\varphi_1})^jd\theta}{(\rho e^{-i\theta}-e^{i\varphi_1})^{n-k+\frac{1}{2}}(\rho e^{-i\theta}-e^{-i\varphi_1})^{\frac{1}{2}}}=0\ (0\leqslant j\leqslant n+1,\ 0\leqslant k\leqslant n-1), \tag{2.6.39}$$

所以可用同样的方法得到

$$\lim_{\rho\to 1+0}F_\rho=\sum_{k=0}^{n-1}\frac{C_k^{n+k-1}}{(2i)^{n+k}}\frac{2}{\varpi_{N-1}}\int_{\substack{xx'=1\\x_N\geqslant 0}}\frac{\dot{x}}{(\sqrt{1-x_1^2})^{n+k}}$$

$$\times\frac{1}{2\pi}\int_0^{2\pi}\frac{(-1)^k(e^{-i\theta}-e^{i\varphi_1})^{k-n}\psi(e^{i\theta},e^{-i\varphi_1})+(-1)^n(e^{-i\theta}-e^{-i\varphi_1})^{k-n}\psi(e^{i\theta},e^{i\varphi_1})}{(e^{-i\theta}-e^{i\varphi_1})^{\frac{1}{2}}(e^{-i\theta}-e^{-i\varphi_1})^{\frac{1}{2}}}d\theta. \tag{2.6.40}$$

由 (2.6.38) 和 (2.6.40) 所表示的都是通常积分，其中 $\varphi_0^{(j)}(e^{ia})=\left.\frac{d^j\varphi_0(e^{i\theta})}{d(e^{i\theta})^j}\right|_{\theta=a}$，$e^{i\varphi_1}$ 和 $\psi(e^{i\theta},e^{i\varphi_1})$，$\psi(e^{i\theta},e^{-i\varphi_1})$ 分别如(2.6.16)，(2.6.30)和(2.6.31)所示．定理证毕．

参　考　文　献

[1] 华罗庚，多复变数函数论中的典型域的调和分析，科学出版社，1958.

[2] 陆启铿，典型流形与典型域，上海科学技术出版社，1963.

[3] 龚　昇、孙继广，多复变数的 Cauchy 型积分(I)超球的 Cauchy 型积分，数学学报，**15** (1965)，431—443.

[4] 龚　昇、孙继广，圆型域中解析函数的积分表示，中国科学技术大学学报，**1** (1965)，219—226.

[5] Какичев, В. А., Граничные свойства интеграла типа Коши многих переменых, *Уч. зап. Шахтинск*, педин-те, **2** (1959), 29—90.

[6] Мусхелишвили, Н. И., Сингулярные интегральные уравнение, Москва, 1962.

第15卷 第6期 数 学 学 报 Vol. 15, No. 6
1965年11月 ACTA MATHEMATICA SINICA Nov., 1965

多复变数的 Cauchy 型积分 (III)
矩陣双曲空間的 Cauchy 型积分*

龔 昇 孙继广
(中国科学技术大学) (中国科学院数学研究所)

§3.1. 引 言

$\mathfrak{R}_I(m, n)$ 表示矩阵双曲空间: $I - Z\bar{Z}' > 0$, $\mathfrak{R}_I^*(m, n)$ 表示空间: $I - Z\bar{Z}' < 0$, Z 是 m 行 n 列(不妨设 $m \leqslant n$) 复元素矩阵. 华罗庚[1]指出: 如果 $\varphi(U)$ 在 $\mathfrak{R}_I(m, n)$ 的特征流形 $\mathfrak{L}_I(m, n)$ 上连续,则 $\mathfrak{R}_I(m, n)$ 的 Cauchy 型积分

$$F(Z) = \frac{1}{V(\mathfrak{L}_I(m, n))} \int_{\mathfrak{L}_I(m,n)} \varphi(U) \det (I - Z\bar{U}')^{-n} \dot{U} \quad Z \in \mathfrak{R}_I(m, n) \tag{3.1.1}$$

和 $\mathfrak{R}_I^*(n) = \mathfrak{R}_I^*(n, n)$ 的 Cauchy 型积分

$$F(Z) = \frac{1}{V(\mathfrak{L}_I(n))} \int_{\mathfrak{L}_I(n)} \varphi(U) \det (I - Z\bar{U}')^{-n} \dot{U} \quad Z \in \mathfrak{R}_I^*(n) \tag{3.1.2}$$

是存在的. 其中 $V(\mathfrak{L}_I(m, n))$ 表示 $\mathfrak{L}_I(m, n)$ 的体积. 同时还指出: 如果 $Z\bar{Z}'$ 的特征根并非都大于1或都小于1, 则常有酉方阵 U, 使 $\det (I - Z\bar{U}') = 0$. 即相应的 Cauchy 核 $\det (I - Z\bar{U}')^{-n}$ 失去意义. 容易看出, 即使 $Z\bar{Z}' > I$, 只要 $m < n$ 而 $n > 2$, 亦必有酉方阵 $U^{(m,n)}$, 使 $\det (I - Z\bar{U}') = 0$. 事实上, Z 可分解为 $Z = U_0^{(m)}(\Lambda^{(m)}O)V_0^{(n)}$, U_0, V_0 皆为酉方阵, $\Lambda = [\lambda_1, \cdots, \lambda_m] > I$. 我们取

$$U = U_0 \begin{pmatrix} I^{(m-2)} & 0 & 0 & 0 & \\ & \vdots & \vdots & \vdots & \\ & 0 & 0 & 0 & \\ 0 & 0 & \sqrt{1-\frac{1}{\lambda_m^2}} & -\frac{1}{\lambda_m} & O^{(m, n-m-1)} \\ 0 & 0 & \frac{1}{\lambda_m} & \sqrt{1-\frac{1}{\lambda_m^2}} & \end{pmatrix} V_0,$$

则立得

$$\det (I - Z\bar{U}') = \det \left([1-\lambda_1, \cdots, 1-\lambda_{m-2}] \dotplus \begin{pmatrix} 1 & 0 \\ -\lambda_m\sqrt{1-\frac{1}{\lambda_m^2}} & 0 \end{pmatrix} \right) = 0.$$

因而, 在 $\mathfrak{R}_I^*(m, n)$ $(m < n)$ 内, Cauchy 核 $\det (I - Z\bar{U}')^{-n}$ 亦无意义. 所以, 要研究的

* 1964年3月8日收到, 1964年5月7日收到修改稿.

是 $\mathfrak{R}_I(m, n)$ $(m \leqslant n)$ 及 $\mathfrak{R}_I^*(n)$ 的 Cauchy 型积分（3.1.1）及（3.1.2）.

我们所得到的主要结果是：如果 $\varphi(U)$ 在 $\mathfrak{L}_I(n)$ 上具有 $N(n)$$\left(n \text{ 为偶数时, } N(n) = \frac{n^2}{4};\ n \text{ 为奇数时, } N(n) = \frac{n^2-1}{4}\right)$阶满足 Lipschitz 条件的偏微商，则 $\mathfrak{R}_I(n)$ 和 $\mathfrak{R}_I^*(n)$ 的 Cauchy 型积分（3.1.1）和（3.1.2）当 $Z = \rho V$ $(0 < \rho < 1$ 或 $\rho > 1)$ 趋于 $\mathfrak{L}_I(n)$ 上的任一点 V 时的极限值是存在的，并且给出了当 $V = I^{(n)}$ 时的极限值的表达式. 用 $\mathfrak{L}_I^{(m-1)}$ 表示 $\overline{\mathfrak{R}_I(m, n)}$ 上使 $I - Z\bar{Z}'$ 的秩为 $m-1$ 的点所成的集合. 我们证明了：如果 $\varphi(U)$ 在 $\mathfrak{L}_I(m, n)$ 上满足 Lipschitz 条件，则 Cauchy 型积分（3.1.1）当 Z 从 $\mathfrak{R}_I(m, n)$ 的內部趋于 $\mathfrak{L}_I^{(m-1)}$ 时的极限值是存在的，并给出了极限值的一般表达式. 由此可以推知：$\mathfrak{R}_I(m, n)$ 上的任一 $\mathfrak{B}$-调和函数在 $\mathfrak{L}_I^{(m-1)}$ 上的值，必可用它的 $\mathfrak{B}$-调和共轭函数在 $\mathfrak{L}_I(m, n)$ 上的值通过积分表出. 当 $m=n$ 时，利用解析变换 $Z^{-1}=W(Z\bar{Z}' > I)$ 即可得到 $\mathfrak{R}_I^*(n)$ 上的 Cauchy 型积分（3.1.2）的相应结果.

前一个结果是本文中 § 3.2 的內容，定理 3.2.1 曾摘要发表[2]，这里将给出详细的证明，从证明过程中可以看出，我们充分利用了矩阵积分的技巧，这是由华罗庚充分发挥，而在作者之一处理酉羣上的富里埃分析时是行之有效的（例如见[3]）.

后一个结果及相应的推论是本文后几节的內容，在这里可以清楚的看到，这些结果的得来是自然的，几乎可以看成我们处理超球的 Cauchy 型积分[5]之后的必然推论.

本文结果曾在中国科学技术大学数学系综合讨论班上报告过. 在工作过程中，自始至终得到我们的导师华罗庚先生的关怀与鼓励. 我们向他表示衷心的感激.

§ 3.2. Cauchy 型积分在 $\mathfrak{L}_I(n)$ 上的极限值

关于 Cauchy 型积分

$$F(Z) = (V(\mathfrak{L}_I(n)))^{-1} \int_{\mathfrak{L}_I(n)} \varphi(U) \det (I - Z\bar{U}')^{-n} \dot{U} \tag{3.2.1}$$

（其中 $Z \in \mathfrak{R}_I(n)$ 或 $\mathfrak{R}_I^*(n)$）在 $\mathfrak{L}_I(n)$ 上的极限值，我们有下面的定理 3.2.1.

首先，我们指出多圆柱上的 Cauchy 型积分的一个结果. 设 $\varphi(\zeta_1, \cdots, \zeta_n)$ 是定义在 $|\zeta_1| = 1, \cdots, |\zeta_n| = 1$ 上的连续函数，如果对于任意二点 $(\xi_1, \cdots, \xi_n)$ 和 $(t_1, \cdots, t_n)$，$|\xi_j| = 1$, $|t_j| = 1$, $j = 1, 2, \cdots, n$，恆有

$$|\varphi(\xi_1, \cdots, \xi_n) - \varphi(t_1, \cdots, t_n)| \leqslant M \cdot \left(\sum_{j=1}^{n} |\xi_j - t_j|^2\right)^{\alpha/2},$$

其中 M 为一仅与 $\varphi(\zeta_1, \cdots, \zeta_n)$ 有关的常数，$0 < \alpha \leqslant 1$，则称函数 $\varphi(\zeta_1, \cdots, \zeta_n)$ 在 $|\zeta_1| = 1, \cdots, |\zeta_n| = 1$ 上满足 Lipschitz 条件. 我们有

引理 3.2.1. 设 $\varphi(\zeta_1, \cdots, \zeta_n)$ 是定义在 $|\zeta_1| = 1, \cdots, |\zeta_n| = 1$ 上的对称函数，并且满足 Lipschitz 条件，则 Cauchy 型积分

$$\Phi_\rho = (2\pi i)^{-n} \int_{|\zeta_1|=1} \cdots \int_{|\zeta_n|=1} \varphi(\zeta_1, \cdots, \zeta_n) \prod_{k=1}^{n} (\zeta_k - \rho)^{-1} d\zeta_1 \cdots d\zeta_n \tag{3.2.2}$$

（其中 $0 < \rho < 1$ 或 $\rho > 1$）的极限值为

$$\lim_{\rho\to 1-0}\Phi_\rho=\sum_{k=0}^{n}C_k^n\left(\frac{1}{2}\right)^{n-k}\text{V. P.}\,(2\pi i)^{-k}\int_{|\zeta_1|=1}\cdots\int_{|\zeta_k|=1}\varphi(\zeta_1,\cdots,\zeta_k,1,\cdots,1)\cdot\prod_{l=1}^{k}(\zeta_l-1)^{-1}d\zeta_1\cdots d\zeta_k \tag{3.2.3}$$

和

$$\lim_{\rho\to 1+0}\Phi_\rho=\sum_{k=0}^{n}C_k^n\left(-\frac{1}{2}\right)^{n-k}\text{V. P.}\,(2\pi i)^{-k}\int_{|\zeta_1|=1}\cdots\int_{|\zeta_k|=1}\varphi(\zeta_1,\cdots,\zeta_k,1,\cdots,1)\cdot\prod_{l=1}^{k}(\zeta_l-1)^{-1}d\zeta_1\cdots d\zeta_k, \tag{3.2.4}$$

其中 V. P. 的意义为 Cauchy 主值，即

$$\text{V. P.}\int_{|\zeta_1|=1}\cdots\int_{|\zeta_k|=1}\phi(\zeta_1,\cdots,\zeta_k)\prod_{l=1}^{k}(\zeta_l-1)^{-1}d\zeta_1\cdots d\zeta_k$$
$$=\lim_{\varepsilon\to 0}\int_{\substack{|\zeta_1|=1\\|\theta_1|>\varepsilon}}\cdots\int_{\substack{|\zeta_k|=1\\|\theta_k|>\varepsilon}}\phi(\zeta_1,\cdots,\zeta_k)\prod_{l=1}^{k}(\zeta_l-1)^{-1}d\zeta_1\cdots d\zeta_k. \tag{3.2.5}$$

公式(3.2.3)和(3.2.4)可由 B. A. Какичев[8] 中的结果导出.

设 $\varphi(U)$ 是在 $\mathfrak{L}_I(n)$ 上的连续函数，如果对于 $\mathfrak{L}_I(n)$ 上任二点 V, W，恒有

$$|\varphi(V)-\varphi(W)|\leqslant M[\operatorname{tr}(V-W)\overline{(V-W)}']^{\alpha/2}, \tag{3.2.6}$$

其中 M 为一仅与 $\varphi(U)$ 有关的常数，$0<\alpha\leqslant 1$，则称 $\varphi(U)$ 在 $\mathfrak{L}_I(n)$ 上满足 Lipschitz 条件. 如果 $\varphi(U)$ 的所有 N 阶偏微商都在 $\mathfrak{L}_I(n)$ 上满足 Lipschitz 条件，则记 $\varphi(U)\in\mathscr{L}(N,\alpha)$.

定理 3.2.1. 在 $\mathfrak{L}_I(n)$ 上任与一函数 $\varphi(U)$：n 为偶数时，设 $\varphi(U)\in\mathscr{L}\left(\frac{n^2}{4},\alpha\right)$；$n$ 为奇数时，设 $\varphi(U)\in\mathscr{L}\left(\frac{n^2-1}{4},\alpha\right)$，则当 $Z=\rho V$ 从 $\mathfrak{R}_I(n)$ 的内部 $(0<\rho<1)$ 以及从 $\mathfrak{R}_I^*(n)$ 的内部 $(\rho>1)$ 趋于 $\mathfrak{L}_I(n)$ 上的任一点 V 时，Cauchy 型积分(3.2.1)的极限值是存在的，并且有

$$\lim_{\rho\to 1-0}F(\rho I)=\sum_{l=1}^{n}\sum_{(\mu,l)}s!\,(n!)^{-1}C_{\mu_{n-1},\cdots,\mu_0}\sum_{k=0}^{l}C_k^l\left(\frac{1}{2}\right)^{l-k}\text{V. P.}\,(2\pi i)^{-k}$$
$$\int_{|\zeta_1|=1}\cdots\int_{|\zeta_k|=1}g_{\mu_{n-1},\cdots,\mu_{n-l}}(\zeta_1,\cdots,\zeta_k,1,\cdots,1)\prod_{j=1}^{k}(\zeta_j-1)^{-1}d\zeta_1\cdots d\zeta_k \tag{3.2.7}$$

和

$$\lim_{\rho\to 1+0}F(\rho I)=\sum_{l=1}^{n}\sum_{(\mu,l)}s!\,(n!)^{-1}C_{\mu_{n-1},\cdots,\mu_0}\sum_{k=0}^{l}C_k^l\left(\frac{-1}{2}\right)^{l-k}\text{V. P.}\,(2\pi i)^{-k}$$
$$\int_{|\zeta_1|=1}\cdots\int_{|\zeta_k|=1}g_{\mu_{n-1},\cdots,\mu_{n-l}}(\zeta_1,\cdots,\zeta_k,1,\cdots,1)\prod_{j=1}^{k}(\zeta_j-1)^{-1}d\zeta_1\cdots d\zeta_k, \tag{3.2.8}$$

其中

$$(\mu, l)\text{表示}\begin{cases}\mu_{n-1} \leqslant \cdots \leqslant \mu_{n-l} \leqslant n-1 < \mu_{n-l-1} \leqslant \cdots \leqslant \mu_0, \\ \mu_{n-1} + \cdots + \mu_0 = n(n-1),\end{cases}$$

及

$$C_{\mu_{n-1},\cdots,\mu_0} = (-1)^{\frac{n(n-1)}{2}} \sum_{\substack{l_{n-1}+k_{n-1}=\mu_{n-1} \\ \cdots\cdots \\ l_0+k_0=\mu_0}} \delta^{n-1,\cdots,1,0}_{l_{n-1},\cdots,l_1,l_0} \delta^{n-1,\cdots,1,0}_{k_{n-1},\cdots,k_1,k_0}, \tag{3.2.9}$$

而 $(l_{n-1}, \cdots, l_1, l_0)$ 与 $(k_{n-1}, \cdots, k_1, k_0)$ 跑遍 $(n-1, \cdots, 1, 0)$ 的所有排列，s 是 $\mu_{n-1}, \cdots, \mu_0$ 中不同値的个数，而

$$g_{\mu_{n-1},\cdots,\mu_{n-l}} = [(n-\mu_{n-1}-1)!\cdots(n-\mu_{n-l}-1)!]^{-1} \cdot \frac{\partial^{(n-1)l-\mu_{n-1}-\cdots-\mu_{n-l}} g^{[\mu]}(\zeta_1, \cdots, \zeta_l)}{\partial\zeta_1^{n-\mu_{n-1}}\cdots\partial\zeta_l^{n-\mu_{n-l}}}, \tag{3.2.10}$$

$$g^{[\mu]}(\zeta_1, \cdots, \zeta_l) = (2\pi i)^{l-n} \int_{|\zeta_{l+1}|=1} \cdots \int_{|\zeta_n|=1} f(\zeta_1, \cdots, \zeta_n)\, p_{l+1}^{\mu_{n-l-1}-n} \cdots p_n^{\mu_0-n}\, d\zeta_{l+1} \cdots d\zeta_n, \tag{3.2.11}$$

此处 $p_j = \zeta_j - 1$, $\mu_{n-j} \geqslant n$, $j = l+1, \cdots, n$,

$$f(\zeta_1, \cdots, \zeta_n) = \varpi_n'^{-1} \int_{[V]} \varphi(V \wedge \bar{V}')[\dot{V}], \tag{3.2.12}$$

其中 $\Lambda = [\zeta_1, \cdots, \zeta_n]$, $\varpi'_n = (2\pi)^{\frac{n(n-1)}{2}}[1!\,2!\cdots(n-1)!]^{-1}$，积分 (3.2.12) 的意义，参见[1].

证.　无妨取 $V = I$.

令　$F_\rho = F(\rho I)$，于是有[1]

$$F(\rho) = (2\pi)^{-n} \int_{2\pi \geqslant \theta_1 \geqslant \cdots \geqslant \theta_n \geqslant 0} \cdots \int f(e^{i\theta_1}, \cdots, e^{i\theta_n}) \prod_{\mu<\nu} |e^{i\theta_\mu} - e^{i\theta_\nu}|^2 \cdot \prod_{l=1}^{n} (1 - \rho e^{-i\theta_l})^{-n}\, d\theta_1\, d\theta_2 \cdots d\theta_n, \tag{3.2.13}$$

易知

$$f(e^{i\theta_1}, \cdots, e^{i\theta_n}) = \varpi_n'^{-1} \int_{[V]} \varphi(V \wedge \bar{V}')[\dot{V}]\ (\Lambda = [e^{i\theta_1}, \cdots, e^{i\theta_n}])$$

是 $e^{i\theta_1}, \cdots, e^{i\theta_n}$ 的对称函数[3]，并且 $f(\zeta_1, \cdots, \zeta_n)$ 在 $|\zeta_1| = 1, \cdots, |\zeta_n| = 1$ 上属于 $\mathscr{L}\left(\frac{n^2}{4}, \alpha\right)$（当 n 为偶数时）或 $\mathscr{L}\left(\frac{n^2-1}{4}, \alpha\right)$（当 n 为奇数时）.

令　$p_{k,\rho} = \zeta_k - \zeta$, $\zeta_k = e^{i\theta_k}$, $k = 1, 2, \cdots, n$ 便得到

$$\begin{aligned} F_\rho &= (n!)^{-1}(2\pi i)^{-n} \int_{|\zeta_1|=1} \cdots \int_{|\zeta_n|=1} f(\zeta_1, \cdots, \zeta_n)\, p_{1,\rho}^{-n} \cdots p_{n,\rho}^{-n} \prod_{\mu \neq \nu} (p_{\mu,\rho} - p_{\nu,\rho})\, d\zeta_1 \cdots d\zeta_n \\ &= \sum_{\mu_{n+1}+\cdots+\mu_0 = n(n-1)} (n!)^{-1}(2\pi i)^{-n} C_{\mu_{n-1},\cdots,\mu_0} \int_{|\zeta_1|=1} \cdots \int_{|\zeta_n|=1} p_{1,\rho}^{-n} \cdots p_{n,\rho}^{-n} \\ &\quad \times p_{1,\rho}^{\mu_{n-1}} \cdots p_{n,\rho}^{\mu_0} f(\zeta_1, \cdots, \zeta_n)\, d\zeta_1 \cdots d\zeta_n \end{aligned}$$

$$= \sum_{(\mu)} (n!)^{-1} s!\, C_{\mu_{n-1},\cdots,\mu_0} (2\pi i)^{-n} \int_{|\zeta_1|=1} \cdots \int_{|\zeta_n|=1} f(\zeta_1, \cdots, \zeta_n) \cdot p_{1,\rho}^{-n+\mu_{n-1}} \cdots p_{n,\rho}^{-n+\mu_0}\, d\zeta_1 \cdots d\zeta_n$$

$$= \sum_{l=1}^{n} \sum_{(\mu,l)} (n!)^{-1} s!\, C_{\mu_{n-1},\cdots,\mu_0} I^{[l]}_{\mu_{n-1},\cdots,\mu_0}(\rho) \tag{3.2.14}$$

其中 $C_{\mu_{n-1},\cdots,\mu_0}$ 为 (3.2.9) 所示，s 是 $\mu_{n-1}, \cdots, \mu_0$ 中不同值的个数，(μ) 表示

$$\begin{cases} \mu_{n-1} \leqslant \cdots \leqslant \mu_0 \\ \mu_{n+1} + \cdots + \mu_0 = n(n-1), \end{cases}$$

而

$$I^{[l]}_{\mu_{n-1},\cdots,\mu_0}(\rho) = (2\pi i)^{-n} \int_{|\zeta_1|=1} \cdots \int_{|\zeta_n|=1} p_{1,\rho}^{\mu_{n-1}-n} \cdots p_{n,\rho}^{\mu_0-n} f(\zeta_1, \cdots, \zeta_n)\, d\zeta_1 \cdots d\zeta_n$$

$$= (2\pi i)^{-l} \int_{|\zeta_1|=1} \cdots \int_{|\zeta_l|=1} g_\rho^{[\mu]}(\zeta_1, \cdots, \zeta_l) p_{1,\rho}^{\mu_{n-1}-n} \cdots p_{l,\rho}^{\mu_{n-l}-n}\, d\zeta_1 \cdots d\zeta_l, \tag{3.2.15}$$

此处

$$g_\rho^{[\mu]}(\zeta_1, \cdots, \zeta_l) = (2\pi i)^{l-n} \int_{|\zeta_{l+1}|=1} \cdots \int_{|\zeta_n|=1} f(\zeta_1, \cdots, \zeta_n)\, p_{l+1,\rho}^{\mu_{n-l-1}-n} \cdots p_{n,\rho}^{\mu_0-n} \cdot d\zeta_{l+1} \cdots d\zeta_n, \tag{3.2.16}$$

连续分部积分可得

$$I^{[l]}_{\mu_{n-1},\cdots,\mu_0}(\rho) = (2\pi i)^{-l} \int_{|\zeta_1|=1} \cdots \int_{|\zeta_l|=1} g_{\mu_{n-1},\cdots,\mu_{n-l},\rho}(\zeta_1, \cdots, \zeta_l)(p_{1,\rho} \cdots p_{l,\rho})^{-1}\, d\zeta_1 \cdots d\zeta_l, \tag{3.2.17}$$

其中

$$g_{\mu_{n-1},\cdots,\mu_{n-l},\rho}(\zeta_1, \cdots, \zeta_l) = [(n-\mu_{n-1}-1)! \cdots (n-\mu_{n-l}-1)!]^{-1} \frac{\partial^{(n-1)l-\mu_{n-1}-\cdots-\mu_{n-l}} g_\rho^{[\mu]}(\zeta_1, \cdots, \zeta_l)}{\partial \zeta_1^{n-\mu_{n-1}-1} \cdots \partial \zeta_l^{n-\mu_{n-l}-1}}. \tag{3.2.18}$$

由引理 3.2.1 易知

$$\lim_{\rho \to 1-0} I^{[l]}_{\mu_{n-1},\cdots,\mu_0}(\rho) = \sum_{k=0}^{l} C_k^l \left(\frac{1}{2}\right)^{l-k} \text{V. P. } (2\pi i)^{-k} \cdot \int_{|\zeta_1|=1} \cdots \int_{|\zeta_k|=1} g_{\mu_{n-1},\cdots,\mu_{n-l}}(\zeta_1, \cdots, \zeta_k, 1, \cdots, 1) \prod_{\nu=1}^{k} (\zeta_\nu - 1)^{-1}\, d\zeta_1 \cdots d\zeta_k, \tag{3.2.19}$$

$$\lim_{\rho \to 1+0} I^{[l]}_{\mu_{n-1},\cdots,\mu_0}(\rho) = \sum_{k=0}^{l} C_k^l \left(-\frac{1}{2}\right)^{l-k} \text{V. P. } (2\pi i)^{-k} \cdot \int_{|\zeta_1|=1} \cdots \int_{|\zeta_k|=1} g_{\mu_{n-1},\cdots,\mu_{n-l}}(\zeta_1, \cdots, \zeta_k, 1, \cdots 1) \cdot \prod_{\nu=1}^{k} (\zeta_\nu - 1)^{-1}\, d\zeta_1 \cdots d\zeta_k, \tag{3.2.20}$$

其中 $g_{\mu_{n-1},\cdots,\mu_{n-l}}$ 如 (3.2.10) 所示。所以在 (3.2.14) 式中令 $\rho \to 1-0$ 和 $1+0$，便得到公式(3.2.7)和(3.2.8)。

必须指出，在公式(3.2.7)和(3.2.8)中所出现的对 $f(\zeta_1,\cdots,\zeta_n)$ 的偏微商不超过 $\frac{n^2}{4}$ 阶(当 n 为偶数时)或 $\frac{n^2-1}{4}$ 阶(当 n 为奇数时). 即在定理对 $\varphi(U)$ 所设的条件下，公式(3.2.7)和(3.2.8)成立.

事实上，只需注意到

$$\prod_{j\neq k}(p_{j,\rho}-p_{k,\rho})\prod_{l=1}^{n}p_{l,\rho}^{-n}=\prod_{j\neq k}(1-p_{k,\rho}p_{j,\rho}^{-1})\prod_{l=1}^{n}p_{l,\rho}^{-1}$$

$$=\prod_{l=1}^{n}p_{l,\rho}^{-1}\prod_{j<k}(2-p_{k,\rho}p_{j,\rho}^{-1}-p_{j,\rho}p_{k,\rho}^{-1})$$

其最高次项有，例如

$$\prod_{l=1}^{n}p_{l,\rho}^{-1}p_{1,\rho}^{1-n}p_{2,\rho}\cdots p_{n,\rho}\cdot p_{2,\rho}^{2-n}p_{3,\rho}\cdots p_{n,\rho}\cdots p_{n-1,\rho}^{-1}p_{n,\rho}$$

$$=\prod_{l=1}^{n}p_{l,\rho}^{-1}\{(p_{2,\rho}p_{3,\rho}^{2}\cdots p_{j-1,\rho}^{j-2}p_{j,\rho}^{j-1}\cdots p_{n-1,\rho}^{n-2}p_{n,\rho}^{n-1})(p_{1,\rho}^{n-1}p_{2,\rho}^{n-2}\cdots p_{j-1,\rho}^{n-j+1}p_{j,\rho}^{n-j}\cdots p_{n-1,\rho})^{-1}\}$$

若 $n=2m$ 或 $2m+1$, 易知当且仅当 $j\geqslant m+1$ 时，上式{ }中的 $p_{j,\rho}$ 的指数为正. 由此可知，相应于此项，必须要求 $f(\zeta_1,\cdots,\zeta_n)$ 所具有的偏微商的阶数为

$$(n-1)+(n-3)+\cdots+[(n-m)-(m-1)]=(n-m)m$$

$$=\begin{cases}\frac{n^2}{4}, & \text{当 } n=2m \text{ 时},\\ \frac{n^2-1}{4}, & \text{当 } n=2m+1 \text{ 时},\end{cases}$$

定理证毕.

§3.3. 一 条 引 理

在 $\mathfrak{L}_I(m,n)$ 上任与一函数 $\varphi(U)$, 如果对于 $\mathfrak{L}_I(m,n)$ 上的任意二点 U_0, V_0, 恒有不等式

$$|\varphi(U_0)-\varphi(V_0)|\leqslant K(\mathrm{tr}\,(U_0-V_0)(\overline{U_0-V_0})')^{\alpha/2},$$

其中 K 为一仅与 $\varphi(U)$ 有关的常数, $0<\alpha\leqslant1$, 则称 $\varphi(U)$ 在 $\mathfrak{L}_I(m,n)$ 上属于 $\mathrm{Lip}\,\alpha$.

我们首先证明

引理 3.3.1. 若 $\varphi(U)$ 在 $\mathfrak{L}_I(m,n)$ 上属于 $\mathrm{Lip}\,\alpha\ (0<\alpha\leqslant1)$, 则

$$g(u)=(V(\mathfrak{L}_I(m-1,n-1)))^{-1}\int_{W\overline{W}'=I^{(m-1)}}\varphi\left(\begin{pmatrix}u\\0\ W\end{pmatrix}\overline{P}'\right)\dot{W}$$

在 $u\bar{u}'=1$ 上亦属于 $\mathrm{Lip}\,\alpha$. 其中 P 为 n 阶酉方阵，使得 $uP=(1,0,\cdots,0)$.

证. 今

$$g_1(u)=\int_{W\overline{W}'=I}\varphi\left(\begin{pmatrix}u\\0\ W\end{pmatrix}\overline{P}'\right)\dot{W},$$

$$g_1(v)=\int_{W\overline{W}'=I}\varphi\left(\begin{pmatrix}v\\0\ W\end{pmatrix}\overline{P}_1'\right)\dot{W},$$

其中 $u=(1,0,\cdots,0)\bar{P}'$, $v=(1,0,\cdots,0)\bar{P}_1'$. P, P_1 皆为酉方阵. 于是

$$|g_1(u)-g_1(v)|=\left|\int_{W\bar{W}'=I}(\varphi((\begin{smallmatrix}1&0\\0&W\end{smallmatrix})\bar{P}')-\varphi((\begin{smallmatrix}1&0\\0&W\end{smallmatrix})\bar{P}_1'))\dot{W}\right|$$

$$\leqslant\int_{W\bar{W}'=I}|\varphi((\begin{smallmatrix}1&0\\0&W\end{smallmatrix})\bar{P}')-\varphi((\begin{smallmatrix}1&0\\0&W\end{smallmatrix})\bar{P}_1')|\dot{W}.$$

因为 $\varphi(U)$在 $\mathfrak{L}_I(m,n)$ 上属于 $\mathrm{Lip}\,\alpha$, 所以

$$|g_1(u)-g_1(v)|\leqslant K\int_{W\bar{W}'=I}[\mathrm{tr}\,((\begin{smallmatrix}1&0\\0&W\end{smallmatrix})(\bar{P}'-\bar{P}_1')(P-P_1)(\begin{smallmatrix}1&0\\0&\bar{W}'\end{smallmatrix}))]^{\alpha/2}\dot{W}$$

$$=K\int_{W\bar{W}'=I}[\mathrm{tr}\,(\bar{P}'-\bar{P}_1')(P-P_1)(\begin{smallmatrix}1&0\\0&\bar{W}'\end{smallmatrix})(\begin{smallmatrix}1&0\\0&W\end{smallmatrix})]^{\alpha/2}\dot{W}$$
$$=K_1[\mathrm{tr}\,(2I-\bar{P}'P_1-\bar{P}_1'P)]^{\alpha/2}$$
$$=K_1[\mathrm{tr}\,(2I-B-\bar{B}')]^{\alpha/2}$$
$$=K_1[\mathrm{tr}(I-B)(\overline{I-B})']^{\alpha/2}.$$

此处 $K_1=K\cdot V(\mathfrak{L}_I(m-1,n-1))$, $B=P_1\bar{P}'$.

显然, $u=vP_1\bar{P}'=vB$. 值得注意的是,这里的酉方阵 B,在满足 $u=vB$ 的前提下,是任意的. 事实上, 可先取一酉方阵 P, 使 $u=(1,0\cdots0)\bar{P}'$, 再任取一酉方阵 B, 只要满足 $u=vB$, 令 $P_1=BP$, 则必有 $v=uB^{-1}=uP\bar{P}_1'=(1,0\cdots0)\bar{P}_1'$.

习知对于 u, v, 存在酉方阵 A, 使

$$vA=(1,0\cdots0)=v_0,$$
$$uA=(a,b,0\cdots0)=u_0,\quad |a|^2+|b|^2=1.$$

令 $C=\bar{A}'BA$, 便有

$$u_0=uA=vBA=v_0\bar{A}'BA=v_0C.$$

显然,此处的酉方阵 C, 在满足 $u_0=v_0C$ 的前提下,也是任意的. 我们取

$$C=\begin{pmatrix}\begin{matrix}a&b\\-\bar{b}&\bar{a}\end{matrix}&\\&I^{(n-2)}\end{pmatrix},$$

于是

$$\mathrm{tr}\,((I-C)(\overline{I-C})')=2(|1-a|^2+|b|^2)=4(1-\mathrm{Re}\,a),$$

而

$$(u-v)(\overline{u-v})'=(u_0-v_0)(\overline{u_0-v_0})'=|1-a|^2+|b|^2=2(1-\mathrm{Re}\,a),$$

所以

$$\mathrm{tr}\,((I-B)(\overline{I-B})')=\mathrm{tr}\,((I-C)(\overline{I-C})')=2(u-v)(\overline{u-v})'.$$

因之,最后得到

$$|g(u)-g(v)|=(V(\mathfrak{L}_I(m-1,n-1)))^{-1}|g_1(u)-g_1(v)|$$
$$=O([(u-v)(\overline{u-v})']^{\alpha/2}),$$

引理得证.

§3.4. $\mathfrak{R}_I(m,n)$ 的 Cauchy 型积分在 $\mathfrak{L}_I^{(m-1)}$ 上的极限值

应用引理 3.3.1, 我们有

定理 3.4.1. 若 $\varphi(U)$ 在 $\mathfrak{L}_I(m, n)$ 上属于 $\mathrm{Lip}\,\alpha\ (0<\alpha\leqslant 1)$，则 Cauchy 型积分

$$F(Z)=(V(\mathfrak{L}_I(m, n)))^{-1}\int_{\mathfrak{L}_I(m,n)}\varphi(U)\det(I-Z\bar{U}')^{-n}\dot{U} \tag{3.4.1}$$

当 $Z=U_0'\begin{pmatrix}\rho & 0\\ 0 & Z_1\end{pmatrix}V_0\in\mathfrak{R}_I(m, n)$ 趋于点 $Q=U_0'\begin{pmatrix}1 & 0\\ 0 & Z_1\end{pmatrix}V_0\in\mathfrak{L}_I^{(m-1)}$ 时，其极限值是存在的，并且等于

$$\begin{aligned}&\mathrm{V.\,P.}\,(V(\mathfrak{L}_I(m, n)))^{-1}\int_{\mathfrak{L}_I(m,n)}\varphi(U)\det(I-Q\bar{U}')^{-n}\dot{U}\\&+\frac{1}{2}\det(I-Z_1\bar{Z}_1')^{-1}(V(\mathfrak{L}_I(m-1, n-1)))^{-1}\cdot\\&\cdot\int_{\mathfrak{L}_I(m-1,n-1)}\varphi\left(U_0'\begin{pmatrix}1 & 0\\ 0 & U_1\end{pmatrix}V_0\right)\det(I-U_1\bar{Z}_1')\det(I-Z_1\bar{U}_1')^{1-n}\dot{U}_1,\end{aligned} \tag{3.4.2}$$

其中 V. P. 之意义为

$$\lim_{\varepsilon\to 0}(V(\mathfrak{L}_I(m, n)))^{-1}\int_{\mathfrak{L}_I(m,n,\varepsilon)}\varphi(U)\det(I-Q\bar{U}')^{-n}\dot{U}, \tag{3.4.3}$$

这里 $\mathfrak{L}_I(m, n, \varepsilon)$ 表示点集 $\{U\,|\,U\in\mathfrak{L}_I(m, n),\ |\det(I-Q\bar{U}')|\geqslant\varepsilon|\det(I-T_0\bar{U}')|\}$.

証. (i) 先设 $Q=\begin{pmatrix}1 & 0\\ 0 & O^{(m-1,n-1)}\end{pmatrix}$，取 $Z=\begin{pmatrix}\rho & 0\\ 0 & O^{(m-1,n-1)}\end{pmatrix}$，$0<\rho<1$，令

$$g(u)=(V(\mathfrak{L}_I(m-1, n-1)))^{-1}\int_{W\bar{W}'=I}\varphi\begin{pmatrix}u\\ (0W)\bar{P}'\end{pmatrix}\dot{W},$$

其中 P 为引理 3.3.1 所述. 则

$$F(Z)=\varpi_{2n-1}^{-1}\int_{u\bar{u}'=1}g(u)(1-\rho\bar{u}')^{-n}\dot{u},$$

此处 $u=(u_1, u_2, \cdots, u_n)$，$\varpi_{2n-1}$ 表示 $u\bar{u}'=1$ 之体积，根据引理 3.3.1，$g(u)$ 在 $u\bar{u}'=1$ 上属于 $\mathrm{Lip}\,\alpha$，应用定理 1.3[5]，便有

$$\begin{aligned}\lim_{Z\to Q}F(Z)&=\lim_{\rho\to 1}\varpi_{2n-1}^{-1}\int_{u\bar{u}'=1}g(u)(1-\rho\bar{u}_1)^{-n}\dot{u}\\&=\mathrm{V.\,P.}\,\varpi_{2n-1}^{-1}\int_{u\bar{u}'=1}g(u)(1-\bar{u}_1)^{-n}\dot{u}+\frac{1}{2}g(1, 0, \cdots, 0)\\&=\lim_{\varepsilon\to 0}\varpi_{2n-1}^{-1}\int_{\substack{u\bar{u}'=1\\|1-\bar{u}_1|\geqslant\varepsilon}}g(u)(1-\bar{u}_1)^{-n}\dot{u}\\&\quad+\frac{1}{2}(V(\mathfrak{L}_I(m-1, n-1)))^{-1}\int_{\mathfrak{L}_I(m-1,n-1)}f\left(\begin{pmatrix}1 & 0\\ 0 & U_1\end{pmatrix}\right)\dot{U}_1\\&=\lim_{\varepsilon\to 0}(V(\mathfrak{L}_I(m, n)))^{-1}\int_{\substack{\mathfrak{L}_I(m,n)\\|\det(I-Q\bar{U}')|\geqslant\varepsilon}}\varphi(U)\det(I-Q\bar{U}')^{-n}\dot{U}\\&\quad+\frac{1}{2}(V(\mathfrak{L}_I(m, n)))^{-1}\int_{\mathfrak{L}_I(m-1,n-1)}\varphi\left(\begin{pmatrix}1 & 0\\ 0 & U_1\end{pmatrix}\right)\dot{U}_1\\&=\mathrm{V.\,P.}\,(V(\mathfrak{L}_I(m, n)))^{-1}\int_{\mathfrak{L}_I(m,n)}\varphi(U)\det(I-Q\bar{U}')^{-n}\dot{U}\\&\quad+\frac{1}{2}(V(\mathfrak{L}_I(m-1, n-1)))^{-1}\int_{\mathfrak{L}_I(m-1,n-1)}\varphi\left(\begin{pmatrix}1 & 0\\ 0 & U_1\end{pmatrix}\right)\dot{U}_1.\end{aligned}$$

(iii) 在 $\mathfrak{L}_I^{(m-1)}$ 上任取一点 $Q = U_0'\begin{pmatrix}1 & 0\\ 0 & Z_1\end{pmatrix}V_0$, $Z_1\bar{Z}_1' < I^{(m-1)}$. 令 $Z = U_0'\begin{pmatrix}\rho & 0\\ 0 & Z_1\end{pmatrix}V_0$ 趋于 Q. 先作 $\mathfrak{R}_I$ 的运动羣 Γ^I 中的变换 $W = \Phi(Z)$: $W = A(Z - T_0)(I - \bar{T}_0'Z)^{-1}D^{-1}$, 将 $T_0 = U_0'\begin{pmatrix}0 & 0\\ 0 & Z_1\end{pmatrix}V_0$ 变至原点, 其中 $A^{(m)} = \begin{pmatrix}1 & 0\\ 0 & A_1\end{pmatrix}\bar{U}_0$, $D^{(n)} = \begin{pmatrix}1 & 0\\ 0 & D_1\end{pmatrix}V_0$, A_1, D_1 分别满足 $\bar{A}_1'A_1 = (I - Z_1\bar{Z}_1')^{-1}$, $\bar{D}_1'D_1 = (I - \bar{Z}_1'Z_1)^{-1}$. 相应地, $W = \Phi(Z)$ 将已与之点 Z 变为 $W = \begin{pmatrix}\rho & 0\\ 0 & 0\end{pmatrix}$, 将 Q 变为 $Q_0 = \begin{pmatrix}1 & 0\\ 0 & 0\end{pmatrix}$, 其逆变换认为 $Z = \Phi^{-1}(W)$, 在 $\mathfrak{L}_I(m, n)$ 上即为 $U = \Phi^{-1}(V)$, 利用熟知的关系式

$$\det(I - W\bar{V}')^{-1}\det(I - Z\bar{U}') = \det(I - T_0\bar{T}_0')^{-1}\det(I - T_0\bar{U}')\det(I - Z\bar{T}_0')$$

及

$$\dot{V} = \det(I - T_0\bar{T}_0')^n|\det(I - T_0\bar{U}')|^{-2n}\dot{U}$$

可得

$$\det(I - W\bar{V}')^{-n}\dot{V} = \det(I - Z\bar{U}')^{-n}(\det(I - Z\bar{T}_0')\det(I - U\bar{T}_0')^{-1})^n\dot{U},$$

于是

$$\lim_{Z\to Q}F(Z) = \lim_{W\to Q_0}\det(I - \Phi^{-1}(W)\bar{T}_0')^{-n}V(\mathfrak{L}_I(m, n))^{-1}\int_{\mathfrak{L}_I(m,n)}g(V)\det(I - W\bar{V}')^{-n}\dot{V},$$

其中 $g(V) = f(\Phi^{-1}(V))(\det(I - \Phi^{-1}(V)\bar{T}_0')^n$. 显然 $g(V)$ 在 $\mathfrak{L}_I(m, n)$ 上属于 Lip α. 由 (i) 便知

$$\lim_{Z\to Q}F(Z) = \det(I - \Phi^{-1}(Q_0)\bar{T}_0')^{-1}\Big[\text{V. P. }V(\mathfrak{L}_I(m, n))^{-1}\int_{\mathfrak{L}_I(m,n)}g(V)\det(I - Q_0\bar{V}')^{-n}\dot{V}$$

$$+\frac{1}{2}V(\mathfrak{L}_I(m-1, n-1))^{-1}\int_{\mathfrak{L}_I(m-1,n-1)}g\left(\begin{pmatrix}1 & 0\\ 0 & V_1\end{pmatrix}\right)\dot{V}_1\Big] = J_1 + J_2,$$

其中

$$J_1 = \lim_{\varepsilon\to 0}V(\mathfrak{L}_I(m, n))^{-1}\int_{\substack{\mathfrak{L}_I(m,n)\\ |\det(I - Q_0\bar{V}')|\geqslant\varepsilon}}\varphi(\Phi^{-1}(V))\det(I - \Phi^{-1}(V)\bar{T}_0')^n\cdot$$

$$\cdot[\det(I - Q_0\bar{V}')\det(I - \Phi^{-1}(Q_0)\bar{T}_0')]^{-n}\dot{V}$$

$$= \lim_{\varepsilon\to 0}V(\mathfrak{L}_I(m, n))^{-1}\int_{\mathfrak{L}_I(m,n,\varepsilon)}\varphi(U)\det(I - Q\bar{U}')^{-n}\dot{U},$$

取之定义为

$$\text{V. P. }V(\mathfrak{L}_I(m, n))^{-1}\int_{\mathfrak{L}_I(m,n)}\varphi(U)\det(I - Q\bar{U}')^{-n}\dot{U},$$

关于积分区域之确定, 用到如下事实: 将所取之 Z, W 及 T_0 代入

$$\det(I - W\bar{V}') = \det(I - Z\bar{U}')\det(I - T_0\bar{T}_0')\det(I - T_0\bar{U}')^{-1}\det(I - Z\bar{T}_0')^{-1},$$

注意到 $\det(I - T_0\bar{T}_0') = \det(I - Z\bar{T}_0')$, 并令 Z 趋于 Q, 相应地, W 趋于 Q_0, 因而对 V 的积分区域 $|\det(I - Q_0\bar{V}')| \geqslant \varepsilon$, 经变换 $V = \Phi(U)$, 则变为对 U 的积分区域 $|\det(I - Q\bar{U}')| \geqslant \varepsilon|\det(I - T_0\bar{U}')|$.

再看 J_2, 由于在变换 $W = \Phi(Z)$ 下, $\mathfrak{L}_I(m, n)$ 上的点 $U_0'\begin{pmatrix}1 & 0\\ 0 & U_1\end{pmatrix}V_0$ 变为 $\begin{pmatrix}1 & 0\\ 0 & V_1\end{pmatrix}$, 且有 $V_1 = A_1(U_1 - Z_1)(I - \bar{Z}_1'U_1)^{-1}D_1^{-1}$, 因而

$$\dot{V}_1 = \det(I - Z_1\bar{Z}_1')^{n-1}|\det(I - Z_1\bar{U}_1')|^{-2(n-1)}\dot{U}_1,$$

所以

$$J_2=(2V(\mathfrak{L}_I(m-1,\,n-1)))^{-1}\int_{\mathfrak{L}_I(m-1,\,n-1)}\varphi\left(\Phi^{-1}\left(\begin{pmatrix}1&0\\0&V_1\end{pmatrix}\right)\right)\cdot$$

$$\cdot\det\left(I-\Phi^{-1}\left(\begin{pmatrix}1&0\\0&V_1\end{pmatrix}\bar{T}_0'\right)\right)^n\det(I-\Phi^{-1}(Q_0)\bar{T}_0')^{-n}\dot{V}_1$$

$$=(2V(\mathfrak{L}_I(m-1,\,n-1)))^{-1}\int_{\mathfrak{L}_I(m-1,\,n-1)}\varphi\left(U_0'\begin{pmatrix}1&0\\0&U_1\end{pmatrix}V_0\right)\det(I-U_1\bar{Z}_1')^n\cdot$$

$$\cdot\det(I-Z_1\bar{Z}_1')^{-n}\det(I-Z_1\bar{Z}_1')^{n-1}|\det(I-Z_1\bar{U}_1')|^{-2(n-1)}\dot{U}_1$$

$$=(2\det(I-Z_1\bar{Z}_1'))^{-1}(V(\mathfrak{L}_I(m-1,\,n-1)))^{-1}\int_{\mathfrak{L}_I(m-1,n-1)}\varphi\left(U_0'\begin{pmatrix}1&0\\0&U_1\end{pmatrix}V_0\right)\cdot$$

$$\cdot\det(I-U_1\bar{Z}_1')\cdot\det(I-Z_1\bar{U}_1')^{1-n}\dot{U}_1$$

综合上面所证便得到公式(3.4.2).

应用定理 3.4.1，我们还可得到

定理 3.4.2. 若 $\varphi(U)$ 在 $\mathfrak{L}_I(n)$ 上属于 $\mathrm{Lip}\,\alpha\ (0<\alpha\leqslant 1)$，则 $\mathfrak{R}_I^*(n)$ 的 Cauchy 型积分

$$F(Z)=(V(\mathfrak{L}_I(n)))^{-1}\int_{\mathfrak{L}_I(n)}\varphi(U)\det(I-Z\bar{U}')^{-n}\dot{U},\tag{3.4.4}$$

当 $Z=U_0'\begin{pmatrix}\rho&0\\0&Z_1\end{pmatrix}V_0\in\mathfrak{R}_I^*(n)\ (\rho>1)$ 趋于 $\mathfrak{L}_I^{*(n-1)}$ 上的点 $Q^*=U_0'\begin{pmatrix}1&0\\0&Z_1\end{pmatrix}V_0$ 时的极限值是存在的，并且等于

$$\mathrm{V.\,P.}\,(V(\mathfrak{L}_I(n)))^{-1}\int_{\mathfrak{L}_I(n)}\varphi(U)\det(I-Q^*\bar{U}')^{-n}\dot{U}$$

$$-\frac{1}{2}\det(I-Z_1\bar{Z}_1')^{-1}(V(\mathfrak{L}_I(n-1)))^{-1}\int_{\mathfrak{L}_I(n-1)}\varphi\left(U_0'\begin{pmatrix}1&0\\0&U_1\end{pmatrix}V_0\right)\det(I-U_1\bar{Z}_1')$$

$$\times\det(I-Z_1\bar{U}_1')^{1-n}\dot{V}_1,\tag{3.4.5}$$

此处 V. P. 之意义为

$$\lim_{\varepsilon\to 0}(V(\mathfrak{L}_I(n)))^{-1}\int_{\mathfrak{L}_I^*(n,\varepsilon)}\varphi(U)\det(I-Q^*\bar{U}')^{-n}\dot{U},\tag{3.4.6}$$

其中 $\mathfrak{L}_I^*(n,\varepsilon)$ 表示点集 $\{U\,|\,U\in\mathfrak{L}_I(n),\ |\det(Q^*-U)|\geqslant\varepsilon|\det(Q^*-Q_0^*U)|\}$，而 $Q_0^*=U_0'\begin{pmatrix}0&0\\0&I^{(n-1)}\end{pmatrix}\bar{U}_0$.

上述定理的证明是容易的，令 $Z^{-1}=W$，相应地 $U^{-1}=V$，记 $\phi(V)=\varphi(V^{-1})\det V^{-n}$，则因 $\dot{U}=\dot{V}$，便得到

$$G(W)=F(W^{-1})=(-1)^{n^2}\det W^n(V(\mathfrak{L}_I(n)))^{-1}\int_{\mathfrak{L}_I(n)}\phi(V)\det(I-W\bar{V}')^{-n}\dot{V}.\tag{3.4.7}$$

由于 $\phi(V)$ 满足定理 3.4.1 中的条件，当 $Z\to Q^*$ 时，$W=\bar{V}_0'\begin{pmatrix}\lambda&0\\0&W_1\end{pmatrix}\bar{U}_0\in\mathfrak{R}_I(n)$ 趋于 $\mathfrak{L}_I^{(n-1)}$ 的点 $Q=Q^{*-1}=\bar{V}_0'\begin{pmatrix}1&0\\0&W_1\end{pmatrix}\bar{U}_0$，其中 $0<\lambda=\dfrac{1}{\rho}<1$，$W_1=Z_1^{-1}$，于是应用定理 3.4.1 以及变换 $Z^{-1}=W$，$U^{-1}=V$，即可得到公式(3.4.5).

§ 3.5. $\mathfrak{R}_I(m,n)$ 内的 $\mathfrak{B}$-调和函数的边界性质

可以应用定理 3.4.1 来研究 $\mathfrak{R}_I(m,n)$ 的 $\mathfrak{B}$-调和函数的某些边界性质. 我们有

定理 3.5.1. 设 $f(Z)=u(Z)+iv(Z)$ 在 $\mathfrak{R}_I(m,n)$ 内解析,在 $\mathfrak{R}_I(m,n)+\mathfrak{L}_I(m,n)$ 上连续, 并且 $u(U)$ 在 $\mathfrak{L}_I(m,n)$ 上属于 $\mathrm{Lip}\,\alpha(0<\alpha\leqslant 1)$, 则对于 $\mathfrak{L}_I^{(m-1)}$ 上任一点 $Q=U_0'\begin{pmatrix}1&0\\0&Z_1\end{pmatrix}V_0$, $Z_1\bar Z_1'<I^{(m-1)}$, 有

$$v(Q)=\mathrm{V.P.}\,(V(\mathfrak{L}_I(m,n)))^{-1}\int_{\mathfrak{L}_I(m,n)}u(U)i^{-1}(H(Q,\bar U)-H(U,\bar Q'))\dot U$$

$$+\frac{1}{2}(\det(I-Z_1\bar Z_1')V(\mathfrak{L}_I(m-1,n-1)))^{-1}\cdot$$

$$\cdot\int_{\mathfrak{L}_I(m-1,n-1)}u\left(U_0'\begin{pmatrix}1&0\\0&U_1\end{pmatrix}V_0\right)i^{-1}(T(Z_1,\bar U_1)-T(U_1,\bar Z_1))\dot U_1+v(0),\quad(3.5.1)$$

其中 $H(Q,\bar U)=\det(I-Q\bar U')^{-n}$,

$$T(Z_1,\bar U_1)=\det(I-U_1\bar Z_1')\det(I-Z_1\bar U_1')^{1-n}.$$

证. 利用 $\mathfrak{R}_I(m,n)$ 内的解析函数的 Schwarz 公式[7]

$$f(Z)=(V(\mathfrak{L}_I(m,n)))^{-1}\int_{\mathfrak{L}_I(m,n)}u(U)S(Z,U)\dot U+iv(0),\quad(3.5.2)$$

其中 Schwarz 核 $S(Z,U)=2\det(I-Z\bar U')^{-n}-1$. 令 $Z=U_0'\begin{pmatrix}\rho&0\\0&Z_1\end{pmatrix}V_0$ 趋于 Q, $\lim\limits_{Z\to Q}f(Z)=f(Q)$ 是存在的[4], 而 (3.5.2) 式之右端的极限值, 可由定理 3.4.1 得出, 于是有

$$f(Q)=\mathrm{V.P.}\,2\cdot(V(\mathfrak{L}_I(m,n)))^{-1}\int_{\mathfrak{L}_I(m,n)}u(U)\det(I-Q\bar U')^{-n}\dot U$$

$$+(\det(I-Z_1\bar Z_1')V(\mathfrak{L}_I(m-1,n-1)))^{-1}\int_{\mathfrak{L}_I(m-1,n-1)}u\left(U_0'\begin{pmatrix}1&0\\0&U_1\end{pmatrix}V_0\right)$$

$$\times\det(I-U_1\bar Z_1')\det(I-Z_1\bar U_1')^{1-n}\dot U_1-u(0)+iv(0),\quad(3.5.3)$$

对上式两端取虚部,便得到公式(3.5.1).

利用变换 $Z^{-1}=W$, 容易得到

定理 3.5.2. 设 $f(Z)=u(Z)+iv(Z)$ 在 $\mathfrak{R}_I^*(n)$ 内解析,在 $\mathfrak{R}_I^*(n)+\mathfrak{L}_I(n)$ 上连续,并且 $u(U)$ 在 $\mathfrak{L}_I(n)$ 上属于 $\mathrm{Lip}\,\alpha\,(0<\alpha\leqslant 1)$, 则对于 $\mathfrak{L}_I^{*(n-1)}$ 上任一点 $Q^*=U_0'\begin{pmatrix}1&0\\0&Z_1\end{pmatrix}V_0$, $Z_1\bar Z_1'>I^{(n-1)}$ 有

$$v(Q^*)=\mathrm{V.P.}\,(V(\mathfrak{L}_I(n)))^{-1}\int_{\mathfrak{L}_I(n)}u(U)i^{-1}(H^*(Q^*,\bar U)-H^*(U,\bar Q^*))\dot U$$

$$-(2\det(I-Z_1\bar Z_1')V(\mathfrak{L}_I(n-1)))^{-1}\int_{\mathfrak{L}_I(n-1)}u\left(U_0'\begin{pmatrix}1&0\\0&U_1\end{pmatrix}V_0\right)i^{-1}(T^*(Z_1,\bar U_1)$$

$$-T^*(U_1,\bar Z_1)\dot U_1+v(Z_\infty),\quad(3.5.4)$$

其中

$$H^*(Q^*,\bar U)=(-1)^{n^2}\det(Q^*\bar U)^n\det(I-Q^*\bar U')^{-n},$$

$$T^*(Z_1, \bar{U}_1) = (-1)^{n^2} \det(Z_1\bar{U}_1) \det(I - U_1\bar{Z}_1') \det(I - Z_1\bar{U}_1')^{1-n}.$$

Z_∞ 表示 Grassmann 流形中，齐次坐标 $(Z_1, Z_2) = (O^{(n)}, I^{(n)})$ 的无穷远点（非齐次坐标 Z 与齐次坐标 (Z_1, Z_2) 的关系为 $Z = Z_1^{-1}Z_2$）.

参　考　文　献

[1] 华罗庚，多复变数函数论中的典型域的调和分析，科学出版社，1958.

[2] 龚　昇，多复变数典型域的 Cauchy 型积分 (I)，中国科学技术大学建校五周年纪念论文集，1963，23—25.

[3] 龚　昇，酉群上的富里埃分析 III，数学学报，**13** (1963)，152—161.

[4] 陆启铿，典型流形与典型域，上海科学技术出版社，1963.

[5] 龚　昇、孙继广，多复变数的 Cauchy 型积分 (I) 超球的 Cauchy 型积分，数学学报，**15** (1965)，431—443.

[6] 龚　昇、孙继广，多复变数的 Cauchy 型积分 (II) Lie 球双曲空间的 Cauchy 型积分，数学学报，**15** (1965)，775—799.

[7] 龚　昇、孙继广，圆型域中解析函数的积分表示，中国科学技术大学学报，**1** (1965)，219—226.

[8] Какичев, В. А., Граничные свойства интеграла типа Коши многих переменых, *Уч. Зап. Шахтинск*, педин-те, **2** (1959), 29—90.

第16卷 第2期 数学学报 Vol. 16, No. 2
1966年3月 ACTA MATHEMATICA SINICA March, 1966

复超球面上的奇异积分方程*

龚 升 孙继广
(中国科学技术大学) (中国科学院数学研究所)

§1. 引 言

一维的奇异积分方程的理论已经相当完整，并有着极为广泛和重要的应用(例如参阅 Н. И. Мусхелишвили[5], И. Н. Векуа[6], Ф. Д. Гахов[7], С. Г. Михлин[8]). 高维的奇异积分方程还未完整. С. Г. Михлин 写了一本关于高维奇异积分与奇异积分方程的书[9]，总结了近年来的发展及其影响，从中可以看出，高维奇异积分方程距离完整还有一定的距离.

首先研究高维的奇异积分方程的是 F. G. Tricomi，他所研究的是在二维欧氏空间中的奇异积分方程. 之后，G. Giraud, С. Г. Михлин, W. J. Trjitzinsky, J. Horwáth, A. P. Calderon, A. Zygmund, И. Н. Векуа, J. J. Kohn, R. T. Seely, A. B. Бицадзе, E. M. Stein 等等都获得不少有趣而重要的结果. 在这些工作中，积分大多是在整个欧氏空间中进行，奇异积分中的奇性是用二点之间的距离来刻划的. 所以要这样做有不少是有着实际的应用的原因. 我们现在从另一个角度来考虑高维的奇异积分方程.

在一维的奇异积分方程中，复变函数的 Cauchy 型积分起着十分重要的作用. 有了 Cauchy 型积分的工具，使得一维的奇异积分方程的处理显得十分简单与深刻. 可是在高维的奇异积分方程的研究中，据作者所看到的，还未有应用 Cauchy 型积分的工具来处理的. 本文就是想利用多复变数的 Cauchy 型积分的工具来处理高维的奇异积分方程及方程组的一个尝试.

本文的主要内容是应用我们以前研究的超球的 Cauchy 型积分[4]来处理复超球面上的奇异积分方程及方程组. 我们对于常系数的奇异积分方程及方程组，得到了比较完整的解决. 从处理过程中看来，使用这个工具，还是比较简便的.

本文的结果曾在中国科学技术大学数学系综合讨论班上报告过. 在工作过程中，自始至终的得到我们的导师华罗庚先生的关怀与鼓励，并给予了不少宝贵的指点，关肇直同志对本文也提供了不少宝贵的指点，我们向他们表示衷心的感谢.

在工作过程中，我们还得到了曾肯成同志，陆洪文同志的帮助，我们也向他们表示感谢.

在本文的§2中，我们证明了含有 Cauchy 核的复合奇性积分公式，在一个复变数时，这就是 Poincerè-Bertrand 公式，在§3中，我们得到了二个实的奇性核：h-核及 B-核，并讨论了它们的一些性质. h-核在一个变数时就是 Hilbert 核. 但 B-核在一个变数时是不

* 1964年6月8日收到.

出现的，这也许是显示了多个变数与一个变数时的根本性不同．在§3和§4中，我们给出了含有 B-核与 h-核的复合奇性积分公式，定理4.1的系1，在一个复变数时，就是Hilbert 公式，在本文中，我们讨论由在复超球面上满足 Lipschitz 条件的复值连续函数所组成的线性空间 $\mathscr{L}^*$ 和由在其上满足 Lipschitz 条件的实值连续函数所组成的线性空间 $\mathscr{L}$，这些核实际上是定义了在 $\mathscr{L}^*$ 上或 $\mathscr{L}$ 上的线性算子．通过线性算子的语言，我们在以下几节中，解了带有 Cauchy 核的常系数的奇异积分方程及方程组，以及带有 B-核和 h-核的常系数的奇异积分方程及方程组．

为了便于行文，我们把[4]中主要结果，简述如下：

在[4]中，我们研究了如下的 Cauchy 型积分

$$\varpi_{2n-1}^{-1}\int_{u\bar{u}'=1}(1-z\bar{u}')^{-n}f(u)\dot{u},$$

这里 $z=(z_1,\cdots,z_n)$，$z\bar{z}'<1$，$u=(u_1,\cdots,u_n)$，ϖ_{2n-1} 表示 $u\bar{u}'=1$ 的体积，$\dot{u}$ 表示 $u\bar{u}'=1$ 的体积元素，在[4]中，我们证明了：若 $f(u)\in \mathrm{Lip}\,\alpha\ (0<\alpha\leqslant 1)$，则当 z 沿着使

$$\frac{[(z-v)\overline{(z-v)'}]^{\frac{1}{2}}}{\min\limits_{u\bar{u}'=1}|1-z\bar{u}'|}<M\ (\text{常数})$$

的路径趋于 v 时，我们有

$$F^+(v)=\lim_{z\to v}\varpi_{2n-1}^{-1}\int_{u\bar{u}'=1}(1-z\bar{u}')^{-n}f(u)\dot{u}=\mathrm{V.P.}\varpi_{2n-1}^{-1}\int_{u\bar{u}'=1}(1-v\bar{u}')^{-n}f(u)\dot{u}+f(v)/2,$$

这里 $\mathrm{V.P.}\int_{u\bar{u}'=1}$ 表示 $\lim\limits_{\delta\to 0}\int_{|1-v\bar{u}'|>\delta}$（[4]中的定理1.3）．

此外，我们还证明了：①若 $f(u)\in\mathrm{Lip}\,\alpha$，则 $F^+(v)\in\mathrm{Lip}\,(\alpha-\delta)$，其中 δ 是小于 α 的任一正数（[4]定理1.4），②若 $\varphi(u,v)$ 为定义在 $u\bar{u}'=1$，$w\bar{w}'=1$ 上的连续函数，作为 u 的函数以及作为 w 的函数都满足 Lipschitz 条件，那末

$$\Phi(v,u)=\varpi_{2n-1}^{-1}\int_{w\bar{w}'=1}(1-v\bar{w}')^{-n}\varphi(u,w)\dot{w}$$

作为 u 的函数及 v 的函数，都满足 Lipschitz 条件（[4]定理1.5）．

§2. 含有 Cauchy 核的复合奇性积分公式

利用§1中所述的诸定理，容易证明下面的含有 Cauchy 核的复合奇性积分公式[5]．

定理2.1.[1)] 若 $\varphi(w)\in\mathrm{Lip}\,\alpha$，$0<\alpha\leqslant 1$，$(w\bar{w}'=1)$ 则我们有

$$\varpi_{2n-1}^{-2}\int_{u\bar{u}'=1}(1-v\bar{u}')^{-n}\dot{u}\int_{w\bar{w}'=1}(1-u\bar{w}')^{-n}\varphi(w)\dot{w}=\varphi(v)/4. \tag{2.1}$$

其中 $v\bar{v}'=1$，积分都是取 Cauchy 主值．

证．由[4]中定理1.1及1.3，我们可以在 Cauchy 主值的意义下，令

$$\varphi_1(v)=\varpi_{2n-1}^{-1}\int_{u\bar{u}'=1}(1-v\bar{u}')^{-n}\varphi(u)\dot{u},$$

1) 在一个变数时，此即著名的 Poincaré-Bertrand 公式．

$$\varphi_2(v)=\varpi_{2n-1}^{-1}\int_{u\bar{u}'=1}(1-v\bar{u}')^{-n}\varphi_1(u)\dot{u}.$$

作 Cauchy 型积分

$$f(z)=\varpi_{2n-1}^{-1}\int_{u\bar{u}'=1}(1-z\bar{u}')^{-n}\varphi(u)\dot{u},$$

$$f_1(z)=\varpi_{2n-1}^{-1}\int_{u\bar{u}'=1}(1-z\bar{u}')^{-n}\varphi_1(u)\dot{u},$$

其中 $z\bar{z}'<1$. 当 z 从 $z\bar{z}'<1$ 中趋于边界时,由[4]中定理 1.3 得到

$$f(v)=\varpi_{2n-1}^{-1}\int_{u\bar{u}'=1}(1-v\bar{u}')^{-n}\varphi(u)\dot{u}+\varphi(v)/2,$$

$$f_1(v)=\varpi_{2n-1}^{-1}\int_{u\bar{u}'=1}(1-v\bar{u}')^{-n}\varphi_1(u)\dot{u}+\varphi_1(v)/2.$$

由 $\varphi_1(v)$ 及 $\varphi_2(v)$ 的定义知

$$\varphi_1(v)=f(v)-\varphi(v)/2,\qquad \varphi_2(v)=f_1(v)-\varphi_1(v)/2,$$

将 $\varphi_1(v)$ 代入 $f_1(z)$ 的表达式中

$$\begin{aligned}f_1(z)&=\varpi_{2n-1}^{-1}\int_{u\bar{u}'=1}(1-z\bar{u}')^{-n}f(u)\dot{u}\\&\quad-(2\varpi_{2n-1})^{-1}\int_{u\bar{u}'=1}(1-z\bar{u}')^{-n}\varphi(u)\dot{u}\\&=f(z)-f(z)/2=f(z)/2,\end{aligned}$$

因而

$$\varphi_2(v)=f(v)/2-(f(v)-\varphi(v)/2)/2=\varphi(v)/4,$$

于是得到了公式(2.1).

公式(2.1)实际上是给出了一组反演公式

系. 若 $\varphi(u)\in \mathrm{Lip}\,\alpha$, 则由下面的(2.2)所定义的 $\psi(v)$ 也满足 Lipschitz 条件,且

$$2\varpi_{2n-1}^{-1}\int_{u\bar{u}'=1}(1-v\bar{u}')^{-n}\varphi(u)\dot{u}=\psi(v),\tag{2.2}$$

$$2\varpi_{2n-1}^{-1}\int_{u\bar{u}'=1}(1-v\bar{u}')^{-n}\psi(u)\dot{u}=\varphi(v)\tag{2.3}$$

相互推导.

§3. B-核与 h-核、B-型积分与 h-型积分

若 $f(z)=U(z)+iV(z)$ 在 $z\bar{z}'<1$ 内解析, 在 $z\bar{z}'\leqslant 1$ 上连续(事实上, 只要求 $f(z)\in H_1$ 已足够[3]),则根据圆型域内的解析函数的 Schwarz 公式[3],有

$$f(z)=\varpi_{2n-1}^{-1}\int_{u\bar{u}'=1}U(u)S(z,u)\dot{u}+iV(0).\tag{3.1}$$

其中 $S(z,u)=2H(z,\bar{u})-1$ 为 Schwarz 核,而 $H(z,\bar{u})=(1-z\bar{u}')^{-n}$ 即为超球的 Cauchy 核[1].

令

$$B(z,u)=\mathrm{Re}\,(S(z,u))=H(z,\bar{u})+H(u,\bar{z})-1,\tag{3.2}$$

$$h(z,u)=\mathrm{Im}\,(S(z,u))=(H(z,\bar{u})-H(u,\bar{z}))/i,\tag{3.3}$$

于是公式(3.1)也可以写成

$$U(z)=\varpi_{2n-1}^{-1}\int_{u\bar{u}'=1}U(u)B(z,u)\dot{u},\tag{3.4}$$

$$V(z)=\varpi_{2n-1}^{-1}\int_{u\bar{u}'=1}U(u)h(z,u)\dot{u}+V(0).\tag{3.5}$$

在 $n=1$ 的情形，核 $B(z,u)$ 与 $h(z,u)$ 分别为 Poisson 核与共轭 Poisson 核，但当 $n>1$ 时，$B(z,u)$ 与 $h(z,u)$ 只能分别称之为 Schwarz 核的实部与虚部，而不能称之为 Poisson 核与共轭 Poisson 核，因为它们已不再具有作为 Poisson 核与共轭 Poisson 核的一些重要性质了。

若 $U(u)\in\mathrm{Lip}\,\alpha(0<\alpha\leqslant 1)$，在(3.5)中令 z 沿着[4]中定理 1.3 中所要求的那样趋于边界上的一点 v，则由[4]中定理 1.3 可得

$$V(v)=\mathrm{V.P.}\varpi_{2n-1}^{-1}\int_{u\bar{u}'=1}U(u)h(v,u)\dot{u}+V(0),\tag{3.6}$$

其中

$$h(v,u)=(H(v,\bar{u})-H(u,\bar{v}))/i\tag{3.7}$$

称为 h-核，当 $n=1$ 时，它就是熟知的 Hilbert 核。

为了确定起见，不妨取 $V(0)=0$，则我们有

$$V(v)=\varpi_{2n-1}^{-1}\int_{u\bar{u}'=1}U(u)h(v,u)\dot{u}.\tag{3.8}$$

此处积分是按照 Cauchy 主值的意义下进行的。由(3.7)及[4]中定理 1.4，易知：若 $z\bar{z}'<1$ 内的 B-调和函数 $U(z)$ 的边界值 $U(u)\in\mathrm{Lip}\,\alpha$，则它的(唯一确定的)共轭 B-调和函数 $V(z)$ 的边界值 $V(u)$（如(3.8)所示）$\in\mathrm{Lip}\,(\alpha-\delta)$，$\delta$ 是小于 α 的任一正数。

定义. 如果在 $u\bar{u}'=1$ 上任与一个可积实值函数 $\varphi(u)$，我们称

$$\Phi(z)=\varpi_{2n-1}^{-1}\int_{u\bar{u}'=1}\varphi(u)B(z,u)\dot{u}\tag{3.9}$$

为 B-型积分，称

$$\Psi(z)=\varpi_{2n-1}^{-1}\int_{u\bar{u}'=1}\varphi(u)h(z,u)\dot{u}\tag{3.10}$$

为 h-型积分。

(3.9)及(3.10)定义了一对在 $z\bar{z}'<1$ 内共轭的 B-调和函数。自然，$\Phi(z)$、$\Psi(z)$ 在边界上的极限值，当 $n>1$ 时未必存在，即使存在了，也未必是 $\varphi(v)$ 及 $\varpi_{2n-1}^{-1}\int_{u\bar{u}'=1}\varphi(u)h(v,u)\dot{u}$。当 $n>1$ 时，$\varphi(v)$ 即使是连续的，或是满足 Lipschitz 条件，或是有更強一些的要求，$\Phi(z)$ 的极限值也不一定是 $\varphi(v)$，这显示出多复变数与单复变数在这时候有着根本性的不同。

应用[4]中定理 1.3，我们可以得到

定理 3.1. 若 $\varphi(u)\in\mathrm{Lip}\,\alpha$，那末，由(3.9)及(3.10)所定义的 $\Phi(z)$ 与 $\Psi(z)$ 在边界上的极限值存在，并且有

$$\Phi(v)=\mathrm{V.P.}\varpi_{2n-1}^{-1}\int_{u\bar{u}'=1}\varphi(u)B(v,u)\dot{u}+\varphi(v),\tag{3.11}$$

及

$$\Psi(v)=\text{V.P.}\,\varpi_{2n-1}^{-1}\int_{u\bar{u}'=1}\varphi(u)h(v,u)\dot{u}. \tag{3.12}$$

(3.11)中的

$$B(v,u)=H(v,\bar{u})+H(u,\bar{v})-1 \tag{3.13}$$

称为 B-核.

証. 事实上,如果 z 从 $z\bar{z}'<1$ 内沿着[4]中定理 1.3 中所要求的那样趋于 v, 则

$$\begin{aligned}\Phi(v)&=\lim_{z\to v}\varpi_{2n-1}^{-1}\int_{u\bar{u}'=1}\varphi(u)(H(z,\bar{u})+H(u,\bar{z})-1)\dot{u}\\&=\text{V.P.}\,\varpi_{2n-1}^{-1}\int_{u\bar{u}'=1}\varphi(u)H(v,\bar{u})\dot{u}+\frac{1}{2}\varphi(v)\\&\quad+\text{V.P.}\,\varpi_{2n-1}^{-1}\int_{u\bar{u}'=1}\varphi(u)H(v,\bar{u})\dot{u}+\frac{1}{2}\varphi(v)-\varpi_{2n-1}^{-1}\int_{u\bar{u}'=1}\varphi(u)\dot{u}\\&=\text{V.P.}\,\varpi_{2n-1}^{-1}\int_{u\bar{u}'=1}\varphi(u)(H(v,\bar{u})+H(u,\bar{v})-1)\dot{u}+\varphi(v)\\&=\text{V.P.}\,\varpi_{2n-1}^{-1}\int_{u\bar{u}'=1}\varphi(u)B(v,u)\dot{u}+\varphi(v).\end{aligned}$$

同理可证(3.12).

从公式(3.11)与(3.12),可以得到一系列重要推论

系 1. 若 $\varphi(u)\in\mathscr{L}$, 则 $\varphi(u)$ 是 $z\bar{z}'<1$ 内某一 B-调和函数的边界值的充分与必要条件是

$$\text{V.P.}\int_{u\bar{u}'=1}\varphi(u)B(v,u)\dot{u}=0. \tag{3.14}$$

公式(3.11)及(3.12)表明: 若 $\varphi(u)\in\mathscr{L}$, 则 $\varphi(v)+\varpi_{2n-1}^{-1}\int_{u\bar{u}'=1}\varphi(u)B(v,u)\dot{u}$ 及 $\varpi_{2n-1}^{-1}\int_{u\bar{u}'=1}\varphi(u)h(v,u)\dot{u}$ 必分别为 $z\bar{z}'<1$ 内的二个 B-调和函数的边界值, 因而利用系 1 可得

系 2. 若 $\varphi(u)\in\mathscr{L}$, 则

$$\varpi_{2n-1}^{-2}\int_{u\bar{u}'=1}B(v,u)\dot{u}\int_{w\bar{w}'=1}\varphi(w)B(u,w)\dot{w}+\varpi_{2n-1}^{-1}\int_{u\bar{u}'=1}\varphi(u)B(v,u)\dot{u}=0, \tag{3.15}$$

$$\varpi_{2n-1}^{-2}\int_{u\bar{u}'=1}B(v,u)\dot{u}\int_{w\bar{w}'=1}\varphi(w)h(u,w)\dot{w}=0. \tag{3.16}$$

同时,利用(3.11)及(3.12)容易得到

系 3. 若 $\varphi(u)\in\mathscr{L}$, 则

$$\varpi_{2n-1}^{-2}\int_{u\bar{u}'=1}h(v,u)\dot{u}\int_{w\bar{w}'=1}\varphi(w)B(u,w)\dot{w}=0. \tag{3.17}$$

事实上,(3.17)的左端等于

$$\begin{aligned}&\varpi_{2n-1}^{-1}\int_{u\bar{u}'=1}(\Phi(u)-\varphi(u))h(v,u)\dot{u}\\&=\varpi_{2n-1}^{-1}\int_{u\bar{u}'=1}\Phi(u)h(v,u)\dot{u}-\varpi_{2n-1}^{-1}\int_{u\bar{u}'=1}\varphi(u)h(v,u)\dot{u}\end{aligned}$$

$$= \Psi(v) - \omega_{2n-1}^{-1}\int_{u\bar{u}'=1} \varphi(u)h(v,u)\dot{u} = 0.$$

系 4. 若 $\varphi(u)\in\mathscr{L}$，则

$$\omega_{2n-1}^{-2}\int_{v\bar{v}'=1}\dot{v}\int_{u\bar{u}'=1}\varphi(u)B(v,u)\dot{u} = 0, \tag{3.18}$$

$$\omega_{2n-1}^{-2}\int_{v\bar{v}'=1}\dot{v}\int_{u\bar{u}'=1}\varphi(u)h(v,u)\dot{u} = 0. \tag{3.19}$$

事实上，(3.18)式之左端等于

$$\omega_{2n-1}^{-1}\int_{v\bar{v}'=1}(\Phi(v)-\varphi(v))\dot{v} = \Phi(0) - \omega_{2n-1}^{-1}\int_{v\bar{v}'=1}\varphi(v)\dot{v}$$

$$= \omega_{2n-1}^{-1}\int_{v\bar{v}'=1}\varphi(v)B(0,v)\dot{v} - \omega_{2n-1}^{-1}\int_{v\bar{v}'=1}\varphi(v)\dot{v} = 0.$$

同理可得(3.19).

在多复变函数论中，一般来说，以下的 Dirichlet 问题

$$\begin{cases}\dfrac{\partial^2 u(z_1,\cdots,z_n)}{\partial z_k\partial\bar{z}_l} = 0 \qquad k,l=1,2,\cdots,n\\ u(z_1,\cdots,z_n)\big|_{\substack{z=u\\ u\bar{u}'=1}} = \varphi(u)\end{cases} \tag{3.20}$$

是无解的[13]. 但从系 1 我们得到

系 5. 若 $\varphi(u)\in\mathscr{L}$，则由(3.20)所定义的 Dirichlet 问题有解的充分与必要条件为

$$\int_{u\bar{u}'=1}\varphi(u)B(v,u)\dot{u} = 0,$$

若这条件满足时，(3.20)有唯一解，并且它的解就是

$$u(z) = \omega_{2n-1}^{-1}\int_{u\bar{u}'=1}\varphi(u)B(z,u)\dot{u}.$$

§4. 含有 h-核的复合奇性积分公式

我们在 §3 中已经得到了一系列含有 B-核与 h-核的复合奇性积分公式. 此外，我们还有

定理 4.1. 若 $\varphi(u)\in\mathscr{L}$，则

$$\omega_{2n-1}^{-2}\int_{u\bar{u}'=1}h(v,u)\dot{u}\int_{w\bar{w}'=1}\varphi(w)h(u,w)\dot{w}$$

$$= -\varphi(v) - \omega_{2n-1}^{-1}\int_{u\bar{u}'=1}\varphi(u)B(v,u)\dot{u} + \omega_{2n-1}^{-1}\int_{u\bar{u}'=1}\varphi(u)\dot{u}. \tag{4.1}$$

证. 由于

$$\frac{2}{(1-v\bar{u}')^n} = B(v,u) + ih(v,u) + 1, \tag{4.2}$$

根据定理 2.1 和 §3 中的结果，便有

$$\varphi(v) = 4\omega_{2n-1}^{-2}\int_{u\bar{u}'=1}(1-v\bar{u}')^{-n}\dot{u}\int_{w\bar{w}'=1}(1-u\bar{w}')^{-n}\varphi(w)\dot{w}$$

$$= \omega_{2n-1}^{-2}\int_{u\bar{u}'=1}(B(v,u)+1+ih(v,u))\dot{u}\int_{w\bar{w}'=1}\varphi(w)(B(u,w)+1+ih(u,w))\dot{w}$$

$$= -\varpi_{2n-1}^{-1}\int_{u\bar{u}'=1}\varphi(u)B(v,u)\dot{u} + \varpi_{2n-1}^{-1}\int_{u\bar{u}'=1}\varphi(u)\dot{u}$$
$$-\varpi_{2n-1}^{-2}\int_{u\bar{u}'=1}h(v,u)\dot{u}\int_{w\bar{w}'=1}\varphi(w)h(u,w)\dot{w},$$

此即公式(4.1).

从定理 4.1,立即可以得出

系 1[1]. 若 $U(z)$ 为 $z\bar{z}'<1$ 内的 B-调和函数,具有边界值 $U(u)$,且 $U(u)\in\mathscr{L}$,则

$$\varpi_{2n-1}^{-2}\int_{u\bar{u}'=1}h(v,u)\dot{u}\int_{w\bar{w}'=1}U(w)h(u,w)\dot{w}$$
$$= -U(v) + \varpi_{2n-1}^{-1}\int_{u\bar{u}'=1}U(u)\dot{u}. \tag{4.3}$$

前已指出,当 $n=1$ 时,核 $h(v,u)$ 就是 Hilbert 核 $h(\tau,\theta)=\operatorname{ctg}\frac{\tau-\theta}{2}$. 易知,$h(\tau,\theta)$ 有如下的重要性质: 设 $\varphi(\theta)$ 是在 $|z|=1$ 上满足 Lipschitz 条件的实值函数,则 $\int_0^{2\pi}\varphi(\theta)h(\tau,\theta)d\theta=0$ 的充分与必要条件是 $\varphi(\theta)=\text{const}$. 但当 $n>1$ 时,如果 $\varphi(u)\in\mathscr{L}$,且 $\int_{u\bar{u}'=1}\varphi(u)h(v,u)\dot{u}=0$,$\varphi(u)$ 未必是常数. 这是因为任取 $\psi(u)\in\mathscr{L}$,令 $\varphi(u)=\varpi_{2n-1}^{-1}\int_{w\bar{w}'=1}\psi(w)B(u,w)\dot{w}$,则必有 $\int_{u\bar{u}'=1}\varphi(u)h(v,u)\dot{u}=0$ (据前节系 3). 例如取 $\psi(u)=u_1\bar{u}_2+\bar{u}_1u_2$,记 $\varpi_{2n-1}^{-1}\int_{u\bar{u}'=1}(u_1\bar{u}_2+\bar{u}_1u_2)\dot{u}=c$,则相应的 $\varphi(u)=c-u_1\bar{u}_2-\bar{u}_1u_2$,它在 $u\bar{u}'=1$ 上并非常数,然而 $\int_{u\bar{u}'=1}\varphi(u)h(v,u)\dot{u}=0$. 不过,我们可以从系 1 得到

系 2. 若 $\varphi(u)\in\mathscr{L}$,并且是 $z\bar{z}'<1$ 内某一 B-调和函数的边界值,则 $\int_{u\bar{u}'=1}\varphi(u)\cdot h(v,u)\dot{u}=0$ 的充分与必要条件是 $\varphi(u)=\text{const}$.

充分性是显然的. 必要性可由

$$0=\varpi_{2n-1}^{-2}\int_{u\bar{u}'=1}h(v,u)\dot{u}\int_{w\bar{w}'=1}\varphi(w)h(u,w)\dot{w}=-\varphi(v)+\varpi_{2n-1}^{-1}\int_{u\bar{u}'=1}\varphi(u)\dot{u}$$

立刻得到.

§5. 算子 H、B、h

在 §1 中,我们已经定义了线性空间 $\mathscr{L}^*$ 和 $\mathscr{L}$. 现在定义算子 H、B、h 和 $\dot{A}$、I、O 如下:

$$H\varphi = 2\varpi_{2n-1}^{-1}\int_{u\bar{u}'=1}\varphi(u)H(v,\bar{u})\dot{u} \quad (\varphi\in\mathscr{L}^*), \tag{5.1}$$

$$B\varphi = \varpi_{2n-1}^{-1}\int_{u\bar{u}'=1}\varphi(u)B(v,u)\dot{u} \quad (\varphi\in\mathscr{L}), \tag{5.2}$$

$$h\varphi = \varpi_{2n-1}^{-1}\int_{u\bar{u}'=1}\varphi(u)h(v,u)\dot{u} \quad (\varphi\in\mathscr{L}), \tag{5.3}$$

---

1) 当 $n=1$ 时,此即著名的 Hilbert 公式.

$$A\varphi = \omega_{2n-1}^{-1}\int_{u\bar{u}'=1}\varphi(u)\dot{u} \qquad (\varphi \in \mathscr{L}^*), \tag{5.4}$$

$$I\varphi = \varphi \qquad (\varphi \in \mathscr{L}^*), \tag{5.5}$$

$$O\varphi = 0 \qquad (\varphi \in \mathscr{L}^*). \tag{5.6}$$

显然，它们是相应的线性空间 $\mathscr{L}^*$ 或 $\mathscr{L}$ 上的线性算子.（参阅[1]）

所有在 $u\bar{u}'=1$ 上满足 Lipschitz 条件，并且是 $z\bar{z}'<1$ 内某一 B-调和函数的边界值的实值函数的全体，显然构成实数域上的一个线性空间，并且是 $\mathscr{L}$ 的一个子空间，记之为 $\mathscr{B}$.

利用线性算子的符号，将 § 2、§ 3 和 § 4 中的结果，列表如下：

$$H^2 = I \qquad (\text{在 } \mathscr{L}^* \text{ 上}) \tag{I}$$

$$B^2 = -B \qquad (\text{在 } \mathscr{L} \text{ 上}) \tag{II}$$

$$h^2 = -I - B + A \qquad (\text{在 } \mathscr{L} \text{ 上}) \tag{III}$$

$$A^2 = A \qquad (\text{在 } \mathscr{L}^* \text{ 上}). \tag{IV}$$

在 $\mathscr{L}$ 上有

$$AB = BA = 0 \tag{V}$$

$$Ah = hA = 0 \tag{VI}$$

$$Bh = hB = 0. \tag{VII}$$

设 $\varphi \in \mathscr{L}$，则

$$B\varphi = 0 \Longleftrightarrow \varphi \in \mathscr{B}, \tag{VIII}$$

设 $\varphi \in \mathscr{B}$，则

$$h\varphi = 0 \Longleftrightarrow \varphi = \text{const}. \tag{IX}$$

在以下几节中，我们将利用这些公式，求解复超球面上的，常系数的奇异积分方程和方程组.

§ 6. 有 Cauchy 核的奇异积分方程

本节考虑方程

$$a\varphi + bH\varphi + K\varphi = f, \tag{6.1}$$

其中 a, b 为复数，且满足 $a^2 - b^2 \neq 0$，已给 $f \in \mathscr{L}^*$，$K\varphi = \omega_{2n-1}^{-1}\int_{u\bar{u}'=1}\varphi(u)K(v,u)\dot{u}$，核 $K(v,u)$ 是对 v、u 皆满足 Lipschitz 条件的复值函数.

6.1. 特征方程.

$$a\varphi + bH\varphi = f \tag{6.2}$$

叫做方程(6.1)的特征奇异方程.

用算子

$$M = aI - bH \tag{6.3}$$

作用于(6.2)式的两端，应用公式(I)，得到

$$(a^2 - b^2)\varphi = af - bHf.$$

易知

$$\varphi = [a/(a^2 - b^2)]f - [b/(a^2 - b^2)]Hf \tag{6.4}$$

是方程(6.2)在 $\mathscr{L}^*$ 内的唯一解.

6.2. 一般方程.

对于方程(6.1),我们可以应用如(6.3)式所示的算子 M,把它归结为一个与之等价的 Fredholm 方程.

事实上,以 M 作用于(6.1),应用公式(I),立得

$$(a^2-b^2)\varphi+aK\varphi-bHK\varphi=af-bHf, \tag{6.5}$$

由于

$$\begin{aligned}HK\varphi&=\varpi_{2n-1}^{-2}\int_{u\bar{u}'=1}(1-v\bar{u}')^{-n}\dot{u}\int_{w\bar{w}'=1}\varphi(w)K(u,w)\dot{w}\\&=\lim_{\varepsilon\to0}\varpi_{2n-1}^{-2}\int_{\substack{u\bar{u}'=1\\|1-v\bar{u}'|>\varepsilon}}(1-v\bar{u}')^{-n}\dot{u}\int_{w\bar{w}'=1}\varphi(w)K(u,w)\dot{w}\\&=\lim_{\varepsilon\to0}\varpi_{2n-1}^{-2}\int_{w\bar{w}'=1}\varphi(w)\dot{w}\int_{\substack{u\bar{u}'=1\\|1-v\bar{u}'|>\varepsilon}}(1-v\bar{u}')^{-n}K(u,w)\dot{u},\end{aligned}$$

应用 Lebesgue 定理,上式即为

$$\varpi_{2n-1}^{-2}\int_{w\bar{w}'=1}\varphi(w)\dot{w}\int_{u\bar{u}'=1}(1-v\bar{u}')^{-n}K(u,w)\dot{u},$$

根据[4]中定理1.5, $\varpi_{2n-1}^{-1}\int_{u\bar{u}'=1}(1-v\bar{u}')^{-n}K(u,w)\dot{u}=K_1(v,w)$ 对 v, w 皆满足 Lipschitz 条件,所以(6.5)是一个 Fredholm 方程,可写之为

$$a_1\varphi+K_1\varphi=F, \tag{6.6}$$

其中 $a_1=a^2-b^2\neq0$, $F=Mf$, $K_1=aK-bHK$.

方程(6.6)与(6.1)是等价的,这只需证明,凡方程(6.6)的解(即(6.5)的解),必满足方程(6.1). 将算子

$$M^*=aI+bH \tag{6.7}$$

作用于(6.5)式两端,应用公式(I),便可得到方程(6.1).

综上所证,便得

定理 6.1. 设在方程(6.1)中, a, b 为复数,满足 $a^2-b^2\neq0$, 已给 $f\in\mathscr{L}^*$, 积分算子 K 的核 $K(v,u)$ 满足 Lipschitz 条件,则

(i) 方程(6.1)的特征方程(6.2)在 $\mathscr{L}^*$ 内有唯一解,如(6.4)式所示;

(ii) 在 $\mathscr{L}^*$ 内,奇异积分方程(6.1)与 Fredholm 方程(6.6)等价.

§7. 有 *B*-核和 *h*-核的奇异积分方程

本节讨论方程

$$L\varphi\equiv a\varphi+bB\varphi+ch\varphi+dA\varphi=f, \tag{7.1}$$

其中 a, b, c, d 为实数,且 $a-b\neq0$, $a^2+c^2\neq0$. 已给 $f\in\mathscr{L}$. 在 $\mathscr{L}$ 内求解方程(7.1)如下:

(i) 当 $a+d\neq0$ 时,易知算子

$$M=\frac{a}{a^2+c^2}I-\frac{c^2+ab}{(a-b)(a^2+c^2)}B-\frac{c}{a^2+c^2}h+\frac{c^2-ad}{(a+d)(a^2+c^2)}A \tag{7.2}$$

是算子

$$L = aI + bB + ch + dA \tag{7.3}$$

的逆算子，即

$$ML = LM = I. \tag{7.4}$$

因而，方程(7.1)在这时有唯一解

$$\varphi = Mf. \tag{7.5}$$

(ii) 当 $a + d = 0$ 时

①如果 $a \neq 0$，方程为

$$a(I - A)\varphi + bB\varphi + ch\varphi = f, \tag{7.6}$$

令

$$(I - A)\varphi = \phi, \tag{7.7}$$

则 ϕ 应满足方程

$$a\phi + bB\phi + ch\phi = f. \tag{7.8}$$

根据(i)中结论，方程(7.8)有唯一解

$$\phi = \left(\frac{a}{a^2 + c^2} I - \frac{c^2 + ab}{(a - b)(a^2 + c^2)} B - \frac{c}{a^2 + c^2} h + \frac{c^2}{a(a^2 + c^2)} A\right) f. \tag{7.9}$$

将(7.9)代入(7.7)中，易知当且仅当 $Af = 0$ 时，方程(7.6)有解，并且解为

$$\varphi = \left(\frac{a}{a^2 + c^2} I - \frac{c^2 + ab}{(a - b)(a^2 + c^2)} B - \frac{c}{a^2 + c^2} h + \frac{c^2}{a(a^2 + c^2)} A\right) f + c_1, \tag{7.10}$$

其中 c_1 为任一实数.

②如果 $a = 0$，方程为

$$bB\varphi + ch\varphi = f \quad (bc \neq 0). \tag{7.11}$$

以算子 B 作用之，得 $B(b\varphi + f) = 0$，因而

$$\varphi = -\frac{1}{b}(f + U) \quad (U \in \mathscr{B}), \tag{7.12}$$

将(7.12)代入(7.11)，知 U 必须满足方程

$$hU = W, \tag{7.13}$$

其中 $W = -\frac{b}{c} f - \frac{b}{c} Bf - hf \in \mathscr{B}$.

用算子 h 作用于(7.13)，得

$$U - AU = -hW. \tag{7.14}$$

记 $AU = d$，以 $U = -hW + d$ 代入(7.13)，利用公式(III)及(IX)可知，当且仅当 $AW = 0$ 时，方程(7.13)在 $\mathscr{B}$ 内有，并且它的解是

$$U = -hW + c_1. \tag{7.15}$$

其中 c_1 为任一实数. 由此可知，当且仅当 $Af = 0$ 时，方程(7.11)在 $\mathscr{L}$ 内有解，解为

$$\begin{aligned}\varphi &= -\frac{1}{b}(f - hW + c_1) \\ &= \left(\frac{1}{b} B - \frac{1}{c} h - \frac{1}{b} A\right) f + c_1',\end{aligned} \tag{7.16}$$

其中 c_1' 为任一实数.

于是，我们得到

定理 7.1. 设在方程(7.1)中，a, b, c, d 为实数，且 $a-b\neq 0$, $a^2+c^2\neq 0$, 已给 $f\in\mathscr{L}$, 那末

(i) 当 $a+d\neq 0$ 时，(7.1)在 $\mathscr{L}$ 内有唯一解，解如(7.5)所示；

(ii) 若 $a+d=0$, 则当且仅当 $Af=0$ 时，(7.1)在 $\mathscr{L}$ 内有解，并且如果 $a\neq 0$, 则解如(7.10)所示，如果 $a=0$, 则解如(7.16)所示.

§8. 一些例外情形

现在讨论在方程(6.2)中 $a^2-b^2=0$ 及方程(7.1)中 $a=b$ 或 $a^2+c^2=0$ 的情形. 也就是讨论下列方程:

$$\varphi\pm H\varphi=f, \tag{8.1}$$

其中 $f\in\mathscr{L}^*$ 已给，在 $\mathscr{L}^*$ 内求解 φ；

$$A\varphi=f, \tag{8.2}$$

$$B\varphi=f, \tag{8.3}$$

$$B\varphi+dA\varphi=f\quad (d\neq 0), \tag{8.4}$$

$$h\varphi=f, \tag{8.5}$$

$$h\varphi+dA\varphi=f\quad (d\neq 0), \tag{8.6}$$

$$\varphi+B\varphi+ch\varphi-A\varphi=f, \tag{8.7}$$

$$\varphi+B\varphi+ch\varphi+dA\varphi=f\quad (d\neq -1), \tag{8.8}$$

在(8.2)—(8.8)诸方程中，已给 $f\in\mathscr{L}$, 在 $\mathscr{L}$ 内求解 φ.

8.1. 如果方程(8.1)在 $\mathscr{L}^*$ 内有解，以算子 $I\mp H$ 作用之，知必须

$$(I\mp H)f=0, \tag{8.9}$$

当 f 满足(8.9)时，任取 $\psi\in\mathscr{L}^*$, 则

$$\varphi=\frac{1}{2}f+(I\mp H)\psi \tag{8.10}$$

必为(8.1)的解.

8.2. 如果方程(8.2)有解，必须 $f=C$ 为常数；当 $f=C$ 时，任取 $\psi\in\mathscr{L}$, 则

$$\varphi=C+h\psi+B\psi \tag{8.11}$$

必为(8.2)的解.

8.3. 如果(8.3)有解，必须

$$hf=0,\qquad f+Bf=0, \tag{8.12}$$

当 f 满足(8.12)时，任取 $v\in\mathscr{B}$, $\psi\in\mathscr{L}$, 则

$$\varphi=-f+v+h\psi \tag{8.13}$$

必为(8.3)的解.

8.4. 如果(8.4)有解，必须

$$hf=0, \tag{8.14}$$

当 f 满足(8.14)时，任取 $\psi\in\mathscr{L}$, 则

$$\varphi=\frac{1}{d}f+\frac{1+d}{d}Bf+h\psi \tag{8.15}$$

必为(8.4)的解.

8.5. 如果(8.5)有解,必须

$$Af = 0, \qquad Bf = 0, \tag{8.16}$$

当 f 满足(8.16)时,任取 $\psi \in \mathscr{L}$,则

$$\varphi = -hf + B\psi \tag{8.17}$$

必为(8.5)的解.

8.6. 如果(8.6)有解,必须 $f \in \mathscr{B}$,当 $f \in \mathscr{B}$ 时,任取 $\psi \in \mathscr{L}$,则

$$\varphi = -hf + \frac{1}{d}Af + B\psi \tag{8.18}$$

必为(8.6)的解.

8.7. 如果(8.7)有解,必须 $f \in \mathscr{B}$. 当 $f \in \mathscr{B}$ 时,任取 $\psi \in \mathscr{L}$,则

$$\varphi = \frac{1}{1+c^2}f - \frac{c}{1+c^2}hf + \frac{c^2}{1+c^2}Af + B\psi \tag{8.19}$$

必为(8.7)的解.

8.8. 如果(8.8)有解,必须 $f \in \mathscr{B}$,当 $f \in \mathscr{B}$ 时,任取 $\psi \in \mathscr{L}$,则

$$\varphi = \frac{1}{1+c^2}f - \frac{c}{1+c^2}hf + \frac{c^2-d}{(1+c^2)(1+d)}Af + B\psi \tag{8.20}$$

必为(8.8)的解.

总之，上述诸方程如果有解，解都是不唯一的. 在以后几节所讨论的方程和方程组中,也要出现一些例外情形,将不再赘述.

§9. 有 B-核和 h-核的奇异积分方程(續)

本节讨论比(6.1)、(7.1)更为一般的方程.

9.1. 首先讨论方程

$$av + chv + KV = W, \tag{9.1}$$

其中 a, c 为实数,已给 $W \in \mathscr{B}$, $KV = \varpi_{2n-1}^{-1}\int_{u\bar{u}'=1} V(u)K(v, u)\dot{u}$. 核 $K(v, u)$ 是对 v, u 皆满足 Lipschitz 条件的实值函数. 我们有

定理 9.1. 如果在 $\mathscr{B}$ 中求解,则方程(9.1)可以化成与之等价的 Fredholm 方程.

证. 以算子

$$M = aI - ch \tag{9.2}$$

作用于(9.1),利用公式(III),得到

$$a_1V + K_1V = W_1, \tag{9.3}$$

其中

$$a_1 = a^2 + c^2,\ W_1 = (aI - ch)W \in \mathscr{B},\ K_1V = -c^2AV + (aI - ch)KV$$
$$= \varpi_{2n-1}^{-1}\int_{u\bar{u}'=1} V(u)K_1(v, u)\dot{u}.$$

应用§6 中 6.2 同样的推理,易知 $K_1(v, u)$ 对 v, u 皆满足 Lipschitz 条件,所以,(9.3)是 Fredholm 方程.

方程(9.3)与(9.1)的等价性是显然的. 因为以算子

$$M^* = aI + ch \tag{9.4}$$

作用于(9.3),立得(9.1).

9.2. 考虑方程

$$a\varphi + bB\varphi + ch\varphi + K\varphi = f, \tag{9.5}$$

其中 a, b, c 为实数,已给 $f \in \mathscr{L}$, $K\varphi = \varpi_{2n-1}^{-1}\int_{u\bar{u}'=1} \varphi(u)K(v, u)\dot{u}$, 核 $K(v, u)$ 对 v, u 皆满足 Lipschitz 条件. 显然,(7.1)是(9.5)的一个特殊情形.

为了讨论的方便,不妨将(9.5)改写为

$$a\varphi + bB\varphi + ch\varphi + dA\varphi + K_1\varphi = f, \tag{9.6}$$

其中 $K_1\varphi = \varpi_{2n-1}^{-1}\int_{u\bar{u}'=1} \varphi(u)K_1(v, u)\dot{u}$, $K_1(v, u)$ 对 v, u 皆满足 Lipschitz 条件. 我们有

定理 9.2. 如果在方程(9.6)中, $a - b \neq 0$, $a + d \neq 0$, $a^2 + c^2 \neq 0$, 并在 $\mathscr{L}$ 内求解 φ, 则方程(9.6)可以化成与之等价的 Fredholm 方程.

証. 用(7.2)所示的算子M作用于(9.6),得到方程

$$\varphi + K_2\varphi = g, \tag{9.7}$$

其中 $g = Mf$, $K_2\varphi = MK_1\varphi = \varpi_{2n-1}^{-1}\int_{u\bar{u}'=1} \varphi(u)K_2(v, u)\dot{u}$, 易知 $K_2(v, u)$ 对 v, u 皆满足 Lipschitz 条件. 所以(9.7)是 Fredholm 方程. 用算子 L (如(7.3)所示)作用于(9.7), 立得(9.6),所以(9.7)与(9.6)等价.

§10. 含有 Cauchy 核的奇异积分方程組

考虑方程组

$$A_1\varphi + HB_1\varphi + K\varphi = f, \tag{10.1}$$

其中 A_1, B_1 皆为 m 阶复数方阵,已给 $f = \begin{pmatrix} f_1 \\ \vdots \\ f_m \end{pmatrix} \in \mathscr{L}^*$(指每个 $f_i \in \mathscr{L}^*$, $i = 1, 2, \cdots, m$),

$K = (K_{ij})_{1\leqslant i,j\leqslant m}$, $K_{ij}\varphi_j = \varpi_{2n-1}^{-1}\int_{u\bar{u}'=1} \varphi_j(u)K_{ij}(v,u)\dot{u}$. 每个核 $K_{ij}(v, u)$ 皆对 v, u 满足 Lipschitz 条件. 我们有

定理 10.1. 如果在方程组(10.1)中, A_1, B_1 满足

$$\det\begin{pmatrix} A_1 & B_1 \\ B_1 & A_1 \end{pmatrix} \neq 0, \tag{10.2}$$

并且在 $\mathscr{L}^*$ 内求解 $\varphi = \begin{pmatrix} \varphi_1 \\ \vdots \\ \varphi_m \end{pmatrix}$, 则方程组(10.1)可以化成与之等价的 Fredholm 方程组.

証. 记

$$\begin{pmatrix} A_1 & B_1 \\ B_1 & A_1 \end{pmatrix}^{-1} = \begin{pmatrix} C_1 & D_1 \\ D_1 & C_1 \end{pmatrix}, \tag{10.3}$$

用算子

$$M = C_1 I + D_1 H \tag{10.4}$$

作用于(10.1)，得到

$$\varphi + K_1\varphi = F, \tag{10.5}$$

其中 $F = Mf \in \mathscr{L}^*$, $K_1\varphi = MK\varphi$, $K_1 = (K_{ij}^{(1)})_{1\leqslant i,j\leqslant m}$. 易知 $K_{ij}^{(1)}\varphi_j = \varpi_{2n-1}^{-1}\int_{u\bar{u}'=1}\varphi_j(u) K_{ij}^{(1)}(v, u)\dot{u}$ 的核 $K_{ij}^{(1)}(v, u)$ 对 v, u 皆满足 Lipschitz 条件，所以(10.5)是 Fredholm 方程组. 用算子 $A_1 I + B_1 H$ 作用于(10.5)，立得(10.1)，所以(10.5)与(10.1)等价.

§11. 有 B-核与 h-核的奇异积分方程組

11.1. 考虑方程组

$$L\varphi \equiv A_1\varphi + BB_1\varphi + hC_1\varphi + AD_1\varphi = f, \tag{11.1}$$

其中 A_1, B_1, C_1, D_1 皆为 m 阶实数方阵，且

$$\left.\begin{aligned} &\det(A_1 - B_1) \neq 0, \\ &\det\begin{pmatrix} A_1 & C_1 \\ -C_1 & A_1 \end{pmatrix} \neq 0, \end{aligned}\right\} \tag{11.2}$$

已给 $f = \begin{pmatrix} f_1 \\ \vdots \\ f_m \end{pmatrix} \in \mathscr{L}$，在 $\mathscr{L}$ 内求解 φ 如下:

(i) 如果 $\det(A_1 + D_1) \neq 0$，令

$$\left.\begin{aligned} &\begin{pmatrix} A_2 & C_2 \\ -C_2 & A_2 \end{pmatrix} = \begin{pmatrix} A_1 & C_1 \\ -C_1 & A_1 \end{pmatrix}^{-1}, \\ &B_2 = (C_2C_1 - A_2B_1)(A_1 - B_1)^{-1}, \\ &D_2 = -(A_2D_1 + C_2C_1)(A_1 + D_1)^{-1}, \end{aligned}\right\} \tag{11.3}$$

则算子

$$M = A_2 I + B_2 B + C_2 h + D_2 A \tag{11.4}$$

是算子

$$L = A_1 I + B_1 B + C_1 h + D_1 A \tag{11.5}$$

的左逆，即

$$ML = I. \tag{11.6}$$

因而，方程(11.1)在这时有唯一解

$$\varphi = Mf. \tag{11.7}$$

(ii) 如果 $\det(A_1 + D_1) = 0$，用算子 B 作用于(11.1)，得到 $B((A_1 - B_1)\varphi - f) = 0$，即

$$\varphi = (A_1 - B_1)^{-1}(f + U) = g + V, \tag{11.8}$$

其中

$$U = \begin{pmatrix} U_1 \\ \vdots \\ U_m \end{pmatrix} \in \mathscr{B},\ g = (A_1 - B_1)^{-1}f,\ V = (A_1 - B_1)^{-1}U \in \mathscr{B}.$$

将(11.8)代入(11.1),知V必须满足方程组

$$A_1V + hC_1V + AD_1V = F, \tag{11.9}$$

其中

$$F = -B_1g - BB_1g - hC_1g - AD_1g \in \mathscr{B}.$$

令

$$M^* = A_2I + C_2h, \tag{11.10}$$

其中 A_2, C_2 满足(11.3). 用 M^* 作用于(11.9),得

$$V + (A_2D_1 + C_2C_1)AV = M^*F, \tag{11.11}$$

其中

$$M^*F = -[(A_2B_1 - C_2C_1)I + A_2B_1B + (A_2C_1 + C_2B_1)h + (A_2D_1 + C_2C_1)A]g.$$

记 $AV = C^* = \begin{pmatrix} C_1^* \\ \vdots \\ C_m^* \end{pmatrix}$ 为一列实数,以

$$V = M^*F - (A_2D_1 + C_2C_1)C^* \tag{11.12}$$

代入(11.9),得到

$$(A_2D_1 + C_2C_1)[(I + A_2D_1 + C_2C_1)C^* + (A_2B_1 + A_2D_1)(A_1 - B_1)^{-1}f] = 0, \tag{11.13}$$

这就表示,如果方程组(11.11)有解,所与之 f 必须使得线性方程组

$$(A_2D_1 + C_2C_1)[(I + A_2D_1 + C_2C_1)x + (A_2B_1 + A_2D_1)(A_1 - B_1)^{-1}f] = 0 \tag{11.14}$$

有解,其中 $x = \begin{pmatrix} x_1 \\ \vdots \\ x_m \end{pmatrix}$ 为一组变元. 反之,如果方程组(11.14)有解,任取其一组解 $x = C^*$,则由(11.12)所示之 V 必为(11.9)的解,从而(11.8)是(11.1)的解. 方程组(11.1)的解的唯一性,可如下得知:即当 $f \equiv 0$ 时,$F \equiv 0$,从(11.11)立知 V 为一组常数,而且必须满足方程

$$(A_2D_1 + C_2C_1)(I + A_2D_1 + C_2C_1)V = 0.$$

于是得到

定理 11.1. 在方程组(11.1)中,如果 A_1, B_1, C_1 满足(11.2),那末

(i) 当 $\det(A_1 + D_1) \neq 0$ 时,(11.1)在 $\mathscr{L}$ 内有唯一解,解如(11.7)所示;

(ii) 若 $\det(A_1 + D_1) = 0$,则当且仅当所与之 f 使得线性方程组(11.14)有解时,(11.1)在 $\mathscr{L}$ 内有解,解的形式为 $\varphi = (A_1 - B_1)^{-1}f + V$,其中 V 如(11.12)所示,C^* 是(11.14)的任一组解.

11.a. 最后,考虑方程组

$$A_1\varphi + BB_1\varphi + hC_1\varphi + AD_1\varphi + K\varphi = f, \tag{11.15}$$

其中 A_1, B_1, C_1, D_1 皆为 m 阶实数方阵,已给 $f = \begin{pmatrix} f_1 \\ \vdots \\ f_m \end{pmatrix} \in \mathscr{L}$, $K = (K_{ij})_{1\leqslant i,j\leqslant m}$, $K_{ij}\varphi_j = \tilde{\omega}_{2n-1}^{-1}\int_{u\bar{u}'=1}\varphi_j(u)K_{ij}(v, u)\dot{u}$,每个 $K_{ij}(v, u)$ 皆对 v, u 满足 Lipschitz 条件. 我们有

定理 11.2. 如果在 $\mathscr{L}$ 内求解方程组(11.15)，则当 A_1, B_1, C_1, D_1 满足

$$\det(A_1 - B_1) \neq 0,\ \det(A_1 + D_1) \neq 0,\ \det\begin{pmatrix} A_1 & C_1 \\ -C_1 & A_1 \end{pmatrix} \neq 0 \tag{11.16}$$

时，可将方程组(11.15)化成与之等价的 Fredholm 方程组.

證. 用(11.4)式所示的算子 M 作用于(11.15)，得到 Fredholm 方程组

$$\varphi + K^*\varphi = F, \tag{11.17}$$

其中 $K^* = MK = (K_{ij}^*)$, $K_{ij}^*\varphi_j = \varpi_{2n-1}^{-1}\int_{u\bar{u}'=1}\varphi_j(u)K_{ij}^*(v, u)\dot{u}$ 的核 $K_{ij}^*(v, u)$ 对 v, u 皆满足 Lipschitz 条件 $(i, j = 1, 2, \cdots, m)$.

由于

$$\begin{aligned}\det(A_2 - B_2) &= \det[A_2 - (C_2C_1 - A_2B_1)(A_1 - B_1)^{-1}] \\ &= \det(A_1 - B_1)^{-1} \neq 0, \\ \det(A_2 + D_2) &= \det[A_2 - (A_2D_1 + C_2C_1)(A_1 + D_1)^{-1}] \\ &= \det(A_1 + D_1)^{-1} \neq 0,\end{aligned}$$

$$\det\begin{pmatrix} A_2 & C_2 \\ -C_2 & A_2 \end{pmatrix} = \det\begin{pmatrix} A_1 & C_1 \\ -C_1 & A_1 \end{pmatrix}^{-1} \neq 0,$$

所以同理可以作出M的左逆算子 M^*，用 M^* 作用于(11.17)，便得到(11.15)，所以(11.17)与(11.15)等价.

系. 如果在 $\mathscr{L}$ 内求解方程组

$$A_1\varphi + BB_1\varphi + hC_1\varphi + K\varphi = f, \tag{11.18}$$

其中 A_1, B_1, C_1 为m阶实数方程阵且满足

$$\det(A_1 - B_1) \neq 0,\ \det\begin{pmatrix} A_1 & C_1 \\ -C_1 & A_1 \end{pmatrix} \neq 0,$$

而 $f = \begin{pmatrix} f_1 \\ \vdots \\ f_m \end{pmatrix} \in \mathscr{L}$, $K = (K_{ij})_{1\leqslant i,j\leqslant m}$, $K_{ij}\varphi_j = \varpi_{2n-1}^{-1}\int_{u\bar{u}'=1}\varphi_j(u)K_{ij}(v, u)\dot{u}$, $K_{ij}(v, u)$ 皆对 v, u 满足 Lipschitz 条件则可将方程组(11.18)化成与之等价的 Fredholm 方程组.

事实上，只要将(11.18)改写成

$$A_1\varphi + BB_1\varphi + hC_1\varphi + AD_1\varphi + K_1\varphi = f,$$

这里 $K_1 = K - AD_1$，若K满足系中的条件，则 K_1 也满足. 选取 D_1 使 $\det(A_1 + D_1) \neq 0$，这是一定能做到的. 于是由定理 11.2，立知系中结论正确.

最后还想指出二点：

(1) 注意到 H, h, 与 B 之间的关系，我们可以将含有 Cauchy 核的常系数奇异积分方程及方程组归化为含有 B-核和 h-核的常系数的奇异积分方程组.

(2) 如果我们把 $\mathscr{L}^*$ 及 $\mathscr{L}$ 看作为定义在特征为零的域上的线性空间，$\mathscr{B}$ 为 $\mathscr{L}$ 的线性子空间，在 $\mathscr{L}^*$ 及 $\mathscr{L}$ 上引进线性算子 H, B, h, A, I, K，且满足关系式如 §5 中(I)—(IX)所示，那末从 §6 以后的这些结果都可以理解为这样一般的线性空间上的结果，只要我们对线性算子K作些必要的要求与说明(请参阅[2]).

参 考 文 献

[1] 华罗庚，多复变数函数论中的典型域的调和分析，科学出版社，1958.

[2] 关肇直，泛函分析讲义，高等教育出版社，1958.

[3] 龚 升、孙继广，圆型域中解析函数的积分表示，中国科学技术大学学报，**1** (1965)，219—226.

[4] 龚 升、孙继广，多复变数的 Cauchy 型积分 (I)，数学学报，**15** (1965)，431—443.

[5] Мусхелишвили, Н. И., Сингулярные интегральные уравнения, ГТТИ, (1962).

[6] Векуа, Н. П., Системы сингулярных интегральных уравнений и некоторые граничные задачи, ГТТИ, (1950) (有中译本,路见可译，上海科学技术出版社).

[7] Гахов, Ф. Д., Краевые задачи, Физматгиз, (1962).

[8] Михлин, С. Г., Сингулярные интегральные уравнение, *Успехи Матем. Наук*, **3** (1948), 29—112.

[9] Михлин, С. Г., Многомерные сингулярные интегралы и интегральные уравнения, *Физматгиз*, (1962).

[10] Tricomi, F. G., Formula di inversione dell' ordine di due integrazioni doppis "con asterisco", *Rend. Accad. Naz. Lincei* 3 *ser.* 6*a*, *fasc.* 9(1926), 535—539.

[11] Tricomi, F. G., Equazioni integrali contenenti il valor principale di un integrale doppis, *Math. Z.*, **27** (1928), 87—133.

[12] Бицадзе, А. В., О двумерных интегралах типа коши, *Сообщ. АН Груз ССР*, **16** (1955), 177—184.

[13] Bergman, S., Kernel function and extended classes in the Theory of functions of complex variables, Colloque aux les Functions de plusieurs variables, Bruxelles, 1953.

Chin. Ann. of Math.
3 (4) 1982

SINGULAR INTEGRALS IN SEVERAL COMPLEX VARIABLES (I)——HENKIN INTEGRALS OF STRICTLY PSEUDOCONVEX DOMAIN

GONG SHENG
(*University of Science and Technology of China, Institute of Applied Mathematics, Academia Sinica*)

SHI JIHUAI
(*University of Science and Technology of China*)

Dedicated to Professor Su Bu-chin on the Occasion of his 80th Birthday and his 50th Year of Educational Work

§0. Introduction

0.1. Cauchy-Fantappiè kernels

In the case of one complex variable, the Cauchy kernel gives rise to singular integrals on boundaries of arbitrary smooth domains. Cauchy kernel defines an analytic function $\mathbf{H}u$ on plane domain Ω by

$$\mathbf{H}u(w)=\int_{z\in b\Omega} H(w,\ z)u(z)d\sigma_z,\ w\in\Omega,$$

where $H(w,\ z)$ is the Cauchy kernel $[2\pi i(z-w)]^{-1}$, $d\sigma_z$ is Lebesgue measure on $b\Omega$. There is the famous Plemelj formula

$$\mathbf{H}u(w)=\frac{1}{2}u(w)+\text{p. v.}\int_{z\in b\Omega} H(w,\ z)u(z)d\sigma_z,\ w\in b\Omega, \tag{0.1}$$

as w approaches $b\Omega$, where $\text{p. v.}\int_{z\in b\Omega}$ is defined by

$$\lim_{\substack{\varepsilon\to 0\\ \varepsilon>0}}\int_{\substack{z\in b\Omega\\ |z-w|>\varepsilon}} H(w,\ z)u(z)d\sigma_z. \tag{0.2}$$

An important point is that the deleted neighborhood around w in (0.2), i. e. $\{z\in b\Omega,\ |z-w|<\varepsilon\}$ is symmetric. If the deleted neighborhood around w is not symmetric, then the limit might fail to exist or the number $\frac{1}{2}$ in (0.1) might have to be modified. There is a corresponding Plemelj formula too as w approaches $b\Omega$ from the exterior of Ω. According to (0.1) and the corresponding formula, we may establish the theory of singular integrals and singular integral equations of one variable. For

Manuscript received April 24, 1981.

example, we can obtain the famous Poincaré-Bertrand formula: If $\varphi\in \mathrm{Lip}\alpha$, $(0<\alpha\leqslant 1)$, then

$$\mathbf{H}^2\varphi=\frac{1}{4}\varphi. \tag{0.3}$$

Here $\mathbf{H}$ is the Cauchy singular integral operator defined by (0.2). Using these formulas, we can solve the singular integral equations with Cauchy kernel, Hilbert kernel and some important boundary value problems. There are many works about this topic([1, 2]).

In several complex variables, there is no perfect analogue of the Cauchy kernel, what come closest to it are certain Cauchy-Fantappié (C-F) kernels

$$K(w, z)=\frac{c_n}{g^n}\omega\wedge dz_1\wedge\cdots\wedge dz_n,\ w\in\Omega,\ z\in b\Omega, \tag{0.4}$$

where

$$c_n=(-1)^{\frac{n(n-1)}{2}}(n-1)!(2\pi i)^{-n}, \tag{0.5}$$

$$g(w, z)=\sum_{i=1}^{n}(z_i-w_i)g_i(w, z), \tag{0.6}$$

$$\omega=g_1\bar\partial g_2\wedge\cdots\wedge\bar\partial g_n+\cdots+(-1)^{n-1}g_n\wedge\bar\partial g_1\wedge\cdots\wedge\bar\partial g_{n-1}. \tag{0.7}$$

These kernels are different from the Cauchy kernel of one complex variable, they depend on the domain Ω. But they can be constructed for a wide class of Ω (smooth, bounded strictly pseudoconvex domains). The very important one of these domains is the ball. Henkin, Ramirez, Stein and Kerzman have constructed some important C-F kernels which are analytic when $w\in\Omega$. These kernels are called Henkin-Ramirez (H-R) kernel or Stein-Kerzman (S-K) kernel. It is well known that the requirement of the analyticity of the C-F kernels is very important.

0.2. Cauchy integral on the sphere

Singular integrals on the boundaries of special domains in $\mathbf{C}^n$ (such as the ball) first appeared and were studied in connection with Szegö kernel. Generally speaking, the Cauchy kernel is different from the Szegö kernel. C-F kernels are in the spirit of the Cauchy kernel and not of the Szegö kernel. However, if Ω is a ball in $\mathbf{C}^n$, the two kernels coincide. In 1965, Kung, S. and Sun, C. K.[3] studied the Cauchy integral on the sphere in $\mathbf{C}^n$. They proved: If $z=(z_1, \cdots, z_n)\in\mathbf{C}^n$, $z\bar z'<1$, $u\bar u'=1$, $f(u)\in\mathrm{Lip}\,p$ $(0<p\leqslant 1)$, then when z approaches v $(v\bar v'=1)$ along the nontangential direction, we have

$$\lim_{z\to v}\frac{1}{\omega_{2n-1}}\int_{u\bar u'=1}\frac{f(u)\dot u}{(1-z\bar u')^n}=\frac{1}{2}f(v)+\text{p. v. }\frac{1}{\omega_{2n-1}}\int_{u\bar u'=1}\frac{f(u)\dot u}{(1-v\bar u')^n}, \tag{0.8}$$

where ω_{2n-1} is the volume of $u\bar u'=1$. The principal value is defined by

$$\text{p. v. }\frac{1}{\omega_{2n-1}}\int_{u\bar u'=1}\frac{f(u)\dot u}{(1-v\bar u')^n}=\lim_{\varepsilon\to 0}\frac{1}{\omega_{2n-1}}\int_{\substack{u\bar u'=1\\ |1-v\bar u'|>\varepsilon}}\frac{f(u)\dot u}{(1-v\bar u')^n}.$$

After that, Koranyi and Vagi[4] obtained the same results by generalized Cayley transformation, this is equivalent to the Plemelj formula of generalized upper half plane

$$D=\left\{z=(z_1, \cdots, z_n):\ \operatorname{Im} z_1-\sum_{k=2}^{n}|z_k|^2>0\right\}.$$

On the other hand, using the Plemelj formulas of the sphere, we may obtain the Plemelj formulas of the matrix hyperbolic space and the Lie sphere hyperbolic space[5,6]. Shi Jihuai[7] also proved the following result

$$\lim_{z\to v}\frac{1}{\omega_{2n-1}}\int_{u\bar{u}'=1}\frac{f(u)\dot{u}}{(1-z\bar{u}')^n}=2^{n-2}f(v)+\text{p. v. }\frac{1}{\omega_{2n-1}}\int_{u\bar{u}'=1(\text{Im})}\frac{f(u)\dot{u}}{(1-v\bar{u}')^n},$$

where p. v. $\frac{1}{\omega_{2n-1}}\int_{u\bar{u}'=1(\text{Im})}$ is defined by $\lim_{\varepsilon\to 0}\frac{1}{\omega_{2n-1}}\int_{\substack{u\bar{u}'=1\\ |\operatorname{Im}(v\bar{u}')|>\varepsilon}}$.

If $u\bar{u}'=1$, $v\bar{v}'=1$, we have the following singular kernels

$$H(v, u)=(1-v\bar{u}')^{-n}\quad\text{(Cauchy kernel)},$$
$$B(v, u)=H(v, u)+H(u, v)-1\quad\text{(B kernel)},$$
$$h(v, u)=\frac{1}{i}(H(v, u)-H(u, v))\quad\text{(Hilbert kernel)}.$$

By these singular kernels, we may get the following singular integral operators on sphere

$$\mathbf{H}\varphi=2\omega_{2n-1}^{-1}\int_{u\bar{u}'=1}\varphi(u)H(v, u)\dot{u},$$

$$\mathbf{B}\varphi=2\omega_{2n-1}^{-1}\int_{u\bar{u}'=1}\varphi(u)B(v, u)\dot{u},$$

$$\mathbf{h}\varphi=2\omega_{2n-1}^{-1}\int_{u\bar{u}'=1}\varphi(u)h(v, u)\dot{u},$$

where the principal value is defined by $\lim_{\substack{\varepsilon\to 0\\ \varepsilon>0}}\int_{\substack{u\bar{u}'=1\\ |1-v\bar{u}'|>\varepsilon}}$. If $\varphi\in \operatorname{Lip}p$ $(0<p\leqslant 1)$, these singular integrals exist. From them we may also obtain the generalized Poincaré-Bertrand formula ($\mathbf{H}^2=I$), the generalized Hilbert formula of Hilbert kernel and solve the linear singular integral equations or the systems of linear singular integral equation with Cauchy kernel, B kernel and Hilbert kernel, etc[8,9]. It is worth mentioning that B kernel does not appear in one varirable. This shows the difference between one and several variables.

0.3. Main results of this paper

Pseudoconvex domains are the domains of holomorphy. For the strictly pseudoconvex domains, we can construct the analytic Henkin-Ramirez (H-R) kernel and Stein-Kerzman (S-K) kernel.

Alt[13] in 1974, Kerzman and Stein[14] in 1978 proved respectively the plemelj formula for the H-R kernel and S-K kernel. They proved: If Ω is C^∞ smooth, bounded strictly pseudoconvex domain, $u\in C^\infty(b\Omega)$, $H(w, z)$ is H-R kernel or S-K kernel

$$\mathbf{H}u(w)=\int_{z\in b\Omega} H(w,\ z)u(z)d\sigma_z,\ w\in\Omega.$$

Then when w approaches $b\Omega$ along the nontangential direction

$$\mathbf{H}u(w)=\frac{1}{2}u(w)+\text{p. v.}\int_{z\in b\Omega} H(w,\ z)u(z)d\sigma_z\quad (w\in b\Omega)\tag{0.9}$$

holds, where p. v. $\int_{z\in b\Omega}$ is defined by $\lim\limits_{\varepsilon\to 0}\int_{z\in b\Omega-B(w,\varepsilon)}$, and

$$B(w,\ \varepsilon)=\{z\in b\Omega,\ |g(z,\ w)|<\varepsilon\}.$$

In the section 1 of this paper, we consider that the deleted neighborhood $B(w,\ \varepsilon)$ around w in the boundary of a strictly pseudoconvex domain is a more general form instead of above form, a corresponding plemelj formula is obtained. This shows the essential difference between one and several variables: There is only one method to define the principal value of the Cauchy integral in the former, but there are infinite many methods in the latter, even if the deleted part is symmetry. Basing upon these results, we discuss the singular integral on the boundary of a strictly pseudoconvex domain.

In the sections 2 and 3, we discuss some special cases of $B\ (w,\ \varepsilon)$ and the theory of the corresponding singular integral equations. In the section 2, we consider the situation that the deleted neighborhood $B(w,\ \varepsilon)$ is an "ellipse" or a "belt" when the "ellipse" is a "disc", we obtain the results of Alt and Kerzman-Stein. We also discover that there is a method of deletion, such that the term $u(w)$ does not appear in the Plemelj formula, in other words, when w approaches the boundary from the interior, the Cauchy integral approaches a special principal value. In the section 3, we consider the situation that the deleted neighborhood is a "rectangle". All these show the variety of the definition of the principal value of the Cauchy integral.

Part results of this paper were announced in [10].

§1. Cauchy integrals on a strictly pseudoconvex domain

1.1 Henkin-Ramirez kernel and Stein-Kerzman kernel

Suppose Ω is a smooth, bounded strictly pseudoconvex domain, λ is a real function, $\lambda\in C^\infty(\overline{\Omega})$, $\lambda(z)<0$, if $z\in\Omega$; $\lambda(z)=0$, if $z\in b\Omega$; $\lambda(z)>0$, if $z\overline{\in}\Omega$; grad $\lambda(z)\neq 0$, if $z\in b\Omega$ and

$$\left(\frac{\partial^2\lambda}{\partial z_i\partial\bar{z}_i}\right)\geqslant CI,\tag{1.1.1}$$

where the constant $C>0$ is independent of $z\in\overline{\Omega}$, I is the identity matrix. For $z\in b\Omega$, $w\in\Omega$ and w near z, set

$$g_i^L(w,\ z)=\frac{\partial\lambda}{\partial z_i}(z)+\frac{1}{2}\sum_{j=1}^{n}\frac{\partial^2\lambda}{\partial z_i\partial z_j}(z)(w_j-z_j),\tag{1.1.2}$$

$$g^L(w, z)=\sum_{i=1}^{n}(z_i-w_i)g_i^L(w, z).$$

Extend the local $g^L(w, z)$ to a global $g(w, z)$: for w near z, set

$$g(w, z)=g^L(w, z)\phi(w, z),$$

where $\phi(w, z)$ is a locally defined function and is holomorphic in w for w near z. Furthermore $\phi(z, z)\neq 0$. The global $g_i(w, z)$ are obtained from the division problem

$$g(w, z)=\sum(z_i-w_i)g_i(w, z). \tag{1.1.3}$$

It may be proved that there are the functions $g_i(w, z)$ which are holomorphic in w such that (1.1.3) holds. Substitute such $g_i(w, z)$ into $g_i(w, z)$ which are needed in the C-F kernel (0.6) of Ω and substitute $g_i(w, z)$ of (1.1.3) into g of (0.4), we obtain the C-F kernel $H(w, z)$ of Ω, this is just the Henkin-Ramirez (H-R) kernel. As for the details of proof, see [11, 15].

Stein and kerzman[14] constructed another kernel $E+C$, called Stein-Kerzman (S-K) kernel, where $E(w, z)\,d\sigma_z$ is a C-F form (0.4), in which $g_i(w, z)$ are totally explicit, but are holomorphic only when w is close to $z\in b\Omega$, namely as w near z, $g_i(w, z)=g_i^L(w, z)$ and $C(w, z)$ is the correction term, such that $H(w, z)=E(w, z)+C(w, z)$ is holomorphic in w globally. Furthermore

$$C(w, z)\in C^\infty(U(\overline{\Omega})\times V(b\Omega)),$$

where $U(\overline{\Omega})$ and $V(b\Omega)$ are the neighborhoods of $\overline{\Omega}$ and $b\Omega$ respectively. That is to say, $C(w, z)$ is infinitely smooth even when $w=z$. $C(w, z)$ is a solution of a $\bar{\partial}$ problem. $C(w, z)$ is not a C-F form, but $E+C$ can reproduce holomorphic functions. As for the details, see [14].

It is known that for H-R kernel and S-K kernel, we both have

$$\int_{z\in b\Omega}|H(w, z)|\,d\sigma_z=\infty,\quad w\in b\Omega, \tag{1.1.4}$$

where $d\sigma_z$ is the Lebesgue element of area on $b\Omega$.

Fix $w\in b\Omega$, without loss of generality, take $w=0$, then we have the following

Theorem 1.1.[12,14] *Suppose Ω is a smooth, bounded strictly pseudoconvex domain, $0\in b\Omega$, $H(w, z)$ is the H-R kernel or S-K kernel. Then near $z=0$, there is a holomorphic local change of variables, such that $b\Omega$ and $H(0, z)$ have the following forms: $z\in b\Omega$ if and only if $z_n=t+i\tau$ satisfies*

$$\tau=|\xi|^2+\varepsilon(\xi, t), \tag{1.1.5}$$

where $z=(z_1, \cdots, z_n)$, $\xi=(z_1, \cdots, z_{n-1})$, the error term ε is of third order, namely

$$|\varepsilon(\xi, t)|=O(\rho^3), \tag{1.1.6}$$

where $\rho^4=|\xi|^4+t^2$. For $H(0, z)$, we have

$$H(0, z)=\frac{\gamma_n}{(|\xi|^2+it)^n}+\phi(z), \tag{1.1.7}$$

where $\phi(z)$ is absolutely integrable in $z\in b\Omega$ and γ_n is a constant.

Obviously $(\xi, t)\in \mathbf{C}^{n-1}\times\mathbf{R}$ and $\tau=|\xi|^2$ is "Heisenberg group" surface S_{n-1}.

(1.1.5) shows that near $z=0$ the $b\Omega$ may be approximated by the "Heisenberg group" surface S_{n-1} and the $\varepsilon(\xi, t)$ of (1.1.6) is the error of approximation.

It is easy to know[12, 13]

$$\int_{\rho<1} |(|\xi|^2+it)|^{-n} d\mu = \infty, \tag{1.1.8}$$

$$\int_{0<a<\rho<b} |(|\xi|^2+it)^{-n} d\mu = 0, \tag{1.1.9}$$

where $d\mu$ is the usual Lebesgue measure $dx_1 dy_1 \cdots dx_{n-1} dy_{n-1} dt$ $(z_j = x_j + iy_j)$. In fact, we may prove

$$\frac{1}{g^n(0, z)} = \frac{1}{(|\xi|^2+it)^n}(1+O(\rho)). \tag{1.1.10}$$

1.2. General plemelj formula

We first prove the following general Plemelj formula.

Theorem 1.2. *Suppose Ω is a smooth, bounded strictly pseudoconvex domain, $H(w, z)$ is the H-R kernel or S-K kernel, $u \in C^\infty(b\Omega)$,*

$$\mathbf{H}u(w) \equiv \int_{z\in b\Omega} H(w, z) u(z) d\sigma_z, \quad w \in \Omega \tag{1.2.1}$$

is the Cauchy integral, where $d\sigma_z$ is the Lebesgue element of area on $b\Omega$. Then $\mathbf{H}u$ is holomorphic in Ω and admits a continous extension up to $\overline{\Omega}$. When $w \in b\Omega$, define

$$\text{p. v.} \int_{z\in b\Omega} H(w, z) u(z) d\sigma_z \equiv \lim_{\substack{\varepsilon\to 0\\ \varepsilon>0}} \int_{z\in b\Omega - D(w,\varepsilon)\cap b\Omega} H(w, z) u(z) d\sigma_z, \tag{1.2.2}$$

here $D(w, \varepsilon)$ is the neighborhood around w and contracts to the point w as $\varepsilon \to 0$. If

$$\lim_{\varepsilon\to 0} \lim_{\delta\to 0} \int_{z\in D(w,\varepsilon)\cap b\Omega} H(w+\delta\nu, z) d\sigma_z = a \tag{1.2.3}$$

exists, where a is a constant and ν is the inner normal to $b\Omega$ at w, $\delta > 0$. Denote the value of (1.2.2) by $\mathbf{H}_a u$ and

$$\mathbf{H}u(w) = \lim_{w_0\to w} \mathbf{H}u(w_0),$$

where w_0 approaches $w \in b\Omega$ along the nontangential direction from the interior of Ω. Then we have the Plemelj formula

$$\mathbf{H}u(w) = au(w) + \mathbf{H}_a u(w). \tag{1.2.4}$$

Obviously, this deduces to the Theorem of Alt[13] and Kerzman-Stein[14] when $D(w, \varepsilon)$ is $\{z \in b\Omega, |g(z, w)| < \varepsilon\}$. In this case, $a = \frac{1}{2}$ and Plemelj formula becomes

$$\mathbf{H}u(w) = \frac{1}{2} u(w) + \mathbf{H}_{\frac{1}{2}} u(w). \tag{1.2.5}$$

We now prove Theorem 1.2.

Keep $w \in b\Omega$ fixed, say $w=0$. Then the condition (1.2.3) becomes

$$\lim_{\varepsilon\to 0} \lim_{\delta\to 0} \int_{z\in D(0,\varepsilon)\cap b\Omega} H(0+\delta\nu, z) d\sigma_z = a, \tag{1.2.6}$$

where ν is the inner normal to $b\Omega$ at $w=0$. Write

$$\lim_{\varepsilon\to 0}\lim_{\delta\to 0}\int_{z\in b\Omega-D(0,\varepsilon)\cap b\Omega} H(0+\delta\nu,\ z)u(z)d\sigma_z$$

$$=\lim_{\varepsilon\to 0}\lim_{\delta\to 0}\int_{z\in b\Omega-D(0,\varepsilon)\cap b\Omega} H(0+\delta\nu,\ z)[u(z)-u(0)]d\sigma_z$$

$$+u(0)\lim_{\varepsilon\to 0}\lim_{\delta\to 0}\int_{z\in b\Omega-D(0,\varepsilon)\cap b\Omega} H(0+\delta\nu,\ z)d\sigma_z,$$

since $u\in C^\infty(b\Omega)$, the above first integral exists[13], and the second integral is equal to $u(0)(1-a)$ *by* (1.2.6). Therefore, for $0\in b\Omega$, we have

$$\text{p. v.}\int_{z\in b\Omega} H(0,\ z)u(z)d\sigma_z=\lim_{\substack{\varepsilon\to 0\\ \varepsilon>0}}\int_{z\in b\Omega-D(0,\varepsilon)\cap b\Omega} H(0,\ z)u(z)d\sigma_z$$

$$=\int_{z\in b\Omega} H(0,\ z)[u(z)-u(0)]d\sigma_z+u(0)(1-a).$$

Hence, for $w\in b\Omega$, we also have

$$\text{p. v.}\int_{z\in b\Omega} H(w,\ z)u(z)d\sigma_z=\int_{z\in b\Omega} H(w,\ z)[u(z)-u(w)]d\sigma_z+u(w)(1-a). \tag{1.2.7}$$

Let $w_0\in\Omega$

$$\int_{z\in b\Omega} H(w_0,\ z)u(z)d\sigma_z=\int_{z\in b\Omega} H(w_0,\ z)[u(z)-u(w)]d\sigma_z$$

$$+u(w)\int_{z\in b\Omega} H(w_0,\ z)d\sigma_z=I_1+I_2.$$

When w_0 approaches w along the nontangential direction, by (1.2.7) we obtain

$$\lim_{w_0\to w} I_1=\int_{z\in b\Omega} H(w,\ z)[u(z)-u(w)]d\sigma_z$$

$$=\text{p. v.}\int_{z\in b\Omega} H(w,\ z)u(z)d\sigma_z-u(w)(1-a),$$

hence

$$\mathbf{H}u(w)=\text{p. v.}\int_{z\in b\Omega} H(w,\ z)u(z)d\sigma_z-u(w)(1-a)+u(w)$$

$$=\text{p. v.}\int_{z\in b\Omega} H(w,\ z)u(z)d\sigma_z+au(w).$$

This ends the proof.

Notice the condition (1.2.3) By Theorem 1.1, there is a holomorphic local change of variables, such that $b\Omega$ and $H(0,\ z)$ can be represented by (1.1.5) and (1.1.7) respectively near $0\in b\Omega$ and (1.2.6) becomes[13, 14]

$$\lim_{\varepsilon\to 0}\lim_{\delta\to 0}\left\{\int_{z\in D(0,\varepsilon)}\gamma_n(|\xi|^2+\delta+it)^{-n}\,d\mu+\int_{z\in D(0,\varepsilon)}\phi_\delta(z)d\mu\right\},$$

where $\phi_\delta(z)$ is absolutely integrable in $z\in b\Omega$ and $\lim_{\delta\to 0}\phi_\delta(z)=\phi(z)$. Since $D(0,\ \varepsilon)$ contracts to a point as $\varepsilon\to 0$, obviously

$$\lim_{\varepsilon\to 0}\lim_{\delta\to 0}\int_{z\in D(0,\varepsilon)}\phi_\delta(z)d\mu=0,$$

therefore the condition (1.2.3) becomes

$$\lim_{\varepsilon\to 0}\lim_{\delta\to 0}\int_{z\in D(0,\varepsilon)}\gamma_n(|\xi|^2+\delta+it)^{-n}d\mu=a.$$

Consider generalized Cayley transformation T

$$z_1=\frac{u_1}{1+u_n},\ \cdots,\ z_{n-1}=\frac{u_{n-1}}{1+u_n},\ z_n=i\frac{1-u_n}{1+u_n}.$$

Since $z_n=t+i\tau$, hence

$$t=\frac{2\operatorname{Im}u_n}{|1+u_n|^2},\quad \tau=\frac{1-|u_n|^2}{|1+u_n|^2},$$

T transforms $|\xi|^2+it+\delta$ into

$$\frac{|u_1|^2+\cdots+|u_{n-1}|^2+2i\operatorname{Im}u_n}{|1+u_n|^2}+\delta.$$

Since T maps upper half plane $\tau>0$, $\tau=|\xi|^2$ onto $uu'=1$, the above can also be written as

$$\frac{1+2i\operatorname{Im}u_n-|u_n|^2}{|1+u_n|^2}+\delta=\frac{(1+u_n)(1-\bar u_n)}{|1+u_n|^2}+\delta=\frac{1-\bar u_n}{1+\bar u_n}+\delta=\frac{(1+\delta)-(1-\delta)\bar u_n}{1+\bar u_n}$$

and $d\mu=\omega_{2n-1}^{-1}(1+\bar u_n)^{-n}\dot u$.

So the condition (1.2.6) becomes

$$\lim_{\varepsilon\to 0}\lim_{\delta\to 0}\frac{1}{\omega_{2n-1}}\int_{z\in D(p_n,\varepsilon)}\frac{\dot u}{(1+\delta)^n\left(1-\dfrac{1-\delta}{1+\delta}\bar u_n\right)^n}=\lim_{\varepsilon\to 0}\lim_{\rho\to 1}\frac{1}{\omega_{2n-1}}\int_{z\in D(p_n,\varepsilon)}\frac{\dot u}{(1-\rho\bar u_n)^n}$$

$$=\lim_{\varepsilon\to 0}\lim_{\rho\to 1}\frac{1}{\omega_{2n-1}}\int_{z\in D(p_n,\varepsilon)}\frac{\dot u}{(1-\rho p_n\bar u')^n}=a, \tag{1.2.8}$$

where $p_n=(0,\cdots,0,1)$, $D(p_n,\varepsilon)=T(D(0,\varepsilon))$ is the neighborhood around p_n, $\rho=\dfrac{1-\delta}{1+\delta}$.

Let us observe the $D(w,\varepsilon)$ in detail. In Theorem 1.1, when changing the variables, we set $g(z,0)=-it+\tau=-iz_n$. Hence, if the definition of $D(w,\varepsilon)$ depends on $g(z,w)$, $D(w,\varepsilon)$ may be written as $D(w,\varepsilon;\operatorname{Re}g,\operatorname{Im}g)$ it becomes $D(0,\varepsilon,\operatorname{Re}g(z,0),\operatorname{Im}g(z,0))$, i. e. $D(0,\varepsilon,\tau,-t)$ after the change of variables and becomes

$$D\left(p_n,\ \varepsilon,\ \frac{1-|u_n|^2}{|1+u_n|^2},\ \frac{-2\operatorname{Im}u_n}{|1+u_n|^2}\right)$$

under the generalized transformation of Cayley.

For example

$$D(w,\varepsilon,\operatorname{Re}g,\operatorname{Im}g)=\{\alpha^2(\operatorname{Re}g)^2+\beta^2(\operatorname{Im}g)^2\leqslant\varepsilon^2\}$$

may be reduced to

$$\{\alpha^2(1-|u_n|^2)^2+4\beta^2(\operatorname{Im}u_n)^2\leqslant\varepsilon^2\};$$

$$D(w,\varepsilon,\operatorname{Re}g,\operatorname{Im}g)=\{|\operatorname{Re}g|<\alpha\varepsilon,\ |\operatorname{Im}g|<\beta\varepsilon,\ \alpha>0,\ \beta>0\}$$

may be reduced to

$$\{1-|u_n|^2<\alpha\varepsilon,\ 2|\operatorname{Im}u_n|<\beta\varepsilon,\ \alpha>0,\ \beta>0\}.$$

In the sections 2 and 3, we shall give some concrete neighborhoods $D(w,\varepsilon)$; calculate the value of a in (1.2.8), one can discover some interesting results.

1.3. Singular integrals on $b\Omega$

When $w\in b\Omega$, $\int_{z\in b\Omega} H(w, z)u(z)d\sigma_z$ is a singular integral.

Let $z\in b\Omega$, $\varphi(z)\in \mathrm{Lip}\,p$, $0<p\leqslant 1$, v, $w\in b\Omega$, set

$$\varphi_1(w)=\int_{b\Omega(a)} H(w, z)\varphi(z)d\sigma_z.\quad \varphi_2(v)=\int_{b\Omega(b)} H(v, z)\varphi_1(z)d\sigma_z,$$

where $\int_{b\Omega(a)}$ stands for the Cauchy principal value, the deleted neighborhood is $D_a(w, \varepsilon)$, such that the corresponding value of (1.2.3) is a.

Consider the Cauchy integrals

$$f(\eta)=\int_{b\Omega} H(\eta, z)\varphi(z)d\sigma_z,\ f_1(\eta)=\int_{b\Omega} H(\eta, z)\varphi_1(z)d\sigma_z,\ \eta\in\Omega,$$

when η approaches $\zeta\in b\Omega$ along the nontangential direction, by (1.2.4)

$$f(\zeta)=\int_{b\Omega(a)} H(\zeta, z)\varphi(z)d\sigma_z+a\varphi(\zeta),\quad f_1(\zeta)=\int_{b\Omega(b)} H(\zeta, z)\varphi_1(z)d\sigma_z+b\varphi_1(\zeta).$$

By the definition of $\varphi_1(\zeta)$; $\varphi_2(\zeta)$, we have

$$\varphi_1(\zeta)=f(\zeta)-a\varphi(\zeta),\ \varphi_2(\zeta)=f_1(\zeta)-b\varphi_1(\zeta),\ \zeta\in b\Omega.$$

Substitute $\varphi_1(\zeta)$ into the expression of $f_1(\eta)$

$$f_1(\eta)=\int_{b\Omega(b)} H(\eta, z)f(z)d\sigma_z-a\int_{b\Omega(b)} H(\eta, z)\varphi(z)d\sigma_z=(1-a)f(\eta),$$

therefore

$$\varphi_2(\zeta)=(1-a)f(\zeta)-b[f(\zeta)-a\varphi(\zeta)]=(1-a-b)f(\zeta)+ab\varphi(\zeta).$$

For a, if there is a neighborhood $D_b(w, \varepsilon)$ around w, such that the corresponding value of (1.2.3) is b and $1-a-b=0$, then $\varphi_2(\zeta)=ab\varphi(\zeta)$, that is to say

$$\mathbf{H}_b\mathbf{H}_a=abI.$$

where I stands for the identity operator. In the same manner, we can also prove

$$\mathbf{H}_a\mathbf{H}_b=abI.$$

Thus we obtain the generalized Poincaré-Bertrand formula.

Theorem 1.3. *Let Ω be a smooth, bounded strictly pseudoconvex domain of $\mathbf{C}^n$, $H(w, z)$ is H-R kernel or S-K kernel, $\varphi\in C^\infty(b\Omega)$. The singular integral operator $\mathbf{H}_a$ is defined by (1.2.2). If for a, there is another definition of Cauchy principal value, such that the value of (1.2.3) is b, and $1-a-b=0$. Suppose the corresponding singular integral operator is $\mathbf{H}_b$, then*

$$\mathbf{H}_a\mathbf{H}_b=\mathbf{H}_b\mathbf{H}_a=abI,\tag{1.3.1}$$

where I stands for the identity operator.

If $ab\neq 0$, (1.3.1) may also be written as: denote $\frac{1}{a}\mathbf{H}_a\varphi=\psi$, then $\frac{1}{b}\mathbf{H}_b\psi=\varphi$; or denote $\frac{1}{b}\mathbf{H}_b\varphi=\psi$, then $\frac{1}{a}\mathbf{H}_a\psi=\varphi$.

It is easy to know by the operator theory, if λ is not a characteristic value of the singular integral operator $\mathbf{H}_a$, $\mathbf{K}$ is a continuous operator, $f\in C^\infty(b\Omega)$, then the singular integral equation on $b\Omega$

$$(-\lambda I+\mathbf{H}_a+\mathbf{K})\varphi=\psi \tag{1.3.2}$$

has exactly one solution, namely it can be normalized to a Fredholm integral equation. When $D(w, \varepsilon)=\{|g(w, z)|<\varepsilon\}$, Ω is a ball, this is just the theory of the singular integral equation on a sphere[8, 9].

When $z\in b\Omega$, $w\in b\Omega$, the H-R kernel or S-K kernel $H(w, z)$ of Ω are both singular kernels. For the case of sphere, one can define B kernel and Hilbert kernel from $H(w, z)$

$$B(w, z)=H(w, z)+H(z, w)-1,$$

$$h(w, z)=\frac{1}{i}[H(w, z)-H(z, w)].$$

From these kernels, we also have

$$\mathbf{B}_a\varphi=\int_{b\Omega(a)} B(w, z)\varphi(z)d\sigma_z,$$

$$\mathbf{h}_a\varphi=\int_{b\Omega(a)} h(w, z)\varphi(z)d\sigma_z,$$

where the integrals denote the Cauchy principal value, the deleted neighborhood around w is $D_\varepsilon(w, \varepsilon)$ such that the value of (1.2.3) is a.

Obviously, if $\mathbf{H}_a$ has an inverse operator, then $\mathbf{B}_a$ and $\mathbf{h}_a$ also have the inverse operators. Therefore, we can also solve the singular integral equations on $b\Omega$

$$(-\mu I+\mathbf{B}_a+\mathbf{K})\varphi=f \tag{1.3.3}$$

and

$$(-\nu I+\mathbf{h}_a+\mathbf{K})\varphi=f, \tag{1.3.4}$$

where μ, ν are not the characteristic value of $\mathbf{B}_a$ and $\mathbf{h}_a$ respectively, $\mathbf{K}$ is a continuous operator, $f\in C^\infty(b\Omega)$.

As in [8, 9], one can solve the systems of singular integral equations similar to (1.3.2), (1.3.3), (1.3.4) and when $D(w, \varepsilon)=\{|g(w, \varepsilon)|<\varepsilon\}$, Ω is a ball, this is just the theory of singular integral equations discussed in [8].

§ 2. The case that the neighborhood is an "ellipse"

2.1. Plemelj formula

In this section, we shall consider some special neighborhoods $D(w, \varepsilon)$, such that the concrete Plemelj formulas are obtained from the general Plemelj formula. We first consider the case that the neighborhood $D_\varepsilon(w, \varepsilon)$ is an "ellipse":

$$\{z\in b\Omega,\ \alpha^2(\operatorname{Re} g)^2+\beta^2(\operatorname{Im} g)^2\leqslant\varepsilon^2\},$$

where $\alpha\geqslant0$, $\beta\geqslant0$, $\alpha+\beta\neq0$. When $\alpha=\beta$, the "ellipse" becomes a, "disc", this is just the case discussed by Alt[13] and Kerzman-Stein[14]. In this case, we shall prove that (1.2.3) becomes

$$\lim_{\varepsilon\to 0}\lim_{\delta\to 0}\int_{z\in D_\varepsilon(w,\varepsilon)} H(w+\delta\nu,\ z)d\sigma_z=\frac{1}{2}\left(\frac{2\beta}{\alpha+\beta}\right)^{n-1}. \tag{2.1.1}$$

That is to say, the value of a in Plemelj formula (1.2.4) is

$$a=\frac{1}{2}\left(\frac{2\beta}{\alpha+\beta}\right)^{n-1}.$$

When $\alpha=\beta$, this is just the Theorem of Alt[13] and Kerzman-Stein[14]. Therefore, this result has generalized the works of Alt and Kerzman-Stein.

Another case $\beta=0$ must be noted. In this case, $a=0$, it follows that the Plemelj formula becomes vary simple

$$\mathbf{H}u(w)=\mathbf{H}_0u(w).$$

In other words, when w_0 approaches $w\in b\Omega$ along the nontangential direction from the interior of Ω

$$\lim_{w_0\to w}\int_{z\in b\Omega} H(w,\ z)u(z)d\sigma_z=\lim_{\substack{\varepsilon\to 0\\ \varepsilon>0}}\int_{z\in b\Omega-D(w,\varepsilon)\cap b\Omega} H(w,\ z)u(z)d\sigma_z.$$

Now $D(w,\ z)$ is a "belt"

$$\{z\in b\Omega,\ |\mathrm{Re}\,g|<\varepsilon\}.$$

Namely, one can find a manner to define the principal value of Cauchy integral, such that this principal value is just the limit value of the Cauchy integral $\mathbf{H}u(w_0)$ when w_0 approaches $w\in b\Omega$ along the nontangential direction from the interior of Ω. It is impossible in one variable that the limit value of the Cauchy integral on $b\Omega$ can be represented by a certain principal value of the Cauchy integral on $b\Omega$.

We now discuss the conditions under which the singular integral operator $\mathbf{H}_a$ has an inverse operator. It has been pointed out in the section **1.3** that if one can find an operator $\mathbf{H}_b$, such that $1-a-b=0$, $ab\neq 0$, then $\frac{1}{b}\mathbf{H}_b$ is the inverse operator of $\frac{1}{a}\mathbf{H}_a$. Obviously, if

$$\frac{1}{2}\left(\frac{2\beta}{\alpha+\beta}\right)^{n-1}<1,$$

i. e.

$$\frac{\alpha}{\beta}>2^{\frac{n-2}{n-1}}-1, \tag{2.1.2}$$

take
$$D_e(w,\ \varepsilon)=\{z\in b\Omega,\ \alpha'^2(\mathrm{Re}\,g)^2+\beta'^2(\mathrm{Im}\,g)^2\leqslant\varepsilon^2\},$$

where $\alpha'\geqslant 0$, $\beta'\geqslant 0$, $\alpha'+\beta'\neq 0$ and

$$\frac{\alpha'}{\beta'}=\left[2^{2-n}-\left(\frac{\alpha+\beta}{\beta}\right)^{1-n}\right]^{-\frac{1}{n-1}}-1,$$

if denote
$$\frac{1}{2}\left(\frac{2\beta'}{\alpha'+\beta'}\right)^{n-1}=b,$$

then $\frac{1}{ab}\mathbf{H}_b$ is the inverse operator of $\mathbf{H}_a$. That is to say, under the condition of (2.1.2), $\mathbf{H}_a$ has an inverse operator. In the same way, we may set up the theory of singnlar integral equations.

2.2. The proof of (2.1.1)

To prnve (2.1.1), it is enough to prove that (1.2.8) holds by § 1.2. Namely

$$\lim_{\varepsilon\to 0}\lim_{\rho\to 1}\frac{1}{\omega_{2n-1}}\int_{u\in D(p_n,\varepsilon)}\frac{\dot{u}}{(1-\rho p_n\bar{u}')^n}=\lim_{\varepsilon\to 0}\lim_{\rho\to 1}\frac{1}{\omega_{2n-1}}\int_{u\in D(p_n,\varepsilon)}\frac{\dot{u}}{(1-\rho\bar{u}_n)^n}=\frac{1}{2}\left(\frac{2\beta}{\alpha+\beta}\right)^{n-1},$$

where $p_n=(0,\cdots,0,1)$, $0<\rho<1$, and

$$D(p_n,\varepsilon)=\{u\bar{u}'=1,\ \alpha^2(1-|u_n|^2)^2+4\beta^2(\operatorname{Im}u_n)^2\leqslant\varepsilon^2\}.$$

Let

$$\tilde{D}(p_n,\varepsilon)=\{u\bar{u}'=1,\ \alpha^2(1-|u_n|^2)^2+4\beta^2(\operatorname{Im}u_n)^2>\varepsilon^2\}.$$

Since

$$\frac{1}{\omega_{2n-1}}\int_{u\in D(p_n,\varepsilon)}\frac{\dot{u}}{(1-\rho\bar{u}_n)^n}+\frac{1}{\omega_{2n-1}}\int_{u\in\tilde{D}(p_n,\varepsilon)}\frac{\dot{u}}{(1-\rho\bar{u}_n)^n}=1,$$

we only need to prove

$$\lim_{\varepsilon\to 0}\lim_{\rho\to 1}\frac{1}{\omega_{2n-1}}\int_{u\in\tilde{D}(p_n,\varepsilon)}\frac{\dot{u}}{(1-\rho\bar{u}_n)^n}=\lim_{\varepsilon\to 0}\frac{1}{\omega_{2n-1}}\int_{u\in\tilde{D}(p_n,\varepsilon)}\frac{\dot{u}}{(1-\bar{u}_n)^n}=1-\frac{1}{2}\left(\frac{2\beta}{\alpha+\beta}\right)^{n-1},\tag{2.2.1}$$

where $\alpha\geqslant 0$, $\beta\geqslant 0$, $\alpha+\beta\neq 0$.

when $\alpha=0$, $\beta>0$, (2.2.1) has been proved in [7].

We now consider the case $\alpha>0$, $\beta>0$.

As has been discussed in [3, 7], let

$$\bar{u}_n=re^{i\theta},\quad v=(u_1,\cdots,u_{n-1}),$$

tnen $\tilde{D}(p_n,\varepsilon)$ may be witten in form

$$\begin{cases} v\bar{v}'=1-r^2,\\ \alpha^2(1-r^2)^2+4\beta^2r^2\sin^2\theta>\varepsilon^2.\end{cases}$$

Set

$$c=\arcsin\frac{\sqrt{\varepsilon^2-\alpha^2(1-r^2)^2}}{2\beta r},$$

we have

$$\frac{1}{\omega_{2n-1}}\int_{u\in\tilde{D}(p_n,\varepsilon)}\frac{\dot{u}}{(1-\bar{u}_n)^n}=\frac{1}{\omega_{2n-1}}\int_{v\bar{v}'<\frac{\varepsilon}{\alpha}}\dot{v}\left\{\int_{-(\pi-c)}^{-c}\frac{d\theta}{(1-re^{i\theta})^n}+\int_c^{\pi-c}\frac{d\theta}{(1-re^{i\theta})^n}\right\}+\frac{1}{\omega_{2n-1}}\int_{v\bar{v}'>\frac{\varepsilon}{\alpha}}\dot{v}\int_{-\pi}^{\pi}\frac{d\theta}{(1-re^{i\theta})^n}=I_1+I_2.\tag{2.2.2}$$

It is known by [3, 7]

$$\int_{-(\pi-c)}^{-c}\frac{d\theta}{(1-re^{i\theta})^n}+\int_c^{\pi-c}\frac{d\theta}{(1-re^{i\theta})^n}=2\operatorname{Im}\left\{\sum_{k=1}^{n-1}\frac{1}{k(1+re^{-ic})^k}-\sum_{k=1}^{n-1}\frac{1}{k(1-re^{ic})^k}+\log\frac{1-re^{ic}}{1+re^{-ic}}\right\}+2(\pi-2c),$$

hence

$$I_1=\frac{2}{\omega_{2n-1}}\operatorname{Im}\left\{\sum_{k=1}^{n-1}J_k-\sum_{k=1}^{n-1}H_k+J_0-H_0\right\}+\frac{2}{\omega_{2n-1}}\int_{v\bar{v}'<\frac{\varepsilon}{\alpha}}(\pi-2c)\dot{v},$$

where $J_0=\int_{v\bar{v}'<\frac{\varepsilon}{\alpha}}\log\frac{1}{1+re^{-ic}}\dot{v}$, $J_k=\int_{vv'<\frac{\varepsilon}{\alpha}}\frac{\dot{v}}{k(1+re^{-ic})^k}$, $k=1,2,\cdots,n-1$,

$$H_0=\int_{v\bar{v}'<\frac{\varepsilon}{\alpha}}\log\frac{1}{1-re^{ic}}\dot{v},\quad H_k=\int_{v\bar{v}'<\frac{\varepsilon}{\alpha}}\frac{\dot{v}}{k(1-re^{ic})^k},\ k=1,2,\cdots,n-1.$$

Since $$\sin c=\frac{\sqrt{\varepsilon^2-\alpha^2(1-r^2)^2}}{2\beta r},\quad \cos c=\frac{\sqrt{4\beta^2r^2-\varepsilon^2+\alpha^2(1-r^2)^2}}{2\beta r},$$

so $$1+re^{-ic}=1+\frac{\sqrt{4\beta^2(1-s^2)+\alpha^2s^4-\varepsilon^2}}{2\beta}-i\frac{\sqrt{\varepsilon^2-\alpha^2s^4}}{2\beta},$$

here $s^2=1-r^2$. Let $v=(x_1, x_2, \cdots, x_{2n-2})$, adopt the spherical polar coorolinates

$$x_1=s\cos\varphi_1,\ x_2=s\sin\varphi_1\cos\varphi_2,\ \cdots,\ x_{2n-2}=s\sin\varphi_1\sin\varphi_2\cdots\sin\varphi_{2n-3},$$

then $$\dot{v}=s^{2n-3}\sin^{2n-4}\varphi_1\sin^{2n-5}\varphi_2\cdots\sin\varphi_{2n-4}dsd\varphi_1\cdots d\varphi_{2n-3}.$$

It follows that

$$J_k=\frac{2\pi^{n-1}}{k\Gamma(n-1)}\int_0^{\sqrt{\frac{\varepsilon}{\alpha}}}\left\{1+\frac{\sqrt{4\beta^2(1-s^2)+\alpha^2s^4-\varepsilon^2}}{2\beta}-i\frac{\sqrt{\varepsilon^2-\alpha^2s^4}}{2\beta}\right\}^{-k}s^{2n-3}ds.$$

Let $\eta=\frac{\varepsilon}{\alpha}$, $s=\sqrt{\eta}\,t$, then J_k is equal to

$$\frac{2\pi^{n-1}(2\beta)^k}{k\Gamma(n-1)}\eta^{n-1}\int_0^1\frac{t^{2n-3}dt}{[2\beta+\sqrt{4\beta^2(1-\eta t^2)+\alpha^2\eta^2(t^4-1)}-i\sqrt{\alpha^2\eta^2(1-t^4)}]^k},$$

since the absolute value of the integrand is bounded, hence

$$\lim_{\varepsilon\to0}J_k=0,\ k=1,2,\cdots,n-1.$$

We may prove $\lim\limits_{\varepsilon\to0}J_0=0$ in the same manner.

As we do for J_k, H_k may be written as

$$H_k=\frac{2\pi^{n-1}(2\beta)^k}{k\Gamma(n-1)}\int_0^1\frac{\eta^{n-1}t^{2n-3}dt}{[2\beta-\sqrt{4\beta^2(1-\eta t^2)+\alpha^2\eta^2(t^4-1)}-i\sqrt{\alpha^2\eta^2(1-t^4)}]^k}.$$

When $k<n-1$, the absolute value of the integrand is not greater than

$$\frac{M}{\beta^2t^4+\alpha^2(1-t^4)-1},$$

where M is an absolute constant, it is integrable in $[0, 1]$, so

$$\lim_{\varepsilon\to0}H_k=0,\ k=1,2,\cdots,n-2.$$

When $k=n-1$, we have

$$H_{n-1}=\frac{2(2\beta\pi)^{n-1}}{\Gamma(n)}\int_0^1\frac{\eta^{n-1}t^{2n-3}dt}{[2\beta-\sqrt{4\beta^2(1-\eta t^2)+\alpha^2\eta^2(t^4-1)}-i\sqrt{\alpha^2\eta^2(1-t^4)}]^{n-1}}$$
$$=\frac{2(2\beta\pi)^{n-1}}{\Gamma(n)}\int_0^1\frac{t^{2n-3}dt}{[\beta t^2-i\alpha\sqrt{1-t^4}+O(\eta)]^{n-1}}.$$

Set $\gamma=\frac{\alpha}{\beta}$, then

$$\lim_{\varepsilon\to0}H_{n-1}=\frac{2^n\pi^{n-1}}{\Gamma(n)}\int_0^1\frac{t^{2n-3}dt}{(t^2-i\gamma\sqrt{1-t^4})^{n-1}}.$$

Let $t^2=\cos\theta$

$$\lim_{\varepsilon\to 0} H_{n-1} = \frac{(2\pi)^{n-1}}{\Gamma(n)} \int_0^{\frac{\pi}{2}} \frac{\cos^{n-2}\theta \sin\theta\, d\theta}{(\cos\theta - i\gamma\sin\theta)^{n-1}}$$

$$= \frac{(2\pi)^{n-1}}{\Gamma(n)(1+\gamma)^{n-1}} \int_0^{\frac{\pi}{2}} \frac{(e^{2i\theta}+1)^{n-2}(e^{2i\theta}-1)}{\left(1+\frac{1-\gamma}{1+\gamma}e^{2i\theta}\right)^{n-1}} d\theta$$

$$= \frac{(2\pi)^{n-1}}{\Gamma(n)(1+\gamma)^{n-1}} \int_0^{\frac{\pi}{2}} \left\{ \frac{1}{i} \sum_{\substack{p=0\\p+q\neq 0}}^{n-2} \sum_{q=0}^{\infty} (-1)^q C_p^{n-2} C_q^{n+q-2} \left(\frac{1-\gamma}{1+\gamma}\right)^q \right.$$

$$\left. \times \left[e^{2i\theta(p+q+1)} - e^{2i\theta(p+q)}\right] + \frac{1}{i}(e^{2i\theta}-1) \right\} d\theta$$

$$= -\frac{(2\pi)^{n-1}}{2\Gamma(n)(1+\gamma)^{n-1}} \sum_{\substack{p=0\\p+q\neq 0}}^{n-2} \sum_{p\neq 0}^{\infty} (-1)^q C_p^{n-2} C_q^{n+q-2} \left(\frac{1-\gamma}{1+\gamma}\right)^q$$

$$\times \left[\frac{(-1)^{p+q+1}-1}{p+q+1} - \frac{(-1)^{p+q}-1}{p+q}\right] + \frac{(2\pi)^{n-1}}{\Gamma(n)(1+\gamma)^{n-1}} \left(1 - \frac{\pi}{2i}\right).$$

It follows that

$$\lim_{\varepsilon\to 0} \operatorname{Im}(H_{n-1}) = \operatorname{Im}(\lim_{\varepsilon\to 0} H_{n-1}) = \frac{2^{n-2}\pi^n}{\Gamma(n)(1+\gamma)^{n-1}}.$$

We now calculate H_0

$$H_0 = \int_{v\bar{v}' < \frac{\varepsilon}{\alpha}} \log \frac{1}{1-re^{ic}} \dot{v} = -\frac{2\pi^{n-1}}{\Gamma(n-1)} \int_0^{\sqrt{\frac{\varepsilon}{\alpha}}} s^{2n-3} \log(1-re^{ic})\, ds.$$

Since $\operatorname{Im}\log(1-re^{ic}) = \arg(1-re^{ic}) = O(1),$

hence when $n>1$

$$\operatorname{Im}(H_0) = O\left(\int_0^{\sqrt{\frac{\varepsilon}{\alpha}}} s^{2n-3}\, ds\right) = O(\varepsilon^{n-1}) = o(1) \quad (\varepsilon\to 0).$$

By the same reason, when $n>1$, we have

$$\int_{v\bar{v}' < \frac{\varepsilon}{\alpha}} (\pi - 2c)\dot{v} = o(1).$$

Sum up the above results, we obtain

$$\lim_{\varepsilon\to 0} I_1 = -\frac{1}{2}\left(\frac{2}{1+\gamma}\right)^{n-1} = -\frac{1}{2}\left(\frac{2\beta}{\alpha+\beta}\right)^{n-1}.$$

As to I_2, by virtue of

$$\int_{-\pi}^{\pi} \frac{d\theta}{(1-re^{i\theta})^n} = 2\operatorname{Im}\left\{\sum_{k=1}^{n-1} \frac{1}{k(1+r)^k} + \sum_{k=1}^{n-1} \frac{1}{k(1-r)^k}\right\} + 2\pi = 2\pi,$$

$$\lim_{\varepsilon\to 0} I_2 = \frac{2\pi}{\omega_{2n-1}} \int_{0<v\bar{v}'<1} \dot{v} = 1.$$

Substitute thses results into (2.2.2), we obtain (2.2.1) immediately.

Finally, we consider the case $\alpha>0$, $\beta=0$.

(2.2.1) now becomes

$$\lim_{\varepsilon\to 0} \frac{1}{\omega_{2n-1}} \int_{u\in \tilde{D}(p_n,\varepsilon)} \frac{\dot{u}}{(1-\bar{u}_n)^n} = 1,$$

where $\tilde{D}(p_n, \varepsilon) = \left\{u\bar{u}' = 1,\ 1 - |u_n|^2 > \frac{\varepsilon}{\alpha}\right\}.$

using the above transformations, one may write $\tilde{D}(p_n, \varepsilon)$ as

$$\begin{cases} vv'=1-r^2, \\ 1-r^2>\dfrac{\varepsilon}{\alpha}, \end{cases}$$

then
$$\frac{1}{\omega_{2n-1}}\int_{u\in\tilde{D}(p_n,\varepsilon)}\frac{\dot{u}}{(1-\bar{u}_n)^n}=\frac{1}{\omega_{2n-1}}\int_{\frac{\varepsilon}{\alpha}<v\bar{v}'<1}\dot{v}\int_{-\pi}^{\pi}\frac{d\theta}{(1-re^{i\theta})^n}.$$

This is just the integral I_2 discussed before, but it is known $\lim_{\varepsilon\to 0} I_2=1$, so (2.2.1) holds when $\alpha>0$, $\beta=0$.

§ 3. The case that the neighborhood is a "rectangle"

3.1. Plemelj formula

Now let the neighborhood $D_R(w, \varepsilon)$ around w be a "rectangle"

$$\{z\in b\Omega,\ |\mathrm{Re}\, g|<\alpha\varepsilon,\ |\mathrm{Im}\, g|<\beta\},$$

here $\alpha>0$, $\beta>0$. In this case, we shall prove that (1.2.3) becomes

$$\lim_{\varepsilon\to 0}\lim_{\delta\to 0}\int_{z\in D_R(w,\varepsilon)} H(w+\delta v, z)d\sigma_z=\frac{2^{n-1}}{\pi}\left\{\frac{\pi}{2}-h_n\left(\mathrm{arc\,tg}\,\frac{\beta}{\alpha}\right)\right\}, \tag{3.1.1}$$

where
$$h_n(x)=\int_0^x \cos^{n-2}t\,\frac{\sin(n-1)t}{\sin t}\,dt,$$

and the value of a in Plemelj formula (1.2.4) is

$$a=\frac{2^{n-1}}{\pi}\left\{\frac{\pi}{2}-h_n\left(\mathrm{arc\,tg}\,\frac{\beta}{\alpha}\right)\right\}. \tag{3.1.2}$$

In particular when $\alpha=\beta$, i. e. the rectangle becomes a square,

$$a=\frac{2^{n-1}}{\pi}\left\{\frac{\pi}{2}-h_n\left(\frac{\pi}{4}\right)\right\}.$$

When $\alpha=\infty$, $D_R(w, \varepsilon)$ becomes

$$\{z\in b\Omega,\ |\mathrm{Im}\, g|<\beta\varepsilon\},$$

this is just the case discussed in the preceding section. By virture of $h_n(0)=0$, so $a=2^{n-2}$, this coincides with the result of the preceding section.

When $\beta=\infty$, $D_R(w, \varepsilon)$ becomes

$$\{z\in b\Omega,\ |\mathrm{Re}\, g|<\alpha\varepsilon\},$$

this is the case discussed in the preceding section too. We shall see in § 3.2

$$h_n\left(\mathrm{arc\,tg}\,\frac{\beta}{\alpha}\right)=\mathrm{arc\,tg}\,\frac{\beta}{\alpha}+\sum_{k=1}^{n-2}\frac{\alpha^k}{k(\alpha^2+\beta^2)^{k/2}}\sin\left(k\,\mathrm{arc\,tg}\,\frac{\beta}{\alpha}\right),$$

therefore, when $\beta\to\infty$ $\mathrm{arc\,tg}\,\frac{\beta}{\alpha}\to\frac{\pi}{2}$, so $h_n\left(\frac{\pi}{2}\right)=\frac{\pi}{2}$. It follows that $a=0$, this coincides with the result of the preceding section too.

As the preceding section, for the singular integal operator $\mathbf{H}_a$, if one can find an operator $\mathbf{H}_b$, such that $1-a-b=0$, $ab\neq 0$ then $\frac{1}{b}\mathbf{H}_b$ is the inverse of $\frac{1}{a}\mathbf{H}_a$.

It is not hard to prove $h_n(x)\geqslant 0$ for arbitrary positive integer n and $x\in\left[0, \frac{\pi}{2}\right]$, and

$$\max_{0<x<\frac{\pi}{2}} h_n(x)=h_n\left(\frac{\pi}{n-1}\right).$$

If α, β satisfy

$$(1-2^{1-n})\pi-h_n\left(\frac{\pi}{n-1}\right)\leqslant h_n\left(\operatorname{arc\,tg}\frac{\beta}{\alpha}\right)\leqslant(1-2^{1-n})\pi,$$

then $\mathbf{H}_a$ (a is defined by (3.1.2)) has an inverse operator. That is to say, we only choose $D_R(w, \varepsilon)$ as

$$\{z\in b\Omega,\ |\operatorname{Re} g|<\alpha'\varepsilon,\ |\operatorname{Im} g|<\beta'\varepsilon\},$$

where $\alpha'>0$, $\beta'>0$, such that

$$h_n\left(\operatorname{arc\,tg}\frac{\beta'}{\alpha'}\right)=(1-2^{1-n})\pi-h_n\left(\operatorname{arc\,tg}\frac{\beta}{\alpha}\right).$$

Set

$$b=\frac{2^{n-1}}{\pi}\left\{\frac{\pi}{2}-h_n\left(\operatorname{arc\,tg}\frac{\beta'}{\alpha'}\right)\right\},$$

then $\frac{1}{b}\mathbf{H}_b$ is the inverse of $\frac{1}{a}\mathbf{H}_a$. It follows that under the condition (3.1.3), $\mathbf{H}_a$ is invertible. We may also discuss the theory of the singular integral equations in the same way.

3.2. The proof of (3.1.1)

To prove (3.1.1), by the argument of § 1.2, we only need to prove that (1.2.8) holds, i. e.

$$\lim_{\varepsilon\to 0}\lim_{\rho\to 1}\frac{1}{\omega_{2n-1}}\int_{u\in D_R(p_n,\varepsilon)}\frac{\dot{u}}{(1-\rho\bar{u}_n)^n}=\frac{2^{n-1}}{\pi}\left\{\frac{\pi}{2}-h_n\left(\operatorname{arc\,tg}\frac{\beta}{\alpha}\right)\right\},$$

where $p_n=(0, \cdots, 0, 1)$, $\alpha>0$, $\beta>0$

$$D_R(p_n, \varepsilon)=\{u\bar{u}'=1,\ 1-|u_n|^2<\alpha\varepsilon,\ 2|\operatorname{Im} u_n|<\beta\varepsilon\}.$$

Let

$$\tilde{D}_R(p_n, \varepsilon)=\{u\bar{u}'=1\}-D_R(p_n, \varepsilon),$$
$$M_\varepsilon=\{u\bar{u}'=1,\ 1-|u_n|^2>\alpha\varepsilon\},$$
$$N_\varepsilon=\{u\bar{u}'=1,\ 2|\operatorname{Im} u_n|>\beta\varepsilon\},$$
$$Q_\varepsilon=\{u\bar{u}'=1,\ 1-|u_n|^2>\alpha\varepsilon,\ 2|\operatorname{Im} u_n|>\beta\varepsilon\}.$$

Thus the result which needs to be proved is equivalent to

$$\lim_{\varepsilon\to 0}\frac{1}{\omega_{2n-1}}\int_{\tilde{D}_R(p_n,\varepsilon)}\frac{\dot{u}}{(1-\bar{u}_n)^n}=1-\frac{2^{n-1}}{\pi}\left\{\frac{\pi}{2}-h_n\left(\operatorname{arc\,tg}\frac{\beta}{\alpha}\right)\right\}. \tag{3.2.1}$$

But

$$\frac{1}{\omega_{2n-1}}\int_{\tilde{D}_R(p_n,\varepsilon)}\frac{\dot{u}}{(1-\bar{u}_n)^n}=\left\{\frac{1}{\omega_{2n-1}}\int_{M_\varepsilon}+\frac{1}{\omega_{2n-1}}\int_{N_\varepsilon}-\frac{1}{\omega_{2n-1}}\int_{Q_\varepsilon}\right\}\frac{\dot{u}}{(1-\bar{u}_n)^n}. \tag{3.2.2}$$

It is known by § 2.2

$$\lim_{\varepsilon\to 0}\frac{1}{\omega_{2n-1}}\int_{M_\varepsilon}\frac{\dot{u}}{(1-\bar{u}_n)^n}=1,\quad \lim_{\varepsilon\to 0}\frac{1}{\omega_{2n-1}}\int_{N_\varepsilon}\frac{\dot{u}}{(1-\bar{u}_n)^n}=1-2^{n-2}. \tag{3.2.3}$$

so we only need to calculate

$$\lim_{\varepsilon\to 0}\frac{1}{\omega_{2n-1}}\int_{Q_\varepsilon}\frac{\dot{u}}{(1-\bar{u}_n)^n}.$$

As the argument in § 2.2, set

$$\bar{u}_n=re^{i\theta},\ v=(u_1,\ \cdots,\ u_{n-1}),$$

then Q_ε may be written as

$$\begin{cases}v\bar{v}'=1-r^2,\\ 1-r^2>\alpha\varepsilon,\ 2r|\sin\theta|>\beta\varepsilon,\end{cases}$$

$$\frac{1}{\omega_{2n-1}}\int_{Q_\varepsilon}\frac{\dot{u}}{(1-\bar{u}_n)^n}=\frac{1}{\omega_{2n-1}}\int_{P_\varepsilon}\dot{v}\left\{\int_{-(\pi-c)}^{-c}+\int_c^{\pi-c}\right\}\frac{d\theta}{(1-re^{i\theta})^n},$$

where $c=\arcsin\dfrac{\beta\varepsilon}{2r}$ and

$$P_\varepsilon=\left\{v=(u_1,\ \cdots,\ u_{n-1}),\ \alpha\varepsilon<v\bar{v}'<1-\left(\frac{\beta\varepsilon}{2}\right)^2\right\}.$$

By [3, 7], the above equality may be written as

$$\frac{1}{\omega_{2n-1}}\int_{Q_\varepsilon}\frac{\dot{u}}{(1-\bar{u}_n)^n}=\frac{2}{\omega_{2n-1}}\operatorname{Im}\left\{\sum_{k=1}^{n-1}J_k-\sum_{k=1}^{n-1}H_k+J_0-H_0\right\}+\frac{1}{\omega_{2n-1}}\int_{P_\varepsilon}(\pi-2c)\dot{v},$$

here $\quad J_0=\displaystyle\int_{P_\varepsilon}\log\frac{1}{1+re^{-ic}}\dot{v},\ J_k=\int_{P_\varepsilon}\frac{\dot{v}}{k(1+re^{-ic})^k},\ k=1,\ 2,\ \cdots,\ n-1,$

$$H_0=\int_{P_\varepsilon}\log\frac{1}{1-re^{ic}}\dot{v},\ H_k=\int_{P_\varepsilon}\frac{\dot{v}}{k(1-re^{ic})^k},\ k=1,\ 2,\ \cdots,\ n-1.$$

Use the spherical polar coordinates

$$\int_{P_\varepsilon}c\dot{v}=\int_{P_\varepsilon}\arcsin\frac{\beta\varepsilon}{2\sqrt{1-v\bar{v}'}}\dot{v}=\frac{2\pi^{n-1}}{\Gamma(n-1)}\int_{\sqrt{\alpha\varepsilon}}^{\sqrt{1-\left(\frac{\beta\varepsilon}{2}\right)^2}}s^{2n-3}\arcsin\frac{\beta\varepsilon}{2\sqrt{1-s^2}}ds$$

$$=\frac{2\pi^{n-1}}{\Gamma(n-1)}\left\{\int_0^1\left[1-\left(\frac{\beta\varepsilon}{2}\right)^2\right]^{n-1}\arcsin\frac{\beta\varepsilon}{2\sqrt{1-\left[1-\left(\frac{\beta\varepsilon}{2}\right)^2\right]t^2}}dt\right.$$

$$\left.-\int_0^1(\alpha\varepsilon)^{n-1}\arcsin\frac{\beta\varepsilon}{2\sqrt{1-\alpha\varepsilon t^2}}dt\right\}.$$

Since the integrands are bounded,

$$\lim_{\varepsilon\to 0}\int_{P_\varepsilon}c\dot{v}=0.$$

It follows that

$$\lim_{\varepsilon\to 0}\frac{2}{\omega_{2n-1}}\int_{P_\varepsilon}(\pi-2c)\dot{v}=1.$$

By virture of $\qquad 1+re^{-ic}=1+\dfrac{1}{2}\sqrt{4r^2-\beta^2\varepsilon^2}-i\dfrac{1}{2}\beta\varepsilon,$

hence

$$J_k=\frac{2\pi^{n-1}}{k\Gamma(n-1)}\int_{\sqrt{\alpha\varepsilon}}^{\sqrt{1-\left(\frac{\beta\varepsilon}{2}\right)^2}}s^{2n-3}\left[1+\frac{1}{2}\sqrt{4(1-s^2)-\beta^2\varepsilon^2}-\frac{i}{2}\beta\varepsilon\right]^{-k}ds$$

$$=\frac{2\pi^{n-1}}{k\Gamma(n-1)}\int_0^1\left[1-\left(\frac{\beta\varepsilon}{2}\right)^2\right]^{n-1}t^{2n-3}\left[1+\frac{1}{2}\sqrt{4\left[1-\left(1-\frac{1}{4}\beta^2\varepsilon^2\right)t^2\right]-\beta^2\varepsilon^2}\right.$$

$$\left.-\frac{i}{2}\beta\varepsilon\right]^{-k}dt-\frac{2\pi^{n-1}}{k\Gamma(n-1)}\int_0^1(\alpha\varepsilon)^{n-1}t^{2n-3}\left[1+\frac{1}{2}\sqrt{4(1-\alpha\varepsilon t^2)-\beta^2\varepsilon^2}-\frac{i}{2}\beta\varepsilon\right]^{-k}dt,$$

the integrands of these two integrals are all bounded, so

$$\lim_{\varepsilon\to 0} J_k=\frac{2\pi^{n-1}}{k\Gamma(n-1)}\int_0^1\frac{t^{2n-3}}{(1+\sqrt{1-t^2})^k}dt,$$

it is a real number, hence

$$\lim_{\varepsilon\to 0}\operatorname{Im}(J_k)=0,\ k=1,\ 2,\ \cdots,\ n-1.$$

We may prove in the same way

$$\lim_{\varepsilon\to 0}\operatorname{Im}(J_0)=0,\ \lim_{\varepsilon\to 0}\operatorname{Im}(H_0)=0.$$

As the calculation of J_k, H_k may be written as

$$\begin{aligned}H_k=&\frac{2\pi^{n-1}}{k\Gamma(n-1)}\left\{\int_0^1\left(1-\frac{1}{4}\beta^2\varepsilon^2\right)^{n-1}t^{2n-3}\left[1-\sqrt{1-\left(1-\frac{1}{4}\beta^2\varepsilon^2\right)t^2-\frac{1}{4}\beta^2\varepsilon^2}\right.\right.\\&\left.\left.-\frac{i}{2}\beta\varepsilon\right]^{-k}dt-\int_0^1(\alpha\varepsilon)^{n-1}t^{2n-3}\left[1-\sqrt{1-\alpha\varepsilon t^2-\frac{1}{4}\beta^2\varepsilon^2}-\frac{i}{2}\beta\varepsilon\right]^{-k}dt\right\}\\=&\frac{2\pi^{n-1}}{k\Gamma(n-1)}(X_k-Y_k).\end{aligned}$$

The absolute value of the integrand of the first integral is not greater than $\dfrac{t^{2n-3}}{(1-\sqrt{1-t^2})^k}$, it is an integrable function on the interval [0, 1] when $k<n-1$, so we have

$$\lim_{\varepsilon\to 0}X_k=\int_0^1\frac{t^{2n-3}}{(1-\sqrt{1-t^2})^k}dt,$$

it is a real number. The absolute value of the integrand of the second integral is $O(\varepsilon^{n-k-1})$, so

$$\lim_{\varepsilon\to 0}Y_k=0,$$

as $k<n-1$. It follows that when $k<n-1$

$$\lim_{\varepsilon\to 0}\operatorname{Im}(H_k)=0.$$

Finally, we calculate H_{n-1}

$$H_{n-1}=\frac{2\pi^{n-1}}{\Gamma(n)}(X_{n-1}-Y_{n-1}),$$

where

$$X_{n-1}=\left(1-\frac{1}{4}\beta^2\varepsilon^2\right)^{n-1}\int_0^1 t^{2n-3}\left[1-\sqrt{\left(1-\frac{1}{4}\beta^2\varepsilon^2\right)(1-t^2)}-\frac{i}{2}\beta\varepsilon\right]^{-(n-1)}dt,$$

$$Y_{n-1}=(\alpha\varepsilon)^{n-1}\int_0^1 t^{2n-3}\left[1-\sqrt{1-\alpha\varepsilon t^2-\frac{1}{4}\beta^2\varepsilon^2}-\frac{i}{2}\beta\varepsilon\right]^{-(n-1)}dt.$$

Let $1-\frac{1}{4}\beta^2\varepsilon^2=\eta^2$, we have

$$X_{n-1}=\eta^{2n-2}\int_0^1 t^{2n-3}[1-\eta\sqrt{1-t^2}-i\sqrt{1-\eta^2}]^{-(n-1)}dt,$$

by [7], we obtain

$$\lim_{\varepsilon\to 0}\operatorname{Im}(X_{n-1})=2^{n-3}\pi.$$

Since $$1-\sqrt{1-\alpha\varepsilon t^2-\frac{1}{4}\beta^2\varepsilon^2}-\frac{i}{2}\beta\varepsilon=\left(\frac{1}{2}\alpha t^2-\frac{i}{2}\beta\right)\varepsilon+O(\varepsilon^2),$$

we have

$$\lim_{\varepsilon\to 0} Y_{n-1}=2^{n-1}\int_0^1\frac{t^{2n-3}dt}{\left(\frac{1}{2}\alpha t^2-\frac{i}{2}\beta\right)^{n-1}}=2^{n-1}\int_0^1\frac{t^{2n-3}dt}{(t^2-i\gamma)^{n-1}}=2^{n-2}\int_0^1\frac{x^{n-2}dx}{(x-i\gamma)^{n-1}}$$

$$=2^{n-2}\left\{\log\frac{1-i\gamma}{-i\gamma}-D_n+\sum_{k=1}^{n-2}C_k^{n-2}\frac{(-1)^k}{k}\right\},$$

where $\gamma=\frac{\beta}{\alpha}$, $D_n=\sum_{k=1}^{n-2}C_k^{n-2}\frac{1}{k}\left(\frac{i\gamma}{1-i\gamma}\right)^k (n\geqslant 3)$. It follows that

$$\lim_{\varepsilon\to 0}\operatorname{Im}(Y_{n-1})=2^{n-2}\left[\operatorname{arc\,tg}\frac{1}{\gamma}-\operatorname{Im}(D_n)\right]\quad(n\geqslant 3)\tag{3.2.4}$$

when $n=2$ $$\lim_{\varepsilon\to 0}\operatorname{Im}(Y_1)=2\operatorname{Im}\left(\int_0^1\frac{t\,dt}{t^2-i\gamma}\right)=\operatorname{arc\,tg}\frac{1}{\gamma}.$$

If define $D_2=0$, then (3.2.4) holds too when $n\geqslant 2$.

We now have

$$\lim_{\varepsilon\to 0}\operatorname{Im}(H_{n-1})=\frac{2^{n-2}\pi^n}{\Gamma(n)}\left\{1-\frac{2}{\pi}\operatorname{arc\,tg}\frac{1}{\gamma}+\frac{2}{\pi}\operatorname{Im}(D_n)\right\},$$

here $n\geqslant 2$. Consequently

$$\lim_{\varepsilon\to 0}\frac{1}{\omega_{2n-1}}\int_{Q_\varepsilon}\frac{\dot{u}}{(1-\bar{u}_n)^n}=1-2^{n-2}\left(1-\frac{2}{\pi}\operatorname{arc\,tg}\frac{1}{\gamma}+\frac{2}{\pi}\operatorname{Im}(D_n)\right).\tag{3.2.5}$$

By (3.2.2), (3.2.3), (3.2.5), we obtain

$$\lim_{\varepsilon\to 0}\frac{1}{\omega_{2n-1}}\int_{D_z(p_n,\varepsilon)}\frac{\dot{u}}{(1-\bar{u}_n)^n}=1-\frac{2^{n-1}}{\pi}\left(\operatorname{arc\,tg}\frac{1}{\gamma}-\operatorname{Im}(D_n)\right).\tag{3.2.6}$$

It is not hard to prove

$$\operatorname{Im}(D_n)=-\operatorname{arc\,tg}\frac{\beta}{\alpha}+h_n\left(\operatorname{arc\,tg}\frac{\beta}{\alpha}\right).$$

Substitute it into (3.2.6), this completes the proof of (3.1.1).

References

[1] Muskhelishvili N. I., Singular integral equations, Translated from Russian, Groningen, 1953.

[2] Zygmund, A., Trigonometric series, *Cambridge University Press*, **2**(1959).

[3] Kung S. and Sun C. K., Integrals of Cauchy type in several complex variables I. *Acta Mathematica Sinica*, **15**(1965), 431—443(In Chinese).

[4] Koranyi, A. and Vagi, S., Singular integrals in homogeneous spaces and some problems of classical analysis, *Ann. Scuola Normale Superiore Pisa*, **25** (1971) 575—648.

[5] Kung S. and Sun C. K., Integrals of Cauchy type in several complex variables III, *Acta Mathematica Sinica*, **15** (1965), 800—811 (In Chinese).

[6] Kung S. and Sun C. K., Integrals of Cauchy type in several complex variables II, *Acta Mathematica Sinica*, **15** (1965), 775—799 (*In Chinese*).

[7] Shi Ji-huai, On the Cauchy type integrals for the hypersphere, *Journal of the China University of the science and technology*, **2**(1980), 1—9(In Chinese).

[8] Kung S. and Sun C. K., Singular integral equations on a complex hyperspher, *Acta Mathematica Sinica*, **16** (1966), 194—210 (*In Chinese*).

[9] Sun C. K., Regularization theorem of the singular integral equations on a complex hypersphere, *Acta Mathematica Sinica*, **20** (1977), 287—290 (In Chinse).

[10] Kung, S., A remark on integrals of Cauchy type in several complex variables, Proceeding of the Ist symposia of partial differential equation and differential geometry, Beijing, 1980.

[11] Ramirez De Arellano, Ein Divisionproblem und Randintegraldarstellungen in der Komplexen Analysis, *Math. Ann.*, **184** (1970), 172—187.

[12] Folland, G. B. and Stein, E. M., Estimates for the $\bar{\partial}$ complex and analysis on the Heisenberg group, *Comm. Pure. Appl. Math.*, **27** (1974), 429—522.

[13] Alt, W., Singuläre Mtegrale mit gemischten hologeneitäten auf Mannigfaltigkeiten und Anwendungen in der Funktionentheorie, *Math. Z.*, **137** (1974), 227—256.

[14] Kerzman, N. and Stein, E. M., The Szegö kernel in terms of Cauchy-Fantappiè kernels, *Duke Math* J., **45** (1978), 197—224.

[15] Henkin, G., Integral representations of functions holomorphic in strictly pseudoconvex domains and some applications, *Math. USSR, Sbornik*, **7** (1969), 597—616.

Chin. Ann. of Math.
4B (3) 1983

SINGULAR INTEGRALS IN SEVERAL COMPLEX VARIABLES (II)——HADAMARD PRINCIPAL VALUE ON A SPHERE

SHI JIHUAI (史济怀) GONG SHENG (龚 昇)
(University of Science and Technology of China)

Abstract

Hadamard introduced the concept of finite parts of divergent integrals. i.e. Hadamard principal value, when he researched the Cauchy problems of the hyperbolic type partial differential equations. In this paper, the authors try to generalize this concept to the singular integrals on a sphere of several complex variables space C^n. The Hadamard principal value of higher order singular integral

$$\frac{1}{\omega_{2n-1}}\int_{u\bar{u}'=1}\frac{f(u)\dot{u}}{(1-v\bar{u}')^{n+\frac{1}{2}}}$$

is defined and the corresponding Plemelj formula is obtained.

§ 0. Introduction

Hadamard introduced the concept of finite parts of divergent integrals, when he researched the Cauchy problems of the hyperbolic type partial differential equations. In the case of one dimension, the integral

$$\int_a^b \frac{f(x)\,dx}{(x-u)^{n+1}} \tag{0.1}$$

is singular, where $a<u<b$, $n\geqslant 0$. As $n=0$, the Cauchy principal value of this integral is defined as

$$\lim_{\varepsilon\to 0^+}\left\{\int_a^{u-\varepsilon}\frac{f(x)}{x-u}\,dx+\int_{u+\varepsilon}^b\frac{f(x)}{x-u}\,dx\right\}$$

if the above limit exists. As $n>0$, the integral (0.1) is divergent. For given $f(x)$, take $g(x)$, such that

$$\lim_{t\to u^-}\int_a^t\frac{f(x)-g(x)}{(x-u)^{n+1}}\,dx$$

exists. Hadamard called the limit value the finite part of this divergent integral. He researched the singular integral

$$\int_a^b\frac{f(x)}{(x-u)^{n+\frac{1}{2}}}\,dx$$

Manuscript received July 20, 1981

and gave many important properties of the finite part. Using these results he also solved the Cauchy problem of hyperbolic type partial differential equations[1].

According to Hadamard's idea, Fox[3] regarded the finite part of divergent singular integral (0.1) as a generalization of Cauchy principal value of singular integral, and called it Hadamard principal value. He defined the Hadamard principal value of the divergent integral (0.1) as follow:

$$P\int_a^b \frac{f(x)}{(x-u)^{n+1}}\,dx=\lim_{\varepsilon\to 0^+}\left\{\int_a^{u-\varepsilon}\frac{f(x)}{(x-u)^{n+1}}\,dx+\int_{u+\varepsilon}^b\frac{f(x)}{(x-u)^{n+1}}\,dx-H_n(u,\ \varepsilon)\right\},$$

where

$$H_0(u,\ \varepsilon)=0,\ H_n(u,\ \varepsilon)=\sum_{i=0}^{n-1}\frac{f^{(i)}(u)}{i!}\left\{\frac{1-(-1)^{n-i}}{(u-i)\varepsilon^{n-i}}\right\}.$$

It is clear, when $n=0$, it is just the Cauchy principal value.

Generalize this idea to the complex plane, Hadamard principal value becomes a generalization of the Cauchy principal value of Cauchy type integral. Fox obtained the corresponding Plemolj formula and solved the boundary value problem and singular integral equations with the aid of these tools.

This paper generalizes the concept of Hadamard principal value to the unit ball ——the simplest and the most essential irreducible domain. We shall give the Hadamard principal value of following singular integral

$$\frac{1}{\omega_{2n-1}}\int_{u\bar{u}'=1}\frac{f(u)\dot{u}}{(1-v\bar{u}')^{n+\frac{1}{2}}}$$

and derive the corresponding Plemolj formula by means of this principal value.

It is not hard to see, these results can be generalized to the strictly pseudoconvex domain Ω. For the Cauchy-Fantapiee kernel

$$K(w,\ z)=C_n g^{-n}\omega\wedge dz_1\wedge\cdots\wedge dz_n,\ w\in\Omega,\ z\in b\Omega,$$

where

$$C_n=(-1)^{\frac{n(n-1)}{2}}(n-1)!(2\pi i)^{-n},$$

$$g(w,\ z)=\sum_{i=1}^{n}(z_i-w_i)g_i(w,\ z),$$

$$\omega=\sum_{j=1}^{n}(-1)^{j-1}g_j\bar{\partial}g_1\wedge\cdots\wedge[\bar{\partial}g_j]\wedge\cdots\wedge\bar{\partial}g_n.$$

we can discuss the Hadamard principal value of singular integral

$$\int_{b\Omega}f(z)C_n g^{-(n+\frac{1}{2})}\omega\wedge dz_1\wedge\cdots\wedge dz_n,\ w\in b\Omega,\ z\in b\Omega.$$

When the kernel is Henkin-Ramirez kernel or Stein-Kerzman kernel, we can give the corresponding Hadamard principal value according to the method of [4].

The notations in this paper are similar to the paper [4].

§1. Some Lemmas

Lemma 1. *Suppose* $c\in\left[0,\ \frac{\pi}{2}\right]$ *we have*

$$J=\int_{-(\pi-c)}^{-c}\frac{d\theta}{(1-re^{i\theta})^{n+\frac{1}{2}}}+\int_{c}^{\pi-c}\frac{d\theta}{(1-re^{i\theta})^{n+\frac{1}{2}}}=2\mathrm{Im}(B_1-B_2),$$

where

$$B_1=\sum_{k=1}^{n}\frac{2}{2k-1}\frac{1}{(1+re^{-ic})^{k-\frac{1}{2}}}+\log(1-\sqrt{1+re^{-ic}})-\log(1+\sqrt{1+re^{-ic}}),$$

$$B_2=\sum_{k=1}^{n}\frac{2}{2k-1}\frac{1}{(1-re^{ic})^{k-\frac{1}{2}}}+\log(1-\sqrt{1-re^{-ic}})-\log(1+\sqrt{1-re^{ic}}).$$

Proof It is clear that

$$J=2\,\mathrm{Re}\int_{c}^{\pi-c}\frac{d\theta}{(1-re^{i\theta})^{n+\frac{1}{2}}},$$

put $z=re^{i\theta}$, then

$$\int_{c}^{\pi-c}\frac{d\theta}{(1-re^{i\theta})^{n+\frac{1}{2}}}=\frac{1}{i}\int_{re^{ic}}^{-re^{-ic}}\frac{dz}{z(1-z)^{n+\frac{1}{2}}}.$$

Substitute

$$\frac{1}{z(1-z)^{n+\frac{1}{2}}}=\sum_{k=1}^{n}\frac{1}{(1-z)^{k+\frac{1}{2}}}+\frac{1}{z(1-z)^{\frac{1}{2}}}$$

into the above equality, we obtain

$$J=2\mathrm{Im}(B_1-B_2).$$

Lemma 2. *Suppose* $\alpha>0,\ \beta>0,\ \gamma=\frac{\alpha}{\beta}$. *Let*

$$A_n=-\frac{2^{n+\frac{1}{2}}\sqrt{\alpha}\,(n-1)}{\pi(2n-1)}\,\mathrm{Im}\left(\int_0^1\frac{x^{n-2}dx}{(x-i\gamma\sqrt{1-x^2})^{n-\frac{1}{2}}}\right)$$

then

$$\lim_{\varepsilon\to 0}\left\{\frac{1}{\omega_{2n-1}}\int_{\sigma_\varepsilon}\frac{\dot{u}}{(1-\bar{u}_n)^{n+\frac{1}{2}}}-\frac{A_n}{\sqrt{\varepsilon}}\right\}=1,$$

where

$$\sigma_\varepsilon=\{u:u\bar{u}'=1,\ \alpha^2(1-|u_n|^2)^2+4\beta^2(\mathrm{Im}\,u_n)^2>\varepsilon^2\}.$$

Proof Let

$$c=\arcsin\frac{\sqrt{\varepsilon^2-\alpha^2(1-r^2)^2}}{2\beta r},$$

$$\frac{1}{\omega_{2n-1}}\int_{\sigma_\varepsilon}\frac{\dot{u}}{(1-\bar{u}_n)^{n+\frac{1}{2}}}=\frac{1}{\omega_{2n-1}}\int_{v\bar{v}'<\frac{\varepsilon}{\alpha}}\dot{v}\left\{\int_{-(\pi-c)}^{-c}\frac{d\theta}{(1-re^{i\theta})^{n+\frac{1}{2}}}+\int_{c}^{\pi-c}\frac{d\theta}{(1-re^{i\theta})^{n+\frac{1}{2}}}\right\}$$

$$+\frac{1}{\omega_{2n-1}}\int_{v\bar{v}'>\frac{\varepsilon}{\alpha}}\dot{v}\int_{-\pi}^{\pi}\frac{d\theta}{(1-re^{i\theta})^{n+\frac{1}{2}}}=I_1+I_2.$$

By Lemma 1, I_1 is equal to

$$\frac{2}{\omega_{2n-1}}\operatorname{Im}\left\{\left(\sum_{k=1}^{n}J_k+J_0-J'_0\right)-\left(\sum_{k=1}^{n}H_k+H_0-H'_0\right)\right\},$$

where

$$J_k=\frac{2}{2k-1}\int_{v\bar{v}'<\frac{\varepsilon}{\alpha}}\frac{\dot{v}}{(1+re^{-ic})^{k-\frac{1}{2}}},\quad J_0=\int_{v\bar{v}'<\frac{\varepsilon}{\alpha}}\log(1-\sqrt{1+re^{-ic}})\dot{v},$$

$$J'_0=\int_{v\bar{v}'<\frac{\varepsilon}{\alpha}}\log(1+\sqrt{1+re^{-ic}})\dot{v},\quad H_k=\frac{2}{2k-1}\int_{v\bar{v}'<\frac{\varepsilon}{\alpha}}\frac{\dot{v}}{(1-re^{ic})^{k-\frac{1}{2}}},$$

$$H_0=\int_{v\bar{v}'<\frac{\varepsilon}{\alpha}}\log(1-\sqrt{1-re^{ic}})\dot{v},\quad H'_0=\int_{v\bar{v}'<\frac{\varepsilon}{\alpha}}\log(1+\sqrt{1-re^{ic}})\dot{v}.$$

Since

$$1+re^{-ic}=1+(2\beta)^{-1}(\sqrt{4\beta^2r^2-\varepsilon^2+\alpha^2(1-r^2)^2}-i\sqrt{\varepsilon^2-\alpha^2(1-r^2)^2}),$$

using the polar coordinate and letting $1-r^2=s^2$, we have

$$J_k=\frac{4\pi^{n-1}}{(2k-1)\Gamma(n-1)}\int_0^{\sqrt{\varepsilon\alpha^{-1}}}s^{2n-3}\{1+(2\beta)^{-1}[\sqrt{4\beta^2(1-s^2)+\alpha^2s^4-\varepsilon^2}$$
$$-i\sqrt{\varepsilon^2-\alpha^2s^4}]\}^{-(k-\frac{1}{2})}ds.$$

Put $\eta=\varepsilon\alpha^{-1}$, $s=\sqrt{\eta}\,t$. Then J_k equals

$$\frac{4\pi^{n-1}(2\beta)^{k-\frac{1}{2}}}{(2k-1)\Gamma(n-1)}\eta^{n-1}\int_0^1 t^{2n-3}[2\beta+\sqrt{4\beta^2(1-\eta t^2)+\alpha^2\eta^2(t^4-1)}$$
$$-i\sqrt{\alpha^2\eta^2(1-t^4)}]^{-(k-\frac{1}{2})}dt.$$

Since the absolute value of the integrand does not exceed $(4\beta^2)^{-1}$, so

$$\lim_{\varepsilon\to 0}J_k=0,\quad (k=1,\ 2,\ \cdots,\ n).$$

By the same reason, we have

$$\lim_{\varepsilon\to 0}\operatorname{Im}(J_0)=\lim_{\varepsilon\to 0}\operatorname{Im}(J'_0)=0.$$

H_k may be written as

$$\frac{4\pi^{n-1}(2\beta)^{k-\frac{1}{2}}}{(2k-1)\Gamma(n-1)}\eta^{n-1}\int_0^1 t^{2n-3}[2\beta-\sqrt{4\beta^2(1-\eta t^2)+\alpha^2\eta^2(t^4-1)}$$
$$-i\sqrt{\alpha^2\eta^2(1-t^4)}]^{-(k-\frac{1}{2})}dt.$$

Let $P=\beta t^2-i\alpha\sqrt{1-t^4}$, and $Q=(4\beta)^{-1}[\alpha^2-(\alpha^2-\beta^2)t^4]$. Then

$$[2\beta-\sqrt{4\beta^2(1-\eta t^2)+\alpha^2\eta^2(t^4-1)}-i\alpha\eta\sqrt{1-t^4}]^{k-\frac{1}{2}}$$
$$=\eta^{k-\frac{1}{2}}\left[P^{k-\frac{1}{2}}+\left(k-\frac{1}{2}\right)QP^{k-\frac{3}{2}}\eta+O(\eta^2)\right],$$

so

$$H_k=\frac{4\pi^{n-1}(2\beta)^{k-\frac{1}{2}}}{(2k-1)\Gamma(n-1)}\eta^{n-k-\frac{1}{2}}\int_0^1 t^{2n-3}\left[P^{k-\frac{1}{2}}+\left(k-\frac{1}{2}\right)QP^{k-\frac{3}{2}}\eta+O(\eta^2)\right]^{-1}dt.$$

When $k<n-\frac{1}{2}$, we have

$$\lim_{\varepsilon\to 0} H_k=0 \quad (k=1,\ 2,\ \cdots,\ n-1).$$

On the other hand, we have

$$\lim_{\varepsilon\to 0}\operatorname{Im}(H_0)=\lim_{\varepsilon\to 0}\operatorname{Im}(H_0')=0.$$

Now we calculate H_n

$$H_n=\frac{4\pi^{n-1}(2\beta)^{n-\frac{1}{2}}}{(2n-1)\Gamma(n-1)}\eta^{-\frac{1}{2}}\int_0^1 t^{2n-3}\left[P^{n-\frac{1}{2}}+\left(n-\frac{1}{2}\right)QP^{n-\frac{3}{2}}\eta+O(\eta^2)\right]^{-1}dt.$$

It is easy to see, A_n may be written as

$$A_n=\frac{-2}{\omega_{2n-1}}\frac{4\pi^{n-1}(2\beta)^{n-\frac{1}{2}}\sqrt{\alpha}}{(2n-1)\Gamma(n-1)}\operatorname{Im}\left(\int_0^1 t^{2n-3}[\beta t^2-i\alpha\sqrt{1-t^4}]^{-(n-\frac{1}{2})}\,dt\right),$$

then

$$-\frac{2}{\omega_{2n-1}}\operatorname{Im}(H_n)-\frac{1}{\sqrt{\varepsilon}}A_n$$

$$=\frac{-2}{\omega_{2n-1}}\frac{4\pi^{n-1}(2\beta)^{n-\frac{1}{2}}}{(2n-1)\Gamma(n-1)\sqrt{\alpha}}\sqrt{\varepsilon}\operatorname{Im}\left\{\int_0^1\frac{-t^{2n-3}\left[\left(n-\frac{1}{2}\right)P^{n-\frac{3}{2}}Q+O(\eta)\right]dt}{P^{n-\frac{1}{2}}\left[P^{n-\frac{1}{2}}+\left(n-\frac{1}{2}\right)P^{n-\frac{3}{2}}Q\eta+O(\eta^2)\right]}\right\}.$$

So we have

$$\lim_{\varepsilon\to 0}\left\{\frac{-2}{\omega_{2n-1}}\operatorname{Im}(H_n)-\frac{1}{\sqrt{\varepsilon}}A_n\right\}=0.$$

For I_2, we notice

$$(1-re^{i\theta})^{-(n+\frac{1}{2})}=\sum_{q=0}^{\infty}\frac{\Gamma\left(n+q+\frac{1}{2}\right)}{\Gamma(q+1)\Gamma\left(n+\frac{1}{2}\right)}r^qe^{iq\theta},$$

so

$$\int_{-\pi}^{\pi}\frac{d\theta}{(1-re^{i\theta})^{n+\frac{1}{2}}}=2\pi,$$

and then $\lim\limits_{\varepsilon\to 0} I_2=1$. Summing up the above, we have

$$\lim_{\varepsilon\to 0}\left\{\frac{1}{\omega_{2n-1}}\int_{\sigma_\varepsilon}\frac{\dot{u}}{(1-\bar{u}_n)^{n+\frac{1}{2}}}-\frac{A_n}{\sqrt{\varepsilon}}\right\}=1.$$

The proof of Lemma 2 is complete.

Lemma 3.

$$\lim_{\varepsilon\to 0}\frac{1}{\omega_{2n-1}}\int_{\sigma_\varepsilon}\frac{\operatorname{Im}(\bar{u}_n)\dot{u}}{(1-\bar{u}_n)^{n+\frac{1}{2}}}=-\frac{2n+1}{4ni},$$

$$\lim_{\varepsilon\to 0}\frac{1}{\omega_{2n-1}}\int_{\sigma_\varepsilon}\frac{\operatorname{Re}(1-\bar{u}_n)}{(1-\bar{u}_n)^{n+\frac{1}{2}}}\dot{u}=\frac{2n-1}{4n}.$$

Proof

$$\frac{1}{\omega_{2n-1}}\int_{\sigma_\varepsilon}\frac{\operatorname{Im}(\bar{u}_n)}{(1-\bar{u}_n)^{n+\frac{1}{2}}}\dot{u}$$

$$=\frac{2i}{\omega_{2n-1}}\int_{v\bar{v}'<\frac{\varepsilon}{\alpha}}\left\{\operatorname{Im}\int_{c}^{\pi-c}\frac{r\sin\theta d\theta}{(1-re^{i\theta})^{n+\frac{1}{2}}}\right\}\dot{v}+\frac{1}{\omega_{2n-1}}\int_{v\bar{v}'>\frac{\varepsilon}{\alpha}}\dot{v}\int_{-\pi}^{\pi}\frac{r\sin\theta d\theta}{(1-re^{i\theta})^{n+\frac{1}{2}}}=I_1+I_2.$$

We calculate I_1 and I_2 respectively. Since

$$\frac{1}{z^2(1-z)^{n+\frac{1}{2}}}=\sum_{k=1}^{n}\frac{n+1-k}{(1-z)^{k+\frac{1}{2}}}+\frac{1}{z^2(1-z)^{\frac{1}{2}}}+\frac{n}{z(1-z)^{\frac{1}{2}}}$$

so

$$\begin{aligned}\int_{c}^{\pi-c}\frac{r\sin\theta d\theta}{(1-re^{i\theta})^{n+\frac{1}{2}}}=&\left\{\frac{r^2-1}{2n-1}\frac{1}{(1+re^{-ic})^{n-\frac{1}{2}}}+\sum_{k=1}^{n-1}\frac{n+1-k}{2k-1}\frac{r^2}{(1+re^{-ic})^{k-\frac{1}{2}}}\right.\\&\left.+\frac{r\sqrt{1+re^{-ic}}}{2e^{-ic}}-\frac{2n+1}{4}r^2\log\frac{1+\sqrt{1+re^{-ic}}}{1-\sqrt{1+re^{-ic}}}\right\}\\&-\left\{\frac{r^2-1}{2n-1}\frac{1}{(1-re^{ic})^{n-\frac{1}{2}}}+\sum_{k=1}^{n-1}\frac{n+1-k}{2k-1}\frac{r^2}{(1-re^{ic})^{k-\frac{1}{2}}}\right.\\&\left.-\frac{r\sqrt{1-re^{ic}}}{2e^{ic}}-\frac{2n+1}{4}r^2\log\frac{1+\sqrt{1-re^{ic}}}{1-\sqrt{1-re^{ic}}}\right\}\end{aligned}\tag{1.1}$$

and

$$I_1=\frac{2i}{\omega_{2n-1}}\left\{\operatorname{Im}\left(M_n+\sum_{k=1}^{n-1}M_k+M_0-M_0'\right)-\operatorname{Im}\left(N_n+\sum_{k=1}^{n-1}N_k-N_0-N_0'\right)\right\}.$$

Where

$$M_n=\frac{1}{2n-1}\int_{v\bar{v}'<\frac{\varepsilon}{\alpha}}(r^2-1)(1+re^{-ic})^{-(n-\frac{1}{2})}\dot{v},$$

$$M_k=\frac{n+1-k}{2k-1}\int_{v\bar{v}'<\frac{\varepsilon}{\alpha}}r^2(1+re^{-ic})^{-(k-\frac{1}{2})}\dot{v}\ (k=1,\ \cdots,\ n-1).$$

$$M_0=\frac{1}{2}\int_{v\bar{v}'<\frac{\varepsilon}{\alpha}}re^{ic}\sqrt{1+re^{-ic}}\dot{v},$$

$$M_0'=\frac{2n+1}{4}\int_{v\bar{v}'<\frac{\varepsilon}{\alpha}}r^2\log\frac{1+\sqrt{1+re^{-ic}}}{1-\sqrt{1+re^{-ic}}}\dot{v},$$

$$N_n=\frac{1}{2n-1}\int_{v\bar{v}'<\frac{\varepsilon}{\alpha}}(r^2-1)(1-re^{ic})^{-(n-\frac{1}{2})}\dot{v},$$

$$N_k=\frac{n+1-k}{2k-1}\int_{v\bar{v}'<\frac{\varepsilon}{\alpha}}r^2(1-re^{ic})^{-(k-\frac{1}{2})}\dot{v}(k=1,\ \cdots,\ n-1),$$

$$N_0=\frac{1}{2}\int_{v\bar{v}'<\frac{\varepsilon}{\alpha}}re^{-ic}\sqrt{1-re^{ic}}\dot{v},$$

$$N_0'=\frac{2n+1}{4}\int_{v\bar{v}'<\frac{\varepsilon}{\alpha}}r^2\log\frac{1+\sqrt{1-re^{ic}}}{1-\sqrt{1-re^{ic}}}\dot{v}.$$

It is easy to prove that

$$\lim_{\varepsilon\to 0}M_k=0\quad(k=1,\ \cdots,\ n).$$

Since the absolute values of M_0, N_0 are bounded, so

$$\lim_{\varepsilon\to 0}M_0=\lim_{\varepsilon\to 0}N_0=0.$$

Clearly

$$\lim_{\varepsilon\to 0}\operatorname{Im}(M'_0)=\lim_{\varepsilon\to 0}\operatorname{Im}(N'_0)=0.$$

Similar to the proof of

$$\lim_{\varepsilon\to 0}H_k=0\quad(k=1,\cdots,n-1)$$

in Lemma 2, we can prove

$$\lim_{\varepsilon\to 0}\operatorname{Im}(N_k)=0\quad(k=1,\cdots,n-1).$$

N_n is equal to

$$\frac{-2\pi^{n-1}(2\beta)^{n-\frac{1}{2}}}{(2n-1)\Gamma(n-1)}\int_0^1\eta^n t^{2n-1}\left[2\beta-\sqrt{4\beta^2(1-\eta t^2)+\alpha^2\eta^2(t^4-1)}-i\sqrt{\alpha^2\eta^2(1-t^4)}\right]^{-\left(n-\frac{1}{2}\right)}dt$$

$$=\frac{-2\pi^{n-1}(2\beta)^{n-\frac{1}{2}}}{(2n-1)\Gamma(n-1)}\sqrt{\eta}\int_0^1 t^{2n-1}\left[\beta t^2-i\alpha\sqrt{1-t^4}+O(\eta)\right]^{-\left(n-\frac{1}{2}\right)}dt,$$

so $\lim_{\varepsilon\to 0}N_n=0$. Hence $\lim_{\varepsilon\to 0}I_1=0$.

Since

$$\int_{-\pi}^{\pi}\frac{r\sin\theta d\theta}{(1-re^{i\theta})^{n+\frac{1}{2}}}=\frac{1}{2i}\sum_{q=0}^{\infty}\frac{\Gamma\left(n+\frac{1}{2}+q\right)}{\Gamma(q+1)\Gamma\left(n+\frac{1}{2}\right)}r^{q+1}\int_{-\pi}^{\pi}\left[e^{i(q+1)\theta}-e^{i(q-1)\theta}\right]d\theta$$

$$=i\left(n+\frac{1}{2}\right)\pi r^2.$$

so

$$\lim_{\varepsilon\to 0}I_2=-\frac{2n+1}{4ni}.$$

Summing up the above, we obtain

$$\lim_{\varepsilon\to 0}\frac{1}{\omega_{2n-1}}\int_{\sigma_\varepsilon}\frac{\operatorname{Im}(\bar{u}_n)}{(1-\bar{u}_n)^{n+\frac{1}{2}}}\dot{u}=-\frac{2n+1}{4ni}.$$

Since $\lim_{\varepsilon\to 0}\frac{1}{\omega_{2n-1}}\int_{\sigma_\varepsilon}\frac{\dot{u}}{(1-\bar{u}_n)^{n-\frac{1}{2}}}=1$, and

$$\frac{1}{\omega_{2n-1}}\int_{\sigma_\varepsilon}\frac{\operatorname{Re}(1-\bar{u}_n)}{(1-\bar{u}_n)^{n+\frac{1}{2}}}\dot{u}=\frac{1}{\omega_{2n-1}}\int_{\sigma_\varepsilon}\frac{\dot{u}}{(1-\bar{u}_n)^{n-\frac{1}{2}}}+\frac{i}{\omega_{2n-1}}\int_{\sigma_\varepsilon}\frac{\operatorname{Im}(\bar{u}_n)}{(1-\bar{u}_n)^{n+\frac{1}{2}}}\dot{u},$$

another equality of this Lemma holds.

Lemma 4.

$$\frac{1}{\omega_{2n-1}}\int_{\sigma_\varepsilon}\frac{\operatorname{Re}(u_k)}{(1-\bar{u}_n)^{n+\frac{1}{2}}}\dot{u}=0,\ \frac{1}{\omega_{2n-1}}\int_{\sigma_\varepsilon}\frac{\operatorname{Im}(u_k)}{(1-\bar{u}_n)^{n+\frac{1}{2}}}\dot{u}=0\ (k=1,\cdots,n-1).$$

Proof We only need to prove the case $k=1$. Set $\bar{u}_n=re^{i\theta}$, $u_1=v_1$, $\cdots$, $u_{n-1}=v_{n-1}$, we have

$$\frac{1}{\omega_{2n-1}}\int_{\sigma_\varepsilon}\frac{\operatorname{Re}u_1}{(1-\bar{u}_n)^{n+\frac{1}{2}}}\dot{u}=\frac{1}{\omega_{2n-1}}\int_{v\bar{v}'<\frac{\varepsilon}{\alpha}}\operatorname{Re}v_1\dot{v}\left\{\int_{-(\pi-c)}^{-c}\frac{d\theta}{(1-re^{i\theta})^{n+\frac{1}{2}}}+\int_c^{\pi-c}\frac{d\theta}{(1-re^{i\theta})^{n+\frac{1}{2}}}\right\}$$

$$+\frac{1}{\omega_{2n-1}}\int_{v\bar{v}'>\frac{\varepsilon}{\alpha}}\operatorname{Re}v_1\dot{v}\int_{-\pi}^{\pi}\frac{d\theta}{(1-re^{i\theta})^{n+\frac{1}{2}}}=I_1+I_2.$$

It is easy to see

$$I_1=\frac{2}{\omega_{2n-1}}\operatorname{Im}\left\{\left(\sum_{k=1}^{n}J_k+J_0-J_0'\right)\cdot\left(\sum_{k=1}^{n}H_k+H_0-H_0'\right)\right\},$$

where

$$J_k=\frac{2}{2k-1}\int_{v\bar{v}'<\frac{s}{\alpha}}(\operatorname{Re}v_1)\left(1+re^{-ic}\right)^{-\left(k-\frac{1}{2}\right)}\dot{v},$$

$$J_0=\int_{v\bar{v}'<\frac{s}{\alpha}}(\operatorname{Re}v_1)\log\left(1-\sqrt{1+re^{-ic}}\right)\dot{v},$$

$$J_0'=\int_{v\bar{v}'<\frac{s}{\alpha}}(\operatorname{Re}v_1)\log\left(1+\sqrt{1+re^{-ic}}\right)\dot{v},$$

$$H_k=\frac{2}{2k-1}\int_{v\bar{v}'<\frac{s}{\alpha}}(\operatorname{Re}v_1)\left(1-re^{ic}\right)^{-\left(k-\frac{1}{2}\right)}\dot{v},$$

$$H_0=\int_{v\bar{v}'<\frac{s}{\alpha}}(\operatorname{Re}v_1)\log\left(1-\sqrt{1-re^{ic}}\right)\dot{v},$$

$$H_0'=\int_{v\bar{v}'<\frac{s}{\alpha}}(\operatorname{Re}v_1)\log\left(1++\sqrt{1-re^{ic}}\right)\dot{v}.$$

Using the spherical polar coordinate and putting $v=(x_1, x_2, \cdots, x_{2n-2})$, we have

$$J_k=\frac{2}{2k-1}\int_0^{\sqrt{s\alpha^{-1}}}\frac{s^{2n-2}ds}{(1+re^{-ic})^{k-\frac{1}{2}}}\int_0^{\pi}\cos\varphi_1\sin^{2n-4}\varphi_1 d\varphi_1\int_0^{\pi}\sin^{2n-5}\varphi_2 d\varphi_2\cdots\int_0^{2\pi}d\varphi_{2n-3}=0$$

$$(k=1, \cdots, n).$$

Similarly $J_0=J_0'=H_k=H_0=H_0'=0$, so $I_1=0$. Accordingly, $I_2=0$ It follows that

$$\frac{1}{\omega_{2n-1}}\int_{\sigma_s}\frac{\operatorname{Re}\dot{u}_1}{(1-\bar{u}_n)^{n+\frac{1}{2}}}\dot{u}=0.$$

We can prove the another equality of this Lemma by the same method.

Lemma 5. *Suppose* $f(u_1, \cdots, u_n)=f(x_1+iy_1, \cdots, x_n+iy_n)$ *is a real function defined on the unit ball* $\bar{B}=\{u: u\bar{u}'\leqslant 1\}$. *If*

$$\frac{\partial f}{\partial x_j}\in \operatorname{Lip}\alpha_j,\ \frac{\partial f}{\partial y_j}\in\operatorname{Lip}\beta_j\ (j=1, \cdots, n),$$

then

(i) *The integrals*

$$I_{x_k}=\omega_{2n-1}^{-1}\int_{u\bar{u}'=1}\Big[f(0, \cdots, 0, x_k+iy_k, \cdots, x_n+iy_n)-f(0, \cdots, 0, iy_k, \cdots, x_n+iy_n)$$
$$-\frac{\partial f}{\partial x_k}(p_n)x_k\Big](1-\bar{u}_n)^{-\left(n+\frac{1}{2}\right)}\dot{u},$$

$$I_{y_k}=\omega_{2n-1}^{-1}\int_{u\bar{u}'=1}\Big[f(0, \cdots, 0, iy_k, x_{k+1}+iy_{k+1}, \cdots, x_n+iy_n)$$
$$-f(0, \cdots, 0, x_{k+1}+iy_{k+1}, \cdots, x_n+iy_n)-\frac{\partial f}{\partial y_k}(p_n)y_k\Big](1-\bar{u}_n)^{-\left(n+\frac{1}{2}\right)}\dot{u},$$

$$I_{x_n}=\omega_{2n-1}^{-1}\int_{u\bar{u}'=1}\Big[f(0, \cdots, 0, x_n)-f(0, \cdots, 0, 1)-\frac{\partial f}{\partial x_n}(p_n)(x_n-1)\Big](1-\bar{u}_n)^{-\left(n+\frac{1}{2}\right)}\dot{u},$$

$$I_{y_n}=\omega_{2n-1}^{-1}\int_{u\bar{u}'=1}\Big[f(0, \cdots, 0, x_n+iy_n)-f(0, \cdots, 0, x_n)-\frac{\partial f}{\partial y_n}(p_n)y_n\Big](1-\bar{u}_n)^{-\left(n+\frac{1}{2}\right)}\dot{u}$$

exist, where $p_n=(0, \cdots, 0, 1)$, $k=1, \cdots, n-1$.

(ii) *Let* $z\in B$. *Replacing* $(1-\bar{u}_n)^{-\left(n+\frac{1}{2}\right)}$ *by* $(1-z\bar{u}')^{-\left(n+\frac{1}{2}\right)}$ *in the above integrals, denote these new integrals by* $I_{x_k}(z)$, $I_{y_k}(z)$, $I_{x_n}(z)$, $I_{y_n}(z)$ *respectively. Then*

$$(K-\lim_{z\to p_n})I_{x_k}(z)=I_{x_k},\quad (K-\lim_{z\to p_n})I_{y_k}(z)=I_{y_k}\quad (k=1, \cdots, n).$$

The concept of K limit was introduced by A. Koranyi[5]: Let

$$D_\alpha(v)=\left\{z{:}z\bar{z}'<1,\ v\bar{v}'=1,\ |1-z\bar{v}'|<\frac{\alpha}{2}(1-z\bar{z}'),\ \alpha>1\right\}.$$

If

$$\lim_{z\to v, z\in D_\alpha(v)} f(z)=\lambda.$$

holds for all $\alpha>1$, we say that f has the K limit λ at the point v.

Proof It is clear that

$$\left|f(0, \cdots, 0, x_k+iy_k, \cdots, x_n+iy_n)-f(0, \cdots, 0, iy_k, \cdots, x_n+iy_n)-\frac{\partial f}{\partial x_k}(p_n)x_k\right|$$
$$\leqslant K|1-\bar{u}_n|^{\frac{1}{2}(\alpha_k+1)},$$

where K is a constant. By Lemma 5.4.3 of [6], when $z\in D_\alpha(p_n)$, $u\bar{u}'=1$, then

$$|1-z\bar{u}'|\geqslant\frac{1}{4\alpha}|1-\bar{u}_n|.$$

Let $g_k(z, u)$ be the integrand of $I_{x_k}(z)$, then

$$|g_k(z, u)|\leqslant M|1-\bar{u}_n|^{-\left(n-\frac{\alpha_k}{2}\right)}.$$

So $\lim\limits_{z\to p_n} g_k(z, u)$ is integrable on $u\bar{u}'=1$, i. e. I_{x_k} exists, and

$$\lim_{z\to p_n}\int_{u\bar{u}'=1} g_k(z, u)\dot{u}=\int_{u\bar{u}'=1}\lim_{z\to p_n} g_k(z, u)\dot{u}.$$

It follows that $(K-\lim\limits_{z\to p_n})I_{x_k}(z)=I_{x_k}$. The proofs of other parts of this lemma are similar.

§2. Hadamard principal value

Now we can prove the following

Theorem 1. *If, with* I_{x_k}, I_{y_k} *defined by Lemma* 5, f *satisfies the same conditions in Lemma* 5, *then we have*

$$\lim_{\varepsilon\to 0}\left\{\frac{1}{\omega_{2n-1}}\int_{\sigma_\varepsilon}\frac{f(u)\dot{u}}{(1-\bar{u}_n)^{n+\frac{1}{2}}}-\frac{f(p_n)}{\sqrt{\varepsilon}}A_n\right\}$$
$$=\sum_{k=1}^{n}(I_{x_k}+I_{y_k})+\frac{1}{2n}\frac{\partial f}{\partial u_n}(p_n)-\frac{\partial f}{\partial \bar{u}_n}(p_n)+f(p_n). \tag{2.1}$$

Proof f can be written as

$$f(u)=\sum_{k=1}^{n-1}\Big[f(0,\cdots,0,x_k+iy_k,\cdots,x_n+iy_n)-f(0,\cdots,0,iy_k,\cdots,x_n+iy_n)$$
$$-\frac{\partial f}{\partial x_k}(p_n)x_k\Big]+\sum_{k=1}^{n-1}\Big[f(0,\cdots,0,iy_k,\cdots,x_n+iy_n)$$
$$-f(0,\cdots,0,x_{k+1}+iy_{k+1},\cdots,x_n+iy_n)-\frac{\partial f}{\partial y_k}(p_n)y_k\Big]$$
$$+\sum_{k=1}^{n-1}\frac{\partial f}{\partial x_k}(p_n)x_k+\sum_{k=1}^{n-1}\frac{\partial f}{\partial y_k}(p_n)y_k$$
$$+\Big[f(0,\cdots,0,x_n+iy_n)-f(0,\cdots,0,x_n)-\frac{\partial f}{\partial y_n}(p_n)y_n\Big]+\frac{\partial f}{\partial y_n}(p_n)y_n$$
$$+\Big[f(0,\cdots,0,x_n)-f(p_n)-\frac{\partial f}{\partial x_n}(p_n)(x_n-1)\Big]+\frac{\partial f}{\partial x_n}(p_n)(x_n-1)+f(p_n),$$

then

$$\frac{1}{\omega_{2n-1}}\int_{\sigma_\varepsilon}\frac{f(u)\dot{u}}{(1-\bar{u}_n)^{n+\frac{1}{2}}}-\frac{f(p_n)}{\sqrt{\varepsilon}}A_n$$
$$=\sum_{k=1}^{n-1}\omega_{2n-1}^{-1}\int_{\sigma_\varepsilon}\Big[f(0,\cdots,0,x_k+iy_k,\cdots,x_n+iy_n)-f(0,\cdots,0,iy_k,\cdots,x_n+iy_n)$$
$$-\frac{\partial f}{\partial x_k}(p_n)x_k\Big](1-\bar{u}_n)^{-(n+\frac{1}{2})}\dot{u}+\sum_{k=1}^{n-1}\omega_{2n-1}^{-1}\int_{\sigma_\varepsilon}\Big[f(0,\cdots,0,iy_k,\cdots,x_n+iy_n)$$
$$-f(0,\cdots,0,x_{k+1}+iy_{k+1},\cdots,y_n+iy_n)-\frac{\partial f}{\partial y_k}(p_n)y_k\Big](1-\bar{u}_n)^{-(n+\frac{1}{2})}\dot{u}$$
$$+\sum_{k=1}^{n-1}\frac{\partial f}{\partial x_k}(p_n)\omega_{2n-1}^{-1}\int_{\sigma_\varepsilon}x_k(1-\bar{u}_n)^{-(n+\frac{1}{2})}\dot{u}$$
$$+\sum_{k=1}^{n-1}\frac{\partial f}{\partial y_k}(p_n)\omega_{2n-1}^{-1}\int_{\sigma_\varepsilon}y_k(1-\bar{u}_n)^{-(n+\frac{1}{2})}\dot{u}$$
$$+\omega_{2n-1}^{-1}\int_{\sigma_\varepsilon}\Big[f(0,\cdots,0,x_n+iy_n)-f(0,\cdots,0,x_n)-\frac{\partial f}{\partial y_n}(p_n)y_n\Big](1-\bar{u}_n)^{-(n+\frac{1}{2})}\dot{u}$$
$$+\frac{\partial f}{\partial y_n}(p_n)\omega_{2n-1}^{-1}\int_{\sigma_\varepsilon}y_k(1-\bar{u}_n)^{-(n+\frac{1}{2})}\dot{u}$$
$$+\omega_{2n-1}^{-1}\int_{\sigma_\varepsilon}\Big[f(0,\cdots,0,x_n)-f(p_n)-\frac{\partial f}{\partial x_n}(p_n)(x_n-1)\Big](1-\bar{u}_n)^{-(n+\frac{1}{2})}\dot{u}$$
$$+\frac{\partial f}{\partial x_n}(p_n)\omega_{2n-1}^{-1}\int_{\sigma_\varepsilon}(x_n-1)(1-\bar{u}_n)^{-(n+\frac{1}{2})}\dot{u}$$
$$+f(p_n)\Big\{\omega_{2n-1}^{-1}\int_{\sigma_\varepsilon}(1-\bar{u}_n)^{-(n+\frac{1}{2})}\dot{u}-A_n\varepsilon^{-\frac{1}{2}}\Big\}.$$

By Lemma 2, 3, 4, 5, we obtain the equality (2.1) immediately.

Theorem 2. *Suppose v is an arbitrary point on the sphere S and f satisfies the same condition as in Lemma 5, then we have*

$$\lim_{\varepsilon\to 0}\Big\{\omega_{2n-1}^{-1}\int_{\sigma_\varepsilon(v)}f(u)(1-v\bar{u}')^{-(n+\frac{1}{2})}\dot{u}-\frac{A_n}{\sqrt{\varepsilon}}f(v)\Big\}$$
$$=\sum_{k=1}^{n}(J_{x_k}+J_{y_k})+\frac{1}{2n}\sum_{j=1}^{n}\frac{\partial f}{\partial u_j}(v)v_j-\sum_{j=1}^{n}\frac{\partial f}{\partial \bar{u}_j}(v)\bar{v}_j+f(v).\tag{2.2}$$

where $\sigma_\varepsilon(v)=\{u:u\bar{u}'=1,\ \alpha^2(1-|v\bar{u}'|^2)^2+4\beta^2(\operatorname{Im}(v\bar{u}'))^2>\varepsilon^2\}$, J_{x_k}, J_{y_k} are the integrals which are obtained by substituting $f(u)$ into $f(u\bar{u}')$ in the integrals I_{x_k}, I_{y_k}, and U is the

$n\times n$ matrix such that $vU=p_n$.

Proof Let $uU=w$. Then the left side of (2.2) is

$$\omega_{2n-1}^{-1}\int_{\sigma_\varepsilon} f(w\bar U')\,(1-\bar w_n)^{-\left(n+\frac{1}{2}\right)}\,\dot w-\frac{A_n}{\sqrt{\varepsilon}}f(p_n\bar U').$$

by Theorem 1, its limit is

$$\sum_{k=1}^{n}(J_{x_k}+J_{y_k})+\frac{1}{2n}\frac{\partial f(w\bar U')}{\partial w_n}(p_n\bar U')-\frac{\partial f(w\bar U')}{\partial \bar w_n}(p_n\bar U')+f(p_n\bar U') \tag{2.3}$$

as $\varepsilon\to 0$. Let $\bar U'=(\alpha_{ij})$. Then $(v_1,\cdots,v_n)=(\alpha_{n1},\cdots,\alpha_{nn})$. and

$$\frac{\partial f(w\bar U')}{\partial w_n}=\sum_{j=1}^{n}\frac{\partial f}{\partial u_j}v_j,\qquad \frac{\partial f(w\bar U')}{\partial \bar w_n}=\sum_{j=1}^{n}\frac{\partial f}{\partial \bar u_j}\bar v_j.$$

From (2.3) we obtain (2.2).

We call the limit (2.2) Hadamard principal value of singular integral

$$\omega_{2n-1}^{-1}\int_{u\bar u'=1}\frac{f(u)\dot u}{(1-v\bar u')^{n+\frac{1}{2}}}$$

and denote it by

$$P\frac{1}{\omega_{2n-1}}\int_{u\bar u'=1}\frac{f(u)\dot u}{(1-v\bar u')^{n+\frac{1}{2}}}=\lim_{\varepsilon\to 0}\left\{\frac{1}{\omega_{2n-1}}\int_{\sigma_\varepsilon(v)}\frac{f(u)\dot u}{(1-v\bar u')^{n+\frac{1}{2}}}-\frac{A_n}{\sqrt{\varepsilon}}f(v)\right\}.$$

Since A_n depends on σ_ε and there are many ways to define σ_ε, so there are many ways to define the Hadamard principal value. But we can see from (2.2) that the Hadamard principal value are the same, it does not depend on the choice of σ_ε.

§ 3. Plemelj formula

Lemma 6.

$$\lim_{z\to p_n}\omega_{2n-1}^{-1}\int_{u\bar u'=1}\frac{x_n-1}{(1-z\bar u')^{n+\frac{1}{2}}}\dot u=\frac{1-2n}{4n},$$

$$\lim_{z\to p_n}\omega_{2n-1}^{-1}\int_{u\bar u'=1}\frac{y_n}{(1-z\bar u')^{n+\frac{1}{2}}}\dot u=-\frac{2n+1}{4n}i.$$

Proof Fix $z\in B$. Suppose $z\bar z'=\rho^2$, there is a unitary matrix U, such that $zU=\rho p_n$. Let $u=w\bar U'$, $\bar U'=(\alpha_{ij})$. Then

$$x_n=\mathrm{Re}\,u_n=\frac{1}{2}\sum_{j=1}^{n}(\alpha_{jn}w_j+\bar\alpha_{jn}\bar w_j),$$

so

$$\frac{1}{\omega_{2n-1}}\int_{u\bar u'=1}\frac{(x_n-1)\dot u}{(1-z\bar u')^{n+\frac{1}{2}}}=\frac{1}{2\omega_{2n-1}}\left\{\alpha_{nn}\int_{w\bar w'=1}\frac{w_n\dot w}{(1-\rho\bar w_n)^{n+\frac{1}{2}}}+\bar\alpha_{nn}\int_{w\bar w'=1}\frac{\bar w_n\dot w}{(1-\rho\bar w_n)^{n+\frac{1}{2}}}\right.$$

$$\left.-2\int_{w\bar w'=1}\frac{\dot w}{(1-\rho\bar w_n)^{n+\frac{1}{2}}}\right\}. \tag{3.1}$$

Since

$$\int_{-\pi}^{\pi}\frac{re^{-i\theta}\,d\theta}{(1-\rho re^{i\theta})^{n+\frac{1}{2}}}=(2n+1)\pi\rho r^2,$$

so

$$\frac{1}{2\omega_{2n-1}}\int_{w\bar{w}'=1}\frac{w_n\dot{w}}{(1-\rho w_n)^{n+\frac{1}{2}}}=\frac{1}{2\omega_{2n-1}}\int_{0<v\bar{v}'<1}\dot{v}\int_{-\pi}^{\pi}\frac{re^{-i\theta}d\theta}{(1-\rho re^{i\theta})^{n+\frac{1}{2}}}=\frac{2n+1}{4n}\rho.$$

Using the same method, we have

$$\int_{w\bar{w}'=1}\frac{\bar{w}_n\dot{w}}{(1-\rho\bar{w}_n)^{n+\frac{1}{2}}}=0,\quad \frac{1}{\omega_{2n-1}}\int_{w\bar{w}'=1}\frac{\dot{w}}{(1-\rho\bar{w}_n)^{n+\frac{1}{2}}}=1.$$

It is not hard to see, when $\rho\to1$, $a_{nn}\to1$, $a_{kn}\to0(k=1,\cdots,n-1)$. From (3.1), we obtain

$$\lim_{z\to p_n}\frac{1}{\omega_{2n-1}}\int_{u\bar{u}'=1}\frac{(x_n-1)\dot{u}}{(1-z\bar{u}')^{n+\frac{1}{2}}}=\frac{1-2n}{4n}.$$

The proof of the other equality is similar.

Lemma 7.

$$\lim_{z\to p_n}\frac{1}{\omega_{2n-1}}\int_{u\bar{u}'=1}\frac{\operatorname{Re}u_k}{(1-z\bar{u}')^{n+\frac{1}{2}}}\dot{u}=0,$$

$$\lim_{z\to p_n}\frac{1}{\omega_{2n-1}}\int_{u\bar{u}'=1}\frac{\operatorname{Im}u_k}{(1-z\bar{u}')^{n+\frac{1}{2}}}\dot{u}=0\quad(k=1,\cdots,n-1).$$

Proof The proof of this lemma is similar to Lemma 6.

Theorem 3. *(Plemelj formula) Suppose f satisfies the condition in Lemma* 5, *then we have*

$$(K-\lim_{z\to p_n})\frac{1}{\omega_{2n-1}}\int_{u\bar{u}'=1}\frac{f(u)\dot{u}}{(1-z\bar{u}')^{n+\frac{1}{2}}}=P\frac{1}{\omega_{2n-1}}\int_{u\bar{u}'=1}\frac{f(u)\dot{u}}{(1-\bar{u}_n)^{n+\frac{1}{2}}}.$$

From this theorem, we can obtain immediately the following

Theorem 4. *Suppose $f=f_1+if_2$ is a complex function defined on the unit ball $\bar{B}$, f_1, f_2 satisfy the conditions in Lemma* 5, *then*

$$(K-\lim_{z\to v})\frac{1}{\omega_{2n-1}}\int_{u\bar{u}'=1}\frac{f(u)\dot{u}}{(1-z\bar{u}')^{n+\frac{1}{2}}}=P\frac{1}{\omega_{2n-1}}\int_{u\bar{u}'=1}\frac{f(u)\dot{u}}{(1-v\bar{u}')^{n+\frac{1}{2}}},$$

where v is an arbitrary point on the sphere S, and

$$P\frac{1}{\omega_{2n-1}}\int_{u\bar{u}'=1}\frac{f(u)\dot{u}}{(1-v\bar{u}')^{n+\frac{1}{2}}}$$
$$=P\frac{1}{\omega_{2n-1}}\int_{u\bar{u}'=1}\frac{f_1(u)\dot{u}}{(1-v\bar{u}')^{n+\frac{1}{2}}}+P\frac{i}{\omega_{2n-1}}\int_{u\bar{u}'=1}\frac{f_2(u)\dot{u}}{(1-v\bar{u}')^{n+\frac{1}{2}}}.$$

References

[1] Hadamard, J., Lectures on Cauchy problem in linear partial differential equations, 1952.

[2] Fox, C., A generalization of the Cauchy principal value, *Canadian J. Math.*, 1957 (9), 110—117.

[3] Kung, S. and Sun, C. K., Integrals of Cauchy type in several complex variables (I), *Acta Mathematica Sinica*, **15** (1965), 431—443 (In Chinese).

[4] Gong Sheng (Sheng Kung) and Shi Jihuai, Singular integrals in several complex variables (1), *Chinese annals of mathematics*, **3** (1982), 483—502.

[5] Koranyi, A., Harmonic functions on Hermitian hyperbolic space, *Tran. Amer. Math. Soc.*, **135** (1969), 507—516.

[6] Rudin, W., Function Theory in the Unit Ball of C^n, Springer-Verlag, New Jork, 1980.

Chin. Ann. of Math.
4B (4) 1983

SINGULAR INTERGRALS IN SEVERAL COMPLEX VARIABLES (III)—CAUCHY INTEGRALS OF CLASSICAL DOMAINS

GONG SHENG (龚 昇)
(University of Science and Technology of China, Institute of Applied Mathematics, Academia Sinica)

SHI JIHUAI (史济怀)
(University of Science and Technology of China)

Abstract

In [1], we discussed the Henkin-Ramirez integrals and Stein-Kerzman integrals of strictly pseudoconvex domains. We found that there are many different ways to define the principal values of H-R integrals or S-K integrals, so that there are many different Plemelj formulas. In the present paper, the similar problems of Cauchy integrals of classical domains are considered. The conclusions similar to [1] and the corresponding Plemelj formulas are obtained. Finally, we discuss the same problems for the Bochner-Martinelli integrals.

§ 0. Introduction

In [1], we discussed the Henkin-Ramiroz (H-R) integrals and Stein-Kerzman (S-K) integrals of strictly psoudoconvex domains and gave the general Plemelj formula. We found that there are many different ways to define the principal values of H-R integrals or S-K integrals, so that there are many different Plemelj formulas.

Can the above situations happen for the other integral representations? This is the problem that will be discussed in the present paper.

Let Ω be a unit ball. Then the S-K kernel of Ω is $(1-z\bar{u}')^{-n}$, and we have the Plemelj formula of the unit ball in [1]. When the deleted neighborhood around the boundary point is an "ellipse" or a "rectangle", the Plemelj formulas can be written as Theorem 1.1 and Theorem 1.2 in the present paper respectively. In § 1, we shall give other proofs for those two theorems. These direct proofs can be easily generalized to a more general situation. This is the content of § 2.

In [4], the Cauchy integral of Lie sphere hyperbolic space was considered. The elements of Lie sphere hyperbolic spaces $R_{\mathrm{IV}}(N)$ are the all complex vectors $z=(z_1, \cdots,$

Manuscript received June 14, 1981.

z_N) satisfying the following conditions

$$\begin{cases}1+|zA_0z'|^2-z\bar{z}'>0\\ 1-|zA_0z'|>0,\end{cases}\qquad A_0=\frac{1}{2}\begin{pmatrix}0&1\\-1&0\end{pmatrix}\dotplus\frac{1}{2}I^{N-2}.$$

By B_{IV} we denote the boundary of R_{IV}, and by L_{IV} the characteristic manifold of R_{IV}.

Let R^*_{IV} be the set of all complex vectors $z=(z_1, \cdots, z_N)$ satisfying

$$\begin{cases}1+|zA_0z'|^2-z\bar{z}'>0,\\ 1-|zA_0z'|<0,\end{cases}$$

L_{IV} is also the characteristic manifold of R^*_{IV}.

If $\varphi(\xi)$ is continuous in L_{IV}, then when $z\in R_{IV}$ or $z\in R^*_{IV}$, the Cauchy integral

$$V(L_{IV})^{-1}\int_{L_{IV}} H_{IV}(z, \xi)\varphi(\xi)\dot{\xi} \tag{0.1}$$

exists, where H_{IV}, $V(L_{IV})$ and $\dot{\xi}$ stand for the Cauchy kernel $(1+zA_0z'\overline{\xi A_0\xi'}-z\bar{\xi}')^{-\frac{N}{2}}$, the volume of L_{IV} and the volume element of L_{IV} respectively.

In [4], the limit values of (0.1), as z approaches $B_{IV}-L_{IV}$ or L_{IV} from the interior of R_{IV}, were obtained. By the same way we can obtain the limit value of (0.1) as z approaches the boundary from the interior of R^*_{IV}. These limit values can be represented by the Cauchy principal values.

In the section 3, using the results of § 1 and § 2, we prove that the limit value of (0.1) can be represented by a lot of distinct Cauchy principal values when z approaches $B_{IV}-L_{IV}$ from the interior of R_{IV}, but it cannot be represented as the above when z approaches L_{IV} from the interior of R_{IV}, because the computation of the limit value of (0.1) in this case must reduce to that of the Cauchy integrals of polydisc, however the polydisc is the topological product of unit discs, it cannot give the various results. It is the same for R^*_{IV}.

In [5, 6], the Cauchy integrals of matrix hyperbolic spaces were considered.

Let $R_I(m, n)$ and $R^*_I(m, n)$ stand for the matrix hyperbolic spaces $I-Z\bar{Z}'>0$ and $I-Z\bar{Z}'<0$ respectively, where Z is the $m\times n (m\leqslant n)$ complex matrix. If $\varphi(U)$ is continuous on the characteristic manifold $L_I(m, n)$ of $R_I(m, n)$, then the Cauchy integrals of $R_I(m, n)$ and $R^*_I(n)=R^*_I(n, n)$

$$V(L_I(m, n))^{-1}\int_{L_I(m,n)}\varphi(U)\det{}^{-n}(I-Z\bar{U}')\dot{U},\ Z\in R_I(m, n), \tag{0.2}$$

$$V(L_I(n))^{-1}\int_{L_I(n)}\varphi(U)\det{}^{-n}(I-Z\bar{U}')\dot{U},\ Z\in R^*_I(n) \tag{0.3}$$

exist. Here $V(L_I(m, n))$ denotes the volume of $L_I(m, n)$.

In [5], the limit values of (0.2) and (0.3), when Z approaches $L_I(n)$ from the interior of $R_I(n)$ and $R^*_I(n)$ respectively, were obtained. By the same reason as above, it cannot give the various definitions of Cauchy principal value. But when Z approaches $L_I^{(m-1)}$ ($L_I^{(m-1)}$ is a part of the boundary of $R_I(m, n)$, it is formed by the matrix satisfying rank$(I-Z\bar{Z}')=m-1$) from the interior of $R_I(m, n)$, by the results in § 1, we can obtain a lot of distinct definitions of Cauchy principal value and Plemelj formulas.

Another important kind of Cauchy-Fantappie kernels is of Bochner-Martinelli kernels. Lu qikeng and Zhong tongde[7] gave the Plemelj formulas of Bochner-Martinelli integrals when z approaches the boundary from the interior or exterior of domain. It is natural to ask whether one can obtain the various Cauchy principal values and Plemelj formulas if one deletes the various neighborhoods as in [1]. The answer is negative, this will be illustrated in § 5.

§ 1. The Cauchy integrals on sphere

Since the ball is a strictly pseudoconvex domain, it is easy to see, the Stein-Kerzman kernel is just $(1-z\bar{u}')^{-n}$. Therefore, the Theorems 1.1 and 1.2 below are the special cases of the Plemelj formulas in § 2.1 and § 3.1 of [1]. We shall give a direct proof here in order to prove the Lemmas 2.1 and 2.2 in § 2.

We first consider the case that the deleted neighborhood is an ellipse.

Lemma 1.1. *Suppose* $u=(u_1, \cdots, u_n)$, $u\bar{u}'=1$ *then*

$$\lim_{\varepsilon\to 0}\omega_{2n-1}^{-1}\int_{\sigma}(1-\bar{u}_1)^{-n}\dot{u}=1-\frac{1}{2}\left(\frac{2\beta}{\alpha+\beta}\right)^{n-1},$$

where $\sigma=\{u\,|\,u\bar{u}'=1,\ \alpha^2(\mathrm{Re}(1-\bar{u}_1))^2+\beta^2(\mathrm{Im}\,\bar{u}_1)^2>\varepsilon^2\}$, $\alpha\geqslant 0$, $\beta\geqslant 0$, $\alpha+\beta\neq 0$.

Proof We first assume that $\alpha>0$, $\beta>0$. Since the case of $\alpha=\beta$ was proved in [2], it is enough to prove it for the case of $\alpha\neq\beta$.

Pick $0<\rho<1$, we have

$$1=\omega_{2n-1}^{-1}\int_{\sigma}(1-\rho\bar{u}_1)^{-n}\dot{u}+\omega_{2n-1}^{-1}\int_{\sigma'}(1-\rho\bar{u}_1)^{-n}\dot{u}=I+I',$$

where $\sigma'=\{u\,|\,u\bar{u}'=1,\ \alpha^2(\mathrm{Re}(1-\bar{u}_1))^2+\beta^2(\mathrm{Im}\,\bar{u}_1)^2\leqslant\varepsilon^2\}$.

Let

$$I'=\omega_{2n-1}^{-1}\int_{\sigma'}(1-\rho\bar{u}_1)^{-n}\dot{u}.$$

Set $\bar{u}_1=re^{i\theta}$, $u_2=v_2$, $\cdots$, $u_n=v_n$, $v=(v_2, \cdots, v_n)$, $\alpha^2=t$, $\beta^2=1-t\,(0<t<1)$, then the points of σ' satisfy

$$t(1-r\cos\theta)^2+(1-t)r^2\sin^2\theta\leqslant\varepsilon^2,\quad v\bar{v}'=1-r^2,$$

so

$$\cos\theta\geqslant\frac{t-\sqrt{(1-t)(t+r^2(1-2t))+(2t-1)\varepsilon^2}}{r(2t-1)}. \tag{1.1}$$

Let

$$a=\arccos\frac{t-\sqrt{(1-t)(t+r^2(1-2t))+(2t-1)\varepsilon^2}}{r(2t-1)},$$

then $-a\leqslant\theta\leqslant a$. By (1.1), we have

$$1\geqslant\frac{t-\sqrt{(1-t)(t+r^2(1-2t))+(2t-1)\varepsilon^2}}{r(2t-1)}.$$

It is not hard to see $1-r^2\leqslant\frac{2\varepsilon}{\sqrt{t}}-\frac{\varepsilon^2}{t}$, thus

$$I'=\omega_{2n-1}^{-1}\int_{E(\varepsilon)}\dot{v}\int_{-a}^{a}(1-\rho re^{i\theta})^{-n}\,d\theta=2\omega_{2n-1}^{-1}\operatorname{Im}\left\{\sum_{p=1}^{n-1}K_p+K_0\right\}+2\omega_{2n-1}^{-1}\int_{E(\varepsilon)}a\dot{v},\quad(1.2)$$

where $E(\varepsilon)=\left\{v=(v_2,\ \cdots,\ v_n)\,|\,v\bar{v}'\leqslant\frac{2\varepsilon}{\sqrt{t}}-\frac{\varepsilon^2}{t}\right\}$,

$$K_p=\int_{E(\varepsilon)}\frac{\dot{v}}{p(1-\rho re^{ia})^p},\quad K_0=\int_{E(\varepsilon)}\log\frac{1}{1-\rho re^{ia}}\dot{v}.$$

Obviously

$$\frac{t-\sqrt{(1-t)(t+r^2(1-2t))+(2t-1)\varepsilon^2}}{r(2t-1)}-1=O(\varepsilon^2),$$

so $a=O(\varepsilon^2)$, and it follows that

$$2\omega_{2n-1}^{-1}\int_{E(\varepsilon)}a\dot{v}=O(\varepsilon^{n+1}).$$

Let

$$Q=\sqrt{(1-t)(t+r^2(1-2t))+(2t-1)\varepsilon^2},$$

then $r\cos a=\frac{t-Q}{2t-1}$, so

$$r\sin a=\frac{\sqrt{r^2(2t-1)t-t-(2t-1)\varepsilon^2+2Qt}}{|2t-1|}.$$

When $p\geqslant1$

$$K_p=\frac{1}{p}\int_{E(\varepsilon)}\left\{1-\rho\frac{t-Q}{2t-1}-i\rho\frac{\sqrt{r^2(2t-1)t-t-(2t-1)\varepsilon^2+2Qt}}{|2t-1|}\right\}^{-p}\dot{v}.$$

Set $v=(x_1,\ x_2,\ \cdots,\ x_{2n-2})$, using the sphero polar coordinates

$$x_1=s\cos\varphi_1,\ x_2=s\sin\varphi_1\cos\varphi_2,\ \cdots,\ x_{2n-2}=s\sin\varphi_1\sin\varphi_2\cdots\sin\varphi_{2n-3},$$
$$\dot{v}=s^{2n-3}\sin^{2n-4}\varphi_1\sin^{2n-5}\varphi_2\cdots\sin\varphi_{2n-4}ds\,d\varphi_1\cdots d\varphi_{2n-3},$$

we have

$$K_p=\frac{2\pi^{n-1}}{p\Gamma(n-1)}\int_0^{\sqrt{\frac{2\varepsilon}{\sqrt{t}}-\frac{\varepsilon^2}{t}}}\left\{1-\rho\frac{t-Q}{2t-1}-i\rho\frac{\sqrt{t(1-s^2)(2t-1)-t-(2t-1)\varepsilon^2+2Qt}}{|2t-1|}\right\}^{-p}s^{2n-3}ds,$$

the absolute value of integrand is

$$\left\{\left(1-\rho\frac{t-Q}{2t-1}\right)^2+\rho^2\frac{t(1-s^2)(2t-1)-t-(2t-1)\varepsilon^2+2Qt}{(2t-1)^2}\right\}^{-\frac{p}{2}}s^{2n-3},$$

and its limit value is

$$|2t-1|^ps^{2n-3}[(t+Q-1)^2+t(1-s^2)(2t-1)-t-(2t-1)\varepsilon^2+2Qt]^{-\frac{p}{2}}$$

as $\rho\to1$. It is not hard to prove that there are $\eta_0>0$, $\delta>0$, such that

$$(t+Q-1)^2+t(1-s^2)(2t-1)-t-(2t-1)\varepsilon^2+2Qt>\eta_0\varepsilon^2\quad(1.3)$$

for $t\in\left(0,\ \sqrt{\frac{2\varepsilon}{\sqrt{t}}-\frac{\varepsilon^2}{t}}\right)$, $\varepsilon\in(0,\ \delta)$.

Thus

$$J_p=\lim_{\rho\to1}K_p=O\left(\varepsilon^{-p}\int_0^{\sqrt{\frac{2\varepsilon}{\sqrt{t}}-\frac{\varepsilon^2}{t}}}s^{2n-3}\,ds\right)=O(\varepsilon^{n-p-1}).$$

When $p<n-1$

$$\lim_{\varepsilon\to0}J_p=\lim_{\varepsilon\to0}\lim_{\rho\to1}K_p=0.$$

By the Lobesgue theorem

$$\begin{aligned}J_{n-1}=&\frac{2\pi^{n-1}}{(n-1)!}\int_0^{\sqrt{\frac{2\varepsilon}{\sqrt{t}}-\frac{\varepsilon^2}{t}}}\left[\frac{t+Q-1}{2t-1}\right.\\&\left.-i\frac{\sqrt{t(2t-1)(1-s^2)-t-(2t-1)\varepsilon^2+2Qt}}{|2t-1|}\right]^{-(n-1)}s^{2n-3}\,ds\\=&\frac{2\pi^{n-1}}{(n-1)!}\int_0^1\left[\frac{t+Q-1}{2t-1}\right.\\&\left.-i\frac{\sqrt{t(2t-1)(1-\eta^2R^2)-t-(2t-1)\varepsilon^2+2Qt}}{|2t-1|}\right]^{-(n-1)}\eta^{2n-2}R^{2n-3}\,dR,\end{aligned}$$

where $\eta=\sqrt{\frac{2\varepsilon}{\sqrt{t}}-\frac{\varepsilon^2}{t}}$, $s=\eta R$.

Since

$$\eta^{-2}\frac{t-1+Q}{2t-1}=\eta^{-2}\frac{t-1+\sqrt{(1-t)^2+(2t-1)(1-t)\eta^2R^2+(2t-1)\varepsilon^2}}{2t-1}=\frac{R^2}{2}+O(\varepsilon)$$

and

$$\begin{aligned}&\frac{1}{\eta^2|2t-1|}\left\{t(2t-1)(1-\eta^2R^2)-t(2t-1)\varepsilon^2-2Qt\right\}^{\frac12}\\&=\left(\frac{t}{1-t}\right)^{\frac12}\left\{\frac1t\frac{1}{\left(\frac{2}{\sqrt{t}}-\frac{\varepsilon}{t}\right)^2}-\frac{R^4}{4}+O(\varepsilon)\right\}^{\frac12}=\frac12\left(\frac{t}{1-t}\right)^{\frac12}(1-R^4)^{\frac12}+O(\varepsilon),\end{aligned}$$

we have

$$\lim_{\varepsilon\to0}J_{n-1}=\frac{2^n\pi^{n-1}}{\Gamma(n)}\int_0^1\frac{R^{2n-3}dR}{\left[R^2-i\left(\frac{t}{1-t}\right)^{\frac12}\sqrt{1-R^4}\right]^{n-1}}=\frac{(2\pi)^{n-1}}{\Gamma(n)}\int_0^1\frac{u^{n-2}du}{(u-ib\sqrt{1-u^2})^{n-1}}.$$

where $b=\left(\frac{t}{1-t}\right)^{\frac12}$. Let $u=\cos\theta$, then

$$\lim_{\varepsilon\to0}J_{n-1}=\frac{(2\pi)^{n-1}}{i\Gamma(n)(1+b)^{n-1}}\int_0^{\frac{\pi}{2}}\frac{(e^{2i\theta}+1)^{n-2}(e^{2i\theta}-1)}{\left(1+\frac{1-b}{1+b}e^{2i\theta}\right)^{n-1}}\,d\theta.$$

Since $b>0$, $-1<\frac{1-b}{1+b}<1$, we have

$$\begin{aligned}\lim_{\varepsilon\to0}J_{n-1}=&\frac{(2\pi)^{n-1}}{\Gamma(n)}\left\{\frac{-1}{2(1+b)^{n-1}}\sum_{p=0}^{n-2}\sum_{q=0}^{\infty}(-1)^qC_{n-2}^pC_{n+q-2}^q\left(\frac{1-b}{1+b}\right)^q\right.\\&\left.\times\left[\frac{(-1)^{p+q+1}-1}{p+q+1}-\frac{(-1)^{p+q}-1}{p+q}\right]-\frac{1}{i(1+b)^{n-1}}\frac{\pi}{2}+\frac{1}{(1+b)^{n-1}}\right\}.\end{aligned}$$

Then

$$\operatorname{Im}(\lim_{\varepsilon\to0} J_{n-1}) = \lim_{\varepsilon\to0} (\operatorname{Im} J_{n-1}) = \frac{\pi^n}{(n-1)!}\,\frac{1}{2}\left(\frac{2}{1+b}\right)^{n-1}.$$

Now we calculate

$$J_0 = \lim_{\rho\to1} K_0 = \int_{E(\varepsilon)} \log\frac{1}{1-re^{i\alpha}}\,\dot{v} = \frac{-2\pi^{n-1}}{\Gamma(n-1)}\int_0^{\sqrt{\frac{2\varepsilon}{t}-\frac{\varepsilon^2}{t}}} s^{2n-3}\log(1-re^{i\alpha})\,ds.$$

Since $\operatorname{Im}\log(1-re^{i\alpha}) = O(1)$, so $J_0 = O(\varepsilon^{n-1})$.

In summary, we have

$$\lim_{\varepsilon\to0}\lim_{\rho\to1} I' = \frac{2}{\omega_{2n-1}}\,\frac{\pi^n}{(n-1)!}\,\frac{1}{2}\left(\frac{2}{1+b}\right)^{n-1} = \frac{1}{2}\left(\frac{2}{1+b}\right)^{n-1}.$$

Substituting $b = \left(\frac{t}{1-t}\right)^{\frac{1}{2}} = \frac{\alpha}{\beta}$ into the above equality, we have proved Lemma 1.1 for the case $\alpha>0$, $\beta>0$.

When $\alpha=0$, $\beta>0$, this was proved in [8].

When $\alpha>0$, $\beta=0$, Lemma 1.1 becomes

$$\lim_{\varepsilon\to0} \omega_{2n-1}^{-1}\int_{\sigma_1} (1-\bar{u}_1)^{-n}\dot{u} = 1, \tag{1.4}$$

where $\sigma_1 = \{u \mid u\bar{u}'=1,\ \operatorname{Re}(1-\bar{u}_1)>\varepsilon\}$.

The proof is similar to the case $\beta>0$. Taking $0<\rho<1$, we have

$$1 = \omega_{2n-1}^{-1}\int_{\sigma_1} (1-\rho\bar{u}_1)^{-n}\dot{u} + \omega_{2n-1}^{-1}\int_{\sigma_1'} (1-\rho\bar{u}_1)^{-n}\dot{u} = I_1 + I_1',$$

where $\sigma_1' = \{u \mid u\bar{u}'=1,\ \operatorname{Re}(1-\bar{u}_1)\leqslant\varepsilon\}$.

Similar to (1.2)

$$I_1' = 2\omega_{2n-1}^{-1}\operatorname{Im}\left\{\sum_{p=1}^{n-1} K_p + K_0\right\} + 2\omega_{2n-1}^{-1}\int_{v\bar{v}'<2\varepsilon-\varepsilon^2} a\dot{v},$$

where $a = \arccos\frac{1-\varepsilon}{r}$. As in the case $\beta>0$, we can prove

$$\int_{v\bar{v}'<2\varepsilon-\varepsilon^2} a\dot{v} = o(1),\ J_p = O(\varepsilon^{n-p-1}),\ J_0 = o(1),\ p=1,\ \cdots,\ n-1.$$

At this case

$$J_{n-1} = \frac{2\pi^{n-1}}{(n-1)!}\int_0^{\sqrt{2\varepsilon-\varepsilon^2}} \frac{s^{2n-3}ds}{(\varepsilon - i\sqrt{2\varepsilon-\varepsilon^2-s^2})^{n-1}}$$

$$= \frac{2\pi^{n-1}}{(n-1)!}\left(\frac{2\varepsilon-\varepsilon^2}{\varepsilon}\right)^{n-1}\int_0^1 \frac{R^{2n-3}dR}{(1-i\eta\varepsilon^{-1}\sqrt{1-R^2})^{n-1}},$$

where $\eta = \sqrt{2\varepsilon-\varepsilon^2}$, $s=\eta R$.

The absolute value of the denominator of the integrand is

$$(1+\eta^2\varepsilon^{-2}(1-R^2))^{\frac{n-1}{2}} = [1+(2\varepsilon-\varepsilon^2)\varepsilon^{-2}(1-R^2)]^{\frac{n-1}{2}} \geqslant 1.$$

By Lebesgue theorem, $\lim_{\varepsilon\to0} J_{n-1}=0$, so $\lim_{\varepsilon\to0} I_1'=0$. The proof of Lemma 1.1 is complete.

Using this Lemma, as in [2], we can prove the following Plemelj formula:

Theorem 1.1. *Suppose f is in Lipα on $u\bar{u}'=1$, v is an arbitrary point on*

$u\bar{u}'=1$, *then*

$$(K\text{-}\lim_{z\to v})\ \omega_{2n-1}^{-1}\int_{u\bar{u}'=1}\frac{f(u)}{(1-z\bar{u}')^n}\dot{u}=\frac{1}{2}\left(\frac{2\beta}{\alpha+\beta}\right)^{n-1}f(v)+p.\ v.\ \frac{1}{\omega_{2n-1}}\int_{u\bar{u}'=1(\gamma)}\frac{f(u)\dot{u}}{(1-v\bar{u}')^n}.$$

Here the Cauchy principal value is defined as

$$p.\ v.\ \frac{1}{\omega_{2n-1}}\int_{u\bar{u}'=1(\gamma)}\frac{f(u)\dot{u}}{(1-v\bar{u}')^n}=\lim_{\varepsilon\to0}\frac{1}{\omega_{2n-1}}\int_{\sigma}\frac{f(u)\dot{u}}{(1-v\bar{u}')^n},\quad \gamma=\frac{\alpha}{\beta},$$

where $\sigma=\{u|u\bar{u}'=1,\ \alpha^2(1-\mathrm{Re}v\bar{u}')^2+\beta^2(\mathrm{Im}v\bar{u}')^2\geqslant\varepsilon^2\}$ (*The definition of K-limit, see* [9])

It is clear that this theorem is a generalization of Theorem 1.3 in [2] (see [2, 3]).

Now we consider the case that the deleted neighborhood is a rectangle.

Theorem. 1.2 *Suppose f satisfies the same condition as in Theorem 1.1, then*

$$(K\text{-}\lim_{z\to v})\frac{1}{\omega_{2n-1}}\int_{u\bar{u}'=1}\frac{f(u)\dot{u}}{(1-z\bar{u}')^n}=bf(v)+p.\ v.\ \frac{1}{\omega_{2n-1}}\int_{u\bar{u}'=1(b)}\frac{f(u)\dot{u}}{(1-v\bar{u}')^n},$$

where $b=\dfrac{2^{n-1}}{\pi}\left\{\dfrac{\pi}{2}-\displaystyle\int_0^{\operatorname{arctg}\frac{\beta}{\alpha}}\cos^{n-2}t\,\frac{\sin(n-1)t}{\sin t}dt\right\}$, *and the Cauchy principal value is defined as*

$$p.v.\ \frac{1}{\omega_{2n-1}}\int_{u\bar{u}'=1(b)}\frac{f(u)\dot{u}}{(1-v\bar{u}')^n}=\lim_{\varepsilon\to0}\frac{1}{\omega_{2n-1}}\int_{D^*(v,\varepsilon)}\frac{f(u)}{(1-v\bar{u}'^n)}\dot{u},$$

$D^*(v,\ \varepsilon)$ *is the complementary set of* $D(v,\ \varepsilon)$ *on* $u\bar{u}'=1$ *and*

$$D(v,\ \varepsilon)=\{u|u\bar{u}'=1,\ 1-\mathrm{Re}v\bar{u}'<\alpha\varepsilon,\ |\mathrm{Im}v\bar{u}'|<\beta\varepsilon\},\ \alpha\geqslant0,\ \beta\geqslant0.$$

To prove Theorem 1.2, it is enough to prove the following

Lemma 1.2.

$$\lim_{\varepsilon\to0}\frac{1}{\omega_{2n-1}}\int_{D^*(v,\varepsilon)}\frac{\dot{u}}{(1-\bar{u}_1)^n}=1-\frac{2^{n-1}}{\pi}\left\{\frac{\pi}{2}-\int_0^{\operatorname{arctg}\frac{\beta}{\alpha}}\cos^{n-2}t\,\frac{\sin(n-1)t}{\sin t}dt\right\}.\qquad(1.5)$$

Proof It is only need to prove

$$\lim_{\varepsilon\to0}\frac{1}{\omega_{2n-1}}\int_{D(v,\varepsilon)}\frac{\dot{u}}{(1-u_1)^n}=\frac{2^{n-1}}{\pi}\left\{\frac{\pi}{2}-\int_0^{\operatorname{arctg}\frac{\beta}{\alpha}}\cos^{n-2}t\,\frac{\sin(n-1)t}{\sin t}dt\right\}.$$

Since

$$D(v,\ \varepsilon)=\{u|u\bar{u}'=1,\ 1-\mathrm{Re}v\bar{u}'<\alpha\varepsilon\}\cup\{u|u\bar{u}'=1,\ |\mathrm{Im}\,v\bar{u}'|<\beta\varepsilon\}$$
$$-\{u|u\bar{u}'=1\}\cup\{u|u\bar{u}'=1,\ 1-\mathrm{Re}v\bar{u}'>\alpha\varepsilon,\ |\mathrm{Im}v\bar{u}'|>\beta\varepsilon\},$$

it is known that (see [8])

$$\lim_{\varepsilon\to0}\omega_{2n-1}^{-1}\int_{\substack{u\bar{u}'=1\\1-\mathrm{Re}u_1<\alpha\varepsilon}}(1-\bar{u}_1)^{-n}\dot{u}=0,\qquad(1.6)$$

$$\lim_{\varepsilon\to0}\omega_{2n-1}^{-1}\int_{\substack{u\bar{u}'=1\\|\mathrm{Im}u_1|<\beta\varepsilon}}(1-\bar{u}_1)^{-n}\dot{u}=2^{n-2}.\qquad(1.7)$$

Thus we only need to prove

$$\lim_{\varepsilon\to0}\omega_{2n-1}^{-1}\int_{\substack{u\bar{u}'=1\\1-\mathrm{Re}u_1>\alpha\varepsilon,\ |\mathrm{Im}u_1|>\beta\varepsilon}}(1-\bar{u}_1)^{-n}\dot{u}=1-\frac{2^{n-1}}{\pi}\int_0^{\operatorname{arctg}\frac{\beta}{\alpha}}\cos^{n-2}t\,\frac{\sin(n-1)t}{\sin t}dt.\qquad(1.8)$$

When $\beta=0$, (1.8) is just (1.6). When $\alpha=0$, by [1]

$$\int_0^{\frac{\pi}{2}}\cos^{n-2}t\,\frac{\sin(n-1)t}{\sin t}\,dt=\frac{\pi}{2}.$$

(1.8) is just (1.7). So it is enough to prove the theorem for the case $\alpha>0$, $\beta>0$.

It is not hard to see that

$$\frac{1}{\omega_{2n-1}}\int_{S(\varepsilon)}\frac{\dot u}{(1-u_1)^n}=\frac{1}{\omega_{2n-1}}\int_{M(\varepsilon)}\dot v\Big(\int_{-(\pi-b)}^{-a}+\int_a^{\pi-b}\Big)\frac{d\theta}{(1-re^{i\theta})^n}$$
$$+\frac{1}{\omega_{2n-1}}\int_{N(\varepsilon)}\dot v\Big(\int_{-(\pi-b)}^{-b}+\int_b^{\pi-b}\Big)\frac{d\theta}{(1-re^{i\theta})^n}=I_1+I_2,$$

where

$$S(\varepsilon)=\{u\,|\,u\bar u'=1,\ 1-\mathrm{Re}\,u_1>\alpha\varepsilon,\ |\mathrm{Im}\,u_1|>\beta\varepsilon\},$$
$$M(\varepsilon)=\{v\,|\,v\bar v'<2\alpha\varepsilon-(\alpha^2+\beta^2)\varepsilon^2\},$$
$$N(\varepsilon)=\{v\,|\,2\alpha\varepsilon-(\alpha^2+\beta^2)\varepsilon^2\leqslant v\bar v'<1-\beta^2\varepsilon^2\}.$$

Since

$$\int_{-(\pi-b)}^{-a}(1-re^{i\theta})^{-n}d\theta+\int_a^{\pi-b}(1-re^{i\theta})^{-n}d\theta=2\,\mathrm{Im}(B_1-B_2)+2(\pi-a-b),$$

where

$$B_1=\sum_{k=1}^{n-1}\frac{1}{k(1+re^{-ib})^k}-\log(1+re^{-ib}),$$
$$B_2=\sum_{k=1}^{n-1}\frac{1}{k(1-re^{ia})^k}-\log(1-re^{ia}),$$

and

$$a=\arccos\frac{1-\alpha\varepsilon}{r},\quad b=\sin\frac{\beta\varepsilon}{r},$$

hence

$$I_1=2\omega_{2n-1}^{-1}\,\mathrm{Im}\int_{M(\varepsilon)}(B_1-B_2)\dot v+2\omega_{2n-1}^{-1}\int_{M(\varepsilon)}(\pi-a-b)\dot v$$
$$=2\omega_{2n-1}^{-1}\,\mathrm{Im}\Big[\Big(\sum_{k=1}^{n-1}J_k-J_0\Big)-\Big(\sum_{k=1}^{n-1}K_k-K_0\Big)\Big]+2\omega_{2n-1}^{-1}\int_{M(\varepsilon)}(\pi-a-b)\dot v,$$

where

$$J_0=\int_{M(\varepsilon)}\log(1+re^{-ib})\dot v,\quad J_k=\frac{1}{k}\int_{M(\varepsilon)}(1+re^{-ib})^{-k}\dot v,\quad k=1,\cdots,n-1,$$
$$K_0=\int_{M(\varepsilon)}\log(1-re^{ia})\dot v,\quad K_k=\frac{1}{k}\int_{M(\varepsilon)}(1-re^{ia})^{-k}\dot v,\quad k=1,\cdots,n-1.$$

Since $1+re^{-ib}=1+\sqrt{r^2-\beta^2\varepsilon^2}-i\beta\varepsilon$, we have

$$J_k=\frac{2\pi^{n-1}}{k\Gamma(n-1)}\int_0^{\sqrt{2\alpha\varepsilon-(\alpha^2+\beta^2)\varepsilon^2}}s^{2n-3}\,(1+\sqrt{r^2-\beta^2\varepsilon^2}-i\beta\varepsilon)^{-k}\,ds.$$

The absolute value of integrand does not exceed s^{2n-3}, so $J_k=O(\varepsilon^{n-1})$ and

$$J_0=\int_{M(\varepsilon)}\log(1+re^{-ib})\dot v=\frac{2\pi^{n-1}}{\Gamma(n-1)}\int_0^{\sqrt{2\alpha\varepsilon-(\alpha^2+\beta^2)\varepsilon^2}}s^{2n-3}\log(1+\sqrt{r^2-\beta^2\varepsilon^2}-i\beta\varepsilon)ds.$$

Thus $\mathrm{Im}\,J_0=O(\varepsilon^{n-1})$. On the other hand, since $1-re^{ia}=\alpha\varepsilon-i\sqrt{r^2-(1-\alpha\varepsilon)^2}$, we have $\mathrm{Im}K_0=O(\varepsilon^{n-1})$. By the inequality

$$|s^{2n-3}(\alpha\varepsilon - i\sqrt{1-s^2-(1-\alpha\varepsilon)^2})^{-k}| \leqslant (\alpha^2+\beta^2)^{-\frac{k}{2}}\varepsilon^{-k}(\sqrt{2\alpha\varepsilon-(\alpha^2+\beta^2)\varepsilon^2})^{2n-3},$$

$$K_k = \frac{2\pi^{n-1}}{k\Gamma(n-1)}\int_0^{\sqrt{2\alpha\varepsilon-(\alpha^2+\beta^2)\varepsilon^2}} s^{2n-3}(\alpha\varepsilon - i\sqrt{1-s^2-(1-\alpha\varepsilon)^2})^{-k}ds = O(\varepsilon^{n-k-1}).$$

Finally

$$K_{n-1} = \frac{2\pi^{n-1}}{\Gamma(n)}\int_0^{\sqrt{2\alpha\varepsilon-(\alpha^2+\beta^2)\varepsilon^2}} s^{2n-3}(\alpha\varepsilon - i\sqrt{1-s^2-(1-\alpha\varepsilon)^2})^{-(n-1)}ds$$
$$= \frac{2\pi^{n-1}}{\Gamma(n)}\left(\frac{\eta^2}{\alpha\varepsilon}\right)^{n-1}\int_0^1 R^{2n-3}\left[1-\frac{i}{\alpha\varepsilon}\sqrt{1-\eta^2R^2-(1-\alpha\varepsilon)^2}\right]^{-(n-1)}dR,$$

where $\eta = \sqrt{2\alpha\varepsilon-(\alpha^2+\beta^2)\varepsilon^2}$, $s=\eta R$. Since

$$\frac{1}{\alpha\varepsilon}\sqrt{1-\eta^2R^2-(1-\alpha\varepsilon)^2} = \left[\frac{1}{\alpha\varepsilon}(1-R^2)(2-\alpha\varepsilon)+\frac{\beta^2R^2}{\alpha^2}\right]^{\frac{1}{2}} = O(\varepsilon^{-\frac{1}{2}}),$$

it follows that $\lim\limits_{\varepsilon\to 0} K_{n-1}=0$. Using the same method, we have $\int_{M(\varepsilon)}(\pi-a-b)\dot{v}=o(1)$. Hence $\lim\limits_{\varepsilon\to 0} I_1=0$.

Now we calculate I_2

$$I_2 = \omega_{2n-1}^{-1}\int_{N(\varepsilon)}\dot{v}\left(\int_{-(\pi-b)}^{-b}+\int_b^{\pi-b}\right)\frac{d\theta}{(1-re^{i\theta})^n}$$
$$= 2\omega_{2n-1}^{-1}\,\mathrm{Im}\left\{\left(\sum_{k=1}^{n-1}P_k-P_0\right)-\left(\sum_{k=1}^{n-1}Q_k-Q_0\right)\right\}+2\omega_{2n-1}^{-1}\int_{N(\varepsilon)}(\pi-2b)\dot{v},$$

where $N(\varepsilon)=\{v\,|\,\eta^2\leqslant v\bar{v}'\leqslant 1-\beta^2\varepsilon^2\}$

$$P_0=\int_{N(\varepsilon)}\log(1+re^{-ib})\dot{v},\quad P_k=\frac{1}{k}\int_{N(\varepsilon)}(1+re^{-ib})^{-k}\dot{v},\ k=1,\ \cdots,\ n-1,$$

$$Q_0=\int_{N(\varepsilon)}\log(1-re^{ib})\dot{v},\quad Q_k=\frac{1}{k}\int_{N(\varepsilon)}(1-re^{ib})^{-k}\dot{v},\ k=1,\ \cdots,\ n-1.$$

It is clear that

$$\int_{N(\varepsilon)}b\dot{v}=o(1)$$

and

$$\lim_{\varepsilon\to 0}\int_{N(\varepsilon)}\pi\dot{v}=\lim_{\varepsilon\to 0}\frac{\pi^{n-1}}{\Gamma(n)}\pi[(\sqrt{1-\beta^2\varepsilon^2})^{2n-2}-\eta^{2n-2}]=\frac{\pi^n}{\Gamma(n)},$$

so

$$\lim_{\varepsilon\to 0}2\omega_{2n-1}^{-1}\int_{N(\varepsilon)}(\pi-2b)\dot{v}=1.$$

On the other hand

$$P_k=\frac{2\pi^{n-1}}{k\Gamma(n-1)}\int_\eta^{\sqrt{1-\beta^2\varepsilon^2}}s^{2n-3}(1+\sqrt{r^2-\beta^2\varepsilon^2}-i\beta\varepsilon)^{-k}ds$$
$$=\frac{2\pi^{n-1}}{k\Gamma(n-1)}\int_0^{\sqrt{1-\beta^2\varepsilon^2}}s^{2n-3}(1+\sqrt{r^2-\beta^2\varepsilon^2}-i\beta\varepsilon)^{-k}ds+o(1)$$
$$=\frac{2\pi^{n-1}}{k\Gamma(n-1)}\int_0^1\gamma^{2n-2}R^{2n-3}(1-\gamma\sqrt{1-R^2}-i\beta\varepsilon)^{-k}dR+o(1),$$

where $\gamma=\sqrt{1-\beta^2\varepsilon^2}$, $s=\gamma R$. When $\varepsilon\to 0$, P_k tends to a real number and $\lim\limits_{\varepsilon\to 0}\mathrm{Im}P_k=0$, $(1\leqslant k\leqslant n-1)$. Similarly, $\mathrm{Im}\,P_0=\mathrm{Im}\,Q_0=o(1)$. Since $1-re^{ib}=1-\sqrt{r^2-\beta^2\varepsilon^2}-i\beta\varepsilon$,

$$Q_k=\frac{2\pi^{n-1}}{k\Gamma(n-1)}\int_\eta^{\sqrt{1-\beta^2\varepsilon^2}} s^{2n-3}(1-\sqrt{1-s^2-\beta^2\varepsilon^2}-i\beta\varepsilon)^{-k}ds.$$

It is clear that

$$\int_0^\eta s^{2n-3}(1-\sqrt{1-s^2-\beta^2\varepsilon^2}-i\beta\varepsilon)^{-k}ds=O(\varepsilon^{n-k-1}),$$

$$\left|\int_0^{\sqrt{1-\beta^2\varepsilon^2}} s^{2n-3}(1-\sqrt{1-s^2-\beta^2\varepsilon^2}-i\beta\varepsilon)^{-k}ds\right|=\left|\int_0^1\frac{\gamma^{2n-2}R^{2n-3}dR}{(1-\gamma\sqrt{1-R^2}-i\beta\varepsilon)^k}\right|$$
$$\leqslant\int_0^1\frac{R^{2n-3}dR}{(1-\sqrt{1-R^2})^k},$$

when $k<n-1$, the integral of right side converges, so

$$\lim_{\varepsilon\to 0}Q_k=\text{real number}.$$

it follows that $\lim\limits_{\varepsilon\to 0}\operatorname{Im}Q_k=0\ (k<n-1)$. Finally

$$Q_{n-1}=\frac{2\pi^{n-1}}{\Gamma(n)}\int_\eta^\gamma s^{2n-3}(1-\sqrt{\gamma^2-s^2}-i\beta\varepsilon)^{-(n-1)}ds=\frac{2\pi^{n-1}}{\Gamma(n)}\left(\int_0^\gamma-\int_0^\eta\right)=T_1-T_2.$$

Note

$$T_1=\frac{2\pi^{n-1}}{\Gamma(n)}\int_0^1\gamma^{2n-2}R^{2n-3}(1-\gamma\sqrt{1-R^2}-i\sqrt{1-\gamma^2})^{-(n-1)}dR$$

so

$$\lim_{\varepsilon\to 0}\operatorname{Im}T_1=\frac{2^{n-2}\pi^n}{\Gamma(n)}$$

and

$$T_2=\frac{2\pi^{n-1}}{\Gamma(n)}\int_0^1\eta^{2n-2}R^{2n-3}(1-\sqrt{\gamma^2-\eta^2R^2}-i\beta\varepsilon)^{-(n-1)}dR.$$

Since

$$(\gamma^2-\eta^2R^2)^{\frac12}=(1-\beta^2\varepsilon^2-\eta^2R^2)^{\frac12}=1-\alpha R^2\varepsilon+O(\varepsilon^2),$$
$$(1-\sqrt{\gamma^2-\eta^2R^2}-i\beta\varepsilon)^{n-1}=(\alpha R^2-i\beta)^{n-1}\varepsilon^{n-1}+O(\varepsilon^n),$$

and $\eta^{2n-2}R^{2n-3}=(2\alpha\varepsilon-(\alpha^2+\beta^2)\varepsilon^2)^{n-1}R^{2n-3}$, hence

$$\lim_{\varepsilon\to 0}T_2=\frac{2\pi^{n-1}}{\Gamma(n)}\int_0^1\frac{(2\alpha)^{n-1}}{(\alpha R^2-i\beta)^{n-1}}R^{2n-3}dR=\frac{(2\pi)^{n-1}}{\Gamma(n)}\int_0^1\frac{u^{n-2}du}{(u-ic)^{n-1}},$$

where $c=\dfrac{\beta}{\alpha}$. An easy computation shows that

$$\operatorname{Im}\left(\int_0^1\frac{u^{n-2}du}{(u-ic)^{n-1}}\right)=\operatorname{arctg}\frac1c-\sum_{k=1}^{n-2}\frac1k C_{n-2}^k\operatorname{Im}\left(\frac{ic}{1-ic}\right)$$
$$=\operatorname{arctg}\frac{\alpha}{\beta}-\sum_{k=1}^{n-2}\frac{(-1)^{k+1}C_{n-2}^k\beta^k}{k(\alpha^2+\beta^2)^k}\sin k\left(\operatorname{arctg}\frac{\alpha}{\beta}\right).$$

and

$$\lim_{\varepsilon\to 0}\operatorname{Im}Q_{n-1}=\frac{2\pi^{n-1}}{\Gamma(n)}\left(2^{n-3}\pi-2^{n-2}\operatorname{arc\,tg}\frac{\alpha}{\beta}+2^{n-2}h_n\right),$$

where $h_n=\sum\limits_{k=1}^{n-2}(-1)^{k+1}\dfrac{C_{n-2}^k\beta^k}{k(\alpha^2+\beta^2)^{k/2}}\sin k\left(\operatorname{arc\,tg}\dfrac{\alpha}{\beta}\right)$. Hence

$$\lim_{\varepsilon\to 0}I_2=1-2^{n-2}\left(1-\frac{2}{\pi}\operatorname{arc\,tg}\frac{\alpha}{\beta}+\frac{2}{\pi}h_n\right)$$

By [1], (1.8) holds.

§ 2. Two Lemmas

It is not hard to generalize Lemmas 1.1 and 1.2 to the following Lemmas 2.1 and 2.2.

Lemma 2.1. *Suppose* $y=(y_1, \cdots, y_N)$, $y\bar{y}'=1$, *then*

$$\lim_{\varepsilon\to 0}\omega_{N-1}^{-1}\int_{\sigma(\varepsilon)}[1-(y_1+iy_2)]^{-\frac{N}{2}}\dot{y}=1-\frac{1}{2}\left(\frac{2\beta}{\alpha+\beta}\right)^{\frac{N}{2}-1},$$

where $\sigma(\varepsilon)=\{y\,|\,y\bar{y}'=1,\ \alpha^2(1-y_1)^2+\beta^2y_2^2\geqslant\varepsilon^2\}$, ω_{N-1}, $\dot{y}$ *stand for the volume and the element of volume of* $y\bar{y}'=1$ *respectively.*

Proof When $N=2n$, this is just Lemma 1.1, so it is enough to prove it when $N=2n+1$. It is known[4] that

$$\omega_{N-1}^{-1}\int_{y\bar{y}'=1}(1-\rho(y_1+iy_2))^{-\frac{N}{2}}\dot{y}=1,$$

so we only need to prove that

$$\lim_{\varepsilon\to 0}\lim_{\rho\to 1}\omega_{N-1}^{-1}\int_{\sigma'(\varepsilon)}(1-\rho(y_1+iy_2))^{-(n+\frac{1}{2})}\dot{y}=\frac{1}{2}\left(\frac{2\beta}{\alpha+\beta}\right)^{n-\frac{1}{2}},$$

where $\sigma'(\varepsilon)=\{y\,|\,y\bar{y}'=1,\ \alpha^2(1-y_1)^2+\beta^2y_2^2\leqslant\varepsilon^2\}$. Let

$$I=\omega_{2n}^{-1}\int_{\sigma'(\varepsilon)}(1-\rho(y_1+iy_2))^{-(n+\frac{1}{2})}\dot{y}$$

and suppose $\alpha>0$, $\beta>0$. Set $y_1+iy_2=re^{i\theta}$, $v=(y_3, \cdots, y_{2n+1})$, then

$$I=\omega_{2n}^{-1}\int_{P(\varepsilon)}\dot{v}\int_{-a}^{a}(1-\rho re^{i\theta})^{-(n+\frac{1}{2})}d\theta,$$

where $P(\varepsilon)=\left\{v\,\middle|\,v\bar{v}'\leqslant\frac{2\varepsilon}{\sqrt{t}}-\frac{\varepsilon^2}{t^2}\right\}$

$$a=\arccos\frac{t-\sqrt{(1-t)(t+r^2(1-2t))+(2t-1)\varepsilon^2}}{r(2t-1)},\quad t=\alpha^2,\ 1-t=\beta^2.$$

Using the same method as that of the proof of Lemma 2.1 in [4], we have

$$\int_{-a}^{a}(1-\rho re^{i\theta})^{-\frac{N}{2}}d\theta=2\,\mathrm{Im}\left\{\sum_{k=1}^{n}\frac{1}{\left(k-\frac{1}{2}\right)(1-\rho re^{ia})^{k-\frac{1}{2}}}+2\log\frac{2}{1+\sqrt{1-\rho re^{ia}}}\right\}+2a,$$

so

$$I=2\omega_{N-1}^{-1}\,\mathrm{Im}\left\{\sum_{k=1}^{n}L_k+L_0\right\}+L_\varepsilon,$$

where

$$L_\varepsilon=2a\omega_{N-1}^{-1}\int_{P(\varepsilon)}\dot{v}=O(\varepsilon^{\frac{N}{2}}),$$

$$L_0=2\int_{P(\varepsilon)}\log\frac{2}{1+\sqrt{1-\rho re^{ia}}}\dot{v},$$

$$L_k=\int_{P(\varepsilon)}\frac{\dot{v}}{\left(k-\frac{1}{2}\right)(1-\rho re^{ia})^{k-\frac{1}{2}}},\quad k=1, \cdots, n.$$

Similar to the proof of Lemma 1.1, it is easy to show

$$\operatorname{Im} L_0=o(1),\ \lim_{\rho\to1}L_k=O(\varepsilon^{n-k}),\ 1\leqslant k\leqslant n-1.$$

Since

$$re^{i\alpha}=\frac{t-Q}{2t-1}+i\frac{\sqrt{r^2(2t-1)t-t-(2t-1)\varepsilon^2+2Qt}}{|2t-1|}$$

where $Q=\sqrt{(1-t)(t+r^2(1-2t))+(2t-1)\varepsilon^2}$, hence

$$J_n=\lim_{\rho\to1}L_n=\frac{2\pi^{n-\frac12}}{\left(n-\frac12\right)\Gamma\left(n-\frac12\right)}\int_0^{\left(\frac{2\varepsilon}{\sqrt t}-\frac{\varepsilon^2}{t}\right)^{\frac12}}\left[\frac{t+Q-1}{2t-1}\right.$$

$$\left.-i\frac{\sqrt{t(2t-1)(1-s^2)-t-(2t-1)\varepsilon^2+2Qt}}{|2t-1|}\right]^{-n+\frac12}s^{2n-2}ds$$

$$=\frac{2\pi^{n-\frac12}}{\left(n-\frac12\right)\Gamma\left(n-\frac12\right)}\int_0^1\left[\frac{t+Q-1}{2t-1}\right.$$

$$\left.-i\frac{\sqrt{t(2t-1)(1-\eta^2R^2)-t-(2t-1)\varepsilon^2+2Qt}}{|2t-1|}\right]^{-n+\frac12}\eta^{2n-1}R^{2n-2}dR,$$

where $\eta=\left(\frac{2\varepsilon}{\sqrt t}-\frac{\varepsilon^2}{t}\right)^{\frac12}$, $s=\eta R$. As the proof in § 1

$$\eta^{-2}\frac{t-1+Q}{2t-1}=\frac{R^2}{2}+O(\varepsilon),$$

$$\eta^{-2}\frac{1}{|2t-1|}\{t(2t-1)(1-\eta^2R^2)-t-(2t-1)\varepsilon^2+2Qt\}^{\frac12}=\frac12\left(\frac{t}{1-t}\right)^{\frac12}(1-R^4)^{\frac12}+O(\varepsilon).$$

Thus

$$J=\lim_{\varepsilon\to0}J_n=\frac{2\pi^{n-\frac12}}{\left(n-\frac12\right)\Gamma\left(n-\frac12\right)}\int_0^1\frac{R^{2n-2}dR}{\left[R^2-i\left(\frac{t}{1-t}\right)^{\frac12}\sqrt{1-R^4}\right]^{n-\frac12}}$$

$$=\frac{(2\pi)^{n-\frac12}}{\Gamma\left(n+\frac12\right)}\int_0^{\frac\pi2}\frac{\cos^{n-\frac32}\theta\sin\theta\,d\theta}{(\cos\theta-ib\sin\theta)^{n-\frac12}}\ \left(b=\sqrt{\frac{t}{1-t}}\right).$$

Note that

$$\frac{\cos^{n-\frac32}\theta\sin\theta}{(\cos\theta-ib\sin\theta)^{n-\frac12}}=\frac1i\,\frac{1}{(1+b)^{n-\frac12}}\,\frac{(e^{2i\theta}+1)^{n-\frac32}(e^{2i\theta}-1)}{(1+de^{2i\theta})^{n-\frac12}}\ \left(d=\frac{1-b}{1+b}\right),$$

$$J=\frac{1}{i(1+b)^{n-\frac12}}\int_0^{\frac\pi2}\frac{(e^{2i\theta}-1)(e^{2i\theta}+1)^{n-\frac32}d\theta}{(1+de^{2i\theta})^{n-\frac12}}$$

$$=\frac{1}{i(1+b)^{n-\frac12}}\lim_{\xi\to1^-}\int_0^{\frac\pi2}\frac{(e^{2i\theta}-1)(1+\xi e^{2i\theta})^{n-\frac32}}{(1+de^{2i\theta})^{n-\frac12}}d\theta$$

$$=\frac{1}{i(1+b)^{n-\frac12}}\lim_{\xi\to1^-}\int_0^{\frac\pi2}\sum_{p=0}^\infty\sum_{q=0}^\infty\frac{\left(n-\frac12+q-1\right)\cdots\left(n-\frac12-p\right)}{p!q!}(-1)^qd^q\xi^pe^{2i(p+q)\theta}(e^{2i\theta}-1)d\theta$$

$$=\text{real number}+i\,\frac\pi2\,\frac{1}{(1+b)^{n-\frac12}}.$$

Then we have

$$\lim_{\varepsilon\to 0}\lim_{\rho\to 1} I=\frac{2}{\omega_{N-1}}\frac{(2\pi)^{n-\frac{1}{2}}}{\Gamma\left(n+\frac{1}{2}\right)}\frac{\pi}{2}\frac{1}{(1+b)^{n-\frac{1}{2}}}=\frac{1}{2}\left(\frac{2\beta}{\alpha+\beta}\right)^{n-\frac{1}{2}}.$$

Lemma 2.1 is proved when $\alpha>0$, $\beta>0$.

When $\alpha=\beta$, it is just Lemma 2.1 of [4].

When $\alpha=0$, $\beta>0$ and $\alpha>0$, $\beta=0$, the proofs are similar, we omit the details.

Lemma 2.2. *Suppose $y=(y_1, \cdots, y_N)$, $y\bar{y}'=1$, then*

$$\lim_{\varepsilon\to 0}\omega_{N-1}^{-1}\int_{D^*(p_1,\varepsilon)}(1-(y_1+iy_2))^{-\frac{N}{2}}\dot{y}=1-\frac{2^{n-1}}{\pi}\left\{\frac{\pi}{2}-\int_0^{\operatorname{arc\,tg}\frac{\beta}{\alpha}}\cos^{n-2}t\,\frac{\sin(n-1)t}{\sin t}\,dt\right\},$$

where $D^*(p_1, \varepsilon)$ is the complementary set of $\{y\,|\,y\bar{y}'=1, 1-y_1<\alpha\varepsilon, |y_2|<\beta\varepsilon\}$ on $y\bar{y}'=1$.

The proof is similar to that of Lemmas 1.2 and 2.1.

§ 3. The Cauchy integrals of Lie sphere hyperbolic spaces

Replacing Lemma 2.1 in [4] by Lemma 2.1 in the last section and using the same method in [4], we can obtain the following results, which are the generalizations of [4]. Because the details of proofs are similar, we only state these theorems and omit their proofs.

Theorem 3.1. *Suppose $\varphi(\xi)$ is in* Lip p $(0<p\leqslant 1)$ *on L_{IV}, then the Cauchy principal value*

$$p.v.V(L_{\mathrm{IV}})^{-1}\int_{L_{\mathrm{IV}}(\gamma)}H_{\mathrm{IV}}(\eta_0,\xi)\varphi(\xi)\dot{\xi}=\lim_{\varepsilon\to 0}V(L_{\mathrm{IV}})^{-1}\int_{G(\varepsilon)}H_{\mathrm{IV}}(\eta_0,\xi)\varphi(\xi)\dot{\xi}$$

exists. Here $\eta_0=e^{i\theta_0}(t, 1, 0, \cdots, 0)P_0^{-1}\Gamma P_0\in B_{\mathrm{IV}}-L_{\mathrm{IV}}$, $P_0=\begin{pmatrix}1 & 1\\ i & -i\end{pmatrix}\dotplus\sqrt{2}\,I^{(N-2)}$ is a real orthogonal matrix, $z_0=e^{i\theta_0}(t, 0, \cdots, 0)P_0^{-1}\Gamma P_0$, $\gamma=\dfrac{\alpha}{\beta}$

$$G(\varepsilon)=\{\xi\,|\,\xi\in L_{\mathrm{IV}},\ \alpha^2[\mathrm{Re}(1+\eta_0A_0\eta_0'\overline{\xi A_0\xi'}-\eta_0\bar{\xi}')(1-\bar{z}_0\xi')]^2$$
$$+\beta^2[\mathrm{Im}(1+\eta_0A_0\eta_0'\overline{\xi A_0\xi'}-\eta_0\bar{\xi}')(1-\bar{z}_0\xi')]^2\geqslant\varepsilon^2\}.$$

Theorem 3.2. *Suppose $\varphi(\xi)$ is in* Lip $p(0<p\leqslant 1)$ *on L_{IV}. Let*

$$F(z)=V(L_{\mathrm{IV}})^{-1}\int_{L_{\mathrm{IV}}}H_{\mathrm{IV}}(z,\xi)\varphi(\xi)\xi,\ z\in R_{\mathrm{IV}}.$$

If z approaches $\eta_0=e^{i\theta_0}(t, 1, 0, \cdots, 0)P_0^{-1}\Gamma P_0\in B_{\mathrm{IV}}-L_{\mathrm{IV}}$ from the interior of R_{IV} satisfying the following condition

$$|1+\eta_0A_0\eta_0'\overline{\xi A_0\xi'}-\eta_0\bar{\xi}'|\,|1+zA_0z'\overline{\xi A_0\xi'}-z\bar{\xi}'|^{-1}\leqslant M\,(M=\text{constant}),$$

then

$$\lim_{z\to\eta_0}F(z)=p.v.V(L_{\mathrm{IV}})^{-1}\int_{L_{\mathrm{IV}}(\gamma)}H_{\mathrm{IV}}(\eta_0,\xi)\varphi(\xi)\dot{\xi}+\frac{1}{2}\left(\frac{2\beta}{\alpha+\beta}\right)^{\frac{N}{2}-1}$$
$$\times\frac{1}{2\pi}\int_0^{2\pi}\varphi\left(e^{i\theta_0}\left(e^{i\theta_0}\frac{1+te^{-i\theta}}{1+te^{i\theta}}, 1, 0, \cdots, 0\right)P_0^{-1}\Gamma P_0\right)\frac{d\theta}{(1+te^{i\theta})^{N/2}},$$

where $p.v.\int_{L_{IV}(\gamma)}$ *is defined in Theorem* 3.1.

In particular, when $\beta=0$, *we have*

$$\lim_{z\to\eta_0}F(z)=p.v.\int_{L_{IV}(\infty)}H_{IV}(\eta_0,\ \xi)\varphi(\xi)\dot{\xi}=\lim_{\varepsilon\to0}V(L_{IV})^{-1}\int_{G'(\varepsilon)}H_{IV}(\eta_0,\ \xi)\varphi(\xi)\dot{\xi},$$

where

$$G'(\varepsilon)=\{\xi\,|\,\xi\in L_{IV},\ \mathrm{Re}\{(1+\eta_0A_0\eta_0'\overline{\xi A_0\xi'}-\eta_0\bar{\xi}')(1-\bar{z}_0\xi')\}\geqslant\varepsilon\}.$$

Namely, the limit value of Cauchy integral can be represented by a certain Cauchy principal value.

Theorem 3.3. *Suppose* $f(z)=u(z)+iv(z)$ *is holomorphic in* R_{IV} *continuous in* $R_{IV}\cup L_{IV}$ *and* $u(\xi)$ *is in* $\mathrm{Lip}\,p(0<p\leqslant1)$ *on* L_{IV}, $\eta=e^{i\theta_0}(t,\ 1,\ 0,\ \cdots,\ 0)P_0^{-1}\Gamma P_0\in B_{IV}-L_{IV}$, *then*

$$v(\eta)=p.v.V(L_{IV})^{-1}\int_{L_{IV}(\gamma)}u(\xi)\frac{1}{i}(H_{IV}(\eta,\ \xi)-H_{IV}(\xi,\ \eta))\dot{\xi}$$

$$+\frac{1}{2}\left(\frac{2\beta}{\alpha+\beta}\right)^{\frac{N}{2}-1}\frac{1}{2\pi}\int_0^{2\pi}u\left(e^{i\theta_0}\left(e^{i\theta}\frac{1+te^{-i\theta}}{1+te^{i\theta}},\ 1,\ 0,\ \cdots,\ 0\right)P_0^{-1}\Gamma P_0\right)$$

$$\times\frac{1}{i}(T(t,\ \theta)-T(t,\ -\theta))d\theta+v(0),$$

where $T(t,\ \theta)=(1+te^{i\theta})^{-\frac{N}{2}}$, $p.v.V(L_{IV})^{-1}\int_{L_{IV}(\gamma)}$ *is defined in Theorem* 3.1.

In particular, when $\beta=0$ *we have*

$$v(\eta)=p.v.V(L_{IV})^{-1}\int_{L_{IV}(\infty)}u(\xi)\frac{1}{i}(H_{IV}(\eta,\ \xi)-H_{IV}(\xi,\ \eta))\dot{\xi}+v(0),$$

where $p.\ v.\ V(L_{IV})^{-1}\int_{S_{IV}(\infty)}$ *is defined as* $\lim_{\varepsilon\to0}V(L_{IV})^{-1}\int_{G'(\varepsilon)}$.

That is to say, under the conditions of Theorem 3.3, v, the imaginary part of f, can be represented by a certain Cauchy principal value of Cauchy integral of u, the real part of f.

Theorem 3.4. *Under the conditions of Theorem* 3.1, *the Cauchy principal value*

$$p.\ v.\ V(L_{IV})^{-1}\int_{L_{IV}(b)}H_{IV}(\eta_0,\ \xi)\varphi(\xi)\dot{\xi}=\lim_{\varepsilon\to0}V(L_{IV})^{-1}\int_{D_b(\eta_0,\varepsilon)}H_{IV}(\eta_0,\ \xi)\varphi(\xi)\dot{\xi}$$

exists, where $D_b(\eta_0,\ \varepsilon)$ *is the complementary set of*

$$\{\xi\,|\,\xi\in L_{IV},\ \mathrm{Re}\{(1+\eta_0A_0\eta_0'\overline{\xi A_0\xi'}-\eta_0\bar{\xi}')(1-\bar{z}_0\bar{\xi}')\}<\alpha\varepsilon\},$$
$$\mathrm{Im}\{(1+\eta_0A_0\eta_0'\overline{\xi A_0\xi'}-\eta_0\bar{\xi}')(1-\bar{z}_0\xi')\}<\beta\varepsilon\}$$

on L_{IV}, $z_0=e^{i\theta_0}(t,\ 0,\ \cdots,\ 0)P_0^{-1}\Gamma P_0$, $b=\dfrac{2^{n-1}}{\pi}\left\{\dfrac{\pi}{2}-\displaystyle\int_0^{\operatorname{arctg}\frac{\beta}{\alpha}}\cos^{n-2}t\,\frac{\sin(n-1)t}{\sin t}\,dt\right\}$.

Theorem 3.5. *Under the conditions of Theorem* 3.2, *we have*

$$\lim_{z\to\eta_0}F(z)=p.\ v.\ V(L_{IV})^{-1}\int_{L_{IV}(b)}H_{IV}(\eta_0,\ \xi)\varphi(\xi)\dot{\xi}$$

$$+b\cdot\frac{1}{2\pi}\int_0^{2\pi}\varphi\left(e^{i\theta_0}\left(e^{i\theta_0}\frac{1+te^{-i\theta}}{1+te^{i\theta}},\ 1,\ 0,\ \cdots,\ 0\right)P_0^{-1}\Gamma P_0\right)\frac{d\theta}{(1+te^{i\theta})^{N/2}}$$

where p. v. $V(L_{IV})^{-1}\int_{L_{IV}(b)}$ *is defined in Theorem* 3.4.

Theorem 3.6. *Under the conditions of Theorem* 3.3, *we have*

$$v(\eta)=\text{p. v. } V(L_{IV})^{-1}\int_{L_{IV}(b)} u(\xi)\frac{1}{i}(H_{IV}(\eta,\xi)-H_{IV}(\xi,\eta))\dot{\xi}$$

$$+b\frac{1}{2\pi}\int_0^{2\pi} u\left(e^{i\theta_0}\left(e^{i\theta}\frac{1+te^{-i\theta}}{1+te^{i\theta}},1,0,\cdots,0\right)P_0^{-1}\Gamma P_0\right)\frac{1}{i}(T(t,\theta)-T(t,-\theta))\,d\theta+v(0),$$

where p. v. $V(L_{IV})^{-1}\int_{L_{IV}(b)}$ *and T are defined in Theorems* 3.4 *and* 3.3 *respectively*

§ 4. The Cauchy integrals of matrix hyperbolic spaces

Replacing Theorem 1.3 in [2] by Theorem 1.1 of § 1, and using the same method as that in the proofs of Theorems 3.4.1, 3.4.2, 3.5.1, and 3.5.2 in [5], we can obtain the following results, which are generalization of the corresponding theorems in [5]. In order to save space, we omit their proofs.

Theorem 4.1. *Suppose $\varphi(U)$ is in* Lipα $(0<\alpha\leqslant 1)$ *on $L_I(m,n)$, then when* $Z=U_0'\begin{pmatrix}\rho & 0\\ 0 & Z_1\end{pmatrix}V_0\in R_I(m,n)$ *approaches* $Q=U_0'\begin{pmatrix}1 & 0\\ 0 & Z_1\end{pmatrix}V_0\in L_I^{(m-1)}$, *the limit value of the Cauchy integral*

$$F(Z)=V(L_I(m,n))^{-1}\int_{L_I(m,n)}\varphi(U)\det(I-Z\overline{U}')^{-n}\dot{U}$$

exists and equals

$$\text{p. v. } V(L_I(m,n))^{-1}\int_{L_I(m,n)(\gamma)}\varphi(U)\det(I-Q\overline{U}')^{-n}\dot{U}$$

$$+\frac{1}{2}\left(\frac{2\beta}{\alpha+\beta}\right)^{n-1}\det(I-Z_1\overline{Z}_1')^{-1}V(L_I(m-1,n-1))^{-1}$$

$$\times\int_{L_I(m-1,n-1)}\varphi\left(U_0'\begin{pmatrix}1 & 0\\ 0 & U_1\end{pmatrix}V_0\right)\frac{\det(I-U_1\overline{Z}_1')}{\det(I-Z_1\overline{U}_1')^{n-1}}\dot{U}_1,$$

where the definition of the principal value is

$$\lim_{\varepsilon\to 0}V(L_I(m,n))^{-1}\int_{D(\varepsilon)}\varphi(U)\det(I-Q\overline{U}')^{-n}\dot{U}$$

and

$$D(\varepsilon)=\{U\,|\,U\in L_I(m,n),\ \alpha^2(\text{Re}\det S_1)^2+\beta^2(\text{Im}\det S_1)^2\geqslant\varepsilon^2\},$$

$$S_1=I-Q\overline{U}'-U\overline{T}_0'+Q\overline{T}_0',\ T_0=U_0'\begin{pmatrix}0 & 0\\ 0 & Z_1\end{pmatrix}V_0,\ \gamma=\frac{\alpha}{\beta}.$$

In particular, when $\beta=0$, we have

$$\lim_{z\to Q}F(Z)=\text{p. v. } V(L_I(m,n))^{-1}\int_{L_I(m,n)(\infty)}\varphi(U)\det(I-Q\overline{U}')^{-n}\dot{U},$$

where the definition of principal value is

$$\lim_{\varepsilon\to 0}V(L_I(m,n))^{-1}\int_{\substack{L_I(m,n)\\ \text{Re}\det S_1>\varepsilon}}\varphi(U)\det(I-Q\overline{U}')^{-n}\dot{U}.$$

Namely, the limit value of Cauchy integral may be represented by a certain Cauchy principal value.

Theorem 4.2. *Suppose $\varphi(U)$ is in* Lipα $(0<\alpha\leqslant 1)$ *on* $L_I(n)$, *then when* $Z=U_0'\begin{pmatrix}\rho & 0\\ 0 & Z_1\end{pmatrix}V_0\in R_I^*(n)$, $(\rho>1)$ *approaches* $Q^*=U_0'\begin{pmatrix}1 & 0\\ 0 & Z_1\end{pmatrix}V_0$, *the limit value of Cauchy integral* $F(Z)=V(L_I(n))^{-1}\int_{L_I(n)}\varphi(U)\det(I-Z\bar{U}')^{-n}\dot{U}$ *exists and equals*

$$p.\ v.\ V(L_I(n))^{-1}\int_{L_I(n)}\varphi(U)\det(I-Q^*\bar{U}')^{-n}\dot{U}$$

$$-\frac{1}{2}\left(\frac{2\beta}{\alpha+\beta}\right)^{n-1}[V(L_I(n-1))\det(I-Z_1\bar{Z}_1')]^{-1}$$

$$\times\int_{L_I(n-1)}\varphi\left(U_0'\begin{pmatrix}1 & 0\\ 0 & U_1\end{pmatrix}V_0\right)\det(I-U_1\bar{Z}_1')\det(I-Z_1\bar{U}_1')^{1-n}\dot{U}_1,$$

where the definition of principal value is

$$\lim_{\varepsilon\to 0}V(L_I(n))^{-1}\int_{D'(\varepsilon)}\varphi(U)\det(I-Q^*\bar{U}')^{-n}\dot{U}$$

and $D'(\varepsilon)=\{U\mid U\in L_I(n),\ \alpha^2(\text{Re}\det S_2)^2+\beta^2(\text{Im}\det S_2)^2\leqslant\varepsilon^2\}$,

$$S_2=Q^*\bar{Q}^{*\prime}+\bar{Q}_0^{\prime *}-U\bar{Q}^{\prime *}-Q^*\bar{U}'\bar{Q}_0^*,\ Q_0^*=U_0'\begin{pmatrix}0 & 0\\ 0 & I^{(n-1)}\end{pmatrix}\bar{U}_0.$$

when $\beta=0$, *we have*

$$\lim_{z\to Q^*}F(Z)=p.\ v.\ V(L_I(n))^{-1}\int_{L_I(n)}\varphi(U)\det(I-Q^*\bar{U}')^{-n}\dot{U},$$

where the difinition of principal value is

$$\lim_{\varepsilon\to 0}V(L_I(n))^{-1}\int_{\substack{L_I(n)\\ \text{Re}\det S_2>\varepsilon}}\varphi(U)\det(I-Q^*\bar{U}')^{-n}\dot{U}.$$

That is to say, in this case, the limit value of Cauchy integral may also be represented by a certain Cauchy principal value.

Theorem 4.3. *Suppose* $f(Z)=u(Z)+iv(Z)$ *is holomorphic in* $R_I(m,n)$ *continuous in* $R_I(m,n)\cup L_I(m,n)$ *and* $u(U)$ *is in* Lip $\alpha(0<\alpha\leqslant 1)$ *on* $L_I(m,n)$, *then for an arbitrary point* $Q=U_0'\begin{pmatrix}1 & 0\\ 0 & Z_1\end{pmatrix}V_0$ *on* $L_I(m,n)$, $Z_1\bar{Z}'<I^{(n-1)}$ *we have*

$$v(Q)=p.\ v.\ V(L_I(m,n))^{-1}\int_{L_I(m,n)(\gamma)}\frac{1}{i}u(U)(H(Q,\bar{U})-H(U,\bar{Q}))\dot{U}$$

$$+\frac{1}{2}\left(\frac{2\beta}{\alpha+\beta}\right)^{n-1}\frac{\det(I-Z_1\bar{Z}_1')^{-1}}{V(L_I(m-1,n-1))}$$

$$\times\int_{L_I(m-1,n-1)}u\left(U_0'\begin{pmatrix}1 & 0\\ 0 & U_1\end{pmatrix}V_0\right)\frac{1}{i}(T(Z_1,\bar{U}_1)-T(U_1,\bar{Z}_1))\dot{U}_1+v(0),$$

where $H(Q,\bar{U})=\det(I-Q\bar{U}')^{-n}$, $T(Z_1,\bar{U}_1)=\det(I-U_1\bar{Z}_1')\det(I-Z_1\bar{U}_1')^{1-n}$.

When $\beta=0$, *we have*

$$v(Q)=p.\ v.\ V(L_I(m,n))^{-1}\int_{L_I(m,n)(\infty)}u(U)\frac{1}{i}(H(Q,\bar{U})-H(U,\bar{Q}))\dot{U}+v(0).$$

Namely v can be represented by a certain Cauchy principal value of u. Here the definition of principal value is the same as that of Theorem 4.1.

Theorem 4.4. *Suppose* $f(Z)=u(Z)+iv(Z)$ *is holomorphic in* $R_I^*(n)$, *continuous on* $R_I^*(n)\cup L_I(n)$, *and* $u(U)$ *is in* Lip $\alpha(0<\alpha\leqslant 1)$ *on* $L_I(n)$, *then for an arbitrary point* $Q^*=U_0'\begin{pmatrix}1&0\\0&Z_1\end{pmatrix}V_0$, $Z_1\bar{Z}_1'>I^{(n-1)}$ *on* $L_I^*(n-1)$, *we have*

$$v(Q^*)=p.v.\ (V(L_I(n)))^{-1}\int_{L_I(n)(\gamma)}u(U)\frac{1}{i}H^*(Q^*,\bar{U})-H^*(U,\bar{Q}^*))\dot{U}$$

$$-\frac{1}{2}\left(\frac{2\beta}{\alpha+\beta}\right)^{n-1}\frac{\det(I-Z_1\bar{Z}_1')^{-1}}{V(L_I(n))}$$

$$\times\int_{L_I(n-1)}u\left(U_0'\begin{pmatrix}1&0\\0&U_1\end{pmatrix}V_0\right)\frac{1}{i}(T^*(Z_1,\bar{U}_1)-T^*(U_1,\bar{Z}_1))\dot{U}_1+v(Z_\infty),$$

where $H^*(Q^*,\bar{U})=(-1)^n\det(Q^*\bar{U})^n\det(I-Q^*\bar{U}')^{-n}$,

$$T^*(Z_1,\bar{U}_1)=(-1)^n\det(Z,\bar{U}_1)\det(I-U_1\bar{Z}_1')\det(I-Z_1\bar{U}_1')^{1-n},$$

Z_∞ *denotes the infinite point of homogenous coordinates* $(Z_1,Z_2)=(O^{(n)},I^{(n)})$ *in the Grassmann manifold.*

When $\beta=0$, *we have*

$$v(Q^*)=p.v.\ V(L_I(n))^{-1}\int_{L_I(n)(\infty)}u(U)\frac{1}{i}(H^*(Q^*,\bar{U})-H^*(U,\bar{Q}^*))\dot{U}+v(Z_\infty),$$

i. e. v may be represented by a certain Cauchy principal value of u. The definition of principal value is as that of Theorem 4.2.

Replacing Theorem 1.3 in [2] by Theorem 1.2 of § 1, and using hte same method as that in the proofs of Theorems 3.4.1, 3.4.2, 3.5.1 and 3.5.2 in [5], we also obtain the analogues of Theorem 4.1, 4.2, 4.3 and 4.4. The details are omited here.

§ 5. On the Bochner-Martinelli integrals

On the Bochner-Martinelli integrals, Lu Qikong and Zhong Tongde[7] proved the following Plemelj formula.

Suppose D is a domain in $\mathbf{R}^{2n}$, $\Omega=\partial D$ is an orientable, C^2 manifold, $f(z)$ is a continuous complex function in Ω and satisfies Lipschtz condition, and $K_{(2n-1)}(z,\zeta)$ is the Bochner-Martinelli kernel. Then, when w approaches $w_0\in\Omega$ from the interior or exterior of D, the limit values of Cauchy integral

$$F(w)=\int_\Omega f(z)K_{(2n-1)}(z,w)$$

are expressed by the following formulas respectively

$$F_i(w_0)=p.v.\int_\Omega f(z)K_{(2n-1)}(z,w_0)+\frac{1}{2}f(w_0),$$

$$F_{\varepsilon}(w_0)=p.\ v.\int_{\Omega}f(z)K_{(2n-1)}(z,\ w_0)-\frac{1}{2}f(w_0).$$

where $p.\ v.\int_{\Omega}$ is defined as $\lim_{\delta\to 0}\int_{\Omega-\Omega\cap S_{\delta}(w_0)}$, and $S_{\varepsilon}\ (w_0)$ is the ball with the center W_0 and raduis ε.

For the Bochner-Martinelli kernel, we can raise the same questions: Are there various definitions for the Cauchy principal value as above? Will the value $\frac{1}{2}$ also change into different values by virture of the different definitions?

The answer is negative, the reason is as follows: In the proof of the above theorem, the deleted neighborhood around w_0 is $\Omega\cap S_{\varepsilon}(w_0)$. Now if the deleted neighborhood around w_0 is $\Omega\cap M_{\varepsilon_1}(w_0)$ rather than $\Omega\cap S_{\varepsilon}(w_0)$ ($M_{\varepsilon_1}(w_0)$ is a neighborhood of w_0, and it contracts to the point w_0 as $\varepsilon_1\to 0$), then we can take ε_1 sufficiently small, such that $\Omega\cap M_{\varepsilon_1}(w_0)\subset\Omega\cap S_{\varepsilon}(w_0)$. Since the Bochner-Martinelli kernel is well defined in $\mathbf{C}^n-\{w_0\}$, and it is a closed form, hence, by Stokes formula, the integral on $\Omega\cap S_{\varepsilon}(w_0)$ is equal to that on $\Omega\cap M_{\varepsilon_1}(w_0)$.

That is to say, no matter what neighborhood is deleted, the results are the same.

The reason why the kernels discussed above have the different Cauchy principal values is that either these kernels are only well defined in the interior or they are not closed forms.

References

[1] Gong Sheng and Shi Jihuai, Singular integrals in several complex variables (I), *Chin. Ann. of Math.*, 3: 4(1982), 483—502.

[2] Kung S. and Sun C. K., Integrals of Cauchy type in several complex variables (I), *Acta Math. Sinica*, 15 (1965), 431—443 (In Chinese).

[3] Koranyi, A. and Vagi, S., Singular integrals in homogeneous spaces and some problems of classical analysis, *Ann. Scuola Normale Superiore Pisa*, 25 (1971), 575—648.

[4] Kungt S. and Sun C. K., Integrals of Cauchys type in several complex variables (II), *Acta Math. Sinica*, 15 (1965), 775—799 (In Chinese).

[5] Kung S., Integrals of Cauchy type of the classical domains in several complex variables (I), Proceeding of the fifth anniversary of the founding of USTC, 1963, 23—25.

[6] Kung S. and Sun C. K., Integrals of Cauchy type in several complex variables (III), *Acta Math. Sinica*, 15 (1965), 800—811. (In Chinese).

[7] Lu Qikeng and Zhong Tongde, Generalizations of Privalof theorem, *Acta Math. Sinica*, 7 (1957), 144—163 (In Chinese).

[8] Shi Jihuai, On the Cauchy type integrals for the hypersphere, *Journal of USTC*, 2 (1980), 1—9 (In Chinese).

[9] Shi Jihuai, and Gong Sheng, Singular integrals in several complex variables (II), *Chin. Ann. of Math.*, 4B: 3(1983).

Chin. Ann. of Math.
5B (1) 1984

SINGULAR INTEGALS IN SEVERAL COMPLEX VARIABLES (IV)——THE DERIVATIVE OF CAUCHY INTEGRAL ON SPHERE

SHI JIHUAI (史济怀) GONG SHENG (龚 升)
(University of Science and Technology of China)

Abstract

The aim of this paper is to study the boundary properties of the derivative of Cauchy integral on the sphere. Using the concept of the Hadamard principal value of singular integral of higher order introduced in [1], the author obtains the corresponding Plemelj formula expressed by Hadamard principal value.

§ 0. Introduction

In the case of one complex variable, if L is a smooth closed curve in $\mathbf{C}$, the Cauchy integral of f is

$$F(z)=\frac{1}{2\pi i}\int_L \frac{f(\zeta)}{\zeta-z}\,d\zeta,$$

its derivative

$$F'(z)=\frac{1}{2\pi i}\int_L \frac{f(\zeta)}{(\zeta-z)^2}\,d\zeta=\frac{1}{2\pi i}\int_L \frac{f'(\zeta)}{\zeta-z}\,d\zeta$$

is also a Cauchy integral. In the case of several complex variables, the derivative of Cauchy integral is no longer a Cauchy integral.

In what follows, B denotes the unit ball in $\mathbf{C}^n$, S denotes the boundary of B, i. e.

$$B=\{z=(z_1,\ \cdots,\ z_n)\,|\,|z_1|^2+\cdots+|z_n|^2<1\},$$
$$S=\{z=(z_1,\ \cdots,\ z_n)\,|\,|z_1|^2+\cdots+|z_n|^2=1\}.$$

Suppose f is integrable on S, the Cauchy integral of f in the ball is

$$F(z)=\frac{1}{\omega_{2n-1}}\int_S \frac{f(u)\dot{u}}{(1-zu')^n}\quad (z\in B),$$

its partial derivative

$$\frac{\partial F(z)}{\partial z}=\frac{n}{\omega_{2n-1}}\int_S \frac{\bar{u}f(u)\dot{u}}{(1-z\bar{u}')^{n+1}} \tag{0.1}$$

is no longer a Cauchy integral, where $\dfrac{\partial F(z)}{\partial z}=\left(\dfrac{\partial F(z)}{\partial z_1},\ \cdots,\ \dfrac{\partial F(z)}{\partial z_n}\right)$. When z

Manuscript received July 20, 1981.

approaches the point v of S, (0.1) is a higher order singular integral.

In [1], using the concept of Hadamaad principal value, we obtain the Plemelj formula of integral

$$\frac{1}{\omega_{2n-1}}\int_S \frac{f(u)\dot{u}}{(1-z\bar{u}')^{n+\frac{1}{2}}} \quad (z\in B).$$

The aim of present paper is to study the limit value of the derivative of Cauchy integral. Let

$$P_j=\{v=(v_1,\ \cdots,\ v_j,\ \cdots,\ v_n)\,|\,v\in S,\ v_j=0\},$$
$$Q_j=\{v=(v_1,\ \cdots,\ v_j,\ \cdots,\ v_n)\,|\,v\in S,\ v_j\neq 0\}.$$

We find that for the points of P_j, the limit value of $\frac{\partial F(z)}{\partial z_j}$ can be represented by the Cauchy principal value and it must be represented by the Hadamard principal value for the points of Q_j. By the uniformity of those two kinds of principal values, it can be represented uniformly by Hadamard principal value.

Using the same method, we can discuss the derivative of H-R integral and S-K integral of strictly pseudoconvex domain.

§ 1. Some lemmas

Lemma 1. *Suppose $z\in B$. Then*

$$\int_S \frac{\bar{u}_j\dot{u}}{(1-z\bar{u}')^m}=0, \tag{1.1}$$

where m is an arbitrary positive number.

Proof Let $z\bar{z}'=\rho^2$. Pick a unitary matrix U, such that $zU=\rho p_n$, where $p_n=(0,\ \cdots,\ 0,\ 1)$. Let $u=w\bar{U}'$ and $\bar{U}'=(\alpha_{lj})$. Then $u_j=\sum_{l=1}^{n}\alpha_{lj}w_l$ and

$$\int_S \bar{u}_j(1-z\bar{u}')^{-m}\dot{u}=\sum_{l=1}^{n}\int_S \bar{\alpha}_{lj}\bar{w}_l(1-\rho\bar{w}_n)^{m}\dot{w}. \tag{1.2}$$

When $l<n$, we have

$$\int_S \bar{w}_l(1-\rho\bar{w}_n)^{-m}\dot{w}=\int_{v\bar{v}'<1}\bar{v}_l\dot{v}\int_{-\pi}^{\pi}(1-\rho re^{i\theta})^{-m}d\theta=2\pi\int_{v\bar{v}'<1}\bar{v}_l\dot{v}=0. \tag{1.3}$$

When $l=n$, since

$$\int_{-\pi}^{\pi} re^{i\theta}(1-\rho re^{i\theta})^{-m}d\theta=\sum_{q=0}^{\infty}\frac{\Gamma(m+q)}{\Gamma(q+1)\Gamma(m)}\rho^q r^{q+1}\int_{-\pi}^{\pi}e^{i(q+1)\theta}d\theta=0,$$

we have

$$\int_S \bar{w}_n(1-\rho\bar{w}_n)^{-m}\dot{w}=0. \tag{1.4}$$

Inserting (1.3), (1.4) into (1.2), we obtain (1.1).

Lemma 2. *Suppose $z\in B$. Then*

$$\frac{1}{\omega_{2n-1}}\int_S \frac{\bar{u}_j x_k}{(1-z\bar{u}')^{n+1}}\dot{u}=\frac{1}{2n}\delta_{kj}, \tag{1.5}$$

$$\frac{1}{\omega_{2n-1}}\int_S \frac{\bar{u}_j y_k}{(1-z\bar{u}')^{n+1}}\dot{u}=\frac{1}{2ni}\delta_{kj}, \tag{1.6}$$

where $j, k=1, \cdots, n$; $x_k=\mathrm{Re}(u_k)$, $y_k=\mathrm{Im}(u_k)$.

Proof Let

$$G(z)=\frac{1}{\omega_{2n-1}}\int_S \frac{x_k}{(1-z\bar{u}')^n}\dot{u}.$$

By Cauchy integral formula and Lemma 1, we have

$$G(z)=\frac{1}{2}\left\{\omega_{2n-1}^{-1}\int_S u_k(1-z\bar{u}')^{-n}\dot{u}+\omega_{2n-1}^{-1}\int_S \bar{u}_k(1-z\bar{u}')^{-n}\dot{u}\right\}=\frac{1}{2}z_k,$$

so $\dfrac{\partial G}{\partial z_j}=\dfrac{1}{2}\delta_{kj}$, i. e. (1.5) holds. The proof of (1.6) is similar.

Lemma 3. *Suppose* $f:\bar{B}\to\mathbf{R}$. *If* $\dfrac{\partial f}{\partial x_j}$, $\dfrac{\partial f}{\partial y_j}$ $(j=1, \cdots, n)$ *are in* Lip α_j *and* Lipβ_j *on* $\bar{B}$ *respectively, then*

(i) *For* $j, k=1, \cdots, n-1$, *the integrals*

$$I_{x_k}^{(j)}=\omega_{2n-1}^{-1}\int_S \bar{u}_j\Big[f(0, \cdots, 0, x_k+iy_k, \cdots, x_n+iy_n)-f(0, \cdots, 0, iy_k, \cdots, x_n+iy_n)$$
$$-\frac{\partial f}{\partial x_k}(p_n)x_k\Big](1-\bar{u}_n)^{-(n+1)}\dot{u},$$

$$I_{y_k}^{(j)}=\omega_{2n-1}^{-1}\int_S \bar{u}_j\Big[f(0, \cdots, 0, iy_k, x_{k+1}+iy_{k+1}, \cdots, x_n+iy_n)$$
$$-f(0, \cdots, 0, x_{k+1}+iy_{k+1}, \cdots, x_n+iy_n)-\frac{\partial f}{\partial y_k}(p_n)y_k\Big](1-\bar{u}_n)^{-(n+1)}\dot{u},$$

$$I_{x_n}^{(j)}=\omega_{2n-1}^{-1}\int_S \bar{u}_j\Big[f(0, \cdots, 0, x_n)-f(0, \cdots, 0, 1)-\frac{\partial f}{\partial x_n}(p_n)(x_n-1)\Big]$$
$$\times(1-\bar{u}_n)^{-(n+1)}\dot{u},$$

$$I_{y_n}^{(j)}=\omega_{2n-1}^{-1}\int_S \bar{u}_j\Big[f(0, \cdots, 0, x_n+iy_n)-f(0, \cdots, 0, x_n)-\frac{\partial f}{\partial y_n}(p_n)y_n\Big]$$
$$\times(1-\bar{u}_n)^{-(n+1)}\dot{u}.$$

exist.

(ii) *Suppose* $z\in B$. Write

$$I_{x_k}^{(j)}(z)=\omega_{2n-1}^{-1}\int_S \bar{u}_j\Big[f(0, \cdots, 0, x_k+iy_k, \cdots, x_n+iy_n)$$
$$-f(0, \cdots, 0, iy_k, \cdots, x_n+iy_n)-\frac{\partial f}{\partial x_k}(p_n)x_k\Big](1-z\bar{u}')^{-(n+1)}\dot{u},$$

$$I_{y_k}^{(j)}(z)=\omega_{2n-1}^{-1}\int_S \bar{u}_j\Big[f(0, \cdots, 0, iy_k, \cdots, x_n+iy_n)$$
$$-f(0, \cdots, 0, x_{k+1}+iy_{k+1}, \cdots, x_n+iy_n)-\frac{\partial f}{\partial y_k}(p_n)y_k\Big](1-z\bar{u}')^{-(n+1)}\dot{u},$$

$$I_{x_n}^{(j)}(z)=\omega_{2n-1}^{-1}\int_S \bar{u}_j\Big[f(0, \cdots, 0, x_n)-f(0, \cdots, 0, 1)-\frac{\partial f}{\partial x_n}(p_n)(x_n-1)\Big]$$
$$\times(1-z\bar{u}')^{-(n+1)}\dot{u},$$

$$I_{y_n}^{(j)}(z)=\omega_{2n-1}^{-1}\int_S \bar{u}_j\Big[f(0, \cdots, 0, x_n+iy_n)-f(0, \cdots 0, x_n)-\frac{\partial f}{\partial y_n}(p_n)y_n\Big]$$
$$\times(1-z\bar{u}')^{-(n+1)}\dot{u},$$

then we have

$$(K-\lim_{z\to p_n}) I^{(j)}_{x_k}(z)=I^{(j)}_{x_k},\quad (K-\lim_{z\to p_n}) I^{(j)}_{y_k}(z)=I^{(j)}_{y_k}\quad (j=1,\ \cdots,\ n-1,\ k=1,\ \cdots,\ n).$$

Proof Note the inequalities

$$|u_j|\leqslant\sqrt{2}\,|1-\bar{v}_n|^{\frac{1}{2}}\quad (j=1,\ \cdots,\ n-1).$$

Using the method of proving Lemma 5 in [1], Lemma 3 follows.

Lemma 4. *For $j=1,\ \cdots,\ n-1$, we have*

$$\int_{\sigma_\varepsilon}\bar{u}_j(1-\bar{u}_n)^{-(n+1)}\dot{u}=0,$$

where $\sigma_\varepsilon=\{u\,|\,u\in S,\ \alpha^2(1-|u_n|^2)^2+4\beta^2(\operatorname{Im}u_n)^2>\varepsilon^2,\ \alpha>0,\ \beta>0\}$.

Proof Let $u_n=re^{i\theta}$, $u_1=v_1$, $\cdots$, $u_{n-1}=v_{n-1}$. Then

$$\int_{\sigma_\varepsilon}\bar{u}_j(1-\bar{u}_n)^{-(n+1)}\dot{u}=\int_{v\bar{v}'<\frac{\varepsilon}{\alpha}}\bar{v}_j\dot{v}\left\{\int_{-(\pi-c)}^{-c}(1-re^{i\theta})^{-(n+1)}d\theta+\int_{c}^{\pi-c}(1-re^{i\theta})^{-(n+1)}d\theta\right\}$$

$$+\int_{v\bar{v}'>\frac{\varepsilon}{\alpha}}\bar{v}_j\dot{v}\int_{-\pi}^{\pi}(1-re^{i\theta})^{-(n+1)}d\theta=I_1+I_2,$$

where $C=\sin^{-1}\dfrac{\sqrt{\varepsilon^2-\alpha^2(1-r^2)^2}}{2\beta r}$. Denote the inner integral of I_1 by $\varphi(r)$. Let $v_j=e^{i\tau}\zeta_i$ and note that $r^2=1-v\bar{v}'$, we have

$$I_1=\int_{v\bar{v}'<\frac{\varepsilon}{\alpha}}\bar{v}_j\varphi(r)\dot{v}=\int_{\zeta\bar{\zeta}'<\frac{\varepsilon}{\alpha}}e^{-i\tau}\bar{\zeta}_j\varphi(r)\dot{\zeta}=e^{-i\tau}I_1,$$

and it follows that $I_1=0$. We can prove $I_2=0$ by the same method. This ends the proof.

Lemma 5. *For $j=1,\ \cdots,\ n-1$, $k=1,\ \cdots,\ n$, we have*

$$\lim_{\varepsilon\to 0}\omega_{2n-1}^{-1}\int_{\sigma_\varepsilon}\bar{u}_jx_k(1-\bar{u}_n)^{-(n+1)}\dot{u}=\frac{1}{2n}\left[1-\frac{1}{2}\left(\frac{2\beta}{\alpha+\beta}\right)^n\right]\delta_{jk},\tag{1.7}$$

$$\lim_{\varepsilon\to 0}\omega_{2n-1}^{-1}\int_{\sigma_\varepsilon}\bar{u}_jy_k(1-\bar{u}_n)^{-(n+1)}\dot{u}=\frac{1}{2ni}\left[1-\frac{1}{2}\left(\frac{2\beta}{\alpha+\beta}\right)^n\right]\delta_{jk}.\tag{1.8}$$

Proof It is clear that

$$\omega_{2n-1}^{-1}\int_{\sigma_\varepsilon}\bar{u}_jx_k(1-\bar{u}_n)^{-(n+1)}\dot{u}=(2\omega_{2n-1})^{-1}\int_{\sigma_\varepsilon}\bar{u}_ju_k(1-\bar{u}_n)^{-(n+1)}\dot{u}+(2\omega_{2n-1})^{-1}$$

$$\int_{\sigma_\varepsilon}\bar{u}_j\bar{u}_k(1-\bar{u}_n)^{-(n+1)}\dot{u}.\tag{1.9}$$

Using the similar method as in proving Lemma 4, we can obtain the following equalities

$$\int_{\sigma_\varepsilon}\bar{u}_j\bar{u}_k(1-\bar{u}_n)^{-(n+1)}\dot{u}=0\quad (j=1,\ \cdots,\ n-1;\ k=1,\ \cdots,\ n),\tag{1.10}$$

$$\int_{\sigma_\varepsilon}\bar{u}_ju_k(1-\bar{u}_n)^{-(n+1)}\dot{u}=0\quad (j\neq k).\tag{1.11}$$

When $j=k<n$, without loss of generality, taking $j=1$, we have

$$(2\omega_{2n-1})^{-1}\int_{\sigma_\varepsilon}|u_1|^2(1-\bar{u}_n)^{-(n+1)}\dot{u}$$

$$=(2\omega_{2n-1})^{-1}\int_{v\bar{v}'<\frac{\varepsilon}{\alpha}}|v_1|^2\dot{v}\left\{\int_{-(\pi-c)}^{-c}(1-re^{i\theta})^{-(n+1)}d\theta+\int_0^{\pi-c}(1-re^{i\theta})^{-(n+1)}d\theta\right\}$$

$$+(2\omega_{2n-1})^{-1}\int_{v\bar{v}'>\frac{\varepsilon}{\alpha}}|v_1|^2\dot{v}\int_{-\pi}^{\pi}(1-re^{i\theta})^{-(n+1)}d\theta=I_1+I_2. \tag{1.12}$$

It is not hard to see that

$$I_1=\omega_{2n-1}^{-1}\operatorname{Im}\left\{\sum_{k=1}^n J_k-\sum_{k=1}^n H_k+J_0-H_0\right\}+\omega_{2n-1}^{-1}\int_{v\bar{v}'<\frac{\varepsilon}{\alpha}}|v_1|^2(\pi-2c)\dot{v},$$

where

$$J_0=-\int_{v\bar{v}'<\frac{\varepsilon}{\alpha}}|v_1|^2\log(1+re^{-ic})\dot{v},\quad J_k=\frac{1}{k}\int_{v\bar{v}'<\frac{\varepsilon}{\alpha}}|v_1|^2(1+re^{-ic})^{-k}\dot{v},$$

$$H_0=-\int_{v\bar{v}'<\frac{\varepsilon}{\alpha}}|v_1|^2\log(1-re^{ic})\dot{v},\quad H_k=\frac{1}{k}\int_{v\bar{v}'<\frac{\varepsilon}{\alpha}}|v_1|^2(1-re^{ic})^{-k}\dot{v}.$$

Suppose $v=(t_1,\cdots,t_{2n-2})$. Using the sphere coordinates, we have

$$J_k=\frac{1}{k}\int_0^\pi\cdots\int_0^\pi\int_0^{2\pi}(\cos^2\varphi_1+\sin^2\varphi_1\cos^2\varphi_2)\sin^{2n-4}\varphi_1\sin^{2n-5}\varphi_2\cdots\sin\varphi_{2n-4}\,d\varphi_1\cdots d\varphi_{2n-3}$$

$$\times\int_0^{\sqrt{\frac{\varepsilon}{\alpha}}}\{1+(2\beta)^{-1}\sqrt{4\beta^2(1-s^2)+\alpha^2s^4-s^2}-i(2\beta)^{-1}\sqrt{s^2-\alpha^2s^4}\}^{-k}s^{2n-1}ds.$$

Set $\eta=\dfrac{\varepsilon}{\alpha}$, $s=\sqrt{\eta}t$. The inner integral becomes

$$(2\beta)^k\eta^n\int_0^1 t^{2n-1}\{2\beta+\sqrt{4\beta^2(1-\eta t^2)+\alpha^2\eta^2(t^4-1)}-i\sqrt{\alpha^2\eta^2(1-t^4)}\}^{-k}dt.$$

Since the integrand is bounded,

$$\lim_{\varepsilon\to0}J_k=0\quad(k=1,\cdots,n).$$

Similarly $\lim\limits_{\varepsilon\to0}J_0=0$.

It is the same as J_k that H_k may be written as

$$H_k=\frac{1}{k}\int_0^\pi\cdots\int_0^\pi\int_0^{2\pi}(\cos^2\varphi_1+\sin^2\varphi_1\cos^2\varphi_2)\sin^{2n-4}\varphi_1\sin^{2n-5}\varphi_2\cdots\sin\varphi_{2n-4}\,d\varphi_1\cdots d\varphi_{2n-3}$$

$$\times(2\beta)^k\eta^n\int_0^1 t^{2n-1}[2\beta-\sqrt{4\beta^2(1-\eta t^2)+\alpha^2\eta^2(t^4-1)}-i\sqrt{\alpha^2\eta^2(1-t^4)}]^{-k}dt.$$

Denote the inner integral by M_k, then

$$M_k=(2\beta)^k\eta^n\int_0^1 t^{2n-1}\eta^{-k}[(\beta t^2-i\alpha\sqrt{1-t^4})+O(\eta)]^{-k}dt.$$

When $k<n$, $\lim\limits_{\varepsilon\to0}M_k=0$, so $\lim\limits_{\varepsilon\to0}H_k=0$. When $k=n$,

$$\lim_{\varepsilon\to0}M_n=(2\beta)^n\int_0^1 t^{2n-1}(\beta t^2-i\alpha\sqrt{1-t^4})^{-n}dt=2^{n-1}\int_0^1 x^{n-1}(x-i\gamma\sqrt{1-x^2})^{-n}dx,$$

it is known in [2] that

$$\operatorname{Im}\left(\int_0^1 x^{n-1}(x-i\gamma\sqrt{1-x^2})^{-n}dx\right)=\frac{\pi}{2}\left(\frac{\beta}{\alpha+\beta}\right)^n.$$

Hence we have

$$\lim_{\varepsilon\to 0}\operatorname{Im}(M_n)=2^{n-2}\pi\left(\frac{\beta}{\alpha+\beta}\right)^n.$$

A direct computation shows

$$\int_0^\pi\cdots\int_0^\pi\int_0^{2\pi}(\cos^2\varphi_1+\sin^2\varphi_1\cos^2\varphi_2)\sin^{2n-4}\varphi_1\sin^{2n-5}\varphi_2\cdots\sin\varphi_{2n-4}\,d\varphi_1\cdots d\varphi_{2n-3}$$
$$=\frac{2\pi^{n-1}}{\Gamma(n)},$$

so

$$\lim_{\varepsilon\to 0}\operatorname{Im}(H_n)=\frac{\pi^n}{2n\Gamma(n)}\left(\frac{2\beta}{\alpha+\beta}\right)^n.$$

It is easy to prove that

$$\lim_{\varepsilon\to 0}\operatorname{Im}H_0=0,\quad \lim_{\varepsilon\to 0}\int_{v\bar{v}'<\frac{\varepsilon}{\alpha}}|v_1|^2(\pi-2c)\dot{v}=0.$$

Substituting these results into the expression of I_1, we have

$$\lim_{\varepsilon\to 0}I_1=-\lim_{\varepsilon\to 0}(\omega_{2n-1})^{-1}\operatorname{Im}(H_n)=-\frac{1}{4n}\left(\frac{2\beta}{\alpha+\beta}\right)^n.$$

On the other hand

$$\lim_{\varepsilon\to 0}I_2=\pi\omega_{2n-1}^{-1}\int_{v\bar{v}'<1}|v_1|^2\dot{v}=\frac{1}{2n}.$$

Substituting these into (1.12), we obtain

$$\lim_{\varepsilon\to 0}(2\omega_{2n-1})^{-1}\int_{\sigma_\varepsilon}|u_1|^2(1-\bar{u}_n)^{-(n+1)}\dot{u}=\frac{1}{2n}\left[1-\frac{1}{2}\left(\frac{2\beta}{\alpha+\beta}\right)^n\right].\tag{1.13}$$

Obviously, replacing u_1 by u_j, $(j=2,\cdots,n-1)$, (1.13) is also true. Substituting (1.10), (1.11) and (1.13) into (1.9), (1.7) is obtained. (1.8) can be proved in a similar way.

§ 2. The limit value of a part of special points

Employing the above Lemmas, we can prove the following

Theorem 1. *Suppose that f satisfies the conditions of Lemma 3, $F(z)$ isthe Cauchy integral of f:*

$$F(z)=\frac{1}{\omega_{2n-1}}\int_S\frac{f(u)\dot{u}}{(1-z\bar{u}')^n}.$$

Then, for $j=1,\cdots,n-1$, we have

$$(K-\lim_{z\to p_n})\frac{\partial F}{\partial z_j}(z)=n\omega_{2n-1}^{-1}\int_{S(\gamma)}\bar{u}_jf(u)(1-\bar{u}_n)^{-(n+1)}\dot{u}+\frac{1}{2}\left(\frac{2\beta}{\alpha+\beta}\right)^n\frac{\partial f}{\partial u_j}(p_n),$$

where $\int_{S(\gamma)}$ denotes the Cauchy principal value, defined as $\lim_{\varepsilon\to 0}\int_{\sigma_\varepsilon}$.

Proof Obviously

$$\frac{\partial F}{\partial z_j}=n\omega_{2n-1}^{-1}\int_S\bar{u}_jf(u)(1-z\bar{u}')^{-(n+1)}\dot{u}.\tag{2.1}$$

According to the expression (2.1) of f in [1], we have

$$\omega_{2n-1}^{-1}\int_S \bar{u}_j f(u)(1-z\bar{u}')^{-(n+1)}\dot{u}$$

$$=\sum_{k=1}^{n}[I_{x_k}^{(j)}(z)+I_{y_k}^{(j)}(z)]-\sum_{k=1}^{n-1}\frac{\partial f}{\partial x_k}(p_n)\omega_{2n-1}^{-1}\int_S \bar{u}_j x_k(1-z\bar{u}')^{-(n+1)}\dot{u}$$

$$+\sum_{k=1}^{n-1}\frac{\partial f}{\partial y_k}(p_n)\omega_{2n-1}^{-1}\int_S \bar{u}_j y_k(1-z\bar{u}')^{-(n+1)}\dot{u}+\frac{\partial f}{\partial y_n}(p_n)\omega_{2n-1}^{-1}\int_S \bar{u}_j y_n(1-z\bar{u}')^{-(n+1)}\dot{u}$$

$$+\frac{\partial f}{\partial x_n}(p_n)\omega_{2n-1}^{-1}\int_S \bar{u}_j(x_n-1)(1-z\bar{u}')^{-(n+1)}\dot{u}+f(p_n)\omega_{2n-1}^{-1}\int_S \bar{u}_j(1-z\bar{u}')^{-(n+1)}\dot{u}.$$

By Lemmas 1, 2, 3, we have

$$(K-\lim_{z\to p_n})\,\omega_{2n-1}^{-1}\int_S \bar{u}_j f(u)(1-z\bar{u}'_n)^{-(n+1)}\dot{u}$$

$$=\sum_{k=1}^{n}(I_{x_k}^{(j)}+I_{y_k}^{(j)})+\frac{1}{2n}\frac{\partial f}{\partial x_j}(p_n)+\frac{1}{2ni}\frac{\partial f}{\partial y_j}(p_n)$$

$$=\sum_{k=1}^{n}(I_{x_k}^{(j)}+I_{y_k}^{(j)})+\frac{1}{n}\frac{\partial f}{\partial u_j}(p_n). \tag{2.2}$$

On the other hand, by Lemmas 3, 4, 5, we also have

$$\lim_{\varepsilon\to 0}\omega_{2n-1}^{-1}\int_{\sigma_\varepsilon}\bar{u}_j f(u)(1-\bar{u}_n)^{-(n+1)}\dot{u}$$

$$=\sum_{k=1}^{n}(I_{y_k}^{(j)}+I_{y_k}^{(j)})+\frac{\partial f}{\partial x_j}(p_n)\frac{1}{2n}\left[1-\frac{1}{2}\left(\frac{2\beta}{\alpha+\beta}\right)^n\right]$$

$$+\frac{\partial f}{\partial y_j}(p_n)\frac{1}{2ni}\left[1-\frac{1}{2}\left(\frac{2\beta}{\alpha+\beta}\right)^n\right]$$

$$=\sum_{k=1}^{n}(I_{x_k}^{(j)}+I_{y_k}^{(j)})+\frac{1}{n}\left[1-\frac{1}{2}\left(\frac{2\beta}{\alpha+\beta}\right)^n\right]\frac{\partial f}{\partial u_j}(p_n),$$

i. e.

$$\sum_{k=1}^{n}(I_{x_k}^{(j)}+I_{y_k}^{(j)})=\omega_{2n-1}^{-1}\int_{S(\gamma)}\bar{u}_j f(u)(1-\bar{u}_n)^{-(n+1)}\dot{u}-\frac{1}{n}\left[1-\frac{1}{2}\left(\frac{2\beta}{\alpha+\beta}\right)^n\right]\frac{\partial f}{\partial u_j}(p_n).$$

Inserting it into (2.2), the desired result is obtained.

Theorem 2. *Suppose f: $\bar{B}\to\mathbf{C}$, where $f_1=\operatorname{Re} f$ and $f_2=\operatorname{Im} f$ satisfy the conditions of Lemma 3 respectively, and $F(z)$ is the Cauchy integral of f. Let*

$$P_k=\{v=(v_1,\cdots,v_k,\cdots,v_n)\,|\,v\in S,\ v_k=0\}.$$

Then

$$(K-\lim_{z\to v})\frac{\partial F}{\partial z_k}(z)=n\omega_{2n-1}^{-1}\int_S \bar{u}_k f(u)(1-v\bar{u}')^{-(n+1)}\dot{u}+\frac{1}{2}\left(\frac{2\beta}{\alpha+\beta}\right)^n\frac{\partial f}{\partial u_k}(v)$$

holds for any $v\in P_k$. Here the integral

$$n\omega_{2n-1}^{-1}\int_S \bar{u}_k f(u)(1-v\bar{u}')^{-(n+1)}\dot{u}$$

is defined as

$$\lim_{\varepsilon\to 0}n\omega_{2n-1}^{-1}\left\{\int_{\sigma_\varepsilon(v)}\bar{u}_k f_1(u)(1-v\bar{u}')^{-(n+1)}\dot{u}+i\int_{\sigma_\varepsilon(v)}\bar{u}_k f_2(u)(1-v\bar{u}')^{-(n+1)}\dot{u}\right\},$$

where

$$\sigma_\varepsilon(v)=\{u\,|\,u\in S,\ \alpha^2(1-|v\bar{u}'|^2)^2+4\beta^2(\operatorname{Im} v\bar{u}')^2>\varepsilon^2\}.$$

Proof Take a unitary matrix U such that $vU=p_n$. Set $\bar{U}'=(\alpha_{ij})$, then $(v_1,\cdots,v_n)=(\alpha_{n1},\cdots,\alpha_{nn})$, so $\alpha_{nk}=0$. Let $uU=w$, $zU=\zeta$, we have

$$\begin{aligned}&(K-\lim_{z\to v})n\omega_{2n-1}^{-1}\int_S \bar{u}_k f_1(u)(1-z\bar{u}')^{-(n+1)}\dot{u}\\&=\sum_{j=1}^{n-1}\bar{\alpha}_{jk}\left\{(K-\lim_{\zeta\to p_n})n\omega_{2n-1}^{-1}\int_S \bar{w}_j f_1(w\bar{U}')(1-\zeta\bar{w}')^{-(n+1)}\dot{w}\right\}\\&=n\omega_{2n-1}^{-1}\int_S \bar{u}_k f_1(u)(1-v\bar{u}')^{-(n+1)}\dot{u}+\frac{1}{2}\left(\frac{2\beta}{\alpha+\beta}\right)^n\sum_{j=1}^{n}\sum_{l=1}^{n}\bar{\alpha}_{jk}\alpha_{jl}\frac{\partial f_1}{\partial u_l}(v)\\&=n\omega_{2n-1}^{-1}\int_S \bar{u}_k f_1(u)(1-v\bar{u}')^{-(n+1)}\dot{u}+\frac{1}{2}\left(\frac{2\beta}{\alpha+\beta}\right)^n\frac{\partial f_1}{\partial u_k}(v).\end{aligned}$$

We have the same equality for $f_2=\mathrm{Im}(f)$. The proof is completed.

Taking $\alpha=\beta$ in the above theorem, we obtain a simple formula

$$(K-\lim_{z\to v})\frac{\partial F}{\partial z_k}(z)=n\omega_{2n-1}^{-1}\int_S \bar{u}_k f(u)(1-v\bar{u}')^{-(n+1)}\dot{u}+\frac{1}{2}\frac{\partial f}{\partial u_k}(v),$$

where $v\in P_k$.

§ 3. Representing the boundary value by means of the Hadamard principal value

The aim of this section is to study the limit value of the derivative of Cauchy integral at any point on the sphere. We need the following Lemmas.

Lemma 6. *Suppose $z\in B$, then*

$$\lim_{z\to p_n}\frac{1}{\omega_{2n-1}}\int_S\frac{\bar{u}_n x_k x_l}{(1-z\bar{u}')^{n+1}}\dot{u}=\frac{1}{2n}\delta_{nk}\delta_{nl},\tag{3.1}$$

$$\lim_{z\to p_n}\frac{1}{\omega_{2n-1}}\int_S\frac{\bar{u}_n x_k y_l}{(1-z\bar{u}')^{n+1}}\dot{u}=\frac{1}{2ni}\delta_{nk}\delta_{nl},\tag{3.2}$$

$$\lim_{z\to p_n}\frac{1}{\omega_{2n-1}}\int_S\frac{\bar{u}_n y_k y_l}{(1-z\bar{u}')^{n+1}}\dot{u}=-\frac{1}{2n}\delta_{nk}\delta_{nl}.\tag{3.3}$$

Proof Let

$$G(z)=\frac{1}{\omega_{2n-1}}\int_S\frac{x_k x_l}{(1-z\bar{u}')^n}\dot{u}.$$

Since $x_k=\frac{1}{2}(u_k+\bar{u}_k)$, we have

$$G(z)=\frac{1}{4\omega_{2n-1}}\int_S\frac{u_k u_l+\bar{u}_k u_l+u_k\bar{u}_l+\bar{u}_k\bar{u}_l}{(1-z\bar{u}')^n}\dot{u}.\tag{3.4}$$

By the Cauchy integral formula,

$$\frac{1}{\omega_{2n-1}}\int_S\frac{u_k u_l}{(1-z\bar{u}')^n}\dot{u}=z_k z_l.$$

Using the method of proving Lemma 1, we have

$$\begin{aligned}\frac{1}{4\omega_{2n-1}}\int_S\frac{\bar{u}_k u_l}{(1-z\bar{u}')^n}\dot{u}&=\frac{1}{4\omega_{2n-1}}\sum_{i=1}^{n}\sum_{j=1}^{n}\alpha_{il}\bar{\alpha}_{jk}\int_S\frac{w_i\bar{w}_j}{(1-\rho\bar{w}_n)^n}\dot{w}\\&=\frac{1}{4\omega_{2n-1}}\sum_{j=1}^{n}\alpha_{jl}\bar{\alpha}_{jk}\int_S\frac{|w_j|^2}{(1-\rho w_n)^n}\dot{w}.\end{aligned}$$

An easy computation shows

$$\int_S \frac{|w_j|^2 \dot{w}}{(1-\rho\bar{w}_n)^n} = \frac{2\pi^n}{\Gamma(n+1)}, \quad (j=1, \cdots, n),$$

so

$$\frac{1}{4\omega_{2n-1}} \int_S \frac{\bar{u}_k u_l \dot{u}}{(1-\bar{z}u')^n} = \frac{1}{4n} \delta_{kl}.$$

It is easy to see that

$$\frac{1}{4\omega_{2n-1}} \int_S \frac{\bar{u}_k \bar{u}_l}{(1-\bar{z}u')^n} \dot{u} = \frac{1}{4\omega_{2n-1}} \sum_{i=1}^n \sum_{j=1}^n \bar{\alpha}_{il} \bar{\alpha}_{jk} \int_S \frac{\bar{w}_i \bar{w}_j}{(1-\rho w_n)^n} \dot{w} = 0$$

Substituting these results into (3.4), we obtain

$$G(z) = \frac{1}{4} z_k z_l + \frac{1}{2n} \delta_{kl}.$$

It follows that

$$\lim_{z \to p_n} \frac{\partial G}{\partial z_n}(z) = \frac{1}{2} \delta_{nk} \delta_{nl}.$$

This is just the equality (3.1). The proofs of (3.2), (3.3) are similar.

Lemma 7.

$$\lim_{\varepsilon\to 0} \frac{1}{\omega_{2n-1}} \int_{\sigma_\varepsilon} \frac{\bar{u}_n(x_n-1)}{(1-u_n)^{n+1}} \dot{u} = \frac{1}{2n} + \frac{1}{2}\left(\frac{2\beta}{\alpha+\beta}\right)^{n-1} \frac{(n-1)\beta}{n(\alpha+\beta)}, \tag{3.5}$$

$$\lim_{\varepsilon\to 0} \frac{1}{\omega_{2n-1}} \int_{\sigma_\varepsilon} \frac{\bar{u}_n y_n}{(1-u_n)^{n+1}} \dot{u} = \frac{1}{i}\left\{\frac{1}{2n} - \frac{1}{2}\left(\frac{2\beta}{\alpha+\beta}\right)^{n-1} \frac{\beta+n\alpha}{n(\alpha+\beta)}\right\}, \tag{3.6}$$

$$\lim_{\varepsilon\to 0} \frac{1}{\omega_{2n-1}} \int_{\sigma_\varepsilon} \frac{\bar{u}_n x_k}{(1-u_n)^{n+1}} \dot{u} = 0 \quad (k=1, \cdots, n-1), \tag{3.7}$$

$$\lim_{\varepsilon\to 0} \frac{1}{\omega_{2n-1}} \int_{\sigma_\varepsilon} \frac{\bar{u}_n y_k}{(1-u_n)^{n+1}} \dot{u} = 0 \quad (k=1, \cdots, n-1). \tag{3.8}$$

Proof First we prove (3.6).

$$\frac{1}{\omega_{2n-1}} \int_{\sigma_\varepsilon} \frac{\bar{u}_n y_n \dot{u}}{(1-u_n)^{n+1}} = \frac{1}{2i\omega_{2n-1}} \int_{v\bar{v}' < \frac{\varepsilon}{\alpha}} \dot{v}\left\{2\,\mathrm{Re} \int_0^{\pi-\varepsilon} \frac{r^2 - r^2 e^{2i\theta}}{(1-re^{i\theta})^{n+1}} d\theta\right\}$$

$$+ \frac{1}{2i\omega_{2n-1}} \int_{v\bar{v}' > \frac{\varepsilon}{\alpha}} \dot{v} \int_{-\pi}^{\pi} \frac{r^2(1-e^{2i\theta})}{(1-re^{i\theta})^{n+1}} d\theta = I_1 + I_2,$$

$$\int_0^{\pi-c} \frac{r^2 - r^2 e^{2i\theta}}{(1-re^{i\theta})^{n+1}} d\theta = \frac{1}{i}\left\{r^2 \int_{re^{ic}}^{-re^{-ic}} \frac{dz}{z(1-z)^{n+1}} - \int_{re^{ic}}^{-re^{-ic}} \frac{z dz}{(1-z)^{n+1}}\right\}$$

$$= \frac{1}{i}\left\{\left[\frac{r^2-1}{n} \frac{1}{(1+re^{-ic})^n} - \frac{r^2+1}{n-1} \frac{1}{(1+re^{-ic})^{n-1}}\right.\right.$$

$$\left. + \sum_{k=1}^{n-2} \frac{r^2}{k} \frac{1}{(1+re^{-ic})^k} + r^2(\log(-re^{-ic}) - \log(1+re^{-ic}))\right]$$

$$- \left[\frac{r^2-1}{n(1-re^{ic})^n} + \frac{r^2+1}{(n-1)(1-re^{ic})^{n-1}} + \sum_{k=1}^{n-2} \frac{r^2}{k(1-re^{ic})^k}\right.$$

$$\left.\left. + r^2 \log(re^{ic}) - r^2 \log(1-re^{ic})\right]\right\} = \frac{1}{i}(B_1 - B_2),$$

then

$$I_1=\frac{1}{i\omega_{2n-1}}\int_{v\bar{v}'<\frac{\varepsilon}{\alpha}}\operatorname{Im}(B_1-B_2)\dot{v}$$
$$=\frac{1}{i\omega_{2n-1}}\operatorname{Im}\left\{\left(\sum_{k=1}^{n}J_k+J_0-J_0'\right)-\left(\sum_{k=1}^{n}H_k+H_0-H_0'\right)\right\},$$

where

$$J_n=\frac{1}{n}\int_{v\bar{v}'<\frac{\varepsilon}{\alpha}}\frac{r^2-1}{(1+re^{-i\sigma})^n}\dot{v},\quad J_{n-1}=\frac{1}{n-1}\int_{v\bar{v}'<\frac{\varepsilon}{\alpha}}\frac{r^2+1}{(1+re^{-i\sigma})^{n-1}}\dot{v},$$

$$J_k=\frac{1}{k}\int_{v\bar{v}'<\frac{1}{\alpha}}\frac{r^2\dot{v}}{(1-re^{-i\sigma})^k},\quad (k=1,\cdots,n-1),$$

$$J_0=\int_{x\bar{v}'<\frac{\varepsilon}{\alpha}}r^2\log(-re^{-i\sigma})\dot{v},\quad J_0'=\int_{v\bar{v}'<\frac{\varepsilon}{\alpha}}r^2\log(1+re^{-i\sigma})\dot{v},$$

$$H_n=\frac{1}{n}\int_{v\bar{v}'<\frac{\varepsilon}{\alpha}}\frac{r^2-1}{(1-re^{i\sigma})^n}\dot{v},\quad H_{n-1}=\frac{1}{n-1}\int_{v\bar{v}'<\frac{\varepsilon}{\alpha}}\frac{r^2+1}{(1-re^{i\sigma})^{n-1}}\dot{v},$$

$$H_k=\frac{1}{k}\int_{v\bar{v}'<\frac{\varepsilon}{\alpha}}\frac{r^2}{(1-re^{i\sigma})^k}\dot{v},\quad (k=1,\cdots,n-2),$$

$$H_0=\int_{v\bar{v}'<\frac{\varepsilon}{\alpha}}r^2\log(re^{i\sigma})\dot{v},\quad H_0'=\int_{v\bar{v}'<\frac{\varepsilon}{\alpha}}r^2\log(1-re^{i\sigma})\dot{v}.$$

Using the method we used in the proof of Lemma 5, it is easy to prove

$$\lim_{\varepsilon\to0}\operatorname{Im}(J_0)=\lim_{\varepsilon\to0}\operatorname{Im}(J_0')=0,\quad \lim_{\varepsilon\to0}J_k=0,\quad (k=1,\cdots,n).$$

On the other hand, we have

$$\lim_{\varepsilon\to0}\operatorname{Im}(H_0)=\lim_{\varepsilon\to0}\operatorname{Im}(H_0')=0,$$

and

$$H_{n-1}=\frac{1}{n-1}\int_{v\bar{v}'<\frac{\varepsilon}{\alpha}}\frac{1+r^2}{(1-re^{i\sigma})^{n-1}}\dot{v}=\frac{2\pi^{n-1}}{(n-1)\Gamma(n-1)}(2\beta)^{n-1}$$
$$\times\int_0^1\frac{(2-\eta t^2)\eta^{n-1}t^{2n-3}dt}{\eta^{n-1}[\beta t^2-i\alpha\sqrt{1-t^4}+O(\eta)]^{n-1}},$$

so

$$\lim_{\varepsilon\to0}\operatorname{Im}(H_{n-1})=\frac{2\pi^{n-1}}{\Gamma(n)}2^{n-1}\operatorname{Im}\left\{\int_0^1\frac{x^{n-2}dx}{(x-i\gamma\sqrt{1-x^2})^{n-1}}\right\}=\frac{\pi^n}{\Gamma(n)}\left(\frac{2\beta}{\alpha+\beta}\right)^{n-1}$$

Using the same method, we may prove

$$\lim_{\varepsilon\to0}\operatorname{Im}(H_n)=-\frac{\pi^n}{n\Gamma(n-1)}\frac{1}{2}\left(\frac{2\beta}{\alpha+\beta}\right)^n,$$
$$\lim_{\varepsilon\to0}H_k=0\quad(k=1,\cdots,n-1).$$

Hence

$$\lim_{\varepsilon\to0}I_1=\frac{1}{i\omega_{2n-1}}\lim_{\varepsilon\to0}\operatorname{Im}(-H_n-H_{n-1})=-\frac{1}{2i}\left(\frac{2\beta}{\alpha+\beta}\right)^{n-1}\frac{\beta+n\alpha}{n(\alpha+\beta)}.$$

The computation of I_2 is easy. Since

$$\int_{-\pi}^{\pi}\frac{r^2(1-e^{2i\theta})}{(1-re^{i\theta})^{n+1}}d\theta=\sum_{q=0}^{\infty}\frac{\Gamma(n+q+1)}{\Gamma(n+1)\Gamma(q+1)}r^{q+2}\int_{-\pi}^{\pi}e^{iq\theta}(1-e^{2i\theta})d\theta=2\pi r^2,$$

we have

$$\lim_{\varepsilon\to 0} I_2=\frac{1}{2i}\frac{2\pi}{\omega_{2n-1}}\frac{2\pi^{n-1}}{\Gamma(n-1)}\int_0^1(1-s^2)s^{2n-3}ds=\frac{1}{2ni}.$$

Thus

$$\lim_{\varepsilon\to 0}\frac{1}{\omega_{2n-1}}\int_{\sigma_\varepsilon}\frac{\bar{u}_n y_n}{(1-\bar{u}_n)^{n+1}}\dot{u}=\frac{1}{i}\left[\frac{1}{2n}-\frac{1}{2}\left(\frac{2\beta}{\alpha+\beta}\right)^{n-1}\frac{\beta+n\alpha}{n(\alpha+\beta)}\right].$$

This is just the equality (3.6).

Employing the equalities

$$\lim_{\varepsilon\to 0}\frac{1}{\omega_{2n-1}}\int_{\sigma_\varepsilon}\frac{\bar{u}_n(x_n-1)}{(1-\bar{u}_n)^{n+1}}\dot{u}=\lim_{\varepsilon\to 0}\frac{i}{\omega_{2n-1}}\int_{\sigma_\varepsilon}\frac{\bar{u}_n y_n}{(1-\bar{u}_n)^{n+1}}\dot{u}-\lim_{\varepsilon\to 0}\frac{1}{\omega_{2n-1}}\int_{\sigma_\varepsilon}\frac{\bar{u}_n\dot{u}}{(1-\bar{u}_n)^{n}}$$

and

$$\lim_{\varepsilon\to 0}\frac{1}{\omega_{2n-1}}\int_{\sigma_\varepsilon}\frac{\bar{u}_n}{(1-\bar{u}_n)^{n}}\dot{u}=-\frac{1}{2}\left(\frac{2\beta}{\alpha+\beta}\right)^{n-1},$$

we obtain (3.5) immediately. Using the same method as in the proof of Lemma 4, we can prove the equalities (3.7), (3.8).

Lemma 8. *For $i,\ j=1,\ \cdots,\ n,\ i+j<2n$, we have*

$$\lim_{\varepsilon\to 0}\frac{1}{\omega_{2n-1}}\int_{\sigma_\varepsilon}\frac{\bar{u}_n x_i x_j}{(1-\bar{u}_n)^{n+1}}\dot{u}=-\frac{1}{4n}\left(\frac{2\beta}{\alpha+\beta}\right)^{n}\delta_{ij},\tag{3.9}$$

$$\lim_{\varepsilon\to 0}\frac{1}{\omega_{2n-1}}\int_{\sigma_\varepsilon}\frac{\bar{u}_n y_i y_j}{(1-\bar{u}_n)^{n+1}}\dot{u}=-\frac{1}{4n}\left(\frac{2\beta}{\alpha+\beta}\right)^{n}\delta_{ij}.\tag{3.10}$$

Proof

$$\frac{1}{\omega_{2n-1}}\int_{\sigma_\varepsilon}\frac{\bar{u}_n x_i x_j}{(1-\bar{u}_n)^{n+1}}\dot{u}=\frac{1}{4\omega_{2n-1}}\int_{\sigma_\varepsilon}\frac{\bar{u}_n(u_iu_j+\bar{u}_iu_j+u_i\bar{u}_j+\bar{u}_i\bar{u}_j)}{(1-\bar{u}_n)^{n+1}}\dot{u}.\tag{3.11}$$

It is easy to prove that the value of above integral is 0, when $i\neq j$. Now suppose $i=j<n$. Without loss of generality, take $i=j=1$. Using the method of proving Lemma 4, we have

$$\int_{\sigma_\varepsilon}\frac{\bar{u}_n u_1^2\dot{u}}{(1-\bar{u}_n)^{n+1}}=\int_{\sigma_\varepsilon}\frac{\bar{u}_n\bar{u}_1^2}{(1-\bar{u}_n)^{n+1}}\dot{u}=0.\tag{3.12}$$

On the other hand

$$\int_{\sigma_\varepsilon}\frac{\bar{u}_n|u_1|^2}{(1-\bar{u}_n)^{n+1}}\dot{u}=\int_{v\bar{v}'<\frac{\varepsilon}{\alpha}}|v_1|^2\dot{v}\left\{2\,\mathrm{Re}\int_{c}^{\pi-c}\frac{re^{i\theta}d\theta}{(1-re^{i\theta})^{n+1}}\right\}$$

$$+\int_{v\bar{v}'>\frac{\varepsilon}{\alpha}}|v_1|^2\dot{v}\int_{-\pi}^{\pi}\frac{re^{i\theta}d\theta}{(1-re^{i\theta})^{n+1}}=I_1+I_2.$$

Clearly $I_2=0$. The inner integral of I_1 equals

$$\frac{2}{n}\mathrm{Im}\{(1+re^{-ic})^{-n}-(1-re^{ic})^{-n}\},$$

so $I_1=\mathrm{Im}(J_n-H_n)$, where

$$J_n=\frac{2}{n}\int_{v\bar{v}'<\frac{\varepsilon}{\alpha}}|v_1|^2(1+re^{-ic})^{-n}\dot{v},\quad H_n=\frac{2}{n}\int_{v\bar{v}'<\frac{\varepsilon}{\alpha}}|v_1|^2(1-re^{ic})^{-n}\dot{v}.$$

It is known in Lemma 5 that

$$\lim_{\varepsilon\to 0}J_n=0,\quad \lim_{\varepsilon\to 0}\mathrm{Im}(H_n)=\frac{\pi^n}{n\Gamma(n)}\left(\frac{2\beta}{\alpha+\beta}\right)^n,$$

i. e.

$$\lim_{\varepsilon\to 0} I_1 = -\frac{\pi^n}{n\Gamma(n)}\left(\frac{2\beta}{\alpha+\beta}\right)^n,$$

so

$$\lim_{\varepsilon\to 0}\int_{\sigma_\varepsilon}\frac{\bar{u}_n|u_1|^2}{(1-\bar{u}_n)^{n+1}}\dot{u} = -\frac{\pi^n}{n\Gamma(n)}\left(\frac{2\beta}{\alpha+\beta}\right)^n. \tag{3.13}$$

Substituting (3. 12) and (3.13) into (3.11), we obtain

$$\lim_{\varepsilon\to 0}\frac{1}{\omega_{2n-1}}\int_{\sigma_\varepsilon}\frac{\bar{u}_n x_1^2}{(1-\bar{u}_n)^{n+1}}\dot{u} = -\frac{1}{4n}\left(\frac{2\beta}{\alpha+\beta}\right)^n.$$

Replacing x_1 by $x_j(j=2, \cdots, n-1)$, the above equality also holds. Thus (3.9) is proved. The proof of (3.10) is similar.

Lemma 9.

$$\lim_{\varepsilon\to 0}\frac{1}{\omega_{2n-1}}\int_{\sigma_\varepsilon}\frac{\bar{u}_n(x_n-1)^2}{(1-\bar{u}_n)^{n+1}}\dot{u} = -\frac{1}{2n}, \tag{3.14}$$

$$\lim_{\varepsilon\to 0}\frac{1}{\omega_{2n-1}}\int_{\sigma_\varepsilon}\frac{\bar{u}_n y_n^2}{(1-\bar{u}_n)^{n+1}}\dot{u} = -\frac{1}{2n}. \tag{3.15}$$

Proof Since

$$\left|\frac{\bar{u}_n(x_n-1)^2}{(1-\bar{u}_n)^{n+1}}\right| \leqslant \frac{1}{|1-\bar{u}_n|^{n-1}},$$

$\int_S\frac{\bar{u}_n(x_n-1)^2}{(1-\bar{u}_n)^{n+1}}\dot{u}$ exists and

$$\lim_{\varepsilon\to 0}\frac{1}{\omega_{2n-1}}\int_{\sigma_\varepsilon}\frac{\bar{u}_n(x_n-1)^2}{(1-\bar{u}_n)^{n+1}}\dot{u} = \frac{1}{\omega_{2n-1}}\int_S\frac{\bar{u}_n(x_n-1)^2}{(1-\bar{u}_n)^{n+1}}\dot{u}. \tag{3.16}$$

On the other hand

$$(K-\lim_{z\to p_n})\frac{1}{\omega_{2n-1}}\int_S\frac{\bar{u}_n(x_n-1)^2}{(1-z\bar{u}')^{n+1}}\dot{u} = \frac{1}{\omega_{2n-1}}\int_S\frac{\bar{u}_n(x_n-1)^2}{(1-\bar{u}_n)^{n+1}}\dot{u}. \tag{3.17}$$

Thus

$$(K-\lim_{z\to p_n})\frac{1}{\omega_{2n-1}}\int_S\frac{\bar{u}_n(x_n-1)^2}{(1-z\bar{u}')^{n+1}}\dot{u} = \lim_{\varepsilon\to 0}\frac{1}{\omega_{2n-1}}\int_{\sigma_\varepsilon}\frac{\bar{u}_n(x_n-1)^2}{(1-\bar{u}_n)^{n+1}}\dot{u}. \tag{3.18}$$

Let

$$G(z) = \frac{1}{\omega_{2n-1}}\int_S\frac{(x_n-1)^2}{(1-z\bar{u}')^n}\dot{u}.$$

It is easy to see that

$$G(z) = \frac{1}{4\omega_{2n-1}}\int_S\frac{u_n^2+\bar{u}_n^2+4-4u_n-4\bar{u}_n+2u_n\bar{u}_n}{(1-z\bar{u}')^n}\dot{u} = \frac{1}{4}\left(z_n^2+4-4z_n+\frac{2}{n}\right),$$

so

$$\lim_{z\to p_n}\frac{1}{\omega_{2n-1}}\int_S\frac{\bar{u}_n(x_n-1)^2}{(1-z\bar{u}')^{n+1}}\dot{u} = -\frac{1}{2n}. \tag{3.19}$$

(3.14) follows from (3.18) and (3.19).

Using the same method, we can prove

$$\lim_{\varepsilon\to 0}\frac{1}{\omega_{2n-1}}\int_{\sigma_\varepsilon}\frac{\bar{u}_n|1-\bar{u}_n|^2}{(1-\bar{u}_n)^{n+1}}\dot{u} = -\frac{1}{n}.$$

Now (3.15) follows from the following equality

$$\frac{1}{\omega_{2n-1}}\int_{\sigma_\varepsilon}\frac{\bar{u}_n y_n^2}{(1-\bar{u}_n)^{n+1}}\dot{u} = \frac{1}{\omega_{2n-1}}\int_{\sigma_\varepsilon}\frac{\bar{u}_n|1-\bar{u}_n|^2}{(1-\bar{u}_n)^{n+1}}\dot{u} - \frac{1}{\omega_{2n-1}}\int_{\sigma_\varepsilon}\frac{\bar{u}_n(x_n-1)^2}{(1-\bar{u}_n)^{n+1}}\dot{u}.$$

Lemma 10.

$$\int_{\sigma_\varepsilon} \frac{\bar{u}_n x_k y_j}{(1-\bar{u}_n)^{n+1}} \dot{u} = 0 \tag{3.20}$$

for $k, j=1, \cdots, n$, $k+j<2n$.

$$\lim_{\varepsilon\to 0}\int_{\sigma_\varepsilon} \frac{\bar{u}_n (x_n-1) y_j}{(1-\bar{u}_n)^{n+1}} \dot{u} = 0 \tag{3.21}$$

for $j=1, \cdots, n$.

Proof Since

$$\int_{\sigma_\varepsilon} \frac{\bar{u}_n x_k y_j \dot{u}}{(1-\bar{u}_n)^{n+1}} = \frac{1}{4i}\int_{\sigma_\varepsilon} \frac{\bar{u}_n (u_k u_j + \bar{u}_k u_j - u_k \bar{u}_j - \bar{u}_k \bar{u}_j)}{(1-\bar{u}_n)^{n+1}} \dot{u},$$

if $k\neq j$, then the above integral is equal to 0, if $k=j<n$, then

$$\int_{\sigma_\varepsilon} \frac{\bar{u}_n x_k y_k}{(1-\bar{u}_n)^{n+1}} \dot{u} = \frac{1}{4i}\int_{\sigma_\varepsilon} \frac{\bar{u}_n (u_k^2 - \bar{u}_k^2)}{(1-\bar{u}_n)^{n+1}} \dot{u}.$$

Using the method used in proving Lemma 4, it is easy to prove that the above integral also equals 0. Hence (3.20) is true.

(3.21) holds evidently for $j=1, \cdots, n-1$. When $j=n$, since

$$\left|\frac{\bar{u}_n (x_n-1) y_n}{(1-\bar{u}_n)^{n+1}}\right| \leqslant \frac{1}{|1-\bar{u}_n|^{n-\frac{1}{2}}},$$

using the same method as in the proof of Lemma 9, we have

$$\lim_{z\to p_n} \frac{1}{\omega_{2n-1}}\int_S \frac{\bar{u}_n (x_n-1) y_n}{(1-z\bar{u}')^{n+1}} \dot{u} = \lim_{\varepsilon\to 0} \frac{1}{\omega_{2n-1}}\int_{\sigma_\varepsilon} \frac{\bar{u}_n (x_n-1) y_n}{(1-\bar{u}_n)^{n+1}} \dot{u}.$$

Thus

$$\lim_{z\to p_n} \frac{1}{\omega_{2n-1}}\int_S \frac{\bar{u}_n (x_n-1) y_n}{(1-z\bar{u}')^{n+1}} \dot{u} = \frac{1}{2ni} - \frac{1}{2ni} = 0,$$

by Lemma 2 and Lemma 6. Hence (3.21) is also ture.

Applying the above Lemmas, we have

Theorem 3. *Suppose* $f: \bar{B}\to \mathbf{R}$ *and* $f\in C^3(\bar{B})$. *Then*

$$\lim_{\varepsilon\to 0} \frac{1}{\omega_{2n-1}}\int_{\sigma_\varepsilon} \frac{\bar{u}_n [f(u) - f(p_n)]}{(1-\bar{u}_n)^{n+1}} \dot{u}$$

exists and equals

$$c_n \frac{\partial f(p_n)}{\partial \bar{u}_n} + \left(\frac{1}{n} - 2c_n d_n + c_n\right)\frac{\partial f(p_n)}{\partial u_n} - \frac{c_{n+1}}{n}\sum_{j=1}^{n-1} \frac{\partial^2 f(p_n)}{\partial u_j \partial \bar{u}_j} - \frac{1}{n}\frac{\partial^2 f(p_n)}{\partial u_n \partial \bar{u}_n} + I,$$

where $c_n = \frac{1}{2}\left(\frac{2\beta}{\alpha+\beta}\right)^{n-1}$, $d_n = \frac{\beta+n\alpha}{n(\alpha+\beta)}$, *I is a number to be determined.*

Proof By the Taylor formula

$$\begin{aligned} f(u_1, \cdots, u_n) &= f(x_1+iy_1, \cdots, x_n+iy_n) \\ &= f(p_n) + \sum_{j=1}^{n-1} \frac{\partial f(p_n)}{\partial x_j} x_j + \frac{\partial f(p_n)}{\partial x_n}(x_n-1) + \sum_{j=1}^{n} \frac{\partial f(p_n)}{\partial y_j} y_j \\ &\quad + \frac{1}{2}\Big\{\sum_{i,j=1}^{n-1} \frac{\partial^2 f(p_n)}{\partial x_i \partial x_j} x_i x_j + 2\sum_{i=1}^{n-1} \frac{\partial^2 f(p_n)}{\partial x_i \partial x_n} x_i (x_n-1) + \frac{\partial^2 f(p_n)}{\partial x_n^2}(x_n-1)^2 \\ &\quad + 2\sum_{i=1}^{n} \frac{\partial^2 f(p_n)}{\partial y_i \partial x_n} y_i (x_n-1) + 2\sum_{i=1}^{n-1}\sum_{j=1}^{n} \frac{\partial^2 f(p_n)}{\partial x_i \partial y_j} x_i y_j + \sum_{i,j=1}^{n} \frac{\partial^2 f(p_n)}{\partial y_i \partial y_j} y_i y_j\Big\} \\ &\quad + R_3(u), \end{aligned} \tag{3.22}$$

where $R_3(u)$ is the remainder term of third order. Obviously

$$R_3(u)=O(|1-\bar{u}_n|^{3/2}),$$

so the integral $\int_S \bar{u}_n R_3(u)(1-\bar{u}_n)^{-(n+1)}\dot{u}$ exists, and we denote its value by I. By (3.22) and the Lemmas 7, 8, 9, and 10, we have

$$\lim_{\varepsilon\to 0}\frac{1}{\omega_{2n-1}}\int_{\sigma_\varepsilon}\frac{\bar{u}_n[f(u)-f(p_n)]}{(1-\bar{u}_n)^{n+1}}\dot{u}$$
$$=\left[\frac{1}{2n}+c_n(1-d_n)\right]\frac{\partial f(p_n)}{\partial x_n}+\frac{1}{i}\left(\frac{1}{2n}-c_n d_n\right)\frac{\partial f(p_n)}{\partial y_n}$$
$$+\frac{1}{2}\left\{\sum_{j=1}^{n-1}\left(\frac{\partial^2 f(p_n)}{\partial x_j^2}+\frac{\partial^2 f(p_n)}{\partial y_j^2}\right)\left(-\frac{c_{n+1}}{2n}\right)-\frac{1}{2n}\left(\frac{\partial^2 f(p_n)}{\partial x_n^2}+\frac{\partial^2 f(p_n)}{\partial y_n^2}\right)\right\}+I,$$

this is just the desired result.

Theorem 4. *Suppose f is as in Theorem 3, the limit*

$$\lim_{\varepsilon\to 0}\frac{1}{\omega_{2n-1}}\int_{\sigma_\varepsilon(v)}\frac{\bar{u}_k[f(u)-f(v)]}{(1-v\bar{u}')^{n+1}}\dot{u},\quad (k=1,\cdots,n)\tag{3.23}$$

exist for any $v\in S$.

Proof Take a unitary matrix U such that $vU=p_n$, let $u=w\bar{U}'$, $\bar{U}'=(\alpha_{ij})$, then

$$\frac{1}{\omega_{2n-1}}\int_{\sigma_\varepsilon(v)}\frac{\bar{u}_k[f(u)-f(v)]}{(1-v\bar{u}')^{n+1}}\dot{u}=\sum_{j=1}^{n-1}\bar{\alpha}_{jk}\frac{1}{\omega_{2n-1}}\int_{\sigma_\varepsilon}\frac{\bar{w}_j[f(w\bar{U}')-f(p_n\bar{U}')]}{(1-w_n)^{n+1}}\dot{w}$$
$$+\bar{\alpha}_{nk}\frac{1}{\omega_{2n-1}}\int_{\sigma_\varepsilon}\frac{\bar{w}_n[f(w\bar{U}')-f(p_n\bar{U}')]}{(1-w_n)^{n+1}}\dot{w}.\tag{3.24}$$

Since $\int_{\sigma_\varepsilon}\frac{\bar{w}_j}{(1-\bar{w}_n)^{n+1}}\dot{w}=0$, we have

$$\lim_{\varepsilon\to 0}\frac{1}{\omega_{2n-1}}\int_{\sigma_\varepsilon}\frac{\bar{w}_j[f(w\bar{U}')-f(p_n\bar{U}'))}{(1-w_n)^{n+1}}\dot{w}=\lim_{\varepsilon\to 0}\frac{1}{\omega_{2n-1}}\int_{\sigma_\varepsilon}\frac{\bar{w}_j f(w\bar{U}')}{(1-\bar{w}_n)^{n+1}}\dot{w}.$$

We have proved the existence of above limit in the proof of Theorem 1. By Theorem 3, tqe limit of the second term of the right hand of (3.24) exists. Thus the theorem is proved.

We call limit (3.23) the Hadamard principal value of singular integral

$$\frac{1}{\omega_{2n-1}}\int_S\frac{\bar{u}_k f(u)\dot{u}}{(1-v\bar{u}')^{n+1}},$$

and denote it by

$$P\frac{1}{\omega_{2n-1}}\int_S\frac{\bar{u}_k f(u)}{(1-v\bar{u}')^{n+1}}\dot{u}=\lim_{\varepsilon\to 0}\frac{1}{\omega_{2n-1}}\int_{\sigma_\varepsilon(v)}\frac{\bar{u}_k[f(u)-f(v)]}{(1-v\bar{u}')^{n+1}}\dot{u}.$$

From Theorem 3, we have the following

Theorem 5. *Suppose f is as in Theorem 3, then*

$$(K-\lim_{z\to p_n})\frac{1}{\omega_{2n-1}}\int_S\frac{\bar{u}_n f(u)}{(1-z\bar{u}')^{n+1}}\dot{u}=P\frac{1}{\omega_{2n-1}}\int_S\frac{\bar{u}_n f(u)}{(1-\bar{u}_n)^{n+1}}\dot{u}-c_n\frac{\partial f(p_n)}{\partial\bar{u}_n}$$
$$-(c_n-2c_nd_n)\frac{\partial f(p_n)}{\partial u_n}+\frac{c_{n+1}}{n}\sum_{j=1}^{n-1}\frac{\partial^2 f(p_n)}{\partial u_j\partial\bar{u}_j}.$$

Proof In the expression (3.22) of f, $R_3(u)=O(|1-\bar{u}_n|^{3/2})$. It is known in [1] that the following inequality

$$|1-z\bar{u}'|\geqslant\frac{1}{4\alpha}|1-\bar{u}_n|$$

holds for $z\in D_\alpha(p_n)$, $u\in S$. So

$$\left|\frac{\bar{u}_nR_3(u)}{(1-z\bar{u}')^{n+1}}\right|\leqslant\frac{K}{|1-\bar{u}_n|^{n-\frac{1}{2}}}.$$

By the Lebesgue theorem,

$$(K-\lim_{z\to p_n})\frac{1}{\omega_{2n-1}}\int_S\frac{\bar{u}_nR_3(u)}{(1-z\bar{u}')^{n+1}}\dot{u}=\frac{1}{\omega_{2n-1}}\int_S\frac{\bar{u}_nR_3(u)}{(1-\bar{u}_n)^{n+1}}\dot{u}=I.$$

Employing the expression (3.22) of f and Lemmas 1, 2 and 6, we find

$$(K-\lim_{z\to p_n})\frac{1}{\omega_{2n-1}}\int_S\frac{\bar{u}_nf(u)\dot{u}}{(1-z\bar{u}')^{n+1}}=I+\frac{1}{n}\frac{\partial f(p_n)}{\partial u_n}-\frac{1}{n}\frac{\partial^2f(p_n)}{\partial u_n\partial\bar{u}_n}\tag{3.25}$$

By Theorem 3,

$$\begin{aligned}I=P\frac{1}{\omega_{2n-1}}\int_S\frac{\bar{u}_nf(u)}{(1-\bar{u}_n)^{n+1}}\dot{u}-c_n\frac{\partial f(p_n)}{\partial\bar{u}_n}-\left(\frac{1}{n}-2c_nd_n+c_n\right)\frac{\partial f(p_n)}{\partial u_n}\\+\frac{c_{n+1}}{n}\sum_{j=1}^{n-1}\frac{\partial^2f(p_n)}{\partial u_j\partial\bar{u}_j}+\frac{1}{n}\frac{\partial^2f(p_n)}{\partial u_n\partial\bar{u}_n}.\end{aligned}\tag{3.26}$$

Inserting (3.26) into (3.25), we obtain the desired result.

We now prove our main result.

Theorem 6. *Suppose f is as in Theorem 3, and $F(z)$ is the Cauchy integral of f, then*

$$\begin{aligned}&(K-\lim_{z\to v})\frac{\partial F(z)}{\partial z}\\&=P\frac{n}{\omega_{2n-1}}\int_S\frac{\bar{u}f(u)\dot{u}}{(1-v\bar{u}')^{n+1}}+\frac{1}{2}\left(\frac{2\beta}{\alpha+\beta}\right)^{n-1}\left\{\frac{\partial f(v)}{\partial u}\left[\frac{2\beta}{\alpha+\beta}I-n\frac{\beta-\alpha}{\beta+\alpha}v'\bar{v}\right]\right.\\&\quad\left.-n\frac{\partial f(v)}{\partial\bar{u}}\bar{v}'\bar{v}+\frac{2\beta}{\alpha+\beta}\bar{v}\left[\mathrm{tr}\left(\frac{\partial^2f(v)}{\partial u\partial\bar{u}}\right)-v\frac{\partial^2f(v)}{\partial u\partial\bar{u}}\bar{v}'\right]\right\}\end{aligned}\tag{3.27}$$

holds for any $v\in S$. Here $\frac{\partial F}{\partial z}=\left(\frac{\partial F}{\partial z_1},\cdots,\frac{\partial F}{\partial z_n}\right)$, $\frac{\partial f}{\partial u}=\left(\frac{\partial f}{\partial u_1},\cdots,\frac{\partial f}{\partial u_n}\right)$,

$$\frac{\partial f}{\partial\bar{u}}=\left(\frac{\partial f}{\partial\bar{u}_1},\cdots,\frac{\partial f}{\partial\bar{u}_n}\right),\quad\frac{\partial^2f}{\partial u\partial\bar{u}}=\begin{pmatrix}\frac{\partial^2f}{\partial u_1\partial\bar{u}_1},&\cdots,&\frac{\partial^2f}{\partial u_1\partial\bar{u}_n}\\ \cdots&\cdots&\cdots\\ \frac{\partial^2f}{\partial u_n\partial\bar{u}_1},&\cdots,&\frac{\partial^2f}{\partial u_n\partial\bar{u}_n}\end{pmatrix}.$$

Proof (3.27) may be written as the component form

$$\begin{aligned}(K-\lim_{z\to v})\frac{\partial F(z)}{\partial z_k}&=P\frac{n}{\omega_{2n-1}}\int_S\frac{\bar{u}_kf(u)}{(1-v\bar{u}')^{n+1}}\dot{u}+\frac{1}{2}\left(\frac{2\beta}{\alpha+\beta}\right)^n\frac{\partial f(v)}{\partial u_k}\\&\quad-\frac{n}{2}\left(\frac{2\beta}{\alpha+\beta}\right)^{n-1}\bar{v}_k\left\{\frac{\beta-\alpha}{\beta+\alpha}\sum_{j=1}^n\frac{\partial f(v)}{\partial u_j}v_j\right.\\&\quad+\sum_{j=1}^n\frac{\partial f(v)}{\partial\bar{u}_j}\bar{v}_j-\frac{1}{n}\frac{2\beta}{\alpha+\beta}\sum_{j=1}^n\frac{\partial^2f(v)}{\partial u_j\partial\bar{u}_j}\\&\quad\left.+\frac{1}{n}\frac{2\beta}{\alpha+\beta}\sum_{i,j=1}^n\frac{\partial^2f(v)}{\partial u_i\partial\bar{u}_j}v_i\bar{v}_j\right\},\quad(k=1,\cdots,n).\end{aligned}\tag{3.28}$$

As before, take a unitary matrix U such that $vU=p_n$. Let $\overline{U}'=(\alpha_{ij})$, then $v_k=\alpha_{nk}$ $(k=1,\cdots,n)$. Set $u=w\overline{U}'$, then

$$(K-\lim_{z\to v})\frac{1}{\omega_{2n-1}}\int_S\frac{\bar{u}_k f(u)\dot{u}}{(1-z\bar{u}')^{n+1}}=\sum_{j=1}^n\bar{a}_{jk}(K-\lim_{\zeta\to p_n})\frac{1}{\omega_{2n-1}}\int_S\frac{\bar{w}_j f(w\bar{U}')}{(1-\zeta\bar{w}')^{n+1}}\dot{w}. \quad (3.29)$$

By Theorems 1 and 5, the right hand side of (3.29) is equal to

$$\sum_{j=1}^{n-1}\bar{a}_{jk}\left\{\frac{1}{\omega_{2n-1}}\int_S\frac{\bar{w}_j f(w\bar{U}')\dot{w}}{(1-w_n)^{n+1}}+\frac{c_{n+1}}{n}\frac{\partial f(p_n\bar{U}')}{\partial w_j}\right\}$$
$$+\bar{a}_{nk}\left\{P\frac{1}{\omega_{2n-1}}\int_S\frac{\bar{w}_n f(w\bar{U}')\dot{w}}{(1-\bar{w}_n)^{n+1}}-\frac{\partial f(p_n\bar{U}')}{\partial w_n}c_n-(c_n-2c_nd_n)\frac{\partial f(p_n\bar{U}')}{\partial \bar{w}_n}\right.$$
$$\left.+\frac{c_{n+1}}{n}\sum_{j=1}^{n-1}\frac{\partial^2 f(p_n\bar{U}')}{\partial w_j\partial\bar{w}_j}\right\}$$
$$=P\frac{1}{\omega_{2n-1}}\int_S\frac{\bar{u}_k f(u)\dot{u}}{(1-v\bar{u}')^{n+1}}+\frac{c_{n+1}}{n}\sum_{l=1}^n\sum_{j=1}^n\bar{a}_{jk}a_{jl}\frac{\partial f(v)}{\partial u_l}$$
$$-\bar{a}_{nk}\left(c_n-2c_nd_n+\frac{c_{n+1}}{n}\right)\sum_{j=1}^n\frac{\partial f(v)}{\partial u_j}v_j-\bar{a}_{nk}c_n\sum_{j=1}^n\frac{\partial f(v)}{\partial\bar{u}_j}\bar{v}_j$$
$$+\frac{c_{n+1}}{n}\bar{a}_{nk}\sum_{j=1}^n\frac{\partial^2 f(v)}{\partial u_j\partial\bar{u}_j}-\frac{c_{n+1}}{n}\bar{a}_{nk}\sum_{l,j=1}^n\frac{\partial^2 f(v)}{\partial u_l\partial\bar{u}_j}v_j\bar{v}_j$$
$$=P\frac{1}{\omega_{2n-1}}\int_S\frac{\bar{u}_k f(u)\dot{u}}{(1-v\bar{u}')^{n-1}}+\frac{c_{n+1}}{n}\frac{\partial f(v)}{\partial u_k}-\bar{a}_{nk}\left\{\left(c_n-2c_nd_n+\frac{c_{n+1}}{n}\right)\sum_{j=1}^n\frac{\partial f(v)}{\partial u_j}v_j\right.$$
$$\left.+c_n\sum_{j=1}^n\frac{\partial f(v)}{\partial\bar{u}_j}\bar{v}_j-\frac{c_{n+1}}{n}\sum_{j=1}^n\frac{\partial^2 f(v)}{\partial u_j\partial\bar{u}_j}+\frac{c_{n+1}}{n}\sum_{l,j=1}^n\frac{\partial^2 f(v)}{\partial u_l\partial\bar{u}_j}v_l\bar{v}_j\right\}.$$

Inserting

$$c_n=\frac{1}{2}\left(\frac{2\beta}{\alpha+\beta}\right)^{n-1},\quad c_n-2c_nd_n+\frac{c_{n-1}}{n}=\frac{1}{2}\left(\frac{2\beta}{\alpha+\beta}\right)^{n-1}\frac{\beta-\alpha}{\beta+\alpha}$$

into the above equality, (3.28) is obtained.

Theorem 7. *Suppose f: $\bar{B}\to\mathbf{C}$, where $f_1=\operatorname{Re}f$ and $f_2=\operatorname{Im}f$ are in $C^3(\bar{B})$, and $F(z)$ is the Cauchy integral of f, then (3.27) also holds for any $v\in S$. Here*

$$P\frac{1}{\omega_{2n-1}}\int_S\frac{\bar{u}_k f(u)\dot{u}}{(1-v\bar{u}')^{n+1}}$$

is defined as

$$P\frac{1}{\omega_{2n-1}}\int_S\frac{\bar{u}_k f_1(u)\dot{u}}{(1-v\bar{u}')^{n+1}}+P\frac{i}{\omega_{2n-1}}\int_S\frac{\bar{u}_k f_2(u)\dot{u}}{(1-v\bar{u}')^{n+1}}.$$

The proof is obvious.

Taking $\alpha=\beta$ in Theorem 7, we have

$$(K-\lim_{z\to v})\frac{\partial F(z)}{\partial z}=P\frac{n}{\omega_{2n-1}}\int_S\frac{\bar{u}f(u)\dot{u}}{(1-v\bar{u}')^{n+1}}$$
$$+\frac{1}{2}\left\{\frac{\partial f(v)}{\partial u}-n\frac{\partial f(v)}{\partial\bar{u}}\bar{v}'\bar{v}+\bar{v}\left[\operatorname{tr}\left(\frac{\partial^2 f(v)}{\partial u\partial\bar{u}}\right)-v\frac{\partial^2 f(v)}{\partial u\partial\bar{u}}\bar{v}'\right]\right\}.$$

References

[1] Shi Jihuai and Gong sheng, Singular integrals in several complex variables (II), *Chin. Ann. of Math.*, **4B**:3 (1983), 307—318.

[2] Gong Sheng and Shi Jihuai, Singular integrals in several complex variables (I), *Chin. Ann. of Math.*, **3** (1982), 483—502.

[3] Gong sheng and Shi Jihuai, Singular integrals in several complex variables (III), *Chin. Ann. of Math.*

[4] Koranyi, A., Harmonic functions on Hermitian hyperbolic space, *Trans. Amer. Math. Soc.*, **135** (1969), 507—516.

[5] Rudin, W., Function theory in the unit ball of C^n, Springer-Verlag, New York, 1980.

第1卷 第2期 中国科学技术大学学报 Vol. 1, No. 2
1965年8月 ZHONGGUO KEXUE JISHU DAXUE XUEBAO August, 1965

圆型域中解析函数的积分表示

龚 昇 孙繼广

提 要

设域 $\mathfrak{R}$ 为 n 个复变数 $z=(z_1, z_2, \cdots, z_n)$ 空间中包有原点的圆型有界单连通域，并且对原点而言是星状的；又设 $\mathfrak{R}$ 的特征流形是圆型、紧致的．本文证明了，凡 $\mathfrak{R}$ 内的解析函数 $f(z)$，如果属于 Hardy 族 H_1，则必可表为 Cauchy 积分和 Schwarz 积分，而且 $f(z)$ 属于 H_1 是 $f(z)$ 可表为 poisson 积分的充分必要条件．本文同时给出了这些结果在正规族方面的某些应用．

§1. Cauchy 积分

域 $\mathfrak{R}$ 为 n 个复变数 $z=(z_1, z_2, \cdots, z_n)$ 的圆型有界单连通域，包有原点，且对原点而言是星状的．$\mathfrak{R}$ 的特征流形为 $\mathfrak{C}$，我们假设 $\mathfrak{C}$ 也是圆型的，而且是紧致的．本文中所考虑的域，都是上述的圆型域．

华罗庚教授[1]早已指出：对于上述域 $\mathfrak{R}$，它的 Cauchy 核是存在的，并且具体的作出了 Cauchy 核，这就使圆型域中的解析函数有了最主要的积分表示．他指出：若 $f(z)$ 在 $\mathfrak{R}$ 的闭包 $\bar{\mathfrak{R}}$ 上解析，则 $f(z)$ 可以用 Cauchy 积分表示．我们称在 $\mathfrak{R}$ 内解析的函数 $f(z)$ 属于 Hardy 族 H_1，若

$$\int_{\mathfrak{C}} |f(r\xi)|\dot{\xi}$$

对 $0 \leqslant r < 1$ 一致有界，这里 $\xi \in \mathfrak{C}$，$\dot{\xi}$ 表示 $\mathfrak{C}$ 的体积元素．我们有

定理 1 若 $f(z) \in H_1$，则 $f(z)$ 可以表为 Cauchy 积分．

证 华罗庚教授[1]证明了：对于上述圆型域 $\mathfrak{R}$，存在一组完整直交函数系 $\psi_m(z)$，它在 $\mathfrak{C}$ 上是就范直交的．于是 $\mathfrak{R}$ 的 Cauchy 核 $H(z, \bar{\xi})$ 等于

$$\sum_{m=0}^{\infty} \psi_m(z)\overline{\psi_m(\xi)}, \tag{1.1}$$

这里 $z \in \mathfrak{R}$，$\xi \in \mathfrak{C}$，而每一个 $\psi_m(z)$ 为齐次多项式．由假设 $f(z) \in H_1$，据 Bochner[4] 的一条定理知道，境界值 $f(\xi)$ 绝对可积．于是作 Cauchy 型积分

$$g(z) = \int_{\mathfrak{C}} f(\xi)H(z, \bar{\xi})\dot{\xi},$$

它是存在的，并且表示一个对 z 的解析函数．由 Cauchy 核的定义，知

$$g(z) = \sum_{m=0}^{\infty} \psi_m(z)\int_{\mathfrak{C}} f(\xi)\overline{\psi_m(\xi)}\dot{\xi},$$

这是因为固定 z 时，级数(1)的收敛对 ξ 是一致的．记

$$C_m = \int_{\mathfrak{C}} f(\xi)\overline{\psi_m(\xi)}\dot{\xi}, \tag{1.2}$$

于是

$$g(z)=\Sigma c_m\psi_m(z).$$

因为 $f(z)$ 在 $\mathfrak{R}$ 内解析,故有 Taylor 展开

$$f(z)=\sum_{m=0}^{\infty}a_m\psi_m(z),$$

取 $z=\gamma\eta$, $\eta\in\mathfrak{C}$, $0<\gamma<1$,由于 ψ_m 为齐次多项式,设其次数为 k_m,则

$$f(\gamma\eta)=\sum_{m=0}^{\infty}a_m\gamma^{k_m}\psi_m(\eta),$$

当 $\gamma<1$ 固定时,它对 η 在 $\mathfrak{C}$ 上的收敛是一致的. 于是

$$\int_{\mathfrak{C}}f(\gamma\xi)\overline{\psi_m(\xi)}\dot{\xi}=\int_{\mathfrak{C}}\sum_{l=0}^{\infty}a_l\gamma^{k_l}\psi_l(\xi)\overline{\psi_m(\xi)}\dot{\xi}=\gamma^{k_m}a_m. \tag{1.3}$$

由(1.2),(1.3)得

$$c_m-a_m\gamma^{k_m}=\int_{\mathfrak{C}}(f(\xi)-f(\gamma\xi))\psi_m(\xi)\dot{\xi},$$

因而

$$|c_m-a_m\gamma^{k_m}|\leqslant M_m\int_{\mathfrak{C}}|f(\xi)-f(\gamma\xi)|\dot{\xi},$$

此处 $M_m=\max\limits_{\xi\in\mathfrak{C}}|\psi_m(\xi)|$, 由于 ψ_m 为齐次多项式, $\mathfrak{C}$ 是紧致的, 故 M_m 是存在的. 令 $\gamma\to 1$,由 Riesz 定理[4]知 $c_m=a_m$,故 $g(z)=f(z)$, 即 $f(z)$ 可以用 Cauchy 积分表达:

$$f(z)=\int_{\mathfrak{C}}f(\xi)H(z,\bar{\xi})\dot{\xi}. \tag{1.4}$$

§ 2. Schwarz 积分

如果 $f(z)\in H_1$, 则由 § 1 知,它可以表成 Cauchy 积分,即(1.4)式成立. 固定 $w\in\mathfrak{R}$ 时, $f(z)H(z,\bar{w})$ 也属于 H_1,故由公式(1.4):

$$\int_{\mathfrak{C}}f(\xi)H(\xi,\bar{w})\dot{\xi}=f(0).$$

取共轭,得

$$\int_{\mathfrak{C}}\overline{f(\xi)}H(w,\bar{\xi})\dot{\xi}=\overline{f(0)}, \tag{2.1}$$

在(2.1)中取 $w=z$,于是

$$\int_{\mathfrak{C}}(f(\xi)+\overline{f(\xi)})H(z,\bar{\xi})\dot{\xi}=f(z)+\overline{f(0)},$$

即

$$\begin{aligned}f(z)&=\int_{\mathfrak{C}}(f(\xi)+\overline{f(\xi)})H(z,\bar{\xi})\dot{\xi}-\frac{1}{2}(f(0)+\overline{f(0)})+i\,\mathrm{Im}\,f(0)=\\&=\int_{\mathfrak{C}}(f(\xi)+\overline{f(\xi)})H(z,\bar{\xi})\dot{\xi}-\frac{1}{2V(\mathfrak{C})}\int_{\mathfrak{C}}(f(\xi)+\overline{f(\xi)})\dot{\xi}+i\,\mathrm{Im}\,f(0),\end{aligned}$$

这里 $V(\mathfrak{C})$ 表示 $\mathfrak{C}$ 的体积. 所以,若记 $f(z)=u(z)+iv(z)$,则有

$$f(z)=\int_{\mathfrak{C}}u(\xi)\left(2H(z,\bar{\xi})-\frac{1}{V(\mathfrak{C})}\right)\dot{\xi}+i\,\mathrm{Im}\,f(0).$$

我们称

$$S(z, \xi) = 2H(z, \bar{\xi}) - \frac{1}{V(\mathfrak{C})} \tag{2.2}$$

为 $\mathfrak{R}$ 的 Schwarz 核. 于是有

定理 2　若 $f(z) \in H_1$,则 $f(z)$ 可以表为 Schwarz 积分:

$$f(z) = \int_{\mathfrak{C}} u(\xi) S(z, \xi) \dot{\xi} + i \operatorname{Im} f(0), \tag{2.3}$$

其中 $u(z) = \operatorname{Re} f(z)$, $S(z, \xi)$ 如(2.2)所示.

1959 年, Какичев[5] 给出了多圆柱的 Schwarz 核

$$\frac{1}{(2\pi)^n}\left[\frac{1}{2^{n-1}} \prod_{m=1}^{n}\left(\frac{\tau_m + z_m}{\tau_m - z_m} + 1\right) - 1\right],$$

其中 $|z_m| < 1$, $|\tau_m| = 1$ $(m = 1, 2, \cdots, n)$. 其实,它也就是

$$\frac{1}{(2\pi)^n}\left(2 \prod_{m=1}^{n} \frac{1}{1 - z_m \bar{\tau}_m} - 1\right) = \frac{2}{(2\pi)^n} \prod_{m=1}^{n} \frac{1}{1 - z_m \bar{\tau}_m} - \frac{1}{(2\pi)^n},$$

而 $\frac{1}{(2\pi)^n} \prod_{m=1}^{n} \frac{1}{1 - z_m \bar{\tau}_m}$ 是熟知的多圆柱的 Cauchy 核, $(2\pi)^n$ 是多圆柱的特征流形的体积.

H. Tornehave[6] 在美国数学评论 24 卷上认为 Какичев 得到多圆柱的 Schwarz 核并非显然的. 现在看来,未必如此.

利用积分表示,容易证明如下的

定理 3　如果 $\mathfrak{M} = \{f(z)\}$ 为域 $\mathfrak{R}$ 内的一个函数族, $\mathfrak{R}$ 除了满足 § 1 中的条件外,并且是可递的. 若在 $\mathfrak{M}$ 的任一函数列中,可以选出一个子函数列,其实部(或虚部)在 $\mathfrak{R}$ 上内闭匀敛,且这个子函数列在 $\mathfrak{R}$ 中的某一点 z_0 上收敛,则 $\mathfrak{M}$ 为正规族.

证明　由于 $\mathfrak{R}$ 是可递的,故不妨设 $z_0 = 0$. 对于任一 $r: 0 < r < 1$,考虑 $\overline{\mathfrak{R}(r)}$($\mathfrak{R}(r)$ 表示以原点为中心,把 $\mathfrak{R}$ 缩小 r 倍所得的域, $\overline{\mathfrak{R}(r)}$ 为其闭包),这是包在 $\mathfrak{R}$ 中的一个闭域. 若 $z \in \overline{\mathfrak{R}(r)}$,取 $r_1: r < r_1 < 1$,令 $z = r_1 z_1$,则 $z_1 \in \mathfrak{R}$. 若 $\{f_k(z)\}$ 为 $\mathfrak{M}$ 的一个函数列中所选出的满足已设条件的子函数列,记 $f_k(z) = f_k(r_1 z_1) = \tilde{f}_k(z_1)$, $u_k(z) = \operatorname{Re}\{\tilde{f}_k(z_1)\} = \tilde{u}_k(z_1)$,则据假设, $\{\tilde{f}_k(z_1)\}$ 在 $\bar{\mathfrak{R}}$ 上解析,其实部 $\{\tilde{u}_k(z_1)\}$ 在 $\mathfrak{C}$ 上一致收敛,且 $\tilde{v}_k^{(0)} = \operatorname{Im}\{\tilde{f}_k(0)\}$ 收敛. 由 Schwarz 公式,我们有

$$\tilde{f}_k(z_1) = \int_{\mathfrak{C}} \tilde{u}_k(\xi) S(z, \xi) \dot{\xi} + i \tilde{v}_k^{(0)},$$

若 $z_1 \in \overline{\mathfrak{R}\left(\frac{r}{r_1}\right)}$,则由华罗庚教授所证明的定理[1]易知

$$|S(z_1, \xi)| = \left| 2H(z_1, \bar{\xi}) - \frac{1}{V(\mathfrak{C})} \right| = \left| 2 \sum_{m=0}^{\infty} \varphi_m(z_1) \overline{\varphi_m(\xi)} - \frac{1}{V(\mathfrak{C})} \right| \leqslant$$

$$\leqslant \frac{1}{V(\mathfrak{C})} + 2 \left| \sum_{m=1}^{\infty} \varphi_m(z_1) \overline{\varphi_m(z)} \right| \leqslant$$

$$\leqslant \frac{1}{V(\mathfrak{C})} + 2 \max_{\zeta \in \mathfrak{C}} \left(\sum_{m=1}^{\infty} |\varphi_m(\zeta)|^2 \gamma_0^{k_m} \right) =$$

$$= \frac{1}{V(\mathfrak{C})} + 2 \max_{\zeta \in \mathfrak{C}} \left(\sum_{m=0}^{\infty} |\psi_m(\zeta)|^2 \gamma_0^{k_m} - \frac{1}{V(\mathfrak{C})} \right) =$$

$$= \frac{1}{V(\mathfrak{C})} \left(\frac{2}{(1-\gamma_0)^n} - 1 \right) = M_0, \quad \left(\gamma_0 = \frac{\gamma}{\gamma_1} \right)$$

由此即得

$$|\tilde{f}_k(z_1) - \tilde{f}_l(z_1)| \leqslant \int_{\mathfrak{C}} |\tilde{u}_k(\xi) - \tilde{u}_l(\xi)| |S(z_1, \xi)| \dot{\xi} + |\tilde{v}_k^{(0)} - \tilde{v}_l^{(0)}|,$$

由假设，任与 $\varepsilon > 0$，存在正整数 K，当 $k, l \geqslant K$ 时，

$$|\tilde{u}_k(\xi) - \tilde{u}_l(\xi)| \leqslant \frac{\varepsilon}{2MV(\mathfrak{C})}$$

对 $\xi \in \mathfrak{C}$ 一致地成立，且

$$|\tilde{v}_k^{(0)} - \tilde{v}_l^{(0)}| < \frac{\varepsilon}{2}.$$

于是当 $z_1 \in \overline{\Re\left(\frac{\gamma}{r_1}\right)}$ 时，$|\tilde{f}_k(z_1) - \tilde{f}_l(z_1)| < \varepsilon$ 一致地成立．即 $\{f_k(z)\}$ 在 $\overline{\Re(r)}$ 上一致收敛．故从 $\mathfrak{M}$ 中任取一个函数列，一定有一个子函数列内闭匀敛，故 $\mathfrak{M}$ 为正规族．

定理 3 以 Какичев 的一个定理[5]作为最简单的特例．

§3. Poisson 积分

华罗庚教授从 Cauchy 核出发，定义

$$P(z, \xi) = \frac{H(z, \bar{\xi}) H(\xi, \bar{z})}{H(z, \bar{z})} \tag{3.1}$$

为 poisson 核，并且称由 $\mathfrak{C}$ 上的连续函数 $u(\xi)$ 所定义的 poisson 积分 $\int_{\mathfrak{C}} u(\xi) P(z, \xi) \dot{\xi}$ 为调和函数．

我们从定理 1 立刻得到

定理 4 若 $f(z)$ 在 $\Re$ 中解析，则 $f(z)$ 可以表为 poisson 积分

$$f(z) = \int_{\mathfrak{C}} f(\xi) P(z, \xi) \dot{\xi} \tag{3.2}$$

的充分必要条件为 $f(z) \in H_1$．

证明 充分性．若 $f(z) \in H_1$，则当 w 固定时，易见 $f(z)H(z, \bar{w}) \in H_1$，故由定理 1，有

$$f(z)H(z, \bar{w}) = \int_{\mathfrak{C}} f(\xi) H(\xi, \bar{w}) H(z, \bar{\xi}) \dot{\xi},$$

特别取 $w = z$，除以 $H(z, \bar{z})$ 即得．

必要性．若 $f(z)$ 可以表为(3.2)，取 $z = \gamma\eta$ $(0 \leqslant \gamma < 1)$，$\eta \in \mathfrak{C}$，则

$$\int_{\mathfrak{C}} |f(\gamma\eta)| \dot{\eta} \leqslant \int_{\mathfrak{C}} \int_{\mathfrak{C}} |f(\xi)| P(\gamma\eta, \xi) \dot{\xi} \dot{\eta}, \tag{3.3}$$

由于 $P(\gamma\eta, \xi) = P(\gamma\xi, \eta)$，故(4.3)式右端为

$$\int_{\mathfrak{C}} \int_{\mathfrak{C}} |f(\xi)| P(\gamma\xi, \eta) \dot{\xi} \dot{\eta} = \int_{\mathfrak{C}} \left(\int_{\mathfrak{C}} P(\gamma\xi, \eta) \dot{\eta} \right) |f(\xi)| \dot{\xi} = \int_{\mathfrak{C}} |f(\xi)| \dot{\xi},$$

所以 $f(z)\in H_1$. 这里的积分次序能够交换是因为 $|f(\xi)|$ 可积[4]，而且当 $0<\gamma<1$ 固定时，$P(\gamma\xi,\eta)$ 对 ξ,η 在 $\mathfrak{C}$ 上是连续的. 这里同时利用了 poisson 核的性质[1]：

$$\int_{\mathfrak{C}} P(z,\xi)\dot{\xi}=1(z\in\mathfrak{R}).$$

在一个复变数时，poisson 核为 Schwarz 核的实部，在现在一般来说是不对的，因为调和函数未必是解析函数的实部. Schwarz 核的实部为

$$B(z,\xi)=H(z,\bar{\xi})+H(\xi,\bar{z})-\frac{1}{V(\mathfrak{C})}{}^{1)},\tag{3.4}$$

虚部为

$$C(z,\xi)=\frac{1}{i}(H(z,\bar{\xi})-H(\xi,\bar{z})).\tag{3.5}$$

从 Schwarz 公式即得：若 $f(z)=u(z)+iv(z)\in H_1$，则

$$u(z)=\int_{\mathfrak{C}} u(\xi)B(z,\xi)\dot{\xi},\tag{3.6}$$

$$v(z)=\int_{\mathfrak{C}} u(\xi)C(z,\xi)\dot{\xi}+\operatorname{Im}f(0).\tag{3.7}$$

从(3.6)，(3.7)可得，若 $f(z)\in H_1$，则

$$f(z)=\int_{\mathfrak{C}} f(\xi)B(z,\xi)\dot{\xi},\tag{3.8}$$

$$f(z)=i\int_{\mathfrak{C}} f(\xi)C(z,\xi)\dot{\xi}+f(0).\tag{3.9}$$

即当 $f(z)\in H_1$ 时，$f(z)$ 可以用 $B(z,\xi)$, $C(z,\xi)$ 为核的积分表示之. B-调和函数也能用这些核来表达，问题在于对 B-调和函数来讲，Dirichlet 问题未必有解，并且容易看出 $B(z,\xi)$、$C(z,\xi)$ 都不是正的，例如：若 $\eta\in\mathfrak{C}$，则 $B(\gamma\eta,-\eta)$ 等于

$$\frac{2}{V(\mathfrak{C})}\sum_{m=0}^{\infty}C_m^{m+n-1}(-\gamma)^m-\frac{1}{V(\mathfrak{C})}=\frac{1}{V(\mathfrak{C})}\left(\frac{2}{(1+\gamma)^n}-1\right),$$

若 $n>1$，且 r 充分接近 1 时，这是小于零的.

§4. 正规族 $E_1(\mathfrak{R}_I)$

令 $\mathfrak{R}_I$ 表示典型域 $\mathfrak{R}_I(m,n)$: $I-Z\bar{Z}'>0$, $Z=Z^{(m,n)}$, $m\leqslant n$. 本节考虑函数族

$$E_1(\mathfrak{R}_I)=\left\{f(Z)\in H_1(\mathfrak{R}_I)\,\middle|\,\frac{1}{V(\mathfrak{C}_I)}\int_{\mathfrak{C}_I}|f(rU)|\dot{U}\leqslant 1\quad 0\leqslant r<1\right\}.\tag{4.1}$$

习知，在一个复变数时，这就是 Смирнов 函数族[7],[8].

本节应用上面的结果，给出 $E_1(\mathfrak{R}_I)$ 内的函数的模、系数和模平均的估计. 对于其他三类典型域，亦可同样讨论之，此处不再赘述.

定理 5　若 $f(Z)\in E_1(\mathfrak{R}_I)$，则

$$|f(Z)|\leqslant\frac{1}{d(I-Z\bar{Z}')^n}.\tag{4.2}$$

---

1) 1958年左右，陆启鏗先生曾对典型域获得过这个核，称之为 Poisson 核，并告诉作者之一，现在看来，应该说是 Schwarz 核的实部.

证明　由于 $f(Z)\in E_1(\mathfrak{R}_I)$，而且当 $Z, W\in\overline{\mathfrak{R}}_I$ 时，

$$\left|\frac{1}{H(Z,\overline{W})}\right| = V(\mathfrak{C}_I)\,|d(I-Z\overline{W}')^n| \leqslant V(\mathfrak{C}_I)d(I+Z\overline{Z}')^{\frac{n}{2}}d(I+W\overline{W}')^{\frac{n}{2}} \leqslant \leqslant 2^{mn}V(\mathfrak{C}_I),$$

所以

$$\frac{f(W)}{H(W,\overline{Z})}\in H_1(\mathfrak{R}_I),\quad W, Z\in\mathfrak{R}_I.$$

据定理1，有

$$\frac{f(W)}{H(W,\overline{Z})} = \int_{\mathfrak{C}}\frac{f(U)}{H(U,\overline{Z})}H(W,\overline{U})\dot{U},$$

特别取 $W=Z$，便有

$$f(Z) = H(Z,\overline{Z})\int_{\mathfrak{C}_I}f(U)\frac{H(Z,\overline{U})}{H(U,\overline{Z})}\dot{U},\ \left(H(Z,\overline{Z}) = \frac{1}{V(\mathfrak{C}_I)d(I-Z\overline{Z}')^n}\right).$$

应用 Riesz 定理的推广[4]，$\dfrac{1}{V(\mathfrak{C}_I)}\displaystyle\int_{\mathfrak{C}_I}|f(U)|\dot{U}\leqslant 1$，即可得到(4.2)。

定理 6　若 $f(Z)=\sum\limits_{m=0}^{\infty}C_m\psi_m(Z)\in E_1(\mathfrak{R}_I)$，$(\{\psi_m(Z)\}_{m=0,1,2,\cdots}$ 为在 $\mathfrak{C}_I$ 上就范正交，而在 $\mathfrak{R}_I$ 内完整正交的函数系，$\psi_m(Z)$ 为 k_m 次齐次多项式[1])则

$$|C_m|\leqslant V(\mathfrak{C}_I)\max_{U\in\mathfrak{C}_I}|\psi_m(U)|. \tag{4.3}$$

证明　取 $Z=rU$，$0<r<1$，则因

$$f(rU)=\sum_{m=0}^{\infty}C_m\psi_m(rU)$$

对 U 在 $\mathfrak{C}_I$ 上一致收敛，所以

$$\int_{\mathfrak{C}_I}f(rU)\overline{\psi_m(U)}\dot{U} = \sum_{v=0}^{\infty}C_v\gamma^{k_v}\int_{\mathfrak{C}_I}\psi_v(U)\overline{\psi_m(U)}\dot{U} = C_m\gamma^{k_m}.$$

于是

$$|C_m|\leqslant\gamma^{-k_m}\int_{\mathfrak{C}_I}|f(rU)|\,|\psi_m(U)|\dot{U}\leqslant\gamma^{-k_m}V(\mathfrak{C}_I)\max_{U\in\mathfrak{C}_I}|\psi_m(U)|,$$

令 $r\to 1$ 即得(4.3)式。

定理 7　若 $f(Z)\in E_1(\mathfrak{R}_I)$，则

$$\int_{\mathfrak{R}_{I_r}}|f(Z)|\dot{Z}\leqslant V(\mathfrak{R}_{I_r}). \tag{4.4}$$

此处 $\mathfrak{R}_{I_r}$ 表示 $\mathfrak{R}_I$ 内 $Z\overline{Z}'<r^2I$ 的区域，$0<r\leqslant 1$. $V(\mathfrak{R}_{I_r})$ 表示 $\mathfrak{R}_{I_r}$ 的体积，$\dot{Z}$ 表示 $\mathfrak{R}_I$ 的体积元素.

证明　令 $P(Z,U)$ 表示 $\mathfrak{R}_I$ 的 poisson 核

$$\frac{d(I-Z\overline{Z}')^n}{V(\mathfrak{C}_I)|d(I-Z\overline{U}')|^{2n}}{}^{[1]}, \tag{4.5}$$

我们先来证明

$$\int_{\mathfrak{R}_{I_r}}P(Z,U)\dot{Z} = \frac{V(\mathfrak{R}_{I_r})}{V(\mathfrak{C}_I)},\quad 0<r\leqslant 1. \tag{4.6}$$

习知[1]，对于任何 $Z\in\Re_I$，都有

$$\int_{\mathfrak{C}_I} P(Z,U)\dot{U}=1, \tag{4.7}$$

于是

$$V(\Re_{I_r})=\int_{\Re_{I_r}}\dot{Z}=\int_{\Re_{I_r}}\dot{Z}\int_{\mathfrak{C}_I}P(Z,U)\dot{U}=\int_{\mathfrak{C}_I}\dot{U}\int_{\Re_{I_r}}P(Z,U)\dot{Z}. \tag{4.8}$$

当 $m=n$ 时，令 $Z\overline{U}'=W$，由于 $W\in\Re_{I_r}$，$P(Z,U)=P(W,I)$，以及 $\dot{Z}=\dot{W}$，所以若记 $\int_{\Re_{I_r}}P(Z,U)\dot{Z}=C_r$，则(4.8)式的右端为

$$\int_{\mathfrak{C}_I}\dot{U}\int_{\Re_{I_r}}P(W,I)\dot{W}=C_r\int_{\mathfrak{C}_I}\dot{U}=C_rV(\mathfrak{C}_I). \tag{4.9}$$

由此便得到(4.6)式。

当 $m<n$ 时，作 n 阶酉方阵 $V=\begin{pmatrix}U^{(m,n)}\\ U_1^{(n-m,n)}\end{pmatrix}$，易知，如果 $Z\in\Re_{I_r}$，则

$$\int_{\mathfrak{C}_I(m,n)}P(Z,U)\dot{U}=\frac{V(\mathfrak{C}_I(m,n))}{V(\mathfrak{C}_I(n,n))}\int_{\mathfrak{C}_I(n,n)}P_1\left(\begin{pmatrix}Z\\0\end{pmatrix},V\right)\dot{V},$$

其中 $P_1(Z_1,V)$ 表示 $\Re_I(n,n)$ 的 poisson 核 $\dfrac{d(I-Z_1Z_1')^n}{V(\mathfrak{C}_I(n,n))|d(I-Z_1\overline{V}')|^{2n}}$，$Z_1\in\Re_I(n,n)$。

于是

$$\begin{aligned}V(\Re_{I_r}(m,n))&=\int_{\Re_{I_r}(m,n)}\dot{Z}\int_{\mathfrak{C}_I(m,n)}P(Z,U)\dot{U}=\\&=\frac{V(\mathfrak{C}_I(m,n))}{V(\mathfrak{C}_I(n,n))}\int_{\Re_{I_r}(m,n)}\dot{Z}\int_{\mathfrak{C}_I(n,n)}P_1\left(\begin{pmatrix}Z\\0\end{pmatrix},V\right)\dot{V}=\\&=\frac{V(\mathfrak{C}_I(m,n))}{V(\mathfrak{C}_I(n,n))}\int_{\mathfrak{C}_I(n,n)}\dot{V}\int_{\Re_I(m,n)}P_1\left(\begin{pmatrix}Z\overline{V}'\\0\end{pmatrix},I\right)\dot{Z}.\end{aligned}$$

令 $Z\overline{V}'=W$，由于 $W\in\Re_{I_r}(m,n)$，$P_1\left(\begin{pmatrix}Z\\0\end{pmatrix},V\right)=P_1\left(\begin{pmatrix}W\\0\end{pmatrix},I\right)$，$\dot{Z}=\dot{W}$，所以

$$\begin{aligned}V(\Re_{I_r}(m,n))&=\frac{V(\mathfrak{C}_I(m,n))}{V(\mathfrak{C}_I(n,n))}\int_{\mathfrak{C}_I(n,n)}\dot{V}\int_{\Re_I(m,n)}P_1\left(\begin{pmatrix}W\\0\end{pmatrix},I\right)\dot{W}=\\&=V(\mathfrak{C}_I(m,n))\int_{\Re_I(m,n)}P_1\left(\begin{pmatrix}W\\0\end{pmatrix},I\right)\dot{W}=\\&=V(\mathfrak{C}_I(m,n))\int_{\Re_{I_r}(m,n)}P(Z,U)\dot{Z},\end{aligned}$$

此即(4.6)式。

现在来证明定理 7．根据定理 4，有

$$f(Z)=\int_{\mathfrak{C}_I}f(U)P(Z,U)\dot{U}.$$

因此，

$$|f(Z)|\leqslant\int_{\mathfrak{C}_I}|f(U)|P(Z,U)\dot{U},$$

于是当 $0<r<1$ 时,

$$\int_{\Re_{I_r}}|f(Z)|\dot{Z}\leqslant\int_{\Re_{I_r}}\dot{Z}\int_{\Re_I}|f(U)|P(Z,U)\dot{U}=\int_{\mathfrak{C}_I}|f(U)|\dot{U}\int_{\Re_{I_r}}P(Z,U)\dot{Z}.$$

这里积分次序可以交换是由于 $|f(U)|$ 在 $\mathfrak{C}_I$ 上可积[4], 而 $P(Z,U)$ 是 $Z\in\overline{\Re}_{I_r}$ 和 $U\in\mathfrak{C}_I$ 的连续函数. 利用(4.6)式,并由于 $f(Z)\in E_1(\Re_I)$,便有

$$\int_{\Re_{I_r}}|f(Z)|\dot{Z}\leqslant\frac{V(\Re_{I_r})}{V(\mathfrak{C}_I)}\int_{\mathfrak{C}_I}|f(U)|\dot{U}\leqslant V(\Re_{I_r}). \tag{4.10}$$

令 $r\to1$,可知当 $r=1$ 时,(4.4)式亦成立. 定理证毕.

至于积分表示的别的应用,将另文发表,例如参阅[2]、[3].

参 考 文 献

[1] 华罗庚, 1958, 多复变数函数论中的典型域的调和分析,科学出版社.

[2] 龚 昇、孙继广,多复变数的 Cauchy 型积分 (I)、(II)、(III), 中国科学, Vol. 14, No. 2,3,4, 1965.

[3] 龚 昇、孙继广,复超球面上的奇异积分方程, 中国科学, Vol. 14, No. 6, 1965.

[4] Bochner, S., 1960. Classes of holomorphic functions of several variables in circular domains, *Pro. Nat. Acad. Sci.*, **46**, 720—723.

[5] Какичев, В. А., 1959, Интеграл Шварца и Формулы Гильберта для аналитических функций многих кмплексных переменных, *Изв. выс. уч. зав. мат.*, 2(9), 80—93.

[6] Tornehave, H., 1962. *Math. Rev.*, 24.

[7] 戈鲁辛, Г. М. 1958, 复变函数的几何理论,科学出版社.

[8] 普里瓦洛夫, И. И. 1956, 解析函数的边界性质,科学出版社.

第2卷 第1期 中国科学技术大学学报 Vol. 2, No. 1
1966年2月 ZHONGGUO KEXUE JISHU DAXUE XUEBAO Feb., 1966

关于多复变数可递域的Poisson公式的一点注記

龔 昇 史济怀

提 要

设$\mathfrak{R}$是一可递域，$\mathfrak{L}$是它的闭包的一部分．设羣G把$\mathfrak{R}$和$\mathfrak{L}$都一一地变为自己．设o是$\mathfrak{R}$中一固定点，G中把$\mathfrak{R}$中的点a变为o点的变换记为

$$y = F(x; a, \cdots).$$

命

$$\Phi(a) = \frac{1}{V(\mathfrak{L})}\int_{\mathfrak{L}} \varphi(\xi) P(\xi; a, \cdots)\dot{\xi},$$

这里$P(\xi; a, \cdots)$是$\mathfrak{R}$的 Poisson 核．

华罗庚教授证明了：如果$\varphi(\xi)$是$\mathfrak{L}$上的连續函数，且

$$(*) \qquad \lim_{a\to\zeta} F^{-1}(\eta; a, \cdots) = \zeta, \qquad \eta\in\mathfrak{L}, \zeta\in\mathfrak{L}$$

成立，那末

$$\lim_{a\to\zeta}\Phi(a) = \varphi(\zeta), \qquad \zeta\in\mathfrak{L}.$$

本文证明了下面的结果：设$\mathfrak{L}$是李羣，如果条件$(*)$和

$$P(\xi, r\zeta, \cdots) = P(\xi\zeta^{-1}, re, \cdots), \qquad \zeta\in\mathfrak{L}, \xi\in\mathfrak{L}$$

成立，这儿e是$\mathfrak{L}$的单位元；那末对于$\varphi(\xi)\in L^p$，有

$$\lim_{r\to 1}\Phi(r\zeta) = \varphi(\zeta) \qquad (L^p).$$

$\mathfrak{R}$为一可递域，$\mathfrak{L}$为其闭包的一部分．设$\mathfrak{L}$是一可定向的微分流形．假定有群G把$\mathfrak{R}$及$\mathfrak{L}$都一一地映为自己．若o为$\mathfrak{R}$中的一定点，G中把$\mathfrak{R}$中的点a变为o的变换为

$$y = F(x; a, \cdots). \tag{1}$$

当然，变换的参数可以不止一个，即G中可以有很多变换把a变为o．变换(1)在$\mathfrak{L}$上有

$$\eta = F(\xi; a, \cdots), \tag{2}$$

这里$\xi, \eta\in\mathfrak{L}$．这时候$\mathfrak{L}$上的体积元素有关系

$$\dot{\eta} = P(\xi; a, \cdots)\dot{\xi}, \tag{3}$$

这里的$P(\xi; a, \cdots)$就称为 Poisson 核，从而得到 Poisson 积分

$$\Phi(a) = \frac{1}{V(\mathfrak{L})}\int_{\mathfrak{L}} \varphi(\xi) P(\xi; a, \cdots)\dot{\xi}, \tag{4}$$

其中$\varphi(\xi)$是$\mathfrak{L}$上的可积函数，$V(\mathfrak{L})$表$\mathfrak{L}$的体积．这就是华罗庚教授[1]关于可递域的 Poisson 积分的看法．由此出发，他证明了：如果$\varphi(\xi)$是$\mathfrak{L}$上的连續函数，且对任一$\zeta\in\mathfrak{L}$，

$$\lim_{a\to\zeta} F^{-1}(\eta; a, \cdots) = \zeta \tag{5}$$

对于 $\eta\in\mathfrak{L}$ 几乎处处成立，则有

$$\lim_{a\to\zeta}\Phi(a)=\varphi(\zeta).$$

有了这个结果，解多复变数典型域的调和方程的 Dirichlet 问题便成为它的简单的特例了[4]．

用经典的方法来研究 L^p 中函数的 Poisson 积分将产生本质性的困难，而华罗庚教授的上述看法却使这项研究有了可能．在本文中，我们给出了两条定理，尽管还没有完全彻底解决问题，但却已使不少最重要的可递域成为其特例．

本文中所讨论的可递域 $\mathfrak{R}$ 具有这样的性质：

$$\{z\,|\,z=r\xi+o,\ \xi\in\mathfrak{L},\ 0\leqslant r<1\}\subset\mathfrak{R},$$

不失普遍性，我们假设 $\mathfrak{R}$ 包有原点，且取 o 为原点．

我们还假定 $\mathfrak{L}$ 是紧致的，以后不再声明．

命

$$L_G^p=\{\varphi(\xi)\,|\,\varphi(\xi)\in L^p,\ \text{且对任一}\ g\in G,\ \varphi(g\xi)\in L^p\}.$$

于是我们有以下的

定理一　若 $\varphi(\xi)\in L_G^p(1\leqslant p<\infty)$，则

$$\lim_{r\to1}\Phi(r\xi)=\varphi(\xi),\ (L^p). \tag{6}$$

但是要求条件(5)及

$$P(\xi;\ r\zeta,\ \cdots)=P(\zeta;\ r\xi,\ \cdots) \tag{7}$$

对于几乎所有的 $\zeta\in\mathfrak{L}$, $\xi\in\mathfrak{L}$, $0\leqslant r<1$ 都成立．

证　由于连续函数类 C 在 L^p 中稠密，故对任一 $\varphi(\xi)\in L_G^p$ 及 $\varepsilon>0$，存在 $\psi(\xi)\in C$，使

$$\|\varphi-\psi\|_p<\varepsilon. \tag{8}$$

命

$$\Psi_r(\zeta)=\Psi(r\zeta)=\frac{1}{V(\mathfrak{L})}\int_{\mathfrak{L}}\psi(\xi)P(\xi;\ r\zeta,\ \cdots)\dot{\xi},$$

于是

$$\|\Phi_r-\varphi\|_p\leqslant\|\Phi_r-\Psi_r\|_p+\|\Psi_r-\psi\|_p+\|\psi-\varphi\|_p, \tag{9}$$

这里 $\Phi_r(\zeta)=\Phi(r\zeta)$．由[1]知，可取 $0<r_0<1$，使当 $r>r_0$ 时，

$$\|\Psi-\psi\|_p<\varepsilon$$

成立．又

$$\|\Phi_r-\Psi_r\|_p=\left(\int_{\mathfrak{L}}|\Phi(r\zeta)-\Psi(r\zeta)|^p\dot{\zeta}\right)^{\frac{1}{p}}=$$

$$=\left(\int_{\mathfrak{L}}\left|\frac{1}{V(\mathfrak{L})}\int_{\mathfrak{L}}(\varphi(\xi)-\psi(\xi))P(\xi;\ r\zeta,\ \cdots)\dot{\xi}\right|^p\dot{\zeta}\right)^{\frac{1}{p}}. \tag{10}$$

记(2)的逆变换为

$$\xi=F^{-1}(\eta;\ a,\ \cdots),$$

则由(3)知，(10)的右边即为

$$\left(\int_{\mathfrak{L}}\left|\frac{1}{V(\mathfrak{L})}\int_{\mathfrak{L}}(\varphi(F^{-1}(\eta; r\zeta, \cdots)) - \psi(F^{-1}(\eta; r\zeta, \cdots)))\dot{\eta}\right|^p\dot{\zeta}\right)^{\frac{1}{p}},$$

由于 $\varphi\in L_G^p$，故上式有意义. 应用 Hölder 不等式，上式不大于

$$\frac{1}{V(\mathfrak{L})^{1-\frac{1}{p}}}\left(\int_{\mathfrak{L}}\int_{\mathfrak{L}}|\varphi(F^{-1}(\eta; r\zeta, \cdots)) - \psi(F^{-1}(\eta; r\zeta, \cdots))|^p\dot{\eta}\dot{\zeta}\right)^{\frac{1}{p}}.$$

作变换(2)，此即

$$\frac{1}{V(\mathfrak{L})^{1-\frac{1}{p}}}\left(\int_{\mathfrak{L}}\dot{\zeta}\int_{\mathfrak{L}}|\varphi(\xi) - \psi(\xi)|^p P(\xi; r\zeta, \cdots)\dot{\xi}\right)^{\frac{1}{p}},$$

由(7)，上式等于

$$\begin{aligned}
&\frac{1}{V(\mathfrak{L})^{1-\frac{1}{p}}}\left(\int_{\mathfrak{L}}|\varphi(\xi) - \psi(\xi)|^p\dot{\xi}\int_{\mathfrak{L}}P(\xi; r\zeta, \cdots)\dot{\zeta}\right)^{\frac{1}{p}}\\
&= \frac{1}{V(\mathfrak{L})^{1-\frac{1}{p}}}\left(\int_{\mathfrak{L}}|\varphi(\xi) - \psi(\xi)|^p\dot{\xi}\int_{\mathfrak{L}}P(\zeta; r\xi, \cdots)\dot{\zeta}\right)^{\frac{1}{p}}\\
&= V(\mathfrak{L})^{\frac{2}{p}-1}\|\varphi - \psi\|_p < \varepsilon V(\mathfrak{L})^{\frac{2}{p}-1}.
\end{aligned}$$

由(8)，(9)，这就证明了定理.

例 1 $\mathfrak{R}$ 为 n 维单位超球 $xx' < 1$，$\mathfrak{L}$ 为超球面 $xx' = 1$，这儿 $x = (x_1, \cdots, x_n)$.

例 2 $\mathfrak{R}$ 为多复变数的四类典型域，$\mathfrak{L}$ 为其相应的特征流形.

例 3 $\mathfrak{R}$ 为多实变数的矩阵双曲空间，$\mathfrak{L}$ 为其特征流形.

例 4 $\mathfrak{R}$ 为实 n 阶对称方阵且满足条件

$$I - XX > 0$$

所成的域，$\mathfrak{L}$ 为实对称正交方阵的全体所组成的流形中的任一连通分支.

例 5 $\mathfrak{R}$ 为实 n 阶斜对称方阵且满足条件

$$I + XX > 0$$

所成的域，$\mathfrak{L}$ 为其特征流形.

上述各例皆满足定理一的条件(参阅[1]，[2]，[4]).

如果对 $\mathfrak{L}$ 略加条件，则(6)可以在 L^p 中成立. 我们有

定理二 若 $\mathfrak{L}$ 是紧致李群，则对任意 $\varphi(\xi)\in L^p$，(6)成立. 但要求条件(5)及

$$P(\xi; r\zeta, \cdots) = P(\xi\zeta^{-1}; re, \cdots) \tag{11}$$

对 $\xi\in\mathfrak{L}$，$\zeta\in\mathfrak{L}$，$0\leqslant r<1$ 几乎处处成立，这里 e 为 $\mathfrak{L}$ 的单位.

证 由(8)，(9)，我们只要证明

$$\lim_{r\to 1}\|\Phi_r - \Psi_r\|_p = 0$$

即可.

任取 $g(\zeta)\in L^q$，这里 L^q 是 L^p 的共轭空间. 考虑积分

$$\int_{\mathfrak{L}}(\Phi(r\zeta) - \Psi(r\zeta))g(\zeta)\dot{\zeta} = \int_{\mathfrak{L}}\left(\frac{1}{V(\mathfrak{L})}\int_{\mathfrak{L}}(\varphi(\xi) - \psi(\xi))P(\xi; r\zeta, \cdots)\dot{\xi}\right)g(\zeta)\dot{\zeta}$$

由条件(11)，上式右边等于

$$\frac{1}{V(\mathfrak{L})}\int_{\mathfrak{L}} g(\zeta)\dot{\zeta}\int_{\mathfrak{L}}(\varphi(\xi)-\psi(\xi))P(\xi\zeta^{-1};re,\cdots)\dot{\xi}.$$

命 $\xi\zeta^{-1}=\eta$，则上式就是

$$\frac{1}{V(\mathfrak{L})}\int_{\mathfrak{L}} P(\eta;re,\cdots)\dot{\eta}\int_{\mathfrak{L}}(\varphi(\eta\zeta)-\psi(\eta\zeta)g(\zeta)\dot{\zeta}.$$

由 Hölder 不等式，上式的绝对值不大于

$$\frac{1}{V(\mathfrak{L})}\int_{\mathfrak{L}}\|\varphi_\eta-\psi_\eta\|_p\|g\|_q P(\eta;re,\cdots)\dot{\eta},$$

这里 $\varphi_\eta=\varphi(\eta\zeta)$，$\psi_\eta=\psi(\eta\zeta)$. 由 Hahn-Banach 定理，我们可以选取 $g(\zeta)\in L^q$，使 $\|g\|_q=1$，且

$$\int_{\mathfrak{L}}(\Phi(r\zeta)-\Psi(r\zeta))g(\zeta)\dot{\zeta}=\|\Phi_r-\Psi_r\|_p,$$

于是得到

$$\|\Phi_r-\Psi_r\|_p\leqslant\frac{1}{V(\mathfrak{L})}\int_{\mathfrak{L}}\|\varphi_\eta-\psi_\eta\|_p P(\eta;re,\cdots)\dot{\eta},$$

但

$$\|\varphi_\eta-\psi_\eta\|_p=\|\varphi-\psi\|_p.$$

由(8)，(9)立得定理.

例 1　$\mathfrak{R}$ 为多圆柱：$\{|z_1|<1,\cdots,|z_n|<1\}$，$\mathfrak{L}$ 为 $\{|z_1|=1,\cdots,|z_n|=1\}$.

例 2　$\mathfrak{R}$ 为由 n 阶复方阵且满足

$$I-Z\bar{Z}'>0$$

的 $Z^{(n)}$ 所组成的域，$\mathfrak{L}$ 为 n 阶酉群.

例 3　$\mathfrak{R}$ 为由 n 阶实方阵且满足

$$I-XX'>0$$

的 $X^{(n)}$ 所组成的域，$\mathfrak{L}$ 为行列式等于 $+1$ 的 n 阶正交群.

以上各例都满足定理二的条件.

从定理二的例 1，例 2 立得以下的

推论 1　如果 $f(\theta_1,\cdots,\theta_n)\in L^p$，则其 Fourier 级数

$$\Sigma a_{\nu_1\cdots\nu_n}e^{i(\nu_1\theta_1+\cdots+\nu_n\theta_n)}$$

在 L^p 中可以 Abel 求和于自己[6].

推论 2　如果 $u(U)$ 是定义在 n 阶酉群 U_n 上的 p 方可积函数，则其 Fourier 级数在 L^p 中可以 Abel 求和于自己[3],[5].

感谢我们的导师华罗庚教授，本文是在他的指导下完成的；殷涌泉，林群及孙继广等同志对本文提供了不少宝贵意见，一并致谢.

林群同志还指出：本文的结果可以推广到 Orlicz 空间.

参　考　文　献

[1] 华罗庚，关于可递域的 Poisson 公式(将发表).

[2] 华罗庚,多复变数函数论中的典型域的调和分析, 1958, 科学出版社.
[3] 华罗庚,紧致羣上的连续函数所成的空间中的一条收敛定理,科学记录, 2 (1958), 341—344.
[4] 华罗庚与陆启铿, Theory of harmonic functions in classical domains, *Scientia Sinica*, 8 (1959), 1031—1094.
[5] 龚 昇,酉羣上的 Fourier 分析 (I), 数学学报 **10** (1960), 239—261.
[6] Zygmund, Trigonometric series (1959).

III. 多复变数的 Bloch 常数

与单复变数不一样，在多复变数的时候，一般来说：Bloch 常数是不存在的．必须要对全纯映照加以适当的限制后，Bloch 常数才会存在．最早看出这一点的也许是 S. Bochner[1]．之后伍鸿熙、陈省身等一大批数学家对此作了很多重要的贡献．[32.1], [28.1], [25], [21] 对有界对称域上的 Bloch 映照族的 Bloch 常数进行了估计，用的方法是先用李群李代数为工具推广 Bonk 定理 [2], 回到单复变数，这依然是著名的 Ahlfors 定理 [3] 等.

这些工作总结在 (4) ([M6]) 中的第一章中.

参　考　文　献

(1) Bochner S. Bloch theorem for real variables, Bull. Amer. Math. Soc., 1946, 52: 715-719.

(2) Bonk M. On Bloch constant, Proc. Amer. Math. Soc., 1990, 110: 889-894.

(3) Ahlfors L V. An extension of Schwarz lemma, Tran. Amer. Math. Soc., 1938, 43: 359-364.

(4) 龚昇, 余其煌，郑学安．布洛赫常数与许瓦尔兹导数．上海：上海科学技术出版社，1998.

The Journal of Geometric Analysis
Volume 4, Number 1, 1994

The Bloch Theorem in Several Complex Variables

By *Carl H. FitzGerald and Sheng Gong*

ABSTRACT. The background theory for the Bloch theorem is generalized to several complex variables. This work involves study of the Bergman kernel functions in order to extend work of Landau and Bonk. The main conclusion is an estimate for Bloch's constant for mappings of domains of the first classical type. In the special case of the n-dimensional ball, the estimate of Bloch's constant coincides with that of Liu.

1. Introduction

Recently Bonk [3] made a numerically small improvement in Ahlfors' fifty-year-old lower bound [1] on Bloch's constant. Bonk's method has inspired further work in one and several variables (Minda [15], Liu & Minda [14], Liu [13]). In particular, Liu extended Bonk's theory to the ball in $\mathbb{C}^n$ and obtained a lower bound on Bloch's constant for a class of mappings defined on the ball in $\mathbb{C}^n$.

Here we extend Bonk's theory to the first classical domain $\mathcal{R}_I$: the set of all matrices in $\mathbb{C}^{mn}$

$$Z = \begin{pmatrix} z_{11} & \cdots & z_{1n} \\ z_{21} & \cdots & z_{2n} \\ \vdots & & \\ z_{m1} & \cdots & z_{mn} \end{pmatrix}$$

such that $I^{(m)} - Z\bar{Z}' > 0$, where $I^{(m)}$ is the identity $m \times m$ matrix and the inequality means that the matrix is positive definite. Then bounds on the Bloch constant are derived for certain families of mappings. This work generalizes most of Liu's paper [13] and uses new methods. Similar results for the other classical domains can be obtained.

Math Subject Classification 32H99.
Key Words and Phrases Block constant, Bergman kernel function, and biholomorphic mappings.
We gratefully acknowledge helpful communications with Professors L. A. Harris, X. Y. Liu, C. D. Minda, T. J. Suffridge, and N. Wallach.

ISSN 1050-6926

The presentation in this paper involves many standard theorems from the theory of several complex variables. Such theorems are stated and reference given the classical book of Hua [10]. A theorem of Minda [15] is also used. Of course, various papers have been helpful for motivation and guidance; in particular, Liu [13] was very useful.

For general background on Bloch's constant for one variable the reader may consult Minda [15]. For several variables, examples show that Bloch's theorem does not hold as widely. In particular, examples have been given by Wu [17] and Harris [9]. Nevertheless, since Solomón Bôchner [2] initiated the study of Bloch's constant in higher dimensions, there have been several other important contributions including H. Wu [17], S. S. Chern [4], L. Harris [9], A. E. Greene and H. Wu [5], K. T. Hahn [6, 7, 8], S. Krantz and D. Ma [11], and R. M. Timoney [16]. In order that there be a positive Bloch constant, it is necessary to restrict the class of mappings considered.

For any rectangular matrix A we let $\|A\|_M$ denote the usual matrix norm of A, that is, $\|A\|_M = \sup|Au|$ where u is a unit vector.

If $F(z) = (F_1(z), F_2(z), \ldots, F_n(z))$ is a holomorphic mapping of a domain in $\mathbb{C}^n$ into $\mathbb{C}^n$, then let $F' = \partial F/\partial z$ denote the Jacobian of F, that is, the matrix $(\partial F_i/\partial z_j)^n_{i,j=1}$. If G is a holomorphic mapping of a domain in $\mathbb{C}^n$ that is contained in the range of F, then $H(z) = G(F(z)) = G \circ F(z)$ is defined and

$$\frac{\partial H}{\partial z} = \frac{\partial G}{\partial w}\frac{\partial F}{\partial z} \quad \text{where} \quad w = F(z).$$

We let Ω denote a bounded homogeneous domain in $\mathbb{C}^n$ that contains the origin. The expression $\mathrm{Aut}(\Omega)$ means the group of holomorphic automorphisms of Ω. And the phrase "a Bloch mapping of Ω" means a holomorphic mapping F of Ω such that the family

$$\mathcal{F}_F = \{g : g(z) = F(\varphi(z)) - F(\varphi(0)) \quad \text{for} \quad \varphi \in \mathrm{Aut}(\Omega)\}$$

is a normal family.

The Bloch norm of a mapping is then defined as follows:

$$\|F\|_B = \sup\left\{\left\|\frac{\partial F \circ \varphi}{\partial z}(0)\right\|_M : \varphi \in \mathrm{Aut}(\Omega)\right\}.$$

A Bloch function of bounded homogeneous domain Ω in $\mathbb{C}^n$ means a holomorphic function f of Ω such that the family

$$\mathcal{F}_f = \{h : h(z) = f(\varphi(z)) - f(\varphi(0)) \quad \text{for} \quad \varphi \in \mathrm{Aut}(\Omega)\}$$

is a normal family.

The Bloch norm of a function is then defined as follows:

$$\|f\|_B = \sup\left\{\left\|\frac{\partial f\circ\varphi}{\partial z}(0)\right\| : \varphi\in \operatorname{Aut}(\Omega)\right\}$$

where $\frac{\partial f}{\partial z} = \left(\frac{\partial f}{\partial z_1},\ldots,\frac{\partial f}{\partial z_n}\right)$ and $\|\ \|$ is the length of the vector.

Timoney [16] proved that a function f of Ω is a Bloch function if and only if $\|f\|_B$ is finite. We observe the following two facts: (1) $F=(f_1,\ldots,f_n)$ is a Bloch mapping of Ω if and only if each f_j, $j=1,2,\ldots,n$, is a Bloch function of Ω; (2) A is a matrix, then $\|A\|_M^2\le \operatorname{tr}(A\bar{A}')\le n\|A\|_M^2$. Thus we have proved the following.

Theorem 1.1. *A mapping F of Ω is a Bloch mapping if and only if $\|F\|_B$ is finite.*

Another important notion involves the Bergman kernel function for the region Ω, which is denoted by $K_\Omega(z,\bar\zeta)$. We introduce a functional on the holomorphic mappings $F:\Omega\to\mathbb{C}^n$:

$$d(F)=\sup_{z\in\Omega}\frac{|\det F'(z)|}{\sqrt{K_\Omega(z,\bar z)/K_\Omega(0,0)}}. \tag{1.1}$$

In the case of one complex variable, and Ω the unit disc, the functional $d(F)$ is the usual Bloch seminorm. Note that, in general, for holomorphic mappings F and G of Ω into $\mathbb{C}^n$, the quantity $d(F+G)$ may fail to be less than or equal to $d(F)+d(G)$. Hence, this functional is not a seminorm.

However, this functional has an important invariance property: if $\varphi\in\operatorname{Aut}(\Omega)$, then $d(F\circ\varphi)=d(F)$ holds for any $\varphi\in\operatorname{Aut}(\Omega)$. This can be easily proven as follows. Since

$$\left|\det\frac{\partial\varphi}{\partial z}\right|^2=\frac{K_\Omega(z,\bar z)}{K_\Omega(\varphi(z),\overline{\varphi(z)})}, \tag{1.2}$$

we have

$$d(F)=\sup\{|\det(F\circ\varphi)'(0)| : \varphi\in\operatorname{Aut}(\Omega)\}.$$

Then $d(F\circ\varphi)=d(F)$ follows immediately. Moreover, we have

$$(d(F))^{\frac{1}{n}}\le\|F\|_B.$$

Let Ω denote a bounded homogeneous domain in $\mathbb{C}^n$ that contains the origin. Let $H(\Omega)$ denote the family of holomorphic mappings on Ω that map Ω into $\mathbb{C}^n$, and let $\mathcal{F}$ be a subfamily of $H(\Omega)$.

For any $F \in \mathcal{F}$, we use $r(a, F)$ to denote the largest nonnegative number r such that there is a simply connected domain R with $a \in R \subset \Omega$ which is mapped biholomorphically onto $B(F(a), r)$ by F, where $B(F(a), r)$ denotes a ball centered at $F(a)$ with radius r. Moreover, we let $r(F) = \sup\{r(a, F) : a \in \Omega\}$, and we define the Bloch constant of the family $\mathcal{F}$ on Ω by $B_{\mathcal{F}} = \inf\{r(F) : F \in \mathcal{F}, \det \frac{\partial F}{\partial z}(0) = 1\}$.

The families of holomorphic mappings on Ω that we will study in this paper are

$$\begin{aligned} \mathcal{B} &= \left\{F \in H(\Omega) : \det \frac{\partial F}{\partial z}(0) = 1, \|F\|_B < \infty\right\}; \\ \mathcal{B}_K &= \{F \in \mathcal{B} : \|F\|_B \le K\},\ 1 \le K < \infty\}; \\ \mathcal{B}_{K,1} &= \{F \in \mathcal{B}_K : d(F) = 1\}. \end{aligned}$$

We let B_K denote the Bloch constant of $\mathcal{B}_K$.

Lemma 1.2. *Suppose Ω is a bounded homogeneous domain in $\mathbb{C}^n$. Given F in $\mathcal{B}_K$ there exists G in $\mathcal{B}_{K,1}$ with $r(F) \ge r(G)$.*

Proof. If $d(F) = 1$, there is nothing to prove. Otherwise, select φ_m in Aut (Ω) with

$$1 < d_m = |\det (F \circ \varphi_m)'(0)|$$

tending to $d(F)$. Set

$$G_m(z) = \frac{F \circ \varphi_m(z) - F \circ \varphi_m(0)}{d_m^{1/n}}.$$

Then $G_m(0) = 0$ and $\det G'_m(0) = 1$. Also, $\|G_m\|_B \le \|F\|_B \le K$ and $r(G_m) \le r(F)$ for all m since $1 < d_m$. Because F is a Bloch mapping, $\{G_m\}$ is normal. Assume G_m converges normally to G—at least some subsequence does. Then $G(0) = 0$, $\det G'(0) = 1$, $\|G\|_B \le K$, and $r(G) \le r(F)$.

The following theorem is a generaliztion of a theorem of Landau [12].

Theorem 1.3. *Suppose Ω is a bounded homogeneous domain in $\mathbb{C}^n$. There exists F in $\mathcal{B}_{K,1}$ with $r(F) = B_k$. Then $B_K > 0$, if $\mathcal{B}_K$ is nonempty.*

Proof. Take F_m in $\mathcal{B}_K$ with $r(F_m) \le B_k + 1/m$. Then there exists G_m in $\mathcal{B}_{K,1}$ with $r(G_m) \le r(F_m) \le B_k + 1/m$; hence $r(G_m)$ tends to B_k. Then $\{G_m - G_m(0)\}$ is normal, and we can assume that it converges normally to F. Then $r(F) = B_k$.

2. Some results on bounded homogeneous domains

Recall two definitions: a domain Ω in $\mathbb{C}^n$ is *circular* if Ω is a domain and if z is in Ω, then $e^{i\varphi}z$ is in Ω for all real numbers φ. Further, a domain Ω is a *complete circular* domain if Ω is a domain in $\mathbb{C}^n$ such that if z is in Ω, $re^{i\varphi}z$ is a point in Ω for all real numbers φ and real numbers $0 \le r \le 1$.

Let $\mathbb{D} = \{\xi : \xi \in \mathbb{C} \text{ and } |\zeta| < 1\}$ and let a' denote the transpose of the vector a.

Theorem 2.1. *Suppose that Ω is a bounded complete circular domain and that a belongs to the closure of Ω. If g is a holomorphic function in $\Omega \cup \{a\}$ that maps $\Omega \cup \{a\}$ into $\mathbb{C}$ and $|g(w)| \le |g(a)|$ for all $w = za$ for $z \in \mathbb{D}$, then $c = \frac{dg}{dw}(a) \cdot a' / g(a)$ is positive and*

$$\operatorname{Re}\left\{\frac{g(ta)}{g(a)}\right\} \ge \frac{(c+1)t-(c-1)}{(c+1)-(c-1)t} \tag{2.1}$$

for $t \in (-1, 1)$.

Proof. Let $h(z) = g(za)/g(a)$ for $z \in \mathbb{D}$. Then $h(z)$ is holomorphic in $\mathbb{D} \cup \{1\}$, $h(\mathbb{D}) \subset \mathbb{D}$, $h(1) = 1$, and $c = \left.\frac{dh(z)}{dz}\right|_{z=1} = h'(1)$. Thus Minda's theorem [15] with h replacing g and $s = a = 1$ gives that c is positive, and the desired inequality holds.

Minda's proof involves a geometric understanding of Julia's lemma. He also discusses the case of equality.

In the several variable situation, equality holds for a wide class of functions.

Theorem 2.2. *Suppose that Ω is a bounded, complete circular, homogeneous domain. Consider $F \in H(\Omega)$ such that $d(F) = 1$ and $\det F'(0) = 1$. Let $\varphi_a(z)$ be a holomorphic automorphism that takes a to the origin. Let $G(z) = F(\varphi_a(z))$ and $g = \det \frac{\partial G}{\partial z}$. Then, for every a in Ω,*

$$c = \left(\frac{dg(a)}{da}\right) \cdot a' / g(a) = \frac{\partial}{\partial a} K_\Omega(a, \bar{a}) \cdot a' / K_\Omega(a, \bar{a})$$

where $\frac{dg(a)}{da}$ means $\left(\frac{\partial g}{\partial z_1}, \frac{\partial g}{\partial z_2}, \ldots, \frac{\partial g}{\partial z_n}\right)$ evaluated at a.

Proof. Fix $a \in \Omega$. We now show that

$$K_\Omega(z, \bar{z}) \le K_\Omega(a, \bar{a}) \tag{2.2}$$

for all $z = \xi a$ for some $\xi \in \mathbb{D}$. Since Ω is a bounded, complete circular domain, there exists (Hua [10]) an expansion of the Bergman kernel function as follows:

$$K_\Omega(z, \bar{\xi}) = \sum_{\nu=0}^{\infty} \varphi_\nu(z)\overline{\varphi_\nu(\zeta)} \tag{2.3}$$

where each $\varphi_\nu(z)$ is a homogeneous polynomial of degree ν for $\nu = 0, 1, 2, \ldots$. Evaluation of equation (2.3) at $z = \xi a$ gives

$$\begin{aligned} K_\Omega(z, \bar{z}) &= K_\Omega(\xi a, \overline{\xi a}) \\ &= \sum_{\nu=0}^{\infty} |\xi|^{2\nu}|\varphi_\nu(a)|^2 \\ &\le \sum_{\nu=0}^{\infty} |\varphi_\nu(a)|^2 \\ &= K_\Omega(a, \bar{a}) \end{aligned}$$

as was to be shown.

Let $\varphi_a(z)$ be a biholomorphic automorphism that takes a to the origin. Consider $G(z) = F(\varphi_a(z))$. Let $g(z) = \det(\partial G/\partial z)$. Since $d(F) = 1$ and since $d(\ )$ is invariant, $d(F) = d(G) = 1$. We have $|\det \frac{\partial G}{\partial z}(z)| \le \sqrt{\frac{K_\Omega(z,\bar{z})}{K_\Omega(0,0)}}$ for $z \in \Omega$. From the hypothesis of Theorem 2.2,

$$\left.\frac{\left|\det \frac{\partial G}{\partial z}(z)\right|}{\sqrt{K_\Omega(z, \bar{z}) / K_\Omega(0, 0)}}\right|_{z=a} = \left.\frac{\left|\det \frac{\partial F}{\partial w}(w)\right|}{\sqrt{K_\Omega(w, \bar{w}) / K_\Omega(0, 0)}}\right|_{w=0} \frac{\left|\det \frac{\partial F}{\partial w}(0)\right|}{1} = 1. \tag{2.4}$$

Thus

$$\begin{aligned} \left|\det \frac{\partial G}{\partial z}(a)\right| &= \sqrt{K_\Omega(a, \bar{a}) / K_\Omega(0, 0)} \ge \sqrt{K_\Omega(z, \bar{z}) / K_\Omega(0, 0)} \\ &\ge \left|\det \frac{\partial G}{\partial z}(z)\right| \end{aligned}$$

for $z = \xi a$ for $\xi \in \mathbb{D}$. Since $g(z) = \det \frac{\partial G}{\partial z}(z)$ satisfies the conditions of g in Theorem 2.1, we have inequality (2.1), where

$$c = \left[\frac{d}{da} \det \frac{\partial G}{\partial z}(a)\right] \cdot a' \Big/ \det \frac{\partial G}{\partial z}(a) \tag{2.5}$$

but we can also find c in terms of the Bergman kernel function. Note that

$$\det \left.\frac{\partial G(z)}{\partial z}\right|_{z=a} = \det \left.\frac{\partial F}{\partial w}\right|_{w=0} \det \left.\frac{\partial \varphi(z)}{\partial z}\right|_{z=a} \tag{2.6}$$

and

$$\left(\frac{d}{dz}\det\frac{\partial G(z)}{\partial z}\right)\bigg|_{z=a} = \left[\left(\frac{d}{dz}\det\frac{\partial F}{\partial w}\right)\det\frac{\partial\varphi}{\partial z}+\left(\det\frac{\partial F}{\partial w}\right)\left(\frac{d}{dz}\det\frac{\partial\varphi}{\partial z}\right)\right]\bigg|_{z=a}$$

$$= \left(\frac{d}{dw}\det\frac{\partial F}{\partial w}\right)\bigg|_{w=0}\frac{\partial\varphi}{\partial z}\bigg|_{z=a}\cdot\det\frac{\partial\varphi}{\partial z}\bigg|_{z=a}$$

$$+\left(\det\frac{\partial F}{\partial w}\right)\bigg|_{w=0}\left(\frac{d}{dz}\det\frac{\partial\varphi}{\partial z}\right)\bigg|_{z=a}.$$

By the assumptions that $\det\frac{\partial F}{\partial w}(0)=1$ and $d(F)=1$,

$$\left|\det\frac{\partial F}{\partial w}\right| = \left|1+\frac{\partial}{\partial w}\left(\det\frac{\partial F}{\partial w}(0)\right)\cdot w'+\cdots\right|$$

$$\le \sqrt{K_\Omega(w,\bar{w})\,/\,K_\Omega(0,0)}\,.$$

Consider the representation of the Bergman kernel in terms of homogeneous polynomials as given in equation (2.3). Clearly $K_\Omega(w,\bar{w})=K_\Omega(0,0)+O(|w|^2)$. We obtain

$$\left|1+\frac{\partial}{\partial w}\left(\det\frac{\partial F}{\partial w}(0)\right)\cdot w'\cdots\right|\le 1+O(|w|^2)$$

as $|w|\to 0$. Thus $\frac{d}{dw}\left(\det\frac{\partial F}{\partial w}\right)\bigg|_{w=0}=0$. Hence

$$\left(\frac{d}{dz}\det\frac{\partial G(z)}{\partial z}\right)\bigg|_{z=a}=\left(\det\frac{\partial F}{\partial w}\right)\bigg|_{w=0}\left(\frac{d}{dz}\det\frac{\partial\varphi}{\partial z}\right)\bigg|_{z=a}. \tag{2.7}$$

From equation (1.2),

$$\left|\det\frac{\partial\varphi}{\partial z}\right|^2=K_\Omega(z,\bar{z})\,/\,K_\Omega(\varphi(z),\overline{\varphi(z)})\,.$$

Taking $\frac{d}{dz}$ we obtain

$$\left(\frac{d}{dz}\det\frac{\partial\varphi}{\partial z}\right)\cdot\overline{\det\frac{\partial\varphi}{\partial z}} = \frac{\frac{\partial}{\partial z}K_\Omega(z,\bar{z})}{K_\Omega(\varphi(z),\overline{\varphi(z)})}-\frac{K_\Omega(z,\bar{z})}{[K_\Omega(\varphi(z),\overline{\varphi(z)})]^2}\frac{\partial}{\partial z}K_\Omega(\varphi(z),\overline{\varphi(z)})$$

$$= \frac{\frac{\partial}{\partial z}K_\Omega(z,\bar{z})}{K_\Omega(\varphi(z),\overline{\varphi(z)})}-\frac{K_\Omega(z,\bar{z})}{[K_\Omega(\varphi(z),\overline{\varphi(z)})]^2}\frac{\partial}{\partial w}K_\Omega(w,\bar{w})\frac{\partial w}{\partial z}.$$

At $z = a$, $w = 0$. It is easy to see that $\left[\frac{\partial}{\partial w} K_\Omega(w, \bar{w})\right]_{w=0} = 0$. We have

$$\left[\left(\frac{d}{dz}\det\frac{\partial\varphi}{\partial z}\right)\cdot\overline{\det\frac{\partial\varphi}{\partial z}}\right]\Bigg|_{z=a} = \frac{\partial}{\partial z}K_\Omega(z,\bar{z})\Bigg|_{z=a} \Bigg/ K_\Omega(0,0)\,. \tag{2.8}$$

Equations (2.5), (2.6), and (2.7) give that

$$\begin{aligned} c &= \left(\frac{d}{dz}\det\frac{\partial G}{\partial z}\right)\Bigg|_{z=a}\cdot a' \Bigg/ \det\frac{\partial G}{\partial z}\Bigg|_{z=a} \\ &= \left(\frac{d}{dz}\det\frac{\partial\varphi}{\partial z}\right)\Bigg|_{z=a}\cdot a' \Bigg/ \left(\det\frac{\partial\varphi}{\partial z}\right)\Bigg|_{z=a} \\ &= \left(\frac{d}{dz}\det\frac{\partial\varphi}{\partial z}\right)\Bigg|_{z=a}\overline{\left(\det\frac{\partial\varphi}{\partial z}\right)}\Bigg|_{z=a}\cdot a' \Bigg/ \left|\det\frac{\partial\varphi}{\partial z}\right|^2_{z=a}. \end{aligned}$$

By equations (1.2) and (2.8), we further find that

$$\begin{aligned} c &= \frac{\frac{\partial}{\partial z}K_\Omega(z,\bar{z})\big|_{z=a}\cdot a'}{K_\Omega(0,0)} \Bigg/ \frac{K_\Omega(a,\bar{a})}{K_\Omega(0,0)} \\ &= \frac{\frac{\partial}{\partial a}K_\Omega(a,\bar{a})\cdot a'}{K_\Omega(a,\bar{a})} \end{aligned}$$

which proves Theorem 2.2.

Moreover, substituting this value of c and $g = \det(\partial G/\partial z)$ into inequality (2.1), we have the following corollary.

Corollary 2.3. *With the same assumptions as in Theorem 2.2,*

$$\operatorname{Re}\left\{\frac{\det\frac{\partial G}{\partial z}(ta)}{\det\frac{\partial G}{\partial z}(a)}\right\}$$

$$\geq \frac{\left[\frac{\partial}{\partial a}K_\Omega(a,\bar{a})\cdot a' + K_\Omega(a,\bar{a})\right]t - \left[\frac{\partial}{\partial a}K_\Omega(a,\bar{a})\cdot a' - K_\Omega(a,\bar{a})\right]}{\left[\frac{\partial}{\partial a}K_\Omega(a,\bar{a})\cdot a' + K_\Omega(a,\bar{a})\right] - \left[\frac{\partial}{\partial a}K_\Omega(a,\bar{a})\cdot a' - K_\Omega(a,\bar{a})\right]t}\,. \tag{2.9}$$

Using $\frac{\partial G}{\partial z}=\frac{\partial F}{\partial w}\frac{\partial w}{\partial z}$ and equation (2.6), we can rewrite equation (2.9) as follows:

$$\mathrm{Re}\left\{\frac{\det\frac{\partial F(w)}{\partial w}\Big|_{w=\varphi(ra)}\det\frac{\partial\varphi(z)}{\partial z}\Big|_{z=ra}}{\det\frac{\partial\varphi(z)}{\partial z}\Big|_{z=a}}\right\}$$

$$\geq\frac{\left[\frac{\partial}{\partial a}K_\Omega(a,\bar{a})\cdot a'+K_\Omega(a,\bar{a})\right]t-\left[\frac{\partial}{\partial a}K_\Omega(a,\bar{a})\cdot a'-K_\Omega(a,\bar{a})\right]}{\left[\frac{\partial}{\partial a}K_\Omega(a,\bar{a})\cdot a'+K_\Omega(a,\bar{a})\right]-\left[\frac{\partial}{\partial a}K_\Omega(a,\bar{a})\cdot a'-K_\Omega(a,\bar{a})\right]t}. \quad (2.10)$$

3. Bonk's Theorem for $\mathcal{R}_I$

Consider the first classical domain $\mathcal{R}_I$. Fix $a=Z_0\in\mathcal{R}_I$. Let $\varphi_a(Z)$ be a holomorphic automorphism of $\mathcal{R}_I$ that takes Z_0 to the origin. We represent $\varphi_a(Z)$ as follows (Hua [10] p. 83):

$$\varphi_a(Z)=A(Z-Z_0)(I-\bar{Z}_0'Z)^{-1}D^{-1}=W \quad (3.1)$$

where

$$\bar{A}'A=(I-Z_0\bar{Z}_0')^{-1} \quad \text{and} \quad \bar{D}'D=(I-\bar{Z}_0'Z_0)^{-1} \quad (3.2)$$

and A is an $m\times m$ matrix and D is an $n\times n$ matrix. Then the differential of the matrix variable W is given by

$$dW=AdZ(I-\bar{Z}_0'Z)^{-1}D^{-1}+A(Z-Z_0)(I-\bar{Z}_0'Z)^{-1}\bar{Z}_0'dZ(I-\bar{Z}_0'Z)^{-1}D^{-1}. \quad (3.3)$$

When $Z=Z_0$, equations (3.2) and (3.3) imply that

$$dW=AdZ\bar{D}' \quad (3.3')$$

and when $Z=Z_0$,

$$\left(\frac{\partial W}{\partial Z}\right)_{Z=Z_0}=A'\otimes\bar{D}',$$

that is, the Jacobian is the Kronecker product of the transpose of A and the conjugate transpose of D. Since the determinant of a matrix is the same as the determinant of the transpose,

$$\det\left(\frac{\partial W}{\partial Z}\right)_{Z=Z_0}=(\det A)^n(\det\bar{D})^m. \quad (3.4)$$

For a more general formula similar to equation (3.3′) we start by simplifying equation (3.3).

$$\begin{aligned} dW &= A[I + (Z - Z_0)(I - \bar{Z}_0'Z)^{-1}\bar{Z}_0']dZ(I - \bar{Z}_0'Z)^{-1}D^{-1} \\ &= A[I + (Z - Z_0)\bar{Z}_0'(I - Z\bar{Z}_0')^{-1}]dZ(I - \bar{Z}_0'Z)^{-1}D^{-1} \\ &= A[(I - Z\bar{Z}_0' + (Z - Z_0)\bar{Z}_0'][I - Z\bar{Z}_0']^{-1}dZ[I - \bar{Z}_0'Z]^{-1}D^{-1} \\ &= A[I - Z_0\bar{Z}_0'][I - Z\bar{Z}_0']^{-1}dZ[I - \bar{Z}_0'Z]^{-1}D^{-1}. \end{aligned}$$

from the inverse of the first equation (3.2), we have

$$\begin{aligned} dW &= AA^{-1}\bar{A}'^{-1}[I - Z\bar{Z}_0']^{-1}dZ[I - \bar{Z}_0'Z]^{-1}D^{-1} \\ &= \bar{A}'^{-1}[I - Z\bar{Z}_0']^{-1}dZ[I - \bar{Z}_0'Z]^{-1}D^{-1}. \end{aligned} \tag{3.5}$$

At $Z = tZ_0$, we obtain a general formula similar to equation (3.3′):

$$dW = \bar{A}'^{-1}(I - tZ_0\bar{Z}_0')^{-1}dZ(I - t\bar{Z}_0'Z_0)^{-1}D^{-1}.$$

Thus

$$\frac{\partial W}{\partial Z}(tZ_0) = [\bar{A}'^{-1}(I - tZ_0\bar{Z}_0')^{-1}]' \otimes [(I - t\bar{Z}_0'Z_0)^{-1}D^{-1}]$$

and

$$\begin{aligned} \det\frac{\partial W}{\partial Z}(tZ_0) &= [\det(I - tZ_0\bar{Z}_0')]^{-n}[\det\bar{A}]^{-n}[\det(I - t\bar{Z}_0'Z_0)]^{-m}[\det D]^{-m} \\ &= \det[I - tZ_0\bar{Z}_0']^{-n-m}[\det\bar{A}]^{-n}[\det D]^{-m}. \end{aligned}$$

From this equation and (3.4), and then (3.2), we have

$$\begin{aligned} \frac{\det\frac{\partial W}{\partial Z}(tZ_0)}{\det\frac{\partial W}{\partial Z}(Z_0)} &= \frac{[\det(I - tZ_0\bar{Z}_0')]^{-m-n}[\det\bar{A}]^{-n}[\det D]^{-m}}{[\det A']^{n}[\det\bar{D}']^{m}} \\ &= \frac{[\det(I - tZ_0\bar{Z}_0')]^{-m-n}}{[\det\bar{A}'A]^{n}[\det\bar{D}'D]^{m}} = \frac{[\det(I - tZ_0\bar{Z}_0)]^{-m-n}}{[\det(I - Z_0\bar{Z}_0')]^{-m-n}} \\ &= \frac{[\det(I - Z_0\bar{Z}_0')]^{m+n}}{[\det(I - tZ_0\bar{Z}_0')]^{m+n}}. \end{aligned}$$

Since $\mathcal{R}_I$ is a domain that satisfies the conditions on the domain in Theorems 2.1 and 2.2, inequality (2.10) becomes

$$\operatorname{Re}\left\{\det \frac{\partial F}{\partial W}(\varphi(tZ_0))\right\} \tag{3.6}$$

$$\geq \frac{\det(I - tZ_0\bar{Z}_0')^{m+n}}{\det(I - Z_0\bar{Z}_0')^{m+n}} \cdot \frac{\left[\frac{\partial}{\partial Z}K_{R_I}(Z,\bar{Z})Z' + K_{R_I}(Z,\bar{Z})\right]t - \left[\frac{\partial}{\partial Z}K_{R_I}(Z,\bar{Z})Z' - K_{R_I}(Z,\bar{Z})\right]}{\left[\frac{\partial}{\partial Z}K_{R_I}(Z,\bar{Z})Z' + K_{R_I}(Z,\bar{Z})\right] - \left[\frac{\partial}{\partial Z}K_{R_I}(Z,\bar{Z})Z' - K_{R_I}(Z,\bar{Z})\right]t}$$

where $Z = (z_{11}, z_{12}, \ldots, z_{1n}, z_{21}, z_{22}, \ldots, z_{2n}, \ldots, z_{m1}, z_{m2}, \ldots, z_{mn})$ and the evaluation of the second factor of the right-hand side of the above inequality is at $Z = Z_0$.

Now we examine

$$\frac{\partial}{\partial Z}K_{R_I}(Z,\bar{Z})Z'\Big|_{Z=Z_0} \quad \text{for} \quad \Omega = \mathcal{R}_I .$$

We know [Hua [10] p. 84] that

$$K_{\mathcal{R}_I}(Z,\bar{W}) = V(\mathcal{R}_I)^{-1}\{\det(I - Z\bar{W}')\}^{-(m+n)}$$

where $V(\mathcal{R}_I)$ is the Euclidean volume of $\mathcal{R}_I$.

$$\frac{\partial}{\partial z_{ij}}K_{\mathcal{R}_I}(Z,\bar{W}) = -(m+n)V(\mathcal{R}_I)^{-1}\{\det(I - Z\bar{W}')\}^{-(m+n+1)} \cdot \frac{\partial}{\partial z_{ij}}\det(I - Z\bar{W}') .$$

Now we examine the last factor.

Consider column vector representations of the matrices Z and W.

$$Z = \begin{pmatrix} v_1 \\ \vdots \\ v_m \end{pmatrix} \quad \text{and} \quad W = \begin{pmatrix} u_1 \\ \vdots \\ u_m \end{pmatrix} .$$

Then

$$I - Z\bar{W}' = I - \begin{pmatrix} v_1 \\ \vdots \\ v_m \end{pmatrix}(\bar{u}_1', \bar{u}_2', \ldots, \bar{u}_m')$$

$$= \begin{pmatrix} 1 - v_1\bar{u}_1', & -v_1\bar{u}_2', & \cdots & -v_1\bar{u}_m' \\ -v_2\bar{u}_1', & 1 - v_2\bar{v}_2', & & -v_2\bar{u}_m' \\ \cdots & \cdots & \cdots & \cdots \\ \cdots & \cdots & \cdots & 1 - v_m\bar{u}_m' \end{pmatrix} .$$

Then z_{ij} is only involved in one row, specifically the ith row:

$$\frac{\partial}{\partial z_{ij}}\det(I - Z\bar{W}') = \begin{vmatrix} 1 - v_1\bar{u}_1' & \cdots & -v_1\bar{u}_m' \\ \cdots & \cdots & \cdots \\ -\bar{w}_{ij} & \cdots & -\bar{w}_{mj} \\ \cdots & \cdots & \cdots \\ -v_m\bar{u}_1' & \cdots & 1 - v_m\bar{u}_m' \end{vmatrix}$$

where only the ith row of the $m \times m$ matrix is of a different pattern.

The case of interest is when $W = Z$. Multiplying by z_{ij} and summing on j, we almost undo the differentiation; only the ii term is altered. The $m \times m$ identity matrix has been changed to have a zero at the ii position. Thus

$$\sum_{j=1}^{n}\frac{\partial}{\partial z_{ij}}\det(I - Z\bar{Z}') = \begin{vmatrix} 1 - v_1\bar{v}_1' & \cdots & -v_1\bar{v}_m' \\ \cdots & \cdots & \cdots \\ -v_i\bar{v}_i' & \cdots & -v_i\bar{v}_m' \\ \cdots & \cdots & \cdots \\ -v_m\bar{v}_1' & \cdots & 1 - v_m\bar{v}_m' \end{vmatrix} = \det\left[\begin{pmatrix} 1 & & 0 \\ & 0 & \\ 0 & & I \end{pmatrix} - Z\bar{Z}'\right].$$

Thus

$$\begin{aligned}\sum_{i=1}^{m}\sum_{j=1}^{n} z_{ij}\frac{\partial}{\partial z_{ij}}\det(I - Z\bar{Z}') \\ = \det\left[\begin{pmatrix} 0 & \\ & I \end{pmatrix} - Z\bar{Z}'\right] + \cdots + \det\left[\begin{pmatrix} I & \\ & 0 \end{pmatrix} - Z\bar{Z}'\right].\end{aligned} \tag{3.7}$$

Let $S(Z)$ denote the sum of m determinants in equation (3.7). Consider the second and third displayed equations after inequality (3.6), and substitute $W = Z$. We can use equation (3.7) to rewrite inequality (3.6) as follows:

$$\operatorname{Re}\left\{\det\frac{\partial F}{\partial w}(\varphi(tZ_0))\right\} \geq \tag{3.8}$$
$$\frac{\det(I - tZ_0\bar{Z}_0')^{m+n}\{[\det(I - Z_0\bar{Z}_0') - (m+n)S(Z_0)]t + [\det(I - Z_0\bar{Z}_0') + (m+n)S(Z_0)]\}}{\det(I - Z_0\bar{Z}_0')^{m+n}\{[\det(I - Z_0\bar{Z}_0') - (m+n)S(Z_0)] + [\det(I - Z_0\bar{Z}_0') + (m+n)S(Z_0)]t\}}.$$

We now examine the value of $\varphi(tZ_0)$. For $Z = tZ_0$ in the representation of φ in equation (3.1), we have

$$\varphi(tZ_0) = (t-1)AZ_0(I - t\bar{Z}_0'Z_0)^{-1}D^{-1} = (t-1)A(I - tZ_0\bar{Z}_0')^{-1}Z_0D^{-1}. \tag{3.9}$$

Recall that any matrix Z in $\mathcal{R}_I$ can be represented as $U\Lambda V$ where U and V are unitary and Λ is diagonal with entries $1 > \lambda_1 \geq \lambda_2 \geq \cdots \geq \lambda_m \geq 0$. Thus it is sufficient to consider

$$Z_0 = -\begin{pmatrix} \lambda_1 & & 0 & \\ & \ddots & & 0 \\ 0 & & \lambda_m & \end{pmatrix}, \qquad 1 > \lambda_1 \geq \lambda_2 \geq \cdots \geq \lambda_m \geq 0. \tag{3.10}$$

Then, from equations (3.2),

$$\bar{A}'A = (I - Z_0\bar{Z}_0')^{-1} = \left(I - \begin{pmatrix} \lambda_1^2 & & 0 \\ & \ddots & \\ 0 & & \lambda_m^2 \end{pmatrix}\right)^{-1} = \begin{pmatrix} \frac{1}{1-\lambda_1^2} & & & 0 \\ & \frac{1}{1-\lambda_2^2} & & \\ & & \ddots & \\ & 0 & & \frac{1}{1-\lambda_m^2} \end{pmatrix}.$$

Thus we can take

$$A = \begin{pmatrix} \frac{1}{\sqrt{1-\lambda_1^2}} & & 0 \\ & \ddots & \\ 0 & & \frac{1}{\sqrt{1-\lambda_m^2}} \end{pmatrix}.$$

And

$$\bar{D}'D = (I - \bar{Z}_0'Z_0)^{-1} = \left(I - \begin{pmatrix} \lambda_1^2 & & & & 0 \\ & \lambda_2^2 & & & \\ & & \ddots & & \\ & & & \lambda_m^2 & \\ 0 & & & & 0 \end{pmatrix}\right)^{-1}$$

$$= \begin{pmatrix} \frac{1}{1-\lambda_1^2} & & & 0 & \\ & \ddots & & & \\ & & \frac{1}{1-\lambda_m^2} & & \\ 0 & & & 1 & \\ & & & & \ddots \end{pmatrix}.$$

We can take

$$D = \begin{pmatrix} \frac{1}{\sqrt{1-\lambda_1^2}} & & 0 & \\ & \ddots & & 0 \\ 0 & & \frac{1}{\sqrt{1-\lambda_m^2}} & \\ 0 & & & I \end{pmatrix}.$$

Now we substitute these values into equation (3.9).

$$\varphi(tZ_0) = (t-1)\begin{pmatrix} \frac{1}{\sqrt{1-\lambda_1^2}} & & 0 \\ & \ddots & \\ 0 & & \frac{1}{\sqrt{1-\lambda_m^2}} \end{pmatrix}\begin{pmatrix} \frac{1}{1-t\lambda_1^2} & & 0 \\ & \ddots & \\ 0 & & \frac{1}{1-t\lambda_m^2} \end{pmatrix}\begin{pmatrix} -\lambda_1 & & 0 & \\ & \ddots & & 0 \\ 0 & & -\lambda_m & \end{pmatrix}$$

$$\times\begin{pmatrix} \sqrt{1-\lambda_1^2} & & 0 \\ & \ddots & \sqrt{1-\lambda_m^2} \\ 0 & & I \end{pmatrix}$$

$$= \begin{pmatrix} \frac{1-t}{(1-t\lambda_1^2)\sqrt{1-\lambda_1^2}} & & 0 \\ & \ddots & \\ 0 & & \frac{1-t}{(1-t\lambda_m^2)\sqrt{1-\lambda_m^2}} \end{pmatrix}\begin{pmatrix} \lambda_1\sqrt{1-\lambda_1^2} & & 0 & \\ & \ddots & & 0 \\ 0 & & \lambda_m\sqrt{1-\lambda_m^2} & \end{pmatrix}$$

$$= \begin{pmatrix} \frac{(1-t)\lambda_1}{1-t\lambda_1^2} & & 0 & \\ & \ddots & & 0 \\ 0 & & \frac{(1-t)\lambda_m}{1-t\lambda_m^2} & \end{pmatrix}$$

$$= \begin{pmatrix} \mu_1 & & 0 & \\ & \ddots & & 0 \\ 0 & & \mu_m & \end{pmatrix}$$

where $\mu_j = \frac{(1-t)\lambda_j}{1-t\lambda_j^2}$ for $j = 1, \ldots, m$. Note that $|\mu_j| < 1$ since φ has range $\mathcal{R}_I$. Now solving for t we find that

$$t = \frac{\lambda_j - \mu_j}{\lambda_j(1-\lambda_j\mu_j)}. \tag{3.11}$$

Since $\mathcal{R}_I$ is a complete circular domain, the range of t so that tZ_0 is in $\mathcal{R}_I$ is at least $[-1, 1]$. From the equation defining the μ_j and the monotonicity as a function of t, it follows that

$$0 \le \mu_j \le \frac{2\lambda_j}{1+\lambda_j^2} \qquad \text{for} \quad j = 1, 2, \ldots, n. \tag{3.12}$$

Consider equation (3.8) at Z_0 as given by equation (3.10). The result is

$$Re\left\{\det\frac{\partial F}{\partial w}(\mu)\right\} \ge \frac{\prod_{j=1}^{m}(1-t\lambda_j^2)^{m+n}}{\prod_{j=1}^{m}(1-\lambda_j^2)^{m+n}} \cdot \frac{N}{E} \tag{3.13}$$

where

$$N = \left\{\prod_{j=1}^{m}(1-\lambda_j^2) + (m+n)\left[\lambda_1^2\prod_{j=2}^{m}(1-\lambda_j^2) + \cdots + \lambda_m^2\prod_{j=1}^{m-1}(1-\lambda_j^2)\right]\right\} t$$
$$+ \left\{\prod_{j=1}^{m}(1-\lambda_j^2) - (m+n)\left[\lambda_1^2\prod_{j=2}^{m}(1-\lambda_j^2) + \cdots + \lambda_m^2\prod_{j=1}^{m-1}(1-\lambda_j^2)\right]\right\},$$

and

$$E = \prod_{j=1}^{m}(1-\lambda_j^2) + (m+n)\left[\lambda_1^2\prod_{j=2}^{m}(1-\lambda_j^2) + \cdots + \lambda_m^2\prod_{j=1}^{m-1}(1-\lambda_j^2)\right]$$
$$+ \left(\prod_{j=1}^{m}(1-\lambda_j^2) - (m+n)\left[\lambda_1^2\prod_{j=2}^{m}(1-\lambda_j^2) + \cdots + \lambda_m^2\prod_{j=1}^{m-1}(1-\lambda_j^2)\right]\right) t.$$

To understand the first factor of the right side of inequality (3.13), we use equation (3.11) to observe that·

$$1 - t\lambda_j^2 = 1 - \frac{\lambda_j - \mu_j}{\lambda_j(1-\lambda_j\mu_j)}\lambda_j^2 = \frac{1-\lambda_j^2}{1-\lambda_j\mu_j}.$$

Thus

$$\frac{\prod_{j=1}^{m}(1-t\lambda_j^2)}{\prod_{j=1}^{m}(1-\lambda_j^2)} = \prod_{j=1}^{m}\frac{1}{1-\lambda_j\mu_j}. \tag{3.14}$$

Next we simplify expressions N and E of inequality (3.13).

$$N = (t+1)\prod_{j=1}^{m}(1-\lambda_j^2) - (m+n)(1-t)\sum_{k=1}^{m}\lambda_k^2\prod_{\substack{j=1\\ j\neq k}}^{m}(1-\lambda_j^2)$$

$$E = (1+t)\prod_{j=1}^{m}(1-\lambda_j^2) + (m+n)(1-t)\sum_{k=1}^{m}\lambda_k^2\prod_{\substack{j=1\\ j\neq k}}^{m}(1-\lambda_j^2).$$

From equation (3.11) we find that

$$1-t=1-\frac{\lambda_j-\mu_j}{\lambda_j-\lambda_j^2\mu_j} = \frac{\mu_j(1-\lambda_j^2)}{\lambda_j(1-\lambda_j\mu_j)} \quad \text{for} \quad j=1,\dots,m, \text{ and}$$

$$(1-t)\sum_{k=1}^{m}\lambda_k^2\prod_{\substack{j=1\\ j\neq k}}^{m}(1-\lambda_j^2) = \sum_{k=1}^{m}\frac{\lambda_k\mu_k}{1-\lambda_k\mu_k}\prod_{j=1}^{m}(1-\lambda_j^2).$$

Thus

$$\frac{N}{E} = \frac{(t+1)-(m+n)\sum_{k=1}^{m}\frac{\lambda_k\mu_k}{1-\lambda_k\mu_k}}{(t+1)+(m+n)\sum_{k=1}^{m}\frac{\lambda_k\mu_k}{1-\lambda_k\mu_k}} = \frac{(t+1)+m(m+n)-(m+n)\sum_{j=1}^{m}\frac{1}{1-\lambda_j\mu_j}}{(t+1)-m(m+n)+(m+n)\sum_{j=1}^{m}\frac{1}{1-\lambda_j\mu_j}}. \tag{3.15}$$

Using equations (3.14) and (3.15), we find that inequality (3.13) becomes

$$\operatorname{Re}\left\{\det\frac{\partial F}{\partial w}(\mu)\right\} \geq \prod_{j=1}^{m}\frac{1}{(1-\lambda_j\mu_j)^{m+n}}\,\frac{(t+1)+m(m+n)-(m+n)\sum_{j=1}^{m}\frac{1}{1-\lambda_j\mu_j}}{(t+1)-m(m+n)+(m+n)\sum_{j=1}^{m}\frac{1}{1-\lambda_j\mu_j}}. \tag{3.16}$$

As the final step, we wish to determine when the right side of inequality (3.16) is positive. We fix $\lambda_1=(m^2+mn+1)^{-1/2}$. We now examine the discussion from equations (3.10) and (3.12). Having fixed $\lambda_1=(m^2+mn+1)^{-1/2}$, we find that inequality (3.12) requires that

$$\mu_1 < \frac{2\sqrt{m^2+mn+1}}{m^2+mn+2}.$$

We now show that all choices of the μ_k's are possible provided only that

$$\frac{2\sqrt{m^2+mn+1}}{m^2+mn+2} > \mu_1 \geq \mu_2 \geq \cdots \geq \mu_m \geq 0.$$

It is necessary to show that there is a corresponding choice of $-1<t<1$ and $1>\lambda_1\geq\lambda_2\geq\cdots\geq\lambda_m\geq 0$. The value of t is determined by equation (3.11) with $j=1$. For fixed λ_1, equation

(3.11) implies that t *strictly* monotonically decreases with μ. As μ_1 goes from 0 to $\frac{2\sqrt{m^2+mn+1}}{m^2+mn+2}$, t varies from $+1$ to -1.

Having fixed μ_1 and λ_1 and thus t, we can now think of equation (3.11) as defining each of the other λ_k in terms of the corresponding μ_k. Dropping the index, we can write equation (3.11) in the following form with t a constant:

$$t = \frac{\lambda - \mu}{\lambda(1 - \lambda\mu)}. \tag{3.17}$$

We think of λ as a function of μ and let λ' denote its derivative with respect to μ. Differentiation of equation (3.17) with respect to μ gives

$$0 = \frac{(\lambda' - 1)[\lambda(1 - \lambda\mu)] - (\lambda - \mu)[\lambda'(1 - \lambda\mu) + \lambda(-\lambda'\mu) + \lambda(-\lambda)]}{[\lambda(1 - \lambda\mu)]^2}.$$

It follows that

$$\lambda'\mu\{1 - 2\lambda\mu + \lambda^2\} = \lambda(1 - \lambda^2).$$

For $0 < \mu \le \mu_1 < 1$, the factor in braces is positive. And, provided $0 < \lambda < 1$,

$$\lambda' = \frac{1}{\mu}\frac{1 - \lambda^2}{1 - 2\lambda\mu + \lambda^2} > 0. \tag{3.18}$$

At $\mu = \mu_1$, we have fixed $\lambda_1 = (m^2+mn+1)^{-1/2}$. As μ decreases from μ_1, the corresponding λ decreases, but remains positive for μ positive according to equation (3.18). Also $\lim\limits_{\mu\to 0^+} \lambda(\mu) = 0$.

Hence for any $\frac{2\sqrt{m^2+mn+1}}{m^2+mn+2} > \mu_1 \ge \mu_2 \ge \cdots \ge \mu_m \ge 0$, there exists corresponding value of t such that $-1 < t < 1$ and $(m^2 + mn + 1)^{-1/2} = \lambda_1 \ge \lambda_2 \ge \cdots \ge \lambda_m \ge 0$.

Now returning to inequality (3.16), we wish to estimate the right-hand side. Since $(m^2 + mn + 1)^{-1/2} = \lambda_1 \ge \lambda_2 \ge \cdots \ge \lambda_m \ge 0$ and $\frac{2\sqrt{m^2+mn+1}}{m^2+mn+2} > \mu_1 \ge \mu_2 \ge \cdots \ge \mu_m \ge 0$ and $-1 < t < 1$, the denominator of the second factor is positive. And also

$$\frac{1}{1 - \lambda_1\mu_1} \ge \cdots \ge \frac{1}{1 - \lambda_m\mu_m} \ge 1.$$

Thus, letting $\gamma = (m^2 + mn + 1)^{1/2}$ and $j = 1$ in equation (3.11), we have that $t = \frac{\gamma(1-\gamma\mu_1)}{\gamma-\mu_1}$.

We let R denote the right side of inequality (3.16). It is to be determined when R is positive.

Fix the value of the index i as 2 or $3 \cdots$ or m and let $x = 1/(1 - \lambda_i \mu_i)$. Then let $A = t + 1 + m(m+n) - (m+n)\sum_{j \neq i} \frac{1}{1-\lambda_j \mu_j}$, and $B = t + 1 - m(m+n) + (m+n)\sum_{j \neq i} \frac{1}{1-\lambda_j \mu_j}$, and $p = m + n$. The isolation of the expression x allows R to be written as follows:

$$R = \prod_{j \neq i} \frac{1}{(1 - \lambda_j \mu_j)^{m+n}} \left(x^p \frac{A - px}{B + px} \right).$$

Let $f(x) = x^p \frac{A-px}{B+px}$. Then $f'(x) = \frac{x^{p-1} p}{(B+px)^2} \{AB + [A(p-1) - B(p+1)]x - p^2 x^2\}$. Substitute the values of A, B, and p into the brace factor; and let $Q = \sum_{j \neq i} 1/(1 - \lambda_j \mu_j)$. Then $f'(x) = \frac{x^{p-1}}{(B+px)^2} p\{-(m+n)^2(x - m + Q)^2 + (t+1)(t+1-2x)\}$. We take $j = i$ in equation (3.11) to obtain $t = \frac{\lambda_i - \mu_i}{\lambda_i - \lambda_i^2 \mu_i}$. Recall that $x = 1/(1 - \lambda_i \mu_i)$. Then

$$t + 1 - 2x = \frac{\lambda_i - \mu_i}{\lambda_i - \lambda_i^2 \mu_i} + 1 - \frac{2}{1 - \lambda_i \mu_i} = \frac{-\mu_i - \lambda_i^2 \mu_i}{\lambda_i (1 - \lambda_i \mu_i)} \leq 0.$$

Hence $f'(x) \leq 0$. Since $\frac{1}{1-\lambda_1 \mu_1} \geq \frac{1}{1-\lambda_i \mu_i}$ for $i = 2, 3, \ldots, m$, if $1/(1 - \lambda_i \mu_i)$ is replaced by $1/(1 - \lambda_1 \mu_1)$ in R for $i = 2, 3, \ldots, m$, then R is not increased. From inequality (3.16), we have

$$\operatorname{Re}\left\{\det \frac{\partial F}{\partial w}(\mu)\right\} \geq \frac{1}{(1 - \lambda_1 \mu_1)^{m(m+n)}} \frac{t + 1 + m(m+n) - m(m+n)\frac{1}{1-\lambda_1 \mu_1}}{t + 1 - m(m+n) + m(m+n)\frac{1}{1-\lambda_1 \mu_1}}.$$

Take $t = \frac{\lambda_1 - \mu_1}{\lambda_1 - \lambda_1^2 \mu_1} = \frac{\sqrt{m^2+mn+1}(1 - \sqrt{m^2+mn+1}\,\mu_1)}{\sqrt{m^2+mn+1} - \mu_1}$, then

$$\begin{aligned} \operatorname{Re}\left\{\det \frac{\partial F}{\partial w}(\mu)\right\} &\geq \frac{1}{(1 - \frac{1}{\gamma}\mu_1)^{m(m+n)}} \frac{\frac{\gamma(1-\gamma\mu_1)}{\gamma - \mu_1} + 1 - (m+n)m\frac{\frac{1}{\gamma}\mu_1}{1-\frac{1}{\gamma}\mu_1}}{\frac{\gamma(1-\gamma\mu_1)}{\gamma - \mu_1} + 1 + (m+n)m\frac{\frac{1}{\gamma}\mu_1}{1-\frac{1}{\gamma}\mu_1}} \\ &= \frac{1}{(1 - \frac{\mu_1}{\gamma})^{m(m+n)}} \frac{\gamma - (m^2 + mn + 1)\mu_1}{\gamma - \mu_1} \end{aligned}$$

where $\gamma = \sqrt{m^2 + mn + 1}$. Thus

$$\operatorname{Re}\left\{\det \frac{\partial F}{\partial w}(\mu)\right\} \geq \frac{1 - \sqrt{m^2 + mn + 1}\,\mu_1}{\left(1 - \frac{\mu_1}{\sqrt{m^2+mn+1}}\right)^{m^2+mn+1}} \tag{3.19}$$

provided that $\mu_1 < \frac{2\sqrt{m^2+mn+1}}{m^2+mn+2}$.

We now generalize inequality (3.19) to nondiagonal points in $\mathcal{R}_I$. Consider $W_0 \in \mathcal{R}_I$. There exist unitary matrices U and V such that for some diagonal matrix μ, $W_0 = U\mu V$. Having fixed U and V, we let $W = UZV$ as Z ranges over $\mathcal{R}_I$. Let $G(Z) = U^{-n}F(UZV)V^{-m}$. Then $\frac{\partial G}{\partial Z} = U^{-n}\frac{\partial F}{\partial W}\frac{\partial W}{\partial Z}V^{-m}$, and

$$\det\frac{\partial G}{\partial Z} = (\det U)^{-n}(\det V)^{-m}\det\frac{\partial F}{\partial W}\det\frac{\partial W}{\partial Z}.$$

Since $dW = UdZV$, $\frac{\partial W}{\partial Z} = U'\otimes V$ and $\det\frac{\partial W}{\partial Z} = (\det U')^n(\det V)^m$. Thus $\det\frac{\partial G}{\partial Z}(Z) = \det\frac{\partial F}{\partial W}(W)$. In particular, when $Z = \mu$, $W = W_0$. Since G satisfies the conditions: $G \in \mathcal{B}$, $d(G) = 1$, $\det\frac{\partial G}{\partial Z}(0) = 1$, inequality (3.19) holds with G replacing F and Z replacing W, that is,

$$\operatorname{Re}\left\{\det\frac{\partial G}{\partial Z}(\mu)\right\} \geq \frac{1-\sqrt{m^2+mn+1}\,\mu_1}{\left(1-\frac{\mu_1}{\sqrt{m^2+mn+1}}\right)^{m^2+mn+1}}$$

holds. Interpreting this inequality in terms of F and W, we have the following general result.

Theorem 3.1. *If $F \in \mathcal{B}$, $d(F) = 1$, $\det\frac{\partial F}{\partial W}(0) = 1$, then*

$$\operatorname{Re}\left\{\det\frac{\partial F}{\partial W}(W)\right\} \geq \frac{1-\sqrt{m^2+mn+1}\,\|W\|_M}{\left(1-\frac{\|W\|_M}{\sqrt{m^2+mn+1}}\right)^{m^2+mn+1}} \tag{3.20}$$

provided $\|W\|_M \leq \frac{2\sqrt{m^2+mn+1}}{m^2+mn+2}$ where $\|W\|_M$ is the matrix norm.

This theorem generalizes Bonk's result [3]. In case $m = 1$, Theorem 3.1 coincides with Liu's extension [13] of Bonk's work.

4. Bloch's constant for $\mathcal{R}_I$

The following lemma will be used in the calculation of a lower bound on Bloch's constant.

Lemma 4.1. *If f is a Bloch mapping in $\mathcal{R}_I$ that maps $\mathcal{R}_I$ into $\mathbb{C}^{mn}$, then*

$$\left\|\frac{\partial f}{\partial Z}(Z)\right\|_M \leq \frac{\|f\|_B}{1-\|Z\|_M^2}. \tag{4.1}$$

Proof. Let φ_Z be a holomorphic automorphism of $\mathcal{R}_I$ that takes Z to the origin. Note that

$$\begin{aligned}\|f'(Z)\|_M &= \|[(f\circ\varphi_Z)'(0)]\cdot[\varphi_Z'(0)]^{-1}\|_M\\ &= \|(f\circ\varphi_Z)'(0)\|_M\cdot\|\varphi_Z'(0)^{-1}\|_M\\ \|f'(Z)\|_M &\le \|f\|_B\cdot\|\varphi_Z'(0)^{-1}\|_M\,. \end{aligned}\tag{4.2}$$

Now we calculate the last factor of inequality (4.2). Fixing Z_0, we can write $\varphi_{Z_0}(Z) = A(Z - Z_0)(I - \bar{Z}_0'Z)^{-1}D^{-1}$. From equation (3.5) we can find dW, and conclude that

$$\frac{\partial W}{\partial Z} = \left[\bar{A}'^{-1}(I - Z\bar{Z}_0')^{-1}\right]' \otimes \left[(I - \bar{Z}_0'Z)^{-1}D^{-1}\right].$$

At $Z = 0$,

$$\frac{\partial W}{\partial Z} = \bar{A}^{-1}\otimes D^{-1}\tag{4.3}$$

and

$$\frac{\overline{\partial W}'}{\partial Z} = A'^{-1}\otimes\bar{D}'^{-1}.$$

Thus

$$\frac{\partial W}{\partial Z}\frac{\overline{\partial W}'}{\partial Z} = \bar{A}^{-1}A'^{-1}\otimes D^{-1}\bar{D}'^{-1}.\tag{4.4}$$

Taking the inverse of equations (3.2),we obtain

$$A^{-1}\bar{A}'^{-1} = I - Z_0\bar{Z}_0' \qquad \text{and} \qquad D^{-1}\bar{D}'^{-1} = I - \bar{Z}_0'Z_0\,.$$

The second of these and the conjugate of the first can be used to simplify equation (4.4):

$$\frac{\partial W}{\partial Z}\frac{\overline{\partial W}'}{\partial Z} = (I - \bar{Z}_0Z_0')\otimes(I - \bar{Z}_0'Z_0)\,.$$

Hence the minimum eigenvalue of $(\partial W/\partial Z)(\overline{\partial W/\partial Z})'$ is $(1-\mu_1^2)^2$ where μ_1^2 is the largest eigenvalue of $Z_0\bar{Z}_0'$. And the maximum eigenvalue of $(\overline{\partial W/\partial Z})'^{-1}(\partial W/\partial Z)^{-1}$ is $(1-\mu_1^2)^{-2}$. Thus

$$\left\|\left(\frac{\partial W}{\partial Z}\right)^{-1}\right\|_M = (1-\mu_1^2)^{-1},$$

that is,

$$\|[\varphi_Z'(0)]^{-1}\|_M = \frac{1}{1-\|Z\|_M^2}.$$

Then inequality (4.2) becomes the desired inequality (4.1).

We can now find a lower bound for the Bloch constant for mappings of $\mathcal{R}_I$. In case $m = 1$, the estimate is the same as Liu's result [13].

Theorem 4.2. *If $F : \mathcal{R}_I \to \mathbb{C}^{mn}$ and $F \in \mathcal{B}(\mathcal{R}_I)$, $d(F) = 1$, $\det\frac{\partial F}{\partial Z}(0) = 1$ and $\|F\|_B \le K$, then*

$$r(0,F) \ge C(K,m,n) = K^{1-mn}\int_0^{\frac{1}{\sqrt{m^2+mn+1}}} \frac{(1-\sqrt{m^2+mn+1}\,t)(1-t^2)^{mn-1}}{\left(1-\frac{t}{\sqrt{m^2+mn+1}}\right)^{m^2+mn+1}}\,dt. \quad (4.5)$$

Proof. Following Liu [13], we consider what causes a ball in the range of F to not have a well-defined inverse mapping of F. For a small ball centered at the origin, since $\det\frac{\partial F}{\partial W}(0) = 1$ and $F(0) = 0$, there is a well-defined inverse of F on the ball. As the ball in the range expands, one of two things happens: either the preimage reaches a point at which $\det\frac{\partial F}{\partial W} = 0$ or it reaches a point of the boundary. Otherwise, one could argue that the ball in the range could be enlarged further.

Thus, from the estimate on the Re $\det\frac{\partial F}{\partial W}$ in Theorem 3.1 in inequality (3.20), we can force $\det\frac{\partial F}{\partial W}$ to be nonzero by requiring that $\|W\|_M < 1/\sqrt{m^2+mn+1}$. Now consider a straight line interval Γ in the range of F. The interval Γ starts at the origin and goes as far as it can without having its preimage run into the boundary of $\mathcal{R}_I$ or a point farther than $1/\sqrt{m^2+mn+1}$ from the origin. We also assume that the direction of Γ is such that it has the minimum length among such intervals. Then

$$\begin{aligned} r(0,F) &\ge \left|\int_\Gamma dW\right| \\ &= \int_\gamma \left|\frac{\partial F}{\partial Z}\frac{dZ'}{|dZ|}\right|\cdot|dZ| \quad \text{where} \quad \gamma = F^{-1}(\Gamma) \\ &\ge \int_\gamma \frac{|\det\frac{\partial F}{\partial Z}|}{\|\frac{\partial F}{\partial Z}\|_M^{mn-1}}|dZ| \\ &\ge \int_0^{\frac{1}{\sqrt{m^2+mn+1}}} \frac{1-\sqrt{m^2+mn+1}\,\|Z\|_M}{\left(1-\frac{\|Z\|_M}{\sqrt{m^2+mn+1}}\right)^{m^2+mn+1}}\cdot\frac{(1-\|Z\|_M^2)^{mn-1}}{K^{mn-1}}\,d|Z| \end{aligned} \quad (4.6)$$

by inequality (3.20) and inequality (4.1).

Let $c = \sqrt{m^2+mn+1}$ and $x = \|Z\|_M$. Now let $g(x)$ equal the integrand in expression (4.6) times K^{mn-1}. Then

$$g(x) = \frac{(1-cx)(1-x^2)^{mn-1}}{\left(1-\frac{x}{c}\right)^{c^2}}.$$

In order to simplify expression (4.6), it is useful to prove that $g'(x) \le 0$ for $0 \le x \le \frac{1}{c}$. Straightforward calculation gives

$$g'(x) = \frac{(1-x^2)^{mn-2}\left\{x(1-x^2)(1-c^2) - 2x\left(1-\frac{x}{c}\right)(1-cx)(mn-1)\right\}}{\left(1-\frac{x}{c}\right)^{c^2+1}}. \tag{4.7}$$

Since the factor in braces in expression (4.7) is nonpositive for $0 \le x \le \frac{1}{c}$, we have $g'(x) \le 0$ for $0 \le x \le \frac{1}{c}$. Since $|Z| \ge \|Z\|_M$, the expression (4.6) can be replaced by one using only $|Z|$. Substitution of t for $|Z|$ allows the integral to be written as

$$\int_0^{\frac{1}{\sqrt{m^2+mn+1}}} \frac{(1-\sqrt{m^2+mn+1}\,t)(1-t^2)^{mn-1}}{\left(1-\frac{t}{\sqrt{m^2+mn+1}}\right)^{m^2+mn+1}}\,dt$$

as was claimed in the theorem.

To obtain an upper bound on Bloch's constant it is sufficient to consider an explicit mapping. Let $K \ge 1$, we consider the mapping

$$W = (w_{ij})_{1\le i,j\le m} = F(Z) = \begin{pmatrix} Kz_{11}, & \dots, & Kz_{1,n-1}, & Kz_{1,n} \\ \dots & \dots & \dots & \dots \\ Kz_{m1}, & \dots, & Kz_{m,n-1}, & K^{1-mn}z_{mn} \end{pmatrix} \tag{4.8}$$

which maps R_I onto

$$I - \begin{pmatrix} \frac{w_{11}}{K}, & \dots, & \frac{w_{1,n-1}}{K}, & \frac{w_{1,n}}{K} \\ \dots & \dots & \dots & \dots \\ \frac{w_{m1}}{K}, & \dots, & \frac{w_{m,n-1}}{K}, & \frac{K^{mn}w_{mn}}{K} \end{pmatrix} \overline{\begin{pmatrix} \frac{w_{11}}{K}, & \dots, & \frac{w_{1,n-1}}{K}, & \frac{w_{1,n}}{K} \\ \dots & \dots & \dots & \dots \\ \frac{w_{m1}}{K}, & \dots, & \frac{w_{m,n-1}}{K}, & \frac{K^{mn}w_{mn}}{K} \end{pmatrix}}' > 0. \tag{4.9}$$

The Jacobian of the mapping (4.8) is

$$\frac{\partial W}{\partial Z} = \begin{pmatrix} K & & & 0 \\ & \ddots & & \\ & & K & \\ 0 & & & K^{1-mn} \end{pmatrix}.$$

This implies that $\det \frac{\partial W}{\partial Z}(0) = 1$.

If $\varphi(Z)$ is the holomorphic automorphism of R_I that maps $Z_0 \in R_I$ to the origin, then

$$\frac{\partial F \circ \varphi}{\partial Z}(0) = \frac{\partial F}{\partial W}\frac{\partial G}{\partial Z}(0) = \begin{pmatrix} K & & & 0 \\ & \ddots & & \\ & & K & \\ 0 & & & K^{1-mn} \end{pmatrix} \bar{A}^{-1} \otimes D^{-1}$$

by (4.3). It is easy to show

$$\left\| \frac{\partial \varphi}{\partial Z}(0) \right\|_M = 1 - \mu_n^2$$

where μ_n^2 is the smallest eigenvalue of $\bar{Z}_0' Z_0$. We immediately get

$$\| F \|_B = \sup \left\{ \left\| \frac{\partial F \circ \varphi}{\partial Z}(0) \right\|_M : \varphi \in \mathrm{Aut}(R_I) \right\} \le K .$$

This means the mapping (4.8) belongs to $\mathcal{B}_K$. The largest radius of the ball centered at origin and containing in the range (4.9) is K^{1-mn}.

Combining this observation, Theorem 4.2 and the Landau's Theorem 1.3, we conclude the following.

Theorem 4.3. *The Bloch constant $B_K(R_I)$ of the family $\mathcal{B}_K(R_I)$ satisfies the following inequalities:*

$$K^{1-mn} \ge B_K(R_I) \ge K^{1-mn} \int_0^{(m^2+mn+1)^{-1/2}} \frac{(1-\sqrt{m^2+mn+1}\,t)(1-t^2)^{mn-1}}{\left(1-\frac{t}{\sqrt{m^2+mn+1}}\right)^{m^2+mn+1}} dt .$$

We note that in the case that $m = 1$, Theorem 4.3 becomes exactly the result found by Liu [13]. And if both $m = 1$ and $n = 1$, Theorem 4.3 becomes the result found by Ahlfors [1].

References

[1] Ahlfors, L. V. An extension of Schwarz's lemma. *Trans. Amer. Math. Soc.* **43**, 359–364 (1938).
[2] Bochner, S. Bloch theorem for real variables. *Bull. Amer. Math. Soc.* **52**, 715–719 (1946).
[3] Bonk, M. On Bloch's constant. *Proc. Amer. Math. Soc.* **110**, 889–894 (1990).
[4] Chern, S. S. On holomorphic mappings of hermitian manifolds of the same dimension. In *Entire Functions and Related Parts of Analysis, Proc. Symposia Pure Math. XI*. AMS, Providence, RI. 1968.
[5] Greene, R. E., and Wu, H. Bloch's theorem for meromorphic functions. *Math. Z.* **116**, 247–257 (1970).
[6] Hahn, K. T. High dimensional generalizations of the Bloch constant and their lower bounds. *Trans. Amer. Math. Soc.* **179**, 263–274 (1973).

[7] Hahn, K. T. Holomorphic mappings of the hyperbolic space into the complex Euclidean space and the Bloch theorem. *Canad. J. Math.* **27**, 446–458 (1975).
[8] Hahn, K. T. Quantitative Bloch's theorem for certain classes of holomorphic mappings of the ball in $P_n\mathbb{C}$. *J. Reine Angew. Math.* **283**, 99–109 (1976).
[9] Harris, L. A. On the size of balls covered by analytic transformations. *Monatshefte für Mathematik*, **83**, 9–23 (1977).
[10] Hua, L. K. Harmonic analysis of functions of several complex variables in the classical domains. *Translations of Mathematical Monographs, Amer. Math. Soc.* **6** (1963).
[11] Krantz, S., and Ma, D. Bloch functions on strongly pseudoconvex domains. *Indiana Univ. Math. J.* **37**, 145–163 (1988).
[12] Landau, E. Über die Bloch'sche Konstante und zwei verwandte Weltkonstanten. *Math. Z.* **30**, 608–634 (1929).
[13] Liu, X. Y. Bloch functions of several complex variables. *Pacific J. Math.*, **152**, 347–363 (1992).
[14] Liu, X. Y., and Minda, D. Distortion theorems for Bloch functions. *Trans. Amer. Math. Soc.*, **333**, 325–338 (1992).
[15] Minda, D. The Bloch and Marden constant. *Proc. CMFT'89, Lecture Notes in Math.*, vol. 1435, pp. 131–142. Springer-Verlag 1990.
[16] Timoney, R. M. Bloch functions in several complex variables. *Bull. London Math. Soc.* **12**, 241–267 (1980). Ibid, II: *J. Reine Angew. Math.* **319**, 1–22 (1980).
[17] Wu, H. Normal families of holomorphic mappings. *Acta Math.* **119**, 193–233 (1967).

Received May 30, 1991; revised September 3, 1991

Department of Mathematics, 0112, University of California, San Diego, La Jolla, CA 92093-0112
Department of Mathematics, 0112, University of California, San Diego, La Jolla, CA 92093-0112 and University of Science and Technology of China, Hefei, Anhui 230026, People's Republic of China

Vol. 36 No. 3 SCIENCE IN CHINA (Series A) March 1993

Bloch Constant of Holomorphic Mappings on Bounded Symmetric Domains

YAN Zhi-Min (严志敏)
(Department of Mathematics, University of California, Berkeley, California 94720, USA)

and GONG Sheng (龚 昇)
(Department of Mathematics, University of Science and Technology of China, Hefei 230026, PRC)

Received June 30, 1992.

Abstract The estimation of the lower bound of the Bloch constant of holomorphic mappings on irreducible bounded symmetric domains is given.

Keywords: bounded symmetric domains, Bloch mappings, Bloch constant.

1 Some Results on Bounded Transitive Domains

First of all, we restate some results on bounded transitive domains which was proved in [1].

Let D be an irreducible bounded transitive domain in C^n which contains the origin, and Aut (D) be the group of holomorphic automorphisms of D. A holomorphic mapping on D is defined as a Bloch mapping if the family of mappings $F_f=\{g \mid g(z)=f(\phi(z))-f(\phi(0)),$ for all $\phi \in \mathrm{Aut}(D)\}$ is a normal family. All the Bloch mappings form a family of Bloch holomorphic mappings. The Bloch norm of a Bloch mapping is defined as

$$\|f\|_B=\sup\left\{\left\|\frac{\partial f\circ\phi}{\partial z}(0)\right\|_M : \phi\in \mathrm{Aut}(D)\right\}, \tag{1.1}$$

where $\| \quad \|_M$ is the norm of the matrix. In [1], we proved the following

Lemma 1.1. *Let f be a holomorphic mapping on D which maps D into C^n and $f(0)=0$. Then f is a Bloch mapping if and only if $\|f\|_B$ is finite.*

Let $F(z)$ be a holomorphic mapping which maps D into C^n. In [1], we introduced a functional $D(F)$ of F

$$D(F)=\sup_{z\in D}\frac{|\det F'(z)|}{\sqrt{K_D(z,\bar z)/K_D(0,0)}}, \tag{1.2}$$

where K_D is the Bergman kernel function of D, and $F'(z)=\dfrac{\partial F}{\partial z}$ is the Jacobian of F. In [1], we proved that $D(F)$ is invariant under the group Aut (D), i.e. $D(F\circ\phi)=D(F)$ holds for any $\phi\in \mathrm{Aut}(D)$.

Denote by $H(D)$ the family of holomorphic mappings which map D into C^n, and let $\mathscr{F}$ be a subfamily of $H(D)$.

For any $F \in \mathscr{F}$, $\lambda \in D$, $r(\lambda, F)$ denotes the largest nonnegative number r such that there exists a domain R in D, $\lambda \in R$, and F biholomorphically maps R onto $B(F(\lambda), r)$, where $B(F(\lambda), r)$ denote a ball centered at $F(\lambda)$ with radius r. Let $r(F) = \sup\{r(\lambda, F), \lambda \in D\}$. Then we define the Bloch constant of the subfamily $\mathscr{F}$ on D as

$$B_{\mathscr{F}} = \inf\left\{r(F): F \in \mathscr{F},\ \det\frac{\partial F}{\partial z}(0) = 1\right\}.$$

In this paper, we consider the following subfamilies of holomorphic mappings:

$$\mathscr{B} = \mathscr{B}(D) = \left\{F \in H(D),\ \det\frac{\partial f}{\partial z}(0) = 1, \|F\|_B < \infty\right\};$$

$$\mathscr{B}_K = \mathscr{B}_K(D) = \{F \in \mathscr{B},\ \|F\|_B \leqslant K,\ 1 \leqslant K < \infty\};$$

$$\mathscr{B}_{K1} = \mathscr{B}_{K1}(D) = \{F \in \mathscr{B}_K, D(F) = 1\}.$$

Denote by B_K the Bloch constant of $\mathscr{B}_k$.

Moreover, in [1] we also proved the following

Lemma 1.2. *Let D be a bounded complete circular transitive domain, let $F \in \mathscr{B}(D)$, let $D(F) = 1$, and $\det\frac{\partial F}{\partial z}(0) = 1$, and let $\lambda \in D$. $\phi_\lambda(z)$ is a holomorphic automorphism which maps λ onto origin, then*

$$\mathrm{Re}\left\{\frac{\det\dfrac{\partial F(w)}{\partial w}\Big|_{w=\phi(t\lambda)}\det\dfrac{\partial \phi_\lambda(z)}{\partial z}\Big|_{z=t\lambda}}{\det\dfrac{\partial \phi_\lambda(z)}{\partial z}\Big|_{z=\lambda}}\right\}$$

$$\geqslant \frac{\left[\dfrac{\partial}{\partial\lambda}K_D(\lambda,\lambda)\cdot\lambda' + K_D(\lambda,\lambda)\right]t - \left[\dfrac{\partial}{\partial\lambda}K_D(\lambda,\lambda)\cdot\lambda' - K_D(\lambda,\lambda)\right]}{\left[\dfrac{\partial}{\partial\lambda}K_D(\lambda,\lambda)\cdot\lambda' + K_D(\lambda,\lambda)\right] - \left[\dfrac{\partial}{\partial\lambda}K_D(\lambda,\lambda)\cdot\lambda' - K_D(\lambda,\lambda)\right]t}, \tag{1.3}$$

where $\dfrac{\partial}{\partial\lambda}K_D(\lambda,\lambda) = \dfrac{\partial}{\partial z}K_D(z,z)|_{z=\lambda}$, $\dfrac{\partial}{\partial z} = \left(\dfrac{\partial}{\partial z_1}, \cdots, \dfrac{\partial}{\partial z_m}\right)$, and $-1 < t < 1$.

Moreover, in [1] we also gave a theorem of Landau type.

Lamma 1.3. *Let D be a bounded transitive domain in C^n which contains the origin. Then there exists a mapping in $\mathscr{B}_{K1}$ such that it is one extremal mapping of the Bloch constant in $\mathscr{B}_K$.*

Hence we can get the estimation of the Bloch constant in $\mathscr{B}_K$ when we get the estimation of the Bloch constant in $\mathscr{B}_{K1}$.

In [1], we got the estimations of the lower and the upper bounds of the Bloch constant of $\mathscr{B}_K(R_I)$ of the classical domain of type I, R_I, when we used the above results.

In this paper, we give the estimation of the lower bound of the Bloch constant of the family of holomorphic mappings $\mathscr{B}_K(D)$ on the irreducible bounded symmetric domains D. When $D=R_I$, it coincides with the result in [1].

2 Some Results of Faraut-Koranyi

In this section, we state some results of Faraut-Koranyi[2].

Let D be an irreducible bounded symmetric domain with the standard Harish-Chandra realization. D contains the origin. Let G be the connected component of Aut (D) and $\mathfrak{G}$ be the Lie algebra of G. Let K be an isotropy group of G which leaves origin fixed and $\mathfrak{k}$ be the maximal compact subalgebra of $\mathfrak{G}$ which corresponds to K. Then $\mathfrak{G}$ is a real simple Lie algebra, and has the Cartan decomposition $\mathfrak{G}=\mathfrak{k}+\mathfrak{P}$. Let $\mathfrak{h}$ be Cartan subalgebra of $\mathfrak{k}$. The $\mathfrak{h}^C$-root of $\mathfrak{G}^C$ and roots of $\mathfrak{k}^C$ are called a compact root. We denote by Φ^+ the set of positive non-compact roots. Denoting by τ the conjugation with respect to the real form $\mathfrak{k}+i\mathfrak{P}$, we consider a basis of root vectors e_α such that $\tau e_\alpha=-e_{-\alpha}$, $[e_\alpha,e_{-\alpha}]=h_\alpha$, $[h_\alpha,e_{\pm\alpha}]=\pm 2e_{\pm\alpha}$. Setting $\mathfrak{P}^\pm=\sum_{\alpha\in\Phi^+}\mathbb{C}e_{\pm\alpha}$, we have $\mathfrak{G}^C=\mathfrak{P}^-+\mathfrak{k}^C+\mathfrak{P}^+$ as a vector space direct sum. We denote by $\gamma_1,\cdots,\gamma_r$ the strong orthogonal roots of Harish-Chandra, where r is the real rank of $\mathfrak{G}$. We simply write $e_j=e_{\gamma_j}$, $h_j=h_{\gamma_j}$, $1\leqslant j\leqslant r$ and $e=\sum_{j=1}^r e_j$. It is known that in Harish-Chandra realization, D is a bounded symmetric domain in $\mathfrak{P}^+$, and the isotropy subgroup K of G at origin acts on D by unitary transformations. We denote by $\mathfrak{h}^-$ the real span of the vectors ih_j, by $\mathfrak{U}^+$ the real span of the e_j, and by $\mathfrak{U}$ the real span of the vectors $i(e_j+\tau e_j)$ $(1\leqslant j\leqslant r)$. The $i\mathfrak{h}^-$-roots of $\mathfrak{G}^C$ are $\pm\frac{1}{2}(\gamma_j\pm\gamma_k)$, $\pm\gamma_j$, $\pm\frac{1}{2}\gamma_j$ $(1\leqslant j,\ k\leqslant r)$ with respective multiplicities a, 1 and $2b$ (independent of j, k). We denote by p the important constant $p=(r-1)a+b+2$ and by $h(z)$ the K-invariant polynomial on $\mathfrak{P}^+$ whose restriction to $\mathfrak{U}^+$ is given by $h\left(\sum_1^r t_je_j\right)=\prod_{j=1}^r(1-t_j^2)$, then the function

$$h(z,w)=\exp\sum_1^n z_j\frac{\partial}{\partial z_j}\exp\sum_1^n \bar{w}_j\frac{\partial}{\partial z_j}h(z) \tag{2.1}$$

is still a polynomial, holomorphic in z, and antiholomorphic in w. Then the Bergman kernel function of D is given by

$$K(z,w)=h(z,w)^{-p}. \tag{2.2}$$

We write $\underline{m}\geqslant 0$ when an r-tple $\underline{m}=(m_1,\cdots,m_r)\in Z^r$ consists of nonnegative integers such that $m_1\geqslant m_2\geqslant\cdots\geqslant m_r$. For a complex number ξ, we define

$$(\xi)_{\underline{m}}=\prod_{j=1}^r\left(\xi-\frac{j-1}{2}a\right)_{m_j}, \tag{2.3}$$

where $(c)_k=\prod_{l=1}^k(c+l-1)$, $c_0=1$. For $1\leqslant j\leqslant r$, we simply write I_j for

$(\underbrace{1,\cdots,1}_{j's},0,\cdots,0)$. Then

$$(-1)_{l_j}=(-1)^j\prod_{l=1}^{j}\left[1+(l-1)\frac{a}{2}\right].$$

Under K, which acts on $\mathfrak{P}^+$ by unitary transformations, the space $P(\mathfrak{P}^+)$ of holomorphic polynomials on $\mathfrak{P}^+$ decomposes into irreducible subspaces as

$$P(\mathfrak{P}^+)=\sum_{\underline{m}\geqslant 0}\mathfrak{P}_{\underline{m}}.$$

The dimension of $\mathfrak{P}_{\underline{m}}$ is given by (cf.[2])

$$d_{\underline{m}}=\left(\prod_{j=1}^{r}\frac{B\left(m_j,\ (r-j)\frac{a}{2}+1+b\right)}{B\left(m_j,\ (r-j)\frac{a}{2}+1\right)}\right)^{-1}\prod_{1\leqslant p<q\leqslant r}\frac{m_p-m_q+\frac{a}{2}(p-q)}{\frac{a}{2}(q-p)}\cdot\frac{B\left(m_p-m_q,\ \frac{a}{2}(q-p-1)+1\right)}{B\left(m_p-m_q,\ \frac{a}{2}(q-p+1)\right)}.\tag{2.4}$$

For $\underline{m}\geqslant 0$, let $K^{\underline{m}}(z,w)$ be the reproducing kernel of $\mathfrak{P}_{\underline{m}}$. Then for all $\xi\in C$ and all $z,\ w\in D$, one has

$$h(z,w)^{-\xi}=\sum_{\underline{m}\geqslant 0}(\xi)_{\underline{m}}K^{\underline{m}}(z,w).\tag{2.5}$$

Every $z\in D$ can be written as

$$z=k\sum_{i=1}^{r}\lambda_ie_i,\quad 0\leqslant\lambda_i<1,1\leqslant i\leqslant r,\ k\in K.$$

We also write $\lambda^2=\sum_{i=1}^{r}\lambda_i^2e_i$ if $\lambda=\sum_{i=1}^{r}\lambda_ie_i$. For $\underline{m}\geqslant 0$, let $\phi_{\underline{m}}$ be the spherical polynomial. Then

$$K^{\underline{m}}(z,z)=\frac{d_{\underline{m}}}{\left(\frac{n}{r}\right)_{\underline{m}}}\phi_{\underline{m}}(\lambda^2)\tag{2.6}$$

if $z=k\sum_{i=1}^{r}\lambda_ie_i$ and n is the dimension of $\mathfrak{P}^+$.

3 Some Lemmas

In order to get the estimations of the two sides of (1.3), we need the following lemmas.

Noting that $(-1)_{\underline{m}}=0$ for all $\underline{m}\geqslant 0$ with $m_1>1$, as a consequence of (2.5), we have .

Lemma 3.1. $h(z,w)=\sum_{j=0}^{r}(-1)_{l_j}K^{l_j}(z,w).$ (3.1)

From (2.6), we have

$$K^{l_j}(z,z)=\frac{d_{l_j}}{\left(\frac{n}{r}\right)_{l_j}}\phi_{l_j}(\lambda^2).$$

Using (2.4), we have, after some computation

$$\frac{d_{l_j}}{\left(\frac{n}{r}\right)_{l_j}}=\frac{\prod_{l=1}^{j}(r-1+1)}{j!\prod_{l=1}^{j}\left(1+(l-1)\frac{a}{2}\right)},\quad j=1,2,\cdots,r. \tag{3.2}$$

We write $z\in\mathfrak{P}^+$ as $z=(z_1,z_2,\cdots,z_n)$ with respect to the basis $\{e_\alpha\}$. Let E denote the differential operator $\sum_{i=1}^{n}z_i\frac{\partial}{\partial z_i}$. We have

Lemma 3.2. $(EK)(z,z)=p\prod_{j=1}^{r}(1-\lambda_j^2)^{-(p+1)}\left[\sum_{i=1}^{r}\lambda_i^2\prod_{j\neq i}^{r}(1-\lambda_j^2)\right],$

where

$$z=k\sum_{i=1}^{r}\lambda_i e_i.$$

Proof. For $j=1,2,\cdots,r$, $K^{l_j}(z,z)$ is a homogeneous polynomial of degree j. It follows from (3.1) and Euler relation that

$$(Eh)(z,z)=\sum_{j=1}^{r}(-1)_{l_j}\cdot jK^{l_j}(z,z).$$

(2.3),(2.6) and (3.2) yield

$$\begin{aligned}(Eh)(z,z)&=\sum_{j=1}^{r}j(-1)^j\prod_{l=1}^{r}\left[1+(l-1)\frac{a}{2}\right]_j\cdot\frac{d_{l_j}}{\left(\frac{n}{r}\right)_{l_j}}\phi_{l_j}(\lambda^2)\\&=\sum_{j=1}^{r}(-1)^j\cdot j\frac{\prod_{l=1}^{j}[r-l+1]}{j!}\phi_{l_j}(\lambda^2)\\&=\sum_{j=1}^{r}(-1)^j\cdot j\binom{r}{j}\phi_{l_j}(\lambda^2).\end{aligned}$$

Hence, by (2.2), we have

$$\begin{aligned}(EK)(z,z)&=E(h(z,z)^{-p})=(-p)h(z,z)^{-p-1}(Eh)(z,z)\\&=(-p)h(z,z)^{-(p+1)}\sum_{j=1}^{r}(-1)^j j\binom{r}{j}\phi_{l_j}(\lambda^2).\end{aligned} \tag{3.3}$$

We denote by $S_j(x_1, \cdots, x_n)(j = 1, \cdots r)$ the elementary symmetric polynomial of degree j. Then

$$\phi_{l_j}(\lambda) = \frac{1}{\binom{r}{j}} S_j(\lambda_1, \cdots, \lambda_r)$$

for each j. Since $h(z,z) = \prod_{j=1}^{r} (1 - \lambda_j^2)$, by(3.3), we get

$$(EK)(z,z) = (-p) \prod_{j=1}^{r} (1 - \lambda_j^2)^{-(p+1)} \left[\sum_{j=1}^{r} (-1)^j j S_j (\lambda_1^2, \cdots, \lambda_r^2) \right].$$

It remains to show that

$$\sum_{j=1}^{r} (-1)^j j S_j(\lambda_1^2, \cdots, \lambda_r^2) = - \sum_{i=1}^{r} \lambda_i^2 \prod_{j \neq i}^{r} (1 - \lambda_j^2). \tag{3.4}$$

It only needs to consider $\frac{d}{ds} \prod_{i=1}^{r} (1 - s\lambda_i)|_{s=1}$. On the one hand, it is

$$\frac{d}{ds}\left(\sum_{i=0}^{r} (-1)^j S_j(\lambda_1, \cdots, \lambda_r) s^j \right)\Bigg|_{s=1} = \sum_{i=1}^{r} (-1)^j j S_j(\lambda_1, \cdots, \lambda_r).$$

On the other hand, it is

$$-\lambda_i \prod_{i=2}^{r} (1 - \lambda_i) - \cdots - \lambda_r \prod_{i=1}^{r-1} (1 - \lambda_i) = - \left[\sum_{i=1}^{r} \lambda_i \prod_{j \neq i}^{r} (1 - \lambda_i) \right].$$

Thus (3.4) is true. This completes the proof of Lemma 3.2.

Let $\lambda = \sum_{i=1}^{r} \lambda_i e_i \in D$, $-1 < \lambda_i < 1$, $i = 1, \cdots, r$. We denote by λ_m the mid-point of a geodesic which connects 0 and λ. Put $\gamma(t) = \sum_{i=1}^{r} (\tanh tc_i) e_i$, $0 \leqslant t \leqslant 1$. Then $\gamma(t)$ is a geodesic in D with respect to the Bergman metric, and $\gamma(0) = 0$, $\gamma(1) = \lambda$ and $\gamma\left(\frac{1}{2}\right) = \lambda_m$. We define ϕ_λ to be the unique holomorphic automorphism such that $\phi_\lambda(\lambda_m) = \lambda_m$, $\phi^2 = id$ and λ_m is the isolated fixed point of ϕ_λ in D.

We choose an element Z in the center of $\mathfrak{k}$ such that the complex structure J on $\mathfrak{P}$ is given by $J = \mathrm{ad}(Z)|_{\mathfrak{P}}$. Then $\exp(\pi Z)$ is in K and acts on D by $-id$.

Lemma 3.3. *If* $\lambda = \sum_{i=1}^{r} \lambda_i e_i \in D$, $-1 < \lambda_i < 1$, $i = 1, \cdots, r$, *then* ϕ_λ *is given by*

$$\phi_\lambda = g \exp \pi Z g^{-1}, \tag{3.5}$$

where $g = \exp \sum_{i=1}^{r} \frac{c_i}{2} X_i$, $c_i = \tanh^{-1} \lambda_i$ *and* $X_i = e_{\gamma_i} + e_{-\gamma_i}$.

Proof. Since $\exp\pi Z = -id$ is the symmetry at 0, it is enough to show that $g^{-1}\lambda_m = 0$. It follows from Lemma 3.5 of [3]. If $g = \exp \sum_{i=1}^{r} t_i X_i, z = \sum_{i=1}^{r} b_i e_i$, then

$$gz = \sum_{i=1}^{r} \frac{b_i \cos ht_i + \sin ht_i}{b_i \sin ht_i + \cos ht_i} e_i. \tag{3.6}$$

Since $g^{-1} = \exp\left(-\sum_{i=1}^{r} \frac{c_i}{2} X_i\right)$, $c_i = \tan h^{-1}\lambda_i, \lambda_m = \gamma\left(\frac{1}{2}\right) = \sum_{i=1}^{r}\left(\tan h \frac{1}{2} c_i\right)e_i$, we have

$$g^{-1}\lambda_m = \sum_{i=1}^{r} \frac{\tan h \frac{1}{2} c_i \cos h\left(-\frac{c_i}{2}\right) + \sin h\left(-\frac{c_i}{2}\right)}{\tan h \frac{1}{2} c_i \sin h\left(-\frac{c_i}{2}\right) + \cos h\left(-\frac{c_i}{2}\right)} e_i = 0.$$

Lemma 3.4. *If* $\lambda = \sum_{i=1}^{r} \lambda_i e_i$, $0 \leqslant \lambda_i < 1, -1 < t < 1$, *then*

$$\phi_\lambda(t\lambda) = \sum_{i=1}^{r} \frac{(1-t)\lambda_i}{1-t\lambda_i^2} e_i. \tag{3.7}$$

Proof. Let $t_i = -\frac{c_i}{2}$, $i = 1, \cdots, r$. By (3.5), we have

$$\phi_\lambda(t\lambda) = g\exp\pi Z g^{-1} \cdot t \sum_{i=1}^{r} \lambda_i e_i$$

$$= \exp\left(-\sum_{i=1}^{r} t_i X_i\right) \exp\pi Z \exp\left(\sum_{i=1}^{r} t_i X_i\right) \cdot t \sum_{i=1}^{r} \lambda_i e_i.$$

By (3.6), it equals

$$\exp\left(-\sum_{i=1}^{r} t_i X_i\right) \exp\pi Y \sum_{i=1}^{r} \frac{t\lambda_i \cos ht_i + \sin ht_i}{t\lambda_i \sin ht_i + \cos ht_i} e_i.$$

Since $\exp e Z = -id$, it equals

$$\exp\left(-\sum_{i=1}^{r} t_i X_i\right) \sum_{i=1}^{r} \left[-\frac{t\lambda_i \cos ht_i + \sin ht_i}{t\lambda_i \sin ht_i + \cos ht_i} e_i\right].$$

By (3.6), it also equals

$$\sum_{i=1}^{r} \left[\frac{-\frac{t\lambda_i \cos ht_i + \sin ht_i}{t\lambda_i \sin ht_i + \cos ht_i} \cos ht_i - \sin ht_i}{\frac{t\lambda_i \cos ht_i + \sin ht_i}{t\lambda_i \sin ht_i + \cos ht_i} \sin ht_i + \cos ht_i}\right] e_i$$

$$= \sum_{i=1}^{r} (-1) \frac{t\lambda_i \cos h^2 t_i + \sin ht_i \cos ht_i + t\lambda_i \sin h^2 t_i + \sin ht_i \cos ht_i}{t\lambda_i \sin ht_i \cos ht_i + \sin h^2 t_i + t\lambda_i \sin ht_i \cos ht_i + \cos h^2 t_i} e_i$$

$$= \sum_{i=1}^{r} (-1) \frac{t\lambda_i \cos h2t_i + \sin h2t_i}{t\lambda_i \sin h2t_i + \cos h2t_i} e_i$$

$$= \sum_{i=1}^{r} (-1) \frac{t\lambda_i + \tan h2t_i}{t\lambda_i \tan h2 t_i + 1} e_i$$

$$= \sum_{i=1}^{r} (-1) \frac{t\lambda_i - \lambda_i}{-t\lambda_i^2 + 1} e_i$$

$$= \sum_{i=1}^{r} \frac{(1-t)\lambda_i}{1 - t\lambda_i^2} e_i.$$

For any $z \in D$, we write $z = k \sum_{i=1}^{r} \mu_i e_i$ with $k \in K$, $\mu_1 \geqslant \mu_0 \geqslant \cdots \geqslant \mu_r \geqslant 0$, and define $\|z\| = \mu_1$. It can be easily shown that $\|kz\| = \|z\|$ holds for every $k \in K$. Let f be a Bloch mapping with Bloch norm $\|f\|_B$, and $\lambda \in D$, $\phi_\lambda(z)$ be the holomorphic automorphism of D which maps λ to origin and origin to λ. Then

$$\frac{\partial f}{\partial z}(\lambda) = \frac{\partial}{\partial z}(f \circ \phi_\lambda)(0)\left(\frac{\partial \phi_\lambda}{\partial z}(0)\right)^{-1},$$

hence

$$\left\|\frac{\partial f}{\partial z}(\lambda)\right\|_M \leqslant \|f\|_B \left\|\left(\frac{\partial \phi_\lambda}{\partial z}(0)\right)^{-1}\right\|_M,$$

where $\| \ \|_M$ is the norm of matrix. By a result of J. Arazy[1)],

$$\frac{\partial \phi_\lambda}{\partial z}[z] = -B[\lambda,\lambda]^{\frac{1}{2}} B(z,\lambda)^{-1}, \tag{3.8}$$

where $B(z,w)$ is the Bergman kernel operator, therefore

$$\left\|\frac{\partial \phi_\lambda}{\partial z}(0)^{-1}\right\|_M = \|B(\lambda,\lambda)^{-\frac{1}{2}}\|_M.$$

It is straightforward that for $\lambda \in D$, we have

$$\|B(\lambda,\lambda)^{-\frac{1}{2}}\|_M = \frac{1}{1 - \|\lambda\|^2}. \tag{3.9}$$

Thus we have proved

Lemma 3.5. *If f is a Bloch mapping of an irreducible bounded symmetric domain D into C^n, then* $\left\|\frac{\partial f}{\partial z}(\lambda)\right\|_M \leqslant \frac{\|f\|_B}{1 - \|\lambda\|^2}$.

Now we calculate the left-hand side of (1.3)

$$\frac{\det \left.\frac{\partial \phi_\lambda}{\partial z}\right|_{z=t\lambda}}{\det \left.\frac{\partial \phi_\lambda}{\partial z}\right|_{z=\lambda}}.$$

1) Personal communication.

It follows from (2.2) and (3.8) that

$$\det \frac{\partial \phi_\lambda}{\partial z} = (-1)^n K(z,\lambda) K(\lambda,\lambda)^{-\frac{1}{2}} = (-1)^n h(z,\lambda)^{-p} h(\lambda,\lambda)^{\frac{p}{2}}. \tag{3.10}$$

If $\lambda = \sum_{i=1}^{r} \lambda_i e_i$, $\lambda_1 \geqslant \cdots \geqslant \lambda_r \geqslant 0$, $-1 < t < 1$, then

$$h(\lambda,\lambda) = \prod_{i=1}^{r} (1 - \lambda_i^2), \tag{3.11}$$

$$h(t\lambda,\lambda) = \prod_{i=1}^{r} (1 - t\lambda_i^2). \tag{3.12}$$

Now by (3.10), (3.11) and (3.12), we have

$$\frac{\det \dfrac{\partial \phi_\lambda}{\partial z}\Big|_{z=t\lambda}}{\det \dfrac{\partial \phi_\lambda}{\partial z}\Big|_{z=\lambda}} = \frac{\prod_{i=1}^{r} (1 - \lambda_i^2)^p}{\prod_{i=1}^{r} (1 - t\lambda_i^2)^p}. \tag{3.13}$$

4 Bloch Constant of Family of Holomorphic Mappings on Bounded Symmetric Domains

Now we can estimate (1.3) when D is a bounded symmetric domain.

By Lemma 3.2, suppose $\lambda = \sum_{i=1}^{r} \lambda_i e_i$, $\lambda_1 \geqslant \lambda_2 \geqslant \cdots \geqslant \lambda_r \geqslant 0$. Then

$$(EK)(\lambda,\lambda) = p \prod_{j=1}^{r} (1-\lambda_j^2)^{-(p+1)} \left[\sum_{i=1}^{r} \lambda_i^2 \prod_{j \neq i}^{r} (1 - \lambda_j^2) \right].$$

Since $K(\lambda,\lambda) = h(\lambda,\lambda)^{-p} = \prod_{j=i}^{r} (1 - \lambda_j^2)^{-p}$, we have

$$(EK)(\lambda,\lambda) + K(\lambda,\lambda) = \prod_{j=1}^{r} (1 - \lambda_j^2)^{-(p+1)} \left\{ \prod_{j=1}^{r} (1 - \lambda_j^2) + p \left[\sum_{i=1}^{r} \lambda_i^2 \prod_{j \neq i}^{r} (1 - \lambda_j^2) \right] \right\}, \tag{4.1}$$

$$(EK)(\lambda,\lambda) - K(\lambda,\lambda) = - \prod_{j=1}^{r} (1 - \lambda_j^2)^{-(p+1)} \left\{ \prod_{j=1}^{r} (1 - \lambda_j^2) - p \left[\sum_{i=1}^{r} \lambda_i^2 \prod_{j \neq i}^{r} (1 - \lambda_j^2) \right] \right\}. \tag{4.2}$$

Substituting (3.13), (4.1) and (4.2) into (1.3), we have

Lemma 4.1. *If $F: D \to C^n$ is a holomorphic mapping on an irreducible* *bounded symmetric domain*, $\lambda=\sum_{i=1}^{r}\lambda_i e_i\in D$, $-1<t<1$, $F\in B(D)$, $D(F)=1$, $\det\frac{\partial F}{\partial z}(0)=1$, $\phi_\lambda(z)$ *is a holomorphic automorphism which maps* λ *onto origin and origin onto* λ, *then* $Re\left\{\det\frac{\partial F}{\partial z}(\phi_\lambda(t\lambda))\right\}$ *is greater than or equal to*

$$\prod_{j=1}^{r}(1-t\lambda_j^2)^p\left\{\prod_{j=1}^{r}(1-\lambda_j^2)+p\left[\sum_{i=1}^{r}\lambda_i^2\prod_{j\neq i}^{r}(1-\lambda_i^2)\right]\right\}t+\left\{\prod_{j=1}^{r}(1-\lambda_j^2)\right.$$
$$\left.-p\left[\sum_{j=1}^{r}\lambda_i^2\prod_{j\neq i}^{r}(1-\lambda_i^2)\right]\right\}\Big/\prod_{j=1}^{r}(1-\lambda_j^2)^p\left\{\prod_{j=1}^{r}(1-\lambda_j^2)+p\left[\sum_{i=1}^{r}\lambda_i^2\prod_{j\neq i}^{r}(1-\lambda_j^2)\right]\right\}$$
$$+\left\{\prod_{j=1}^{r}(1-\lambda_j^2)-p\left[\sum_{i=1}^{r}\lambda_i^2\prod_{j\neq i}^{r}(1-\lambda_i^2)\right]\right\}t. \tag{4.3}$$

Let

$$\mu_i=\frac{(1-t)\lambda_i}{1-t\lambda_i^2} \tag{4.4}$$

when $-1<t<1$. Then $\phi_\lambda(t\lambda)=\sum_{i=1}^{r}\mu_i e_i=\mu$, $1>\mu_1\geqslant\cdots\geqslant\mu_r\geqslant 0$ and (4.3) becomes

$$Re\left\{\det\frac{\partial F}{\partial z}(\mu)\right\}\geqslant\prod_{j=1}^{r}\frac{1}{(1-\lambda_j\mu_j)^p}\cdot\frac{(t+1)-p}{(t+1)+p}. \tag{4.5}$$

We have

$$(t+1)-p\sum_{i=1}^{r}\frac{\lambda_i\mu_i}{1-\lambda_i\mu_i}=\frac{1}{r}\sum_{i=1}^{r}\left(\frac{\lambda_i-\mu_i}{\lambda_i(1-\lambda_i\mu_i)}+1\right)-p\sum_{i=1}^{r}\frac{\lambda_i\mu_i}{1-\lambda_i\mu_i}$$
$$=\sum_{i=1}^{r}\frac{2\lambda_i-\mu_i-(pr+1)\lambda_i^2\mu_i}{r\lambda_i(1-\lambda_i\mu_i)}.$$

If we take $\lambda_1=(pr+1)^{-\frac{1}{2}}$, then $0<\mu_1<\frac{2\lambda_1}{1+\lambda_1^2}=\frac{2(pr+1)^{-\frac{1}{2}}}{pr+2}$. Conversely, if $\mu_1,\cdots,\mu_r$ are given, satisfying the following conditions

$$\frac{2(pr+1)^{\frac{1}{2}}}{pr+2}>\mu_1\geqslant\mu_2\geqslant\cdots\geqslant 0,$$

then we can choose t such that $\lambda_1=(pr+1)^{-\frac{1}{2}}$, $\lambda_2,\cdots,\lambda_r$ satisfies $-1<t<1$, and $(pr+1)^{-\frac{1}{2}}=\lambda_1\geqslant\lambda_2\geqslant\cdots\geqslant\lambda_r\geqslant 0$. And λ_i, μ_i satisfy relation (4.4). Obviously, we only need to take

$$t=\frac{(pr+1)^{\frac{1}{2}}(1-\mu_1(pr+1)^{\frac{1}{2}})}{(pr+1)^{\frac{1}{2}}-\mu_1}. \tag{4.6}$$

Then t satisfies $-1<t<1$ when $0<\mu_1<\frac{2(pr+1)^{\frac{1}{2}}}{pr+2}$. By (4.4), we get $\lambda_2,\cdots,$

λ_r when t is chosen. It remains to show that $\lambda_1 \geqslant \lambda_2 \geqslant \cdots \geqslant \lambda_r \geqslant 0$. We drop the index in (4.4), then the relation between λ and μ is

$$t = \frac{\lambda - \mu}{\lambda(1 - \lambda\mu)}, \tag{4.7}$$

where t is a constant. Thus λ is a function of μ. Differentiating (4.7) with respect to μ, we have

$$\lambda' = \frac{1 - \lambda^2}{\mu(1 - 2\lambda\mu + \lambda^2)}.$$

Hence $\lambda' \geqslant 0$ since $0 \leqslant \mu \leqslant 1, 0 \leqslant \lambda < 1$, i.e. λ is the monotonic function of μ. Thus $\lambda_1 \geqslant \lambda_2 \geqslant \cdots \geqslant \lambda_r \geqslant 0$ if $\mu_1 \geqslant \mu_2 \geqslant \cdots \geqslant \mu_r \geqslant 0$.

The denominator of (4.5) is positive, and

$$\frac{\lambda_1\mu_1}{1 - \lambda_1\mu_1} \geqslant \cdots \geqslant \frac{\lambda_r\mu_r}{1 - \lambda_r\mu_r} \geqslant 0.$$

From the above inequalities, (4.6) and (4.5), we have

Lemma 4.2. *Let D be an irreducible bounded symmetric domain in C^n, $F \in \mathscr{B}(D)$,*

$$D(F) = 1,\ \det\frac{\partial F}{\partial z}(0) = 1, \mu = \sum_{i=1}^{r} \mu_i e_i \in D,\ \mu_1 \geqslant \mu_2 \geqslant \cdots \geqslant \mu_r \geqslant 0. \text{ Then}$$

$$Re\left\{\det\frac{\partial F}{\partial z}(\mu)\right\} \geqslant \frac{1 - (pr+1)^{\frac{1}{2}}\mu_1}{\left(1 - \dfrac{\mu_1}{(pr+1)^{\frac{1}{2}}}\right)^{pr+1}} \tag{4.8}$$

holds provided that $\mu_1 < \dfrac{2(pr+1)^{\frac{1}{2}}}{pr+2}$.

Proof. The right-hand side can be rewritten as

$$\prod_{j=1}^{r} \frac{1}{(1 - \lambda_j\mu_j)^p} \cdot \frac{(t+1) + pr - p\sum_{j=1}^{r}\dfrac{1}{1 - \lambda_j\mu_j}}{(t+1) - pr + p\sum_{j=1}^{r}\dfrac{1}{1 - \lambda_j\mu_j}}. \tag{4.9}$$

Fix an i, and let $x = \dfrac{1}{1 - \lambda_i\mu_i}$, $A = t + 1 + pr - p\sum_{j\neq i}\dfrac{1}{1 - \lambda_j\mu_j}$, $B = t + 1 - pr + p\sum_{j\neq i}\dfrac{1}{1 - \lambda_j\mu_j}$. Then (4.9) can be expressed as

$$\prod_{j\neq i} \frac{1}{(1 - \lambda_j\mu_j)^p} x^p \frac{A - px}{B + px}.$$

Let $f(x) = x^p \dfrac{A - px}{B + px}$. Then

$$f'(x) = \frac{px^{p-1}}{(B + px)^2}[AB + ((p-1)A - (p+1)B)x - p^2x^2].$$

Let $Q = \sum_{j \neq i} \frac{1}{1 - \lambda_j \mu_j}$. Then

$$
\begin{aligned}
AB &+ ((p-1)A - (p+1)B)x - p^2x^2 = (t+1+pr-pQ)(t+1-pr+pQ) \\
&+ ((p-1)(t+1+pr-pQ) - (p+1)(t+1-pr+pQ))x - p^2x^2 \\
&= (t+1)^2 - (r-Q)^2p + 2p^2(r-Q)x - 2(t+1)x - p^2x^2 \\
&= (t+1)(t+1-2x) - p^2(r-Q-x)^2.
\end{aligned}
$$

From (4.4), we have $t = \frac{\lambda_i - \mu_i}{\lambda_i - \lambda_i^2 \mu_i}$, then

$$
t + 1 - 2x = \frac{\lambda_i - \mu_i}{\lambda_i - \lambda_i \mu_i^2} + 1 - \frac{2}{1 - \lambda_i \mu_i} = -\frac{\mu_i(1+\lambda_i^2)}{\lambda_i(1-\lambda_i\mu_i)} \leqslant 0,
$$

and hence $f'(x) \leqslant 0$. Since $\frac{1}{1-\lambda_i\mu_i} \leqslant \frac{1}{1-\lambda_1\mu_1}$, (4.9) is non-increasing if we replace $\frac{1}{1-\lambda_i\mu_i}$ by $\frac{1}{1-\lambda_1\mu_1}$. (4.9) is greater than or equal to

$$
\frac{1}{(1-\lambda_1\mu_1)^{pr}} \frac{t+1+pr-pr\dfrac{1}{1-\lambda_1\mu_1}}{t+1-pr+pr\dfrac{1}{1-\lambda_1\mu_1}}.
$$

We get (4.8) when we substitute $t = \frac{\lambda_1 - \mu_1}{\lambda_1 - \lambda_1^2\mu_1}$ and $\lambda_1 = (1+pr)^{-\frac{1}{2}}$ in the above expression.

If $z \in D$, then z can be written as $z = k\sum_{i=1}^{r} \mu_i e_i, 1 > \mu_1 \geqslant \cdots \geqslant \mu_r \geqslant 0, k \in K$.

Fix k. Let $\mu = \sum_{i=1}^{r} \mu_i e_i$. Then $z = k\mu$. Let $G(\mu) = F(k\mu)\left(\frac{\partial z}{\partial \mu}\right)^{-1}$. Then $\frac{\partial z}{\partial \mu}$ is independent of μ, and hence, $\frac{\partial G}{\partial \mu} = \frac{\partial F}{\partial z}$. Thus $\det \frac{\partial G}{\partial \mu} = \det \frac{\partial F}{\partial z}$. If $F \in \mathscr{B}(D)$, $D(F) = 1$, $\det \frac{\partial F}{\partial \mu}(0) = 1$, then G satisfies these conditions, i.e. $G \in \mathscr{B}(D), D(G) = 1$, $\det \frac{\partial G}{\partial \mu}(0) = 1$. By Lemma 4.2, we have

Lemma 4.3 (Bonk's Distortion Theorem). *Let D be an irreducible bounded symmetric domain in C^n, $F \in \mathscr{B}(D), D(F) = 1$, $\det \frac{\partial F}{\partial z}(0) = 1$, $z = k\sum_{i=1}^{r} \mu_i e_i \in D, k \in K$. Then*

$$
Re\left\{\det \frac{\partial F}{\partial z}(z)\right\} \geqslant \frac{1-(pr+1)^{\frac{1}{2}}\|z\|}{\left(1-\dfrac{\|z\|}{\sqrt{pr+1}}\right)^{pr+1}} \tag{4.10}
$$

holds provided thar $\|z\| < \frac{2(pr+1)^{\frac{1}{2}}}{pr+2}$, where $\|z\| = \mu_1$.

Lemma 4.3 is the generalization of Bonk's theorem on the irreducible bounded domain. Then we have the following theorem for the estimation of the lower bound of the Bloch constant of $B_K(D)$.

Theorem 4.1. *The Bloch constant* $\mathscr{B}_K$ *of* $B_K(D)$ *satisfies the following inequality*

$$B_K \geqslant K^{1-n}\int_0^{\frac{1}{\sqrt{pr+1}}}\frac{(1-\sqrt{pr+1}t)(1-t^2)^{n-1}}{\left(1-\dfrac{t}{\sqrt{pr+1}}\right)^{pr+1}}dt. \tag{4.11}$$

Proof. Let $F\in\mathscr{B}_K(D)$. We consider what causes a ball in the range of F not having a well-defined inverse delete of F. For a small ball centered at the origin, since $\det\dfrac{\partial F}{\partial z}(0)=1$ and $F(0)=0$, there is a well-defined inverse of F on the ball. When the ball in the range expands, either the preimage reaches a point at which $\det\dfrac{\partial F}{\partial z}=0$ or it reaches a point of the boundary. Otherwise, one could argue that the ball in the range could be enlarged further. Thus from the estimate on the Re $\det\dfrac{\partial F}{\partial z}$ in inequality (4.10), we can force $\det\dfrac{\partial F}{\partial z}$ to be non-zero by requiring $\|z\|<\dfrac{1}{\sqrt{pr+1}}$. Now consider a straight line interval Γ in the range of F. The interval Γ starts at the origin and goes as far as it can without having its preimage run into the boundary of D or $\det\dfrac{\partial F}{\partial z}=0$. Then

$$r(0,F)\geqslant\left|\int_\Gamma dw\right|=\int_\Gamma|dw|=\int_\gamma\left|\frac{\partial F}{\partial z}\frac{dz}{|dz|}\right|\cdot|dz|,$$

where $\gamma=F^{-1}(\Gamma)$. Obviously

$$\int_\gamma\left|\frac{\partial F}{\partial z}\frac{dz'}{|dz|}\right|\cdot|dz|\geqslant\int_\gamma\frac{\left|\det\dfrac{\partial F}{\partial z}\right|}{\left\|\dfrac{\partial F}{\partial z}\right\|_M^{n-1}}dz.$$

The right-hand side of the above inequality is greater than or equal to

$$\int_0^{\frac{1}{\sqrt{pr+1}}}\frac{(1-\sqrt{pr+1}\|z\|)}{\left(1+\dfrac{\|z\|}{\sqrt{pr+1}}\right)^{pr+1}}\frac{(1-\|z\|^2)^{n-1}}{K^{n-1}}d|z| \tag{4.12}$$

by Lemma 3.5 and Lemma 4.3.

Let $c=\sqrt{pr+1}$, $x=\|z\|$, $g(x)=\dfrac{(1-cx)(1-x^2)^{n-1}}{\left(1-\dfrac{x}{c}\right)^{c^2}}$. Then

$$g'(x) = \frac{(1-x^2)^{n-2}}{\left(1-\frac{x}{c}\right)^{c^2+1}}\left[(1-c^2)x(1-x^2) - 2(n-1)x\left(1-\frac{x}{c}\right)(1-cx)\right]$$

$$= \frac{(1-x^2)^{n-2}}{\left(1-\frac{x}{c}\right)^{c^2+1}}\left[-prx(1-x^2) - 2(n-1)x\left(1-\frac{x}{c}\right)(1-cx)\right] \leqslant 0$$

when $x \leqslant \frac{1}{c}$. Since $|z| \geqslant \|z\|$, (4.12) is greater than or equal to

$$\int_0^{\frac{1}{\sqrt{pr+1}}} \frac{(1-\sqrt{pr+1}\,t)(1-t^2)^{n-1}}{\left(1-\frac{t}{\sqrt{pr+1}}\right)^{pr+1} K^{n-1}}\,dt.$$

We complete the proof of Theorem 4.1.

We already know that [5,6], for $R_1: I - U\bar{Z}' > 0$, $Z = Z^{(m,n)}$, $1 \leqslant m \leqslant n$, then $p = n+m, r = m$. Substituting these values in (4.11), the Bloch constant $B_K(R_1)$ is greater than or equal to

$$K^{1-mn}\int_0^{\frac{1}{\sqrt{mn+m^2+1}}} \frac{(1-\sqrt{mn+m^2+1}\,t)(1-t^2)^{mn-1}}{\left(1-\frac{t}{\sqrt{mn+m^2+1}}\right)^{mn+m^2+1}}\,dt.$$

This is the result in [1].

For R_{II}: $I - Z\bar{Z} > 0$, $Z = Z' = Z^{(n)}$, then $p = n+1$, $r = n$, and $(pr+1)^{\frac{1}{2}} = (n^2+n+1)^{\frac{1}{2}}$. Substituting these values in (4.11), the Bloch constant $B_K(R_{II})$ is greater than or equal to

$$K^{-\frac{1}{2}(n-1)(n+2)}\int_0^{\frac{1}{\sqrt{n^2+n+1}}} \frac{(1-\sqrt{n^2+n+1}\,t)(1-t^2)^{\frac{1}{2}(n-1)(n+2)}}{\left(1-\frac{t}{\sqrt{n^2+n+1}}\right)^{n^2+n+1}}\,dt.$$

For R_{III}: $I + Z\bar{Z} > 0$, $Z = -Z' = Z^{(n)}$, then $p = 2n-2$, $r = \left[\frac{n}{2}\right]$, and $(pr+1)^{\frac{1}{2}} = (n^2-n+1)^{\frac{1}{2}}$ when n is even; $(pr+1)^{\frac{1}{2}} = (n^2-2n+2)^{\frac{1}{2}}$ when n is odd. Substituting these values in (4.11), the Bloch constant $B_K(R_{III})$ is greater than or equal to

$$K^{-\frac{1}{2}(n+1)(n-2)}\int_0^{\frac{1}{\sqrt{n^2-n+1}}} \frac{(1-\sqrt{n^2-n+1}\,t)(1-t^2)^{\frac{1}{2}(n+1)(n-2)}}{\left(1-\frac{t}{\sqrt{n^2-n+1}}\right)^{n^2-n+1}}\,dt$$

when n is even, or

$$K^{-\frac{1}{2}(n+1)(n-2)}\int_0^{\frac{1}{\sqrt{n^2-2n+2}}} \frac{(1-\sqrt{n^2-2n+\alpha}\,t)(1-t^2)^{\frac{1}{2}(n+1)(n-2)}}{\left(1-\frac{t}{\sqrt{n^2-2n+2}}\right)^{n^2-2n+2}}\,dt$$

when n is odd.

For R_{IV} : $1+|zz'|^2-2z\bar{z}'>0, 1-|zz'|>0$, then $p=n$, $r=2$, and $(pr+1)^{\frac{1}{2}}=(2n+1)^{\frac{1}{2}}$. Substituting these values in (4.11), the Bloch constant $B_K(R_{IV})$ is greater than or equal to

$$K^{1-n}\int_0^{\frac{1}{\sqrt{2n+1}}}\frac{(1-\sqrt{2n+1}\,t)(1-t^2)^{n-1}}{\left(1-\frac{t}{\sqrt{2n+1}}\right)^{2n+1}}dt.$$

For the exceptional domain R_V of dimension 16, $p=12$, $r=2$, and $(pr+1)^{\frac{1}{2}}=5$. Substituting these values in (4.11), the Bloch constant $B_K(R_V)$ is greater than or equal to

$$K^{-15}\int_0^{\frac{1}{5}}\frac{(1-5t)(1-t^2)^{15}}{\left(1-\frac{t}{5}\right)^{25}}dt=0.02553342\cdots K^{-15}.$$

For the exceptional domain R_{VI} of dimension 27, $p=18$, $r=3$, and $(pr+1)^{\frac{1}{2}}=7$. Substituting these values in (4.11), the Bloch constant $B_K(R_{IV})$ is greater than or equal to

$$K^{-26}\int_0^{\frac{1}{7}}\frac{(1-7t)(1-t^2)^{26}}{\left(1-\frac{t}{7}\right)^{49}}dt=0.01318193\cdots K^{-26}.$$

References

1 Gong, S. & FitzGerald, C. H., *Chinese Science Bulletin*, 1992, **37**(5):406.
2 Faraut, J. & Koranyi, A., *Functional Analysis* 1990, **88**(1):64.
3 Upmeier, H., *Tran. Amer. Math. Soc.*, 1983, **280**:221.
4 Koranyi, A. & Wolf, J. A., *Ann. of Math.*, 1965, **81**:265.
5 Arazy, J. & Fisher, S. D., Invariant Hilbert spaces of analytic functions on bounded symmetric domains, *preprint*.
6 Hua, L. K., *Amer. Math. Soc.*, Providence, RI, 1963.

The Locally Biholomorphic Bloch and Marden Constants in Several Complex Variables

C. H. FitzGerald
and S. Gong

Abstract. This work continues a program to generalize one variable geometric function theory to several complex variables. The first mappings studied are required to be locally biholomorphic on spaces of matrices into $\mathbb{C}^n$. It is required that these mappings have a limited rate of growth as the boundary of the domain is approached. (Otherwise the Bloch constant may not exist.) These requirements specify classes of mappings. For each class of mappings there is a positive number r such that, for each mapping f in the class, there is a subdomain which f carries onto a ball of radius r in a one-to-one fashion. The supremum of such numbers r is the Bloch constant for the class of mappings. For a wide range of classes, a positive lower bound is found for the Bloch constant.
For the last portion of the paper, the mappings are required to be holomorphic, not necessarily locally biholomorphic. For each class of mappings, there is a positive number ρ such that, for each mapping f, there is a hyperbolic ball of radius ρ in the domain of f on which f is one-to-one. The supremum of such numbers ρ is the Marden constant for the class of mappings. From the earlier results of the paper, a positive lower bound of the Marden constant is obtained.

1 Background

This paper continues an effort to extend geometric function theory of one variable to several complex variables. In a previous paper [2] we extended the theory of the Bloch constant of holomorphic mappings to the classical domain $\mathcal{R}_I$: $\{Z : Z$ is an $m\times n$ complex matrix and $I-Z\overline{Z}'$ is positive definite, where $\overline{Z}'$ is the transpose of the complex conjugate$\}$. Here we use several of those results to discuss the locally biholomorphic Bloch constant and the Marden constant.

There are several definitions which need to be made. Most are the same as used in [2]. Let Ω be a bounded, homogeneous domain containing the origin in the space of several complex variables. Let $Aut(\Omega) = \{\varphi : \varphi$ is a biholomorphic mapping of Ω onto $\Omega\}$. Let $H(\Omega)$ denote the family of all holomorphic mappings which map Ω into $\mathbb{C}^n$.

To say that a mapping $F \in H(\Omega)$ is a *Bloch mapping* of Ω means that $\{g : g(z) = F(\varphi(z)) - F(\varphi(0))$ for some $\varphi \in Aut(\Omega)\}$ is a normal family on Ω.

The *Bloch norm* of a mapping is given by

$$\|F\|_B = \sup\{\|\frac{\partial F\circ\varphi}{\partial z}(0)\|_M : \varphi \in Aut(\Omega)\}$$

Computational Methods and Function Theory (CMFT'94)
R. M. Ali, St. Ruscheweyh and E. B. Saff (Eds.), pp. 147–158

ISBN 981-02-2129-0

where $\|A\|_M$ denotes the matrix norm of A, that is, the nonnegative square root of the maximum eigenvalue of $A\overline{A}'$. Let $d(F) = \sup_{z\in\Omega} \frac{|\det F'(z)|}{\sqrt{K_\Omega(z,\overline{z})/K_\Omega(0,0)}}$, where $K_\Omega(z,\overline{z})$ is the Bergman kernel function of Ω and $\det F'(z)$ is the determinant of the complex Jacobian of $F(z)$. Then d is a functional. In general, it is not a semi-norm. It is invariant in the sense that for each $\varphi \in Aut(\Omega)$, $d(F\circ\varphi) = d(F)$. It follows that

$$d(F) = \sup\{|\det(F\circ\varphi)'(0)| : \varphi \in Aut(\Omega)\}.$$

Certain families of mappings are denoted with script capital letters:

$$\begin{aligned} \mathcal{B} &= \{F \in H(\Omega) : \det \frac{\partial F}{\partial z}(0) = 1 \text{ and } \|F\|_{\mathcal{B}} < +\infty\}, \\ \mathcal{B}_K &= \{F \in \mathcal{B} : \|F\|_{\mathcal{B}} \le K\} \text{ for } 1 \le K < +\infty, \\ \mathcal{B}_{K,1} &= \{F \in \mathcal{B}_K : d(F) = 1\}. \end{aligned}$$

Various numbers are associated with those families: B_K is the Bloch constant of $\mathcal{B}_K$; specifically it is the supremum of the radii r such that, for all $F \in \mathcal{B}_K$, there is a ball of radius r on which there is a well-defined branch of F^{-1}. Let $r(a, F)$ be the largest nonnegative number r such that there is a domain $\mathcal{D}$ with $a \in \mathcal{D} \subset \Omega$ which is mapped by F biholomorphically onto the ball of radius r centered at $F(a)$, that is, onto $B(F(a), r)$.

Let $r(F) = \sup_{a\in\Omega} r(a, F)$. Then $B_K = \inf r(F)$ where the infimum is taken over $F \in \mathcal{B}_K$.

In fact, not all of $\mathcal{B}_K$ needs to be considered. We showed [2] that there exists $F \in \mathcal{B}_{K,1}$ with $r(F) = B_K$. (This is a Landau type theorem cf. Landau [4].)

In [2] we studied the Bloch constant for the family of holomorphic mappings in several variables. In this paper we examine the Bloch constant for the family of locally biholomorphic mappings on Ω. The following definitions will make this more specific.

A superscript zero will indicate that the mappings considered are locally biholomorphic:

$$\begin{aligned} \mathcal{B}^0 &= \{F \in \mathcal{B} : \det \frac{\partial F}{\partial z}(z) \ne 0 \text{ for } z \in \Omega\}, \\ \mathcal{B}^0_K &= \{F \in \mathcal{B}^0 : \|F\|_{\mathcal{B}} \le K\} \text{ for } 1 \le K < +\infty, \\ \mathcal{B}^0_{K,1} &= \{F \in \mathcal{B}^0_K : d(F) = 1\}. \end{aligned}$$

Then B^0_K is the Bloch constant of $\mathcal{B}^0_K$, that is, the supremum of the radii r such that, for all $F \in \mathcal{B}^0_K$ there is a ball of radius r on which F^{-1} can be defined. We call B^0_K the locally biholomorphic Bloch constant. Landau's observation holds, that is, there exists $F \in \mathcal{B}^0_{K,1}$ with $r(F) = B^0_K$.

The main part of this paper is devoted to finding a lower bound on the locally biholomorphic Bloch constant for $\mathcal{B}^0_K$ where the domain is $\mathcal{R}_I$.

The paper concludes with an estimation of the Marden constant for a certain class of mappings of $\mathcal{R}_I$. The Marden constant is the largest number r_0 such that every mapping in the class is biholomorphic on a ball of hyperbolic radius r_0 in the domain.

2 General theorems

The following theorem is a version of Julia's lemma. In [2] we extended Minda's version [7] to several variables. Here we state a version which uses the additional assumption that $g(z)$ is nonzero. The proof is based on the work of Liu and Minda [6].

Theorem 1 *Suppose that Ω is a bounded, complete circular domain and that the point a belongs to the closure of Ω. Let $\mathbf{D}$ denote the open unit disk in $\mathbb{C}$. If g is a nonzero, holomorphic function in $\Omega \cup \{a\}$ which maps $\Omega \cup \{a\}$ into $\mathbb{C}$ and $|g(w)| \leq |g(a)|$ for all $w = za$ for $z \in \mathbf{D}$, then $c = [\frac{dg}{dw}(a) \cdot a']/g(a)$ is positive and*

$$\left|\frac{g(xa)}{g(a)}\right| \geq \exp\left\{-2c\frac{1-x}{1+x}\right\} \quad \text{for } x \in (-1,1) \tag{2.1}$$

Proof: Let $h(z) = \frac{g(za)}{g(a)}$ for $z \in \mathbf{D}$. Then $h(z)$ is holomorphic in $\mathbf{D} \cup \{1\}$, $h(\mathbf{D}) \subset \mathbf{D}$, $h(1) = 1$ and $c = \left.\frac{dh(z)}{dz}\right|_{z=1} = h'(1)$. Since $h(z)$ is nonzero and $\mathbf{D}$ is simply connected, a well-defined choice of $H(z) = -\log h(z)$ can be made. Then $Re\, H(z) > 0$ for $z \in \mathbf{D}$ and $H(1) = 0$.

$$\begin{aligned} c &= \left.\frac{dh}{dz}\right|_{z=1} = h'(1) = \left[\frac{dg}{dw}(a)\cdot a'\right]/g(a) \\ &= [\exp(-H(z))]\left[-\frac{dH(z)}{dz}\right]\Big|_{z=1} \\ &= -\left.\frac{dH(z)}{dz}\right|_{z=1} > 0. \end{aligned}$$

Note that $|h(z)| = \exp\{-Re\, H(z)\}$. Consider Julia's lemma ([6], p. 327). The function $H(z)$ satisfies all the conditions for the half-plane version and for the corollary. For any $x \in (-1,1)$, $Re\,(H(x)) \leq 2d\frac{1-x}{1+x}$ where $d = -\left.\frac{dH(z)}{dz}\right|_{z=1} = c > 0$. We conclude that $|h(z)| = \left|\frac{g(xa)}{g(a)}\right| \geq \exp\{-2d\frac{1-x}{1+x}\} = \exp\{-2c\frac{1-x}{1+x}\}$ as claimed.

The following theorem is Theorem 2.2 in [2].

Theorem 2 *Suppose that Ω is a bounded, complete circular, homogeneous domain. Consider a holomorphic mapping F of Ω into $\mathbb{C}^n$ such that $d(F) = 1$ and*

$\det F'(0) = 1$. Let $\varphi_a(z)$ be a biholomorphic automorphism that takes a to the origin. Let $G(z) = F(\varphi_a(z))$ and $g = \det \frac{\partial G}{\partial z}$. Then, for every a in Ω,

$$\begin{aligned} c &= \left(\frac{dg(a)}{da}\right) \cdot a'/g(a) \\ &= \frac{\partial}{\partial a} K_\Omega(a, \bar{a}) \cdot a'/K_\Omega(a, \bar{a}). \end{aligned} \tag{2.2}$$

3 Locally biholomorphic mappings of $\mathcal{R}_I$

In addition to the hypothesis of Theorem 2.2, we add that F is locally biholomorphic. Then $G(z) = F(\varphi_a(z))$ is locally biholomorphic and $g(z) = \det \frac{\partial G}{\partial z}$ is nonzero. Theorem 2.1 then applies to g (cf. [2]) so that equation (2.2) makes equation (2.1) become

$$\begin{aligned} \left|\frac{g(xa)}{g(a)}\right| &\geq \exp\left\{-2\left(\left[\frac{dg(a)}{da} \cdot a'\right]/g(a)\right)\frac{1-x}{1+x}\right\} \\ &= \left\{-2\,\frac{\frac{\partial}{\partial a} K_\Omega(a, \bar{a}) \cdot a'}{K_\Omega(a, \bar{a})}\,\frac{1-x}{1+x}\right\} \end{aligned} \tag{3.1}$$

for $x \in (-1, 1)$.

We have considerable information about g. Note that $g(z) = \det \frac{\partial G(z)}{\partial z}$ and $\frac{\partial G(z)}{\partial z} = \frac{\partial F}{\partial w}\frac{\partial w}{\partial z}$ where $w = \varphi_a(z)$. Also $\varphi_a(a) = 0$ and $\det F'(0) = 1$. Thus the left side of inequality (3.1) can be modified so that the inequality implies that

$$\left|\frac{\det \frac{\partial F(w)}{\partial w}\Big|_{w=\varphi(xa)} \det \frac{\partial \varphi}{\partial z}\Big|_{z=xa}}{\det \frac{\partial \varphi}{\partial z}\Big|_{z=a}}\right| \geq \exp\left\{\frac{-2\,\frac{\partial}{\partial a} K_\Omega(a, \bar{a}) \cdot a'}{K_\Omega(a, \bar{a})}\,\frac{1-x}{1+x}\right\}. \tag{3.2}$$

Now the domain is specialized to Ω being a classical domain of first type $\mathcal{R}_I$. Points will be indicated with capital letters Z or W if they are to be thought of as $m \times n$ matrices or with lower case letters z or w if they are to be thought of as mn vectors. The first part of §3 of [2] studies $\varphi_a(Z)$ in such a case. Just before equation (3.6) of [2] it is noted that

$$\frac{\det \frac{\partial \varphi}{\partial z}\Big|_{Z=xa}}{\det \frac{\partial \varphi}{\partial z}\Big|_{Z=a}} = \left(\frac{\det(I - Z_0\overline{Z}_0')}{\det(I - xZ_0\overline{Z}_0')}\right)^{m+n} > 0$$

where $Z_0 = a$. Substitution of this equation in inequality (3.2) gives

$$\begin{aligned} &\left|\det \frac{\partial F}{\partial w}(\varphi(xZ_0))\right| \geq \\ &\left[\frac{\det(I - xZ_0\overline{Z}_0')}{\det(I - Z_0\overline{Z}_0')}\right]^{m+n} \exp\left\{\frac{-2\,\frac{\partial}{\partial z} K(Z_0, \overline{Z}_0) \cdot z_0'}{K(Z_0, \overline{Z}_0)}\,\frac{1-x}{1+x}\right\}. \end{aligned} \tag{3.3}$$

A more explicit expression can be found for the terms involving $K(Z_0, \overline{Z}_0)$ because we are assuming that Ω is a classical domain of the first type. Recall ([3], p. 84) that $K_{\mathcal{R}_I}(Z, \overline{W}) = V(\mathcal{R}_I)^{-1}\{\det(I - Z\overline{W}')\}^{-(m+n)}$ where $V(\mathcal{R}_I)$ is the Euclidean volume of $\mathcal{R}_I$. Note ([2], eq. 3.7) that

$$\sum_{i=1}^{m}\sum_{j=1}^{n} z_{ij}\frac{\partial}{\partial z_{ij}}\det(I - Z\overline{Z}') =$$

$$\det\left[\begin{pmatrix} 0 & \\ & I \end{pmatrix} - Z\overline{Z}'\right] + \cdots + \det\left[\begin{pmatrix} I & \\ & 0 \end{pmatrix} - Z\overline{Z}'\right]. \tag{3.4}$$

where det[ ] means the determinant of the matrix inside [ ]. From the formula for K_{R_I} it follows that

$$\frac{\partial}{\partial z_{ij}}K_{\mathcal{R}_I}(Z, \overline{W}) =$$

$$-(m+n)V(\mathcal{R}_I)^{-1}\{\det(I - Z\overline{W}')\}^{-(m+n+1)}\frac{\partial}{\partial z_{ij}}\det(I - Z\overline{W}').$$

Thus equality (3.4) implies that

$$\sum_{i=1}^{m}\sum_{j=1}^{n} z_{ij}\frac{\partial}{\partial z_{ij}}K_{\mathcal{R}_I}(Z, \overline{Z}) = -(m+n)V(\mathcal{R}_I)^{-1}\{\det(I - Z\overline{W}')\}^{-(m+n+1)}S(Z)$$

and

$$\frac{\frac{\partial}{\partial z_0}K_{R_I}(Z_0, \overline{Z}_0)\cdot z_0'}{K_{R_I}(Z_0, \overline{Z}_0)} = \frac{-(m+n)}{\det(I - Z_0\overline{Z}_0')}S(Z_0).$$

Inequality (3.3) becomes

$$\left|\det\frac{\partial F}{\partial w}(\varphi(xZ_0))\right| \geq$$

$$\left(\frac{\det(I - xZ_0\overline{Z}_0')}{\det(I - Z_0\overline{Z}_0')}\right)^{m+n}\exp\left\{\frac{2(m+n)S(Z_0)}{\det(I - Z_0\overline{Z}_0')}\frac{1-x}{1+x}\right\}. \tag{3.5}$$

Each point in Z in $\mathcal{R}_I$ can be written in the form $Z = U\Lambda V$ where U and V are unitary and Λ is diagonal with entries $1 > \lambda_1 \geq \lambda_2 \geq \ldots \geq \lambda_m \geq 0$. First we consider the case Z_0 is diagonal:

$$Z_0 = -\begin{pmatrix} \lambda_1 & & 0 \\ & \ddots & 0 \\ 0 & & \lambda_m \end{pmatrix}, \quad 1 > \lambda_1 \geq \lambda_2 \geq \ldots \geq \lambda_m \geq 0.$$

As observed in equation (3.9) of [2], for $-1 < x < 1$,

$$\begin{aligned}\varphi(xZ_0) &= (x-1)AZ_0(I - x\overline{Z}_0'Z_0)^{-1}D^{-1} \\ &= (x-1)A(I - xZ_0\overline{Z}_0')Z_0D^{-1}\end{aligned}$$

where we can take

$$A = \begin{pmatrix} \frac{1}{\sqrt{1-\lambda_1^2}} & & 0 \\ & \ddots & \\ 0 & & \frac{1}{\sqrt{1-\lambda_m^2}} \end{pmatrix} \quad \text{and} \quad D = \begin{pmatrix} \frac{1}{\sqrt{1-\lambda_1^2}} & 0 & & \\ 0 & \ddots & & 0 \\ & & \frac{1}{\sqrt{1-\lambda_m^2}} & \\ & 0 & & I \end{pmatrix}.$$

Then $\varphi(xZ_0) = \begin{pmatrix} \mu_1 & & 0 \\ & \ddots & 0 \\ 0 & & \mu_m \end{pmatrix}$ where $\mu_j = \frac{(1-x)\lambda_j}{1-x\lambda_j^2}$ for $j = 1, \ldots, m$.
Thus we have equations (3.11) and inequalities (3.12) of [2]

$$x = \frac{\lambda_j - \mu_j}{\lambda_j(1-\lambda_j\mu_j)} \quad \text{for } j = 1, \ldots, m$$

and

$$0 \le \mu_j \le \frac{2\lambda_j}{1+\lambda_j^2} \quad \text{for } j = 1, \ldots, m.$$

We will simplify inequality (3.5). Since

$$\det(I - Z_0\overline{Z}_0') = \prod_{j=1}^{m}(1-\lambda_j^2)$$

and

$$S(Z_0) = \sum_{k=1}^{m}\prod_{j\neq k}(1-\lambda_j^2)(-\lambda_k),$$

$$\frac{S(Z_0)}{\det(I - Z_0\overline{Z}_0')} = -\sum_{k=1}^{m}\frac{\lambda_k^2}{1-\lambda_k^2}.$$

Also, equation (3.14) of [2] is easily obtained:

$$\frac{\prod_{j=1}^{m}(1-x\lambda_j^2)}{\prod_{j=1}^{m}(1-\lambda_j^2)} = \prod_{j=1}^{m}\frac{1}{1-\lambda_j\mu_j}.$$

Let $\mu = \varphi(xZ_0)$. Inequality (3.5) becomes

$$\left|\det\frac{\partial F}{\partial w}(\mu)\right| \ge \prod_{j=1}^{m}\frac{1}{(1-\lambda_j\mu_j)^{m+n}}\exp\left\{\frac{-2(m+n)(1-x)}{1+x}\sum_{k=1}^{m}\frac{\lambda_k^2}{1-\lambda_k^2}\right\}. \tag{3.6}$$

It is easily checked that

$$\frac{(1-x)\lambda_k^2}{1-\lambda_k^2} = \frac{\lambda_k\mu_k}{1-\lambda_\mu\mu_k} \quad \text{for } k=1,2,\ldots,m. \tag{3.7}$$

Inequality (3.6) becomes

$$\begin{aligned}\left|\det\frac{\partial F}{\partial w}(\mu)\right| &\geq \prod_{j=1}^{m}\frac{1}{(1-\lambda_j\mu_j)^{m+n}}\exp\left\{\frac{-2(m+n)}{1+x}\sum_{k=1}^{m}\frac{\lambda_k\mu_k}{1-\lambda_k\mu_k}\right\}\\ &= \prod_{j=1}^{m}\left[\frac{1}{1-\lambda_j\mu_j}\exp\left\{\frac{-2}{1+x}\frac{\lambda_j\mu_j}{1-\lambda_j\mu_j}\right\}\right]^{m+n}.\end{aligned} \tag{3.8}$$

For each $j=1,\ldots,m$, the factor in the product can be thought of as

$$f(y) = \frac{1}{1-y}\exp\left(\frac{-2}{1+x}\frac{y}{1-y}\right)$$

where $y=\lambda_j\mu_j$ raised to the $m+n$ power. Note that $0\leq y<1$ and $-1<x<1$,

$$\begin{aligned}f'(y) &= \frac{1}{(1-y)^3(1+x)}\left[\exp\left\{\frac{-2}{1+x}\frac{y}{1-y}\right\}\right][(1+x)(1-y)-2]\\ &< 0.\end{aligned}$$

Thus $f(y)$ is a decreasing, positive function. Since the λ_k's are less than one and decreasing, equation (3.7) implies that the products $\lambda_j\mu_j$ are decreasing with j. Inequality (3.8) still holds if each $\lambda_j\mu_j$ is replaced with $\lambda_1\mu_1$.

$$\left|\det\frac{\partial F}{\partial w}(\mu)\right| \geq \frac{1}{(1-\lambda_1\mu_1)^{m(m+n)}}\exp\left\{\frac{-2(m(m+n))}{1+x}\frac{\lambda_1\mu_1}{1-\lambda_1\mu_1}\right\}. \tag{3.9}$$

We now want to express the right side as a function of μ only as the left side is. Recall that the values of x, $\lambda_1,\ldots\lambda_m$ and $\mu_1,\ldots,\mu_m$ are related by

$$x = \frac{\lambda_j-\mu_j}{\lambda_j-\lambda_j^2\mu_j} \quad \text{for } j=1,\ldots,m. \tag{3.10}$$

When x is given and $Z_0 = \begin{pmatrix}\lambda_1 & & 0\\ & \ddots & 0\\ 0 & & \lambda_m\end{pmatrix}$, equation (3.10) determines the values of $\mu_1,\ldots,\mu_m$. On the other hand, we now show that if the μ_j's are given, we can find x and λ_j's which satisfy equation (3.10) and also satisfy the other conditions: $x\in(-1,1)$, $1>\lambda_1\geq\ldots\geq\lambda_m\geq 0$ and $1>\mu_1\geq\mu_2\geq\ldots\geq\mu_m\geq 0$. When the μ_j's are given with $1>\mu_1\geq\mu_2\geq\ldots\geq\mu_m\geq 0$, we take $\lambda_1=1-\epsilon>\mu_1$ for some small positive ϵ. Then equation (3.10) with $j=1$ gives

$$x = \frac{1-\epsilon-\mu_1}{1-\epsilon-(1-\epsilon)^2\mu_1}$$

and $x \in (0,1)$. Then the values of $\lambda_2, \dots, \lambda_m$ can be found from equation (3.10). The positive choice should be made in the quadratic formula. And, if $\mu_k = 0$, then $\lambda_k = 0$. Now inequality (3.9) can be written using only information from μ

$$\begin{aligned}\left|\det \frac{\partial F}{\partial w}(\mu)\right| &\geq \frac{1}{(1-(1-\epsilon)\mu_1)^{m(m+n)}} \\ &\times \exp\left\{\frac{-m(m+n)(1-\epsilon-(1-\epsilon)^2\mu_1)(1-\epsilon)\mu_1}{[2(1-\epsilon)-\mu_1-(1-\epsilon)^2\mu_1](1-(1-\epsilon)\mu_1)}\right\}.\end{aligned}$$

Consider the limit on $\epsilon \to 0^+$.

$$\left|\det \frac{\partial F}{\partial w}(\mu)\right| \geq \frac{1}{(1-\mu_1)^{m(m+n)}} \exp\left\{\frac{-m(m+n)\mu_1}{1-\mu_1}\right\}. \tag{3.11}$$

Note that $\|\mu\|_M = \mu_1$ and $\|U\mu V\|_M = \|\mu\|_M$ for any unitary matrices U and V. Thus the right side of inequality (3.11) can be expressed in an invariant form.

The analysis of the left side of inequality (3.11) is the same as that toward the end of §3 of [2]. To generalize the inequality to a nondiagonal point W_0 in $\mathcal{R}_I$, find unitary matrices U and V such that $W_0 = U\mu V$ where μ is diagonal. Fix U and V. Consider $W = UZV$ as Z ranges over $\mathcal{R}_I$. Let $G(Z) = U^{-n}F(UZV)V^{-m}$. Then $\frac{\partial G}{\partial z} = U^{-n}\frac{\partial F}{\partial w}\frac{\partial w}{\partial z}V^{-m}$.

$$\det \frac{\partial G}{\partial z} = (\det U)^{-n}(\det V)^{-m} \det \frac{\partial F}{\partial w} \det \frac{\partial w}{\partial z}.$$

Since $\frac{\partial w}{\partial z} = U' \otimes V$, $\det \frac{\partial w}{\partial z} = (\det U')^n(\det V)^m$. Hence $\left|\det \frac{\partial w}{\partial z}(Z)\right| = \left|\det \frac{\partial F}{\partial W}(W)\right|$. In particular, for $Z = \mu$, $W = W_0$ and $\left|\det \frac{\partial G}{\partial z}(\mu)\right| = \left|\det \frac{\partial F}{\partial W}(W_0)\right|$. Thus we have shown the following inequality.

Theorem 3 *If $\Omega = \mathcal{R}_I$ and $F \in \mathcal{B}^0$, $d(F) = 1$, $\det \frac{\partial F}{\partial w}(0) = 1$, then*

$$\left|\det \frac{\partial F}{\partial w}(W)\right| \geq (1-\|W\|_M)^{-m(m+n)} \exp\left\{\frac{-m(m+n)\|W\|_M}{1-\|W\|_M}\right\} \tag{3.12}$$

where $\|W\|_M$ is the matrix norm of the matrix W.

This theorem generalizes Theorem 1 of [6]. In case $m = 1$, Theorem 3.1 coincides with a theorem of Liu [5].

4 The locally biholomorphic Bloch constant

The following lemma is Lemma 4.1 of [2].

Lemma 1 *If f is a Bloch mapping in $\mathcal{R}_I$ which maps $\mathcal{R}_I$ into $\mathbb{C}^{mn}$, then*

$$\left\|\frac{\partial f}{\partial z}(Z)\right\|_M \le \frac{\|f\|_B}{1-\|Z\|_M^2}.$$

The analogous result to Theorem 4.1 of [2] is now presented.

Theorem 4 *If $F:\mathcal{R}_I\to\mathbb{C}^{mn}$ and $F\in\mathcal{B}^0(\mathcal{R}_I)$, $d(F)=1$, $\det\frac{\partial F}{\partial z}(0)=1$, and $\|F\|_B\le K$, then*

$$r(0,F)\ge C(K,m,n)=K^{1-mn}\int_0^1\frac{(1+t)^{mn-1}}{(1-t)^{m^2+1}}\exp\{\frac{-m(m+n)t}{1-t}\}\,dt.$$

Proof: Following Liu [5], we consider what causes a ball in the range of F centered at $F(0)$ to not have a well-defined inverse mapping F. Since $\det\frac{\partial F}{\partial z}(0)=1$, there is a well-defined inverse of F on a small ball centered at $F(0)$. Since $\det\frac{\partial F}{\partial z}\neq 0$ on $\mathcal{R}_I$, the ball can be increased in radius until the preimage hits the boundary of $\mathcal{R}_I$.

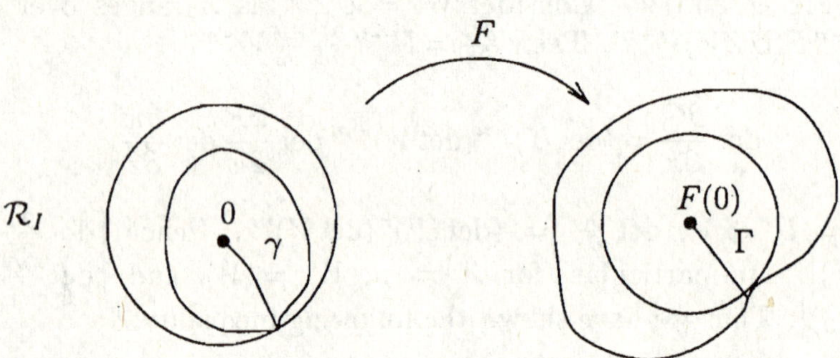

Consider a straight line interval Γ in the range of F. The interval starts at $F(0)$ with $F^{-1}(F(0))=0$ and goes until the image of F^{-1} runs into the boundary of $\mathcal{R}_I$. We assume that Γ has the minimal length among such intervals. Let $\gamma=F^{-1}(\Gamma)$. Then

$$\begin{aligned} r(0,F) &\ge \left|\int_\Gamma dw\right| \\ &= \int_\gamma\left|\frac{\partial F}{\partial z}\frac{dz'}{|dz|}\right|\cdot|dz| \\ &\ge \int_\gamma\frac{\left|\det\frac{\partial F}{\partial z}\right|}{\left\|\frac{\partial F}{\partial z}\right\|_M^{mn-1}}|dz|. \end{aligned}$$

By Lemma 4.1, and the assumed bound $\|F\|_B \le K$,

$$r(0,F) \ge \int_\gamma \frac{\left|\det \frac{\partial F}{\partial z}\right|}{K^{mn-1}}(1-\|Z\|_M^2)^{mn-1}\,|dz|.$$

By inequality (3.12) and the fact that $|dz| \ge d|z|$,

$$\begin{aligned} & r(0,F) \\ \ge\ & \int_\gamma (1-\|Z\|_M)^{-m^2-mn}\left(\exp\left\{\frac{-m(m_n)\|Z\|_M}{1-\|Z\|_M}\right\}\right)\frac{(1-\|Z\|_M^2)^{mn-1}}{K^{mn-1}}\,d|z| \\ \ge\ & K^{1-mn}\int_\gamma (1+\|Z\|_M)^{mn-1}(1-\|Z\|_M)^{1-m^2} \\ & \times \exp\left\{\frac{-m(m+n)\|Z_M\|}{1-\|Z\|_M}\right\} d|z|. \end{aligned} \tag{4.1}$$

To understand the dependence of the left side of inequality (4.1) on $\|Z\|_M$, we consider $\ell(x) = (1+x)^{mn-1}(1-x)^{-1-m^2}\exp\left\{\frac{-m(m+n)x}{1-x}\right\}$. The function $\ell(x)$ is decreasing since

$$\begin{aligned} \ell'(x) &= (1+x)^{-m^2-3}(1-x)^{mn-2}[(-2m^2-2) \\ &\quad +(-2mn-m^2+2)x+(m^2+mn)x^2] \\ &\quad \times \exp\left\{\frac{-m(m+n)x}{1-x}\right\} < 0. \end{aligned}$$

Since $|z| \ge \|Z\|_M$, inequality (4.1) implies that

$$r(0,F) \ge \int_0^1 (1+t)^{mn-1}(1-t)^{-1-m^2}\exp\left\{\frac{-m(m+n)t}{1-t}\right\} dt.$$

Thus the second inequality in the following theorem has been proved.

Theorem 5 *The Bloch constant $B_K^0(\mathcal{R}_I)$ of the family $\mathcal{B}_K^0(\mathcal{R}_I)$ satisfies the following inequalities*

$$\begin{aligned} & K^{1-mn} \ge B_K^0(\mathcal{R}_I) \ge \\ & \quad K^{1-mn}\int_0^1 (1+t)^{mn-1}(1-t)^{-m^2-1}\exp\left\{\frac{-m(m+n)t}{1-t}\right\} dt. \end{aligned} \tag{4.2}$$

The first inequality in inequalities (4.2) follows from an example:

$$F(Z) = \begin{pmatrix} Kz_{1,1} & \cdots & Kz_{1,n-1} & Kz_{1,n} \\ \cdots & \cdots & \cdots & \cdots \\ Kz_{m,1} & \cdots & Kz_{m,n-1} & K^{1-mn}z_{m,n} \end{pmatrix}$$

We note that in case $m = 1$, Theorem 4.2 becomes exactly the result found by Liu [5]. For both $m = 1$ and $n = 1$, Theorem 4.2 becomes the result found in Ahlfors [1] and Pommerenke [8].

5 The Marden constant for $\mathcal{R}_I$

Let Ω be $\mathcal{R}_I$. Consider a fixed real number K greater than or equal one. Let $\hat{\mathcal{B}}(K)$ be the set of holomorphic mappings F of $\mathcal{R}_I$ into $\mathbb{C}^{mn}$ such that

(1) $\det \frac{\partial F}{\partial z}(0) = 1$,

(2) $d(F) = 1$, and

(3) for each $\varphi \in Aut(\mathcal{R}_I)$, each root of the characteristic equation of $\frac{\partial F \circ \varphi}{\partial z}(0)$ has magnitude between $\frac{1}{K}$ and K.

In particular, the mappings in $\hat{\mathcal{B}}(K)$ are locally biholomorphic.

Let $a = K^{1-mn} \int_0^1 (1+t)^{mn-1}(1-t)^{-m^2-1} \exp\left\{\frac{-m(m+n)t}{1-t}\right\} dt$. Then, by Theorem 4.2, if $F \in \hat{\mathcal{B}}(K)$, then the range of F contains $B(F(0), a)$, a ball in $\mathbb{C}^{mn}$ centered at $F(0)$ of radius a, such that there is a well-defined mapping F^{-1} defined on $B(F(0), a)$ with $F^{-1}(F(0)) = 0$. Recall $B(F(0), a) = \{w \in \mathbb{C}^{mn} : a^2 - [W - F(0)][W - F(0)]' > 0\}$.

Let $X = a^{-1}(W - F(0))$. And let $G(X) = a^{-1}F^{-1}(aX + F(0))$. Then $\frac{\partial G}{\partial x} = a^{-1}\frac{\partial F^{-1}}{\partial x} = \frac{\partial F^{-1}(aX+F(0))}{\partial(ax)}$. Hence, $G \in B(K)$ by the lower bounds on the roots of the characteristic equation for $\frac{\partial F \circ \varphi}{\partial z}(0)$ where the domain considered is the unit ball for $X \in \mathbb{C}^{mn}$. By Liu's estimate of the Bloch constant for such mappings [5], a ball of radius $b = K^{1-mn} \int_0^1 \frac{(1+t)^{mn-1}}{(1-t)^2} \exp\left\{\frac{-(mn+1)t}{1-t}\right\} dt$ about the origin is in the range of $G(X)$. (Note that a and b are not the same because for a we are using our result on $\mathcal{R}_I$ but for b we have only a ball in $\mathbb{C}^{mn}$ thus the same theorem is used in the formula m is 1 and n is mn.) Since $G(X) = a^{-1}F^{-1}(aX + F(0))$, the ball of radius ab centered at the origin is in range of F^{-1} restricted to the ball $B(F(0), a)$. Hence a lower bound on the Marden constant has been found.

Theorem 6 *If $F \in \hat{\mathcal{B}}(K)$ for $\Omega = \mathcal{R}_I$, then F is one-to-one on a ball of hyperbolic radius* $2 \arctan ab = \log \frac{1+ab}{1-ab}$.

We believe that this lower bound of the Marden constant is much smaller than the actual value of the Marden constant.

6 Summary

The work done in [2] has been extended to apply to locally biholomorphic mappings. In particular, bounds have been obtained for the Bloch constant for this class of mappings on $\mathcal{R}_I$. Also a lower bound has been derived for the Marden constant for a class of mappings of $\mathcal{R}_I$.

Acknowledgement. We gratefully acknowledge helpful conversations and preprints from Dr. M. Bonk, Dr. X. Y. Liu and Dr. D. Minda.

References

[1] Ahlfors, L. V., An extension of Schwarz's lemma, *Trans. Amer. Math. Soc.* **43** (1938), 359–364.

[2] FitzGerald, C. H., and S. Gong, The Bloch theorem in several complex variables, *J. of Geometric Analysis* **4** (1994), 35–58.

[3] Hua, L. K., *Harmonic Analysis of Functions of Several Complex Variables in Classical Domains*, Translations of Mathematical Monographs, 6, American Mathematical Society, 1963.

[4] Landau, E., Über die Blochsche Konstante und zwei verwandte Weltkonstanten, *Math. Z.* **30** (1929), 608–634.

[5] Liu, X. Y., Bloch functions of several complex variables, *Pacific J. of Math.* **152** (1992), 347–363.

[6] Liu, X. Y., and D. Minda, Distortion theorems for Bloch functions, *Trans. Amer. Math. Soc.* **333** (1992), 325–338.

[7] Minda, D., The Bloch and Marden constants, in *Computational Methods and Function Theory, Proceedings, Valparaíso 1989*. St. Ruscheweyh, E. B. Saff, L. C. Salinas, R. S. Varga (eds). Lecture Notes in Mathematics 1435, pp. 131–142, Springer-Verlag, 1990.

[8] Pommerenke, Ch., On Bloch functions, *J. London Math. Soc.* (2) **2** (1970), 689–693.

C. H. FitzGerald
Department of Mathematics
University of California, San Diego, La Jolla, CA 92093-0112
email: cfitzgerald@ucsd.edu

S. Gong
Department of Mathematics
University of California, San Diego, La Jolla, CA 92093-0112
and
University of Science and Technology of China,
Hefei, Anhui 230026, P. R. of China

Chin. Ann. of Math.
17B: 3(1996),271-278.

THE BLOCH CONSTANT OF LOCALLY BIHOLOMORPHIC MAPPINGS ON BOUNDED SYMMETRIC DOMAINS

GONG SHENG*

Abstract

The estimations of the bounds of the Bloch constant of locally biholomorphic mappings on irreducible bounded symmetric domains are given. When the domain is a unit circle, the estimation of the lower bounds is just the famous one-half estimation.

Keywords Bounded symmetric domain, Bloch mapping, Bloch constant, Locally biholomorphic mapping
1991 MR Subject Classification 32H05
Chinese Library Classification O174.5

§1. Introduction

In [1], we studied the bounds of Bloch constant of holomorphic mappings on irreducible bounded symmetric domains. This paper is a successive paper of [1]. We study the bounds of the Bloch constant of locally biholomorphic mappings on irreducible bounded symmetric domains. All the notations of this paper are the same as those we used in [1].

Let D be a domain, which contains the origin. A holomorphic mapping on D is defined as a Bloch mapping if the family of mappings

$$F_f = \{g : g(z) = f(\phi(z)) - f(\phi(0)), \quad \text{for all } \phi \in \mathrm{Aut}(D)\},$$

is a normal family. The Bloch norm of a Bloch mapping is defined as

$$||f||_B = \sup\left\{ \left\| \frac{\partial f \circ \phi}{\partial z}(0) \right\|_M : \phi \in \mathrm{Aut}(D) \right\},$$

where $||\quad||_M$ is the norm of the matrix.

Denote by $H(D)$ the family of holomorphic mappings which maps D into $\mathbf{C}^n$. If D contains the origin, we introduced a functional $d(F)$ of $F \in H(D)$ by ([2, formula (1.1)])

$$d(F) = \sup_{z \in D} \frac{|\det F'(z)|}{\sqrt{K(z,\bar{z})/K(0,0)}},$$

where K is the Bergman kernel function of D, $F'(z) = \frac{\partial F}{\partial z}$ is the Jacobian of F. $d(F)$ is invariant under the group of holomorphic automorphisms $\mathrm{Aut}(D)$ of D.

Manuscript received May 30, 1994.
*Department of Mathematics, University of Science and Technology of China, Hefei 230026, China..

We consider the following subfamilies of $H(D)$:

$$\mathcal{B} = \mathcal{B}(D) = \Big\{F \in H(D) : \det \frac{\partial F}{\partial z}(0) = 1, \quad ||F||_{\mathcal{B}} < \infty\Big\};$$

$$\mathcal{B}^0 = \mathcal{B}^0(D) = \Big\{F \in \mathcal{B} : \det \frac{\partial F}{\partial z}(z) \neq 0, \quad z \in D\Big\};$$

$$\mathcal{B}^0_K = \mathcal{B}^0_K(D) = \{F \in \mathcal{B}^0 : ||F||_{\mathcal{B}} \leq K, \quad 1 \leq K < \infty\};$$

and

$$\mathcal{B}^0_{K,1} = \mathcal{B}^0_{K,1}(D) = \{F \in \mathcal{B}^0_K : D(F) = 1\}.$$

In [2], we proved a theorem of Landau type for the Bloch constant. Similarly, we may prove the following result of Landau type for the Bloch constant for the family of locally biholomorphic mappings.

Lemma 1.1. *Let D be a bounded homogeneous domain in $\mathbf{C}^n$ which contains the origin. Then there exists a mapping in $\mathcal{B}^0_{K,1}$ such that it is an extremal mapping of the Bloch constant in $\mathcal{B}_K$.*

The proof of Lemma 1.1 is similar to the proof of Theorem 1.3 in [2], and we omit it.

Lemma 1.2. *Suppose that D is a bounded complete circular domain and that a belongs to the closure of D. If g is a non-zero holomorphic function in $D \cup \{a\}$ which maps $D \cup \{a\}$ into $\mathbf{C}$, and $|g(w)| \leq |g(a)|$ for all $w = za$ when $z \in \Delta$ (Δ is the unit disc in $\mathbf{C}$), then*

$$c = \frac{dg}{dw}(a) \cdot a'/g(a)$$

is positive and

$$\Big|\frac{g(xa)}{g(a)}\Big| \geq \exp\Big\{-2c\frac{1-x}{1+x}\Big\} \tag{1.1}$$

for $x \in (-1,1)$.

Proof. Let $h(z) = \frac{g(za)}{g(a)}$ for $z \in D$. Then $h(z)$ is holomorphic in $\Delta \cup \{1\}$,

$$h(\Delta) \subset \Delta, \quad h(1) = 1, \quad \text{and} \quad c = \frac{dh(z)}{dz}|_{z=1} = h'(1).$$

Since $g(z)$, and hence $h(z)$, is a non-zero functions in $D \cup \{a\}$, there exists a holomorphic function $H(z)$ on Δ, such that $h(z) = \exp(-H(z))$ and $\Re eH(z) > 0$, when $z \in \Delta$. $H(z)$ maps Δ into $\Re eH > 0$, and $H(1) = 0$. By elementary geometric observation, we have

$$\begin{aligned} c &= \frac{dh(z)}{dz}\Big|_{z=1} = h'(1) \\ &= \frac{dg}{dw}(a) \cdot a'/g(a) \\ &= \exp(-H(z))\Big[-\frac{dH(z)}{dz}\Big]\Big|_{z=1} \\ &= -\frac{dH(z)}{dz}\Big|_{z=1} > 0. \end{aligned}$$

$H(z)$ satisfies all conditions for the Julia lamma of half-plane version, and by its corollary ([3, p.327]) we have, for any $x \in (-1,1)$,

$$\Re e(H(x)) \leq 2c\frac{1-x}{1+x}.$$

Thus

$$|h(x)| = \left|\frac{g(xa)}{g(a)}\right| = \exp\{-\Re H(x)\} \geq \exp\Big\{-2c\frac{1-x}{1+x}\Big\}.$$

Suppose D is a bounded, complete circular, homogeneous domain. Suppose $F \in H(D)$ and satisfies $d(F)=1$ and $\det F'(0)=1$. Let $\phi_\lambda(z) \in \mathrm{Aut}(D)$, that takes $\lambda \in D$ to the origin. Let $G(z) = F(\phi_\lambda(z))$ and $g = \det \frac{\partial G}{\partial z}$. Then for every λ in D,

$$c = \Big(\frac{dg(\lambda)}{d\lambda}\Big)\cdot \lambda'/g(\lambda) = \frac{\partial}{\partial\lambda}K(\lambda,\bar\lambda)\cdot\lambda'/K(\lambda,\bar\lambda)$$

by Theorem 2.2 of [2]. Substituting this result into (1.1) of Lemma 1.2, we have

$$\left|\frac{g(x\lambda)}{g(\lambda)}\right| \geq \exp\Big\{\frac{-2\frac{\partial}{\partial\lambda}K(\lambda,\bar\lambda)\cdot\lambda'}{K(\lambda,\bar\lambda)}\cdot\frac{1-x}{1+x}\Big\},$$

where $x\in(-1,1)$. We get

$$\left|\frac{\det\frac{\partial F(w)}{\partial w}|_{w=\phi_\lambda(x\lambda)}\det\frac{\partial\phi_a}{\partial z}|_{z=x\lambda}}{\det\frac{\partial\phi_\lambda}{\partial z}|_{z=\lambda}}\right| \geq \exp\Big\{\frac{-2\frac{\partial}{\partial\lambda}K(\lambda,\bar\lambda)\cdot\lambda'}{K(\lambda,\bar\lambda)}\cdot\frac{1-x}{1+x}\Big\}. \qquad (1.2)$$

§2. Distortion Theorem

Let D be an irreducible bounded symmetric domain with the standard Harish-Chandra realization. D contains the origin. Let G be the connected component of $\mathrm{Aut}(D)$ and $\mathcal{G}$ be the Lie algebra of G. Let K be an isotropy group of G which leaves the origin fixed and $\mathcal{K}$ be the maximal compact subalgebra of $\mathcal{G}$ which corresponds to K. Then $\mathcal{G}$ has Cartan decomposition of $\mathcal{G}=\mathcal{K}+\mathcal{P}$. Let $\mathcal{H}$ be Cartan subalgebra of $\mathcal{K}$. We denote by Φ^+ the set of positive non-compact roots. Denote by τ the conjugation with respect to the real form $\mathcal{K}+i\mathcal{P}$. We consider a basis of root vectors e_α such that

$$\tau e_\alpha = -e_{-\alpha},\quad [e_\alpha, e_{-\alpha}] = h_\alpha,\quad \text{and}\quad [h_\alpha, e_{\pm\alpha}] = \pm 2e_{\pm\alpha}.$$

Setting $\mathcal{P}^\pm = \sum\limits_{\alpha\in\Phi^+} ce_{\pm\alpha}$, we have $\mathcal{G}^c = \mathcal{P}^- + \mathcal{K}^c + \mathcal{P}^+$ as a vector space direct sum. We denote by $\gamma_1,\cdots,\gamma_r$ the strong orthogonal roots of Harish-Chandra, where r is the real rank of $\mathcal{G}$. We simply write $e_j = e_{\gamma_j}$, $h_j = h_{\gamma_j}$, $1\leq j\leq r$, and $e=\sum\limits_{j=1}^r e_j$. It is known that, in Harish-Chandra realization, D is a bounded symmetric domain in $\mathcal{P}^+$ and the isotropy subgroup K of G at the origin acts on D by unitary transformations. We denote by $\mathbf{h}^-$ the real span of the vector ih_j, by $\mathcal{U}^+$ the real span of e_j, and by $\mathcal{U}$ the real span of the vectors $i(e_j+\tau e_j)$ $(1\leq j\leq r)$. The $i\mathbf{h}^-$-roots of $\mathcal{G}^c$ are $\pm\frac12(\gamma_j+\gamma_k)$, $\pm\gamma_j$, $\pm\frac12\gamma_j$ $(1\leq j,k\leq r)$ with respect multiplicities a, 1 and $2b$ (independent of j, k). We denote by p the important constant, $p=(r-1)a+b+2$.

Every $z\in D$ can be written as

$$z = k\sum_{i=1}^r \lambda_i e_i,\quad 0\leq\lambda_i<1,\quad 1\leq i\leq r,\quad k\in K.$$

We consider

$$z=\lambda=\sum_{i=1}^r\lambda_i e_i,\quad 1>\lambda_1\geq\lambda_2\geq\cdots\geq\lambda_r\geq 0,$$

at first. Let E denote the differential operator $\sum\limits_{i=1}^{n} z_i \frac{\partial}{\partial z_i}$. Then

$$(EK)(z,\bar{z}) = p\prod_{j=1}^{r}(1-\lambda_j^2)^{-(p+1)}\Big[\sum_{i=1}^{r}\lambda_i^2\prod_{j\neq i}^{r}(1-\lambda_j^2)\Big], \tag{2.1}$$

by [1, Lemma 3.2] and

$$K(z,\bar{z}) = \prod_{j=1}^{r}(1-\lambda_j^2)^{-p}. \tag{2.2}$$

By the formula (3.13) of [1], we know

$$\frac{\det \frac{\partial \phi_\lambda}{\partial z}|_{z=x\lambda}}{\det \frac{\partial \phi_\lambda}{\partial z}|_{z=\lambda}} = \prod_{i=1}^{r}\frac{(1-\lambda_i^2)^p}{(1-x\lambda_i^2)^p}. \tag{2.3}$$

Substituting (2.1), (2.2) and (2.3) into (1.2), we have

Lemma 2.1. *Suppose F: $D \to \mathbf{C}^n$ is a holomorphic mapping on an irreducible bounded symmetric domain D, with $F \in B^0(D)$, $d(F) = 1$, $\det \frac{\partial F}{\partial z}(0) = 1$ and*

$$\lambda = \sum_{i=1}^{r}\lambda_i e_i \in D, \quad -1 < x < 1.$$

If $\phi_\lambda(z) \in \mathrm{Aut}(D)$ which maps λ to the origin, then

$$\Big|\det \frac{\partial F}{\partial w}(\phi_\lambda(x\lambda))\Big| \geq \prod_{j=1}^{r}\Big(\frac{1-x\lambda_j^2}{1-\lambda_j^2}\Big)^p \exp\Big\{\frac{-2p(1-x)}{1+x}\sum_{i=1}^{r}\frac{\lambda_i^2}{1-\lambda_i^2}\Big\}. \tag{2.4}$$

By [1, Lemma 3.4] we know

$$\phi_\lambda(x\lambda) = \sum_{i=1}^{r}\frac{(1-x)\lambda_i}{1-x\lambda_i^2}e_i.$$

Let

$$\mu_i = \frac{(1-x)\lambda_i}{1-x\lambda_i^2}, \quad i = 1,\cdots,r. \tag{2.5}$$

Then $\phi_\lambda(x\lambda) = \sum\limits_{i=1}^{r}\mu_i e_i = \mu$ and $1 > \mu_1 \geq \cdots \geq \mu_r \geq 0$. By the formula (3.14) of [2], we know

$$\prod_{j=1}^{r}\frac{1-x\lambda_j^2}{1-\lambda_j^2} = \prod_{j=1}^{r}\frac{1}{1-\lambda_j\mu_j}.$$

Thus (2.4) becomes

$$\Big|\det \frac{\partial F}{\partial w}(\mu)\Big| \geq \prod_{j=1}^{r}\frac{1}{(1-\lambda_j\mu_j)^p}\exp\Big\{\frac{-2p(1-x)}{1+x}\sum_{i=1}^{r}\frac{\lambda_i^2}{1-\lambda_i^2}\Big\}.$$

By (2.5), we know

$$1-x = \frac{\mu_j(1-\lambda_j^2)}{\lambda_j(1-\lambda_j\mu_j)}, \quad j = 1,\cdots,r.$$

It implies

$$\frac{(1-x)\lambda_j^2}{1-\lambda_j^2} = \frac{\lambda_j\mu_j}{1-\lambda_j\mu_j}.$$

Then (2.4) becomes

$$\left|\det\frac{\partial F}{\partial z}(\mu)\right| \geq \prod_{j=1}^{r}\left[\frac{1}{1-\lambda_j\mu_j}\exp\left\{\frac{-2}{1+x}\cdot\frac{\lambda_j\mu_j}{1-\lambda_j\mu_j}\right\}\right]^{p}. \tag{2.6}$$

Fix x, and consider the function

$$f(Y)=\frac{1}{1-Y}\exp\left\{\frac{-2}{1+x}\cdot\frac{Y}{1-Y}\right\}.$$

Then

$$f'(Y)=\frac{(1+x)(1-Y)-2}{(1-Y)^3(1+x)}\exp\left\{\frac{-2}{1+x}\cdot\frac{Y}{1-Y}\right\}<0,$$

when $0\leq Y<1$, $-1<x<1$. This means $f(Y)$ is a decreasing function. By (2.5), it is easy to verify $\lambda_1\mu_1\geq\cdots\geq\lambda_r\mu_r$. Using the decreasing property of $f(Y)$ to (2.6), we have

$$\left|\det\frac{\partial F}{\partial w}(\mu)\right| \geq \frac{1}{(1-\lambda_1\mu_1)^{rp}}\exp\left\{\frac{-2pr}{1+x}\cdot\frac{\lambda_1\mu_1}{1-\lambda_1\mu_1}\right\}. \tag{2.7}$$

The x, λ_j, μ_j, $j=1,\cdots,r$, are related by (2.5). When $z=\lambda=\sum\limits_{i=1}^{r}\lambda_ie_i$ and x are given, μ is decided by (2.5). Conversely, if μ is given, we may find x and λ_j, $j=1,\cdots,n$, such that they satisfy the relation (2.5) and $x\in(-1,1)$, $1>\lambda_1\geq\cdots\geq\lambda_r\geq 0$, $1>\mu_1\geq\cdots\geq\mu_r\geq 0$.

Now suppose μ is given. We may take a small positive number ϵ such that $1-\epsilon>\mu_1$, and then let

$$\lambda_1=1-\epsilon>\mu_1 \quad\text{and}\quad x=\frac{1-\epsilon-\mu_1}{1-\epsilon-(1-\epsilon)^2\mu_1},$$

and then decide $\lambda_2,\cdots,\lambda_r$ by (2.5). Substituting the value of λ_1 and x into (2.7), we have

$$\left|\det\frac{\partial F}{\partial w}(\mu)\right| \geq \frac{1}{(1-(1-\epsilon)\mu_1)^{rp}}\times\exp\left\{\frac{-2pr(1-\epsilon-(1-\epsilon)^2\mu_1)(1-\epsilon)\mu_1}{[2(1-\epsilon)-\mu_1-(1-\epsilon)^2\mu_1](1-(1-\epsilon)\mu_1)}\right\}.$$

Letting $\epsilon\to 0$, we obtain

$$\left|\det\frac{\partial F}{\partial w}(\mu)\right| \geq \frac{1}{(1-\mu)^{rp}}\exp\left\{\frac{-pr\mu_1}{1-\mu_1}\right\}. \tag{2.8}$$

Lemma 2.2. *If $F\in\mathcal{B}^0(D)$, $d(F)=1$, $\det\frac{\partial F}{\partial w}(0)=1$, and*

$$\mu=\sum_{i=1}^{r}\mu_ie_i\in D,\quad 1>\mu_1\geq\cdots\geq\mu_r\geq 0,$$

then (2.8) *holds.*

Let $z\in D$. Then z can be written as

$$z=k\sum_{i=1}^{r}\mu_ie_i,\quad 1>\mu_1\geq\cdots\geq\mu_r\geq 0,\quad k\in K.$$

Fix k, and let $\mu=\sum\limits_{i=1}^{r}\mu_ie_i$. Then $z=k\mu$. Let $G(\mu)=F(k\mu)\left(\frac{\partial z}{\partial\mu}\right)^{-1}$. Then $\frac{\partial z}{\partial\mu}$ is independent of μ, and hence $\frac{\partial G}{\partial\mu}=\frac{\partial F}{\partial z}$. Thus

$$\det\frac{\partial G}{\partial\mu}=\det\frac{\partial F}{\partial z}.$$

If $F\in\mathcal{B}^0(D)$, $d(F)=1$, and $\det\frac{\partial F}{\partial\mu}(0)=1$, then G satisfies the same conditions, i.e.,

$$G\in\mathcal{B}^0(D),\quad d(F)=1,\quad\text{and}\quad\det\frac{\partial G}{\partial\mu}(0)=1.$$

By Lemma 2.2, we have

Theorem 2.1(Distortion Theorem). *Let D be an irreducible bounded symmetric domain in $\mathbf{C}^n$, $F \in \mathcal{B}^0(D)$, $d(F) = 1$, $\det \frac{\partial F}{\partial z}(0) = 1$, and*

$$z = k \sum_{i=1}^{r} \mu_i e_i \in D, \quad k \in K.$$

Then

$$\left|\det \frac{\partial F}{\partial z}(z)\right| \geq \frac{1}{(1-\|z\|)^{rp}} \exp\left\{\frac{-pr\|z\|}{1-\|z\|}\right\} \tag{2.9}$$

holds, where $\|z\| = \mu_1$.

When D is the unit ball in $\mathbf{C}^n$, Theorem 2.1 coincides with a theorem of Liu[4].

§3. Bloch Constant of Locally Biholomorphic Mappings

Using Theorem 2.1, we may prove

Theorem 3.1. *The Bloch constant B_K^0 of $\mathcal{B}_K^0(D)$ satisfies the following inequality:*

$$K^{1-n} \geq B_K^0 \geq K^{1-n} \int_0^1 \frac{(1-t^2)^{n-1}}{(1-t)^{pr}} \exp\left\{\frac{-rpt}{1-t}\right\} dt. \tag{3.1}$$

In order to prove Theorem 3.1, we need

Lemma 3.1. *If $F \in B(D)$, then*

$$\left\|\frac{\partial F}{\partial z}(z)\right\|_M \leq \frac{\|F\|_B}{1-\|z\|^2}. \tag{3.2}$$

Proof of Theorem 3.1. The inequality of the left-hand side of (3.1) was proved in [1].

Let $F \in \mathcal{B}_K^0(D)$. We consider what causes a ball in the range of F not having a well-defined inverse deletion of F. For a small ball centered at the origin, since $\det \frac{\partial F}{\partial z}(0) = 1$, and $F(0) = 0$, there is a well-defined inverse of F on the ball. When the ball in the range expands, either the preimage reaches a point at which $\det \frac{\partial F}{\partial z} = 0$ or it reaches a point of the boundary. Otherwise, one could argue that the ball in range could be enlarged further. Then, from the estimate on $\left|\det \frac{\partial F}{\partial z}\right|$ in inequality (2.9), we can force $\det \frac{\partial F}{\partial z}$ to be non-zero. Now consider a straight line interval Γ in the range of F. The interval Γ starts at the origin and goes as far as it can with its preimage not running into the boundary of D or $\det \frac{\partial F}{\partial z} = 0$. If $r(0, F)$ denotes the largest nonnegative number r such that there exists a domain R in D, and F biholomorphically maps R onto a ball centered at the origin with radius r, then

$$r(0,F) \geq \left|\int_\Gamma dw\right| = \int_\Gamma |dw| = \int_\gamma \left|\frac{\partial F}{\partial z} \cdot \frac{dz'}{|dz|}\right| \cdot |dz|,$$

where $\gamma = F^{-1}(\Gamma)$. Obviously

$$\int_\gamma \left|\frac{\partial F}{\partial z} \cdot \frac{dz'}{|dz|}\right| \cdot |dz| \geq \int_\gamma \frac{\left|\det \frac{\partial F}{\partial z}\right|}{\left\|\frac{\partial F}{\partial z}\right\|_M^{n-1}} |dz|.$$

The right-hand side of the preceding inequality is greater than or equal to

$$\int_0^1 \frac{1}{(1-\|z\|)^{rp}} \exp\left\{\frac{-pr\|z\|}{1-\|z\|}\right\} \frac{(1-\|z\|^2)^{n-1}}{K^{n-1}} d|z|$$

by (2.9) and (3.2). Let

$$l(x) = (1-x)^{-c} \exp\left\{\frac{-cx}{1-x}\right\} (1-x^2)^{n-1}.$$

Then

$$l'(x) = -l(x)\frac{c(1-x)x + 2(n-1)x}{(1-x)^2(1+x)} \le 0$$

when $0 \le x < 1$, $c > 0$, and $n \ge 1$. Thus

$$r(0, F) \ge K^{1-n} \int_0^1 \frac{(1-t^2)^{n-1}}{(1-t)^{rp}} \exp\left\{\frac{-prt}{1-t}\right\} dt$$

since $|z| \ge \|z\|$. We prove (3.1).

We already know that[5,6]: For $\mathcal{R}_I$: $I - Z\bar{Z}' > 0$, $Z = Z^{(m,n)}$, $1 \le m \le n$, then $p = n + m$, $r = m$. Substituting these values into (3.1), we see that the Bloch constant $B_K^0(\mathcal{R}_I)$ is greater than or equal to

$$K^{1-mn} \int_0^1 \frac{(1+t)^{mn-1}}{(1-t)^{m^2+1}} \exp\left\{\frac{-m(m+n)t}{1-t}\right\} dt.$$

This result coincides with the result in [7], and when $m = 1$, it coinçides with the result of Liu[4].

For $\mathcal{R}_{II}$: $I - Z\bar{Z}' > 0$, $Z = Z' = Z^{(n)}$, then $p = n + 1$, $r = n$. Substituting these values into (3.1), we see that the Bloch constant $B_K^0(\mathcal{R}_{II})$ is greater than or equal to

$$K^{-\frac{1}{2}(n-1)(n+2)} \int_0^1 \frac{(1+t)^{\frac{1}{2}(n-1)(n+2)}}{(1-t)^{\frac{1}{2}(n^2+n+2)}} \exp\left\{\frac{-n(n+1)t}{1-t}\right\} dt.$$

For $\mathcal{R}_{III}$: $I + Z\bar{Z}' > 0$, $Z = -Z' = Z^{(n)}$, then $p = 2n - 2$, $r = \left[\frac{n}{2}\right]$, and then $pr = n(n-1)$ when n is even; $pr = (n-1)^2$ when n is odd. Substituting these values into (3.1), we see that the Bloch constant $B_K^0(\mathcal{R}_{III})$ is greater than or equal to

$$K^{-\frac{1}{2}(n+1)(n-2)} \int_0^1 \frac{(1+t)^{\frac{1}{2}(n+1)(n-2)}}{(1-t)^{\frac{1}{2}(n^2-n+2)}} \exp\left\{\frac{-n(n-1)t}{1-t}\right\}$$

when n is even; or

$$K^{-\frac{1}{2}(n+1)(n-2)} \int_0^1 \frac{(1+t)^{\frac{1}{2}(n+1)(n-2)}}{(1-t)^{\frac{1}{2}(n^2-3n+4)}} \exp\left\{\frac{-(n-1)^2 t}{1-t}\right\} dt$$

when n is odd.

For $\mathcal{R}_{IV}$: $1 + |z\bar{z}| - 2z\bar{z}' > 0$, $1 - |z\bar{z}| > 0$, then $p = n$, $r = 2$. Substituting these values into (3.1), we see that the Bloch constant $B_K^0(\mathcal{R}_{IV})$ is greater than or equal to

$$K^{1-n} \int_0^1 \frac{(1+t)^{n-1}}{(1-t)^{n+1}} \exp\left\{\frac{-2nt}{1-t}\right\} dt.$$

For the exceptional domain $\mathcal{R}_V$ of dimension 16, then $p = 12$, $r = 2$. Substituting these values into (3.1), we see that the Bloch constant $B_K^0(\mathcal{R}_V)$ is greater than or equal to

$$K^{-15} \int_0^1 \frac{(1+t)^{15}}{(1-t)^9} \exp\left\{\frac{-24t}{1-t}\right\} dt.$$

For the exceptional domain $\mathcal{R}_{VI}$ of dimension 27, then $p = 18$, $r = 3$. Substituting these values into (3.1), we see that the Bloch constant $B_K^0(\mathcal{R})$ is greater than or equal to .

$$K^{-26} \int_0^1 \frac{(1+t)^{26}}{(1-t)^{28}} \exp\left\{\frac{-54t}{1-t}\right\} dt.$$

REFERENCES

[1] Yan Zhimin & Gong Sheng, Bloch costant of holomorphic mappings on bounded symmetric domains, *Science in China (Series A)* **36** (1993), 285-299.

[2] FitzGerald Carl H. & Gong Sheng, The bloch theorem in several complex variables, *The Journal of Geometric Analysis*, **4** (1994), 36-58.

[3] Liu Xiangyang & Minda Divid, Distortion theorem for bloch functions, *Tran. Amer. Math. Soc.*, **333** (1992), 325-338.

[4] Liu Xiangyang; Bloch functions of several complex variables, *Pacific J. Math.*, **152**(1992), 347-363.

[5] Arazy, J. & Fisher, S.d., Invariant Hilbert spaces of analytic functions on bounded symmetric domains, Topics in operator theory, Ernst D. Hellinger memorial volume, 67-91, Operator theory Adv. Appl. 48, Birkhäuser, Basel, 1990.

[6] Hua Luokeng, Harmonic analysis of functions of several complex variables in the classical domains, *Transl. Amer. Math. Soc.*, **6**, (1963).

[7] FitzGerald Carl, H. & Gong Sheng, The locally biholomorphic Bloch and Marden constants in several complex variables, Computational methods and function theory 1994 (R. M. Ali. St. Ruscheweyh and E. B. Saff, eds.), Series in approximations and decompositions, Vol. 5, 1995, World Scientific Publishing Co., 147-158.

第 7 卷 第 4 期 数學學報 Vol. 7, No. 4
1957 年 12 月 ACTA MATHEMATICA SINICA Dec., 1957

多連通區域上的 Bloch 常數*

龔 昇
(中國科學院數學研究所)

本文可以看作[1]文的一個應用，在這裏將提供一個關於多連通區域的 Bloch 常數的估計的普遍方法．並且特別對於二連通區域的 Bloch 常數進行了具體的估計．

我們知道任何一個 z 平面上的二連通區域 G，它的每一個境界都不止含有一點．一定存在唯一的 R 使 G 與圓環 $R<|z|<1$ 相互共形映照．我們稱 $\|G\|=\dfrac{-\log R}{2\pi}$ 爲 G 的模[2]．以 S_R 表示具有模 $\dfrac{-\log R}{2\pi}$ 的 G 的全體，那末 S_R 中任何二個區域都可以相互共形映照． 設 F_R 是在圓環 $R<|z|<1$ 上的單葉的解析函數的全體．H_R 是在圓環 $R<|z|<1$ 上的解析函數的全體．我們不妨假設 $\dfrac{df}{dz}$ 在 $z=\sqrt{R}$ 的値爲 1．

記 $S_{FR}=\left\{w=f(z)\in F_R,\ \dfrac{df}{dz}\Big|_{z=\sqrt{R}}=1;\ R<|z|<1\right\}$，$S_{HR}=\left\{w=f(z)\in H_R,\ \dfrac{df}{dz}\Big|_{z=\sqrt{R}}=1;\ R<|z|<1\right\}$．

若 $G\in S_{FR}$，且在 G 中所能容納的最大圓的半徑爲 d_G 的話，那末稱

$$L_R=\min_{G\in S_{FR}} d_G$$

爲圓環 $R<|z|<1$ 的 Bloch-Landau 常數．

若 $G\in S_{HR}$，且在 G 中所能容納的最大圓的半徑爲 c_G 的話，那末稱

$$B_R=\min_{G\in S_{HR}} c_G$$

爲圓環 $R<|z|<1$ 的 Bloch 常數．

同樣的，我們可以定義一般的多連通區域的 Bloch-Landau 常數和 Bloch 常數．

本文首先證明了如下的

定理 1.

$$L_R\geqslant\frac{\pi\sqrt{R(1-A)}}{2\pi\sqrt{k(2A+1)}-2(1-A)},$$

$$B_R\geqslant\frac{\pi\sqrt{3R(1-A)}}{4\pi\sqrt{k(2A+1)}-2(1-A)},$$

* 1956 年 6 月 21 日收到．

這裏 k^2 為 Jacobi 橢圓函数的模，$k^2 = \frac{\pi^4}{x^4}\left[\sum_{\nu=1}^{\infty}\frac{\nu^2(\nu-1)R^2}{1-R^{2\nu}} + \frac{1}{4\ln\frac{1}{R}}\right]$,

$A = \frac{1}{x}\int_0^x dn^2 u du$, $\sqrt{\frac{2x}{\pi}} = 1 + 2R + 2R^4 + 2R^9 + \cdots$.

至於一般的多連通區域的 L_R, B_R 的估計將在以後給出.

我們知道

$$\begin{cases}\varphi_\nu = \sqrt{\frac{\nu+1}{\pi}}\frac{z^\nu}{\sqrt{1-R^{2(\nu+1)}}}, \nu = 0, 1, \pm 2, \pm 3, \cdots \\ \varphi_{-1} = \frac{z^{-1}}{\sqrt{2\pi\log\frac{1}{R}}}\end{cases}$$

是 $R<|z|<1$ 上的一組完整正常直交系，因之，$R<|z|<1$ 的核函數是

$$K(z,\bar{\zeta}) = \frac{1}{\pi}\sum_{\substack{\nu=-\infty\\ \nu\neq -1}}^{\infty}\frac{\nu+1}{1-R^{2(\nu+1)}}z^\nu\bar{\zeta}^\nu + \frac{z^{-1}\bar{\zeta}^{-1}}{2\pi\log\frac{1}{R}} =$$

$$= \frac{1}{\pi z\bar{\zeta}}\left[\sum_{\nu=1}^{\infty}\frac{\nu R^\nu}{1-R^{2\nu}}\left(\frac{z^\nu\bar{\zeta}^\nu}{R^\nu} + \frac{R^\nu}{z^\nu\bar{\zeta}^\nu}\right) + \frac{1}{2\log\frac{1}{R}}\right] \tag{1}$$

或者可以把他寫成 Jacobi 橢圓函數的形式[3]

$$K(z,\bar{z}) = \frac{x}{\pi^3 z\bar{\zeta}}\left[x dn^2\left(\frac{x}{\pi i}\ln\frac{z\bar{\zeta}}{R}\right) - \int_0^x dn^2 u du\right]. \tag{2}$$

此處 x 定義如前，因之，

$$K(z,\bar{z}) = \frac{x^2}{\pi^3 r^2}\left[dn^2\left(\frac{x}{\pi i}\ln\frac{r^2}{R}\right) - \frac{1}{x}\int_0^x dn^2 v dv\right],$$

此處 $r = |z|$.

由於 $\frac{\partial^2}{\partial z\partial\bar{z}}\log\frac{1}{r^2} = 0$，所以

$$T = \frac{\partial^2}{\partial z\partial\bar{z}}\log K(z,\bar{z}) =$$

$$= \frac{\partial^2}{\partial z\partial\bar{z}}\log\left[dn^2\left(\frac{x}{\pi i}\ln\frac{r^2}{R}\right) - \frac{1}{x}\int_0^r dn^2 v dv\right] =$$

$$= \frac{2k^2x^2}{\pi^2 r^2(dn^2 u - A)^2}[(dn^2 u - A)(cn^2 u - sn^2 u dn^2 u) + 2Ak^2 sn^2 u cn^2 u].$$

此處 $u = \frac{x}{\pi i}\ln\frac{r^2}{R}$, k^2 為 Jacobi 橢圓函數的模，可以將 $\frac{\partial^2 K(z,\bar{z})}{\partial z\partial\bar{z}}$ 分別用(1)，(2)算出，以 $z=\sqrt{R}$ 代入即得 $T_{\sqrt{R}} = \frac{2k^2x^2}{\pi^2 R(1-A)}$. 同時,由(1)可以看出,當 z 從 $R<|z|<1$

的内部趨於 $|z|=R$ 及 $|z|=1$ 上的時候，T 是趨於無窮的．

由 $d\sigma^2 = T\,|dz|^2$ 所定義的 Gauss 曲率為

$$\tau = \frac{-1}{T}\frac{\partial^2}{\partial z\partial \bar z}\log T = \frac{-1}{T}\Big[-2T + \frac{\partial^2}{\partial z\partial \bar z}\ln\{(dn^2u - A)\cdot$$

$$\cdot(cn^2u - sn^2u\,dn^2u) + 2Ak^2 sn^2u\,cn^2u\}\Big],$$

經過冗長的計算，可以得出，在 $z=\sqrt{R}$ 處的 Gauss 曲率等於

$$-\frac{2}{k^2}(1-A)+1+2A.$$

我們所以取 $z=\sqrt{R}$，是因為 τ 在 $z=\sqrt{R}$ 處為最小，參閱[4]．

若 $G\in S_{FR}$，在 G 上作度量

$$ds^2 = \frac{-2}{\tau_{\sqrt{R}}}\frac{|dw|^2}{4\rho^2\left(\log\dfrac{c}{\rho}\right)^2},$$

這裏 c 是比 eL'_R 大的常數，L'_R 是 G 所能容納的最大圓的半徑，$\rho(w)$ 是以點 w 為中心的 G 所能容納的最大圓的半徑．此圓與 G 的境界切於 a 點．作圓 $|w-a|=\rho$，且定義度量

$$ds^{*2} = \frac{-2}{\tau_{\sqrt{R}}}\frac{|dw|^2}{4|w-a|^2(\log c/|w-a|)^2}.$$

因為在點 w 處 $|w-a|=\rho$，而在 w 的近旁的點 w'，$|w'-a|\geqslant\rho(w')$，所以

$$ds^{*2}=ds^2,\qquad \text{在點 } w \text{ 處成立},$$

$$ds^{*2}\leqslant ds^2,\qquad \text{在點 } w \text{ 的近旁的點 } w' \text{ 處成立},$$

因之，ds^{*2} 是 ds^2 的支持度量，容易證明由 ds^{*2} 所定義的曲率是 $\tau_{\sqrt{R}}$，所以 [1] 中的定理所要求的條件全部滿足，因之，

$$\left|\frac{df(z)}{dz}\right|^2 \leqslant T\Big/\frac{-2}{\tau_{\sqrt{R}}}\cdot\frac{1}{4\rho^2(\log c/\rho)^2},$$

取 $z=\sqrt{R}$，得

$$1\leqslant T_{\sqrt{R}}\Big/\frac{-2}{\tau_{\sqrt{R}}}\cdot\frac{1}{4\rho^2_{\sqrt{R}}(\log c/\rho_{\sqrt{R}})^2}.$$

因為 $\dfrac{-2}{\tau_{\sqrt{R}}}\cdot\dfrac{1}{4\rho^2(\log c/\rho)^2}$ 是 ρ 的減少函數，因之

$$1\leqslant T_{\sqrt{R}}\Big/\frac{-2}{\tau_{\sqrt{R}}4L'^2_R(\log c/L'_R)^2},$$

讓 $c\to eL'_R$，得到

$$L'^2_R \geqslant \frac{1}{2(-\tau_{\sqrt{R}})T_{\sqrt{R}}}.$$

因為 L_R 是 L'_R 的下界，且 $T_{\sqrt{R}} = \dfrac{2k^2x^2}{\pi^2R(1-A)}$，$\tau_{\sqrt{R}} = 1+2A-\dfrac{2}{R^2}(1-A)$，所以我

們有

$$L'_R \geqslant \frac{\pi\sqrt{R(1-A)}}{2x\sqrt{k}(2A+1)-2(1-A)}.$$

若 $G \in S_{HR}$，在G上作度量

$$ds_1^2 = \frac{-2}{\tau_{\sqrt{R}}}\cdot\frac{A^2|dw|^2}{4\rho(A^2-\rho)^2}$$

這裏 A 是滿足 $A^2 > B'_R$ 的常數，B'_R 是 G 所能容納的最大圓的半徑，$\rho(w)$ 定義如前，在支點 a 的附近，我們有 $\rho = |w-a|$，若 a 是 n 階支點，則 $w_1 = (w-a)^{\frac{1}{n}}$ 是單值化變數，而從 $\lambda_1|dw_1| = \lambda|dw|$ 所得的 $\lambda_1 = n\rho^{\frac{1}{2}-\frac{1}{n}}/2(A^2-\rho)$. 立刻可以看出，當 $n=2$ 時，λ_1 是正則的，而當 $n>2$ 時 λ_1 在 a 處爲零．作圓 $|w-a|=\rho$，定義度量

$$ds_1^{*2} = \frac{-2}{\tau_{\sqrt{R}}}\frac{A^2|dw|^2}{4|w-a|(A^2-|w-a|)^2}.$$

容易看出 ds_1^{*2} 是 ds_1^2 的支持度量，而 ds^{*2} 所定義的曲率是 $\tau_{\sqrt{R}}$，所以由 [1] 中的定理，我們有

$$\left|\frac{df(z)}{dz}\right|^2 < T \Big/ \frac{-2}{\tau_{\sqrt{R}}}\frac{A^2}{4\rho(A^2-\rho)^2}.$$

取 $z=\sqrt{R}$，得

$$1 \leqslant T_{\sqrt{R}} \Big/ \frac{-2}{\tau_{\sqrt{R}}}\frac{A^2}{4\rho_{\sqrt{R}}(A^2-\rho_{\sqrt{R}})^2}.$$

因爲 $\dfrac{-2}{\tau_{\sqrt{R}}}\cdot\dfrac{A^2}{4\rho(A^2-\rho)^2}$ 當 $\rho < A^2/3$ 時是 ρ 的減少函數，因之，當 $B'_R < A^2/3$ 時，

$$1 \leqslant T_{\sqrt{R}} \Big/ \frac{-2A^2}{\tau_{\sqrt{R}}4B'_R(A^2-B_R)^2}.$$

讓 $A^2 \to 3B'_R$，得到

$$B'^2_R \geqslant \frac{-3}{8T_{\sqrt{R}}\tau_{\sqrt{R}}},$$

以 $T_{\sqrt{R}}$, $\tau_{\sqrt{R}}$ 的値代入即得定理 1.

由上述的證明中，可以看出，這個方法不僅對於二連通區域有效，而且對一般的多連通區域全都有效．可以概述如下：

設 D 是 z 平面上的一個多連通區域，其 Bergman 核爲 K，作其度量

$$d\sigma^2 = \frac{\partial^2 \log K}{\partial z\partial\bar{z}}|dz|^2 = T|dz|^2,$$

由此得出在每一點 z 處的 Gauss 曲率 $\tau(z)$，若 (i) $T\to\infty$ 當 z 由 D 的內部趨於邊界 (ii) 在 D 中有一點 z，使 $\tau(z)\leqslant 0$，(這二個條件照我想全是可以免去的，將在另文中

加以討論）若 $f(z)$ 將 D 映照到 D_f，且 $\frac{\partial f}{\partial z}$ 在 D 中之某一點 z 處其值爲 1．我們在 D_f 中作一度量

$$ds^2 = g(\rho^2)\,|dw|^2,$$

這裏 $\rho = \rho(w)$ 是由點 w 到 D 的邊界的距離，$g(x)$ 滿足以下幾個條件

（i）$g(x) > 0$，　若 $x > 0$；

（ii）$g'(x) < 0$，　當 $0 \leqslant x <$ 適當大的常數 A．

若 w 到 D_f 的邊界的最近點是 a，作度量

$$ds^{*2} = g(|w-a|^2)\,|dw|^2,$$

由 f 及 ρ 的定義，可知 ds^{*2} 是 ds^2 的支持度量，取適當的 g 使由 ds^{*2} 定義的曲率不大於 $\tau(z)$（這是常可以辦得到的）．於是由 [1] 中的引理得到

$$\left|\frac{dw}{dz}\right|^2 < T/g(\rho^2),$$

取 $z = z_0$，則

$$g(\rho^2(z_0)) < T(z_0),$$

由 g 的定義得到

$$g(B^2) < T(z_0),$$

讓 $\frac{1}{g(x)} = \varphi(x)$，則

$$\frac{1}{T(z_0)} < \varphi(B^2),$$

由於 $\varphi(x)$ 是 x 的增加函數，當 $x > 0$，所以

$$B^2 \geqslant \varphi^{-1}\left(\frac{1}{T(z_0)}\right).$$

這樣就得到多連通區域的 Bloch-Landan 或 Bloch 常數的估計了．

若多連通區域 G 是簡單域，在 $z = \xi$ 處其曲率爲最小且爲非正的，記作 $\tau(\xi)$，則我們可以取

$$ds^2 = \frac{-2}{\tau(\xi)} \frac{|dw|^2}{4\rho^2\left(\log\frac{c}{\rho}\right)^2}$$

即取 $g(x) = \frac{-2}{\tau(\xi)} \frac{1}{x\left(\log\frac{c}{x}\right)^2}$，　$(c > eL')$

這裏 L' 是 G_f 所能容納的最大圓的半徑，這樣的 $g(x)$ 全滿足所要求的條件，於是可以得到

$$L^2 > \frac{1}{2(-\tau(\xi))T(z_0)},$$

這裏L是域G的 Bloch-Landau 常数.

同様可以證明：若B是域G的 Bloch 常数,則

$$B^2 > \frac{3}{-8\tau(\xi)T(z_0)}.$$

因之,我們有

定理 2.

$$L > \sqrt{\frac{1}{-2\tau(\xi)T(z_0)}},$$

$$B > \sqrt{\frac{3}{-8\tau(\xi)T(z_0)}}.$$

最後要附帶提到一點,假如我們將 [5] 中定理B略加變化的話,我們可以得到多連通區域的 Schottky 定理.

定理 3. 若$f(z)$在有界多連通區域R(不妨假設它包有原點)中解析,且不取 0 和 1,若R的 Bergman 度量爲 $ds^2 = T(z)|dz|^2$,並且 $T(z) \to \infty$ 當z從R中趨於∂R,那末

$$\log|f(z)| \leq (7 + \log^+|f(0)|) \exp\left\{2\int_0^{|z|}\sqrt{T(z)}\,|dz|\right\}$$

在$z \in R$中成立.

参考文献

[1] 龔 昇，多複變數 Einstein 空間中的 Schwarz 引理，數學學報，3 卷 4 期，1957.

[2] Carathéodory, C., Conformal representation, Cambr. Univ. Press, 1932.

[3] Куфарев, П. П., Über des zwei fach zusammenhanganden Minimalgebiet, Изв. научно-исся. ин-та математики и механики при Томском гос. университете, т. 1, 1935—1937.

[4] Фукс, Б. А., Неевклидова геометрия в теории конформных и псевдоконформных отображений, 1951.

[5] Ahlfors, L. V., An extension of Schwarz's Lemma, *Tran. Amer. Math. Soc.*, **43**, 359—364, 1938.

ON THE BLOCH CONSTANT FOR MULTIPLY CONNECTED REGIONS

KUNG SUN

(*Institute of Mathematics, Academia Sinica*)

ABSTRACT

Let D be a bounded multiply connected region in z-plane. The curvature of Bergman metric of this region takes the minimum value τ_a at $z = a$.

Let $w = f(z)$ be an analytic function in D with $\left.\frac{\partial f}{\partial z}\right|_{z=b} = 1$, which maps D onto D_f in w-plane.

The object of the present note is to prove that D_f must contain a circle with radius $\sqrt{\frac{3}{-8\tau_a T_b}}$ where T_b is the value of the Bergman metric T of the region D at $z = b$. Further, if the function $f(z)$ is simple in D, then D_f must contain a circle with radius $\sqrt{\frac{1}{-2\tau_a T_b}}$.

IV. 多复变数的 Schwarz 导数

有众多的不同的观点来认识 Schwarz 导数，于是人们用各种不同的观点来定义与研究高维空间的 Schwarz 导数.

[29.1], [17], [16.2], [15.2], [14] 是从 Ahlfors 的观点 (1), 即： Schwarz 导数是交比的无穷小形式，以及 Thurston 的观点 (2), 即： Schwarz 导数是度量曲线或流形与平坦之间的偏离，这两种观点出发来研究典型域上的 Schwarz 导数， [20.1] 及 [18.1] 是研究 Kähler 流形上的 Schwarz 导数. [23.4] 是另一种方法来定义与研究 $\mathbb{C}^n$ 上的 Schwarz 导数.

这些定义的 Schwarz 导数大多满足: 1. 在线性变换下的不变性， 2. 映射的 Schwarz 导数为零当且仅当映射为线性变换. 有的还给出了相应的 Nehari 定理 (3) .

这些工作总结在 (4) ([M8]) 中的第二章中.

参 考 文 献

(1) Ahlfors L V. Cross ratio and Schwarzian derivative in $\mathbb{R}^n$, Complex Analysis, Edited by Hersch J. and Huber A, Basel, 1-15.

(2) Thurston W F. Zippers and univalent functions. The Bieberbach Conjecture, A. Baernstein II, et al. ed. Amer. Math. Soc., 1986, 185-197.

(3) Nehari Z. Some criteria of univalence, Proc. Amer. Math. Soc., 1954, 5:700-704.

(4) 龚昇, 余其煌，郑学安. 布洛赫常数与许瓦尔兹导数. 上海：上海科学技术出版社，1998.

Vol. 36 No. 5 SCIENCE IN CHINA (Series A) May 1993

The Schwarzian Derivative in Several Complex Variables

GONG Sheng (龚 昇)
(*Department of Mathematics, University of Science and Technology of China, Hefei 230026, PRC*)
and Carl H. FitzGerald
(*Department of Mathematics, University of California, San Diego, La Jolla, CA 92093, USA*)

Received December 29, 1991.

Abstract The Schwarzian derivative of holomorphic mappings on some classical domains is defined. Some Schwarzian properties are studied.

Keywords: Schwarzian derivative, classical domains.

1 Cross Ratio

In one complex variable, the cross ratio of four points is invariant when the points are mapped by a fractional linear transformation. It is commonly accepted that the Schwarzian derivative is an infinitesimal version of the cross ratio, but it is seldom made explicit. The Schwarzian derivative of a function is invariant if the function is composed of a fractional linear map of its range. The importance of cross ratio in complex analysis is well known. For the Schwarzian, the original application of the Schwarzian to finding mapping onto circular triangle is still interesting. The relation between Schwarzian and the solution of a certain differential equation is still studied. Several univalence criteria in spirit of Nehari[1] have been obtained. These criteria involve some bound on the Schwarzian of a function in a domain which then implies that the function is one-to-one. A differential geometry interpretation of Schwarzian has been given. It shows that the Schwarzian is a type of curvature of a curve in the projective space CP^1 (Flanders[2]).

We will introduce appropriate generalization of the Schwarzian in several complex variables. It is more complicated in that it has an additional matrix parameter that gives the direction of differentiation. We generalize the invariance properties of Schwarzian to a weaker form of invariance. Furthermore, the mappings are found for which the Schwarzian is identically zero.

Finally we will note a new invariant in several complex variables that does not exist in one complex variable. It involves the first derivative of mapping in either of different directions. It is simpler than the Schwarzian, and, in a sence, it replaces the Schwarzian for differentiation in several directions.

In this paper, we will consider points which are square matrices in the classical domain of type one. It is not difficult to extend the results in this paper to the classical domains of type two, three and the corresponding classical domains in real cases.

For reference and completeness we will start with the cross ratio. The basic invariance of the cross ratio was discoverde by Loo-keng Hua[3].

Let

$$Z = Z^{(n,n)} = \begin{bmatrix} z_{11} \cdots z_{1n} \\ \cdots\cdots\cdots \\ z_{n1} \cdots z_{nn} \end{bmatrix} \in C^{n\times n}$$

denote a square matrix, let $\bar{Z}'$ denote the conjugate transpose of Z, and let I denote the $n \times n$ identity matrix. Then the classical domain of type one R_I is a domain in $C^{n\times n}$ such that

$$I - Z\bar{Z}' > 0,$$

where > 0 means positive definite. Let G_I denote the group of holomorphic automorphisms of R_I. Such transformations can be expressed as

$$W = (AZ + B)(CZ + D)^{-1} = (\bar{A}' + Z\bar{B}')^{-1}(\bar{C}' + Z\bar{D}'), \tag{1.1}$$

where A, B, C, D satisfy the conditions

$$\bar{A}'A - \bar{C}'C = I,\ \bar{A}'B = \bar{C}'D,\ \bar{B}'B - \bar{D}'D = -I,$$
$$A\bar{A}' - B\bar{B}' = I,\ A\bar{C}' = B\bar{D}',\ C\bar{C}' - D\bar{D}' = -I. \tag{1.2}$$

Let Z_k for $k = 1, 2, 3, 4$ be four points in R_I, and let W_k for $k = 1, 2, 3, 4$ be their respective images under (1.1) so that $W_k = W(Z_k)$ for $k = 1,2,3, 4$. Then by analogy to the one complex variable definition, we define the cross ratio by (Hua[3])

$$\begin{aligned}[W] &= (W_1, W_2, W_3, W_4) \\ &= (W_1 - W_3)(W_2 - W_3)^{-1}(W_2 - W_4)(W_1 - W_4)^{-1}.\end{aligned} \tag{1.3}$$

From (1.1) and (1.2), it can be shown that

$$W_1 - W_3 = (\bar{A}' + Z_1\bar{B}')^{-1}(Z_1 - Z_3)(CZ_3 + D)^{-1}. \tag{1.4}$$

Similarly,

$$(W_2 - W_3)^{-1} = (CZ_3 + D)(Z_2 - Z_3)^{-1}(\bar{A}' + Z_2\bar{B}'), \tag{1.5}$$

$$W_2 - W_4 = (\bar{A}' + Z_2\bar{B})^{-1}(Z_2 - Z_4)(CZ_4 + D)^{-1}, \tag{1.6}$$

$$(W_1 - W_4)^{-1} = (CZ_4 + D)(Z_1 - Z_4)^{-1}(\bar{A}' + Z_1\bar{B}'). \tag{1.7}$$

Substituting (1.4)—(1.7) into (1.3), we have

$$[W] = (\bar{A}' + Z_1\bar{B}')^{-1}[Z](\bar{A}' + Z_1\bar{B}'). \tag{1.8}$$

Eq. (1.8) shows that the cross ratio is invariant up to similarity.

2 Definition of the Schwarzian

In one complex variable, the Schwarzian is a combination of derivatives which is invariant if the function is composed of a fractional linear transformation. The original idea of the method by which Schwarz found a satisfactory combination of derivatives is easily indicated (Caratheodory[4]). He considered a function $u(z)$, and $w = (au+b)/(cu+d)$ for constants a, b, c, d. Taking the first few derivatives and eliminating the four constants, he obtained an invariant expression.

Unfortunately, Schwarz's approach does not appear useful in our situation. When the products of matrices such as (1.1) are differentiated, the resulting expression becomes very complicated because the factors cannot be commuted.

The approach we take is straightforward from the point of view that the Schwarzian derivative is an infinitesimal version of cross ratio, but involves considerable computation. Indeed, even in one complex variable, note has been taken of the difficulties[5,6].

The one variable equality suggests our approach in several variables. Consider $z_k(k=1,2,3,4)$ points close together and on a straight line. For $k=2,3,4$, we think of expressing the value of w at z_k as a power series expansion about z_1. We insert these expressions for the w_k' s in Eq. (1.3) and allow z_k' s to coalesce at z_1. The appropriate expression for the Schwarzian will become clear.

In order to carry out this calculation we introduce some notations. Let

$$\Lambda = \begin{pmatrix} \lambda_{11}, \lambda_{12}, \cdots, \lambda_{1n} \\ \cdots\cdots\cdots \\ \lambda_{n1}, \lambda_{n2}, \cdots, \lambda_{nn} \end{pmatrix},$$

$\lambda = (\lambda_{11}, \lambda_{12}, \cdots, \lambda_{1n}, \cdots, \lambda_{n1}, \cdots, \lambda_{nn})$, and let $\lambda^{(2)}$, $\lambda^{(3)}, \cdots, \lambda^{(k)}$ be the second, third, $\cdots$, kth Kronecker products of λ.

The directional derivative of $W = W(Z)$ in the direction Λ is easily expressed.

$$DW = D_\Lambda W = \lambda\left(\frac{\partial}{\partial z}\right)' W$$

$$= \begin{pmatrix} \sum\limits_{i,j} \lambda_{ij} \dfrac{\partial}{\partial z_{ij}} w_{11}, & \sum\limits_{i,j} \lambda_{ij} \dfrac{\partial}{\partial z_{ij}} w_{12}, \cdots, & \sum\limits_{i,j} \lambda_{ij} \dfrac{\partial}{\partial z_{ij}} w_{1n} \\ & \cdots\cdots\cdots & \\ \sum\limits_{i,j} \lambda_{ij} \dfrac{\partial}{\partial z_{ij}} w_{n1}, & \sum\limits_{i,j} \lambda_{ij} \dfrac{\partial}{\partial z_{ij}} w_{n2}, \cdots, & \sum\limits_{i,j} \lambda_{ij} \dfrac{\partial}{\partial z_{ij}} w_{nn} \end{pmatrix},$$

where $\dfrac{\partial}{\partial z} = \left(\dfrac{\partial}{\partial z_{11}}, \cdots, \dfrac{\partial}{\partial z_{1n}}, \cdots, \dfrac{\partial}{\partial z_{n1}}, \cdots, \dfrac{\partial}{\partial z_{nn}}\right)$. The second and third derivatives in that direction are given by

$$D^2 W = D_\Lambda^2 W = \lambda^{(2)}\left(\frac{\partial}{\partial z}\right)^{(2)'} W,$$

$$D^3 W = D_\Lambda^3 W = \lambda^{(3)}\left(\frac{\partial}{\partial z}\right)^{(3)'} W.$$

We can also express these derivatives as differentiation with respect to one complex variable,

$$D_\Lambda W(Z) = \frac{d}{d\xi} W(Z+\Lambda\xi)|_{\xi=0}, \quad D_\Lambda^2 W(Z) = \left(\frac{d}{d\xi}\right)^2 W(Z+\Lambda\xi)|_{\xi=0}$$

etc, where $\xi \in C^1$.

Now we examine the expansion of $W(Z_1 + \Lambda\xi)$ where $\xi \in C^1$. Note that

$$\begin{aligned} w_{11}(Z_1 + \Lambda\xi) &= w_{11}(Z_1) + \sum_{i,j} \frac{\partial w_{11}(Z)}{\partial z_{ij}}\Big|_{Z=Z_1} \lambda_{ij}\xi \\ &\quad + \frac{1}{2} \sum_{i,j,k,l} \frac{\partial^2 w(Z)}{\partial z_{ij}\partial z_{kl}}\Big|_{Z=Z_1} \lambda_{ij}\lambda_{kl}\xi^2 + \cdots \\ &= w_{11}(Z_1) + Dw_{11}(Z_1)\xi + \frac{1}{2} D^2 w_{11}(Z_1)\xi^2 + \frac{1}{6} D^3 w_{11}(Z_1)\xi^3 + \cdots. \end{aligned}$$

Similarly, the other functions w_{ij} can be expanded. Thus we have

$$\begin{aligned} W(Z_1 + \Lambda\xi) &= W(Z_1) + DW(Z_1)\xi + \frac{1}{2} D^2 W(Z_1)\xi^2 + \frac{1}{6} D^3 W(Z_1)\xi^3 + \cdots \\ &= W(Z_1) + A_1\xi + \frac{1}{2} A_2\xi^2 + \frac{1}{6} A_3\xi^3 + \cdots, \end{aligned} \tag{2.1}$$

where $A_k = D^k W(Z_1)$ for $k = 1, 2, 3, \cdots$.

Then each of $W_1 = W(Z)$, $W_2 = W(Z + \Lambda\xi)$, $W_3 = W(Z + 2\Lambda\xi)$ and

$$W_4 = W(Z + 3\Lambda\xi)$$

can be similarly expanded. Therefore

$$\begin{aligned} W_1 - W_3 &= -2A_1\xi - 2A_2\xi^2 - \frac{4}{3} A_3\xi^3 - \cdots \\ &= -2\left(I + A_2A_1^{-1}\xi + \frac{2}{3} A_3A_1^{-1}\xi^2 + \cdots\right) A_1\xi, \end{aligned}$$

$$\begin{aligned} W_2 - W_3 &= -A_1\xi - \frac{2}{3} A_2\xi^2 - \frac{7}{6} A_3\xi^3 - \cdots \\ &= -\left(I + \frac{3}{2} A_2A_1^{-1}\xi + \frac{7}{6} A_3A_1^{-1}\xi^2 + \cdots\right) A_1\xi, \end{aligned}$$

$$\begin{aligned} W_2 - W_4 &= -2A_1\xi - 4A_2\xi^2 - \frac{13}{3} A_3\xi^3 - \cdots \\ &= -2\left(I + 2A_2A_1^{-1}\xi + \frac{13}{6} A_3A_1^{-1}\xi^2 + \cdots\right) A_1\xi, \end{aligned}$$

and

$$\begin{aligned} W_1 - W_4 &= -3A_1\xi - \frac{9}{2} A_2\xi^2 - \frac{9}{2} A_3\xi^3 - \cdots \\ &= -3\left(I + \frac{3}{2} A_2A_1^{-1}\xi + \frac{3}{2} A_2A_1^{-1}\xi^2 + \cdots\right) A_1\xi. \end{aligned}$$

Hence, for small ξ, the cross ratio $[W] = (W_1 - W_3)(W_2 - W_3)^{-1}(W_2 - W_4)$

$$(W_1 - W_4)^{-1} = \frac{4}{3}\left(I + A_2A_1^{-1}\xi + \frac{2}{3}A_3A_1^{-1}\xi^2 + \cdots\right)\left(I + \frac{3}{2}A_2A_1^{-1}\xi\right.$$

$$\left. + \frac{7}{6}A_3A_1^{-1}\xi^2 + \cdots\right)^{-1}\left(I + 2A_2A_1^{-1}\xi + \frac{13}{6}A_3A_1^{-1}\xi^2 + \cdots\right)\left(I + \frac{3}{2}A_2A_1^{-1}\xi\right.$$

$$\left. + \frac{3}{2}A_3A_1^{-1}\xi^2 + \cdots\right)^{-1} = \frac{4}{3}\left(I + \left(\frac{1}{6}A_3A_1^{-1} - \frac{1}{4}A_2A_1^{-1}A_2A_1^{-1}\right)\xi^2 + \cdots\right). \tag{2.2}$$

The coefficient of ξ^2 in Eq. (2.2) will be identified as the Schwarzian.

Now we assume that the range of W is in R_I and examine the invariance of that expression. Let $X = (AW + B)(CW + D)^{-1}$, that is, X and W are related by a holomorphic automorphism of R_I. By Eq. (1.8), we have

$$[X] = (\bar{A}' + W\bar{B}')^{-1}[W](\bar{A}' + W\bar{B}').$$

Let $A_k(X)$, $A_k(W)$ denote the directional derivatives of order k of X and W respectively. Then Eq. (2.2) implies that

$$[X] = \frac{4}{3}\left(I + \left(\frac{1}{6}A_3(X)A_1^{-1}(X) - \frac{1}{4}A_2(X)A_1^{-1}(X)A_2(X)A_1^{-1}(X)\right)\xi^2 + \cdots\right)$$

$$= (\bar{A}' + W\bar{B}')^{-1}\frac{4}{3}\left(I + \left(\frac{1}{6}A_3(W)A_1^{-1}(W)\right.\right.$$

$$\left.\left. - \frac{1}{4}A_2(W)A_1^{-1}(W)A_2(W)A_1^{-1}(W)\right)\xi^2 + \cdots\right)(\bar{A}' + W\bar{B}').$$

By identifying the coefficients of ξ^2 and multiplying by a scalar, we find that

$$A_3(X)A_1^{-1}(X) - \frac{3}{2}A_2(X)A_1^{-1}(X)A_2(X)A_1^{-1}(X) = (\bar{A}' + W\bar{B}')^{-1}$$

$$\cdot\left(A_3(W)A_1^{-1}(W) - \frac{3}{2}A_2(W)A_1^{-1}(W)A_2(W)A_1^{-1}(W)\right)(\bar{A}' + W\bar{B}'). \tag{2.3}$$

Hence the expression on the left side of (2.3) is invariant up to similarity under the group of holomorphic automorphism of R_I.

Thus we define

$$A_3(W)A_1^{-1}(W) - \frac{3}{2}A_2(W)A_1^{-1}(W)A_2(W)A_1^{-1}(W)$$

$$= (D_A^3W)(D_AW)^{-1} - \frac{3}{2}(D_A^2W)(D_AW)^{-1}(D_A^2W)(D_AW)^{-1}$$

as our generalization of Schwarzian in several complex variables, and denote it by $\{W;Z\}_A$. And we note that it agrees with the standard definition if $n = 1$.

As an alternative expression for our Schwarzian,

$$\{W;Z\}_A = D_A((D_A^2W)(D_AW)^{-1}) - \frac{1}{2}((D_A^2W)(D_AW)^{-1})^2, \qquad (2.4)$$

where A is any non-zero $n\times n$ complex matrix and W depends on the variable Z.

In general, if we start from the points $W_1 = W(Z_1)$, $W_2 = W(Z_1+\alpha A\xi)$, $W_3 = W(Z_1+\beta A\xi)$ and $W_4 = W(Z_1+\gamma A\xi)$ where α, β, γ are distinct non-zero complex numbers, then after long calculation, we get

$$[W] = (0,\alpha,\beta,\gamma)\left(I + \frac{1}{6}\{W,Z\}_{A\xi}\xi^2 + \cdots\right), \qquad (2.5)$$

where $(0,\alpha,\beta,\gamma)$ are the cross ratio of the numbers $0,\alpha,\beta,\gamma$. Formula (2.2) is a special case of (2.5) for $\alpha=1$, $\beta=2$ and $\gamma=3$.

3 Some Algebraic Lemmas

In one complex variable, it is easy to solve the third order differential equation $\{w;z\}=0$, that is,

$$\frac{w'''}{w'} - \frac{3}{2}\left(\frac{w''}{w'}\right)^2 = 0. \qquad (3.1)$$

By standard substitution and elementary integration, one can show that $w(z)$ satisfies Eq. (3.1) if and only if $w(z)$ is a fractional linear transformation: $w(z) = (az+b)/(cz+d)$ where $ad\text{-}bc$ is non-zero.

In several complex variables, we can prove the following theorem.

Theorem 1. *If $W=W(Z)$ is a holomorphic $n\times n$ matrix mapping of an $n\times n$ matrix variable $Z\in R_I$, then*

$$\{W;Z\}_A = 0 \qquad (3.2)$$

for all Z in R_I and all non-zero $n\times n$ matrices A (when the left side of (3.2) exists) if and only if

$$W = W(Z) = (w_{ij}(Z))_{1\le i,j\le n}, \qquad (3.3)$$

where

$$w_{ij} = c_{ij} + tr(C_{ij}Z(I-L(Z))^{-1}), \text{ or } w_{ij} = c_{ij} + tr(C_{ij}(I-L(Z))^{-1}Z), \qquad (3.4)$$

$i,j=1,2,\cdots,n$; c_{ij} are constants, C_{ij} are $n\times n$ constant matrices, and $L(Z)$ is $n\times n$ matrix. Each entry of $L(Z)$ is a linear homogeneous function of z_{ij}, $i,j=1,2,\cdots,n$.

If Eq. (3.3) holds, then a straightforward calculation shows that (3.2) is true. The difficult part is that (2.2) implies (3.3). It needs some algebraic lemmas.

Lemma 1. *If Z is a matrix and all the entries are independent complex variables, then* $\det Z$ *is irreducible.*

Proof. Suppose that $\det Z$ is reducible, that is, $\det Z = PQ$ for some non-constant polynomials P and Q. If all but one entry of Z hold fixed, $\det Z$ is linear in remaining variable. For each entry, one and only one of P and Q depends on that variable. Since each is assumed to be non-constant, some entries appear in P and others in Q. Consider a variable on which P depends, and consider the row on which that variable appears. Think of the entries along that row as variables while all other matrix entries hold constant in general position. Expand $\det Z$ along that row. The resulting polynomial is of the first degree with each variable having a constant coefficient. Then, if all but one entry of the row hold constant, the zero of $\det Z = 0$ in that variable depends on the entry where each of the other row entries is fixed; hence every entry on that row is a variable on which P depends. The same argument is applied to columns. Thus each column from the origin row must consist of entries on which P depends. The polynomial P depends on every entry of Z and Q depends on none of them. This contradicts the fact that Q is to be non-constant.

Lemma 2. *Suppose that Z is a matrix of complex variables. Let W be the adjoint of Z and let R_k denote the kth row of Z. Then each entry of R_kW is divisible by the determinant of Z.*

Proof. Of course $R_k W$ is a row vector. The kth entry is the determinant of Z. The mth entry, for m not equal to k, is equal to the determinant of the matrix obtained from Z by replacing the mth row of Z by a row identical to the kth row. Thus that entry is zero.

Theorem 2. *Let $Z = (z_{ij})$ be an $n \times n$ matrix of complex variables. Let W be the adjoint of Z. Consider a row vector A of n entries, with each entry being a homogeneous polynomial of degree two in the variables z_{ij}. And consider a column vector B of n entries, with each entry being a homogeneous of degree two in the variables z_{ij}. If AWB is divisible by the polynomial* $\det Z$, *then either each entry of the vector AW is divisible by* $\det Z$, *or each entry of the vector WB is divisible by* $\det Z$. *Consequently, there exists either a row vector L with first order entries in the z_{ij} variables such that $A = LZ$ or a column vector L with first order entries such that $B = ZL$.*

Proof. There is a minor of Z that has a non-zero determinant. By exchanging rows and columns if necessary, it can be assumed to be the n, n minor of Z; then its determinant is the n, n entry of W. Let the determinant of that minor be p. Since it is of lower degree than the irreducible polynomial $\det Z$, it is sufficient to show that either pAW or WBp has entries which are divisible by $\det Z$.

Let R_{n-1k} denote the vector having the same first $n-1$ terms as the kth row of Z. Consider the equation

$$\sum_{k=1}^{n-1} x_k R_{n-1k} = pA_{n-1}. \tag{3.5}$$

Since the determinant of the $(n-1) \times (n-1)$ matrix $\begin{pmatrix} R_{n-11} \\ \cdots\cdots \\ R_{n-1n-1} \end{pmatrix}$ is p and is non-zero, by Cramer's rule, there is a set of polynomials x_k which solves Eq. (3.5).

By Lemma 2,

$$\left(pA-\sum_{k=1}^{n-1}x_kpR_k\right)W=pAW+\text{a term with entries divisible by detZ.} \quad (3.6)$$

On the other hand, by the choice of x_k's, the vector in the parentheses has its first $n-1$ entries equal to zero. The nth entry equals a polynomial times p; let the entry be represented by pa_n.

A similar discussion can be given concerning the column vector B. Then $pAWBp$ being divisible by $\det Z$ implies that

$$\left(pa_n-p\sum_{k=1}^{n}x_kz_{kn}\right)p\left(b_np-\sum_{k=1}^{n}y_kz_{nk}p\right) \quad (3.7)$$

is divisible by $\det Z$.

First we assume that the irreducible polynomial $\det Z$ divides the factor represented by the expression in the first parentheses of expression (3.7). Then the entries of $\left(pA-p\sum_{k=1}^{n}x_kR_k\right)$ are divisible by $\det Z$ and each entry of

$$\left(pA-p\sum_{k=1}^{n}x_kR_k\right)W$$

is divisible by det Z. But from (3.6), the previous expression equals pAW + a term divisible by $\det Z$. Hence pAW has entries divisible by $\det Z$. Since p and $\det Z$ have no common factors $AW=\alpha\det Z$ for some vector α with polynomial entries, $A=\alpha Z$. But the degree of the polynomial entries of α must be one since the degree of each entry of Z is one and the degree of each entry of A is two; thus α is a row vector L with entries of degree one as described in the theorem and $A=LZ$.

A similar analysis for the factor in the second parentheses of (3.7) being divisible by $\det Z$ gives the conclusion that there is a column vector L with the first degree entries such that $B=ZL$.

4 Proof of Theorem 1

Now we prove the "only if" part of Theorem 1.

Let $U=(D_A^2W)(D_AW)^{-1}$. Then

$$D_AU-\frac{1}{2}U^2=0, \quad (4.1)$$

since

$$\{W;Z\}_A=D_A((D_A^2W)(D_AW)^{-1})-\frac{1}{2}[(D_A^2W)(D_AW)^{-1}]^2=0.$$

(4.1) is equivalent to

$$D_A(U^{-1})=-U^{-1}(D_AU)U^{-1}=-\frac{1}{2}I.$$

Obviously, $D_A\left(-\frac{1}{2}A^{-1}Z\right)=-\frac{1}{2}I$, hence $D_A\left(U^{-1}+\frac{1}{2}A^{-1}Z\right)=0$. We have

$$U^{-1}+\frac{1}{2}A^{-1}Z=A^{-1}C_1(Z,A), \tag{4.2}$$

where $C_1(Z,A)$ is a matrix for which each entry is function of Z and satisfies $D_AC_1(Z,A)=0$, and $C_1(Z,A)$ is a constant matrix if $Z=\xi A$ where $\xi\in C^1$.

From (4.2), we get

$$U=\left(-\frac{1}{2}Z+C_1\right)^{-1}A. \tag{4.3}$$

Let $V=D_AW$. Then

$$(D_AV)V^{-1}=U=\left(-\frac{1}{2}Z+C_1\right)^{-1}A=-(Z+C_0)^{-1}2A,$$

where $C_0=-2C_1$. Let $V_0=(Z+C_0)^{-1}A(Z+C_0)^{-1}$. Then

$$D_AV_0=-2(Z+C_0)^{-1}AV_0,$$

and hence $(D_AV)V^{-1}-(D_AV_0)V_0^{-1}=0$,

$$\begin{aligned}D_A(V^{-1}V_0)&=-V^{-1}(D_AV)V^{-1}V_0+V^{-1}D_AV_0\\&=-V^{-1}[(D_AV)V^{-1}-(D_AV_0)V_0^{-1}]V_0=0.\end{aligned}$$

We have

$$V^{-1}V_0=C_2(Z,A)^{-1}, \tag{4.4}$$

where $C_2(Z,A)$ is a matrix for which each entry is function of Z and A, and satisfies $D_AC_2(Z,A)=0$, and $C_2(Z,A)$ is a constant matrix if $Z=\xi A$ where $\xi\in C^1$. From (4.4), we get

$$V=D_AW=(Z+C_0)^{-1}A(Z+C_0)^{-1}C_2. \tag{4.5}$$

Let $W_0=-(Z+C_0)^{-1}C_2$. Then $D_AW_0=(Z+C_0)^{-1}A(Z+C_0)^{-1}C_2$ and

$$D_A(W-W_0)=0.$$

We have

$$W(Z)=-(Z+C_0)^{-1}C_2+C_3, \tag{4.6}$$

where $C_3=C_3(Z,A)$ is a matrix for which each entry is function of Z and A, and satisfies $D_AC_3(Z,A)=0$, and $C_3(Z,A)$ is a constant matrix if $Z=\xi A$ where $\xi\in C^1$.

Suppose that $\{W;Z\}_A=0$ for all invertible A, and for all Z in the neighborhood N of the origin. The Jacobian of W must be non-zero in N in order that the Schwarzian is meaningful for all invertible A, and for $Z\in N$. By translation and multiplication by $J_W(0)^{-1}$, we can assume that in N,

$$W(Z)=Z+C_2(Z)+C_3(Z)+\cdots, \tag{4.7}$$

where for each $k = 2, 3, \cdots$, each entry of the matrix $C_k(Z)$ is a homogeneous polynomial of degree k.

Fix a complex line $Z = \xi A$ for an invertible A. We have already proved that in a line W is linear fractional. Thus by (4.6),

$$W(Z) = Z(I - L_1(Z))^{-1}, \tag{4.8}$$

or

$$W(Z) = (I - L_1(Z))^{-1}Z, \tag{4.9}$$

where $L_1(Z)$ is homogeneous linear.

Assume (4.8) holds, we expand (4.8) as

$$W(Z) = Z + ZL_1(Z) + ZL_1(Z)L_1(Z) + ZL_1(Z)L_1(Z)L_1(Z) + \cdots$$

at the neighborhood of the origin.

From (4.7), for all line ξA with invertible A, we have

$$C_2(Z) = ZL_1(Z),$$

$$C_3(Z) = ZL_1(Z)L_1(Z),$$

thus

$$C_3(Z) = C_2(Z)Z^{-1}C_2(Z), \tag{4.10}$$

etc. By continuity, (4.10) holds for all points of N, then by Theorem 2, we have

$$C_2(Z) = ZL(Z) \qquad \text{for all } Z \in N, \tag{4.11}$$

or

$$C_2(Z) = L(Z)Z \qquad \text{for all } Z \in N, \tag{4.12}$$

where $L(Z)$ is homogeneous linear in each coordinate.

Suppose (4.11) holds. Then (4.10) implies

$$C_3(Z) = ZL(Z)Z^{-1}ZL(Z) = ZL(Z)L(Z),$$

$$\cdots\cdots\cdots$$

$$C_k(Z) = Z\underbrace{L(Z)L(Z)\cdots L(Z)}_{k-1 \text{ product}}.$$

Finally, we have $W(Z) = Z(I - L(Z))^{-1}$ for all of N. Similarly, suppose (4.12) holds, then we have $W(Z) = (I - L(Z))^{-1}Z$ for all of N. Thus we prove the theorem in the case $W(0) = 0$, $J_W(0) = I$ when (4.8) holds.

If (4.9) holds, we can use the same argument to prove the theorem in the case $W(0) = 0$, $J_W(0) = I$. The general case of the theorem follows immediately.

5 A New Invariant

Let $W = W(Z)$ be a mapping in R_I. The range of W is in R_I, and

$$X = (AW + B)(CW + D)^{-1},$$

that is, X and W are related by a holomorphic automorphism of R_I. Let Λ and Π be two $n\times n$ matrices. Then

$$D_\Lambda X = A(D_\Lambda W)(CW+D)^{-1} - (AW+B)(CW+D)^{-1}C(D_\Lambda W)(CW+D)^{-1}$$
$$= (\bar{A}'+W\bar{B}')^{-1}(D_\Lambda W)(CW+D)^{-1}$$

and

$$D_\Pi X = (\bar{A}'+W\bar{B}')^{-1}(D_\Pi W)(CW+D)^{-1}.$$

Let $[X;\ Z]_{\Lambda,\Pi} = D_\Lambda X(D_\Pi X)^{-1}$, and let $[W;\ Z]_{\Lambda,\Pi} = D_\Lambda W(D_\Pi W)^{-1}$. If $(D_\Pi X)^{-1}$ and $(D_\Pi W)^{-1}$ exist, then

$$[X;Z]_{\Lambda,\Pi} = (\bar{A}'+W\bar{B}')^{-1}[W;Z]_{\Lambda,\Pi}(\bar{A}'+W\bar{B}').$$

Hence $[W;\ Z]_{\Lambda,\Pi}$ is invariant up to similarity under the group of holomorphic automorphisms of R_I. This is a new invariant in several complex variables that does not exist in one complex variable.

Finally, we like to point out that, during the last few years there are many interesting extensions of the Schwarzian derivatives into high dimensions. For examples, K. Carne[7], B. Osgood and D. Stowe[8,9] extended the Schwarzian derivatives in the viewpoint of the conformal mappings between Riemannian manifolds. Weiqi Gao[10] extended Flanders' result to high dimensions.

The authors gratefully acknowledge helpful communications with Profs. Avinash Suthaye and Ted Suffridge.

References

1 Nehari, Z., *Bull. Amer. Math. Soc.*, 1949, **55**:545.
2 Flanders, H., *J. Diff. Geometry*, 1970, **4**:515.
3 Hua Lookeng, *Tran. Amer. Math. Soc.*, 1945, **57**:441.
4 Caratheodory, C., *Funktionentheorie*, Basel, Birthauser, 1950.
5 Ahlfors, L. V., *Complex Analysis* (eds. Hersch, J. & Huber, A.), Basel, 1980, pp. 1—15.
6 Ahlfors, L. V., *Ordway Professorship Lecture in Mathematics*, University of Minnesota, 1981.
7 Carne, K., *J. Reine Angew. Math.*, 1990, **408**:10.
8 Osgood, B. & Stowe, D., *Duke Math. J.*, 1922, **67**:57.
9 Osgood, B. & Stowe, D., *Comment. Math. Helv.*, 1990, **65**:234.
10 Gao Weiqi, Schwarzian curvatures of holomorphic curves, preprint, 1992.

Chin. Ann. of Math.
19B: 1(1998),1-8.

THE SCHWARZIAN DERIVATIVE IN SEVERAL COMPLEX VARIABLES (II)****

GONG SHENG* YU QIHUANG** ZHENG XUEAN***

Abstract

The Schwarzian derivative of holomorphic mapping on classical domain $\mathbb{R}_I$ is zero iff it is linear fractional.

Keywords Schwarzian derivative, Classical domain, Linear fractional mapping
1991 MR Subject Classification 32H02
Chinese Library Classification O174.56

§1.

Let $Z = (z_{ij})_{1\le i,j\le n} \in \mathbb{C}^{n\times n}$ denote a square complex matrix. The classical domain of type one R_I is a domain in $\mathbb{C}^{n\times n}$ such that $I - Z\bar{Z}' > 0$, where I denotes the $n\times n$ idendity matrix, $\bar{Z}'$ denotes the conjugate transpose of Z, >0 means positive definite. Let $\Lambda = (\lambda_{ij})_{1\le i,j\le n} \in \mathbb{C}^{n\times n}, \lambda = (\lambda_{11},\cdots,\lambda_{1n},\cdots,\lambda_{n1},\cdots,\lambda_{nn}) \in \mathbb{C}^{n^2}, \lambda^{(k)}$ is the k-th Kronecker product of λ. Let $W = W(Z)$ be a holomorphic $n\times n$ matrix mapping of $n\times n$ matrix variable Z. We denote the k-th directional derivative of a holomorphic mapping $W = W(Z)$ in the direction Λ by

$$D^k W(Z) = D^k_\Lambda W(Z) = \lambda^{(k)} \left(\frac{\partial}{\partial Z}\right)' W,$$

where

$$\frac{\partial}{\partial Z} = \left(\frac{\partial}{\partial z_{11}},\cdots\frac{\partial}{\partial z_{1n}},\cdots,\frac{\partial}{\partial z_{n1}},\cdots,\frac{\partial}{\partial z_{nn}}\right).$$

In [1], Gong and FitzGerald defined the Schwarzian derivative of $W = W(Z)$ along the direction Λ by

$$\{W;Z\}_\Lambda = (D^3_\Lambda W)(D_\Lambda W)^{-1} - \frac{3}{2}(D^2_\Lambda W)(D_\Lambda W)^{-1}(D^2_\Lambda W)(D_\Lambda)^{-1},$$

and proved that

$$\{W;Z\}_\Lambda = 0 \tag{1.1}$$

Manuscript received August 30, 1996.

*Department of Mathematics, University of Science and Technology of China, Hefei 230026, China.
**Department of Mathematics, Institute of Applied Mathematics, Academia Sinica, Beijing 100080, China.
***Department of Mathematics, Beijing Normal University, Beijing 100875, China.

****Project supported by the National Natural Science Foundation of China.

for all $Z \in R_I$ and all non-zero matrix $\Lambda \in \mathbb{C}^{n\times n}$ (when $\{W;Z\}_\Lambda$ exists), if and only if

$$W = W(Z) = W(0)+(I-L(Z))^{-1}D_Z W(0) \text{ or } W(Z) = W(0)+D_Z W(0)(I-L(Z))^{-1} \quad (1.2)$$

where $L(Z)$ is an $n\times n$ matrix, each entry of $L(Z)$ is a linear homogeneous polynomial of $z_{ij}, i,j = 1,2,\cdots,n$, $I-L(Z)$ is non-singular when $Z \in R_I$.

In this note we prove that (1.1) may be replaced by $\{W;Z\}_Z = 0$ where $Z \in R_I$.

Theorem 1.1. *Let $W = W(Z) : R_I \to \mathbb{C}^{n\times n}$ be a holomorphic mapping with $J_W(0)$ non-singular. Then*

$$\{W;Z\}_Z = 0 \quad (1.3)$$

for all Z in R_I (when $\{W;Z\}_Z$ exists) if and only if $W = W(Z)$ is defined by (1.2). Moreover, $W(Z)$ is biholomorphic in R_I.

As a consequance, we have

Corollary 1.1. *Let $W = W(Z) : R_I \to \mathbb{C}^{n\times n}$ be a holomorphic mapping with $J_W(0)$ non-singular. Then $\{W;Z\}_Z = 0$ implies $\{W;Z\}_\Lambda = 0$ for all non-zero Λ when $\{W;Z\}_\Lambda$ exists.*

Denote $\{W;Z\}_Z$ by $\{W;Z\}$. By Theorem 1.1, it is reasonable to define it as the Schwarzian derivative of $W(Z)$ at Z.

Of course Theorem 1.1 is still true if we replace R_I by any domain in $\mathbb{C}^{n\times n}$.

§2.

Now we are going to prove Theorem 1.1.

A straighforward calculation shows that (1.3) is true if $W = W(Z)$ is defined by (1.2). The difficult part is that (1.3) implies $W = W(Z)$ is defined by (1.2).

We expand $W(Z)$ at $Z = 0$

$$W(Z) = W(0) + D_Z W(0) + C_2(Z) + C_3(Z) + C_4(Z) + \cdots,$$

where $C_2, C_3, C_4, \cdots$ are $n \times n$ matrices, all entries of $C_2(Z), C_3(Z), C_4(Z), \cdots$ are homogeneous polynomials of entries of Z of degree 2, 3, 4,..., respectively.

Fix $Z \in R_I$, if $Z, D_Z W(0)$ are non-singular. Let $Q(t) = W(tZ), t \in [0,1]$. Then

$$\begin{aligned}\frac{dQ}{dt} &= D_Z W(tZ) = \frac{1}{t} D_{tZ} W(tZ)\\ &= D_Z W(0) + 2tC_2(Z) + 3t^2 C_3(Z) + \cdots,\end{aligned}$$

and

$$\begin{aligned}\frac{d^2Q}{dt^2} &= D_Z^2 W(tZ) = \frac{1}{t^2} D_{tZ}^2 W(tZ)\\ &= 2C_2(Z) + 6tC_3(Z) + \cdots.\end{aligned}$$

If we take $\epsilon > 0$ sufficiently small, such that $\frac{dQ(t)}{dt}$ is non-singular when $t \in [0,\epsilon]$, then we may define

$$A(t) = \frac{d^2Q}{dt^2}\left(\frac{dQ}{dt}\right)^{-1} = D_Z^2 W(tZ)(D_Z W(tZ))^{-1}$$

for $t \in [0,\epsilon]$. It is easy to verify that

$$\frac{dA}{dt} - \frac{1}{2}A^2 = 0,$$

since $\{W; Z\} = 0$.

Consider a C^∞ mapping $G(t) : [0, \epsilon] \to \mathbb{C}^{n\times n}$ which is the solution of the following initial value problem

$$\begin{cases} \dfrac{dG}{dt} = -\dfrac{1}{2}GA, \\ G(0) = I. \end{cases} \tag{2.1}$$

It is known that the solution exists and is unique (cf. [2]).

By (2.1), we have

$$\frac{d^2G}{dt^2} = -\frac{1}{2}G\left(\frac{dA}{dt} - \frac{1}{2}A^2\right) = 0.$$

That means $\frac{dG}{dt}$ is independent of t. Thus

$$\frac{dG(t)}{dt} = -\frac{1}{2}G(t)A(t) = -\frac{1}{2}G(0)A(0) = -\frac{1}{2}A(0). \tag{2.2}$$

Integrating both sides of (2.2) with respect to t from 0 to t, we have

$$G(t) - G(0) = \frac{t}{2}A(0);$$

that is,

$$G(t) = I - \frac{t}{2}A(0). \tag{2.3}$$

Substituting it into (2.2), we obtain

$$-\frac{1}{2}\left(I - \frac{t}{2}A(0)\right)A(t) = -\frac{1}{2}A(0).$$

Hence

$$\left(I - \frac{t}{2}A(0)\right)A(t) = A(0).$$

By the definition of $A(t)$, the preceding equation is

$$\left(I - \frac{t}{2}A(0)\right)\frac{d^2Q(t)}{dt^2} = A(0)\frac{dQ(t)}{dt}.$$

It is the same as

$$\frac{d}{dt}\left[\left(I - \frac{t}{2}A(0)\right)\frac{dQ(t)}{dt}\right] = \frac{1}{2}A(0)\frac{dQ(t)}{dt}.$$

Integrating both sides of the preceding equation with respect to t from 0 to t, we have

$$\left(I - \frac{t}{2}A(0)\right)\frac{dQ}{dt} - D_ZW(0) = \frac{1}{2}A(0)(Q(t) - W(0)),$$

since $Q(0) = W(0)$, $\frac{dQ}{dt}(0) = D_ZW(0)$. Thus

$$\frac{d}{dt}\left[(I - \frac{t}{2}A(0))Q(t)\right] = D_ZW(0) - \frac{1}{2}A(0)W(0)$$

holds. Integrating both sides of the preceding equation with respect to t from 0 to t, we have

$$\left(I - \frac{t}{2}A(0)\right)Q(t) - W(0) = t\left(D_ZW(0) - \frac{1}{2}A(0)W(0)\right). \tag{2.4}$$

By (2.3), $G(t)$ is non-singular when t is sufficiently small, solving $Q(t)$ from (2.4),

$$Q(t) = W(tZ) = t\left(I - \frac{t}{2}A(0)\right)^{-1}D_ZW(0) + W(0).$$

Expanding $W(tZ)$ and $t(I-\frac{t}{2}A(0))^{-1}D_ZW(0)+W(0)$ with respect to t at a neighborhood of $t=0$, we have

$$W(tZ)=W(0)+tD_ZW(0)+t^2C_2(Z)+t^3C_3(Z)+t^4C_4(Z)+\cdots, \tag{2.5}$$

where $C_2(Z),C_3(Z),C_4(Z),\cdots$ are $n\times n$ matrices, all entries of $C_2(Z),C_3(Z),C_4(Z),\cdots$ are homogeneous polynomails of entries of Z of degree 2, 3, 4, $\cdots$ respectively.

On the other hand, in a neighborhood of $t=0$, we have

$$\begin{aligned}&W(0)+t\Big(I-\frac{t}{2}A(0)\Big)^{-1}D_ZW(0)\\&=W(0)+tD_ZW(0)+\frac{t^2}{2}A(0)D_ZW(0)+\frac{t^3}{4}A^2(0)D_ZW(0)+\cdots.\end{aligned} \tag{2.6}$$

Comparing the corresponding coefficeints of t^2,t^3 in (2.5) and (2.6), we get

$$C_2(Z)=\frac{1}{2}A(0)D_ZW(0),\qquad C_3(Z)=\frac{1}{4}A^2(0)D_ZW(0).$$

Since $D_ZW(0)$ is non-singular, we have

$$C_3(Z)=\frac{1}{4}A(0)D_ZW(0)(D_ZW(0))^{-1}A(0)D_ZW(0)=C_2(Z)(D_ZW(0))^{-1}C_2(Z). \tag{2.7}$$

Let $\xi=(\xi_{ij})_{1\le i,j\le n}=D_ZW(0)$. Because $J_W(0)$ is non-singular, we may express $z_{ij},i,j=1,2,\cdots,n$ as homogeneous polynomials of the entries of ξ of degree one; that is,

$$Z=P(\xi)=(p_{ij}(\xi))_{1\le i,j\le n},$$

each $p_{ij},i,j=1,2,\cdots,n$, is a homogeneous polynomial of the entries of ξ of degree one. By (2.7), we have

$$C_3(P(\xi))=C_2(P(\xi))\xi^{-1}C_2(P(\xi)).$$

Let $C_3(P(\xi))=B_3(\xi)$, $C_2(P(\xi))=B_2(\xi)$. Then all entries of $B_2(\xi)$, $B_3(\xi)$ are homogeneous polynomials of the entries of ξ of deree 2, degree 3, respectively. From $B_3(\xi)=B_2(\xi)\xi^{-1}B_2(\xi)$, and Theorem 2 of [1], we have $B_2(\xi)=\xi L_0(\xi)$ or $B_2(\xi)=L_0(\xi)\xi$, where $L_0(\xi)$ is an $n\times n$ matrix, each entry is a homogeneous polynomial of the entries of degree one. Thus we have $C_2(Z)=D_ZW(0)L_0(D_ZW(0))$ or $C_2(Z)=L_0(D_ZW(0))D_ZW(0)$. We get

$$C_2(Z)=D_ZW(0)L(Z),\quad\text{or}\quad C_2(Z)=L(Z)D_ZW(0).$$

If $L(Z)+L_0(D_ZW(0))$, obviously each entry of $L(Z)$ is a homogeneous polynomial of the entries of Z of degree one.

If $C_2(Z)=D_ZW(0)L(Z)$, then

$$\begin{aligned}C_3(Z)&=C_2(Z)(D_ZW(0))^{-1}C_2(Z)\\&=D_ZW(0)L(Z)(D_ZW(0))^{-1}D_ZW(0)L(Z)\\&=D_ZW(0)(L(Z))^2.\end{aligned}$$

Similarly, we have $C_4(Z)=D_ZW(0)(L(Z))^3,\cdots$. Substituting all these results into (2.5), we obtain

$$\begin{aligned}W(tZ)&=W(0)+tD_ZW(0)+t^2D_ZW(0)L(Z)+t^3D_ZW(0)(L(Z))^2\\&\quad+t^4D_ZW(0)(L(Z))^3+\cdots\\&=tD_ZW(0)(I-tL(Z))^{-1}.\end{aligned} \tag{2.8}$$

Similarly, in the case $C_2(Z) = L(Z)D_ZW(0)$, we obtain

$$W(tZ) = W(0) + t(I - tL(Z))^{-1}D_ZW(0). \tag{2.9}$$

Thus (2.8) or (2.9) holds true at a neighborhood of $t = 0$. No doubt, $W(tZ)$ is an analytic function of t when $tZ \in R_I$. It implies (2.8) or (2.9) holds true when $tZ \in \mathbb{R}_I$.

Let $t \to 1$. We have

$$W(Z) = W(0) + (I - L(Z))^{-1}D_ZW(0)$$

or $W(Z) = W(0) + D_ZW(0)(I - L(Z))^{-1}$ when $D_ZW(0)$ is non-singular. Since $W(Z)$ is holomorphic in $\mathbb{R}_I$, we know that $I - L(Z)$ is non-singular when Z is non-singular.

If $D_ZW(0)$ is non-singular at a point $Z \in R_I$, then there exists a neighborhood of Z, such that $D_ZW(0)$ is non-singular at this neighborhood.

By the uniqueness theorem of holomorphic mapping, we conclude that

$$W(Z) = W(0) + (I - L(Z))^{-1}D_ZW(0)$$

or $W(Z) = W(0) + D_ZW(0)(I_L(Z))^{-1}$ for all $Z \in \mathbb{R}_I$, and hence $I - L(Z)$ is non-singular when $Z \in \mathbb{R}_I$. Thus we have proved that $W = W(Z)$ is defined by (1.2) if (1.3) is true in the case that $J_W(0)$ is non-singular.

§3.

Now, we prove that $W = W(Z)$ which is defined by (1.2) is biholomorphic in $\mathbb{R}_I$.

If it is not biholomorphic, then there exist two points $Z_1, Z_2 \in R_I$, such that $W_1 = W(Z_1) = W_2 = W(Z_2)$. We consider $W(Z) = W(0) + D_ZW(0)(I - L(Z))^{-1}$ at first. R_I is a convex domain, $(1-t)Z_1 + tZ_2 \in R_I$, if $t \in [0,1]$. Let

$$\begin{aligned} Q(t) &= D_{(1-t)Z_1+tZ_2}W(0)(I - L((1-t)Z_1 + tZ_2))^{-1} - D_ZW(0)(I - L(Z_1))^{-1} \\ &= (D_{Z_1}W(0) + t(D_{Z_2}W(0) - D_{Z_1}W(0))[I - L(Z_1) - t(L(Z_2) - L(Z_1))]^{-1} \\ &\quad - D_{Z_1}W(0)(I - L(Z_1))^{-1}. \end{aligned}$$

Then

$$Q(0) = Q(1) = 0,$$

and

$$\begin{aligned} \frac{dQ}{dt} &= \{(D_{Z_2}W(0) - D_{Z_1}W(0) + (D_{Z_1}W(0) + t(D_{Z_2}W(0) - D_{Z_1}W(0)))[I - L(Z_1) \\ &\quad - t(L(Z_2) - L(Z_1))]^{-1}(L(Z_2) - L(Z_1))\}[I - L(Z_1) - t(L(Z_2) - L(Z_1))]^{-1}, \end{aligned} \tag{3.1}$$

and

$$\begin{aligned} \frac{d^2Q}{dt^2} &= 2\{(D_{Z_2}W(0) - D_{Z_1}W(0) + (D_{Z_1}W(0) + t(D_{Z_2}W(0) - D_{Z_1}W(0)))[I - L(Z_1) \\ &\quad - t(L(Z_2) - L(Z_1))]^{-1}(L(Z_2) - L(Z_1))\}[I - L(Z_1) - t(L(Z_2) - L(Z_1))]^{-1} \\ &\quad \times (L(Z_2) - L(Z_1))[I - L(Z_1) - t(L(Z_2) - L(Z_1))]^{-1} \\ &= 2\frac{dQ}{dt}(L(Z_2) - L(Z_1))[I - L(Z_1) - t(L(Z_2) - L(Z_1))]^{-1}. \end{aligned} \tag{3.2}$$

Let

$$A(t) = 2(L(Z_2) - L(Z_1))[I - L(Z_1) - t(L(Z_2) - L(Z_1))]^{-1}. \tag{3.3}$$

Then

$$\frac{dA(t)}{dt}=2(L(Z_2)-L(Z_1))[I-L(Z_1)-tL(L(Z_2)-L(Z_1)]^{-1}$$
$$\times(L(Z_2)-L(Z_1))[I-L(Z_1)-t(L(Z_2)-L(Z_1)]^{-1}.$$

It follows that

$$\frac{dA(t)}{dt}-\frac{1}{2}A^2(t)=0, \tag{3.4}$$

when $t\in[0,1]$. If $G(t),t\in[0,1]$ is the solution of the initial value problem

$$\begin{cases}\dfrac{dG}{dt}=-\dfrac{1}{2}AG,\\ G(0)=I,\end{cases} \tag{3.5}$$

where $A(t)$ is defined by (3.3), we have

$$\frac{d^2(QG)}{dt^2}=\frac{d^2Q}{dt^2}G+2\frac{dQ}{dt}\frac{dG}{dt}+Q\frac{d^2G}{dt^2}.$$

By (3.3), (3.4) and (3.5),

$$\frac{d^2G}{dt^2}=-\frac{1}{2}\frac{dA}{dt}G-\frac{1}{2}A\frac{dG}{dt}=-\frac{1}{2}\frac{dA}{dt}G+\frac{1}{4}A^2G=0.$$

By (3.2), (3.5),

$$2\frac{dQ}{dt}\frac{dG}{dt}+\frac{d^2Q}{dt^2}G=-\frac{dQ}{dt}AG+\frac{d^2Q}{dt^2}G=0.$$

We have

$$\frac{d^2(QG)}{dt^2}=0 \tag{3.6}$$

when $t\in[0,1]$.

Since $Q(t)$ is an analytic function of t when $t\in[0,1]$, $G(t)\in C^\infty$, (3.6) is true for $t\in[0,1]$.

The matrix $QG\bar{G}'\bar{Q}'$ is a semi-positive definite Hermitian matrix, and hence

$$\operatorname{tr}(QG\bar{G}'\bar{Q}')\geq 0.$$

By (3.6),

$$\frac{d^2}{dt^2}\operatorname{tr}(QG\bar{G}'\bar{Q}')=\operatorname{tr}\left(\frac{d^2}{dt^2}(QG\bar{G}'\bar{Q}')\right)=\operatorname{tr}\left(\frac{d(QG)}{dt}\frac{d(\overline{QG}')}{dt}\right)\geq 0$$

for $t\in[0,1]$. That means $\operatorname{tr}(QG\bar{G}'\bar{Q}')$, as a function of t in [0,1], is a concave contineous function. We know that $QG\bar{G}'\bar{Q}'=0$, at $t=0$ and $t=1$. It implies $\operatorname{tr}(QG\bar{G}'\bar{Q}')\equiv 0$, when $t\in[0,1]$, that is, $QG\bar{G}'\bar{Q}'\equiv 0, t\in[0,1]$. Hence $QG\equiv 0, t\in[0,1]$. There is a neighborhood N of $t=0$ such that $G(t)$ is non-singular for $t\in N$, since $G(0)=I$, It follows that $Q(t)\equiv 0$, when $t\in N$. It is impossible.

We have proved that $W(Z)=W(0)+D_ZW(0)(I-L(Z))^{-1}$ is biholomorphic in $\mathbb{R}_I$.

Using the similar argument we can prove $W(Z)=W(0)+(I-L(Z))^{-1}D_ZW(0)$ is biholpmorphic in $\mathbb{R}_I$.

§4.

Finally, we would like to make the following remark.

Let us consider the mappings

$$W(Z) = (A(Z) + B)(C(Z) + D)^{-1} \quad \text{or} \quad W(Z) = (C(Z) + D)^{-1}(A(Z) + B), \tag{4.1}$$

where $A(Z), B, C(Z), D \in \mathbb{C}^{n\times n}$, all entries of $A(Z), C(Z)$ are the homogeneous polynomials of entries of Z of degree one, and $C(Z) + D$ is non-singular when $Z \in \mathbb{R}_I$.

We consider $W = W(Z) = (C(Z) + D)^{-1}(A(Z) + B)$ at first. Fix $0 \neq \Lambda \in \mathbb{C}^{n\times n}$. We have

$$D_\Lambda W(Z) = (C(Z) + D)^{-1}\left[A(\Lambda) - C(\Lambda)(C(Z) + D)^{-1}(A(Z) + B)\right], \tag{4.2}$$

$$\begin{aligned} D_\Lambda^2 W(Z) &= -2(C(Z) + D)^{-1}C(\Lambda)(C(Z) + D)^{-1}[A(\Lambda) - C(\Lambda)(C(Z) + D)^{-1}(A(Z) + B)] \\ &= -2(C(Z) + D)^{-1}C(\Lambda)D_\Lambda W(Z), \end{aligned}$$

and

$$\begin{aligned} D_\Lambda^3 W(Z) &= 6(C(Z) + D)^{-1}C(\Lambda)(C(Z) + D)^{-1}C(\Lambda)(C(Z) + D)^{-1} \\ &\quad \times [A(\Lambda) - C(\Lambda)(C(Z) + D)^{-1}(A(Z) + B)] \\ &= 6(C(Z) + D)^{-1}C(\Lambda)(C(Z) + D)^{-1}C(\Lambda)D_\Lambda W(Z). \end{aligned}$$

It follows that

$$\{W; Z\}_\Lambda = 0$$

for $Z \in \mathbb{R}_I$ except a lower dimension manifold of Z such that $A(\Lambda) - C(\Lambda)(C(Z) + D)^{-1}(A(Z) + B)$ is singular.

By (4.2), we have

$$D_Z W(0) = D^{-1}\left[A(Z) - C(Z)D^{-1}B\right].$$

Using the identity of matrix

$$\begin{pmatrix} A & C \\ B & D \end{pmatrix}\begin{pmatrix} I & 0 \\ 0 & D^{-1} \end{pmatrix}\begin{pmatrix} I & 0 \\ -B & I \end{pmatrix} = \begin{pmatrix} A - CD^{-1}B & CD^{-1} \\ 0 & I \end{pmatrix},$$

we have $\det D_Z W(0) \not\equiv 0$ if and only if $\det\begin{pmatrix} A & B \\ C & D \end{pmatrix} \not\equiv 0$.

Similarly, if $W = W(Z) = (A(Z) + B)(C(Z) + D)^{-1}$, fixing $0 \neq \Lambda \in \mathbb{C}^{n\times n}$, then we have

$$\{W; Z\}_\Lambda = 0$$

for $Z \in \mathbb{R}_I$, except a lower dimension manifold of Z, and $\det D_Z W(0) \not\equiv 0$ if and only if $\det\begin{pmatrix} A & B \\ C & D \end{pmatrix} \not\equiv 0$. Thus we may rewrite Theorem 1.1 as

Theorem 4.1. *The assumptions are the same as Theorem 1.1. Then $\{W; Z\} = 0$ holds for $Z \in R_I$ if and only if*

$$W = W(Z) = (A(Z) + B)(C(Z) + D)^{-1} \quad or \quad W = W(Z) = (C(Z) + D)^{-1}(A(Z) + B),$$

where $A(Z), B, C(Z), D \in \mathbb{C}^{n\times n}$, all entries of $A(Z), C(Z)$ are homogeneous polynomials of the entries of Z of degree one. Moreover, $C(Z) + D$ is non-singular in $\mathbb{R}_I$, and $\det\begin{pmatrix} A(Z) & B \\ C(Z) & D \end{pmatrix} \not\equiv 0$ in $\mathbb{R}_I$. The mappings $W = W(Z)$ which was defined by (4.1) are biholomorphic in $\mathbb{R}_I$.

Actually, mappings (1.2) and (4.1) are equivalent to each other. For example,

$$\begin{aligned} W = W(Z) &= (A(Z)+B)(C(Z)+D)^{-1} \\ &= BD^{-1} + (A(Z) - BD^{-1}C(Z))D^{-1}(I+C(Z))^{-1}; \end{aligned}$$

comparing it with (1.2), we get

$$W(0) = BD^{-1}, \quad L(Z) = -C(Z), \quad D_Z W(0) = A(Z)D^{-1} - BD^{-1}C(Z)D^{-1}.$$

Hence $\det D_Z W(0) \not\equiv 0$ and $\det\begin{pmatrix} A(Z) & B \\ C(Z) & D \end{pmatrix} \not\equiv 0$ are equivalent to each other.

Conversely,

$$\begin{aligned} W = W(Z) &= W(0) + D_Z W(0)(I - L(Z))^{-1} \\ &= [W(0)(I - L(Z) + D_Z W(0)](I - L(Z))^{-1}; \end{aligned}$$

comparing it with (4.1), we get

$$A(Z) = D_Z W(0) - W(0)L(Z), \quad B = W(0), \quad C(Z) = -L(Z), \quad D = I.$$

Hence

$$\det\begin{pmatrix} A & B \\ C & D \end{pmatrix} = \det\begin{pmatrix} D_Z W(0) - W(0)L(Z) & W(0) \\ -L(Z) & I \end{pmatrix} = \det\begin{pmatrix} D_Z W(0) & 0 \\ -L(Z) & I \end{pmatrix}.$$

Thus $\det\begin{pmatrix} A & B \\ C & D \end{pmatrix} \not\equiv 0$ and $\det D_Z W(0) \not\equiv 0$ are equivalent to each other.

The conclusion is also true when

$$W(Z) = (C(Z)+D)^{-1}(A(Z)+B) \quad \text{and} \quad W(Z) = W(0) + (I - L(Z))^{-1} D_Z W(0).$$

Theorem 4.1 is also true if we replace $\mathbb{R}_I$ by a convex domain in $\mathbb{C}^{n\times n}$.

REFERENCES

[1] Gong Sheng & FitzGerald, C. H., The Schwarzian derivative in several complex variables, *Science in China (Series A)*, **36**(1993), 513–523.

[2] Birkhoff, G. & Rota, G. C., Ordinary differential equations, Gim and Co., 1962.

Vol. 41 No. 2 SCIENCE IN CHINA (Series A) February 1998

The Schwarzian derivative in several complex variables (Ⅲ) *

GONG Sheng (龚 昇),

(Department of Mathematics, University of Science and Technology of China, Hefei 230026, China)

ZHENG Xuean (郑学安)

(Department of Mathematics, Beijing Normal University, Beijing 100875, China)

and YU Qihuang (余其煌)

(Institute of Applied Mathematics, Chinese Academy of Sciences, Beijing 100080, China)

Received August 28, 1997.

Abstract In the point view of Lie group, the cross ratio and Schwarzian derivative in $\mathbb{C}^n$ are defined and discussed, especially the Schwarzian derivative of holomorphic mappings on the domains in matrix space $\mathbb{C}^{m\times n}$ is defined and discussed. It is proved that it is invariant up to similarity under the group of holomorphic automorphism of the Grassmann manifold $\mathbb{C}G(m, n)$. And it is also proved that the Schwarzian derivative equals zero if and only if the mapping is linearly fractional.

Keywords: cross ratio, Schwarzian derivative, regular trajectory.

1 Cross ratio, Schwarzian derivative

Let $Z = Z^{(m,n)} = (z_{ij})_{1\leqslant i\leqslant m,\ 1\leqslant j\leqslant n}$ denote the rectangular complex matrix. The classical domain of type one $R_I^{(m,n)}$ is a domain in $\mathbb{C}^{m\times n}$ such that $I - Z\bar{Z}^{\mathrm{T}} > 0$, where I denotes the $m\times m$ identity matrix, $\bar{Z}^{\mathrm{T}}$ denotes the conjugate transpose of Z, and >0 means positive definite. In ref. [1] and another paper[1)], we considered the Schwarzian derivative of holomorphic mappings on $R_I^{(m,n)}$ when $m = n$. The conclusions in ref. [1] hold true in general if we replace $R_I^{(m,n)}$ by any simply connected domain in $\mathbb{C}^{n\times n}$. The main point of view of ref. [1] is that the Schwarzian derivative is the infinitesimal form of the cross ratio. In this paper, we still use this point of view to define and to discuss the cross ratio and Schwarzian derivative in $\mathbb{C}^n$, especially the Schwarzian derivative of holomorphic mappings on domains in matrix space $\mathbb{C}^{m\times n}$. Of course, it is more difficult compared with $\mathbb{C}^{n\times n}$. We need to use the Lie group to consider how to define the cross ratio of four points such that it is invariant up to similarity under the group of holomorphic automorphism. If we can do that, we may define the Schwarzian derivative.

Regarding a point in $\mathbb{C}^n$ as a column vector, the elements of complex general linear group $GL(n, \mathbb{C})$ act from left. Let G be a subgroup of $GL(n, \mathbb{C})$. For every $X\in\mathbb{C}^n$, $G(X)$ denotes the trajectory of G in $\mathbb{C}^n$ which passes point X. We decompose $\mathbb{C}^n$ as the union of all trajectories which are mutually non-intersected. We call $X\in\mathbb{C}^n$ a regular point with respect to G if $G(X)$ is holomorphic homeomorphism of G. Otherwise, we call X a singular point, and $G(X)$ a singular

* Project supported by the National Natural Science Foundation of China.

1) Gong Sheng, Yu Qihuang, Zheng Xuean, The Schwarzian derivative in several complex variables (Ⅱ), *Chinese Annuals of Math.*, to appear.

trajectory. All regular points with respect to G in $\mathbb{C}^n$ form a set, denoted by $\mathbb{C}^n(G)$.

Definition 1. Let G be a complex linear Lie subgroup of $GL(n, \mathbb{C})$, let $\mathbb{C}^n(G)$ be the set of all regular points with respect to G in $\mathbb{C}^n$, and let $\mathbb{C}^n(G)$ be a dense open set in $\mathbb{C}^n$. Let Z_1, Z_2, Z_3 and Z_4 be any four points in $\mathbb{C}^n$.

1) If $Z_i - Z_j$, $(i,j) = (1,3), (2,3), (1,4), (2,4)$ belong to the same regular trajectory $G(X)$. Then we say that the cross ratio $[Z_1, Z_2, Z_3, Z_4]$ of four points Z_1, Z_2, Z_3 and Z_4 exists. Let $Z_i - Z_j = A_{ij}X$, $A_{ij} \in G$. The cross ratio of Z_1, Z_2, Z_3 and Z_4 is defined as

$$[Z_1, Z_2, Z_3, Z_4] = A_{13}A_{23}^{-1}A_{24}A_{14}^{-1}. \tag{1}$$

Obviously, definition 1 is independent of the choice of X.

2) If $Z_i - Z_j, (i,j) = (1,3), (2,3), (1,4), (2,4)$ belong to two or more different regular trajectories, then we say that the cross ratio $[Z_1, Z_2, Z_3, Z_4]$ of four points Z_1, Z_2, Z_3, Z_4 does not exist.

When $n = 1$, definition 1 defines the cross ratio in original sense. We may take $G = GL(1, \mathbb{C})$, then $\mathbb{C}(G)$ is a dense open set in $\mathbb{C}$. When $n \geqslant 2$, there are many choices of G, and the regular trajectory is not unique.

We may complete definition 1 in the following way. Let $\bar{G}$ be the closure of G in $GL(n, \mathbb{C})$. We may admit $A_{13}, A_{24} \in \bar{G}$. Adding a set of points of infinity in $\bar{G}$, we may allow A_{23}^{-1}, A_{14}^{-1} to belong to this set. Since $\mathbb{C}^n(G)$ is dense in $\mathbb{C}^n$, then the cross ratio for any four points in $\mathbb{C}^n$ may exist and be given by (1), or may not exist.

Moreover, if there exists a regular trajectory $G(X) = \mathbb{C}^n(G)$ which is a dense open set in $\mathbb{C}^n$, then case 2) in definition 1 is impossible. For example, the cross ratio defined in ref. [1] is this case.

With definition 1, we may define the infinitesimal form of the cross ratio of four points as the Schwarzian derivative.

Definition 2. Let $\Omega \in \mathbb{C}^n$ be a domain. Let $W = W(Z): \Omega \subset \mathbb{C}^n$ be a holomorphic mapping in Ω, $Z \in \Omega$. The Jacobian of W at point Z, $J_W(Z)$, is non-singular. Let $0 \neq \lambda \in \mathbb{C}^n$, $D_\lambda W(Z)$ is a regular point with respect to G. Let

$$l = \{Z + \mu\lambda \mid u \in \mathbb{C}, \| u\lambda \| < \varepsilon\},$$

where $\| \cdot \|$ means the length of a vector. If there exists $\varepsilon > 0$ such that for any four distinct points $Z_i, i = 1, 2, 3, 4$, on l, the cross ratio $[W(Z_1), W(Z_2), W(Z_3), W(Z_4)]$ which is defined by definition 1 exists, then we say that the Schwarzian derivative of W at point Z in the direction λ, $\{W; Z\}_\lambda$, exists, and we may express it as

$$\{W; Z\}_\lambda = \lim_{t \to 0} \frac{6([0, \alpha, \beta, \gamma]I_n - [W(Z_1), W(Z_2), W(Z_3), W(Z_4)])}{t^2 \alpha(\beta - \gamma)[0, \alpha, \beta, \gamma]}, \tag{2}$$

where $Z_1 = Z, Z_2 = Z + \alpha t\lambda, Z_3 = Z + \beta t\lambda, Z_4 = Z + \gamma t\lambda$, $\alpha, \beta, \gamma, t \in \mathbb{C}$. It is easy to verify that (2) is independent of the choice of α, β and γ. If there does not exist such $\varepsilon > 0$, then we say that the Schwarzian derivative of W at point Z in the direction λ, $\{W; Z\}_\lambda$ does not exist. For point Z where $J_W(Z)$ is singular, or for the direction λ, $D_\lambda W(Z)$ is a singular point with respect to G, when $J_W(Z)$ is nonsingular, we do not define the Schwarzian derivative of W at Z along λ in these cases. In particular, fix $Z_0 \in \Omega$, for any $Z \in \Omega$, $Z \neq Z_0$. $\{W; Z\}_{Z - Z_0}$ is defined as the Schwarzian derivative of W at Z with respect to Z_0. If $0 \in \Omega$, and $Z_0 = 0$, we denote $\{W; Z\}_Z$ by $\{W; Z\}$, and define it as the Schwarzian derivative of W at Z.

For every $n \geqslant 2$, there exist some $G \subset GL(n, \mathbb{C})$ such that definitions 1 and 2 make sense. If $\{W; Z\}_\lambda$ exists, it can be expressed by directional derivatives $D_\lambda^k W(Z)$, $k = 1, 2, 3$. For example, let $n = l_1 s_1 + l_2 s_2 + \cdots + l_p s_p$, $l_j \leqslant s_j$, $j = 1, 2, \cdots, p$. All l_j, s_j are positive integers. Then we may take $G = GL(l_1, \mathbb{C}) \otimes I_{s_1} \oplus GL(l_2, \mathbb{C}) \otimes I_{s_2} \oplus \cdots \oplus GL(l_p, \mathbb{C}) \otimes I_{s_p}$. Now we explain this idea by the cross ratio of four points and Schwarzian derivative in matrix space $\mathbb{C}^{m \times n}$.

We consider $\mathbb{C}^{mn}$ $(m \leqslant n)$. Obviously $\mathbb{C}^{mn}$ is isomorphic to $\mathbb{C}^{m \times n}$. Take the complex Lie subgroup $G = GL(m, \mathbb{C}) \otimes I_n$ in $GL(mn, \mathbb{C})$. Under the unique isomorphism between $\mathbb{C}^{mn}$ and $\mathbb{C}^{m \times n}$, G acting on $\mathbb{C}^{mn}$ is corresponding to $GL(m, \mathbb{C})$ acting on $\mathbb{C}^{m \times n}$, i.e. the left multiplies an $m \times m$ matrix on $\mathbb{C}^{m \times n}$.

When $m < n$, all regular points with respect to $GL(m, \mathbb{C})$ in $\mathbb{C}^{m \times n}$ are the points in $\mathbb{C}^{m \times n}$ with rank m. There are infinitely many regular trajectories.

Let $Z_1, Z_2, Z_3, Z_4 \in \mathbb{C}^{m \times n}$. Supposing that the cross ratio (1) of the four points exists, (1) becomes

$$[Z_1, Z_2, Z_3, Z_4] = (Z_1 - Z_3)(Z_2 - Z_3)^{-1}(Z_2 - Z_4)(Z_1 - Z_4)^{-1}, \tag{3}$$

where $Z^{-1} = \bar{Z}'(Z\bar{Z}')^{-1}$ is a generalized inverse matrix of Z.

If the cross ratio of four points Z_1, Z_2, Z_3, Z_4 by definition 1 does not exist, the right side of (3) still makes sense. But we cannot use it as the definition of the cross ratio of four points Z_1, Z_2, Z_3, Z_4. We will prove that (Theorem 1) the existence or nonexistence of the cross ratio of four points by definition 1 is the invariant property under the group of holomorphic automorphism of the complex dual space, the Grassmann manifold $\mathbb{C}G(m, n)$ of the classical domain of type one $R_I^{(m,n)}$. If the cross ratio exists, then it is invariant up to similarity under the group of holomorphic automorphism. If we use (3) as the definition of the cross ratio of four points, this is not true. Thus we cannot use (3) as the definition of the cross ratio of four points in general (except $m = n$).

2 Invariant property

Theorem 1. *Let G be $G = GL(m, \mathbb{C}) \otimes I_n$ in definition 1 for $\mathbb{C}^{mn}$. Suppose that the cross ratio of four points in $\mathbb{C}^{mn}$ exists by definition 1, i.e. the cross ratio of the corresponding four points in $\mathbb{C}^{m \times n}$ exists. Then the cross ratio of the four images by the holomorphic automorphism mapping of the complex Grassmann manifold $\mathbb{C}G(m, n)$ exists, and is similar to the original cross ratio; that is, if the cross ratio of four points in $\mathbb{C}^{mn}$ exists, then it is invariant up to similarity under the group of holomorphic automorphism of $\mathbb{C}G(m, n)$.*

Conversely, if the cross ratio of four points does not exist, then the cross ratio of the four image points by the holomorphic automorphism mapping does not exist; that is, the existence and nonexistence of the cross ratio of four points are the invariant property under the group of holomorphic automorphism of $\mathbb{C}G(m, n)$.

Proof. We may regard $\mathbb{C}G(m, n)$ as a compact complex manifold which is $\mathbb{C}^{m \times n}$ by adding points of infinity. In $\mathbb{C}^{m \times n}$ we may express any holomorphic automorphism of $\mathbb{C}G(m, n)$ as

$$W(Z) = (AZ + B)(CZ + D)^{-1}, \tag{4}$$

where $A \in \mathbb{C}^{m \times m}$, $B \in \mathbb{C}^{m \times n}$, $C \in \mathbb{C}^{n \times m}$, $D \in \mathbb{C}^{n \times n}$, $Z \in \mathbb{C}^{m \times n}$, $W \in \mathbb{C}^{m \times n}$, $(m \leqslant n)$, and $\det\begin{pmatrix} A & B \\ C & D \end{pmatrix} = 1$.

If $\det D \neq 0$, then

$$\begin{pmatrix} A & B \\ C & D \end{pmatrix} \begin{bmatrix} I_m & 0 \\ 0 & D^{-1} \end{bmatrix} \begin{bmatrix} I_m & 0 \\ -C & I_n \end{bmatrix} = \begin{bmatrix} A - BD^{-1}C & BD^{-1} \\ 0 & I_n \end{bmatrix}.$$

Thus $\det(A - BD^{-1}C) = \det D^{-1}$.

If $\det D = 0$, we may find $Z_0 \in \mathbb{C}^{m \times n}$ such that $\det(CZ_0 + D) \neq 0$. Actually, all the points Z_0 for which $\det(CZ_0 + D) \neq 0$ form a dense open set in $\mathbb{C}^{m \times n}$. Letting $\tilde{D} = CZ_0 + D$, $\tilde{B} = AZ_0 + B$, $Z = \xi + Z_0$, we have $(AZ + B)(CZ + D)^{-1} = (A\xi + \tilde{B})(C\xi + \tilde{D})^{-1}$, and

$$\begin{pmatrix} A & \tilde{B} \\ C & \tilde{D} \end{pmatrix} = \begin{bmatrix} A & AZ_0 + B \\ C & CZ_0 + D \end{bmatrix} = \begin{pmatrix} A & B \\ C & D \end{pmatrix} \begin{bmatrix} I_m & Z_0 \\ 0 & I_n \end{bmatrix}.$$

Thus $\det \begin{pmatrix} A & \tilde{B} \\ C & \tilde{D} \end{pmatrix} = 1$.

Let $Z_i = \xi_i + Z_0$, $i = 1, 2, 3, 4$. Then

$$[Z_1, Z_2, Z_3, Z_4] = [\xi_1, \xi_2, \xi_3, \xi_4].$$

It tells us that we only need to prove the theorem under the assumption $\det D \neq 0$.

Obviously,

$$\begin{aligned} &(AZ + B)(CZ + D)^{-1} - BD^{-1} \\ &= (AZ + B)(D^{-1}CZ + I_n)^{-1}D^{-1} - BD^{-1} \\ &= [AZ + B - B(D^{-1}CZ + I_n)](D^{-1}CZ + I_n)^{-1}D^{-1} \\ &= (A - BD^{-1}C)Z(D^{-1}CZ + I_n)^{-1}D^{-1}, \end{aligned} \tag{5}$$

and

$$Z(D^{-1}CZ + I_n)^{-1} = (ZD^{-1}C + I_m)^{-1}Z \tag{6}$$

hold true. Let $W_i = W(Z_i)$, $i = 1, 2, 3, 4$. Then

$$\begin{aligned} W_i - W_j &= (W_i - BD^{-1}) - (W_j - BD^{-1}) \\ &= (A - BD^{-1}C)(Z_iD^{-1}C + I_m)^{-1}Z_iD^{-1} - (A - BD^{-1}C)Z_j(D^{-1}CZ_j + I_n)^{-1}D^{-1} \\ &= (A - BD^{-1}C)(Z_iD^{-1}C + I_m)^{-1}[Z_i(D^{-1}CZ_j + I_n) \\ &\quad - (Z_iD^{-1}C + I_m)Z_j](D^{-1}CZ_j + I_n)^{-1}D^{-1} \\ &= (A - BD^{-1}C)(Z_iD^{-1}C + I_m)^{-1}(Z_i - Z_j)(D^{-1}CZ_j + I_n)^{-1}D^{-1} \\ &= P_i(Z_i - Z_j)Q_j \end{aligned} \tag{7}$$

by (4)—(6), where $P_i = (A - BD^{-1}C)(Z_iD^{-1}C + I_m)^{-1}$, $Q_j = (CZ_j + D)^{-1}$; P_i and Q_j exist; otherwise, we only need to make some translation of Z, and the cross ratio is invariant under translation. When $m = n$, the generalized inverse matrix is an inverse matrix. We have

$$\begin{aligned} [W_1, W_2, W_3, W_4] &= P_1(Z_1 - Z_3)Q_3(P_2(Z_2 - Z_3)Q_3)^{-1} \\ &\quad \times P_2(Z_2 - Z_4)Q_4(P_1(Z_1 - Z_4)Q_4)^{-1} = P_1[Z_1, Z_2, Z_3, Z_4]P_1^{-1} \end{aligned} \tag{8}$$

by (7). When $m < n$, if the cross ratio of four points Z_1, Z_2, Z_3, Z_4 exists, then there exists a regular point $X \in \mathbb{C}^{m \times n}$ such that $Z_i - Z_j = A_{ij}X$, $A_{ij} \in G$, where $i, j = 1, 2, 3, 4$. From (7), we have $W_i - W_j = P_iA_{ij}XQ_j$, $i, j = 1, 2, 3, 4$. The pair $W_1 - W_3$ and $W_2 - W_3$; $W_1 - W_4$ and $W_2 - W_4$ belong to the same trajectory since $P_iA_{ij} \in G$. We will be able to prove the existence of

$[W_1, W_2, W_3, W_4]$ if we can show that XQ_3 and XQ_4 belong to the same regular trajectory.

By the definition of Q_j, $j=3,4$, we have

$$\begin{aligned}XQ_3 - XQ_4 &= X(CZ_3+D)^{-1} - X(CZ_4+D)^{-1}\\ &= X(CZ_3+D)^{-1}[CZ_4+D-CZ_3-D](CZ_4+D)^{-1}\\ &= XQ_3C(Z_4-Z_3)Q_4 = XQ_3C(A_{13}-A_{14})XQ_4.\end{aligned}$$

Hence

$$XQ_3 = [I + XQ_3C(A_{13}-A_{14})]XQ_4 \in G(XQ_4),$$

i.e. XQ_3 and XQ_4 belong to the same regular trajectory. Thus the cross ratio of W_1, W_2, W_3, W_4 exists if the cross ratio of Z_1, Z_2, Z_3, Z_4 exists.

Moreover, by (3) and (7), we have

$$\begin{aligned}[W_1, W_2, W_3, W_4] &= (W_1-W_3)(W_2-W_3)^{-1}(W_2-W_4)(W_1-W_4)^{-1}\\ &= P_1A_{13}XQ_3(P_2A_{23}XQ_3)^{\mathrm{T}}[P_2A_{23}XQ_3(P_2A_{23}XQ_3)^{\mathrm{T}}]^{-1}\\ &\quad\times P_2A_{24}XQ_4(P_1A_{14}XQ_4)^{\mathrm{T}}[P_1A_{14}XQ_4(P_1A_{14}XQ_4)^{\mathrm{T}}]^{-1}\\ &= P_1A_{13}(XQ_3Q_3^{\mathrm{T}}X^{\mathrm{T}})A_{23}^{\mathrm{T}}P_2^{\mathrm{T}}(P_2^{\mathrm{T}})^{-1}(A_{23}^{\mathrm{T}})^{-1}(XQ_3Q_3^{\mathrm{T}}X^{\mathrm{T}})^{-1}A_{23}^{-1}P_2^{-1}\\ &\quad\times P_2A_{24}(XQ_4Q_4^{\mathrm{T}}X^{\mathrm{T}})A_{14}^{\mathrm{T}}P_1^{\mathrm{T}}(P_1^{\mathrm{T}})^{-1}(A_{14}^{\mathrm{T}})^{-1}(XQ_4Q_4^{\mathrm{T}}X^{\mathrm{T}})^{-1}A_{14}^{-1}P_1^{-1}\\ &= P_1A_{13}A_{23}^{-1}A_{24}A_{14}^{-1}P_1^{-1} = P_1[Z_1, Z_2, Z_3, Z_4]P_1^{-1}.\end{aligned}\tag{9}$$

Equalities (8) and (9) show that when $m \leqslant n$, if the cross ratio of the four points Z_1, Z_2, Z_3, Z_4 exists, then the cross ratio of four points W_1, W_2, W_3, W_4 under transformation (4) exists, and these two cross ratios are similar, i.e. the cross ratio is invariant up to similarity under transformation (4).

Conversely, if the cross ratio of four points Z_1, Z_2, Z_3, Z_4 does not exist, we will prove that the cross ratio of the four image points W_1, W_2, W_3, W_4 under mapping (4) does not exist.

From (7), we know that $W_1 - W_3$ and $W_2 - W_3$ do not belong to the same trajectory if $Z_1 - Z_3$ and $Z_2 - Z_3$ do not belong to the same trajectory. Similarly, if $Z_1 - Z_4$ and $Z_2 - Z_4$ do not belong to the same trajectory, then $W_1 - W_4$ and $W_2 - W_4$ do not belong to the same trajectory either. We only need to consider the case where $Z_1 - Z_3$ and $Z_2 - Z_3$ belong to the same trajectory, $Z_1 - Z_4$ and $Z_2 - Z_4$ belong to the same trajectory but $Z_1 - Z_3$ and $Z_1 - Z_4$ do not belong to the same trajectory. In this case, we have

$$Z_1 - Z_3 = A_{13}X,\ Z_2 - Z_3 = A_{23}X,\ Z_1 - Z_4 = A_{14}Y,\ Z_2 - Z_4 = A_{24}Y,$$

where X, Y are regular points, and X, Y do not belong to the same trajectory.

From (7), we have

$$\begin{aligned}W_1 - W_3 &= P_1A_{13}XQ_3, & W_2 - W_3 &= P_2A_{23}XQ_3,\\ W_1 - W_4 &= P_1A_{14}YQ_4, & W_2 - W_4 &= P_2A_{24}YQ_4.\end{aligned}$$

We want to prove that XQ_3 and YQ_4 do not belong to the same trajectory.

Obviously,

$$Q_3^{-1} - Q_4^{-1} = CZ_3 + D - CZ_4 - D = C(Z_3 - Z_4).$$

Hence

$$Q_3^{-1} = Q_4^{-1} + C(Z_3 - Z_4),$$

which implies

$$YQ_4Q_3^{-1} = Y[I_n + Q_4C(Z_3 - Z_4)] = Y + YQ_4C[Z_1 - Z_4 - (Z_1 - Z_3)]$$
$$= Y + YQ_4C(A_{14}Y - A_{13}X) = (I_m + YQ_4CA_{14})Y - (YQ_4CA_{13})X. \quad (10)$$

If YQ_4 and XQ_3 belong to the same trajectory, then there exists $\alpha \in G(m,\mathbb{C})$ such that $\alpha YQ_4 = XQ_3$, i.e. $\alpha YQ_4Q_3^{-1} = X$. Multiplying both sides of (10) by α from left, we have

$$X = \alpha(I_m + YQ_4CA_{14})Y - \alpha(YQ_4CA_{13})X,$$

i.e.

$$(I_m + \alpha YQ_4CA_{13})X = \alpha(I_m + YQ_4CA_{14})Y.$$

That means X and Y belong to the same trajectory. We get a contradiction. Thus XQ_3 and YQ_4 do not belong to the same trajectory if X and Y do not belong to the same regular trajectory. Hence the cross ratio of the four points W_1, W_2, W_3, W_4 does not exist; that is, the existence of the cross ratio of the four points is invariant under mapping (4).

We have proved Theorem 1.

From Theorem 1 and definition 2, we have the following theorem.

Theorem 2. *Let $\Omega \subset \mathbb{C}^{m\times n}$ be a domain, and let $W = W(Z): \Omega \to \mathbb{C}^{m\times n}$ be a holomorphic mapping. If the Schwarzian derivative of W at $Z \in \Omega$ in direction $\lambda \in \mathbb{C}^{m\times n}$, $\{W;Z\}_\lambda$ exists, then $\{W;Z\}_\lambda$ is invariant up to similarity under the group of holomorphic automorphism of $\mathbb{C}G(m,n)$.*

Theorems 1 and 2 generalize the corresponding results in ref. [1] not only from $m = n$ to $m \leqslant n$, but also from the group of holomorphic automorphism of R_I to $\mathbb{C}G(m,n)$. When $m = n = 1$, Theorems 1 and 2 coincide with the classical results in one complex variable.

3 Evaluation of Schwarzian derivative

Let $\Omega \subset \mathbb{C}^{m\times n}$ be a domain, and let $W = W(Z): \Omega \to \mathbb{C}^{m\times n}$ be a holomorphic mapping. We define the Schwarzian derivative $\{W;Z\}_\lambda$ according to definition 2 when $G = GL(m,\mathbb{C}) \otimes I_n$.

Let

$$\{W;Z\}_\lambda^0 = D_\lambda^3 W(Z)(D_\lambda W(Z))^{-1} - \frac{3}{2}(D_\lambda^2 W(Z)(D_\lambda W(Z))^{-1})^2. \quad (11)$$

Then

1) $\{W;Z\}_\lambda^0$ exists and is holomorphic for Z and λ, when $J_W(Z)$ is nonsingular and $(D_\lambda W(Z))^{-1}$ exists.

2)

$$\{W;Z\}_\lambda = \{W;Z\}_\lambda^0, \quad (12)$$

when $m = n$, and $J_W(Z)$, $D_\lambda W(Z)$ are nonsingular. Hence the Schwarzian derivatives are the same by definition 1 and by ref. [1] when $m = n$.

3) In general, $\{W;Z\}_\lambda$ and $\{W;Z\}_\lambda^0$ are different when $m < n$. $\{W;Z\}_\lambda$ is invariant up to similarity under the group of holomorphic automorphism of $\mathbb{C}G(m,n)$, but $\{W;Z\}_\lambda^0$ is not. If the Schwarzian derivatives $\{W;Z\}_\lambda$ and $(D_\lambda W(Z))^{-1}$ exist, then

$$\{W;Z\}_\lambda = \{W;Z\}_\lambda^0.$$

1) is obvious, we check 2) and 3) as follows.

Since $W(Z)$ is holomorphic on Ω, we expand W at point Z as a power series of t when $|t|$ is sufficiently small,

$$W(Z+\alpha t\lambda)=W(Z)+\alpha tB_1+\frac{\alpha^2t^2}{2}B_2+\frac{\alpha^3t^3}{6}B_3+\cdots,$$

where $B_k=D_\lambda^kW(Z)$, $k=1,2,3,\cdots$, are the directional derivatives of W at Z in direction λ. Thus, when $|t|$ is sufficiently small,

$$\begin{aligned}
W_1-W_3&=-\beta t\left(B_1+\frac{\beta t}{2}B_2+\frac{\beta^2t^2}{6}B_3+\cdots\right),\\
W_2-W_3&=(\alpha-\beta)t\left(B_1+\frac{\alpha+\beta}{2}tB_2+\frac{\alpha^2+\alpha\beta+\beta^2}{6}t^2B_3+\cdots\right),\\
W_2-W_4&=(\alpha-\gamma)t\left(B_1+\frac{\alpha+\gamma}{2}tB_2+\frac{\alpha^2+\alpha\gamma+\gamma^2}{6}t^2B_3+\cdots\right),\\
W_1-W_4&=-\gamma t\left(B_1+\frac{\gamma t}{2}B_2+\frac{\gamma^2t^2}{6}B_3+\cdots\right).
\end{aligned}\tag{13}$$

We have

$$\begin{aligned}
&(W_1-W_3)(W_2-W_3)^{\mathrm T}\\
&=-\beta(\alpha-\beta)t^2\Big|I_m+\frac{\beta t}{2}B_2B_1^{-1}+\frac{(\alpha+\beta)t}{2}B_1B_2^{\mathrm T}(B_1B_1^{\mathrm T})^{-1}+\frac{\beta^2t^2}{6}B_3B_1^{-1}\\
&\quad+\frac{\beta(\alpha+\beta)}{4}t^2(B_2B_2^{\mathrm T})(B_1B_1^{\mathrm T})^{-1}+\frac{\alpha^2+\alpha\beta+\beta^2}{6}t^2B_1B_3^{\mathrm T}(B_1B_1^{\mathrm T})^{-1}+\cdots\Big|B_1B_1^{\mathrm T},\\
&[(W_2-W_3)(W_2-W_3)^{\mathrm T}]^{-1}\\
&=(\alpha-\beta)^{-2}t^{-2}(B_1B_1^{\mathrm T})^{-1}\Big|I_m-\frac{\alpha+\beta}{2}t[B_2B_1^{-1}+B_1B_2^{\mathrm T}(B_1B_1^{\mathrm T})^{-1}]\\
&\quad-\left(\frac{\alpha+\beta}{2}\right)^2t^2B_2B_2^{\mathrm T}(B_1B_1^{\mathrm T})^{-1}-\frac{\alpha^2+\alpha\beta+\beta^2}{6}t^2[B_3B_1^{-1}+B_1B_3^{\mathrm T}(B_1B_1^{\mathrm T})^{-1}]\\
&\quad+\left(\frac{\alpha+\beta}{2}\right)^2t^2[B_2B_1^{-1}+B_1B_2^{\mathrm T}(B_1B_1^{\mathrm T})^{-1}]^2+\cdots\Big|,
\end{aligned}$$

and

$$\begin{aligned}
&(W_2-W_4)(W_1-W_4)^{\mathrm T}\\
&=-\gamma(\alpha-\gamma)t^2\Big|I_m+\frac{\alpha+\gamma}{2}tB_2B_1^{-1}+\frac{\gamma}{2}tB_1B_2^{\mathrm T}(B_1B_1^{\mathrm T})^{-1}\\
&\quad+\frac{\gamma(\alpha+\gamma)}{4}t^2B_2B_2^{\mathrm T}(B_1B_1^{\mathrm T})^{-1}+\frac{\alpha^2+\alpha\gamma+\gamma^2}{6}t^2B_3B_1^{-1}+\frac{\gamma^2}{6}t^2B_1B_3^{\mathrm T}(B_1B_1^{\mathrm T})^{-1}+\cdots\Big|B_1B_1^{\mathrm T},\\
&[(W_1-W_4)(W_1-W_4)^{\mathrm T}]^{-1}\\
&=\gamma^2t^2(B_1B_1^{\mathrm T})^{-1}\Big|I_m-\frac{\gamma}{2}t[B_2B_1^{-1}+B_1B_2^{\mathrm T}(B_1B_1^{\mathrm T})^{-1}]-\frac{\gamma^2}{4}t^2B_2B_2^{\mathrm T}(B_1B_1^{\mathrm T})^{-1}\\
&\quad-\frac{\gamma^2}{6}t^2[B_3B_1^{-1}+B_1B_3^{\mathrm T}(B_1B_1^{\mathrm T})^{-1}]+\frac{\gamma^2}{4}t^2[B_2B_1^{-1}+B_1B_2^{\mathrm T}(B_1B_1^{\mathrm T})^{-1}]^2+\cdots\Big|.
\end{aligned}$$

Substitute these four formulas into the following equations:

$$\begin{aligned}
[W_1,W_2,W_3,W_4]&=A_{13}A_{23}^{-1}A_{24}A_{14}^{-1}\\
&=(W_1-W_3)(W_2-W_3)^{-1}(W_2-W_4)(W_1-W_4)^{-1}.
\end{aligned}$$

After lengthy calculation, we have

$$[W_1,W_2,W_3,W_4]=[0,\alpha,\beta,\gamma]\Big[I_m-\frac{\alpha(\beta-\gamma)}{6}\left(B_3B_1^{-1}-\frac{3}{2}(B_2B_1^{-1})^2\right)t^2$$

$$+\frac{\alpha(\gamma-\alpha-\beta)}{4}B_2(I-B_1B_1^{-1})B_2^{\mathrm{T}}(B_1B_1^{\mathrm{T}})^{-1}t^2+\cdots\Big]. \tag{14}$$

Since the cross ratio of W_1, W_2, W_3, W_4 exists, we obtain

$$\begin{aligned}W(Z_2)-W(Z_3)&=A_{23}A_{13}^{-1}A_{13}X\\&=A_{23}A_{13}^{-1}(W(Z_1)-W(Z_3))=A(W(Z_1)-W(Z_3)),\end{aligned} \tag{15}$$

where $A=A_{23}A_{13}^{-1}=A(t,Z,\lambda)\in\mathbb{C}^{m\times m}$. Obviously, A is a holomorphic mapping of t at $t=0$ when $\det(D_\lambda W(Z)D_\lambda W(Z)^{\mathrm{T}})\neq0$. We expand A at $t=0$ as a power series of t,

$$A(t,Z,\lambda)=A_0(Z,\lambda)+tA_1(Z,\lambda)+t^2A_2(Z,\lambda)+\cdots. \tag{16}$$

Substituting (13) and (16) into (15), we have

$$\begin{aligned}&(\alpha-\beta)t\left(B_1+\frac{\alpha+\beta}{2}tB_2+\frac{\alpha^2+\alpha\beta+\beta^2}{6}t^2B_3+\cdots\right)\\&=(A_0+tA_1+t^2A_2+\cdots)(-\beta t)\left(B_1+\frac{1}{2}\beta tB_2+\frac{1}{6}\beta^2t^2B_3+\cdots\right).\end{aligned}$$

Comparing the coefficients t^k, $t=1,2,3,\cdots$, on both sides of the equation, we obtain

$$(\alpha-\beta)B_1=-\beta A_0B_1,\quad \frac{\alpha^2-\beta^2}{2}B_2=\left(A_1B_1+\frac{1}{2}\beta A_0B_2\right)(-\beta),\cdots.$$

Then $B_2=CB_1$ holds, where $C=C(Z,\lambda)\in\mathbb{C}^{m\times m}$, i.e.

$$\mathrm{D}_\lambda^2W(Z)=C(Z,\lambda)\mathrm{D}_\lambda W(Z). \tag{17}$$

From (17), we obtain

$$B_2(I_m-B_1^{-1}B_1)B_2^{\mathrm{T}}(B_1B_1^{\mathrm{T}})^{-1}=CB_1B_1^{\mathrm{T}}C^{\mathrm{T}}(B_1B_1^{\mathrm{T}})^{-1}-CB_1B_1^{-1}B_1B_1^{\mathrm{T}}C^{\mathrm{T}}(B_1B_1^{\mathrm{T}})^{-1}=0.$$

Substituting this result into (14), we have

$$[W_1,W_2,W_3,W_4]=[0,\alpha,\beta,\gamma]\left[I_m-\frac{\alpha(\beta-\gamma)}{6}\left(B_3B_1^{-1}-\frac{3}{2}(B_2B_1^{-1})^2\right)t^2+\cdots\right].$$

Thus the Schwarzian derivative which is defined by definition 2 is (11) if it exists. Only in the case where (2) makes sense are (11) and (2) equivalent. We can use (11) as the definition of Schwarzian derivative.

4 Schwarzian derivative and linear fractional transformation

Now we will prove the following result.

Theorem 3. *Let $\Omega\subset\mathbb{C}^{m\times n}$ be a domain, and let $W=W(Z):\Omega\to\mathbb{C}^{m\times n}$ be a holomorphic mapping. $Z_0\in\Omega$. The Jacobian of W at Z_0, $J_W(Z_0)$ being nonsingular. Then $\{W;Z\}_{Z-Z_0}\equiv0$ on a dense open set Σ in Ω if and only if*

1) $W(Z)=W(Z_0)+(I-L(Z_0)-L(Z))^{-1}\mathrm{D}_{Z-Z_0}W(Z_0)$, *or* $W(Z)=W(Z_0)+\mathrm{D}_{Z-Z_0}W(Z_0)(I-L(Z_0)-L(Z))^{-1}$, *when* $m=n$;

2) $W(Z)=W(Z_0)+(I-L(Z_0)-L(Z))^{-1}\mathrm{D}_{Z-Z_0}W(Z_0)$, *when* $m<n$;

where $L(Z)$ is an $m\times m$ matrix, each entry of $L(Z)$ is a linear homogeneous polynomial of the entries of Z, and $I-L(Z_0)-L(Z)$ is nonsingular when $Z\in\Omega$.

We may prove that $W(Z)$ is biholomorphic on Ω, and 1) and 2) can be written as

1) $W(Z)=(C(Z)+D)^{-1}(A(Z)+B)$, or $W(Z)=(A(Z)+B)(C(Z)+D)^{-1}$, when $m=n$;

2) $W(Z)=(C(Z)+D)^{-1}(A(Z)+B)$, when $m<n$;

where $C(Z),D\in\mathbb{C}^{m\times m}$, $A(Z),B\in\mathbb{C}^{m\times n}$; each entry of $A(Z)$ and $C(Z)$ is a linear homogeneous polynomial of the entries of Z; $C(Z)+D$ is nonsingular when $Z\in\Omega$; $J_W(Z_0)$ is nonsingular.

The family of the mappings in Theorem 3, 1) and 2) is the group of holomorphic automorphism of $\mathbb{C}G(1,n)=\mathbb{C}P^n$ when $1=m\leqslant n$. There exists a subfamily of the family of the mappings in Theorem 3, 1) and 2) which is the group of holomorphic automorphism of $\mathbb{C}G(m,n)$. Thus when $1=m\leqslant n$, the Schwarzian derivatives completely characterize the space $\mathbb{C}G(1,n)=\mathbb{C}P^n$, and when $2\leqslant m\leqslant n$, the Schwarzian derivatives essentially characterize the the space $\mathbb{C}G(m,n)$, showing the difference between the Hermitian symmetric space of rank 1 and rank $\geqslant 2$ in the point of view of geometry.

We need the following two lemmas.

Lemma 1. *Let* $V\in\mathbb{C}^{m\times n}$, $m<n$. Then

1) $\det VV^{\mathrm{T}}$ *is the product of two linear independent linear homogeneous polynomials when* $m=1$, $n=2$;

2) $\det VV^{\mathrm{T}}$ *is irreducible in other cases.*

Proof. If $\det VV^{\mathrm{T}}$ is reducible, then

$$\det VV^{\mathrm{T}} = L_1(V)L_2(V), \tag{18}$$

where $L_1(V)$ is a homogeneous polynomial of entries of V of degree $(2m-k)$, $1\leqslant k<2m$. Let β be an $n\times n$ diagonal matrix, there are m entries equal to one, and the rest are equal to zero on the diagonal. Then $V\beta$ is an $m\times n$ matrix, in which $n-m$ columns are zero. We delete all these zero columns, and get an $m\times m$ matrix. We denote it by V_β. Replacing V by $V\beta$ in (18), we have

$$(\det V_\beta)^2 = L_1(V\beta)L_2(V\beta).$$

From Lemma 2.1 in ref. [1], we know that $\det V_\beta$ is irreducible. So

$$L_1(V\beta) = A_\beta\det V_\beta,\quad L_2(V\beta) = B_\beta\det V_\beta, \tag{19}$$

where A_β and B_β are constants, and $A_\beta B_\beta=1$. (19) holds true for all V_β.

We know that

$$\det VV^{\mathrm{T}} = \sum_\beta(\det V_\beta)^2. \tag{20}$$

Since all $\det V_\beta$ are linear independent homogeneous polynomials of degree m, we may expand it as a basis of the homogeneous polynomials of the entries of V of degree m; that is,

$$L_1(V) = \sum_\beta a_\beta\det V_\beta + \text{other terms},\quad L_2(V) = \sum_\beta b_\beta\det V_\beta + \text{other terms},$$

where a_β and b_β are constants, and $a_\beta b_\beta=1$. From (18) and (19), we know

$$L_1(V) = \sum_\beta a_\beta\det V_\beta,\quad L_2(V) = \sum_\beta b_\beta\det V_\beta.$$

If $\det VV^{\mathrm{T}}$ is reducible when $m<n$, then

$$\sum_\beta(\det V_\beta)^2 = \Big(\sum_\beta a_\beta\det V_\beta\Big)\Big(\sum_\gamma b_\gamma\det V_\gamma\Big)$$

by (20). The total number of β is C_n^m. The previous equality holds provided

$$\begin{cases} a_\beta b_\beta = 1, & \beta = 1,2,\cdots,C_n^m,\\ a_\beta b_\gamma + a_\gamma b_\beta = 0, & \beta<\gamma;\beta,\gamma = 1,2,\cdots,C_n^m. \end{cases}$$

We have $2C_n^m$ variables and $\frac{1}{2}C_n^m(C_n^m+1)$ equations. The system of equations has a solution unless

$$2C_n^m \geqslant \frac{1}{2}C_n^m(C_n^m+1),$$

i.e. $3 \geqslant C_n^m$. It holds only in the cases of $m=1$, $n=2$; $m=1$, $n=3$; and $m=2$, $n=3$. When $m=1$, $n=3$ or $m=2$, $n=3$, the system of equations becomes

$$\begin{cases} a_1b_1=1,\ a_2b_2=1,\ a_3b_3=1,\\ a_1b_2+a_2b_1=0,\ a_1b_3+a_3b_1=0,\ a_2b_3+a_3b_2=0. \end{cases}$$

From $a_1b_1=1$, $a_2b_2=1$ and $a_1b_2+a_2b_1=0$ we have $a_1^2+a_2^2=0$, i.e. $a_1=\pm ia_2$. Similarly, we have $a_2=\pm ia_3$, $a_3=\pm ia_1$. These three equations contradict one another. This means that in these cases the system of equations has no solution.

We have proved Lemma 1.

From Lemma 1, we have the following lemma.

Lemma 2. *Let $V\in\mathbb{C}^{m\times n}$, $m<n$, and let A, B be two $m\times m$ matrices. All the entries of A and B are homogeneous polynomials of the entries of V of degree 3. $(VV^T)^*$ is the algebraic complement of VV^T. If each entry $A(VV^T)^*B$ is divisible by $\det VV^T$, then $A=L(V)\cdot VV^T$ or $B=VV^TR(V)$, where $L(V)$, $R(V)$ are $m\times m$ matrices and entries of $L(V)$ and $R(V)$ are linear homogenous polynomials of the entries of V.*

Proof of Lemma 2 is similar to the proof of Theorem 2 in ref. [1]. We omit the details of the proof.

Now we are going to prove Theorem 3.

Part 1) of Theorem 3 was proved previously[1)]. We only need to prove part 2) of Theorem 3. We first prove the necessity; that is, if $\{W;Z\}_{Z-Z_0}\equiv 0$ for all $Z\in\Sigma$, then $W(Z)$ is a a mapping in 2), when $Z_0\in\Omega$, $J_W(Z_0)$ is nonsingular. Without loss of generality, we may assume $Z_0=0$. Thus $\{W;Z\}\equiv 0$ when $Z\in\Sigma$. It is easy to verify that it is equivalent to $\{W;Z\}_{uZ}\equiv 0$, where $v\in\mathbb{C}$, $v\neq 0$. It is equivalent to $\{W;uZ\}_Z\equiv 0$ also, where $uZ\in\Sigma$, $u\neq 0$. Letting $u\to 0$, we have $\{W;0\}_Z\equiv 0$. $|_Z^0$ Since $J_W(0)$ is nonsingular, we may choose $Z\in\Omega$ such that $D_ZW(0)$ is nonsingular.

By (11), we have

$$D_Z^3W(0)(D_ZW(0))^{-1}-\frac{3}{2}(D_Z^2W(0))(D_ZW(0))^{-1}(D_Z^2W(0))(D_ZW(0))^{-1}\equiv 0,$$

i.e.

$$\begin{aligned} &D_Z^3W(0)(D_ZW(0))^T(D_ZW(0)(D_ZW(0))^T)^{-1}\\ &\quad-\frac{3}{2}[D_Z^2W(0)(D_ZW(0))^T[D_ZW(0)(D_ZW(0))^T]^{-1}]^2\equiv 0. \end{aligned}$$

Let $V=D_ZW(0)$. Then the previous equation becomes

$$D_Z^3W(0)V^T=\frac{3}{2}D_Z^2W(0)V^T(VV^T)^*D_Z^2W(0)V^T(\det VV^T)^{-1},$$

where $(VV^T)^*$ is the algebraic complement of VV^T. Thus all entries of $D_Z^2W(0)V^T(VV^T)^*D_Z^2W(0)V^T$ are divisible by $\det VV^T$. Each entry of $D_Z^2W(0)V^T$ is a homogeneous polynomial of entries of Z of degree 3. Hence every entry of $D_Z^2W(0)V^T$ is a homogeneous polynomial of

1) See footnote 1) on page 158.

entries of V of degree 3. By Lemma 2, $D_Z^2W(0)V^T$ is $L(V)VV^T$, or $VV^T R(V)$, where every entry of $L(V)$, $R(V)$ is a linear homogeneous polynomial of the entries of V.

From (17), we know that if we fix $Z_0 \in \Omega$, then for any $0 \neq \lambda \in \mathbb{C}^{m\times n}$, there exists a $\epsilon(\lambda) > 0$ such that

$$D_\lambda^2 W(Z_0 + t\lambda) = c(Z_0 + t\lambda, \lambda) D_\lambda W(Z_0 + t\lambda) \tag{21}$$

holds for $|t| < \epsilon(\lambda)$. Letting $\lambda = Z$, $Z_0 = 0$, $t = u$ in (21), we obtain

$$D_Z^2 W(uZ) = C(uZ, Z) D_Z W(uZ). \tag{22}$$

For every $Z \in \mathbb{C}^{m\times n}$ for which $(D_Z W(0))^{-1}$ makes sense, there exists a neighborhood at $u = 0$ which is dependent on Z. In this neighborhood $C(uZ, Z)$ is a holomorphic mapping of u. Differentiating both sides of (22), we have

$$D_Z^3 W(uZ) = D_Z C(uZ, Z) D_Z W(uZ) + C^2(uZ, Z) D_Z W(uZ).$$

Substituting the previous equation and (22) into $\{W, uZ\}_Z \equiv 0$, we have

$$\begin{aligned} & D_Z^3 W(uZ)(D_Z W(uZ))^{-1} - \frac{3}{2}[D_Z^2 W(uZ)(D_Z W(uZ))^{-1}]^2 \\ &= D_Z C(uZ, Z) + C^2(uZ, Z) - \frac{3}{2} C^2(uZ, Z) = 0, \end{aligned}$$

i.e.

$$D_Z C(uZ, Z) = \frac{1}{2} C^2(uZ, Z). \tag{23}$$

Using mathematical induction, we may prove that for any $Z \in \mathbb{C}^{m\times n}$, $uZ \in \Omega$ in a neighborhood of $u = 0$, equalities

$$\left(\frac{d}{du}\right)^k C(uZ, Z) = 2^{-k} k! (C(uZ, Z))^{k+1}, \quad k = 1, 2, \cdots \tag{24}$$

hold. When $k = 1$, (24) is (23).

If (24) holds true when $k = l$, then

$$\begin{aligned} \left(\frac{d}{du}\right)^{l+1} C(uZ, Z) &= \frac{d}{du}[2^{-l} l! (C(uZ, Z))^{l+1}] \\ &= 2^{-l} l! (l+1)(C(uZ, Z))^l \times \frac{1}{2}(C(uZ, Z))^2 \\ &= 2^{-(l+1)}(l+1)!(C(uZ, Z))^{l+2} \end{aligned}$$

by (24). Thus (24) holds for $k = 1, 2, \cdots$.

Using mathematical induction again, we may prove that for any $Z \in \mathbb{C}^{m\times n}$, $uZ \in \Omega$ in a neighborhood of $u = 0$, equalities

$$\left(\frac{d}{du}\right)^{k+1} W(uZ) = 2^{-k}(k+1)!(C(uZ, Z))^k D_Z W(uZ), \quad k = 1, 2, \cdots \tag{25}$$

hold. When $k = 1$, (25) is (22).

If (25) holds true when $k = l$, then

$$\begin{aligned} \left(\frac{d}{du}\right)^{l+2} W(uZ) &= \frac{d}{du}[2^{-l}(l+1)!(C(uZ, Z))^l D_Z W(uZ)] \\ &= 2^{-l}(l+1)!\Big[l(C(uZ, Z))^{l-1} \frac{d}{du} C(uZ, Z) D_Z W(uZ) + \\ &\quad (C(uZ, Z))^l D_Z^2 W(uZ) \Big] \end{aligned}$$

$$= 2^{-l}(l+1)!\left[l(C(uZ,Z))^{l-1}\frac{1}{2}(C(uZ,Z))^2 D_Z W(uZ) + (C(uZ,Z))^l \times C(uZ,Z)D_Z W(uZ)\right]$$

$$= 2^{-(l+1)}(l+2)!(C(uZ,Z))^{l+1}D_Z W(uZ,Z)$$

by (23) and (25).

Thus (25) holds true for $k=1,2,\cdots$.

Letting $u\to 0$ in (25), we have

$$D_Z^{k+1}W(0) = 2^{-k}(k+1)!(C(0,Z))^k D_Z W(0),\ k=1,2,\cdots. \tag{26}$$

If $D_Z^2 W(0)V^{\mathrm{T}} = L(V)VV^{\mathrm{T}}$ holds true,

$$D_Z^2 W(0)(D_Z W(0))^{\mathrm{T}} = L(D_Z W(0))D_Z W(0)(D_Z W(0))^{\mathrm{T}} \tag{27}$$

holds. When $k=1$ at (26), it becomes

$$D_Z^2 W(0) = C(0,Z)D_Z W(0). \tag{28}$$

Multiplying by $(D_Z W(0))^{\mathrm{T}}$ both sides of (28), we obtain

$$D_Z^2 W(0)(D_Z W(0))^{\mathrm{T}} = C(0,Z)D_Z W(0)(D_Z W(0))^{\mathrm{T}}.$$

From the above equality and (27), we have

$$L(D_Z W(0))D_Z W(0)(D_Z W(0))^{\mathrm{T}} = C(0,Z)D_Z W(0)(D_Z W(0))^{\mathrm{T}}.$$

Thus $L(D_Z W(0)) = C(0,Z)$ holds for almost all $Z\in\mathbb{C}^{m\times n}$. We know that every entry of $L(V)$ is a linear homogeneous polynomial of the entries of V, and every entry of $V=D_Z W(0)$ is a linear homogeneous polynomial of entries of Z. Every entry of $L(D_Z W(0)) = C(0,Z)$ is the continuous function of Z. Thus for all $Z\in\mathbb{C}^{m\times n}$, every entry of $C(0,Z)$ is a linear homogeneous polynomial of the entries of Z.

Since $W(Z)$ is holomorphic on Ω, at $W=0$ we may expand $W(Z)$ as

$$W(Z) = W(0) + \sum_{k=1}^{\infty}\frac{1}{k!}D_Z^{(k)}W(0).$$

Substituting (26) into the left side of the previous equation,

$$W(Z) = W(0) + \sum_{k=1}^{\infty}\left(\frac{1}{2}\right)^{k-1}(C_0(Z))^{k-1}D_Z W(0), \tag{29}$$

where $C_0(Z) = C(0,Z)$, we have

$$W(Z) = W(0) + \left(I - \frac{1}{2}C_0(Z)\right)^{-1} D_Z W(0).$$

We may express it as $W(Z) = (C(Z)+D)^{-1}(A(Z)+B)$, where $A(Z), B\in\mathbb{C}^{m\times n}$, $C(Z)$, $D\in\mathbb{C}^{m\times m}$, and each entry of $A(Z), C(Z)$ is a linear homogeneous polynomial of the entries of Z.

Another possibility is $D_Z^2 W(0)V^{\mathrm{T}} = VV^{\mathrm{T}}R(V)$. We will prove that we can reach the same conclusion under the assumptions of Theorem 3.

We know that

$$C(0,Z) = D_Z^2 W(0)V^{-1} = D_Z^2 W(0)V^{\mathrm{T}}(VV^{\mathrm{T}})^{-1} = D_Z^2 W(0)V^{\mathrm{T}}(VV^{\mathrm{T}})^{*}\det(VV^{\mathrm{T}})^{-1}$$

by (28). Substituting it into (28), we have

$$D_Z^2 W(0) = D_Z^2 W(0)V^{\mathrm{T}}(VV^{\mathrm{T}})^{*}V\det(VV^{\mathrm{T}})^{-1}.$$

Every entry of the matrix on the left-hand side of the above equality is a homogeneous polynomial of the entries of Z of degree 2. It is a homogeneous polynomial of the entries of V of degree 2.

Thus, every entry of the matrix $D_Z^2W(0)V^T(VV^T)^*V$ on the right-hand side of the previous equation is divisible by $\det(VV^T)$. If H is an $m\times m$ matrix, all entries of H are the homogeneous polynomial of the entries of V of degree 2, then every entry of $D_Z^2W(0)V^T(VV^T)^*VH$ is divisible by $\det(VV^T)$.

If $D_Z^2W(0)V^T = VV^TR(V)$ holds, then

$$D_Z^2W(0)V^T(VV^T)^*D_Z^2W(0)V^T = D_Z^2W(0)V^T(VV^T)^*VV^TR(V).$$

We already know that every entry on the left-hand side of the above equality is divisible by det (VV^T), hence every entry on the right-hand side of the equality is divisible by $\det(VV^T)$. Thus every entry of matrix

$$D_Z^2W(0)V^T(VV^T)^*VV^TR(V) + D_Z^2W(0)V^T(VV^T)^*VH$$
$$= D_Z^2W(0)V^T(VV^T)^*V[V^TR(V) + H]$$

is divisible by $\det(VV^T)$. Let $H = BV^TT(V) - V^TR(V)$, where $B=(b_{ij})_{1\leqslant i,j\leqslant n}$ is an $n\times n$ matrix, b_{ij}, $1\leqslant i,j\leqslant n$ are constants, $\det B\neq 0$, $b_{ij}\neq 0$ when $i\neq j$, $T(V)$ is an $m\times m$ matrix, and each entry is linear homogeneous polynomial of the entries of V. Hence each entry of $D_Z^2W(0)V^T(VV^T)^*VBV^TT(V)$ is divisible by $\det VV^T$. By Lemma 2, we have

$$D_Z^2W(0)V^T = L_1(V)VV^T \quad \text{or} \quad VBV^TT(V) = VV^TS(V),$$

where $L_1(V)$ and $S(V)$ are $m\times m$ matrices; each entry is a linear homogenous polynomial of the entries of V. If $VBV^TT(V) = VV^TS(V)$ is true, then

$$(VV^T)^{-1}VBV^TT(V) = S(V).$$

Each entry of V is an independent variable. Let

$$V = \begin{pmatrix} v_{11} & \cdots & v_{1n} \\ \cdots & \cdots & \cdots \\ v_{m1} & \cdots & v_{mn} \end{pmatrix}, \text{ and } T(V) = \begin{pmatrix} v_{11} & \cdots & v_{1m} \\ \cdots & \cdots & \cdots \\ v_{m1} & \cdots & v_{mm} \end{pmatrix}.$$

Then

$$(VV^T)^{-1}VBV^T\begin{pmatrix} v_{11} & \cdots & v_{1n} \\ \cdots & \cdots & \cdots \\ v_{m1} & \cdots & v_{mn} \end{pmatrix} = S\begin{pmatrix} v_{11} & \cdots & v_{1n} \\ \cdots & \cdots & \cdots \\ v_{m1} & \cdots & v_{mn} \end{pmatrix}$$

holds for all V. Let $v_{ii}\neq 0$, $i=1,2,\cdots,m$, $v_{ij}=0$, when $i\neq j$. We have

$$\begin{pmatrix} v_{11}^{-1} & \cdots & 0 & 0 & \cdots & 0 \\ \vdots & \ddots & \vdots & \vdots & \ddots & \vdots \\ 0 & \cdots & v_{mm}^{-1} & 0 & \cdots & 0 \end{pmatrix}\begin{pmatrix} b_{11} & \cdots & b_{1n} \\ \cdots & \cdots & \cdots \\ b_{n1} & \cdots & b_{nn} \end{pmatrix}\begin{pmatrix} v_{11}^2 & \cdots & 0 \\ & \ddots & \vdots \\ 0 & \cdots & v_{mm}^2 \\ 0 & \cdots & 0 \\ \vdots & \ddots & \vdots \\ 0 & \cdots & 0 \end{pmatrix}$$

$$= S\begin{pmatrix} v_{11} & \cdots & 0 & 0 & \cdots & 0 \\ \vdots & \ddots & \vdots & \vdots & \ddots & \vdots \\ 0 & \cdots & v_{mm} & 0 & \cdots & 0 \end{pmatrix}. \tag{30}$$

The left side of the above equation is $(v_{ii}^{-1}b_{ij}v_{jj}^2)_{1\leqslant i,j\leqslant m}$. When $m\geqslant 2$, we may take $b_{ij}\neq 0$ and when $i\neq j$, $v_{ii}^{-1}b_{ij}v_{ij}^2$ is a rational fraction function of V. It is not a polynomial. The entry on the right side of (30) is a linear homogenous polynomial of the entries of V. We get a contradiction. Thus we only need to consider the case $D_Z^2W(0)V^T = L(V)VV^T$. When $m=1$, VV^T is a complex number, $VV^{T}R(V)=R(V)VV^{T}$. It belongs to the case of $D_Z^2W(0)V^{T}=L(V)VV^{T}$ also.

We have proved the necessity part of Theorem 3.

We may prove the sufficiency part of Theorem 3 by direct evaluation of the Schwarzian derivative. To prove that the mappings in 1) or 2) are biholomorphic on Ω is similar to the proof of the corresponding results[1)]. We omit the details of the proof.

Moreover, direct evaluation shows that $\{W;Z\}_{\lambda}\equiv 0$ for any $\lambda\in\mathbb{C}^{m\times n}$ when W is a mapping in 1) or 2).

Corollary 1. *The assumptions are the same in Theorem 3. Then* $\{W;Z\}_{Z-Z_0}\equiv 0$ *implies* $\{W;Z\}_{\lambda}\equiv 0$ *for any* $Z\in\Omega, 0\neq\lambda\in\mathbb{C}^{m\times n}$.

Reference

1 Gong Sheng, FitzGerald, C. H., The Schwarzian derivative in several complex variables, *Science in China*, Ser. A, 1993, 36: 513.

1) See footnote 1) on page 158.

Vol. 41 No. 8 SCIENCE IN CHINA (Series A) August 1998

The Schwarzian derivative in several complex variables IV*

GONG Sheng (龚 昇),
(Department of Mathematics, University of Science and Technology of China, Hefei 230026, China)
YU Qihuang (余其煌)
(Institute of Applied Mathematics, Chinese Academy of Sciences, Beijing 100080, China)
and ZHENG Xuean (郑学安)
(Department of Mathematics, Beijing Normal University, Beijing 100875, China)

Received August 28, 1997

Abstract Using two different Lie groups, two different Schwarzian derivatives of holomorphic mappings on domains in $\mathbb{C}^n$ are defined and discussed. The necessary and sufficient conditions for annihilation of these two different Schwarzian derivatives are given.

Keywords: cross ratio, Schwarzian derivative, linear fractional mapping.

1 Two Schwarzian derivatives

In previous papers[1-3] we defined the Schwarzian derivative of holomorphic mapping on classical domains and matrix spaces as the infinitesimal form of the cross ratio. In this paper we discuss the Schwarzian derivative on domains in $\mathbb{C}^n$ by the same idea. Starting from two different Lie groups, we may define two different cross ratios, and then we have two different Schwarzian derivatives. These two Schwarzian derivatives are not equivalent to each other.

In sec. 2, we define the cross ratio of four points in $\mathbb{C}^n$ by $GL(n,\mathbb{C})$, and then we have a Schwarzian derivative (8). This is a matrix. We call it a matrix Schwarzian derivative. We will prove that the matrix Schwarzian derivative is invariant up to similarity under the group of holomorphic automorphism of $\mathbb{CP}^n$ (Theorem 1). We will also prove that the matrix Schwarzian derivative of holomorphic mapping is equal to zero if and only if the mapping is linear fractional (Theorem 2).

In ref. [3], we studied the Schwarzian derivative of holomorphic mappings on domains in $\mathbb{C}^{m\times n}(m\leqslant n)$. When $m=1$, we have the Schwarzian derivative of holomorphic mappings on domains in $\mathbb{C}^n$. It is a scalar. Of course, it is different from the matrix Schwarzian derivative. Besides the properties of the Schwarzian derivative we already know in ref. [3], in sec. 3, we will prove that in a neighborhood of a fixed point, if the Schwarzian derivative of holomorphic mapping exists at any point in the neighborhood and in any direction, then the mapping is linear fractional, and hence the Schwarzian derivative is equal to zero (Theorem 3)..

In ref. [4], we defined and discussed a kind of the Schwarzian derivative of holomorphic mapping on the domain $\mathbb{C}^n$ which was defined by two directions λ and μ. When $\lambda=\mu$,

* Project supported by the National Natural Science Foundation of China.

it coincides with the Schwarzian derivative which was defined in ref. [3]. Thus the Schwarzian derivative in ref. [4] is a generalization of Schwarzian derivative in ref. [3]. Theorem 3 tells us that this generalization is natural and desirable.

2 Matrix Schwarzian derivative

Let $\Omega\subset\mathbb{C}^n$ be a domain, and let $W=W(Z):\Omega\to\mathbb{C}^n$ be a holomorphic mapping. If $Z\in\Omega$, and the Jacobian, $J_W(Z)$ is nonsingular. Let $l=\{Z+u\lambda: u\in\mathbb{C}, \|u\lambda\|<\varepsilon\}$ be a complex line segment which passes point Z in the direction λ, where $0\neq\lambda\in\mathbb{C}^n$, $\varepsilon>0$ and $\|\cdot\|$ means the length of vector. Let $Z_i=Z+u_i\lambda\in l$, $i=1,2,3,4$ and let $Z_i\in\Omega$, $i=1,2,3,4$; when ε is sufficiently small. If there exists a column vector $X\in\mathbb{C}^n$, $W(Z_i)-W(Z_j)=A_{ij}X$, where $A_{ij}\in GL(n,\mathbb{C})$, $X\neq 0$, $W=W(Z)$ is a column vector, then the cross ratio of four points $W(Z_i)$, $i=1,2,3,4$ exists and

$$[W(Z_1),\ W(Z_2),\ W(Z_3),\ W(Z_4)] = A_{13}A_{23}^{-1}A_{24}A_{14}^{-1}$$

by Definition 1 in ref. [3]. Hence we have a kind of Schwarzian derivative of W by Definition 2 in reference [3].

We may expand W at point Z into power series. Fix λ. There exists $\varepsilon>0$,

$$W(Z+t\lambda) = W(Z)+\sum_{k=1}^{\infty}\frac{t^k}{k!}D_\lambda^k W(Z) \tag{1}$$

when $|t|<\varepsilon$, where $D_\lambda^k W(Z)$ is the kth-directional derivative of $W=W(Z)$ at point Z in direction λ.

Let $\lambda=(\lambda_1,\lambda_2,\cdots,\lambda_n)^{\mathrm T}$, $W=(w_1,w_2,\cdots,w_n)^{\mathrm T}$ be column vectors. Then

$$D_\lambda^k W(Z) = P^{(k)}(Z,\lambda)\lambda, \tag{2}$$

where $P^{(k)}(Z,\lambda)=(p_{ij}^{(k)}(Z,\lambda))_{1\leqslant i,j\leqslant n}$, and

$$p_{ij}^{(k)}(Z,\lambda) = \sum_{l_2,\cdots,l_k}\lambda_{l_2}\cdots\lambda_{l_k}\frac{\partial^k w_i(Z)}{\partial z_j\partial z_{l_2}\cdots\partial z_{l_k}}.$$

Obviously, we have $D_\lambda W(Z)=J_W(Z)\lambda$, and

$$D_\lambda^k W(Z) = \left(\frac{\mathrm d}{\mathrm dt}\right)^k W(Z+t\lambda)\big|_{t=0},\quad D_\lambda P^{(k)}(Z,\lambda) = P^{(k+1)}(Z,\lambda).$$

Substituting (2) into (1), we obtain

$$W(Z+t\lambda) = W(Z)+\sum_{k=1}^{\infty}\frac{t^k}{k!}P^{(k)}(Z,\lambda)\lambda.$$

We may express $W(Z_i)-W(Z_j)$ as

$$\begin{aligned}W(Z_i)-W(Z_j) &= \sum_{k=1}^{\infty}\frac{u_i^k-u_j^k}{k!}P^{(k)}(Z,\lambda)\lambda\\ &= \sum_{k=1}^{\infty}\frac{u_i^k-u_j^k}{k!}P^{(k)}(Z,\lambda)(J_W(Z))^{-1}J_W(Z)\lambda\\ &= A_{ij}J_W(Z)\lambda,\end{aligned} \tag{3}$$

where

$$A_{ij} = \sum_{k=1}^{\infty}\frac{u_i^k-u_j^k}{k!}P^{(k)}(Z,\lambda)(J_W(Z))^{-1}, \tag{4}$$

or

$$W(Z_i)-W(Z_j)=\sum_{k=1}^{\infty}\frac{u_i^k-u_j^k}{k!}(J_W(Z))J_W(Z)^{-1}P^{(k)}(Z,\lambda)\lambda=J_W(Z)B_{ij}\lambda,$$

where

$$B_{ij}=\sum_{k=1}^{\infty}\frac{u_i^k-u_j^k}{k!}(J_W(Z))^{-1}P^{(k)}(Z,\lambda). \tag{5}$$

We may take $X=J_W(Z)\lambda$ in (3). Since $J_W(Z)$ is nonsingular, and $\lambda\neq 0$, we have $X\neq 0$. Hence the cross ratio of four points $W(Z_i)$, $i=1,2,3,4$, can be defined as

$$[W(Z_1),W(Z_2),W(Z_3),W(Z_4)]=A_{13}A_{23}^{-1}A_{24}A_{14}^{-1},$$

where A_{ij} is defined by (4). It tells us that if $Z_1,Z_2,Z_3,Z_4\in l$, are distinct from one another, and $J_W(Z)$ is nonsingular, then the cross ratio of $W(Z_1)$, $W(Z_2)$, $W(Z_3)$, $W(Z_4)$ always exists.

From (4) and (5), we have $J_W(Z)B_{ij}(J_W(Z))^{-1}=A_{ij}$, and

$$[W(Z_1),W(Z_2),W(Z_3),W(Z_4)]=J_W(Z)B_{13}B_{23}^{-1}B_{24}B_{14}^{-1}(J_W(Z))^{-1}.$$

Let

$$[W(Z_1),W(Z_2),W(Z_3),W(Z_4)]^{\sim}=B_{13}B_{23}^{-1}B_{24}B_{14}^{-1}. \tag{6}$$

Then

$$\begin{aligned}&[W(Z_1),W(Z_2),W(Z_3),W(Z_4)]^{\sim}\\&=(J_W(Z))^{-1}[W(Z_1),W(Z_2),W(Z_3),W(Z_4)]J_W(Z).\end{aligned} \tag{7}$$

Now we evaluate the Schwarzian derivative of W at point Z in the direction λ by definition 2 in reference [3].

Let $u_i=v_is$, s, $v_i\in\mathbb{C}$, $i=1,2,3,4$. Then

$$\begin{aligned}A_{ij}&=\sum_{k=1}^{\infty}\frac{v_i^k-v_j^k}{k!}P^{(k)}(Z,\lambda)(J_W(Z))^{-1}s^k\\&=(v_i-v_j)s\left(I+\frac{v_i+v_j}{2!}P^{(2)}(Z,\lambda)(J_W(Z))^{-1}s\right.\\&\quad\left.+\frac{v_i^2+v_iv_2+v_j^2}{3!}P^{(3)}(Z,\lambda)(J_W(Z))^{-1}s^2+\cdots\right).\end{aligned}$$

Substituting it into formula (2) in ref. [3], after lengthy calculation, we have

$$\{W;Z\}_{\lambda,M}=P^{(3)}(Z,\lambda)(J_W(Z))^{-1}-\frac{3}{2}(P^{(2)}(Z,\lambda)(J_W(Z))^{-1})^2. \tag{8}$$

We call $\{W;Z\}_{\lambda,M}$ matrix Schwarzian derivative of $W=W(Z)$ at point Z in directioin λ. Similarly, if $Z_0\in\Omega$, and $J_W(Z_0)$ is nonsingular, then we may define the matrix Schwarzian derivative of W at point Z with respect to point Z_0 as $\{W;Z\}_{Z-Z_0,M}$. When $0\in\Omega$, $Z_0=0$, we denote $\{W;Z\}_{Z,M}$ by $\{W;Z\}_M$, and call it the matrix Schwarzian derivative of $W=W(Z)$ at Z.

If we use (6) as another definition of cross ratio, then we have another matrix Schwarzian derivative $\{W;Z\}_{\lambda,M}^{\sim}$. From (7), we know that

$$\{W;Z\}_{\lambda,M}^{\sim}=(J_W(Z))^{-1}P^{(3)}-\frac{3}{2}((J_W(Z))^{-1}P^{(2)})^2.$$

It follows that

$$\{W;Z\}_{\lambda,M}^{\sim}=(J_W(Z))^{-1}\{W;Z\}_{\lambda,M}J_W(Z),$$

i.e. these two matrix Schwarzian derivatives are similar.

Thus for any four points $Z_i\in l$, $i=1,2,3,4$, the cross ratio of these four points always exists, and we may define the matrix Schwarzian derivative (8) for W at point Z in direction $\lambda(\lambda \neq 0)$. We will prove the following theorems.

Theorem 1. *Let $\Omega \in \mathbb{C}^n$ be a domain, let $W = W(Z): \Omega \to \mathbb{C}^n$ be a holomorphic mapping on Ω, $Z \in \Omega$, and let $J_W(Z)$ be nonsingular. Then the matrix Schwarzian derivative $\{W; Z\}_{\lambda, M}$ which is defined by (8) is invariant up to similarity under the group of holomorphic automorphism of $\mathbb{C}P^n$.*

Theorem 2. *Let $\Omega \subset \mathbb{C}^n$ be a domain, let $W = W(Z): \Omega \to \mathbb{C}^n$ be a holomorphic mapping on Ω, $Z_0 \in \Omega$, and let Jacobian $J_W(Z_0)$ be nonsingular. Then $\{W; Z\}_{Z-Z_0, M} \equiv 0$ for all $Z \in \Omega$ if and only if*

$$W(Z) = W(Z_0) + \left(I_n - \frac{1}{2}P^{(2)}(Z_0, Z)(J_W(Z_0))^{-1}\right)^{-1} J_W(Z_0)(Z - Z_0), \tag{9}$$

where $W = W(Z)$, Z are column vectors, $P^{(2)}(Z_0, Z) = (p_{ij}^{(2)}(Z_0, Z))$, $p_{ij}^{(2)}(Z_0, Z) = \sum_l z_l \cdot \dfrac{\partial^2 W_i(Z_0)}{\partial z_j \partial z_l}$ is linear homogeneous polynomial of the entries of Z. $W(Z)$ is biholomorphic on Ω when Ω is convex.

Of course, we can write (9) as $W(Z) = (C(Z) + D)^{-1}(A(Z) + B)$, where $C(Z)$, $D \in \mathbb{C}^{n\times n}$, $A(Z)$, $B \in \mathbb{C}^{n\times 1}$. The entries of $A(Z)$, $C(Z)$ are linear homogeneous polynomials of the entries of Z, $C(Z) + D$ is nonsingular on Ω, and $J_W(Z_0)$ is nonsingular.

The proof of Theorem 1 is similar to the proof of Theorem 2 in ref. [3]. We omit the details.

Now we prove Theorem 2. We prove the necessity part of Theorem 2 at first.

Without loss of generality, we may assume $Z_0 = 0$. Otherwise we make a translation. Since $J_W(0)$ is nonsingular, there exists $0 \neq Z \in \mathbb{C}^n$, and $\varepsilon > 0$ such that $J_W(tZ)$ is nonsingular, when $|t| < \varepsilon \|Z\|^{-1}$, where $\|X\|$ is the length of the vector X.

When $Z_0 = 0$, $\{W; Z\}_{Z-Z_0, M} = 0$ becomes $\{W; Z\}_{Z, M} = 0$. By the definition of $\{W; Z\}_{\lambda, M}$, we know that $\{W; Z\}_{Z, M} = 0$ is equivalent to $\{W; tZ\}_{Z, M} = 0$, when $0 \neq Z \in \mathbb{C}^n$, $tZ \in \Omega$, and $J_W(tZ)$ is nonsingular. By (8),

$$P^{(3)}(tZ, X)J_W(tZ)^{-1} - \frac{3}{2}(P^{(2)}(tZ, Z)(J_W(tZ)^{-1})^2 = 0 \tag{10}$$

holds when $0 < |t| < \varepsilon \|Z\|^{-1}$. The left side of (10) is a holomorphic mapping of t when $|t| < \varepsilon \|Z\|^{-1}$. Let $t \to 0$. (10) becomes $\{W; 0\}_{Z, M} = 0$ for all $Z \in \mathbb{C}^n$.

By (10), we have

$$P^{(3)}(tZ, Z) = 3!\left(\frac{1}{2}P^{(2)}(tZ, Z)(J_W(tZ))^{-1}\right)^2 J_W(tZ). \tag{11}$$

It is a holomorphic mapping of t when $|t| < \varepsilon \|Z\|^{-1}$. Moreover, we know that

$$\begin{cases} \dfrac{d}{dt}P^{(k)}(tZ, Z) = P^{(k+2)}(tZ, Z), \quad k = 1, 2, \cdots, \\ P^{(1)}(tZ, Z) = J_W(tZ). \end{cases}$$

Differentiating both sides of (11) with respect to t, we may prove that

$$P^{(k)}(tZ, Z) = k!\left(\frac{1}{2}P^{(2)}(tZ, Z)(J_W(tZ))^{-1}\right)^{k-1} J_W(tZ) \tag{12}$$

holds for $k = 1, 2, \cdots$ by mathematical induction.

Obviously, (12) is true when $k=1,2,3$. If (12) is true when $k=l$, then

$$\begin{aligned}
P^{(l+1)}(tZ,Z) &= \frac{d}{dt}P^{(l)}(tZ,Z) = \frac{d}{dt}\left[l!\left(\frac{1}{2}P^{(2)}(tZ,Z)(J_W(tZ))^{-1}\right)^{l-1}J_W(tZ)\right] \\
&= l!\left(\frac{1}{2}\right)^{l-1}[(l-1)(P^{(2)}(tZ,Z)(J_W(tZ))^{-1})^{l-2}\{P^{(3)}(tZ,Z)(J_W(tZ))^{-1} \\
&\quad - P^{(2)}(tZ,Z)(J_W(tZ))^{-1}P^{(2)}(tZ,Z)(J_W(tZ))^{-1}\}J_W(tZ) \\
&\quad + (P^{(2)}(tZ,Z)J_W(tZ)^{-1})^{l-1}P^{(2)}(tZ,Z)] \\
&= l!\left(\frac{1}{2}\right)^{l-1}(P^{(2)}(tZ,Z)(J_W(tZ))^{-1})^{l-2}\{(l-1)[P^{(3)}(tZ,Z)-P^{(2)}(tZ,Z) \\
&\quad \cdot J_W(tZ)^{-1}P^{(2)}(tZ,Z)] + P^{(2)}(tZ,Z)(J_W(tZ)^{-1}P^{(2)}(tZ,Z)\} \\
&= l!\left(\frac{1}{2}\right)^{l-1}(P^{(2)}(tZ,Z)(J_W(tZ))^{-1})^{l-2}\left(\frac{l+1}{2}\right)P^{(2)}(tZ,Z) \\
&\quad \cdot (J_W(tZ))^{-1}P^{(2)}(tZ,Z) \\
&= (l+1)!\left(\frac{1}{2}\right)^{l}(P^{(2)}(tZ,Z)(J_W(tZ))^{-1})^{l}J_W(tZ)
\end{aligned}$$

by (12). Hence (12) is true when $k=l+1$. We have proved that (12) is true for $k=1,2,\cdots$. Letting $t\to 0$ in (12), we know that

$$P^{(k)}(0,Z) = k!\left(\frac{1}{2}P^{(2)}(0,Z)(J_W(0))^{-1}\right)^{k-1}J_W(0) \tag{13}$$

holds for $k=1,2,3,\cdots$.

Taking $Z=0$, $\lambda=Z$ in (1), there exists $\varepsilon>0$ such that when $|t|<\varepsilon$

$$W(tZ) = W(0) + \sum_{k=1}^{\infty}\frac{t^k}{k!}D_Z^kW(0)$$

holds true.

$$W(tZ) = W(0) + \sum_{k=1}^{\infty}\frac{t^k}{k!}P^k(0,Z)Z$$

holds true when $|t|<\varepsilon$ by (2). Inserting (13) in the left side of the previous equality,

$$W(tZ) = W(0) + \left(I - \frac{t}{2}P^{(2)}(0,Z)(J_W(0))^{-1}\right)^{-1}J_W(0)Z$$

holds when $|t|<\varepsilon$.

When $tZ\in\Omega$, both sides of the previous equality are holomorphic mappings of t. By the uniqeness theorem of the holomorphic mapping of one complex variable, the equality holds for all t when $tZ\in\Omega$. Letting $t=1$, we see that

$$W(Z) = W(0) + \left(I - \frac{1}{2}P^{(2)}(0,Z)(J_W(0))^{-1}\right)^{-1}J_W(0)Z$$

holds true at a neighborhood of $Z=0$.

By the uniqueness theorem of the holomorphic mapping of several complex variables, the previous equality holds true for all $Z\in\Omega$.

We have proved the necessity part of Theorem 2.

We may use similar argument as we used in ref. [2] to prove that (9) is biholomorphic on Ω when Ω is convex. We omit the details.

Now we prove the sufficiency part of Theorem 2.

Letting $Z_0=0$ and $Z=t\lambda$ in (9), we have

$$W(t\lambda) = W(0) + t\left(I_n - \frac{t}{2}P^{(2)}(0,\lambda)(J_W(0))^{-1}\right)^{-1} J_W(0)\lambda.$$

Let $F(Z+\xi) = W(\xi)$. Then the previous equation becomes

$$F(Z+t\lambda) = F(Z) + t\left(I_n - \frac{t}{2}P_F^{(2)}(Z,\lambda)(J_F(Z))^{-1}\right)^{-1} J_F(Z)\lambda,$$

where $P_F^{(2)}(Z,\lambda)$ means that the mapping in $P^{(2)}$ is F. Expanding F at Z into a power series of t, we have

$$F(Z+t\lambda) = F(Z) + \sum_{k=1}^{\infty}\frac{t^k}{k!}P_F^{(k)}(Z,\lambda)\lambda.$$

Comparing the coefficients of t^k, $k=1,2,3,\cdots$, we obtain

$$P_F^{(k)}(Z,\lambda) = k!\,Z^{1-k}(P_F^{(2)}(Z,\lambda)(J_F(Z))^{-1})^{k-1}J_F(Z).$$

Thus $\{F;Z\}_{\lambda,M}=0$; it is equivalent to $\{W;Z\}_{\lambda,M}=0$.

We have proved the sufficiency part of Theorem 2.

As a consequence of Theorem 2, we have the following.

Corollary 1. *Assumptions are the same as Theorem 2. $\{W;Z\}_{Z-Z_0,M}=0$ holds true on $\Omega\setminus\{Z_0\}$ and $\{W,Z\}_{\lambda,M}=0$ holds for all $Z\in\Omega$, and $0\neq\lambda\in\mathbb{C}^n$ also. Moreover, they are equivalent to each other.*

We can write $\{W;Z\}_{\lambda,M}$ as

$$\{W;Z\}_{\lambda,M} = \sum_{i,j}\lambda_i\lambda_j\frac{\partial^2 J_W(Z)}{\partial z_i\partial z_j}J_W^{-1}(Z) - \frac{3}{2}\left(\sum_i\lambda_i\frac{\partial J_W(Z)}{\partial z_i}J_W^{-1}(Z)\right)^2$$

by (8) where $Z=(z_1,z_2,\cdots,z_n)$, $\lambda=(\lambda_1,\lambda_2,\cdots,\lambda_n)$.

Corollary 2. *Assumptions are the same as Theorem 2. Moreover, if there exists a function of $Z-Z_0$, $a(Z-Z_0)$ such that*

$$\sum_i(z_i-z_i^0)\frac{\partial J_W(Z_0)}{\partial z_i}(Z-Z_0) = a(Z-Z_0)J_W(0)(Z-Z_0)$$

holds where $Z_0=(z_{01},z_{02},\cdots,z_{0n})^T$, $Z=(z_1,z_2,\cdots,z_n)^T\in\Omega$, then

$$W(Z) = W(Z_0) + \frac{J_W(0)(Z-Z_0)}{1-b(Z-Z_0)},$$

where $b=(b_1,b_2,\cdots,b_n)\in\mathbb{C}^n$ is a constant row vector, and W is a column vector.

Proof. The assumptions of Corollary 2 mean $P^{(2)}(0,Z-Z_0)=a(Z-Z_0)J_W(Z_0)$. Since every entry of $P^{(2)}(0,Z-Z_0)$ is a linear homogeneous polynomial of the entries of $Z-Z_0$, $a(Z-Z_0)$ is a linear homogeneous polynomial of the entries of $Z-Z_0$; that is, $a(Z-Z_0)=b(Z-Z_0)$, where $b=(b_1,b_2,\cdots,b_n)\in\mathbb{C}^n$. Substituting it into (9), we have proved Corollary 2.

3 Another Schwarzian derivative

We recall the definition of Schwarzian derivative in ref. [3], where $Z\in\mathbb{C}^{m\times n}$, $G=GL(m,\mathbb{C})\otimes I_n$, In ref. [3], we proved that $\{W;Z\}_\lambda$ is invariant up to similarity under the group of holomorphic automorphism of $\mathbb{C}G(m,n)$ (Theorem 2 in ref. [3]). We also proved that if Ω is a domain in $\mathbb{C}^{m\times n}$, $W=W(Z);\Omega\to\mathbb{C}^{m\times n}$ is a holomorphic mapping on Ω, $Z_0\in\Omega$, and $J_W(Z_0)$ is nonsingular, then $\{W;Z\}_{Z-Z_0}=0$ holds true in a neighborhood of Z_0, if and only if W is a linear fractional mapping (Theorem 3). When $m=1$, then $Z\in\mathbb{C}^n$, $G=GL(1,\mathbb{C})\otimes I_n$. Let Ω

$\subset \mathbb{C}^n$ be a domain, and let $W = W(Z)$: $\Omega \to \mathbb{C}^n$ be a holomorphic mapping on Ω. For $Z \in \Omega$, if there exist $\alpha, \beta, \gamma \in \mathbb{C}, \lambda \in \mathbb{C}^n$ such that the cross ratio of four points $W(Z_i)$, $i = 1,2,3,4$ exists and the limit (2) in ref. [3] exists, where $Z_1 = Z$, $Z_2 = Z + t\alpha\lambda$, $Z_3 = Z + t\beta\lambda$, $Z_4 = Z + t\gamma\lambda$, and the Schwarzian derivative of W at point Z in direction λ, $\{W;Z\}_\lambda$ exists, then it can be expressed as

$$\{W;Z\}_\lambda^\circ = D_\lambda^3 W(Z)(D_\lambda W(Z))^{-1} - \frac{3}{2}(D_\lambda^2 W(Z)(D_\lambda W(Z))^{-1})^2, \tag{14}$$

where the inverse of matrix is the generalized inverse matrix, and W, Z, λ are row vectors. Since $D_\lambda W(Z) = \lambda J_W(Z)^T$, (14) can be rewritten as

$$\{W;Z\}_\lambda^\circ = (\lambda\lambda^T) D_\lambda^3 W(Z)(J_W(Z)^T)^{-1} - \frac{3}{2}(D_\lambda^2 W(Z)(J_W(Z)^T)^{-1}\lambda^T)^2(\lambda\lambda^T)^{-2}.$$

Hence

$$(\lambda\lambda^T)^2\{W;Z\}_\lambda^\circ = D_\lambda^3 W(Z)(J_W(Z)^T)^{-1}\lambda^T(\lambda\lambda^T) - \frac{3}{2}(D_\lambda^2 W(Z)(J_W(Z)^T)^{-1}\lambda^T)^2.$$

We have

$$\begin{aligned}
&(\lambda\lambda^T)^2\{W;Z\}_\lambda^\circ \\
&= \sum_{i,j}\lambda_i\lambda_j \frac{\partial J_W(Z)^T}{\partial z_i \partial z_j}(J_W(Z)^T)^{-1}\lambda^T(\lambda\lambda^T) - \frac{3}{2}\left(\sum_i \lambda_i \frac{\partial J_W(Z)^T}{\partial z_i}(J_W(Z)^T)^{-1}\lambda^T\right)^2 \\
&= (\lambda\lambda^T)\lambda(J_W(Z))^{-1}\left(\sum_{i,j}\lambda_i\lambda_j \frac{\partial^2 J_W(Z)}{\partial z_i \partial z_j}\right)\lambda^T - \frac{3}{2}\left(\lambda(J_W(Z))^{-1}\left(\sum_i \lambda_i \frac{\partial J_W(Z)}{\partial z_i}\right)\lambda^T\right)^2 \\
&= \lambda(J_W(Z))^{-1}\left[(\lambda\lambda^T)\sum_{i,j}\lambda_i\lambda_j \frac{\partial^2 J_W(Z)}{\partial z_i \partial z_j} - \frac{3}{2}\sum_{i,j}\lambda_i\lambda_j \frac{\partial J_W(Z)}{\partial z_i}\lambda\lambda^T(J_W(Z))^{-1}\frac{\partial J_W(Z)}{\partial z_j}\right]\lambda^T
\end{aligned} \tag{15}$$

by $D_\lambda^2 W(Z) = D_\lambda(\lambda J_W(Z)^T) = \sum_i \lambda\lambda_i \frac{\partial J_W(Z)^T}{\partial z_i}$, $D_\lambda^3 W(Z) = \sum_{i,j}\lambda\lambda_i\lambda_j \frac{\partial^2 J_W(Z)^T}{\partial z_i \partial z_j}$.

When $m = 1$, Theorems 2 and 3 in ref. [3] may be stated as follows: $\{W;Z\}_\lambda$ is invariant up to similarity under the group of holomorphic automorphism of $\mathbb{C}P^n$. Let $\Omega \subset \mathbb{C}^n$ be a domain, let $W = W(Z)$: $\Omega \to \mathbb{C}^n$ be a holomorphic mapping, and suppose that $0 \in \Omega$ and $\{W;Z\}$ exist at a neighborhood of Z. Then $\{W;Z\} \equiv 0$ holds true on Ω if and only if

$$W(Z) = W(0) + \frac{Z J_W(0)^T}{1 - bZ^T}, \tag{16}$$

where $b = (b_1, b_2, \cdots, b_n) \in \mathbb{C}^n$ is a constant vector.

In this section, we will prove the following Theorem.

Theorem 3. *Let $\Omega \subset \mathbb{C}^n$ be a domain, and let $W = W(Z)$: $\Omega \to \mathbb{C}^n$ be a holomorphic mapping. If there exists $Z_0 \in \Omega$, $J_W(Z_0)$ is nonsingular, and $\{W;Z\}_\lambda$ exists at a neighborhood of Z_0, then $W(Z)$ can be expressed as (16), where $0 \neq \lambda \in \mathbb{C}^n$. Hence $\{W;Z\}_\lambda = 0$ for any $Z \in \Omega$, $0 \neq \lambda \in \mathbb{C}^n$.*

Proof. Without loss of generality, we assume $Z_0 = 0$, $W(0) = 0$, $J_W(0) = I$. Otherwise we make a translations and normalize the mapping. When r is sufficiently small, $B_r = \{Z \in \mathbb{C}^n: Z\bar{Z}^T < r\} \subset \Omega$, contained in the neighborhood of $Z = 0$ under the assumption of Theorem 3. By (22) in ref. [3], there exists a function $a(tZ, Z)$ such that

$$\sum_{i,j}z_i z_j \frac{\partial^2 w_\alpha(tZ)}{\partial z_i \partial z_j} = a(tZ, Z)\sum_i z_i \frac{w_\alpha(tZ)}{\partial z_i}, \quad \alpha = 1, \cdots, n$$

holds when $Z \in B_r$, where $W=(w_1, \cdots, w_n)$. This equality is the same as

$$\frac{d^2}{dt^2} w_\alpha(tZ) = a(tZ, Z) \frac{d}{dt} w_\alpha(tZ).$$

Hence, for any $\alpha, \beta = 1, \cdots, n$, we have

$$\frac{\frac{d^2}{dt^2} w_\alpha(tZ)}{\frac{d}{dt} w_\alpha(tZ)} = \frac{\frac{d^2}{dt^2} w_\beta(tZ)}{\frac{d}{dt} w_\beta(tZ)} = a(tZ, Z).$$

Integrating the left equality with respect to t from 0 to t, we have

$$\frac{\frac{d}{dt} w_\alpha(tZ)}{\frac{d}{dt} w_\alpha(0)} = \frac{\frac{d}{dt} w_\beta(tZ)}{\frac{d}{dt} w_\beta(0)}.$$

We know that $\frac{d}{dt} w_\alpha(tZ) \mid_{t=0} = \sum z_i \frac{\partial w_\alpha(0)}{\partial z_i}$; since $J_W(0) = I$, we have

$$\frac{\frac{d}{dt} w_\alpha(tZ)}{z_\alpha} = \frac{\frac{d}{dt} w_\beta(tZ)}{z_\beta}.$$

Integrating both sides of the above equality, we obtain

$$\frac{w_\alpha(tZ)}{z_\alpha} = \frac{w_\beta(tZ)}{z_\beta},$$

since $W(0)=0$. Let $t=1$. It is $\frac{w_\alpha(Z)}{z_\alpha} = \frac{w_\beta(Z)}{z_\beta}$. Denoting $\frac{w_\alpha(Z)}{z_\alpha} = \widetilde{w}$, $\alpha = 1, \cdots, n$, we have

$$w_\alpha(Z) = z_\alpha \widetilde{w}, \quad \alpha = 1, \cdots, n.$$

Differentiating both sides of the equality with respect to z_i, we have

$$\frac{\partial w_\alpha(Z)}{\partial z_i} = \delta_i^\alpha \widetilde{w} + \frac{\partial \widetilde{w}}{\partial z_i} z_\alpha.$$

We know that $\frac{\partial w_\alpha}{\partial z_i}(0) = \delta_i^\alpha$. Let $Z=0$ in the previous equality. It is $\delta_i^\alpha = \delta_i^\alpha \widetilde{w}(0)$. Thus $\widetilde{w}(0) = 1$.

If $\{W; Z\}_\lambda$ exists at a neighborhood of $Z=0$ in any direction $0 \neq \lambda \in \mathbb{C}^n$, from (17) in ref. [3], there exists $a(Z, \lambda)$ such that

$$\sum_{i,j} \lambda_i \lambda_j \frac{\partial w_\alpha(Z)}{\partial z_i \partial z_j} = a(Z, \lambda) \sum_i \lambda_i \frac{\partial w_\alpha(Z)}{\partial z_i}, \quad \alpha = 1, \cdots, n$$

holds, where $\lambda = (\lambda_1, \cdots, \lambda_n)$. Substituting $w_\alpha = z_\alpha \widetilde{w}$ in the previous equality, we have

$$2\lambda_\alpha \sum_i \lambda_i \frac{\partial \widetilde{w}(Z)}{\partial z_i} + z_\alpha \sum_{i,j} \lambda_i \lambda_j \frac{\partial^2 \widetilde{w}(Z)}{\partial z_i \partial z_j} = a(Z, \lambda)\left(\lambda_\alpha \widetilde{w} + z_\alpha \sum_i \lambda \frac{\partial \widetilde{w}(Z)}{\partial z_i}\right).$$

Hence

$$a(Z, \lambda) = \left(2\lambda_\alpha \sum_i \lambda_i \frac{\partial \widetilde{w}}{\partial z_i} + z_\alpha \sum_{i,j} \lambda_i \lambda_j \frac{\partial^2 \widetilde{w}(Z)}{\partial z_i \partial z_j}\right)\left(\lambda_\alpha \widetilde{w} + z_\alpha \sum_i \lambda_i \frac{\partial \widetilde{w}}{\partial z_i}\right)^{-1}.$$

Denoting the right-hand side of this equality by Q_α we have $Q_\alpha = Q_\beta$, $\alpha, \beta = 1, \cdots, n$, i.e.

$$(z_\alpha \lambda_\beta - z_\beta \lambda_\alpha)\left(\widetilde{w}(Z) \sum_{i,j} \lambda_i \lambda_j \frac{\partial^2 \widetilde{w}(Z)}{\partial z_i \partial z_j} - 2\left(\sum_i \lambda \frac{\partial \widetilde{w}}{\partial z_i}\right)^2\right) = 0.$$

If Z and λ are two different directions, then

$$\tilde{w}(Z)\sum_{i,j}\lambda_i\lambda_j\frac{\partial^2\tilde{w}(Z)}{\partial z_i\partial z_j}=2\left(\sum_i\lambda_i\frac{\partial\tilde{w}(Z)}{\partial z_i}\right)^2 \tag{17}$$

holds. For any fixed Z, we may choose a sequence $\lambda^{(j)}$ such that $\lambda^{(j)}$ is different from Z, and $\lim_{j\to\infty}\lambda^{(j)}=Z$. Both sides of (17) are continuous functions of λ_i, $i=1,2,3,\cdots,n$. Using the limiting process, (17) holds for any Z and λ. Let $\lambda=Z$

$$\tilde{w}(tZ)\sum_{i,j}z_iz_j\frac{\partial^2\tilde{w}(Z)}{\partial z_i\partial z_j}=2\left(\sum_i z_i\frac{\partial\tilde{w}(tZ)}{\partial z_i}\right)^2.$$

Replacing Z by tZ in $0\leqslant t\leqslant 1$, we have

$$\tilde{w}(tZ)\frac{\mathrm{d}^2}{\mathrm{d}t^2}\tilde{w}(tZ)=2\left(\frac{\mathrm{d}}{\mathrm{d}t}\tilde{w}(tZ)\right)^2,$$

i.e.

$$\frac{\frac{\mathrm{d}^2}{\mathrm{d}t^2}\tilde{w}(tZ)}{\frac{\mathrm{d}}{\mathrm{d}t}\tilde{w}(tZ)}=2\frac{\frac{\mathrm{d}}{\mathrm{d}t}\tilde{w}(tZ)}{\tilde{w}(tZ)}.$$

Integrating both sides of this equality with respect to t from 0 to t, we have

$$\frac{\frac{\mathrm{d}}{\mathrm{d}t}\tilde{w}(tZ)}{Zb^{\mathrm{T}}}=(\tilde{w}(tZ))^2,$$

where $b=(b_1,b_2,\cdots,b_n)\in\mathbb{C}^n$, since $\frac{\mathrm{d}}{\mathrm{d}t}\tilde{w}(tZ)\mid_{t=0}=\sum_i z_i\frac{\partial\tilde{w}}{\partial z_i}(0)=Zb^{\mathrm{T}}$, $\tilde{w}(0)=1$.

Integrating both sides of the equality $\frac{\frac{\mathrm{d}}{\mathrm{d}t}\tilde{w}(tZ)}{(\tilde{w}(tZ))^2}=Zb^{\mathrm{T}}$ with respect to t from 0 to t, we obtain

$$1-\frac{1}{\tilde{w}(tZ)}=tZb^{\mathrm{T}}.$$

Letting $t=1$, we have $\tilde{w}(Z)=\frac{1}{1-Zb^{\mathrm{T}}}$. By the definition of $\tilde{w}$, we have $w_a(Z)=\frac{z_a}{1-Zb^{\mathrm{T}}}$, when $Z\in B_r$, $a=1,2,\cdots,n$. $w_a(Z)$ is a holomorphic function of z when $Z\in\Omega$. Thus $w_a(Z)=\frac{z_a}{1-Zb^{\mathrm{T}}}$ holds true for any $Z\in\Omega$.

We have proved that if $\{W;Z\}_\lambda$ exists at a neighborhood of $Z=0$ in any direction $0\neq\lambda\in\mathbb{C}^n$, then W can be expressed as (16).

Now we prove that if $W(Z)$ can be expressed as (16), then $\{W;Z\}_\lambda=0$ holds for any $0\neq\lambda\in\mathbb{C}^n$.

(16) can be written as

$$W(Z)=\left(w_1(0)+\frac{1}{1-Zb^{\mathrm{T}}}\sum_i z_i\frac{\partial w_1}{\partial z_i}(0),\cdots,w_n(0)+\frac{1}{1-Zb^{\mathrm{T}}}\sum_i z_i\frac{\partial w_n}{\partial z_i}(0)\right).$$

It can be verified that

$$J_W(Z)=\left(\frac{I_n}{1-Zb^{\mathrm{T}}}+\frac{Zb^{\mathrm{T}}}{(1-Zb^{\mathrm{T}})^2}\right),$$

$$\frac{\partial J_W(Z)}{\partial z_i} = J_W(0)\left[\frac{b_i I_n}{(1-Zb^T)^2} + \frac{2b_i Z^T b}{(1-Zb^T)^3} + \frac{e_i^T b}{(1-Zb^T)^2}\right],$$

and

$$\frac{\partial^2 J_W(Z)}{\partial z_i \partial z_j} = J_W(0)\left[\frac{2b_i b_j I_n}{(1-Zb^T)^3} + \frac{2b_i e_j^T b}{(1-Zb^T)^3} + \frac{6b_i b_j Z^T b}{(1-Zb^T)^4} + \frac{2b_j e_i^T b}{(1-Zb^T)^3}\right]$$

hold true where e_i, e_j are unit vectors of the ith and jth coordinates, respectively. Hence

$$J_W(Z)^{-1} = (1-Zb^T)(I - Z^T b)(J_W(0))^{-1},$$

$$\lambda(J_W(Z))^{-1}\left(\sum_{i,j}\lambda_i\lambda_j \frac{\partial^2 J_W(Z)}{\partial z_i \partial z_j}\right) = \frac{b(\lambda b^T)^2}{(1-Zb^T)^2}\lambda \tag{18}$$

and

$$\lambda(J_W(Z))^{-1}\left(\sum_i \lambda_i \frac{\partial J_W(Z)}{\partial z_i}\right) = \frac{2\lambda b^T}{1-Zb^T} \tag{19}$$

hold. Substituting (18) and (19) into (15), we have $\{W;Z\}_\lambda^0 = 0$, and $\{W;Z\}_\lambda = 0$.

We have proved Theorem 3.

On the right-hand side of (15), we replace λ^T by ν^T, where $\nu = (\nu_1, \cdots, \nu_n) \in \mathbb{C}^n$, and denote it by $\{W;Z\}_{\lambda,\nu}$,

$$\{W;Z\}_{\lambda,\nu} = (\lambda\nu^T)\lambda(J_W(Z))^{-1}\left(\sum_{i,j}\lambda_i\lambda_j \frac{\partial^2 J_W(Z)}{\partial z_i \partial z_j}\right)\nu^T - \frac{3}{2}\left[\lambda(J_W(Z))^{-1}\left(\sum_i \lambda_i \frac{\partial J_W(Z)}{\partial z_i}\right)\nu^T\right]^2. \tag{20}$$

(20) is the Schwarzian derivative of W at point Z in direction λ, ν which was defined in reference [4].

Let $\Omega \subset \mathbb{C}^n$ be a domain, let $0 \in \Omega$, $W = W(Z): \Omega \to \mathbb{C}^n$ be a locally biholomorphic mapping and let $F(Z) = F(0) + \frac{ZJ_F(0)^T}{1-bZ^T}$. In ref. [4], we proved that if $F \circ W$ and $W \circ F$ make sense on Ω, then

$$\{W \circ F, Z\}_{\lambda, bJ_W(Z)} = \{W;Z\}_{\lambda, bJ_W(Z)}$$

and

$$\{W \circ F, Z\}_{\lambda,\nu} = \{W;Z\}_{\lambda \overline{J_W(Z)}^T, \nu J_W^{-1}(Z)}$$

hold for any non-zero λ, $\nu \in \mathbb{C}^n$.

4 Comparison of two kinds of Schwarzian derivatives

We conclude the paper by comparing two kinds of Schwarzian derivatives which we discussed in this paper.

Let $\Omega \subset \mathbb{C}^n$ be a domain, $W = W(Z): \Omega \to \mathbb{C}^n$ be a holomorphic mapping. Starting from two different Lie groups, we have two different Schwarzian derivatives. Letting $G = GL(1,\mathbb{C}) \otimes I_n$, we obtain the Schwarzian derivative of W at Z, $\{W;Z\}$, which is a scalar function. Letting $G = GL(n,\mathbb{C})$, we obtain the matrix Schwarzian derivative of W at Z, $\{W;Z\}_M$, which is an $n \times n$ matrix function. In ref. [3] and this paper we have proved that the Schwarzian derivative $\{W;Z\}$ and the matrix Schwarzian derivative $\{W;Z\}_M$ are invariant up to similarity under the group of holomorphic automorphism $\mathbb{CP}^n$. We also proved that if $0 \in \Omega$, $\{W;Z\}$ exists at a neighborhood N of $Z=0$, and $\{W;Z\} = 0$ (one equation) holds on N, then $W(Z)$ is in the form (16). Moreover, if Ω is convex, then it is biholomorphic on Ω. The converse of it also holds true. We have also proved that if $\{W;Z\}_M = 0$(n^2 equation) holds on N, then $W(Z)$ is in the form (9) (if $Z_0 = 0$). If Ω is convex, then it is biholomorphic on Ω. The converse of it also holds true. Thus $\{W;Z\}$ and $\{W;Z\}_M$ characterize $\mathbb{C}P^n$.

If $0 \neq \lambda \in \mathbb{C}^n$, we defined the Schwarzian derivatives $\{W;Z\}_\lambda$ and $\{W;Z\}_{\lambda;M}$. Theorem 3 tells us that if $W(Z)$ is not in the form (16), then there at least exists a direction $\lambda \neq 0$, $\{W;Z\}_\lambda$ does not exist, but $\{W;Z\}_{\lambda,M}$ exists for any direction $\lambda \neq 0$. This is the difference between these two Schwarzian derivatives.

References

1 Gong Sheng, Fitz Gerald, C. H., The Schwarzian derivative in several complex variables, *Science in China*, Ser A, 1993, 36: 513.

2 Gong Sheng, Yu Qihuang, Zheng Xuean, The Schwarzian derivative in several complex variables (Ⅱ), *Chinese Ann. of Math.*, Ser. B, 1998, 41(1):8.

3 Gong Sheng, Zheng Xuean, Yu Qihuang, The Schwarzian derivative in several complex variables (Ⅲ), *Science in China*, Ser. A, 1998, 41(2): 158.

4 Yu Qihuang, Gong Sheng, The Schwarzian derivative in $\mathbb{C}^n$, *Chinese J. of Contemporary Math.*, 1995, 16: 301.

第42卷第2期
1999年3月

数 学 学 报
ACTA MATHEMATICA SINICA

Vol.42, No. 2
march, 1999

多复变数的 Schwarz 导数 (V)

龚 升

(中国科学技术大学数学系 合肥 230026)
(Email: gongs@public.intercom.com.cn)

郑学安

(北京师范大学数学系 北京 100875)
(Email: zhengxa@bnu.edu.cn)

余其煌

(中国科学院应用数学研究所 北京 100080)
(Email: yqh@public.intercom.com.cn)

摘 要 本文从 Thurston 的观点出发，用二阶逼近来定义与讨论矩阵空间 $\mathbb{C}^{m\times n}$ $(m\le n)$ 中的域上全纯映照的 Schwarz 导数及高阶 Schwarz 导数，证明：如果它们存在的话，那么它们是在 $R_I(m,n)$ 的紧对偶空间 $\mathbb{C}G(m,n)$ 的全纯自同构群下的相似不变量. 并证明：这样得到的 Schwarz 导数与前几文 [1-4] 中由 Ahlfors 的观点得到的 Schwarz 导数是相一致的. 此外，还应用这种观点定义与讨论了 $\mathbb{C}^N$ 中的域上全纯映照的 Schwarz 导数.

关键词 Möbius 变换，二阶逼近，Schwarz 导数，高阶 Schwarz 导数
MR(1991) 主题分类 32H02
中图分类 O174.56

The Schwarzian Derivative in Several Complex Variables (V)

Gong Sheng
(Department of Mathematics, University of Science and Technology of China, Hefei 230026, P. R. China)
(Email: gongs@public.intercom.com.cn)

Zheng Xuean
(Department of Mathematics, Beijing Normal University, Beijing 100875, P. R. China)
(Email: zhengxa@bnu.edu.cn)

Yu Qihuang
(Institute of Applied Mathematics, Academia Sinica, Beijing 100080, P. R. China)
(Email: yqh@public.intercom.com.cn)

Abstract From point view of Thurston we define and discuss the Schwarzian derivative and Schwarzian derivative of high order of holomorphic mappings on domains of the matrix space $\mathbb{C}^{m\times n}$ $(m\le n)$ by using the approximation of second order. We prove

收稿日期: 1997-08-12, 接受日期: 1998-01-08
国家自然科学基金资助项目

that if the Schwarzian derivatives exist, then they are the invariant up to similarity under the group of holomorphic automophism of the compact dual space of $R_I(m,n)$. Also we prove that the Schwarzian in this paper is same as the Schwarzian in [1-4] which obtained by the point view of Ahlfors. Moreover, we define and discuss the Schwarzian of holomorphic mappings on domains of $\mathbb{C}^N$.

Keywords Möbius translation, Approximation of second order, Schwarzian derivative, Schwarzian derivative of high order

MR(1991) Subject Classification 32H02

Chinese Library Classification O174.56

1 引言

有各种不同的观点与途径来得到复分析中的 Schwarz 导数. 例如: Schwarz, Ahlfors,Thurston 以及 Deligne 等等. 于是有从各种不同的观点与途径来推广 Schwarz 导数到高维空间中去的文献, 在前四文 [1-4] 中, 龚升, FitzGerald, 余其煌, 郑学安等人, 按照 Ahlfors 的观点: Schwarz 导数是交比的无穷小形式, 将 Schwarz 导数推广到高维空间, 尤其是矩阵空间. Thurston[5] 在 1985 年曾经说过: "Schwarz 导数很像一种曲率, 在微分几何中, 各种不同的曲率是用来度量曲线或流形与平坦之间的偏离, 而 Schwarz 导数是用来度量共形映照与 Möbius 变换之间的偏离." 他是根据 Mario 与 Sarvas[6] 的工作而说的这番话的. 按照他的观点, Osgood 与 Stove[7,8] 以及 Carne[9] 讨论了 Riemann 流形之间的共形映照的 Schwarz 导数. Nolzon 与 Tamanoi[10,11] 讨论了 $\mathbb{C}^n$ 上全纯映照的 Schwarz 导数等等.

本文就是按照 Thurston 的观点来讨论矩阵空间 $\mathbb{C}^{m\times n}$ $(m \leq n)$ 中的域上的局部双全纯映照的 Schwarz 导数. 证明了: 由这种观点导出的 Schwarz 导数与从 Ahlfors 的观点导出的 Schwarz 导数是相一致的, 如果这样的 Schwarz 导数存在的话. 此外, 从 Thurston 的观点, 还可以定义高阶 Schwarz 导数. 而且可以证明: 这些高阶 Schwarz 导数如同 Schwarz 导数那样, 在 R_I 的紧对偶空间的全纯自同构作用下是相似不变量. 要注意的是, 当 $m < n$ 时, 这时候由 Thurston 观点出发, 未必能定义 Schwarz 导数, 但是如果能定义 Schwarz 导数的话, 那么 Schwarz 导数与高阶 Schwarz 导数也都具有相似不变性. 这时候, 要求能定义 Schwarz 导数的条件, 相当于文 [3] 中要求四点交比存在的条件. 要有这样的条件应该说是当然的. 应用上述的这些结果以及 Waring-Hilbert 问题的解, 可以定义在 $\mathbb{C}^N$ 上全纯映照的 Schwarz 导数及高阶 Schwarz 导数, 并可以证明: 它们也具有相似不变性. 此外, 应用文 [3] 中的方法, 还可以证明: 这样定义的 Schwarz 导数等于零当且仅当是对应某种意义下的线性分式变换.

如前指出, Thurston 的观点是: Schwarz 导数是用来衡量全纯映照与 Möbius 变换之间的偏离的. 为了刻划两者之间的偏离, 要引入二阶逼近的概念. 若 $\Omega \subset \mathbb{C}^n$ 为域. 映照 $f(Z):\Omega \to \mathbb{C}^n$ 为 Ω 上局部双全纯映照, $0 \in \Omega$, 对每一点 $Z_0 \in \Omega$, 若在 $\mathbb{C}^n$ 的 Möbius 变换群中能找到一个元素 W_{Z_0}, 使得 f 在 Z_0 点与 W_{Z_0} 在一定点 (例如原点) 的值、沿 Z_0 方向的一阶、二阶方向导数的值全相等, 则称 W_{Z_0} 为 f 在点 Z_0 的二阶逼近. 而 $W_{Z_0}^{-1} \circ f$ 即为 f 与 Möbius 变换之间的偏离, 而其主项为 Schwarz 导数.

2 Schwarz 导数与高阶 Schwarz 导数

在这一节中，从 Thurston 的观点来定义 Schwarz 导数及高阶 Schwarz 导数.

先来考虑 $\mathbb{C}^{m\times n}$, $m<n$ 的情形，即讨论 $\mathbb{C}^{m\times n}$ 中的域 Ω. $f(Z):\Omega\to\mathbb{C}^{m\times n}$ 为 Ω 上局部双全纯映照. 若 $Z_0\in\Omega$, 且秩为 m, 寻找 $f(Z)$ 在 Z_0 点的二阶逼近，即寻找 Möbius 变换

$$W_{Z_0}(Z)=A(Z_0)+B(Z_0)Z(I-C(Z_0)Z)^{-1}, \tag{1}$$

这里 $A(Z_0)\in\mathbb{C}^{m\times n}$, $B(Z_0)\in\mathbb{C}^{m\times m}$, $C(Z_0)\in\mathbb{C}^{n\times m}$ 使得

$$f(Z_0)=W_{Z_0}(0);\quad D_{Z_0}f(Z_0)=D_{Z_0}W_{Z_0}(0);\quad D^2_{Z_0}f(Z_0)=D^2_{Z_0}W_{Z_0}(0). \tag{2}$$

成立. 条件 (2) 导出

$$A(Z_0)=f(Z_0);\quad B(Z_0)Z_0=D_{Z_0}f(Z_0);\quad 2B(Z_0)Z_0C(Z_0)Z_0=D^2_{Z_0}f(Z_0).$$

由于 Z_0 的秩为 m, f 在 Ω 上局部双全纯，故 $D_{Z_0}f(Z_0)$ 的秩为 m. 将 $B(Z_0)$ 的元素看作独立变量，于是有 m^2 个独立变量，而 $B(Z_0)Z_0=D_{Z_0}f(Z_0)$ 有 mn 个方程，由于 $m<n$, 故方程个数大于独立变量的个数，一般来说，这样的方程是无解的. 也就是说，一般来说，不存在 (1) 这样的 Möbius 变换，使之满足 (2). 因之，不能用 (1) 来对 $f(Z)$ 进行二阶逼近.

来考虑另一种 Möbius 变换

$$W_{Z_0}(Z)=A(Z_0)+(I-ZC(Z_0))^{-1}ZB(Z_0), \tag{3}$$

这里 $A(Z_0)\in\mathbb{C}^{m\times n}$, $B(Z_0)\in\mathbb{C}^{n\times n}$, $C(Z_0)\in\mathbb{C}^{n\times m}$, 使得 (2) 成立，这时 (2) 导出

$$A(Z_0)=f(Z_0);\quad Z_0B(Z_0)=D_{Z_0}f(Z_0);\quad 2Z_0C(Z_0)Z_0B(Z_0)=D^2_{Z_0}f(Z_0). \tag{4}$$

将 $B(Z_0)$ 的元素看作独立变量，于是有 n^2 个独立变量，而 $Z_0B(Z_0)=D_{Z_0}f(Z_0)$ 有 mn 个方程. 由于 $m<n$, 故方程的个数小于独立变量个数. $D_{Z_0}f(Z_0)$ 的秩为 m. 因此, 存在无穷多个秩为 n 的 $B(Z_0)$, 满足 $Z_0B(Z_0)=D_{Z_0}f(Z_0)$. 取定了 $B(Z_0)$ 之后，考虑方程 $2Z_0C(Z_0)Z_0B(Z_0)=D^2_{Z_0}f(Z_0)$, 即 $2Z_0C(Z_0)D_{Z_0}f(Z_0)=D^2_{Z_0}f(Z_0)$. 将 $C(Z_0)\in\mathbb{C}^{n\times m}$ 的元素看作独立变量，于是有 mn 个方程，但是方程未必一定有解. 如果方程无解，则不能用 (3) 来对 $f(Z)$ 进行二阶逼近. 如果方程有解，则能用 (3) 来对 $f(Z)$ 进行二阶逼近，这时候可用二阶逼近来定义 Schwarz 导数及高阶 Schwarz 导数，并可证明，它们在 $R_I(m,n)$ 的紧对偶空间 $\mathbb{C}G(m,n)$ 的全纯自同构群作用下是相似不变的. 而且这时得到的 Schwarz 导数与文 [3] 中得到的 Schwarz 导数是相同的. 也就是说：在 $\mathbb{C}^{m\times n}$, $m<n$ 的情形，用 Thurston 的观点得到的 Schwarz 导数 (如果存在的话) 与用 Ahlfors 的观点得到的 Schwarz 导数 (如果存在的话) 是相一致的.

现在在假设方程 $2Z_0C(Z_0)D_{Z_0}f(Z_0)=D^2_{Z_0}f(Z_0)$ 有解的情形下来定义 f 在 Z_0 点的 Schwarz 导数及高阶 Schwarz 导数，并证明其相似不变性.

这时可以定义

$$W^{-1}_{Z_0}\circ f(Z)B(D_{Z_0}f(Z_0))^{-1}$$

$$= (f(Z) - A)(Cf(Z) - CA + B)^{-1}B(D_{Z_0}f(Z_0))^{-1}$$
$$= (f(Z) - f(Z_0))(B^{-1}C(f(Z) - f(Z_0)) + I)^{-1}(D_{Z_0}f(Z_0))^{-1}. \tag{5}$$

这里 $(D_{Z_0}f(Z_0))^{-1}$ 为 $D_{Z_0}f(Z_0)$ 广义逆. 取 $Z = Z_0 + tZ_0$, 则当 $|t|$ 充分小时, 有展开式

$$W_{Z_0}^{-1} \circ f(Z + tZ_0)B(D_{Z_0}f(Z_0))^{-1} = \sum_{k=0}^{\infty} S_k[f](Z_0)\frac{t^k}{k!}, \tag{6}$$

定义 $S_k[f](Z_0)$ 为 f 在 Z_0 点的 k 阶 Schwarz 导数, $k = 0, 1, 2, \ldots$, 而定义 3 阶 Schwarz 导数为 Schwarz 导数, 记作 $S[f](Z_0)$.

在 $\mathbb{C}^{mn}$, 当 $m = n$, 即 $\mathbb{C}^{n\times n}$ 的情形, 情况要简单的多. 这时方程 (2) 及 (4) 都有唯一解. 因此, 二种逼近都存在. 如用 (3) 进行二阶逼近, 就可用 (6) 来定义 Schwarz 导数及高阶 Schwarz 导数; 如用 (1) 来进行二阶逼近, 就用后面的 (25) 来定义 *-Schwarz 导数及高阶 *-Schwarz 导数, 这将在第四节中进行讨论.

3 相似不变性

在这一节中将证明:在上一节中所定义的 Schwarz 导数及高阶 Schwarz 导数, 都是在$R_I(m,n)$ 的紧对偶空间 Grassman 流形 $\mathbb{C}G(m,n)$ 的全纯自同构群作用下的相似不变量. 仍然假设 $m < n$.

$R_I(m,n)$ 的紧对偶空间 Grassman 流形 $\mathbb{C}G(m,n)$ 的全纯自同构群由

$$T(Z) = (PZ + Q)(RZ + S)^{-1}$$

且 $\det\begin{pmatrix} P & Q \\ R & S \end{pmatrix} = 1$ 所组成, 这里 $P \in \mathbb{C}^{m\times m}$, $Q \in \mathbb{C}^{m\times n}$, $R \in \mathbb{C}^{n\times m}$ 及 $S \in \mathbb{C}^{n\times n}$.

首先一个问题是: 对于 $(T\circ f)(Z)$, 在 Z_0 点的二阶逼近是否存在, 即能否找到

$$\widetilde{W}_{Z_0}(Z) = A_1(Z_0) + (I - ZC_1(Z_0))^{-1}ZB_1(Z_0), \tag{7}$$

这里 $A_1(Z_0) \in \mathbb{C}^{m\times n}$, $B_1(Z_0) \in \mathbb{C}^{n\times n}$, $C_1(Z_0) \in \mathbb{C}^{n\times m}$, 使得

$$(T\circ f)(Z_0) = \widetilde{W}_{Z_0}(0);\quad D_{Z_0}(T\circ f)(Z_0) = D_{Z_0}\widetilde{W}_{Z_0}(0);\quad D^2_{Z_0}(T\circ f)(Z_0) = D^2_{Z_0}\widetilde{W}_{Z_0}(0)$$

成立, 即

$$A_1 = (T\circ f)(Z_0),\quad Z_0B_1 = D_{Z_0}(T\circ f)(Z_0),\quad Z_0C_1Z_0B_1 = \frac{1}{2}D^2_{Z_0}(T\circ f)(Z_0),$$

的解 A_1, B_1, C_1 存在. 显然, 解 A_1, B_1 是存在的, 问题是解 C_1 是否存在. 下面要证明: 如果解 C 存在, 则解 C_1 也存在. 这时, (7) 成为 $(T\circ f)(Z)$ 在 Z_0 点的二阶逼近. 于是: 当 $|t|$ 充分小时, 有

$$\widetilde{W}_{Z_0}\circ(T\circ f)(Z_0 + tZ_0)B_1[D_{Z_0}(T\circ f)(Z_0)]^{-1} = \sum_{k=0}^{\infty} S_k[T\circ f](Z_0)\frac{t^k}{k!}. \tag{8}$$

而 $S_k[T\circ f](Z_0)$ 为 $(T\circ f)(Z)$ 在点 Z_0 处的 k 阶 Schwarz 导数, $k = 0, 1, \ldots$.

由于

$$(T\circ f)(Z)=(Pf(Z)+Q)(Rf(Z)+S)^{-1},$$

对上式求导，就有

$$\begin{aligned}D_{Z_0}(T\circ f)(Z)=&P(D_{Z_0}f(Z))(Rf(Z)+S)^{-1}\\&-(Pf(Z)+Q)(Rf(Z)+S)^{-1}R(D_{Z_0}f(Z))(Rf(Z)+S)^{-1}.\end{aligned}\tag{9}$$

要简化 (9) 式，就要用到以下的恒等式，这时不妨假设 S 非异 (如 S 奇异，则可用文 [3] 中的办法处理).

$$\begin{aligned}(PZ+Q)(RZ+S)^{-1}&=(PZ+Q)(S^{-1}RZ+I)^{-1}S^{-1}\\&=[(P-QS^{-1}R)Z+Q(S^{-1}RZ+I)](S^{-1}RZ+I)^{-1}S^{-1}\\&=(P-QS^{-1}R)Z(S^{-1}RZ+I)^{-1}S^{-1}+QS^{-1}\\&=(P-QS^{-1}R)(ZS^{-1}R+I)^{-1}ZS^{-1}+QS^{-1}.\end{aligned}\tag{10}$$

由文 [3] 知：如果 S 非异，则 $P-QS^{-1}R$ 也是非异的.

应用 (10) 于 (9), 得到

$$\begin{aligned}&D_{Z_0}(T\circ f)(Z)\\&=P(D_{Z_0}f(Z))(Rf(Z)+S)^{-1}-(P-QS^{-1}R)(f(Z)S^{-1}R+I)^{-1}f(Z)S^{-1}R(D_{Z_0}f(Z))\\&\quad\cdot(Rf(Z)+S)^{-1}-QS^{-1}R(D_{Z_0}f(Z))(Rf(Z)+S)^{-1}\\&=(P-QS^{-1}R)(D_{Z_0}f(Z))(Rf(Z)+S)^{-1}-(P-QS^{-1}R)(f(Z)S^{-1}R+I)^{-1}f(Z)\\&\quad\cdot S^{-1}R(D_{Z_0}f(Z))(Rf(Z)+S)^{-1}\\&=(P-QS^{-1}R)(f(Z)S^{-1}R+I)^{-1}(f(Z)S^{-1}R+I-f(Z)S^{-1}R)(D_{Z_0}f(Z))(Rf(Z)+S)^{-1}\\&=(P-QS^{-1}R)(f(Z)S^{-1}R+I)^{-1}(D_{Z_0}f(Z))(Rf(Z)+S)^{-1}.\end{aligned}\tag{11}$$

对 (11) 求导，可得

$$\begin{aligned}&D^2_{Z_0}(T\circ f)(Z)\\&=(P-QS^{-1}R)(f(Z)S^{-1}R+I)^{-1}(D^2_{Z_0}f(Z))(Rf(Z)-S)^{-1}-(P-QS^{-1}R)\\&\quad\cdot(f(Z)S^{-1}R+I)^{-1}(D_{Z_0}f(Z))S^{-1}R(f(Z)S^{-1}R+I)^{-1}(D_{Z_0}f(Z))(Rf(Z)+S)^{-1}\\&\quad-(P-QS^{-1}R)(f(Z)S^{-1}R+I)^{-1}(D_{Z_0}f(Z))(Rf(Z)+S)^{-1}R(D_{Z_0}f(Z))(Rf(Z)+S)^{-1}\\&=(P-QS^{-1}R)(f(Z)S^{-1}R+I)^{-1}(D^2_{Z_0}f(Z))(Rf(Z)+S)^{-1}-(P-QS^{-1}R)\\&\quad\cdot(f(Z)S^{-1}R+I)^{-1}(D_{Z_0}f(Z))(Rf(Z)+S)^{-1}\{(Rf(Z)+S)S^{-1}R+R(f(Z)S^{-1}R+I)\}\\&\quad\cdot(f(Z)S^{-1}R+I)^{-1}(D_{Z_0}f(Z))(Rf(Z)+S)^{-1}\\&=(P-QS^{-1}R)(f(Z)S^{-1}R+I)^{-1}(D^2_{Z_0}f(Z))(Rf(Z)+S)^{-1}-2(P-QS^{-1}R)\\&\quad\cdot(f(Z)S^{-1}R+I)^{-1}(D_{Z_0}f(Z))(Rf(Z)+S)^{-1}R(D_{Z_0}f(Z))(Rf(Z)+S)^{-1}.\end{aligned}\tag{12}$$

记

$$H=(P-QS^{-1}R)\big(f(Z_0)S^{-1}R+I\big)^{-1},\tag{13}$$

则 (11) 及 (12) 可以写成

$$D_{Z_0}(T\circ f)(Z_0)=H(D_{Z_0}f(Z_0))(Rf(Z_0)+S)^{-1},\tag{14}$$

$$\begin{aligned}D^2_{Z_0}(T\circ f)(Z_0)&=H\big(D^2_{Z_0}f(Z_0)\big)(Rf(Z_0))+S)^{-1}-2H(D_{Z_0}f(Z_0))\\&\quad\cdot(Rf(Z_0)+S)^{-1}R(D_{Z_0}f(Z_0))(Rf(Z_0)+S)^{-1}.\end{aligned}\tag{15}$$

由 (4), (13), (14) 及 (15), 可得

$$\begin{aligned}&\frac{1}{2}D^2_{Z_0}(T\circ f)(Z_0)\\&=HZ_0CZ_0B(Rf(Z_0)+S)^{-1}-HZ_0B(Rf(Z_0)+S)^{-1}RZ_0B(Rf(Z_0)+S)^{-1}\\&=(HZ_0CH^{-1}-HZ_0B(Rf(Z_0)+S)^{-1}RH^{-1})(D_{Z_0}(T\circ f)(Z_0)).\end{aligned}$$

另一方面，由

$$\frac{1}{2}D^2_{Z_0}(T\circ f)(Z_0)=Z_0C_1Z_0B_1=Z_0C_1(D_{Z_0}(T\circ f)(Z_0)),$$

可知，可取下列方程的解 C_1,

$$Z_0C_1=HZ_0\big(C-B(Rf(Z_0)+S)^{-1}R\big)H^{-1}.\tag{16}$$

将 C_1 的元素看作独立变量，于是有 mn 个变量， m^2 个方程，而 Z_0 的秩为 m, 故方程有解，且解不唯一. 因此，如果 $f(Z)$ 在 Z_0 点有二阶逼近，则 $T\circ f(Z)$ 在 Z_0 点也有二阶逼近 (7).

由 (13), (14) 及 (16) 就有

$$Z_0C_1=Z_0B_1B_1^{-1}C_1=D_{Z_0}(T\circ f)(Z_0)B_1^{-1}C_1,$$

及

$$\begin{aligned}Z_0C_1&=HZ_0B(Rf(Z_0)+S)^{-1}(Rf(Z_0)+S)B^{-1}CH^{-1}-HZ_0B(Rf(Z_0)+S)^{-1}RH^{-1}.\\&=D_{Z_0}(T\circ f)(Z_0)\big[(Rf(Z_0)+S)B^{-1}CH^{-1}-RH^{-1}\big].\end{aligned}$$

因之，可以选取 C_1, 为

$$C_1=B_1\big((Rf(Z_0)+S)B^{-1}CH^{-1}-RH^{-1}\big).\tag{17}$$

当 B,C 及 B_1 确定后， C_1 是唯一确定，且 (17) 的右边只依赖于 T 及 $f(Z_0)$.

现在就可以来计算 $W_{Z_0}^{-1}\circ T\circ f(Z)$.

由 (7), 可得

$$\begin{aligned}&\overline{W}_{Z_0}^{-1}\circ(T\circ f)(Z)\\&=((T\circ f)(Z)-(T\circ f)(Z_0))\big(I+B^{-1}C_1^{-1}((T\circ f)(Z)-(T\circ f)(Z_0))\big)^{-1}B_1^{-1}.\end{aligned}\tag{18}$$

由 T 的定义，

$$\begin{aligned}&((T\circ f)(Z)-(T\circ f)(Z_0))(Rf(Z)+S)\\&=[Pf(Z)+Q-(T\circ f)(Z_0)(Rf(Z)+S)]\\&=P(f(Z)-f(Z_0))-(T\circ f)(Z_0)R(f(Z)-f(Z_0))\\&=(P-(T\circ f)(Z_0)R)(f(Z)-f(Z_0)).\end{aligned}$$

由 (10), (13) 知

$$\begin{aligned}P-(T\circ f)(Z_0)R&=P-(Pf(Z_0)+Q)(Rf(Z_0)+S)^{-1}R\\&=P-(P-QS^{-1}R)\big(f(Z_0)S^{-1}R+I\big)^{-1}f(Z_0)S^{-1}R-QS^{-1}R\\&=(P-QS^{-1}R)\big[I-\big(f(Z_0)s^{-1}R+I\big)^{-1}f(Z_0)S^{-1}R\big]\\&=(P-QS^{-1}R)\big(I+f(Z_0)S^{-1}R\big)^{-1}\\&=H.\end{aligned}$$

因此，

$$((T\circ f)(Z)-(T\circ f)(Z_0))(Rf(Z)+S)=H(f(Z)-f(Z_0)). \tag{19}$$

但是由 (17), (19)

$$\begin{aligned}&(Rf(Z)+S)^{-1}\big(I+B_1^{-1}C_1((T\circ f)(Z)-(T\circ f)(Z_0))\big)^{-1}\\&=[Rf(Z)+S+B_1^{-1}C_1((T\circ f)(Z)-(T\circ f)(Z_0))(Rf(Z)+S)]^{-1}\\&=\big[Rf(Z)+S+B_1^{-1}C_1H(f(Z)-f(Z_0))\big]^{-1}\\&=\big[Rf(Z_0)+S+\big(R+B_1^{-1}C_1H\big)(f(Z)-f(Z_0))\big]^{-1}\\&=\big[Rf(Z_0)+S+(Rf(Z_0)+S)B^{-1}C(f(Z)-f(Z_0))\big]^{-1}\\&=\big[I+B^{-1}C(f(Z)-f(Z_0))\big]^{-1}(Rf(Z_0)+S)^{-1}.\end{aligned} \tag{20}$$

将 (19)、(20) 代入 (18)，得到

$$\begin{aligned}&\overline{W}_{Z_0}^{-1}\circ(T\circ f)(Z)\\&=H(f(Z)-f(Z_0))\big[I+B^{-1}C(f(Z)-f(Z_0))\big]^{-1}(Rf(Z_0)+S)^{-1}B_1^{-1}.\end{aligned}$$

再由 (14)

$$\begin{aligned}&\overline{W}_{Z_0}^{-1}\circ(T\circ f)(Z)B_1(D_{Z_0}(T\circ f)(Z_0))^{-1}\\&=H(f(Z)-f(Z_0))[I+B^{-1}C(f(Z)-f(Z_0))]^{-1}(Rf(Z_0)+S)^{-1}\\&\quad\cdot\big[H(D_{Z_0}f(Z_0))(Rf(Z_0)+S)^{-1}\big]^{-1}\\&=H\big(W_{Z_0}^{-1}\circ f(Z)\big)B(Rf(Z_0)+S)^{-1}\big[H(D_{Z_0}f(Z_0))(Rf(Z_0)+S)^{-1}\big]^{-1}.\end{aligned} \tag{21}$$

如假设，当 $|t|$ 充分小时，(4) 中以 Z_0+tZ_0 代替 Z_0 时恒有解，则有

$$2(Z_0+tZ_0)C(Z_0+tZ_0)D_{Z_0+tZ_0}f(Z_0+tZ_0)=D^2_{Z_0+tZ_0}f(Z_0+tZ_0).$$

上式两边在 $t=0$ 处展开，比较 t^k 的系数，$k=2,3,\ldots$, 就可以得到：

$$D^k_{Z_0}f(Z_0)=E_k(Z_0)D_{Z_0}f(Z_0),\quad k=2,3,\ldots, \tag{22}$$

这里 $E_k\in\mathbb{C}^{m\times m}$.

由于

$$\begin{aligned}&W^{-1}_{Z_0}\circ f(Z_0+tZ_0)B\\&=t\left(D_{Z_0}f(Z_0)+D^2_{Z_0}f(Z_0)\frac{t}{2}+D^3_{Z_0}f(Z_0)\frac{t^2}{6}+\cdots\right)\\&\cdot\left(I+B^{-1}CD_{Z_0}f(Z_0)t+B^{-1}CD^2_{Z_0}\frac{t^2}{2}+B^{-1}CD^3_{Z_0}f(Z_0)\frac{t^3}{6}+\cdots\right).\end{aligned}$$

故由 (22), $W^{-1}_{Z_0}\circ f(Z)B$ 可以写成 $U(Z_0)D_{Z_0}f(Z_0)$, 这里 $U(Z_0)\in\mathbb{C}^{m\times m}$. 因之,

$$W^{-1}_{Z_0}\circ f(Z)B=W^{-1}_{Z_0}f(Z)B(D_{Z_0}f(Z_0))^{-1}(D_{Z_0}f(Z_0)), \tag{23}$$

在 (21) 中，令 $Z=Z_0+tZ_0$, 并应用 (23) 就有

$$\begin{aligned}&\widetilde{W}^{-1}_{Z_0}\circ(T\circ f)(Z_0+tZ_0)B_1(D_{Z_0}(T\circ f)(Z_0))^{-1}\\&=H(W^{-1}_{Z_0}\circ f(Z_0+tZ_0))BD^{-1}_{Z_0}f(Z_0)H^{-1}HD_{Z_0}f(Z_0)(Rf(Z_0)+S)^{-1}\\&\quad\cdot[H(D_{Z_0}f(Z_0))(Rf(Z_0)+S)^{-1}]^{-1}\\&=H[W^{-1}_{Z_0}\circ f(Z_0+tZ_0)BD^{-1}_{Z_0}f(Z_0)]H^{-1}.\end{aligned}$$

由 (6), 就可以得到

$$S_k[T\circ f](Z_0)=HS_k[f](Z_0)H^{-1},\quad k=0,1,2,3,\ldots.$$

现在来具体计算由 (6) 所定义的 $S_k[F](Z_0)$.

显然有：$S_0[f](Z_0)=0$, $S_1[f](Z_0)=I$. 而

$$\begin{aligned}S_2[f](Z_0)&=(D^2_{Z_0}f(Z_0)-2D_{Z_0}f(Z_0)B^{-1}CD_{Z_0}f(Z_0))B^{-1}B(D_{Z_0}f(Z_0))^{-1}\\&=D^2_{Z_0}f(Z_0)(D_{Z_0}f(Z_0))^{-1}-2D_{Z_0}f(Z_0)B^{-1}C.\end{aligned}$$

由 (4) 可得

$$S_2[f](Z_0)=2Z_0C(Z_0)D_{Z_0}f(Z_0)(D_{Z_0}f(Z_0))^{-1}-2Z_0BB^{-1}C=0.$$

而再次应用 (4) 及 $S_2[f](Z_0)=0$, 可得

$$S_3[f](Z_0)=(D^3_{Z_0}f(Z_0)-3D^2_{Z_0}f(Z_0)B^{-1}CD_{Z_0}f(Z_0)+6D_{Z_0}f(Z_0)\cdot(B^{-1}CD_{Z_0}f(Z_0))^2$$

$$
\begin{aligned}
&\quad - 3D_{Z_0}f(Z_0)B^{-1}CD^2_{Z_0}f(Z_0))B^{-1}B(D_{Z_0}f(Z_0))^{-1}\\
&= D^3_{Z_0}f(Z_0)(D_{Z_0}f(Z_0))^{-1} - 3D^2_{Z_0}f(Z_0)B^{-1}C + 6D_{Z_0}f(Z_0)B^{-1}CD_{Z_0}f(Z_0)B^{-1}C\\
&\quad - 3D_{Z_0}f(Z_0)B^{-1}CD^2_{Z_0}f(Z_0)(D_{Z_0}f(Z_0))^{-1}\\
&= D^3_{Z_0}f(Z_0)(D_{Z_0}f(Z_0))^{-1} - 6Z_0CZ_0BB^{-1}C + 6Z_0BB^{-1}CZ_0BB^{-1}C - 6Z_0CZ_0C\\
&= D^3_{Z_0}f(Z_0)(D_{Z_0}f(Z_0))^{-1} - \frac{3}{2}(D^2_{Z_0}f(Z_0)(D_{Z_0}f(Z_0))^{-1})^2\\
&= \{f; Z_0\},
\end{aligned}
\tag{24}
$$

这里 $\{f; Z_0\}$ 就是文 [3] 中由 Ahlfors 观点得到的 Schwarz 导数 (如果存在的话).

因之，当 $m < n$ 时，从 Thurston 观点得到的 Schwarz 导数 $S[f](Z_0)$ (如果可以定义的话) 与从 Ahlfors 观点所得到的 Schwarz 导数 $\{f; Z_0\}$ (如果存在的话) 是相一致的. 用同样的方法，可以得出由 (6) 所定义的高阶 Schwarz 导数 $S_k[f](Z_0)$ $k = 4, 5, \ldots$ 的具体表达式. 这里不一一具体计算了.

于是就有

定理 1 若 $\Omega \subset \mathbb{C}^{m\times n}$ $(m < n)$ 为域，$f: \Omega \to \mathbb{C}^{m\times n}$ 为 Ω 上局部双全纯映照. $Z_0 \in \Omega$ 且非异，若存在正数 ϵ 充分小，使得 $|t| < \epsilon$ 时，方程 (4) 中以 $Z_0 + tZ_0$ 代替 Z_0 时，恒有解. 则由 (6) 所定义的 k 阶 Schwarz 导数，是在 $R_I(m, n)$ 的紧对偶空间 $\mathbb{C}G(m, n)$ 的全纯自同构群作用下的相似不变量，$k = 0, 1, 2, \ldots$ 且 $S_3[f](Z_0) = S[f](Z_0) = \{f; Z_0\}$. 这里 $\{f; Z_0\}$ 为文 [3] 中所定义的 Schwarz 导数.

4 $\mathbb{C}^{n\times n}$ 的情形

以上讨论了 $\mathbb{C}^{m\times n}$ $(m < n)$ 的情形. 如果 $m = n$, 即 $\mathbb{C}^{n\times n}$ 的情形，则比起 $m < n$ 的情形要简单得多. 这时候方程 (2), (4) 都有唯一解. 因此，也可以像上一节中那样进行计算，而定理 1 中的条件已无必要. 因之，对于 $\Omega \subset \mathbb{C}^{n\times n}$ 上的局部双全纯映照 $f: \Omega \to \mathbb{C}$ $(n \times n)$, 在非异点 $Z_0 \in \Omega$, 即可以用 (1) 来进行二阶逼近，也可以用 (3) 来进行二阶逼近，从前面的计算，可得，如用 (3) 来进行二阶逼近，这时候 $Z_0^{-1} = B(D_{Z_0}f(Z_0))^{-1}$, 故由 (6) 所得到的 Schwarz 导数 $S[f](Z_0) = S_3[f](Z_0)$ 就是在文 [3] 中由 Ahlfors 观点所得到的 Schwarz 导数 $\{f; Z_0\}$. 即这时候，从二种不同观点得到的 Schwarz 导数是相一致的.

如果用 (1) 作为 $f(Z)$ 在 Z_0 点的二阶逼近，定义

$$
Z_0^{-1}W_0^{-1} \circ f(Z_0 + tZ_0) = \sum_{k=0}^{\infty} S_k^*[f](Z_0)\frac{t^k}{k!}. \tag{25}
$$

中的 $S_k^*[f](Z_0)$ 为 f 在 Z_0 点的 k 阶 $*$-Schwarz 导数，$k = 0, 1, 2, \ldots$, 特别称 3 阶 $*$-Schwarz 导数为 Schwarz 导数，记作 $S^*[f](Z_0)$ 同上面一样，可以证明 $S_k^*[f](Z_0)$, $k = 0, 1, 2, \ldots$ 都在 $R_I(n, n)$ 的紧对偶空间 $\mathbb{C}G(n, n)$ 的全纯自同构群作用下的相似不变量.

具体计算可得，

$$
\begin{aligned}
&S_0^*[f](Z_0) = 0, \quad S_1^*[f](Z_0) = I, \quad S_2^*[f](Z_0) = 0,\\
&S_3[f](Z_0) = S^*[f](Z_0) = (D_{Z_0}f(Z_0))^{-1}\{f; Z_0\}(D_{Z_0}f(Z_0)). \ldots.
\end{aligned}
$$

即由 (25) 所定义的 Schwarz 导数 $S^*[f](Z_0)$ 与文 [3] 中 Ahlfors 观点定义的 Schwarz 导数 $\{f;Z_0\}$ 是相似的.

于是有

定理 2 若 $\Omega\in\mathbb{C}^{n\times n}$ 为域, $f:\Omega\to\mathbb{C}^{n\times n}$ 为局部双全纯映照, $Z_0\in\Omega$ 且非异. 则由 (6) 所定义的 k 阶 Schwarz 导数及由 (25) 所定义的 k 阶 $*$-Schwarz 导数, 都是在 $R_I(n,n)$ 的紧对偶空间 $\mathbb{C}G(n,n)$ 的全纯自同构群作用下的相似不变量. 特别, Schwarz 导数 $S[f](Z_0)=S_3[f](Z_0)=\{f;Z_0\}$, 以及 $*$-Schwarz 导数 $S^*[f](Z_0)=S^*[f](Z_0)=(D_{Z_0}f(Z_0))^{-1}\{f;Z_0\}\ (D_{Z0}f(Z_0))$, 即 $S^*[f](Z_0)$ 与 $\{f;Z_0\}$ 是相似的. 这里 $\{f;Z_0\}$ 为文 [3] 中所定义的 Schwarz 导数.

5 $\mathbb{C}^N$ 的情形

从上面对 $\mathbb{C}^{n\times n}$ 中域上局部双全纯映照的 Schwarz 导数及高阶 Schwarz 导数的定义及讨论, 可以导出另一种对 $\mathbb{C}^N$ 种的域上双全纯映照的 Schwarz 导数及高阶 Schwarz 导数的定义.

熟知, 对于任意正整数 N, 一定可以写成不超过四个平方数之和, 即一定存在一个或二个或三个或四个正整数 n_1,n_2,n_3,n_4, 使得 $N=n_1^2+n_2^2+n_3^2+n_4^2$ (参考华罗庚 [12]). 总可以将 $\mathbb{C}^N$ 看成一个矩阵空间. 例如, 若 N 可以写成四个平方数之和时, $\mathbb{C}^N$ 可看成下面的矩阵空间:

$$\mathbb{C}^N=\mathbb{C}^{n_1\times n_1}\times\mathbb{C}^{n_2\times n_2}\times\mathbb{C}^{n_3\times n_3}\times\mathbb{C}^{n_4\times n_4}.$$

而 $\mathbb{C}^N$ 中的一点 Z 可以表成如下的准对角阵:

$$Z=\begin{pmatrix}Z_1&0&0&0\\0&Z_2&0&0\\0&0&Z_3&0\\0&0&0&Z_4\end{pmatrix}=\mathrm{diag}\,(Z_1,Z_2,Z_3,Z_4).\tag{26}$$

这里 $Z_i\in\mathbb{C}^{n_i\times n_i}$, $i=1,2,3,4$. 记 G_i 为 $R_I(n_i,n_i)$ 的紧对偶空间 $\mathbb{C}G(n_i,n_i)$ 的全纯自同构群, $i=1,2,3,4$. 令

$$G=G_1\times G_2\times G_3\times G_4,$$

则 G 是 $R_I(n_1,n_1)\times R_I(n_2,n_2)\times R_I(n_3,n_3)\times R_I(n_4,n_4)$ 的紧对偶空间的全纯自同构群. 若

$$T=\begin{pmatrix}P&Q\\R&S\end{pmatrix}\in G.\tag{27}$$

则有 $\det T\neq0$, 且 P,Q,R 及 $S\in\mathbb{C}^{n_1\times n_1}\times\mathbb{C}^{n_2\times n_2}\times\mathbb{C}^{n_3\times n_3}\times\mathbb{C}^{n_4\times n_4}$, 即

$$P=\mathrm{diag}\,(P_1,P_2,P_3,P_4),\quad Q=\mathrm{diag}\,(Q_1,Q_2,Q_3,Q_4),$$
$$R=\mathrm{diag}\,(R_1,R_2,R_3,R_4),\quad S=\mathrm{diag}\,(R_1,R_2,R_3,R_4).$$

这里 $\begin{pmatrix}P_i&Q_i\\R_i&S_i\end{pmatrix}\in G_i$, $i=1,2,3,4$ 且 $P_i,Q_i,R_i,S_i\in\mathbb{C}^{n_i\times n_i}$, $i=1,2,3,4$.

由 (27) 定义的 T 作用于由 (26) 定义的 Z 为

$$T(Z)=(PZ+Q)(RZ+S)^{-1}=\mathrm{diag}\,(T_1(Z),T_2(Z),T_3(Z),T_4(Z))$$

$$= \mathrm{diag}\,\big((P_1Z_1+Q_1)(R_1Z_1+S_1)^{-1},(P_2Z_2+Q_2)(R_2Z_2+S_2)^{-1},$$
$$(P_3Z_3+Q_3)(R_3Z_3+S_3)^{-1},(P_4Z_4+Q_4)(R_4Z_4+S_4)^{-1}\big).$$

其中 $T_k=(P_kZ_k+Q_k)(R_kZ_k+S_k^{-1})$, $k=1,2,3,4$. 于是 G 仍可视为 $\mathbb{C}^N$ 上的一个 Möbius 变换群. $T\in G$ 称为 Möbius 变换.

若 $\Omega\subset\mathbb{C}^N$ 为域，$0\in\Omega$, $f(Z):\Omega\to\mathbb{C}^N$ 为 Ω 上的局部双全纯映照，将 Z,f 表成 (26) 的形式，且取 $Z_0\in\mathbb{C}^N$, 使得用 (26) 表示时，矩阵非异. 于是可以考虑 $f(Z)$ 在 Z_0 点的二阶逼近，即考虑

$$W_{Z_0}(Z)=A+(I-ZC)^{-1}ZB.$$

且要求 A,B,C 满足：

$$A=f(Z_0),\quad Z_0B=D_{Z_0}f(Z_0),\quad Z_0CZ_0B=\frac{1}{2}D_{Z_0}^2f(Z_0).$$

这里 $A,B,C,D,Z\in\mathbb{C}^{n_1\times n_1}\times\cdots\times\mathbb{C}^{n_4\times n_4}$. 由于 $f(Z)$ 在 Ω 上是局部双全纯映照，故 $J_f(Z_0)$ 非异. 于是就可以考虑

$$\begin{aligned}W_{Z_0}^{-1}\circ f(Z)&=(f(Z)-A)(Cf(Z)-CA-B)^{-1}\\&=(f(Z)-f(Z_0))\big(I+B^{-1}C(f(Z)-f(Z_0))\big)^{-1}B^{-1}.\end{aligned}$$

由于 Z_0 非异，故有：当 $|t|$ 充分小时，

$$W_{Z_0}^{-1}\circ f(Z_0+tZ_0)Z_0=\sum_{k=0}^{\infty}S_k[f](Z_0)\frac{t^k}{k!}.\tag{28}$$

$S_k[f]f(Z_0)$ 定义为 $f(z)$ 在点 Z_0 处的 k 阶 Schwarz 导数，$k=0,1,2,\ldots$. 而定义 $S_3[f](Z_0)$ 为 $f(Z)$ 在点 Z_0 处的 Schwarz 导数. 由前面的讨论可知，这样定义的 $S_k[f](Z_0)$ 是在 G 作用下的相似不变量.

当然，这样定义的 k 阶 Schwarz 导数是依赖于 $N=n_1^2+n_2^2+n_3^2+n_4^2$ 的分拆. 如果分拆不止一种，则这样定义的 k 阶 Schwarz 导数也不止一种. 直接计算，由 (28) 所定义的 $S_k[f](Z_0)$ 为：

$$S_0[f](Z_0)=0,\quad S_1[f](Z_0)=I,\quad S_2[f](Z_0)=0,$$

以及

$$S[f](Z_0)=S_3[f](Z_0)=D_{Z_0}^3f(Z_0)(D_{Z_0})^{-1}-\frac{3}{2}\big(D_{Z_0}^2f(Z_0)(D_{Z_0}f(Z_0))^{-1}\big)^2,$$
$$\cdots\cdots\cdots\cdots$$

于是可以利用文 [3] 中的方法证明 $S[f](Z)=0$ 对所有 $Z\in\Omega$ 都成立，当且仅当

$$f(Z)=\mathrm{diag}\,(F_1(Z),F_2(Z),F_3(Z),F_4(Z)),$$

这里 $F_i(Z)=(A_i(Z)+B_i)(C_i(Z)+D_i)^{-1}$ 或是 $(C_i(Z)+D_i)^{-1}(A_i(Z)+B_i)$, 而 $A_i,B_i,C_i,D_i\in\mathbb{C}^{n_i\times n_i}$, $A_i(Z)$ 及 $C_i(Z)$ 为 Z 的元素的一次齐次式，$i=1,2,3,4$. 如果 N 不能表成四个平方数

之和，则 N 可表成一个平方数，二个平方数之和或三个平方数之和. 则上述讨论依然可以进行，只是 $\mathbb{C}^N=\mathbb{C}^{n_1\times n_1}$ (已讨论过)，$\mathbb{C}^N=\mathbb{C}^{n_1\times n_1}\times\mathbb{C}^{n_2^n\times n_2}$，或是 $\mathbb{C}^N=\mathbb{C}^{n_1\times n_1}\times\mathbb{C}^{n_2^2\times n_2}\times\mathbb{C}^{n_3\times n_3}$. 可以得到相应的结果.

更一般地，如将 N 分拆成 $N=m_1\times n_1+\cdots+m_s\times n_s$，则 $\mathbb{C}^N$ 表成 $\mathbb{C}^{m_1\times n_1}\times\cdots\times\mathbb{C}^{m_s\times n_s}$，而 $\mathbb{C}^N$ 中的一点 Z 可以表成

$$Z=\begin{pmatrix} Z_1 & & 0 \\ & \ddots & \\ 0 & & Z_s \end{pmatrix},$$

这里 $Z_i\in\mathbb{C}^{m_i\times n_i}$，$i=1,2,\ldots,s$，而上述结果的讨论一样可以进行，并可得到相应的结果. 不在此一一叙述了.

参 考 文 献

1 Gong S, FitzGerald C H. The Schwarzian derivatives in several complex variables. Science in China, Series A, 1993, 36(5): 513-523

2 Gong S, Yu Q H, Zheng X A. The Schwarzian derivatives in several complex variables (II). Chinese Annals of Math, 1998, 19B(1): 1-8

3 龚升, 郑学安, 余其煌. 多复变数的 Schwarz 导数 (III). 中国科学, A 辑, 1997, 27(11): 454-461

4 龚升, 余其煌, 郑学安. 多复变数的 Schwarz 导数 (IV). 中国科学, 1998, 28(1): 14-23

5 Thurston W P. Zippers and univalent Functions. The Bieberbach Conjucture, A Baernstein II et al ed, Amer Math Soc, 1986, 185-192

6 Matio O, Sarvas J. Injectivity theorems in plane and space. Ann Acad Sci Fenn. A1, Math (1978/1979), 253-257

7 Osgood B, Stove D. The Schwarzian derivative and conformal mappings of Riemannian manifolds. Duke Math J, 1992, 67(1): 57-99

8 Osgood B, Stove D. A generalization of Nehari's univalence criterion. Comment Math Helv, 1990, 65(2): 234-242

9 Carne K. The Shwarzian derivative for conformal maps. J Raine Angew Math, 1990, 408(1): 10-33

10 Molzon R, Tamanoi H. Generalized Schwarzian in several variables and Möbius invariant differential operators. IHES, preprint, M/94/64, Dec 1994

11 Tamanoi H. Higher Schwarzian operators and combinations of Schwarzian derviative. IHES, preprint, M/94/67, Dec 1994.

12 华罗庚. 数论导引. 科学出版社, 1975

Vol. 38 No. 9 SCIENCE IN CHINA (Series A) September 1995

The Schwarzian derivative in Kähler manifolds*

GONG Sheng (龚 昇)

(Department of Mathematics, University of Science and Technology of China, Hefei 230026, China)

and YU Qihuang (余其煌)

(Institute of Applied Mathematics, Chinese Academy of Sciences, Beijing 100080, China)

Received February 2, 1995

Abstract Let f be a holomorphic immersion which maps a Kähler manifold into a Kähler manifold of the same dimension. A Schwarzian derivative S_f of f is proposed. It is proved that: i) if $S_f=0$ and $S_g=0$, then $S_{f\cdot g}=0$; ii) if the real part of the Schwarzian derivative of f on a convex domain in the Käler manifold is bounded above, then f is an embedding. The upper bound is related to the holomorphic sectional curvature of the domain. This second theorem is an extension of Nehari's criterion of univalence.

Keywords: Schwarzian derivative, Kähler manifold, holomorphic mapping.

The Schwarzian derivative of a holomorphic function plays a very important role in one complex variable. With the expectation that a similar concept would have importance in several complex variables, various generalizations have been proposed and studied. Certain of these generalizations are related to our approach. We divide them according to the class of mappings on which the Schwarzian derivative is defined:

i) Osgood, Stove[1, 2] and Carne[3] extended the Schwarzian derivative to apply it to conformal mappings between Riemannian manifolds.

ii) FitzGerald, Gong[4] and Gong, Yu[1)] extended the Schwarzian derivative to apply it to holomorphic mappings in higher dimension Euclidean space.

Motivated by this work, we define a Schwarzian derivative on holomorphic immersions between two Kähler manifolds. The two theorems of the paper are properties of the Schwarzian related to the properties of the standard Schwarzian. The first theorem examines the Schwarzian of the composition of mappings. The second is a generalization of Nehari's univalence criterion.

The summation convention will be used throughout this paper.

---

* Project supported by the National Natural Science Foundation of China.

1) Yu Qihuang, Gong Sheng, The Schwarzian derivative in $\mathbb{C}^n$, to appear in *Chinese Annals of Math*.

1 Definition and theorems

Let M, N, together with $ds_M^2=2g_{i\bar{j}}dz^i\overline{dz^j}$, $ds_N^2=2h_{\alpha\bar{\beta}}dw^\alpha\overline{dw^\beta}$ respectively, be two Kähler manifolds of the same dimension, whose corresponding (1, 0) connections are ∇ and $\widetilde{\nabla}$ respectively. Let f: $M\to N$ be a holomorphic immersion. Suppose that $\chi(M)$ is the set of $\mathbb{C}^\infty$ vector fields of (1, 0) type defined on M. We define the derivative of f and the derivatives of higher order of f by

$$Df:\ \chi(M)\to\chi(N),$$

$$Df(X)=f_*X, \tag{1}$$

where f_* is the differential of f, i.e.

$$f_*:\ T_xM\to T_{f(x)}N,$$

$$f_*\xi^i\frac{\partial}{\partial z^i}=\xi^i\frac{\partial w^\alpha}{\partial z^i}\frac{\partial}{\partial w^\alpha}, \tag{2}$$

$$D^2f:\ \chi(M)\times\chi(M)\to\chi(N), \tag{3}$$

$$D^2f(X,\ Y)=\widetilde{\nabla}_{f_*X}f^*Y-f^*\nabla_XY, \tag{4}$$

$$D^3f:\ \chi(M)\times\chi(M)\times\chi(M)\to\chi(N), \tag{5}$$

$$D^3f(X,\ Y,\ Z)=\widetilde{\nabla}_{f_*X}D^2f(Y,\ Z)-D^2f(Y,\ \nabla_XZ)-D^2f(\nabla_XY,\ Z). \tag{6}$$

Obviously,

$$D^2f(Y,\ X)=D^2f(X,\ Y), \tag{7}$$

and

$$D^3f(X,\ Y,\ Z)=D^3f(X,\ Z,\ Y). \tag{8}$$

Hence D^2f is bilinear and symmetric with respect to X, Y, and D^3f is trilinear with respect to X, Y, Z, and symmetric with respect to Y, Z.

Since f: $M\to N$ is an immersion, for every $z\in M$, there exists a neighborhood U_z of z, such that $f|_{U_z}$ is injective and its inverse $f|_{U_z}^{-1}$ is holomorphic. Since there will be no confusion, we use f^{-1} instead of $f|_{U_z}^{-1}$. Similarly, we will use f_*^{-1} to mean the differential of $f|_{U_z}^{-1}$.

Now, we define the Schwarzian derivative Sf_z of the immersion f: $M\to N$ by

$$S_f=Sf=Sf_z:\ \chi(M)\times\chi(M)\to\chi(N),$$

where

$$Sf(X,\ Y)=Sf_z(X,\ Y)$$

$$=\langle X,\ Y\rangle_M f_*^{-1}D^3f(X,\ X,\ X)-\frac{3}{2}\langle f_*^{-1}D^2f(X,\ X),\ Y\rangle_M f_*^{-1}D^2f(X,\ X). \tag{9}$$

Clearly $Sf_z(X, Y)$ is linear with respect to Y for all Kähler manifolds M and N.

If $N=M=\mathbb{C}^n$, and M, N have the Euclidean metric, let

$$X=\xi^i\frac{\partial}{\partial z^i},\quad Y=\eta^i\frac{\partial}{\partial z^i}.$$

Then (9) becomes

$$2Sf_z(X,\ Y)=\xi^i\eta^{-i}\xi^j\xi^k\xi^l\frac{\partial^3 w^\alpha}{\partial z^i\partial z^j\partial z^l}\frac{\partial z^m}{\partial\omega^\alpha}\frac{\partial}{\partial z^m}$$

$$-\frac{3}{2}\xi^i\xi^j\frac{\partial^2 w^\alpha}{\partial z^i\partial z^j}\frac{\partial z^k}{\partial\omega^\alpha}\eta^{-k}\times\xi^l\xi^m\frac{\partial^2 w^\beta}{\partial z^l\partial z^m}\frac{\partial z^s}{\partial w^\beta}\frac{\partial}{\partial z^s}.\tag{10}$$

When $n=1$, (10) is equal to

$$2Sf_z(X,\ Y)=\xi^4\bar{\eta}\left(\frac{f'''}{f'}-\frac{3}{2}\left(\frac{f''}{f'}\right)^2\right)\frac{\partial}{\partial z}.\tag{11}$$

The coefficient of the $\dfrac{\partial}{\partial z}$ is the standard Schwarzian derivative of holomorphic functions of one complex variable timing an explicit nonzero factor.

Now let $M=B^n$ with Bergman metric $ds_M^2=4\dfrac{(1-|z|^2)\delta_{ij}+\bar{z}^iz^j}{(1-|z|^2)^2}dz^id\bar{z}^j$, where B^n is the unit ball in $\mathbb{C}^n$, and $N=\mathbb{C}^n$. Then the Schwarzian derivative of the holomorphic immersion f: $M\to N$ can be written as

$$Sf_z(X,\ Y)=\langle X,\ Y\rangle_{B^n}\xi^i\xi^j\xi^k\frac{\partial^3 w^\alpha}{\partial z^i\partial z^j\partial z^k}\frac{\partial z^l}{\partial w^\alpha}\frac{\partial}{\partial z^l}$$

$$-\frac{3}{2}\xi^i\xi^j\frac{\partial^2 w^\alpha}{\partial z^i\partial z^j}\frac{\partial z^l}{\partial w^\alpha}\bar{\eta}^s\left\langle\frac{\partial}{\partial z^l},\ \frac{\partial}{\partial z^s}\right\rangle_{B_n}\xi^k\xi^m\frac{\partial^2 w^\beta}{\partial z^k\partial z^m}\frac{\partial z^s}{\partial w^\beta}\frac{\partial}{\partial z^s}-\frac{3\xi^i\bar{z}^i}{1-|z|^2}\times$$

$$\left[\langle X,\ Y\rangle_{B^n}\xi^i\xi^j\frac{\partial^2 w^\alpha}{\partial z^i\partial z^j}\frac{\partial z^k}{\partial w^\alpha}\frac{\partial}{\partial z^k}-\xi^i\xi^j\frac{\partial^2 w^\alpha}{\partial z^i\partial z^j}\frac{\partial z^l}{\partial w^\alpha}\bar{\eta}^s\left\langle\frac{\partial}{\partial z^l},\ \frac{\partial}{\partial z^s}\right\rangle_{B^n}X\right].\tag{12}$$

In particular, if $n=1$, (12) is equal to

$$Sf_z(X,\ Y)=\frac{2\xi^4\bar{\eta}}{(1-|z|^2)^2}\left(\frac{f'''}{f'}-\frac{3}{2}\left(\frac{f''}{f'}\right)^2\right)\frac{\partial}{\partial z}.\tag{13}$$

The coefficient of the $\dfrac{\partial}{\partial z}$ is the Schwarzian derivative of holomorphic functions of one complex variable timing an explicit nonzero factor which is different from the previous case.

In this paper, we shall prove the following two theorems.

Theorem 1. *Let M, N, H be Kähler manifolds with complex dimension of n, and f: $M \to N$, g: $N \to H$ be holomorphic immersions. If the equation*

$$Sf_z(X, Y)=0, \quad Sg_w(\tilde{X}, \tilde{Y})=0$$

holds for every $z \in M$, and every X, $Y \in \chi(M)$; and for every $w \in N$, and every $\tilde{X}$, $\tilde{Y} \in \chi(N)$, then for every $z \in M$, and every X, $Y \in \chi(M)$,

$$Sg \circ f_z(X, Y)=0.$$

As a consequence of Theorem 1, we have

Corollary 1. *Let M be a Kähler manifolds,* $\mathrm{Aut}(M)$ *be the holomorphic automorphism group of M, and*

$$\text{Möb}(M)=\{f \in \mathrm{Aut}(M):\ Sf_z(X, Y) \equiv 0,\ \forall z \in M,\ \forall X,\ Y \in \chi(M)\}.$$

Then Möb (M) is the subgroup of $\mathrm{Aut}(M)$.

Theorem 2 (Injective Criterion). *Let M be a complete Kähler manifold of n-dimension, $\Omega \subseteq M$ be a convex domain, whose holomorphic sectional curvature is bounded above by K. And suppose that there exists a number δ, $0<\delta \leqslant \infty$, such that the distance of every two points of Ω is less than δ, and f: $\Omega \to \mathbb{C}^n$ is a holomorphic immersion. If the inequality*

$$\mathrm{Re}\frac{\langle Sf_z(X, Y), X\rangle}{\langle X, X\rangle^2\langle X, Y\rangle} \leqslant \frac{4\pi^2}{\delta^2}-K$$

holds for every $z \in \Omega$, X, $Y \in \chi(\Omega)$, then f is an embedding.

2 Proof of Theorem 1

Let M, N, H be Kähler manifolds, with ∇, $\tilde{\nabla}$, $\tilde{\tilde{\nabla}}$ as corresponding connections of (1, 0)-type respectively. Suppose that

$$\begin{array}{ccccc} M & \xrightarrow{f} & N & \xrightarrow{g} & H \\ z & \mapsto & w & \mapsto & \xi \end{array}$$

are holomophic immersions.

We now calculate the derivatives of the composition.

By definition (4), if X, Y, $Z \in \chi(M)$, then

$$D^2 g \circ f(X, Y)=\tilde{\tilde{\nabla}}_{g_* f_* X} g \circ f_* Y-g \circ f_* \nabla_X Y$$

$$
\begin{aligned}
&=\tilde{\tilde{\nabla}}_{g\cdot f\cdot X} g\cdot f_* Y - g_*\tilde{\nabla}_{f\cdot X} f_* Y + g_*\tilde{\nabla}_{f\cdot X} f_* Y - g_*(f_*\nabla_X Y)\\
&= D^2 g(f_* X,\ f_* Y) + g_* D^2 f(X,\ Y).
\end{aligned}
\tag{14}
$$

Using (14) and definition (6), we get

$$
\begin{aligned}
& D^3 g\circ f(X,\ Y,\ Z)\\
&=\tilde{\tilde{\nabla}}_{g_*(f_* X)} D^2 g\circ f(Y,\ Z) - D^2 g\circ f(Y,\ \nabla_X Z) - D^2 g\circ f(\nabla_X Y,\ Z)\\
&=\tilde{\tilde{\nabla}}_{g\cdot f\cdot X}(D^2 g(f_* Y,\ f_* Z) + g_* D^2 f(Y,\ Z)) - D^2 g(f_* Y,\ f_*\nabla_X Z) - g_* D^2 f(Y,\ \nabla_X Z)\\
&\quad - D^2 g(f_*\nabla_X Y,\ f_* Z) - g_* D^2 f(\nabla_X Y,\ Z)\\
&=\tilde{\tilde{\nabla}}_{g_*(f_* X)} D^2 g(f_* Y,\ f_* Z) - D^2 g(f_* Y,\ \tilde{\nabla}_{f_* X} f_* Z) - D^2 g(\tilde{\nabla}_{f_* X} f_* Y,\ f_* Z)\\
&\quad + D^2 g(f_* Y,\ \tilde{\nabla}_{f_* X} f_* Z) - D^2 g(f_* Y,\ f_*\nabla_X Z) + D^2 g(\tilde{\nabla}_{f_* X} f_* Y,\ f_* Z)\\
&\quad - D^2 g(f_*\nabla_X Y,\ f_* Z) + \tilde{\tilde{\nabla}}_{g\cdot f\cdot X} g_* D^2 f(Y,\ Z) - g_* D^2 f(Y,\ \nabla_X Z)\\
&\quad - g_* D^2 f(\nabla_X Y,\ Z)\\
&= D^3 g(f_* X,\ f_* Y,\ f_* Z) + D^2 g(f_* Y,\ D^2 f(X,\ Z)) + D^2 g(f_* Z,\ D^2 f(X,\ Y))\\
&\quad + \tilde{\tilde{\nabla}}_{g\cdot f\cdot X} g_* D^2 f(Y,\ Z) - g_*\tilde{\nabla}_{f_* X} D^2 f(Y,\ Z) + g_*\tilde{\nabla}_{f_* X} D^2 f(Y,\ Z)\\
&\quad - g_* D^2 f(Y,\ \nabla_X Z) - g_* D^2 f(\nabla_X Y,\ Z)\\
&= D^3 g(f_* X,\ f_* Y,\ f_* Z) + D^2 g(f_* Y,\ D^2 f(X,\ Z)) + D^2 g(f_* Z,\ D^2 f(X,\ Y))\\
&\quad + D^2 g(f_* X,\ D^2 f(Y,\ Z)) + g_*\tilde{\nabla}_{f_* X} D^2 f(Y,\ Z) - g_* D^2 f(Y,\ \nabla_X Z)\\
&\quad - g_* D^2 f(\nabla_X Y,\ Z)\\
&= D^3 g(f_* X,\ f_* Y,\ f_* Z) + g_* D^3 f(X,\ Y,\ Z) + D^2 g(f_* X,\ D^2 f(Y,\ Z))\\
&\quad + D^2 g(f_* Y,\ D^2 f(Z,\ X)) + D^2 g(f_* Z,\ D^2 f(X,\ Y)).
\end{aligned}
\tag{15}
$$

Substituting (14) and (15) into the expression of $Sg\circ f_z$, we have

$$
\begin{aligned}
Sg\circ f_z(X,\ Y) &= \langle X,\ Y\rangle_M f_*^{-1} g_*^{-1}[D^3 g(f_* X,\ f_* X,\ f_* X) + g_* D^3 f(X,\ X,\ X)\\
&\quad + 3D^2 g(f_* X,\ D^2 f(X,\ X))] - \frac{3}{2}\langle f_*^{-1} g_*^{-1}[D^2 g(f_* X,\ f_* X) + g_* D^2 f(X,\ X)];\ Y\rangle\\
&\quad \times f_*^{-1} g_*^{-1}[D^2 g(f_* X,\ f_* X) + g_* D^2 f(X,\ X)].
\end{aligned}
\tag{16}
$$

Now we are going to prove Theorem 1.

Let X be fixed, and $\langle X,\ Y\rangle_M=0$. Since $Sf_z(X,\ Y)=0$, we have

$$\langle f_*^{-1} D^2 f(X,\ X),\ Y\rangle_M = 0,$$

so that

$$D^2 f(X,\ X) = a f_* X, \tag{17}$$

where a is a complex-valued function of z which is related to X.

By (17), for any $Y\in\chi(M)$, we get

$$\langle f_*^{-1} D^2 f(X,\ X),\ Y\rangle_M f_*^{-1} D^2 f(X,\ X) = \langle X,\ Y\rangle_M a^2 X. \tag{18}$$

By assumption

$$0=sf_z(X, Y)=\langle X, Y\rangle_M f_*^{-1}D^3f(X, X, X)$$
$$-\frac{3}{2}\langle f_*^{-1}D^2f(X, X), Y\rangle f_*^{-1}D^2f(X, X)$$

and from (18), we get

$$f_*^{-1}D^3f(X, X, X)=\frac{3}{2}a^2X. \tag{19}$$

Similarly we can verify that

$$D^2g(f_*X, f_*X)=bg_*f_*X \tag{20}$$

and

$$g_*^{-1}D^3g(f_*X, f_*X, f_*X)=\frac{3}{2}b^2f_*X. \tag{21}$$

Substituting (17)—(21) into (16), we find

$$Sg\cdot f_z(X, Y)=\langle X, Y\rangle\left[\frac{3}{2}b^2X+\frac{3}{2}a^2X+3f_*^{-1}g_*^{-1}D^2g(f_*X, af_*X)\right]$$
$$-\frac{3}{2}\langle(a+b)X, Y\rangle(a+b)X$$
$$=\langle X, Y\rangle\left[\frac{3}{2}(a^2+b^2)X+3af_*^{-1}g_*^{-1}D^2g(f_*X, f_*X)\right.$$
$$\left.-\frac{3}{2}(a+b)^2X\right]$$
$$=\langle X, Y\rangle\left[\frac{3}{2}(a^2+b^2)+3ab-\frac{3}{2}(a+b)^2\right]X$$
$$=0.$$

Thus we have proved Theorem 1.

In order to prove the corollary, it is sufficient to prove $f^{-1}\in \text{Möb}(M)$. By (14), we know

$$0=D^2f^{-1}\cdot f(X, X)=D^2f^{-1}(f_*X, f_*X)+f_*^{-1}D^2f(X, X)$$
$$=D^2f^{-1}(f_*X, f_*X)+f_*^{-1}af_*X=D^2f^{-1}(f_*X, f_*X)+aX,$$

and hence we have the equation

$$D^2f^{-1}(f_*X, f_*X)=-aX. \tag{22}$$

By (15), (19) and (22), we have

$$0=D^3f^{-1}\cdot f(X, X, X)=D^3f^{-1}(f_*X, f_*X, f_*X)$$
$$+f_*^{-1}D^3f(X, X, X)+3D^2f^{-1}(f_*X, D^2f(X, X))$$
$$=D^3f^{-1}(f_*X, f_*X, f_*X)+\frac{3}{2}a^2X+3D^2f^{-1}(f_*X, af_*X)$$
$$=D^3f^{-1}(f_*X, f_*X, f_*X)+\frac{3}{2}a^2X-3a^2X,$$

i.e.

$$D^3 f^{-1}(f_* X, f_* X, f_* X) = \frac{3}{2} a^2 X.$$

Substituting these results into the expression of Sf_z^{-1}, we get

$$Sf_z^{-1}(X, Y) = 0.$$

We have proved the corollary.

3 Proof of Theorem 2

In order to prove Theorem 2, we need the following lemma.

Lemma 1. *Let $p(t)$ be a real continuous function on $[0, T]$ $(T>0)$, and the differential inequality*

$$\frac{d^2 u}{dt^2} + p(t) u \leqslant 0 \tag{23}$$

have a positive solution $u(t)>0$, where $t \in [0, T]$. Let q be a complexed-valued function q: $[0, T] \to \mathbb{C}$ such that

$$\mathrm{Re} q(t) \leqslant p(t), \qquad t \in [0, T]. \tag{24}$$

Then every nontrivial solution of the differential equation

$$\frac{d^2 w}{dt^2} + q(t) w = 0 \tag{25}$$

has at most one zero point in $[0, T]$.

Proof. Let w be a nontrivial solution with at least two zero points t_1, t_2, and $0 \leqslant t_1 < t_2 \leqslant T$. Since $u>0$, then from (23)—(25), we know

$$\frac{d}{dt} \mathrm{Re}\left[\frac{dw}{dt} \bar{w} - \frac{1}{u} \frac{du}{dt} |w|^2 \right] \geqslant |w|^2 (p(t) - \mathrm{Re} q(t)) + \left| \frac{dw}{dt} - \frac{w}{u} \frac{du}{dt} \right|^2. \tag{26}$$

Integrating the inequality (26) from t_1 to t_2, we get

$$0 \geqslant \int_{t_1}^{t_2} |w|^2 (p(t) - \mathrm{Re} q(t)) dt + \int_{t_1}^{t_2} \left| \frac{dw}{dt} - \frac{w}{u} \frac{du}{dt} \right|^2 dt \geqslant 0.$$

Since each integral equals zero, it follows that

$$\frac{dw}{dt} - \frac{w}{u} \frac{du}{dt} = 0, \quad \text{a.e. } t \in (t_1, t_2);$$

that is,

$$\frac{d}{dt}\left(\frac{w}{u}\right)=\frac{1}{u}\left(\frac{dw}{dt}-\frac{w}{u}\frac{dw}{dt}\right)=0, \quad \text{a.e.} \quad t\in(t_1, t_2).$$

Then we have $w(t)=cu(t)$, where $t\in(t_1, t_2)$ and c is a non-zero constant. By the continuity of $\dot{u}$, we get $w(t)=cu(t)$, for $t\in[t_1, t_2]$. While $w(t_2)=0$, which contradicts $u(t_2)\neq 0$, we have completed the proof.

Now we are going to prove Theorem 2.

Let

$$ds^2_M=2g_{i\bar{j}}dz^i d\bar{z}^j. \tag{27}$$

In the proof, we will use the symbol $\langle,\rangle$ instead of the symbol $\langle,\rangle_M$. First, we assert that for the differential vector field X of (1, 0) type in a neighborhood of z, there always exists a $\mathbb{C}^\infty$ complex-valued function a defined in a neighborhood of z which is related to z such that

$$D^2f(X, X)=af_*X. \tag{28}$$

In fact, for the fixed X, let $X_1, X_2, \cdots, X_n$ be the orthonormal basis of the inner product space composed of all the (1, 0) vectors, and $X_1=\frac{X}{|X|}$. Then we have

$$f_*^{-1}D^2f(X, X)=a^\alpha X_\alpha. \tag{29}$$

Let

$$Y_k=\frac{1}{k}X_1-\bar{a}^1\left(\sum_{\alpha=2}^n \bar{a}^\alpha X_\alpha\right). \tag{30}$$

Then

$$\begin{aligned}
&\mathrm{Re}\frac{\langle Sf_z(X, Y_k), X\rangle}{\langle X, X\rangle^2\langle X, Y_k\rangle}\\
&=\mathrm{Re}\left\{\frac{1}{\langle X, X\rangle^2}\langle f_*^{-1}D^3f(X, X, X), Y_k\rangle\right\}\\
&\quad-\frac{3}{2}\mathrm{Re}\left\{\frac{k}{\langle X, X\rangle^2}\left(\frac{a^1}{k}-\bar{a}^1\sum_{\alpha=2}^n|\bar{a}^\alpha|^2\right)a^1\right\}\\
&=\mathrm{Re}\left\{\frac{1}{\langle X, X\rangle^2}\langle f_*^{-1}D^3f(X, X, X), Y_k\rangle\right\}\\
&\quad-\frac{3}{2}\mathrm{Re}\left\{\frac{(a^1)^2}{\langle X, X\rangle^2}-\frac{k}{\langle X, X\rangle^2}|a^1|^2\sum_{\alpha=2}^n|a^\alpha|^2\right\}.
\end{aligned}$$

Each of $a^2, a^3, \cdots, a^n$ is 0. Suppose that it was not true. As k tends to $+\infty$, the right-hand side of the previous equation tends to $+\infty$. This would contradict the assumption of Theorem 2.

Suppose that the theorem is not true, i.e. there exists $z_1, z_2 \in \Omega$ such that $z_1 \neq z_2$ and $f(z_1)=f(z_2)$. Since M is complete and Ω is a convex domain, there exists a minimum normal geodesic that connects z_1 and z_2,

$$\gamma:[0, T] \to \Omega, \ T \leqslant \delta.$$

Let

$$X_0 = \dot{\gamma}(t) = \gamma_* \frac{\partial}{\partial t} \tag{31}$$

be the real tangent vector (along γ) of $\gamma(t)$, and

$$X = X_0 - iJX_0, \ \overline{X} = X_0 + iJX_0, \tag{32}$$

$$X_0 = \frac{1}{2}(X+\overline{X}), \tag{33}$$

where J is the complex structure of M, and X is the (1, 0) vector field along $\dot{\gamma}(t)$. Also we shall denote any $\mathbb{C}^\infty$ extension of X in the neighborhood of $\gamma(t)$ by X. By the previous assertion, we have

$$D^2 f(X, X) = af_* X. \tag{34}$$

Let us calculate $D^3 f(X, X, X)$. By definition, we know

$$D^3 f(X, X, X) = \widetilde{\nabla}_{f_* X} D^2 f(X, X) - 2D^2 f(X, \nabla_X X), \tag{35}$$

where $\widetilde{\nabla}$ is the flat connection of $\mathbb{C}^n$. From (34), we get

$$\begin{aligned} \widetilde{\nabla}_{f_* X} D^2 f(X, X) &= \widetilde{\nabla}_{f_* X} af_* X = (Xa) f_* X + a\widetilde{\nabla}_{f_* X} f_* X \\ &= (Xa) f_* X + a(\widetilde{\nabla}_{f_* X} f_* X - f_* \nabla_X X) + af_* \nabla_X X \\ &= (Xa) f_* X + af_* \nabla_X X + a^2 f_* X, \end{aligned} \tag{36}$$

and

$$\frac{da}{dt} = \gamma_* \frac{\partial}{\partial t} a = X_0 a = \frac{1}{2}(X+\overline{X})a = \frac{1}{2}Xa + \frac{1}{2}\overline{X}a,$$

$$Xa = 2\frac{da}{dt} - \overline{X}a. \tag{37}$$

Combining (37) with (36), we have

$$\widetilde{\nabla}_{f_* X} D^2 f(X, X) = \left(2\frac{da}{dt} - \overline{X}a + a^2\right) f_* X + af_* \nabla_X X. \tag{38}$$

Moreover, recalling that $\gamma(t)$ is the normal geodesic curve, i.e. $\nabla_{X_0} X_0 = 0$, we have

$$0 = \nabla_{\frac{1}{2}(X+\overline{X})} \frac{1}{2}(X+\overline{X}) = \frac{1}{4}(\nabla_X X + \nabla_{\overline{X}} X + \nabla_X \overline{X} + \nabla_{\overline{X}} \overline{X}).$$

Noticing that $\nabla_X X + \nabla_{\overline{X}} X$ is a (1, 0) vector, while $\nabla_X \overline{X} + \nabla_{\overline{X}} \overline{X}$ is a (0. 1) vector, we find

$$\nabla_X X+\nabla_{\overline{X}} X=0,$$

or

$$\nabla_X X=-\nabla_{\overline{X}} X. \tag{39}$$

By (39), this implies

$$-D^2 f(X, \nabla_X X)=D^2 f(X, \nabla_{\overline{X}} X). \tag{40}$$

In order to calculate $D^2 f(X, \nabla_{\overline{X}} X)$, we shall adopt the local coordinates, with $X=\xi^i \frac{\partial}{\partial z^i}$, $\overline{X}=\overline{\xi}^i \frac{\partial}{\partial \overline{z}^i}$, and then

$$D^2 f(X, \nabla_{\overline{X}} X)=D^2 f\left(\xi^i \frac{\partial}{\partial z^i}, \nabla_{\overline{\xi}^j \frac{\partial}{\partial \overline{z}^j}} \xi^k \frac{\partial}{\partial z^k}\right)=\xi^i \overline{\xi}^j D^2 f\left(\frac{\partial}{\partial z^i}, \nabla_{\frac{\partial}{\partial \overline{z}^j}} \xi^k \frac{\partial}{\partial z^k}\right)$$

$$=\xi^i \overline{\xi}^j D^2 f\left(\frac{\partial}{\partial z^i}, \frac{\partial \xi^k}{\partial \overline{z}^j} \frac{\partial}{\partial z^k}+\xi^k \nabla_{\frac{\partial}{\partial \overline{z}^j}} \frac{\partial}{\partial z^k}\right). \tag{41}$$

Since ∇ is a (1, 0) connection, i.e. $\nabla_{\frac{\partial}{\partial \overline{z}^j}} \frac{\partial}{\partial z^k}=0$, we get

$$D^2 f(X, \nabla_{\overline{X}} X)=\xi^i \overline{\xi}^j \frac{\partial \xi^k}{\partial \overline{z}^j} D^2 f\left(\frac{\partial}{\partial z^i}, \frac{\partial}{\partial z^k}\right). \tag{42}$$

Substituting (38), (42) into (35), we obtain the equation

$$D^3 f(X, X, X)=\left(2 \frac{\mathrm{d} a}{\mathrm{d} t}-\overline{X} a+a^2\right) f_* X+a f_* \nabla_X X+2 \xi^i \overline{\xi}^j \frac{\partial \xi^k}{\partial \overline{z}^j} D^2 f\left(\frac{\partial}{\partial z^i}, \frac{\partial}{\partial z^k}\right).$$

It follows that

$$f_*^{-1} D^3 f(X, X, X)=\left(2 \frac{\mathrm{d} a}{\mathrm{d} t}-\overline{X} a+a^2\right) X+a \nabla_X X+2 \xi^i \overline{\xi}^j \frac{\partial \xi^k}{\partial \overline{z}^j} f_*^{-1} D^2 f\left(\frac{\partial}{\partial z^i}, \frac{\partial}{\partial z^k}\right). \tag{43}$$

From the definition of $D^2 f$ and (34),

$$a X=f_*^{-1} \widetilde{\nabla}_{f_* X} f_* X-\nabla_X X=f_*^{-1} D^2 f(X, X). \tag{44}$$

We co-differentiate the left-hand side of (44) along X to obtain

$$\nabla_{\overline{X}} a X=(\overline{X} a) X+a \nabla_{\overline{X}} X=(\overline{X} a) X-a \nabla_X X \tag{45}$$

by (39). We co-differentiate the right-hand side of (44) along X to obtain

$$\nabla_{\overline{X}} f_*^{-1} \widetilde{\nabla}_{f_* X} f_* X-\nabla_{\overline{X}} \nabla_X X=\nabla_{\overline{\xi}^j \frac{\partial}{\partial \overline{z}^j}} f_*^{-1} D^2 f\left(\xi^i \frac{\partial}{\partial z^i}, \xi^k \frac{\partial}{\partial z^k}\right)$$

$$=\overline{\xi}^j \widetilde{\nabla}_{\frac{\partial}{\partial \overline{z}^j}} f_*^{-1} \xi^i \xi^k D^2 f\left(\frac{\partial}{\partial z^i}, \frac{\partial}{\partial z^k}\right)$$

$$=\overline{\xi}^j\frac{\partial(\xi^i\xi^k)}{\partial\overline{z}^j}f_*^{-1}D^2f\left(\frac{\partial}{\partial z^i},\ \frac{\partial}{\partial z^k}\right)+\xi^i\overline{\xi}^j\xi^k\nabla_{\frac{\partial}{\partial\overline{z}^j}}f_*^{-1}D^2f\left(\frac{\partial}{\partial z^i},\ \frac{\partial}{\partial z^k}\right)$$

$$=2\xi^i\overline{\xi}^j\frac{\partial\xi^k}{\partial\overline{z}^j}f_*^{-1}D^2f\left(\frac{\partial}{\partial z^i},\ \frac{\partial}{\partial z^k}\right)+\xi^i\overline{\xi}^j\xi^k\nabla_{\frac{\partial}{\partial\overline{z}^j}}f_*^{-1}D^2f\left(\frac{\partial}{\partial z^i},\ \frac{\partial}{\partial z^k}\right). \tag{46}$$

We note that $\nabla_{\frac{\partial}{\partial\overline{z}^j}}\frac{\partial}{\partial z^k}=\Gamma^l_{ik}\frac{\partial}{\partial z^l}$, where Γ^l_{ik} are the coefficients of the connection. Also f is holomorphic. And ∇ is of (1, 0) type. From these facts

$$\xi^i\overline{\xi}^j\xi^k\nabla_{\frac{\partial}{\partial\overline{z}^j}}f_*^{-1}D^2f\left(\frac{\partial}{\partial z^i},\ \frac{\partial}{\partial z^k}\right)$$

$$=\xi^i\overline{\xi}^j\xi^k\nabla_{\frac{\partial}{\partial\overline{z}^j}}f_*^{-1}\left(\widetilde{\nabla}_{f_*\frac{\partial}{\partial z^i}}f_*\frac{\partial}{\partial z^k}-f_*\nabla_{\frac{\partial}{\partial z^i}}\frac{\partial}{\partial z^k}\right)$$

$$=\xi^i\overline{\xi}^j\xi^k\nabla_{\frac{\partial}{\partial\overline{z}^j}}f_*^{-1}\left(\frac{\partial\omega^\alpha}{\partial z^i\partial z^k}\frac{\partial}{\partial\omega^\alpha}-f_*\Gamma^l_{ki}\frac{\partial}{\partial z^l}\right)$$

$$=\xi^i\overline{\xi}^j\xi^k\left(\nabla_{\frac{\partial}{\partial\overline{z}^j}}\frac{\partial\omega^\alpha}{\partial z^i\partial z^k}\frac{\partial z^l}{\partial\omega^\alpha}\frac{\partial}{\partial z^l}-\nabla_{\frac{\partial}{\partial\overline{z}^j}}\Gamma^l_{ki}\frac{\partial}{\partial z^l}\right)$$

$$=\xi^i\overline{\xi}^j\xi^kS^l_{ik\overline{j}}\frac{\partial}{\partial z^l}, \tag{47}$$

where $S^l_{ik\overline{j}}$ is the curvature tensor.

Combining (47) with (46), we find

$$\nabla_{\overline{X}}f_*^{-1}D^2f(X,\ X)=2\xi^i\overline{\xi}^j\frac{\partial\xi^k}{\partial\overline{z}^j}f_*^{-1}D^2f\left(\frac{\partial}{\partial z^i},\ \frac{\partial}{\partial z^k}\right)+\xi^i\overline{\xi}^j\xi^kS^l_{ik\overline{j}}\frac{\partial}{\partial z^l}. \tag{48}$$

Comparing (45) with (48), we get

$$(\overline{X}a)X-a\nabla_XX=2\xi^i\overline{\xi}^j\frac{\partial\xi^k}{\partial\overline{z}^j}f_*^{-1}D^2f\left(\frac{\partial}{\partial z^i},\ \frac{\partial}{\partial z^k}\right)+\xi^i\overline{\xi}^j\xi^kS^l_{ik\overline{j}}\frac{\partial}{\partial z^l},$$

thus

$$\langle(\overline{X}a)X,\ X\rangle=\left\langle2\xi^i\overline{\xi}^j\frac{\partial\xi^k}{\partial\overline{z}^j}f_*^{-1}D^2f\left(\frac{\partial}{\partial z^j},\ \frac{\partial}{\partial z^k}\right),\ X\right\rangle$$

$$+a\langle\nabla_XX,\ X\rangle+\xi^i\overline{\xi}^j\xi^k\overline{\xi}^lS_{i\overline{j}k\overline{l}}. \tag{49}$$

Since $\gamma(t)$ is a normal geodesic curve, we have

$$I=\langle X_0,\ X_0\rangle=\frac{1}{4}\langle X,\ X\rangle+\frac{1}{4}\langle\overline{X},\ \overline{X}\rangle=\frac{1}{2}\langle X,\ X\rangle,$$

and hence

$$2=\langle X, X\rangle=\xi^i\bar{\xi}^j g_{i\bar{j}}. \tag{50}$$

From (49) and (50), we know

$$2(\bar{X}a)=\langle L, X\rangle+a\langle\nabla_X X, X\rangle+\xi^i\bar{\xi}^j\xi^k\bar{\xi}^l S_{i\bar{j}k\bar{l}}, \tag{51}$$

where

$$L=2\xi^i\bar{\xi}^j\frac{\partial\xi^k}{\partial\bar{z}^j}f_*^{-1}D^2f\left(\frac{\partial}{\partial z^i}, \frac{\partial}{\partial z^k}\right). \tag{52}$$

By (43) and (34),

$$\begin{aligned}Sf_z(X, Y)&=\langle X, Y\rangle\left[\left(2\frac{da}{dt}-\bar{X}a+a^2\right)X+a\nabla_X X+L\right]-\frac{3}{2}\langle X, Y\rangle a^2X\\&=\langle X, Y\rangle\left[\left(2\frac{da}{dt}-\frac{1}{2}a^2\right)X-(\bar{X}a)X+a\nabla_X X+L\right].\end{aligned} \tag{53}$$

Thus, it follows that

$$\langle Sf_z(X, Y), X\rangle=\langle X, Y\rangle\left[2\left(2\frac{da}{dt}-\frac{1}{2}a^2\right)-2(\bar{X}a)+\langle a\nabla_X X+L, X\rangle\right]. \tag{54}$$

By (51) and (54), it is easy to verify that

$$\begin{aligned}&\langle Sf_z(X, Y), X\rangle\\&=\langle X, Y\rangle\left[2\left(2\frac{da}{dt}-\frac{1}{2}a^2\right)-\langle L, X\rangle-a\langle\nabla_X X, X\rangle\right]\\&\quad-\langle X, Y\rangle[\xi^i\bar{\xi}^j\xi^k\bar{\xi}^l S_{i\bar{j}k\bar{l}}+a\langle\nabla_X X, X\rangle+\langle L, X\rangle]\\&=\langle X, Y\rangle\left[2\left(2\frac{da}{dt}-\frac{1}{2}a^2\right)-\xi^i\bar{\xi}^j\xi^k\bar{\xi}^l S_{i\bar{j}k\bar{l}}\right].\end{aligned} \tag{55}$$

By the assumption of Theorem 2 that the holomorphic sectional curvature is bounded above by K, we know

$$\frac{\xi^i\bar{\xi}^j\xi^k\bar{\xi}^l S_{i\bar{j}k\bar{l}}}{(\xi^i\bar{\xi}^j g_{i\bar{j}})^2}\leqslant K,$$

and by (50), the following inequality is true:

$$\xi^i\bar{\xi}^j\xi^k\bar{\xi}^l S_{i\bar{j}k\bar{l}}\leqslant 4K. \tag{56}$$

By the assumption of Theorem 2 of the real part of the Schwarzian being bounded above, we have

$$\frac{4\pi^2}{\delta^2}-K\geqslant\operatorname{Re}\frac{\langle Sf_z(X, Y), X\rangle}{\langle X, X\rangle^2\langle X, Y\rangle}=\operatorname{Re}\frac{2\left(2\frac{da}{dt}-\frac{1}{2}a^2\right)-\xi^i\bar{\xi}^j\xi^k\bar{\xi}^l S_{i\bar{j}k\bar{l}}}{\langle X, X\rangle^2}$$

$$=\frac{1}{2}\mathrm{Re}\left(2\frac{\mathrm{d}a}{\mathrm{d}t}-\frac{1}{2}a^2\right)-\frac{1}{4}\xi^i\overline{\xi^j}\xi^k\overline{\xi^l}S_{i\bar{j}k\bar{l}}\geqslant\frac{1}{2}\mathrm{Re}\left(2\frac{\mathrm{d}a}{\mathrm{d}t}-\frac{1}{2}a^2\right)-K,$$

and hence

$$\frac{1}{8}\mathrm{Re}\left(2\frac{\mathrm{d}a}{\mathrm{d}t}-\frac{1}{2}a^2\right)\leqslant\frac{\pi^2}{\delta^2}. \tag{57}$$

Since $[0, T]$ is compact, from (57) we know that $T<\delta$ and that there exists $\varepsilon>0$, such that

$$\frac{1}{8}\mathrm{Re}\left(2\frac{\mathrm{d}a}{\mathrm{d}t}-\frac{1}{2}a^2\right)\leqslant\frac{\pi^2}{(T+2\varepsilon)^2}. \tag{58}$$

Now we set

$$g^\alpha(t)=f^\alpha\circ\gamma(t)-f^\alpha\circ\gamma(0),\ \alpha=1,\ 2,\cdots,n. \tag{59}$$

Then

$$g^\alpha(0)=g^\alpha(T)=0, \tag{60}$$

and

$$\frac{\mathrm{d}g^\alpha}{\mathrm{d}t}=X_0g^\alpha=\frac{1}{2}(X+\overline{X})g^\alpha=\frac{1}{2}Xg^\alpha=\frac{1}{2}f_*Xg^\alpha,$$

because f is a holomorphic mapping, and

$$\frac{\mathrm{d}^2g^\alpha}{\mathrm{d}t^2}=X_0(X_0g^\alpha)=X_0\left(\frac{1}{2}Xg^\alpha\right)=\frac{1}{4}(X+\overline{X})(Xg^\alpha)=\frac{1}{4}X(Xg^\alpha)+\frac{1}{4}\overline{X}(Xg^\alpha). \tag{61}$$

On the other hand,

$$\begin{aligned}D^2f(X,\ X)g^\alpha&=(\widetilde{\nabla}_{f_*X}f_*X)g^\alpha-(f_*\nabla_XX)g^\alpha\\&=(\widetilde{\nabla}_{f_*X}f_*X)g^\alpha+(f_*\nabla_{\overline{X}}X)g^\alpha=X(Xg^\alpha)+\overline{X}(Xg^\alpha).\end{aligned} \tag{62}$$

Comparing (61) with (62) we see that

$$\frac{\mathrm{d}^2g^\alpha}{\mathrm{d}t^2}=\frac{1}{4}D^2f(X,\ X)g^\alpha=\frac{1}{4}a(f_*Xg^\alpha), \tag{63}$$

so that

$$\frac{\mathrm{d}^2g^\alpha}{\mathrm{d}t^2}-\frac{a\mathrm{d}g^\alpha}{2\mathrm{d}t}=0. \tag{64}$$

Let

$$v^\alpha=g^\alpha\mathrm{e}^{-\frac{1}{4}\int_0^t a(s)\mathrm{d}s}.$$

Then (64) is the same as

$$\frac{\mathrm{d}^2v^\alpha}{\mathrm{d}t^2}+\frac{1}{8}\left(2\frac{\mathrm{d}a}{\mathrm{d}t}-\frac{1}{2}a^2\right)v^\alpha=0. \tag{65}$$

Of course, $u=\sin\frac{\varepsilon+t}{T+2\varepsilon}\pi$ is a solution of the equation

$$\frac{d^2u}{dt^2}+\frac{\pi^2}{(T+2\varepsilon)^2}u=0, \quad 0\leqslant t\leqslant\pi$$

and satisfies

$$u(t)>0, \quad 0\leqslant t\leqslant\pi.$$

Then by Lemma 1, we have that $v^{\alpha}(t)$ must be the trivial solution,

$$v^{\alpha}(t)\equiv 0, \quad \alpha=1, 2,\cdots n.$$

That is,

$$f\cdot\gamma(t)=\text{const},$$

which contradicts the fact that f is an immersion. We have completed the proof.

The following Corollary 2 is just the Nehari's theorem.

Corellary 2. *Assume that f: $D=\{z\in\mathbb{C}: |z|<1\}\to\mathbb{C}$, and f is locally biholomorphic. If the inequality*

$$\text{(a) } \operatorname{Re}\frac{\xi^2}{|\xi|^2}\left(\frac{f'''}{f'}-\frac{3}{2}\left(\frac{f''}{f'}\right)^2\right)\leqslant\frac{\pi^2}{2}, \text{ or}$$

$$\text{(b) } \operatorname{Re}\frac{\xi^2}{|\xi|^2}\left(\frac{f'''}{f'}-\frac{3}{2}\left(\frac{f''}{f'}\right)^2\right)\leqslant\frac{2}{(1-|z|^2)^2}$$

holds, then f is biholomorphic.

Proof. (a) We apply the Euclidean metric in D, then $K=0$ and $\delta=2$. Noticing $\langle\frac{\partial}{\partial z}, \frac{\partial}{\partial z}\rangle=\frac{1}{2}$, we have

$$\operatorname{Re}\frac{\langle Sf_z(X, Y), X\rangle}{\langle X, X\rangle^2\langle X, Y\rangle}=\operatorname{Re}\frac{\frac{1}{4}\xi^3|\xi|^2\bar{\eta}}{\frac{1}{8}|\xi|^4\xi\bar{\eta}}\left(\frac{f'''}{f'}-\frac{3}{2}\left(\frac{f''}{f'}\right)^2\right)\leqslant\frac{4\pi^2}{2^2}.$$

By Theorem 2, f is a holomorphic embedding, i.e. biholomorphic.

(b) We adopt the Poincaré metric $ds^2=\frac{4dzd\bar{z}}{(1-|z|^2)^2}$, then $\delta=\infty$, $K=-1$, and $\langle\frac{\partial}{\partial z}, \frac{\partial}{\partial z}\rangle=\frac{2}{(1-|z|^2)^2}$, then

$$\operatorname{Re}\frac{\langle Sf_z(X,\ Y),\ X\rangle}{\langle X,\ X\rangle^4\langle X,\ Y\rangle}=\operatorname{Re}\frac{\dfrac{4}{(1-|z|^2)^4}\xi^3|\xi|^2\bar{\eta}}{\dfrac{8}{(1-|z|^2)^6}|\xi|^4\xi\bar{\eta}}\left(\frac{f'''}{f'}-\frac{3}{2}\left(\frac{f''}{f'}\right)^2\right)\leqslant 1.$$

By Theorem 2, f is biholomorphic.

We have proved Corollary 2.

Remark 1. Condition (a) is equivalent to the following condition

$$\text{(c)}\quad\left|\left(\frac{f'''}{f'}-\frac{3}{2}\left(\frac{f''}{f'}\right)^2\right)\right|\leqslant\frac{\pi^2}{2}.$$

1) $(c)\Rightarrow(a)$. Actually,

$$\operatorname{Re}\frac{\xi^2}{|\xi|^2}\left(\frac{f'''}{f'}-\frac{3}{2}\left(\frac{f''}{f'}\right)^2\right)\leqslant\left|\frac{\xi^2}{|\xi|^2}\left(\frac{f'''}{f'}-\frac{3}{2}\left(\frac{f''}{f'}\right)^2\right)\right|\leqslant\left|\frac{f'''}{f'}-\frac{3}{2}\left(\frac{f''}{f'}\right)^2\right|\leqslant\frac{\pi^2}{2}.$$

2) $(a)\Rightarrow(c)$. If not, i.e. there exists z_0 such that $\left|\frac{f'''}{f'}-\frac{3}{2}\left(\frac{f''}{f'}\right)^2\right|>\frac{\pi^2}{2}$. Since (a) is true for all ξ, if

$$\xi^2=\overline{\frac{f'''}{f'}(z_0)-\frac{3}{2}\left(\frac{f''}{f'}(z_0)\right)^2},$$

then

$$\frac{\pi^2}{2}\geqslant\frac{\left|\dfrac{f'''}{f'}(z_0)-\dfrac{3}{2}\left(\dfrac{f''}{f'}(z_0)\right)^2\right|^2}{\left|\dfrac{f'''}{f'}(z_0)-\dfrac{3}{2}\left(\dfrac{f''}{f'}(z_0)\right)^2\right|}=\left|\frac{f'''}{f'}(z_0)-\frac{3}{2}\left(\frac{f''}{f'}(z_0)\right)^2\right|,$$

which is a contradiction.

Remark 2. In a similar fashion, we assert that condition (b) and the following condition (d) are equivalent.

$$\text{(d)}\quad\left|\frac{f'''}{f'}(z)-\frac{3}{2}\left(\frac{f''}{f'}(z)\right)^2\right|\leqslant\frac{2}{(1-|z|^2)^2}.$$

Corollary 3. *Assume that f: $D=\{z\in\mathbb{C}:\ |z|<1\}\to\mathbb{C}$ and f is locally biholomorphic. If the inequality*

$$\left|\left(\frac{f'''}{f'}-\frac{3}{2}\left(\frac{f''}{f'}\right)^2\right)\right|\leqslant\frac{6}{(1+|z|^2)^2}$$

holds, then f is biholomorphic.

Proof. Consider D as a convex domain in CP^1 with metric $\frac{4dzd\bar{z}}{(1+|z|^2)^2}$, the cuvature $k=1$, and $\delta=\pi$. By Theorem 2, if the following inequality holds, then Corollary 3 will be true:

$$\mathrm{Re}\frac{\langle Sf_z(X,Y),X\rangle}{\langle X,X\rangle^2\langle X,Y\rangle}=\mathrm{Re}\frac{\frac{4}{(1+|z|^2)^4}\xi^3|\xi|^2\bar{\eta}}{\frac{8}{(1+|z|^2)^6}|\xi|^4\xi\bar{\eta}}\left(\frac{f'''}{f'}-\frac{3}{2}\left(\frac{f''}{f'}\right)^2\right)\leqslant 3,$$

$$\mathrm{Re}\frac{\xi^3|\xi|^2\bar{\eta}}{|\xi|^4\xi\bar{\eta}}\left(\frac{f'''}{f'}-\frac{3}{2}\left(\frac{f''}{f'}\right)^2\right)\leqslant\frac{6}{(1+|z|^2)^2}. \tag{66}$$

In fact, Corollary 3 follows from inequality (66).

$$\mathrm{Re}\frac{\xi^3|\xi|^2\bar{\eta}}{|\xi|^4\xi\bar{\eta}}\left(\frac{f'''}{f'}-\frac{3}{2}\left(\frac{f''}{f'}\right)^2\right)\leqslant\left|\left(\frac{f'''}{f'}-\frac{3}{2}\left(\frac{f''}{f'}\right)^2\right)\right|\leqslant\frac{6}{(1+|z|^2)^2}.$$

References

1 Osgood, B., Stove, D., The Schwarzian derivative and conformal mapping of Riemannian manifolds, *Duke Math. J.*, 1992, 67: 57.

2 Osgood, B., Stove, D., A generalization of Nehari's univalence criterion, *Comment. Math. Helv.*, 1990, 65: 234.

3 Carne, K., The Schwarzian derivative for conformal mappings, *J. reine angew. Math.*, 1990, 408: 10.

4 Gong Sheng, FitzGerald, C. H., The Schwarzian derivative in several complex variables, *Science in China*, Ser. A, 1993, 36: 513.

Vol. 40 No. 1 SCIENCE IN CHINA (Series A) January 1997

Schwarzian derivative in Kähler manifolds (II) *

GONG Sheng (龚 升),

(Department of Mathematics, University of Science and Technology of China, Hefei 230026, China)

YU Qihuang (余其煌) and NA Jisheng (那吉生)

(Institute of Applied Mathematics, Chinese Academy of Sciences, Beijing 100080, China)

Received July 4, 1996

Abstract Two injectivity theorems are proved which are for objective manifold $N=\mathbb{CP}^n$ and $N=B^n$ (unit ball), both equipped with canonical metric. Thus, the injectivity theorems are established in three cases of objective manifolds, i.e. with different holomorphic sectional curvatures -1, 0 and 1 respectively.

Keywords: Schwarzian derivative, immersion, embedding.

This paper is a continuation to ref. [1]. Let $f: M\to N$ be a holomorphic mapping between Kähler manifolds M and N of the same dimension. In ref. [1] we defined the differential $\mathrm{D}f$ of f, higher-order differentials $\mathrm{D}^2 f$, $\mathrm{D}^3 f$ respectively, and with them gave a definition of the Schwarzian derivative for f. Then we proved the invariance property (Theorem 1 of ref. [1]) and injectivity theorem (Theorem 2 of ref. [1]) for the Schwarzian derivative. In Theorem 2 of ref. [1], the objective manifold $N=\mathbb{C}^n$ has flat metric. In this paper, other two injectivity theorems are proved. Theorem 1 is for objective manifold $N=\mathbb{CP}^n$ with canonical metric, and Theorem 2 is for $N=B^n$ (unit ball) with canonical metric. Thus, the injectivity theorems are established in three cases of objective manifolds, i.e., with different holomorphic sectional curvatures -1, 0 and 1.

All symbols in this paper coincide with those in ref. [1], and summation convention is still adopted. The main theorems of this paper are as follows.

Theorem 1. *Let M be a complete Kähler manifold of dimension n, and $\Omega\subseteq M$ be a convex domain, whose holomophic sectional curvature has an upper bound κ and diameter δ, $0<\delta\leqslant\infty$. Let $f: \Omega\to\mathbb{CP}^n$ be a holomorphic immersion and inequality*

$$\operatorname{Re}\frac{\langle Sf_z(X,Y),X\rangle}{\langle X,X\rangle^2\langle X,Y\rangle}\leqslant\frac{4\pi^2}{\delta^2}-\kappa+\frac{\langle f_*X,f_*X\rangle}{2\langle X,X\rangle}$$

holds for all $z\in\Omega$ and X, $Y\in\chi(M)$, where $\mathbb{CP}^n$ is the complex projective space with canonical metric. Then f is an embedding.

Theorem 2. *Let M be a complete Kähler manifold of dimension n, $\Omega\subseteq M$ be a convex do-*

* Project supported by the National Natural Science Foundation of China and partially supported by the Laboratory of Management, Decision and Information System, the Chinese Academy of Sciences.

main, whose holomorphic sectional curvature has an upper bound κ *and diameter* $\delta, 0<\delta\leqslant\infty$. *Let* $f: \Omega\to B^n$ *be a holomorphic immersion, and the inequality*

$$\operatorname{Re}\frac{\langle Sf_z(X,Y),X\rangle}{\langle X,X\rangle^2\langle X,Y\rangle}\leqslant\frac{4\pi^2}{\delta^2}-\kappa-\frac{\langle f_*X,f_*X\rangle}{2\langle X,X\rangle}$$

holds for all $z\in\Omega$ *and* $X,\ Y\in\chi(M)$, *where* B^n *is the unit ball in* $\mathbb{C}^n$ *with canonical metric. Then* f *is an embedding.*

1 Several lemmas

In ref. [1], we already proved the following

Lemma 1. *Let* M, N *be Kähler manifolds of the same dimension,* $f: M\to N$ *be a holomorphic immersion and for all* $z\in M$, $X,\ Y\in\chi(M)$ *the expression*

$$\operatorname{Re}\frac{\langle Sf_z(X,Y),X\rangle}{\langle X,X\rangle^2\langle X,Y\rangle}$$

has an upper bound. Then there exists a C^∞ *complex-valued function* a *of* M *which is defined in a neighbourhood of* z, *such that* $D^2f(X,X)=af_*X$.

Lemma 2. *Suppose* M *is an n-dimensional Kähler manifold,* N *is an n-dimensional Kähler manifold with constant holomorphic sectional curvature* $(=c)$. *Let* $\gamma(t)$ *be a normal geodesic on* M, $X_0=\gamma_*\frac{d}{dt}$ *is its real tangent vector. And* $X=X_0-iJX_0$, $\bar X=X_0+iJX_0$, *where* J *is the complex structure of* M. *Let* $f: M\to N$ *be a holomorphic immersion which satisfies the boundedness assumption of Lemma 1. Then for all* $Y\in\chi(M)$,

$$\frac{\langle Sf_z(X,Y),X\rangle}{\langle X,X\rangle^2\langle X,Y\rangle}=\frac{1}{2}\left(\frac{2da}{dt}-\frac{a^2}{2}\right)+c\,\frac{\langle f_*X,f_*X\rangle}{\langle X,X\rangle}-\frac{R^M(X,JX,X,JX)}{\langle X,X\rangle^2},$$

where a *satisfies* $D^2f(X,X)=af_*X$, *and* R^M *is the curvature operator of* M.

Proof. By Lemma 1, we have

$$\widetilde{\nabla}_{f_*X}D^2f(X,X)-f_*\nabla_Xf_*^{-1}D^2f(X,X)=D^2f(X,f_*^{-1}D^2f(X,X))$$
$$=D^2f(X,f_*^{-1}af_*X)=aD^2f(X,X)=a^2f_*X,\tag{1}$$

and

$$Sf_z(X,Y)=\langle X,Y\rangle f_*^{-1}D^3f(X,X,X)-\frac{3}{2}\langle f_*^{-1}D^2f(X,X),Y\rangle f_*^{-1}D^2f(X,X)$$

$$=\langle X,Y\rangle f_*^{-1}[\widetilde{\nabla}_{f_*X}D^2f(X,X)-2D^2f(X,\nabla_XX)]-\frac{3}{2}a^2\langle X,Y\rangle X$$

$$=\langle X,Y\rangle\{f_*^{-1}[\widetilde{\nabla}_{f_*X}D^2f(X,X)-f_*\nabla_Xf_*^{-1}D^2f(X,X)$$

$$+ f_*\nabla_X f_*^{-1}\mathrm{D}^2 f(X, X) - 2\mathrm{D}^2 f(X, \nabla_X X)] - \frac{3}{2}a^2 X\}. \tag{2}$$

Substituting eq. (1) into eq. (2), we get

$$\begin{aligned} Sf_z(X, Y) &= \langle X, Y\rangle \left\{ -\frac{a^2}{2}X + \nabla_X f_*^{-1}\mathrm{D}^2 f(X, X) - 2f_*^{-1}\mathrm{D}^2 f(X, \nabla_X X) \right\} \\ &= \langle X, Y\rangle \left\{ -\frac{a^2}{2}X + \nabla_X aX - 2f_*^{-1}\mathrm{D}^2 f(X, \nabla_X X) \right\} \\ &= \langle X, Y\rangle \left\{ \left(Xa - \frac{a^2}{2} \right) X + a\nabla_X X - 2f_*^{-1}\mathrm{D}^2 f(X, \nabla_X X) \right\}. \end{aligned} \tag{3}$$

Letting $\xi^i \frac{\partial}{\partial z^i} = X$, and co-differentiating both sides of the equality $aX = f_*^{-1}\mathrm{D}^2 f(X, X)$ with respect to $\overline{X}$, we obtain

$$\begin{aligned} \nabla_{\overline{X}} aX &= (\overline{X}a)X + a\nabla_{\overline{X}} X = (\overline{X}a)X - a\nabla_X X = \nabla_{\overline{X}} f_*^{-1}\mathrm{D}^2 f(X, X) \\ &= \nabla_{\overline{X}} f_*^{-1}\mathrm{D}^2 f\left(\xi^i \frac{\partial}{\partial z^i}, \xi^k \frac{\partial}{\partial z^k} \right) = \nabla_{\overline{X}} \xi^i \xi^k f_*^{-1}\mathrm{D}^2 f\left(\frac{\partial}{\partial z^i}, \frac{\partial}{\partial z^k} \right) \\ &= (\overline{X}\xi^i\xi^k) f_*^{-1}\mathrm{D}^2 f\left(\frac{\partial}{\partial z^i}, \frac{\partial}{\partial z^k} \right) + \xi^i\xi^k \nabla_{\overline{X}} f_*^{-1}\mathrm{D}^2 f\left(\frac{\partial}{\partial z^i}, \frac{\partial}{\partial z^k} \right). \end{aligned} \tag{4}$$

Since $z(t)$ is a normal geodesic, $\nabla_X X = -\nabla_{\overline{X}} X$ (ref. [1], eq. (39)). Using the property: ∇ is (1,0)-connection, we know

$$\begin{aligned} &(\overline{X}\xi^i\xi^k) f_*^{-1}\mathrm{D}^2 f\left(\frac{\partial}{\partial z^i}, \frac{\partial}{\partial z^k} \right) = 2\xi^i(\overline{X}\xi^k) f_*^{-1}\mathrm{D}^2 f\left(\frac{\partial}{\partial z^i}, \frac{\partial}{\partial z^k} \right) \\ &= 2f_*^{-1}\mathrm{D}^2 f\left(\xi^i \frac{\partial}{\partial z^i}, (\overline{X}\xi^k)\frac{\partial}{\partial z^k} \right) = 2f_*^{-1}\mathrm{D}^2 f(X, \nabla_{\overline{X}} X) = -2f_*^{-1}\mathrm{D}^2 f(X, \nabla_X X). \end{aligned} \tag{5}$$

Substituting eq. (5) into eq. (4), we have

$$(\overline{X}a)X - a\nabla_X X = \xi^i\xi^k \nabla_{\overline{X}} f_*^{-1}\mathrm{D}^2 f\left(\frac{\partial}{\partial z^i}, \frac{\partial}{\partial z^k} \right) - 2f_*^{-1}\mathrm{D}^2 f(X, \nabla_X X). \tag{6}$$

Moreover, we have

$$\begin{aligned} &\xi^i\xi^k \nabla_{\overline{X}} f_*^{-1}\mathrm{D}^2 f\left(\frac{\partial}{\partial z^i}, \frac{\partial}{\partial z^k} \right) \\ &= \xi^i\xi^k \nabla_{\overline{\xi}\frac{\partial}{\partial \overline{z}^j}} f_*^{-1}\left[\widetilde{\nabla}_{f_*\frac{\partial}{\partial z^i}} f_* \frac{\partial}{\partial z^k} - \nabla_{\frac{\partial}{\partial z^i}} \frac{\partial}{\partial z^k} \right] \\ &= \xi^i \overline{\xi}^j \xi^k \nabla_{\frac{\partial}{\partial \overline{z}^j}} \left[f_*^{-1} \frac{\partial^2 w^\gamma}{\partial z^i \partial z^k} \frac{\partial}{\partial w^\gamma} + f_*^{-1} \frac{\partial w^\alpha}{\partial z^i} \frac{\partial w^\gamma}{\partial z^k} \widetilde{\nabla}_{\frac{\partial}{\partial w^\alpha}} \frac{\partial}{\partial w^\gamma} - \Gamma_{ik}^l \frac{\partial}{\partial z^l} \right] \\ &= \xi^i \overline{\xi}^j \xi^k \left[\frac{\partial w^\alpha}{\partial z^i} \frac{\partial w^\gamma}{\partial z^k} \frac{\partial \widetilde{\Gamma}_{\alpha\gamma}^\delta}{\partial \overline{w}^\beta} \frac{\partial \overline{w}^\beta}{\partial \overline{z}^j} \frac{\partial z^l}{\partial w^\delta} \frac{\partial}{\partial z^l} - \frac{\partial \Gamma_{ik}^l}{\partial \overline{z}^j} \frac{\partial}{\partial z^l} \right] \\ &= \xi^i \overline{\xi}^j \xi^k \left[S_{ikj}^l \frac{\partial}{\partial z^l} - \frac{\partial w^\alpha}{\partial z^i} \frac{\partial \overline{w}^\beta}{\partial \overline{z}^j} \frac{\partial w^\gamma}{\partial z^k} \widetilde{S}_{\alpha\gamma\overline{\beta}}^\delta \frac{\partial z^l}{\partial w^\delta} \frac{\partial}{\partial z^l} \right]. \end{aligned} \tag{7}$$

N is a Kähler manifold of constant holomorphic sectional curvature c, whose curvature tensor is

$$\widetilde{S}^{\delta}_{\alpha\gamma\bar{\beta}} = \frac{c}{2}(\delta_{\alpha\delta}\tilde{g}_{\gamma\bar{\beta}} + \delta_{\gamma\delta}\tilde{g}_{\alpha\bar{\beta}}). \tag{8}$$

Substituting eq. (8) into eq. (7), we obtain

$$\begin{aligned}\xi^i\xi^k \nabla_{\bar{X}} f_*^{-1} D^2 f\left(\frac{\partial}{\partial z^i}, \frac{\partial}{\partial z^k}\right) &= \xi^i\bar{\xi}^j\xi^k S^l_{ik\bar{j}} \frac{\partial}{\partial z^l} - c\xi^k\bar{\xi}^j\xi^l \frac{\partial w^\gamma}{\partial z^k}\frac{\partial \bar{w}^\beta}{\partial \bar{z}^j} g_{\gamma\bar{\beta}} \frac{\partial}{\partial z^l} \\ &= \xi^i\bar{\xi}^j\xi^k S^l_{ik\bar{j}} \frac{\partial}{\partial z^l} - c\langle f_* X, f_* X\rangle X. \end{aligned} \tag{9}$$

Inserting eq. (9) into eq. (6) and taking the inner product, it follows that

$$\begin{aligned}(\bar{X}a)\langle X, X\rangle &= \xi^i\bar{\xi}^j\xi^k\bar{\xi}^l S_{i\bar{j}k\bar{l}} - c\langle f_* X, f_* X\rangle\langle X, X\rangle \\ &\quad + a\langle \nabla_X X, X\rangle - 2\langle f_* D^2 f(X, \nabla_X X), X\rangle. \end{aligned} \tag{10}$$

Using $Xa = 2\frac{\mathrm{d}a}{\mathrm{d}t} - \bar{X}a$ (ref. [1], eq. (37)) in eq. (3) and taking the inner product, we get

$$\begin{aligned}&\frac{\langle Sf_z(X, Y), X\rangle}{\langle X, X\rangle^2\langle X, Y\rangle} \\ &= \frac{2\frac{\mathrm{d}a}{\mathrm{d}t} - \bar{X}a - \frac{a^2}{2}}{\langle X, X\rangle} + \frac{\langle a\nabla_X X, X\rangle}{\langle X, X\rangle^2} - \frac{\langle 2f_*^{-1}D^2 f(X, \nabla_X X), X\rangle}{\langle X, X\rangle^2}. \end{aligned} \tag{11}$$

Because $\gamma(t)$ is a normal geodesic, it implies $\langle X, X\rangle = 2$. Substituting eq. (10) into eq. (11), we obtain

$$\begin{aligned}\frac{\langle Sf_z(X, Y), X\rangle}{\langle X, X\rangle^2\langle X, Y\rangle} &= \frac{2\frac{\mathrm{d}a}{\mathrm{d}t} - \frac{a^2}{2}}{\langle X, X\rangle} - \frac{\xi^i\bar{\xi}^j\xi^k\bar{\xi}^l S_{i\bar{j}k\bar{l}} - c\langle f_* X, f_* X\rangle\langle X, X\rangle}{\langle X, X\rangle^2} \\ &\quad - \frac{a\langle \nabla_X X, X\rangle - 2\langle f_*^{-1}D^2 f(X, \nabla_X X), X\rangle}{\langle X, X\rangle} \\ &\quad + \frac{a\langle \nabla_X X, X\rangle}{\langle X, X\rangle^2} - \frac{2\langle f_*^{-1}D^2 f(X, \nabla_X X), X\rangle}{\langle X, X\rangle} \\ &= \frac{\mathrm{d}a}{\mathrm{d}t} - \frac{a^2}{4} - \frac{R^M(X, JX, X, JX)}{\langle X, X\rangle^2} + c\frac{\langle f_* X, f_* X\rangle}{\langle X, X\rangle}. \end{aligned}$$

Thus, Lemma 2 is proved.

2 Proof of theorems

If Theorem 1 is not true, i.e. supposing f is not an embedding, there exist two points z_1, $z_2 \in \Omega$, such that $z_1 \neq z_2$, $f(z_1) = f(z_2)$. Let $\gamma: [0, T] \to M$ be a normal geodesic, $\gamma(0) = z_1$, $\gamma(T) = z_2$. Let τ be an arbitrary holomorphic function defined in a neighbourhood of $f \circ \gamma(t)$ and $h = \tau - \tau(0)$. For convenience, denoting $h \circ f \circ \gamma$ by h, then $h(0) = h(T) = 0$. We have

$$\frac{\mathrm{d}h}{\mathrm{d}t} = X_0 h = \frac{1}{2}(X + \bar{X})h = \frac{1}{2}Xh \tag{12}$$

and

$$\frac{d^2h}{dt^2} = X_0(X_0h) = \frac{1}{2}(X+\bar{X})\frac{1}{2}Xh = \frac{1}{4}[X(Xh)+\bar{X}(Xh)], \tag{13}$$

where $X_0 = \gamma_* \frac{d}{dt}$, $X = X_0 - iJX_0$, $\bar{X} = X_0 + iJX_0$, J is the complex structure of M.

Let $X = \xi^i \frac{\partial}{\partial z^i}$, U be the coordinate neighbourhood of $f(\gamma)$, with coordinates $(w^1, w^2, \cdots, w^n)$. Then we get

$$\begin{aligned} D^2f(X,X)h &= (\widetilde{\nabla}_{f_*X}f_*X - f_*\nabla_XX)h = (\widetilde{\nabla}_{f_*X}f_*X + f_*\nabla_{\bar{X}}X)h \\ &= X(Xh) + \bar{X}(Xh) + \xi^i\xi^j\frac{\partial w^\alpha}{\partial z^i}\frac{\partial w^\beta}{\partial z^j}\widetilde{\Gamma}^\gamma_{\alpha\beta}\frac{\partial h}{\partial w^\gamma}, \end{aligned} \tag{14}$$

where $\widetilde{\Gamma}^\gamma_{\alpha\beta}$ are coefficients of connection of $\mathbb{C}P^n$. Comparing eq.(13) with eq.(14), we obtain

$$D^2f(X,X)h = 4\frac{d^2h}{dt^2} + \xi^i\xi^j\frac{\partial w^\alpha}{\partial z^i}\frac{\partial w^\beta}{\partial z^j}\widetilde{\Gamma}^\gamma_{\alpha\beta}\frac{\partial h}{\partial w^\gamma}. \tag{15}$$

By Lemma 1 and eq.(13), we have

$$D^2f(X,X)h = 2a\frac{dh}{dt}. \tag{16}$$

Inserting eq.(16) into eq.(15), we have

$$2a\frac{dh}{dt} - 4\frac{d^2h}{dt^2} = \xi^i\xi^j\frac{\partial w^\alpha}{\partial z^i}\frac{\partial w^\beta}{\partial z^j}\widetilde{\Gamma}^\gamma_{\alpha\beta}\frac{\partial h}{\partial w^\gamma}. \tag{17}$$

But it is known that

$$\widetilde{\Gamma}^\gamma_{\alpha\beta} = -\frac{1}{1+\rho^2}(\bar{w}^\beta\delta_{\gamma\alpha} + \bar{w}^\alpha\delta_{\gamma\beta}), \tag{18}$$

where $\rho^2 = \sum_{\alpha=1}^{n} w^\alpha\bar{w}^\alpha$.

Substituting eq.(18) into eq.(17), we get

$$\frac{d^2h}{dt^2} - \frac{1}{2}\left[a + \frac{2(X\rho^2)}{1+\rho^2}\right]\frac{dh}{dt} = 0. \tag{19}$$

Let $b = a + \frac{2(X\rho^2)}{1+\rho^2}$. Then eq.(19) can be written as

$$\frac{d^2h}{dt^2} - \frac{b}{2}\frac{dh}{dt} = 0. \tag{20}$$

If we set

$$v = h\mathrm{e}^{-\frac{1}{4}\int_0^t b(s)\mathrm{d}s},$$

then

$$\frac{\mathrm{d}^2 v}{\mathrm{d}t^2} + \frac{1}{8}\left(2\frac{\mathrm{d}b}{\mathrm{d}t} - \frac{b^2}{2}\right)v = 0. \tag{21}$$

It is easy to compute

$$2\frac{\mathrm{d}b}{\mathrm{d}t} - \frac{b^2}{2} = 2\frac{\mathrm{d}a}{\mathrm{d}t} - \frac{a^2}{2} - \frac{4(X\rho^2)^2 + 2(\bar{X}\rho^2)(X\rho^2)}{(1+\rho^2)^2} + \frac{2X(X\rho^2) + 2\bar{X}(X\rho^2)}{1+\rho^2} - \frac{2\mathrm{D}^2 f(X,X)\rho^2}{1+\rho^2}. \tag{22}$$

Now, we compute $\mathrm{D}^2 f(X,X)\rho^2$ as follows:

$$\begin{aligned}\mathrm{D}^2 f(X,X)\rho^2 &= (\widetilde{\nabla}_{f_*X} f_* X - f_*\nabla_X X)\rho^2 = (\widetilde{\nabla}_{f_*X} f_* X + f_*\nabla_{\bar{X}} X)\rho^2 \\ &= (\nabla_{f_*X} f_* X)\rho^2 + \bar{X}(X\rho^2) - \xi^i\bar{\xi}^j\frac{\partial w^\alpha}{\partial z^i}\frac{\partial \bar{w}^\beta}{\partial \bar{z}^j}\delta_{\alpha\beta} \\ &= X(X\rho^2) + \bar{X}(X\rho^2) + \xi^i\xi^k\frac{\partial w^\alpha}{\partial z^i}\frac{\partial w^\gamma}{\partial z^k}\tilde{\Gamma}^\beta_{\alpha\gamma}\bar{w}^\beta - \xi^i\bar{\xi}^j\frac{\partial w^\alpha}{\partial z^i}\frac{\partial \bar{w}^\beta}{\partial \bar{z}^j}\delta_{\alpha\beta} \\ &= X(X\rho^2) + \bar{X}(X\rho^2) - \frac{2(X\rho^2)^2}{1+\rho^2} - \xi^i\bar{\xi}^j\frac{\partial w^\alpha}{\partial z^i}\frac{\partial \bar{w}^\beta}{\partial \bar{z}^j}\delta_{\alpha\beta}.\end{aligned} \tag{23}$$

On the other hand, we have

$$\begin{aligned}\langle f_*X, f_*X\rangle &= \xi^i\bar{\xi}^j\frac{\partial w^\alpha}{\partial z^i}\frac{\partial \bar{w}^\beta}{\partial \bar{z}^j}g_{\alpha\bar{\beta}} = 4\xi^i\bar{\xi}^j\frac{\partial w^\alpha}{\partial z^i}\frac{\partial \bar{w}^\beta}{\partial \bar{z}^j}\frac{(1+\rho^2)\delta_{\alpha\beta} - \bar{w}^\alpha w^\beta}{(1+\rho^2)^2} \\ &= \frac{4}{1+\rho^2}\xi^i\bar{\xi}^j\frac{\partial w^\alpha}{\partial z^i}\frac{\partial \bar{w}^\beta}{\partial \bar{z}^j}\delta_{\alpha\beta} - \frac{4}{(1+\rho^2)^2}(X\rho^2)(\bar{X}\rho^2).\end{aligned}$$

Thus

$$\xi^i\bar{\xi}^j\frac{\partial w^\alpha}{\partial z^i}\frac{\partial \bar{w}^\beta}{\partial \bar{z}^j}\delta_{\alpha\beta} = \frac{1+\rho^2}{4}\langle f_*X, f_*X\rangle + \frac{(X\rho^2)(\bar{X}\rho^2)}{1+\rho^2}. \tag{24}$$

Substituting eq. (24) into eq. (23), we obtain

$$\mathrm{D}^2 f(X,X)\rho^2 = X(X\rho^2) + \bar{X}(X\rho^2) - \frac{2(X\rho^2)^2}{1+\rho^2} - \frac{1+\rho^2}{4}\langle f_*X, f_*X\rangle - \frac{(X\rho^2)(\bar{X}\rho^2)}{1+\rho^2}. \tag{25}$$

Substituting eq. (25) into eq. (22), we obtain

$$2\frac{db}{dt}-\frac{b^2}{2}=2\frac{da}{dt}-\frac{a^2}{2}+\frac{1}{2}\langle f_*X, f_*X\rangle. \tag{26}$$

Substituting eq. (26) into eq. (21), we obtain

$$\frac{d^2v}{dt^2}+\frac{1}{8}\left(2\frac{da}{dt}-\frac{a^2}{2}+\frac{1}{2}\langle f_*X, f_*X\rangle\right)v=0.$$

By the assumption and Lemma 2, noting that $c=1$, we have

$$\frac{4\pi^2}{\delta^2}-\kappa+\frac{\langle f_*X, f_*X\rangle}{2\langle X, X\rangle}\geqslant \mathrm{Re}\frac{\langle Sf_z(X,Y),X\rangle}{\langle X,X\rangle^2\langle X,Y\rangle}$$
$$=\mathrm{Re}\left[\frac{1}{2}\left(\frac{2da}{dt}-\frac{a^2}{2}\right)\right]+\frac{\langle f_*X, f_*X\rangle}{\langle X,X\rangle}-\frac{R^M(X,JX,X,JX)}{\langle X,X\rangle^2}$$
$$\geqslant \mathrm{Re}\left[\frac{1}{2}\left(\frac{2da}{dt}-\frac{a^2}{2}\right)\right]+\frac{\langle f_*X, f_*X\rangle}{\langle X,X\rangle}-\kappa.$$

Thus,

$$\frac{4\pi^2}{\delta^2}\geqslant \mathrm{Re}\left[\frac{1}{2}\left(2\frac{da}{dt}-\frac{a^2}{2}\right)+\frac{\langle f_*X, f_*X\rangle}{2\langle X,X\rangle}\right].$$

Noting $\langle X, X\rangle=2$, we obtain

$$\frac{\pi^2}{\delta}\geqslant\frac{1}{8}\mathrm{Re}\left(2\frac{da}{dt}-\frac{a^2}{2}+\frac{1}{2}\langle f_*X, f_*X\rangle\right).$$

Obviously, $u=\sin\frac{\varepsilon+t}{T+2\varepsilon}$, $t\in[0,T]$ is a positive solution of the equation

$$\frac{d^2u}{dt^2}+\frac{\pi^2}{(T+2\varepsilon)^2}u=0.$$

Since $T<\delta$, we may choose ε sufficiently small such that $T+2\varepsilon<\delta$. Thus we have

$$\frac{\pi^2}{(T+2\varepsilon)^2}>\frac{\pi^2}{\delta^2}\geqslant\frac{1}{8}\mathrm{Re}\left(2\frac{da}{dt}-\frac{a^2}{2}+\frac{1}{2}\langle f_*X, f_*X\rangle\right).$$

By Lemma 1 of ref. [1], we have $v\equiv$constant. But $v=he^{\frac{1}{4}\int_0^t b(s)ds}$, h is an arbitrary holomorphic function, so $f\circ\gamma(t)\equiv$constant, $t\in[0,T]$. That is $f\circ\gamma(t)$ is just a point, $\forall t\in[0,T]$. It contradicts the assumption: f is immersion. Theorem 1 is proved.

The proof of Theorem 2 is similar, so we omit the details.

Reference

1 Gong, S., Yu, Q., Schwarzian derivative of Kähler manifolds, *Science in China*, Ser. A, 1995, 38: 1033.

Chinese Journal of Contemporary Mathematics
Volume 16, Number 3, 1995

THE SCHWARZIAN DERIVATIVE IN C^n

Yu Qihuang* and Gong Sheng**

**Institute of Applied Mathematics, Academia Sinica, Beijing, 100080, China.*
***Department of Mathematics, University of Science and Technology of China, Hefei, Anhui, 230026, China.*
Received June 8, 1993. Revised April 5, 1994.
Project supported by the National Natural Science Foundation of China.

A Schwarzian derivative S_f of a holomorphic function f defined in a domain in C^n is proposed and the following properties are proved: (1) $S_f = 0$ if and only if f is a linear fractional transformation; (2) S_f is invariant in some sense under composition of f with a linear fractional function; (3) the univalence criterion of Nehari is generalized.

Keywords Schwarzian derivative, Fractional transformation, univalence criterion.

1991 MR Subject Classification 30A, 32H.

§1. INTRODUCTION

The Schwarzian derivative of holomorphic functions is an important topic in the function theory of one complex variable. The Schwarzian derivative of a holomorphic function $f(z)$ of one complex variable z is defined as

$$S_f = \frac{f'''(z)}{f'(z)} - \frac{3}{2}\left(\frac{f''(z)}{f'(z)}\right)^2$$

if $f'(z) \neq 0$. The basic properties of the Schwarzian derivative of a holomorphic function of one complex variable are: 1) $S_f = 0$ if and only if f is a linear fractional transformation. 2) the Schwarzian derivative is invariant if f is composed with a linear fractional function. In addition, there are many important applications of the Schwarzain derivative, in particular, there is Nehari's univalence criterion.

Recently, high dimensional generalizations of the Schwarzian derivative of holomorphic functions have been given. For example, Osgood and Stowe[1] and Carne[3] extended the Schwarzian derivative from the point of view of conformal mappings between Riemannian manifolds. Osgood and Stowe[2] generalized Nehari's univalence criterion by a bound on the Schwarzian derivative. Weiqi Gao[4] extended Flanders' result to high dimensions. Flanders[5] pointed out that the Schwarzian derivative of a holomorphic function of one complex variable is some kind of curvature of a curve in the complex projective space CP^1. C. H. FitzGerald and S. Gong[6] extended the Schwarzian derivative of holomorphic mappings on some classical domains from the point of view that the Schwarzian derivative is an infinitesimal version of cross ratio.

The classical domain of type I is a domain R_I of $m \times n$ matrices with complex entries satisfying the condition $I^{(m)} - Z\overline{Z}' > 0$. Here $I^{(m)}$ is the identity matrix of order m, and $\overline{Z}'$ is the complex conjugate of the transposed matrix Z', and $H > 0$ means that H is a Hermitian positive definite matrix. We are particularly interested in square matrices, that is the case with $m = n$.

Let $\Lambda = (\lambda_{ij})_{1 \leq i,j \leq n}$ be an $n \times n$ matrix with complex entries. If $W = W(Z)$ is a holomorphic mapping on R_I which maps R_I into $\mathbf{C}^{n\times n}$, then the directional derivative of $W = W(Z)$ in the direction Λ is defined as

$$D_\Lambda W = \frac{d}{dt} W(Z + t\Lambda)|_{t=0},$$

where $t \in \mathbf{C}$. The second and third order derivatives in that direction are given by

$$D_\Lambda^2 W = D_\Lambda(D_\Lambda W), \qquad D_\Lambda^3 W = D_\Lambda(D_\Lambda^2 W).$$

C. H. FitzGerald and S. Gong[6] defined the Schwarzian derivative of $W = W(Z)$ in the direction Λ as

$$\{W; Z\}_\Lambda = (D_\Lambda^3 W)(D_\Lambda W)^{-1} - \frac{3}{2}((D_\Lambda^2 W)(D_\Lambda W)^{-1})^2 \tag{1.1}$$

when $(D_\Lambda W)^{-1}$ exists. Then $\{W; Z\}_\Lambda$ is invariant up to similarity under the group of holomorphic automorphism of R_I, and $\{W; Z\}_\Lambda = 0$ for all Z in R_I and all non-zero Λ (when the left hand side of (1) exists) if and only if

$$W = W(Z) = (w_{ij}(Z))_{1 \leq i,j \leq n},$$

where

$$w_{ij} = c_{ij} + \operatorname{tr}(C_{ij} Z(I - L(Z))^{-1})$$

or

$$w_{ij} = c_{ij} + \operatorname{tr}(C_{ij}(I - L(Z))^{-1} Z), \quad i, j = 1, 2, \cdots n.$$

and c_{ij} are constants, C_{ij} are $n \times n$ constant matrices, $L(Z)$ is $n \times n$ matrix, each entry of $L(Z)$ is a linear homogeneous function of z_{ij}, $i, j = 1, 2, \cdots n$.

We may define the Schwarzian derivatives of holomorphic mappings on classical domains of type II and III, and also for the corresponding classical domains on real fields in a similar way

How to define the Schwarzian derivative of holomorphic mappings on a vector space in C^n? In this note, we propose a Schwarzian derivative of holomorphic mappings on vector space in C^n, which is very close to the definition of FitzGerald and Gong for the Schwarzian derivative of holomorphic mappings on matrix spaces.

Using the definition of Schwarzian derivative in this note, we can generalize the property 1) under the condition that the domain is starlike, generalize the property 2) without this restriction, and generalize Nehari's univalence criterion on convex domains.

§2. DEFINITION OF SCHWARZIAN

Let $\Omega \subset C^n$ be a domain, and $f : \Omega \to C^n$ be a locally biholomorphic mapping on Ω. If $z = (z_1, \cdots, z_n) \in \Omega, \lambda = (\lambda_1, \cdots, \lambda_n) \in C^n$ and $\mu = (\mu_1, \cdots, \mu_n) \in C^n$, we define the Schwarzian derivative of f at the point z along the directions λ, μ as

$$\{f; z; \lambda, \mu\} = \langle\lambda, \mu\rangle\langle D_\lambda^3 f \cdot J_f^{-1}, \mu\rangle - \frac{3}{2}\langle D_\lambda^2 f \cdot J_f^{-1}, \mu\rangle^2. \tag{2.1}$$

where $\langle\ ,\ \rangle$ is the Hermitian inner product in C^n, $D_\lambda f$ is the directional derivative of f along λ, that is,

$$D_\lambda f = D_\lambda(f_1, \cdots, f_n) = \left(\sum_i \lambda_i \frac{\partial f_1}{\partial z_i}, \cdots, \sum_i \lambda_i \frac{\partial f_n}{\partial z_i}\right) = \lambda J_f,$$

where J_f is the transpose of the Jacobian of f with respect to z, that is,

$$J_f = \begin{pmatrix} \frac{\partial f_1}{\partial z_1} & \cdots & \frac{\partial f_n}{\partial z_1} \\ & \cdots & \\ \frac{\partial f_1}{\partial z_n} & \cdots & \frac{\partial f_n}{\partial z_n} \end{pmatrix}.$$

Compare with (1.1) and (2.1), the expression looks very similar.

It is easy to verify

$$D_\lambda^2 f = D_\lambda(D_\lambda f) = \sum_i \lambda\lambda_i \frac{\partial J_f}{\partial z_i} \tag{2.2}$$

and

$$D_\lambda^3 f = D_\lambda(D_\lambda^2 f) = \sum_{i,j} \lambda\lambda_i\lambda_j \frac{\partial^2 J_f}{\partial z_i \partial z_j}. \tag{2.3}$$

By (2.2) and (2.3), we can rewrite (2.1) as

$$\{f; z; \lambda, \mu\} = (\lambda\bar{\mu}')\lambda \sum_{i,j} \lambda_i\lambda_j \frac{\partial^2 J_f}{\partial z_i \partial z_j} J_f^{-1}\bar{\mu}' - \frac{3}{2}\left(\lambda \sum_i \lambda_i \frac{\partial J_f}{\partial z_i} J_f^{-1}\bar{\mu}'\right)^2. \tag{2.4}$$

Now we study the necessary and sufficient conditions that the Schwarzian derivative of a holomorphic mapping is equal to zero along every λ and every $\mu \in C^n$ and every $z \in \Omega$. We have the following.

Theorem 2.1. Let $\Omega \subset C^n$ be a domain starlike with respect to the origin, and $f : \Omega \to C^n$ be a local biholomorphic mapping. Then $\{f; z; \lambda, \mu\} = 0$ for every $z \in \Omega$ and every $\lambda, \mu \in C^n$ if and only if

$$f(z) = f(0) + \frac{zJ_f(0)}{1 - zb'}, \tag{2.5}$$

where $b = (b_1, \cdots, b_n) \in \mathbf{C}^n$ is a constant vector, and $1 - z\bar{b}' \neq 0$ when $z \in \Omega$.

Actually, (2.5) is a very general form of a linear fractional mapping which is equivalent to the linear fractional mapping

$$f(z) = \frac{a + zB}{c + zd'}, \tag{2.6}$$

where a, d are arbitrary n-dimensional constant vectors, c is an arbitrary non-zero constant. B is an arbitrary $n \times n$ constant matrix, and $c + zd' \neq 0$ when $z \in \Omega$.

§3. PROOF OF THEOREM 2.1

Since Ω is a starlike domain with respect to the origin, $\{f; tz; z, \mu\}$ is well defined for every $z \in \Omega$, $0 \leq t \leq 1$, and

$$\begin{aligned}\{f; tz; z, \mu\} = (z\bar{\mu}')z \sum_{i,j} z_i z_j \frac{\partial^2 J_f}{\partial z_i \partial z_j}(tz) J_f^{-1}(tz)\bar{\mu}' \\ - \frac{3}{2}\left(z \sum_i z_i \frac{\partial J_f}{\partial z_i}(tz) J_f^{-1}(tz)\bar{\mu}'\right)^2,\end{aligned} \tag{3.1}$$

where

$$J_f(tz) = J_f(w)|_{w=tz}, \quad \frac{\partial J_f}{\partial z_i}(tz) = \frac{\partial J_f(w)}{\partial w_i}|_{w=tz}$$

and

$$\frac{\partial^2 J_f}{\partial z_i \partial z_j}(tz) = \left.\frac{\partial^2 J_f(w)}{\partial w_i \partial w_j}\right|_{w=tz}.$$

If $\{f; z; \lambda, \mu\} = 0$ for every $z \in \Omega$ and every $\lambda, \mu \in \mathbf{C}^n$, then $\{f; tz; z, \mu\} = 0$. We denote the vector subspace which is generated by z in $\mathbf{C}^n$ by L_z, and the orthogonal complement of L_z by $L_z^{\perp}$.

If $\mu \in L_z^{\perp}$, then from (3.1) and $\{f; tz; z, \mu\} = 0$, we have

$$z \sum_i z_i \frac{\partial J_f}{\partial z_i}(tz) J_f^{-1}(tz)\bar{\mu}' = 0$$

for all $\mu \in L_z^{\perp}$. This implies

$$z \sum_i z_i \frac{\partial J_f}{\partial z_i}(tz) J_f^{-1}(tz) \in L_z.$$

Thus there exists a function $a(t, z)$ such that

$$\sum_i z_i \frac{\partial J_f}{\partial z_i}(tz) J_f^{-1}(tz) = a(t, z)z. \tag{3.2}$$

Obviously, $a(t, z)$ is a holomorphic function of z, and a C^∞ function of t.

Differentiating (3.2) with respect to t, we have

$$\begin{aligned}\frac{\partial}{\partial t}a(t,z)z &= z\frac{\partial}{\partial t}\left(\sum_i z_i\frac{\partial J_f}{\partial z_i}(tz)J_f^{-1}(tz)\right)\\ &= z\sum_i\left(z_i\frac{\partial}{\partial t}\left(\frac{\partial J_f}{\partial z_i}(tz)\right)J_f^{-1}(tz)\right.\\ &\quad\left.-z_i\frac{\partial J_f}{\partial z_i}(tz)J_f^{-1}(tz)\frac{\partial}{\partial t}J_f(tz)J_f^{-1}(tz)\right).\end{aligned} \tag{3.3}$$

But we know that

$$\frac{\partial}{\partial t}J_f(tz)=\sum_i\frac{\partial J_f}{\partial z_i}(tz)z_i \tag{3.4}$$

and

$$\frac{\partial}{\partial t}\left(\frac{\partial J_f}{\partial z_i}(tz)\right)=\sum_j\frac{\partial^2 J_f}{\partial z_i\partial z_j}(tz)z_j. \tag{3.5}$$

Substituting (3.4), (3.5) into (3.6), and using (3.2), we get

$$\begin{aligned}&z\sum_{i,j}z_iz_j\frac{\partial^2 J_f}{\partial z_i\partial z_j}(tz)J_f^{-1}(tz)\\ &=\frac{\partial}{\partial t}a(t,z)z+z\sum_{i,j}z_i\frac{\partial J_f}{\partial z_i}(tz)J_f^{-1}(tz)z_j\frac{\partial J_f}{\partial z_i}(tz)J_f^{-1}(tz)\\ &=\frac{\partial}{\partial t}a(t,z)z+a^2(t,z)z.\end{aligned} \tag{3.6}$$

By assumption, $\{f;tz;z,\mu\}=0$ and it follows from (3.1),

$$(z\bar{\mu}')z\sum_{i,j}z_iz_j\frac{\partial^2 J_f}{\partial z_i\partial z_j}(tz)J_f^{-1}(tz)\bar{\mu}'-\frac{3}{2}\left(z\sum_i z_i\frac{\partial J_f}{\partial z_i}(tz)J_f^{-1}(tz)\bar{\mu}'\right)^2=0 \tag{3.7}$$

which holds for every $z\in\Omega$ and $\mu\in C^n$. Substituting (3.2) and (3.6) into (3.7), we obtain an equation of $a(t,z)$:

$$(z\bar{\mu}')^2\left(\frac{\partial}{\partial t}a(t,z)+a^2(t,z)-\frac{3}{2}a^2(t,z)\right)=0.$$

that is,

$$\frac{\partial}{\partial t}a(t,z)-\frac{1}{2}a^2(t,z)=0. \tag{3.8}$$

Integrating the equation

$$\frac{1}{a^2(t,z)}\frac{\partial}{\partial t}a(t,z)=\frac{1}{2}$$

with respect to t from $t=0$ to $t=t$, we obtain

$$\frac{1}{a(0,z)}-\frac{1}{a(t,z)}=\frac{1}{2}t.$$

Then $a(t,z)$ can be expressed as

$$a(t,z)=\frac{2a(0,z)}{2-a(0,z)t}. \tag{3.9}$$

Substituting (3.9) into (3.2), and then multiplying $J_f(tz)$ on both sides, we get

$$z\sum_i z_i\frac{\partial J_f}{\partial z_i}(tz)=\frac{2a(0,z)}{2-a(0,z)t}zJ_f(tz).$$

The left-hand side of the previous equality is $\frac{\partial}{\partial t}(zJ_f(tz))$ by (3.4), thus

$$\frac{\partial}{\partial t}(zJ_f(tz))=\frac{2a(0,z)}{2-a(0,z)t}(zJ_f(tz)). \tag{3.10}$$

If the α-th component of $zJ_f(tz)$ is $(zJ_f(tz))_\alpha$, then the equation of the α-th component of (3.10) is

$$\frac{\partial}{\partial t}(zJ_f(tz))_\alpha=\frac{2a(0,z)}{2-a(0,z)t}(zJ_f(tz))_\alpha,\quad \alpha=1,2,\cdots,n.$$

We solve this equation and get

$$\frac{(zJ_f(tz))_\alpha}{(zJ_f(0))_\alpha}=\left(\frac{2}{2-a(0,z)t}\right)^2,\quad \alpha=1,2,\cdots,n.$$

The vector form of these equalities is

$$zJ_f(tz)=\frac{4}{(2-a(0,z)t)^2}zJ_f(0).$$

This is

$$\frac{\partial}{\partial t}f(tz)=\frac{4}{(2-a(0,z)t)^2}zJ_f(0)$$

since $zJ_f(tz)=\frac{\partial}{\partial t}f(tz)$. Integrating this equation with respect to t from $t=0$ to $t=t$, we have

$$f(tz)-f(0)=\frac{2t}{2-a(0,z)t}zJ_f(0), \tag{3.11}$$

then let $t=1$, it is

$$f(z)=f(0)+\frac{2}{2-a(0,z)}zJ_f(0),$$

and replace z by tz , it becomes

$$f(tz)=f(0)+\frac{2t}{2-a(0,tz)}zJ_f(0). \tag{3.12}$$

Comparing (3.11) and (3.12), we have $a(0,z)t=a(0,tz)$, it implies $a(0,z)$ is a linear function of z, because $a(0,z)$ is a holomorphic function of z. Thus $a(0,z)=zb_0'$, where $b_0\in C^n$, and (2.5) is proved if we let $\frac{1}{2}b_0=\bar{b}$.

The function (2.5) satisfies the equation $\{f; z; \lambda, \mu\} = 0$ for any $z \in \Omega$ and $\lambda, \mu \in \mathbf{C}^n$ by direct evaluation. It does not use the starlike condition of the domain.

It is easy to verify directly that: if $f(z) = f(0) + \dfrac{zJ_f(0)}{1 - z\bar{b}'}$, then

$$J_f(z) = \left(\frac{1}{1 - z\bar{b}'} I + \frac{\bar{b}'z}{(1 - z\bar{b}')^2}\right) J_f(0),$$

$$\frac{\partial J_f(z)}{\partial z_i} = \left[\frac{\bar{b}_i}{(1 - z\bar{b}')^2} I + \frac{2\bar{b}_i\bar{b}'z}{(1 - z\bar{b}')^3} + \frac{\bar{b}'e_i}{(1 - z\bar{b}')^2}\right] J_f(0)$$

and

$$\frac{\partial^2 J_f(z)}{\partial z_i \partial z_j} = \left[\frac{2\bar{b}_i\bar{b}_j}{(1 - z\bar{b}')^3} I + \frac{2b_i\bar{b}'e_j}{(1 - z\bar{b}')^3} + \frac{6\bar{b}_i\bar{b}_j\bar{b}'z}{(1 - z\bar{b}')^4} + \frac{2\bar{b}_j\bar{b}'e_i}{(1 - z\bar{b}')^3}\right] J_f(0)$$

hold, where e_i, e_j are the unit vectors at i-th and j-th coordinates, respectively.

From these equalities, we have

$$\lambda \sum_i \lambda_i \frac{\partial J_f(z)}{\partial z_i} J_f^{-1}(z) = \frac{2\lambda\bar{b}'}{1 - z\bar{b}'} \lambda \tag{3.13}$$

and

$$\lambda \sum_{i,j} \lambda_i \lambda_j \frac{\partial^2 J_f(z)}{\partial z_i \partial z_j} J_f^{-1}(z) = \frac{6(\lambda\bar{b}')^2}{(1 - z\bar{b}')^2} \lambda. \tag{3.14}$$

Substituting (3.13) and (3.14) into (2.4), we get $\{f; z; \lambda, \mu\} = 0$ for every $z \in \Omega$ and every $\lambda, \mu \in \mathbf{C}^n$.

§4. AN INVARIANT THEOREM

The following Theorem 4.1 is a generalization of the invariant property 2) of the Schwarzian derivative of holomorphic function of one variable in the introduction.

Theorem 4.1. If $\Omega \subset \mathbf{C}^n$ is a domain, $F : \Omega \to \mathbf{C}^n$ is a local biholomorphic mapping, and

$$f(z) = f(0) + \frac{zJ_f(0)}{1 - z\bar{b}'}$$

is defined by (2.5), then

$$\{f \circ F; z; \lambda, \overline{bJ_F(z)'}\} = \{F; z; \lambda, \overline{bJ_F(z)'}\} \tag{4.1}$$

holds for any f which is holomorphic on the image of Ω by F.

Moreover, the equality

$$\{F \circ f; z; \lambda, \mu\} = \{F; f(z); \lambda J_f(z), \overline{\mu J_f^{-1}(z)'}\} \tag{4.2}$$

holds for any f which is defined on a domain $D \subset \mathbf{C}^n$, f is holomorphic on D, and the image of D by f is contained in Ω.

We prove (4.1) at first.

Let $w=F(z)$ and $G(z)=f\circ F(z)$, then

$$J_G(z)=J_F(z)J_f(w).$$

By direct calculation, we have

$$\frac{\partial J_G(z)}{\partial z_i}=\frac{\partial J_F(z)}{\partial z_i}J_f(w)+J_F(z)\sum_\alpha\frac{\partial J_f(w)}{\partial w_\alpha}\frac{\partial w_\alpha}{\partial z_i},$$

$$\begin{aligned}&\lambda\sum_i\lambda_i\frac{\partial J_G(z)}{\partial z_i}J_G^{-1}(z)\bar{\mu}'\\&=\lambda\sum_i\lambda_i\frac{\partial J_F(z)}{\partial z_i}J_F^{-1}(z)\bar{\mu}'\\&\quad+\lambda J_F(z)\sum_\alpha(\lambda J_F(z))_\alpha\frac{\partial J_f(w)}{\partial w_\alpha}J_f^{-1}(w)J_F^{-1}(z)\bar{\mu}'.\end{aligned}$$

$$\begin{aligned}\frac{\partial^2 J_G(z)}{\partial z_i\partial z_j}&=\frac{\partial^2 J_F(z)}{\partial z_i\partial z_j}J_f(w)+\frac{\partial J_F(z)}{\partial z_i}\sum_\alpha\frac{\partial J_f(w)}{\partial w_\alpha}\frac{\partial w_\alpha}{\partial z_j}\\&\quad+\frac{\partial J_F(z)}{\partial z_j}\sum_\alpha\frac{\partial J_f(w)}{\partial w_\alpha}\frac{\partial w_\alpha}{\partial z_i}+J_F(z)\sum_{\alpha,\beta}\frac{\partial^2 J_f(w)}{\partial w_\alpha\partial w_\beta}\frac{\partial w_\alpha}{\partial z_i}\frac{\partial w_\beta}{\partial z_j}\\&\quad+J_F(z)\sum_\alpha\frac{\partial J_f(w)}{\partial w_\alpha}\frac{\partial^2 w_\alpha}{\partial z_i\partial z_j},\end{aligned}\tag{4.3}$$

and

$$\begin{aligned}&\lambda\sum_{i,j}\lambda_i\lambda_j\frac{\partial^2 J_G(z)}{\partial z_i\partial z_j}J_G^{-1}(z)\bar{\mu}'\\&=\lambda\sum_{i,j}\lambda_i\lambda_j\frac{\partial^2 J_F(z)}{\partial z_i\partial z_j}J_F^{-1}(z)\bar{\mu}'\\&\quad+2\lambda\sum_i\lambda_i\frac{\partial J_F(z)}{\partial z_i}\sum_\alpha\frac{\partial J_f(w)}{\partial w_\alpha}(\lambda J_F(z))_\alpha J_f^{-1}(w)J_F^{-1}(z)\bar{\mu}'\\&\quad+\lambda J_F(z)\sum_{\alpha,\beta}\frac{\partial^2 J_f(w)}{\partial w_\alpha\partial w_\beta}(\lambda J_F(z))_\alpha(\lambda J_F(z))_\beta J_f^{-1}(w)J_F^{-1}(z)\bar{\mu}'\\&\quad+\lambda J_F(z)\sum_\alpha\frac{\partial J_f(w)}{\partial w_\alpha}\sum_{i,j}\lambda_i\lambda_j\frac{\partial^2 w_\alpha}{\partial z_i\partial z_j}J_f^{-1}(w)J_F^{-1}(z)\bar{\mu}',\end{aligned}\tag{4.4}$$

where $(\lambda J_F(z))_\alpha$ is the α-th component of $\lambda J_F(z)$.

From (4.3) and (4.4), we get

$$\{f\circ F;z;\lambda,\mu\}=\mathrm{I}+\mathrm{II}+\mathrm{III}+\mathrm{IV}+\mathrm{V}+\mathrm{VI}+\mathrm{VII}.\tag{4.5}$$

where

$$I = (\lambda\bar{\mu}')\lambda\sum_{i,j}\lambda_i\lambda_j\frac{\partial^2 J_F(z)}{\partial z_i\partial z_j}J_F^{-1}(z)\bar{\mu}',$$

$$II = 2(\lambda\bar{\mu}')\lambda\sum_i\lambda_i\frac{\partial J_F(z)}{\partial z_i}\sum_\alpha\frac{\partial J_f(w)}{\partial w_\alpha}(\lambda J_F(z))_\alpha J_f^{-1}(w)J_F^{-1}(z)\bar{\mu}',$$

$$III = (\lambda\bar{\mu}')\lambda J_F(z)\sum_{\alpha,\beta}\frac{\partial^2 J_f(w)}{\partial w_\alpha\partial w_\beta}(\lambda J_F(z))_\alpha(\lambda J_F(z))_\beta J_f^{-1}(w)J_F^{-1}(z)\bar{\mu}',$$

$$IV = (\lambda\bar{\mu}')\lambda J_F(z)\sum_\alpha\frac{\partial J_f(w)}{\partial w_\alpha}\sum_{i,j}\lambda_i\lambda_j\frac{\partial^2 w_\alpha}{\partial z_i\partial z_j}J_f^{-1}(w)J_F^{-1}(z)\bar{\mu}',$$

$$V = -\frac{3}{2}(\lambda\sum_i\lambda_i\frac{\partial J_F(z)}{\partial z_i}J_F^{-1}(z)\bar{\mu}')^2,$$

$$VI = -3\lambda\sum_{i,\alpha}\lambda_i\frac{\partial J_F(z)}{\partial z_i}J_F^{-1}(z)\bar{\mu}'\lambda J_F(z)(\lambda J_F(z))_\alpha\frac{\partial J_f(w)}{\partial w_\alpha}J_f^{-1}(w)J_F^{-1}(z)\bar{\mu}',$$

$$VII = -\frac{3}{2}(\lambda J_F(z)\sum_\alpha(\lambda J_F(z))_\alpha\frac{\partial J_f(w)}{\partial w_\alpha}J_f^{-1}(w)J_F^{-1}(z)\bar{\mu}')^2.$$

Obviously,

$$I + V = \{F; z; \lambda, \mu\} \tag{4.6}$$

and

$$III + VII = \{f; w; \lambda J_F(z), \mu\overline{J_F^{-1}(z)'}\}. \tag{4.7}$$

It is easy to verify

$$\sum_\alpha(\lambda J_F(z))_\alpha\frac{\partial J_f(w)}{\partial w_\alpha} = \frac{\lambda J_F(z)\bar{b}'}{1 - w\bar{b}'}J_f(w) + \frac{\bar{b}'\lambda J_F(z)J_f(w)}{1 - w\bar{b}'}. \tag{4.8}$$

Using (4.8), we can simplify II and VI as

$$\begin{aligned} II = {} & \frac{2(\lambda\bar{\mu}')(\lambda J_F(z)\bar{b}')}{1 - w\bar{b}'}\lambda\sum_i\lambda_i\frac{\partial J_F(z)}{\partial z_i}J_F^{-1}(z)\bar{\mu}' \\ & + \frac{2(\lambda\bar{\mu}')^2}{1 - w\bar{b}'}\lambda\sum_i\frac{\partial J_F(z)}{\partial z_i}\bar{b}' \end{aligned} \tag{4.9}$$

and

$$VI = -\frac{6(\lambda\bar{\mu}')(\lambda J_F(z)\bar{b}')}{1 - w\bar{b}'}\lambda\sum_i\lambda_i\frac{\partial J_F(z)}{\partial z_i}J_F^{-1}(z)\bar{\mu}'. \tag{4.10}$$

We know that

$$\lambda\sum_i\lambda_i\frac{\partial J_F(z)}{\partial z_i} = \sum_{i,j}\lambda_i\lambda_j\frac{\partial^2 w}{\partial z_i\partial z_j}$$

and

$$\sum_\alpha \frac{\partial J_f(w)}{\partial w_\alpha}\left(\lambda \sum_i \lambda_i \frac{\partial J_F(z)}{\partial z_i}\right)_\alpha$$
$$= \lambda \sum_i \lambda_i \frac{\partial J_F(z)}{\partial z_i} \bar{b}'(1 - w\bar{b}')^{-1} J_f(w) + \frac{\bar{b}'\lambda}{1 - w\bar{b}'} \sum_i \lambda_i \frac{\partial J_F(z)}{\partial z_i} J_f(w). \tag{4.11}$$

By (4.11), we can simplify IV as

$$\mathrm{IV} = \frac{(\lambda\bar{\mu}')^2}{1 - w\bar{b}'} \lambda \sum_i \lambda_i \frac{\partial J_F(z)}{\partial z_i} \bar{b}'$$
$$+ \frac{(\lambda\bar{\mu}')\lambda J_F(z)\bar{b}'}{1 - w\bar{b}'} \lambda \sum_i \lambda_i \frac{\partial J_F(z)}{\partial z_i} J_F^{-1}(z)\bar{\mu}'. \tag{4.12}$$

From (4.9), (4.10) and (4.12), the sum of II, IV and VI is equal to

$$\mathrm{VIII} = 3\frac{\lambda\bar{\mu}'}{1 - w\bar{b}'} \lambda \sum_i \lambda_i \frac{\partial J_F(z)}{\partial z_i} [\lambda\bar{\mu}'\bar{b}' - \lambda J_F \bar{b}' J_F^{-1}(z)\bar{\mu}']. \tag{4.13}$$

By (4.5), (4.6), (4.7) and (4.13), we get

$$\{f \circ F; z; \lambda, \beta\} = \{F; z; \lambda, \mu\} + \{f; w; \lambda J_F(z), \mu \overline{J_F^{-1}(z)'}\} + \mathrm{VIII}. \tag{4.14}$$

By Theorem 2.1,

$$\{f; w; \lambda J_F(z), \mu \overline{J_F^{-1}(z)'}\} = 0$$

for every $z \in \Omega$ and $\lambda, \beta \in \mathbf{C}^n$. If we take $\mu = b\overline{J_F(z)'}$, then VIII $= 0$ by (4.13). Combining all these results, we have proved (4.1).

The proof of (4.2) is similar to the proof of (4.1). Let $w = f(z)$ and $H(z) = F \circ f(z)$, then $J_H(z) = J_f(z) J_F(w)$. By a similar argument as we used before, it is easy to verify

$$\{F \circ f; z; \lambda, \mu\} = \{f; z; \lambda, \mu\} + \{F; w; \lambda J_f(z), \mu \overline{J_F^{-1}(z)'}\} + \mathrm{II}' + \mathrm{IV}' + \mathrm{VI}', \tag{4.15}$$

where

$$\mathrm{II}' = 2(\lambda\bar{\mu}')\lambda \sum_i \frac{\partial J_f(z)}{\partial z_i} \sum_\alpha \frac{\partial J_F(w)}{\partial w_\alpha} (\lambda J_f(z))_\alpha J_F^{-1}(w) J_f^{-1}(z)\bar{\mu}',$$

$$\mathrm{IV}' = (\lambda\bar{\mu}')\lambda J_f(z) \sum_\alpha \frac{\partial J_F(w)}{\partial w_\alpha} \sum_{i,j} \lambda_i \lambda_j \frac{\partial^2 w_\alpha}{\partial z_i \partial z_j} J_F^{-1}(w) J_f^{-1}(z)\bar{\mu}'.$$

and

$$\mathrm{VI}' = -3\lambda \sum_i \lambda_i \frac{\partial J_f(z)}{\partial z_i} J_f^{-1}(z)\bar{\mu}' \lambda J_f(z) \sum_\alpha (\lambda J_f(z))_\alpha \frac{\partial J_F(w)}{\partial w_\alpha} J_F^{-1}(w) J_f^{-1}(z)\bar{\mu}'.$$

Using (3.13), we can simplify II′ and VI′ as

$$\mathrm{II}' = \frac{4(\lambda\bar{\mu}')(\lambda\bar{b}')}{1 - z\bar{b}'} \lambda J_f(z) \sum_\alpha (\lambda J_f(z))_\alpha \frac{\partial J_F(w)}{\partial w_\alpha} J_F^{-1}(w) J_f^{-1}(z)\bar{\mu}' \tag{4.16}$$

and

$$\mathrm{VI}' = \frac{-6(\lambda\bar{\mu}')(\lambda\bar{b}')}{1-z\bar{b}'}\lambda J_f(z)\sum_\alpha(\lambda J_f(z))_\alpha\frac{\partial J_F(w)}{\partial w_\alpha}J_F^{-1}(w)J_f^{-1}(z)\bar{\mu}'. \tag{4.17}$$

From the definition of f in the theorem,

$$\sum_{i,j}\lambda_i\lambda_j\frac{\partial^2 f_\alpha}{\partial z_i\partial z_j} = \left(\lambda\sum_i\lambda_i\frac{\partial J_f(z)}{\partial z_i}\right)_\alpha = \frac{2\lambda\bar{b}'}{1-z\bar{b}'}(\lambda J_f(z))_\alpha$$

and

$$\sum_\alpha\frac{\partial J_F(w)}{\partial w_\alpha}\sum_{i,j}\lambda_i\lambda_j\frac{\partial^2 w_\alpha}{\partial z_i\partial z_j} = \frac{2\lambda\bar{b}'}{1-z\bar{b}'}\sum_\alpha(\lambda J_f(z))_\alpha\frac{\partial J_F(w)}{\partial w_\alpha}. \tag{4.18}$$

From (4.18), we can simplify IV' as

$$\mathrm{IV}' = \frac{2(\lambda\bar{\mu}')(\lambda\bar{b}')}{1-z\bar{b}'}\lambda J_f(z)\sum_\alpha(\lambda J_f(z))_\alpha\frac{\partial J_F(w)}{\partial w_\alpha}J_F^{-1}(w)J_f^{-1}(z)\bar{\mu}'. \tag{4.19}$$

By (4.16), (4.17) and (4.20), we have

$$\mathrm{II}' + \mathrm{IV}' + \mathrm{VI}' = 0. \tag{4.20}$$

By Theorem 2.1, $\{f; z; \lambda, \mu\} = 0$. Combining (4.15) and (4.20), we have proved (4.2).

This completes the proof of Theorem 4.1.

§ 5. A UNIVALENCE CRITERION

The following Theorem 5.1 is a generalization of Nehari's theorem of the univalency criterion.

Theorem 5.1. If Ω is a bounded convex domain in $\mathbf{C}^n$, whose diameter is d, and $f: \Omega - \mathbf{C}^n$ is a locally biholomorphic mapping. If for every λ and $\mu \in \mathbf{C}^n$,

$$|\{f; z; \lambda, \mu\}| \le \frac{\pi^2}{2}|\lambda\bar{\mu}'|^2 \tag{5.1}$$

holds true, then f is a biholomorphic mapping on Ω.

We need the following.

Lemma 5.1. For any real-valued continuous function $p(t)$ on $[0,1]$, if there exists a function $u(t)$ which is continuous on $[0,1]$, is positive for $t \in (0,1]$, and satisfies the differential inequality

$$\frac{d^2u}{dt^2} + p(t)u(t) \le 0 \tag{5.2}$$

for $t \in (0,1)$. If $q(t)$ is a complex-valued continuous function on $[0,1]$, such that

$$\Re e q(t) \le p(t) \tag{5.3}$$

when $t \in [0,1]$, then every nontrivial solution of the differential equation

$$\frac{d^2w}{dt^2} + q(t)w(t) = 0 \tag{5.4}$$

has at most one zero at $[0, 1]$.

Proof. Suppose that w is a nontrivial solution of (5.4) and has at least two zeros in $[0, 1]$. Let t_1 and t_2 be two such zeros with $0 \leq t_1 \leq t_2 \leq 1$. From (5.2), (5.4) and $u(t) > 0$ when $t \in (0, 1)$, we have

$$\frac{d}{dt}\Re e\left[\frac{dw}{dt}\bar{w} - \frac{1}{u}\frac{du}{dt}|w|^2\right] \geq |w|^2(p(t) - \Re eq(t)) + \left|\frac{dw}{dt} - \frac{w}{u}\frac{du}{dt}\right|^2. \tag{5.5}$$

Integrating both sides of (5.5) with respect to t from t_1 to t_2, we find that

$$0 \geq \int_{t_1}^{t_2} |w|^2(p(t) - \Re eq(t))dt + \int_{t_1}^{t_2}\left|\frac{dw}{dt} - \frac{w}{u}\frac{du}{dt}\right|^2 dt \geq 0.$$

It follows from (5.3),

$$\frac{dw}{dt} - \frac{w}{u}\frac{du}{dt} = 0$$

for $t \in (0, 1)$ almost everywhere, that is,

$$\frac{d}{dt}\left(\frac{w}{u}\right) = \frac{1}{u}\left(\frac{dw}{dt} - \frac{w}{u}\frac{du}{dt}\right) = 0$$

for $t \in (0, 1)$ almost everywhere. Thus $w(t) = cu(t)$ for $t \in (0, 1)$ almost everywhere, and c is a non-zero constant. By the continuity of $u(t)$ and $w(t)$, we get $w(t) = cu(t)$ for $t \in [0, 1]$. But $w(t_2) = 0$ on one hand, and $cu(t_2) \neq 0$ on the other hand. This contradiction proves the lemma.

§ 6. PROOF OF THEOREM 5.1

For every $\lambda \in \mathbf{C}^n$, $\lambda \neq 0$ and every $\mu \in L_\lambda^\perp$, we have $\{f; z; \lambda, \mu\} = 0$ for any $z \in \Omega$, if condition (5.1) is satisfied. It follows that

$$\lambda \sum_i \lambda_i \frac{\partial J_f}{\partial z_i} J_f^{-1} \mu' = 0$$

holds for any $z \in \Omega$, and any $\mu \in L_\lambda^\perp$. This implies

$$\lambda \sum_i \lambda_i \frac{\partial J_f}{\partial z_i} J_f^{-1} = a_\lambda(z)\lambda, \tag{6.1}$$

where $a_\lambda(z)$ is a holomorphic function of z in Ω, and parameterized with λ.

If Theorem 5.1 is not true, then there exist a pair of points $\xi \in \Omega$ and $\eta \in \Omega$ such that $f(\xi) = f(\eta)$. The function

$$g(t) = f(t\xi + (1-t)\eta) - f(\xi)$$

is well defined on $[0, 1]$ because Ω is convex. Then $g(0) = g(1) = 0$, and

$$\frac{dg(t)}{dt} = \left(\frac{dg_1}{dt}, \cdots, \frac{dg_n}{dt}\right) = \lambda J_f(t\xi + (1-t)\eta), \tag{6.2}$$

$$\frac{d^2g(t)}{dt^2} = \lambda \sum_i \lambda_i \frac{\partial J_f}{\partial z_i}(t\xi + (1-t)\eta). \tag{6.3}$$

where $\lambda = \xi - \eta$. From (6.1) and (6.3), we get

$$\frac{d^2g(t)}{dt^2} J^{-1}(t\xi + (1-t)\eta) = a_\lambda(t\xi + (1-t)\eta)\lambda.$$

Operating $\sum_j \lambda_j \frac{\partial}{\partial z_j}$ on (6.1), and using (6.1) again, we obtain

$$\lambda \sum_{i,j} \lambda_i \lambda_j \frac{\partial^2 J_f(z)}{\partial z_i \partial_j} J_f^{-1}(z) = \lambda \sum_j \lambda_j \frac{\partial a}{\partial z_j} + a_\lambda^2(z)\lambda. \tag{6.4}$$

Substitution of (6.1) and (6.4) into (2.4) shows that the Schwarzian derivative $\{f; z; \lambda, \mu\}$ is equal to $(\lambda\bar{\mu}')^2 \left(\sum_j \lambda_j \frac{\partial a_\lambda}{\partial z_j} - \frac{1}{2}a_\lambda^2\right)$ for every $\lambda \in C^n$ and $\mu \in C^n$. The hypothesis (5.1) becomes

$$\left|\sum_j \lambda_j \frac{\partial a_\lambda(z)}{\partial z_j} - \frac{1}{2}a_\lambda^2(z)\right| \leq \frac{\pi^2}{2}.$$

Consider $z = t\xi + (1-t)\eta$. Since

$$\frac{da_\lambda}{dt}(t\xi + (1-t)\eta) = \sum_i \lambda_i \frac{\partial a_\lambda}{\partial z_i}(t\xi + (1-t)\eta)$$

we have

$$\left|\frac{da_\lambda}{dt}(t\xi + (1-t)\eta) - \frac{1}{2}a_\lambda^2(t\xi + (1-t)\eta)\right| \leq \frac{\pi^2}{2}. \tag{6.5}$$

The equalities (6.1), (6.2) and (6.3) imply

$$\frac{d^2g(t)}{dt} - a_\lambda(z)\frac{dg(t)}{dt} = 0. \tag{6.6}$$

Let

$$v(t) = g(t)\exp\left\{-\frac{1}{2}\int_0^t b_\lambda(\tau)d\tau\right\}, \tag{6.7}$$

where $b_\lambda(\tau) = a_\lambda(\tau\xi + (1-\tau)\eta)$, then $v(0) = v(1) = 0$ since $g(0) = g(1) = 0$, and

$$\frac{dv(t)}{dt} = \frac{dg(t)}{dt}\exp\left\{-\frac{1}{2}\int_0^t b_\lambda(\tau)d\tau\right\} - \frac{1}{2}g(t)b_\lambda(t)\exp\left\{-\frac{1}{2}\int_0^t b_\lambda(\tau)d\tau\right\},$$

$$\frac{d^2v(t)}{dt^2} = \left(\frac{d^2g(t)}{dt^2} - b_\lambda(t)\frac{dg(t)}{dt}\right)\exp\left\{-\frac{1}{2}\int_0^t b_\lambda(\tau)d\tau\right\}$$
$$-\frac{1}{2}\frac{db_\lambda(t)}{dt}v(t) + \frac{1}{4}b_\lambda^2(t)v(t).$$

That is

$$\frac{d^2v(t)}{dt^2}+\frac{1}{2}\left(\frac{db_\lambda(t)}{dt}-\frac{1}{2}b_\lambda^2(t)\right)v(t)=0$$

by (6.6). Each component of this equation is

$$\frac{d^2v_\alpha(t)}{dt^2}+\frac{1}{2}\left(\frac{db_\lambda(t)}{dt}-\frac{1}{2}b_\lambda(t)\right)v_\alpha(t)=0$$

for $\alpha=1,2,\cdots,n$.

If we take

$$q(t)=\frac{1}{2}\left(\frac{da_\lambda}{dt}(t\xi+(1-t)\eta)-\frac{1}{2}a_\lambda^2(t\xi+(1-t)\eta)\right)$$

and $p(t)=\dfrac{\pi^2}{4}$, then (6.5) implies $\Re e\ q(t)\le p(t)$. Obviously

$$\frac{d^2u}{dt^2}+\frac{\pi^2}{4}u(t)\le 0$$

has a positive solution $u(t)=\sin\dfrac{\pi t}{2}$ for $t\in(0,1]$. By Lemma 5.1, $v_\alpha(t\xi+(1-t)\eta)$ cannot have two zeros when $t\in[0,1]$. With this contradiction, Theorem 5.1 is proved.

Acknowledgment. This paper was written while the second author was at the Department of Mathematics of University of California, San Diego as a visiting professor. He would like to express his gratitute to the Department and Professor C. H. FitzGerald.

REFERENCES

1. Osgood, B. & Stove, D., "The Schwarzian derivative and conformal mappings of Riemannian manifolds," *Duke Math. J.*, **67**, pp. 57-99, 1992.
2. Osgood, B. & Stove D., "A generalization of Nehari's univalence criterion," *Comment. Math. Helv.*, **65**, pp. 234-242, 1990.
3. Carne, K., "The Schwarzian derivative for conformal maps," *J. Reine Angew. Math.*, **408**, pp. 10-33, 1990.
4. Gao Weiqi, "Schwarzian curvatures of holomorphic curves," Complex Variables (to appear).
5. Flanders, H., "The Schwarzian derivative as a curvature," *J. Diff. Geom.*, **4**, pp. 515-519, 1970.
6. FitzGerald, C. H. & Gong Sheng, "The Schwarzian derivative in several complex variables," *Science in China (Series A)*, **36**, pp. 513-523, 1993.

V. 一类 Reinhardt 域的 Bergman 核函数

自从 C. Fefferman [1] 给出了强拟凸域上的 Bergman 核函数的渐近表达式之后，有很多有关 Bergman 核函数的讨论，尤其是弱拟凸域上的 Bergman 核函数的渐进表达.

[45.1], [22.1], [11] 是对一类 Reinhardt 域 $D = \{z \in \mathbb{C}^n \mid \|z\|_\alpha = \sum_{j=1}^{n} |z_j|^{2/\alpha_j} < 1\}$ $\alpha_j > 0$, $j = 1, 2, \cdots, n$ 及其更为一般的域的 Bergman 核函数的渐进表达及性质进行讨论.

参 考 文 献

(1) Fefferman C. The Bergman Kernel and biholomorphic mappings of psendo convex domains, Invent. Math., 1974, 26: 1-65.

数学年刊
8A (2) 1987

关于 Reinhardt 域（一）

龚 昇
（中国科技大学，中国科学院应用数学研究所）

提 要

本文讨论了Reinhardt域$D=\{z\in\mathbb{C}^n:\sum_{j=1}^{n}|z_j|^{\frac{2}{\alpha_j}}<1\}$, $1\geqslant\alpha_j>0$, $j=1,\cdots,n$. 当α_j不全为1时为弱拟凸域. 首先给出了其Bergman核的渐近展开，指出了核函数的奇性与Levi形式的退化关系. 其次指出A. Bonanie与N. Lohone给出的Szegö核是错误的，并给出了Szegö核的估计. 最后讨论了一类域的几何性质，给出了Bergman度量，Ricci曲率及无向曲率的明确表达式，显示了强拟凸点与弱拟凸点的几何性质的不同.

§1. 引 言

在多复变数函数论中，经过多年来，特别是近二十年来的深入研究，人们对强拟凸域有了十分深刻的了解，例如见文[1]，但是对于弱拟凸域的研究还了解不多. 弱拟凸域与强拟凸域是有本质上的不同的. 弱拟凸域有多少模型至今还不清楚.

为了对弱拟凸域有所了解，选择其典型的区域进行深入的研究，并与强拟凸域进行比较，这是十分必要的，这样做，对一般的弱拟凸域的了解会有所帮助，例如见文[2, 3, 4].

与强拟凸域的模型——超球最接近的是Reinhardt域

$$D=\left\{z\in\mathbb{C}^n:\sum_{j=1}^{n}|z_j|^{2/\alpha_j}<1\right\},\ 1\geqslant\alpha_j>0,\ j=1,\ \cdots,\ n. \tag{1.1}$$

显然，当α_j不全为1时，这是弱拟凸域. 这项研究是对这类域的一些性质进行系统的讨论.

对于强拟凸域，C. Fefferman[5]对Bergman核$K(z, z)$的表达式有著名的结果. 对于这类弱拟凸域D如何？在§2中我们给出D的$K(z, z)$的渐近展开，指出了核函数的奇性与Levi形式的退化之间的关系，这包含了J. P. D'Angelo[3]的结果. 1977年，A. Bonanie与N. Lohone[4]给出了D的Szegö核，在§3中指出，他们的结果是错误的，并给出了D的Szegö核的估计式. 为了讨论D的几何性质，我们讨论了最简单的$\alpha_1=1$, $\alpha_2=\alpha$, $0<\alpha<1$, $n=2$的情形，而所得的结果，不难推广到$\alpha_1=\cdots=\alpha_{n-1}=1$, $\alpha_n=\alpha$, $0<\alpha<1$的情形. 在§4中，先给出了其Bergman度量. 在§5中给出了在每点的Ricci曲率的确切表达式，§6中给出了每点的无向曲率。我们知道，当区域为超球时，Ricci曲率为常量-1，无向曲率为常量$-n$. 而当区域为D时，Ricci曲率与无向曲率不再是常量，这依赖于α, z及dz. 在§5, §6中得到的结果显示了强拟凸点与弱拟凸点的几何性质的不同.

本文1984年10月18日收到.

考虑这样的域是 E. M. Stein 建议于我的，为此向他表示感谢．在撰写此文时，史济怀与我进行了有益的讨论，也一并致谢．

§2. Bergman 核的渐近表示

若 U 为强拟凸域，$K(z, w)$ 为其 Bergman 核，在 ∂U 的附近，$K(z, z)$ 的表示式有 C. Fefferman 的著名定理：在 ∂U 附近有光滑函数 f 及 g，使得

$$K(z, z)=f(z)(-r(z))^{-n-1}+g(z)\log(-r(z)), \tag{2.1}$$

这里 $r(z)$ 为 U 的定义函数．

对于弱拟凸域，特别在弱拟凸点，其 Bergman 核相应的表示是什么？如 D 由(1.1)所定义，J. P. D'Angelo 在文[3]中给出了 D 的 Bergman 核 $K(z, w)$ 的级数表达式为

$$\frac{1}{\alpha_1\cdots\alpha_n\pi^n}\sum_{m_1,\cdots,m_n\geq 0}\frac{\Gamma\left(\sum_{j=1}^n\alpha_j(m_j+1)+1\right)}{\Gamma(\alpha_1m_1+\alpha_1)\cdots\Gamma(\alpha_nm_n+\alpha_n)}(z_1\bar{w}_1)^{m_1}\cdots(z_n\bar{w}_n)^{m_n}. \tag{2.2}$$

在 ∂D 附近，其渐近表示是什么？在文[4]中，他给出了，当 $\alpha_1=\cdots=\alpha_{n-1}=1$, $\alpha_n=\alpha$, $0<\alpha<1$ 的情形，此时(2.2)可表为

$$\frac{n!\alpha^{n-1}}{\pi^n}\sum_{k=1}^n d_k\left(1-\sum_{i=1}^{n-1}z_i\bar{w}_i\right)^{-n+k\alpha}\left[\left(1-\sum_{i=1}^{n-1}z_i\bar{w}_i\right)^{\alpha}-z_n\bar{w}_n\right]^{-k-1}, \tag{2.3}$$

这里 d_k 为 $\frac{1}{\alpha}$ 的 $n-k$ 次多项式，当 $\alpha=1$, $k\neq n$ 时，

$$d_k=0;\ d_n=1,\ d_{n-1}=\frac{(n-1)(1-\alpha)}{2\alpha}.$$

由此可以推出：当 $\alpha_1=\cdots=\alpha_{n-1}=1$, $\alpha_n=\alpha$ 时，

$$\lim_{\substack{r\to 0\\ z_n\text{ 固定}}}\frac{(-r(z))^{n+1}K(z, z)}{\det(r_{z_i\bar{z}_j}(z))}=\frac{n!}{\pi^n}, \tag{2.4}$$

这里 $r(z)$ 为 D 的定义函数，$(r_{z_i\bar{z}_j}(z))=\left(\frac{\partial^2 r}{\partial z_i\partial\bar{z}_j}\right)_{1<i,j<n}$.

对于一般的情形，文[4]中证明了：当 $z_1=\cdots=z_k=0$ 时，则存在正的常数 c_1 及 c_2 及非负多项式 f_k，使得

$$\begin{aligned}c_1\det(r_{z_i\bar{z}_j}(k, z))(-r(k, z))^{-n-1+k-\sum_{i=1}^k d_i}&\leqslant K(z, z)\\ &\leqslant c_2(\det(r_{z_i\bar{z}_j}(k, z)+f_k(z))\cdot(-r(k, z))^{-n-1+k-\sum_{i=1}^k d_i},\end{aligned} \tag{2.5}$$

这里 $r(k, z)=\sum_{l=k+1}^n|z_l|^{2/\alpha_l}-1$, $(r_{z_i\bar{z}_j}(k, z))=\left(\frac{\partial^2 r(k, z)}{\partial z_i\partial\bar{z}_j}\right)_{k+1<i,j<n}$.

本节将分别各种情形，给出(2.2)的渐近表示式，并以(2.4)，(2.5)为其推论．

事实上，(2.2)可表为

$$\frac{1}{\alpha_1\cdots\alpha_n\pi^n}\int_0^\infty e^{-t}t^{\alpha_1+\cdots+\alpha_n}E_{\alpha_1}(t^{\alpha_1}z_1\bar{w}_1)\cdots E_{\alpha_n}(t^{\alpha_n}z_n\bar{w}_n)dt, \tag{2.6}$$

这里

$$E_\alpha(u)=\sum_{m=0}^\infty\frac{u^m}{\Gamma(\alpha m+\alpha)}$$

为 Mittag-Leffler 函数，当 $|\operatorname{Arg}u|<\frac{7\pi\alpha}{8}$ 时，有

$$\left| E_\alpha(u) - \frac{1}{\alpha} u^{\frac{1}{\alpha}-1} \exp u^{\frac{1}{\alpha}} \right| \leqslant c, \tag{2.7}$$

这里 c 为依赖于 α 的常数(见文[6, 7]). 由此可得: 当 $z_1 \neq 0, \cdots, z_n \neq 0$ 时, $K(z, z)$ 不大于

$$\frac{1}{\alpha_1 \cdots \alpha_n \pi^n} \int_0^\infty e^{-t} t^{\alpha_1+\cdots+\alpha_n} \left(\frac{1}{\alpha_1} t^{1-\alpha_1} |z_1|^{\frac{2}{\alpha_1}-2} \exp(t|z_1|^{\frac{2}{\alpha_1}}) + c_1 \right) \times \cdots$$
$$\times \left(\frac{1}{\alpha_n} t^{1-\alpha_n} |z_n|^{\frac{2}{\alpha_n}-2} \exp(t|z_n|^{\frac{2}{\alpha_n}}) + c_n \right) dt,$$

这里 c_j 为依赖于 α_j 的常数, $j=1, 2, \cdots, n$. 经过计算, 上式等于

$$\frac{\Gamma(n+1)}{\pi^n} \left(1 - \sum_{j=1}^n |z_j|^{\frac{2}{\alpha_j}}\right)^{-n-1} \prod_{j=1}^n \left(\frac{|z_j|^{\frac{1}{\alpha_j}-1}}{\alpha_j} \right)^2 + \sum_{k=1}^n \frac{c_k \Gamma(n+\alpha_k)}{\alpha_k \pi^n \left(1 - \sum_{j\neq k} |z_j|^{2/\alpha_j}\right)^{n+\alpha_k}}$$
$$\cdot \prod_{j\neq k} \left(\frac{|z_j|^{\frac{1}{\alpha_j}-1}}{\alpha_j} \right)^2 + \sum_{k,l=1}^n \frac{c_k c_l \Gamma(n+\alpha_k+\alpha_l-1)}{\alpha_k \alpha_l \pi^n \left(1 - \sum_{j\neq k,l} |z_j|^{2/\alpha_j}\right)^{n+\alpha_k+\alpha_l-1}} \prod_{j\neq k,l} \left(\frac{|z_j|^{\frac{1}{\alpha_j}-1}}{\alpha_j} \right)^2$$
$$+ \cdots + \frac{c_1 \cdots c_n \Gamma(\alpha_1 + \cdots + \alpha_n + 1)}{\alpha_1 \cdots \alpha_n \pi^n}.$$

如果记 $r^{[k_1,\cdots,k_l]} = \sum_{j\neq k_1,\cdots,k_l} |z_j|^{\frac{2}{\alpha_j}} - 1$, 则上式可写成

$$\frac{\Gamma(n+1)\det(r_{z_i\bar{z}_j})}{\pi^n(-r(z))^{n+1}} + \sum_{l=1}^n \sum_{k_1<\cdots<k_l} \frac{c_{k_1}\cdots c_{k_l} \Gamma(n+1-l+\alpha_{k_1}+\cdots+\alpha_{k_l}) \det(r^{[k_1,\cdots,k_l]}_{z_i\bar{z}_j})}{\alpha_{k_1}\cdots\alpha_{k_l} \pi^n (-r^{[k_1,\cdots,k_l]})^{n+1-l+\alpha_{k_1}+\cdots+\alpha_{k_l}}},$$

这也可写成

$$\frac{\Gamma(n+1)\det(r_{z_i\bar{z}_j})}{\pi^n(-r(z))^{n+1}}$$
$$\cdot \left[1 + \sum_{l=1}^n \sum_{k_1<\cdots<k_l} \frac{c_{k_1}\cdots c_{k_l} \Gamma(n+1-l+\alpha_{k_1}+\cdots+\alpha_{k_l}) (-r(z))^{n+1} \det(r^{[k_1,\cdots,k_l]}_{z_i\bar{z}_j})}{\alpha_{k_1}\cdots\alpha_{k_l} \Gamma(n+1) (-r^{[k_1,\cdots,k_l]})^{n+1-l+\alpha_{k_1}+\cdots+\alpha_{k_l}} \det(r_{z_i\bar{z}_j})} \right]. \tag{2.8}$$

可以证明: 当 $z_1 \neq 0, \cdots, z_n \neq 0$, $r(z) \to 0$ 时,

$$\frac{(-r(z))^{n+1} \det(r^{[k_1,\cdots,k_l]}_{z_i\bar{z}_j})}{(-r^{[k_1,\cdots,k_l]})^{n+1-l+\alpha_{k_1}+\cdots+\alpha_{k_l}} \det(r_{z_i\bar{z}_j})} \to 0. \tag{2.9}$$

现在以 $l=1$ 为例来证(2.9). 这时(2.9)的左边为

$$\left(1 - \frac{|z_k|^{2/\alpha_k}}{1 - \sum_{j\neq k} |z_j|^{2/\alpha_j}}\right)^{n+1} \left(\frac{1 - \sum_{j\neq k} |z_j|^{2/\alpha_j}}{|z_k|^{2/\alpha_k}} \right)^{1-\alpha_k} = \frac{(1-Q_k)^{n+1}}{Q_k^{1-\alpha_k}},$$

此处 $Q_k = |z_k|^{2/\alpha_k} / (1 - \sum_{j\neq k} |z_j|^{2/\alpha_j})$. 当 z 在 D 内沿着 $z_1 \neq 0, \cdots, z_n \neq 0$ 的途径趋于边界时, $Q_k \to 1$, 故(2.9)成立. 同样可证 $l>1$ 的情形.

再利用(2.7), 可以证明 $K(z, z)$ 不小于

$$\frac{\Gamma(n+1)\det(r_{z_i\bar{z}_j})}{\pi^n(-r(z))^{n+1}}$$
$$+ \sum_{l=1}^n (-1)^l \sum_{k_1<\cdots<k_l} \frac{c_{k_1}\cdots c_{k_l} \Gamma(n+1-l+\alpha_{k_1}+\cdots+\alpha_{k_l}) \det(r^{[k_1,\cdots,k_l]}_{z_i\bar{z}_j})}{\alpha_{k_1}\cdots\alpha_{k_l} \pi^n (-r^{[k_1,\cdots,k_l]})^{n+1-l+\alpha_{k_1}+\cdots+\alpha_{k_l}}} \tag{2.10}$$

由于 ∂D 上任何一点均可沿 $z_1 \neq 0, \cdots, z_n \neq 0$ 的途径趋于这点, 故得

$$\lim_{\substack{r(z)\to 0\\ z_1\neq 0,\cdots,z_n\neq 0}} \frac{K(z,z)(-r(z))^{n+1}}{\det(r_{z_i\bar{z}_j})} = \frac{\Gamma(n+1)}{\pi^n} \tag{2.11}$$

对一般的 D 都成立,即使趋向的点是弱拟凸点也成立.

由于 $E_1(u)=e^u$,所以如果 $\alpha_1=\cdots=\alpha_k=1$,则(2.8)与(2.10)可更简单些,即 $K(z,z)$ 不大于

$$\frac{\Gamma(n+1)\det(r_{z_i\bar{z}_j})}{\pi^n(-r(z))^{n+1}} + \sum_{l=1}^{n-k}\sum_{\substack{k=1\\ s_1<\cdots<s_l}}^{n} \frac{c_{s_1}\cdots c_{s_l}\Gamma(n+1-l+\alpha_{s_1}+\cdots+\alpha_{s_l})\det(r_{v_i\bar{z}_j}^{[s_1,\cdots,s_l]})}{\alpha_{s_1}\cdots\alpha_{s_l}\pi^n(-r^{[s_1,\cdots,s_l]})^{n+1-l+\alpha_{l_1}+\cdots+\alpha_{s_l}}}.$$

同样 $K(z,z)$ 不小于

$$\frac{\Gamma(n+1)\det(r_{z_i\bar{z}_j})}{\pi^n(-r(z))^{n+1}} + \sum_{l=1}^{n-k}(-1)^l\sum_{\substack{k+1\\ s_1<\cdots<s_l}}^{n} \frac{c_{s_1}\cdots c_{s_l}\Gamma(n+1-l+\alpha_{s_1}+\cdots+\alpha_{s_l})\det(r_{z_i\bar{z}_j}^{[s_1,\cdots,s_l]})}{\alpha_{s_1}\cdots\alpha_{s_l}\pi^n(-r^{[s_1,\cdots,s_l]})^{n+1-l+\alpha_{s_1}+\cdots+\alpha_{s_l}}}.$$

在这种情形,当 z 在 D 内沿着 $z_{k+1}\neq 0,\cdots,z_n\neq 0$ 的途径趋于边界时,可得

$$\lim_{\substack{r(z)\to 0\\ z_{k+1}\neq 0,\cdots,z_n\neq 0}} \frac{K(z,z)(-r(z))^{n+1}}{\det(r_{z_i\bar{z}_j})} = \frac{\Gamma(n+1)}{\pi^n},$$

即使这点是弱拟凸的也成立. 显然(2.4)不过是上述结论的一个特例.

再来考虑 $z_1=\cdots=z_k=0$ 的情形,这时(2.2)成为

$$\frac{1}{\alpha_1\cdots\alpha_n\pi^n}\sum_{m_{k+1},\cdots,m_n\geq 0}\frac{\Gamma\left(\alpha_1+\cdots+\alpha_k+\sum\limits_{j=k+1}^{n}\alpha_j(m_j+1)+1\right)}{\Gamma(\alpha_1)\cdots\Gamma(\alpha_k)\Gamma(\alpha_{k+1}(m_{k+1}+1))\cdots\Gamma(\alpha_n(m_n+1))}\times$$
$$\times(z_{k+1}\bar{w}_{k+1})^{m_{k+1}}\cdots(z_n\bar{w}_n)^{m_n},$$

这可表为

$$\frac{1}{\Gamma(\alpha_1+1)\cdots\Gamma(\alpha_n+1)\alpha_{k+1}\cdots\alpha_n\pi^n}\int_0^\infty e^{-t}t^{\alpha_1+\cdots+\alpha_n}E_{\alpha_{k+1}}(t^{\alpha_{k+1}}z_{k+1}\bar{w}_{k+1})\cdots E_{\alpha_n}(t^{\alpha_n}z_n\bar{w}_n)dt.$$

故 $K(z,z)$ 不大于

$$\frac{1}{\Gamma(\alpha_1+1)\cdots\Gamma(\alpha_k+1)\alpha_{k+1}\cdots\alpha_n\pi^n}\int_0^\infty e^{-t}t^{\alpha_1+\cdots+\alpha_n}\times$$
$$\times\left(\frac{1}{\alpha_{k+1}}t^{1-\alpha_{k+1}}|z_{k+1}|^{\frac{2}{\alpha_{k+1}}-2}\exp\left(t|z_{k+1}|^{\frac{2}{\alpha_{k+1}}}\right)+c_{k+1}\right)\times\cdots$$
$$\times\left(\frac{1}{\alpha_n}t^{1-\alpha_n}|z_n|^{\frac{2}{\alpha_n}-2}\exp\left(t|z_n|^{\frac{2}{\alpha_n}}\right)+c_n\right)dt$$

这等于

$$\frac{|z_{k+1}|^{\frac{2}{\alpha_{k+1}}-2}\cdots|z_n|^{\frac{2}{\alpha_n}-2}\Gamma(n-k+1+\alpha_1+\cdots+\alpha_k)}{\Gamma(\alpha_1+1)\cdots\Gamma(\alpha_k+1)\alpha_{k+1}^2\cdots\alpha_n^2\pi^n\left(1-\sum\limits_{k+1}^{n}|z_j|^{\frac{2}{\alpha_j}}\right)^{n-k+1+\alpha_1+\cdots+\alpha_k}}$$
$$+\sum_{l=k+1}^{n}c_l\frac{|z_{k+1}|^{\frac{2}{\alpha_{k+1}}-2}\cdots|z_n|^{\frac{2}{\alpha_n}-2}\alpha_l\Gamma(n-k+\alpha_1+\cdots+\alpha_k+\alpha_l)}{\Gamma(\alpha_1+1)\cdots\Gamma(\alpha_k+1)\alpha_{k+1}^2\cdots\alpha_n^2\pi^n|z_l|^{\frac{2}{\alpha_l}-2}\times\left\{\left(1-\sum\limits_{\substack{k+1\\ j\neq l}}^{n}|z_j|^{\frac{2}{\alpha_j}}\right)^{n-k+\alpha_1+\cdots+\alpha_k+\alpha_l}\right\}}$$
$$+\cdots+\frac{c_{k+1}\cdots c_n\Gamma(\alpha_1+\cdots+\alpha_n+1)}{\Gamma(\alpha_1+1)\cdots\Gamma(\alpha_k+1)\alpha_{k+1}\cdots\alpha_n\pi^n}.$$

如果记 $r^{[l]}(k, z)=\sum_{\substack{k+1\\ j\neq l}}^{n}|z_j|^{\frac{2}{\alpha_j}}-1,\ r^{[l,s]}(k, z)=\sum_{\substack{k+1\\ j\neq l,s}}^{n}|z_j|^{\frac{2}{\alpha_j}}-1,\ \cdots,$

则上式可写成

$$\frac{\Gamma(n-k+1+\alpha_1+\cdots+\alpha_k)\det(r_{z_i\bar{z}_j}(k, z))}{\pi^n\Gamma(\alpha_1+1)\cdots\Gamma(\alpha_k+1)(-r(k, z))^{n-k+1+\alpha_1+\cdots+\alpha_k}}$$
$$+\sum_{l=k+1}^{n}\frac{c_l\Gamma(n-k+\alpha_1+\cdots+\alpha_k+\alpha_l)\det(r^{[l]}_{z_i\bar{z}_j}(k, z))}{\alpha_l\pi^n\Gamma(\alpha_1+1)\cdots\Gamma(\alpha_k+1)(-r^{[l]}(k, z))^{n-k+\alpha_1+\cdots+\alpha_k+\alpha_l}}$$
$$+\cdots+\frac{c_{k+1}\cdots c_n\Gamma(\alpha_1+\cdots+\alpha_n+1)}{\Gamma(\alpha_1+1)\cdots\Gamma(\alpha_k+1)\alpha_{k+1}\cdots\alpha_n\pi^n},$$

同样可以证明 $K(z, z)$ 不小于

$$\frac{\Gamma(n-k+1+\alpha_1+\cdots+\alpha_k)\det(r_{z_i\bar{z}_j}(k, z))}{\pi^n\Gamma(\alpha_1+1)\cdots\Gamma(\alpha_k+1)(-r(k, z))^{n-k+1+\alpha_1+\cdots+\alpha_k}}$$
$$-\sum_{l=k+1}^{n}\frac{c_l\Gamma(n-k+\alpha_1+\cdots+\alpha_k+\alpha_l)\det(r^{[l]}_{z_i\bar{z}_j}(k, z))}{\alpha_l\pi^n\Gamma(\alpha_1+1)\cdots\Gamma(\alpha_k+1)(-r^{[l]}(k, z))^{n-k+\alpha_1+\cdots+\alpha_k+\alpha_l}}$$
$$+\cdots+(-1)^{n-k}\frac{c_{k+1}\cdots c_n}{\alpha_{k+1}\cdots\alpha_n\Gamma(\alpha_1+1)\cdots\Gamma(\alpha_k+1)\pi^n}.$$

于是(2.5)可以写成更精确的形式

$$\frac{\Gamma(n-k+1+\alpha_1+\cdots+\alpha_k)\det(r_{z_i\bar{z}_j}(k, z))}{\pi^n\Gamma(\alpha_1+1)\cdots\Gamma(\alpha_k+1)(-r(k, z))^{n-k+1+\alpha_1+\cdots+\alpha_k}}$$
$$+\sum_{s=1}^{n-k}(-1)^s\sum_{\substack{k+1\\ l_1<\cdots<l_s}}^{n}\frac{\begin{pmatrix}c_{l_1}\cdots c_{l_s}\Gamma(n-k+1-s+\alpha_1+\cdots+\alpha_k+\alpha_{l_1}+\\ +\cdots+\alpha_{l_s})\det(r^{[l_1,\cdots,l_s]}_{z_i\bar{z}_j}(k, z))\end{pmatrix}}{\begin{pmatrix}\alpha_{l_1}\cdots\alpha_{l_s}\pi^n\Gamma(\alpha_1+1)\cdots\Gamma(\alpha_k+1)\times\\ \times(-r^{[l_1,\cdots,l_s]}(k, z))^{n-k-s+1+\alpha_1+\cdots+\alpha_k+\alpha_{l_1}+\cdots+\alpha_{l_s}}\end{pmatrix}}$$
$$\leqslant K(z, z)\leqslant\frac{\Gamma(n-k+1+\alpha_1+\cdots+\alpha_k)\det(r_{z_i\bar{z}_j}(k, z))}{\pi^n\Gamma(\alpha_1+1)\cdots\Gamma(\alpha_k+1)(-r(k, z))^{n-k+1+\alpha_1+\cdots+\alpha_k}}$$
$$+\sum_{s=1}^{n-k}\sum_{\substack{k+1\\ l_1<\cdots<l_s}}^{n}\frac{\begin{pmatrix}c_{l_1}\cdots c_{l_s}\Gamma(n-k+1-s+\alpha_1+\cdots+\alpha_k+\alpha_{l_1}+\\ +\cdots+\alpha_{l_s})\det(r^{[l_1,\cdots,l_s]}_{z_i\bar{z}_j}(k, z))\end{pmatrix}}{\begin{pmatrix}\alpha_{l_1}\cdots\alpha_{l_s}\Gamma(\alpha_1+1)\cdots\Gamma(\alpha_k+1)\times\\ \times(-r^{[l_1,\cdots,l_s]}(k, z))^{n-k-s+1+\alpha_1+\cdots+\alpha_k+\alpha_{l_1}+\cdots+\alpha_{l_s}}\end{pmatrix}}$$

由于 Levi 形式与 $(\det r_{z_i\bar{z}_j})$ 只差一个不为零的函数，故上述估计说明了域 D 的 Levi 形式的退化性质与 $K(z, z)$ 的奇性之间的关系.

此外，如果由(1.1)所定义的 D 中一部分 $\alpha_j(1\leqslant j\leqslant n)$ 为 1，也可以用前面同样方法，得到 $K(z, z)$ 的估计式，这里从略．并可证明：当 D 由(1.1)所定义，$1\geqslant\alpha_j>0, j=1,\cdots n$，只要 z 沿着 $z_{k+1}\neq0, \cdots, z_n\neq0$ 趋于边界时，则有

$$\lim_{r\to0}\frac{K(z, z)(-r(k, z))^{n-k+1+\alpha_1+\cdots+\alpha_k}}{\det(r_{z_i\bar{z}_j}(k, z))}=\frac{\Gamma(n-k+1+\alpha_1+\cdots+\alpha_k)}{\pi^n\Gamma(\alpha_1+1)\cdots\Gamma(\alpha_k+1)}.$$

于是归纳起来，可得

1° 若 $D=\left\{z\in\mathbb{C}^n, \sum_{j=1}^{n}|z_j|^{\frac{2}{\alpha_j}}<1\right\}, 1>\alpha_j>0,\ j=1,\ \cdots,\ n$，则当 $z_1\neq0, \cdots, z_n\neq0$ 时，

$$\frac{\Gamma(n+1)\det(r_{z_i\bar{z}_j}(z))}{\pi^n(-r(z))^{n+1}}$$
$$+\sum_{l=1}^{n}(-1)^l\sum_{\substack{1\\ k_1<\cdots<k_l}}^{n}\frac{c_{k_1}\cdots c_{k_l}\Gamma(n+1-l+\alpha_{k_1}+\cdots+\alpha_{k_l})\det(r^{[k_1,\cdots,k_l]}_{z_i\bar{z}_j}(z))}{\alpha_{k_1}\cdots\alpha_{k_l}\pi^n(-r^{[k_1,\cdots,k_l]}(z))^{n+1-l+\alpha_{k_1}+\cdots+\alpha_{k_l}}}$$

$$\leqslant K(z,z)\leqslant\frac{\Gamma(n+1)\det(r_{z_i\bar z_j}(z))}{\pi^n(-r(z))^{n+1}}$$

$$+\sum_{l=1}^{n}\sum_{\substack{1\\k_1<\cdots<k_l}}^{n}\frac{c_{k_1}\cdots c_{k_l}\Gamma(n+1-l+\alpha_{k_1}+\cdots+\alpha_{k_l})\det(r^{[k_1,\cdots,k_l]}_{z_i\bar z_j}(z))}{\alpha_{k_1}\cdots\alpha_{k_l}\pi^n(-r^{[k_1,\cdots,k_l]}(z))^{n+1-l+\alpha_{k_1}+\cdots+\alpha_{k_l}}},$$

当 $z_1=0,\cdots,z_k=0$ 时,

$$\frac{\Gamma(n-k+1+\alpha_1+\cdots+\alpha_k)\det(r_{z_i\bar z_j}(k,z))}{\pi^n\Gamma(\alpha_1+1)\cdots\Gamma(\alpha_k+1)(-r(k,z))^{n-k+1+\alpha_1+\cdots+\alpha_k}}+\sum_{s=1}^{n-k}(-1)^s$$

$$\times\sum_{\substack{k+1\\l_1<\cdots<l_s}}^{n}\frac{c_{l_1}\cdots c_{l_s}\Gamma(n-k+1-s+\alpha_1+\cdots+\alpha_k+\alpha_{l_1}+\cdots+\alpha_{l_s})\det(r^{[l_1,\cdots,l_s]}_{z_i\bar z_j}(k,z))}{\alpha_{l_1}\cdots\alpha_{l_s}\pi^n\Gamma(\alpha_1+1)\cdots\Gamma(\alpha_k+1)(-r^{[l_1,\cdots,l_s]}(k,z))^{n-k-s+1+\alpha_1+\cdots+\alpha_{l_1}+\cdots+\alpha_{l_s}}}$$

$$\leqslant K(z,z)\leqslant\frac{\Gamma(n-k+1+\alpha_1+\cdots+\alpha_k)\det(r_{z_i\bar z_j}(k,z))}{\pi^n\Gamma(\alpha_1+1)\cdots\Gamma(\alpha_k+1)(-r(k,z))^{n-k+1+\alpha_1+\cdots+\alpha_n}}$$

$$+\sum_{s=1}^{n-k}\sum_{\substack{k+1\\l_1<\cdots<l_s}}^{n}\frac{\left(\begin{array}{c}c_{l_1}\cdots c_{l_s}\Gamma(n-k+1-s+\alpha_1+\cdots+\alpha_k+\alpha_{l_1}+\\+\cdots+\alpha_{l_s})\det(r^{[l_1,\cdots,l_s]}_{z_i\bar z_j}(k,z))\end{array}\right)}{\left(\begin{array}{c}\alpha_{l_1}\cdots\alpha_{l_s}\pi^n\Gamma(\alpha_1+1)\cdots\Gamma(\alpha_k+1)\times\\\times(-r^{[l_1,\cdots,l_s]}(k,z))^{n-k-s+1+\alpha_1+\cdots+\alpha_k+\alpha_{l_1}+\cdots+\alpha_{l_s}}\end{array}\right)},$$

这里 $c_j(j=1,\cdots,n)$ 为与 α_j 有关的常数,

$$r(k,z)=\sum_{j=k+1}^{n}|z_j|^{\frac{2}{\alpha_j}}-1,\quad r(0,z)=r(z),\quad r^{[l_1,\cdots,l_s]}(k,z)=\sum_{\substack{j=k+1\\j\neq l_1,\cdots,l_s}}^{n}|z_j|^{\frac{2}{\alpha_j}}-1,$$

以及
$$(r^{[l_1,\cdots,l_s]}_{z_i\bar z_j}(k,z))=\left(\frac{\partial^2 r^{[l_1,\cdots,l_s]}(k,z)}{\partial z_i\partial\bar z_j}\right)_{k+1\leqslant i,j\leqslant n},$$

$$k=0,\cdots,n-1;\ s=1,\cdots,n-k.$$

2° 若 $D=\left\{z\in\mathbb{C}^n,\ \sum_{j=1}^{k}|z_j|^2+\sum_{k+1}^{n}|z_j|^{\frac{2}{\alpha_j}}<1\right\}$, $1>\alpha_j>0$, $j=k+1,\cdots,n$, 则当 $z_{k+1}\neq0,\cdots,z_n\neq0$ 时,

$$\frac{\Gamma(n+1)\det(r_{z_i\bar z_j}(z))}{\pi^n(-r(z))^{n+1}}$$

$$+\sum_{l=1}^{n-k}(-1)^l\sum_{\substack{k+1\\s_1<\cdots<s_l}}^{n}\frac{c_{s_1}\cdots c_{s_l}\Gamma(n+1-l+\alpha_{s_1}+\cdots+\alpha_{s_l})\det(r^{[s_1,\cdots,s_l]}_{z_i\bar z_j}(z))}{\alpha_{s_1}\cdots\alpha_{s_l}\pi^n(-r^{[s_1,\cdots,s_l]}(z))^{n+1-l+\alpha_{s_1}+\cdots+\alpha_{s_l}}}$$

$$\leqslant K(z,z)\leqslant\frac{\Gamma(n+1)\det(r_{z_i\bar z_j}(z))}{\pi^n(-r(z))^{n+1}}$$

$$+\sum_{l=1}^{n-k}\sum_{\substack{k+1\\s_1<\cdots<s_l}}^{n}\frac{c_{s_1}\cdots c_{s_l}\Gamma(n+1-l+\alpha_{s_1}+\cdots+\alpha_{s_l})\det(r^{[s_1,\cdots,s_l]}_{z_i\bar z_j}(z))}{\alpha_{s_1}\cdots\alpha_{s_l}\pi^n(-r^{[s_1,\cdots,s_l]}(z))^{n+1-l+\alpha_{s_1}+\cdots+\alpha_{s_l}}}.$$

特别当 $n=k$ 时,就有

$$K(z,z)=\frac{\Gamma(n+1)\det(r_{z_i\bar z_j}(z))}{\pi^n(-r(z))^{n+1}}.$$

3° 若 $D=\left\{z\in\mathbb{C}^n,\sum_{j=1}^{n}|z_j|^{\frac{2}{\alpha_j}}<1\right\}$, $1>\alpha_j>0$, $j=1,\cdots,n$, 且 z 沿着 $z_1\neq0,\cdots,z_n\neq0$ 趋于边界; 或是 $D=\left\{z\in\mathbb{C}^n,\sum_{j=1}^{k}|z_j|^2+\sum_{k+1}^{n}|z_j|^{\frac{2}{\alpha_j}}<1\right\}$, $1>\alpha_j>0$, $j=k+1,\cdots,n$, 且 z 沿着 $z_{k+1}\neq0,\cdots,z_n\neq0$ 趋于边界; 则

$$\lim_{r\to0}\frac{K(z,z)(-r(z))^{n+1}}{\det(r_{z_i\bar z_j}(z))}=\frac{\Gamma(n+1)}{\pi^n}.$$

若 $D=\left\{z\in\mathbb{C}^n, \sum_{j=1}^{n}|z_j|^{\frac{2}{\alpha_j}}<1\right\}$, $1\geqslant\alpha_j>0$, $j=1,\cdots,n$, 当 z 沿着 $z_1=0,\cdots,z_k=0$, $z_{k+1}\neq0,\cdots,z_n\neq0$ 趋于边界时,

$$\lim_{r\to0}\frac{K(z,z)(-r(k,z))^{n-k+1+\alpha_1+\cdots+\alpha_k}}{\det(r_{z_i\bar{z}_j}(k,z))}=\frac{\Gamma(n-k+1+\alpha_1+\cdots+\alpha_k)}{\pi^n\Gamma(\alpha_1+1)\cdots\Gamma(\alpha_k+1)}.$$

从此可以看出:对于强拟凸域 $|z_1|^2+\cdots+|z_n|^2<1$ 来讲, $K(z,z)$ 当 z 从内部趋于边界时,不论是何种途径,得到的结果是一样的,但是对于弱拟凸域 $|z_1|^{2/\alpha_1}+\cdots+|z_n|^{2/\alpha_n}<1$ 来讲, $K(z,z)$ 当 z 从内部趋于边界时,对那些 $\alpha_j\neq1$ 的 z_j 来讲,当 $z_j=0$ 与 $z_j\neq0$ 所得到的结果是不一样的,也就是说对于弱拟凸点来讲, z 从内部趋于边界时,选取 $z_j=0$ 的途径与选取 $z_j\neq0$ 的途径 $K(z,z)$ 所得到的结果是不一样的. 这反映了 Levi 形式的退化性质与 $K(z,z)$ 的奇性之间的关系. 从这点上来讲,强拟凸域与弱拟凸域是不一样的.

§3. Szegö 核

若 D 由 (1.1) 所定义, A. Bonanio 与 N. Lohoné 在文 [4] 中给出了 D 的 Szegö 核 $S(z,\zeta)$ 为

$$I(z,\zeta)=\sum_{m_1,\cdots,m_n\geqslant0}\frac{\Gamma(\alpha_1m_1+\cdots+\alpha_nm_n+\alpha_1+\cdots+\alpha_n)}{\Gamma(\alpha_1m_1+\alpha_1)\cdots\Gamma(\alpha_nm_n+\alpha_n)}(z_1\bar{\zeta}_1)^{m_1}\cdots(z_n\bar{\zeta}_n)^{m_n},\quad(3.1)$$

这里 $z\in D$, $\zeta\in\partial D$.

首先指出这个结果是不对的.

显然, $\partial D=\left\{z\in\mathbb{C}^n, \sum_{j=1}^{n}|z_j|^{\frac{2}{\alpha_j}}=1\right\}$, 在 ∂D 上参数化, $z_j=r_j^{\alpha_j}e^{i\theta_j}$, $j=1,\cdots,n$, 于是

$$\partial D=\left\{z=(z_1,\cdots,z_n)=(r_1^{\alpha_1}e^{i\theta_1},\cdots,r_n^{\alpha_n}e^{i\theta_n})\in\mathbb{C}^n, \sum_{j=1}^{n}r_j^2=1\right\},$$

∂D 上距离元素为

$$ds^2=\sum_{j=1}^{n}r_j^{2\alpha_j-2}(\alpha_j^2dr_j^2+r_j^2d\theta_j^2),$$

∂D 上的体积元素为

$$d\mu(z)=A^{\frac{1}{2}}r_1^{\alpha_1}\cdots r_n^{\alpha_n}d\theta_1\cdots d\theta_n,$$

这里 $$A=\det\left[\begin{pmatrix}\alpha_1^2r_1^{2\alpha_1-2}&&\\&\ddots&\\&&\alpha_{n-1}^2r_{n-1}^{2\alpha_{n-1}-2}\end{pmatrix}+\alpha_n^2r_n^{2\alpha_n-4}\begin{pmatrix}r_1\\\vdots\\r_{n-1}\end{pmatrix}(r_1,\cdots,r_{n-1})\right].$$

容易证明 $$A=\frac{\alpha_1^2\cdots\alpha_{n-1}^2\alpha_n^2r_1^{2\alpha_1-2}\cdots r_n^{2\alpha_n-2}}{r_n^2}\left(\frac{r_1^{4-2\alpha_1}}{\alpha_1^2}+\cdots+\frac{r_n^{4-2\alpha_n}}{\alpha_n^2}\right).$$

故 $$d\mu(z)=\alpha_1\cdots\alpha_nr_1^{2\alpha_1-1}\cdots r_n^{2\alpha_n-1}d\theta_1\cdots d\theta_nd\sigma(r)\left(\frac{r_1^{4-2\alpha_1}}{\alpha_1^2}+\cdots+\frac{r_n^{4-2\alpha_n}}{\alpha_n^2}\right)^{\frac{1}{2}},$$

这里 $d\sigma(r)$ 为单位球面 $r_1^2+\cdots+r_n^2=1$ 上的体积元素. 由于 $\{z_1^{m_1},\cdots,z_n^{m_n}, m_1\geqslant0,\cdots,m_n\geqslant0\}$ 为 D 中全纯函数的完备正交系,在 ∂D 上也是正交的,记

$$\frac{1}{V(\partial D)}\int_{\partial D}|z_1^{m_1}\cdots z_n^{m_n}|^2d\mu(z)=a_{m_1\cdots m_n},\ m_1\geqslant0,\cdots,m_n\geqslant0,\quad(3.2)$$

这里 $V(\partial D)=\int_{r_1^2+\cdots+r_n^2=1}\alpha_1\cdots\alpha_n r_1^{2\alpha_1-1}\cdots r_n^{2\alpha_n-1}\left(\frac{r_1^{4-2\alpha_1}}{\alpha_1^2}+\cdots+\frac{r_n^{4-2\alpha_n}}{\alpha_n^2}\right)^{\frac{1}{2}}d\sigma(r)$ 为 ∂D 的体积，则 D 的 Szegö 核 $S(z, \zeta)$ 为

$$\sum_{m_1,\cdots,m_n\geq 0} a_{m_1\cdots m_n}^{-1}(z_1\bar\zeta_1)^{m_1}\cdots(z_n\bar\zeta_n)^{m_n}. \tag{3.3}$$

由于文[4]中计算 $d\mu$ 时少了 $\left(\frac{r_1^{4-2\alpha_1}}{\alpha_1^2}+\cdots+\frac{r_n^{4-2\alpha_n}}{\alpha_n^2}\right)^{\frac{1}{2}}$ 的因子，所以得到了 $S(z, \zeta)$ 等于由(3.1)所定义的 $I(z, \zeta)$ 的错误结果．实际上，(3.2)在一般情况下是难于计算出来的．

记

$$H(r)=\frac{r_1^{4-2\alpha_1}}{\alpha_1^2}+\cdots+\frac{r_n^{4-2\alpha_n}}{\alpha_n^2} \tag{3.4}$$

在约束条件 $r_1^2+\cdots+r_n^2=1$ 之下，进行估计．当 $0<\alpha_j<1$, $j=1, 2, \cdots, n$ 时，可用 Lagrange 乘子法，容易证明

$$H(r)\geqslant\sum_{j=1}^n\frac{\beta}{2-\alpha_j}\left(\frac{\beta\alpha_j^2}{2-\alpha_j}\right)^{\frac{1}{1-\alpha_j}}=\beta-\beta\sum_{j=1}^n\frac{1-\alpha_j}{2-\alpha_j}\left(\frac{\beta\alpha_j}{2-\alpha_j}\right)^{\frac{1}{1-\alpha_j}}=h(\alpha), \tag{3.5}$$

这里 β 为

$$\sum_{j=1}^n\left(\frac{\beta\alpha_j^2}{2-\alpha_j}\right)^{\frac{1}{1-\alpha_j}}=1 \tag{3.6}$$

的正实数解．当 $\alpha_1=\cdots=\alpha_k=1$, $0<\alpha_j<1$, $j=k+1, \cdots, n$ 时，容易证明

$$H(r)\geqslant 1-\sum_{j=k+1}^n\frac{1-\alpha_j}{2-\alpha_j}\left(\frac{\alpha_j}{2-\alpha_j}\right)^{\frac{1}{1-\alpha_j}}=h_k(\alpha). \tag{3.7}$$

另一方面，显然有

$$H(r)\leqslant(\min_{1\leqslant j\leqslant n}\alpha_j)^{-2}. \tag{3.8}$$

由此立即可以得到，若 $z=(z_1, \cdots, z_n)=(r_1\zeta_1, \cdots, r_n\zeta_n)$, $0<r_j<1$, $j=1, 2, \cdots, n$, $\alpha_1=\cdots=\alpha_k=1$, $0<\alpha_j<1$, $j=k+1, \cdots, n$, 则

$$(\min_{1\leqslant j\leqslant n}\alpha_j)I(z, \zeta)\leqslant S(z, \zeta)\leqslant\frac{I(z, \zeta)}{h_k(\alpha)}, \tag{3.9}$$

$h_k(\alpha)$ 由(3.7)所定义，$h_0(\alpha)=h(\alpha)$ 由(3.5)所定义．

特别当 $k=n-1$，即 $\alpha_1=\cdots=\alpha_{n-1}=1$, $\alpha_n=\alpha$, $0<\alpha<1$，则 $h_{n-1}=1-\frac{1-\alpha}{2-\alpha}$，于是 $I(z, \zeta)$ 成为

$$\sum_{k=1}^{n-1}d_k(g(z, \zeta)^\alpha-z_k\bar\zeta_k)^{-k-1}(g(z, \zeta))^{-n+1+\alpha k}. \tag{3.10}$$

此处 $g(z, \zeta)=1-\sum_{j=1}^{n-1}z_j\bar\zeta_j$, d_k 为 α 的 k 次多项式，当 $k\neq n-1$, $\alpha=1$ 时，$d_k=0$; $d_{n-1}=(n-1)!\alpha^{n-1}$．所以，在这种情形，当 $z=(z_1, \cdots, z_n)=(r_1\zeta_1, \cdots, r_n\zeta_n)$, $0\leqslant r_j<1$, $1\leqslant j\leqslant n$ 时，

$$\alpha\sum_{k=1}^{n-1}d_k(g(z, \zeta)^\alpha-z_n\bar\zeta_n)^{-k-1}(g(z, \zeta))^{-n+1+\alpha k}\leqslant S(z, \zeta)$$
$$\leqslant\left(1-\frac{1-\alpha}{2-\alpha}\left(\frac{\alpha^2}{2-\alpha}\right)^{\frac{1}{1-\alpha}}\right)^{-\frac{1}{2}}\cdot\sum_{k=1}^{n-1}d_k(g(z, \zeta)^\alpha-z_n\bar\zeta_n)^{-k-1}(g(z, \zeta)^{-n+1+\alpha k}.$$

现在来证明(3.10)．由于 $\alpha_1=\cdots=\alpha_{n-1}=1$, $\alpha_n=\alpha$, 故 $I(z,\zeta)$等于

$$\sum_{m_1,\cdots,m_n\geq 0}\frac{\Gamma(m_1+\cdots+m_{n-1}+\alpha m_n+\alpha+n-1)}{\Gamma(m_1+1)\cdots\Gamma(m_{n-1}+\Gamma(\alpha m_n+2)1)}(z_1\bar\zeta_1)^{m_1}\cdots(z_n\bar\zeta_n)^{m_n},$$

利用公式 $\sum_0^\infty\frac{\Gamma(m+\alpha)}{\Gamma(m+1)}x^m=\Gamma(\alpha)(1-x)^{-\alpha}$, 上式等于

$$\sum_{m_2,\cdots,m_n\geq 0}\frac{\Gamma(m_2+\cdots+m_{n-1}+\alpha m_n+\alpha+n-1)}{\Gamma(m_2+1)\cdots\Gamma(m_{n-1}+1)\Gamma(\alpha m_n+\alpha)}(1-z_1\bar\zeta_1)^{-(m_2+\cdots+m_{n-1}+\alpha m_n+\alpha+n-1)}$$
$$\times(z_2\bar\zeta_2)^{m_2}\cdots(z_n\bar\zeta_n)^{m_n},$$

反复应用得到

$$\sum_{m\geq 0}(\alpha m+\alpha)\cdots(\alpha m+\alpha+n-2)(g(z,\zeta))^{-n+1-\alpha}(z_n\bar\zeta_n g(z,\zeta)^{-\alpha})^m.$$

由于$(\alpha m+\alpha)\cdots(\alpha_m+\alpha+n-2)$为$m$的 $n-1$ 次多项式，故 $I(z,\zeta)$为

$$\sum_{k=1}^{n-1}d_k(1-z_n\bar\zeta_n g(z,\zeta)^{-\alpha})^{-k-1}(g(z,\zeta))^{-n+1-\alpha},$$

此即(3.10)．显然可以证明：d_k 为 α 的 k 次多项式，当 $k\neq n-1$, $\alpha=1$ 时，

$$d_k=0;\quad d_{n-1}=(n-1)!\alpha^{n-1}.$$

§4. Bergman 度量

有了 Bergman 核，就可以考虑 Bergman 度量，对于由(1.1)所定义的一般的域 D，来讨论 Bergman 度量是有困难的，因为这时 Bergman 核只有级数表达式，而难于写出更一般的明确表达式．但是在 $\alpha_1=\cdots=\alpha_{n-1}=1$, $\alpha_n=\alpha$, $0<\alpha<1$ 的情形，D 的 Bergman 核有表达式(2.3)．不失一般性，我们只讨论 $n=2$ 的情形，因为在 $n=2$ 情形下所得到的一些结论都不难可以推广到一般的 n 的情形．也就是我们讨论的域为

$$D=\{z\in\mathbb{C}^n,\ |z_1|^2+|z_2|^{2p}<1\},\ p>1. \tag{4.1}$$

由(2.3)，D 的 Bergman 核 $K(z,z)$为

$$\frac{p+1}{\pi^2 p}[(1-|z_1|^2)^{\frac{1}{p}}-|z_2|^2]^{-3}(1-|z_1|^2)^{-2+\frac{1}{p}}\left[(1-|z_1|^2)^{\frac{1}{p}}-\frac{p-1}{p+1}|z_2|^2\right],$$

这可写成

$$\frac{p+1}{\pi^2 p}(1-Q)^{-3}(1-qQ)(1-|z_1|^2)^{-2-\frac{1}{p}}, \tag{4.2}$$

这里 $Q=\frac{|z_2|^2}{(1-|z_1|^2)^{\frac{1}{p}}}$, $q=\frac{p-1}{p+1}$. 故 Bergman 度量为

$$d\bar d\log K(z,z)=-3d\bar d\log(1-Q)+d\bar d\log(1-qQ)-\left(2+\frac{1}{p}\right)d\bar d\log(1-|z_1|^2),$$

所以

$$d\bar d\log K=\left(\frac{3}{1-Q}-\frac{q}{1-qQ}\right)d\bar dQ+\left(\frac{3}{(1-Q)^2}-\frac{q^2}{(1-qQ)^2}\right)dQ\overline{dQ}$$
$$+\left(2+\frac{1}{p}\right)\frac{|dz_1|^2}{(1-|z_1|^2)^2},$$

不难直接从 Q 的定义验证

$$d\bar dQ=\frac{Q|dz_1|^2}{p(1-|z_1|^2)^2}+\frac{dQ\overline{dQ}}{Q}, \tag{4.3}$$

因之 Bergman 度量 $ds^2=d\bar{d}\log K$ 又可写成

$$\left(\frac{3Q}{p(1-Q)}-\frac{qQ}{p(1-qQ)}+2+\frac{1}{p}\right)\frac{|dz_1|^2}{(1-|z_1|^2)^2}$$
$$+\left(\frac{3}{Q(1-Q)^2}-\frac{q}{Q(1-qQ)^2}\right)dQ\overline{dQ}. \tag{4.4}$$

具体写出来，令

$$ds^2=dzH\overline{dz}',\ dz=(dz_1,\ dz_2),\ H=\begin{pmatrix}h_{11} & h_{12}\\ h_{21} & h_{22}\end{pmatrix},$$

由于 $\dfrac{dQ\,dQ}{Q}=\dfrac{1}{p^2}dz\begin{pmatrix}|z_1|^2Q(1-|z_1|^2)^{-2}, & p\bar{z}_1z_2(1-|z_1|^2)^{-1-\frac{1}{p}}\\ pz_1\bar{z}_2(1-|z_1|^2)^{-1-\frac{1}{p}}, & p^2(1-|z_1|^2)^{-\frac{1}{p}}\end{pmatrix}\overline{dz}'$. 所以

$$\begin{cases}h_{11}=\dfrac{2p+1+|z_2|^2T+|z_1|^2|z_2|^2M}{p(1-|z_1|^2)^2},\\ h_{12}=\dfrac{\bar{z}_1z_2M}{1-|z_1|^2},\\ h_{21}=\dfrac{z_1\bar{z}_2M}{1-|z_1|^2},\\ h_{22}=pM,\end{cases} \tag{4.5}$$

这里

$$\begin{cases}T=\dfrac{3}{(1-|z_1|^2)^{\frac{1}{p}}-|z_2|^2}-\dfrac{p-1}{(p+1)(1-|z_1|^2)^{\frac{1}{p}}-(p-1)|z_2|^2},\\ M=\dfrac{(1-|z_1|^2)^{\frac{1}{p}}}{p}\left[\dfrac{3}{((1-|z_1|^2)^{\frac{1}{p}}-|z_2|^2)^2}-\dfrac{p^2-1}{((p+1)(1-|z_1|^2)^{\frac{1}{p}}-(p-1)|z_2|^2)^2}\right].\end{cases} \tag{4.6}$$

容易证明

$$h=\det H=\frac{M(2p+1+|z_2|^2T)}{(1-|z_1|^2)^2}$$
$$=\left(2-\frac{1}{p}+\frac{3}{p(1-Q)}-\frac{1}{p(1-qQ)}\right)\left(\frac{3}{(1-Q)^2}-\frac{q}{(1-qQ)^2}\right)\frac{1}{(1-|z_1|^2)^{\frac{1}{p}+2}}. \tag{4.7}$$

由此可得

$$H^{-1}=\begin{pmatrix}h^{11} & h^{12}\\ h^{21} & h^{22}\end{pmatrix}=\begin{pmatrix}\dfrac{p(1-|z_1|^2)^2}{2p+1+|z_2|^2T}, & \dfrac{-\bar{z}_1z_2(1-|z_1|^2)}{2p+1+|z_2|^2T}\\ \dfrac{-z_1\bar{z}_2(1-|z_1|^2)}{2p+1+|z_2|^2T}, & \dfrac{2p+1+|z_2|^2(T+|z_1|^2M)}{p(2p+1+|z_2|^2T)M}\end{pmatrix}. \tag{4.8}$$

当然，上述这些式子的计算很复杂，从略.

§5. Ricci 曲 率

为了计算由(4.1)所定义的域 D 的 Ricci 曲率，我们先将(4.7)式改写成

$$h(z,z)=K(z,z)\cdot\frac{\pi^2}{p+1}((2p-1)(1-Q)(1-qQ)+3(1-qQ)-(1-Q))$$
$$\cdot(3(1-qQ)^2-q(1-Q)^2(1-qQ)^{-4},$$

这就是

$$\frac{\pi^2 K(z, z)}{(p+1)^2}[2p+1-4(p-1)Q+(2p-1)qQ^2][2(p+2)-4(p-1)Q+(p-1)(3q-1)Q^2](1-qQ)^{-4}.$$

因之

$$\log h=\log K+\log\frac{\pi^2}{(p+1)^2}+\log(2p+1-4(p-1)Q+(2p-1)qQ^2)-4\log(1-qQ)+\log(2(p+2)-4(p-1)Q+(p-1)(3q-1)Q^2).$$

分解因子

$$\begin{cases}A=2p+1+4(p-1)Q+(2p-1)qQ^2=(2p+1)(1-\gamma_1 Q)(1-\gamma_2 Q),\\B=p+2-2(p-1)Q+q(p-2)Q^2=(p+2)(1-\gamma_3 Q)(1-\gamma_4 Q).\end{cases}\tag{5.1}$$

容易证明，当 $z\in D$ 时，$A>0$, $B>0$. 这时

$$\log h=\log K+\log\frac{2\pi^2(2p+1)(p+2)}{(p+1)^2}+\sum_{j=1}^{4}\log(1-\gamma_j Q)-4\log(1-qQ),\tag{5.2}$$

这里

$$\begin{cases}\gamma_1,\ \gamma_2=\dfrac{q^{\frac{1}{2}}(2(p^2-1)^{\frac{1}{2}}\pm\sqrt{3}i)}{2p+1},\\[2ex]\gamma_3,\ \gamma_4=\dfrac{q^{\frac{1}{2}}((p^2-1)^{\frac{1}{2}}\pm\sqrt{3})}{p+2}.\end{cases}\tag{5.3}$$

由于 $$d\bar{d}\log(1-\beta Q)=\frac{-\beta}{(1-\beta Q)^2}[(1-\beta Q)d\bar{d}Q+\beta\, dQ\,\bar{d}Q],$$

由(4.3)，上式右边等于

$$\frac{-\beta}{(1-\beta Q)^2}\left[(1-\beta Q)\frac{Q|dz_1|^2}{p(1-|z_1|^2)^2}+\frac{dQ\,\bar{d}Q}{Q}\right].\tag{5.4}$$

于是由(5.2)及(5.4)得到 $d\bar{d}\log h$ 等于

$$d\bar{d}\log K+\left(\frac{4qQ}{p(1-qQ)}-\sum_{j=1}^{4}\frac{\gamma_j Q}{p(1-\gamma_j Q)}\right)\frac{|dz_1|^2}{(1-|z_1|^2)^2}+\left(\frac{4q}{Q(1-qQ)^2}-\sum_{j=1}^{4}\frac{\gamma_j}{Q(1-\gamma_j Q)^2}\right)dQ\,\bar{d}Q.\tag{5.5}$$

经过计算，可得

$$\frac{4qQ}{p(1-qQ)}-\sum_{j=1}^{4}\frac{\gamma_j Q}{p(1-\gamma_j Q)}=\frac{2q(p-1)Q(1-Q)^3}{p(1-qQ)AB}>0.\tag{5.6}$$

经过计算，可以证明

$$\begin{aligned}&\frac{4q}{(1-qQ)^2}-\sum_{j=1}^{4}\frac{\gamma_j}{(1-\gamma_j Q)^2}\\&=\frac{2q(p-1)(1-Q)^2(1-qQ^2)}{(1-qQ)^2AB}-\frac{36q(p^2-1)(1-Q)^2(1-qQ)^2Q}{A^2B^2}\\&=\frac{2q(p-1)(1-Q)^2}{(1-qQ)^2AB}\left[1-qQ^2-\frac{18(p+1)(1-qQ)^4Q}{AB}\right],\end{aligned}\tag{5.7}$$

将(5.6), (5.7)代入(5.5)得到: $d\bar{d}\log h$ 等于

$$
\begin{aligned}
&d\bar{d}\log K+\frac{2q(p-1)Q(1-Q)^3|dz_1|^2}{p(1-qQ)AB(1-|z_1|^2)^2}+\frac{2q(p-1)(1-Q)^2}{(1-qQ)^2AB}\\
&\quad\times\left[1-qQ^2-\frac{18(p+1)(1-qQ)^4Q}{AB}\right]\frac{dQ\,\overline{dQ}}{Q}\\
&=d\bar{d}\log K+\frac{2q(p-1)(1-Q)^2}{(1-qQ)^2AB}\left[\frac{Q(1-Q)(1-qQ)|dz_1|^2}{p(1-|z_1|^2)^2}\right.\\
&\quad\left.+\left(1-qQ^2-\frac{18(p+1)(1-qQ)^4Q}{AB}\right)\frac{dQ\,\overline{dQ}}{Q}\right].
\end{aligned}
$$

于是 Ricci 曲率 R 为

$$
\begin{aligned}
&-1-\frac{2q(p-1)(1-Q)^2}{d\bar{d}\log K(1-qQ)^2AB}\left[\frac{Q(1-Q)(1-qQ)|dz_1|^2}{p(1-|z_1|^2)^2}\right.\\
&\quad\left.+\left(1-qQ^2-\frac{18(p+1)(1-qQ)^4Q}{AB}\right)\frac{dQ\,\overline{dQ}}{Q}\right],
\end{aligned}\tag{5.8}
$$

由(4.4)，$d\bar{d}\log K$ 为

$$
\begin{aligned}
&\left(\frac{3Q}{p(1-Q)}-\frac{qQ}{p(1-qQ)}+2+\frac{1}{p}\right)\frac{|dz_1|^2}{(1-|z_1|^2)^2}\\
&\quad+\left(\frac{3}{(1-Q)^2}-\frac{q}{(1-qQ)^2}\right)\frac{dQ\,\overline{dQ}}{Q}.
\end{aligned}
$$

归纳起来，得到：如 D 由(4.1)所定义，则当 $z\in D$ 时，对 dz 的 Ricci 曲率 $R(z, dz, p)$ 为(5.8)，这里 A, B 由(5.1)所定义，Q, q 由(4.2)所定义.

由此可得如下有趣的推论：

1° 固定 $p>0$ 及点 $(z_1, 0)\in D$，任给 dz，有

$$
Q=0,\ d\bar{d}\log K=\left(2+\frac{1}{p}\right)\frac{|dz_1|^2}{(1-|z_1|^2)^2}+(3-q)\frac{|dz_2|^2}{(1-|z_1|^2)^{\frac{1}{p}}},
$$

于是 $R((z_1, 0), dz, p)$ 等于

$$
-1-\frac{2p(p-1)^2(1-|z_1|^2)^2|dz_2|^2}{\left\{\begin{array}{l}(2p+1)(p+2)[(2p+1)(p+1)(1-|z_1|^2)^{\frac{1}{p}}|dz_1|^2+\\ \quad+2p(p+2)(1-|z_1|^2)^2|dz_2|^2\end{array}\right\}}.\tag{5.9}
$$

所以当 $(z_1, 0)\in D$ 时，不论 dz 取何方向，不论 p 取何值，只要 $p>0$，总有

$$
R((z_1, 0), dz, p)\leqslant-1.\tag{5.10}
$$

由(5.9)还可以看出，对于弱拟凸点 $(e^{i\theta}, 0)$，当 dz 固定，且 $dz_2\neq0$，$(z_1, 0)$ 从 D 内趋于 $(e^{i\theta}, 0)\in\partial D$ 时，$R((z_1, 0), dz, p)$ 单调增加到 -1. 此外，当 $dz_2\neq0$ 时，(5.10)中等号成立，当且仅当 $p=1$，即 D 为强拟凸域超球.

2° 当 p 固定，$z=(z_1, z_2)$ 在 D 中沿着 $z_2\neq0$ 的途径趋于 ∂D，由于此时 $Q\to1$，所以有

$$
R(z, dz, p)=-1+\frac{(p^2-1)^2(-r(z))^4}{36(1-|z_1|^2)^{4/p}}+O((-r(z))^5).
$$

由此可得，当 z 沿着 $z_2\neq0$ 的途径趋于边界时，在十分靠近边界时，$R(z, dz, p)$ 是单调下降到 -1. 因之，对于 ∂D 上的弱拟凸点 $(e^{i\theta}, 0)$ 来讲，作任意小的一个邻域 N，则在 $N\cap D$ 中，有点使 R 大于 -1，小于 -1 及等于 -1，不论 dz 是如何选取. 但是对于 ∂D 上的其他的点，这些点都是强拟凸点，对其中的任意一点，总可选取一个小的邻域 N，使

在 $N\cap D$ 中的每一点，其 Ricci 曲率都 >-1，不论 dz 取什么方向. 这是强，弱拟凸点的差别.

3° 此外，从(5.8)容易看出，当 z 属于区域 E

$$\{z\in D,\ AB(1-qQ^2)-18(p+1)(1-qQ)^4Q\geqslant 0\}$$

时，$R(z, dz, p)\leqslant -1$ 永远成立，等号成立当且仅当 $p=1$. 可以证明，在 ∂D 上只有点 $(e^{i\theta}, 0)$ 满足

$$AB(1-qQ^2)-18(p+1)(1-qQ)^4Q>0,$$

而在 ∂D 上其余的点都满足

$$AB(1-qQ^2)-18(p+1)(1-qQ)^4Q<0.$$

所以只有弱拟凸点 $(e^{i\theta}, 0)$ 可以有沿着在 E 内的途径，即 $R<-1$ 的途径趋于 $(e^{i\theta}, 0)$ 使 R 取极大值 -1，也可以有沿着在 E 外的途径，即 $R>-1$ 的途径趋于 $(e^{i\theta}, 0)$ 使 R 取极小值 -1. 而在 ∂D 上的其余的点都不能这样做.

4° 固定 $z\in D$ 及 dz，容易证明：当 $p\to 1$ 时，$R(z, dz, p)$ 等于

$$-1-\frac{(p-1)^2(1-Q_0)^4}{27}\left(\frac{Q_0(1-Q_0)|dz_1|^2}{(1-|z_1|^2)^2}+\frac{(1-4Q_0)dQ_0\overline{dQ_0}}{Q_0}\right)\times$$
$$\times\left(\frac{(1-Q_0)|dz_1|^2}{(1-|z_1|^2)^2}+\frac{dQ_0\overline{dQ_0}}{Q_0}\right)^{-1}+O((p-1)^3),$$

这里 $Q_0=|z_2|^2/(1-|z_1|^2)$. 于是得到：在区域 $1-|z_1|^2-4|z_2|^2\geqslant 0$ 中取定 $z, dz, R(z, dz, p)$ 作为 p 的函数，当 $p=1$ 时取极大值 -1，即为强拟凸域超球时，$R(z, dz, p)$ 取极大值 -1.

5° 附带还可看出：当 $z_2=0$ 时，

$$\lim_{\substack{z\to\partial D\\ z_2=0}}\lim_{p\to 1}\frac{R+1}{(p-1)^2(1-|z_1|^2)^{2-\frac{1}{p}}}=\lim_{p\to 1}\lim_{\substack{z\to\partial D\\ z_2=0}}\frac{R+1}{(p-1)^2(1-|z_1|^2)^{2-\frac{1}{p}}}$$
$$=\frac{-2|dz_2|^2}{81|dz_1|^2},$$

当 $z_2\neq 0$ 时，

$$\lim_{\substack{z\to\partial D\\ z_2\neq 0}}\lim_{p\to 1}\frac{R+1}{(p-1)^2(-r(z))^4}=\lim_{p\to 1}\lim_{\substack{z\to\partial D\\ z_2\neq 0}}\frac{R+1}{(p-1)^2(-r(z))^4}=\frac{1}{9(1-|z_1|^2)^4}$$

极限值一正一负，这也反映了强弱拟凸点的不同

§6. 无向曲率

若记由(4.1)所定义的域 D 的 Ricci 曲率为

$$R=dz\,S\,\overline{dz}'/dz\,H\,\overline{dz}',$$

则
$$S=-H-\frac{2q(p-1)(1-Q)^2}{(1-qQ)^2AB}\left[\frac{Q(1-Q)(1-qQ)}{p(1-|z_1|^2)^2}\begin{pmatrix}1&0\\0&0\end{pmatrix}\right.$$
$$\left.+\left(1-qQ^2-\frac{18(p+1)(1-qQ)^4Q}{AB}\right)\cdot\begin{pmatrix}\dfrac{|z_1|^2Q}{p^2(1-|z_1|^2)^2}, & \dfrac{\bar z_1z_2}{p(1-|z_1|^2)^{\frac{1}{p}+1}}\\[2ex] \dfrac{z_1\bar z_2}{p(1-|z_1|^2)^{\frac{1}{p}+1}}, & \dfrac{1}{(1-|z_1|^2)^{\frac{1}{p}}}\end{pmatrix}\right]$$

由(4.8)

$$H^{-1}S=-I-\frac{2q(p-1)(1-Q)^2}{(1-qQ)^2AB}\left[\frac{Q(1-Q)(1-qQ)}{p(1-|z_1|^2)^2}\begin{pmatrix}\frac{p(1-|z_1|^2)^2}{2p+1+|z_2|^2T}, & 0\\ \frac{-z_1\bar{z}_2(1-|z_1|^2)}{2p+1+|z_2|^2T}, & 0\end{pmatrix}\right.$$

$$\left.+\left(1-qQ^2-\frac{18(p+1)(1-qQ)^4Q}{AB}\right)\begin{pmatrix}u_{11} & u_{12}\\ u_{21} & u_{22}\end{pmatrix}\right],$$

此处 $u_{11}=u_{12}=0$,而

$$u_{21}=z_1\bar{z}_2/(p^2(1-|z_1|^2)^{1+\frac{1}{p}}M),\ u_{22}=1/(p(1-|z_1|^2)^{\frac{1}{p}}M),$$

这里 M 由(4.6)所定义.

因之无向曲率 $V=\operatorname{tr}(H^{-1}S)$ 等于

$$-2-\frac{2q(p-1)(1-Q)^4}{AB}\left[\frac{Q}{A}+\frac{p+1}{2B}\left(1-qQ^2-\frac{18(p+1)(1-qQ)^4Q}{AB}\right)\right] \quad (6.1)$$

由此可得如下的推论

1° 在点 $(z_1,0)\in D$,则

$$V((z_1,0),p)=-2-\frac{(p-1)^2}{(2p+1)(p+2)^2},$$

即 V 为不依赖于 z_1 的常量,且永远 $\leqslant -2$,等号成立当且仅当 $p=1$,固定 z_1,将 V 看作 p 的函数,则当 $p=1$ 时,V 取极大值 -2.

2° 当 z 沿着 $z_2\neq 0$ 途径趋于 ∂D 时,则

$$V=-2+\frac{(p^2-1)^2(-r(z))^4}{54(1-|z_1|^2)^{4/p}}+O((-r(z))^5),$$

所以在弱拟凸点 $(e^{i\theta},0)$ 的任意小的邻域 N 中,总存在 $z\in D\cap N$,使 V 取值大于,小于及等于 -2. 而在边界的其余的强拟凸点上,总可以找到一个充分小的邻域 N,使所有在 $N\cap D$ 中的点上,$V>-2$ 成立,$V=-2$ 只是在 ∂D 上.

3° 此外,若记

$$f(z,p)=(1-qQ^2)AB+BQ-18(p+1)(1-qQ)^4Q,$$

从(6.1)立即可以得到,当 $p\neq 1$ 时,当 z 在区域 D_1:$f(z,p)>0$ 中,则 $V<-2$;当 z 在区域 D_2:$f(z,p)<0$ 中,则 $V>-2$;当 z 满足 $f(z,p)=0$,则 $V=-2$. 而在 ∂D 上的点,只有 $(e^{i\theta},0)$ 是 D_1 中,其余的全在 D_2 中,所以对于弱拟凸点 $(e^{i\theta},0)$,在 $D\cap D_1$ 中存在着途径趋于 $(e^{i\theta},0)$,在这途径上 V 全是 <-2 的;在 $D\cap D_2$ 中存在着途径趋于 $(e^{i\theta},0)$,在这途径上 V 全是 >-2 的. 而在 ∂D 上的其余的点,则不能.

4° 固定 z,$z_2\neq 0$,若 $p\to 1$,则

$$V(z,p)=-2-\frac{(p-1)^2(-r(z))^4(1-|z_1|^2-3|z_2|^2)}{27(1-|z_1|^2)^5}+O((p-1)^3).$$

记 $D_3=\{z\in D,\ 1-|z_1|^2-3|z_2|^2>0\}$,则当 $z\in D_3$,$p=1$ 时 $V(z,p)$ 取极大值 -2,当 $z\in D$,$z\overline{\in}\bar{D}_3$,$p=1$ 时 $V(z,p)$ 取极小值 -2.

5° 附带得到:当 $z_2=0$ 时,

$$\lim_{p\to 1}\frac{V((z_1,0),p)+2}{(p-1)^2}=\frac{-1}{27};$$

当 $z_2\neq 0$ 时,

$$\lim_{p\to 1}\lim_{\substack{r(z)\to 0\\ z_1\neq 0}}\frac{V+2}{(-r(z))^4(p-1)^2}=\lim_{\substack{r(z)\to 0\\ z_1\neq 0}}\lim_{p\to 1}\frac{V+2}{(-r(z))^4(p-1)^2}$$

$$=\frac{2}{27(1-|z_1|^2)^4}$$

极限值一正一负，这也反映了强弱拟凸点的不同.

最后重复指出：§4，§5，§6 中所进行的讨论，不难立即可推广到域$\{z\in\mathbb{C}^n, 1-|z_1|^2-\cdots-|z_{n-1}|^2-|z_n|^{2p}>0\}(p>1)$的情形.

参 考 文 献

[1] Beals, M. Fefferman, C. and Grossman, R., Strictly Pseudoconvex Domains in $\mathbb{C}^n$, *Bull. Amer. Math. Soc.*, **8** (1983), 125—322.

[2] Range, R. M., On Hölder Estimotes for $\bar{\partial}u=f$ on Weakly Pseudoconvex Domains, *Proc. Conf. Complex Anal. Cortona*, (1976—77), *Sc. Norm. Sup. Pisa*, (1979).

[3] D'Angelo, J. P., A Note on the Bergman Kernel, *Duke Math. J.*, **45** (1978), 259—265.

[4] Bonanie, A., Lohone, N., Noyaux de Szegö de Certains Dom aines de C^n et Integlites L^p, *C. R.*, **235A** (1977), 699—702.

[5] Fefferman, C., The Bergman Kernel and Biholomorphic Mappings of Pseudoconvex Domains, *Invent. Math.*, **26** (1974), 1—65.

[6] Valiron, G., Théorie des Fonctions, Masson, Paris, 1966.

[7] Erdélyi, A. etc., Higher Transcendental Functions, Vol. III (1955), McGraw Hill.

TRANSACTIONS OF THE
AMERICAN MATHEMATICAL SOCIETY
Volume 348, Number 5, May 1996

THE BERGMAN KERNEL FUNCTION OF SOME REINHARDT DOMAINS

SHENG GONG AND XUEAN ZHENG

ABSTRACT. The boundary behavior of the Bergman Kernel function of some Reinhardt domains is studied. Upper and lower bounds for the Bergman kernel function are found at the diagonal points $(z, \bar{z})$. Let D be the Reinhardt domain

$$D = \left\{ z \in \mathbf{C}^n \mid \|z\|_\alpha = \sum_{j=1}^{n} |z_j|^{2/\alpha_j} < 1 \right\}$$

where $\alpha_j > 0$, $j = 1, 2, \ldots, n$; and let $K(z, \bar{w})$ be the Bergman kernel function of D. Then there exist two positive constants m and M and a function F such that

$$mF(z, \bar{z}) \leq K(z, \bar{z}) \leq MF(z, \bar{z})$$

holds for every $z \in D$. Here

$$F(z, \bar{z}) = (-r(z))^{-n-1} \prod_{j=1}^{n} (-r(z) + |z_j|^{2/\alpha_j})^{1-\alpha_j}$$

and $r(z) = \|z\|_\alpha - 1$ is the defining function for D. The constants m and M depend only on $\alpha = (\alpha_1, \ldots, \alpha_n)$ and n, not on z.

1. INTRODUCTION

The Bergman kernel function $K(z, \bar{w})$, $z, w \in \Omega$ for a domain $\Omega \subset \mathbf{C}^n$ is the kernel of the Bergman projection operator, the operator projecting $L^2(\Omega)$ onto its holomorphic subspace. In this paper, we consider the Reinhardt domain

$$D = \left\{ z \in \mathbf{C}^n \mid \|z\|_\alpha = \sum_{j=1}^{n} |z_j|^{2/\alpha_j} < 1 \right\} \tag{1}$$

where $\alpha_j > 0$, $j = 1, 2, \ldots, n$. The purpose of this paper is to give an estimate of the kernel function $K(z, \bar{z})$ of D in a "small constant-large constant" sense. Precisely, we prove the following

Theorem 1. *There exist two positive constants m and M, which only depend on n and α_j, $j = 1, 2, \ldots, n$, such that*

$$mF(z, \bar{z}) \leq K(z, \bar{z}) \leq MF(z, \bar{z}) \tag{2}$$

Received by the editors October 13, 1994.
1991 *Mathematics Subject Classification*. Primary 32H10.

holds for every $z \in D$, where

$$F(z,\bar{z}) = (-r(z))^{-n-1}\prod_{j=1}^{n}(-r(z)+|z_j|^{2/\alpha_j})^{1-\alpha_j} \tag{3}$$

and $r(z) = \|z\|_\alpha - 1$ is the defining function for D.

The study of the boundary behavior of $K_\Omega(z,\bar{w})$ for a domain $\Omega \subset \mathbf{C}^n$ is quite old, going back to the original inquires of Bergman [1]. Both Fefferman [6] and later, Boutet de Monvel and Sjöstrand [2] obtained the asymptotic expansion of $K_\Omega(z,\bar{w})$, when $\Omega \Subset \mathbf{C}^n$ is strongly pseudo-convex. For domains in $\mathbf{C}^2$, Catlin [3] has given a precise description of the behavior of $K_\Omega(z,\bar{z})$ near finite-type points in $\partial\Omega$. Sharp estimates for $K_\Omega(z,\bar{w})$ for this class of domains were obtained by McNeal [7] and Nagel et al. [9]. For a "decoupled" class of $\Omega \Subset \mathbf{C}^n$, McNeal [8] described the exact estimates on $K_\Omega(z,\bar{w})$ for z near a point of finite type in $\partial\Omega$. For the Reinhardt domain (1), D'Angelo [4] gives the series form of the Bergman kernel function $K(z,w)$ as

$$\begin{aligned} K(z,\bar{w}) &= (\alpha_1\alpha_2\cdots\alpha_n\pi^n)^{-1} \\ &\times \sum_{m_1,m_2,\ldots,m_n\geq 0}\frac{\Gamma(\sum_{j=1}^{n}\alpha_j(m_j+1)+1)}{\Gamma(\alpha_1 m_1+\alpha_1)\cdots\Gamma(\alpha_n m_n+\alpha_n)}(z_1\bar{w}_1)^{m_1}\cdots(z_n\bar{w}_n)^{m_n}. \end{aligned} \tag{4}$$

We will use equation (4) to obtain the estimate of $K(z,\bar{z})$ given in inequalities (2).

It should be noted that D'Angelo [4, 5] has studied $K(z,\bar{z})$ for certain domains. He showed

$$K(z,\bar{z}) = \sum_{k=0}^{l+1} c_k\left(1-\sum_{j=1}^{l}|z_j|^2\right)^{-l-1+kp}\left(\left(1-\sum_{j=1}^{l}|z_j|^2\right)^{p}-\sum_{j=l+1}^{n}|z_j|^2\right)^{-n+l-k}$$

when the domain is $\Omega = \{z\in\mathbf{C}^n \mid \sum_{j=1}^{l}|z_j|^2 + (\sum_{j=l+1}^{n}|z_j|^2)^{1/p} < 1\}$, where $1\leq l < n$, p is a positive real number, and the constants c_k depend on k, l, n and p only.

The sketch of the proof is now indicated. We will informally use the word "comparable" to mean the two functions or sequences of coefficients are related by inequalities such as those between K and F given in inequalities (2). The proof can be outlined as follows: We start with the expansion of the Bergman kernel function in (4). Then we show that the coefficients are comparable to certain coefficients A that are more tractable. The Main Lemma shows that these coefficients A are comparable to certain coefficients B. It follows that the coefficients in equation (4) are comparable to the coefficients B. Multiplying by the powers of z and summing, we find that the kernel function is comparable to a function F which has an expansion with the coefficients B.

Most of the lemmas are combinatorial. The Main Lemma which relates the coefficients A and B is stated in section 3. Its proof is carried out in sections 4 and 5. The proof of the Theorem is given in sections 6 and 7.

2. Preliminary lemmas

We start with some combinatorial lemmas proved using generating function. The routine proofs are not included.

Let $\mathbf{R}^n_+$ denote the set of vectors $x = (x_1, \ldots, x_n)$ in $\mathbf{R}^n$, such that all the components $x_j, j = 1, 2, \ldots, n$, are non-negative, and let $\sigma(x) = \sum_{i=1}^n x_i$. As usual we let e_j denote the unit vector of the j-th coordinate in $\mathbf{R}^n$, and let $x(j) = (x_1, \ldots, x_{j-1}, 0, x_{j+1}, \ldots, x_n)$, $\sigma(x(j)) = \sum_{i \neq j} x_i$ if $x = (x_1, \ldots, x_n)$.

Let a, t be real numbers, $|t| < 1$, and

$$(5) \qquad (1-t)^{-a} = \sum_{k=0}^{\infty} b(a,k) t^k$$

Lemma 1.

$$(6) \qquad b(a,k) = \frac{\Gamma(k+a)}{\Gamma(a)\Gamma(k+1)}, \quad \text{when } a \neq 0,\ a > -1;$$

and $b(0,0) = 1, b(0,k) = 0$ *when* $k \neq 0$.

Let $u = (u_1, \ldots, u_n) \in \mathbf{R}^n_+$, $\sigma(u) = \sum_{i=1}^n u_i < 1$, and

$$(7) \qquad (1-\sigma(u))^{-a} = \sum_{m \in \mathbf{Z}^n_+} C(a,m) u^m$$

where $m = (m_1, \ldots, m_n)$, $u^m = u_1^{m_1} \cdots u_n^{m_n}$.

Lemma 2.

$$(8) \qquad C(a,m) = \frac{\Gamma(a+\sigma(m))}{\Gamma(a)\prod_{j=1}^n \Gamma(1+m_j)}, \quad \text{when } a \neq 0,\ a > -1;$$

and $C(0,0) = 1$, $C(0,m) = 0$ *when* $m \neq 0 = (0, \ldots, 0)$.

Lemma 3. *Let* a_j, $j = 1, \ldots, s$, *be real numbers,* $k \in \mathbf{Z}_+$, *then*

$$(9) \qquad b(a_1 + \cdots + a_s, k) = \sum_{\sigma(l)=k} b(a_1, l_1) b(a_2, l_2) \cdots b(a_s, l_s)$$

where $b(a,k)$ *is defined by* (5), $l = (l_1, \ldots, l_s) \in \mathbf{Z}^s_+$, *and* $\sigma(l) = \sum_{j=1}^n l_j$.

Lemma 4. *Let* a_j, $j = 1, \ldots, s$, *be real numbers,* $l \in \mathbf{Z}^n_+$, *then*

$$(10) \qquad C(a_1 + \cdots + a_s, l) = \sum_{m+p+\cdots+q=l} C(a_1, m) C(a_2, p) \cdots C(a_s, q)$$

where $C(a,l)$ *is defined by* (7) *and* $l, m, \ldots, q \in \mathbf{Z}^n_+$.

Lemma 5. *If* $a > -1$, $m = (m_1, \ldots, m_n) \in \mathbf{Z}^n_+$, $p = (p_1, \ldots, p_n) \in \mathbf{Z}^n_+, q = (q_1, \ldots, q_n) \in \mathbf{Z}^n_+$ *and* $m = p + q, p_j q_j = 0$ *for* $j = 1, 2, \ldots, n$, *then*

$$(11) \qquad C(a,m) = C(a,p) C(a + \sigma(p), q)$$

where $C(a,l)$ *is defined by* (8).

Consider the special case that

$$p = m(j) = (m_1, \ldots, m_{j-1}, 0, m_{j+1}, \ldots, m_n),$$
$$q = m_j e_j = (0, \ldots, 0, m_j, 0, \ldots, 0).$$

Then (11) becomes

$$C(a,m) = C(a, m(j)) C(a + \sigma(m(j)), m_j e_j) = C(a, m(j)) b(a + \sigma(m(j)), m_j)$$

since

$$C(a + \sigma(m(j)), m_j e_j) = b(a + \sigma(m(j)), m_j).$$

If we take $p = m_j e_j$, $q = m(j)$, then (11) becomes

$$C(a,m) = b(a,m_j)C(a+m_j, m(j)).$$

Thus we have

$$C(a,m) = C(a,m(j))b(a+\sigma(m(j)), m_j) = C(a+m_j, m(j))b(a,m_j). \tag{12}$$

In particular, if $j = n$, then (12) becomes

$$C(a,m) = C(a,m(n))b(a+\sigma(m(n)), m_n) = C(a+m_n, m(n))b(a,m_n). \tag{13}$$

Lemma 6. *If a, m are two real numbers, $m \geq 1$, and $m + a > 0$, then there exist two positive numbers c and C, which are independent of m, and only depend on a, such that*

$$c\Gamma(m)m^a < \Gamma(m+a) < C\Gamma(m)m^a. \tag{14}$$

Proof. When m is a positive integer, then (14) is just the consequence of Euler-Gauss formula

$$\Gamma(a) = \lim_{m\to\infty} m^a \frac{1\cdot 2\cdots(m-1)}{a\cdot(a+1)\cdots(a+m)}.$$

When m is not a positive integer, we may prove it by using the Stirling's formula: For real $x > 0$,

$$\Gamma(x) = \sqrt{2\pi}x^{x-1/2}e^{-x}e^{\theta(x)/12x}$$

with $0 < \theta(x) < 1$. Using the Stirling formula, we have

$$\frac{\Gamma(m+a)}{\Gamma(m)m^a} = \left(1+\frac{a}{m}\right)^{m+a-1/2} e^{-a}\exp\left\{\frac{\theta_1(m+a)}{12(m+a)} - \frac{\theta_2(m)}{12m}\right\}$$

where $0 < \theta_1 < 1, 0 < \theta_2 < 1$. It is easy to verify that the right hand side of the previous equality is bounded above by $\exp\{1 - a + \frac{1}{12(1+a)}\}\max[1,(1+a)^{a-1/2}]$, and bounded below by $\exp\{-\frac{1}{12} - a\}\min[1,e^a]\min[1,(1+a)^{a-1/2}]$. Thus we prove (14).

Actually, we have $\lim_{m\to\infty}\frac{\Gamma(m+a)}{\Gamma(m)m^a} = 1$, for any positive real number sequence $\{m\}$. □

3. The Main Lemma

Lemma 7 (Main Lemma). *Let $u = (u_1,\ldots,u_n) \in \mathbf{R}^n_+, \sigma(u) < 1$, $a > 0$, $\varepsilon = (\varepsilon_1,\varepsilon_2,\ldots,\varepsilon_n) \in \mathbf{R}^n$, $\varepsilon_j > -1$, $j = 1,2,\ldots,n$. Further, if all ε_j, $j = 1,2,\ldots,n$, are non-positive, we assume that $a+\sum_{j=1}^n \varepsilon_j > 1$. If some of the ε_j, $j = 1,2,\ldots,n$, are non-negative and the others are non-positive, then we assume that $a+\sum_{j=1}^n \varepsilon_j > \sum_{j=1}^n\{\varepsilon_j\} + 1$, where $\{\varepsilon_j\}$ means the least integer which is equal to or greater than ε_j. Let*

$$F_n^{a,\varepsilon}(u) = (1-\sigma(u))^{-a}\prod_{j=1}^n(1+u_j-\sigma(u))^{-\varepsilon_j}, \tag{15}$$

and let the Taylor expansion of $F_n^{a,\varepsilon}(u)$ be

$$F_n^{a,\varepsilon}(u) = \sum_{s\in\mathbf{Z}^n_+} B_{s,n}^{a,\varepsilon}u^s \tag{16}$$

where $s = (s_1, s_2, \ldots, s_n) \in \mathbf{Z}_+^n$, $u^s = u_1^{s_1} \cdots u_n^{s_n}$, *then there exist two positive constants* m_n *and* M_n, *which are independent of* s, *and depend on* a, n, *and* ε *only, such that*

$$m_n A_{s,n}^{a,\varepsilon} \le B_{s,n}^{a,\varepsilon} \le M_n A_{s,n}^{a,\varepsilon} \tag{17}$$

where

$$A_{s,n}^{a,\varepsilon} = \frac{\Gamma(a+\sigma(\varepsilon)+\sigma(s))}{\Gamma(a+\sigma(\varepsilon))\prod_{j=1}^n \Gamma(1+s_j)(1+s_j)^{\varepsilon_j}} = \frac{C(a+\sigma(\varepsilon), s)}{\prod_{j=1}^n (1+s_j)^{\varepsilon_j}}, \tag{18}$$

and $\sigma(\varepsilon) = \sum_{j=1}^n \varepsilon_j$, $\sigma(s) = \sum_{j=1}^n s_j$.

We prove the Main Lemma by induction for n.

When $n = 1$,

$$F_1^{a,\varepsilon}(u) = (1-u_1)^{-a} = \sum b(a, m_1) u_1^{m_1},$$

that is, $B_{m,1}^{a,\varepsilon} = b(a, m_1)$, $(m = m_1)$; and

$$A_{m,1}^{a,\varepsilon} = \frac{\Gamma(a+\varepsilon_1+m_1)}{\Gamma(a+\varepsilon_1)\Gamma(1+m_1)(1+m_1)^{\varepsilon_1}} = b(a, m_1) I_1$$

where

$$I_1 = \frac{\Gamma(a+\varepsilon_1+m_1)\Gamma(a)}{\Gamma(a+\varepsilon_1)\Gamma(a+m_1)(1+m_1)^{\varepsilon_1}}.$$

Obviously (17) is true when $m_1 = 0$. If $m_1 \ne 0$, then by Lemma 6, there exist c and C, which depends only on a and ε, such that

$$I_1 \le C \frac{\Gamma(a)}{\Gamma(a+\varepsilon_1)} \left(\frac{m_1}{1+m_1}\right)^{\varepsilon_1} < C \frac{\Gamma(a)}{\Gamma(a+\varepsilon_1)}$$

if $\varepsilon_1 \ge 0$; and

$$I_1 \le C 2^{-\varepsilon_1} \frac{\Gamma(a)}{\Gamma(a+\varepsilon_1)}$$

if $\varepsilon_1 < 0$. Similarly,

$$I_1 > c \frac{\Gamma(a)}{\Gamma(a+\varepsilon_1)}$$

if $\varepsilon_1 < 0$; and

$$I_1 > c 2^{-\varepsilon_1} \frac{\Gamma(a)}{\Gamma(a+\varepsilon_1)}$$

if $\varepsilon_1 \ge 0$. We have proved (17) in the case $n = 1$.

In order to complete the induction process, we will need the following lemmas.

Lemma 8. *The assumptions are as in the Main Lemma. Moreover, we assume for a fixed* k, $1 \le k \le n$, $\varepsilon_k > 0$, *then*

$$B_{s,n}^{a,\varepsilon} = \sum_{l(k)+\eta(k)=s(k)} C(\varepsilon_k, \eta(k)) b(a+\sigma(\varepsilon(k))+\sigma(l(k)), s_k) B_{l(k),n-1}^{a,\varepsilon(k)}; \tag{19}$$

if we assume $\varepsilon_k = 0$, *then*

$$B_{s,n}^{a,\varepsilon} = b(a+\sigma(\varepsilon)+\sigma(s(k)), s_k) B_{s(k),n-1}^{a,\varepsilon(k)}; \tag{20}$$

where $l=(l_1,\ldots,l_n)\in\mathbf{Z}_+^n$, $\eta=(\eta_1,\ldots,\eta_n)\in\mathbf{Z}_+^n$, $(s_1,\ldots,s_n)\in\mathbf{Z}_+^n$ *and* $B^{a,\varepsilon(k)}_{s(k),n-1}$ *is the coefficient of the Taylor expansion of the function*

$$F^{a,\varepsilon(k)}_{n-1}(u(k))=(1-\sigma(u(k))^{-a}\prod_{j\neq k}(1-\sigma(u(k))+u_j)^{-\varepsilon_j}. \tag{21}$$

Lemma 9. *The assumptions are as in the Main Lemma. Fix any* k, $1\le k\le n$, *then*

$$\begin{aligned}B^{a,\varepsilon}_{s,n}=&\sum_{l+\xi=s}b(a_k,\xi_k)C(a_k+\varepsilon_k+\xi_k,\xi(k))\\&\times b(a-a_k+\sigma(\varepsilon(k))+\sigma(l(k)),l_k)B^{a-a_k,\varepsilon(k)}_{l(k),n-1}\end{aligned} \tag{22}$$

where $a=a_1+a_2+\cdots+a_n$, *such that* $a_j>0$, $a_j+\varepsilon_j>0$, $j=1,2,\ldots,n$, *and* $l=(l_1,l_2,\ldots,l_n)\in\mathbf{Z}_+^n,\xi=(\xi_1,\xi_2,\ldots,\xi_n)\in\mathbf{Z}_+^n$; $s=(s_1,s_2,\ldots,s_n)\in\mathbf{Z}_+^n$ *and* $B^{a-a_k,\varepsilon(k)}_{l(k),n-1}$ *is the coefficient of the Taylor expansion of the function*

$$F^{a-a_k,\varepsilon(k)}_{n-1}(u(k))=(1-\sigma(u(k)))^{a_k-a}\prod_{j\neq k}(1-\sigma(u(k))+u_j)^{-\varepsilon_j}.$$

Lemma 10. *Assumptions are as in the Main Lemma, then all the coefficients* $B^{a,\varepsilon}_{s,n}$ *in* (16) *are non-negative.*

Proof of Lemma 8. In the case $\varepsilon_k>0$, we can expand each factor in (15) by power series, then

$$\begin{aligned}F^{a,\varepsilon}_n(u)=&\sum_{m,\beta(1),\ldots,\gamma(n-1),\delta(n)\in\mathbf{Z}_+^n}C(a,m)C(\varepsilon_1,\beta(1))\cdots\\&\qquad C(\varepsilon_{n-1},\gamma(n-1))C(\varepsilon_n,\delta(n))u^{m+\beta(1)+\cdots+\delta(n)}\\=&\sum_{s\in\mathbf{Z}_+^n}\sum_{m+\beta(1)+\cdots+\gamma(n-1)+\delta(n)=s}C(a,m)C(\varepsilon_1,\beta(1))\cdots C(\varepsilon_n,\delta(n))u^s.\end{aligned}$$

Comparing with (16), we get

$$B^{a,\varepsilon}_{s,n}=\sum_{m+\beta(1)+\cdots+\delta(n)=s}C(a,m)C(\varepsilon_1,\beta(1))\cdots C(\varepsilon_n,\delta(n)).$$

Using Lemma 5, we can decompose each factor (except the factor $C(\varepsilon_k,\xi(k))$) on the right-hand side of this equality as

$$\begin{aligned}C(a,m)&=C(a,m(k))b(a+\sigma(m(k)),m_k),\\C(\varepsilon_1,\beta(1))&=C(\varepsilon_1,\beta(1,k))b(\varepsilon_1+\sigma(\beta(1,k)),\beta_k),\\&\cdots\cdots\cdots\\C(\varepsilon_n,\delta(n))&=C(\varepsilon_n,\delta(k,n))b(\varepsilon_n+\sigma(\delta(k,n)),\delta_k).\end{aligned}$$

Then we have

$$\begin{aligned}
B_{s,n}^{a,\varepsilon} = & \sum_{m+\beta(1)+\cdots+\delta(n)=s} C(\varepsilon_k,\xi(k))C(a,m(k))C(\varepsilon_1,\beta(1,k))\cdots \\
& [C(\varepsilon_k,\xi(k,k))]\cdots C(\varepsilon_n,\delta(k,n))b(a+\sigma(m(k)),m_k) \\
& \times b(\varepsilon_1+\sigma(\beta(1,k)),\beta_k)\cdots[b(\varepsilon_k+\sigma(\xi(k,k)),\beta_k)]\cdots b(\varepsilon_n+\sigma(\delta(k,n)),\delta_n) \\
= & \sum_{l(k)+\xi(k)=s(k)} C(\varepsilon_k,\xi(k)) \sum_{m(k)+\beta(1,k)+\cdots+\delta(k,n)=l(k)} C(a,m(k)) \\
& \times C(\varepsilon_1,\beta(1,k))\cdots[C(\varepsilon_k,\xi(k,k))]\cdots C(\varepsilon_n,\delta(k,n)) \\
& \times \sum_{m_k+\beta_k+\cdots+\delta_k=s_k} b(a+\sigma(m(k)),m_k)b(\varepsilon_1+\sigma(\beta(1,k)),\beta_k)\cdots \\
& [b(\varepsilon_k+\sigma(\xi(k,k)),\beta_k)]\cdots b(\varepsilon_n+\sigma(\delta(k,n)),\delta_n).
\end{aligned}$$

where [ ] means that the term inside [ ] is deleted from the product By Lemma 3, the last sum of the right hand side of this equality is equal to $b(a+\sigma(\varepsilon(k))+\sigma(l(k)),s_k)$, and the second sum is just the coefficient $B_{l(k),n-1}^{a,\varepsilon(k)}$ of the Taylor expansion of the function (21).

In the case $\varepsilon_k=0$, we have

$$B_{s,n}^{a,\varepsilon} = \sum_{m+\beta(1)+\cdots+\delta(n)=s} C(a,m)C(\varepsilon_1,\beta(1))\cdots[C(\varepsilon_k,\xi(k))]\cdots C(\varepsilon_n,\delta(n)),$$

and

$$\begin{aligned}
B_{s,n}^{a,\varepsilon} = & \sum_{m(k)+\beta(1,k)+\cdots+\delta(k,n)=l(k)} C(a,m(k))C(\varepsilon_1,\beta(1,k))\cdots \\
& [C(\varepsilon_k,\xi(k,k))]\cdots C(\varepsilon_n,\delta(k,n)) \\
& \times \sum_{m_k+\beta_k+\cdots+\delta_k=s_k} b(a+\sigma(m(k)),m_k)b(\varepsilon_1+\sigma(\beta(1,k)),\beta_k)\cdots \\
& [b(\varepsilon_k+\sigma(\xi(k,k)),\xi_k)]\cdots b(\varepsilon_n+\sigma(\delta(k,n)),\delta_n).
\end{aligned}$$

We get (20).

We have proved Lemma 8. □

Proof of Lemma 9. We can express $F_n^{a,\varepsilon}(u)$ as

$$F_n^{a,\varepsilon} = \prod_{j=1}^{n}[(1-\sigma(u))^{-a_j}(1-\sigma(u(j)))^{-\varepsilon_j}]. \tag{23}$$

Since $\sigma(u)=\sigma(u(j))+u_j$, we get

$$\begin{aligned}
&(1-\sigma(u))^{-a_j}(1-\sigma(u(j)))^{-\varepsilon_j} \\
&\quad = (1-u_j(1-\sigma(u(j)))^{-1})^{-a_j}(1-\sigma(u(j)))^{-\varepsilon_j-a_j}.
\end{aligned}$$

The assumption $\sigma(u)<1$ implies $u_j(1-\sigma(u(j)))^{-1}<1$,

$$(1-u_j(1-\sigma(u(j)))^{-1})^{-a_j} = \sum_{k=0}^{\infty} b(a_j,k)u_j^k(1-\sigma(u(j)))^{-k}$$

and

$$(1-\sigma(u))^{-a_j}(1-\sigma(u(j)))^{-\varepsilon_j}=\sum_{k=0}^{\infty}b(a_j,k)u_j^k(1-\sigma(u(j)))^{-\varepsilon_j-a_j-k}$$

$$(24)\qquad =\sum_{k=0}^{\infty}b(a_j,k)u_j^k\sum_{m(j)\in\mathbf{Z}_+^n}C(\varepsilon_j+a_j+k,m(j))u^{m(j)}$$

$$=\sum_{m\in\mathbf{Z}_+^n}b(a_j,m_j)C(a_j+\varepsilon_j+m_j,m(j))u^m$$

where $\sum_{m(j)\in\mathbf{Z}_+^n}$ means the sum taking for all lattice points in $\mathbf{Z}_+^n$, where the jth coordinate is zero.

Thus we can expand every factor in (23) by (24) as

$$F_n^{a,\varepsilon}(u)=\sum_{\beta,\gamma,\dots,\delta,\eta\in\mathbf{Z}_+^n}b(a_1,\beta_1)C(a_1+\varepsilon_1+\beta_1,\beta(1))\cdots$$
$$b(a_n,\eta_n)C(a_n+\varepsilon_n+\eta_n,\eta(n))u^{\beta+\gamma+\cdots+\delta+\eta}$$
$$=\sum_{s\in\mathbf{Z}_+^n}\sum_{\beta+\gamma+\cdots+\delta+\eta=s}b(a_1,\beta_1)C(a_1+\varepsilon_1+\beta_1,\beta(1))\cdots$$
$$b(a_n,\eta_n)C(a_n+\varepsilon_n+\eta_n,\eta(n))u^s.$$

Comparing this formula with (16), we get

$$(25)\qquad B_{s,n}^{a,\varepsilon}=\sum_{\beta+\gamma+\cdots+\delta+\eta=s}b(a_1,\beta_1)C(a_1+\varepsilon_1+\beta_1,\beta(1))\cdots$$
$$b(a_n,\eta_n)C(a_n+\varepsilon_n+\eta_n,\eta(n)).$$

Let

$$\beta(1,k)=(0,\beta_2,\dots,\beta_{k-1},0,\beta_{k+1},\dots,\beta_n),$$
$$\gamma(2,k)=(\gamma_1,0,\gamma_3,\dots,\gamma_{k-1},0,\gamma_{k+1},\dots,\gamma_n),$$

etc., then we may decompose $C(a_1+\varepsilon_1+\beta_1,\beta(1))$, etc., by Lemma 5 as,

$$C(a_1+\varepsilon_1+\beta_1,\beta(1))=C(a_1+\varepsilon_1+\beta_1,\beta(1,k))b(a_1+\varepsilon_1+\sigma(\beta(k)),\beta_k),$$
$$C(a_2+\varepsilon_2+\gamma_2,\gamma(2))=C(a_2+\varepsilon_2+\gamma(2,k))b(a_2+\varepsilon_2+\sigma(\gamma(k)),\gamma_k),$$
$$\cdots\cdots\cdots$$
$$C(a_n+\varepsilon_n+\eta_n,\eta(n))=C(a_n+\varepsilon+\eta_n,\eta(k,n))b(a_n+\varepsilon+\sigma(\eta(k)),\eta_k)$$

except the term $C(a_k+\varepsilon_k+\xi_k,\xi(k))$.

Substituting the above equalities into (25), and let $l = s - \xi$, we get

$$\begin{aligned}
B^{a,\varepsilon}_{s,n} = & \sum_{l+\xi=s} b(a_k,\xi_k)C(a_k+\varepsilon_k+\xi_k,\xi(k)) \sum_{\beta+\gamma+\cdots+\eta=l} b(a_1,\beta_1) \\
& \times C(a_1+\varepsilon_1+\beta_1,\beta(1,k))b(a_2,\gamma_2)C(a_2+\varepsilon_2+\gamma_2,\gamma(2,k))\cdots \\
& [b(a_k,\xi_k)C(a_k+\varepsilon_k+\xi_k,\xi(k,k))]\cdots b(a_n,\eta_n)C(a_n+\varepsilon_n+\eta_n,\eta(k,n)) \\
& \times b(a_1+\varepsilon_1+\sigma(\beta(k)),\beta_k)b(a_2+\varepsilon+\sigma(\gamma(k)),\gamma_k) \\
& \cdots[b(a_k+\varepsilon_k+\sigma(\xi(k)),\xi_k)]\cdots b(a_n+\varepsilon_n+\sigma(\eta(k)),\eta_k) \\
= & \sum_{l+\xi=s} b(a_k,\xi_k)C(a_k+\varepsilon_k+\xi_k,\xi(k))b(a-a_k+\sigma(\varepsilon(k))+\sigma(l(k)),l_k) \\
& \times \sum_{\beta(k)+\gamma(k)+\cdots+\eta(k)=l(k)} b(a_1,\beta_1)C(a_1+\varepsilon+\beta_1,\beta(1,k))\cdots \\
& [b(a_k,\xi_k)C(a_k+\varepsilon_k+\xi_k,\xi(k,k))]\cdots b(a_k,\eta_k)C(a_n+\varepsilon_n+\eta_n,\eta(k,n))
\end{aligned}$$

by Lemma 3, where [ ] means that the term inside [ ] is deleted from the product. The last sum of the right-hand side of the previous equality is just the coefficient $B^{a-a_k,\varepsilon(k)}_{l(k),n-1}$ of the Taylor expansion of the function

$$\begin{aligned}
F^{a-a_k,\varepsilon(k)}_{n-1} &= (1-\sigma(u(k)))^{a_k-a}\prod_{j\neq k}(1+u_j-\sigma(u(k)))^{-\varepsilon_j} \\
&= \sum_{l(k)\in \mathbf{Z}^n_+} B^{a-a_k,\varepsilon(k)}_{l(k),n-1}u^{l(k)}.
\end{aligned}$$

We have proved Lemma 9. □

Proof of Lemma 10. By the assumptions of the Main Lemma, we may choose $a_j > 0$, $j=1,2,\ldots,n$, such that $a = a_1+\cdots+a_n$, and $a_j+\varepsilon_j>0$, $j=1,2,\ldots,n$, then $B^{a,\varepsilon}_{s,n}\geq 0$ for all s follows by (25). □

4. Proof of the Main Lemma, Part 1

Fix $s=(s_1,s_2,\ldots,s_n)\in \mathbf{Z}^n_+$, and let $s_k = \max_{1\leq j\leq n} s_j$. We may assume $s_k\geq 1$, otherwise $s=0$, then the Main Lemma is true obviously. For that k, there are three possibilities of ε_k: (1) $\varepsilon_k>0$; (2) $\varepsilon_k=0$ and (3) $\varepsilon_k<0$.

In this section, we prove the Main Lemma in the cases (1) $\varepsilon_k>0$ holds or (2) $\varepsilon_k=0$ holds.

If $\varepsilon_k>0$, by the conditions of the Main Lemma, all the factors in the right-hand side of (19) are non-negative.

The induction process is as follows: to prove (17) is true if we assume

$$m_{n-1}(j)A^{a,\varepsilon(j)}_{s(j),n-1}\leq B^{a,\varepsilon(j)}_{s(j),n-1}\leq M_{n-1}(j)A^{a,\varepsilon(j)}_{s(j),n-1}$$

is true for $j=1,2,\ldots,n$ where $m_{n-1}(j)$ and $M_{n-1}(j)$ are two positive constants which are independent of $s(j)$, and only dependent on a and ε, and

$$A^{a,\varepsilon(j)}_{l(j),n-1} = C(a+\sigma(\varepsilon(j)),l(j))\prod_{p\neq j}(1+l_p)^{-\varepsilon_p}.$$

Let $m_{n-1}=\min_{1\leq j\leq n} m_{n-1}(j)$ and $M_{n-1}=\max_{1\leq j\leq n} M_{n-1}(j)$, then

$$m_{n-1}A^{a,\varepsilon(j)}_{l(j),n-1}\leq B^{a,\varepsilon(k)}_{l(j),n-1}\leq M_{n-1}A^{a,\varepsilon(j)}_{l(j),n-1}.$$

In particular, if we take $j = k$, we have

$$m_{n-1}A^{a,\varepsilon(k)}_{l(k),n-1} \le B^{a,\varepsilon(k)}_{l(k),n-1} \le M_{n-1}A^{a,\varepsilon(k)}_{l(k),n-1},$$

where

$$A^{a,\varepsilon(k)}_{l(k),n-1} = C(a+\sigma(\varepsilon(k)), l(k)) \prod_{j\neq k}(1+l_j)^{-\varepsilon_j}. \tag{26}$$

Let

$$I^{a,\varepsilon}_s = \sum_{l(k)+\eta(k)=s(k)} C(\varepsilon_k, \eta(k))b(a+\sigma(\varepsilon(k))+\sigma(l(k)), s_k)A^{a,\varepsilon(k)}_{l(k),n-1},$$

then

$$m_{n-1}I^{a,\varepsilon}_s \le B^{a,\varepsilon}_{s,n} \le M_{n-1}I^{a,\varepsilon}_s$$

by (19).

We estimate $I^{a,\varepsilon}_s$.

Fix l_k. We define the set E as

$$E = \{l(k) = (l_1, \dots, l_{k-1}, 0, l_{k+1}, \dots, l_n) \in \mathbb{Z}^n_+,\ 0 \le l_j \le s_j,\ j \neq k\}.$$

We define the mapping $\phi : E \to E$ as

$$\phi(l(k)) = \bar{l}(k) = (\bar{l}_1, \dots, \bar{l}_{k-1}, 0, \bar{l}_{k+1}, \dots, \bar{l}_n)$$

where $\bar{l}_j = l_j$, if $\frac{1}{2}s_j \le l_j \le s_j$; $\bar{l}_j = s_j - l_j$, if $0 \le l_j < \frac{1}{2}s_j$; and define $\bar{\eta}(k)$ as $\bar{l}(k) + \bar{\eta}(k) = l(k) + \eta(k) = s(k)$.

It is easily verified that

$$\prod_{j\neq k}\Gamma(1+\bar{\eta}_j)\Gamma(1+\bar{l}_j) = \prod_{j\neq k}\Gamma(1+\eta_j)\Gamma(1+l_j). \tag{27}$$

Consider

$$I \equiv \frac{C(\varepsilon_k, \bar{\eta}(k))b(a+\sigma(\varepsilon(k))+\sigma(\bar{l}(k)), s_k)A^{a,\varepsilon(k)}_{\bar{l}(k),n-1}}{C(\varepsilon_k, \eta(k))b(a+\sigma(\varepsilon(k))+\sigma(l(k)), s_k)A^{a,\varepsilon(k)}_{l(k),n-1}}.$$

By (26) and (27), we have

$$I = \frac{\Gamma(\varepsilon_k+\sigma(\bar{\eta}(k)))\Gamma(a+\sigma(\varepsilon(k))+\sigma(\bar{l}(k))+s_k)}{\Gamma(\varepsilon_k+\sigma(\eta(k)))\Gamma(a+\sigma(\varepsilon(k))+\sigma(l(k))+s_k)} \prod_{j\neq k}\left(\frac{1+l_j}{1+\bar{l}_j}\right)^{\varepsilon_j}.$$

If all $\bar{l}_j = l_j$, $j \neq k$, then $I = 1$. If at least one of $\bar{l}_j \neq l_j$, then $d = \sigma(\bar{l}(k)) - \sigma(l(k)) = \sigma(\eta(k)) - \sigma(\bar{\eta}(k)) \ge 1$. Let $d_j = \bar{l}_j - l_j$, $j \neq k$, then

$$I = \prod_{\nu=0}^{d-1}\frac{a+\sigma(\varepsilon(k))+s_k+\sigma(l(k))+\nu}{\varepsilon_k+\sigma(\bar{\eta}(k))+\nu}\prod_{j\neq k}\left(\frac{1+l_j}{1+l_j+d_j}\right)^{\varepsilon_j}.$$

The reciprocal of the first product in the previous equality is

$$\begin{aligned}
&\prod_{\nu=0}^{d-1}\frac{\varepsilon_k+\sigma(\eta(k))+\nu}{a+\sigma(\varepsilon(k))+s_k+\sigma(l(k))+\nu}\\
&=\prod_{\nu=0}^{d-1}\left(1+\frac{\varepsilon_k+\sigma(\eta(k))-a-\sigma(\varepsilon(k))-s_k-\sigma(l(k))}{a+\sigma(\varepsilon(k))+s_k+\sigma(l(k))+\nu}\right)\\
&\le\prod_{\nu=0}^{d-1}\exp\left\{\frac{\varepsilon_k+\sigma(\eta(k))-a-\sigma(\varepsilon(k))-s_k-\sigma(l(k))}{a+\sigma(\varepsilon(k))+s_k+\sigma(l(k))+\nu}\right\}
\end{aligned}$$

since $e^x\ge 1+x$ when $x\ge -1$. Then

$$I\ge\exp\left(\sum_{\nu=0}^{d-1}\frac{a+\sigma(\varepsilon(k))+s_k+\sigma(l(k))-\sigma(\bar\eta(k))-\varepsilon_k}{a+\sigma(\varepsilon(k))+s_k+\sigma(l(k))+\nu}\right)\prod_{j\ne k}\left(\frac{1+l_j}{1+l_j+d_j}\right)^{\varepsilon_j}.$$

When ε is given and $s_k=\max_{1\le j\le n}s_j$, there are only a finite number of $s\in\mathbf{Z}_+^n$ such that $s_k\le 2\varepsilon_k$. In this situation, we only have a finite number of $B_{s,n}^{a,\varepsilon}$ and $A_{s,n}^{a,\varepsilon}$, thus we just need to take

$$M_n=\max_{s_k=\max s_j,\,s_k\le 2\varepsilon_k}\frac{B_{s,n}^{a,\varepsilon}}{A_{s,n}^{a,\varepsilon}},\qquad m_n=\min_{s_k=\max s_j,\,s_k\le 2\varepsilon_k}\frac{B_{s,n}^{a,\varepsilon}}{A_{s,n}^{a,\varepsilon}}.$$

It only remains for us to consider the case $s_k\ge 2\varepsilon_k$.

From the definition of the mapping ϕ, we know $\bar\eta_j\le l_j$, $j\ne k$, and hence $\sigma(l(k))\ge\sigma(\bar\eta(k))$. Then

$$\begin{aligned}
&\frac{a+\sigma(\varepsilon(k))+s_k+\sigma(l(k))-\sigma(\bar\eta(k))-\varepsilon_k}{a+\sigma(\varepsilon(k))+s_k+\sigma(l(k))+\nu}\\
&>\frac{s_k+\sigma(l(k))-\sigma(\bar\eta(k))-\varepsilon_k}{s_k+\sigma(l(k))+\nu}>\frac{\frac12 s_k}{s_k+\sigma(l(k))+\nu}
\end{aligned}$$

since $s_k\ge 2\varepsilon_k$. We have

$$I>\exp\left\{\sum_{\nu=0}^{d-1}\frac{s_k}{2(s_k+\sigma(l(k))+\nu)}\right\}\prod_{j\ne k}\left(\frac{1+l_j}{1+l_j+d_j}\right)^{\varepsilon_j},$$

thus

$$I>\exp\left\{\sum_{\nu=0}^{d-1}\frac{s_k}{2(s_k+\sigma(l(k))+\nu)}\right\}\prod_{j\ne k}\left(\frac{1}{1+d_j}\right)^{|\varepsilon_j|},$$

since

$$\left(\frac{1+l_j}{1+l_j+d_j}\right)^{\varepsilon_j}\ge\left(\frac{1}{1+d_j}\right)^{|\varepsilon_j|}.$$

Moreover, since $0\le\sigma(l(k))\le\sigma(s(k))\le(n-1)s_k$,

$$\nu\le d\le\sigma(\bar l(k))-\sigma(l(k))\le\sigma(\bar l(k))\le(s(k))\le(n-1)s_k,$$

we have

$$\frac{s_k}{2(s_k+\sigma(l(k))+\nu)}\ge\frac{s_k}{2(s_k+(n-1)s_k+(n-1)s_k)}=\frac{1}{4n-2}.$$

Therefore, we obtain

$$I > \left(\frac{1}{1+d}\right)^{\sum_{j\neq k}|\varepsilon_j|} \exp\left(\frac{d}{4n-2}\right) \equiv f(d).$$

The function $f(d)$ takes the minimum value at $d = (4n-2)\sum_{j\neq k}|\varepsilon_j| - 1$, hence

$$I > e^{-1}\left((4n-2)\sum_{j\neq k}|\varepsilon_j|\right)^{-\sum_{j\neq k}|\varepsilon_j|} \equiv a_0,$$

where a_0 is a constant which only depends on ε and n. From the definition of I, the following inequality

$$\begin{aligned} &C(\varepsilon_k, \eta(k))b(a+\sigma(\varepsilon(k))+\sigma(l(k)), s_k)A^{a,\varepsilon(k)}_{l(k),n-1} \\ &\leq a_0^{-1}C(\varepsilon_k, \eta(k))b(a+\sigma(\varepsilon(k))+\sigma(\bar{l}(k)), s_k)A^{a,\varepsilon(k)}_{\bar{l}(k),n-1} \end{aligned}$$

holds. According to the definition of the set E and the mapping ϕ, at most 2^{n-1} different $l(k)$ map to the same $\bar{l}(k)$ by ϕ. Let $\bar{E} = \phi(E)$, and

$$J = \sum_{l(k)+\eta(k)=s(k),\, l(k)\in\bar{E}} C(\varepsilon_k, \eta(k))b(a+\sigma(\varepsilon(k))+\sigma(l(k)), s_k)A^{a,\varepsilon(k)}_{l(k),n-1},$$

then

$$J \leq I^{a,\varepsilon}_{s,n},$$

and

$$\begin{aligned} I^{a,\varepsilon}_{s,n} &= \sum_{l(k)+\eta(k)=s(k)} C(\varepsilon_k, \eta(k))b(a+\sigma(\varepsilon(k))+\sigma(l(k)), s_k)A^{a,\varepsilon(k)}_{l(k),n-1} \\ &\leq 2^{n-1}a_0^{-1} \sum_{l(k)+\eta(k)=s(k),\, l(k)\in\bar{E}} C(\varepsilon_k, \eta(k)) \\ &\quad \times b(a+\sigma(\varepsilon(k))+\sigma(l(k)), s_k)A^{a,\varepsilon(k)}_{l(k),n-1} \\ &= 2^{n-1}a_0^{-1}J. \end{aligned}$$

Thus

$$J \leq I^{a,\varepsilon}_s \leq 2^{n-1}a_0^{-1}J.$$

Hence, we only need to estimate J.

Since $l \in \bar{E}$, $\frac{1}{2}s_j \leq l_j \leq s_j$, $j \neq k$, we have

$$2^{-|\varepsilon_j|}(1+s_j)^{-\varepsilon_j} \leq (1+l_j)^{-\varepsilon_j} \leq 2^{|\varepsilon_j|}(1+s_j)^{-\varepsilon_j} \tag{28}$$

for $j \neq k$. By (26) and (12),

$$J = \sum_{l(k)+\eta(k)=s(k),l(k)\in\bar{E}} C(\varepsilon_k,\eta(k))b(a+\sigma(\varepsilon(k))+\sigma(l(k)),s_k) \times C(a+\sigma(\varepsilon(k)),l(k))\prod_{j\neq k}(1+l_j)^{-\varepsilon_j}$$

$$= \sum_{l(k)+\eta(k)=s(k),l(k)\in\bar{E}} C(\varepsilon_k,\eta(k))C(a+\sigma(\varepsilon(k))+s_k,l(k)) \times b(a+\sigma(\varepsilon(k)),s_k)\prod_{j\neq k}(1+l_j)^{-\varepsilon_j}.$$

By (28), we get $2^{-\sum_{j\neq k}|\varepsilon_j|}L \le J \le 2^{\sum_{j\neq k}|\varepsilon_j|}L$, where

$$L = b(a+\sigma(\varepsilon(k)),s_k)\prod_{j\neq k}(1+s_j)^{-\varepsilon_j} \times \sum_{l(k)+\eta(k)=s(k),l(k)\in\bar{E}} C(\varepsilon_k,\eta(k))C(a+\sigma(\varepsilon(k))+s_k,l(k)).$$

Using Lemma 4, we have

$$L \le \prod_{j\neq k}(1+s_j)^{-\varepsilon_j}b(a+\sigma(\varepsilon(k)),s_k)C(a+\sigma(\varepsilon)+s_k,s(k)).$$

Using Lemma 6, we have

$$\frac{b(a+\sigma(\varepsilon(k)),s_k)(1+s_k)^{\varepsilon_k}}{b(a+\sigma(\varepsilon),s_k)} = \frac{\Gamma(a+\sigma(\varepsilon(k))+s_k)\Gamma(a+\sigma(\varepsilon(k))+\varepsilon_k)(1+s_k)^{\varepsilon_k}}{\Gamma(a+\sigma(\varepsilon(k)))\Gamma(a+\sigma(\varepsilon(k))+\varepsilon_k+s_k)}$$

$$\le C\frac{\Gamma(a+\sigma(\varepsilon))}{\Gamma(a+\sigma(\varepsilon(k)))}\left(\frac{1+s_k}{a+\sigma(\varepsilon)+s_k}\right)^{\varepsilon_k}$$

$$\le C\frac{\Gamma(a+\sigma(\varepsilon))}{\Gamma(a+\sigma(\varepsilon(k)))}\max(1,(a+\sigma(\varepsilon))^{-\varepsilon_k}) = C_1.$$

Of course, the constant C_1 depends only on a and ε. Thus

$$J \le 2^{\sum_{j\neq k}|\varepsilon_j|}C_1\prod_{j=1}^{n}(1+s_j)^{-\varepsilon_j}C(a+\sigma(\varepsilon)+s_k,s(k))b(a+\sigma(\varepsilon),s_k)$$

$$= 2^{\sum_{j\neq k}|\varepsilon_j|}C_1C(a+\sigma(\varepsilon),s)\prod_{j=1}^{n}(1+s_j)^{-\varepsilon_j} = 2^{\sum_{j\neq k}|\varepsilon_j|}C_1A_{s,n}^{a,\varepsilon}$$

by (12) and (18)

We have proved the right-hand inequality of (17) when $\varepsilon_k > 0$.

Now we are going to prove the left-hand inequality of (17) when $\varepsilon_k > 0$.

From the definition of I,

$$I = I_0\prod_{j\neq k}\left(\frac{1+\bar{l}_j}{1+l_j}\right)^{-\varepsilon_j}$$

where I_0 is equal to

$$\frac{C(\varepsilon_k,\bar{\eta}(k))b(a+\sigma(\varepsilon(k))+\sigma(\bar{l}(k)),s_k)C(a+\sigma(\varepsilon(k)),\bar{l}(k))}{C(\varepsilon_k,\eta(k))b(a+\sigma(\epsilon(k))+\sigma(l(k)),s_k)C(a+\sigma(\varepsilon(k)),l(k))}.$$

Then

$$\begin{aligned}I_0&=\frac{\Gamma(\varepsilon_k+\sigma(\bar{\eta}(k)))\Gamma(a+\sigma(\varepsilon(k))+\sigma(\bar{l}(k)))}{\Gamma(\varepsilon_k+\sigma(\eta(k)))\Gamma(a+\sigma(\varepsilon(k))+s_k+\sigma(l(k)))}\\&=\prod_{\nu=0}^{d-1}\frac{a+\sigma(\varepsilon(k))+s_k+\sigma(l(k))+\nu}{\varepsilon_k+\sigma(\bar{\eta}(k))+\nu}\\&\geq\exp\left\{\sum_{\nu=0}^{d-1}\frac{a+\sigma(\varepsilon(k))+s_k+\sigma(l(k))-\sigma(\bar{\eta}(k))-\varepsilon_k}{a+\sigma(\varepsilon(k))+s_k+\sigma(l(k))+\nu}\right\}\\&\geq\exp\left\{\sum_{\nu=0}^{d-1}\frac{s_k}{2(s_k+\sigma(l(k))+\nu)}\right\}>\exp\frac{d}{4n-2}\geq 1\end{aligned}$$

since $s_k\geq 2\varepsilon_k$ and $e^x\leq(1-x)^{-1}$ when $1>x\geq 0$.

Applying this inequality to the J, we obtain

$$\begin{aligned}J=&\sum_{l(k)+\eta(k)=s(k),l(k)\in E}C(\varepsilon_k,\eta(k))C(a+\sigma(\varepsilon(k))+s_k,l(k))\\&\times b(a+\sigma(\varepsilon(k)),s_k)\prod_{j\neq k}(1+l_j)^{-\varepsilon_j}\\\geq&\,2^{-\sum_{j\neq k}|\varepsilon_j|}\sum_{l(k)+\eta(k)=s(k),l(k)\in E}C(\varepsilon_k,\eta(k))\\&\times b(a+\sigma(\varepsilon(k))+\sigma(l(k)),s_k)C(a+\sigma(\varepsilon(k)),l(k))\prod_{j\neq k}(1+s_j)^{-\varepsilon_j}\\\geq&\,2^{1-n-\sum_{j\neq k}|\varepsilon_j|}\prod_{j\neq k}(1+s_j)^{-\varepsilon_j}b(a+\sigma(\varepsilon(k)),s_k)\\&\times\sum_{l(k)+\eta(k)=s(k)}C(\varepsilon_k,\eta(k))C(a+\sigma(\varepsilon(k))+s_k,l(k))\\=&\,2^{1-n-\sum_{j\neq k}|\varepsilon_j|}\prod_{j\neq k}(1+s_j)^{-\varepsilon_j}b(a+\sigma(\varepsilon(k)),s_k)\\&\times C(a+\sigma(\varepsilon)+s_k,s(k))\end{aligned}$$

by Lemma 4.

Using Lemma 6, we have

$$\begin{aligned}&\frac{b(a+\sigma(\varepsilon(k)),s_k)(1+s_k)^{\varepsilon_k}}{b(a+\sigma(\varepsilon),s_k)}\\&\quad=\frac{\Gamma(a+\sigma(\varepsilon(k))+\varepsilon_k)\Gamma(a+\sigma(\varepsilon(k))+s_k)(1+s_k)^{\varepsilon_k}}{\Gamma(a+\sigma(\varepsilon(k)))\Gamma(a+\sigma(\varepsilon(k))+\varepsilon_k+s_k)}\\&\quad\geq C\frac{\Gamma(a+\sigma(\varepsilon))}{\Gamma(a+\sigma(\varepsilon(k)))}\left(\frac{1+s_k}{a+\sigma(\varepsilon)+s_k}\right)^{\varepsilon_k}\\&\quad\geq C\frac{\Gamma(a+\sigma(\varepsilon))}{\Gamma(a+\sigma(\varepsilon(k)))}\min\left\{1,\left(\frac{1+2\varepsilon_k}{a+\sigma(\varepsilon)+2\varepsilon_k}\right)\right\}\equiv c_1\end{aligned}$$

since $s_k > 2\varepsilon_k$, and c_1 is a constant which only depends on a and ε. Thus

$$J \geq 2^{1-n-\sum_{j\neq k}|\varepsilon_k|} c_1 \prod_{j=1}^{n} (1+s_j)^{-\varepsilon_j} C(a+\sigma(\varepsilon)+s_k, s(k)) b(a+\sigma(\varepsilon), s_k)$$

$$= 2^{1-n-\sum_{j\neq k}|\varepsilon_j|} c_1 C(a+\sigma(\varepsilon), s) \prod_{j=1}^{n} (1+s_j)^{-\varepsilon_j}$$

$$= 2^{1-n-\sum_{j\neq k}|\varepsilon_j|} c_1 A^{a,\varepsilon}_{s,n}$$

by (12) and (18).

We have proved the left-hand inequality of (17) when $\varepsilon_k > 0$.

We now consider the case 2) $\varepsilon_k = 0$. The proof of (17) is easy. By (20), we know

$$B^{a,\varepsilon}_{s,n} = b(a+\sigma(\varepsilon(k))+\sigma(s(k)), s_k) B^{a,\varepsilon(k)}_{s(k),n-1}.$$

By induction hypothesis, there exist two constants m_{n-1} and M_{n-1}, which are independent of s, such that

$$m_{n-1} A^{a,\varepsilon(k)}_{s(k),n-1} \leq B^{a,\varepsilon(k)}_{s(k),n-1} \leq M_{n-1} A^{a,\varepsilon(k)}_{s(k),n-1}.$$

Thus

$$B^{a,\varepsilon}_{s,n} \leq M_{n-1} b(a+\sigma(\varepsilon(k))+\sigma(s(k)), s_k) A^{a,\varepsilon(k)}_{s(k),n-1}$$

$$= M_{n-1} b(a+\sigma(\varepsilon(k))+\sigma(s(k)), s_k) C(a+\sigma(\varepsilon(k)), s(k)) \prod_{j\neq k} (1+s_j)^{-\varepsilon_j}$$

$$= M_{n-1} C(a+\sigma(\varepsilon), s) \prod_{j=1}^{n} (1+s_j)^{-\varepsilon_j} = M_{n-1} A^{a,\varepsilon}_{s,n}$$

since $\varepsilon_k = 0$.

Similarly, we can prove $B^{a,\varepsilon}_{s,n} \geq m_{n-1} A^{a,\varepsilon}_{s,n}$.

5. Proof of the Main Lemma, Part 2

In this section we prove the Main Lemma in the final case 3) $\varepsilon_k < 0$.

We take $a_k = -\varepsilon_k$ at (22). By Lemma 10, we know $B^{a-a_k,\varepsilon(k)}_{l(k),n-1} \geq 0$. All the factors in each term of the right-hand side of the equality (22) are non-negative.

The induction process is as follows: to prove (17) is true if we assume

$$m'_{n-1}(j) A^{a-a_k,\varepsilon(j)}_{l(j),n-1} \leq B^{a-a_k,\varepsilon(j)}_{l(j),n-1} \leq M'_{n-1}(j) A^{a-a_k,\varepsilon(j)}_{l(j),n-1}$$

is true for $j = 1, 2, \ldots, n$, where $m'_{n-1}(j)$ and $M'_{n-1}(j)$ are two positive constants, which are independent of $s(j)$, and only depend on a and $\varepsilon(j)$, and

$$A^{a-a_j,\varepsilon(j)}_{l(j),n-1} = C(a-a_j+\sigma(\varepsilon(j)), l(j)) \prod_{j\neq k} (1+l_j)^{-\varepsilon_j}.$$

Let

$$m'_{n-1} = \min_{1\leq j\leq n} m'_{n-1}(j), \quad M'_{n-1} = \max_{1\leq j\leq n} M'_{n-1}(j),$$

then

$$m'_{n-1} A^{a-a_j,\varepsilon(j)}_{l(j),n-1} \leq B^{a-a_k,\varepsilon(j)}_{l(j),n-1} \leq M'_{n-1} A^{a-a_j,\varepsilon(j)}_{l(j),n-1}.$$

In particular, if we take $j=k$, then

$$m'_{n-1}A^{a-a_k,\varepsilon(k)}_{l(k),n-1}\le B^{a-a_k,\varepsilon(k)}_{l(k),n-1}\le M'_{n-1}A^{a-a,\varepsilon(k)}_{l(k),n-1}.$$

Let

$$P^{a,\varepsilon}_s=\sum_{l+\xi=s}C(\xi_k,\xi(k))b(a_k,\xi_k)b(a-a_k+\sigma(\varepsilon(k))+\sigma(l(k)),l_k)A^{a-a_k,\varepsilon(k)}_{l(k),n-1},$$

then

$$m'_{n-1}P^{a,\varepsilon}_s\le B^{a,\varepsilon}_{s,n}\le M'_{n-1}P^{a,\varepsilon}_s$$

hold by (22).

We estimate $P^{a,\varepsilon}_s$.

Partition the sum $P^{a,\varepsilon}_s$ into two parts,

$$\begin{aligned}P^{a,\varepsilon}_s=\sum_{l+\xi=s}&=\sum_{l_k+\xi_k=s_k,l_k\ge\frac34 s_k}\ \sum_{l(k)+\xi(k)=s(k)}\\&+\sum_{l_k+\xi_k=s_k,l_k<\frac34 s_k}\ \sum_{l(k)+\xi(k)=s(k)}=P_1+P_2.\end{aligned}$$

Let

$$Q^{a,\varepsilon}_{l,\xi}=C(\xi_k,\xi(k))b(a-a_k+\sigma(\varepsilon(k))+\sigma(l(k)),l_k)A^{a-a_k,\varepsilon(k)}_{l(k),n-1},$$

then

$$\begin{aligned}Q^{a,\varepsilon}_{l,\xi}&=C(\xi_k,\xi(k))b(a-a_k+\sigma(\varepsilon(k))+\sigma(l(k)),l_k)C(a-a_k+\sigma(\varepsilon(k)),l(k))\\&\times\prod_{j\ne k}(1+l_j)^{-\varepsilon_j}=C(\xi_k,\xi(k))C(a-a_k+\sigma(\varepsilon(k)),l)\prod_{j\ne k}(1+l_j)^{-\varepsilon_j}\end{aligned}$$

by (12), and

$$P^{a,\varepsilon}_s=\sum_{l+\xi=s}b(a_k,\xi_k)Q^{a,\varepsilon}_{l,\xi}=P_1+P_2.$$

Let

$$E_0=\{l=(l_1,\dots,l_n)\in\mathbf{Z}^n_+,0\le l_j\le s_j,\ j=1,\dots,n\},$$

and define the mapping ϕ_0: $E_0\to E_0$, to $\phi_0(l)=\bar l=(\bar l_1,\dots,\bar l_n)$ where $\bar l_j=l_j$, if $\frac12 s_j\le l_j\le s_j$; $\bar l_j=s_j-l_j$, if $0\le l_j<\frac12 s_j$, $j=1,\dots,n$; and define $\bar\xi$ as $\bar\xi+\bar l=\xi+l=s$, $\bar E_0=\phi_0(E_0)$.

Consider

$$G=\frac{Q^{a,\varepsilon}_{\bar l,\bar\xi}}{Q^{a,\varepsilon}_{l,\xi}}=\frac{C(\bar\xi_k,\bar\xi(k))C(a-a_k+\sigma(\varepsilon(k)),\bar l)}{C(\xi_k,\xi(k))C(a-a_k+\sigma(\varepsilon(k)),l)}\prod_{j\ne k}\left(\frac{1+l_j}{1+\bar l_j}\right)^{\varepsilon_j}.$$

When $l_k\ge\frac34 s_k$, we have $\bar l_k=l_k$, $\bar\xi_k=\xi_k$, and then

$$\begin{aligned}G&=\frac{\Gamma(\sigma(\bar\xi)\Gamma(a-a_k+\sigma(\varepsilon(k))+\sigma(\bar l))}{\Gamma(\sigma(\xi))\Gamma(a-a_k+\sigma(\varepsilon(k))+\sigma(l))}\prod_{j\ne k}\left(\frac{1+l_j}{1+\bar l_j}\right)^{\varepsilon_j}\\&=\prod_{\nu=0}^{d-1}\frac{a-a_k+\sigma(\varepsilon(k))+\sigma(l)+\nu}{\sigma(\bar\xi)+\nu}\prod_{j\ne k}\left(\frac{1+l_j}{1+l_j+d_j}\right)^{\varepsilon_j}\end{aligned}\tag{29}$$

where $d=\sigma(\bar l(k))-\sigma(l(k))$, $d_j=\bar l_j-l_j$, $j\ne k$.

By the definition of ϕ_0, we have $\sigma(l(k)) \geq \sigma(\bar{\xi}(k))$, and $l_k - \bar{\xi}_k \geq \frac{1}{2}s_k$ since $l_k \geq \frac{3}{4}s_k$. By the hypothesis of the Main Lemma and $a_k + \varepsilon_k = 0$, we have $a - a_k + \sigma(\varepsilon(k)) > 0$. The first product in the right-hand side of (29) is equal to

$$\prod_{\nu=0}^{d-1}\left(1+\frac{a-a_k+\sigma(\varepsilon(k))+\sigma(l)-\sigma(\bar{\xi})}{\sigma(\bar{\xi})+\nu}\right) \geq \prod_{\nu=0}^{d-1}\left(1+\frac{\frac{1}{2}s_k}{\sigma(\bar{\xi})+\nu}\right)$$

since $a - a_k + \sigma(\varepsilon(k)) > 0$, and

$$\sigma(l) - \sigma(\bar{\xi}) = l_k - \bar{\xi}_k + \sigma(l(k)) - \sigma(\bar{\xi}(k)) > l_k - \bar{\xi}_k \geq \tfrac{1}{2}s_k.$$

Moreover,

$$\sigma(\bar{\xi}) + \nu \leq ns_k + (n-1)s_k \leq (2n-1)s_k$$

and

$$\left(\frac{1+l_j}{1+l_j+d_j}\right)^{\varepsilon_j} \geq (1+d_j)^{-|\varepsilon_j|} \geq (1+d)^{-|\varepsilon_j|},$$

hence

$$G \geq \prod_{\nu=0}^{d-1}\left(\frac{4n-1}{4n-2}\right)\prod_{j\neq k}\left(\frac{1}{1+d}\right)^{|\varepsilon_j|}$$
$$= \left(\frac{4n-1}{4n-2}\right)^d \left(\frac{1}{1+d}\right)^{\sum_{j\neq k}|\varepsilon_j|} \equiv f.$$

Then function f take its minimum value at $d = (\log\frac{4n-1}{4n-2})^{-1}\sum_{j\neq k}|\varepsilon_j| - 1$, thus

$$G \geq \left(\frac{4n-1}{4n-2}\right)^{\sum_{j\neq k}|\varepsilon_j|(\log(\frac{4n-1}{4n-2}))^{-1}-1}$$
$$\times \left(\log\left(\frac{4n-1}{4n-2}\right)\right)^{\sum_{j\neq k}|\varepsilon_j|}\left(\sum_{j\neq k}|\varepsilon_j|\right)^{-\sum_{j\neq k}|\varepsilon_j|} \equiv a_0'$$

if $\sum_{j\neq k}|\varepsilon_j| \neq 0$; and $G \geq 1$, if $\sum_{j\neq k}|\varepsilon_j| = 0$. Let $a_1' = \min(1, a_0')$, then $G \geq a_1'$. Obviously, a_1' is a constant which only depends on a and ε.

Using the inequality $G \geq a_1'$ and (29), we have

$$P_1 = \sum_{l_k+\xi_k=s_k, l_k\geq\frac{3}{4}s_k}\ \sum_{l(k)+\xi(k)=s(k)} b(a_k,\xi_k)Q_{l,\xi}^{a,\varepsilon}$$
$$\leq \frac{2^{n-1}}{a_1'}\sum_{l_k+\xi_k=s_k, l_k\geq\frac{3}{4}s_k}\ \sum_{l(k)+\xi(k)=s(k), l\in\bar{E}_0} b(a_k,\xi_k)Q_{l,\xi}^{a,\varepsilon}.$$

By the definition of $Q_{l,\xi}^{a,\varepsilon}$, and we observe (28) holds true if $l \in \bar{E}_0$. Then

$$2^{-\sum_{j\neq k}|\varepsilon_j|}Q_0 \leq Q_{l,\xi}^{a,\varepsilon} \leq 2^{\sum_{j\neq k}|\varepsilon_j|}Q_0$$

where

$$Q_0 = C(\xi_k, \xi(k))C(a - a_k + \sigma(\varepsilon(k)), l)\prod_{j\neq k}(1+s_j)^{-\varepsilon_j}.$$

The right-hand inequality implies

$$
\begin{aligned}
P_1 &\le C_2 \sum_{l_k+\xi_k=s_k, l_k\ge \frac{3}{4}s_k} \sum_{l(k)+\xi(k)=s(k), l\in E_0} b(a_k,\xi_k)C(\xi_k,\xi(k)) \\
&\quad \times C(a-a_k+\sigma(\varepsilon(k)),l)\prod_{j\ne k}(1+s_j)^{-\varepsilon_j} \\
&\le C_2\prod_{j\ne k}(1+s_j)^{-\varepsilon_j}\sum_{l+\xi=s} b(a_k,\xi_k)C(\xi_k,\xi(k))C(a-a_k+\sigma(\varepsilon(k)),l) \\
&= C_2\prod_{j\ne k}(1+s_j)^{-\varepsilon_j}\sum_{l_k+\xi_k=s_k} b(a_k,\xi_k)b(a-a_k+\sigma(\varepsilon(k)),l_k) \\
&\quad \times \sum_{l(k)+\xi(k)=s(k)} C(\xi_k,\xi(k))C(a-a_k+\sigma(\varepsilon(k))+l_k,l(k)) \\
&= C_2\prod_{j\ne k}(1+s_j)^{-\varepsilon_j}\sum_{l_k+\xi_k=s_k} b(a_k,\xi_k)b(a-a_k+\sigma(\varepsilon(k)),l_k) \\
&\quad \times C(a+\sigma(\varepsilon)+s_k,s(k)) \\
&= C_2\prod_{j\ne k}(1+s_j)^{-\varepsilon_j} b(a+\sigma(\varepsilon(k)),s_k)C(a+\sigma(\varepsilon)+s_k,s(k))
\end{aligned}
$$

where $C_2 = 2^{n-1+\sum_{j\ne k}|\varepsilon_j|}/a_1'$.

Using Lemma 6, and a similar process as we used in section 4, we may prove

$$
\frac{b(a+\sigma(\varepsilon(k)),s_k)(1+s_k)^{\varepsilon_k}}{b(a+\sigma(\varepsilon),s_k)} \le C_3
$$

where C_3 is a constant, which only depends on a and ε. So we have

$$
\begin{aligned}
P_1 &\le C_2C_3\prod_{j=1}^{n}(1+s_j)^{-\varepsilon_j} b(a+\sigma(\varepsilon),s_k)C(a+\sigma(\varepsilon)+s_k,s(k)) \\
&= C_2C_3A^{a,\varepsilon}_{s,n}.
\end{aligned}
$$

Now we estimate the upper bound of P_2.

P_2 is defined as

$$
\begin{aligned}
P_2 = &\sum_{l_k+\xi_k=s_k, l_k<\frac{3}{4}s_k} \sum_{l(k)+\xi(k)=s(k)} b(a_k,\xi_k)C(\xi_k,\xi(k)) \\
&\times C(a-a_k+\sigma(\varepsilon(k)),l)\prod_{j\ne k}(1+l_j)^{-\varepsilon_j}.
\end{aligned}
$$

It is equal to

$$
\begin{aligned}
&\sum_{l_k+\xi_k=s_k, l_k<\frac{3}{4}s_k} \sum_{l(k)+\xi(k)=s(k)} \frac{C(\xi_k,\xi(k))}{c(a_k+\xi_k,\xi(k))} b(a-a_k+\sigma(\varepsilon(k))+\sigma(l(k)),l_k) \\
&\quad \times b(a_k+\sigma(\xi(k)),\xi_k)C(a_k,\xi(k))C(a-a_k+\sigma(\varepsilon(k)),l(k))\prod_{j\ne k}(1+l_j)^{-\varepsilon_j}
\end{aligned}
$$

since

$$
b(a_k,\xi_k)C(a_k+\xi_k,\xi(k)) = b(a_k+\sigma(\xi(k)),\xi_k)C(a_k,\xi(k))
$$

by (12). Let $r_j = \{\varepsilon_j\}$, $j = 1, \ldots, n$, where $\{x\}$ means the least integer X, such that $X \geq x$, and $r = (r_1, \ldots, r_n)$, $\bar{\varepsilon}_j = \varepsilon_j - \{\varepsilon_j\}$, then $-1 < \bar{\varepsilon}_j \leq 0$, $j \neq k$, and $\bar{\varepsilon}_k = \varepsilon_k$ since $-1 < \varepsilon_k \leq 0$.

Using Lemma 6, we have

$$\frac{\Gamma(1+l_j+r_j)(1+l_j+r_j)^{\bar{\varepsilon}_j}}{\Gamma(1+l_j)(1+l_j)^{\varepsilon_j}} \leq C_{1j}\frac{(1+l_j)^{r_j}(1+l_j+r_j)}{(1+l_j)^{\varepsilon_j}}$$

$$= C_{1j}\left(\frac{1+l_j+r_j}{1+l_j}\right)^{\varepsilon_j} \leq C_{1j},$$

and

$$\frac{\Gamma(1+l_j+r_j)(1+l_j+r_j)^{\bar{\varepsilon}_j}}{\Gamma(1+l_j)(1+l_j)^{\varepsilon_j}} \geq c_{3j}\left(\frac{1+l_j+r_j}{1+l_j}\right)^{\varepsilon_j}$$

$$\geq c_{3j}(1+r_j)^{\varepsilon_j} = c_{1j}$$

where C_{1j}, c_{3j} and c_{1j} are constants, which only depend on ε_j. Let

$$C_4 = \max_{j\neq k} C_{1j}, \quad c_4 = \min_{j\neq k} c_{1j}.$$

Thus we have

$$c_4(1+l_j+r_j)^{-\bar{\varepsilon}_j}\frac{\Gamma(1+l_j)}{\Gamma(1+l_j+r_j)} \leq (1+l_j)^{-\varepsilon_j}$$

$$\leq C_4(1+l_j+r_j)^{-\bar{\varepsilon}_j}\frac{\Gamma(1+l_j)}{\Gamma(1+l_j+r_j)}.$$

Let $q = l + r$, then

$$C(a - a_k + \sigma(\varepsilon(k)), l)\prod_{j\neq k}(1+l_j)^{-\varepsilon_j}$$

$$\leq C_4\frac{\Gamma(a+\sigma(\varepsilon)+\sigma(l))}{\Gamma(a+\sigma(\varepsilon))}\prod\frac{(1+q_j)^{-\bar{\varepsilon}_j}}{\Gamma(1+q_j)}$$

$$= C_5\frac{\Gamma(a+\sigma(\varepsilon)+\sigma(q)-\sigma(r))}{\Gamma(a+\sigma(\varepsilon)-\sigma(r))\prod\Gamma(1+q_j)}\prod_{j\neq k}(1+q_j)^{-\bar{\varepsilon}_j}$$

$$= C_5C(a+\sigma(\varepsilon)-\sigma(r), q)\prod_{j\neq k}(1+q_j)^{-\bar{\varepsilon}_j}$$

and

$$C(a+\sigma(\varepsilon), l)\prod_{j\neq k}(1+l_j)^{-\varepsilon_j} \geq c_5C(a+\sigma(\varepsilon)-\sigma(r), q)\prod_{j\neq k}(1+q_j)^{-\bar{\varepsilon}_j} \tag{30}$$

where

$$C_5 = C_4\frac{\Gamma(a+\sigma(\varepsilon)-\sigma(r))}{\Gamma(a+\sigma(\varepsilon))}, \quad c_5 = c_4\frac{\Gamma(a+\sigma(\varepsilon)-\sigma(r))}{\Gamma(a+\sigma(\varepsilon))}.$$

Define the mapping $\phi^*: E \to E$, by $l(k) \to \hat{l}(k) = \phi^*(l(k))$, where $\hat{l}_j = l_j$, if $\frac{1}{2}(s_j - r_j) < l_j \leq s_j$; and $\hat{l}_j = s_j - l_j$, if $0 \leq l_j \leq \frac{1}{2}(s_j - r_j)$.

Let $E^* = \phi^*(E)$. Define $\xi, \hat{l}$ and $\hat{\xi}$ by $l + \xi = s$, $\hat{l} = \phi^*(l)$ and $\hat{l} + \hat{\xi} = l + \xi = s$. Consider

$$R = \frac{C(a_k,\bar{\xi})C(a+\sigma(\varepsilon),\bar{l})}{C(a_k,\xi)C(a+\sigma(\varepsilon),l)} \prod_{j\neq k}\left(\frac{1+l_j}{1+\bar{l}_j}\right)^{\varepsilon_j}.$$

Then

$$R \geq \frac{c_5}{C_5}\frac{C(a_k,\bar{\xi})C(a+\sigma(\varepsilon)-\sigma(r),\bar{q})}{C(a_k,\xi)C(a+\sigma(\varepsilon)-\sigma(r),q)} \prod_{j\neq k}\left(\frac{1+\bar{q}_j}{1+q_j}\right)^{-\varepsilon_j}$$

where

$$\bar{q}+\bar{\xi} = q+\xi = l+r+\xi = s+r = \bar{l}+r+\bar{\xi}.$$

By the definition of ϕ^* and $\bar{q}$, we have $q_j \leq \bar{q}_j$, hence

$$\prod_{j\neq k}\left(\frac{1+\bar{q}_j}{1+q_j}\right)^{-\varepsilon_j} \geq 1,$$

and

$$R \geq c_6\frac{C(a_k,\bar{\xi})C(a+\sigma(\varepsilon)-\sigma(r),\bar{q})}{C(a_k,\xi)C(a+\sigma(\varepsilon)-\sigma(r),q)}$$

$$= c_6\frac{\Gamma(\sigma(\bar{\xi})+a_k)\Gamma(a+\sigma(\varepsilon)-\sigma(r)+\sigma(\bar{q}))}{\Gamma(\sigma(\xi)+a_k)\Gamma(a+\sigma(\varepsilon)-\sigma(r)+\sigma(q))}$$

$$\geq c_6\frac{\Gamma(\sigma(\bar{\xi}+a_k))\Gamma(a+\sigma(\varepsilon)-\sigma(r)+\sigma(\bar{q}))}{\Gamma(\sigma(\xi)+a_k)\Gamma(a+\sigma(\varepsilon)-\sigma(r)+\sigma(q))}$$

$$= c_6\prod_{\nu=1}^{d-1}\frac{a+\sigma(\varepsilon)-\sigma(r)+\sigma(q)+\nu}{a_k+\sigma(\bar{\xi})+\nu} \geq c_6$$

since $\sigma(q) \geq \sigma(\bar{\xi})$ and $a+\sigma(\varepsilon) > \sigma(r)+a_k$, where $c_6 = c_5/C_5$. Thus

$$P_2 \leq c_6^{-1}2^{n-1}b_1^{-1} \sum_{l_k+\xi_k=s_k,\, l_k<\frac{3}{4}s_k} \sum_{l(k)+\xi(k)=s(k),\, l(k)\in E^*} b(a_k,\xi_k)$$

$$\times\, b(a+\sigma(\varepsilon)+\sigma(l(k)),l_k)C(\xi_k,\xi(k))C(a+\sigma(\varepsilon),l(k))\prod_{j\neq k}(1+l_j)^{-\varepsilon_j}$$

where

$$0 < b_1 \leq C(\xi_k,\xi(k))(C(a_k+\xi_k,\xi(k))^{-1} \leq 1.$$

Since $l(k) \in E^*$, $s_j \geq l_j > \frac{1}{2}(s_j - r_j)$, we can find a constant $p_j > 0$, which only depends on ε_j, such that $(1+l_j)^{-\varepsilon_j} \leq p_j^{|\varepsilon_j|}(1+s_j)^{-\varepsilon_j}$ hold for all but a finite number of values of s_j. Therefore,

$$\begin{aligned}
P_2 &\le C_6 \prod_{j\neq k}(1+s_j)^{-\varepsilon_j} \sum_{l_k+\xi_k=s_k, l_k<\frac{3}{4}s_k} \sum_{l(k)+\xi(k)=s(k), l(k)\in E^*} b(a_k,\xi_k) \\
&\quad \times b(a+\sigma(\varepsilon)+\sigma(l(k)), l_k)C(\xi_k,\xi(k))C(a+\sigma(\varepsilon), l(k)) \\
&\le C_6 \prod_{j\neq k}(1+s_j)^{-\varepsilon_j} \sum_{l_k+\xi_k=s_k} \sum_{l(k)+\xi(k)=s(k)} b(a_k,\xi_k)b(a+\sigma(\varepsilon) \\
&\quad +\sigma(l(k), l_k)C(\xi_k,\xi(k))C(a+\sigma(\varepsilon), l(k)) \\
&= C_6 \prod_{j\neq k}(1+s_j)^{-\varepsilon_j} \sum_{l_k+\xi_k=s_k} \sum_{l(k)+\xi(k)=s(k)} b(a_k,\xi_k) \\
&\quad \times C(a+\sigma(\varepsilon)+l_k, l(k))b(a+\sigma(\varepsilon), l_k)C(\xi_k,\xi(k)) \\
&= C_6 \prod_{j\neq k}(1+s_j)^{-\varepsilon_j} \sum_{l_k+\xi_k=s_k} b(a_k,\xi_k)b(a+\sigma(\varepsilon), l_k) \\
&\quad \times C(a+\sigma(\varepsilon)+s_k, s(k)) \\
&= C_6 \prod_{j\neq k}(1+s_j)^{-\varepsilon_j} b(a+\sigma(\varepsilon(k)), s_k)C(a+\sigma(\varepsilon)+s_k, s(k)) \\
&\le C_6C_3 \prod_{j=1}^{n}(1+s_j)^{-\varepsilon_j} b(a+\sigma(\varepsilon), s_k)C(a+\sigma(\varepsilon)+s_k, s(k)) \\
&= C_6C_3A^{a,\varepsilon}_{s,n}
\end{aligned}$$

where $C_6 = c_6^{-1}2^{n-1}b_1^{-1}\prod_{j\neq k}p_j^{|\varepsilon_j|}$.

We have proved the right-hand inequality of (17) when $\varepsilon_k < 0$.

Now we are going to prove the left-hand inequality of (17) when $\varepsilon_k < 0$. We just need to estimate the lower bound of $P^{a,\varepsilon}_s$ when $\varepsilon_k < 0$.

We already know

$$\begin{aligned}
P^{a,\varepsilon}_s &= \sum_{l+\xi=s} b(a_k,\xi_k)Q^{a,\varepsilon}_{l,\xi} \\
&= \sum_{l_k+\xi_k=s_k} \sum_{l(k)+\xi(k)=s(k)} b(a_k,\xi_k)C(\xi_k,\xi(k)) \\
&\quad \times C(a+\sigma(\varepsilon), l)\prod_{j\neq k}(1+l_j)^{-\varepsilon_j} \\
&\ge \sum_{l_k+\xi_k=s_k} \sum_{l(k)+\xi(k)=s(k), l(k)\in \tilde{E}} b(a_k,\xi_k)C(\xi_k,\xi(k)) \\
&\quad \times C(a+\sigma(\varepsilon), l)\prod_{j\neq k}(1+l_j)^{-\varepsilon_j}
\end{aligned}$$

where $\tilde{E} = \phi(E)$, and ϕ, E are defined in section 4. If we use inequality (28) and the equality

$$C(a+\sigma(\varepsilon), l) = b(a+\sigma(\varepsilon), l_k)C(a+\sigma(\varepsilon)+l_k, l(k)),$$

we have

$$P_s^{a,\varepsilon} \ge 2^{-\sum_{j\ne k}|\varepsilon_j|}\prod_{j\ne k}(1+s_j)^{-\varepsilon_j}\sum_{l_k+\xi_k=s_k} b(a_k,\xi_k)$$
$$\times\, b(a+\sigma(\varepsilon),l_k)\sum_{l(k)+\xi(k)=s(k),l(k)\in E} C(\xi_k,\xi(k))C(a+\sigma(\varepsilon)+l_k,l(k)).$$

Let $\bar{l}(k)=\phi(l(k)),\bar{l}(k)+\bar{\xi}(k)=l(k)+\xi(k)=s(k)$, and consider

$$\begin{aligned}H&=\frac{C(\xi_k,\bar{\xi}(k))C(a+\sigma(\varepsilon),\bar{l})}{C(\xi_k,\xi(k))C(a+\sigma(\varepsilon),l)}\\&=\frac{\Gamma(\sigma(\bar{\xi}))\Gamma(a+\sigma(\varepsilon)+\sigma(\bar{l}))}{\Gamma(\sigma(\xi))\Gamma(a+\sigma(\varepsilon)+\sigma(l))}\\&=\prod_{\nu=0}^{d-1}\frac{a+\sigma(\varepsilon)+\sigma(l)+\nu}{\sigma(\bar{\xi})+\nu}\end{aligned}$$

where $d=\sigma(\bar{l}(k))-\sigma(l(k))$. Then $H\ge 1$ since $a+\sigma(\varepsilon)>0$, and $\sigma(l)\ge\sigma(\bar{\xi})$. Hence

$$\begin{aligned}P_s^{a,\varepsilon}&\ge c_7\prod_{j\ne k}(1+s_j)^{-\varepsilon_j}\sum_{l_k+\xi_k=s_k} b(a_k,\xi_k)\\&\quad\times b(a+\sigma(\varepsilon),l_k)\sum_{l(k)+\xi(k)=s(k)} C(\xi_k,\xi(k))C(a+\sigma(\varepsilon)+l_k,l(k))\\&=c_7\prod_{j\ne k}(1+s_j)^{-\varepsilon_j}\sum_{l_k+\xi_k=s_k} b(a_k,\xi_k)\\&\quad\times b(a+\sigma(\varepsilon),l_k)C(a+\sigma(\varepsilon)+s_k,s(k))\\&=c_7\prod_{j\ne k}(1+s_j)^{-\varepsilon_j}b(a+\sigma(\varepsilon(k)),s_k)C(a+\sigma(\varepsilon)+s_k,s(k))\\&\ge c_7c_1\prod_{j=1}^{n}(1+s_j)^{-\varepsilon_j}b(a+\sigma(\varepsilon),s_k)C(a+\sigma(\varepsilon)+s_k,s(k))\\&=c_7c_1A_{s,n}^{a\varepsilon}\end{aligned}$$

where $c_7=2^{1-n-\sum_{j\ne k}|\varepsilon_j|}$ and c_1 is defined in section 4.

We have proved the left-hand inequality of (17) when $\varepsilon_k<0$. □

6. Proof of the Theorem, Part 1

Let D be defined as in (1), and its Bergman kernel function be

$$K(z,\bar{w})=\sum_{m\in\mathbf{Z}_+^n} C_m z^m\bar{w}^m$$

where $m=(m_1,\dots,m_n),z=(z_1,\dots,z_n)\in D,w=(w_1,\dots,w_n)\in D$, $z^m=z_1^{m_1}\cdots z_n^{m_n}$ and $w^m=w_1^{m_1}\cdots w_n^{m_n}$. We know

$$C_m=\frac{1}{\pi^n(\prod_{j=1}^n\alpha_j)}\frac{\Gamma(\sigma(\alpha m)+\sigma(\alpha)+1)}{\prod_{j=1}^n\Gamma(\alpha_jm_j+\alpha_j)}\tag{31}$$

by (4).

Lemma 11. *There exist two constants c_8 and C_8, which are independent of m, and depend only on α and n, such that*

$$c_8 B^{n+1,\alpha-1}_{[\alpha m],n} \prod_{j=1}^{n} \left(\frac{1+\sigma(\alpha)+\sigma([\alpha m])}{1+[\alpha_j m_j]}\right)^{\varepsilon_j} \le C_m$$
$$\le C_8 B^{n+1,\alpha-1}_{[\alpha m],n} \prod_{j=1}^{n} \left(\frac{1+\sigma(\alpha)+\sigma([\alpha m])}{1+[\alpha_j m_j]}\right)^{\varepsilon_j}$$

where $[x]$ means the largest integer X, such that $X \le x$, $\varepsilon_j = \alpha_j m_j - [\alpha_j m_j]$, and $\sigma([\alpha m]) = \sum_{j=1}^{n} [\alpha_j m_j]$, and hence $0 \le \varepsilon_j < 1$.

Proof. By Lemma 6, there exist c'_8, c''_8, C'_8 and C''_8, which are independent of m, and only depend on α and n, such that

$$c'_8 \Gamma(\sigma([\alpha m]) + \sigma(\alpha) + 1)(\sigma([\alpha m]) + \sigma(\alpha) + 1)^{\sigma(\varepsilon)}$$
$$\le \Gamma(\sigma([\alpha m]) + \sigma(\varepsilon) + \sigma(\alpha) + 1)$$
$$\le C'_8 \Gamma(\sigma([\alpha m]) + \sigma(\alpha) + 1)(\sigma([\alpha m]) + \sigma(\alpha) + 1)^{\sigma(\varepsilon)};$$

and

$$c''_8 \Gamma([\alpha_j m_j] + 1)([\alpha_j m_j] + 1)^{\varepsilon_j + \alpha_j - 1} \le \Gamma([\alpha_j m_j] + \varepsilon_j + \alpha_j)$$
$$\le C''_8 \Gamma([\alpha_j m_j] + 1)([\alpha_j m_j] + 1)^{\varepsilon_j + \alpha_j - 1}.$$

Thus we have

$$\frac{c'_8}{C''_8} \frac{1}{\pi^n (\prod_{j+1}^{n} \alpha_j)} \frac{\Gamma(\sigma([\alpha m]) + \sigma(\alpha) + 1)(\sigma([\alpha m]) + \sigma(\alpha) + 1)^{\sigma(\varepsilon)}}{\prod_{j=1}^{n} \Gamma([\alpha_j m_j] + 1)([\alpha_j m_j] + 1)^{\varepsilon_j + \alpha_j - 1}} \le C_m$$
$$\le \frac{C'_8}{c''_8} \frac{1}{\pi^n (\prod_{j=1}^{n} \alpha_j)} \frac{\Gamma(\sigma([\alpha m]) + \sigma(\alpha) + 1)(\sigma([\alpha m]) + \sigma(\alpha) + 1)^{\sigma(\varepsilon)}}{\prod_{j=1}^{n} \Gamma([\alpha_j m_j] + 1)([\alpha_j m_j] + 1)^{\varepsilon_j + \alpha_j - 1}}.$$

By the Main Lemma, we know

$$m_n A^{n+1,\alpha-1}_{[\alpha m],n} \le B^{n+1,\alpha-1}_{[\alpha m],n} \le M_n A^{n+1,\alpha-1}_{[\alpha m],n},$$

where $\alpha - 1 = (\alpha_1 - 1, \ldots, \alpha_n - 1)$, m_n and M_n are two constants, which are independent of m, and only depend on α, n. According to the definition (18),

$$A^{n+1,\alpha-1}_{[\alpha m],n} = \frac{\Gamma(\sigma([\alpha m]) + \sigma(\alpha) + 1)}{\Gamma(1+\sigma(\alpha)) \prod_{j=1}^{n} \Gamma(1 + [\alpha_j m_j])(1 + [\alpha_j m_j])^{\alpha_j - 1}},$$

we have

$$m_n \frac{\Gamma(\sigma([\alpha m]) + \sigma(\alpha) + 1)(\sigma([\alpha m]) + \sigma(\alpha) + 1)^{\sigma(\varepsilon)}}{\Gamma(1+\sigma(\alpha)) \prod_{j=1}^{n} \Gamma(1 + [\alpha_j m_j])(1 + [\alpha_j m_j])^{\varepsilon_j + \alpha_j - 1}}$$
$$\le B^{n+1,\alpha-1}_{[\alpha m],n} \prod_{j=1}^{n} \left(\frac{1+\sigma(\alpha)+\sigma([\alpha m])}{1+[\alpha_j m_j]}\right)^{\varepsilon_j}$$
$$\le M_n \frac{\Gamma(\sigma([\alpha m]) + \sigma(\alpha) + 1)(\sigma([\alpha m]) + \sigma(\alpha) + 1)^{\sigma(\varepsilon)}}{\Gamma(1+\sigma(\alpha)) \prod_{j=1}^{n} \Gamma(1 + [\alpha_j m_j])(1 + [\alpha_j m_j])(1 + [\alpha_j m_j])^{\varepsilon_j + \alpha_j - 1}}.$$

Thus we prove the lemma. □

Lemma 12. *If $\delta = (\delta_1, \dots, \delta_n)$, $\delta_j = 0$ or 1, $j = 1, 2, \dots, n$, then there exist two positive constants c_9 and C_9, which are independent of m, and only depend on α and n, such that*

$$c_9 B^{n+1,\alpha-1}_{[\alpha m],n} \prod_{j=1}^{n} \left(\frac{1+\sigma(\alpha)+\sigma([\alpha m])}{1+[\alpha_j m_j]} \right)^{\delta_j} \le B^{n+1,\alpha-1+\delta}_{[\alpha m],n}$$

$$\le C_9 B^{n+1,\alpha-1}_{[\alpha m],n} \prod_{j=1}^{n} \left(\frac{1+\sigma(\alpha)+\sigma([\alpha m])}{1+[\alpha_j m_j]} \right)^{\delta_j}$$

where $\alpha - 1 + \delta = (\alpha_1 - 1 + \delta_1, \dots, \alpha_n - 1 + \delta_n)$.

Proof. By the Main Lemma, there exist two positive constants c_9' and C_9', which are independent of m and δ, and only depend on α and n, such that

$$c_9' A^{n+1,\alpha-1+\delta}_{[\alpha m],n} \le B^{n+1,\alpha-1+\delta}_{[\alpha m],n} \le C_9' A^{n+1,\alpha-1+\delta}_{[\alpha m],n}.$$

According to the definition (18),

$$A^{n+1,\alpha-1+\delta}_{[\alpha m],n} = \frac{\Gamma(1+\sigma(\alpha)+\sigma([\alpha m])+\sigma(\delta))}{\Gamma(1+\sigma(\alpha)+\sigma(\delta)) \prod_{j=1}^{n} \Gamma(1+[\alpha_j m_j])(1+[\alpha_j m_j])^{\alpha_j-1+\delta_j}},$$

where $\sigma(\delta) = \sum_{j=1}^{n} \delta_j$. By Lemma 6, there exist two positive constants c_9'' and C_9'', which are independent of m and δ, and only depend on α and n, such that

$$c_9'' \frac{\Gamma(\sigma([\alpha m])+\sigma(\alpha)+1)(\sigma([\alpha m])+\sigma(\alpha)+1)^{\sigma(\delta)}}{\Gamma(1+\sigma(\alpha))(1+\sigma(\delta))^{\sigma(\delta)} \prod_{j=1}^{n} \Gamma(1+[\alpha_j m_j])(1+[\alpha_j m_j])^{\alpha_j-1+\delta_j}}$$

$$\le A^{n+1,\alpha-1+\delta}_{[\alpha m],n}$$

$$\le C_9'' \frac{\Gamma(\sigma([\alpha m])+\sigma(\alpha)+1)(\sigma([\alpha m])+\sigma(\alpha)+1)^{\sigma(\delta)}}{\Gamma(1+\sigma(\alpha))(1+\sigma(\alpha))^{\sigma(\delta)} \prod_{j=1}^{n} \Gamma(1+[\alpha_j m_j])(1+[\alpha_j m_j])^{\alpha_j-1+\delta_j}}.$$

That is

$$c_9'' B^{n+1,\alpha-1}_{[\alpha m],n} \prod_{j=1}^{n} \left(\frac{\sigma([\alpha m])+\sigma(\alpha)+1}{[\alpha_j m_j]+1} \right)^{\delta_j} (1+\sigma(\alpha))^{-\sigma(\delta)} \le A^{n+1,\alpha-1+\delta}_{[\alpha m],n}$$

$$\le C_9'' B^{n+1,\alpha-1}_{[\alpha m],n} \prod_{j=1}^{n} \left(\frac{\sigma([\alpha m])+\sigma(\alpha)+1}{[\alpha_j m_j]+1} \right)^{\delta_j} (1+\sigma(\alpha))^{-\sigma(\delta)}.$$

Thus we have proved the lemma. □

From (4), we know

$$K(z,\bar{z}) = \sum_{z \in \mathbf{Z}^n_+} C_m |z|^{2m}$$

where C_m is given by (31), and $|z|^{2m} = |z_1|^{2m_1} \cdots |z_n|^{2m_n}$. Let $u_j = |z_j|^{2/\alpha_j}$, $j = 1, 2, \dots, n$. $\|z\|_\alpha = \sum_{j=1}^{n} |z_j|^{2/\alpha_j}$, then $D = \{z \in \mathbf{C}^n \mid \|z\|_\alpha < 1\}$ and

$$K(z,\bar{z}) = \sum_{m \in \mathbf{Z}^n_+} C_m u^{m\alpha}$$

where $u^{m\alpha} = u_1^{m_1\alpha_1} \cdots u_n^{m_n\alpha_n}$.

Let $K_\alpha(u) = \sum_{m\in\mathbf{Z}_+^n} C_m u^{m\alpha}$, then

$$K_\alpha(u) = \sum_{m\in\mathbf{Z}_+^n} C_m u_1^{[\alpha_1 m_1]} \cdots u_n^{[\alpha_n m_n]} u_1^{\varepsilon_1} \cdots u_n^{\varepsilon_n}$$

where $\varepsilon_j = \alpha_j m_j - [\alpha_j m_j]$ and hence $1 < \varepsilon_j \geq 0$. By Lemma 11,

$$\begin{aligned} K_\alpha(u) &\leq C_8 \sum_{m\in\mathbf{Z}_+^n} B_{[\alpha m].n}^{n+1,\alpha-1} \prod_{j=1}^n \left(\frac{1+\sigma(\alpha)+\sigma([\alpha m])}{1+[\alpha_j m_j]} u_j\right)^{\varepsilon_j} u_1^{[\alpha_1 m_1]} \cdots u_n^{[\alpha_n m_n]} \\ &\leq C_8 \sum_{m\in\mathbf{Z}_+^n} B_{[\alpha m].n}^{n+1,\alpha-1} \prod_{j=1}^n \left(1+\frac{1+\sigma(\alpha)+\sigma([\alpha m])}{1+[\alpha_j m_j]} u_j\right) u_1^{[\alpha_1 m_1]} \cdots u_n^{[\alpha_n m_n]} \end{aligned}$$

since $1 + x > x^a$ if $x > 0$ and $0 \leq a \leq 1$.

Let

$$\prod_{j=1}^n (1+x_j) = \sum_{\delta\in Q} x^\delta = \sum_{\delta\in Q} x_1^{\delta_1} \cdots x_n^{\delta_n}$$

where $\delta = (\delta_1, \ldots, \delta_n)$, $\delta_j = 0$ or 1, $j = 1, 2, \ldots, n$, and Q is a set of lattice points in $\mathbf{R}_+^n$, which contains all the lattice points whose components are only 0 or 1, then

$$\begin{aligned} K_\alpha(u) \leq C_8 &\sum_{m\in\mathbf{Z}_+^n} B_{[\alpha m].n}^{n+1,\alpha-1} u_1^{[\alpha_1 m_1]} \cdots u_n^{[\alpha_n m_n]} \\ &\times \sum_{\delta\in Q} \prod_{j=1}^n \left(\frac{1+\sigma(\alpha)+\sigma([\alpha m])}{1+[\alpha_j m_j]}\right)^{\delta_j} u^{\delta_j}. \end{aligned}$$

By Lemma 12, we have

$$K_\alpha(u) \leq \frac{C_8}{c_9} \sum_{\delta\in Q} u^\delta \sum_{m\in\mathbf{Z}_+^n} B_{[\alpha m].n}^{n+1,\alpha-1+\delta} u_1^{[\alpha_1 m_1]} \cdots u_n^{[\alpha_n m_n]}$$

where $u^\delta = u_1^{\delta_1} \cdots u_n^{\delta_n}$.

Let $h = (h_1, \ldots, h_n) \in \mathbf{Z}_+^n$ be given. Then as an equation of m_j, $[\alpha_j m_j] = h_j$ at most has $\{\frac{1}{\alpha_j}\}$ solutions, $j = 1, 2, \ldots, n$, where $\{x\}$ means the least integer X such that $x \leq X$. Thus

$$K_\alpha(u) \leq \frac{C_8}{c_9} \prod_{j=1}^n \left\{\frac{1}{\alpha_j}\right\} \sum_{\delta\in Q} u^\delta \sum_{h\in\mathbf{Z}_+^n} B_{h.n}^{n+1,\alpha-1+\delta} u_1^{h_1} \cdots u_n^{h_n}.$$

By the definition of F, (15) and (16), we have

$$\begin{aligned} \sum_{h\in\mathbf{Z}_+^n} B_{h.n}^{n+1,\alpha-1+\delta} u_1^{h_1} \cdots u_n^{h_n} &= F_n^{n+1,\alpha-1+\delta}(u) \\ &= (1-\sigma(u))^{-n-1} \prod_{j=1}^n (1+u_j-\sigma(u))^{-\alpha_j+1-\delta_j}. \end{aligned}$$

Since $\frac{u_j}{1+u_j-\sigma(u)} < 1$ and $\delta_j = 0$ or 1, we have

$$\begin{aligned}
K_\alpha(u) &\le C_{10} \sum_{\delta\in Q} u^\delta F_n^{n+1,\alpha-1+\delta}(u) \\
&= C_{10} \sum_{\delta\in Q} (1-\sigma(u))^{-n-1} \prod_{j=1}^{n} \left(\frac{u_j}{1+u_j-\sigma(u)}\right)^{\delta_j} \\
&\quad \times \prod_{j=1}^{n} (1+u_j-\sigma(u))^{1-\alpha_j} \\
&= C_{10} \sum_{\delta\in Q} F_n^{n+1,\alpha-1}(u) \prod_{j=1}^{n} \left(\frac{u_j}{1+u_j-\sigma(u)}\right)^{\delta_j} \\
&\le 2^n C_{10} F_n^{n+1,\alpha-1}(u) = 2^n C_{10} F(z,\bar z)
\end{aligned}$$

where

$$C_{10} = \frac{C_8}{c_9} \prod_{j=1}^{n} \left\{\frac{1}{\alpha_j}\right\}.$$

We have proved the right-hand inequality of (2).

7. Proof of Theorem, Part 2

The proof of the left-hand inequality of (2) is little bit difficult.

For any $\delta = (\delta_1, \ldots, \delta_n) \in Q$, we define

$$\phi_\delta(m) = (\phi_{\delta,1}(m_1), \ldots, \phi_{\delta,n}(m_n))$$

as $\phi_{\delta,j}(m_j) = [m_j\alpha_j]$ when $\delta_j = 1$; and $\phi_{\delta,j}(m_j) = \{m_j\alpha_j\}$ when $\delta_j = 0$; $j = 1, 2, \ldots, n$; where Q is defined in section 6.

For any $h \in \mathbf{Z}_+^n$, $m \in \mathbf{Z}_+^n$, and $\delta \in Q$, we define $N_\delta(m,h) = 1$, if $h = \phi_\delta(m)$; and $N_\delta(m,h) = 0$ if $h \ne \phi_\delta(m)$, then we have

(1) for a fixed $m \in \mathbf{Z}_+^n$,

$$\sum_{h\in \mathbf{Z}_+^n} N_\delta(m,h) = 1; \tag{32}$$

(2) for a fixed $h \in \mathbf{Z}_+^n$,

$$0 \le \sum_{m\in\mathbf{Z}_+^n} N_\delta(m,h) \equiv N_\delta(h) \le \prod_{j=1}^{n} \{\frac{1}{\alpha_j}\}$$

since each of the equations $[\alpha_j m_j] = h_j$ and $\{\alpha_j m_j\} = h_j$ has at most $\{1/\alpha_j\}$ integer solutions for m_j.

For any $m \in \mathbf{Z}_+^n$ and $\delta \in Q$, we may decompose αm as

$$\begin{aligned}
\alpha m &= (\alpha_1 m_1, \ldots, \alpha_n m_n) \\
&= (\phi_{\delta,1}(m_1) + \varepsilon_1(\delta_1, m_1), \ldots, \phi_{\delta,n}(m_n) + \varepsilon_n(\delta_n, m_n)),
\end{aligned}$$

where

$$\varepsilon_j(\delta_j, m_j) = \alpha_j m_j - \phi_{\delta,j}(m_j) = \alpha_j m_j - [\alpha_j m_j]$$

if $\delta_j = 1$; and

$$\varepsilon_j(\delta_j, m_j) = \alpha_j m_j - \phi_{\delta,j}(m_j) = \alpha_j m_j - \{\alpha_j m_j\}$$

if $\delta_j = 0$; $j = 1, 2, \ldots, n$.

Let $\phi_{\delta,j}(m_j) = h_j, \varepsilon_j = \varepsilon_j(\delta_j, m_j)$, then

$$\alpha_j m_j = \phi_{\delta,j}(m_j) + \varepsilon_j(\delta_j, m_j) = h_j + \varepsilon_j.$$

We may extend Lemma 11 as

Lemma 13. *If $\phi_\delta(h) = m$, then there exist two positive constants c_{11} and C_{11}, which are independent of m and δ, and only depend on α and n, such that*

$$c_{11} B_{h,n}^{n+1,\alpha-1} \prod_{j=1}^{n} \left(\frac{1+\sigma(\alpha)+\sigma(h)}{h_j+1}\right)^{\varepsilon_j} \le C_m$$

$$\le C_{11} B_{h,n}^{n+1,\alpha-1} \prod_{j=1}^{n} \left(\frac{1+\sigma(\alpha)+\sigma(h)}{h_j+1}\right)^{\varepsilon_j}.$$

Since the proof of Lemma 13 is similar to that of Lemma 11, we omit it.

By (32), we have

$$K_\alpha(u) = \sum_{m\in \mathbf{Z}_+^n} \sum_{h\in \mathbf{Z}_+^n} N_\delta(m,h) C_m u_1^{h_1} \cdots u_n^{h_n} u_1^{\varepsilon_1} \cdots u_n^{\varepsilon_n}.$$

Using Lemma 13, we find that

$$K_\alpha \ge c_{11} \sum_{m,h\in \mathbf{Z}_+^n} N_\delta(m,h) B_{h,n}^{n+1,\alpha-1} \prod_{j=1}^{n} \left(\frac{\sigma(h)+\sigma(\alpha)+1}{h_j+1}\right)^{\varepsilon_j} u^h$$

holds for any $\delta \in Q$. Since the total number of lattice points in Q is 2^n, we have

$$K_\alpha(u) \ge 2^{-n} c_{11} \sum_{h\in \mathbf{Z}_+^n} B_{h,n}^{n+1,\alpha-1} u^h \sum_{\delta\in Q}$$

$$\times \left(\sum_{m\in \mathbf{Z}_+^n} N_\delta(m,h) \prod_{j=1}^{n} \left(\frac{\sigma(h)+\sigma(\alpha)+1}{h_j+1} u_j\right)^{\varepsilon_j(\delta_j,m_j)}\right).$$

We consider the last bracket on the right-hand side of the previous inequality. For a fixed $h \in \mathbf{Z}_+^n$, and a fixed $\delta \in Q$, as an equation in m, $\phi_\delta(m) = h$ may have solutions, that means, $h \in \phi_\delta(\mathbf{Z}_+^n)$; or may not have a solution, that means, h does not belong to $\phi_\delta(\mathbf{Z}_+^n)$. We define $\bar{N}_\delta(h)$ as the characteristic function of $\phi_\delta(\mathbf{Z}_+^n)$ when δ is fixed, that is, $\bar{N}_\delta(h) = 1$ if $h \in \phi_\delta(\mathbf{Z}_+^n)$; and $\bar{N}_\delta(h) = 0$, otherwise, Then

$$K_\alpha(u) > 2^{-n} c_{11} \sum_{h\in \mathbf{Z}_+^N} B_{h,n}^{n+1,\alpha-1} \sum_{\delta\in Q} \bar{N}_\delta(h) \prod_{j=1}^{n} \left(\frac{\sigma(h)+\sigma(\alpha)+1}{h_j+1} u_j\right)^{\varepsilon_j(\delta_j,h_j)}$$

where $\varepsilon_j(\delta_j, h_j) = \varepsilon_j(\delta_j, m_j)$ if m_j is a solution of $h_j = \phi_{\delta,j}(m_j)$, $j = 1, 2, \ldots, n$.

Obviously, $\phi(\mathbf{Z}_+^n)$ has the following properties:

(1) If $h \in \phi_\delta(\mathbf{Z}_+^n)$, then $h(j) = (h_1, \ldots, h_{j-1}, 0, h_{j+1}, \ldots, h_n) \in \phi_\delta(\mathbf{Z}_+^n)$;

(2) If $h \in \phi_\delta(\mathbf{Z}_+^n)$, then there exists an integer k_j, $1 \le k_j \le \{\alpha_j\}$, such that $h + k_j e_j \in \phi_\delta(\mathbf{Z}_+^n)$.

Fix $h(1) \in \phi_\delta(\mathbf{Z}_+^n)$, then there exists an integer sequence

$$n_{1,0} = 0 < n_{1,1} < n_{1,2} < \cdots < n_{1,k} < \cdots, \qquad k = 0, 1, 2, \ldots,$$

which is independent of the choice of $h(1)$, such that $h(1) + n_{1,k} e_1 \in \phi_\delta(\mathbf{Z}_+^n)$, and $1 \le n_{1,k+1} - n_{1,k} \le \{\alpha_1\}$. Moreover, if $n_{1,k+1} \ge n_{1,k} + 2$ holds, then $h(1) + (n_{1,k} + s)e_1$ does not belong to $\phi_\delta(\mathbf{Z}_+^n)$, when $1 \le s \le n_{1,k+1} - n_{1,k} - 1$.

Fix $\delta \in Q$, and consider a subset $D_1(\delta) \subset \mathbf{Z}_+^n$ as

$$D_1(\delta) = \{h(1) + ke_1 \in \mathbf{Z}_+^n, \text{ for all } h(1) \in \phi_\delta(\mathbf{Z}_+^n),\ k = 0, 1, \ldots\},$$

then $D_1(\delta) \supset \phi_\delta(\mathbf{Z}_+^n)$, and

$$K_\alpha(u) > c_{11} 2^{-n} \sum_{\delta \in Q} \sum_{h \in D_1(\delta)} \bar{N}_\delta(h) B_{h,n}^{n+1,\alpha-1} u^h \prod_{j=1}^n \left(\frac{\sigma(h)+\sigma(\alpha)+1}{h_j+1} u_j \right)^{\varepsilon_j(\delta_j,h_j)}.$$

Let $\chi = h(1)+n_{1,k}e_1 \in \phi_\delta(\mathbf{Z}_+^n)$, $\tau = h(1)+n_{1,k+1}e_1 \in \phi_\delta(\mathbf{Z}_+^n)$ and $l = n_{1,k+1}-n_{1,k}$, $1 \le s \le n_{1,k+1} - n_{1,k} - 1 = l-1$ if $l \ge 2$, then

(33)

$$\begin{aligned}
&\frac{s}{l} B_{\tau,n}^{n+1,\alpha-1} u^\tau \prod_{j=1}^n \left(\frac{\sigma(\tau)+\sigma(\alpha)+1}{\tau_j+1} u_j \right)^{\varepsilon_j(\delta_j,\tau_j)} \\
&\quad + \frac{l-s}{l} B_{\chi,n}^{n+1,\alpha-1} u^\chi \prod_{j=1}^n \left(\frac{\sigma(\chi)+\sigma(\alpha)+1}{\chi_j+1} u_j \right)^{\varepsilon_j(\delta_j,\chi_j)} \\
&\quad \ge (B_{\tau,n}^{n+1,\alpha-1})^{s/l} (B_{\chi,n}^{n+1,\alpha-1})^{(l-s)/l} u^{(\tau+\varepsilon(\delta,\tau))s/l+(\chi+\varepsilon(\delta,\chi))^{(l-s)/s}} \\
&\quad \times \prod_{j=1}^n \left(\frac{\sigma(\tau)+\sigma(\alpha)+1}{\tau_j+1} \right)^{s/l\varepsilon_j(\delta_j,\tau_j)} \left(\frac{\sigma(\chi)+\sigma(\alpha)+1}{\chi+1} \right)^{(l-s)/s\varepsilon_j(\delta_j,\chi_j)},
\end{aligned}$$

since the arithmetic mean is greater or equal to the geometric mean.

According to the Main Lemma, there exist two constants c_{12} and C_{12}, which are independent of τ and χ, and only depend on n and α, such that

$$c_{12} A_{\tau,n}^{n+1,\alpha-1} \le B_{\tau,n}^{n+1,\alpha-1} \le C_{12} A_{\tau,n}^{n+1,\alpha-1},$$

and

$$c_{12} A_{\chi,n}^{n+1,\alpha-1} \le B_{\chi,n}^{n+1,\alpha-1} \le C_{12} A_{\tau,n}^{n+1,\alpha-1}.$$

Let $h_1 = n_{1,k} + s$, then

$$\begin{aligned}
A_{\tau,n}^{n+1,\alpha-1} &= \frac{\Gamma(\sigma(\alpha)+\sigma(h(1))+n_{1,k+1}+1)}{\Gamma(\sigma(\alpha)+1)\prod_{j=2}^n \Gamma(1+h_j)(1+h_j)^{\alpha_j-1}\Gamma(1+n_{1,k+1})(1+n_{1,k+1})^{\alpha_1-1}} \\
&= \frac{\Gamma(\sigma(\alpha)+\sigma(h)+l-s+1)}{\Gamma(\sigma(\alpha)+1)\prod_{j=2}^n \Gamma(1+h_j)(1+h_j)^{\alpha_j-1}\Gamma(h_1+l-s+1)(h_1+l-s+1)^{\alpha_1-1}} \\
&= A_{h,n}^{n+1,\alpha-1} \frac{(\sigma(\alpha)+\sigma(h)+1)\cdots(\sigma(\alpha)+\sigma(h)+l-s)(1+h_1)^{\alpha_1-1}}{(h_1+1)\cdots(h_1+l-s)(h_1+l-s+1)^{\alpha_1-1}},
\end{aligned}$$

and

$$\begin{aligned}
A_{\chi,n}^{n+1,\alpha-1} &= \frac{\Gamma(\sigma(\alpha)+\sigma(h(1))+n_{1,k}+1)}{\Gamma(\sigma(\alpha)+1)\prod_{j=2}^n \Gamma(1+h_j)(1+h_j)^{\alpha_j-1}\Gamma(1+n_{1,k})(1+n_{1,k})^{\alpha_1-1}} \\
&= \frac{\Gamma(\sigma(\alpha)+\sigma(h)+1)}{\Gamma(\sigma(\alpha)+1)\prod_{j=2}^n \Gamma(1+h_j)(1+h_j)^{\alpha_j-1}\Gamma(1+h_1)(1+n_{1,k})^{\alpha_1-1}} \\
&\quad \times \frac{(1+n_{1,k})\cdots(s+n_{1,k})}{(\sigma(\alpha)+\sigma(h(1))+n_{1,k}+1)\cdots(\sigma(\alpha)+\sigma(h(1))+n_{1,k}+s)} \\
&= A_{h,n}^{n+1,\alpha-1} \frac{(1+n_{1,k})\cdots(s+n_{1,k})(1+h_1)^{\alpha_1-1}}{(\sigma(\alpha)+\sigma(h(1))+n_{1,k}+1)\cdots(\sigma(\alpha)+\sigma(h(1))+n_{1,k}+s)(1+n_{1,k})^{\alpha_1-1}}.
\end{aligned}$$

Thus

$$(A_{\tau,n}^{n+1,\alpha-1})^{s/l}(A_{\chi,n}^{n+1,\alpha-1})^{(l-s)/l} = A_{h,n}^{n+1,\alpha-1}(1+h_1)^{\alpha_1-1}$$
$$\times\left(\frac{\sigma(\alpha)+\sigma(h)+1)\cdots(\sigma(\alpha)+\sigma(h)+l-s)}{(h_1+1)\cdots(h_1+l-s)(h_1+l-s+1)^{\alpha_1-1}}\right)^{s/l}$$
$$\times\left(\frac{(1+n_{1,k})\cdots(s+n_{1,k})}{(\sigma(\alpha)+\sigma(h(1))+n_{1,k}+1)\cdots(\sigma(\alpha)+\sigma(h(1))+n_{1,k}+s)(1+n_{1,k})^{\alpha_1-1}}\right)^{(l-s)/l}$$
$$= A_{h,n}^{n+1,\alpha-1}I_1I_2I_3,$$

where

$$I_1 = \frac{(1+h_1)^{\alpha_1-1}}{((h_1+l-s+1)^{s/l}(h_1-s+1)^{(l-s)/l})^{\alpha_1-1}},$$
$$I_2 = \frac{((\sigma(\alpha)+\sigma(h)+1)\cdots(\sigma(\alpha)+\sigma(h)+l-s))^{s/l}}{((\sigma(\alpha)+\sigma(h)-s+1)\cdots(\sigma(\alpha)+\sigma(h)))^{(l-s)/l}}$$

and

$$I_3 = \frac{(h_1(h_1-1)\cdots(h_1-s+1))^{(l-s)/l}}{((h_1+1)\cdots(h_1+l-s))^{s/l}}.$$

Obviously, I_1, I_2 and I_3 are bounded above and below since $l \le \{\alpha_1\}$ and $s \le l-1 \le \{\alpha_1\}-1$. Thus there exist two positive constants c_{14} and C_{14}, which are independent of τ, χ and h, and only depend on α and n, such that

$$c_{14}B_{h,n}^{n+1,\alpha-1} \le (B_{\tau,n}^{n+1,\alpha-1})^{s/l}(B_{\chi,n}^{n+1,\alpha-1})^{(l-s)/l} \le C_{14}B_{h,n}^{n+1,\alpha-1}$$

where $1 \le s \le n_{1,k+1}-n_{1,k}-1$.

Moreover, we have

$$\tau\frac{s}{l}+\chi\frac{l-s}{l} = (h(1)+n_{1,k+1}e_1)\frac{s}{l}+(h(1)+n_{1,k}e_1)\frac{l-s}{l}$$
$$= h(1)+(n_{1,k}+s)e_1 = h,$$

and

$$\frac{\sigma(\tau)+\sigma(\alpha)+1}{\tau_j+1} = \frac{\sigma(h)+\sigma(\alpha)+1+l-s}{h_j+1}$$

if $j \ne 1$, then there exist two positive constants c'_{15} and C'_{15}, which are independent of h, and only depend on α and n, such that

$$c'_{15}\frac{\sigma(h)+\sigma(\alpha)+1}{h_j+1} \le \frac{\sigma(\tau)+\sigma(\alpha)+1}{\tau_j+1} \le C'_{15}\frac{\sigma(h)+\sigma(\alpha)+1}{h_j+1}$$

when $j \ne 1$, since $1 \le l-s \le \{\alpha_1\}-1$. When $j=1$, we have

$$\frac{\sigma(\tau)+\sigma(\alpha)+1}{\tau_1+1} = \frac{n_{1,k+1}+\sigma(h(1))+\sigma(\alpha)+1}{n_{1,k+1}+1} = \frac{\sigma(h)+\sigma(\alpha)+1+l-s}{h_1+1+l-s},$$

then there exist two positive constants c''_{15} and C''_{15}, which are independent of h and τ, and only depend on α and n, such that

$$c''_{15}\frac{\sigma(h)+\sigma(\alpha)+1}{h_1+1} \le \frac{\sigma(\tau)+\sigma(\alpha)+1}{\tau_1+1} \le C''_{15}\frac{\sigma(h)+\sigma(\alpha)+1}{h_1+1}.$$

Thus

$$c_{15}\frac{\sigma(h)+\sigma(\alpha)+1}{h_j+1} \le \frac{\sigma(\tau)+\sigma(\alpha)+1}{\tau_j+1} \le C_{15}\frac{\sigma(h)+\sigma(\alpha)+1}{h_j+1}$$

hold for $j = 1,2,\ldots,n$, where $c_{15} = \min(c'_{15}, c''_{15})$, and $C_{15} = \max(C'_{15}, C''_{15})$. Similarly, the inequalities

$$c_{16}\frac{\sigma(h)+\sigma(\alpha)+1}{h_j+1} \le \frac{\sigma(\chi)+\sigma(\alpha)+1}{\chi_j+1} \le C_{16}\frac{\sigma(h)+\sigma(\alpha)+1}{h_j+1}$$

holds for $j = 1,2,\ldots,n$, where c_{16} and C_{16} are two positive constants, which are independent of h, and only depend on α and n.

Combining all these results, we get the right-hand side of (33) is not less than

$$c_{17}B^{n+1,\alpha-1}_{h,n}u^h\prod_{j=2}^{n}\left(\frac{\sigma(h)+\sigma(\alpha)+1}{h_j+1}u_j\right)^{\varepsilon_j(\delta_j,h_j)} \times\left(\frac{\sigma(h)+\sigma(\alpha)+1}{h_1+1}u_1\right)^{\varepsilon'_1(\delta_1,h_1)}$$

where $c_{17} = c_{14}c_{15}^{s/l}c_{16}^{(l-s)/l}$, and $\varepsilon'_1(\delta_1,h_1) = \frac{s}{l}\varepsilon_1(\delta_1,\tau_1) + \frac{l-s}{l}\varepsilon_1(\delta_1,\chi_1)$.

Let

$$T(y) = B^{n+1,\alpha-1}_{y,n}u^y\prod_{j=1}^{n}\left(\frac{\sigma(y)+\sigma(\alpha)+1}{y_j+1}u_j\right)^{\varepsilon_j(\delta_j,y_i)}$$

where $y = \tau, \chi$ or h. Then

$$\begin{aligned}
T(\tau)+T(\chi) &= \frac{2}{l-1}\sum_{s=1}^{l-1}\frac{1}{l}(sT(\tau)+(l-s)T(\chi))\\
&= \frac{1}{2}(T(\tau)+T(\chi)) + \frac{1}{4}(T(\tau)+T(\chi)) + \frac{1}{2(l-1)}\sum_{s=1}^{l-1}\frac{1}{l}(sT(\tau)+(l-s)T(\chi))\\
&\ge \frac{1}{2}(T(\tau)+T(\chi)) + \frac{1}{4}(T(\tau)+T(\chi))\\
&\quad + \frac{c_{17}}{2(l-1)}\sum_{s=1}^{l-1}T(h)\left(\frac{\sigma(h)+\sigma(\alpha)+1}{h_1+1}u_1\right)^{\varepsilon'_1(\delta_1,h_1)-\varepsilon_1(\delta_1,h_1)}\\
&\ge \frac{1}{2}(T(\tau)+T(\chi)) + c_{18}\sum_{s=0}^{l}T(h)\left(\frac{\sigma(h)+\sigma(\alpha)+1}{h_1+1}u_1\right)^{\varepsilon'_1(\delta_1,h_1)-\varepsilon_1(\delta_1,h_1)}
\end{aligned}$$

where $c_{18} = \min(\frac{1}{4}, \frac{c_{17}}{2([\alpha_1]-1)})$. Thus, we have

$$K_\alpha(u) > 2^{-n}c_{11}c_{18}\sum_{\delta\in Q}\sum_{h\in \mathbf{Z}^n_+}\bar{N}_\delta(1,h)B^{n+1,\alpha-1}_{h,n}u^h \times\prod_{j=2}^{n}\left(\frac{\sigma(h)+\sigma(\alpha)+1}{h_j+1}u_j\right)^{\varepsilon_j(\delta_j,h_j)}\left(\frac{\sigma(h)+\sigma(\alpha)+1}{h_1+1}u_1\right)^{\varepsilon'_1(\delta_1,h_1)} \tag{34}$$

where $\bar{N}_\delta(1,h)$ is the characteristic function of $D_1(\delta)$, that is, $\bar{N}_\delta(1,h) = 1$, if $h \in D_1$; $\bar{N}_\delta(1,h) = 0$, if h does not belong to D_1.

If $\delta_1 = 1$, then there exist two positive integers m_1 and m'_1, such that $[\alpha_1 m'_1] = n_{1,k+1}$, and $[\alpha_1 m_1] = n_{1,k}$. Hence

$$\varepsilon_1(1,\tau_1) = \alpha_1 m'_1 - [\alpha_1 m'_1] = \alpha_1 m'_1 - n_{1,k+1} \ge 0,$$

and

$$\varepsilon_1(1,\chi_1) = \alpha_1 m_1 - [\alpha_1 m_1] = \alpha_1 m_1 - n_{1,k} \geq 0.$$

Since by the definition of ε_1',

$$\varepsilon_1'(\delta_1, h_1) = \frac{s}{l}\varepsilon_1(\delta_1, \tau_1) + \frac{l-s}{l}\varepsilon_1(\delta_1, \chi_1),$$

we have $\varepsilon_1'(1, h_1) \geq 0$. Similarly, we can prove that $\varepsilon_1'(0, h_1) \leq 0$.

Fix $h(2) \in D_1(\delta)$, then there exists a sequence of integers

$$n_{2,0} = 0 < n_{2,1} < \cdots < n_{2,k} < \cdots,$$

which is independent of the choice of $h(2) \in D_1(\delta)$, such that $h(2)+n_{2,k}e_2 \in D_1(\delta)$, and $1 \leq n_{2,k+1} - n_{2,k} \leq \{\alpha_2\}$ and $h(2) + (n_{2,k}+s)$ does not belong to $D_1(\delta)$ when $1 \leq s \leq n_{2,k+1} - n_{2,k} - 1$ and $n_{2,k+1} \geq n_{2,k} + 2$.

Define $D_2 = D_2(\delta)$ by

$$D_2(\delta) = \{h(2) + ke_2 \in \mathbf{Z}_+^n, \text{ for all } h(2) \in D_1(\delta),\ k = 0, 1, 2, \ldots\}.$$

Then

$$D_2(\delta) \supset D_1(\delta) \supset \phi_\delta(\mathbf{Z}_+^n),$$

and $D_2(\delta)$ has the following properties:

(1) $h(j) \in D_2(\delta)$, $j = 1, 2, \ldots, n$ if $h \in D_2(\delta)$, in particular, $h(1,2) \in D_2(\delta)$;

(2) $h(1,2) + ke_1 + se_2 \in D_2(\delta)$, $k, s = 0, 1, 2, \ldots$, if $h(1,2) \in D_2(\delta)$;

(3) If $h \in D_2(\delta)$, then there exists k_j, $1 \leq k_j \leq \{\alpha_j\}$, such that $h + k_j e_j \in D_2(\delta)$, $j = 3, 4, \ldots, n$.

Using the same method as we used to prove (34), we can prove

$$K_\alpha(u) > c_{19} \sum_{\delta \in Q} \sum_{h \in \mathbf{Z}_+^n} \bar{N}_\delta(2,h) B_{h,n}^{n+1,\alpha-1} u^h \prod_{j=3}^{n} \left(\frac{\sigma(h)+\sigma(\alpha)+1}{h_j+1} u_j\right)^{\varepsilon_j(\delta_j, h_j)}$$
$$\times \left(\frac{\sigma(h)+\sigma(\alpha)+1}{h_1+1} u_1\right)^{\varepsilon_1'(\delta_1, h_1)} \left(\frac{\sigma(h)+\sigma(\alpha)+1}{h_2+1} u_2\right)^{\varepsilon_2'(\delta_2, h_2)}$$

where c_{19} is a positive constant, which only depends on α and n, and $\bar{N}_\delta(2,h)$ is the characteristic function of $D_2(\delta)$, that is, $\bar{N}_\delta(2,h) = 1$ if $h \in D_2$; $\bar{N}_\delta(2,h) = 0$ if h does not belong to D_2; and $\varepsilon_2'(\delta_2, h_2)$ has the following property:

$$\varepsilon_2'(1, h_2) \geq 0, \quad \varepsilon_2'(0, h_2) \leq 0.$$

We can repeat this process again and again, finally, we have

$$K_\alpha(u) > c_{20} \sum_{\delta \in Q} \sum_{h \in \mathbf{Z}_+^n} \bar{N}_\delta(n,h) B_{h,n}^{n+1,\alpha-1} u^h$$
$$\times \prod_{j=1}^{n} \left(\frac{\sigma(h)+\sigma(\alpha)+1}{h_j+1} u_j\right)^{\varepsilon_j'(\delta_j, h_j)}$$

where c_{20} is a positive constant, which only depends on α and n, and $\bar{N}_\delta(n,h)$ is the characteristic function of $D_n(\delta)$, by the definition of $D_j(\delta)$, $j = 1, 2, \ldots, n$, we have $D_n(\delta) = \mathbf{Z}_+^n$, thus $\bar{N}_\delta(n,h) \equiv 1$, when $h \in \mathbf{Z}_+^n$; and $\varepsilon_j'(\delta_j, h_j)$ has the properties:

$$\varepsilon_j'(1,h_j) \geq 0, \quad \varepsilon_j'(0,h_j) \leq 0, \qquad j = 1,2,\ldots,n.$$

Thus

$$K_\alpha(u) > c_{20} \sum_{h \in \mathbf{Z}_+^n} B_{h,n}^{n+1,\alpha-1} u^h \sum_{s \in Q} \prod_{j=1}^{n} \left(\frac{\sigma(h)+\sigma(\alpha)+1}{h_j+1} u_j \right)^{\varepsilon_j'(\delta_j,h_j)}.$$

For a fixed $h \in \mathbf{Z}_+^n$, we may take $\delta = (\delta_1, \ldots, \delta_n) \in Q$, such that

$$\delta_j = 0 \quad \text{if } \frac{\sigma(h)+\sigma(\alpha)+1}{h_j+1} u_j \leq 1;$$

$$\delta_j = 1 \quad \text{if } \frac{\sigma(h)+\sigma(\alpha)+1}{h_j+1} u_j \geq 1;$$

then

$$\prod_{j=1}^{n} \left(\frac{\sigma(h)+\sigma(\alpha)+1}{h_j+1} u_j \right)^{\varepsilon_j'(\delta_j,h_j)} \geq 1,$$

and hence

$$\sum_{\delta \in Q} \prod_{j=1}^{n} \left(\frac{\sigma(h)+\sigma(\alpha)+1}{h_j+1} u_j \right)^{\varepsilon_j'(\delta_j,h_j)} \geq 1.$$

Finally we obtain

$$K_\alpha(u) > c_{20} \sum_{h \in \mathbf{Z}_+^n} B_{h,n}^{n+1,\alpha-1} u^h = c_{20} F_n^{n+1,\alpha-1}(u) = c_{20} F(z,\bar{z}).$$

We have completed the proof of the Theorem. □

ACKNOWLEDGEMENTS

This paper was written while the first author was at the Department of Mathematics of the University of California, San Diego, as a visiting professor. The first author is grateful to the department for their support, and also would like to express gratitude to Professor John D'Angelo for helpful and important communications.

REFERENCES

1. B. Bergman, *Über die Kernfunktion eines Bereiches und ihr Verhalten am Rande*, J. Reine Angew. Math. **169** (1933), 1–42.
2. L. Boutet de Monvel and J. Sjöstrand, *Sur la singularité des noyaux de Bergman et de Szegö*, Astérisque 34–45, Soc. Math. de Paris, France, 1976, pp. 123–164. MR **58**:28684
3. D. Catlin, *Estimates of invariant metrics on pseudoconvex domains of dimension two*, Math. Z. **200** (1989), 429–466. MR **90e**:32029
4. J. P. D'Angelo, *A note on the Bergman kernel*, Duke Math. J. **45** (1978), 259–266. MR **57**:12906
5. ———, *An explicit computation of the Bergman kernel function*, J. Geom. Anal. **4** (1994), 23–34. MR **95a**:32039
6. C. Fefferman, *The Bergman kernel and biholomorphic mappings of pseudoconvex domains*, Invent. Math. **26** (1974), 1–65. MR **50**:2562
7. J. McNeal, *Boundary behavior of the Bergman kernel function in* $\mathbf{C}^2$, Duke Math. J. **58** (1989), 499–512. MR **91c**:32017

8. ———, *Local geometry of decoupled pseudoconvex domain*, Aspekte der Math. **E17** (1990), 223–230. MR **92g**:32033
9. A. Nagel, J. P. Rosay, E. M. Stein, and S. Wainger, *Estimates for the Bergman and Szegö kernels in* $\mathbf{C}^2$, Ann. of Math. **129** (1989), 113–149. MR **90g**:32028

Department of Mathematics, University of Science and Technology of China, Hefei, Anhui, 230026, People's Republic of China

Department of Mathematics, University of California, San Diego, La Jolla, California 92093

Department of Mathematics, Anhui University, Hefei, Anhui, 230039, People's Republic of China

Vol. 43 No. 5 SCIENCE IN CHINA (Series A) May 2000

The Bergman kernel function of some Reinhardt domains (II)

GONG Sheng (龚 昇)[1] & ZHENG Xue'an (郑学安)[2]

1. Department of Mathematics, University of Science and Technology of China, Hefei 230026, China;
2. Department of Mathematics, Beijing Normal University, Beijing 100875, China
Correspondence should be addressed to Gong Sheng (email: gongs@ustc.edu.cn)

Received May 17, 1999

Abstract The boundary behavior of the Bergman kernel function of a kind of Reinhardt domain is studied. Upper and lower bounds for the Bergman kernel function are found at the diagonal points $(Z, \overline{Z})$. Let Ω be the Reinhardt domain

$$\Omega = \left\{ Z = (Z_1, Z_2, \cdots, Z_n) \in \mathbb{C}^N,\ Z_j \in \mathbb{C}^{N_j},\ j = 1,2,\cdots,n,\ \middle|\ \| Z \|_\alpha = \sum_{j=1}^{n} \| Z_j \|^{\frac{2}{\alpha_j}} < 1 \right\},$$

where $\alpha_j > 0$, $j = 1,2,\cdots,n$, $N = N_1 + N_2 + \cdots + N_n$, $\| Z_j \|$ is the standard Euclidean norm in $\mathbb{C}^{N_j}$, $j = 1,2,\cdots,n$; and let $K(Z, \overline{W})$ be the Bergman kernel function of Ω. Then there exist two positive constants m and M, and a function F such that

$$mF(Z,\overline{Z}) \leqslant K(Z,\overline{Z}) \leqslant MF(Z,\overline{Z})$$

holds for every $Z \in \Omega$. Here

$$F(Z,\overline{Z}) = (-r(Z))^{-N-1} \prod_{j=1}^{n} \left(-r(Z) + \| Z_j \|^{\frac{2}{\alpha_j}}\right)^{(1-\alpha_j)N_j}$$

and $r(Z) = \| Z \|_\alpha - 1$ is the defining function of Ω. The constants m and M depend only on $\alpha = (\alpha_1, \cdots, \alpha_n)$ and $N_1, N_2, \cdots, N_n$, not on Z.

This result extends some previous known results.

Keywords: Reinhardt domain, Bergman kernel function, boundary behavior.

Let $\Omega \subset \mathbb{C}^n$ be a domain in $\mathbb{C}^n$. The Bergman kernel function $K_\Omega(z, \overline{w})$, $z, w \in \Omega$, for Ω is the kernel of the Bergman projection operator, the operator projecting $L^2(\Omega)$ onto its holomorphic subspaces $A^2(\Omega)$. The Bergman kernel can hardly be calculated explicitly, unless the domain Ω has a great deal of symmetry, so that a useful orthonormal basis for $A^2(\Omega)$ can be determined, for example, the classical domain[1]. The boundary behavior of $K_\Omega(z, \overline{w})$ for a domain $\Omega \subset \mathbb{C}^n$ is an old problem[2]. It has become a branch and a hot topic in several complex variables since the famous paper of Fefferman[3] for the boundary behaviors of the Bergman kernel function of strictly pseudo-convex domains was published in 1974. The interest in explicit or asymptotic formula of the Bergman kernel function of the weak pseudo-convex domains was motivated by the surprising effectiveness of these formulas. For example, Mostow and Siu[4] gave a counter-example to the important conjecture that the universal covering of a compact Kähler manifold of negative sectional curvature should be biholomorphic to the ball. In their counter-example, the explicit calculation of the Bergman kernel function and metric of the weak pseudo-convex domain $\{z = (z_1, z_2) \in \mathbb{C}^2 \mid |z_1|^{14} + |z_2|^2 < 1\}$ plays an essential role.

We[5] have studied an important kind of Reinhardt domain:

$$D = \{z = (z_1, \cdots, z_n) \in \mathbb{C}^n \mid \| z \|_\alpha = \sum_{j=1}^{n} |z_j|^{\frac{2}{\alpha_j}} < 1\}, \tag{1}$$

where $\alpha_j > 0$, $j = 1, 2, \cdots, n$ are positive real numbers. The estimate of the Bergman kernel function $K_D(z, \bar{z})$ of D in a "small constant-large constant" sense was given. Precisely, we proved the following result.

There exist two positive constants m_0 and M_0 which depend only on n and α_j, $j = 1, 2, \cdots, n$, such that

$$m_0 F_0(z, \bar{z}) \leqslant K_D(z, \bar{z}) \leqslant M_0 F_0(z, \bar{z}) \tag{2}$$

holds for every $z \in D$, where

$$F_0(z, \bar{z}) = (-r_0(z))^{-n-1} \prod_{j=1}^{n} (-r_0(z) + |z_j|^{\frac{2}{\alpha_j}})^{1-\alpha_j} \tag{3}$$

and $r_0(z) = \| z \|_\alpha - 1$ is the defining function of D.

As we know, D'Angelo may be the first mathematician who studied the Bergman kernel function of such Reinhardt domains. In 1978, he considered the domain of D which is defined by (1) when $\alpha_1 = \cdots = \alpha_{n-1} = 1$ and α_n^{-1} is a positive integer. In this case, he[6] gave the explicit formula of the Bergman kernel function as

$$K_D(z, \bar{w}) = \sum_{k=1}^{n} c_k \frac{\left(1 - \sum_{j=1}^{n-1} z_j \bar{w}_j\right)^{-n+\alpha_n k}}{\left[\left(1 - \sum_{j=1}^{n-1} z_j \bar{w}_j\right)^{\alpha_n} - z_n \bar{w}_n\right]^{k+1}}, \tag{4}$$

where c_k, $k = 1, 2, \cdots, n$ are constants which depend on n, k and α_n only. He also gave the series expansion of the Bergman kernel function of the domain D defined by (1).

In 1996, by using the Appell's multivariable hypergeometric functions, Francsics and Hanges[7] gave the explicit formula of $K_D(z, \bar{w})$ of D when all $\alpha_j^{-1} = p_j$, $j = 1, \cdots, n$ are positive integers. They proved that in this case

$$K_D(z, \bar{w}) = \frac{p_1 \cdots p_n}{\pi^n} \sum_{q_1=0}^{p_1-1} \cdots \sum_{q_n=0}^{p_n-1} \frac{\Gamma\left(1 + \sum_{j=1}^{n} \frac{q_j + 1}{p_j}\right)}{\prod_{j=1}^{n} \Gamma\left(\frac{q_j + 1}{p_j}\right)}$$

$$\times (z\bar{w})^q F_A^{(n)}\left(1 + \sum_{j=1}^{n} \frac{q_j + 1}{p_j}, \mathbf{1}, \frac{\mathbb{Q} + \mathbf{1}}{\mathbb{P}}, (z\bar{w})^p\right) \tag{5}$$

for all $(z, w) \in D \times D$, where $(z\bar{w})^q = (z_1\bar{w}_1)^{q_1} \cdots (z_n\bar{w}_n)^{q_n}$ if $q = (q_1, \cdots, q_n)$, $z = (z_1, \cdots, z_n)$, $w = (w_1, \cdots, w_n)$; and $\mathbf{1} = (1, \cdots, 1)$, $\frac{\mathbb{Q} + \mathbf{1}}{\mathbb{P}} = \left(\frac{q_1 + 1}{p_1}, \cdots, \frac{q_n + 1}{p_n}\right)$ and $F_A^{(n)}$ is the Appell's multivariable hypergeometric function

$$F_A^{(n)}(\alpha, \beta, \gamma, x) = \frac{1}{\Gamma(\alpha)} \prod_{j=1}^{n} \frac{\Gamma(\gamma_j)}{\Gamma(\beta_j)} \sum_{m \geqslant 0} \Gamma(\alpha + \sigma(m)) \left(\prod_{j=1}^{n} \frac{\Gamma(\beta_j + m_j)}{\Gamma(\gamma_j + m_j) m_j!} x_j^{m_j}\right), \tag{6}$$

where $m = (m_1, \cdots, m_n) \in \mathbb{R}_+^n$, $\sigma(m) = \sum_{j=1}^{n} m_j$, $\alpha \in \mathbb{R}$, $\beta = (\beta_1, \cdots, \beta_n) \in \mathbb{R}^n$, $\gamma = (\gamma_1, \cdots, \gamma_n) \in \mathbb{R}^n$.

They also gave the explicit formula of $K_D(z,\bar{w})$ of D when α_j^{-1}, $j=1,\cdots,n-1$ are positive integers, and $\alpha_n>0$ is an arbitrary positive real number. In this case, they gave the Bergman kernel function as

$$K_D(z,\bar{w}) = \frac{1}{\pi^n\prod_{j=1}^{n}\alpha_j}h_0(z\bar{w}),$$

where the function h_0 is given by

$$h_0(x) = \frac{\partial^{n-1}}{\partial x_1\cdots\partial x_{n-1}}\left(\sum_{j_1=0}^{\alpha_1-1}\cdots\sum_{j_{n-1}=0}^{\alpha_{n-1}-1}\frac{\left(1-\sum_{l=1}^{n-1}\omega_l^{j_l}x_l^{\frac{1}{\alpha_l}}\right)^{\alpha_n-1}}{\left[\left(1-\sum_{l=1}^{n-1}\omega_l^{j_l}x_l^{\frac{1}{\alpha_l}}\right)^{\alpha_n}-x_n\right]^2}\right)$$

and $\omega_j=e^{2\pi/\alpha_j}$, $j=1,\cdots,n-1$.

In the same paper, using the properties of Appell's multivariable hypergeometric functions, Francsics and Hanges[7] proved that (5) coincides with (4) when $p_1=\cdots=p_{n-1}=1$ and p_n is a positive integer.

In 1997, Francsics and Hanges[8] used the multivariable hypergeometric function again to study the boundary behavior of the Bergman kernel function of D which is defined by (1) when $2/\alpha_j$, $j=1,2,\cdots,n$ are rational numbers.

Of course, all the domains considered by D'Angelo[6] and Francsics and Hanges[7,8] are special cases of the domain D which was defined by (1).

In 1994, D'Angelo[9] considered another kind of Reinhardt domain:

$$\Omega_0=\left\{z=(z_1,\cdots,z_N)\in\mathbb{C}^N\,\middle|\,\sum_{j=1}^{l}|z_j|^2+\left(\sum_{j=l+1}^{N}|z_j|^2\right)^{\frac{1}{p}}<1\right\}, \tag{7}$$

where $1\leqslant l\leqslant N-1$ and $p>0$ is a positive real number. This is not a "decoupled" domain. He gave the explicit formula of the Bergman kernel functions $K_{\Omega_0}(z,\bar{z})$ of the domains Ω_0 as

$$K_{\Omega_0}(z,\bar{z})=\sum_{k=0}^{l+1}c_k\left(1-\sum_{j=1}^{l}|z_j|^2\right)^{-l-1+kp}\left(\left(1-\sum_{j=1}^{l}|z_j|^2\right)^p-\sum_{j=l+1}^{N}|z_j|^2\right)^{-N+l-k}, \tag{8}$$

where c_k, $k=1,\cdots,l+1$ are constants which depend only on k, l, N and p.

Obviously, the results of refs. [5,9] are not included by each other.

Francsics and Hanges[7] considered a more general kind of Reinhardt domain:

$$\Omega_1=\{Z=(Z_1,Z_2,\cdots,Z_n)\in\mathbb{C}^N,\ Z_j\in\mathbb{C}^{N_j},$$
$$j=1,\cdots,n\mid \|Z\|_\alpha=\sum_{j=1}^{n}\|Z_j\|^{2p_j}<1\},$$

where $N_1,N_2,\cdots,N_n$, $p_1,p_2,\cdots,p_n$ are positive integers, $N=N_1+N_2+\cdots+N_n$, $\|Z_j\|$ is the standard Euclidean norm of $\mathbb{C}^{N_j}$, $j=1,\cdots,n$.

They gave the explicit formula of the Bergman kernel function $K_{\Omega_1}(Z,\overline{W})$ of Ω_1 as

$$K_{\Omega_1}(Z,\overline{W})=\frac{\prod_{j=1}^{n}p_j}{\pi^n}h(Z_1\overline{W}_1',\cdots,Z_n\overline{W}_n')$$

when $(Z,W)\in\Omega\times\Omega$, where

$$h(y) = \frac{\partial^{N_1-1}}{\partial y_1^{N_1-1}} \cdots \frac{\partial^{N_n-1}}{\partial y_n^{N_n-1}} \left(\sum_{q_1=0}^{p_1-1} \cdots \sum_{q_n=0}^{p_n-1} y^q \frac{\Gamma\left(1+\sum_{j=1}^{n} \frac{q_j+1}{p_j}\right)}{\prod_{j=1}^{n} \Gamma\left(\frac{q_j+1}{p_j}\right)} \right.$$

$$\left. \times F_A\left(1+\sum_{j=1}^{n} \frac{q_j+1}{p_j}, 1, \frac{\mathbb{Q}+1}{\mathbb{P}}, y^p\right)\right).$$

Here the meaning of all notations was mentioned as above and $y = (y_1, y_2, \cdots, y_n)$.

Francsics and Hanges[7] considered the boundary behavior of the Bergman kernel function of the domain

$$\Omega_2 = \Big\{ z = (z_1, z_2, \cdots, z_{n+m})$$

$$\in \mathbb{C}^{n+m} \mid |z_1|^{2/q_1} + \cdots + |z_n|^{2/q_n} + (|z_{n+1}|^2 + \cdots + |z_{n+m}|^2)^p < 1 \Big\},$$

where $q_1, \cdots, q_n$ are integers, and p is an arbitrary positive real number.

Our goal in this paper is to consider the boundary behavior of the Bergman kernel function of a more general kind of Reinherdt domain:

$$\Omega = \{ Z = (Z_1, Z_2, \cdots, Z_n) \in \mathbb{C}^N, \ Z_j \in \mathbb{C}^{N_j},$$

$$j = 1, \cdots, n \mid \| Z \|_\alpha = \sum_{j=1}^{n} | Z_j |^{2/\alpha_j} < 1 \}, \tag{9}$$

where $N_1, N_2, \cdots, N_n$ are positive integers, $N = N_1 + N_2 + \cdots + N_n$, $\alpha_j > 0$, $j = 1, 2, \cdots, n$ are positive **real numbers**. $\| Z_j \|$ is the standard Euclidean norm in $\mathbb{C}^{N_j}$, $j = 1, 2, \cdots, n$.

When all $N_j = 1$, $j = 1, \cdots, n$, $\Omega = D$. Thus D is a special case of Ω. If at least one of N_j, $j = 1, \cdots, n$ is greater than 1 and the corresponding $\alpha_j \neq 1$, then Ω is not a "decoupled" domain. Ω_1 is a special case of Ω when all $\frac{1}{\alpha_j}$, $j = 1, 2, \cdots, n$ are positive integers, and Ω_2 is the special case of Ω also.

Using the method as in ref. [5], we may give an estimate of the Bergman kernel function at diagonal points, i.e. $K_\Omega(Z, \bar{Z})$ of Ω in a "small constant-large constant" sense again.

1 Main result

Theorem 1. There exist two positive constants m and M which depend only on N_j, α_j, $j = 1, 2, \cdots, n$, such that

$$mF(Z, \bar{Z}) \leqslant K_\Omega(Z, \bar{Z}) \leqslant MF(Z, \bar{Z}) \tag{10}$$

holds for every $Z \in \Omega$, where

$$F(Z, \bar{Z}) = (-r(Z))^{-N-1} \prod_{j=1}^{n} (-r(Z) + \| Z_j \|^{2/\alpha_j})^{(1-\alpha_j)N_j}, \tag{11}$$

and $r(Z) = \| Z \|_\alpha - 1$ is the defining function for Ω.

Of course, Theorem 1 extends the result of ref. [5] because $\Omega = D$ when $N_1 = \cdots = N_n = 1$, $N = n$, and then the function $F(Z, \bar{Z})$ in (11) is the same as function $F_0(z, \bar{z})$ in (3).

Now, we show that the result of D'Angelo[9] coincides with Theorem 1 when $\Omega = \Omega_0$.

If $n = 2$, $N = N_1 + N_2$, $N_1 = l$, $N_2 = N - l$, $\alpha_1 = 1$, $\alpha_2 = p$, then $\Omega = \Omega_0$, and the function $F(Z, \bar{Z})$ defined by (11) becomes

$$F(z,\bar{z}) = \left(1 - \sum_{j=1}^{l} |z_j|^2 - \left(\sum_{j=l+1}^{N} |z_j|^2\right)^{\frac{1}{p}}\right)^{-N-1}\left(1 - \sum_{j=1}^{l} |z_j|^2\right)^{(1-p)(N-l)}. \quad (12)$$

By (7), we may define $\theta(z)$ as

$$\left(\sum_{j=l+1}^{N} |z_j|^2\right)^{\frac{1}{p}} = \theta\left(1 - \sum_{j=1}^{l} |z_j|^2\right). \quad (13)$$

Then $0 \leqslant \theta < 1$. Substituting (13) into (12) and (8), we have

$$F(z,\bar{z}) = (1-\theta)^{-N-1}\left(1 - \sum_{j=1}^{l} |z_j|^2\right)^{-1-l-pN+pl},$$

and

$$K_{\Omega_0}(z,\bar{z}) = \left(\sum_{k=0}^{l+1} c_k(1-\theta^p)^{-N+l-k}\right)\left(1 - \sum_{j=1}^{l} |z_j|^2\right)^{-1-l-pN+pl}.$$

Thus

$$K_{\Omega_0}(z,\bar{z}) = F(z,\bar{z})\sum_{k=0}^{l+1} c_k\left(\frac{1-\theta^p}{1-\theta}\right)^{-N-1}(1-\theta^p)^{l-k+1}, \ 0 \leqslant \theta < 1.$$

Obviously, there exist two positive constants $M > m > 0$, which depend on N, l and p only such that

$$m \leqslant \sum_{k=0}^{l+1} c_k\left(\frac{1-\theta^p}{1-\theta}\right)^{-N-1}(1-\theta^p)^{l-k+1} \leqslant M.$$

Thus D'Angelo's result[9] coincides with Theorem 1 when $\Omega = \Omega_0$.

In ref. [7], Francsics and Hanges gave the series expansion of the Bergman kernel function $K_\Omega(Z,\overline{W})$ of Ω by

$$K_\Omega(Z,\overline{W}) = \frac{1}{\pi^N\left(\prod_{j=1}^{n} \alpha_j\right)}\sum_{s\in\mathbb{Z}_+^n} C_s(Z_1\overline{W}_1')^{s_1}(Z_2\overline{W}_2')^{s_2}\cdots(Z_n\overline{W}_n')^{s_n}, \quad (14)$$

where $s = (s_1, \cdots, s_n)$,

$$C_s = \frac{\Gamma\left(1 + \sum_{j=1}^{n} \alpha_j(s_j + N_j)\right)}{\prod_{j=1}^{n} \Gamma(\alpha_j(s_j + N_j))}\prod_{j=1}^{n}\frac{\Gamma(N_j + s_j)}{\Gamma(1 + s_j)}. \quad (15)$$

Recall that the series expansion of the Bergman kernel function $K_D(z,\bar{w})$ of D defined by (1) is

$$K_D(z,\bar{w}) = \frac{1}{\pi^n\left(\prod_{j=1}^{n} \alpha_j\right)}\sum_{s\in\mathbb{Z}_+^n} D_s(z_1\bar{w}_1)^{s_1}(z_2\bar{w}_2)^{s_2}\cdots(z_n\bar{w}_n)^{s_n},$$

where $s = (s_1, \cdots, s_n)$,

$$D_s = \frac{\Gamma\left(1 + \sum_{j=1}^{n} \alpha_j(s_j + 1)\right)}{\prod_{j=1}^{n} \Gamma(\alpha_j(s_j + 1))}.$$

If we let $x = (x_1, x_2, \cdots, x_n) \in \mathbb{C}^n$, and

$$\varphi_D(x) = \frac{1}{\pi^n\left(\prod_{j=1}^{n}\alpha_j\right)}\sum_{s\in\mathbb{Z}_+^n} D_s x_1^{s_1}x_2^{s_2}\cdots x_n^{s_n},$$

then it is easy to verify that

$$K_\Omega(Z,\overline{W}) = \pi^{n-N}\frac{\partial^{N_1-1}}{\partial x_1^{N_1-1}}\cdots\frac{\partial^{N_n-1}}{\partial x_n^{N_n-1}}\varphi_D(x)\Bigg|_{x=(Z_1\overline{W}_1',\cdots,Z_n\overline{W}_n')}. \tag{16}$$

Even we already know the asymptotic formula of $K_D(z,\bar{z})$ by (2) and (3), it is impossible to get the asymptotic formula of $K_\Omega(Z,\bar{Z})$ directly from (16). However, we may prove Theorem 1 by using the method and idea which we used in ref. [5].

We state the following notations and lemmas.

Let a, t be real numbers, $|t|<1$, $a>-1$, and

$$(1-t)^{-a} = \sum_{k=0}^{\infty} b(a,k)t^k,$$

where

$$b(a,k) = \frac{\Gamma(k+a)}{\Gamma(a)\Gamma(k+1)}.$$

Let $u=(u_1,\cdots,u_n)\in\mathbb{R}_+^n$, $\sigma(u)=\sum_{i=1}^{n}u_i<1$, $a>-1$, and

$$(1-\sigma(u))^{-a} = \sum_{m\in\mathbb{Z}_+^n} C(a,m)u^m,$$

where $m=(m_1,m_2,\cdots,m_n)$, $u^m=u_1^{m_1}u_2^{m_2}\cdots u_n^{m_n}$, and

$$C(a,m) = \frac{\Gamma(a+\sigma(m))}{\Gamma(a)\prod_{j=1}^{n}\Gamma(1+m_j)},\quad \sigma(m)=\sum_{j=1}^{n}m_j. \tag{17}$$

Lemma 1. If a, m are two real numbers, $m\geqslant 1$ and $m+a>0$, then there exist two positive numbers c and C which are independent of m, and only depend on a, such that

$$c\Gamma(m)m^a < \Gamma(m+a) < C\Gamma(m)m^a.$$

This is Lemma 6 in ref. [5].

Lemma 2 (Main Lemma). Let $u=(u_1,\cdots,u_n)\in\mathbb{R}^n$, $|u_j|<1$, $j=1,2,\cdots,n$, and $\sigma(u)<1$. Let $a>0$, $\varepsilon=(\varepsilon_1,\cdots,\varepsilon_n)\in\mathbb{R}^n$, $\varepsilon_j>-N_j$, $N_j\in\mathbb{Z}_+$, $j=1,2,\cdots,n$, and

$$a+\sigma(\varepsilon) > \sum_{j=1}^{n}\{\varepsilon_j\}+1,$$

where $\{\varepsilon_j\}$ means the least integer which is equal to or greater than ε_j, $\sigma(\varepsilon)=\sum_{j=1}^{n}\varepsilon_j$. Let

$$F_n^{a,\varepsilon}(u) = (1-\sigma(u))^{-a}\prod_{j=1}^{n}(1+u_j-\sigma(u))^{-\varepsilon_j},$$

and let the Taylor expansion of $F_n^{a,\varepsilon}(u)$ be

$$F_n^{a,\varepsilon}(u) = \sum_{s\in\mathbb{Z}_+^n} B_{s,n}^{a,\varepsilon}u^s,$$

where $s=(s_1,s_2,\cdots,s_n)\in\mathbb{Z}_+^n$, $u^s=u_1^{s_1},\cdots,u_n^{s_n}$. Then there exist two positive constants m_n and M_n, which are independent of s, and depend on a; n and ε only, such that

$$m_n A_{s,n}^{a,\varepsilon} \leqslant B_{s,n}^{a,\varepsilon} \leqslant M_n A_{s,n}^{a,\varepsilon},$$

where

$$A_{s,n}^{a,\varepsilon} = \frac{C(a + \sigma(\varepsilon), s)}{\prod_{j=1}^{n} (1 + s_j)^{\varepsilon_j}},$$

and $C(a + \sigma(\varepsilon), s)$ is defined by (17).

When $N_j = 1, j = 1, \cdots, n$, Lemma 2 is the same as Lemma 7 in ref. [5]. The proof of Lemma 7 in ref. [5] was difficult and lengthy. We may use a similar argument to prove Lemma 2 in this paper. Of course, it needs Lemma 1 to Lemma 10 (except Lemma 7) in ref. [5] to prove Lemma 2. We omit the details of the proof of Lemma 2, because the proof is lengthy and similar to the proof of Lemma 7 in ref. [5].

2 Proof of theorems

Let Ω be a set in $\mathbb{R}^n$ or $\mathbb{C}^n$, and let $f_1(x)$, $f_2(x)$ be two non-negative real valued functions which are defined on Ω $(x \in \Omega)$. We use the notation

$$f_1(x) \cong f_2(x) \qquad \text{for every } x \in \Omega$$

to denote that there exist two positive real numbers m and M which are independent of $x \in \Omega$, such that

$$m f_1(x) \leqslant f_2(x) \leqslant M f_1(x)$$

holds for all $x \in \Omega$.

By (14), we know that the series expansion of Bergman kernel function of Ω is

$$K_\Omega(Z, \bar{Z}) = \frac{1}{\pi^N\left(\prod_{j=1}^{n} a_j\right)} \sum_{s \in \mathbb{Z}_+^n} C_s \| Z_1 \|^{2s_1} \| Z_2 \|^{2s_2} \cdots \| Z_n \|^{2s_n},$$

where C_s is defined by (15).

By Lemma 1,

$$\begin{aligned} \Gamma(\alpha_j(s_j + N_j)) &= \Gamma(1 + (\alpha_j - 1)N_j + \alpha_j s_j + N_j - 1) \\ &\cong \Gamma(1 + (\alpha_j - 1)N_j + \alpha_j s_j) \cdot (1 + (\alpha_j - 1)N_j + \alpha_j s_j)^{N_j - 1}, \end{aligned}$$

and

$$\Gamma(N_j + s_j) = \Gamma(1 + s_j + N_j - 1) \cong \Gamma(1 + s_j)(1 + s_j)^{N_j - 1},$$

we have

$$\frac{\Gamma(N_j + s_j)}{\Gamma(\alpha_j(s_j + N_j))\Gamma(1 + s_j)} \cong \frac{1}{\Gamma(1 + (\alpha_j - 1)N_j + \alpha_j s_j)} \left(\frac{1 + s_j}{1 + (\alpha_j - 1)N_j + \alpha_j s_j}\right)^{N_j - 1}.$$

Obviously, there exist two positive real numbers $C_1 > c_1 > 0$, which depend on α_j and N_j, $j = 1, \cdots, n$, only, and independent of s_j, $j = 1, 2, \cdots, n$ such that

$$c_1 < \left(\frac{1 + s_j}{1 + (\alpha_j - 1)N_j + \alpha_j s_j}\right)^{N_j - 1} < C_1.$$

Thus

$$\frac{\Gamma(N_j + s_j)}{\Gamma(\alpha_j(s_j + N_j))\Gamma(1 + s_j)} \cong \frac{1}{\Gamma(1 + (\alpha_j - 1)N_j + \alpha_j s_j)},$$

and hence

$$C_s \cong \tilde{C}_s,$$

where

$$\tilde{C}_s = \frac{\Gamma\left(1 + \sum_{j=1}^{n} \alpha_j (s_j + N_j)\right)}{\prod_{j=1}^{n} \Gamma(1 + (\alpha_j - 1) N_j + \alpha_j s_j)}.$$

Now we are going to prove a theorem which is more general than Theorem 1.

Theorem 2. Let Ω be defined as (9), $G(Z,\bar{Z})$ be a function defined on Ω with the series expansion

$$G(Z,\bar{Z}) = \sum_{m \in \mathbb{Z}_+^N} E_m Z^m \bar{Z}^m, \tag{18}$$

where $Z = (Z_1, \cdots, Z_n) = (z_1, z_2, \cdots, z_N) \in \mathbb{C}^N$, $Z_j \in \mathbb{C}^{N_j}$, $j = 1,2,\cdots,n$, $N = N_1 + \cdots + N_n$, $m = (m^{(1)}, \cdots, m^{(n)}) = (m_1, m_2, \cdots, m_N) \in \mathbb{Z}_+^N$, $m^{(j)} \in \mathbb{Z}_+^{N_j}$, $j = 1,2,\cdots,n$, $Z^m = z_1^{m_1} z_2^{m_2} \cdots z_N^{m_N}$. If every coefficient E_m, $m \in \mathbb{Z}_+^N$ satisfies the following condition: There exist two positive real numbers $C_2 > c_2 > 0$, which depend on $\alpha = (\alpha_1, \cdots, \alpha_n)$, $\alpha_j > 0$, $j = 1,2,\cdots,n$, $\varepsilon = (\varepsilon_1, \cdots, \varepsilon_n) \in \mathbb{R}^n$, $N_j \in \mathbb{Z}_+$, $\varepsilon_j > -N_j$, $j = 1,2,\cdots,n$, only and independent of $m \in \mathbb{Z}_+^N$, such that

$$c_2 E_m \leqslant \frac{\Gamma\left(a + \sigma(\varepsilon) + \sum_{j=1}^{n} \alpha_j \sigma(m^{(j)})\right)}{\prod_{j=1}^{n} (1 + \alpha_j \sigma(m^{(j)}))(1 + \sigma(m^{(j)}))^{\varepsilon_j}} \prod_{j=1}^{n} C(1, m^{(j)}) \leqslant C_2 E_m \tag{19}$$

holds, where a and ε satisfy the conditions in Lemma 2, then there exist two positive real numbers $M_1 > m_1 > 0$, which depend on $\alpha = (\alpha_1, \cdots, \alpha_n)$, $\varepsilon = (\varepsilon_1, \cdots, \varepsilon_n)$ and $N_1, \cdots, N_n$ only and independent of $Z \in \Omega$ such that

$$m_1 G(Z,\bar{Z}) \leqslant (-r(Z))^{-a} \prod_{j=1}^{n} (-r(Z) + \| Z_j \|^{\frac{2}{\alpha_j}})^{-\varepsilon_j} \leqslant M_1 G(Z,\bar{Z}) \tag{20}$$

holds for every $z \in \Omega$.

In particular, if $a = N + 1$, $\varepsilon_j = (\alpha_j - 1) N_j$, then Theorem 2 is Theorem 1.

Proof of Theorem 2.

Let

$$E_m^{(1)} = \frac{\Gamma\left(a + \sigma(\varepsilon) + \sum_{j=1}^{n} \alpha_j \sigma(m^{(j)})\right)}{\prod_{j=1}^{n} \Gamma(1 + \alpha_j \sigma(m^{(j)}))(1 + \sigma(m^{(j)}))^{\varepsilon_j}} \prod_{j=1}^{n} C(1, m^{(j)})$$

and

$$G_1(Z,\bar{Z}) = \sum_{m \in \mathbb{Z}_+^N} E_m^{(1)} Z^m \bar{Z}^m.$$

Then, by (19), we have

$$G_1(Z,\bar{Z}) \cong G(Z,\bar{Z})$$

where $z \in \Omega$. Thus we have proved (20) if we can prove it when we replace G with G_1 in it.

Let $m = (m^{(1)}, \cdots, m^{(n)})$, $m^{(j)} \in \mathbb{Z}_+^{N_j}$, $s_j = \sigma(m^{(j)})$, $j = 1,2,\cdots,n$ and $s = s(m) = (s_1, s_2, \cdots, s_n)$. Then

$$\frac{\Gamma\left(a+\sigma(\varepsilon)+\sum_{j=1}^{n}\alpha_j\rho(m^{(j)})\right)}{\prod_{j=1}^{n}\Gamma(1+\alpha_j\rho(m^{(j)}))(1+\sigma(m^{(j)}))^{\varepsilon_j}}=\frac{\Gamma\left(a+\sigma(\varepsilon)+\sum_{j=1}^{n}\alpha_js_j\right)}{\prod_{j=1}^{n}\Gamma(1+\alpha_js_j)(1+s_j)^{\varepsilon_j}}. \tag{21}$$

We denote the right-hand side of (21) by $H^{a,\varepsilon}_{s(m)}=H^{a,\varepsilon}_{s}$. Then $E^{(1)}_m$ and $G_1(Z,\bar{Z})$ may be expressed as

$$E^{(1)}_m=H^{a,\varepsilon}_{s(m)}\prod_{j=1}^{n}C(1,m^{(j)})=H^{a,\varepsilon}_{s}\prod_{j=1}^{n}C(1,m^{(j)}),$$
$$G_1(Z,\bar{Z})=\sum_{s\in\mathbb{Z}^n_+}H^{a,\varepsilon}_{s}\parallel Z\parallel^{2s},$$

where $\parallel Z\parallel^{2s}=\parallel Z_1\parallel^{2s_1}\cdots\parallel Z_n\parallel^{2s_n}$. Moreover, if we let $u_j=\parallel Z_j\parallel^{2/\alpha_j}$, then we may express $G_1(Z,\bar{Z})$ as

$$G_1(Z,\bar{Z})\equiv G_1(u)=\sum_{s\in\mathbb{Z}^n_+}H^{a,\varepsilon}_{s}u_1^{\alpha_1s_1}u_2^{\alpha_2s_2}\cdots u_n^{\alpha_ns_n}.$$

Let Q be a set in $\mathbb{Z}^n_+$ defined as $Q=\{\delta=(\delta_1,\delta_2,\cdots,\delta_n)\in\mathbb{Z}^n_+$, where $\delta_j=0$ or 1, $j=1,2,\cdots,n\}$. We define the mapping $\phi_\delta:\mathbb{Z}^n_+\to\mathbb{Z}^n_+$ as follows. For any $s=(s_1,\cdots,s_n)\in\mathbb{Z}^n_+$, we define $\phi_\delta(s)$ as

$$\phi_\delta(s)=(\phi_{\delta,1}(s_1),\phi_{\delta,2}(s_2),\cdots,\phi_{\delta,n}(s_n)),$$

where

$$\begin{cases}\phi_{\delta,j}(s_j)=[\alpha_js_j], & \text{if } \delta_j=1;\\ \phi_{\delta,j}(s_j)=\{\alpha_js_j\}, & \text{if } \delta_j=0;\end{cases}$$

where $[x]$ is the largest integer which is equal to or less than x, and $\{x\}$ is the least integer which is equal to or greater than x.

We define the function $N_\delta(s,h)$ of $s,h=(h_1,\cdots,h_n)\in\mathbb{Z}^n_+$ as

$$\begin{cases}N_\delta(s,h)=1, & \text{if } h=\phi_\delta(s);\\ N_\delta(s,h)=0, & \text{if } h\neq\phi_\delta(s).\end{cases}$$

Obviously, we have

$$\sum_{h\in\mathbb{Z}^n_+}N_\delta(s,h)=1, \tag{22}$$

and

$$0\leqslant\sum_{s\in\mathbb{Z}^n_+}N_\delta(s,h)\equiv N_\delta(h)\leqslant\prod_{j=1}^{n}\left\{\frac{1}{\alpha_j}\right\}. \tag{23}$$

Let

$$\eta_j(\delta,s,h)=\alpha_js_j-\phi_{\delta,j}(s_j)\text{ when } h=\phi_\delta(s). \tag{24}$$

Then for any $\delta\in Q$, and every $Z\in\Omega$, we have

$$G_1(u)\equiv G(Z,\bar{Z})=\sum_{s\in\mathbb{Z}^n_+}\sum_{h\in\mathbb{Z}^n_+}N_\delta(s,h)H^{a,\varepsilon}_{s}u_1^{\alpha_1s_1}\cdots u_n^{\alpha_ns_n}$$
$$\cong\sum_{s,h\in\mathbb{Z}^n_+}N_\delta(s,h)\frac{\Gamma(a+\sigma(\varepsilon)+\sigma(h))}{\prod_{j=1}^{n}\Gamma(1+h_j)(1+s_j)^{\varepsilon_j}}\prod_{j=1}^{n}\left(\frac{a+\sigma(\varepsilon)+\sigma(h)}{1+h_j}u_j\right)^{\eta_j(\delta,s,h)}u^h \tag{25}$$

by (22), (24) and Lemma 1, where $u_j = \| Z_j \|^{2/\alpha_j}$, $j = 1, \cdots, n$. We denote the right-hand side of (25) as $G_\delta(u)$.

Taking $\delta = \delta_0 = (1,1,\cdots,1) \in Q$, i.e. all the elements of δ are equal to 1 at $G_\delta(u)$, then $\eta_j(\delta, s, h) \geqslant 0$ in (25), and using the inequality $1 + x \geqslant x^q$ if $x \geqslant 0$ and $0 \leqslant q \leqslant 1$, we have

$$G_{\delta_0}(u) \leqslant \sum_{h \in \mathbb{Z}_+^n} \frac{\Gamma(a + \sigma(\varepsilon) + \sigma(h))}{\prod_{j=1}^n \Gamma(1 + h_j)(1 + h_j)^{\varepsilon_j}} N_{\delta_0}(h) \prod_{j=1}^n \left(1 + \frac{a + \sigma(\varepsilon) + \sigma(h)}{1 + h_j} u_j\right) u^h$$

$$\leqslant \prod_{j=1}^n \left\{\frac{1}{\alpha_j}\right\} \sum_{\delta \in Q} u^\delta \sum_{h \in \mathbb{Z}_+^n} \frac{\Gamma(a + \sigma(\varepsilon) + \sigma(h))}{\prod_{j=1}^n \Gamma(1 + h_j)(1 + h_j)^{\varepsilon_j}} \prod_{j=1}^n \left(\frac{a + \sigma(\varepsilon) + \sigma(h)}{1 + h_j}\right)^{\delta_j} u^h \tag{26}$$

by (23). We denote the right-hand side of inequality (26) by $J(u)$.

Using Lemma 1 again, we have

$$\frac{\Gamma(a + \sigma(\varepsilon) + \sigma(h))}{\prod_{j=1}^n \Gamma(1 + h_j)(1 + h_j)^{\varepsilon_j}} \prod_{j=1}^n \left(\frac{a + \sigma(\varepsilon) + \sigma(h)}{1 + h_j}\right)^{\delta_j} \cong \frac{\Gamma(a + \sigma(\varepsilon) + \sigma(h) + \sigma(\delta))}{\prod_{j=1}^n \Gamma(1 + h_j)(1 + h_j)^{\varepsilon_j + \delta_j}}$$

for every $h \in \mathbb{Z}_+^n$, and every $\delta \in Q$. Thus

$$J(u) \cong \sum_{\delta \in Q} u^\delta \sum_{h \in \mathbb{Z}_+^n} \frac{\Gamma(a + \sigma(\varepsilon) + \sigma(h) + \sigma(\delta))}{\Gamma(a + \sigma(\varepsilon) + \sigma(\delta)) \prod_{j=1}^n \Gamma(1 + h_j)(1 + h_j)^{\varepsilon_j + \delta_j}} u^h.$$

Using Lemma 2, we have

$$\begin{aligned} J(u) &\cong \sum_{\delta \in Q} u^\delta F_n^{a, \varepsilon + \delta}(u) \\ &= \sum_{\delta \in Q} u^\delta (1 - \sigma(u))^{-a} \prod_{j=1}^n (1 + u_j - \sigma(u))^{-\varepsilon_j - \delta_j} \\ &\leqslant 2^n (1 - \sigma(u))^{-a} \prod_{j=1}^n (1 + u_j - \sigma(u))^{-\varepsilon_j}. \end{aligned}$$

Since $G_1(Z, \bar{Z}) \cong G(Z, \bar{Z})$, we conclude that there exists a positive real number C_3, such that

$$G(Z, \bar{Z}) \leqslant C_3(-r(Z))^{-a} \prod_{j=1}^n (-r(Z) + \| Z \|^{\frac{2}{\alpha_j}})^{-\varepsilon_j}.$$

We have proved the left-hand side of inequality (20).

Now we are going to prove the right-hand side of inequality (20).

Fix a $\delta \in Q$, we may divide $\mathbb{R}_+^n$ as the union of non-overlapping closed boxes Ω_+^δ, $t = 1, 2, \cdots$, i.e. $\mathbb{R}_+^n = \bigcup_{t=1}^\infty \Omega_t^\delta$ by the lattice set $\phi_\delta(\mathbb{Z}_+^n)$, every vertex of Ω_t^δ, $t = 1, 2, \cdots$, belongs to $\phi_\delta(\mathbb{Z}_+^n)$. Let $h_{t_0, j}$, $j = 1, 2, \cdots, 2^n$ be all vertices of a box $\Omega_{t_0}^\delta$. Then for every $h_{t_0, j}$, there exists at least one $s \in \mathbb{Z}_+^n$, such that $\phi_\varepsilon(s) = h_{t_0, j}$. We denote the corresponding $\eta_l(\delta, s, h_{t_0, j})$ by $\eta_l(\delta, h_{t_0, j})$. If there exists a lattice point h which is inside the box but is not a vertex of the box, then we choose 2^n non-negative real numbers $Q_{j, t_0} \geqslant 0$, such that $\sum_{j=1}^{2^n} Q_{j, t_0} = 1$. Using the property that the arithmatic mean is greater than or equal to the geometrics mean, we have

$$\sum_{j=1}^{2^n} Q_{j,t_0} T^{a,\varepsilon}_{h_{t_0,j}} \prod_{l=1}^{n} \left(\frac{a+\sigma(\varepsilon)+\sigma(h)}{1+h^{(l)}_{t_0,j}} u_l \right)^{\eta_l(\delta, h_{t_0,j})} u^{h_{t_0,j}}$$

$$\geqslant c_3 T^{a,\varepsilon}_h \prod_{j=1}^{n} \left(\frac{a+\sigma(\varepsilon)+\sigma(h)}{1+h_j} u_j \right)^{\tilde{\eta}_j(\delta,h)} u^h, \tag{27}$$

where $T^{a,\varepsilon}_h = \frac{\Gamma(a+\sigma(\varepsilon)+\sigma(h))}{\prod_{j=1}^{n}\Gamma(1+h_j)(1+h_j)^{\varepsilon_j}}$, $h_{t_0,j} = (h^{(1)}_{t_0,j}, \cdots, h^{(n)}_{t_0,j})$, $h = (h_1, \cdots, h_n)$, $c_3 > 0$ is a positive real number which is independent of $h \in \mathbb{Z}^n_+$ and u, $|\tilde{\eta}_j(\delta,h)| < 1$; and $\tilde{\eta}_j(\delta, h) \geqslant 0$ if $\delta_j = 1$; and $\tilde{\eta}_j(\delta,h) \leqslant 0$ if $\delta_j = 0$.

The total number of lattice points contained in Ω^δ_t is not greater than $\prod_{j=1}^{n}\left(1+\left\{\frac{1}{\alpha_j}\right\}\right)$. Thus the $G_\delta(u)$ which is defined by the right-hand side of (25) has the following inequality

$$G_\delta(u) = \sum_{s,h\in\mathbb{Z}^n_+} \frac{\Gamma(a+\sigma(\varepsilon)+\sigma(h))}{\prod_{j=1}^{n}\Gamma(1+h_j)(1+s_j)^{\varepsilon_j}} N_\delta(s,h) \prod_{j=1}^{n}\left(\frac{a+\sigma(\varepsilon)+\sigma(h)}{1+h_j}u_j\right)^{\eta_j(\delta,h)} u^h$$

$$\geqslant c_4 \prod_{j=1}^{n}\left(1+\left\{\frac{1}{\alpha_j}\right\}\right)^{-1} \sum_{h\in\mathbb{Z}^n_+} \frac{\Gamma(a+\sigma(\varepsilon)+\sigma(h))}{\prod_{j=1}^{n}\Gamma(1+h_j)(1+h_j)^{\varepsilon_j}}$$

$$\times \prod_{j=1}^{n}\left(\frac{a+\sigma(\varepsilon)+\sigma(h)}{1+h_j}u_j\right)^{\tilde{\eta}_j(\delta,h)} u^h \tag{28}$$

by (27) where $c_4 > 0$ is a positive real number which is independent of $h \in \mathbb{Z}^n_+$ and u.

From (25) and (28), we have

$$G_1(u) = G_1(Z,\bar{Z}) \geqslant c_5 \sum_{\delta\in Q}\sum_{h\in\mathbb{Z}^n_+} \frac{\Gamma(a+\sigma(\varepsilon)+\sigma(h))}{\prod_{j=1}^{n}\Gamma(1+h_j)(1+h_j)^{\varepsilon_j}} \prod_{j=1}^{n}\left(\frac{a+\sigma(\varepsilon)+\sigma(h)}{1+h_j}u_j\right)^{\tilde{\eta}_j(\delta,h)} u^h$$

$$= c_5 \sum_{h\in\mathbb{Z}^n_+} \frac{\Gamma(a+\sigma(\varepsilon)+\sigma(h))}{\prod_{j=1}^{n}\Gamma(1+h_j)(1+h_j)^{\varepsilon_j}} \sum_{\delta\in Q}\prod_{j=1}^{n}\left(\frac{a+\sigma(\varepsilon)+\sigma(h)}{1+h_j}u_j\right)^{\tilde{\eta}_j(\delta,h)} u^h, \tag{29}$$

where $c_5 > 0$ is a positive real number which is independent of $h \in \mathbb{Z}^n_+$ and u.

We may prove that

$$\sum_{\delta\in Q}\prod_{j=1}^{n}\left(\frac{a+\sigma(\varepsilon)+\sigma(h)}{1+h_j}u_j\right)^{\tilde{\eta}_j(\delta,h)} \geqslant 1. \tag{30}$$

For fixed u and h, there exists a $\delta^{(0)} \in Q$, where $\delta^{(0)}_j = 0$ if $\frac{a+\sigma(\varepsilon)+\sigma(h)}{1+h_j}u_j < 1$, and $\delta^{(0)}_j = 1$ if $\frac{a+\sigma(\varepsilon)+\sigma(h)}{1+h_j}u_j \geqslant 1$. Then, for this $\delta^{(0)}$, we have

$$\prod_{j=1}^{n}\left(\frac{a+\sigma(\varepsilon)+\sigma(h)}{1+h_j}u_j\right)^{\tilde{\eta}_j(\delta^{(0)},h)} \geqslant 1.$$

Thus (30) is true. From (29) and (30), we obtain

$$G_1(u) = G_1(z,\bar{z}) \geqslant c_5 \sum_{h\in\mathbb{Z}_+^n} \frac{\Gamma(a+\sigma(\varepsilon)+\sigma(h))}{\prod_{j=1}^{n}\Gamma(1+h_j)(1+h_j)^{\varepsilon_j}} u^h \cong F_n^{a,\varepsilon}(u).$$

We have proved the right-hand side of inequality (20).

Acknowledgements This work was supported by the National Natural Science Foundation of China (Grant Nos. 19671076 and 19731004).

References

1. Hua, L.K., Harmonic Analysis of Functions of Several Complex Variables in the Classical Domains, Transl. Amer. Math. Soc., Vol. 6, Providence, Rhode Island: AMS, 1963.
2. Bergman, S., Über die Kernfunction eines Bereiches und ihr Verhalten am Rande, J. Reine Angew. Math., 1933, 169: 1.
3. Fefferman, S., The Bergman kernel and biholomorphic mappings of strictly pseudoconvex comains, Invent. Math., 1974, 26: 1.
4. Mostow, D., Siu, Y.T., A compact Kähler surface of negative curvature not covered by the ball, Ann. of Math., 1980, 112: 321.
5. Gong, S., Zheng, X.A., The Bergman kernel function of some Reinhardt domains, Tran. Amer. Math. Soc., 1996, 348: 1771.
6. D'Angelo, J.P., A note on the Bergman kernel, Duke Math. J., 1978, 45: 259.
7. Francsics, G., Hanges, H., The Bergman kernel of complex ovals and multivariable hypergeometric functions, Jour. of Functional Analysis, 1996, 142: 494.
8. Francsics, G., Hanges, N., Asymptotic behavior of the Bergman kernel and hypergeometric functions, Contemporary Math. 205, Providence, Rhode Island: AMS, 1997, 79—92.
9. D'Angelo, J.P., An explicit computation of the Bergman kernel function, J. Geom. Anal., 1994, 4: 23.

VI. 多复变数的凸映照与星形映照

单复变数几何函数论有丰硕的成果，但到多复变数时，大多可举出反例说明其不再成立．在 1933 年， H. Cartan(1) 就指出这一点，如要得到一些正面的结果，必须对映照加以一定的限制．凸映照与星形映照这是最自然而简单的限制．之后的确有这方面的种种研究．从 1988 年起，我们系统地对此进行研究，最早的两篇文章是 [40.1] 与 [41.1]．之后有一系列的研究与文章，对凸映照与星形映照的判别准则，增长定理、掩盖定理、偏差定理以及其他问题进行讨论，这些结果总结在 (2) ([M1] 英文版 [M2]) 中．

这项工作以及其他一些多复变数几何函数论的工作获得 2002 年中国科学院自然科学一等奖，获奖人：龚昇、余其煌、郑学安、王世坤、刘太顺．

参 考 文 献

(1) Cartan H. Sur la possibilité d'étendre aux functions de plusieurs variables complexes la theorie des functions univalents, Lecons sur les Functions Univalents ou Multivalents by P. Montel. Gauthier-Villar, 1933: 129-155.

(2) 龚昇. 多复变数的凸映照与星形映照．2版．北京：科学出版社，2003.

Preface of *Convex and Starlike Mappings in Several Complex Variables*

This interesting book deals with the theory of convex and starlike biholomorphic mappings in several complex variables. The underlying theme is the extension to several complex variables of geometric aspects of the classical theory of univalent functions. Because the author's introduction provides an excellent overview of the content of the book, I will not duplicate the effort here. Rather, I will place the book into historical context.

The theory of univalent functions long has been an important part of the study of holomorphic functions of one complex variable. The roots of the subject go back to the famous Riemann Mapping Theorem which asserts that a simply connected region Ω which is a proper subset of the complex plane C is biholomorphically equivalent to the open unit disk Δ. That is, there is a univalent function (holomorphic bijection) $f:\Delta \to \Omega$. In the early part of this century work began to focus on the class S of normalized ($f(0) = 0$ and $f'(0) = 1$) univalent functions defined on the unit disk. The restriction to univalent functions defined on the unit disk is justified by the Riemann Mapping Theorem. The subject contains many beautiful results that were obtained by fundamental techniques developed by many mathematicians including Koebe, Bieberbach, Loewner, Goluzin, Grunsky, and Schiffer. The best-known aspect of univalent function theory is the so-called Bieberbach conjecture which was proved by de Branges in 1984.

A particularly elegant branch of complex analysis is geometric function theory in which the objective is to understand the relationship between geometric properties of the range of a holomorphic function and analytic properties of the function. Geometric function theory combines with univalent function theory in the study of image $f(\Delta)$ is a convex subset of C. A function f in S is starlike (with respect to the origin) if for each point $w \in f(\Delta)$ the straight line segment between O and w is contained in $f(\Delta)$. A good understanding of these two families was achieved by exploiting the geometry of the image.

It is natural to inquire whether the theory of univalent functions can extend to several complex variables. In the context of several complex variables one speaks of biholomorphic mappings rather than univalent functions as in one variable. This issue was explicitly raised by Henri Cartan in 1931. In several complex variables there is no analog of the Riemann Mapping Theorem, so there is not a single standard region like Δ in which to consider biholomorphic mappings. Cartan pointed out that some classical results of univalent function theory did not have analogs in several complex variables. At the same time he suggested that one should investigate the important geometrically defined subclasses of convex and starlike biholomorphic mappings. Even though parts of the classical theory of univalent functions have no analogs in several complex variables, Cartan suggested that geometric restrictions on the range of biholomorphic mappings might lead to an interesting theory.

As a matter of fact, there was little work in the geometric directions suggested by Cartan until the 1970's when a number of results dealing with the convex and starlike biholomorphic mappings appeared. At the time there was no systematic development of the subject. The situation changed dramatically in the late 1980's with a burst of new activity. A number of elegant parallels with the classical theory of starlike and convex univalent functions have been emerged. Even though the field is still evolving, the theory of convex and starlike biholomorphic mappings has now advanced to the stage that a unified presentation of the known results is needed. This book gathers together and presents in a unified manner the current state of affairs for convex and starlike biholomorphic mappings in several complex variables.

Because Professor Sheng Gong has been and continues to be one of the main contributors to the development of the theory of convex and starlike biholomorphic mappings, it is natural that he should prepare a monograph on the subject. In fact, the majority of the results presented are due to Professor Gong, his co-workers and his students. At the same time, open problems remain and the interested reader will the majority of the results presented are due to Professor Gong, his co-workers and his students. At the same time, open problems remain and the interested reader will find suggestions for future research.

Unfortunately, there has been a sharp division between researchers in one complex variable and those in several complex variables. The theory of convex and starlike mappings is an ideal meeting ground for the two groups. It is an area of several complex variables which contains many parallels with the classical theory while requiring the tools of several complex variables. This book provides a bridge between the two groups. It will serve as an excellent introduction to geometric function theory in several complex variables to workers in univalent function theory who have limited background in several complex variables. For instance, even though a result may hold in more generality, Professor Sheng Gong often first presents a proof in the special case of the unit ball or polydisk. In this special setting the proofs are more accessible. Then either a more general result is established, or the relevant literature is cited. The presentation is essentially self-contained. Professor Sheng Gong deserves the thanks of the geometric function theory community for writing an informative, up-to-date monograph that will certainly foster more work in the area.

David Minda

Cincinnati, OH

1997

PACIFIC JOURNAL OF MATHEMATICS
Vol. 150, No. 1, 1991

THE GROWTH AND 1/4-THEOREMS FOR STARLIKE MAPPINGS IN C^n

ROGER W. BARNARD, CARL H. FITZGERALD, AND SHENG GONG

Certain geometric function theory results are obtained for holomorphic mappings on the unit ball. Specifically, the mappings studied are one-to-one onto domains that are starlike with respect to the origin. For such a mapping $f(z)$, sharp estimates are derived for $|f(z)|$ in terms of $|z|$. Also, a generalization of the Koebe covering theorem is proved. As a corollary of the work, a new proof is given that, in $\mathbf{C}^n$ for $n \geq 2$, a ball and a polydisc are not biholomorphically equivalent.

More than fifty years ago, Henri Cartan [2, cf. 1, 3] suggested that geometric function theory of one complex variable should be extended to biholomorphic mappings of n complex variables. In particular, he cited the special classes of starlike functions and of convex functions as appropriate topics for generalization. In noting some of the difficulties of generalization, he pointed out the Growth Theorem for univalent functions as one that would not extend to the polydisc (nor to the ball). Also he observed that for normalized biholomorphic mappings there is no neighborhood about the origin that is always covered by the range of a biholomorphic mapping of the polydisc (or the ball), that is, there is no Koebe 1/4-Theorem for these mappings.

We will show that, for the class of normalized one-to-one starlike mappings on the ball, versions of both the Growth Theorem and the Koebe 1/4-Theorem hold. Specifically, we will find sharp upper and lower bounds on the magnitude of such a mapping in terms of the magnitude of the variable; and, we will show that the ball of radius 1/4 centered at the origin is always covered.

As a corollary of the results we will demonstrate a theorem first shown by Poincaré: there is no biholomorphic mapping of the unit ball in $\mathbf{C}^n$ onto the unit polydisc in case $n \geq 2$.

1. Preliminaries. Let points in $\mathbf{C}^n$ be denoted by column vectors $z = (z_1, z_2, \ldots, z_n)'$, where the prime indicates transpose. The origin $(0, 0, \ldots, 0)'$ is sometimes denoted by 0. The norm of z is $|z| = (\overline{z}'z)^{1/2} = \langle z, z\rangle = (|z_1|^2 + |z_2|^2 + \cdots + |z_n|^2)^{1/2}$, and the ball

$B_r = \{z: z \in \mathbf{C}^n \text{ and } |z| < r\}$ with B denoting the unit ball B_1. Mappings f in $\mathbf{C}^n$ are given as column vectors $f = (f_1, f_2, \ldots, f_n)'$ where each function f_k is a function into $\mathbf{C}$. The Jacobian of a holomorphic mapping f is the matrix with ij entry $\partial f_i/\partial z_j$ and it is denoted by J or J_f or $J_f(z)$. A holomorphic mapping f is normalized means that $f(0) = 0$ and $J_f(0) = I$, the identity matrix. A holomorphic mapping f is starlike with respect to the origin means that f is a normalized, one-to-one mapping with an image that is starlike with respect to the origin, that is, each point in the range can be joined to the origin by a straight line interval without leaving the range of f.

The following theorem gives an analytical characterization of starlike mappings of the open ball. Since this theorem will be basic to our work, we indicate a proof of the necessity of the analytical condition in the case that the mapping can be extended holomorphically to the closure of the ball. This argument is intended to facilitate the intuitive understanding of Suffridge's result, Theorem 1.1.

For z_0 on the boundary ∂B of B, the point $f(z_0)$ is a point on the boundary of the range. A vector V starting at $f(z_0)$ and extending in the direction that $f(z_0)$ is from the origin will lie in the complement of the range of f on the open ball. (If V reached a point which came from the interior of the ball, then the starlike property would imply that the point $f(z_0)$ arose from an interior point. But then a neighborhood of $f(z_0)$ would have to be covered by the image of a strictly interior neighborhood of the ball. Since f is one-to-one, the value $f(z_0)$ could not have been reached as a boundary value.)

Now consider the vector V under the inverse of the Jacobian: $J_f(z_0)^{-1}V$. If this vector is placed so that its base is at z_0, then it must be in the complement of the ball. (Note that the linear mapping $J_f(z_0)^{-1}$ may fail to be conformal.) Since $J_f(z_0)^{-1}V$ is in the complement of the ball, the real part of its inner product with z_0 must be nonnegative. That is,

$$\operatorname{Re}\langle J_f(z_0)^{-1}V, z_0\rangle = |J_f(z_0)^{-1}V| \cdot |z_0| \cos\theta,$$

where θ is the angle between the two entries of the inner product, is nonnegative. Furthermore, since V is a real, positive scalar times $f(z_0)$, the necessity of the analytical condition in the following theorem is obtained for points on ∂B. The condition (1.1) then follows for the points of B by the maximum principle.

Theorem 1.1 (*Suffridge* [5]). *Suppose that $f(z)$ is a normalized holomorphic mapping on the unit ball B and that $J_f(z)$ is nonzero in B. Then f is a one-to-one mapping that takes the ball onto a domain that is starlike with respect to the origin if and only if the normalized mapping*

$$w(z) = J_f(z)^{-1} f(z) \tag{1.1}$$

has the property that

$$\operatorname{Re} \overline{z}' w(z) \geq 0 \quad \textit{for every point } z \textit{ in the ball } B. \tag{1.2}$$

An easy corollary will be of use later in our work: If f is starlike with respect to the origin, then, when f is restricted to a smaller ball centered at the origin, it is still a starlike mapping.

The following lemma will be important in working with condition (1.2). A proof is included in order that this work be as self-contained as possible.

Lemma 1.2 (*Pfaltzgraff* [4]). *Let $w(z)$ be a normalized mapping such that*

$$\operatorname{Re} \sum_{j=1}^{n} \overline{z}_j w_j \geq 0 \quad \textit{on } B. \tag{1.3}$$

Then

$$\frac{1-|z|}{|z|(1+|z|)} \cos \Theta \leq \frac{1}{|w(z)|} \leq \frac{1+|z|}{|z|(1-|z|)} \cos \Theta \tag{1.4}$$

where Θ is the angle between z and $w(z)$.

Proof. Fix $\zeta \in \partial B$. Let ξ be a scalar complex number with $0 < |\xi| < 1$. By hypothesis,

$$\operatorname{Re} \sum_{j=1}^{n} \overline{\zeta}_j \overline{\xi} w_j(\zeta\xi) \geq 0.$$

Rewriting the sum, we obtain

$$\operatorname{Re} \sum_{j=1}^{n} \overline{\zeta}_j \zeta_j \overline{\xi} \xi \frac{w_j(\zeta\xi)}{\zeta_j \xi} \geq 0.$$

Since $\overline{\xi}\xi > 0$, we have

$$\operatorname{Re} \sum_{j=1}^{n} \overline{\zeta}_j \frac{w_j(\zeta\xi)}{\xi} \geq 0. \tag{1.5}$$

We define $P(\xi)$ to be $\sum_{j=1}^{n}\overline{\zeta}_j(w_j(\zeta\xi)/\xi)$. Then $P(\xi)$ is an analytic function of a one dimensional complex variable. From the normalization of w, it is clear that we can extend P to be defined at zero and then $P(0)=1$. From inequality (1.5), $P(\xi)$ has positive real part on the unit disc $\{\xi\colon |\xi|<1\}$; and the estimates

$$\frac{1-|\xi|}{1+|\xi|}\leq |\operatorname{Re} P(\xi)|\leq \frac{1+|\xi|}{1-|\xi|}$$

hold on the unit disc.

If the expression $\operatorname{Re}\sum_{j=1}^{n}\overline{\zeta}_j(w_j(\zeta\xi)/\xi)$ is considered as an inner product, we see this real expression equals $(|w(\zeta\xi)|/|\xi|)|\zeta|\cos\theta$ where θ is the angle between $w(\zeta\xi)/\xi$ and ζ. This angle θ is easily seen to be the same as the angle Θ in the conclusion of the theorem; just note that the angle θ is unchanged if each vector is rotated by multiplying by the complex scalar ξ and then think of z as being represented by $\zeta\xi$. Also we use that $|\zeta|=1$ so that $|z|=|\xi|$. The following inequalities result:

$$\frac{1-|z|}{1+|z|}\leq \frac{|w(z)|}{|z|}\cos\Theta\leq \frac{1+|z|}{1-|z|}.$$

The conclusion of the lemma follows by using these inequalities to estimate $1/|w(z)|$.

2. The Growth Theorem for starlike functions. We can now state the Growth Theorem for starlike mappings:

THEOREM 2.1. *Let $f=(f_1, f_2, \ldots, f_n)'$ be a starlike mapping of the unit ball B in $\mathbf{C}^n$. For any point $z\in B$,*

$$\frac{|z|}{(1+|z|)^2}\leq |f(z)|\leq \frac{|z|}{(1-|z|)^2}.$$

Furthermore the estimates are sharp.

Proof. Consider a number $0<r<1$. Let z_1 be a point such that $|z_1|=r$ and $|f(z_1)|=$ maximal magnitude of f on $\{z\colon |z|=r\}$. Since f is starlike, there is a ray from the origin to $f(z_1)$ which is contained in the image of $\{z\colon |z|\leq r\}$. Hence the preimage of this ray is in the closure of the ball B_r. Indeed, as one moves along the preimage starting at the origin, the magnitude of z is monotonically

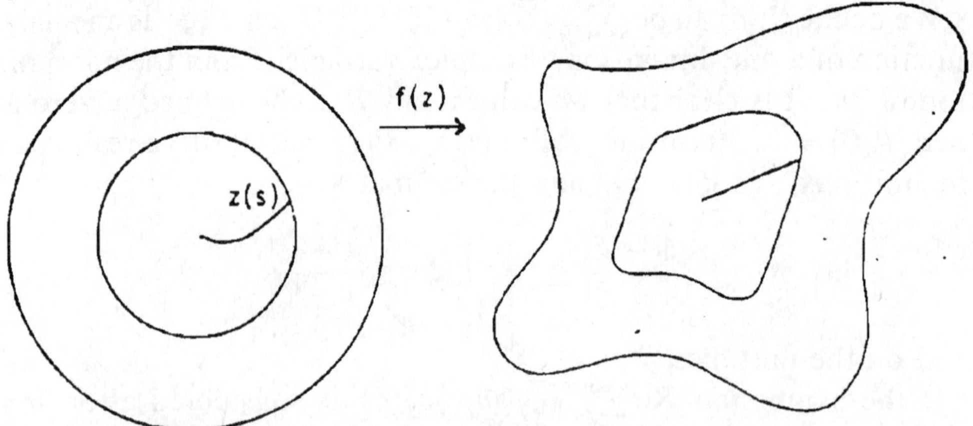

increasing. Since this preimage curve is analytic, we can certainly think of it as $z(s)$ parameterized with arc length.

$$\frac{df(z(s))}{ds} = \sum_{j=1}^{n} \frac{\partial f}{\partial z_j}\frac{dz_j}{ds} = J\frac{dz}{ds}, \tag{2.1}$$
$$\frac{df(z(s))}{ds} = \lambda(z(s))f(z(s)),$$

where λ is a positive, scalar-valued function, and the existence of λ follows from the fact that the image of the arc $z(s)$ is a straight line extending in the direction of $f(z(s))$.

Since f is starlike, we can apply equation (1.1) of Suffridge's theorem to obtain

$$\frac{df(z(s))}{ds} = \lambda J w.$$

We can solve for dz/ds by applying J^{-1}. Since λ is a scalar, we obtain

$$\frac{dz}{ds} = \lambda w.$$

Taking the magnitude of both sides of this equation, and recalling that s is the arc length parameter, we note that

$$1 = \lambda|w|, \quad \text{that is ,} \quad \lambda = \frac{1}{|w|}. \tag{2.2}$$

With this expression for λ, the equation (2.1) can be expressed as a vector equation as follows:

$$\frac{df(z(s))}{ds} = \frac{1}{|w(s)|}f(z(s)).$$

Our estimation of the magnitude of f will be facilitated by introducing a scalar function: $g(s) = |f(z(s))|2 = \overline{f(z(s))'}f(z(s))$. Then we have

$$\frac{dg}{ds} = 2\operatorname{Re}\left\{\overline{f(z(s))'}\frac{df(z(s))}{ds}\right\}$$
$$= 2\frac{1}{|w|}\operatorname{Re}\{\overline{f(z(s))'}f(z(s))\}$$
$$= 2\frac{1}{|w|}g(z(s)).$$

Hence we can write the equation of differentials:

$$\frac{dg}{g} = 2\frac{1}{|w|}ds. \tag{2.3}$$

Thus, from the upper estimate of $|w|$ given by Lemma 1.2, we have

$$\frac{d\log g}{ds} \le 2\frac{1+|z(s)|}{|z(s)|(1-|z(s)|)}\cos\theta(s), \tag{2.4}$$

where $\theta(s)$ is the angle between $z(s)$ and $dz(s)/ds$. From geometrical considerations, it is easy to obtain the general formula $d|z(s)|/ds = \cos\theta(s)$ provided $z(s)$ is not at the origin. Then from formula (2.4) we find that

$$\log g(s_1) - \log g(s_0) \le \int_{s_0}^{s_1} 2\frac{(1+|z(s)|)}{|z(s)|(1-|z(s)|)}\cos\theta(s)\,ds$$
$$= \int_{|z(s_0)|}^{|z(s_1)|} 2\frac{1+|z|}{|z|(1-|z|)}\,d|z|$$
$$= \int_{|z(s_0)|}^{|z(s_1)|}\left(\frac{2}{|z|} + \frac{4}{1-|z|}\right)d|z|$$
$$= 2\log|z(s_1)| - 4\log(1-|z(s_1)|)$$
$$- [2\log|z(s_0)| - 4\log(1-|z(s_0)|)].$$

We recall that f is normalized so that its Jacobian at the origin is equal to the identity matrix. Hence, if we let s_0 be a small positive number ε, then $|z(s_0)| = \varepsilon + o(\varepsilon)$ and $g(s_0) = \varepsilon^2 + o(\varepsilon^2)$. Thus

$$g(s_1) \le \left\{\frac{|z(s_1)|^2}{(1-|z(s_1)|)^4}\right\}\frac{\varepsilon^2 + o(\varepsilon^2)}{\varepsilon^2 + o(\varepsilon^2)}.$$

If we consider the limit as ε tends to zero, and pick s_1 so that $z(s_1)$ is $z_1 = z$, and use the definition of g, we obtain

$$|f(z)| \le \frac{|z|}{(1-|z|)^2} \quad \text{as claimed in the theorem.}$$

The other inequality of the theorem is easily proved. Again we use equality (2.3), but we use the lower bound on $1/|w(z)|$ given by Lemma 1.2. The rest of the argument follows as before.

The inequalities in Theorem 2.1 are best possible. To demonstrate this fact we define

$$f(z)=\left(\frac{z_1}{(1-z_1)^2}, \frac{z_2}{(1-z_2)^2}\right)'.$$

If the variable z is restricted to be of the form $(z_1, 0)'$, then equality can be obtained in both inequalities of Theorem 2.1. That f satisfies the hypothesis that it be normalized at the origin is clear from power series expansions. That it is one-to-one in the ball is obvious since each coordinate function of f is a univalent function of one variable: thus given $(w_1, w_2)'$ the value of z_1 can be obtained by applying the inverse of the function in the first coordinate to w_1 and the value of z_2 is obtained by a similar process from w_2. Finally, that f takes the ball onto a starlike domain follows from the fact that each coordinate function of f is a starlike function of one variable. For any point in the range of f, consider the straight line from that point to the origin. The projection of that line onto the plane of points $(*, 0)'$ is also a straight line to the origin. Its preimage under the first coordinate function is a curve that moves to the origin in a monotonic fashion, that is, the straight line distance to the origin decreases monotonically. Hence if we start at a point in the range of f and move along the straight line to the origin, the points considered are always in the range of f on the unit ball.

Now, it is clear that there are many examples showing that the estimates of Theorem 2.1 are sharp. For example, the second coordinate function of f can be replaced by any one-to-one analytic function f_2 of the single variable z_2 onto a domain that is starlike with respect to the origin provided f_2 takes the origin to the origin and has derivative at the origin equal to one.

3. The covering theorem for starlike mappings. From Theorem 2.1 it is easy to prove the following analogue of Koebe's 1/4-Theorem:

COROLLARY 3.1. *If $f(z)$ is a holomorphic mapping on the unit ball in C^n and is starlike with respect to the origin, then the image of f contains a ball of radius $1/4$ centered at the origin. The value $1/4$ is best possible.*

Proof. Consider the image of a ball B_r of radius $0 < r < 1$. The image is an open set with boundary being the image of the boundary of B_r. By Theorem 2.1, the image of the ∂B_r is at least $r/(1+r)^2$ away from the origin. By letting r tend to 1 from below, we complete the proof of the corollary.

Also it is easy to see that the estimate is sharp. The examples given at the end of the previous section show that the point $w_1 = -1/4$, $w_2 = 0$ can fail to be in the range.

If k-fold symmetry of f is assumed, that is, the image of f is unchanged when multiplied by the scalar complex number $\exp(2\pi i/k)$ for some positive integer k, then Theorem 2.1 and Corollary 3.1 can be strengthened:

COROLLARY 3.2. *If f is holomorphic in the unit ball and is starlike with a k-fold symmetric image for $k \geq 1$, then*

$$\frac{|z|}{(1+|z|^k)^{2/k}} \leq |f(z)| \leq \frac{|z|}{(1-|z|^k)^{2/k}}.$$

Then the image of the ball under f contains a ball of radius $2^{-2/k}$ and these estimates are best possible.

Proof. The geometric hypothesis that the image of the normalized holomorphic mapping is k-fold symmetric has a related analytical property of $f(z)$. Specifically, $\exp(-2\pi i/k)f(\exp(2\pi i/k)z) = f(z)$ for all z in the ball. This latter expression of symmetry is obviously satisfied by the mapping $w(z)$ of Theorem 1.1. Consequently, the positive real part function $P(\xi)$ in Lemma 1.1 has the symmetry property that $P(\exp(2\pi i/k)z) = P(z)$. Then the estimate for this function is that

$$\frac{1-|\xi|^{1/k}}{1+|\xi|^{1/k}} \leq \operatorname{Re} P(\xi) \leq \frac{1+|\xi|^{1/k}}{1-|\xi|^{1/k}}.$$

Following the rest of the proof of Theorem 2.1 gives the result claimed. Many mappings show that the result is sharp, for example,

$$f(z) = \left(\frac{z_1}{(1-z_1^k)^{2/k}}, \frac{z_2}{(1-z_2^k)^{2/k}}\right)'.$$

Finally, the estimate for the Koebe ball is found the same way as before. The example just described shows that the result obtained is sharp.

This corollary implies some restrictions on which symmetrical domains can be mapped onto each other by biholomorphic mappings. We recall that a domain is balanced means it is starlike with respect to the origin and is circular; a domain is circular means that if w belongs to the domain, then ξw also belongs to the domain for every complex scalar ξ of magnitude 1.

COROLLARY 3.3. *The only balanced domain which is the image of the unit ball under a normalized biholomorphic mapping is the unit ball.*

Proof. Consider a normalized, biholomorphic mapping of the ball onto a balanced domain D. Since the image domain is k-fold symmetric for all positive integers k, the estimates obtained in the previous corollary imply that $|f(z)| = |z|$ for all z in B. By the normalization of f, it follows that $f(z) = z$ for all z in B; hence the domain D is the unit ball B.

As an easy consequence of this corollary, we deduce a well-known, classic observation.

COROLLARY 3.4 (*Poincaré* 1907, *Cartan* 1932). *For $n \geq 2$, the unit ball in $\mathbf{C}^n$ is not biholomorphically equivalent to a polydisc in $\mathbf{C}^n$.*

Proof. Suppose there were such a mapping f from B onto a polydisc D. Using a holomorphic affine transformation in the range space, we can require D to be of a certain form: $D = \{w: |w_k| \leq r_k$ for $k = 1, 2, \ldots, n\}$ for some positive numbers $r_1, r_2, \ldots, r_n$. By an application of a Möbius transformation in the domain, we can assume that f takes the origin into the origin. Also, by using a unitary transformation of the ball, we can assume that the Jacobian of f at the origin is diagonal. By rescaling the complex coordinates in the image space, it can be assumed that f is normalized. Since the polydisc, positioned as described, is balanced, a contradiction to the previous corollary has been reached for $n \geq 2$. Hence there cannot be a biholomorphic mapping of B onto a polydisc.

4. Acknowledgments. This paper was written while the first and third authors were visiting the University of California at San Diego.

The authors are grateful for that research opportunity. The authors also appreciate helpful communications with Professor T. J. Suffridge.

References

[1] R. W. Barnard, C. H. FitzGerald, and Sheng Gong, *A distortion theorem for biholomorphic mappings in* $\mathbf{C}^2$, (to appear).

[2] H. Cartan, *Sur la possibilité d'étendre aux fonctions de plusieurs variables complexes la théorie des fonctions univalentes*, pages 129–155. Note added to P. Montel, *Leçons sur les Fonctions Univalentes ou Multivalentes*, Gauthier-Villars, Paris, 1933.

[3] P. Duren and W. Rudin, *Distortion in several variables*, Complex Analysis; Theory and Application, **5** (1986), 323–326.

[4] J. A. Pfaltzgraff, *Subordination chains and univalence of holomorphic mappings in* $\mathbf{C}^n$, Math. Ann., **210** (1974), 55–68.

[5] T. J. Suffridge, *The principle of subordination applied to functions of several variables*, Pacific J. Math., **33** (1970), 241–248.

Received March 20, 1989.

Texas Tech University
Lubbock, TX 79409

University of California, San Diego
La Jolla, CA 92093-0112

and

University of Science and Technology of China
Hefei, Anhui 230026
People's Republic of China

On Holomorphic Starlike Mappings of Several Complex Variables

Gong Sheng

(University of Science and Technology of China, Hefei, Anhui, 230026, China.)

Wang Shikun

(Institute of Applied Mathematics, Academia Sinica, Beijing, 100080, China,)

Yu Qihuang

(Institute of Mathematics, Academia Sinica, Beijing, 100080, China)

Abstract. In this paper some theorems are established to show that locally biholomorphic mappings are globally biholomorphic on complete Carathéodory domains and r-domains in $\mathbb{C}^n$.

1. Introduction

Suppose that Ω is a domain of a complex n-space $\mathbb{C}^n$, $n \geq 2$ and f is a locally biholomorphic mapping from Ω to $\mathbb{C}^n$. One of the interesting problems in several complex variables is to determine the conditions under which f is biholomorphic. The study of this subject is concerned closely with holomorphic equivalence between two domains. In [7] Pinchuk discussed strictly pseudoconvex domains. In [2] the authors discussed a class of Reinhardt domains in $\mathbb{C}^n$. In the present paper we are concerned with the two classes of domains of $\mathbb{C}^n$ as follows.

The first class of domains is the domain on which there is a continuous exhaust function $r(z)$ from Ω to $\mathbb{R}^+ \cup \{0\}$, with a unique zero point. We call it the r-domain.

The second class of domains is the complete Carathéodory domains. Recall that the Carathéodory quasidistance between two points p, q in a domain $\Omega \subset \mathbb{C}^n$ is defined by

$$C_\Omega(p,q) = \frac{1}{2} \sup_{h \in \Theta_p} \ln \frac{1+|h(q)|}{1-|h(q)|},$$

where $\Theta_p = \Theta_p(\Omega, D) = \{h \in \Theta(\Omega) \mid |h| < 1,\ h(p) = 0\}$, and $\Theta(\Omega)$ is the set of all holomorphic functions on Ω. The quasidistance is really a distance for bounded domains in $\mathbb{C}^n$ [6]. The so-called complete Carathéodory domains mean that the subset defined by $\{Z \in \Omega \mid C_\Omega(p, Z) < a\}$ is a relatively compact set in Ω for any $p \in \Omega$ and $a \in \mathbb{R}^+$.

In this paper we will prove two main theorems as follows:

Theorem 1. *Suppose that Ω is an r-domain and f is a locally biholomorphic mapping from Ω to $\mathbb{C}^n$, satisfying the following conditions:*

1. $r(p) = 0,\quad f(p) = 0,$ *p is a fixed point in Ω,* $\quad r(Z) > 0,\quad Z \in \Omega \setminus \{p\}$.
2. *For any $Z \in \Omega$, there is the biholomorphic neighborhood U_Z of Z such that if, for any fixed $t \in [0,1]$, $Z(t)$ satisfies $f(Z(t)) = (1-t)f(Z) \in f(U_Z)$ and $Z(t) \in U_Z$, then we have $r(Z(t)) \le r(Z)$.*

Then f is globally biholomorphic on Ω and $\{f(\Omega)\}$ is starlike with respect to the origin in $\mathbb{C}^n$.

Remark 1. Starlike here means that if $Z \in \Omega$ then $tf(Z)$ falls in $\{f(\Omega)\}$, $0 \le t \le 1$.

Remark 2. Assume $r(Z)$ is a function of class C^1. Then the condition (2) in the theorem can be written as follows:

$$< f^{-1*}|_{U_Z} dr, d\rho >|_{W=f(Z)} \ge 0$$

where $\rho = (\sum_{i=1}^n |w_i|^2)^{\frac{1}{2}}$, $w_i (i = 1, 2, \cdots, n)$ are the components of W.

Theorem 2. *Let Ω be a complete Carathéodory domain with $0 \in \Omega$, and let $f : \Omega \to \mathbb{C}^n$ be a locally biholomorphic mapping. Then the sufficient and necessary conditions for f to be a globally biholomorphic are that for any $Z \in \Omega$ and any fixed $t, 0 \le t \le 1$, if $Z(t) \in U_Z$ which is a biholomorphic neighborhood of Z with respect to $f(Z)$ then $C_\Omega(0, Z) \ge C_\Omega(0, Z(t))$ where $Z(t)$ satisfies $f(Z(t)) = (1-t)f(Z)$.*

2. Some Lemmas

Assume that Ω is an r-domain or complete Carathéodory domain. Let f be a locally biholomorphic mapping from Ω to $\mathbb{C}^n$. For simplicity, we fix some notations. Throughout the paper we denote the components of W, Z by small letters w_i, z_i and U_Z usually represents the neighborhood of Z in which f is biholomorphic. And denote

$$\mathcal{E}_a = \{Z \in \Omega \mid r(Z) < a,\ a \in \mathbb{R}^+\},$$

$$\Delta(t, Z) = \{(1-t)f(Z) \mid t \in [0,1], Z \in \Omega\},$$

$$B_\epsilon(Z) = \{W \in \Omega \mid \sum_{i=1}^{n} \mid w_i - z_i\mid^2 < \epsilon\}, \quad Z \in \mathbb{C}^n.$$

In order to prove Theorem 1 we need two Lemmas:

Lemma 1. *Let Ω and f be defined as in Theorem 1. Let f be biholomorphic on $\mathcal{E}_a$. Then $\{f(\overline{\mathcal{E}}_a)\}$ is starlike with respect to the origin in $\mathbb{C}^n$.*

Lemma 2. *Let Ω and f be defined as in Theorem 1. Let f be biholomorphic on $\mathcal{E}_a$. Then $f : \overline{\mathcal{E}}_a \to f(\overline{\mathcal{E}}_a)$ is a diffeomorphic mapping.*

Proof of Lemma 1. First, we prove that a positive number ϵ exists such that

$$(1-t)f(Z_0) \subset f(\overline{\mathcal{E}}_a), \quad \epsilon > t > 0, \text{for any} \quad Z_0 \in \overline{\mathcal{E}}_a.$$

If the conclusion fails, we can find a sequence $\{t_i\}_{i=1}^{\infty}$ converging to zero such that $(1-t_i)f(Z_0) \notin f(\overline{\mathcal{E}}_a)$, for all $i = 1, 2, \cdots$. We know $f(U_{Z_0})$ is an open set (see [4]). Thus we can find a positive δ such that $B_\delta(f(Z_0))$ is contained in $f(U_{Z_0})$. When i becomes sufficiently large we have

$$W_i = (1-t_i)f(Z_0) \in B_\delta(f(Z_0)).$$

Hence there is a point Z_k in the biholomorphic neighborhood U_{Z_0} satisfying $f(Z_k) = W_k = (1-t_k)f(Z_0)$ and $r(Z_k) > a \geq r(Z_0)$ due to the fact that $f(Z_k) = (1-t_k)f(Z_0) \notin \{f(\overline{\mathcal{E}}_a)\}$. This contradicts the condition (2) in Theorem 1.

Second, we will show that if a sequence $\{\tilde{t}_i\}_{i=1}^{\infty}$, satisfying $(1-\tilde{t}_i)f(Z_0) \in f(\overline{\mathcal{E}}_a)$, converges to $\tilde{t}$ then $(1-\tilde{t})f(Z_0) \in f(\overline{\mathcal{E}}_a)$.

Take $\tilde{Z}_i$ in $\overline{\mathcal{E}}_a$ such that $f(\tilde{Z}_i) = (1-\tilde{t}_i)f(Z_0)$. Since $r(Z)$ is a continuous exhaust function, $\overline{\mathcal{E}}_a$ is a compact set in Ω. So the sequence $\{\tilde{Z}_i\}$ has a limit point $\tilde{Z}$ in $\overline{\mathcal{E}}_a$. Let $\widetilde{W} = f(\tilde{Z})$. We have

$$\begin{aligned}|\widetilde{W} - (1-\tilde{t})f(Z_0)| &= |\widetilde{W} - f(\tilde{Z}_i) + f(\tilde{Z}_i) - (1-\tilde{t})f(Z_0)| \\ &\leq |\widetilde{W} - f(\tilde{Z}_i)| + |\tilde{t}_i - \tilde{t}||f(Z_0)|.\end{aligned}$$

Letting $i \to +\infty$, we have

$$|\widetilde{W} - (1-\tilde{t})f(Z_0)| = 0, \quad i.e. \quad f(\tilde{Z}) = (1-\tilde{t})f(Z_0).$$

If the lemma is not true, then there is at least a point Z_0 in $\overline{\mathcal{E}}_a \subset \Omega$ such that the set

$$T = \{t \in [0,1] \mid (1-t)f(Z_0) \notin f(\overline{\mathcal{E}}_a)\} \tag{1}$$

is nonempty. Take $t^* = \inf\{t | t \in T\}$. Obviously, $t^* < 1$. We consider the two cases below.

1. If $t^* \in T$, then $t^* \neq 0$. We have

$$(1-t)f(Z_0) \in f(\overline{\mathcal{E}}_a), \tag{2}$$

for any $0 \leq t < t^*$. However, this also shows that $t^* \notin T$ because of the second step of previous proof.

2. If $t^* \notin T$, then $(1-t^*)f(Z_0) \in f(\overline{\mathcal{E}}_a)$. Consider the point Z^* in $\overline{\mathcal{E}}_a$ such that $f(Z^*) = (1-t^*)f(Z_0)$. Using the proof in the first step we can find a positive ϵ such that

$$(1-s)f(Z^*) \in f(\overline{\mathcal{E}}_a) \tag{3}$$

for any $\epsilon > s > 0$. So

$$(1-s)(1-t^*)f(Z_0) \in f(\overline{\mathcal{E}}_a) \tag{4}$$

for $\epsilon > s > 0$. That is

$$(1-(s+t^*-st^*))f(Z_0) \in f(\overline{\mathcal{E}}_a),$$

for $\epsilon > s > 0$. Since $t^* = \inf\{t|t \in T\}$ and we can take a sequence $\{\epsilon_i\}_{i=1}^{\infty}$ such that $t^* + \epsilon_i \in T$ and $\epsilon_i \to 0$ if $i \to +\infty$. Then let $s_i = \epsilon_i/(1-t^*)$. We have

$$s_i + t^* - s_i t^* = t^* + \epsilon_i. \tag{5}$$

When i is sufficiently large we have $\epsilon > s_i > 0$ because of $\lim_{i\to+\infty} s_i = 0$. This implies $t^* + s_i \notin T$ if i sufficiently large. We complete the proof of Lemma1.

Proof of Lemma 2. In fact, we need only prove that the map f is one to one in $\overline{\mathcal{E}}_a$.

If the conclusion above is not true, then there are at least two distinct points Z_1 and Z_2 in $\overline{\mathcal{E}}_a$ such that $f(Z_1) = f(Z_2) = W$. Obviously, there is at least one point between Z_1 and Z_2, which falls in $\partial\overline{\mathcal{E}}_a$, the boundary part of $\overline{\mathcal{E}}_a$.

We first show that $\{f^{-1}(0)\} \cap \overline{\mathcal{E}}_a = \{p\}$.

If there is another point $Z \neq p$ in $\overline{\mathcal{E}}_a$ such that $f(Z) = 0$ then $Z \in \partial\mathcal{E}_a$ due to f being biholomorphic on $\mathcal{E}_a$. Let U_Z be a biholomorphic neighborhood of Z with respect to f. Since U_Z is an open set to contain the boundary point Z of $\mathcal{E}_a$, there must be a point $\tilde{Z}$ in U_Z

such that $r(\widetilde{Z}) < r(Z) = a$. On the other hand, U_Z can be also regarded as a biholomorphic neighborhood of $\widetilde{Z}$. Obviously, $Z = Z(1)$ satisfies $f(Z) = (1-t)f(\widetilde{Z})|_{t=1} \in f(U_Z)$. So, from the condition (2) in Theorem 1, we get $r(Z) \leq r(\widetilde{Z})$. That is impossible.

Secondly, if there are two distinct points Z_1 and Z_2 such that

$$f(Z_1) = f(Z_2) = W,$$

then we can prove that there must be a positive t_o in [0, 1], such that the set $\{f^{-1}(t_0W)\}\cap\overline{\mathcal{E}}_a$ is a single point set. In fact, since f is locally biholomorphic we can find the neighbourhoods U_i of $Z_i (i = 1, 2)$ in which f is biholomorphic. By the open mapping theorem $f(U_i) = V_i$ are also open sets in $\mathbb{C}^n$. It is easy to see $W \in V_i$. So $V_1 \cap V_2$ is nonempty and open. Hence an open ball $B_\epsilon(W)$ contained in $V_1 \cap V_2$ exists. Moreover, we have $tW \subset V_1 \cap V_2$ for any $t \geq 1-\epsilon$. So $\{f^{-1}((1-t)W)\} \cap \overline{\mathcal{E}}_a$ is nonempty from the condition (2) in Theorem 1. If $\{f^{-1}((1-\epsilon)W)\}\cap\overline{\mathcal{E}}_a$ only contains one point then we take $t_0 = 1-\epsilon$. If $\{f^{-1}((1-\epsilon)W)\}\cap\overline{\mathcal{E}}_a$ contains at least two points then we take $t_1 = 1-\epsilon$, $W_1 = t_1W$. Following the process to get t_1 we obtain t_2 and $t_1 > t_2$ so on. Repeating the operating we get a monotone sequence $1 > t_1 > t_2 > \cdots \geq 0$ which is a bounded one. Thus

$$\lim_{i\to+\infty} t_i = t_0 \geq 0. \tag{6}$$

If $t_0 = 0$, we obtain that f is not biholomorphic at the point p. Hence $t_0 > 0$. Denote by Z_0 the single point $\{f^{-1}(t_0W)\} \cap \overline{\mathcal{E}}_a$.

Next we will show that there is a continuous curve l_1 jointing Z_0 and Z_1 in $\overline{\mathcal{E}}_a$ such that

$$f(l_1) = \{tW \mid t_0 \leq t \leq 1\}, \tag{7}$$

where t_0, Z_0 are defined as above. Let s be the point in $[t_0, 1]$, satisfying the condition that there is a continuous curve l in $\overline{\mathcal{E}}_a$ of which a boundary point is Z_1 and $f(l) = \{tW \mid s \leq t \leq 1\}$. All the points are denoted by S.S is a subset of [0, 1]. We know $f(U_{Z_1})$ is an open set. Take a small positive ϵ such that $\overline{B_\epsilon(f(Z_1))} \subset f(U_{Z_1})$. If the intersection point of the segment $\Delta(t, W)$ and the circle $\partial B_\epsilon(f(Z_1))$ is sW then $s < 1$ and $f^{-1}|_{U_{Z_1}}(\{tW|s \leq t \leq 1\})$ is continuous in U_{Z_1}. From the condition (2) in Theorem 1 we get $f^{-1}|_{U_{Z_1}}(\{tW|s \leq t \leq 1\}) \subset \overline{\mathcal{E}}_a$. Hence the set S is not empty. Let $s^* = \inf\{S\}$. We claim $s^* = t_0$. Based on the discussion above, in fact, $t_0 \leq s^* < 1$. If $s^* > t_0$ then take $Y_1 \in \overline{\mathcal{E}}_a$ such that $f(Y_1) = s^*W$. Repeating the above discussion we can find $s^{**} < s^*$ and $s^{**} \in S$. This contradicts the definition of s^*. Similarly, we can find the continuous curve l_2 jointing Z_0 and Z_2, satisfying $f(l_2) = \{tW \mid t_0 \leq t \leq 1\}$.

Now let us complete the proof of Lemma 2. Obviously, $l_1 \neq l_2$ because of $Z_1 \neq Z_2$. Let $t^* = \sup\{t \mid t \in H\}$, where $H = \{t \in [0,1] \mid tW \in f(l_1 \cap l_2)\}$. If the inverse image of t^*W is Z^*, then $Z^* \in l_1 \cap l_2 \subset \overline{\mathcal{E}}_a$ and $f(Z^*) = t^*W, t^* \geq t_0$. We have $f^{-1}(tW) \cap l_1 \neq f^{-1}(tW) \cap l_2$, for any $t > t^*$. This implies that f is not biholomorphic at Z^*. We complete the proof of Lemma 2.

3. Proof of Theorem 1

We denote

$$\mathcal{E}_a = \{Z \in \Omega \mid r(Z) < a, a \in \mathbb{R}^+\},$$
$$\mathcal{A} = \{a \in \mathbb{R}^+ \mid f \text{ is biholomorphic on } \mathcal{E}_a\}.$$

We first prove the set $\mathcal{A}$ is nonempty. This is to prove there is a positive number ϵ such that f is biholomorphic in $\mathcal{E}_\epsilon$. If it is not true then a positive sequence $\{\epsilon_i\}_{i=1}^{\infty}$ exists, satisfying $\epsilon_i \to 0$ when $i \to \infty$, and $\mathcal{E}_{\epsilon_i} \setminus U_p \neq \emptyset$, where U_p is a biholomorphic neighborhood of the point p with respect to f. Since the set $\mathcal{E}_{\epsilon_i} \setminus U_p$ is nonempty, we take $Z_i \in \mathcal{E}_{\epsilon_i} \setminus U_p$. The set $\{Z_i\}_{i=1}^{\infty}$ is a subset of $\overline{\mathcal{E}}_1$. According to the definition of exhaust function $\overline{\mathcal{E}}_1$ is a compact set. So there is a convergent subsequence with $\{Z_{i_k}\}$ with $Z_{i_k} \notin U_p$. However, by the continuity of $r(Z)$,

$$r(Z_0) = \lim_{i_k \to \infty} r(Z_{i_k}) = 0. \tag{8}$$

From the assumption (1) in Theorem 1, $Z_0 = p$. Therefore, $\lim_{i_k \to \infty} Z_{i_k} = p \in U_p$, which means $Z_{i_k} \in U_p$ when i_k becomes sufficiently large. Hence $\mathcal{A}$ is not empty.

Next we show $\mathcal{A}$ is a closed set. Assume $\tilde{a}$ is a limit point of $\mathcal{A}$. In fact, if $a_1 \in \mathcal{A}$ then $a \in \mathcal{A}$ for all $a \leq a_1$. Therefore, it suffices to prove that if $\tilde{a} > a$ and a falls in $\mathcal{A}$ then $\tilde{a}$ is also in $\mathcal{A}$. That is to prove f is biholomorphic in $\mathcal{E}_{\tilde{a}}$. If it is not true then there must be two distinct points Z_1, Z_2 in $\mathcal{E}_{\tilde{a}}$ such that $f(Z_1) = f(Z_2)$. Since $r(Z_1) < \tilde{a}$, $r(Z_2) < \tilde{a}$, we can find an a^* which satisfies

$$r(Z_1) < a^* < \tilde{a}, \qquad r(Z_2) < a^* < \tilde{a}. \tag{9}$$

The formulas imply $Z_1, Z_2 \in \overline{\mathcal{E}}_{a^*}$. But $a^* \in \mathcal{A}$. Thus $f(Z_1) - f(Z_2) \neq 0$ which contradicts the assumption of $f(Z_1) = f(Z_2)$. This proves $\mathcal{A}$ is a closed set.

Finally, we prove the set $\mathcal{A}$ is an open set. If $\mathcal{A}$ is not open then there is at least one positive number a_0 in $\mathcal{A}$ such that $a \notin \mathcal{A}$ for any $a > a_0$. Hence we can choose two sequences of points $Z(i)$ and $Y(i)$ satisfying

$$Z(i) \neq Y(i), \qquad f(Z(i)) = f(Y(i)), \qquad i = 1, 2, \cdots \tag{10}$$

and

$$0 < r(Z_i) - a_0 < \frac{1}{i}, \qquad 0 < r(Y_i) - a_0 < \frac{1}{i}, \quad i = 1, 2, \cdots. \tag{11}$$

It is easily seen that $Z(i)$, $Y(i)$ are bounded sequences. So there are convergent subsequences $Z(i_k)$ and $Y(i_k)$ of $Z(i)$ and $Y(i)$. Denote

$$\lim_{k\to\infty} Z(i_k) = Z_0, \qquad \lim_{k\to\infty} Y(i_k) = Y_0. \tag{12}$$

From (10) and (11), we have

$$f(Z_0) = f(Y_0). \tag{13}$$

Since the exhaust function is continuous,

$$\lim_{k\to\infty} r(X_{i_k}) = r(Z_0) = a_0, \tag{14}$$

$$\lim_{k\to\infty} r(Y_{i_k}) = r(Y_0) = a_0. \tag{15}$$

Lemma 2 implies $Z_0 = Y_0$ and (10) implies f is not biholomorphic at the point Z_0. So A is an open set.

The proof of preceding second step also candidates that A is a connected set. Since A is open, closed and nonempty, $A = \mathbb{R}^+$. That is, f is a biholomorphic mapping in Ω. By Lemma 1, $\{f(\Omega)\}$ must be starlike.

4. Proof of Theorem 2

A fundamental property of Carathéodory distance tells us that if f is a holomorphic mapping from a domain Ω_1 to another domain Ω_2 in $\mathbb{C}^n$, then

$$C_{\Omega_1}(Z, Y) \geq C_{\Omega_2}(f(Z), f(Y))$$

holds for any $Z, Y \in \Omega_1$.

Using the above property we can prove Theroem 2. Theorem 1 implies the sufficiency in Theorem 2. We need only prove the necessary condition. If it fails, then there are a point $Z^* \in \Omega$ and a positive number $s \in [0, 1]$ with $sf(Z^*) = f(Z(s))$ such that

$$C_\Omega(0, Z(s)) > C_\Omega(0, Z^*). \tag{16}$$

We define a holomorphic mapping g from Ω to itself by $g(Z) = f^{-1}(sf(Z))$. Obviously, $g(0) = 0, g(Z^*) = Z(s)$. By the property of Carathéodory distance, we have

$$C_\Omega(0, Z^*) \leq C_\Omega(0, Z(s)). \tag{17}$$

(17) contradicts (16). We complete the proof of Theorem 2.

The four types of classical domains are defined as follows [3]:

$$\begin{aligned}
R_I(m,n) &= \{Z \in \mathbb{C}^{m\times n} \mid I - Z\bar{Z}^T > 0\},\\
R_{II}(m,m) &= \{Z \in \mathbb{C}^{m\times m} \mid Z = Z^T, I - Z\bar{Z}^T > 0\},\\
R_{III}(m,m) &= \{Z \in \mathbb{C}^{m\times m} \mid Z = -Z^T, I + Z\bar{Z}^T > 0, m \geq 2\},\\
R_{IV}(n) &= \{Z \in \mathbb{C}^n \mid 1 - 2Z\overline{Z}^T + |ZZ^T|^2 > 0, ZZ^T < 1, n \geq 3\}.
\end{aligned}$$

And the Reinhardt domains B_p are defined as follows:

$$B_p = \{Z \in \mathbb{C}^n \mid \sum_{i=1}^{n} |z_i|^p < 1\}.$$

Theorem 3. *let Ω be any one of the four types of classical domains or B_p. We have*

$$C_\Omega(0, Z) = \frac{1}{2}\log\frac{1 + \|Z\|_\Omega}{1 - \|Z\|_\Omega},$$

for all $Z \in \Omega$, where $\|Z\|_\Omega$ are positive square roots of the largest eigenvalue of the matrix $Z\bar{Z}^T$ when $\Omega = R_K(K = I, II, III)$, $(Z\bar{Z}^T + \sqrt{(Z\bar{Z}^T)^2 - |ZZ^T|^2})^{1/2}$ when $\Omega = K_{IV}$, and $\sum_{i=1}^{n} |z_i|^p$ when $\Omega = B_p$.

Remark. The same results for classical domains were obtained by Wang Daming (see references in [8]).

Proof. Here we give the proof for B_p. Fix a $Z_0 \in B_p$. Let $Z = tZ_0/\|Z_0\|_p$ where $t \in D$. Sicne

$$\|Z\|_p = \left(\sum_{i=1}^{n} \frac{|t|^p |z_{0i}|^p}{\|Z_0\|_p}\right)^{\frac{1}{p}} = |t| < 1,$$

we have $Z \in B_p$. For any $f \in \Theta(B_p, D)$ with $f(0) = 0$ we define an analytic function g from D to D by

$$g(t) = f(\frac{tZ_0}{\|Z_0\|_p}).$$

By the Schwartz Lemma,

$$|g(t)| \leq |t|$$

holds. In particular, if we take $t = \|Z_0\|_p$ then

$$|f(Z_0)| \leq \|Z_0\|_p$$

holds for any $Z_0 \in B_p$. That is,

$$C_{B_p}(0, Z) \leq \frac{1}{2} \log \frac{1 + \|Z_0\|_p}{1 - \|Z_0\|_p}. \tag{18}$$

On the other hand, we can construct a new function by

$$f(Z) = \sum_{i=1}^{n} a_i z_i,$$

where

$$a_i = \begin{cases} z_{0i}^{\frac{p}{2}-1} \overline{z}_{0i}^{\frac{p}{2}} / \|Z_0\|_p^{p-1}, & \text{if } z_{0i} \neq 0, \\ 0, & \text{if } z_{0i} = 0. \end{cases}$$

The linear function is holomorphic in B_p and $f(0) = 0$. Because

$$|f(Z)| = \left|\sum_{i=1}^{n} a_i z_i\right| \leq \sum_{i=1}^{n} |a_i||z_i| \leq \left(\sum |z_i|^p\right)^{\frac{1}{p}} \left(\sum (a_i)^{\frac{p}{p-1}}\right)^{\frac{p-1}{p}} \tag{19}$$

and

$$\sum_{i=1}^{n} |a_i|^{\frac{p}{p-1}} = \frac{1}{\|Z_0\|_p^p} \sum_{i=1}^{n} |z_{0i}|^p = 1,$$

we obtain

$$|f(Z)| \leq \|Z\|_p < 1,$$

i.e. f is holomorphic from B_p to D with $f(0) = 0$. It is clear that

$$f(Z_0) = \sum a_i z_{0i} = \frac{1}{\|Z_0\|_p^{p-1}} \sum |z_{0i}|^p = \|Z_0\|_p.$$

That is

$$C_{B_p}(0, Z_0) \geq \frac{1}{2} \log \frac{1 + \|Z_0\|_p}{1 - \|Z_0\|_p}. \tag{20}$$

Combining (18) with (20), we complete the proof of Theorem 3.

From Theorem 2 and Theorem 3 we get

Corollary 1. *Let Ω be R_K, any one of the four types of classical domains, or B_p. Then the sufficient and necessary conditions for a locally biholomorphic mapping f from Ω to $\mathbb{C}^{\dim \Omega}$ with $f(0) = 0$ to be globally biholomorphic are that for any $Z \in \Omega$ and any fixed*

t, $0 \le t \le 1$, *if* $Z(t) \in U_Z$ *which is a biholomorphic neighborhood of* $Z \in \Omega$ *with respect to* $f(Z)$ *then* $\|Z(t)\|_\Omega \le \|Z\|_\Omega$ *where* $Z(t)$ *satisfies* $f(Z(t)) = (1-t)f(Z)$.

From the result that strong pseudoconvex domains in $\mathbb{C}^n$ are complete Carathéodory domains [4], we have

Corollary 2. *Let* Ω *be a strong pseudoconvex domain in* $\mathbb{C}^n$ *and let* f *be a locally biholomorphic mapping on* Ω *with* $f(0) = 0$. *Then the sufficient and necessary conditions for* f *to be a globally biholomorphic are that for any* $Z \in \Omega$ *and any fixed* t, $0 \le t \le 1$, *if* $Z(t) \in U_Z$ *which is a biholomorphic neighborhood of* $Z \in \Omega$ *with respect to* $f(Z)$ *then* $C_\Omega(0, Z) \ge C_\Omega(0, Z(t))$ *where* $Z(t)$ *satisfies* $f(Z(t)) = (1-t)f(Z)$.

In [5] Kobayashi proved that generalized analytic polyhedrons are complete domains with respect to the Carathéodory distance. Thus we have

Corollary 3. *Let* Ω *be a generalized analytic polyhedron. Let* f *be a local biholomorphic mapping from* Ω *to* $\mathbb{C}^n$. *Then the sufficient and necessary conditions for* f *to be a globally biholomorphic are that for any* $Z \in \Omega$ *and any fixed* t, $0 \le t \le 1$, *if* $Z(t) \in U_Z$ *which is a biholomorphic neighborhood of* $Z \in \Omega$ *with respect to* $f(Z)$ *then* $C_\Omega(0, Z) \ge C_\Omega(0, Z(t))$ *where* $Z(t)$ *satisfies* $f(Z(t)) = (1-t)f(Z)$.

This work was supported by the Natural Science Foundation of China.

References

[1] H. Grauert and K. Fritzshe, *Several complex variables*, Springer-Verlag Press, 1976.

[2] S. Gong, S. K. Wang and Q. H. Yu, *A necessary and sufficient condition that biholomorphic mapping are starlike on a class of Reinhardt domains*, to appear in Chinese Annals of Mathematics.

[3] L. K. Hua, *Harmonic analysis of functions of several complex variables in classical domains*, Amer. Math. Soc., Transl. Math. Monograph, 1963.

[4] D. Kim, *Complete domains with respect to Carathéodory distance*, II, Proc. Amer. Math. Soc., 53(1975), 101-142.

[5] S. Kobayashi, *Hyperbolic manifolds and holomorphic mappings*, New York, Dekker, 1970.

[6] Q. K. Lu, *Introduction to several complex variables*, Chinese Science Press, 1961 (in Chinese).

[7] S. I. Pinchuk, *On proper holomorphic maps of strictly pseudoconvex domains*, Sib. Mat. Zh 15, 909-917 (1974, Russian), Zbl. 289, 32011. English Transl.: Sib. Math. Notes, 15(1974), 116-120 .

[8] Q. K. Lu, *A study of the theory of functions of several complex variables in China during the last four decades (1949-1989)*, to be appear in the book "Contemporary geometry", ed. by H. Wu.

第42卷第1期 数 学 学 报 Vol.42, No. 1
1999年1月 ACTA MATHEMATICA SINICA Jan., 1999

有界星形域上的全纯映照成为星形映照的判别准则

龚 升
(中国科学技术大学数学系 合肥 230026)
(Email: gongs@public.intercom.com.cn)

王世坤* 余其煌
(中国科学院应用数学所 北京 100080)
(Email: yqh@public.intercom.com.cn)
*(Email: xyswsk@pku.edu.cn)

摘 要 本文讨论从有界星形圆形域和 Caratheodory 完备域到 C^n 局部双全纯映照成为双全纯星形映照的充分必要条件.

关键词 星形圆形域, r- 域, Minkowski 泛函, 星形映照
MR(1991) 主题分类 32H02
中图分类 O174.56

Critieria of Biholomorphic Starlike Mappings

Gong Sheng
(Department of Mathematics, University of Science and Technology of China, Hefei 230026, *P. R. China)*
(Email: gongs@public.intercom.com.cn)
Wang Shiku* Yu Qihuang
(Institute of Applied Mathematics, Chinese Academy of Sciences, Beijing 100080, *P. R. China)*
(Email: yqh@public.intercom.com.cn)
* *(Email: xyswsk@pku.edu.cn)*

Abstract In this paper the sufficient and necessary condition that holomorphic mappings are starlike on bounded starlike circle domain is discussed.

Keywords Starlike circular domain, r-domain, Minkowski functional, Starlike mapping
MR(1991) Subject Classification 32H02
Chinese Library Classification O174.56

多复变数的域上的全纯映照成为双全纯星形映照的判别准则已有不少结果, 如文献 [1–6] 等 (参阅文献 [7] 第六章及 Chapter I). 最早的结果也许是 Suffridge T. J.[1] 给出的关于 $B_p = \{Z = (z_1, z_2, \cdots, z_n) \in \mathbb{C}^n : \sum_{j=1}^n |z_j|^p < 1, p \geq 1\}$ 上局部双全纯映照成为双全纯星形映照的判别准则; 在文献 [3] 中龚 升, 王世坤, 余其煌给出了域 $D_p = \{Z = (z_1, z_2, \cdots, z_n) \in \mathbb{C}^n :$

收稿日期: 1997-08-27, 接受日期: 1998-01-08
国家自然科学基金资助项目

$\sum_{j=1}^{n}|z_j|^{p_j}<1,\ 2p_n>p_1\geq p_2\geq\cdots\geq p_n>1\}$ 上局部双全纯映照成为双全纯星形映照的判别准则. 但此时条件 $2p_n>p_1\geq p_2\geq\cdots\geq p_n>1$ 不能去掉. 后来陈正 [5] 以及林运泳与洪毅 [6] 给出了有界凸圆形域上局部双全纯映照成为双全纯星形映照的判别准则.

定理 1 若 $\Omega\subset\mathbb{C}^n$ 为有界凸圆形域, 其 Minkowski 泛函 $p(Z)=\inf\{c>0,\frac{Z}{c}\in\Omega\}$, 除去一个低维子流形外是一阶可微的, $0\in\Omega$. 若 $f(Z):\to\Omega\subset\mathbb{C}^n$ 为正规化局部双全纯映照, 则 $f(z)$ 成为 Ω 上正规化双全纯 (相对于原点) 星形映照当且仅当

$$\Re e\Big\{\frac{\partial p(Z)}{\partial Z}J_f^{-1}(Z)f(Z)\Big\}\geq 0, \tag{1}$$

对所有的 $Z\in\Omega$ 都成立, 这里 $f(Z)=(f_1(Z),f_2(Z),\cdots,f_n(Z))'$ 为列向量, $\frac{\partial p}{\partial Z}=(\frac{\partial p}{\partial z_1},\frac{\partial p}{\partial z_2},\cdots,\frac{\partial p}{\partial z_n})$, $Z=(z_1,z_2,\cdots,z_n)$, $J_f(Z)$ 为 f 在 Z 点的 Jacobian.

显然, 域 $D=\{Z=(z_1,z_2,\cdots,z_n)\in\mathbb{C}^n:\sum_{j=1}^{n}:|z_j|^{p_j}<1,\ p_1\geq p_2\geq\cdots\geq p_n>1\}$ 是有界凸圆形域, 从而除掉了余其煌, 王世坤, 龚 升的结果 [3] 中的条件 $2p_n>p_1\geq p_2\geq\cdots\geq p_n>1$.

本文首先指出, 定理 1 可以推广为:

定理 2 若 $0\in\Omega$ 为 $\mathbb{C}^n$ 中的有界星形圆形域, 其余假设如定理 1, 则 $f(z)$ 成为 Ω 上正规化双全纯星形映照当且仅当 (1) 成立.

有界星形圆形域比有界凸圆形域多得多, 例如: $B=\{Z=(z_1,z_2,\cdots,z_n)\in\mathbb{C}_n:\sum_{j=1}^{n}|z_j|^p<1,p>0\}$, 当 $p\geq1$ 时, B 是有界凸圆形域; 当 $0<p<1$ 时, B 是有界星形圆形域, 而不是有界凸圆形域.

定理 2 的必要部分已由刘太顺与任广斌 [8] 所证明, 而充分性部分为下列定理 3 的一个简单推论. 定理 3 由龚 升, 王世坤, 余其煌 [4] 所证.

定理 3 若 $\Omega\in\mathbb{C}^n$ 为 r- 域, f 为 Ω 到 $\mathbb{C}^n$ 的局部双全纯映照, 满足:

1. 存在定义在 Ω 上的穷竭 (exhaustion) 函数 $r(Z):\Omega\to\mathbb{R}\cup\{0\}$, 且 $r(q)=0$, $f(q)=0$, q 为 Ω 中的一固定点, 而当 $Z\in\Omega\backslash\{q\}$ 时, $r(Z)>0$;

2. 对于任意 $Z\in\Omega$, 存在 Z 的邻域 U_Z, 使得 f 在 U_Z 上是双全纯的, 而且对任意的 $t\in[0,1]$, $Z(t)$ 为 U_Z 中满足 $f(Z(t))=(1-t)f(Z)\in f(U_Z)$ 的点, 就有 $r(Z(t))\leq r(Z)$, 那么 f 在 Ω 上是相对于原点而言的双全纯星形映照.

如果 $r(Z)\in C^1(\Omega)$, 则条件 2 可以写为,

$2'$.
$$\langle f^{-1*}dr,d\rho\rangle\,|_{W=f(Z)}\geq 0, \tag{2}$$

这里 $\rho(W)=\big(\sum_{j=1}^{n}|w_j|^2\big)^{\frac{1}{2}}$, $W=(w_1,w_2,\cdots,w_n)$, f^{-1*} 表示 f^{-1} 在 U_Z 上的拉回 (pull back), $\langle\cdot,\cdot\rangle$ 表示 $\mathbb{C}^n$ 中的 Hermite 内积.

定理 2 的证明 只要取 $r(Z)=\frac{p(Z)}{1-p(Z)}$, $q=0$, 则 $r(Z)$ 为 Ω 上的连续穷竭函数, 于是, 定理 2 中的 Ω 成为 r- 域, 而

$$d\rho^2=d\sum_{j=1}^{n}W\overline{W}'=\sum_{j=1}^{n}(w_jd\overline{w_j}+\overline{w_j}dw_j);$$

$$f^{-1*}dr=\frac{\sum\limits_{j,k=1}^{n}\Big(\frac{\partial p}{\partial z_k}\frac{\partial z_k}{\partial w_j}dw_j+\frac{\partial p}{\partial \overline{z_k}}\frac{\partial\overline{z_k}}{\partial\overline{w_j}}d\overline{w_j}\Big)}{(1-p)^2}.$$

所以

$$\langle f^{-1*}dr, d\rho\rangle\,|_{W=f(Z)}=\frac{\Re e\frac{\partial P(Z)}{\partial Z}J_f^{-1}(Z)f(Z)}{\rho(1-p(Z))^2},$$

由 (1) 式和定理 3, 我们立即推出定理 2 的充分性部分.

此外，还有

定理 4 若 $\Omega\in\mathbb{C}^n$ 为 r- 域， $f:\Omega\to\mathbb{C}^n$ 为 Ω 上的双全纯星形映照，满足

1. $r(q)=0$, $f(q)=0$, q 为 Ω 中固定的一点，且当 $Z\in\Omega\backslash\{q\}$ 时， $r(Z)>0$;

2. 对于任意在 Ω 上的全纯映照 $g:\Omega\to\Omega$, 且 $f(q)=q$, 有收缩性质 (contraction property), 即

$$r(Z)\geq r(f(Z)), \tag{3}$$

对于任意的 $Z\in\Omega$ 都成立，则对任意 $Z\in\Omega$, $0\leq t\leq 1$, $f(Z(t))=(1-t)f(Z)$, 不等式

$$r(Z)\geq r(Z(t)), \tag{4}$$

对于任意的 $Z\in\Omega$ 都成立．若除去一个低维子流形外， $r(Z)\in C^1(\Omega)$ 在 Ω 上成立，则 (1) 成立.

证明 若 $Z\in\Omega$, $0\leq\lambda\leq 1$, $Z_\lambda=f^{-1}(\lambda f(Z))$, 则由 (3), $r(Z)\geq r(f(Z_\lambda))$, 取 $\lambda=1-t$, 即得 (4)．若除去一个低维子流形外， $r(Z)\in C^1(\Omega)$ 在 Ω 上成立，则对所有的 $\lambda\in(0,1)$, 有 $\frac{r(Z)-r(f(z_\lambda))}{1-\lambda}\geq 0$. 于是

$$\frac{dr(z_\lambda)}{d\lambda}\Big|_{\lambda=1}=\sum_{i,j}\left(\frac{dr}{dz_i}\frac{dz_i}{dw_j}w_j+\frac{dr}{d\overline{z_i}}\frac{d\overline{z_i}}{d\overline{w_j}}\overline{w_j}\right)\geq 0, \tag{5}$$

这里 $f=W=(w_1,w_2,\cdots,w_n)$. 但是,

$$f^{-1*}dr=2\Re e\left(\sum_{i,j}\frac{\partial r}{\partial z_i}\frac{\partial z_i}{\partial w_j}dw_j\right),$$

$$d\rho=\Re e\left(\frac{1}{\rho}\sum_k\overline{w_k}dw_k\right),$$

所以

$$\langle f^{-1*}dr, d\rho\rangle\,|_{w=f(Z)}=\Re e\left\{\frac{4}{\rho}\sum_{i,j}\frac{\partial r}{\partial z_i}\frac{\partial z_i}{\partial w_j}w_j\right\}.$$

由 (5) 即得 (2)．这就证明了定理 3.

定理 3 给出 r- 域上局部双全纯映照成为双全纯星形映照的充分条件，而定理 4 给出 r- 域上全纯映照称为双全纯映照的必要条件．应用这二条定理，立即可得定理 1, 这是因为定理 3 给出了充分条件．由刘太顺与任广斌 [9] 已证：在有界凸圆形域上，定理 4 中的条件成立，即：若 $g:\Omega\to\Omega$ 为 Ω 上的全纯映照， $g(0)=0$, 则 $p(Z)\geq p(g(Z))$, 这里的 $p(Z)$ 为 Ω 的 Minkowski 泛函．于是由定理 4, 就得必要条件．因此，定理 1 就成为定理 3 及定理 4 的直接推论.

定理 3 与 4, 以及定理 2, 从某种意义上讲，基本上解决了多复变数的域上的局部双全纯映照成为双全纯星形映照的判别准则．例如：将定理 3 及定理 4 应用于 Carathèodory 完备域上，由于 Carathèodory 度量有收缩性质，所以有

定理 5 若 $\Omega \subset \mathbb{C}^n$ 为 Caratheodory 完备域, $0 \in \Omega$, $f : \Omega \to \mathbb{C}^n$ 为局部双全纯映照, 则 f 为 Ω 上相对原点而言是双全纯星形映照当且仅当: $C_\Omega(0, Z) \geq C(0, Z(t))$ 对所有的 $Z \in \Omega$ 都成立, 这里 C_Ω 为 Caratheodory 度量, $f(Z(t)) = (1-t)f(Z), 0 \leq t \leq 1, Z(t) \in U_Z, U_Z$ 为 Z 的邻域, f 在 U_Z 上双全纯. 若 $C_\Omega(0, Z) \in C^1(\Omega)$ 在 Ω 上成立, 则 f 在 Ω 上相对于原点而言是双全纯星形映照, 当且仅当

$$\langle f^{-1*} dC_\Omega(0, Z), d\rho \rangle \mid_{W=f(Z)} \geq 0,$$

这里 $\rho = \left(\sum_{i=1}^{n} |w_i|^2\right)^{\frac{1}{2}}$, $W = f = (w_1, w_2, \cdots, w_n)$.

已知 [10] 强拟凸域, 一般解析多面体 (general analytic polyhedron), 典型域及 B_p 等都是 Caratheodory 完备域所以定理 5 仍有广泛意义.

参 考 文 献

1 Suffridge T J. The principle of subordination applied to function of several variables. Pacific Jour of Math, 1970, 33:(1) 241–248

2 Kikuchi K. Starlike and convex mappings in several complex variables. Pacific Jour of Math, 1973, 44(2): 569–580

3 Gong S, Wang S K, Yu O II. A necessary and sufficient condition that biholomorphic mappings are starlike on Reinhardt domains. Chinese Annals of Math, 1992, 13B(1): 95–104

4 Gong S, Wang S K, Yu Q H. On holomorphic starlike mappings of several complex variables. Lecture note in contemporary mathematics. Institute of Mathematics Academia Sinica, Science Press, 1991, 2: 91–103

5 陈正. $\mathbb{C}^n$ 中一类有界严格平衡域上的双全纯映照及星形映照的刻划. 数学年刊, 1995, 16A(2): 230–237

6 林运泳, 洪毅. Banach 空间中的全纯映射的若干性质. 数学学报, 1995, 38(2): 234–241

7 龚升. 多复变数的凸映照与星形映照. 纯粹数学与应用数学专著第 34 号. 科学出版社, 1995
Gong Sheng, Convex and starlike mappings in several complex varialtcs. Science Press / Kluwer Academic Publishers, 1998.

8 Liu T S, Ren G B. The growth and 1/4-covering theorems for the normalized starlike mappings on bounded starlike and circular domains. Chinese Annals of Math. to appear

9 Liu T S, Ren G B. The grouth theorems of convex mappings on bounded convex circular domains. Science in China, (Series A) 1998, 41(2): 123–130

10 Poletskii E A, Shabat B V. Invaiant metrices. Several Complex Veriables III, Khenkin G M ed. Springer-Verlag, 1989: 63–112

Chin. Ann. of Math.
11B:1(1990), 100—104.

THE GROWTH AND 1/4-THEOREMS FOR STARLIKE MAPPING IN B^p†

YU QIHUANG(余其煌)* WANG SHIKUN(王世坤)** GONG SHENG(龚 升)***

(*Dedicated to the Tenth Anniversary of CAM*)

Abstract

The Growth and 1/4-Theorems for the class of normalized biholomorphic starlike mappings on $B^p=\{Z=(Z_1, \cdots, Z_n)\in C^n|\sum_{i=1}^{n}|Z_i|^p<1\}$ for any $p>1$ are given.

Let $B^p=\{Z=(Z_1, \cdots, Z_n)\in C^n|\sum_{i=1}^{n}|Z_i|^p<1\}$ for any $p>1$, which is a class of Reinhardt domains. When p=2, B^2 is the ball in C^n. In [1], Carl H. FitzGerald, Gong Sheng and Roger W. Barnard have established the Growth and 1/4-Theorems for normalized biholomorphic starlike mappings in C^n, which generalized some results of geometric function theory of one complex variable to several complex variables. The main purpose of this paper is to establish the corresponding results for B^n, $p>1$. A holomorphic mapping f from B^p to C^n is normalized if $f(0)=0$ and its complex Jacobian at origin is identity matrix, i. e $\left\{\frac{\partial f_i}{\partial Z_j}\right\}\Big|_{z=0}=\delta_{ij}$. The normalized holomorphic mapping is starlike in C^n means that it is a biholomorphic mapping with a starlike image $f(B^p)$ with respect to the origin in C^n. A necessary and sufficient condition for a normalized holomorphic mapping to be starlike was given by F. J. Suffiridge[2]. Here we rewrite the Suffridge's theorem without proof in terms of differential geometry as the following:

Theorem 1. *Let f: $B^p\to C^n$ be a normalized holomorphic mapping. Then it is starlike if and only if*

$$<f^{-1*}du\ dp>|_{w(z)}\geq 0 \quad (\textit{for any } Z\in B^p\setminus\{0\}),$$

where $\langle , \rangle$ is the inner product, w is coordinate of $f(Z)$ in C^n, ρ is the distance from the orign in C^n, $u=\sum_{i=1}^{n}|Z_i|^p$.

We can prove the following

Manuscript received April 19, 1989.

* Institute of Applied Mathematics, Academia Sinica, Beijing, China.

** Institute of Mathematics, Academia Sinica, Beijing, China.

*** Departments of Mathematics, University of Science and Technology of China, Hefei, Anhui, China.

† Projects Supported by National Natural Science Fondation of China.

Theorem 2. *Let $f: B^p \to C^n$ be a normalized biholomarphic starlike mapping. Then*

$$\frac{2pu(1-u^{\frac{1}{p}})}{1+u^{\frac{1}{p}}} \le \langle f^{-1*}du,\ d\rho^2\rangle \le \frac{2pu(1+u^{\frac{1}{p}})}{1-u^{\frac{1}{p}}},$$

where $u=\sum_{i=1}^{n}|Z_i|^p$, $\rho = dist\ (0,\ w(Z))$.

Proof By using the Suffridge's theorem and noting $\langle dw^i,\ d\bar{w}^j\rangle = 2\delta_{ij}$, we have

$$\begin{aligned} 0 \le \langle f^{-1*}du,\ d\rho^2\rangle &= \langle f^{-1*}(\partial u+\bar\partial u),\ \partial\rho^2+\bar\partial\rho^2\rangle \\ &= \langle f^{-1*}\partial u,\ \bar\partial\rho^2\rangle + \langle f^{-1*}\bar\partial u,\ \partial\rho^2\rangle \\ &= 2\mathrm{Re}\langle f^{-1*}\partial u,\ \bar\partial\rho^2\rangle \\ &= 2p\mathrm{Re}\sum_{i,j=1}^{n}\frac{|Z_i|^p}{Z_i}\cdot\frac{\partial Z_i}{\partial w_j}\cdot w_j, \end{aligned} \tag{1}$$

$$\text{i. e.}\quad \mathrm{Re}\sum_{i,j=1}^{n}\frac{|Z_i|^p}{Z_i}\cdot\frac{\partial Z_i}{\partial w_j}\cdot w_j \ge 0. \tag{2}$$

Let t be a complex number with $|t|<1$ and, for any fixed $Z\in B^p$, $Z^0=Z/[u(Z)]^{\frac{1}{p}}$. Then

$$\sum_{i}^{n}|tZ_i^0|^p = |t|^p < 1.$$

That means $tZ^0\in B^p$. Thus tZ^0 can be instead of Z in (2), and the inequality (2) is

$$\mathrm{Re}\sum_{i,j=1}^{n}\frac{|tZ_i^0|^p}{tZ_i^0}\cdot\frac{\partial Z_i}{\partial w_j}\cdot w_j(tZ^0)\ge 0,$$

i. e.

$$\mathrm{Re}\sum_{i,j=1}^{n}\frac{|Z_i^0|^p}{Z_i^*}\cdot\frac{\partial Z_i}{\partial w_j}\cdot\frac{w_j}{t}(tZ^0)\ge 0. \tag{3}$$

Now we define a new function of one complex variable t.

$$A(t)=\sum_{i,j=1}^{n}\frac{|Z_i^0|^p}{tZ_i^0}\cdot\frac{\partial Z_i}{\partial w_j}\cdot w_j(tZ^0). \tag{4}$$

It is clear that $A(t)$ is holomorphic in the unit disc $D=\{t\in C^1 \mid |t|<1\}$ when $t\neq 0$ and $\mathrm{Re}A(t)\ge 0$. Because of the normalized condition of f, i. e. $\left.\frac{\partial w_i}{\partial Z_k}\right|_{z=0}=\delta_{ik}$, we have

$$\lim_{t\to 0}\frac{w_j(tZ^0)}{t}=\sum_{k=1}^{n}\frac{\partial w_j}{\partial Z_k}\cdot\left.\frac{\partial Z_k}{\partial t}\right|_{t=0}=\sum_{k=1}^{n}\left.\frac{\partial w_j}{\partial Z_k}\right|_{t=0}Z_k^0=Z_j^0,$$

which implies $A(t)$ is also holomorphic at $t=0$ and $A(0)=1$, i. e. $A(t)$ is normalized holomorphic in the unit disc. So an argument of one complex variable can be used to obtain

$$\frac{1-|t|}{1+|t|}\le \mathrm{Re}\{A(t)\}\le\frac{1+|t|}{1-|t|} \tag{5}$$

for any $t\in D$. In particular, when $t=u(Z)^{\frac{1}{p}}$,

$$\frac{1-u^{\frac{1}{p}}(Z)}{1+u^{\frac{1}{p}}(Z)}\le \mathrm{Re}\sum_{i,j}^{n}\frac{|Z_i^0|^p}{Z_i}\cdot\frac{\partial Z_i}{\partial w_j}\cdot w_j(Z)\le\frac{1+u^{\frac{1}{p}}(Z)}{1-u^{\frac{1}{p}}(Z)}, \tag{6}$$

i. e.

$$\frac{u(Z)(1-u^{\frac{1}{p}}(Z))}{1+u^{\frac{1}{p}}(Z)} \leq \operatorname{Re} \sum_{i,j}^{n} \frac{|Z_i|^p}{Z_i} \cdot \frac{\partial Z_i}{\partial w_j} \cdot w_j(Z) \leq \frac{u(Z)(1+u^{\frac{1}{p}}(Z))}{1-u^{\frac{1}{p}}(Z)}. \tag{7}$$

Substituting (1) to (7), we obtain

$$\frac{2pu(1-u^{\frac{1}{p}})}{1+u^{\frac{1}{p}}} \leq \langle f^{-1*}du,\ d\rho^2\rangle \leq \frac{2pu(1+u^{\frac{1}{p}})}{1-u^{\frac{1}{p}}}. \tag{8}$$

Theorem 3. *Let $p>1$ and $f: B^p \to C^n$ be a normalized biholomarphic starlike mapping. Then*

$$\frac{|Z|}{(1-u^{\frac{1}{p}})^2} \geq \rho \geq \frac{|Z|}{(1+u^{\frac{1}{p}})^2},$$

where $|Z| = \left(\sum_{i=1}^{n} |Z_i|^2\right)^{\frac{1}{2}}$, $u = \sum_{i=1}^{n} |Z_i|^p$, $\rho = dist(0,\ w(Z))$.

Proof By Theorem 2

$$2\rho\langle f^{-1*}du,\ d\rho\rangle \leq \frac{2pu(1+u^{\frac{1}{p}})}{1-u^{\frac{1}{p}}}. \tag{9}$$

In fact,

$$\langle f^{-1*}du,\ d\rho\rangle = \langle d(u\circ f^{-1}),\ d\rho\rangle = \frac{\partial u}{\partial \rho}$$

is the directional derivative of u along the direction $d\rho$. So (9) is

$$\frac{\partial u}{\partial \rho} \cdot \frac{1-u^{\frac{1}{p}}}{pu(1+u^{\frac{1}{p}})} \leq \frac{1}{\rho}, \tag{10}$$

for any $Z \in B^p \setminus \{0\}$.

Fixed a $Z \in B^p \setminus \{0\}$ and a positive s, the segment joining the origin O and the point $Q=f(Z)$ in C^n will be denoted by L and the plane determined by the three points O, Q and $Q_1=f(sZ)$ in C^n will be denoted by Σ. When the positive number s is small enough the small circle $B(s) = \{W=(W_1, \cdots, W_n) \in C^n \mid |W| = |f(sZ)|\} \cap \Sigma$ must lies in the image $f(B^p)$ and intersects the segment L at S. Now let us constract a path l on the plane Σ. It comsists of the short arc on the circle $B(s)$ from Q_1 to S and the segment joining the intersectional point S and Q. By the difinition of l, ρ is increasing on the path l. Integrating (10) along the path l from Q_1 to S, then to Q, we get

$$\int_{Q_1}^{Q} \frac{d\rho}{\rho} \geq \frac{1}{p} \int_{sZ}^{Z} \frac{1-u^{\frac{1}{p}}}{u(1+u^{\frac{1}{p}})} du. \tag{11}$$

For the right hand side of (11),

$$\frac{1}{p}\int_{sZ}^{Z} \frac{1-u^{\frac{1}{p}}}{u(1+u)^{\frac{1}{p}}} du = [\ln u^{\frac{1}{p}} - 2\ln(1+u^{\frac{1}{p}})]_{sZ}^{Z}.$$

Therefore

$$\ln\rho(Q) - \ln\rho(Q_1) \geq \ln\frac{u^{\frac{1}{p}}(Z)}{[1+u^{\frac{1}{p}}(Z)]^2} + \ln[1+u^{\frac{1}{p}}(sZ)]^2 - \ln u^{\frac{1}{p}}(sZ).$$

Thus

$$\rho(Q)\geq\frac{u^{\frac{1}{p}}(Z)}{[1+u^{\frac{1}{p}}(Z)]^2}\cdot[1+u^{\frac{1}{p}}(sZ)]^2\cdot\frac{\rho(Q_1)}{u^{\frac{1}{p}}(sZ)}. \tag{12}$$

Note that $Q_1=Q(s)=f(sZ)$ and the normalized condition of f,

$$\rho(Q_1)=\left(\sum_{i=1}^{n}|w_i(sZ)|^2\right)^{\frac{1}{2}}=\left(\sum_{i=1}^{n}|sZ_i+O(s^2)|^2\right)^{\frac{1}{2}}$$

$$=\left(\sum_{i=1}^{n}s^2|Z_i|^2+O(s^3)\right)^{\frac{1}{2}}=s\left(\sum_{i=1}^{n}|Z_i|^2+O(s)\right)^{\frac{1}{2}}.$$

But

$$u^{\frac{1}{p}}(sZ)=\left[\sum_{i=1}^{n}|sZ_i|^p\right]^{\frac{1}{p}}=s\left[\sum_{i=1}^{n}|Z_i|^p\right]^{\frac{1}{p}}.$$

So

$$\frac{\rho(Q_1)}{u^{\frac{1}{p}}(sZ)}=\frac{\left(\sum_{i=1}^{n}|Z_i|^2\right)^{\frac{1}{2}}+O(s)}{\left(\sum_{i=1}^{n}|Z_i|^p\right)^{\frac{1}{p}}}.$$

Let $s\to0$; then (12) becomes

$$\rho(Q)\geq\frac{u^{\frac{1}{p}}(Z)}{(1+u^{\frac{1}{p}}(Z))^2}\cdot\frac{(Z\bar{Z})^{\frac{1}{2}}}{u^{\frac{1}{p}}(Z)}=\frac{|Z|}{(1+u^{\frac{1}{p}}(Z))^2}. \tag{13}$$

Using the left side inequality in Theorem 2 and the procedure for getting (13), we can prove the inequality of the left hand side of the Theorem 3.

For any $i=1, 2, \cdots\cdots, n$,

$$\frac{\partial}{\partial|Z_i|^p}(u-|Z|^p)=1-\left(\frac{|Z_i|}{|Z|}\right)^{2-p}$$

such that

$$\begin{cases}\frac{\partial}{\partial|Z_i|^p}(u-|Z|^p)\geqq0, \text{ for } 1<p<2\\ \frac{\partial}{\partial|Z_i|^p}(u-|Z|^p)\leq0, \text{ for } p\geq2.\end{cases}$$

Hence

$$\begin{aligned}&u^{\frac{1}{p}}\geq|Z|, \text{ when } 1<p<2,\\ &u^{\frac{1}{p}}\leq|Z|, \text{ wpen } p\geq2.\end{aligned} \tag{14}$$

Using Hörder's inequality, we have

$$\begin{aligned}&|Z|\geq n^{\frac{p-2}{2p}}u^{\frac{1}{p}}, \text{ when } 1<p<2,\\ &|Z|\leq n^{\frac{p-2}{2p}}u^{\frac{1}{p}}, \text{ when } p\geq2.\end{aligned} \tag{15}$$

Substituting (14)(15) into Theorem 3 we have

Coroliary. *Suppose that f: $B^p\to C^n$ is a starlike and normalized mapping Then*

$$\frac{u^{\frac{1}{p}}}{(1-u^{\frac{1}{p}})^2}\geq\rho\geq\frac{p-2}{n}\cdot\frac{u^{\frac{1}{p}}}{(1+u^{\frac{1}{p}})^2},\ for\ 2>p>1,$$

$$n^{\frac{p-2}{2p}}\cdot\frac{u^{\frac{1}{p}}}{(1-u^{\frac{1}{p}})^2}\geq\rho\geq\frac{u^{\frac{1}{p}}}{(1+u^{\frac{1}{p}})^2},\ for\ p\geq 2.$$

From the corollary above we immediately obtain the Koebe-type 1/4 Theorem for the class of normalized biholomorphic starlike mappings on B^p in C^n for any $p>1$.

Theorem 4. *Let f: $B^p\to C^n$ is a normalized biholomorphic starlike mapping. Then the image of f contains a ball centered at the origin in C^n with radius* $1/4K$, *where*

$$K=\begin{cases} n^{1/p-1/2}, & for\ 1<p<2,\\ 1, & for\ p\geq 2.\end{cases}$$

References

[1] Fitz Gerald, Cral H., Gong Sheng(龚 升) and Barnard, Roger W., The Growth and 1/4-Theorems for starlike mappings in C^n, *Chinese Science Bulletin*, **34**: 3(1989), 161—162.

[2] Suffridge, T. J., The principal of subordination applied to function of several variables, *Pacific. J. Math.*, **33**(1970), 241-248.

Chin. Ann. of Math.
14B: 1(1993),93-104.

THE GROWTH THEOREM FOR BIHOLOMORPHIC MAPPINGS IN SEVERAL COMPLEX VARIABLES***

GONG SHENG*   WANG SHIKUN**   YU QIHUANG**

Abstract

The Growth Theorem for normalized biholomorphic starlike and convex mappings on Reinhardt domains and classical domains are established.

§1. The Growth Theorem for Starlike Mappings on Reinhardt Domains

There are fruitful results on the geometrical function theory of one complex variable. But there exist a lot of counter-examples to show that the corresponding reseults in several complex variables are not true. In 1933, H.Cartan[1] already pointed out that the growth theorem of biholomorphic mappings on the unit ball of C^n is not true. In 1988, Carl H.Fitzgerald, Sheng Gong and Roger W.Barnard[2] gave the first affirmative result about the growth theorem in several complex variables. They proved the following result.

Let $B^n \equiv \{Z = (z_1, z_2, \cdots, z_n) \in C^n | \ |Z| < 1\}$ be the unit ball in C^n, where $|Z| = (\sum\limits_{i=1} |z_i|^2)^{1/2}$, and

$$f(Z) = (f_1(Z), f_2(Z), \cdots, f_n(Z)) = Z + (ZA^{(1)}Z^T, \cdots, ZA^{(n)}Z^T) + \cdots \tag{1.1}$$

be the normalized biholomorphic starlike mappings on B^n, where $A^{(j)} = (a_{kl}^{(j)})_{1\le l,k\le n}$, $j = 1, 2, \cdots, n$, are constant matrices, starlike mapping means the image of B^n by $f(Z)$ is starlike with respect to origin, and normalized mapping means $f(0) = 0$, $J_f(0) = I$, where J_f is the Jacobian of f, I is the identity matrix. Then

$$\frac{|Z|}{(1+|Z|)^2} \le |f(Z)| \le \frac{|Z|}{(1-|Z|)^2} \tag{1.2}$$

holds for every $z \in B^n$. The estimation is precise, but the extremal mapping is not unique. Obviously, $f(B^n) \supset \frac{1}{4}B^n$, that is, the Koebe constant of this family of mappings on B^n is $\frac{1}{4}$.

In 1989, Qihuang Yu, Shikun Wang and Sheng Gong[3] extended the Growth Theorem of normalized biholomorphic starlike mappings on ball to the Reinhardt domain B_p where $B_p \equiv \{Z \in C^n | \ \|Z\|_p = (\sum\limits_{i=1}^{n} |z_i|^p)^{\frac{1}{p}} < 1, p > 1\}$. We proved the following result.

Manuscript received April 3, 1990.
*Department of Mathematics, University of Science and Technology of China, Hefei, Anhui, 230026, China.
**Institute of Applied Mathematics, Academia Sinica, Beijing, 100080, China.
***Project supported by the National Natural Science Foundation of China and the Open Foundation of Institute of Mathematics, Acadimia Sinica .

Let (1.1) be the normalized biholomorphic starlike mappings on B_p. Then

$$\frac{|Z|}{(1+\|Z\|_p)^2} \le |f(Z)| \le \frac{|Z|}{(1-\|Z\|_p)^2} \tag{1.3}$$

holds for every $Z \in B_p$. The estimation is precise, but the extremal mapping is not unique. As a consequence of (1.3), $f(B^p) \supset \frac{1}{4K}B^p$ where $K=1$ if $p \ge 2$, $K = n^{\frac{2-p}{2p}}$ if $2 \ge p > 1$. That is, the Koebe constant of this family of mappings on B_p is $\frac{1}{4K}$.

On the other hand, using Loewner Chain's method, John A. Pfeltzgraff[4] proved that: Let (1.1) be the normalized biholomorphic starlike mappings on B_p. Then

$$\frac{\|Z\|_p}{(1+\|Z\|_p)^2} \le \|f(Z)\|_p \le \frac{\|Z\|_p}{(1-\|Z\|_p)^2} \tag{1.4}$$

holds for every $Z \in B_p$. The estimation is precise, but the extremal mapping is not unique. As a consequence of (1.4), $f(B^p) \supset \frac{1}{4}B^p$. That is, the Koebe constant of this family of mappings on B_p is $\frac{1}{4}$ in the sense of p-norm.

Now we would like to point out that (1.3) and (1.4) are equivalent. That means they can imply each other.

Let $\rho(W) = \mathrm{tr}W\overline{W}^T$ where $W = f(Z)$, $Z \in B_p$. Then it is easy to verify

$$d\rho(W) = \frac{1}{2\rho}\sum_{i=1}^{n}(d\omega_i\overline{\omega_i} + \overline{d\omega_i}\omega_i).$$

and

$$d\|W\|_p = \frac{1}{2}\|W\|_p\sum_{i=1}^{n}\left(\frac{d\omega_i}{\omega_i} + \frac{\overline{d\omega_i}}{\overline{\omega_i}}\right),$$

where $\|W\|_p = (\sum_{i=1}^{n}|\omega_i|^p)^{\frac{1}{p}}$, $p > 1$. We have $< d\|W\|_p, d\rho >= \frac{\|W\|_p}{\rho}$, where $<,>$ is the inner product of the cotangent space of C^n at the point W. On the other hand, $< d\|W\|_p, d\rho >= \frac{d\|W\|_p}{d\rho}$. We obtain

$$\frac{d\|W\|_p}{\|W\|_p} = \frac{d\rho}{\rho}. \tag{1.5}$$

Then we integrate (1.5) on both sides from $W = f(\epsilon Z)$ to $W = f(Z)$ where ϵ is a small positive number. We get

$$\frac{\|f(Z)\|_p}{\|f(\epsilon Z)\|_p} = \frac{|f(Z)|}{|f(\epsilon Z)|}. \tag{1.6}$$

Now we can prove that (1.3) imply (1.4). From (1.6) and (1.3), the following inequality holds

$$\frac{\|f(\epsilon Z)\|_p(1-\epsilon\|Z\|_p)^2}{(1+\|Z\|_p)^2\epsilon} \le \|f(Z)\|_p \le \frac{\|f(\epsilon Z)\|_p(1+\epsilon\|Z\|_p)^2}{(1-\|Z\|_p)^2\epsilon}. \tag{1.7}$$

Then letting $\epsilon \to 0$, we get (1.4) since $\lim\limits_{\epsilon\to 0}\frac{\|f(\epsilon Z)\|_p}{\epsilon} = \|Z\|_p$ due to the normalized condition of f.

Similarly, using (1.6) and (1.4), we can prove (1.3).

Note that the methods used to prove (1.3) in [3] and (1.4) in [4] are essentially different, but (1.3) and (1.4) can imply each other. In other words, they are two different methods to prove both (1.3) and (1.4) from [3] and [4].

§2. The Growth Theorems for Starlike Mappings on Classical Domains

As a part of his doctor thesis, in 1989, Taishun Liu[5] extended the result of [2] in another direction. He got the growth theorems of normalized biholomorphic starlike mapping on classical domains. The four types of classical domains are defined as follows[6]:

$$\mathcal{R}_I(m,n)=\{Z\in C^{m\times n}|I-Z\overline{Z}^T>0\},$$

$$\mathcal{R}_{II}(n)=\{Z\in C^{n\times n}|I-Z\overline{Z}^T>0,Z=Z^T\},$$

$$\mathcal{R}_{III}(n)=\{Z\in C^{n\times n}|I-Z\overline{Z}^T>0,Z=Z^T\},$$

$$\mathcal{R}_{IV}(n)=\{Z\in C^n|1-2Z\overline{Z}^T+|ZZ^T|^2>0,|ZZ^T|<1\}.$$

He proved that: Let $f(Z)$ be the normalized biholomorphic starlike mappings on classical domains $\mathcal{R}_K$, $(K=I,II,III,IV)$ which map $\mathcal{R}_K$ to $C^{\dim \mathcal{R}_K}$. Then

$$\frac{\|Z\|_K}{(1+\|Z\|_K)^2}\le\|f(Z)\|_K\le\frac{\|Z\|_K}{(1-\|Z\|_K)^2} \tag{2.1}$$

holds for every $Z\in\mathcal{R}_K$, where $\|\ \|_K$ is the norm of $\mathcal{R}_K$ (cf. [7]), that is, $\|Z\|_K$ is a positive square root of the largest eigenvalue $\lambda(Z)$ of $Z\overline{Z}^T$ for $K=I,II,III$ and $\|Z\|_{IV}=(|Z|^2+(|Z|^4-|Z\overline{Z}^T|^2)^{\frac{1}{2}}$. As the corollary of (2.1), he got $f(\mathcal{R}_K)\supset\frac{1}{4}\mathcal{R}_K$, $K=I,II,III,IV$.

All these estimation are precise, and the extremal mapping is not unique.

In this paper, we establish another growth theorem of normalized biholomorphic starlike mapping on classical domains.

Theorem. *Let $W=f(Z)$ be the normalized biholomorphic starlike mappings on classical domains $\mathcal{R}_K$, $(K=I,II,III,IV)$, which map $\mathcal{R}_K$ to $C^{\dim \mathcal{R}_K}$. Then*

$$\frac{|Z|}{(1+\|Z\|_K)^2}\le|f(Z)|_K\le\frac{|Z|}{(1-\|Z\|_K)^2} \tag{2.2}$$

holds for every $Z\in\mathcal{R}_K$.

The estimations are precise, but the extremal mapping is not unique.

Finally, we prove that (2.1) and (2.2) are equivalent, that is, (2.1) and (2.2) can imply each other. In other words, we have two different methods, one is Taishun Liu's in [5] and the other is that we use in this paper, they can prove both (2.1) and (2.2). Actually, in the case of the domain is $\mathcal{R}_{IV}$, Taishun Liu already proved that (2.1) implies (2.2), and it is easy to prove that (2.2) implies (2.1) in the same way as he used. So, in this paper, we only need to prove the theorem in the cases that the domains are $\mathcal{R}_I$, $\mathcal{R}_{II}$ and $\mathcal{R}_{III}$, and then prove that (2.1) and (2.2) are equivalent in these domains.

§3. Some Lemmas

To prove the theorem, we need some lemmas.

Lemma 3.1. *Let $\Phi(t)$ be the holomorphic mappings which map $\mathcal{R}_K$ into $\mathcal{R}_K$ and $\Phi(0)=0$. Then*

$$\|\Phi(Z)\|_K\le\|Z\|_K,\quad K=I,II,III,IV,$$

holds for every $Z\in\mathcal{R}_K$.

This lemma can be deduced from Schwartz Lemma of the normal linear space by Lawrence A. Harris[8]. Using the Schwartz Lemma, we have

Lemma 3.2. *Let $W = f(Z)$ be the normalized biholomorphic starlike mappings on $\mathcal{R}_K$ which map $\mathcal{R}_K$ to $C^{\dim \mathcal{R}_K}$. Then*

$$\|f^{-1}(rW)\|_K \leq \|f^{-1}(W)\|_K$$

holds for every $Z \in \mathcal{R}_K$ and every $r \in [0,1]$.

Proof. Let $f(\mathcal{R}_K)$ denote the image of $\mathcal{R}_K$ under f. For a fixed $Z \in \mathcal{R}_K$, the subset of $\mathcal{R}_K$ is defined as

$$\mathcal{E}_a = \{Y \in \mathcal{R}_K | \ \|Y\|_K < \|Z\|_K = a\}.$$

Obviously, $\mathcal{E}_a$ is an open set. Because a holomorphic mapping is an open mapping, $f(\mathcal{E}_a)$ is also an open set and $\{f(\overline{\mathcal{E}_a})\} = \overline{\{f(\mathcal{E}_a)\}}$. By the starlike hypothesis for the mapping f, where $W = f(Z)$, we have $rW \in f(\mathcal{R}_K)$. That means $f^{-1}(rW) \in \mathcal{R}_K$ for all $0 \leq r \leq 1$. We claim that

$$rW \in f(\overline{\mathcal{E}_a}). \tag{3.1}$$

If it is true, then $f^{-1}(rW) \in \overline{\mathcal{E}_a}$, which implies the lemma holds. To show that (3.1) is true we suppose there is such an $r_0 < 1$ that $r_0 W \notin \{f(\overline{\mathcal{E}_a})\}$, i.e., $\|f^{-1}(r_0 W)\|_K > a$. We define a new mapping $K(Z)$ from $\mathcal{R}_K$ to $\mathcal{R}_K$ by $K(Z) = f^{-1}(r_0 f(Z))$. Since f is a biholomorphic starlike mapping, $K(Z)$ is well defined and is holomorphic with $K(0) = 0$. By Lemma 3.1, we have

$$\|Z\|_K \geq \|K(Z)\|_K = \|f^{-1}(r_0 W)\|_K > a.$$

This contradicts $\|f^{-1}(W)\|_K = \|Z\|_K = a$. Thus (3.1) is true. The lemma is proved.

Let $W = f(Z)$ be the biholomorphic starlike mapping on $\mathcal{R}_K$. Then $g(W) = \|f^{-1}(W)\|_K$ is a continuous function of each entry of W. For a fixed $W^0 \in f(\mathcal{R}_K)$, let $\sigma(W^0) \equiv \{W = rW^0, 0 \leq r \leq 1\}$, this is a closed set. Thus $g(W)$ is absolute continuous on $W \in \sigma(W^0)$, and then $g(W)$ is differentiable with respect to W almost everywhere. We have

Lemma 3.3. *Let $W = f(Z)$ be the biholomorphic starlike mappings on $\mathcal{R}_K$ which map $\mathcal{R}_K$ to $C^{\dim \mathcal{R}_K}$. Then $\|f^{-1}(W)\|_K$ is differentiable with respect to W almost everywhere on the $\sigma(W^0) \equiv \{W = rW^0, 0 \leq r \leq 1\}$ where $W^0 = f(Z^0)$ is a fixed point in $f(\mathcal{R}_K)$ and $K = I, II, III, IV$.*

§4. The Proof of the Growth Theorem

For $\mathcal{R}_{IV}$, (2.2) was proved by Taishun Liu[5]. Here we only give the proof of the theorem in the case that the domain is $\mathcal{R}_I$, and the proofs of the theorem are similar in the cases that the domains are $\mathcal{R}_{II}$ and $\mathcal{R}_{III}$.

Let $W \in f(\mathcal{R}_I(m,n))$ and denote $\rho^2(W\overline{W}^T)$, $\lambda(Z) = \|Z\|_I^2$. When $\lambda \circ f^{-1}$ is constrained on the segment $\sigma(W^0)$ joining the origin and $W^0 = f(Z^0)$, for a fixed $Z^0 \in \mathcal{R}_I(m,n)$, the directional derivative of $\lambda \circ f^{-1}$ along the direction $d\rho$, that is $\frac{d(\lambda \circ f^{-1})}{d\rho}$, exists almost everywhere on the segment $\sigma(W^0)$ by Lemma 3.3. By the definition of $\lambda(Z)$, $\lambda(Z)$ satisfies the following equation

$$\det(\lambda(Z)I - Z\overline{Z}^T) = 0.$$

That is,

$$\sum_{k=0}^{m}(-1)^k \mathrm{tr}^{(k)} Z\overline{Z}^T \lambda^{m-k} = 0, \tag{4.1}$$

where $\mathrm{tr}^{(k)} Z\overline{Z}^T$ is the sum of all the $k \times k$ principal minors of $Z\overline{Z}^T$. Differentiating (4.1) on both sides, we have

$$\sum_{k=0}^{m-1}(-1)^k (m-k)\mathrm{tr}^{(k)} Z\overline{Z}^T \lambda^{m-k-1} d\lambda + \sum_{k=1}^{m}(-1)^k d(\mathrm{tr}^{(k)} Z\overline{Z}^T)\lambda^{m-k} = 0.$$

Thus $d\lambda = \frac{P}{Q}$ where

$$P = \sum_{k=1}^{m}(-1)^{k+1} d(\mathrm{tr}^{(k)} Z\overline{Z}^T)\lambda^{m-k}, \tag{4.2}$$

$$Q = \sum_{k=0}^{m-1}(-1)^k (m-k)\mathrm{tr}^{(k)} Z\overline{Z}^T \lambda^{m-k-1}. \tag{4.3}$$

Obviously, P is equal to

$$2\mathrm{Re}\sum_{k=1}^{m}(-1)^{k+1}\lambda^{m-k}\sum_{i,j}\frac{\partial}{\partial \omega_{ij}}(\mathrm{tr}^{(k)} Z\overline{Z}^T) d\omega_{ij}. \tag{4.4}$$

Since $d\rho^2 = 2\mathrm{Re}(\sum_{i,j}\omega_{ij}\overline{d\omega_{ij}})$, we obtain

$$< d\lambda \circ f^{-1}, d\rho^2 > |_{W=f(Z)} = \frac{P_1}{Q} \tag{4.5}$$

where $<,>|_W$ is the inner product of contangent space of $C^{m\times n}$ at the point W, and

$$P_1 = 4\mathrm{Re}\sum_{k=1}^{m}(-1)^{k+1}\lambda^{m-k}\sum_{i,j}\frac{\partial}{\partial \omega_{ij}}(\mathrm{tr}^{(k)} Z\overline{Z}^T)\omega_{ij}. \tag{4.6}$$

It is easy to evaluate, for $1, 2, \cdots, m$,

$$\begin{aligned}
&\frac{\partial}{\partial \omega_{ij}}(\mathrm{tr}^{(k)} Z\overline{Z}^T)\\
&=\frac{\partial}{\partial \omega_{ij}}\sum_{l_1<\cdots<l_k}\begin{vmatrix}\sum_{s=1}^{k} z_{l_1 s}\bar{z}_{l_1 s}, & \cdots, & \sum_{s=1}^{k} z_{l_1 s}\bar{z}_{l_k s}\\ \vdots & \cdots, & \vdots \\ \sum_{s=1}^{k} z_{l_k s}\bar{z}_{l_1 s}, & \cdots, & \sum_{s=1}^{k} z_{l_k s}\bar{z}_{l_k s}\end{vmatrix}\\
&=\frac{\partial}{\partial \omega_{ij}}\sum_{l_1<\cdots<l_k}\sum_{p_1,\cdots,p_k}\delta^{1,\cdots,k}_{p_1,\cdots,p_k}\left(\sum_s z_{l_1 s}\bar{z}_{l_{p_1} s}\right)\left(\sum_s z_{l_2 s}\bar{z}_{l_{p_2} s}\right)\cdots\left(\sum_s z_{l_k s}\bar{z}_{l_{p_k} s}\right)\\
&=\sum_{l_1<\cdots<l_k}\sum_{p_1,\cdots,p_k}\delta^{1,\cdots,k}_{p_1,\cdots,p_k}\sum_{q=1}^{k}\left(\prod_{j=1,j\neq q}^{k}\left(\sum_s z_{l_j s}\bar{z}_{l_{p_j} s}\right)\left(\sum_s \frac{\partial z_{l_q s}}{\partial \omega_{ij}}\bar{z}_{l_{p_q} s}\right)\right),
\end{aligned} \tag{4.7}$$

where $(p_1, p_2, \cdots, p_k)$ is a permutation of $(1, 2, \cdots, k)$ and $\delta^{1,2,\cdots,k}_{p_1,\cdots,p_k} = 1$ if this permutation is even, $= -1$ if this permutation is odd, and the second summation in the right hand side of (4.7) is over all the permutations of $(1,2,\cdots,k)$, and $l_1,l_2,\cdots,l_k$ are K integers of $1,2,\cdots,m$, $1\le k\le m$.

Substituting (4.7) to (4.6), we have

$$P_1=4\mathrm{Re}\{\sum_{k=1}^{m}(-1)^{k+1}\lambda^{m-k}\sum_{i,j}\omega_{ij}$$

$$\cdot\sum_{l_1<\cdots<l_k}\sum_{p_1,\cdots,p_k}\delta^{1,\cdots,k}_{p_1,\cdots,p_k}\sum_{g=1}^{k}(\prod_{i=1,j\neq q}^{k}(\sum_s z_{l_js}\bar z_{l_{p_j}s})(\sum_s\frac{\partial z_{l_qs}}{\partial\omega_{ij}}\bar z_{l_{p_q}s}))\},\tag{4.8}$$

for a $Z\in\mathcal{R}_I(m,n)$. Let $Y=\frac{tZ}{\|Z\|_I}$ where t is a complex number with $|t|<1$. Then

$$I-Y\overline{Y}^T=I-\frac{(t)^2}{\lambda(Z)}\cdot Z\overline{Z}^T>0.$$

That means $Y=\frac{tZ}{\|Z\|_I}\in\mathcal{R}_I(m,n)$, and obviously $\|Y\|_I=|t|$, $\lambda(Y)=|t|^2$. Let $W^0=f(Y)$. By Lemma 3.2,

$$\lambda(f^{-1}(W_0))-\lambda(f^{-1}((1-r)W_0)\ge 0$$

for every $1\ge r\ge 0$, and

$$\frac{\lambda(f^{-1}(W_0))-\lambda(f^{-1}((1-r)W_0))}{r}\ge 0$$

for every $1\ge r>0$. By the definition of directional derivative

$$\left.\frac{d(\lambda\circ f^{-1})}{d\rho}\right|_{W=W_0}\ge 0,\tag{4.9}$$

if the directional derivative at $W=W^0$ exists. But by Lemma 3.3, it exists almost everywhere on the segment $\dot\sigma(W^0)$. On the other hand

$$\langle d(\lambda\circ f^{-1}),d\rho^2\rangle|_{W=W_0}=2\rho\left.\frac{d(\lambda\circ f^{-1})}{d\rho}\right|_{W=W_0},$$

hence (4.9) is equivalent to

$$\langle d(\lambda\circ f^{-1}),d\rho^2\rangle|_{W=W_0}\ge 0.$$

By (4.3),(4.5),(4.8) and (4.9), we have

$$\frac{P_2}{Q_2}\ge 0,$$

where

$$P_2=4\mathrm{Re}\left\{\sum_{k=1}^{m}\frac{(-1)^{k+1}|t|^{2(m-k)}|t|^{2k}}{t\|Z\|_I^{2k-1}}\sum_{i,j}\omega_{ij}\right|_{W=f(\frac{tZ}{\|Z\|_I})}\sum_{l_1<\cdots<l_k}$$

$$\left.\sum_{p_1,\cdots,p_k}\sum_{q=1}^{k}\delta^{1,\cdots,k}_{p_1,\cdots,p_k}\left[\prod_{j=1,j\neq q}^{k}(\sum_s z_{l_js}\bar z_{l_{p_j}s})(\sum_s\left.\frac{\partial z_{l_qs}}{\partial\omega_{ij}}\right|_{W=f(\frac{tZ}{\|Z\|_I})}\bar z_{l_{p_q}s})\right]\right\}.$$

and

$$Q_2=\sum_{k=0}^{m-1}\frac{(-1)^k(m-k)|t|^{2(m-1)}}{\lambda^{(k)}(Z)}\mathrm{tr}^{(k)}Z\overline{Z}^T.$$

It is easy to observe that we can rewrite P_2/Q_2 as $4|t|^2\mathrm{Re}\frac{P_3}{Q_3}$, where

$$P_3(t) = \|Z\|_I \sum_{k=1}^{m} \frac{(-1)^{k+1}\lambda(Z)^{m-k}}{t} \sum_{i,j}\omega_{ij}\Big|_{W=f(\frac{tZ}{\|Z\|_I})} \cdot \sum_{l_1<\cdots<l_k}$$

$$\sum_{p_1,\cdots,p_k} \delta^{1,\cdots,k}_{p_1,\cdots,p_k} \sum_{q=1}^{k}\left\{ \prod_{j=1,j\neq q}^{k} (\sum_s z_{l_j s}\bar z_{l_{p_j} s})(\sum_s \frac{\partial}{\partial \omega_{ij}} z_{l_q s}\Big|_{W=f(\frac{tZ}{\|Z\|_I})} \bar z_{l_{p_q} s})\right\} \tag{4.10}$$

and

$$Q_3 = \sum_{k=0}^{m-1}(-1)^k(m-k)\mathrm{tr}^{(k)} Z\overline{Z}^T\lambda(Z)^{m-k}. \tag{4.11}$$

By (4.9), $\mathrm{Re}\frac{P_3}{Q_3} \geq 0$. Obviously, $C(t) = \frac{P_3(t)}{Q_3}$ is an analytic function of t when $0 < |t| < 1$ and $\mathrm{Re}C(t) \geq 0$. We have

$$\begin{aligned}
&\lim_{t\to 0}\frac{1}{t}\omega_{ij}\left(\frac{tZ}{\|Z\|_I}\right)\\
&= \frac{\partial \omega_{ij}}{\partial t}\Big|_{t=0} = \sum_{k,l}\frac{\partial\omega_{ij}}{\partial z_{kl}}\frac{\partial z_{kl}}{\partial t}\\
&= \sum_{k,l}\delta^{ij}_{kl}\frac{z_{kl}}{\|Z\|_I} = \frac{z_{ij}}{\|Z\|_I}
\end{aligned} \tag{4.12}$$

where δ^{ij}_{kl} are Kronecker symbols and

$$\lim_{t\to 0}\frac{\partial z_{kl}}{\omega_{ij}}\Big|_{W=f(\frac{tZ}{\|Z\|_I})} = \delta^{ij}_{kl}. \tag{4.13}$$

Using (4.12) and (4.13), we get

$$\begin{aligned}
&\lim_{t\to 0} P_3(t)\\
&=\|Z\|_I\sum_{k=1}^{m}(-1)^{k+1}\lambda(Z)^{m-k}\sum_{ij}\frac{z_{ij}}{\|Z\|_I}\sum_{l_1<\cdots<l_k}\sum_{p_1,\cdots,p_k}\delta^{1,\cdots,k}_{p_1,\cdots,p_k}\cdot\\
&\quad\sum_{q=1}^{k}\left\{\prod_{j=1,j\neq q}^{k}(\sum_s z_{l_j s}\bar z_{l_{p_j} s})(\delta^{ij}_{l_q s}\bar z_{l_{p_q} s})\right\}\\
&=\sum_{k=1}^{m}(-1)^{k+1}\lambda(Z)^{m-k}\sum_{l_1<\cdots<l_k}\sum_{p_1,\cdots,p_k}\delta^{1,\cdots,k}_{p_1,\cdots,p_k}\cdot\\
&\quad\sum_{q=1}^{k}\left\{\prod_{j=1,j\neq q}^{k}(\sum_s z_{l_j s}\bar z_{l_{p_j} s})(\sum_s z_{l_q s}\bar z_{l_{p_q} s})\right\}\\
&=\sum_{k=1}^{m}(-1)^{k+1}\lambda(Z)^{m-k}\sum_{l_1<\cdots<l_k}\sum_{p_1,\cdots,p_k}\delta^{1,\cdots,k}_{p_1,\cdots,p_k}k\left\{\prod_{j=1}^{k}(\sum_s z_{l_j s}\bar z_{l_{p_j} s})\right\}\\
&=\sum_{k=1}^{m}(-1)^{k+1}\lambda(Z)^{m-k}k\mathrm{tr}^{(m-k)}Z\overline{Z}^T.
\end{aligned}$$

Thus

$$\lim_{t\to 0}\frac{P_3}{Q_3}=\frac{\sum\limits_{k=1}^{m}(-1)^{k+1}\lambda(Z)^{m-k}\mathrm{tr}^{(k)}Z\overline{Z}^T}{\sum\limits_{k=0}^{m-1}(m-k)(-1)^k\lambda(Z)^{m-k}\mathrm{tr}^{(k)}Z\overline{Z}^T}.$$

This is equal to 1 by (4.1).

The analytic function $C(t)$ of one complex variable t where t is in the unit disc of C satisfies: $C(0)=1$ and $\mathrm{Re}C(t)\ge 0$. By a theorem of analytic function with real part in the unit disc, we have

$$\frac{1-|t|}{1+|t|}\ge \mathrm{Re}C(t)\ge\frac{1+|t|}{1-|t|}. \tag{4.14}$$

Taking $t=\|Z\|_I$ in (4.14), we get

$$\frac{1-\|Z\|_I}{1+\|Z\|_I}\ge \mathrm{Re}C(\|Z\|_I)\ge\frac{1+\|Z\|_I}{1-\|Z\|_I}.$$

But it is easy to verify

$$C(\|Z\|_I)=\frac{1}{4\lambda(Z)}\langle d\lambda\circ f^{-1},d\rho^2\rangle|_{W=f(Z)}=\frac{\rho}{2\lambda(Z)}\cdot\frac{d\lambda\circ f^{-1}}{d\rho}\bigg|_{W=f(Z)}.$$

Thus the inequality

$$\frac{2\lambda(Z)(1-\sqrt{\lambda(Z)})}{(1+\sqrt{\lambda(Z)})}\le\langle d(\lambda\circ f^{-1}),d\rho^2\rangle|_{W=f(Z)}\le\frac{2\lambda(Z)(1+\sqrt{\lambda(Z)})}{(1-\sqrt{\lambda(Z)})} \tag{4.15}$$

holds almost everywhere on the segment $\sigma(W)$.

We take a sufficient small positive number $s\le 1$ such that the small hyperball

$$B(s)=\{W\in C^{m\times n}|\ \mathrm{tr}W\overline{W}^T=\mathrm{tr}(f(sZ)\overline{f(sZ)}^T)\}$$

intersects $\sigma(Z)$. The intersection point is

$$Q=\sqrt{\frac{\mathrm{tr}(f(sZ)\overline{f(sZ)}^T}{\mathrm{tr}W\overline{W}^T}}\cdot W.$$

Obviously $\rho(Q)=\rho(f(sZ))$ and (4.15) is equivalent to the following two inequalities

$$\frac{(1-\sqrt{\lambda(Z)})}{2\lambda(Z)(1+\sqrt{\lambda(Z)})}\cdot\frac{d\lambda(Z)}{d\rho}\bigg|_{Z=f^{-1}(W)}\le\frac{1}{\rho} \tag{4.16}$$

and

$$\frac{(1+\sqrt{\lambda(Z)})}{2\lambda(Z)(1-\sqrt{\lambda(Z)})}\cdot\frac{d\lambda(Z)}{d\rho}\bigg|_{Z=f^{-1}(W)}\ge\frac{1}{\rho}. \tag{4.17}$$

Integrating (4.16) with respect to $d\rho$ along the path $\sigma(W)$ from Q to W gives

$$\ln\rho(W)\ge\ln\frac{\sqrt{\lambda(Z)}(1+\sqrt{\lambda\circ f^{-1}(Q)})^2}{(1+\sqrt{\lambda(Z)})^2}+\ln\frac{\rho(Q)}{\sqrt{\lambda\circ f^{-1}(Q)}}. \tag{4.18}$$

It follows by the normalized assumption for f that

$$\rho^2(Q)=\rho^2(f(sZ))=\mathrm{tr}(f(sZ)\overline{f(sZ)}^T=s^2(\mathrm{tr}Z\overline{Z}^T+O(s)). \tag{4.19}$$

The normalized conditions shows f^{-1} is also normalized. That is $\partial z_{ij}/\partial \omega_{kl} = \delta_{kl}^{ij}$. Then we have

$$f^{-1}(Q) = f^{-1}\left(\sqrt{\frac{\operatorname{tr}(f(sZ)\overline{f(sZ)}^T)}{\operatorname{tr}W\overline{W}^T}}\cdot W\right)$$
$$= f^{-1}\left(s\left(\sqrt{\frac{\operatorname{tr}Z\overline{Z}^T}{\operatorname{tr}W\overline{W}^T}}+O(s)\right)W\right)$$
$$= s\left(\sqrt{\frac{\operatorname{tr}Z\overline{Z}^T}{\operatorname{tr}W\overline{W}^T}}W+O(s)\right). \tag{4.20}$$

Thus we have

$$\lambda\circ f^{-1}(Q) = s^2(\operatorname{tr}Z\overline{Z}^T + O(s)). \tag{4.21}$$

Substituting (4.19),(4.21) into (4.18) and letting $s\to 0$ in (4.18), we get

$$|f(Z)| = \rho(W) \geq \frac{\sqrt{\operatorname{tr}Z\overline{Z}^T}}{(1+\sqrt{\lambda(Z)})^2} = \frac{|Z|}{1+\|Z\|_I)^2} \tag{4.22}$$

holds almost everywhere.

We have almost proved the left hand side inequality of (2.2) is true almost everywhere. Based on the fact that $\rho(f(Z))$ and $\lambda(Z)$ are continuous functions of Z on $\mathcal{R}_I(m,n)$ and the measure of the excluded set is zero, we see that the left hand side inequality of (2.2) holds for all $z\in\mathcal{R}_I(m,n)$. Similarly, we can prove the right hand side inequality of (2.2) holds for all $z\in\mathcal{R}_I(m,n)$.

Using the same method, we also can prove the theorem in the case that the domains are $\mathcal{R}_K$, $K = II, III, IV$. We omit the details of the proof. Obviously, the method used by Taishun Liu[4] to prove (2.1) are different from the one used by us in this paper.

The following mappings show (2.2) is precise, and the extremal mappings are not unique.

$$\mathcal{R}_I:\ \frac{1}{(1-z_n)}Z,\quad Z[I-\begin{pmatrix}Z\\0\end{pmatrix})]^{-2};$$
$$\mathcal{R}_{II}:\ \frac{1}{(1-z_n)}Z, Z=Z^T;\quad Z[I-Z]^{-2}, Z=Z^T;$$
$$\mathcal{R}_{IV}:\ \frac{1}{(1-z_n)}Z, Z=-Z^T;\quad Z[I-AZ]^{-2}, Z=Z^T;$$

where $\begin{pmatrix}Z\\0\end{pmatrix}$ is a square matrix and

$$A=\begin{pmatrix}0&1\\-1&0\end{pmatrix}\dotplus\cdots\dotplus\begin{pmatrix}0&1\\-1&0\end{pmatrix}\text{ if } n \text{ is even,}$$
$$A=\begin{pmatrix}0&1\\-1&0\end{pmatrix}\dotplus\cdots\dotplus\begin{pmatrix}0&1\\-1&0\end{pmatrix}\dotplus 0\text{ if } n \text{ is odd.}$$

§5. The Equivalence of Two Growth Theorems for Starlike Mappings on Classical Domains

Now we prove (2.1) and (2.2) are equivalent. As did in the previous section, we only prove this conclusion in the case that the domains is $\mathcal{R}_I$. And the proofs for the other two cases are similar.

If $\lambda(W)$ is the largest eigenvalue of $W\overline{W}^T$, then $\|W\|_I^2 = \lambda(W)$, where $W = f(Z)$, $Z \in \mathcal{R}$. $\lambda(Z)$ satisfies the equation

$$\det(\lambda(W)I - W\overline{W}^T) = 0,$$

which is equivalent to

$$\sum_{k=0}^{m}(-1)^k\lambda^{m-k}\mathrm{tr}^{(k)}W\overline{W}^T = 0, \tag{5.1}$$

where $\mathrm{tr}^{(0)}W\overline{W}^T = \mathrm{tr}W\overline{W}^T$ and $\mathrm{tr}^{(m)}W\overline{W}^T = \det W\overline{W}^T$. Differentiating on both sides of (5.1), we have

$$\left\{\sum_{k=0}^{m-1}(-1)^k(m-k)\lambda^{m-k-1}\mathrm{tr}^{(k)}W\overline{W}^T\right\}d\lambda + \sum_{k=1}^{m}(-1)^k\lambda^{m-k}d\mathrm{tr}^{(k)}W\overline{W}^T = 0. \tag{5.2}$$

We can easily verify

$$\langle d(\mathrm{tr}^{(k)}W\overline{W}^T), d\rho\rangle = \frac{2k}{\rho}\mathrm{tr}^{(k)}W\overline{W}^T \tag{5.3}$$

because of $d\rho \doteq \frac{1}{2\rho}\sum_{ij}(\omega_{ij}d\bar{\omega}_{ij} + \overline{\omega_{ij}}d\omega_{ij})$. From (5.2) and (5.3), we get

$$\left\{\sum_{k=0}^{m-1}(-1)^k(m-k)\lambda^{m-k-1}\mathrm{tr}^{(k)}W\overline{W}^T\right\}\langle d\lambda, d\rho\rangle = \frac{-2}{\rho}\sum_{k=1}^{m}(-1)^k k\lambda^{m-k}\mathrm{tr}^{(k)}W\overline{W}^T. \tag{5.4}$$

Using (5.1), we have

$$\begin{aligned}
&\sum_{k=0}^{m-1}(-1)^k(m-k)\lambda^{m-k-1}\mathrm{tr}^{(k)}W\overline{W}^T\\
=&\frac{m}{\lambda}(-1)^{m+1}\det W\overline{W}^T - \left\{\sum_{k=1}^{m-1}(-1)^k k\lambda^{m-k-1}\mathrm{tr}^{(k)}W\overline{W}^T\right\}\\
=&-\frac{1}{\lambda}\sum_{k=1}^{m}(-1)^k k\lambda^{m-k}\mathrm{tr}^{(k)}W\overline{W}^T.
\end{aligned}$$

Substituting the previous equality into (5.4), we obtain

$$\langle d\lambda, d\rho\rangle = \frac{2\lambda}{\rho}.$$

That is $\langle d\|Z\|_I, d\rho\rangle = \frac{\|Z\|_I}{\rho}$. Thus

$$\frac{d\|Z\|_I}{\|Z\|_I} = \frac{d\rho}{\rho} \tag{5.5}$$

holds.

As we did in proving that (1.3) and (1.4) are equivalent, we integrate (5.5) on both sides from $W = f(\epsilon Z)$ to $W = f(Z)$ where ϵ is a small positive number, and then let $\epsilon \to 0$.

According to the definition of $\|Z\|_{I,\rho}$ and the normalization conditions of f, we immediately see that (2.1) and (2.2) are equivalent in the case that the domain is $\mathcal{R}_I$.

Similarly, we can prove

$$\frac{d\|Z\|_K}{\|Z\|_K}=\frac{d\rho}{\rho} \tag{5.6}$$

where $\rho=(\sum_{ij}\omega_{ij}\overline{\omega_{ij}})^{1/2}$, $W=f(Z)$, $Z\in\mathcal{R}_K$ if $K=II$ or $K=III$.

Using (5.6) we can also prove that (2.1) and (2.2) are equivalent in the cases that the domains are $\mathcal{R}_{II}$ and $\mathcal{R}_{III}$.

In the case that the domain is $\mathcal{R}_{IV}$, Taishun Liu[5] already proved that (2.1) implies (2.2), but it is easy to prove that (2.2) implies (2.1) if we use the same process.

§6. The Growth Theorems of Convex Mappings on Several Complex Variables

In 1989, Ted Suffridge, Caroly Thomas and Taishun Liu, using three different methods, independently proved the following Growth Theorem of normalized convex biholomorphic mappings on the unit ball in C^n.

Let $f(Z)$ be the normalized convex biholomorphic mapping on unit ball B^n in C^N. Then

$$\frac{|Z|}{1+|Z|}\le|f(Z)|\le\frac{|Z|}{1-|Z|} \tag{6.1}$$

holds for every $X\in B^n$. The estimation is precise, but the extremal mapping is not unique. As a consequence, $f(B^n)\supset\frac{1}{2}B^n$, that is, the Koebe constant of this family of mappings is $\frac{1}{2}$.

In 1989, Taishun Liu and Sheng Gong[9] proved one side of Growth Theorem of normalized convex biholomorphic mappings on the Reinhardt domains B_p, $B_p\equiv\{Z\in C^n|\ \|Z\|_p<1, p\ge 1\}$, in p-norm.

Let $f(Z)$ be the normalized convex biholomorphic mapping on the Reinhardt domains B_p. Then

$$\|f(Z)\|_p\le\frac{\|Z\|_p}{1-\|Z\|_p} \tag{6.2}$$

holds for all $Z\in B_p$. The estimation is precise, but the extremal mapping is not unique.

Now we can use equality (1.5) to get

$$|f(Z)|\le\frac{|Z|}{1-\|Z\|_p} \tag{6.3}$$

from (6.2), and conversely, we can get (6.2) from (6.3).

As the other part of his doctor thesis, Taishun Liu[5] got the Growth Theorems of normalized convex biholomorphic mappings on classical domains.

Let $f(Z)$ be the normalized convex biholomorphic mapping on classical domains $\mathcal{R}_K(I,II,III,IV)$. Then

$$\frac{\|Z\|_K}{1+\|Z\|_K}\le\|f(Z)\|_K\le\frac{\|Z\|_K}{1-\|Z\|_K} \tag{6.4}$$

holds for every $Z \in \mathcal{R}_K$. The estimation is precise, but the extremal mapping is not unique. As a consequence, $f(\mathcal{R}_K) \supset \frac{1}{2}\mathcal{R}_K$, that is, the Koebe constant of this family of mappings is $\frac{1}{2}$.

Now we can use equalities (5.5) and (5.6) to get

$$\frac{|Z|}{1+\|Z\|} \le |f(Z)| \le \frac{|Z|}{1-\|Z\|} \tag{6.5}$$

from (6.4) if $K = I, II, III$ and in the case that the domain is $\mathcal{R}_{IV}$, Taishun Liu already proved (6.4). Conversely, we can get (6.4) from (6.5).

References

[1] Cartan, H., Sur La possibilityé d'étendre aux fonctions de plusieurs variable complexes la theoarié de fonction univalentes, Note added to P. Montel, Lecons sur les fonctions Univalentes ou multivalents,, Gauthier-Villars, Paris, 1933.

[2] FitzGerald, Carl H., Gong Sheng & Barnard, Roger W., The Growth and 1/4-theorems for starlike mappings in C^n, *Chinese Science Bulletin*, **34**:3 (1988), 161-162.

[3] Yu Qihuang, Wang Shikun & Gong Sheng, The Growth and 1/4-theorems for starlike mappings in B^p, *Chinese Annals of Mathematics*, **11B**:1 (1990), 100-104.

[4] Pfaltzgraff, J.A., Personal communication.

[5] Liu Taishun, The Growth and 1/4-theorems for starlike mappings on classical domains, Doctor Thesis, 1989.

[6] Hua, L.K., Harmonic analysis of functions of several complex variables in the classical domain, Science Press, Peking, 1958, Amer. Math. Soc, Transl. Math. Monograph, 1963.

[7] Harris, L.A., Bounded symmetric homogeneous domains in infinite-dimensional space, *Lecture Notes in Math.*, **364**, Springer, Berlin, 1974, 13-40.

[8] Harris, L.A., Schwartz's lemma in normal linear spaces, *Proc. Natl. Acd. Sci. U.S.A.*, **64(4)**:9 (1969), 1014-1017.

[9] Liu Taishun & Gong Sheng, The Growth theorem for convex mappings in B_p, (to appear in *Chinese Quarterly Journal*).

TRANSACTIONS OF THE
AMERICAN MATHEMATICAL SOCIETY
Volume 344, Number 2, August 1994

A DISTORTION THEOREM FOR BIHOLOMORPHIC MAPPINGS IN $\mathbb{C}^2$

ROGER W. BARNARD, CARL H. FITZGERALD, AND SHENG GONG

ABSTRACT. Let J_f be the Jacobian of a normalized biholomorphic mapping f from the unit ball B^2 into $\mathbb{C}^2$. An expression for the $\log\det J_f$ is determined by considering the series expansion for the renormalized mappings F obtained from f under the group of holomorphic automorphisms of B^2. This expression is used to determine a bound for $|\det J_f|$ and $|\arg\det J_f|$ for f in a compact family X of normalized biholomorphic mappings from B^2 into $\mathbb{C}^2$ in terms of a bound $C(X)$ of a certain combination of second-order coefficients. Estimates are found for $C(X)$ for the specific family X of normalized convex mappings from B^2 into $\mathbb{C}^2$.

1. INTRODUCTION

Distortion theorems for families of univalent functions have been studied at least since 1907 when Köbe discovered his classical "Verzerrungsatz": the distortion theorem for the class of univalent functions defined on the unit disc in the complex plane $\mathbb{C}$. Köbe's theorem gives explicit upper and lower bounds on $|f'(z)|$ in terms of $|z|$ for any function f that is one-to-one and analytic on the unit disc and normalized by $f(0) = 0$ and $f'(0) = 1$. The term distortion arises from the geometric interpretation of $|f'(z)|$ as the infinitesimal magnification factor of arc length and the interpretation of the square of $|f'(z)|$ as the infinitesimal magnification factor of area.

In P. Montel's book on univalent function theory, Henri Cartan wrote an appendix titled "Sur la possibilité d'éntendre aux fonctions de plusieurs variables complexes la théorie des fonctions univalentes" in which he called for a number of generalizations of properties of univalent functions in one variable to biholomorphic mappings in several variables. He specifically cited the special classes of starlike and convex mappings as appropriate topics for generalization. (We have followed his suggestions concerning starlike mappings elsewhere [1].) Cartan indicated particular interest in the properties of the determinant of the complex Jacobian of univalent mappings (the square of the magnitude of the determinant of the complex Jacobian is the infinitesimal magnification factor of volume in C^n). He stated a "théorème présumé" that the magnitude of the determinant of the Jacobian of a normalized biholomorphic mapping would have a finite upper and a positive lower bound depending only on $|z| = r < 1$ and indicated the real interest and merit in determining these bounds. That his conjecture did not hold was probably known for some time. A discussion of

Received by the editors March 13, 1989 and, in revised form, November 2, 1993.
1991 *Mathematics Subject Classification*. Primary 32H99.

Cartan's note and a simple class of counterexamples to his conjecture appears in [4]. We include the following counterexample. For any positive integer k, let $f(z) = (f_1(z), f_2(z))$ with

$$(1) \quad \begin{cases} f_1(z) = z_1, \\ f_2(z) = z_2(1 - z_1)^{-k} = z_2 + kz_2z_1 + \cdots. \end{cases}$$

Then f is a normalized biholomorphic function on the unit ball B^2 in $\mathbb{C}^2$, that is, $f(0) = 0$, and the Jacobian of f at the origin is the identity matrix. The Jacobian of f is given by

$$J_f = \begin{pmatrix} 1 & 0 \\ \dfrac{kz_2}{(1-z_1)^{k+1}} & \dfrac{1}{(1-z_1)^k} \end{pmatrix}.$$

Thus $|\det J_f| = |1 - z_1|^{-k}$ giving

$$\max_{|z|\le r} |\det J_f| = (1-r)^{-k} \to \infty \quad \text{as } k \to \infty$$

and

$$\min_{|z|\le r} |\det J_f| = (1+r)^{-k} \to 0 \quad \text{as } k \to \infty.$$

We note that the counterexample also applies for z in the polydisc $\{(z_1, z_2): |z_1| < 1, |z_2| < 1\}$ as well as the unit ball. We also note from (1) that there is no bound on the magnitude of second-order coefficients of normalized univalent functions in the polydisc or in B^2. The lack of a bound on these coefficients shows that the usual proofs of one variable of Köbe's distortion theorem do not extend to $\mathbb{C}^n$ for $n \ge 2$. In one dimension, the usual proofs use the bound of the magnitude of the second coefficient for normalized univalent functions and the invariance of the full class of univalent functions under Möbius transformations.

That Cartan's conjectured distortion theorem does not hold in the class of normalized biholomorphic mappings in $\mathbb{C}^n$, $n \ge 2$, suggests the problem of finding a subset for which the conjecture does hold. We are thus led to consider the class of normalized convex mappings which Cartan had already suggested for study. For this class the coefficients of second-order terms are of bounded magnitude. With considerable alterations, we will extend the one variable proof to obtain a distortion theorem in $\mathbb{C}^2$.

As mentioned, the invariance of the full class of univalent functions under Möbius transformations has been a powerful tool in obtaining theorems and other results concerning the subclass of normalized univalent functions. Also in several variables, invariance has played an important role. In [7, p. 278], W. Rudin introduced the term "$\mathcal{M}$-invariant space" to mean a space of holomorphic mappings on B^2 in $\mathbb{C}^2$, which are invariant under holomorphic automorphisms of the ball B^2 in $\mathbb{C}^2$. That is, if a mapping f on the ball belongs to the space, then f composed with ψ belongs to the space for every ψ in $\mathrm{Aut}(B^2)$. Our main theorem will be for compact subsets of M-invariant spaces of biholomorphic mappings on the ball B^2 in $\mathbb{C}^2$. The M-invariance of convexity is used to obtain estimates in the subclass of normalized convex mappings of B^2.

Characterizations of convex mappings in $\mathbb{C}^n$ have been obtained by Suffridge in [8] and later by Kikuchi in [6]. Suffridge characterizes convex mappings both on the unit ball B^n as well as the polydisc, $E = \{z \in \mathbb{C}^n : |z_1| < 1, |z_2| < 1, \dots, |z_n| < 1\}$. He also proved the rather surprising result that a holomorphic mapping from the polydisc E, a reducible domain, maps onto a convex domain in $\mathbb{C}^n$ if and only if there exist univalent functions f_k $(1 \le k \le n)$ from the unit disc in the plane onto convex domains in the plane such that $f(z) = T(f_1(z_2), f_2(z_2), \dots, f_n(z_n))$ where T is a nonsingular linear transformation. Hence the standard distortion results follow in a straightforward manner in this case. However, in the unit ball B^2, an irreducible domain, this type of reduction does not occur and thus our interest in the ball for this class. Indeed, E. Cartan has shown that every bounded, transitive domain in $\mathbb{C}^2$ is biholomorphically equivalent to either a polydisc or the ball; thus, once we obtain our estimates for the case of the ball, the problem is essentially solved for all bounded transitive domains in $\mathbb{C}^2$.

2. Main results

We now indicate the technique for estimation of the magnitude of the logarithm of the determinant of the Jacobian of a mapping f at a. The rate of change of this magnitude with respect to the components of the vector a is bound by certain combinations of the second-order coefficient of a particular new mapping F. This new mapping is obtained by composing f with a holomorphic automorphism of the ball and then renormalizing. We are then able to give explicit bounds for the logarithm of the determinant of the Jacobian in terms of bounds on this combination of the second-order coefficients, provided these bounds exist. In §4 we determine an estimate for these bounds for the class K of normalized convex mappings on B^2. In the final section we make a conjecture as to sharp bounds and give a supporting example. We concentrate our considerations to mappings from B^2 into $\mathbb{C}^2$ noting that extensions to B^n can be made in principle albeit with even more formable computations.

For an explicit statement of our main result we let points in $\mathbb{C}^2$ be denoted by row vectors $z = (z_1, z_2)$ for z_1, z_2 in $\mathbb{C}$. A prime on a vector will indicate the transpose. The origin $(0, 0)$ is sometimes denoted by 0. Mappings f in $\mathbb{C}^2$ are given as row vectors $f = (f_1, f_2)$ where each function f_i is a function in $\mathbb{C}^2$. The Jacobian of a holomorphic mapping f is the matrix with its ijth entry $\partial f_i/\partial z_j$ and is denoted by J or J_f. A holomorphic mapping f is normalized means that $f(0) = 0$ and $J_f(0) = I$ the identity matrix. A mapping f can be normalized by application of a complex affine transformation: $[f(z) - f(0)]J'^{-1}$ where J is evaluated at the origin. Let f be defined by $f(z) = (f_1(z), f_2(z))$ with

$$(2)\qquad \begin{aligned} f_1(z) &= z_1 + d^{(1)}_{2,0} z_1^2 + d^{(1)}_{1,1} z_1 z_2 + d^{(1)}_{0,2} z_2^2 + \cdots, \\ f_2(z) &= z_2 + d^{(2)}_{2,0} z_1^2 + d^{(2)}_{1,1} z_1 z_2 + d^{(2)}_{0,2} z_2^2 + \cdots. \end{aligned}$$

Applying the result in [5] we let $\phi_a(w)$ be the Möbius transformation of B^2 onto B^2 taking $a = (a_1, a_2)$ into the origin where ϕ_a will be explicitly given in (7); then for $z = (z_1, z_2)$ fixed in B^2 and $(a_1, a_2) = (\rho z_1, \rho z_2)$, $0 < \rho \le 1$, we define G by $G(w) = f[\phi_a(w)]$ and renormalize so that the resulting normalized mapping F has the expansion

$$F(w) = (w_1, w_2) + (w_1^2, w_1 w_2, w_2^2)D + \cdots$$

where

$$D = \begin{pmatrix} d_{2,0}^{(1)}(\rho), & d_{1,1}^{(1)}(\rho), & d_{0,2}^{(1)}(\rho) \\ d_{2,0}^{(2)}(\rho), & d_{1,1}^{(2)}(\rho), & d_{0,2}^{(2)}(\rho) \end{pmatrix}'. \tag{3}$$

Then letting J denote the Jacobian of f at z we have

(4)

$$\log\det J = \log\frac{1}{(1 - z\overline{z}')^{3/2}} - 2\int_0^1 \frac{z_1(d_{2,0}^{(1)} + (1/2)d_{1,1}^{(2)}) + z_2(d_{0,2}^{(2)} + z_2(d_{0,2}^{(2)} + (1/2)d_{1,1}^{(1)})}{1 - \rho^2 z\overline{z}'}\,d\rho$$

where we have dropped the indicated dependency of the d's on ρ. Then we use a unitary transformation and (4) to obtain the main result:

Theorem. *Consider an M-invariant space S of biholomorphic mappings on the unit ball B^2 into $\mathbb{C}^2$. Let X be the mappings obtained from mappings of S by normalizing by a complex affine transformation. Suppose that X is compact. With the notation of equations* (2), *define $C(X)$ by*

$$C(X) = \sup\{|d_{2,0}^{(1)} + (1/2)d_{1,1}^{(2)}| : f \in X\}. \tag{5}$$

For any mapping f in X, let J denote its Jacobian at z. Then

$$|\log[(1 - z\overline{z}')]^{3/2}\det J| \le C(X)\log\left(\frac{1+|z|}{1-|z|}\right).$$

As a consequence we have

$$\frac{(1-|z|)^{C(X)-3/2}}{(1+|z|)^{C(X)+3/2}} < |\det J| < \frac{(1+|z|)^{C(X)-3/2}}{(1-|z|)^{C(X)+3/2}} \tag{6}$$

and

$$|\arg\det J| \le C(X)\log\left(\frac{1+|z|}{1-|z|}\right)$$

where the branch of the argument is chosen to be zero at the origin.

In §4 we determine that for $X = K$, the class of normalized convex functions in B^2,

$$C(K) < 1.761.$$

In §5 we conjecture that $C(K) = 3/2$, give a supporting example, and suggest the corresponding sharp bounds in (6) for $\mathbb{C}^n$, for each $n \ge 2$.

3. Biholomorphic mappings under the group of biholomorphic automorphisms

In this section we show how the expression (4) is obtained.

In [5] Sheng Gong and Z. Yan proved a very general result involving the covariant derivative of the Bergman metric which gives a closed form expansion for the composition of a holomorphic mapping with an automorphism of a bounded domain in $\mathbb{C}^n$. For our case the mapping is from B^2 to $\mathbb{C}^2$. We carry out the initial calculations for a (not necessarily normalized) mapping $g: B^2 \to \mathbb{C}^2$. Let $z = (z_1, z_2)$, $w = (w_1, w_2)$, and $a = (a_1, a_2) \in B^2$. (Since equation (4) is obviously true for $z = (0, 0)$, we can assume that $z \neq (0, 0)$ and $a \neq (0, 0)$ in the following calculations.) Then a Möbius transformation $\phi_a(w): B^2 \to B^2$ taking a into the origin is given by

$$z = \phi_a(w) = \frac{a - P_a z - sQ_a z}{1 - z\bar{a}'}$$

where P_a is the orthogonal projection of $\mathbb{C}^2$ onto the subspace $[a]$ generated by a, $Q_a = I - P_a$, I the identity matrix, so that Q_a is the projection onto the orthogonal complement of $[a]$ and $s = \sqrt{1 - a\bar{a}'}$ with $\bar{a}'$ denoting the conjugate transpose of the row vector a. More explicitly we can write

$$\phi_a(w) = \frac{a - w}{1 - w\bar{a}'} A \tag{7}$$

where A is the matrix defined by

$$A = \frac{\bar{a}'a + s(a\bar{a}'I - \bar{a}'a)}{a\bar{a}'}. \tag{8}$$

This gives

$$\begin{aligned} z_1 &= \frac{a_1 a\bar{a}' - a_1 w\bar{a}' - s(a\bar{a}'w_1 - w\bar{a}'a_1)}{a\bar{a}'(1 - w\bar{a}')}, \\ z_2 &= \frac{a_2 a\bar{a}' - a_2 w\bar{a}' - s(a\bar{a}'w_2 - w\bar{a}'a_2)}{a\bar{a}'(1 - w\bar{a}')}. \end{aligned} \tag{9}$$

From the mapping g which is not necessary normalized, we define a composed $G: B^2 \to \mathbb{C}^2$ by

$$G(w) = g(z) = g[\phi_a(w)] = g\left[\frac{a - w}{1 - w\bar{a}'} A\right].$$

We now compute the initial power series expansion of $G(w)$.

$$\begin{cases} \dfrac{\partial G}{\partial w_1} = \dfrac{\partial g}{\partial z_1}\dfrac{\partial z_1}{\partial w_1} + \dfrac{\partial g}{\partial z_2}\dfrac{\partial z_2}{\partial w_1}, \\ \dfrac{\partial G}{\partial w_2} = \dfrac{\partial g}{\partial z_1}\dfrac{\partial z_1}{\partial w_2} + \dfrac{\partial g}{\partial z_2}\dfrac{\partial z_2}{\partial w_2}. \end{cases} \tag{10}$$

By (9) we have

$$\begin{aligned} \frac{\partial z_1}{\partial w_1} &= \frac{s(-|a_1|^2 s - |a_2|^2 + a\bar{a}'w_2\bar{a}_2)}{a\bar{a}'(1 - w\bar{a}')^2}, \\ \frac{\partial z_2}{\partial w_1} &= \frac{s(\bar{a}_1 a_2 - \bar{a}_1 a_2 s - a\bar{a}'w_2\bar{a}_1)}{a\bar{a}'(1 - w\bar{a}')^2}, \\ \frac{\partial z_1}{\partial w_2} &= \frac{s(\bar{a}_2 a_1 - \bar{a}_2 a_1 s - a\bar{a}'w_1\bar{a}_2)}{a\bar{a}'(1 - w\bar{a}')^2}, \\ \frac{\partial z_2}{\partial w_2} &= \frac{s(-|a_2|^2 s - |a_1|^2 + a\bar{a}'w_1\bar{a}')}{a\bar{a}'(1 - w\bar{a}')^2}. \end{aligned} \tag{11}$$

Moreover,

(12)

$$\frac{\partial^2 G}{\partial w_1^2}=\frac{\partial^2 g}{\partial z_1^2}\left(\frac{\partial z_1}{\partial w_1}\right)^2+2\frac{\partial^2 g}{\partial z_1\partial z_2}\frac{\partial z_1}{\partial w_1}\frac{\partial z_2}{\partial w_1}+\frac{\partial^2 g}{\partial z_2^2}\left(\frac{\partial z_2}{\partial w_1}\right)^2+\frac{\partial g}{\partial z_1}\frac{\partial^2 z_1}{\partial w_1^2}+\frac{\partial g}{\partial z_2}\frac{\partial^2 z_2}{\partial w_1^2},$$

$$\frac{\partial^2 G}{\partial w_1\partial w_2}=\frac{\partial^2 g}{\partial z_1^2}\frac{\partial z_1}{\partial w_2}\frac{\partial z_1}{\partial w_1}+\frac{\partial^2 g}{\partial z_1\partial z_2}\left(\frac{\partial z_1}{\partial w_1}\frac{\partial z_2}{\partial w_2}+\frac{\partial z_1}{\partial w_2}\frac{\partial z_2}{\partial w_1}\right)+\frac{\partial^2 g}{\partial z_2^2}\frac{\partial z_2}{\partial w_2}\frac{\partial z_2}{\partial w_1}+\frac{\partial^2 z_1}{\partial w_1\partial w_2}\frac{\partial g}{\partial z_1}+\frac{\partial^2 z_2}{\partial w_1\partial w_2}\frac{\partial g}{\partial z_2},$$

$$\frac{\partial^2 G}{\partial w_2^2}=\frac{\partial^2 g}{\partial z_1^2}\left(\frac{\partial z_1}{\partial w_2}\right)^2+2\frac{\partial^2 z}{\partial z_1\partial z_2}\frac{\partial z_1}{\partial w_2}\frac{\partial z_2}{\partial w_2}+\frac{\partial^2 g}{\partial z_2^2}\left(\frac{\partial z_2}{\partial w_2}\right)^2+\frac{\partial g}{\partial z_1}\frac{\partial^2 z_1}{\partial w_2^2}+\frac{\partial g}{\partial z_2}\frac{\partial^2 z_2}{\partial w_2^2}.$$

From (11) we get

(13)

$$\frac{\partial^2 z_1}{\partial w_1^2}=\frac{2s\overline{a}_1(-|a_1|^2 s+w\overline{a}'a\overline{a}'-|a_2|^2-a\overline{a}'w_2\overline{a}_2)}{a\overline{a}'(1-w\overline{a}')^3},$$

$$\frac{\partial^2 z_1}{\partial w_1\partial w_2}=\frac{s\overline{a}_2(|a_1|^2-|a_2|^2+a\overline{a}'w_2\overline{a}_2-a\overline{a}'w_1\overline{a}_1-2|a_1|^2 s)}{a\overline{a}'(1-w\overline{a}')^3},$$

$$\frac{\partial^2 z_1}{\partial w_2^2}=\frac{2s\overline{a}_2^2(a_1-a_1 s-a\overline{a}'w_1)}{a\overline{a}'(1-w\overline{a}')^3},$$

$$\frac{\partial^2 z_2}{\partial w_1^2}=\frac{2s\overline{a}_1^2(a_2-a_2 s-a\overline{a}'w_2)}{a\overline{a}'(1-w\overline{a}')^3},$$

$$\frac{\partial^2 z_1}{\partial w_1\partial w_2}=\frac{s\overline{a}_1(|a_2|^2-|a_1|^2+a\overline{a}'(w_1\overline{a}'-w_2\overline{a}_2)-2|a_2|^2 s)}{a\overline{a}'(1-w\overline{a}')^3},$$

$$\frac{\partial^2 z_2}{\partial w_2^2}=\frac{2s\overline{a}_2(-|a_2|^2 s+w\overline{a}'a\overline{a}'-|a_1|^2-a\overline{a}'w_2\overline{a}_2)}{a\overline{a}'(1-w\overline{a}')^3}.$$

We expand $G(w)$ at $w=0$,

(14)

$$G(w)=G(0)+\left.\frac{\partial G}{\partial w_1}\right|_{w=0}w_1+\left.\frac{\partial G}{\partial w_2}\right|_{w=0}w_2+\frac{1}{2}\left.\frac{\partial^2 G}{\partial w_1^2}\right|_{w=0}w_1^2+\left.\frac{\partial^2 G}{\partial w_1\partial w_2}\right|_{w=0}w_1w_2+\frac{1}{2}\left.\frac{\partial^2 G}{\partial w_2^2}\right|_{w=0}w_2^2+\cdots.$$

When $w=0$ then $z=a$ and we let

(15)

$$c=s/a\overline{a}',\qquad \alpha=s|a_1|^2+|a_2|^2,$$
$$\beta=-\overline{a}_1a_2(1-s),\qquad \gamma=s|a_2|^2+|a_1|^2.$$

Then (11) and (13) imply

(16)

$$\left.\frac{\partial z_1}{\partial w_1}\right|_{w=0}=-c\alpha,\qquad \left.\frac{\partial z_1}{\partial w_2}\right|_{w=0}=-c\overline{\beta},$$
$$\left.\frac{\partial z_2}{\partial w_1}\right|_{w=0}=-c\beta,\qquad \left.\frac{\partial z_2}{\partial w_2}\right|_{w=0}=-c\gamma$$

and

$$(17)\quad \begin{cases} \left.\dfrac{\partial^2 z_1}{\partial w_1^2}\right|_{w=0} = -2c\overline{a}_1\alpha, & \left.\dfrac{\partial^2 z_1}{\partial w_1\partial w_2}\right|_{w=0} = c(-\overline{a}_2\alpha - \overline{a}_1\overline{\beta}), \\ \left.\dfrac{\partial^2 z_1}{\partial w_2^2}\right|_{w=0} = -2c\overline{a}_2\overline{\beta}, & \left.\dfrac{\partial^2 z_2}{\partial w_1^2}\right|_{w=0} = -2c\overline{a}_1\beta, \\ \left.\dfrac{\partial^2 z_2}{\partial w_1\partial w_2}\right|_{w=0} = c(-\overline{a}_1\gamma - \overline{a}_2\overline{\beta}), & \left.\dfrac{\partial^2 z_2}{\partial w_2^2}\right|_{w=0} = -2c\overline{a}_2\gamma. \end{cases}$$

By (10), (12), (15), and (16) and by letting

$$\frac{\partial g}{\partial a_1} = \left.\frac{\partial g(z)}{\partial z_1}\right|_{z=a}, \qquad \frac{\partial^2 g}{\partial a_1^2} = \left.\frac{\partial^2 g(z)}{\partial z_1^2}\right|_{z=a}, \text{ etc.},$$

we have

(18)

$$\left.\frac{\partial G}{\partial w_1}\right|_{w=0} = -c\left\{\alpha\frac{\partial g}{\partial a_1} + \beta\frac{\partial g}{\partial a_2}\right\}, \qquad \left.\frac{\partial G}{\partial w_2}\right|_{w=0} = -c\left\{\overline{\beta}\frac{\partial g}{\partial a_1} + \gamma\frac{\partial g}{\partial a_2}\right\},$$

$$\left.\frac{\partial^2 G}{\partial w_1^2}\right|_{w=0} = c^2\left[\alpha^2\frac{\partial^2 g}{\partial a_1^2} + 2\alpha\beta\frac{\partial^2 g}{\partial a_1\partial a_2} + \beta^2\frac{\partial^2 g}{\partial a_2^2}\right] - 2c\overline{a}_1\left[\alpha\frac{\partial g}{\partial a_1} + \beta\frac{\partial g}{\partial a_2}\right],$$

(19)

$$\left.\frac{\partial^2 G}{\partial w_1\partial w_2}\right|_{w=0} = c^2\left[\alpha\overline{\beta}\frac{\partial^2 g}{\partial a_1^2} + (\alpha\gamma + |\beta|^2)\frac{\partial^2 g}{\partial a_1\partial a_2} + \beta\gamma\frac{\partial^2 g}{\alpha a_2^2}\right] - c\left[(\overline{a}_1\overline{\beta} + \overline{a}_2\alpha)\frac{\partial g}{\partial a_1} + (\overline{a}_2\beta + \overline{a}_1\gamma)\frac{\partial g}{\partial a_2}\right],$$

$$\left.\frac{\partial^2 G}{\partial w_2^2}\right|_{w=0} = c^2\left[\gamma^2\frac{\partial^2 g}{\partial a_2^2} + 2\gamma\overline{\beta}\frac{\partial^2 g}{\partial a_1\partial a_2} + \overline{\beta}^2\frac{\partial^2 g}{\partial a_2^2}\right] - 2c\left[\gamma\frac{\partial g}{\partial a_2} + \overline{\beta}\frac{\partial g}{\partial a_1}\right].$$

Substitution of (18) and (19) into equation (14) gives

$$\begin{aligned} g(z) = G(w) = g(a) &+ c\left[-\alpha\frac{\partial g}{\partial a_1} - \beta\frac{\partial g}{\partial a_2}\right]w_1 + c\left[-\overline{\beta}\frac{\partial g}{\partial a_1} - \gamma\frac{\partial g}{\partial a_2}\right]w_2 \\ &+ \frac{1}{2}\left[c^2\left(\alpha^2\frac{\partial^2 g}{\partial a_1^2} + 2\alpha\beta\frac{\partial^2 g}{\partial a_1\partial a_2} + \beta^2\frac{\partial^2 g}{\partial a_2^2}\right)\right. \\ &\qquad \left. + 2c\overline{a}_1\left(-\alpha\frac{\partial g}{\partial a_1} - \beta\frac{\partial g}{\partial a_2}\right)\right]w_1^2 \\ &+ \left\{c^2\left[\alpha\overline{\beta}\frac{\partial^2 g}{\partial a_1^2} + (\alpha\gamma + |\beta|^2)\frac{\partial^2 g}{\partial a_1\partial a_2} + \beta\gamma\frac{\partial^2 g}{\partial a_2^2}\right]\right. \\ &\qquad \left. + c\left[(-\overline{a}_1\overline{\beta} - \overline{a}_2\alpha)\frac{\partial g}{\partial a_1} + (-\overline{a}_2\beta - \overline{a}_1\gamma)\frac{\partial g}{\partial a_2}\right]\right\}w_1w_2 \\ &+ \frac{1}{2}\left[c^2\left(\alpha^2\frac{\partial^2 g}{\partial a_2^2} + 2\overline{\beta}\gamma\frac{\partial^2 g}{\partial a_1\partial a_2} + \overline{\beta}^2\frac{\partial^2 g}{\partial a_1^2}\right)\right. \\ &\qquad \left. + 2c\overline{a}_2\left(-\gamma\frac{\partial g}{\partial a_2} - \overline{\beta}\frac{\partial g}{\partial a_1}\right)\right]w_2^2 + \cdots. \end{aligned} \tag{20}$$

Now we consider the special case in which $g = f$, our normalized mapping. We continue to use G with components G_1, G_2 for the composed mapping so that

$$G(w) = f\left(\frac{a-w}{1-w\overline{a}'}A\right) = (G_1(w), G_2(w)).$$

Then by (20) with c, α, β, γ still defined as in (15), we have

$$\begin{aligned} G(w) - G(0) &= (f_1(z) - f_1(a), f_2(z) - f_2(a)) \\ &= c(w_1, w_2)\begin{pmatrix} -\alpha\frac{\partial f_1}{\partial a_1} - \beta\frac{\partial f_1}{\partial a_2}, & -\alpha\frac{\partial f_2}{\partial a_1} - \beta\frac{\partial f_2}{\partial a_2} \\ -\overline{\beta}\frac{\partial f_2}{\partial a_1} - \gamma\frac{\partial f_1}{\partial a_2}, & -\overline{\beta}\frac{\partial f_1}{\partial a_1} - \gamma\frac{\partial f_2}{\partial a_2} \end{pmatrix} + \cdots \qquad (21) \\ &= c(w_1, w_2)\begin{pmatrix} -\alpha, & -\beta \\ -\overline{\beta}, & -\gamma \end{pmatrix} J' + \frac{1}{2}(w_1^2, w_1w_2, w_2^2)\left(\frac{\partial^2 G}{\partial w^2}\right)_{w=0} + \cdots \end{aligned}$$

where J is the Jacobian of $f(z)$ at $z = a$ and

$$\left(\frac{\partial^2 G}{\partial w^2}\right)_{w=0} = \begin{pmatrix} \frac{\partial^2 G_1}{\partial w_1^2}, & \frac{\partial^2 G_2}{\partial w_1^2} \\ 2\frac{\partial^2 G_1}{\partial w_1 \partial w_2}, & 2\frac{\partial^2 G_2}{\partial w_1 \partial w_2} \\ \frac{\partial^2 G_1}{\partial w_2^2}, & \frac{\partial^2 G_2}{\partial w_2^2} \end{pmatrix}_{w=0}. \qquad (22)$$

By (19), the right-hand side of (22) equals

$$c^2 M\left(\frac{\partial^2 f}{\partial a^2}\right) - 2cNJ' \qquad (23)$$

where

$$M = \begin{pmatrix} \alpha^2, & 2\alpha\beta, & \beta^2 \\ 2\alpha\overline{\beta}, & 2(\alpha\gamma + |\beta|^2), & 2\beta\gamma \\ \overline{\beta}^2, & 2\overline{\beta}\gamma, & \gamma^2 \end{pmatrix} = \begin{pmatrix} \frac{\partial^2 f_1}{\partial a_1^2}, & \frac{\partial^2 f_2}{\partial a_1^2} \\ \frac{\partial^2 f_1}{\partial a_1 \partial a_2}, & \frac{\partial^2 f_2}{\partial a_1 \partial a_2} \\ \frac{\partial^2 f_1}{\partial a_2^2}, & \frac{\partial^2 f_2}{\partial_2^2} \end{pmatrix}, \qquad (24)$$

$$N = \begin{pmatrix} \overline{a}_1\alpha, & \overline{a}_1\beta \\ \overline{a}_2 + \overline{a}_1\beta, & \overline{a}_1\gamma + \overline{a}_2\beta \\ \overline{a}_2\overline{\beta}, & \overline{a}_2\gamma \end{pmatrix}.$$

Hence we have

$$\begin{aligned} G(w) - G(0) &= c(w_1, w_2)\begin{pmatrix} -\alpha, & -\beta \\ -\overline{\beta}, & -\gamma \end{pmatrix} J' \qquad (25) \\ &\quad + \frac{c^2}{2}(w_1^2, w_1w_2, w_2^2)M\left(\frac{\partial^2 f}{\partial a^2}\right) - c(w_1^2, w_1w_2, w_2^2)NJ' + \cdots. \end{aligned}$$

Next we normalize this mapping. This step will complete the plan indicated after equation (2) of transforming the function f by a Möbius transformation and then renormalizing. Multiplying (25) on the right by $\frac{1}{c}J'^{-1}\begin{pmatrix} -\alpha, & -\beta \\ -\overline{\beta}, & -\gamma \end{pmatrix}^{-1}$, we have the normalized mapping $F(w)$ given by

$$F(w) = \frac{1}{c}(G(w) - G(0))J'^{-1}\begin{pmatrix} -\alpha, & -\beta \\ -\bar{\beta}, & -\gamma \end{pmatrix}^{-1}$$

$$(26') \qquad = (w_1, w_2) + \frac{c}{2}(w_1^2, w_1 w_2, w_2^2)M\left(\frac{\partial^2 f}{\partial a^2}\right)J'^{-1}\begin{pmatrix} -\alpha, & -\beta \\ -\bar{\beta}, & -\gamma \end{pmatrix}^{-1}$$

$$- (w_1^2, w_1 w_2, w_2^2)N\begin{pmatrix} -\alpha, & -\beta \\ -\bar{\beta}, & -\gamma \end{pmatrix}^{-1} + \cdots$$

$$= (w_1, w_2) + (w_1^2, w_1 w_2, w_2^2)D + \cdots$$

with

$$D = \begin{pmatrix} d_{2,0}^{(1)}, & d_{2,0}^{(2)} \\ d_{1,1}^{(1)}, & d_{1,1}^{(2)} \\ d_{0,2}^{(1)}, & d_{0,2}^{(2)} \end{pmatrix}$$

where each component of D depends on $a = (a_1, a_2)$.

We now will express $(\partial^2 f/\partial a^2)J'^{-1}$ in terms of the coefficients of the second-order terms and the matrices depending on the point a. From (26) we obtain

$$(27) \qquad D = \frac{c}{2}M\left(\frac{\partial^2 f}{\partial a^2}\right)J'^{-1}\begin{pmatrix} -\alpha, & -\beta \\ -\bar{\beta}, & -\gamma \end{pmatrix}^{-1} - N\begin{pmatrix} -\alpha, & -\beta \\ -\bar{\beta}, & -\gamma \end{pmatrix}^{-1}.$$

Multiplication of (27) on the left by $M^{-1}2/c$ and on the right by $\begin{pmatrix} -\alpha, & -\beta \\ -\bar{\beta}, & -\gamma \end{pmatrix}$ gives

$$\frac{2}{c}M^{-1}D\begin{pmatrix} -\alpha, & -\beta \\ -\bar{\beta}, & -\gamma \end{pmatrix} = \left(\frac{\partial^2 f}{a^2}\right)J'^{-1} - \frac{2}{c}M^{-1}N$$

which we write as

$$(28) \qquad \left(\frac{\partial^2 f}{\partial a^2}\right)J'^{-1} = \frac{2}{c}M^{-1}D\begin{pmatrix} -\alpha, & -\beta \\ -\bar{\beta}, & -\gamma \end{pmatrix} + \frac{2}{c}M^{-1}N.$$

A direct computation of the inverse of M shows that

$$M^{-1} = \frac{1}{(a\bar{a}')^4(1 - a\bar{a}')}\begin{pmatrix} \gamma^2, & -\beta\gamma, & \beta^2 \\ -\gamma\bar{\beta}, & -\frac{1}{2}(\alpha\gamma + |\beta|^2), & -\alpha\beta \\ \bar{\beta}^2, & -\alpha\bar{\beta}, & \alpha^2 \end{pmatrix}.$$

From (15), it follows that

$$(29) \qquad \alpha\gamma - |\beta|^2 = s(a\bar{a}')^2.$$

Using (29), one obtains

$$(30) \qquad M^{-1}N = \frac{1}{(a\bar{a}')^2\sqrt{1 - a\bar{a}'}}\begin{pmatrix} \bar{a}_2\beta - \bar{a}_1\gamma, & 0 \\ \frac{1}{2}(\bar{a}_1\bar{\beta} - \bar{a}_2\alpha), & \frac{1}{2}(\bar{a}_2\beta - \bar{a}_1\gamma) \\ 0, & \bar{a}_1\bar{\beta} - \bar{a}_2\alpha \end{pmatrix}.$$

Thus

(31)

$$\left(\frac{\partial^2 f}{\partial a^2}\right)J'^{-1} = \frac{1}{\det J}\begin{pmatrix} \frac{\partial^2 f_1}{\partial a_1^2}\frac{\partial f_2}{\partial a_2} - \frac{\partial^2 f_2}{\partial a_1^2}\frac{\partial f_1}{\partial a_2}, & -\frac{\partial^2 f_1}{\partial a_1^2}\frac{\partial f_2}{\partial a_1} + \frac{\partial^2 f_2}{\partial a_1^2}\frac{\partial f_1}{\partial a_1} \\ \frac{\partial^2 f_1}{\partial a_1\partial a_2}\frac{\partial f_2}{\partial a_2} - \frac{\partial^2 f_1}{\partial a_1\partial a_2}\frac{\partial f_1}{\partial a_2}, & -\frac{\partial^2 f_1}{\partial a_2\partial a_1}\frac{\partial f_2}{\partial a_1} + \frac{\partial^2 f_2}{\partial a_1\partial a_2}\frac{\partial f_1}{\partial a_1} \\ \frac{\partial^2 f_1}{\partial a_2^2}\frac{\partial f_2}{\partial a_2} - \frac{\partial^2 f_2}{\partial a_1^2}\frac{\partial f_1}{\partial a_2}, & -\frac{\partial^2 f_1}{\partial a_2^2}\frac{\partial f_2}{\partial a_1} + \frac{\partial^2 f_2}{\partial a_2^2}\frac{\partial f_1}{\partial a_1} \end{pmatrix}$$

and

$$M^{-1}D\begin{pmatrix}-\alpha, & -\beta\\ -\overline{\beta}, & -\gamma\end{pmatrix}=\frac{1}{(a\overline{a}')^4(1-a\overline{a}')}\begin{pmatrix}l_{11}, & l_{12}\\ l_{21}, & l_{22}\\ l_{31}, & l_{32}\end{pmatrix} \tag{32}$$

where

$$l_{11}=-\alpha\gamma^2d^{(1)}_{2,0}+\alpha\beta\gamma d^{(1)}_{1,1}-\alpha\beta^2d^{(1)}_{0,2}-\overline{\beta}\gamma^2d^{(2)}_{2,0}+|\beta^2|\gamma d^{(2)}_{1,1}-\beta|\beta^2|d^{(2)}_{0,2},$$

$$l_{12}=-\beta\gamma^2d^{(1)}_{2,0}+\beta^2\gamma d^{(1)}_{1,1}-\beta^3d^{(1)}_{0,2}-\gamma^3d^{(2)}_{2,0}+\beta\gamma^2d^{(2)}_{1,1}-\beta^2\gamma d^{(2)}_{0,2},$$

$$l_{21}=\alpha\overline{\beta}\gamma d^{(1)}_{2,0}-\frac{\alpha}{2}(\alpha\gamma+|\beta^2|)d^{(1)}_{1,1}+\alpha^2\beta d^{(1)}_{0,2}+\gamma\overline{\beta}^2d^{(2)}_{2,0}$$
$$-\frac{\beta}{2}(\alpha\gamma+|\beta|^2)d^{(2)}_{1,1}+\alpha|\beta|^2d^{(2)}_{0,2},$$

$$l_{22}=\gamma|\beta^2|d^{(1)}_{2,0}-\frac{\beta}{2}(\alpha\gamma+|\beta^2|)d^{(1)}_{1,1}+\alpha\beta^2d^{(1)}_{0,2}+\gamma^2\overline{\beta}d^{(2)}_{2,0}$$
$$-\frac{\gamma}{2}(\alpha\gamma+|\beta|^2|)d^{(2)}_{1,1}+\alpha\beta\gamma d^{(2)}_{0,2},$$

$$l_{31}=-\alpha\overline{\beta}^2d^{(1)}_{2,0}+\alpha^2\overline{\beta}d^{(1)}_{1,1}-\alpha^3d^{(1)}_{0,2}-\overline{\beta}^3d^{(2)}_{2,0}-\alpha\overline{\beta}^2d^{(2)}_{1,1}-\alpha^2\overline{\beta}d^{(2)}_{0,2},$$

$$l_{32}=-\overline{\beta}|\beta^2|d^{(1)}_{2,0}+\alpha|\beta^2|d^{(1)}_{1,1}-\alpha^2\beta d^{(1)}_{0,2}-\overline{\beta}^2\gamma d^{(2)}_{2,0}+\alpha\overline{\beta}\gamma d^{(2)}_{1,1}-\alpha^2\gamma d^{(2)}_{0,2}.$$

We now observe that

$$\frac{\partial}{\partial a_1}\log\det J=\frac{1}{\det J}\frac{\partial}{\partial a_1}\left[\frac{\partial f_1}{\partial a_1}\frac{\partial f_2}{\partial a_2}-\frac{\partial f_2}{\partial a_1}\frac{\partial f_1}{\partial a_2}\right]$$
$$=\frac{1}{\det J}\left[\frac{\partial^2 f_1}{\partial a_1^2}\frac{\partial f_2}{\partial a_2}-\frac{\partial^2 f_2}{\partial a_1^2}\frac{\partial f_1}{\partial a_2}-\frac{\partial^2 f_2}{\partial a_1\partial a_2}\frac{\partial f_2}{\partial a_1}+\frac{\partial^2 f_2}{\partial a_1\partial a_2}\frac{\partial f_1}{\partial a_1}\right].$$

The last expression equals the sum of two entries of the right side of (31), specifically the sum of the (1, 1) entry and the (2, 2) entry. Since (31) and (28) restate the same matrix, we can use the right side of (28) as expressed by (30), (32), and (33). Adding entries (1, 1) and (2, 2) we obtain

$$\frac{\partial}{\partial a_1}\log\det J=\frac{2}{(a\overline{a}')^3(1-a\overline{a}')^{3/2}}$$
$$\times\Big(-\alpha\gamma^2d^{(1)}_{2,0}+\alpha\beta\gamma d^{(1)}_{1,1}-\alpha\beta^2d^{(1)}_{0,2}-\overline{\beta}\gamma^2d^{(2)}_{2,0}+|\beta^2|\gamma d^{(2)}_{1,1}$$
$$-\beta|\beta^2|d^{(2)}_{0,2}+\gamma|\beta|^2d^{(1)}_{2,0}-\frac{\beta}{2}(\alpha\gamma+|\beta^2|)d^{(1)}_{1,1}$$
$$+\alpha\beta^2d^{(1)}_{0,2}+\gamma^2\overline{\beta}d^{(2)}_{2,0}-\frac{\gamma}{2}(\alpha\gamma+|\beta|^2)d^{(2)}_{1,1}+\alpha\beta\gamma d^{(2)}_{0,2}\Big)$$
$$+\frac{2}{a\overline{a}'(1-a\overline{a}')}\left[\overline{a}_1\gamma-\overline{a}_2\beta+\frac{1}{2}(\overline{a}_1\gamma-\overline{a}_2\beta)\right].$$

When the last expression is simplified with the use of (29), we find that

$$\frac{\partial}{\partial a_1}\log\det J=\frac{3\overline{a}_1}{1-a\overline{a}'}+\frac{2}{a\overline{a}'(1-a\overline{a}')}\left(-\gamma d^{(1)}_{2,0}+\frac{\beta}{2}d^{(1)}_{1,1}-\frac{\gamma}{2}d^{(2)}_{1,1}+\beta d^{(2)}_{0,2}\right). \tag{34}$$

The partial with respect to a_2 can be obtained in a similar fashion. In this case, the (2, 1) entry and (3, 2) entries are added as they arise in (31) and then (28). Again using (30), (32), (33) and simplifying using (29), we find that (35)

$$\frac{\partial}{\partial a_2}\log\det J=\frac{3\overline{a}_2}{1-a\overline{a}'}+\frac{2}{a\overline{a}'(1-a\overline{a}')}\left(\overline{\beta}d^{(1)}_{2,0}-\frac{\alpha}{2}d^{(1)}_{1,1}+\frac{\overline{\beta}}{2}d^{(2)}_{1,1}-\alpha d^{(2)}_{0,2}\right).$$

Since $\det J$ is holomorphic in a_1 and a_2, $\frac{\partial}{\partial \overline{a}_1}\log\det J=\frac{\partial}{\partial \overline{a}_2}\log\det J=0$. Thus equations (34) and (35) give the total differential:

$$\begin{aligned}d\log\det J&=\frac{\partial}{\partial a_1}(\log\det J)da_1+\frac{\partial}{\partial a_2}(\log\det J)da_2\\&=\frac{3(\overline{a}_1da_1+\overline{a}_2da_2)}{1-a\overline{a}'}+\frac{2}{a\overline{a}'(1-a\overline{a}')}\\&\quad\times\left[(-\gamma da_1+\overline{\beta}da_2)\left(d^{(1)}_{2,0}+\frac{1}{2}d^{(2)}_{1,1}\right)+(\beta da_1-\alpha da_2)\left(d^{(1)}_{0,2}+\frac{1}{2}d^{(1)}_{1,1}\right)\right].\end{aligned}$$

By taking the straight line path from $(0,0)$ to the fixed point (z_1,z_2) in B^2 with the point on the line being given by $\rho z=\rho(z_1,z_2)=(\rho z_1,\rho z_2)$, $0\le\rho\le 1$, i.e., $a_1=\rho z_1$, $a_2=\rho z_2(z\overline{z}'=|z_1|^2+|z_2|^2$, $\sqrt{z\overline{z}'}=|z|)$, we obtain

$$\frac{d}{d\rho}\log\det J=\frac{\partial}{\partial a_1}(\log\det J)\frac{da_1}{d\rho}+\frac{\partial}{\partial a_2}(\log\det J)\frac{da_2}{d\rho}.$$

Thus

$$\frac{d}{d\rho}(\log\det J)d\rho=\frac{\partial}{\partial a_1}(\log\det J)da_1+\frac{\partial}{\partial a_2}(\log\det J)da_2.$$

Hence,

$$\begin{aligned}\log\det J&=\int_0^1\frac{d}{d\rho}(\log\det J)\,d\rho\\&=\int_{0,0}^{(z_1,z_2)}\frac{\partial}{\partial a_1}(\log\det J)\,da_1+\frac{\partial}{\partial a_2}(\log\det J)da_2\\&=3\int_{0,0}^{(z_1,z_2)}\frac{(\overline{a}_1da_1+\overline{a}_2da_2)}{(1-a\overline{a}')}\\&\quad+2\int_{(0,0)}^{(z_1,z_2)}\frac{(d^{(1)}_{2,0}+\frac{1}{2}d^{(2)}_{1,1}(-\gamma da_1+\overline{\beta}da_2)+(d^{(2)}_{0,2}+\frac{1}{2}d^{(1)}_{1,1})(\beta da_1-\alpha da_2)}{a\overline{a}'(1-a\overline{a}')}.\end{aligned}$$

Using (15) we obtain

$$\begin{aligned}\log\det J&=3\int_0^1\frac{\rho(z\overline{z}')}{1-\rho^2z\overline{z}'}d\rho\\&\quad-2\int_0^1\rho^2z\overline{z}'\frac{[z_1(d^{(1)}_{2,0}+\frac{1}{2}d^{(2)}_{1,1})+z_2(d^{(2)}_{0,2}+\frac{1}{2}d^{(1)}_{1,1})]}{\rho^2z\overline{z}'(1-\rho^2z\overline{z}')}d\rho\\&=-\frac{3}{2}\log(1-z\overline{z}')-2\int_0^1\frac{z_1(d^{(1)}_{2,0}+\frac{1}{2}d^{(2)}_{1,1})+z_2(d^{(2)}_{0,2}+\frac{1}{2}d^{(1)}_{1,1})}{1-\rho^2z\overline{z}'}d\rho.\end{aligned}$$

Thus, we have the desired expression (36)

$$\log\det J=\log\frac{1}{(1-z\overline{z}')^{3/2}}-2\int_0^1\frac{z_1(d^{(1)}_{2,0}+\frac{1}{2}d^{(2)}_{1,1})+z_2(d^{(2)}_{0,2}+\frac{1}{2}d^{(1)}_{1,1})}{1-\rho^2z\overline{z}'}d\rho.$$

To determine an upper bound for the integrand on the right-hand side of (36) we make a rotation of B^2, i.e., a transformation of the domain by a unitary matrix U. Let $\xi = (\xi_1, \xi_2) \in B^2$ and

$$U = \begin{pmatrix} U_{11}, & U_{12} \\ U_{21}, & U_{22} \end{pmatrix}$$

be a unitary matrix. Consider $\zeta = \xi U$. The series of $f(\zeta)$ has the form

$$f(\zeta) = \zeta + (\zeta_1^2, \zeta_1\zeta_2, \zeta_2^2) \begin{pmatrix} d_{2,0}^{(1)}, & d_{2,0}^{(2)} \\ d_{1,1}^{(1)}, & d_{1,1}^{(2)} \\ d_{0,2}^{(1)}, & d_{0,2}^{(2)} \end{pmatrix} + \cdots.$$

We define g by $g(\xi) = f(\xi U)$. Then $g(\xi)\overline{U}'$ will be a new normalized convex mapping with coefficients of second-order terms being defined as follows:

$$g(\xi)\overline{U}' = \xi + (\xi_1^2, \xi_1\xi_2, \xi_2^2) \begin{pmatrix} D_{2,0}^{(1)}, & D_{2,0}^{(2)} \\ D_{1,1}^{(1)}, & D_{1,1}^{(2)} \\ D_{0,2}^{(1)}, & D_{0,2}^{(2)} \end{pmatrix} + \cdots.$$

Substitution of $\zeta = \xi U$ shows the coefficient matrices are related in the following way:

$$\begin{pmatrix} D_{2,0}^{(1)}, & D_{2,0}^{(2)} \\ D_{1,1}^{(1)}, & D_{1,1}^{(2)} \\ D_{0,2}^{(1)}, & D_{0,2}^{(2)} \end{pmatrix} = \begin{pmatrix} U_{11}^2, & U_{11}U_{12}, & U_{12}^2 \\ 2U_{11}U_{21}, & U_{11}U_{22} + U_{21}U_{12}, & 2U_{12}U_{22} \\ U_{21}^2, & U_{21}U_{22}, & U_{22}^2 \end{pmatrix}$$

$$\times \begin{pmatrix} d_{2,0}^{(1)}, & d_{2,0}^{(2)} \\ d_{1,1}^{(1)}, & d_{1,1}^{(2)} \\ d_{0,2}^{(1)}, & d_{0,2}^{(2)} \end{pmatrix} \overline{U}'.$$

From this equation and the fact that $\overline{U}'U = I$, we obtain

$$D_{2,0}^{(1)} + \frac{1}{2}D_{1,1}^{(2)} = \left(d_{2,0}^{(1)} + \frac{1}{2}d_{1,1}^{(2)}\right) U_{11} + \left(d_{0,2}^{(2)} + \frac{1}{2}d_{1,1}^{(1)}\right) U_{12}. \tag{37}$$

Since

$$\begin{bmatrix} z_1/|z| & z_2/|z| \\ -\overline{z}_2/|z| & \overline{z}_1/|z| \end{bmatrix}$$

is a unitary matrix, we can take $U_{11} = z_1/|z|$ and $U_{12} = z_2/|z|$. The hypothesis of the theorem gives that the class S is M-invariant and that X comes from A by normalization of the mappings of S. From the definition of $C(X)$ given in (5), we then have

$$C(X) \geq \{|D_{2,0}^{(1)} + (1/2)D_{1,1}^{(2)}|\}. \tag{38}$$

Using (38) and (36) in (37), we find that

$$|\log\det J(1 - z\overline{z}')] \leq 2C(X)\int_0^1 \frac{|z|\,d\rho}{1 - \rho z\overline{z}'} = C(X)\log\frac{1+|z|}{1-|z|}. \tag{39}$$

From (39), the other conclusions of the theorem follow:

$$\frac{(1-|z|)^{C(X)-3/2}}{(1+|z|)^{C(X)+3/2}} \leq |\det J| \leq \frac{(1+|z|)^{C(X)-3/2}}{(1-|z|)^{C(X)+3/2}} \tag{40}$$

and

$$|\arg\det J| \le C(X)\log\left(\frac{1+|z|}{1-|z|}\right).$$

4. Estimates for the second-order coefficients for convex mappings

We now obtain an explicit estimate of $C(X)$ for a class Cartan suggested studying: the convex mappings. We use the characterization for convex mappings obtained by Kikuchi in [6]. He showed that f given by (2) is convex in B^2 if and only if

$$\operatorname{Re}\left\{1-\bar{z}\left(\frac{df}{dz}\right)^{-1}\frac{d^2f}{dz^2}(\alpha^2)'\right\}>0 \tag{41}$$

holds for every unit vector $\alpha=(\alpha_1,\alpha_2)$ which satisfies

$$\operatorname{Re}\{\bar{z}\alpha'\}=0,$$

where $df/dz=J$, $\alpha^2=(\alpha_1^2,\alpha_1\alpha_2,\alpha_2\alpha_1,\alpha_2^2)$ and

$$\frac{d^2f}{dz^2}=\begin{pmatrix}\dfrac{\partial^2 f_1}{\partial z_1^2} & \dfrac{\partial^2 f_1}{\partial z_1\partial z_2} & \dfrac{\partial^2 f_1}{\partial z_2\partial z_1} & \dfrac{\partial^2 f_1}{\partial z_2^2}\\[2ex] \dfrac{\partial^2 f_2}{\partial z_1^2} & \dfrac{\partial^2 f_2}{\partial z_1\partial z_2} & \dfrac{\partial^2 f_2}{\partial z_2\partial z_1} & \dfrac{\partial^2 f_2}{\partial z_2^2}\end{pmatrix}.$$

Using the expansion of f in (2), (41) is equivalent to

$$\operatorname{Re}\left\{1-\bar{z}\begin{pmatrix}2d_{2,0}^{(1)} & d_{1,1}^{(1)} & d_{1,1}^{(1)} & 2d_{0,2}^{(1)}\\ 2d_{2,0}^{(2)} & d_{1,1}^{(2)} & d_{1,1}^{(2)} & 2d_{0,2}^{(2)}\end{pmatrix}\begin{pmatrix}\alpha_1^2\\ \alpha_1\alpha_2\\ \alpha_2\alpha_1\\ \alpha_2^2\end{pmatrix}+\cdots\right\}>0. \tag{42}$$

If we let b_1 and b_2 be positive real numbers and $\alpha=(\alpha_1,\alpha_2)$ where

$$\alpha_1=\frac{ib_1z_1}{\sqrt{b_1^2|z_1|^2+b_2^2|z_2|^2}},\qquad \alpha_2=\frac{ib_2z_2}{\sqrt{b_1^2|z_1|^2+b_2^2|z_2|^2}},$$

then $\operatorname{Re}\{\bar{z}\alpha'\}=0$. We find by (42) that

$$\begin{aligned}\operatorname{Re}\Bigg\{1+&\left(\frac{1}{b_1^2|z_1|^2+b_2^2|z_2|^2}\right)\bar{z}\\ &\times\begin{pmatrix}2d_{2,0}^{(1)} & d_{1,1}^{(1)} & d_{1,1}^{(1)} & 2d_{0,2}^{(1)}\\ 2d_{2,0}^{(2)} & d_{1,1}^{(2)} & d_{1,1}^{(2)} & 2d_{0,2}^{(1)}\end{pmatrix}\begin{pmatrix}b_1^2z_1^2\\ b_1b_2z_1z_2\\ b_2b_1z_1z_2\\ b_2^2z_2^2\end{pmatrix}+\cdots\Bigg\}>0.\end{aligned} \tag{43}$$

We represent $z=(z_1,z_2)$ as $\xi\eta$ where ξ is a complex scalar of magnitude less than one and $\eta=(\eta_1,\eta_2)$ satisfies

$$|\eta_1|^2+|\eta_2|^2=1. \tag{44}$$

Then (43) becomes

$$\text{Re}\left\{1+\frac{\xi(\overline{\eta}_1,\overline{\eta}_2)}{b_1^2|\eta_1|^2+b_2^2|\eta_2|^2}\right. \tag{45}$$
$$\left.\times\begin{pmatrix}2d_{2,0}^{(1)} & d_{1,1}^{(1)} & d_{1,1}^{(1)} & 2d_{0,2}^{(1)}\\ 2d_{2,0}^{(2)} & d_{1,1}^{(2)} & d_{1,1}^{(2)} & 2d_{0,2}^{(2)}\end{pmatrix}\begin{pmatrix}b_1^2\eta_1^2\\ b_1b_2\eta_1\eta_2\\ b_2b_1\eta_1\eta_2\\ b_2^2\eta_2^2\end{pmatrix}+\cdots\right\}>0.$$

The left-hand side of (45) is the real part of an analytic function of one complex variable ξ in the open unit disk. Consider the series expansion about the origin. The coefficient of ξ can be represented with a probability measure ν on $0\le t\le 2\pi$ as follows:

$$\frac{(\overline{\eta}_1,\overline{\eta}_2)}{b_1^2|\eta_1|^2+b_2^2|\eta_2|^2}\begin{pmatrix}2b_1^2\eta_1^2d_{2,0}^{(1)}+2b_1b_2\eta_1\eta_2d_{2,1}^{(1)}+2b_2^2\eta_2^2d_{0,2}^{(1)}\\ 2b_1^2\eta_1^2d_{2,0}^{(2)}+2b_1b_2\eta_1\eta_2d_{1,1}^{(2)}+2b_2^2\eta_2^2d_{0,2}^{(2)}\end{pmatrix}$$
$$=2\int_0^{2\pi}e^{-it}\,d\nu(t).$$

Thus we have

$$b_1^2\eta_1|\eta_1|^2d_{2,0}^{(1)}+b_1b_2|\eta_1|^2\eta_2d_{1,1}^{(1)}+b_2^2\overline{\eta}_1\eta_2^2d_{0,2}^{(1)}+b_1^2\eta_1^2\overline{\eta}_2d_{2,0}^{(2)}$$
$$+b_1b_2\eta_1|\eta_2|^2d_{1,1}^{(2)}+b_2^2\eta_2|\eta_2|^2d_{0,2}^{(2)}=(b_1^2|\eta_1|^2+b_2^2|\eta_2|^2)\int_0^{2\pi}e^{-it}\,d\nu(t). \tag{46}$$

We note here that using Suffridge's characterization [8] leads to this same equation. Let $\eta_1=|\eta_1|e^{i\theta_1}$ and $\eta_2=|\eta_2|e^{i\theta_2}$. Multiplying (46) by $\overline{\eta}_1$ and integrating with respect to $d\theta_1/2\pi$ we find that

$$|b_1^2|\eta_1|^4d_{2,0}^{(1)}+b_1b_2|\eta_1|^2|\eta_2|^2d_{1,1}^{(2)}|\le(b_1^2|\eta_1|^2+b_2^2|\eta_2|^2)|\eta_1|. \tag{47}$$

Multiplying (46) by $\overline{\eta}_2$ and integrating with respect to $d\theta_2/2\pi$ we find that

$$|b_1b_2|\eta_1|^2|\eta_2|^2d_{1,1}^{(1)}+b_2^2|\eta_2|^4d_{0,2}^{(2)}|\le(b_1^2|\eta_1|^2+b_2^2|\eta_2|^2)|\eta_2|. \tag{48}$$

From symmetry and the discussion after (37) we need only consider (47) which becomes

$$|d_{2,0}^{(1)}+(b_2/b_1)|\eta_2/\eta_1|^2d_{1,1}^{(2)}|\le\frac{1}{|\eta_1|}[1+(b_2/b_1)^2|\eta_2/\eta_1|^2]. \tag{49}$$

Let $(b_2/b_1)|\eta_2/\eta_1|^2=c$ and $|\eta_1|=x_1$. A straightforward argument determines the minimum of $\frac{1}{x}+c^2\frac{x}{1-x^2}$ for $0<x<1$. This minimum gives an explicit, but involved, upper bound $T(c)$ for $|d_{2,0}^{(1)}+cd_{1,1}^{(2)}|$ in terms of c. For our purpose we only need $T(c)$ for $c=1/2$. For

$$(b_2/b_1)|\eta_2/\eta_1|^2=1/2,$$

equation (44) shows that (49) becomes

$$|d_{2,0}^{(1)}+(1/2)d_{1,1}^{(2)}|\le\frac{1}{|\eta_1|}+\frac{1}{4}\frac{|\eta_1|}{1-|\eta_1|^2}.$$

Since the function

$$\frac{1}{x}+\frac{1}{4}\frac{x}{1-x^2}$$

is minimized at $x=\sqrt{(9-\sqrt{33})/6}$, we obtain the estimate

$$|d^{(1)}_{2,0}+(1/2)d^{(2)}_{1,1}|<1.761. \tag{50}$$

In the notation of the theorem, if we let $X=K$, the class of normalized convex functions on B^2, then (50) shows that

$$C(K)<1.761. \tag{51}$$

The theorem implies that for $z\neq 0$,

$$|\log[(\det J)(1-z\bar{z}')^{3/2}]|<1.761\log\left(\frac{1+|z|}{1-|z|}\right), \tag{52}$$

$$\frac{(1-|z|)^{0.261}}{(1+|z|)^{3.261}}<|\det J|<\frac{(1+|z|)^{0.261}}{(1-|z|)^{3.261}}, \tag{53}$$

and

$$|\arg\det J|<1.761\log\left(\frac{1+|z|}{1-|z|}\right).$$

5. A CONJECTURE AND AN EXAMPLE

Our work suggests the conjecture that if f is a normalized convex mapping on B^n, then

$$|\log[\det J(1-z\bar{z}')^{(n+1)/2}]|\leq\frac{n+1}{2}\log\left(\frac{1+|z|}{1-|z|}\right). \tag{54}$$

Inequality (54) would have as consequences that

$$\frac{1}{(1+|z|)^{n+1}}\leq|\det J|\leq\frac{1}{(1-|z|)^{n+1}} \tag{55}$$

and

$$|\arg\det J|\leq\frac{n+1}{2}\log\left(\frac{1+|z|}{1-|z|}\right). \tag{56}$$

For $n=1$, it is known that (54) is sharp and (55) is sharp but (56) is not sharp. Hence, for $n>1$, we do not expect (56) to be sharp.

For $n=2$, the conjecture (54) is of the same form and close to the proved result (52). The consequence (55) of the conjecture would be

$$(1+|z|)^{-3}\leq|\det J|\leq(1-|z|)^{-3} \tag{57}$$

which is of a simpler form but still close to the proved inequality (53).

Still considering $n=2$, another implication of the conjectured inequality (54) would be an inequality for $C(K)$. If, as z tends to zero, we consider the first-order terms in the identity (36) in light of the discussion after (37), then the equation (54) would imply that $C(K)\leq 3/2$. The example to be discussed proves that $C(K)\geq 3/2$. Hence, if the conjecture holds, then

$$C(K)=3/2. \tag{58}$$

Conversely, if (58) holds, the theorem implies the conjecture (54) for $n = 2$.

Finally we remark that for n a positive integer, the following example shows that if conjecture (54) is true, it is sharp. Similarly, the example shows that if (55) holds, then it is sharp. We first consider the case $n = 2$. Define $f = (f_1, f_2)$ by

$$(f_1, f_2) = \left(\frac{z_1}{1-z_1}, \frac{z_2}{1-z_1}\right) = (z_1 + z_1^2 + \cdots, z_2 + z_1 z_2 + \cdots). \tag{59}$$

Then $d_{2,0}^{(1)} + (1/2)d_{1,1}^{(2)} = 3/2$ and the Jacobian J for this function is given by

$$J = \begin{pmatrix} \dfrac{1}{(1-z_1)^2} & 0 \\ \dfrac{z_2}{(1-z_1)^2} & \dfrac{1}{(1-z_1)} \end{pmatrix}. \tag{60}$$

We will give two proofs that the mapping f is convex. We generated the mapping as a limit of renormalized Möbius transformations of the identity mapping. Since the full class of convex mappings is M-invariant and the subset of normalized convex mappings is compact, the limit function exists and is convex. Consider the identity mapping (z_1, z_2) on B^2 and the point $a = (a_1, 0)$ in B^2 with $0 < a_1 < 1$. If we let $\phi_a(z)$ be the Möbius transformation taking a into the origin, then from (7) ϕ_a is given by

$$\phi_a(z) = \frac{a - z}{1 - z\bar{a}'} A,$$

where in this case equation (8) shows that

$$A = \begin{pmatrix} 1 & 0 \\ 0 & \sqrt{1-a_1^2} \end{pmatrix}.$$

Then $\phi_a(z)$ can be calculated from these equations or from (9) with $w = z$ and $a = (a_1, 0)$. We find that

$$\phi_a(z) = \left(\frac{a_1 - z_1}{1 - a_1 z_1}, \frac{-z_2\sqrt{1-a_1^2}}{1 - a_1 z_1}\right).$$

The Jacobian is easily found:

$$J_\phi(0) = \begin{pmatrix} a_1^2 - 1 & 0 \\ 0 & -\sqrt{1-a_1^2} \end{pmatrix}.$$

Thus the renormalized mapping is given by

$$[\phi_a(z) - \phi_a(0)]J_\phi'^{-1}(0) = \left(\frac{a_1 - z_1}{1 - z_1 a_1} - a_1, \frac{-z_2\sqrt{1-a_1^2}}{1 - z_1 a_1}\right) \begin{pmatrix} \dfrac{-1}{1-a_1^2} & 0 \\ 0 & \dfrac{-1}{\sqrt{1-a_1^2}} \end{pmatrix}$$

$$= \left(\frac{z_1}{1 - a_1 z_1}, \frac{z_2}{1 - a_1 z_1}\right)$$

which approaches (59) as a_1 approaches 1 as claimed. By construction it is clear that the resulting function is convex. In the general case $n \geq 2$ a similar construction gives the convex mapping $f = (f_1, f_2, \ldots, f_n)$ defined by

$$(f_1(z), f_2(z), \ldots, f_n(z)) = \left(\frac{z_1}{1-z_1}, \frac{z_2}{1-z_1}, \frac{z_3}{1-z_1}, \ldots, \frac{z_n}{1-z_1}\right).$$

The Jacobian of this example is triangular; thus the determinant of the Jacobian is easily calculated. The result is

$$\det J_f(z) = \frac{z_1}{(1-z_1)^{n+1}}.$$

Consequently, if the conjectured inequalities (57) are true, they are sharp.

Returning to the case $n = 2$, we give a second proof that f given by (59) is convex. We consider the Kikuchi criterion (41) with

$$\alpha = i(A_1 z_1, A_2 z_2)/[A_1^2|z_1|^2 + A_2^2|z_2|^2]^{1/2}$$

where A_1 and A_2 are possible. The left side of (41) becomes the real part of

$$\left\{A_1^2|z_1|^2 + A_2^2|z_2|^2 + (\bar{z}_1, \bar{z}_2)\begin{pmatrix}(1-z_1)^2, & 0\\ -z_2(1-z_1), & 1-z_1\end{pmatrix}\right.$$

$$\left.\times\begin{pmatrix}\frac{2}{(1-z_1)^3}, & 0, & 0, & 0\\ \frac{2z_2}{(1-z_1)^3}, & \frac{1}{(1-z_1)^2}, & \frac{1}{(1-z_1)^2}, & 0\end{pmatrix}\begin{pmatrix}A_1^2 z_1^2\\ A_1A_2z_1z_2\\ A_2A_1z_2z_1\\ A_2^2z_2^2\end{pmatrix}\right\}.$$

Hence,

$$\operatorname{Re}\left\{A_1^2|z_1|^2 + A_2^2|z_2|^2 + \frac{2A_1^2z_1|z_1|^2 + 2A_1A_2z_1|z_2|^2}{1-z_1}\right\} \geq 0. \tag{61}$$

The left side of (61) equals

$$\operatorname{Re}\left\{A_1^2|z_1|^2\left(\frac{1+z_1}{1-z_1}\right) + A_2^2|z_2|^2 - A_1A_2|z_2|^2 + A_1A_2|z_2|^2\left(\frac{1+z_1}{1-z_1}\right)\right\}. \tag{62}$$

Using the method of completing the square and the fact that for $|z_1| < 1$

$$\operatorname{Re}\left\{\frac{1+z_1}{1-z_1}\right\} \geq \frac{1-|z_1|}{1+|z_1|},$$

one can show that the nonnegativity of (62) follows from the nonnegativity of

$$A_1^2|z_1|^2\left[\frac{1-(|z_1|^2+|z_2|^2)}{1+|z_1|}\right] + |z_2|(1+|z_1|)\left(A_2 - \frac{|z_1|}{1+|z_1|}A_1\right)^2. \tag{63}$$

The expression (63) is nonnegative for all z such that $|z_1|^2 + |z_2|^2 \leq 1$ and the result follows.

For this convex mapping, the inequality (41) for the Kikuchi criterion is sharp at essentially every point of the boundary of B^2. Consider $z = (z_1, z_2)$ on the boundary. The first term of (63) is zero. And, for any positive choice of A_1, the choice of $A_2 = A_1|z_1|/(1+|z_1|)$ makes the second term of (63) equal to zero. Except for the point $z = (1, 0)$, the steps can be reversed to see that equality holds in the Kikuchi criterion (41). Furthermore, the exceptional point (1, 0) is carried to infinity by the mapping.

This example was essentially known. Consider the Cayley transform [7, p. 31] of B^n onto an "upper half plane" in $\mathbb{C}^n$. By composing with an appropriate unitary transformation of the domain and a complex affine trnasformation of the range, a Cayley transform can be changed into a normalized mapping which is exactly our example.

6. Acknowledgments

This paper was written while the first and third authors were visiting the University of California, San Diego. The authors are grateful for that research opportunity. The authors also appreciate helpful communication with Professor T. J. Suffridge.

References

1. R. W. Barnard, C. H. Fitzgerald, and Sheng Gong, *The growth and 1/4-theorems for starlike functions in* $\mathbb{C}^n$, Pacific J. Math. **150** (1991), 13–22.
2. H. Cartan, *Sur la possibilité d'étendre sux fonctions de plusieurs variables complexes la théorie des fonctions univalentes*, Note added to P. Montel, *Leçons sur les fonctions univalentes on multivalentes*, Gauthier-Villars, Paris, 1993, pp. 129–155.
3. P. Duren, *Univalent functions*, #259, Springer-Verlag, New York, 1983.
4. P. Duren and W. Rudin, *Distortion in several variables*, Complex Analysis: Theory & Appl. **5** (1986), 323–326.
5. S. Gong and Z. M. Yan, *A remark on Möbius transformations* (III), Chinese Quart. J. **1** (1987), 33–40.
6. K. Kikuchi, *Starlike and convex mappings in several complex variables*, Pacific J. Math. 44 (1973), 569–580.
7. W. Rudin, *Function theory in the unit ball of* $\mathbb{C}^n$, Grundlehren Math. Wiss., Springer, Berlin, 1980.
8. T. J. Suffridge, *The principle of subordination applied to functions of several variables*, Pacific J. Math. **33** (1970), 241–248.
9. ——, *Starlikeness, convexity and other geometric properties of holomorphic maps in higher dimensions*, Lecture Notes in Math., vol. 599, Springer-Verlag, New York, 1976, pp. 146–159.

Department of Mathematics, Texas Tech University, Lubbock, Texas 79409-1042
E-mail address: narwb@ttacs1.ttu.edu

Department of Mathematics, University of California, San Diego, La Jolla, California 92093-0112
E-mail address: cfitzgerald@ucsd.edu

Department of Mathematics, University of Science and Technology of China, Hefei (Anhui) 230026, Peoples Republic of China

PACIFIC JOURNAL OF MATHEMATICS
Vol. 161, No. 2, 1993

BIHOLOMORPHIC CONVEX MAPPINGS OF BALL IN $\mathbb{C}^n$

SHENG GONG, SHIKUN WANG, AND QIHUANG YU

Biholomorphic convex mappings from unit ball in $\mathbb{C}^n$ into $\mathbb{C}^n$ are studied in this paper. A Schwartz type lemma for the class of mappings and a necessary and sufficient condition under which a holomorphic mapping is biholomorphic and convex are established. The results are used to describe some characteristics of the image of the class of mappings.

1. Introduction. Since Loebe discovered as early as 1907 his "Verzerrugsatz", classical distortion theorems for families of univalent functions defined the unit disc in the complex plane $\mathbb{C}$ have developed systematically in depth and scope. For several variables, H. Cartan showed his interest in the field and conjectured [**1**] that the magnitude of the determinant of the complex Jacobian of a normalized biholomorphic mapping on the unit polydisc in $\mathbb{C}^2$ should have a finite upper and a positive lower bound. But, it was pointed out in [**2**] that the conjecture is not correct. That is the distortion theorem for general biholomorphic mappings in several variables does not hold. This suggests one has to find some of their subclasses for which the distortion theorem can still hold. In 1988 S. Gong, C. H. FitzGerald and R. W. Bernard [**3**] obtained first time the upper and lower bound of the magnitude of the determinant of the Jacobian of a normalized, convex and biholomorphic mapping from the unit ball $B = \{Z = (z^1, z^2) \in \mathbb{C}^2 | |Z| < 1\}$ into $\mathbb{C}^2$. After that Taishun Liu [**4**] generalized the result to the unit ball in $\mathbb{C}^n$ and X. A. Zheng [**5**] to the bounded symmetric domains.

It is worth pointing out that the estimate of the magnitude of the determinant of the Jacobian of a holomorphic mapping is one of the generalization of distortion theorems of one variable. The estimation of eigenvalues of the Hermitian matrix, the product of the complex Jacobian of a holomorphic mapping and its transpose conjugate, should also be considered to be another form of generalization of distortion theorems of one complex variable. In this contribution we will discuss the type of distortion theorems and its application to biholomorphic convex mappings.

The paper is arranged as follows: In §2 we will prove a Schwartz type lemma of biholomorphic convex mappings from the unit ball in $\mathbb{C}^n$ into $\mathbb{C}^n$. The lemma describes the type of distortion theorems given by us. In §3 we will give a necessary and sufficient condition under which a holomorphic mapping is biholomorphic and convex. In [6, 7] two necessary and sufficient conditions which are equivalent were also given. However, in the section we would quote a counterexample by Taishun Liu in [4] to illustrate that the conditions are not sufficient. In §4, as an application of the above results, we will study the geometric characteristics of the image of biholomorphic convex mappings.

In this paper summation convention is used.

2. A Schwartz type lemma of biholomorphic convex mappings. B denotes the unit ball in $\mathbb{C}^n$, that is

$$B=\left\{Z=(z^1,\dots,z^n)\in\mathbb{C}^n \mid |Z|^2=\sum_{i=1}^{n}|z^i|^2<1\right\}.$$

When $n=1$ replace B by the symbol D. The Jacobian of a mapping f from B to $\mathbb{C}^n$ is denoted by J_f. So-called normalized holomorphic mapping means the holomorphic one with $f(0)=0$ and $J_f(0)=I$, I is the unit matrix.

In this section we demonstrate a Schwartz type lemma of biholomorphic convex mappings from B into $\mathbb{C}^n$.

THEOREM 1. *Let $f\colon B\to\mathbb{C}^n$ be a biholomorphic convex mapping. Then*

$$\left(\frac{1+|Z|}{1-|Z|}\right)^2 g_{\alpha\bar\beta}\,dz^\alpha\,d\bar z^\beta \ge f^*\delta_{ij}\,dw^i\,d\bar w^j \ge \left(\frac{1-|Z|}{1+|Z|}\right)^2 g_{\alpha\bar\beta}\,dz^\alpha\,d\bar z^\beta$$

holds, where

$$g_{\alpha\bar\beta}=\frac{(1-|Z|^2)\delta_{\alpha\beta}+\bar z^\alpha z^\beta}{(1-|Z|^2)^2}$$

is the Bergman metric of the unit ball B in $\mathbb{C}^n$ and the inequalities are sharp.

The proof of the theorem depends on the following two lemmas.

LEMMA 1. *Suppose that $F\colon B\to\mathbb{C}^n$, $W=F(\xi)$ is a biholomorphic convex mapping and $\xi=F^{-1}(W)$ is the inverse mapping of $F(\xi)$.*

Then

$$4\delta_{\alpha\beta}|a|\left.\frac{\partial\xi^\alpha}{\partial w^i}\frac{\partial\overline{\xi}^\beta}{\partial\overline{w}^j}\right|_{\xi=0}v^i\overline{v}^j$$

$$\geq 2\left|\operatorname{Re}\left(\delta_{\alpha\beta}\left.\frac{\partial^2\xi^\alpha}{\partial w^k\partial w^i}\frac{\partial\overline{\xi}^\beta}{\partial\overline{w}^j}\frac{\partial w^k}{\partial z^l}\right|_{\xi=0}a^l v^i\overline{v}^j\right)\right|$$

holds for any $a=(a^1,\ldots,a^n)\in B$ *and* $V=(v^1,\ldots,v^n)\in\mathbb{C}^n$, *where* w^i $(i=1,2,\ldots,n)$ *are components of* W *and* ξ^α $(\alpha=1,2,\ldots,n)$ *ones of* ξ.

LEMMA 2. *Suppose that* $f\colon B\to\mathbb{C}^n$, $W=f(Z)$, *is a biholomorphic convex mapping with* $f(0)=0$ *and* $Z=f^{-1}(W)$ *is the inverse mapping of* $f(Z)$. *Then*

$$4|Z|g_{\alpha\overline{\beta}}\frac{\partial z^\alpha}{\partial w^i}\frac{\partial\overline{z}^\beta}{\partial\overline{w}^j}v^i\overline{v}^j$$

$$\geq\left|2\operatorname{Re}\left\{g_{\beta\overline{\gamma}}\frac{\partial z^\alpha}{\partial w^i}\frac{\partial\overline{z}^\gamma}{\partial\overline{w}^j}\overline{z}^\alpha z^\beta v^i\overline{v}^j+|Z|^2 g_{\alpha\overline{\beta}}\frac{\partial z^\alpha}{\partial w^i}\frac{\partial\overline{z}^\beta}{\partial\overline{w}^j}v^i\overline{v}^j\right.\right.$$

$$\left.\left.+(1-|Z|^2)g_{\alpha\overline{\beta}}\frac{\partial^2 z^\alpha}{\partial w^k\partial w^i}\frac{\partial\overline{z}^\beta}{\partial\overline{w}^j}\frac{\partial w^k}{\partial z^\gamma}z^\gamma v^i\overline{v}^j\right\}\right|$$

holds for any $Z\in B$, $V=(v^1,\ldots,v^n)\in\mathbb{C}^n$.

Proof of Lemma 1. Since $F(B)$ is a convex set we have

$$\lambda F(\xi)+(1-\lambda)F(\xi_0)\in F(B)$$

for any $\lambda\in[0,1]$ and $\xi,\xi_0\in B$. Fix ξ_0 and define a holomorphic mapping ξ_λ from B to B by

$$\xi_\lambda=F^{-1}(\lambda F(\xi)+(1-\lambda)F(\xi_0)),\qquad \xi\in B.$$

From Schwartz's lemma of the unit ball in $\mathbb{C}^n$, it follows that

$$g_{\alpha\overline{\beta}}(\xi)v^\alpha\overline{v}^\beta\geq g_{\alpha\overline{\beta}}(\xi_\lambda)\frac{\partial\xi^\alpha_\lambda}{\partial\xi^k}\frac{\partial\overline{\xi}^\beta_\lambda}{\partial\overline{\xi}^l}v^k\overline{v}^l$$

or

$$(1)\quad g_{\alpha\overline{\beta}}(\xi)v^\alpha\overline{v}^\beta\geq\lambda^2 g_{\alpha\overline{\beta}}(\xi_\lambda)\frac{\partial\xi^\alpha_\lambda}{\partial w^i}(\xi_\lambda)\frac{\partial w^i}{\partial\xi^k}(\xi)\frac{\partial\overline{\xi}^\beta_\lambda}{\partial\overline{w}^j}(\xi_\lambda)\frac{\partial\overline{w}^j}{\partial\overline{\xi}^l}(\xi)v^k\overline{v}^l.$$

The inequality (1) can be rewritten as

$$g_{\alpha\overline{\beta}}(\xi)\frac{\partial\xi^\alpha}{\partial w^i}\frac{\partial\overline{\xi}^\beta}{\partial\overline{w}^j}\bigg|_{\xi} v^i\overline{v}^j - \lambda^2 g_{\alpha\overline{\beta}}(\xi_\lambda)\frac{\partial\xi^\alpha_\lambda}{\partial w^i}\frac{\partial\overline{\xi}^\beta_\lambda}{\partial\overline{w}^j}\bigg|_{\xi_\lambda} v^i\overline{v}^j \geq 0.$$

For $\lambda\in[0,1)$ we have

$$\lim_{\lambda\to 1}\left[\frac{g_{\alpha\overline{\beta}}(\xi)\frac{\partial\xi^\alpha}{\partial w^i}\frac{\partial\overline{\xi}^\beta}{\partial\overline{w}^j}\big|_{\xi} v^i\overline{v}^j - \lambda^2 g_{\partial\overline{\beta}}(\xi_\lambda)\frac{\partial\xi^\alpha_\lambda}{\partial w^i}\frac{\partial\overline{\xi}^\beta_\lambda}{\partial\overline{w}^j}\big|_{\xi_\lambda} v^i\overline{v}^j}{1-\lambda}\right]\geq 0.$$

Straightforward calculation yields

$$\begin{aligned}
0 &\leq \lim_{\lambda\to 1}\frac{d}{d\lambda}\left[\lambda^2 g_{\alpha\overline{\beta}}(\xi_\lambda)\frac{\partial\xi^\alpha_\lambda}{\partial w^i}\frac{\partial\overline{\xi}^\beta_\lambda}{\partial\overline{w}^j}\bigg|_{\xi_\lambda} v^i\overline{v}^j\right]\\
&= \lim_{\lambda\to 1} 2\left\{\lambda g_{\alpha\overline{\beta}}\frac{\partial\xi^\alpha_\lambda}{\partial w^i}\frac{\partial\overline{\xi}^\beta_\lambda}{\partial\overline{w}j}\bigg|_{\xi_\lambda} v^i\overline{v}^j\right.\\
&\qquad + \mathrm{Re}\left[\lambda^2 g_{\alpha\overline{\beta}}\frac{\partial^2\xi^\alpha_\lambda}{\partial w^k\partial w^i}\frac{\partial\overline{\xi}^\beta_\lambda}{\partial\overline{w}^j}\bigg|_{\xi_\lambda}(w^k(\xi)-w^k(\xi_0))v^i\overline{v}^j\right.\\
&\qquad \left.\left.+\lambda^2\frac{\partial\xi^\alpha_\lambda}{\partial w^i}\frac{\partial\overline{\xi}^\beta_\lambda}{\partial\overline{w}^j}\bigg|_{\xi_\lambda} v^i\overline{v}^j\frac{\partial g_{\alpha\overline{\beta}}}{\partial\xi^\gamma_\lambda}\frac{\partial\xi^\gamma}{\partial w^k}\bigg|_{\xi_\lambda}(w^k(\xi)-w^k(\xi_0))\right]\right\}.
\end{aligned}$$

Noting that $\xi_\lambda\to\xi$ as $\lambda\to 1$, we obtain at the point ξ

$$\begin{aligned}
(2)\quad & 2g_{\alpha\overline{\beta}}\frac{\partial\xi^\alpha_\lambda}{\partial w^i}\frac{\partial\overline{\xi}^\beta}{\partial\overline{w}^j}v^i\overline{v}^j\\
&+2\,\mathrm{Re}\left\{g_{\alpha\overline{\beta}}\frac{\partial^2\xi^\alpha}{\partial w^k\partial w^i}\frac{\partial\overline{\xi}^\beta}{\partial\overline{w}^j}(w^k(\xi)-w^k(\xi_0))v^i\overline{v}^j\right.\\
&\qquad\left.+\frac{\partial\xi^\alpha}{\partial w^i}\frac{\partial\overline{\xi}^\beta}{\partial\overline{w}^j}v^i\overline{v}^j\frac{\partial g_{\alpha\overline{\beta}}}{\partial\xi^\gamma}\frac{\partial\xi^\gamma}{\partial w^k}(w^k(\xi)-w^k(\xi_0))\right\}\geq 0.
\end{aligned}$$

The inequality (2) holds for any $\xi,\xi_0\in B$. Now fix ξ and take $\xi_0=\frac{t\xi}{|\xi|}$ in (2) where $t\in D$, D is the unit disc in $\mathbb{C}$. We then obtain a nonnegative harmonic function defined on the unit disc by

$$H(t) = 2g_{\alpha\overline{\beta}}\frac{\partial \xi^\alpha}{\partial w^i}\frac{\partial \overline{\xi}^\beta_\lambda}{\partial \overline{w}^j}v^i\overline{v}^j$$
$$+ 2\operatorname{Re}\left\{g_{\alpha\overline{\beta}}\frac{\partial^2 \xi^\alpha}{\partial w^k \partial w^i}\frac{\partial \overline{\xi}^\beta}{\partial \overline{w}^j}\left(w^k(\xi) - w^k\left(\frac{t\xi}{|\xi|}\right)\right)v^i\overline{v}^j\right.$$
$$\left. + \frac{\partial \xi^\alpha}{\partial w^i}\frac{\partial \overline{\xi}^\beta}{\partial \overline{w}^j}v^i\overline{v}^j\frac{\partial g_{\alpha\overline{\beta}}}{\partial \xi^\gamma}\frac{\partial \xi^\gamma}{\partial w^k}\left(2^k(\xi) - w^k\left(\frac{t\xi}{|\xi|}\right)\right)\right\}.$$

It follows from Harnack's inequality that

$$\frac{1-|t|}{1+|t|}H(0) \le H(t) \le H(0)\frac{1+|t|}{1-|t|}.$$

Now taking $t = |\xi|$, we obtain

$$\frac{1-|\xi|}{1+|\xi|}H(0) \le H(|\xi|) \le H(0)\frac{1+|\xi|}{1-|\xi|}$$

or

(3) $$\frac{1-|\xi|}{1+|\xi|}H(|\xi|) \le H(0) \le H(|\xi|)\frac{1+|\xi|}{1-|\xi|},$$

where

$$H(|\xi|) = 2g_{\alpha\overline{\beta}}\frac{\partial \xi^\alpha}{\partial w^i}\frac{\partial \overline{\xi}^\beta}{\partial \overline{w}^j}v^i\overline{v}^j$$

and

$$H(0) = 2g_{\alpha\overline{\beta}}\frac{\partial \xi^\alpha}{\partial w^i}\frac{\partial \overline{\xi}^\beta}{\partial \overline{w}^j}v^i\overline{v}^j$$
$$+ 2\operatorname{Re}\left\{g_{\alpha\overline{\beta}}\frac{\partial^2 \xi^\alpha}{\partial w^k \partial w^i}\frac{\partial \overline{\xi}^\beta}{\partial \overline{w}^j}(\xi)(w^k(\xi) - w^k(0))v^i\overline{v}^j\right.$$
$$\left. + \frac{\partial \xi^\alpha}{\partial w^i}\frac{\partial \overline{\xi}^\beta}{\partial \overline{w}^j}v^i\overline{v}^j\frac{\partial g_{\alpha\overline{\beta}}}{\partial \xi^\gamma}\frac{\partial \xi^\gamma}{\partial w^k}(\xi)(w^k(\xi) - w(0))\right\}.$$

The inequalities (3) can be rewritten as

$$\frac{4|\xi|}{1-|\xi|}g_{\alpha\overline{\beta}}\frac{\partial \xi^\alpha}{\partial w^i}\frac{\partial \overline{\xi}^\beta}{\partial \overline{w}^j}v^i\overline{v}^j$$
$$\ge 2\operatorname{Re}\left\{g_{\alpha\overline{\beta}}\frac{\partial^2 \xi^\alpha}{\partial w^k \partial w^i}\frac{\partial \overline{\xi}^\beta}{\partial \overline{w}^j}v^i\overline{v}^j(w^k(\xi) - w^k(0))\right.$$
$$\left. + \frac{\partial \xi^\alpha}{\partial w^i}\frac{\partial \overline{\xi}^\beta}{\partial \overline{w}^j}v^i\overline{v}^j\frac{\partial g_{\alpha\overline{\beta}}}{\partial \xi^\gamma}\frac{\partial \xi^\gamma}{\partial w^k}(w^k(\xi) - w^k(0))\right\}$$
$$\ge \frac{-4|\xi|}{1+|\xi|}g_{\alpha\overline{\beta}}\frac{\partial \xi^\alpha}{\partial w^i}\frac{\partial \overline{\xi}^\beta}{\partial \overline{w}^j}v^i\overline{v}^j.$$

Taking $\xi = \nu a$, $\nu \in (0, 1)$, and letting $\nu \to 0$ in the inequalities above, we have

$$(4)\qquad 4|a|\delta_{\alpha\bar{\beta}}\frac{\partial \xi^\alpha}{\partial w^i}\frac{\partial \bar{\xi}^\beta}{\partial \overline{w}^j}v^i\bar{v}^j\bigg|_{\xi=0} \geq \left|2\operatorname{Re}\left(\delta_{\alpha\bar{\beta}}\frac{\partial^2 \xi^\alpha}{\partial w^k \partial w^i}\frac{\partial \bar{\xi}^\beta}{\partial \overline{w}^j}\frac{\partial w^k}{\partial \xi^l}a^l v^i \bar{v}^j\right)\right|\Bigg|_{\xi=0}.$$

Lemma 1 is completely proved.

Proof of Lemma 2. Let $\varphi_a\colon B \to B$ be a holomorphic automorphism defined by

$$\varphi_a(\xi) = \frac{a-\xi}{1-\xi\bar{a}'}A,$$

where $A = sI + \frac{\bar{a}'a}{1+s}$, $s = \sqrt{1-|a|^2}$, $a \in B$. Denote $Z = \varphi_a(\xi)$. The inverse mapping of φ_a is as follows:

$$\xi = \varphi^{-1}(Z) = \frac{a-Z}{1-Z\bar{a}'}A.$$

Let $F(\xi)$ in Lemma 1 be $f(\varphi_a(\xi))$. Straightforward calculation gives

$$(5)\qquad \begin{aligned}\frac{\partial^2 \xi^\alpha}{\partial w^k \partial w^i}\bigg|_{\xi=0} &= \frac{\partial^2 \xi^\alpha}{\partial z^\mu \partial z^\nu}\frac{\partial z^\mu}{\partial w^k}\frac{\partial z^\nu}{\partial w^i} + \frac{\partial \xi^\alpha}{\partial z^\nu}\frac{\partial^2 z^\nu}{\partial w^k \partial w^i}\\ &= \frac{\bar{a}^\mu A^\alpha_\nu + \bar{a}^\nu A^\alpha_\mu}{(1-|a|^2)^2}\frac{\partial z^\nu}{\partial w^k}\frac{\partial z^\mu}{\partial w^i} - \frac{A^\alpha_\mu}{1-|a|^2}\frac{\partial^2 z^\mu}{\partial w^k \partial w^i},\\ \frac{\partial \bar{\xi}^\beta}{\partial \overline{w}^j}\bigg|_{\xi=0} &= -\frac{\overline{A}^\beta_\sigma}{1-|a|^2}\frac{\partial \bar{z}^\sigma}{\partial \overline{w}^j},\\ \frac{\partial w^k}{\partial \xi^l}\bigg|_{\xi=0} &= (\bar{a}^l a^\tau - A^\tau_l)\frac{\partial w^k}{\partial z^\tau}.\end{aligned}$$

Substituting (5) into the right side in (4) we have at $\xi = 0$ $(Z = a)$

$$(6)\qquad \begin{aligned}&2\operatorname{Re}\left(\delta_{\alpha\bar{\beta}}\frac{\partial^2 \xi^\alpha}{\partial w^k \partial w^i}\frac{\partial \bar{\xi}^\beta}{\partial \overline{w}^j}\frac{\partial w^k}{\partial \xi^l}\bigg|_{\xi=0} a^l v^i \bar{v}^j\right)\\ &= -2\operatorname{Re}\left\{(\bar{a}^\mu g_{\nu\bar{\sigma}} + \bar{a}^\nu g_{\mu\bar{\sigma}})a^\nu \frac{\partial z^\mu}{\partial w^i}\frac{\partial \bar{z}^\sigma}{\partial \overline{w}^j}v^i\bar{v}^j\right.\\ &\qquad \left. + (1-|a|^2)g_{\mu\bar{\sigma}}\frac{\partial^2 z^\mu}{\partial w^k \partial w^j}\frac{\partial \bar{z}^\sigma}{\partial \overline{w}^\gamma}\frac{\partial w^k}{\partial z^\tau}v^i\bar{v}^j\right\}.\end{aligned}$$

Notice that the metric $(g_{\alpha\overline{\beta}})_{1\le\alpha,\beta\le n}$ on B is invariant under its holomorphic automorphism. That is to say

$$(7)\qquad \delta_{\alpha\beta}\frac{\partial\xi^\alpha}{\partial w^i}\frac{\partial\overline{\xi}^\beta}{\partial\overline{w}^j}\bigg|_{\xi=0}v^i\overline{v}^j = g_{\alpha\overline{\beta}}(a)\frac{\partial z^\alpha}{\partial w^i}\frac{\partial\overline{z}^\beta}{\partial\overline{w}^j}\bigg|_{Z=a}v^i\overline{v}^j.$$

Combining (6), (7) with (4), we obtain at $Z=a$.

$$\begin{aligned}(8)\quad & 4|a|g_{\alpha\overline{\beta}}\frac{\partial z^\alpha}{\partial w^i}\frac{\partial\overline{z}^\beta}{\partial\overline{w}^j}\cdot v^i\overline{v}^j\\ &\ge\bigg|2\,\mathrm{Re}\bigg\{\overline{a}^\alpha a^\beta g_{\beta\overline{\gamma}}\frac{\partial z^\alpha}{\partial w^i}\frac{\partial\overline{z}^\gamma}{\partial\overline{w}^j}\overline{z}^\alpha z^\beta v^i\overline{v}^j\\ &\qquad+|a|^2 g_{\alpha\overline{\beta}}\frac{\partial z^\alpha}{\partial w^i}\frac{\partial\overline{z}^\beta}{\partial\overline{w}^j}v^i\overline{v}^j\\ &\qquad+(1-|a|^2)g_{\alpha\overline{\beta}}\frac{\partial^2 z^\alpha}{\partial w^k\partial w^i}\frac{\partial\overline{z}^\beta}{\partial\overline{w}^j}\frac{\partial w^k}{\partial z^\gamma}a^\gamma v^i\overline{v}^j\bigg\}\bigg|.\end{aligned}$$

The inequality (8) is valid for any $a\in B$. we complete the proof of Lemma 2.

Proof of Theorem 1. From the inequality (8) it follows that

$$\begin{aligned}(9)\quad & 2|Z|(2-|Z|)g_{\alpha\overline{\beta}}\frac{\partial\overline{z}^\alpha}{\partial w^i}\frac{\partial\overline{z}^\beta}{\partial\overline{w}^j}v^i\overline{v}^j\\ &\ge 2\,\mathrm{Re}\bigg\{\overline{z}^\alpha z^\beta g_{\beta\overline{\gamma}}\frac{\partial z^\alpha}{\partial w^i}\frac{\partial\overline{z}^\gamma}{\partial\overline{w}^j}v^i\overline{v}^j\\ &\qquad+(1-|Z|^2)g_{\alpha\overline{\beta}}\frac{\partial^2 z^\alpha}{\partial w^k\partial w^i}\frac{\partial\overline{z}^\beta}{\partial\overline{w}^j}\frac{\partial w^k}{\partial z^\gamma}z^\gamma v^i\overline{v}^j\bigg\},\end{aligned}$$

for any $V\in\mathbf{C}^n$. Using the definition of $g_{\beta\overline{\gamma}}$ we have

$$(10)\qquad z^\beta g_{\beta\overline{\gamma}}=\frac{z^\gamma}{(1-|Z|^2)^2}$$

and it is easy to see that the second term of the right side in (9)

$$(11)\quad \begin{aligned}&(1-|Z|^2)g_{\alpha\bar\beta}\frac{\partial^2 z^\alpha}{\partial w^k\partial w^i}\frac{\partial\bar\xi^\beta}{\partial\overline{w}^j}\frac{\partial w^k}{\partial z^\gamma}z^\gamma v^i\overline{v}^j\\ &=z^\gamma\frac{\partial}{\partial z^\gamma}\left\{(1-|Z|^2)g_{\alpha\beta}\frac{\partial z^\alpha}{\partial w^i}\frac{\partial\overline{z}_\beta}{\partial\overline{w}^j}v^i\overline{v}^j\right\}\\ &\quad+|Z|^2 g_{\alpha\bar\beta}\frac{\partial z^\alpha}{\partial w^i}\frac{\partial\overline{z}^\beta}{\partial\overline{w}^j}v^i\overline{v}^j\\ &\quad-\frac{\partial z^\alpha}{\partial w^i}\frac{\partial\overline{z}^\beta}{\partial\overline{w}^j}v^i\overline{v}^j\left[|Z|^2 g_{\alpha\bar\beta}+\frac{\overline{z}^\beta z^\alpha}{(1-|Z|^2)^2}\right]\\ &=z^\gamma\frac{\partial}{\partial z^\gamma}\left[(1-|Z|^2)g_{\alpha\bar\beta}\frac{\partial z^\alpha}{\partial w^i}\frac{\partial\overline{z}^\beta}{\partial\overline{w}^j}v^i\overline{v}^j\right]\\ &\quad-\frac{\overline{z}^\alpha z^\beta}{(1-|Z|^2)^2}\frac{\partial z^\alpha}{\partial w^i}\frac{\partial\overline{z}^\beta}{\partial\overline{w}^j}v^i\overline{v}^j.\end{aligned}$$

Substituting (10), (11) into (9) immediately gives

$$(12)\quad \begin{aligned}&2|Z|(2-|Z|)g_{\alpha\bar\beta}\frac{\partial z^\alpha}{\partial w^i}\frac{\partial\overline{z}^\beta}{\partial\overline{w}^j}v^i\overline{v}^j\\ &\geq 2\,\mathrm{Re}\left\{z^\gamma\frac{\partial}{\partial z^\gamma}\left[(1-|Z|^2)g_{\alpha\bar\beta}\frac{\partial z^\alpha}{\partial w^i}\frac{\partial\overline{z}^\beta}{\partial\overline{w}^j}v^i\overline{v}^j\right]\right\}.\end{aligned}$$

Noting the Euclidean inner product $\langle dz^\alpha, dz^\beta\rangle = 2\delta_{\alpha\bar\beta}$, $\langle dz^\alpha, d\overline{z}^\beta\rangle = 0$, we easily see that the right side in (12)

$$(13)\quad \begin{aligned}&2\,\mathrm{Re}\left\{z^\gamma\frac{\partial}{\partial z^\gamma}\left[(1-|Z|^2)g_{\alpha\bar\beta}\frac{\partial z^\alpha}{\partial w^i}\frac{\partial\overline{z}^\beta}{\partial\overline{w}^j}v^i\overline{v}^j\right]\right\}\\ &=\frac{1}{2}\left\langle d\left[(1-|Z|^2)g_{\alpha\bar\beta}\frac{\partial z^\alpha}{\partial w^i}\frac{\partial\overline{z}^\beta}{\partial\overline{w}^j}v^i\overline{v}^j\right],\ d|Z|^2\right\rangle.\end{aligned}$$

By substituting (13) into (12), it follows that

$$(14)\quad \begin{aligned}&2|Z|(2-|Z|)g_{\alpha\bar\beta}\frac{\partial z^\alpha}{\partial w^i}\frac{\partial\overline{z}^\beta}{\partial\overline{w}^j}v^i\overline{v}^j\\ &\geq\frac{1}{2}\left\langle d\left[(1-|Z|^2)g_{\alpha\bar\beta}\frac{\partial z^\alpha}{\partial w^i}\frac{\partial\overline{z}^\beta}{\partial\overline{w}^j}v^i\overline{v}^j\right],\ d|Z|^2\right\rangle.\end{aligned}$$

The inequality (14) can be written as

$$(15)\quad \frac{2(2-|Z|)}{1-|Z|^2}\geq\left\langle d\left\{\lg\left[(1-|Z|^2)g_{\alpha\bar\beta}\frac{\partial z^\alpha}{\partial w^i}\frac{\partial\overline{z}^\beta}{\partial\overline{w}^j}v^i\overline{v}^j\right]\right\},\ d|Z|\right\rangle.$$

Hence, integrating both sides of (15) one obtains

$$\int_0^{|Z|} \frac{2(2-r)}{1-r^2}\,dr \geq \lg\left[(1-|Z|^2)g_{\alpha\overline{\beta}}\frac{\partial z^\alpha}{\partial w^i}\frac{\partial \overline{z}^\beta}{\partial \overline{w}^j}v^i\overline{v}^j\right] - \lg|V|^2.$$

So

$$\frac{(1+|Z|)^3}{(1-|Z|)}|V|^2 \geq (1-|Z|^2)g_{\alpha\overline{\beta}}\frac{\partial z^\alpha}{\partial w^i}\frac{\partial \overline{z}^\beta}{\partial \overline{w}^j}v^i\overline{v}^j. \tag{16}$$

(16) implies that for any $Z \in B$ the following matrix

$$\frac{(1+|Z|)^3}{1-|Z|}I - (1-|Z|^2)\left(g_{\alpha\overline{\beta}}\frac{\partial z^\alpha}{\partial w^i}\frac{\partial \overline{z}^\beta}{\partial \overline{w}^j}\right)_{1\leq i,j\leq n}$$

is positive semi-definite. So we get the following matrix inequality

$$\left(\frac{1-|Z|}{1+|Z|}\right)^2 (g_{\alpha\overline{\beta}})_{1\leq\alpha,\beta\leq n} \leq \left(\delta_{ij}\frac{\partial w^i}{\partial z^\alpha}\frac{\partial \overline{w}^j}{\partial \overline{z}^\beta}\right)_{1\leq\alpha,\beta\leq n}. \tag{17}$$

Similarly, we can prove

$$\left(\frac{1-|Z|}{1+|Z|}\right)^2 (g_{\alpha\overline{\beta}})_{1\leq\alpha,\beta\leq n} \geq \left(\delta_{ij}\frac{\partial w^i}{\partial z^\alpha}\frac{\partial \overline{w}^j}{\partial \overline{z}^\beta}\right)_{1\leq\alpha,\beta\leq n}. \tag{18}$$

Combining (17) with (18), we complete the proof of the inequalities in Theorem 1.

Now let us give an extreme example to illustrate that the inequalities in Theorem 1 are sharp. Set $n = 2$ and let

$$W = f(Z) = f^1(z^1, z^2), \quad f^1(z^1, z^2) = \left(\frac{z^1}{1-z^1}, \frac{z^2}{1-z^1}\right).$$

It is easy to check that the mapping is normalized, convex and biholomorphic and its Jacobian is

$$J_f(Z) = \begin{pmatrix} \dfrac{1}{(1-z^1)^2} & 0 \\ \dfrac{z^2}{(1-z^1)^2} & \dfrac{1}{1-z^1} \end{pmatrix}.$$

Take $V = (1, 0)$. Then

$$VJ_f(Z) = V\left(\frac{\partial W}{\partial Z}\right) = \left(\frac{1}{(1-z^1)^2}, 0\right)$$

and

$$\begin{aligned} &VJ_f(Z)\overline{J_f(Z)}'\overline{V}' \\ &= \frac{1}{(1-z^1-\overline{z}^1+|z^1|^2)^2} = \frac{1}{(1-2r_1\cos\theta_1+r_1^2)^2}, \end{aligned} \tag{19}$$

where $z^1 = r_1 e^{i\theta_1}$. On the other hand, we have

$$\left(\frac{1-|Z|}{1+|Z|}\right)^2 V(g_{\alpha\overline{\beta}})_{\alpha,\beta=1,2}\overline{V}' = \frac{1-|z^2|^2}{(1+|Z|)^4}, \tag{20}$$

$$\left(\frac{1+|Z|}{1-|Z|}\right)^2 V(g_{\alpha\overline{\beta}})_{\alpha,\beta=1,2}\overline{V}' = \frac{1-|z^2|^2}{(1-|Z|)^4}. \tag{21}$$

If $z^2 = 0$, $z^1 = r_1$ $(\theta_1 = 0)$, we obtain from (19) and (20)

$$VJ_f(Z)\overline{J_{f(Z)}}'\overline{V}' = \left(\frac{1+|Z|}{1-|Z|}\right)^2 V(g_{\alpha\overline{\beta}})_{\alpha,\beta=1,2}\overline{V}'. \tag{22}$$

If $z^2 = 0$, $z^1 = -r_1$ $(\theta_1 = \pi)$, then from (19) and (21) we get

$$VJ_f(Z)\overline{J_f(Z)}'\overline{V}' = \left(\frac{1-|Z|}{1+|Z|}\right)^2 V(g_{\alpha\overline{\beta}})_{\alpha,\beta=1,2}\overline{V}'. \tag{23}$$

The two inequalities (22) and (23) imply that the inequalities in Theorem 1 are sharp.

3. Necessary and sufficient conditions of biholomorphic convex mappings. Denote $B_r = \{Z \in \mathbb{C}^n \mid |Z| < r < 1\}$ and the boundary of B_r by S_r. We will demonstrate the following theorem.

THEOREM 2. *If $f: B \to \mathbb{C}^n$ is a holomorphic mapping with $f(0) = 0$ and $|\det J_f(Z)| > 0$, then the following statements are equivalent.*

(i) *f is a biholomorphic convex mapping.*

(ii) *If $Z \in B$ and $b = (b^1, \dots, b^n) \in \mathbb{C}^n$ satisfy $\mathrm{Re}(b^\alpha \overline{z}^\alpha) = 0$, then*

$$|b|^2 + \mathrm{Re}\left(b^\beta b^\gamma \frac{\partial w^i}{\partial z^\beta}\frac{\partial \overline{w}^j}{\partial z^\gamma} z^\alpha \frac{\partial^2 z^\alpha}{\partial w^i \partial w^j}\right) \geq 0. \tag{24}$$

Proof. First prove (ii) from (i).

Since $f(B)$ is a convex set, the mapping

$$Z_\lambda = f^{-1}(\lambda f(Z_1) + (1-\lambda) f(Z_2)),$$

$Z_1, Z_2 \in B$ and $0 \leq \lambda \leq 1$, defines a holomorphic mapping from $B \times B \to B$. Denote Carathéodory distance of the two points P and Q of a domain Ω in $\mathbb{C}^n$ by $C_\Omega(P, Q)$. Writing $F(Z_1, Z_2) = Z_\lambda$ we see [8]

$$\begin{aligned} C_B(F(Z_1, Z_2), 0) &\leq C_{B\times B}((Z_1, Z_2), (0, 0)) \\ &= \max\{C_B(Z_1, 0), C_B(Z_2, 0)\} \end{aligned}$$

by the contraction property of Carathéodory distance for holomorphic mappings. Recalling the Carathéodory distance on the unit ball in $\mathbb{C}^n$

$$C_B(Z, 0) = \frac{1}{2}\lg\frac{1+|Z|}{1-|Z|}, \qquad Z \in B,$$

then one has

$$|Z_\lambda| \le \max\{|Z_1|, |Z_2|\}.$$

It implies that for any $\lambda \in (0, 1)$ and $Z_1, Z_2 \in B_r$, $Z_\lambda \in B_r$. Thus $f(B_r)$ is also a convex set. Since, moreover, f is biholomorphic the image $f(S_r)$ (or $\partial f(B_r)$) of S_r under the mapping f is a real convex hypersurface. The second fundamental form S of the hypersurface is positive semi-definite. In other words, for any tangent vector U to $f(S_r)$, we have

$$S(U, U) \ge 0. \tag{25}$$

Note that $f(S_r)$ is an isothermal surface of the function $r^2 = |Z|^2 \circ f^{-1}$ and

$$\operatorname{grad} r^2 = \frac{\partial r^2}{\partial w^i}\frac{\partial}{\partial \overline{w}^i} + \frac{\partial r^2}{\partial \overline{w}^i}\frac{\partial}{\partial w^i} = \overline{z}^\alpha\frac{\partial z^\alpha}{\partial w^i}\frac{\partial}{\partial \overline{w}^i} + z^\alpha\frac{\partial \overline{z}^\alpha}{\partial \overline{w}^i}\frac{\partial}{\partial w^i}$$

is a normal vector to $f(S_r)$. Denote the normal vector by ν. $\frac{\nu}{|\nu|}$ is a unit normal vector. Assume that $U = a^i\frac{\partial}{\partial w^i} + \overline{a}^i\frac{\partial}{\partial \overline{w}^i}$ with

$$\operatorname{Re}(a^i\overline{z}^\alpha(\partial z^\alpha/\partial w^i)) = 0.$$

Then we have

$$S(U, U) = \left\langle \widetilde{D}_U\frac{\nu}{|\nu|}, U\right\rangle,$$

where $\widetilde{D}$ is the connection in R^{2n} (directional derivative). Via $\langle \nu, U\rangle = 0$ we have

$$\left\langle \widetilde{D}_U\frac{\nu}{|\nu|}, U\right\rangle = \frac{1}{|\nu|}\langle \widetilde{D}_U\nu, U\rangle.$$

Hence, the above inequality (25) can be rewritten as

$$\frac{1}{|\nu|}\left(a^i\overline{a}^j\frac{\partial^2 r^2}{\partial w^i \partial \overline{w}^j} + \operatorname{Re}\left(a^i a^j\frac{\partial^2 r^2}{\partial w^i \partial w^j}\right)\right) \ge 0. \tag{26}$$

Now set $b = a\frac{\partial Z}{\partial w}$ or $a^i = b^\beta(\partial w^i/\partial z^\beta)$. From $\operatorname{Re}(a^i \overline{z}^\beta(\partial z^\beta/\partial w^i)) = 0$, it follows that $\operatorname{Re}(b^\beta\overline{z}^\beta) = 0$. Substituting $a^i = b^\beta(\partial w^i/\partial z^\beta)$ into (26) we can get (24).

Now prove (i) from the condition (ii). We divide the proof into three steps.

In the first step, we show that if the mapping f is injective on B_μ $(0<\mu<1)$, then $f(B_\mu)$ is a convex set. In fact, since f is an injection on B_μ, $f(S_r)$ is a real hypersurface in $\mathbb{C}^n$ for any $r<\mu$. If $U=a^i(\partial/\partial w^i)+\bar{a}^i(\partial/\partial\overline{w}^i)$ is a real tangent vector to $f(S_r)$, the second fundamental form is

$$(27)\quad \begin{aligned} S(U,U)&=\frac{1}{|\nu|}\left(a^i\bar{a}^j\frac{\partial^2 r^2}{\partial w^i\partial\overline{w}^j}+\operatorname{Re}\left(a^ia^j\frac{\partial^2 r^2}{\partial w^i\partial w^j}\right)\right)\\ &=\frac{1}{|\nu|}\left(a^i\bar{a}^j\frac{\partial z^\alpha}{\partial w^i}\frac{\partial\bar{z}^\alpha}{\partial\overline{w}^j}+\operatorname{Re}\left(a^ia^jz^\alpha\frac{\partial^2 z^\beta}{\partial w^i\partial w^j}\right)\right).\end{aligned}$$

Set $b=a\frac{\partial Z}{\partial W}$. We have $\operatorname{Re}(b^\beta\bar{z}^\beta)=0$. Recalling the condition (ii) we get

$$S(U,U)\geq 0.$$

Hence, $f(S_r)$ is a real convex hypersurface for any $r<\mu$, and then we conclude that $f(B_\mu)$ is a convex set.

In the second step, we prove if the mapping f is injective on B_μ, then f is also injective on $\overline{B}_\mu$. Since $f(B_\mu)$ is a convex set $f(B_\mu)$ is starlike with respect to the origin. Therefore, we have

$$\langle dr^2|_{B_\mu},\ d\rho^2\rangle\geq 0$$

thanks to the necessary condition for starlike mappings [9], where $\rho^2=|W|^2$. If the statement is not true then there are at least two distinct points $X, Y\in\overline{B}_\mu$ such that $f(X)=f(Y)$. We know, for all $0\leq s\leq 1$, $sf(X)$, $sf(Y)\in f(\overline{B}_\mu)$. Because f is a holomorphic immersion one can obtain the curve $X(s)$ with $X(1)=X$ which falls in B by the method of analytic continuation such that $f(X(s))=sf(X)$. That is $X(s)=f^{-1}(sf(X))$ is a univalent component of the inverse images of the segment $sf(X)$. Since

$$\frac{dr^2(X(s))}{ds}=\frac{1}{s}\langle dr^2(X(s)),\,d\rho\rangle|_{W=sf(X)}\geq 0$$

for $0\leq s\leq 1$, we have

$$r^2(X(s))\leq r^2(X(1))=r^2(X)=\mu.$$

Thus

$$X(s)\in\overline{B}_\mu.$$

Suppose that $Y(s)$ is another univalent component of the inverse images of the segment $sf(X)$, but $Y(1)=Y$. A similar discussion shows $Y(s)\in\overline{B}_\mu$. Let

$$\mathscr{R}=\{s\in[0,1]|X(s)=Y(s)\}.$$

If the set $\mathscr{R}$ is nonempty then the supremum s^* of $\mathscr{R}$ exists. Because $\mathscr{R}$ is a closed set $s^* \in \mathscr{R}$ and $s^* < 1$ due to $X(1) \neq Y(1)$. This implies that for any small positive ε satisfying $s^* < s^* + \varepsilon \leq 1$ we have $X(s^* + \varepsilon) \neq Y(s^* + \varepsilon)$. But

$$f(X(s^* + \varepsilon)) = f(Y(s^* + \varepsilon)) = (s^* + \varepsilon)f(X).$$

This is contrary to f being locally biholomorphic at $X(s^*)$. If $\mathscr{R}$ is empty then $X(0) \neq Y(0)$. In other words, at this case we have at least two points $X(0)$ and $Y(0)$ such that $f(X(0)) = f(Y(0)) = 0$. Since f is an injection on B_μ, at least one of the two points $X(0)$ and $Y(0)$ must be a boundary point of $\overline{B}_\mu$. Suppose that $X(0) \in \partial \overline{B}_\mu$. Let $B(X(0), \delta)$ be the open ball central at $X(0)$ with radius δ, which is so small that $B(X(0), \delta) \cap \mathscr{U}_0$ is empty, where $\mathscr{U}_0 \subset B_\mu$ is the neighborhood of the origin in B such that f is biholomorphic on it. Because of the open mapping theorem $f(B(X(0), \delta) \cap B_\mu)$ is an open set and $f(\mathscr{U}_0)$ an open set including the origin of $\mathbb{C}^n$. So the origin is also a boundary point of the open set $f(B(X(0), \delta) \cap B_\mu)$ and then $\{f(B(X(0), \delta) \cap B_\mu)\} \cap \{f(\mathscr{U}_0)\}$ is not empty. This implies that for any $W \in \{f(B(X(0), \delta) \cap B_\mu)\} \cap \{f(\mathscr{U}_0)\}$ it has two distinguished inverse images on B_μ. It is impossible due to the fact that f is injective on B_μ. So, f is also an injection on $\overline{B}_\mu$.

In the third step, we demonstrate $\mathscr{A} = [0, 1]$ where $\mathscr{A} = \{t \in [0, 1] | f$ is injective on $B_t\}$. Since $f(0) = 0$, $J_f(0) \neq 0$, the set $\mathscr{A}$ is nonempty. We claim that $\mathscr{A}$ is a closed set. In fact, if $0 < t_1 \in \mathscr{A}$ then all $t \leq t_1$ fall in $\mathscr{A}$. Therefore, to prove $\mathscr{A}$ is a closed set it is sufficient only to prove that if $t^* > t$ and all t fall in $\mathscr{A}$ then t^* is also in $\mathscr{A}$. If the result is not true, there exist at least two distinct points X_1, X_2 in B_{t^*} such that $f(X_1) = f(X_2)$. Because of $|X_1| < t^*$, $|X_2| < t^*$, one can find a t^{**} satisfying $|X_1| < t^{**} < t^*$, $|X_2| < t^{**} < t^*$. The formulas above imply $X_1, X_2 \in B_{t^{**}}$. But $t^{**} \in \mathscr{A}$. This yields $f(X_1) \neq f(X_2)$. It is impossible. So $\mathscr{A}$ is a closed set. Finally, we need to prove the set $\mathscr{A}$ is also an open set. For the end, only to verify that if f is a one to one mapping on B_t then there is a positive ε small enough such that f is also a one to one mapping on $B_{t+\varepsilon}$. If not, there is a sequence $\varepsilon_n > 0$ with $\lim_{n\to\infty} \varepsilon_n = 0$ such that one can find two sequences $\{X_n\}, \{Y_n\}$ satisfying the following conditions:

$$X_n, Y_n \in B_{t+\varepsilon_n}, \quad X_n \neq Y_n, \quad f(X_n) = f(Y_n)$$

for all $n = 1, 2, \ldots$. Obviously,

$$\lim_{n\to\infty} |Y_n| = \lim_{n\to\infty} |Y_n| = t.$$

It is easy to see that $\{X_n\}$, $\{Y_n\}$ are bounded sequences. So there exist two subsequences $\{X_{n_k}\}$, $\{Y_{n_k}\}$ of $\{X_n\}$, $\{Y_n\}$ which converge to X and Y respectively, that is

$$\lim_{k\to\infty} X_{n_k} = X, \quad \lim_{k\to\infty} Y_{n_k} = Y.$$

The two subsequences satisfy

$$f(X_{n_k}) = f(Y_{n_k}), \qquad X_{n_k} \neq Y_{n_k}, \qquad \text{for } k = 1, 2, \ldots,$$

and

$$\lim_{k\to\infty} |X_{n_k}| = \lim_{k\to\infty} |Y_{n_k}| = t.$$

Obviously, $X, Y \in \partial B_t$. If $X \neq Y$ that is contrary to the result in the second step. If $X = Y$ it implies that there are two points denoted still by $\{X_{n_k}\}$, $\{Y_{n_k}\}$ in any neighborhood of $X = Y$ such that $X_{n_k} \neq Y_{n_k}$, but $f(X_{n_k}) = f(Y_{n_k})$. It is obviously impossible due to the assumption of f being local biholomorphic.

Combining the above argument we obtain $\mathscr{A} = [0, 1]$. Hence f is biholomorphic and f_t is convex from the argument of the first step.

It is worth noting when $n = 1$, the inequality (24) has the following form:

$$|b|^2 + \mathrm{Re}\left(b^2\left(\frac{dw}{dz}\right)^2 \overline{z}\frac{d^2z}{dw^2}\right) \geq 0, \qquad \mathrm{Re}(b\overline{z}) = 0.$$

Because of

$$\frac{d^2w}{dz^2}\left(\frac{dz}{dw}\right)^2 = -\frac{dw}{dz}\frac{d^2z}{dw^2},$$

we have

$$(28)\ |b|^2 - \mathrm{Re}\left(\overline{z}b^2\frac{dw}{dz}\left(\frac{dz}{dw}\right)^2\frac{d^2w}{dz^2}\right) = |b|^2 - \mathrm{Re}\left(\overline{z}b^2\frac{d^2w}{dz^2}\frac{dz}{dw}\right)$$

$$= |b|^2 + \mathrm{Re}\left(\overline{z}b^2\frac{f''}{f'}\right) \geq 0.$$

Furthermore, from $\mathrm{Re}(b\overline{z}) = b\overline{z} - \overline{b}z = 0$, $\overline{z}b = -\overline{b}z$, the inequality (28) can be rewritten as

$$|b|^2\left(1 + \mathrm{Re}\,\frac{zf''}{f'}\right) \geq 0.$$

It just is the necessary and sufficient condition of univalent convex functions defined on the unit disc in the complex plane $\mathbb{C}$.

In [6] K. Kikuchi stated the following theorem: Let f be a holomorphic mapping from the ball B into $\mathbb{C}^n$ with $f(0) = 0$ and $\det J_f(Z) \neq 0$, $Z \in B$. Then f is biholomorphic and convex if and only if

$$\operatorname{Re}[|ZA|^2 + \overline{Z} J_f^{-1}(Z)\tilde{A}] \geq 0, \tag{29}$$

where

$$A = \begin{pmatrix} a_1 & 0 & \cdots & 0 \\ 0 & a_2 & \cdots & 0 \\ \vdots & \vdots & \ddots & \vdots \\ 0 & 0 & \cdots & a_n \end{pmatrix}, \quad a_j \geq 0, \quad j = 1, 2, \ldots, n,$$

$$\tilde{A} = \begin{pmatrix} ZA\dfrac{\partial^2 f_1}{\partial Z' \partial Z}(Z)A'Z' \\ \cdots\cdots\cdots\cdots\cdots\cdots \\ ZA\dfrac{\partial^2 f_n}{\partial Z' \partial Z}(Z)A'Z' \end{pmatrix}$$

with $ZA \neq 0$. The necessary and sufficient condition is equivalent to the one given by Suffridge in [7].

We quote a counter-example given by Taishun Liu in [4] to show that the condition (29) is not sufficient to conclude the section. Set

$$f(Z) = \left(\frac{z_1}{1 - z_1}, \frac{z_2}{1 - z_2}, \ldots, \frac{z_n}{1 - z_n}\right).$$

Clearly, the mapping is biholomorphic and a simple calculation gives

$$J_f^{-1}(Z) = \begin{pmatrix} (1 - z_1)^2 & 0 & \cdots & 0 \\ 0 & (1 - z_2)^2 & \cdots & 0 \\ \vdots & \vdots & \ddots & \vdots \\ 0 & 0 & \cdots & (1 - z_n)^2 \end{pmatrix} \tag{30}$$

and $\partial^2 f_j / \partial Z' \partial Z$ is the matrix of which the unique non-zero element is $\frac{1}{(1-z_j)^3}$ at j column and j row.

Putting (30) into the left side of (29), we have

$$\operatorname{Re}[|ZA|^2 + \overline{Z} J_f^{-1}(Z)\tilde{A}] = \operatorname{Re}\left[\sum_{j=1}^{n} a_j^2 |z_j|^2 \left(\frac{1 + \overline{z}_j}{1 + z_j}\right)\right] > 0.$$

Hence, according to the result by Kikuchi, the mapping

$$f(Z) = \left(\frac{z_1}{1 - z_1}, \frac{z_2}{1 - z_2}, \ldots, \frac{z_n}{1 - z_n}\right)$$

should be convex. That is to say for $0 < r < 1$

$$\frac{1}{n}\sum_{k=1}^{n} f(re_k) \in \{f(B)\}$$

where e_k is the unit vector of which kth component equals to 1. So

$$f^{-1}\left[\frac{1}{n}\sum_{k=1}^{n} f(re_k)\right] \in B.$$

But one has

$$f^{-1}\left[\frac{1}{n}\sum_{k=1}^{n} f(re_k)\right] = \left(\frac{r}{r+n(1-r)}, \dots, \frac{r}{r+n(1-r)}\right). \tag{31}$$

When r is close to 1 the right side of (31) does not belong to B. Therefore, the mapping f is not convex.

4. Characteristics of the image of biholomorphic convex mappings. As an application of the theorems in §§1 and 2, we now consider characteristics of the image of biholomorphic convex mappings from B to $\mathbb{C}^n$.

THEOREM 3. *Suppose that $f: B \to \mathbb{C}^n$ is a normalized biholomorphic convex mapping. Then*

(1) $$\frac{(1-\mu)^5}{\mu(1+\mu)^3} \le \text{principal curvature of } f(S_\mu) \le \frac{(1+\mu)^4}{\mu(1-\mu)^4},$$

(2) $$\omega_{2n-1}\frac{1-\mu}{2\mu(1+\mu)}\int_0^\mu \frac{(1-t)^{n+1}t^{2n-1}}{(1+t)^{3n+3}}\,dt \le \text{volume of } f(S_\mu)$$
$$\le n\omega_{2n-1}\frac{1+\mu}{2\mu(1-\mu)}\int_0^\mu \frac{(1+t)^{n+1}t^{2n-1}}{(1-t)^{3n+3}}\,dt,$$

where ω_{2n-1} is the volume of the unit sphere in R^{2n}.

Proof. Fix $Z \in B - \{0\}$ and $b \in \mathbb{C}^n$ fulfilling $\operatorname{Re}(b^\beta \bar{z}^\beta) = 0$. Let $\tilde{Z} = \frac{tZ}{|Z|}$ and $\tilde{b} = \frac{tb}{|Z|}$, $t \in D$. Then $\tilde{Z} \in B$ and $\operatorname{Re}(\tilde{b}^\beta \bar{\tilde{z}}^\beta) = 0$. Since $W = f(Z)$ is a normalized biholomorphic convex mapping it follows from Theorem 2 that

$$\frac{|t|^2|b|^2}{|Z|^2} + \operatorname{Re}\left\{|t|^2\frac{t\bar{z}^\alpha b^\beta b^\gamma}{|Z|^3}\left(\frac{\partial w^i}{\partial z^\beta}\frac{\partial w^j}{\partial z^\gamma}\frac{\partial^2 z^\alpha}{\partial w^i \partial w^j}\right)\bigg|_{\tilde{Z}}\right\} \ge 0$$

or

$$\operatorname{Re}\left\{|b|^2 + \frac{t\bar{z}^\alpha b^\beta b^\gamma}{|Z|}\left(\frac{\partial w^i}{\partial z^\beta}\frac{\partial w^j}{\partial z^\gamma}\frac{\partial^2 z^\alpha}{\partial w^i \partial w^j}\right)\bigg|_{\tilde{Z}}\right\} \ge 0. \tag{32}$$

The left side of (32) is a real-valued and nonnegative harmonic function. From Harnack's inequality we have

$$(33)\ |b|^2\left(\frac{1-|t|}{1+|t|}\right)\le |b|^2+\operatorname{Re}\left\{\frac{t\overline{z}^\alpha b^\beta b^\gamma}{|Z|}\left(\frac{\partial w^i}{\partial z^\beta}\frac{\partial w^j}{\partial z^\gamma}\frac{\partial^2 z^\alpha}{\partial w^i\partial w^j}\right)\Big|_{\widetilde{Z}}\right\}\le |b|^2\left(\frac{1+|t|}{1-|t|}\right).$$

If we take $t=|Z|$, the formulae (33) becomes as follows:

$$|b|^2\frac{1-|Z|}{1+|Z|}\le |b|^2+\operatorname{Re}\left\{\overline{z}^\alpha b^\beta b^\gamma\left(\frac{\partial w^i}{\partial z^\beta}\frac{\partial w^j}{\partial z^\gamma}\frac{\partial^2 z^\alpha}{\partial w^i\partial w^j}\right)\right\}\le |b|^2\frac{1+|Z|}{1-|Z|}$$

for any $Z\in B-\{0\}$ with $\operatorname{Re}(b^\beta\overline{z}^\beta)=0$. Now taking $b=a(\frac{\partial Z}{\partial W})$ and noting (27) we have

$$(34)\qquad a^i\frac{\partial z^\beta}{\partial w^i}\frac{\partial\overline{z}^\beta}{\partial\overline{w}^j}\overline{a}^j\left(\frac{1-|Z|}{1+|Z|}\right)\le|\nu|S(U,U)\le a^i\frac{\partial z^\beta}{\partial w^i}\frac{\partial\overline{z}^\beta}{\partial\overline{w}^j}\overline{a}^j\left(\frac{1+|Z|}{1-|Z|}\right),$$

where

$$|\nu|=\sqrt{\overline{z}^\alpha\frac{\partial z^\alpha}{\partial w^i}\frac{\partial\overline{z}^\beta}{\partial\overline{w}^i}z^\beta},\qquad U=a^i\frac{\partial}{\partial w^i}+\overline{a}^i\frac{\partial}{\partial\overline{w}^i}.$$

From Theorem 1, it follows that

$$\delta_{ij}\ge\left(\frac{1-|Z|}{1+|Z|}\right)^2\left[\frac{1}{1-|Z|^2}\frac{\partial z^\alpha}{\partial\overline{w}^i}\frac{\partial\overline{z}^\alpha}{\partial\overline{w}^j}+\frac{1}{(1-|Z|^2)^2}\overline{z}^\alpha\frac{\partial z^\alpha}{\partial\overline{w}^i}\frac{\partial\overline{z}^\beta}{\partial\overline{w}^j}z^\beta\right],$$

$$\delta_{ij}\le\left(\frac{1+|Z|}{1-|Z|}\right)^2\left[\frac{1}{1-|Z|^2}\frac{\partial z^\alpha}{\partial\overline{w}^i}\frac{\partial\overline{z}^\alpha}{\partial\overline{w}^j}+\frac{1}{(1-|Z|^2)^2}\overline{z}^\alpha\frac{\partial z^\alpha}{\partial\overline{w}^i}\frac{\partial\overline{z}^\beta}{\partial\overline{w}^j}z^\beta\right].$$

Therefore, the inequalities

$$(35)\quad |a|^2\ge\frac{1}{(1+|Z|)^4}\left[(1-|Z|^2)a^i\frac{\partial z^\alpha}{\partial w^i}\frac{\partial\overline{z}^\alpha}{\partial\overline{w}^j}\overline{a}^j+\left|a^i\overline{z}^\alpha\frac{\partial z^\alpha}{\partial w^i}\right|^2\right],$$

$$|a|^2\le\frac{1}{(1-|Z|)^4}\left[(1-|Z|^2)a^i\frac{\partial z^\alpha}{\partial w^i}\frac{\partial\overline{z}^\alpha}{\partial\overline{w}^j}\overline{a}^j+\left|a^i\overline{z}^\alpha\frac{\partial z^\alpha}{\partial w^i}\right|^2\right]$$

hold for any $Z\in B$. By the Schwartz inequality

$$0\le\left|a^i\overline{z}^\alpha\frac{\partial z^\alpha}{\partial w^i}\right|^2\le|Z|^2\left|a^i\frac{\partial z^\alpha}{\partial w^i}\frac{\partial\overline{z}^\alpha}{\partial\overline{w}^j}\overline{a}^j\right|$$

(34) can be read as

$$\frac{1-|Z|}{(1+|Z|)^3}a^i\frac{\partial z^\alpha}{\partial w^i}\frac{\partial \overline{z}^\alpha}{\partial \overline{w}^j}\overline{a}^j \le |a|^2 \le \frac{1}{(1-|Z|)^4}a^i\frac{\partial z^\alpha}{\partial w^i}\frac{\partial \overline{z}^\alpha}{\partial \overline{w}^j}\overline{a}^j. \tag{36}$$

Substituting (36) into (34), we obtain

$$|\nu|S(U,U) \ge \frac{(1-|Z|)^5}{1+|Z|}|U|^2, \tag{37}$$

$$|\nu|S(U,U) \le \left(\frac{1+|Z|}{1-|Z|}\right)^4|U|^2.$$

On the other hand, again using Theorem 1 we get

$$\left(\frac{1+|Z|}{1-|Z|}\right)^2(g_{\alpha\overline{\beta}}) \ge \left(\frac{\partial w^i}{\partial z^\alpha}\frac{\partial \overline{w}^i}{\partial \overline{z}^\beta}\right) \ge \left(\frac{1-|Z|}{1+|Z|}\right)^2(g_{\alpha\overline{\beta}}).$$

This yields

$$\left(\frac{1-|Z|}{1+|Z|}\right)^2(g^{\alpha\overline{\beta}}) \le \left(\frac{\partial z^\alpha}{\partial w^i}\frac{\partial \overline{z}^\beta}{\partial \overline{w}^i}\right) \le \left(\frac{1+|Z|}{1-|Z|}\right)^2(g^{\overline{\beta}\alpha}),$$

where $(g^{\overline{\beta}\alpha})$ is the inverse matrix of $(g_{\alpha\overline{\beta}})$. Hence

$$\frac{(1-|Z|)^3}{1+|Z|}(\delta_{\beta\alpha}-\overline{z}^\beta z^\alpha)\overline{z}^\beta z^\alpha \le \overline{z}^\alpha\frac{\partial z^\alpha}{\partial w^i}\frac{\partial \overline{z}^\beta}{\partial \overline{w}^i}z^\beta$$

$$\le \frac{(1+|Z|)^3}{1-|Z|}(\delta_{\beta\alpha}-\overline{z}^\beta z^\alpha)\overline{z}^\beta z^\alpha.$$

So, we obtain

$$|Z|^2(1-|Z|)^4 \le |\nu|^2 \le |Z|^2(1+|Z|)^4. \tag{38}$$

Putting (38) into (37), we have

$$\frac{(1-\mu)^5}{\mu(1+\mu)^3} \le \text{principal curvature of } f(S_\mu) \le \frac{(1+\mu)^4}{\mu(1-\mu)^4}.$$

To yield the volume of the hypersurface $f(S_\mu)$ of $(2n-1)$-dimension, we consider the function $r^2=|Z|^2\circ f^{-1}$. Denote $\Delta = 4\partial^2/\partial w^i\partial\overline{w}^i$. Then we have

$$\int_{f(B_\mu)}\frac{\Delta r^2}{1-r^2}dW = -\int_{f(B_\mu)}\frac{|\nabla r^2|}{(1-r^2)^2}dW + \frac{1}{1-\mu^2}\int_{\partial f(B_\mu)}\frac{\partial r^2}{\partial n}d\sigma, \tag{39}$$

where dW is the volume element of $\mathbb{C}^n = \mathscr{R}^{2n}$, $d\sigma$ that of $\partial f(B_\mu)$ and $n = \frac{\nu}{|\nu|}$ the unit normal vector. Because of

$$\frac{\partial r^2}{\partial n} = \frac{\nu}{|\nu|} r^2 = \frac{2}{|\nu|}\left(\overline{z}^\alpha \frac{\partial z^\alpha}{\partial w^i}\frac{\partial \overline{z}^\beta}{\partial \overline{w}^i} z^\beta\right) = 2|\nu|$$

(39) becomes

$$\text{(40)} \qquad \int_{f(B_\mu)} \frac{(1-r^2)\Delta r^2 + |\nabla r^2|^2}{(1-r^2)^2}\, dW = \frac{2}{1-\mu^2}\int_{f(S_\mu)} |\nu|\, d\sigma.$$

On the other hand, we have

$$\Delta r^2 = 4\frac{\partial z^\alpha}{\partial w^i}\frac{\partial \overline{z}^\beta}{\partial \overline{w}^i}, \qquad |\nabla r^2|^2 = 4\overline{z}^\alpha \frac{\partial z^\alpha}{\partial w^i}\frac{\partial \overline{z}^\beta}{\partial \overline{w}^i} z^\beta.$$

Hence

$$\text{(41)} \qquad \frac{(1-r^2)\Delta r^2 + |\nabla r^2|^2}{(1-r^2)^2} = g_{\alpha\overline{\beta}}\frac{\partial z^\alpha}{\partial w^i}\frac{\partial \overline{z}^\beta}{\partial \overline{w}^i}.$$

Putting (41) into (40) we obtain

$$\text{(42)} \qquad \int_{f(B_\mu)} g_{\alpha\overline{\beta}}\frac{\partial z^\alpha}{\partial w^i}\frac{\partial \overline{z}^\beta}{\partial \overline{w}^i}\, dW = \frac{2}{1-\mu^2}\int_{f(S_\mu)} |\nu|\, d\sigma.$$

From Theorem 1 we have

$$\text{(43)} \qquad \left(\frac{1+|Z|}{1-|Z|}\right)^2 g_{\alpha\overline{\beta}}\frac{\partial z^\alpha}{\partial w^i}\frac{\partial \overline{z}^\beta}{\partial \overline{w}^i} \geq n \geq \left(\frac{1-|Z|}{1+|Z|}\right)^2 g_{\alpha\overline{\beta}}\frac{\partial z^\alpha}{\partial w^i}\frac{\partial \overline{z}^\beta}{\partial \overline{w}^i}.$$

Substituting (43) into (42), we obtain

$$\text{(44)} \qquad n\int_{f(B_\mu)} \left(\frac{1-r}{1+r}\right)^2 dW \leq \frac{2}{1-\mu^2}\int_{f(S_\mu)} |\nu|\, d\sigma \leq n\int_{f(B_\mu)} \left(\frac{1+r}{1-r}\right)^2 dW.$$

Combining (39) with (44), we obtain

$$\frac{n(1-\mu)}{2\mu(1+\mu)}\int_{f(B_\mu)} \left(\frac{1-r}{1+r}\right)^2 dW \leq \int_{f(S_\mu)} d\sigma \leq \frac{n(1+\mu)}{2\mu(1-\mu)}\int_{f(B_\mu)} \left(\frac{1+r}{1-r}\right)^2 dW$$

or

$$\int_{f(S_\mu)} d\sigma \geq \frac{n(1-\mu)}{2\mu(1+\mu)} \int_{B_\mu} \left(\frac{1-|Z|}{1+|Z|}\right)^2 |\det J_f(Z)|^2 \, dZ \,, \tag{45}$$

$$\int_{f(S_\mu)} d\sigma \leq \frac{n(1+\mu)}{2\mu(1-\mu)} \int_{B_\mu} \left(\frac{1+|Z|}{1-|Z|}\right)^2 |\det J_f(Z)|^2 \, dZ \,,$$

where dZ is the volume element of $\mathbb{C}^n = \mathscr{R}^{2n}$. Finally, again using Theorem 1 we obtain

$$\frac{(1+|Z|)^{n-1}}{(1-|Z|)^{3n+1}} \geq |\det J_f(Z)|^2 \geq \frac{(1-|Z|)^{n-1}}{(1+|Z|)^{3n+1}}. \tag{46}$$

Putting (46) into (45) concludes Theorem 3.

References

[1] H. Cartan, *Sur la Possibilité d'étendre aux Fonctions de Plusieurs Variables Complexes la Theorié des Fonctions Univalentes*, Note to P. Montel, *Leçons sur les fonctions univalentes ou multivalentes*, Gauthier-Villars, Paris, 1933, 129–155.

[2] P. Duren and W. Rudin, *Distortion in several variables*, Complex Analysis, Theory and Application, **5** (1986), 323–326.

[3] C. H. FitzGerald, S. Gong and R. W. Bernard, *Distortion theorem for biholomorphic mappings in* $\mathbb{C}^2$, Chinese Sci. Bull., **5** (1990), 353–356.

[4] Taishun Liu, *The Growth Theorems, Covering Theorems and Distortion Theorems in Several Complex Variables*, Doctoral thesis, 1989 (in Chinese).

[5] S. Gong and X. A. Zheng, *Distortion theorems for biholomorphic mappings in transitive domains* (II), Chinese Sci. Bull., **5** (1990), 330–332 (in Chinese).

[6] K. Kikuchi, *Starlike and convex mappings in several complex variables*, Pacific J. Math., **2** (1973), 569–880.

[7] T. J. Suffridge, *The principle of subordination applied to functions of several variables*, Pacific J. Math., **33** (1970), 241–248.

[8] S. I. Pinchuk, *Holomorphic Maps in* $\mathbb{C}^n$ *and the Problem of Holomorphic Equivalence*, Several Complex Variables III, G. M. Khenkin (Ed.), Springer-Verlag, 1989.

[9] S. Gong, S. K. Wang and Q. Yu, *The growth and* 1/4*-theorems for starlike mappings in* B^p, Chinese Ann. Math., **11B 1** (1990), 100–104.

Received October 29, 1991 and in revised form April 29, 1992. Supported by National Natural Science Foundation of China.

Academia Sinica
P.O. Box 2734
Beijing 100080 China

Distortion Theorem for Biholomorphic Mappings in Transitive Domains (Ⅰ)

Gong Sheng
(*University of Science and Technology of China, Hefei*)
Zheng Xuean
(*Department of Mathematics, Anhui University, Hefei*)

1. Introduction

There are fruitful results in geometrical function theory of one complex variable. But through some counter-examples one can show that the corresponding results in several complex variables are not true. For the classical distortion theorem in one complex veriable, H. Cartan[1] conjectured that it is true for the biholomorphic mappings in the unit ball B^n of $\mathbb{C}^n(n\geqslant 2)$. Unfortunately, this conjecture is not true. The first affirmative result was given in [2].

Let $f(z)=(f_1(z), f_2(z))$, $z=(z_1,z_2)$, be the normalized biholomorphic mapping in B^2, i. e.,

$$\begin{cases} f_1(z)=z_1+d^1_{2,0}z_1^2+d^1_{1,1}z_1z_2+d^1_{0,2}z_2^2+\cdots \\ f_2(z)=z_2+d^2_{2,0}z_1^2+d^2_{1,1}z_1z_2+d^2_{0,2}z_2^2+\cdots \end{cases} \tag{1}$$

Let S be a subclass of the class of normalized biholomorphic mappings, which satisfies the following condition. If $f(z)\in S$, and under the holomorphic automorphism transformation of B^2 we normalized it again, then we get mapping F. If $F\in S$ is true, we say S is an $\mathscr{A}$-invariant subclass.

In [2], we proved that if $f\in S$, and S is an $\mathscr{A}$-invariant subclass, then

$$|\log((1-z\bar{z}')^{3/2}\det J_f(z))|\leqslant C(S)\log\frac{1+|z|}{1-|z|}. \tag{2}$$

where $C(S)=\sup\{|d^1_{2,0}+\frac{1}{2}d^2_{1,1}|\,f\in S\}$, and $J_f(z)$ is the Jacobain of $f(z)$ at z. By (2) we have

$$\frac{(1+|z|)^{C(S)-\frac{3}{2}}}{(1+|z|)^{C(S)+\frac{3}{2}}}\leqslant|\det J_f(z)|\leqslant\frac{(1+|z|)^{C(S)-\frac{3}{2}}}{(1-|z|)^{C(S)+\frac{3}{2}}}. \tag{3}$$

and

$$|\arg \det J_f(z)| \leqslant C(S) \log \frac{1+|z|}{1-|z|} \tag{4}$$

When S is a class of convex mappings, we can prove $C(S) < 1.761$, and conjecture $C(S) = \frac{3}{2}$. The convex mapping $\left(\frac{z_1}{1-z_1}, \frac{z_2}{1-z_1}\right)$ is one mapping attending this value.

If f is a normalized biholomorphic mapping on the unit ball B^n of $\mathbb{C}^n$, $f(z) \in S$, and S is an $\mathscr{A}$-invariant subclass,

$$f(z) = z + (zA^1z', \cdots, zA^nz') + \cdots$$

where $A^i = (a_{jk}^{(i)})_{1 \leqslant j,k \leqslant n}$, $i = 1, \cdots, n$, then

$$|\log((1-zz)^{\frac{n+1}{2}} \det J_f(z))| \leqslant C(S) \log \frac{1+|z|}{1-|z|}. \tag{5}$$

where $C(S) = \operatorname{Sup}\{|\sum_{i=1}^{n} a_{i1}^i|, f \in S\}$. If S is convex, then $C(S) \leqslant 1 + \frac{\sqrt{2}(n-1)}{2}$. This result and (5) was proved by Liu [3]. In [2], we conjectured that $C(S) = \frac{n+1}{2}$, when S is the subclass of convex mappings. The mapping $\left(\frac{z_1}{1-z_1}, \frac{z_2}{1-z_1}, \cdots, \frac{z_n}{1-z_1}\right)$ is a mapping attending this value.

In this paper, we will consider the distortion theorem of normalized biholomorphic mappings on transitive domains.

2. Main results

Let $M \subset \mathbb{C}^n$ be a transitive domain, bounded or unbounded. Let G be the Lie group consisting of some holomorphic automorphism of M and acting transitively on M. Moreover, let K be the isotropy subgroup of G which fixes the point $m \in M$. If $\psi_z \in G$ which transits the point $z \in M$ to m, i. e. $\psi_z(Z) = m$. Denote the Jacobian of ψ_z by ψ_z'. If M is unbounded, we assume $|\det \psi_k'(m)| = 1$ for all $k \in K$ and define $K(z, \bar{z}) = c \det \psi_z(z) \det \psi_z'(z)$ where c is a constant. If M is bounded, $K(z, \bar{z})$ is the Berman kernel function of M. We denote $K(m, \bar{m})^{-1} \left(\frac{\partial}{\partial z_p} K(z, \bar{z})\right)_{z=m}$ by C_p.

If ψ_z is the inverse of ψ_z, then $\psi_z \in G$ and $\psi_z(m) = z$. For any point z in M, we can connect z and m by an analytic curve $\sigma(\rho) = (\sigma_1(\rho), \cdots \sigma_n(\rho))(\sigma \leqslant \rho \leqslant 1)$ such that $\sigma(0) = m$, $\sigma(1) = z$. Moreover, if M is starlike with respect to m, then we can take $\sigma(\rho) = m + \rho(z - m)$.

We can prove the following

Theorem 1. *If f is a biholomorphic mapping which maps the transitive domain m into $\mathbb{C}^n$. Let $\sigma(\rho)=a=(a_1,\cdots a_n)$, $F(w)=F_a(w)=(f(\rho_a(w))-f(z))J_f(a)^{-1}\varphi_a'(m)^{-1}$, $d_{ij}(\rho)=(d_{ij}^{(1)}(\rho),\cdots,d_{ij}^{(n)}(\rho))=\frac{1}{2}\left.\frac{\partial^2 F(w)}{\partial w_i \partial w_j}\right|_{w=m}$ $(1\leqslant i,j\leqslant n)$, $\psi_a'=(u_{pl})$, $J_f(z)$ be the Jacobian of f at z, then*

$$\log\det J_f(z)-\log\det J_f(m)=\frac{1}{2}(\log K(z,\bar{z})-\log K(m,\bar{m}))+$$

$$+\sum_{l,k=1}^{n}\int_0^1 \frac{da_k}{d\rho}u_{kl}(\rho)(2\sum_{j=1}^{n}d_{lj}^{(j)}(\rho)-C_l)d\rho+$$

$$iIm\left\{\int_0^1\sum_{j=1}^{n}\frac{da_j}{d\rho}\frac{\partial}{\partial a_j}K(a,\bar{a})\cdot K(a,\bar{a})^{-1}d\rho\right\}. \tag{6}$$

we give the following corollaries.

Corollary 1. *Assumptions are the same as theorem 1. If m is starlike with respect to m, then (6) becomes*

$$\log\det J_f(z)-\log\det J_f(m)=\frac{1}{2}(\log K(z,\bar{z})-\log K(m,\bar{m}))+$$

$$+\sum_{l,k=1}^{n}\int_0^1(z_k-m_k)u_{kl}(\rho)(2\sum_{j=1}^{n}d_{lj}(\rho)-C_l)d\rho+$$

$$iIm\left\{\int_0^1 K(a,a)^{-1}\sum_{k=1}^{n}(z_k-m_k)\frac{\partial}{\partial a_k}K(a,a)d\rho\right\}. \tag{7}$$

Corollary 2. *Assumptions are the same as corollary 1. If M is bounded, then (7) becomes*

$$\log\det J_f(z)-\log\det J_f(m)=\frac{1}{2}(\log K(z,\bar{z})-\log\det K(m,\bar{m}))+$$

$$+2\sum_{\delta,K,l}\int_0^1(z_k-m_k)u_{kl}(\rho)d_{lj}^{(j)}(\rho)d\rho. \tag{8}$$

Let $M\subset C^n$ be a bounded symmetrical domain containing the origin. M is the standard realization of G/K, where G is a semi-simple connected noncompact Lie group with finite center and K is a maximal compact subgroup of G which keeps 0 fixed.

Let $\mathscr{G}_0$ be the Lie algebra of G, and $\mathscr{L}_0$ be the subalgebra of $\mathscr{G}_0$ corresponding to K. Then $\mathscr{G}_0=\mathscr{L}_0+\mathscr{P}_0$. Suppose $\mathscr{B}_0$ is a maximal Abelian subalgebra in $\mathscr{P}$. Then we can choese a basis of $\mathscr{B}_0$, X_1, $\cdots$, X_q, where $q=\dim \mathscr{B}_0=\text{rank}M$, and for any $X\in \mathscr{B}_0$, there exists a unique expression $X=x_1X_1+\cdots+x_qX_q$. Every $z\in M$ corresponds to a unique coset $\exp Ad(k)X\cdot K$ for some $k\in K$, $X\in\mathscr{B}_0$. Because K acts on M just as a subgroup of unitary group $U(n)$ acting on $\mathbb{C}^n$, we have

$$z=\xi(\exp Ad(k)X\cdot K)=k\,\xi\,(\exp X\cdot K)=\Lambda\widetilde{K},$$

where ξ is the mapping, which maps G/K onto M, $\Lambda=\xi\,(\exp X\cdot K)\subset M$, $\widetilde{k})=(\widetilde{k}_{ij})\in U(n)$.

We say a biholomorphic mapping f is normalized if $f(0)=0$, $J_f(0)=I$. If S is the subclass of classes of normalized biholomorphic mappings which satisfy the following condition: If $f(z)\in S$, and under the holomorphic automorphism transformation of M, we normalize it again, then we get F. If $F\in S$ is true, then we say S is an $\mathscr{A}$-invariant subclass.

Then by (3), we have

Theorem 2. *Assumptions are the same as above. If f is a normalized biholomorphic mapping which maps M into $\mathbb{C}^n$, $f\in S$, and S is an $\mathscr{A}$-invariant subclass, $K(0,0)=1$, then*

$$\left|\log\frac{\det J_f(z)}{\sqrt{K(z,z)}}\right|\leqslant C(S)\sum_{p=1}^{q}\log\frac{1+|\tanh x_p|}{1-|\tanh x_p|}\tag{9}$$

where $C(S)=\text{Sup}\left\{\left|\sum_{i,j}k_{li}\,d_{ij}^{(j)}\right|,\ f\in S,\ l=1,\ \cdots,\ n\right\}$.

From (9), we have

$$\sqrt{K(z,\bar{z})}\left(\frac{1-|\tanh x_p|}{1+|\tanh x_p|}\right)^{C(S)}\leqslant|\det J_f(z)|\leqslant\sqrt{K(z,\bar{z})}\left(\frac{1-|\tanh x_p|}{1+|\tanh x_p|}\right)^{C(S)}\tag{10}$$

If S is the class of normalized convex mappings, then we have $C(S)\leqslant m$. It will be proved in another paper.

In another paper, we will give the explicit forms of the distortion theorems for the classical domains and the nonsymmetrical transitive domains.

3. The Proof of Theorem 1.

Let $a\in M$. Then there exists a neighborhood of a such that f has the expansion

$$f(z)=f(a)+\sum_{k=1}^{\infty}\frac{1}{k!}\sum_{i_1,\cdots,i_k=1}^{n}\left(\left.\frac{\partial^k f(z)}{\partial z_{i_1}\cdots\partial z_{i_k}}\right|_{z=a}\right)(z_{i_1}-a_{i_1})\cdots(z_{i_k}-a_{i_k})$$

at this neighborhood. Let $g(w)=f(\varphi_a(w))$. Then g is a biholomorphic mapping which maps M into $\mathbb{C}^n$, and at a neighborhood of $w=m$, $g(w)$ has the expansion

$$g(w)=f(a)+\sum_{k=1}^{\infty}\frac{1}{k!}\sum_{i_1,\cdots,i_k=1}^{n}\left(\frac{\partial^k g(w)}{\partial w_{i_1}\cdots\partial w_{i_k}}\bigg|_{w=m}\right)(w_{i_1}-m_{i_1})\cdots(w_{i_k}-a_{i_k})$$

$$=f(a)+(w-m)\varphi_a{}'(m)J_f(a)+\frac{1}{2}\sum_{i,j=1}^{n}\left(\frac{\partial^2 g(n)}{\partial w_i\partial w_j}\bigg|_{w=m}\right)(w_i-m_i)(w_j-m_j)+\cdots \tag{11}$$

Denoting $\varphi_a=(\varphi_{a,1},\cdots\varphi_{a,n})$, we have

$$\frac{\partial^2 g(w)}{\partial w_i\partial w_j}=\frac{\partial}{\partial w_i}\left(\frac{\partial g}{\partial w_j}\right)=\frac{\partial}{\partial w_i}\left(\sum_{p=1}^{n}\frac{\partial\psi_{a,p}}{\partial w_j}\frac{\partial f}{\partial z_p}\right)=$$

$$=\sum_{p=1}^{n}\frac{\partial^2\varphi_{a,p}}{\partial w_i\partial w_j}\frac{\partial f}{\partial z_p}+\sum_{p=1}^{n}\frac{\partial\varphi_{a,p}}{\partial w_j}\frac{\partial^2 f}{\partial w_i\partial z_p}$$

$$=\sum_{p=1}^{n}\frac{\partial^2\varphi_{a,p}}{\partial w_i\partial w_j}\frac{\partial f}{\partial z_p}+\sum_{p,q=1}^{n}\frac{\partial\varphi_{a,p}}{\partial w_j}\frac{\partial\varphi_{a,q}}{\partial w_i}\frac{\partial^2 f}{\partial z_p\partial z_q}. \tag{12}$$

Normalizing $g(w)$, we get

$$F(w)=(g(w)-f(a))J_f(a)^{-1}\varphi_a{}'(m)^{-1}=(F_1(w),\ \cdots,\ F_n(w)).$$

The expansion of $F(w)$ at $w=m$ is

$$F(w)=w-m-\sum_{i,j}^{n}d_{ij}(w_i-m_i)(w_j-m_j)+\cdots \tag{13}$$

where $d_{ij}=-\frac{1}{2}\frac{\partial^2 F(w)}{\partial w_i\partial w_j}\bigg|_{w=m}$. On the other hand, taking $w=m$ in (12), and substituting it into (11), we have

$$F(w)=w-m+\frac{1}{2}\sum_{i,j,p,q=1}^{n}(w_i-m_i)(w_j-m_j)\left(\frac{\partial\varphi_{a,p}}{\partial w_j}\frac{\partial\varphi_{a,q}}{\partial w_i}\frac{\partial^2 f}{\partial z_p\partial z_q}\bigg|_{w=m}\right)\cdot$$

$$J_f(a)^{-1}\varphi_a{}'(m)^{-1}+\frac{1}{2}\sum_{i,j,p=1}^{n}(w_i-m_i)(w_j-m_j)\left(\frac{\partial^2\varphi_{a,p}}{\partial w_i\partial w_j}\frac{\partial f}{\partial z_p}\bigg|_{w=m}\right)J_f(a)^{-1}\varphi_a{}'(m)$$

$$+\cdots. \tag{14}$$

Obviously, $d_{ij}=d_{ji}$. Comparing (13) with (14), we get

$$4d_{ij}=\sum_{p,q=1}^{n}\left[\left(\frac{\partial\varphi_{a,p}}{\partial w_j}\frac{\partial\varphi_{a,q}}{\partial w_i}+\frac{\partial\varphi_{a,q}}{\partial w_j}\frac{\partial\varphi_{a,p}}{\partial w_i}\right)\frac{\partial^2 f}{\partial z_p\partial z_q}\bigg|_{w=m}\right]J_f(a)^{-1}\varphi_a{}'(m)^{-1}$$

$$+2\sum_{p=1}^{n}\left(\frac{\partial^2\varphi_{a,p}}{\partial w_i\partial w_j}\frac{\partial f}{\partial z_p}\bigg|_{w=m}\right)J_f(a)^{-1}\varphi_a{}'(m)^{-1}. \tag{15}$$

Moreover, $\frac{\partial f}{\partial z_p}(a)J_f(a)^{-1}=e_p=(0,\ \cdots,\ 0,\ 1,\ 0,\ \cdots,\ 0)$, i.e. at the position p, it is 1; otherwise it is zero. So we have

$$\sum_{p=1}^{n}\frac{\partial^2\varphi_{a,p}(m)}{\partial w_i\partial w_j}\frac{\partial f(a)}{\partial z_p}J_f(a)^{-1}=\sum_{p=1}^{n}\frac{\partial^2\varphi_{a,p}(m)}{\partial w_i\partial w_j}e_p=\frac{\partial^2\varphi_a(m)}{\partial w_i\partial w_j}. \tag{16}$$

Multiplying $\frac{1}{2}u_{ri}u_{sj}$ on both sides of (15), taking sum for i, j, and then multiplying $\varphi_a'(m)$, we have

$$2\sum_{i,j=1}^{n}u_{ri}u_{sj}d_{ij}\varphi_a'(m)=\frac{1}{2}\sum_{i,j,p,q}u_{ri}u_{sj}\left(\frac{\partial\varphi_{a,p}(m)}{\partial w_i}\frac{\partial\varphi_{a,q}(m)}{\partial w_j}+\right.$$

$$\left.+\frac{\partial\varphi_{a,p}(m)}{\partial w_j}\frac{\partial\varphi_{a,q}(m)}{\partial w_i}\right)\frac{\partial^2 f(a)}{\partial z_p\partial z_q}J_f(a)^{-1}+\sum_{i,j}u_{ri}u_{sj}\frac{\partial^2\varphi_a(m)}{\partial w_i\partial w_j}$$

by (16). Since $\varphi_a'(a)=(u_{ij}(a))$ and $\varphi_a'(m)=\left(\frac{\partial\varphi_{a,p}(m)}{\partial w_j}\right)_{1\leqslant p,q\leqslant n}$, $\varphi_a'(a)\varphi_a'(m)=I$, we have

$$2\sum_{i,j=1}^{n}u_{ri}u_{sj}d_{ij}\varphi_a'(m)=\frac{1}{2}\sum_{p,q=1}^{n}(\delta_{rp}\delta_{sq}+\delta_{sp}\delta_{rq})\frac{\partial^2 f(a)}{\partial z_p\partial z_q}J(a)^{-1}+$$

$$+\sum_{i,j=1}^{n}u_{ri}u_{sj}\frac{\partial^2\varphi_a(m)}{\partial w_i\partial w_j}.$$

If $r\neq s$, $(p,q)=(r,s)$ or $(p,q)=(s,r)$, then $\delta_{rp}\delta_{sq}+\delta_{sp}\delta_{rq}=1$, otherwise $\delta_{rp}\delta_{sq}+\delta_{sp}\delta_{rq}=0$. If $r=s$, $p=q=r=s$, then

$$\delta_{rp}\delta_{sq}+\delta_{sp}\delta_{rq}=2;\ \text{otherwise}\ \delta_{rp}\delta_{sq}+\delta_{sp}\delta_{rq}=0.$$

Thus we get

$$2\sum_{i,j}^{n}u_{ri}u_{sj}d_{ij}\varphi_a'(m)=\frac{\partial^2 f(a)}{\partial z_r\partial z_s}J_f(a)+\sum_{i,j=1}^{n}u_{ri}u_{sj}\frac{\partial^2\varphi_a(m)}{\partial w_i\partial w_j}$$

i. e.

$$\frac{\partial^2 f(a)}{\partial z_r\partial z_s}J_f(a)^{-1}=2\sum_{i,j=1}^{n}u_{ri}u_{sj}d_{ij}\varphi_a'(m)-\sum_{i,j=1}^{n}u_{ri}u_{sj}\frac{\partial^2\varphi_a(m)}{\partial w_i\partial w_j}, \tag{17}$$

Computing $\frac{\partial}{\partial z_p}$ logdet $J_f(z)$, we have

$$\frac{\partial}{\partial z_p}\text{logdet}\,J_f=(\det J_f)^{-1}\frac{\partial}{\partial z_p}\det J_f=(\det J_f)^{-1}\sum_{k=1}^{n}\sum_{(i_1,\dots,i_n)}$$

$$\delta_{i_1,i_2,\dots,i_n}^{1,2,\dots,n}\frac{\partial f_{i_1}}{\partial z_1}\cdots\frac{\partial^2 f_{i_k}}{\partial z_p\partial z_k}\cdots\frac{\partial f_{i_n}}{\partial z_n}=\sum_{i,k=1}^{n}\frac{\partial^2 f_i}{\partial z_p\partial z_k}J_{ik}(\det J_f)^{-1}=$$

$$=\mathrm{tr}\left(\left(\frac{\partial}{\partial z_p}J_f\right)J_f^{-1}\right)$$

where J_{ik} is the algebraic minor of element (i, k) of J. The qth row vector of $\left(\frac{\partial}{\partial z_p} J_f\right) J_f^{-1}\Big|_{z=a}$ is $\frac{\partial^2 f(a)}{\partial z_p \partial z_q} J_f(a)^{-1}$. Let $\varphi_a'(m)=(v_{ij})$, $d_{ij}=(d_{ij}^{(1)}, \cdots, d_{ij}^{(n)})$ and $d_{ij}^{(r)}=\frac{1}{2}\left.\frac{\partial^2 F_1(w)}{\partial w_i \partial w_j}\right|_{w=m}$ by (17). Then we have

$$\frac{\partial}{\partial z_p}\log\det J_f\Big|_{z=a}=\sum_{q,r}\frac{\partial^2 f(a)}{\partial z_p\partial z_q}J_f(a)^{-1}=2\sum_{i,j,q,r}u_{pi}u_{qj}d_{ij}^{(r)}v_{rq}$$

$$-\sum_{i,j,k}u_{pi}u_{kj}\frac{\partial^2\varphi_{a,k}(m)}{\partial w_i\partial w_j}=2\sum_{i,j=1}^{n}u_{pi}d_{ij}^{(j)}-\sum_{i,j,h=1}^{n}u_{pi}u_{kj}\frac{\partial^2\varphi_{a,k}(m)}{\partial w_i\partial w_j}. \tag{18}$$

The Bergman kernel function has the property

$$K(w,\overline{w})=\det\varphi_a'(w)K(\varphi_a(w),\overline{\varphi_a(w)})\overline{\det\varphi_a'(w)}. \tag{19}$$

We have

$$\frac{\partial}{\partial w_p}K(w,\overline{w})=K(a,\overline{a})\overline{\det\varphi_a'(m)}\frac{\partial}{\partial w_p}\det\varphi_a'(w)\Big|_{w=m}+$$

$$\det\varphi_a'(m)\overline{\det\varphi_a'(m)}\frac{\partial}{\partial w_p}K(\varphi_a(w),\overline{\varphi_a(w)})\Big|_{w=m}.$$

Dividing (19) on both sides, one can get

$$C_p=\frac{1}{K(m,\overline{m})}\frac{\partial}{\partial w_p}K(w,\overline{w})\Big|_{w=m}=\frac{1}{\det\varphi_a'(m)}\frac{\partial}{\partial w_p}\det\varphi_a'(w)\Big|_{w=m}+$$

$$+\frac{1}{K(a,\overline{a})}\frac{\partial}{\partial w_p}K(\varphi_a(w),\overline{\varphi_a(w)})\Big|_{w=m}=\sum_{k,j=1}^{n}\frac{\partial^2\varphi_{a,k}(m)}{\partial w_p\partial w_j}u_{kj}$$

$$+\sum_{k=1}^{n}\left(\frac{\partial\varphi_{a,k}}{\partial w_p}\frac{\partial}{\partial z_k}K(z,\overline{z})\Big|_{w=m,\,z=a}\right)K(a,\overline{a})^{-1}.$$

Multiplying u_{rp} on both sides, and taking the sum of p from 1 to n, we have

$$\sum_{p=1}^{n}u_{rp}c_p=\sum_{p,k,j=1}^{n}u_{rp}u_{kj}\frac{\partial^2\varphi_{a,k}}{\partial w_p\partial w_j}+\sum_{p,k=1}^{n}\frac{\partial\varphi_{e,k}}{\partial w_p}u_{rp}\frac{\partial}{\partial z_k}K(z,\overline{z})\Big|_{z=a}$$

$$K(a,a)^{-1}=\sum_{p,K,j=1}^{n}u_{rp}u_{kj}\frac{\partial^2\varphi_{a,k}}{\partial w_p\partial w_j}+\left(\frac{\partial}{\partial z_r}K(z,\overline{z})\Big|_{z=a}\right)K(a,\overline{a}).$$

Substituting it into (18), we have

$$\frac{\partial}{\partial z_p}\log\det J_f(z)\Big|_{z=a}=\sum_{i=1}^{n}u_{pi}(2\sum_{j=1}^{n}d_{ij}^{(j)}-c_i)+\left(\frac{\partial}{\partial z_p}K(z,z)\Big|_{z=a}\right)K(a,a). \tag{20}$$

Since $a=(a_1, \cdots, a_n)=a(\rho)$ is an analytic curve in M, $a(0)=m$, $a(1)=z$, $(0\leqslant\rho\leqslant 1)$, we have

$$\frac{d}{d\rho}\log\det J_f(a(\rho))=\sum_{p=1}^{n}\frac{da_p}{d\rho}\frac{\partial}{\partial a_p}\log\det J_f(a)=$$

$$=\sum_{i,p=1}^{n}\frac{da_p}{d\rho}u_{pi}\,(2\sum_{j=1}^{n}d_{ij}^{(j)}-c_i)+\sum_{p=1}^{n}\frac{da_p}{d\rho}\,(\frac{\partial}{\partial a_p}K(a,\bar{a}))K(a,\bar{a})^{-1} \tag{21}$$

by (20). But $\frac{\partial}{\partial\bar{a}_p}K(a,\bar{a})=\overline{\frac{\partial}{\partial a_p}K(a,\bar{a})}$, and

$$\frac{d}{d\rho}K(a,\bar{a})=\sum_{p=1}^{n}\frac{da_p}{d\rho}\frac{\partial}{\partial a_p}K(a,\bar{a})+\sum_{p=1}^{n}\overline{\frac{da_p}{d\rho}}\frac{\partial}{\partial\bar{a}_p}K(a,\bar{a})$$

$$=2\mathrm{Re}\sum_{p=1}^{n}\frac{da_p}{d\rho}\frac{\partial}{\partial a_p}K(a,\bar{a}),$$

and so (21) become

$$\frac{d}{d\rho}\log\det J_f(a)=\sum_{i,p=1}^{n}\frac{da_p}{d\rho}u_{pi}\,(2\sum_{j=1}^{n}d_{ij}^{(j)}-c_i)+\frac{1}{2}\frac{d}{d\rho}\log K(a,\bar{a})$$

$$+\sqrt{-1}\,\mathrm{Im}\sum_{p=1}^{n}\frac{da_p}{d\rho}\,(\frac{\partial}{\partial a_p}K(a,\bar{a}))K(a,\bar{a}).$$

Integrating the preceding equality with respect to ρ from 0 to 1, we obtain (6).

If M is starlike with respect to m, then $a(\rho)=m+\rho(z-m)$ and we get (7) immediately.

If M is bounded and starlike with respect to m, then the kernel function can be expressed as $K(z,\bar{z})=\sum_{\gamma}\eta_\nu(z-m)\overline{\eta_\nu(z-m)}$ where η_ν are homogeneous polynomials. Therefore $c_i=0$, $i=1,\cdots,n$, and $\sum_{p=1}^{n}(z_i-m_i)\frac{\partial}{\partial w_p}K(w,\bar{w})\Big|_{w=\rho(z-m)+m}$ is real. Thus (8) followed.

4. The Proof of Theorem 2.

If ξ maps G/K onto M, such that $\xi\,(\exp Y\cdot K)=z\in M$; where $Y\in\mathscr{P}_0$, M is the standard realization of G/K by ξ, and $\pi_q=\xi(\exp\mathscr{B}_0, K)$ is a flat submanifold of M. Let

$$\Lambda=(\lambda_1,\cdots,\lambda_s,i\lambda_{s+1},\cdots,i\lambda_q,0,\cdots,0)\in\pi_q,\ \lambda_i\in\mathbb{R},\ i=1,\cdots,q.$$

For $X\in\mathscr{B}_0$, $X=x_1X_1+\cdots+x_qX_q$ can be expressed as

$$\Lambda=\xi\,(\exp X\cdot K)=(\tanh x_1,\cdots,\tanh x_q,0,\cdots,0)\,B\in\pi_q$$

where $B=\begin{pmatrix} B_1 & 0 \\ 0 & I \end{pmatrix}=(b_{ij})$, $B_1=B_1^{(q)}$. Since $\exp\mathscr{B}_0\cdot K$ is flat at G/K, B is a complex constant matrix. Since ξ is a standard realized mapping, we have $|\lambda_j|<1$, $j=1,\cdots,q$ and $\sum_{i=1}^{n}|b_{ij}|=1$.

Obviously $\psi_\Lambda(z)=\xi(\exp-X\cdot\exp Y\cdot K)$. If $Y=y_1Y_1+\cdots+y_qY_q\in\mathscr{B}_0$, $\Lambda_1=\xi(\exp Y\cdot K)$, then

$$\psi_\Lambda(\Lambda_1)=\xi(\exp-X\cdot\exp Y\cdot K)=\xi(\exp((y_1-x_1)X+\cdots+(y_q-x_q)X_q)\cdot K).$$

Let $F(y)=F(y_1,\cdots,y_q)=(\tanh y_1,\cdots,\tanh y_q,0,\cdots,0)$. Thus

$$\psi_\Lambda(\Lambda_1)=F(y-x)B,\ \Lambda_1=F(y)B. \tag{22}$$

Since $\psi_\Lambda(\pi_q)=\pi_q$, (22) holds at π_q. Evaluating the Jacobain of ψ_Λ at π_q, on the one hand,

$$\left.\frac{\partial\psi_\Lambda}{\partial y}\right|_{y=x}=\frac{\partial}{\partial y}F(y-x)\Big|_{y=x}B=(I,0)B=(B_1,0). \tag{23}$$

On the other hand

$$\left.\frac{\partial\psi_\Lambda}{\partial y}\right|_{y=x}=\left.\frac{\partial\Lambda_1}{\partial y}\right|_{y=x}\left.\frac{\partial\psi_\Lambda}{\partial\Lambda_1}\right|_{\Lambda_1=\Lambda}=\left(\left.\frac{\partial F}{\partial y}\right|_{y=x}\right)B\left(\left.\frac{\partial\psi_\Lambda}{\partial\Lambda_1}\right|_{\Lambda_1=\Lambda}\right). \tag{24}$$

Since $\left.\frac{\partial F}{\partial y}\right|_{y=x}=(*,0)$, we still use $\left.\frac{\partial F}{\partial y}\right|_{y=x}$ to denote $*$, where $\left.\frac{\partial F}{\partial y}\right|_{y=x}$ is a $q\times q$ matrix. Thus (24) become

$$\left(\left.\frac{\partial F}{\partial y}\right|_{y=x},0\right)\begin{pmatrix} B_1, & 0 \\ 0, & I \end{pmatrix}\left(\frac{\partial\psi_\Lambda}{\partial\Lambda_1}\right)_{\Lambda_1=\Lambda}=\left(\left.\frac{\partial F}{\partial y}\right|_{y=x}B,0\right)\left(\frac{\partial\psi_\Lambda}{\partial\Lambda_1}\right)_{\Lambda_1=\Lambda}.$$

Comparing (23) with (24), we have

$$(B_1,0)=\left(\left.\frac{\partial F}{\partial y}\right|_{y=x}B_1,0\right)\left(\left.\frac{\partial\psi_\Lambda}{\partial\Lambda_1}\right|_{\Lambda_1=\Lambda}\right).$$

Thus

$$B_1=\left(\left.\frac{\partial F}{\partial y}\right|_{y=x}\right)B_1\left(\left.\frac{\partial\psi_\Lambda}{\partial\Lambda_1}\right|_{\Lambda_1=\Lambda}\right),$$

and so we have

$$\psi_\Lambda'(\Lambda)=B^{-1}\begin{pmatrix}\left(\left.\frac{\partial F}{\partial y}\right|_{y=x}\right)^{-1} & 0 \\ * & * \end{pmatrix}B. \tag{25}$$

Let $\Pi_q=\{t_1,\cdots,t_s,it_{s+1},\cdots,it_q,0,\cdots,0\}$, $\pi t_j\in\mathbb{R}$ $j=1,\cdots,q$. ψ_Λ preserves π_q, and $\psi_\Lambda'(\Lambda)$ preserves the tangent space Π_q of π_q. Obviously

$$\left(\frac{\partial F}{\partial y}\bigg|_{y=x}\right)^{-1}=\begin{pmatrix}(1-\tanh^2 x_1)^{-1} & & 0\\ & \ddots & \\ 0 & & (1-\tanh^2 x_q)^{-1}\end{pmatrix}.$$

and

$$\begin{aligned}\Lambda\psi_\Lambda'(\Lambda)&=(\tanh x_1,\cdots,\tanh x_q,0,\cdots,0)B\psi_\Lambda'(\Lambda)\\&=\left(\frac{\tanh x_1}{1-\tanh^2 x_1},\cdots,\frac{\tanh x_q}{1-\tanh^2 x_q},0,\cdots,0\right)B\end{aligned}$$

by (25). Let Λ be as above, $0\leqslant\rho\leqslant 1$,

$$\begin{aligned}\rho\Lambda&=(\tanh s_1,\cdots,\tanh s_q,0,\cdots,0)B\\&=(\rho\tanh t_1,\cdots,\rho\tanh t_q,0,\cdots,0)B.\end{aligned}$$

Thus we have

$$\psi_{\rho\Lambda}'(\rho\Lambda)=B^{-1}\begin{pmatrix}(1-\tanh^2 s_1)^{-1} & & 0 & \\ & \ddots & & \\ 0 & & (1-\tanh^2 s_q)^{-1} & 0\\ & * & & *\end{pmatrix}B$$

$$=B^{-1}\begin{pmatrix}(1-\rho^2\tanh^2 x_1)^{-1} & & 0 & \\ & \ddots & & \\ 0 & & (1-\rho^2\tanh^2 x_q)^{-1} & 0\\ & * & & *\end{pmatrix}B$$

and then

$$\Lambda\psi_{\rho\Lambda}'(\rho\Lambda)=\left(\frac{\tanh x_1}{1-\rho^2\tanh^2 x_1},\cdots,\frac{\tanh x_q}{1-\rho^2\tanh^2 x_q},0,\cdots,0\right)B.$$

For any $Y\in\mathscr{P}_0$, there exist $X\in\mathscr{B}_0$ and $k\in K$ such that $Y=\mathrm{Ad}(k)X$, and then $z=\xi(\exp Y\cdot K)=\xi(\exp(\mathrm{Ad}(k)X)\cdot K)=k\xi(\exp X\cdot K)=k\Lambda=\Lambda\widetilde{k}$. K acting on M is just like a subgroup of unitary group $U(n)$ acting on $\mathbb{C}^n$, *thus* $\widetilde{k}=(k_{ij})\in U(n)$, and hence

$$\psi_z'(z)=\widetilde{k}^{-1}\psi_\Lambda'(\Lambda)\widetilde{k},\ z\psi_z'(z)=\Lambda\psi_\Lambda'(\Lambda)\widetilde{k},\ z\psi_{\rho z}'(\rho z)=\Lambda\psi_{\rho\Lambda}'(f\Lambda)\widetilde{k}$$

Because $X=x_1X_1+\cdots+x_qX_q$, $\Lambda=\xi(\exp X\cdot K)$ $z=\xi$ $(\exp\mathrm{Ad}(k)X\cdot K)$, we have

$$z\psi_{\rho z}'(\rho z)=\Lambda\psi_{\rho\Lambda}'(\rho\Lambda)\widetilde{k}=\left(\frac{\tanh x_1}{1-\rho^2\tanh^2 x_1},\cdots,\frac{\tanh x_q}{1-\rho^2\tanh^2 x_q},0,\cdots,0\right)B\widetilde{k}$$

$$\sum_{p=1}^{n}z_pu_{pl}(\rho z)=\sum_{p,l}b_{pl}\frac{\tanh x_p}{1-\rho^2\tanh^2 x_p}k_{li}.$$

If $m=0$, $J_f(0)=I$, $K(0,0)=1$, then

$$\log\det J_f(z)=\log\sqrt{K(z,\bar{z})}+\int_0^1 2\sum\frac{\tanh x_p}{1-\rho^2\tanh^2 x_p}b_{pl}k_{li}d_{ij}^{(j)}(\rho)d\rho.$$

Let $C(S)=\mathrm{Sup}\left\{\left|\sum_{i,j}k_{li}d_{ij}^{(j)}\right|,f\in S,l=1,\cdots,n\right\}$.

We have the estimation

$$\left|\log\frac{\det J_f(z)}{\sqrt{K(z,\bar{z})}}\right|\leqslant C(S)\int_0^1 2\sum_{l,p}\frac{|\tanh x_p|}{1-\rho^2\tanh^2 x_p}|b_{pl}|\,d\rho.$$

since $\sum_{l=1}^{n}|b_{pl}|=1$, the right hand side is equal to

$$C(S)\int_0^1 2\sum_p\frac{|\tanh x_p|}{1-\rho^2\tanh^2 x_p}\,d\rho=C(S)\sum_p\log\frac{1+|\tanh x_p|}{1-|\tanh x_p|},$$

and finally we get (9).

References

[1] H. Cartan, Sur la possibilité d etendre aux fonctions de plusieurs variables complexes la theorie des fonctions univalentes, page 129—155, Note added to P. Montel, Lecons sur les Fonctions Univalentes ou Multivalentes, Gauthier-Villars, Paris, 1933.

[2] Sheng Gong, C. H. FitzGerald and R. W. Barnard, *Chinese Science Bulletin*, 34(1989), 81—83.

[3] Tashen Liu, (to appear at Chinese Annals of Mathematics).

Chin. Ann. of Math.
16B: 2(1995),203-212.

DISTORTION THEOREM FOR BIHOLOMORPHIC MAPPINGS IN TRANSITIVE DOMAINS (IV)

ZHENG XUEAN* GONG SHENG**

Abstract

The distortion theorem for biholomorphic starlike mappings (with respect to origin) in bounded symmetric domains are given. The distortion theorem for locally biholomorphic convex mappings in bounded symmetric domains are given also.

Keywords Biholomorphic mapping, Transitive domian, Distortion theorem.
1991 MR Subject Classification 32H.

§1. Introduction

Let $M \subset \mathbf{C}^n$ be a bounded symmetric domain. Let G be a Lie group consisting of some holomorphic automorphisms of M and acting transitively on M, K be the isotropy subgroup of G which fixes the point $m \in M$. Then M is a bounded realization of symmetric space G/K. Let ξ be the holomorphic diffeomorphism of G/K onto M. Moreover, let $M_0 \subset \mathbf{C}^n$ be the Harish-Chandra canonical realization of G/K, and ξ_0 be the holomorphic diffeomorphism of G/K onto M_0. We assume $\xi_0\xi^{-1}(m) = 0$. Let $K_M(z,\bar{z})$, $K_{M_0}(z,\bar{z})$ be the Bergman kernel functions of M and M_0 respectively. Let $\Psi_z \in G$ which transits the point $z \in M$ to m, i.e., $\Psi_z(z) = m$, and φ_z be the inverse of Ψ_z, i.e., $\varphi_z \in G$ and $\varphi_z(m) = z$.

Let $\mathcal{G}$ be the Lie algebra of $G, \mathcal{K}$ be the subalgebra of $\mathcal{G}$ corresponding to K. Then we have Cartan decomposition of $\mathcal{G} = \mathcal{K} + \mathcal{P}$. Suppose that $\mathcal{A}$ is a maximal Abelian subspace in $\mathcal{P}$ and A is the analytic subgroup in G corresponding to $\mathcal{A}$ in $\mathcal{G}$. We can choose a basis of $\mathcal{A}$, say $X_1, \cdots X_q$, where $q = \dim\mathcal{A} =$rank G/K, and for any $X \in \mathcal{A}$ there exists a unique expression $X = x_1X_1 + \cdots + x_qX_q$. For any $z \in M_0$, there exist $k \in K$ and $X \in \mathcal{A}$, such that

$$z = \xi_0 (k \exp X \cdot O) = (\tanh x_1, \cdots, \tanh x_q, 0, \cdots, 0)\,\tilde{k}, \tag{1.1}$$

where O is the identity coset in G/K, and $k \in K \to \tilde{k}' \in U_n$ is the unitary representation of K and U_n is the unitary group.

First we will prove the following theorem.

Theorem 1.1. *Let $M \subset \mathbf{C}^n$ be a bounded symmetric domain and $S(M)$ denote the family of normalized biholomorphic starlike mappings (with respect to origin) from M into*

Manuscript received March 19, 1993.
*Department of Mathematics, Anhui University, Hefei 230039, China.
**Deparatment of Mathematics, University of Science and Technology of China, Hefei 230039, China.

$\mathbf{C}^n$. Then there exists a positive constant $C(S(M))$, such that for any $f \in S(M)$ the following inequality

$$\sqrt{\frac{K_M(z,\bar{z})}{K_M(m,\bar{m})}}\liminf_{\theta\to b\in\partial M_0}\left(\frac{K_{M_0}(\theta,\bar{\theta})}{K_{M_0}(\eta,\bar{\eta})}\right)^{C(S(M))}$$

$$\leq|\det J_f(t)|\leq\sqrt{\frac{K_M(z,\bar{z})}{K_M(m,\bar{m})}}\limsup_{\theta\to b\in\partial M_0}\left(\frac{K_{M_0}(\theta,\bar{\theta})}{K_{M_0}(\eta,\bar{\eta})}\right)^{C(S(M))}\tag{1.2}$$

holds, where $z=\xi(ka\cdot O)$, $\theta=\xi_0(a_1\cdot O)$, $\eta=\xi_0(a_1a\cdot O)$, $k\in K, a, a_1\in A$. Furthermore, $C(S(M))$ is bounded by

$$\frac{1}{2}\leq C(S(M))\leq\frac{7n-2}{4}.\tag{1.3}$$

A holomorphic mapping on M is locally biholomorphic if $\det J_f(z)\neq 0$ for all $z\in M$.

Let $d(a,z)$ be the Bergman distance between a and z, and let $D(a,\rho)$ denote

$$\{w\in M\mid d(a,w)<\rho\}.$$

Suppose that f is a locally biholomorphic mapping on M, and denote

$$\sup\{r\mid f\text{ is biholomorphic on }D(z,r)\}$$

by $\rho(z,f)$, and $\inf\{\rho(z,f)|\ z\in M\}$ by $\rho(f)$.

We call f a uniformly locally biholomorphic mapping if $\rho(f)>0$. Moreover, if these exist $0<r\leq\rho(f)$ such that f is a convex mapping on $D(z,r)$, then we call f a uniformly locally biholomorphic convex mapping.

Suppose that $\Omega\subset M$ is an open set in $\mathbf{C}^n$, $m\in\Omega$. Let $f: M\to\mathbf{C}^n$ be a locally biholomorphic mapping. If for any $z\in M$, $f\circ\varphi_z$ is a biholomorphic convex mapping on Ω, then we call f an Ω−uniformly locally biholomorphic convex mapping.

If Ω is a geodesic ball centered at m, then the definition of Ω−uniformly locally biholomorphic convex mapping is coincide with the definition of uniformly locally biholomorphic convex mapping.

The family of Ω−uniformly locally biholomorphic convex mappings is an invariant family (cf. [1]).

We will prove the following theorem.

Theorem 1.2. *Let $M\subset\mathbf{C}^n$ be a bounded symmetric domain, and $f: M\to\mathbf{C}^n$ be a normalized Ω−uniformly locally biholomorphic convex mapping. Suppose that $\tilde{\Omega}=\xi_0\cdot\xi^{-1}(\Omega)$, $\tilde{\Omega}$ contains the largest ball centered at the origin with a radius r, and is in the smallest ball centered at the origin with a radius R. Then*

$$\sqrt{\frac{K_M(z,\bar{z})}{K_M(m,\bar{m})}}\prod_{j=1}^{q}\left(\frac{1-|\tan hx_j|}{1+|\tan hx_j|}\right)^{A_j}$$

$$\leq|\det J_f(z)|\leq\sqrt{\frac{K_M(z,\bar{z})}{K_M(m,\bar{m})}}\prod_{j=1}^{q}\left(\frac{1-|tanhx_j|}{1+|tanhx_j|}\right)^{A_j},\tag{1.4}$$

where $x_j, i=1,\cdots,q$, are as defined in (1.1), and $A_j, i=1,\cdots,q$, are constants which *satisfy the inequalities*

$$1 \le \frac{1}{4}\sum_{p=q+1}^{2n} |\alpha_p(X_j)| + \frac{1}{2} \le A_j \le nr^{-2}R, \tag{1.5}$$

where $\alpha_{q+1},\cdots,\alpha_{2n}$ *are the positive roots (counting the multiplicity) of the adjoint representations of* $\mathcal{A}$ *at* $\mathcal{G}$.

§2. On the Determinant of Jacobian

In [1], we studied transitive domain $M \subset \mathbf{C}^n$, including both bounded and unbounded. Denote the Jacobian of Ψ_z by J_{Ψ_z}. If M is unbounded, we assume $|\det J_{\Psi_k}(m)| = 1$ for all $k \in K$ and define

$$K_M(z,\bar z) = c\det J_{\Psi_z}(z)\overline{\det J_{\Psi_z}(z)},$$

where c is a constant. If M is bounded, $K_M(z,\bar z)$ is the Bergman kernel function of M. We denote $K(m,\bar m)^{-1}\left(\frac{\partial}{\partial z}K(z,\bar z)\right)_{z=m}$ by Cp.

For any point z in M, we can connect z and m by an analytic curve $a(\rho) = (a_1(\rho),\cdots, a_n(\rho))$, $(0 \le \rho \le 1)$ such that $a(0) = m$, $a(1) = z$. If M is starlike with respect to m, we can take $a(\rho) = m + \rho(z - m)$.

In [1], we proved

Theorem A. *Assume that f is a biholomorphic mapping on transitive domain M which maps M into $\mathbf{C}^n$. Let* $a(\rho) = (a_1(\rho),\cdots,a_n(\rho))$, $F(w) = (f(\varphi_a(w)) - f(a))\,(J_f(a)^{-1})'\,(J_{\varphi_a}(m)^{-1})'$,

$$d_{ij}(\rho) = \left(d_{ij}^{(1)}(\rho),\cdots,d_{ij}^{(n)}(\rho)\right) = \frac{1}{2}\frac{\partial^2 F(w)}{\partial w_i \partial w_j}\bigg|_{w=m}, \quad 1 \le i,j \le n,$$

$(J_{\Psi_a}(a))' = (u_{p\ell}(a))$. *Then*

$$\begin{aligned}\log\det J_f(z) - \log\det J_f(m) &= \frac{1}{2}(\log K_M(z,\bar z) - \log K_M(m,\bar m)) \\ &\quad + \sum_{\ell,k=1}^{n}\int_0^1 \frac{da_k}{d\rho}u_{k\ell}(\rho)\left(2\sum_{j=1}^{n} d_{\ell j}^{(j)}(\rho) - c_\ell\right)d\rho \\ &\quad + i\mathrm{Im}\left\{\sum_{j=1}^{n}\frac{da_j}{d\rho}\frac{\partial}{\partial a_j}k(a,\bar a)\cdot K(a,\bar a)^{-1}d\rho\right\}.\end{aligned} \tag{2.1}$$

Actually, Theorem A holds even if M is a locally biholomorphic mapping or f is a holomorphic mapping at the point where $\det J_f(z) \ne 0$ (assume $\det J_f(m) \ne 0$).

We give another version of Theorem A.

Theorem 2.1. *Assumptions are the same as Theorem A and M is a bounded symmetric* *domain. Then*

$$\log\det J_f(z)-\log\det J_f(m)=\frac{1}{2}\left(\log K_M(z,\bar z)-\log k_M(m,\bar m)\right)$$
$$+\sum_{p=1}^{q}\sum_{\ell=1}^{n}x_p\int_0^1 k_{p\ell}\left(2\sum_{j=1}^{n}d_{\ell_j}^{(j)}(\rho)-c_\ell\right)d\rho$$
$$+i\mathrm{Im}\left\{\int_0^1\sum_{j=1}^{n}\frac{da_j}{d\rho}\frac{\partial}{\partial a_j}K(a,\bar a)\cdot K(a,\bar a)d\rho\right\},\qquad(2.2)$$

where $(k_{p\ell})$ is a unitary matrix.

Proof. in [3], we already proved that $\frac{\det J_f(z)}{\sqrt{K_M(z,\bar z)/K_M(m,\bar m)}}$ is a biholomorphic invariant. We have

$$\frac{\det J_f(z)}{\sqrt{K_M(z,\bar z)/K_M(m,\bar m)}}=\frac{\det J_{\tilde f}\left(\xi_0\circ\xi^{-1}(z)\right)}{\sqrt{K_{M_0}\left(\xi_0\circ\xi^{-1}(z),\overline{\xi_0\circ\xi^{-1}(z)}\right)/K_{M_0}(0,0)}},$$

where $\tilde f$ is the normalization of $f\circ\xi\circ\xi_0^{-1}$. We just need to estimate

$$2\sum_{i,k,\ell=1}^{n}\int_0^1\frac{da_k}{d\rho}u_{k\ell}(\rho)d_{\ell j}^{(i)}(\rho)dp$$

on M_0 only.

Let $M_{0,R}=\left\{(x,y)\in\mathbf{R}^{2n}\,|z=x+iy\in M_0\subset\mathbf{C}^n\right\}$, and the real mapping $\xi_{0,R}$ corresponding to ξ_0 be

$$\xi_{0,R}\left(k\exp\left(\sum_{j=1}^{q}x_iX_j\right)\cdot O\right)=(\tanh x_1,\cdots\tanh x_q,0,\cdots,0)\,P(k)\in M_{0,R}.$$

Let $g\in G$, Ψg denote the holomorphic automorphism corresponding to g. If $g=k_0\exp Y$, $k_0\in K$, $Y\in\mathcal{P}$, then $g(\rho)=k_0\exp\rho Y=k_0\exp Adk(\rho X)$ where $k\in K$, $X\in\mathcal{A}$. Then

$$\xi_0(g(\rho)\cdot O)=a(\rho)\in M_0,\quad\Psi_{g(\rho)}=\Psi_{a(\rho)},$$
$$J_{\Psi g(\rho)}\left(\xi_0(g(\rho)\cdot O)\right)=J_{\Psi_a}(a)=(u_{k\ell}(a))',$$
$$\Psi_{g(\rho)}(z)=\Psi_{g(\rho)}\left(\xi_0(g(1)\cdot O)\right)=\xi_0\left(g(\rho)^{-1}g(1)\cdot O\right)$$

and $\Psi_{g(\rho)}\left(\xi_0(g(\rho)\cdot O)\right)$, where $z=\xi_0(g(1)\cdot O)$, i.e., $a(1)=z$.

Let $a(\rho)=x(\rho)+iy(\rho)\in M_0$. Then $(x(\rho),y(\rho))\in M_{0,R}$. Obviously, $k_0\exp\rho Y\exp tY\cdot O$ is a curve in G/K which contains the point $k_0\exp\rho Y\cdot O$. $g(\rho)^{-1}$ transfers this curve as $\exp tY\cdot O$ which contains the point 0. The corresponding curves in M_0 are the curves $\xi_0\left(k_0\exp\rho Y\exp tY\cdot 0\right)$ and $\xi_0\left(k_0\exp tY\cdot O\right)$, these two curves contain the point $\xi_0(\exp tY\cdot 0)$ and $\xi_0(e\cdot O)=0$ respectively. Thus $\Psi_{g(\rho)}$ maps the curve $a(\rho+t)$ to the curve $\xi_0(\exp tY\cdot 0)$, $\frac{da(\rho)}{d\rho}$ is the tangent of the curve $a(\rho+t)$ at the point $a(\rho)$ and $\frac{d\xi_0(\exp tY\cdot O)}{dt}|_{t=0}$ is the tangent of the curve $\xi_0(\exp tY\cdot O)$ at the point 0. We have

$$\frac{da(\rho)}{dp}\left(J_{\Psi_{g(\rho)}}\left(\xi_0(g(\rho)\cdot O)\right)\right)'=\frac{d\xi_0(\exp tY\cdot O)}{dt}\bigg|_{t=0}.$$

It is easily to verify

$$\frac{d\xi_0(\exp tY\cdot O)}{dt}\bigg|_{t=0}=(x_1,\cdots,x_q,0,\cdots,0)\,\tilde k,$$

where $\tilde{k}=(k_{j\ell})$ is a unitary matrix. Noticing

$$\left(J_{\Psi_{g(\rho)}}(\xi_0(g(\rho)\cdot O))\right)'=\left(J_{\Psi_{a(\rho)}}(a(\rho))\right)'=(u_{k\ell}),$$

we get on M_0,

$$\sum_{k=1}^{n}\frac{da_k(\rho)}{d\rho}u_{k\ell}(\rho)=\sum_{j=1}^{q}x_jk_{j\ell}. \tag{2.3}$$

By (2.3), we obtain (2.2) immediately.

§3. Distortion of Biholomorphic Mapping

Moreover, we can prove the following distortion theorem.

Theorem 3.1. *If the assumptions of Theorem 2.1 are satisfied, then*

$$\sqrt{\frac{K_M(z,\bar z)}{K_M(m,\bar m)}}\liminf_{\theta\to b\in\partial M_0}\left(\frac{K_{M_0}(\theta,\bar\theta)}{K_{M_0}(\eta,\bar\eta)}\right)^{c(f,M_0)}$$
$$\le|\det J_f(z)|\le\sqrt{\frac{K_M(z,\bar z)}{K_M(m,\bar m)}}\limsup_{\theta\to b\in\partial M_0}\left(\frac{K_{M_0}(\theta,\bar\theta)}{K_{M_0}(\eta,\bar\eta)}\right)^{c(f,M_0)} \tag{3.1}$$

holds, where $z=\xi(ka\cdot O)$, $\theta=\xi_0(a_1\cdot O)$, $\eta=\xi_0(a_1a\cdot O)$, $k\in K$, $a,a_1\subset A$,

$$c(f,M_0)=\sup\left\{\left|2\int_0^1\sum_{j=1}^{n}d_{\ell j}^{(j)}(\rho)d\rho\right|\Bigg/\left(\frac12\sum_{i=q+1}^{2n}|\alpha_j(X_\ell)|+1\right),\ \ell=1,\cdots,q,\ z\in M_0\right\},$$

$d_{\ell j}(\rho)$ *are given in Theorem A, and* $\alpha_{q+1},\cdots,\alpha_{2n}$ *are the positive roots (counting the multiplicity) of the adjoint representations of* $\mathcal{A}$ *at* $\mathcal{G}$.

Proof. By Theorem 2.1, formula (2.2) and $\det J_f(m)=1$, we have

$$\log|\det J_f(z)|=\frac12\log\frac{K_M(z,\bar z)}{K_M(m,\bar m)}+\mathrm{Re}\sum_{l,p=1}^{n}x_p\int_0^1k_{p\ell}\left(2\sum_{j=1}^{n}d_{\ell_j}^{(j)}(\rho)-c_\ell\right)d\rho.$$

Since $\frac{\det J_f(z)}{\sqrt{K_M(z,\bar z)/K_M(m,\bar m)}}$ is a biholomorphic invariant, we just need to estimate

$$\sum_{\ell,p=1}^{n}x_p\int_0^1k_{p\ell}\left(2\sum_{j=1}^{k}d_{\ell j}^{(j)}(\rho)-c_\ell\right)d\rho$$

on M_0 only. But at $M_0, c_\ell=0$ we have

$$\left|\frac{\det J_f(z)}{\sqrt{K_M(z,\bar z)/K_M(m,\bar m)}}\right|=e^{\mathrm{Re}\left\{\sum\limits_{j,\ell,p=1}^{n}2x_p\int_0^1k_{p\ell}d_{\ell j}^{(j)}(\rho)d\rho\right\}}.$$

Since $e^{2x_j}=\frac{1+\tanh x_j}{1-\tanh x_j}$, the previous equality is equivalent to

$$\left|\frac{\det J_f(z)}{\sqrt{K_M(z,\bar z)/K_M(m,\bar m)}}\right|=\prod_{p=1}^{q}\left(\frac{1+\tanh x_p}{1-\tanh x_p}\right)^{\sum\limits_{j,\ell=1}^{n}\mathrm{Re}\int_0^1k_{p\ell}d_{\ell j}^{(j)}(\rho)d\rho} \tag{3.2}$$

On the other hand, we already proved that (cf. [3])

$$\frac{K_{M_0}(z,\bar z)}{K_{M_0}(0,0)}=\prod_{j=1}^{q}\frac{1}{1-\lambda_j^2}\prod_{j=q+1}^{2n}\frac{e^{\alpha_j(X)}-e^{-\alpha_j(X)}}{2\alpha_j(\tanh X)}, \tag{3.3}$$

where $\lambda_j = \tanh x_j, j = 1, \cdots, q$,

$$\tanh X = \tanh\left(\sum_{j=1}^{q} x_j X_j\right) = \sum_{j=1}^{q} (\tanh x_j)\, X_j = \sum_{j=1}^{q} \lambda_j X_j.$$

By the Dynkin diagram of restricted root systems of the adjoint representation of $\mathcal{A}$ at real semi-simple Lie algebra of $\mathcal{G}$, there are only the following possibilities for $\alpha_j(X)$ (cf. [4]): for $X = x_1X_1 + \cdots + x_qX_q \in \mathcal{A}$, we have

1) $\alpha_j(X) = x_p + x_\ell,\ p \neq \ell,\ 1 \le p, \ell \le q$;

2) $\alpha_j(X) = x_p - x_\ell,\ p \neq \ell,\ 1 \le p, \ell \le q$;

3) $\alpha_j(X) = x_p$;

4) $\alpha_j(X) = 2x_p$.

For the case (1),

$$\alpha_j(X_i) = \begin{cases} 1, & \text{when } i = p \text{ or } \ell; \\ 0, & \text{otherwise;} \end{cases}$$

then we have

$$\frac{e^{\alpha_j(X)} - e^{-\alpha_j(X)}}{\alpha_j(\tanh X)} = \frac{e^{x_p + x_\ell} - e^{-x_p - x_\ell}}{2\,(\lambda_p + \lambda_\ell)}$$
$$= \left(\frac{1}{1-\lambda_p^2}\right)^{\frac12}\left(\frac{1}{1-\lambda_\ell^2}\right)^{\frac12} = \prod_{p=1}^{q}\left(\frac{1}{1-\lambda_p^2}\right)^{\frac12|\alpha_j(X_p)|}.$$

Similarly for the cases 2), 3), 4), we also have

$$\frac{e^{\alpha_j(X)} - e^{-\alpha_j(X)}}{2\alpha_j(\tanh X)} = \prod_{p=1}^{q}\left(\frac{1}{1-\lambda_p^2}\right)^{\frac12|\alpha_j(X_p)|}.$$

From (3.3), we get

$$\frac{K_{M_0}(z,\bar z)}{K_{M_0}(0,0)} = \prod_{p=1}^{q}\frac{1}{1-\lambda_p^2}\prod_{j=q+1}^{2n}\prod_{p=1}^{q}\left(\frac{1}{1-\lambda_p^2}\right)^{\frac12|\alpha_j(X_p)|}$$
$$= \prod_{p=1}^{q}\left(\frac{1}{1-\lambda_p^2}\right)^{\frac12\sum\limits_{j=q+1}^{2n}|\alpha_j(X_p)|+1}. \tag{3.4}$$

Let

$$z = \xi_0(ka\cdot O) = \xi_0\left(k\exp\sum_{j=1}^{q} x_jX_j\cdot O\right), \quad a = \exp\sum_{j=1}^{q} x_jX_j \in A;$$

$$\theta = \xi_0(a_1\cdot O) = \xi_0\left(\exp\sum_{j=1}^{q} y_jX_j\cdot O\right), \quad a_1 = \exp\sum_{j=1}^{q} y_jX_j \in A;$$

$$\eta = \xi_0(a_1a\cdot O) = \xi_0\left(\exp\left(\sum_{j=1}^{q}(x_j+y_j)\,X_j\right)\cdot O\right).$$

Then by (3.4) we have

$$\frac{K_{M_0}(\theta,\bar\theta)}{K_{M_0}(\eta,\bar\eta)} = \prod_{p=1}^{q}\left(\frac{1-\nu_p^2}{1-\mu_p^2}\right)^{\frac12\sum\limits_{j=q+1}^{2n}|\alpha_j(X_p)|+1},$$

where $\nu_p = \tanh(x_p + y_p)$, $\mu_p = \tanh y_p$. Obviously,

$$\sup_{\theta \to b \in \partial M_0} \lim \frac{K_{M_0}(\theta,\bar\theta)}{K_{M_0}(\eta,\bar\eta)} = \prod_{p=1}^{q} \left(\frac{1+|\tanh x_p|}{1-|\tanh x_p|}\right)^{\frac{1}{2}\prod_{j=q+1}^{2n}|\alpha_j(X_p)|+1} \tag{3.5}$$

and

$$\inf_{\theta \to b \in \partial M_0} \lim \frac{K_{M_0}(\theta,\bar\theta)}{K_{M_0}(\eta,\bar\eta)} = \prod_{p=1}^{q} \left(\frac{1-|\tanh x_p|}{1+|\tanh x_p|}\right)^{\frac{1}{2}\prod_{j=q+1}^{2n}|\alpha_j(X_p)|+1}. \tag{3.6}$$

From (3.2), (3.5) and (3.6), we have (3.1).

§4. Proof of Theorem 1.1

We now give a proof of Theorem 1.1 based on Theorem 3.1. Gong Sheng[5] proved the following result.

If M_0 is the bounded symmetric domain, it is the Harish-Chandra canonical realization of symmetric space G/K. If $\varphi(z)$ is defined as: $\varphi(z) = r$, if $z = rz_0$, $z_0 \in \partial M_0$, and f is a normalized biholomorphic starlike (with respect to origin) mapping on M_0 which maps M_0 into $\mathbf{C}^n$, then

$$|f(z)| \le \frac{\sqrt{q}\varphi(z)}{(1-\varphi(z))^2}, \tag{4.1}$$

where $q = \text{rank } G/K$.

Let $\{\phi_\nu(z)\}$ be the othonormal system on M_0. Then the Bergman kernel function $K_{M_0}(z,\bar\xi) = \sum\limits_{\nu=0}^{\infty} \phi_\nu(z)\overline{\phi_\nu(\xi)}$ and

$$F(z) = f(rz) = \int_{M_0} f(r\xi) K_{M_0}(z,\bar\xi)\, d\xi = \sum_{\nu=0}^{\infty} c_\nu(r)\phi_\nu(z),$$

where $c_\nu(r) = \int_{M_0} f(r\xi)\overline{\phi_\nu(\xi)}d\xi$. By Parseval's equality, $\sum\limits_{\nu=0}^{\infty} |c_\nu(r)|^2 = \int_{M_0} |f(r\xi)|^2 d\xi$. Let $w = rz$, $f(z) = (f_1(z), \cdots, f_n(z))$, $c_\nu(r) = \left(c_\nu^{(1)}(r), \cdots, c_\nu^{(n)}(r)\right)$. Then

$$\frac{\partial F(z)}{\partial z} = r\frac{\partial f}{\partial w}(w) = \sum_{\nu=0}^{\infty} c_\nu(r)\frac{\partial \phi_\nu(z)}{\partial z},$$

and $r\frac{\partial f_\ell}{\partial w_j}(w) = \sum\limits_{\nu=0}^{\infty} c_\nu^{(\ell)}(r)\frac{\partial \phi_\nu(z)}{\partial z_j}$. By the Schwarz inequality,

$$\begin{aligned} r\left|\frac{\partial f_\ell}{\partial w_j}(w)\right| &\le \|f_\ell(r\xi)\|_2 \left(\sum_{\nu=0}^{\infty} \frac{\partial}{\partial z_j}\phi_\nu(z)\frac{\partial}{\partial \bar z_j}\overline{\phi_\nu(z)}\right)^{\frac{1}{2}} \\ &= \|f_\ell(r\xi)\|_2 \left(\frac{\partial^2}{\partial z_j \partial \bar z_j} K_{M_0}(z,\bar z)\right)^{\frac{1}{2}}. \end{aligned} \tag{4.2}$$

By (3.4), it is easy to verify that

$$\frac{\partial}{\partial \bar z_j} K_{M_0}(z,\bar z) = O\left(K_{M_0}(z,\bar z)\prod_{\ell=1}^{q}\left(\frac{1}{2}\sum_{j=q+1}^{2n}|\alpha_j(x_\ell)|+1\right)\frac{2\lambda_\ell}{1-\lambda_\ell^2}\bar k_{j\ell}\right)$$

where $z = (\lambda_1, \cdots, \lambda_q, 0, \cdots, 0)\tilde{k}$, $\tilde{k} = (k_{j\ell})$. Similarly, we get

$$\frac{\partial^2 K_{M_0}(z,\bar{z})}{\partial z_j \partial \bar{z}_j} = O\left(K_{M_0}(z,\bar{z}) \prod_{p=1}^{q} \frac{1}{\left(1-\lambda_p^2\right)^2}\right). \tag{4.3}$$

On the other hand, by (4.1),

$$\begin{aligned} \|f(r\xi)\|_2^2 &= \int_{M_0} |f(r\xi)|^2 \, d\xi \le q \int_{M_0} \frac{(\varphi(r\xi))^2}{(1-\varphi(r\xi))^4} d\xi \\ &\le c \int_{-1}^{1} \cdots \int_{-1}^{1} \frac{(\varphi(r\lambda))^2 \, d\lambda}{(1-\varphi(r\lambda))^4}, \end{aligned} \tag{4.4}$$

where $\lambda = (\lambda_1, \cdots, \lambda_q, 0, \cdots, 0)$, $d\lambda = d\lambda_1, \cdots, d\lambda q$.

By the definition of φ, the right hand side of (4.4) is $O\left((1-r)^{-3}\right)$.

Let $z = rz_0$, $z_0 \in \partial M_0$, $w = rz = (\eta_1, \cdots, \eta_q, 0, \cdots, 0)\tilde{k}$. Then

$$\varphi(w) = r^2 = \sup\{|\eta_j| \, j = 1, \cdots, q\}$$

and

$$\frac{1}{1-r^4} \le \frac{1}{1-\sup\{\eta_j^2\}} \le \prod_{j=1}^{q} \frac{1}{1-\eta_j^2}.$$

We have

$$\|f(r\xi)\|_2 = O\left(\left(\prod_{j=1}^{q} \frac{1}{1-\eta_j^2}\right)^{\frac{3}{2}}\right). \tag{4.5}$$

From (4.2), (4.3) and (4.5), we have

$$\left|\frac{\partial f_\ell}{\partial w_j}(w)\right| = O\left(\left(\prod_{j=1}^{q} \frac{1}{1-\eta_j^2}\right)^{\frac{3}{2}} (K_{M_0}(z,\bar{z}))^{\frac{1}{2}} \prod_{p=1}^{q} \frac{1}{1-\lambda_p^2}\right).$$

But $\frac{1}{1-\lambda_p^2} = O\left(\frac{1}{1-\eta_p^2}\right)$, so we get

$$|\det J_f(w)| = O\left(K_{M_0}(z,\bar{z})^{\frac{n}{2}} \left(\prod_{j=1}^{q} \frac{1}{1-\eta_j^2}\right)^{\frac{5n}{2}}\right).$$

Since $\frac{1}{2} \sum_{j=q+1}^{2n} |\alpha_j(x_p)| + 1 \ge 2$, (3.4) implies

$$\frac{K_{M_0}(w,\bar{w})}{K_{M_0}(0,0)} \ge \left(\prod_{p=1}^{q} \frac{1}{1-\eta_p^2}\right)^2$$

and $|\det J_f(w)| = O\left(K_{M_0}(w,\bar{w})^{\frac{7n}{4}}\right)$. Since

$$K_{M_0}(w,\bar{w}) = O\left(\sup_{\theta \to b \in \partial M} \lim \frac{K_{M_0}(\theta,\bar{\theta})}{K_{M_0}(\eta,\bar{\eta})}\right),$$

we have

$$|\det J_f(w)| = O\left(\sqrt{\frac{K_{M_0}(w,\bar{w})}{K_{M_0}(0,0)}} \sup_{\theta \to b \in \partial M} \lim \left(\frac{K_{M_0}(\theta,\bar{\theta})}{K_{M_0}(\eta,\bar{\eta})}\right)^{\frac{7n-2}{4}}\right).$$

This argument proves the right hand side of inequality (1.3). The left hand side of inequality (1.3) is obviously, because otherwise the left hand side of (1.2) will tend to infinity for every point as z tends to the boundary of M_0.

§5. A Theorem on Bounded Domain

Theorem 5.1. *Let $U \subset \mathbf{C}^n$ be a bounded domain and B_d denote a ball centered at origin with radius d. If $B_r \subset U \subset B_R$ let S_U denote the family of normalized biholomorphic convex mapping on U which maps U into $\mathbf{C}^n$. If $f \in S_U$ and $f(z) = z + \sum_{i,j=1}^{n} d_{ij}z_iz_j + \cdots$, then*

$$|d_{ij}| \le Rr^{-2}, \quad i,j = 1,2,\cdots,n. \tag{5.1}$$

Proof. The mapping $\Phi(z) = f^{-1}\left(\frac{1}{2}f(zV) + f(z)\right)$ maps B_r into U and $\Phi(0) = 0$ if $V \in U_n$, $f \in S_U$.

If $\Phi(z) = zJ_\phi(0) + \sum \Phi_{ij}z_iz_j + \cdots$, then

$$\begin{aligned} f(\Phi(z)) &= \Phi(z) + \sum d_{ij}(\Phi(z))_i(\Phi(z)_j + \cdots \\ &= zJ_\phi(0) + \sum \Phi_{ij}z_iz_j + \sum d_{ij}(zJ_\phi(0)_i(zJ_\phi(0))_j + \cdots. \end{aligned}$$

Since $f(zV) = zV + \sum d_{ij}(zV)_i(zV)_j + \cdots$ and $f(\Phi(z)) = \frac{1}{2}(f(zV) + f(z))$, we have

$$\begin{aligned} &zJ_\phi(0) + \sum \Phi_{ij}z_iz_j + \sum d_{ij}(zJ_\phi(0))_i(zJ_\phi(0))_j + \cdots \\ &= \frac{z}{2}(V + I) + \frac{1}{2}\left(\sum d_{ij}(zV)_i(zV)_j + \sum d_{ij}z_iz_j\right) + \cdots \end{aligned}$$

if $z \in B_r$. Thus, $J_\phi(0) = \frac{1}{2}(V + I)$ and $\sum \phi_{ij}z_iz_j = \sum d_{ij}\left(\frac{z}{2}(I - V)\right)_i\left(\frac{z}{2}(I + V)\right)_j$. The mapping

$$\Phi(z) = z \cdot \frac{1}{2}(I + V) + \sum d_{ij}\left(\frac{z}{2}(I - V)\right)_i\left(\frac{z}{2}(I - V)\right)_j + \cdots \tag{5.2}$$

is a biholomorphic mapping on B_r which maps B_r into U, and $|\Phi(z)| < R$ if $z \in B_r$.

Let $\xi = \frac{1}{2}z(I + V)$, $\eta = \frac{1}{2}z(I - V)$. Then $|\xi|^2 + |\eta|^2 = |z|^2 < r^2$ and $\mathrm{Re}\xi\bar{\eta}' = 0$.

Let $\xi = 0, \eta = r_1e^{i\theta}e_j$, where $0 < r_1 < r$ and e_j is the unit vector such that all the components are zero except the j-th component is 1. Then by (5.2),

$$\Phi(z) = r_1^2e^{2i\theta}d_{ij} + \cdots \in U \subset B_R. \tag{5.3}$$

Multiplying (5.3) by $e^{-2i\theta}$, integrating with respect to θ from 0 to 2π and dividing by 2π, we get

$$r_1^2d_{jj} \in B_R, \qquad r_1 < r. \tag{5.4}$$

Let $\xi = 0$, $\eta = r_1e^{it}e^{i\theta}e_k + r_1e^{-it}e^{i\theta}e_j, k \neq j$, where $2r_1^2 \le r^2$, and e_k is the unit vector such that all the components are zero except the k-th component is 1. Then by (5.2),

$$\Phi(z) = r_1^2e^{2i\theta}\left(e^{2it}d_{kk} + 2d_{jk} + e^{-2it}d_{jj}\right) + \cdots \in U \subset B_R. \tag{5.5}$$

Multiplying (5.5) by $e^{-2i\theta}$, integrating with respect to θ from 0 to 2π, dividing by 2π, and then integrating with respect to t from 0 to 2π and dividing by 2π, we get

$$2r_1^2d_{jk} \in B_R, \quad 2r_1^2 < r^2. \tag{5.6}$$

Letting $r_1 \to r$ at (5.4), $2r_1^2 \to r^2$ at (5.6), we get (5.1).

Corollary 5.1. *Assumptions are the same as Theorem 5.1, and U is the unit ball in $\mathbf{C}^n$. Then*

$$|d_{ij}| \leq 1 \tag{5.7}$$

for all $i, j = 1, 2, \cdots, n$.

This estimation is precise, because the mapping $F(z) = \left(\frac{z_1}{1-z_1}, \cdots, \frac{z_n}{1-z_1}\right)$ is a normalized biholomorphic convex mapping on the unit ball in $\mathbf{C}^n$; this mapping makes equality hold in (5.7).

§6. Proof of Theorem 1.2

We start from formula (3.2). As we have already pointed out in [2],

$$\sup\left\{\left|\sum_{j\cdot\ell=1}^{n} k_{p\ell} d_{\ell j}^{(j)}\right|, p = 1, \cdots, n,\ f \in S_{M_0}\right\}$$

is the same as

$$\sup\left\{\left|\sum_{j=1}^{n} d_{\ell j}^{(j)}\right|, \ell = 1, \cdots, n,\ f \in S_{M_0}\right\}$$

if $(k_{p\ell}) = \tilde{k}$, and $S\,(M_0)$ is an invariant family. By Theorem 5.1,

$$\sup\left\{\left|\sum_{j=1}^{n} d_{\ell j}^{(j)}\right|, \ell = 1, \cdots, n,\ f \in S_{M_0}\right\} \leq nr^{-2}R$$

since $S\,(M_0)$ is the family of normalized Ω-uniformly locally biholomorphic convex mappings. We get the right hand side inequality of (1.5).

In order to prove the left hand side inequality of (1.5), we consider $\frac{\det J_{\tilde{f}}(\tilde{z})}{\sqrt{K_{M_0}(\tilde{z},\bar{\tilde{z}})/K_{M_0}(0,0)}}$, where $\tilde{f}$ is the normalization of $f \circ \xi \circ \xi_0^{-1}$ and $\tilde{z} = \xi_0 \circ \xi^{-1}(z) \in M_0$. By (3.4), we have

$$\prod_{\ell=1}^{q} \frac{(1 - |\tanh x_\ell|)^{A_\ell - \frac{1}{4}\sum\limits_{i=q+1}^{2n} |\alpha_j(X_\ell)| - \frac{1}{2}}}{(1 + |\tanh x_\ell|)^{A_\ell + \frac{1}{4}\sum\limits_{i=q+1}^{2n} |\alpha_j(X_\ell)| + \frac{1}{2}}} \leq \left|\det J_{\tilde{f}}(\tilde{z})\right|. \tag{5.8}$$

Since $\det J_{\tilde{f}}(\tilde{z})$ is a holomorphic function on M_0 and $\det J_{\tilde{f}}(\tilde{z}) \neq 0$ for any $\tilde{z} \in M_0$, $\left|\det J_{\tilde{f}}(\tilde{z})\right|$ attends its minimum value on the boundary of M_0. Each $A_\ell - \frac{1}{4}\sum\limits_{j=q+1}^{2n} |\alpha_j(X_\ell)| - \frac{1}{2}$ must be ≥ 0, otherwise the left hand side of (5.8) would tend to infinity as $\tilde{z}$ approaches the boundary. This proves the left hand side inequality of (1.5).

References

[1] Gong Sheng & Zheng Xuean, Distortion theorem for biholomorphic mappings in transitive domains I, International symposium in memory of L.K. Hua, Vol. II, Springer-Verlag, 1991, 111-122.

[2] Zheng Xuean & Gong Sheng, Distortion theorem for biholomorphic mappings in transitive domains II, *Chin. Ann. of Math.*, **13B**:4 (1992), 471-484.

[3] Zheng Xuean & Gong Sheng, Distortion theorem for biholomorphic mappings in transitive domains III, *Chin. Ann of Math.*, **14B**:3 (1993), 367-386.

[4] Helgason, S., Differential geometry, Lie groups, and symmetric spaces, Academic Press, 1978.

[5] Gong Sheng, Convex and Starlike mappings in several complex variables, Science Press, Beijing, China (to appear).

Chin. Ann. of Math.
20B: 3(1999),297-304.

DISTORTION THEOREMS FOR BIHOLOMORPHIC CONVEX MAPPINGS ON BOUNDED CONVEX CIRCULAR DOMAINS**

GONG SHENG* LIU TAISHUN*

Abstract

In terms of Carathéodory metric and Kobayashi metric, distortion theorems for biholomorphic convex mappings on bounded circular convex domains are given.

Keywords Distortion theorem, Convex mappins, Infinitesimal form of Carathéodory metric, Infinitesimal form of Kobayashi-Royden metric, Minkowski functional

1991 MR Subject Classification 32H02
Chinese Library Classification O174.56

§1. Two Different Distortion Theorems

Various distortion theorems for families of univalent functions have been studied since as early as 1907 when Köbe discovered his "Verzerrungssatz", the distortion theorem for the class of univalent functions defined on the unit disk in the complex plane **C**.

Let $\Omega \subset \mathbf{C}^n$ be a domain and $f(z) = (f_1(z), \cdots, f_n(z))$ be a biholomorphic mapping on Ω which maps Ω to $\mathbf{C}^n$. There are many counter-examples to show that $|\det J_f(z)|$ and $J_f(z)\overline{J_f(z)}'$ have no finite upper bound and no non-zero lower bound, where $J_f(z)$ is the Jacobian of f at point z.

Let $\Omega \subset \mathbf{C}^n$ be a domain and $m \in \Omega, f(z) = (f_1(z), \cdots, f_n(z))$ be a holomorphic mapping on Ω which maps Ω into $\mathbf{C}^n$. We say f is normalized at point m if $f(m) = 0$ and $J_f(m) = I$, where I is the $n \times n$ identity matrix. If $0 \in \Omega$ and $m = 0$, then we say that f is normalized.

Let $\Omega \subset \mathbf{C}^n$ be a bounded homogeneous domain and $0 \in \Omega, \mathrm{Aut}(\Omega)$ be the group of holomorphic automorphism of $\Omega, \mathcal{P}$ be a family of normalized holomorphic mapping on Ω which maps Ω into $\mathbf{C}^n$. For any $f \in \mathcal{P}, \varphi \in \mathrm{Aut}(\Omega)$, we normalize $f(\varphi(z))$ and obtain a normalized holomorphic mapping $F(z)$ on Ω. We say $\mathcal{P}$ is a linear-invariant family if $F(z) \in \mathcal{P}$ for any $f \in \mathcal{P}$ and $\varphi \in \mathrm{Aut}(\Omega)$.

The first affirmative result about the estimations of $|\det J_f(z)|$ for normalized holomorphic mappings was given by Barnard, FitzGerald and Gong[1] in the case $n = 2$ at first, and then Liu Taishun[2] extended it to the general case.

Manuscript received May 25, 1998. Revised December 8, 1998.

*Department of Mathematics, University of Science and Technology of China, Hefei 230026, China.
E-mail: gongs@public.intercom.com.cn

**Project supported by the National Natural Science Foundation of China.

Let

$$f(z)=(f_1(z),\cdots,f_n(z))=z+(aA^1z',\cdots,zA^nz')+\cdots\in \mathbf{S},$$

where $A^i=(a^i_{j,k})_{1\le j,k\le n}, i=1,\cdots,n$ and $\mathbf{S}$ is a linear-invarint family of holomorphic mappings on the unit ball

$$B^n=\left\{(z_1,\cdots,z_n)\in \mathbf{C}^n:\sum_{i=1}^n|z_i|^2<1\right\}$$

in $\mathbf{C}^n$. Then

$$\frac{(1-|z|)^{C(S)-\frac{n+1}{2}}}{(1+|z|)^{C(S)+\frac{n+1}{2}}}\le|\det J_f(z)|\le\frac{(1+|z|)^{C(S)-\frac{n+1}{2}}}{(1-|z|)^{C(S)+\frac{n+1}{2}}},$$

where $C(S)=\sup\left\{\left|\sum\limits_{i=1}^n a^i_{i1}\right|:f\in \mathbf{S}\right\}$. In particular, if $\mathbf{S}$ is the family of normalized biholomorphic convex mappings on $\mathbf{B}^n$, then $\frac{n+1}{2}\le C(S)\le 1+\frac{\sqrt{2}(n-1)}{2}$.

Moreover, Barnard, FitzGerald and Gong[1] conjectured that $C(S)=\frac{n+1}{2}$ for the family of normalized biholomorphic convex mappings on $\mathbf{B}^n$. Recently, Pfaltzgraff and Suffridge[3] gave a counter-example to show that the conjecture is not true. To find the exact value of $C(S)$ for the family of normalized biholomorphic convex mappings remains as an open problem. Even we do not know the precise value of $C(S)$ for the family of biholomorphic convex mappings on B^n, but we had the following precise estimations of $J_f(z)\overline{J_f(z)}'$ for the family of normalized biholomorphic convex mappings on B^n which was given by Gong, Wang ang Yu[4] in 1993.

Theorem 1.1. *Let $f:\mathbf{B}^n\to\mathbf{C}^n$ be a normalized biholomorphic convex mapping on $\mathbf{B}^n$ in $\mathbf{C}^n$. Then*

$$\left(\frac{1-|z|}{1+|z|}\right)^2G\le J_f(z)\overline{J_f(z)}'\le\left(\frac{1+|z|}{1-|z|}\right)^2G \tag{1.1}$$

holds for every $z=(z_1,\cdots,z_n)\in\mathbf{B}^n$ where

$$G=(g_{i,j})_{1\le i,j\le n}=\left(\frac{(1-|z|^2)\delta_{i,j}+\bar z_iz_j}{(1-|z|^2)^2}\right)_{1\le i,j\le n}$$

is the matrix of the Bergman matric of $\mathbf{B}^n$ in $\mathbf{C}^n$. The estimations are precise.

§2. Main Results

In the note, we prove the following main result which extends Theorem 1.1.

Theorem 2.1. *Let $\Omega\subset\mathbf{C}^n$ be a bounded convex circular domain with $0\in\Omega$, and $p(z)(z\in\Omega)$ be its Minkowski functional. Let $f(z):\Omega\to\mathbf{C}^n$ be a normalized biholomorphic convex mapping on Ω. Then for every $z\in\Omega$ and vector $\xi\in\mathbf{C}^n$, the inequalities*

$$\frac{1-p(z)}{1+p(z)}F^\Omega(z,\xi)\le p(J_f(z)\xi)\le\frac{1+p(z)}{1-p(z)}F^\Omega(z,\xi) \tag{2.1}$$

hold, where

$$F^\Omega(z,\xi)=F^\Omega_C(z,\xi)=F^\Omega_K(z,\xi),$$

$F^\Omega_C(z,\xi)$ and $F^\Omega_K(z,\xi)$ are the infinitesimal form of Carathéodory metric and the infinitesimal form of Kobayashi-Royden metric of Ω respectively.

When Ω is the unit ball $\mathbf{B}^n$ in $\mathbf{C}^n$, we have

$$(F^{\Omega}(z,\xi))^2 = (F^{\Omega}_C(z,\xi))^2 = (F^{\Omega}_K(z,\xi))^2 = \frac{|\xi|^2}{1-|z|^2} + \frac{\xi\bar{z}'z\bar{\xi}'}{(1-|z|^2)^2}.$$

(2.1) can be written as (1.1). Theorem 2.1 extends Theorem 1.1.

The original proof of Theorem 1.1 in [4] is based on differential geometry and the detail of the proof is a little bit lengthy, but the proof of Theorem 2.1 is shorter and simple.

Moreover, there are two consequences of the main result.

Theorem 2.2. *Let $\Omega \subset \mathbf{C}^n$ be a bounded convex circular domain with $0 \in \Omega$, and $p(z)(z \in \Omega)$ be its Minkowski functional. Let $f(z) : \Omega \to \mathbf{C}^n$ be a normalized convex biholomorphic mapping on Ω. Then the inequalities*

$$\frac{p(z)}{(1+p(z))^2} \le p(J_f(z)z) \le \frac{p(z)}{(1-p(z))^2} \tag{2.2}$$

hold, where z is a column vector.

When Ω is the unit disk Δ in $\mathbf{C}$, (2.2) is the classical growth theorem of normalized starlike function in Δ due to Alexander theorem: $zf'(z)$ is starlike if and only if $f(z)$ is convex. But the Alexander theorem is not true for several complex variables case in general. Liu Taishun and Ren Guangbin[5] already proved the growth theorem for normalized starlike biholomorphic mappings on bounded starlike circular domains as follows.

Let $\Omega \subset \mathbf{C}^n$ be a bounded starlike circular domain with $0 \in \Omega$, and $p(z)(z \in \Omega)$ be its Minkowski functional. Let $g(z) : \Omega \to \mathbf{C}^n$ be a normalized starlike biholomorphic mapping on Ω. Then for any $z \in \Omega$, the inequalities

$$\frac{p(z)}{(1+p(z))^2} \le p(g(z)) \le \frac{p(z)}{(1-p(z))^2} \tag{2.3}$$

hold.

As we already mentioned, the family of normalized starlike biholomorphic mappings and the family of $J_f(z)z$ (z is a column vector) where $f(z)$ is convex are essential different in the case of several complex case. Thus the inequalities (2.2) and (2.3) are two different inequalities.

As another consequence of Theorem 2.1, we have the estimations of the modulus of the eigenvalues of the Jacobian of the convex mapping.

Theorem 2.3. *Let $\Omega \subset \mathbf{C}^n$ be a bounded convex circular domain with $0 \in \Omega$, and $p(z)(z \in \Omega)$ be its Minkowski functional. Let $f(z) : \Omega \to \mathbf{C}^n$ be a normalized convex biholomophic mapping on Ω, and $\lambda_1(z), \cdots, \lambda_n(z)$ be the eigenvalues of $J_f(z)$ with $|\lambda_1(z)| \ge |\lambda_2(z)| \ge \cdots \ge |\lambda_n(z)|$. Then the inequalities*

$$\frac{(1-p(z))}{(1+p(z))^2} \le |\lambda_n(z)| \le \cdots \le |\lambda_2(z)| \le |\lambda_1(z)| \le \frac{1+p(z)}{(1-p(z))^2} \tag{2.4}$$

hold.

§3. Lemmas

We need the following lemmas.

Lemma 3.1. *Let $\Omega \subset \mathbf{C}^n$ be a bounded convex circular domain with $0 \in \Omega$, and $p(z)(z \in$*

Ω) be its Minkowski functional. Then

$$F^{\Omega}(z,\zeta)=F_K^{\Omega}(z,\zeta)=F_C^{\Omega}(z,\zeta)=\frac{1}{1-(p(z))^2} \tag{3.1}$$

holds, where $\zeta=\frac{z}{p(z)}\in\partial\Omega$ is a column vector.

Proof. According to the definition of the infinitesimal form of Kobayashi-Royden metric F_K^{Ω} of Ω,

$$F_K^{\Omega}(z,\xi)=\inf\left\{\frac{|\alpha|}{1-|\lambda|^2}:\exists\varphi\in H(\Delta,\Omega),\exists\lambda\in\Delta,\ \varphi(\lambda)=z,\alpha\varphi'(\lambda)=\xi\right\}, \tag{3.2}$$

where $H(\Delta,\Omega)$ is the family of holomophic mappings on Δ which maps Δ into Ω.

Fix $z\in\Omega$ and define $\varphi(w)=w\zeta$, where $w\in\Delta$. Then $\varphi\in H(\Delta,\Omega)$ since Ω is a bounded convex circular domain. Let $\lambda=p(z)$, then $\varphi(\lambda)=z$. Let $\alpha=1$, then $\alpha\varphi'(\lambda)=\zeta$. By the definition (3.2), we have

$$F_K^{\Omega}(z,\zeta)\le\frac{1}{1-(p(z))^2}. \tag{3.3}$$

On the other hand, according to the definition of the infinitesimal form of Carathéodory metric F_C^{Ω} of Ω,

$$F_C^{\Omega}(z,\zeta)=\sup\{|J_{\varphi}(z)\zeta|:\exists\varphi\in H(\Omega,\Delta),\ \varphi(z)=0\}, \tag{3.4}$$

where $H(\Omega,\Delta)$ is the family of holomorphic mappings in Ω which maps Ω into Δ.

Fix $z\in\Omega$. There exists a continuous linear functional T_z on the Banach space $\mathbf{C}^n$ with the Minkowski functional of Ω as semi-norm, $T_z:\mathbf{C}^n\to\mathbf{C}$, such that $\|T_z\|\le1$ and $T_z(z)=p(z)$, where $\|\ \|$ is the norm of the dual space of $\mathbf{C}^n$.

Let $\psi_{p(z)}(\lambda)$ be a holomorphic automorphism of Δ which maps $p(z)$ to 0, i.e.,

$$\psi_{p(z)}(\lambda)=\frac{p(z)-\lambda}{1-p(z)\lambda}.$$

Let

$$\varphi(w)=\psi_{p(z)}\circ T_z(w).$$

Then $\varphi\in H(\Omega,\Delta)$ and $\varphi(z)=0$. Thus

$$J_{\varphi}(z)\zeta=J_{\psi_{p(z)}}(p(z))J_{T_z}(\zeta)=\frac{-1}{1-(p(z))^2}T_z(\zeta)=\frac{-1}{1-(p(z))^2}$$

since $T_z(\zeta)=1$. By the definition (3.4), we have

$$F_C^{\Omega}(z,\zeta)\ge\frac{1}{1-(p(z))^2}.$$

By the Theorem of Lempert[11], we know that

$$F_C^{\Omega}(z,\xi)=F_K^{\Omega}(z,\xi)$$

when Ω is convex. By (3.3) and (3.5), we obtain (3.1).

From Lemma 3.1 we have

Lemma 3.2. *Let $\Omega\subset\mathbf{C}^n$ be a bounded convex circular domain with $0\in\Omega$, and $p(z)(z\in\Omega)$ be its Minkowski functional. Then*

$$F^{\Omega}(0,\xi)=F_K^{\Omega}(0,\xi)=F_C^{\Omega}(0,\xi)=p(\xi), \tag{3.6}$$

where $\xi\in\mathbf{C}^n$ is a column vector.

Proof. Let $\zeta = \frac{\xi}{p(\xi)} \in \partial\Omega$ and $0 < t < 1$. Then

$$F^{\Omega}(t\zeta, \zeta) = \frac{1}{1 - (p(t\zeta))^2}$$

by Lemma 3.1. Letting $t \to 0$, we have

$$F^{\Omega}(0, \zeta) = 1$$

by the continuous properties of F^{Ω} and $p(z)$ at point 0.

By the definition of F^{Ω}, we obtain

$$F^{\Omega}(0, \zeta) = F^{\Omega}(0, p(\xi)\zeta) = p(\xi)F^{\Omega}(0, \zeta) = p(\xi).$$

Lemma 3.2 is a known result (see [12]).

Lemma 3.3. *Let $\Omega \subset \mathbf{C}^n$ be a bounded convex circular domain with $0 \in \Omega$, and $p(z)(z \in \Omega)$ be its Minkowski functional. Then for every column vector $z \in \Omega$ and every column vector $\xi \in \partial\Omega$,*

$$\frac{1}{1 + p(z)} \leq F^{\Omega}(z, \xi) \leq \frac{1}{1 - p(z)} \tag{3.7}$$

holds, where $F^{\Omega} = F^{\Omega}_K = F^{\Omega}_C$.

Proof. For fixed z and ξ, the mapping

$$\varphi(w) = (\varphi_1(w), \cdots, \varphi_n(w))' = z + w(1 - p(z))\xi$$

is a holomorphic mapping on Δ where $'$ means transpose.

Obviously, $\varphi(0) = z$ and

$$\varphi'(0) = (\varphi_1'(0), \cdots, \varphi_n'(0))' = (1 - p(z))\xi.$$

Moreover,

$$\begin{aligned} p(\varphi(w)) &= p(z + w(1 - p(z))\xi) \leq p(z) + p(w(1 - p(z))\xi) \\ &= p(z) + |w|(1 - p(z))p(\xi) < 1 \end{aligned}$$

since $|w| < 1$ and $p(\xi) = 1$. Thus $\varphi(w) \in \Omega$. This means $\varphi \in H(\Delta, \Omega)$.

By the definition (3.2) of the infinitesimal form of Kobayashi-Royden metric, and letting $\lambda = 0, \alpha = \frac{1}{1-p(z)}$, we have

$$F^{\Omega}_K(z, \xi) \leq \frac{1}{1 - p(z)}.$$

For fixed ξ, there exists a continuous linear functional T_ξ on Banach space $\mathbf{C}^n$ with the Minkowski functional of Ω as semi-norm, $T_\xi : \mathbf{C}^n \to \mathbf{C}$, such that $\|T_\xi\| \leq 1$ and $T_\xi(\xi) = p(\xi) = 1$, where $\| \ \|$ is the norm of the dual space of $\mathbf{C}^n$. Then the function

$$\psi(w) = \frac{T_\xi(w - z)}{1 + p(z)}$$

is a holomorphic function on Ω.

Obviously, $\psi(z) = 0$ since $T_\xi(0) = 0$ and

$$|\psi(w)| = \frac{|T_\xi(w - z)|}{1 + p(z)} \leq \frac{\|T_\xi\| \cdot p(w - z)}{1 + p(z)} \leq \frac{p(w) + p(z)}{1 + p(z)} < 1.$$

Thus $\psi(w) \in \Delta$. This means $\psi \in H(\Omega, \Delta)$. By the definition (3.4) of the infinitesimal form of Carathéodory metric and

$$J_{\psi}(z)\xi = \frac{T_{\xi}(\xi)}{1+p(z)} = \frac{1}{1+p(z)},$$

we have

$$F_C^{\Omega}(z,\xi) \geq \frac{1}{1+p(z)}.$$

By the Theorem of Lempert[11], we know that

$$F_C^{\Omega}(z,\xi) = F_K^{\Omega}(z,\xi)$$

when Ω is convex. We obtain (3.7).

§4. Proof of Theorems

Proof of Theorem 2.1. We prove the right side inequality of (2.1) at first.

For any fixed $z \in \Omega$, we consider the line segment which starts from $f(z)$, passes through the origin, and then meets a point P on the boundary of $f(\Omega)$. There exists a point $z^* \in \partial\Omega$, such that $f(z^*) = P$. Since $f(z)$, 0 and $f(z^*)$ lie on a line segment, and 0 is a point between $f(z)$ and $f(z^*)$, there exists a real number $\lambda \in (0,1)$, such that

$$\lambda f(z) + (1-\lambda) f(z^*) = 0.$$

From the definition of $p(z)$, we have

$$\lambda p(f(z)) = (1-\lambda) p(f(z^*)).$$

Liu and Ren[6] proved that if $f(z)$ is a normalized biholomorphic convex mapping on a bounded convex circular domain in $\mathbf{C}^n$ with $0 \in \Omega$, then the inequalities of the growth theorem

$$\frac{p(z)}{1+p(z)} \leq p(f(z)) \leq \frac{p(z)}{1-p(z)}$$

hold for $z \in \Omega$. We have

$$\frac{\lambda p(z)}{1-p(z)} \geq \lambda p(f(z)) = (1-\lambda) p(f(z^*)) \geq \frac{1}{2}(1-\lambda).$$

It yields

$$\lambda \geq \frac{1-p(z)}{1+p(z)}. \tag{4.1}$$

Let

$$h(w) = f^{-1}[\lambda f(w) + (1-\lambda) f(z^*).$$

$h(w)$ is well defined when $w \in \Omega$ since f is a convex mapping. Moreover, $h(w)$ is biholomorphic on $\Omega, h(\Omega) \subset \Omega$ and $h(z) = 0$ since f is a normalized biholomorphic mapping.

Since $h(z)$ is a biholomorphic mapping on Ω, Carathéodory metric is an invariant metric (see [7, 8, 9, 10]), we have the equality

$$F_C^{h(\Omega)}(0, J_h(z)\xi) = F_C^{\Omega}(z,\xi).$$

Since $h(\Omega) \subset \Omega$, we have the inequality of Carathéodory metrics of inner mapping,

$$F_C^{h(\Omega)}(0, J_h(z)\xi) \geq F_C^{\Omega}(0, J_h(z)\xi). \tag{4.2}$$

Hence

$$F_C^{\Omega}(0, J_h(z)\xi) \le F_C^{\Omega}(z,\xi).$$

From (4.2) and according to the definition of h and F_C^{Ω}, we have $J_h(z) = \lambda J_f(z)$ and

$$\lambda F_C^{\Omega}(0, J_f(z)\xi) \le F_C^{\Omega}(z,\xi). \tag{4.3}$$

Combining the inequalities (4.1) and (4.3), we obtain the right side inequality of (2.1).

Now we prove the left side inequality of (2.1).

For any fixed $z \in \Omega$, we consider a line segment which starts from the origin, passes through $f(z)$, and then meets a point Q in the boundary of $f(\Omega)$. There exists a point $\tilde{z} \in \partial\Omega$, such that $f(\tilde{z}) = Q$. Since $0, f(z)$ and $f(\tilde{z})$ lie on a line segment, and $f(z)$ is a point between 0 and $f(\tilde{z})$, there exists a real number $\mu \in (0,1)$, such that $f(z) = \mu f(\tilde{z})$. Then

$$z = \tilde{z}(\mu) = f^{-1}(\mu f(\tilde{z})).$$

In [6], Liu and Ren proved that if $z(t) = f^{-1}(tf(z)), 0 \le t \le 1$, then

$$\frac{p(z)}{1+p(z)} \le \frac{p(z(t))}{t(1+p(z(t)))}$$

holds. We have

$$\frac{1}{2} = \frac{p(\tilde{z})}{1+p(\tilde{z})} \le \frac{p(\tilde{z}(\mu))}{\mu(1+p(\tilde{z}(\mu)))} = \frac{p(z)}{\mu(1+p(z))}.$$

It implies

$$\mu \le \frac{2p(z)}{1+p(z)}, \quad 1-\mu \ge \frac{1-p(z)}{1+p(z)}. \tag{4.4}$$

Let

$$H(w) = f^{-1}[(1-\mu)f(w) + \mu f(\tilde{z})].$$

$H(w)$ is well defined when $w \in \Omega$ since f is a convex mapping. Moreover, $H(w)$ is biholomorphic on $\Omega, H(\Omega) \subset \Omega, H(0) = z$ since f is a normalized biholomorphic mapping.

Since $H(z)$ is biholomorphic on Ω, the Carathéodory metric is an invariant metric, we have the equality

$$F_C^{H(\Omega)}(z, J_H(0)\eta) = F_C^{\Omega}(0,\eta)$$

which holds for any $z \in \Omega, \eta \in \mathbf{C}^n$. Since $H(\Omega) \subset \Omega$, we have the inequality of Carathéodory metrics of the inner mapping,

$$F_C^{H(\Omega)}(z, J_H(0)\eta) \ge F_C^{\Omega}(z, J_H(0)\eta).$$

Hence

$$F_C^{\Omega}(z, J_H(0)\eta) \le F_C^{\Omega}(0,\eta). \tag{4.5}$$

From (4.5) and according to the definition of H and F_C^{Ω}, we have $J_H(0) = J_f^{-1}(z)(1-\mu)$ and

$$(1-\mu)F_C^{\Omega}(z, J_f^{-1}(z)\eta) \le F_C^{\Omega}(0,\eta). \tag{4.6}$$

Let $\eta = J_f(z)\xi$ in (4.6), it becomes

$$(1-\mu)F_C^{\Omega}(z,\xi) \le F_C^{\Omega}(0, J(z)\xi). \tag{4.7}$$

Combining (4.4) and (4.7), we obtain the left side inequality of (2.1).

The Theorem of Lempet[11] tells us that

$$F_C^\Omega(z,\xi)=F_K^\Omega(z,\xi)$$

when Ω is convex.

We have proved the inequality (2.1) by Lemma 3.2.

By Theorem 2.1 and Lemma 3.1, we have Theorem 2.2.

Proof of Theorem 2.3. By Theorem 2.1 and Lemma 3.3, we have

$$\frac{1-p(z)}{(1+p(z))^2}\le p(J_f(z)\xi)\le\frac{1+p(z)}{(1-p(z))^2}$$

when $\xi\in\partial\Omega$ is a column vector. If $\lambda(z)$ is the eigenvalue of $J_f(z)$, then

$$J_f(z)\eta=\lambda(z)\eta,$$

where η is the eigenvector of $\lambda(z)$. Let $\xi=\frac{\eta}{p(\eta)}$. Then $\xi\in\partial\Omega$ and

$$p(J_f(z)\xi)=p(\lambda(z)\xi)=|\lambda(z)|p(\xi)=|\lambda(z)|.$$

Hence

$$\frac{1-p(z)}{(1+p(z))^2}\le|\lambda(z)|\le\frac{1+p(z)}{(1-p(z))^2}$$

hold for all eigenvalues of $J_f(z)$. Thus we have proved (2.4).

References

[1] Barnard, R. M., FitzGerald, C. H. & Gong Sheng, Distortion theorem for biholomorphic mappings in $\mathbf{C}^n$, *Tran. Amer. Math. Soc.*, **344**(1994), 907–924.

[2] Liu Taishun, The distortion theorems, growth theorems and covering theorems for biholomorphic mappings on classical domains, University of Science and Technology of Science of China, Doctor Thesis, 1989.

[3] Pfaltzgraff, J. A. & Suffridge, T. S., Linear invariant, order and convex mappings in $\mathbf{C}^n$, Research report, 96–6, 1996, Univ. of Kentucky.

[4] Gong Sheng, Wang Shikun & Yu Qihuang, Biholomorphic convex mappings of ball in $\mathbf{C}^n$, *Pacific Jour. of Math.*, **161**(1993), 287–306.

[5] Liu Taishun & Ren Guangbin, The growth and 1/4-covering theorems for the normalized starlike mappings on bounded starlike circular domains, *Chin. Ann. of Math.*, **19B**:4(1998), 401–408.

[6] Liu Taishun & Ren Guangbin, The growth theorem of convex mappings on bounded convex circular domains, *Science in China (series A)*, **41**(1998).

[7] Yau ShingTung & Schoen, Richard, Differential geometry, Science Press, 1991.

[8] Krantz, S. G., Function theory of several complex variables, 2nd edition, Wadsworth & Brooks/col, 1992.

[9] Poletskii, E. A. & Shabat, B V., Invariant metrices, Several complex variables III, G. M. Khenkin (Ed.) Springer-Verlag, 1989, pp.63–11.

[10] Kobayashi, S., Hyperbolic manifolds and holomorphic mappings, Dekker, New York, 1970.

[11] Lempert, L., La metrique Kobayashi et las representation des domains sur la boule, *Bull. Soc. Math. France*, **109**(1981), 427–474.

[12] Jarnicki, M. & Pflug, P., Invariant distances and metrics in complex analysis, de Gruyter, Berlin-New York, 1993.

ASIAN J. MATH.
Vol. 4, No. 4, pp. 795–816, December 2000

THE DISTORTION THEOREMS OF LINEAR INVARIANT FAMILY ON THE UNIT BALL*

SHENG GONG† AND QIHUANG YU‡

1. Introduction. It is well know that the distortion theorem of biholomorphic mapping in any domain in $\mathbb{C}^n$ does not exist. It exists only when the mapping class has more restrictions.

The first affirmative result was established by Barnard, FitzGerald and Gong[1] in 1988(The paper was published at 1994). They proved the following result.

A linear invarient family, hereafter denoted LIF, on the unit ball $\boldsymbol{B}^n$ in $\mathbb{C}^n$ is a family $\mathcal{M}$ of localy biholomorphic mappings $f : \boldsymbol{B}^n \to \mathbb{C}^n$ such that if $f \in \mathcal{M}$ then (i)$f(0) = 0, J_f(0) = I$, where J_f is the Jacobi matrix of f and (ii)Λ_φ of $f \in \mathcal{M}$ for $\varphi \in Aut(\boldsymbol{B}^n)$, the holomorphic automorphisms of $\boldsymbol{B}^n$. Here $\Lambda_\varphi(f)$ is the Koebe transform of f, $\Lambda_\varphi(f) = (J_\varphi(0))^{-1}(J_f(\varphi(0))^{-1}[f(\varphi(z)) - f(\varphi(0))]$.

Let $\mathcal{S}$ be a linear invariant family, on the unit ball $\boldsymbol{B}^2$ in $\mathbb{C}^2$. If $f(z) = (f_1(z), f_2(z)) \in \mathcal{S}$, $z = (z_1, z_2) \in \boldsymbol{B}^2$, and

$$\begin{cases} f_1(z) & = z_1 + d_{2,0}^{(1)} z_1^2 + d_{1,1}^{(1)} z_1 z_2 + d_{0,2}^{(1)} z_2^2 + \cdots\cdots, \\ f_2(z) & = z_2 + d_{2,0}^{(2)} z_1^2 + d_{1,1}^{(2)} z_1 z_2 + d_{0,2}^{(2)} z_2^2 + \cdots\cdots, \end{cases}$$

then

$$\left| log\left[(1 - z\bar{z}')^{\frac{3}{2}} det J_f(z)\right]\right| \le C(\mathcal{S}) log \frac{1+|z|}{1-|z|} \tag{1.1}$$

holds, where $C(\mathcal{S}) = Sup\left\{ |d_{2,0}^{(1)} + \frac{1}{2} d_{1,1}^{(2)}| \;\middle|\; f \in \mathcal{S} \right\}$, $J_f(z)$ is the Jacobi matrix of f at z.

From (1.1), it follows the following distortion theorem

$$\frac{(1-|z|)^{C(\mathcal{S})-\frac{3}{2}}}{(1+|z|)^{C(\mathcal{S})+\frac{3}{2}}} \le |det J_f(z)| \le \frac{(1+|z|)^{C(\mathcal{S})-\frac{3}{2}}}{(1-|z|)^{C(\mathcal{S})+\frac{3}{2}}}. \tag{1.2}$$

In particular, if $\mathcal{S}$ is the family of normalized biholomorphic convex mapping, then one can prove $1.5 \le C(\mathcal{S}) \le 1.761$.

In 1989, at his doctor thesis, Liu[11] extended this result to $\mathbb{C}^n$. He proved that: Let $\mathcal{S}$ be a LIF on the unit ball $\boldsymbol{B}^n$ in $\mathbb{C}^n$. If

$$f(z) = (f_1(z), f_2(z), \cdots\cdots, f_n(z)) = z + (zA^{(1)}z', \cdots\cdots, zA^{(n)}z') + \cdots\cdots \in \mathcal{S}, \tag{1.3}$$

where $A^{(i)} = (a_{jk}^{(i)})_{1 \le j,k \le n}$, $z \in \boldsymbol{B}^n$, then

$$\left| log\left[(1 - z\bar{z}')^{\frac{n+1}{2}} det\, J_f(z)\right]\right| \le log \frac{1+|z|}{1-|z|}, \tag{1.4}$$

---

*Received June 20, 2000; accepted for publication October 23, 2000. Suppoted by National Science Foundation of China.

†Department of Mathematics, University of Science and Technology of China, Hefei, Anhui, 230026, China (gongs@sina.com).

‡Institute of Applied Mathematics, Academia Sinica, Beijing, 100080, China.

where

$$C(\mathcal{S}) = Sup\{|\sum_{i=1}^{n} a_{i1}^{(i)}||f \in \mathcal{S}\}. \tag{1.5}$$

As a consequence, we have

$$\frac{(1-|z|)^{C(\mathcal{S})-\frac{n+1}{2}}}{(1+|z|)^{C(\mathcal{S})+\frac{n+1}{2}}} \le |detJ_f(z)| \le \frac{(1+|z|)^{C(\mathcal{S})-\frac{n+1}{2}}}{(1-|z|)^{C(\mathcal{S})+\frac{n+1}{2}}}. \tag{1.6}$$

He also proved that if $\mathcal{S}$ is the family of normalized biholomorphic convex mappings, then

$$\frac{n+1}{2} \le C(\mathcal{S}) \le 1 + \frac{\sqrt{2}(n-1)}{2}. \tag{1.7}$$

It was conjectured by Barnard, FitzGerald and Gong[1] that $C(\mathcal{S}) = \frac{n+1}{2}$, where $\mathcal{S}$ is the family of biholomorphic convex mappings.

In 1997, Pfaltzgraff [12] proved (1.6) again. The proof of Pfaltzgraff is same as Liu's proof in his doctor thesis.

In 1999, Pfaltzgraff and Suffridge [13] gave a counterexample to disprove the conjecture : $C(\mathcal{S}) = \frac{n+1}{2}$ when $\mathcal{S}$ is the family of normalized convex mappings. Until now, nobody know the exact value of $C(\mathcal{S})$ when $\mathcal{S}$ is the family of normalized convex mappings, even nobody conjecture it.

In 1990, Gong and Zheng [8] extended (1.4) and (1.6) to the bounded symmetric domains.

Let $M \subset \mathbb{C}^n$ be a bounded symmetric domain containing the origin. Let G be the group of holomorphic automorphism of M, and let K be isotropy subgroup in G which fixes the origin. Let $\mathcal{G}$ be Lie algebla of G, $\mathcal{K}$ be the maximal compact subalgebla of $\mathcal{G}$ corresponding to K. Then $\mathcal{G}$ has the Cartan decomposition.

$$\mathcal{G} = \mathcal{K} + \mathcal{P}.$$

Suppose $\mathcal{A}$ is a maximal Abelian subalgebla in $\mathcal{P}$ and A is the corresponding analytic subgroup in G. Then we have the Iwasawa decomposition $G = KAN$. For every $z \in M$, there exist $X \in \mathcal{A}$ and $k \in K$ such that

$$z = \xi(ka \cdot O) = \xi(expAd(k)X \cdot O)$$

where $O = eK$ is the identity coset in G/K and ξ is the holomorphic diffeomorphism from G/K onto M. If $X_1, \cdots\cdots, X_q$ form a basis of $\mathcal{A}$, then $X = \sum_{j=1}^{q} x_j X_j$, where $q = dim\mathcal{A} = rankM$. If $\mathcal{S}$ is a LIF on M, and $f(z) \in \mathcal{S}$, then

$$|log\frac{detJ_f(z)}{\sqrt{K(z,\overline{z})/K(0,0)}}| \le C(\mathcal{S}) \sum_{p=1}^{q} log\frac{1+|tanh\, x_p|}{1-|tanh\, x_p|} \tag{1.8}$$

holds, where $K(z,\overline{z})$ is the Bergman kernel function on M, $C(\mathcal{S})$ is defined by (1.5) when $f(z)$ has expression (1.3).

In 1993, Gong Wang and Yu [6] established the following result. If $f(z)$ is a normalized convex mapping on the unit ball $\boldsymbol{B}^n$ in $\mathbb{C}^n$, then

$$\left(\frac{1-|z|}{1+|z|}\right)^2 G \le J_f(z)\overline{J(z)}' \le \left(\frac{1+|z|}{1-|z|}\right)^2 G \tag{1.9}$$

holds, where

$$G = (g_{ij})_{1\le i,j\le n} = \left(\frac{(1-|z|^2)\delta_{ij} + \overline{z_i}z_j}{(1-|z|^2)^2}\right)_{1\le i,j\le n} \tag{1.10}$$

is the matrix of the Bergman metic of B^n.

From (1.9), we immediately have the following consequences:

$$\frac{(1-|z|)^{n-1}}{(1+|z|)^{3n+1}} \le det(J_f\overline{J_f}') \le \frac{(1+|z|)^{n-1}}{(1-|z|)^{3n+1}}. \tag{1.11}$$

Comparing (1.11) with (1.6) and (1.7), we can see that (1.6) and (1.7) is better then (1.11).

$$\frac{1}{(1+|z|)^2} \le \|J_f(z)\| \le \frac{1}{(1-|z|)^2}, \tag{1.12}$$

where $\|J_f(z)\|$ is the norm of the matrix $J_f(z)$.

$$\frac{1-|z|}{(1+|z|)^3} \le \Lambda_n^2 \le \frac{1+|z|}{(1-|z|)^3}, \tag{1.13}$$

where Λ_n^2 is the smallest eigenvalue of $J_f(z)\overline{J_f(z)}'$.

(1.14)

$$\frac{1}{(1+|z|)^4} + \frac{(n-1)(1-|z|)}{(1+|z|)^3} \le Tr(J_f(z)\overline{J_f(z)}') \le \frac{1}{(1-|z|)^4} + \frac{(n-1)(1+|z|)}{(1-|z|)^3}.$$

In 1999, Gong and Liu [5] extended (1.9) to bounded convex circular domains.

Let Ω be a bounded convex circular domain in $\mathbb{C}^n$,$0 \in \Omega$, its Minkowski functional $p(z) \in C^1$ on $\overline{\Omega}$ except for a lower dimensional manifold. Let $f(z)$ be a normalized biholomorphic convex mapping on Ω, then

$$\frac{1-p(z)}{1+p(z)}F(z,\xi) \le p(J_f(z)\xi) \le \frac{1+p(z)}{1-p(z)}F(z,\xi) \tag{1.15}$$

holds,where $\xi \in \mathbb{C}^n$ is a column vector, $F(z,\xi) = F_c(z,\xi) = F_k(z,\xi)$, $F_c(z,\xi)$ is the Carathéodory metric in infinitesimal form and $F_k(z,\xi)$ is the Kobayashi-Royden metric in infinitesimal form.(By Lempert Theorem [10] , F_c and F_k are same when the domain is convex.) When Ω is unit ball B^n in $\mathbb{C}^n$, then [10]

$$F(z,\xi) = F_c(z,\xi) = F_k(z,\xi) = \frac{|\xi|^2}{1-|z|^2} + \frac{\xi\overline{z}'z\overline{\xi}'}{(1-|z|^2)}.$$

In this case, (1.15) becomes (1.9).

When $\xi = z$ in (1.15), we have

$$\frac{p(z)}{(1+p(z))^2} \le p(J_f(z)z) \le \frac{p(z)}{(1-p(z))^2} \tag{1.16}$$

where $z \in \mathbb{C}^n$ is a column vector. It implies

$$\frac{1-p(z)}{(1+p(z))^2} \le |\lambda_n(z)| \le \cdots\cdots \le |\lambda_1(z)| \le \frac{1+p(z)}{(1-p(z))^2} \tag{1.17}$$

where $|\lambda_1(z)|^2 \geq \cdots\cdots \geq |\lambda_n(z)|^2$ are eigenvalues of $J_f(z)\overline{J_f(z)}'$.

In the case that Ω is the unit ball $\boldsymbol{B}^n$ in $\mathbb{C}^n$, we compare (1.17) with (1.12) and (1.13), we may find that (1.12) and (1.13) are better than (1.17).

Let $f(z)$ be a normalized holomorphic mapping on a domain Ω in $\mathbb{C}^n$, and $0 \in \Omega$. We may expand $f(z)$ at $z = 0$ as

$$f(z) = z + \sum_{j=2}^{\infty} \frac{1}{j!} D^j f(0)(z, \cdots\cdots, z), \tag{1.18}$$

where D^j is the Frechet derivative of order j, $j = 1, 2, \cdots\cdots$. If Ω is a bounded symmetric domain in $\mathbb{C}^n$, and $\mathcal{S}$ is a LIF, then we may find that $C(\mathcal{S})$ defined by (1.5) can express as

$$C(\mathcal{S}) = Sup\left\{\frac{1}{2}\left|Tr(D^2 f(0)u)\right| \;\middle|\; u\bar{u}' = 1, f \in \mathcal{S}\right\}. \tag{1.19}$$

Recently, Pfaltzgraff and Suffridge [14] define another order. If $\mathcal{S}$ is a LIF on a domain Ω which containing origin, they defined

$$\alpha = \|ord\|\mathcal{S} = Sup\left\{\frac{1}{2}\|D^2 f(0)\| \;\middle|\; f \in \mathcal{S}\right\}. \tag{1.20}$$

They proved that if Ω is the unit ball in $\mathbb{C}^n$, $\mathcal{S}$ is a LIF, $f \in \mathcal{S}$, then

$$\frac{(1-|z|)^{\alpha-1}}{(1+|z|)^{\alpha+1}} \leq \|Df(z)\| \leq \frac{(1+|z|)^{\alpha-1}}{(1-|z|)^{\alpha+1}}, \tag{1.21}$$

where α is the order of $\mathcal{S}$ which is defined by (1.20).

Hamada and Kohr [9] used same method as Pfaltzgraff and Suffridge used in [14] to prove the similar result when Ω is the unit polydisk in $\mathbb{C}^n$.

There are many examples of LIF at [14]. When LIF is the family of normalized convex mapping on $\boldsymbol{B}^n$, then $\alpha = 1$, (1.21) becomes (1.12). All normalized localy biholomorphic mappings on $\mathbb{C}^n$ form a LIF, then $\alpha = \infty$ for this family.

As we mentioned above, we gave the estimations of $|detJ_f(z)|$ by (1.6) and (1.7), and the estimations of $\|J_f(z)\|$ by (1.21) already, it is natural to ask the estimations of $Tr(J_f(z)\overline{J_f(z)}')$. In the Section 2, we will give the estimations of $Tr(J_f(z)\overline{J_f(z)}')$ for LIF on the unit ball $\boldsymbol{B}^n$ in $\mathbb{C}^n$ by (1.20). When LIF is the family of normalized convex mappings, it is coincided with (1.14). In Section 3, by using the method as we used in Section 2, we may prove (1.6) by a different approach. As another application of (1.9), we will give the estimations of the covariant derivative of biholomorphic convex mappings on the unit ball in Section 4.

2. The estimations of the trace of Jacobi matrix of mappings in a $\boldsymbol{B}^n$. In this section, all vectors are column vectors.We consider the LIF on the unit ball $\boldsymbol{B}^n$ in $\mathbb{C}^n$. We already know that [14]:

(1) $\|ord\|\mathcal{S} \geq 1$ for every LIF $\mathcal{S}$. $\|ord\|\mathcal{S} = 1$ if $\mathcal{S}$ is the family of normalized biholomorphic convex mappings on $\boldsymbol{B}^n$.

(2) If $\mathcal{S}$ is a LIF and $\|ord\|\mathcal{S} = \alpha < \infty$, then $\mathcal{S}$ is a normal family.

(3) If $\varphi_a(z) \in Aut(\boldsymbol{B}^n)$, which maps 0 to $a \in \boldsymbol{B}^n$, and

$$f(z;a) = (D\varphi_a(0))^{-1}(Df(\varphi_a(0))^{-1}[f(\varphi_a(z)) - f(\varphi_a(0))], \tag{2.1}$$

where $f \in \mathcal{S}$, a LIF. Then the definition of $\|ord\|\mathcal{S}$ can be expressed as

$$\|ord\|\mathcal{S} = Sup\left\{\frac{1}{2}\left|D^2 f(0;a)(w,w)\right| \,\middle|\, f \in \mathcal{S}, |w| = 1, |a| < 1\right\}. \tag{2.2}$$

(4) $C(\mathcal{S}) \le n\|ord\|\mathcal{S}$, when $\mathcal{S}$ is a LIF on $\boldsymbol{B}^n$.

We have the estimations of $|detJ_f(z)|$ by (1.6) and the estimations of $\|J_f(z)\|$ by (1.21) already. In this section, we will give the estimations of $Tr(J_f(z)\overline{J_f(z)}')$ when f belongs to a LIF $\mathcal{S}$ with $\|ord\|\mathcal{S} = \alpha$ on the unit ball $\boldsymbol{B}^n$ in $\mathbb{C}^n$.

THEOREM 2.1 *If $\mathcal{S}$ is a LIF on the unit ball $\boldsymbol{B}^n$ in $\mathbb{C}^n$ with $\|ord\|\mathcal{S} = \alpha$, then*

$$\begin{aligned} &\frac{(1-|z|)^{2\alpha-2}}{(1+|z|)^{2\alpha+2}} + (n-1)\frac{(1-|z|)^{2\alpha-1}}{(1+|z|)^{2\alpha+1}} \\ \le\ & Tr(J_f(z)\overline{J_f(z)}') \le \frac{(1+|z|)^{2\alpha-2}}{(1-|z|)^{2\alpha+2}} + (n-1)\frac{(1+|z|)^{2\alpha-1}}{(1-|z|)^{2\alpha+1}} \end{aligned} \tag{2.3}$$

for all $f \in \mathcal{S}$ and $z \in \boldsymbol{B}^n$.

The inequalities (1.21) are the estimations of the largest eigenvalue of $J_f(z)\overline{J_f(z)}'$. Theorem 2.1 tells us that there exists one largest eigenvalue only in $J_f(z)\overline{J_f(z)}'$, all other eigenvalues have lower order comparing with this largest eigenvalue if the largest eigenvalue of $J_f(z)\overline{J_f(z)}'$ attend the estimating order.

When $\mathcal{S}$ is the family of normalized biholomorphic convex mappings, then $\|ord\|\mathcal{S} = 1$, (2.3) is coincide with (1.14).

If $\mathcal{S}$ is a LIF on the unit polydisc on $\mathbb{C}^n$, then we can get the similar conclusion, but we omit to state the results and its proof. The proof is similar with the proof of Theorem 2.1 in below.

The method of the proof of Theorem 2.1 is different but relatyed with the proof of (1.21) by Pfaltzgraff and Suffridge [14]. In purpose to prove Theorm 2.1, we need the following lemmas.

LEMMA 2.2 *Fix $z_0 = r_0u_0 \in \boldsymbol{B}^n$, $u_0 \in \partial\boldsymbol{B}^n$. If F is the extremal mapping in $\mathcal{S}$ for the following maximum problem:*

$$Tr\left(\overline{J_F(z_0)}'J_F(z_0)\right) \ge Tr\left(\overline{J_f(z_0)}'J_f(z_0)\right)$$

holds for any $f \in \mathcal{S}$, then

$$Tr\left(\overline{J_F(z_0)}'J_F(z_0)\right) = max\left\{Tr\left(\overline{J_F(z)}'J_F(z)\right) \,\middle|\, |z| = r_0\right\}. \tag{2.4}$$

Proof. If (2.4) were false, then there would be a point $z_1 \ne z_0, |z_1| = r_0$ such that

$$Tr\left(\overline{J_F(z_1)}'J_F(z_1)\right) > Tr\left(\overline{J_F(z_0)}'J_F(z_0)\right).$$

There exists an unitary matrix U such that $Uz_0 = z_1$ where z_0, z_1 are column vectors. Consider $F_U(z) = U^{-1}F(Uz) \in \mathcal{S}$, then $J_{F_U}(z) = U^{-1}J_F(Uz)U$, and consequently

$$Tr\left(\overline{J_{F_U}(z)}'J_{F_U}(z)\right) = Tr\left(\overline{J_F(Uz)}'J_F(Uz)\right),$$

which yieds the contradiction

$$Tr\left(\overline{J_{F_U}(z_0)}'J_{F_U}(z_0)\right) = Tr\left(\overline{J_F(z_1)}'J_F(z_1)\right) > Tr\left(\overline{J_F(z_0)}'J_F(z_0)\right).$$

LEMMA 2.3 *Fix $z_0 = r_0u_0 \in \boldsymbol{B}^n, u_0 \in \partial\boldsymbol{B}^n$. If F is the extremal mapping in $\mathcal{S}$ for the maximum problem in Lemma 2.2, then $Tr[D^2F(z_0)z_0\overline{J_F(z_0)}']$ is real, equivalently, $Tr[D^2F(z_0)u_0\overline{J_F(z_0)}']$ is real.*

Proof. Let t be a small real number, then

$$J_F(z_0e^{it}) = J_F(z_0 + itz_0 + O(t^2)) = J_F(z_0) + itD^2F(z_0)z_0 + O(t^2).$$

We have

(2.5)

$$\overline{J_F(z_0e^{it})}'J_F(z_0e^{it}) = \overline{J_F(z_0)}'J_F(z_0) - it\overline{D^2F(z_0)z_0}'J_F(z_0) + itD^2F(z_0)z_0\overline{J_F(z_0)}' + O(t^2).$$

By Lemma 2.2,

$$\begin{aligned} 0 \ &\geq Tr\left(\overline{J_F(z_0e^{it})}'J_F(z_0e^{it})\right) - Tr\left(\overline{J_F(z_0)}'J_F(z_0)\right) \\ &= itTr\left(D^2F(z_0)z_0\overline{J_F(z_0)}' - J_F(z_0)\overline{D^2F(z_0)z_0}'\right) + O(t^2) \\ &= 2\Re eTr\left(itD^2F(z_0)z_0\overline{J_F(z_0)}'\right) + O(t^2). \end{aligned}$$

Dividing by $|t|$ and considering both $t \to 0^+$ and $t \to 0^-$, we have

$$Im\left\{Tr(D^2F(z_0)z_0\overline{J_F(z_0)}'\right\} = 0.$$

LEMMA 2.4. *Fix $z_0 = r_0u_0 \in \boldsymbol{B}^n, u_0 \in \partial\boldsymbol{B}^n$. If F is the extremal mapping in $\mathcal{S}$ for the maximum problem in Lemma 2.2, then*

(2.6)

$$\begin{aligned} &Tr\{\overline{J_F(z_0)}'D^2F(z_0)\nu\} - Tr\{\overline{J_F(z_0)}'D^2F(z_0)z_0\overline{z_0}'\nu\} \\ - \ &Tr\{\overline{z_0}'\overline{J_F(z_0)}'J_F(z_0)\nu\} - Tr\{J_F(z_0)\overline{J_F(z_0)}'\overline{z_0}'\nu\} - Tr\{J_F(z_0)\overline{J_F(z_0)}'D^2F(0)\nu\} \\ = \ &0 \end{aligned}$$

holds for any $\nu \in \partial\boldsymbol{B}^n$, where z_0, ν are column vectors.

Proof. Any $f \in \mathcal{S}$ can expand as (1.18), it can express as

$$f(z) = z + A_2(z,z) + A_3(z,z,z) + \cdots\cdots$$

also. The mapping defined by (2.1) belongs to $\mathcal{S}$ and can be expanded as [14]

$$f(z;a) = f(z) + Df(z)(a - \overline{a}'zz) - a - 2A_2(a, f(z)) + O(|a|^2),$$

where $|a|$ is sufficiently small. Let $DF(z;a)$ be the Fréchet derivative of $F(z;a)$ with respect to z, then [14]

$$DF(z;a) = J_F(z) + D^2F(z)(a - \overline{a}'zz) - J_F(z)z\overline{a}' - \overline{a}'zJ_F(z) - D^2F(0)aJ_F(z) + O(|a|^2),$$

when $|a|$ is sufficiently small, and

(2.7)

$$\begin{aligned}
&\overline{DF(Z;a)}'DF(z;a)\\
=\ & \overline{J_F(z)}'J_F(z)+[D^2F(z)(a-\overline{a}'zz)-J_F(z)z\overline{a}'-\overline{a}'zJ_F(z)-D^2F(0)aJ_F(z)]\overline{J_F(z)}'\\
+\ & J_F(z)\left[\overline{D^2F(z)a}'-z\overline{a}'\overline{D^2F(z)z}'-a\overline{z}'\overline{J_F(z)}'-\overline{J_F(Z)}'\overline{z}'a-\overline{J_F(z)}'\overline{D^2F(0)a}'\right]\\
+\ & O(|a|^2).
\end{aligned}$$

By the extremal property of F at point z_0, we have

$$\begin{aligned}
rl0 &\geq Tr\left\{\overline{DF(z_0;a)}'DF(z_0;a)\right\}-Tr\left(\overline{J_F(z_0)}'J_F(z_0)\right)\\
&=2\Re e\left[Tr[D^2F(z)a\overline{J_F(z)}']-Tr[\overline{a}'zD^2F(z)z\overline{J_F(z)}']\right.\\
&\left.-Tr[J_F(z)a\overline{z}'\overline{J_F(z)}']-Tr[\overline{z}'aJ_F(z)\overline{J_F(z)}']-Tr[D^2F(0)aJ_F(z)\overline{J_F(z)}']\right]+O(|a|^2).
\end{aligned}$$

By Lemma 2.3, we know that $Tr[D^2F(z_0)z_0\overline{J_F(z_0)}']$ is real, thus

$$\Re e[Tr\{\overline{a}'z_0D^2F(z_0)z_0\overline{J_F(z_0)}'\}]=\Re e[Tr\{\overline{z_0}'aD^2F(z_0)z_0\overline{J_F(z_0)}'\}].$$

Substituting it into the previous inequality, we have

$$\begin{aligned}
0\geq \Re e&\left[Tr[D^2F(z_0)a\overline{J_F(z_0)}']-Tr[\overline{z_0}'aD^2F(z_0)z_0\overline{J_F(z_0)}']\right.\\
&-Tr[J_F(z_0)a\overline{z_0}'\overline{J_F(z_0)}']-Tr[\overline{z_0}'aJ_F(z_0)\overline{J_F(z_0)}']\\
&\left.-Tr[D^2F(0)aJ_F(z_0)\overline{J_F(z_0)}']\right]+O(|a|^2).
\end{aligned}$$

Dividing $|a|$, and denoting $\nu=\frac{a}{|a|}$, then let $|a|\to 0$, note that a can be replaced by $e^{it}a$ with arbitrary real t. We have proved (2.6).

LEMMA 2.5. *Fix $z_0=r_0u_0\in \boldsymbol{B}^n, u_0\in\partial\boldsymbol{B}^n$. If F is the extremal mapping in $\mathcal{S}$ for the maximum problem in Lemma 2.2, then*

$$\begin{aligned}
&\frac{Tr\left(\overline{J_F(z)}'J_F(z)\right)-Tr\left(\overline{J_F(z_0)}'J_F(z_0)\right)}{2(r-r_0)}\\
(2.8)\qquad =\ &\frac{r_0}{1-r_0^2}\left[Tr(J_F(z_0)\overline{J_F(z_0)}')+Tr(\overline{u_0}'\overline{J_F(z_0)}'J_F(z_0)u_0)\right]\\
+\ &\frac{1}{1-r_0^2}Tr(J_F(z_0)\overline{J_F(z_0)}'D^2F(0)u_0)+O(|r-r_0|),
\end{aligned}$$

where $z=ru_0$, and r is a real number near r_0.

Proof. We expand $J_F(z)$ at point z_0,

$$J_F(z)=J_F(z_0+z-z_0)=J_F(z_0)+D^2F(z_0)(z-z_0)+O(|z-z_0|^2).$$

Thus

$$\begin{aligned}
(2.9)\qquad \overline{J_F(z)}'J_F(z) &= \overline{J_F(z_0)}'J_F(z_0)+J_F(z_0)\overline{D^2F(z_0)(z-z_0)}'\\
&+D^2F(z_0)(z-z_0)\overline{J_F(z_0)}'+O(|z-z_0|^2).
\end{aligned}$$

We have

$$Tr\left(\overline{J_F(z)}'J_F(z)\right) - Tr\left(\overline{J_F(z_0)}'J_F(z_0)\right)$$
$$= 2\Re e\left[Tr(D^2F(z_0)(z-z_0)\overline{J_F(z_0)}')\right] + O(|z-z_0|^2).$$

Dividing $2(r-r_0)$ on both sides, it is

$$\frac{Tr\left(\overline{J_F(z)}'J_F(z)\right) - Tr\left(\overline{J_F(z_0)}'J_F(z_0)\right)}{2(r-r_0)}$$
$$= \Re e\left[Tr(D^2F(z_0)u_0\overline{J_F(z_0)}')\right] + O(|r-r_0|).$$

By Lemma 2.3, we know that

$$\Re e\left[Tr(D^2F(z_0)u_0\overline{J_F(z_0)}')\right] = Tr(D^2F(z_0)u_0\overline{J_F(z_0)}').$$

Thus

(2.10)
$$\frac{Tr\left(\overline{J_F(z)}'J_F(z)\right) - Tr\left(\overline{J_F(z_0)}'J_F(z_0)\right)}{2(r-r_0)} = Tr(D^2F(z_0)u_0\overline{J_F(z_0)}') + O(|r-r_0|).$$

Let $\nu = u_0$ in Lemma 2.4, (2.6) become

$$\begin{aligned}
& Tr\{\overline{J_F(z_0)}'D^2F(z_0)u_0\} - r_0^2 Tr\{\overline{J_F(z_0)}'D^2F(z_0)u_0\} \\
- \;& Tr\{\overline{z_0}'\overline{J_F(z_0)}'J_F(z_0)u_0\} - Tr\{J_F(z_0)\overline{J_F(z_0)}'u_0\} - Tr\{J_F(z_0)\overline{J_F(z_0)}'D^2F(0)u_0\} \\
= \;& 0.
\end{aligned}$$

Hence

$$\begin{aligned}
& Tr(D^2F(z_0)u_0\overline{J_F(z_0)}') \\
= \;& \frac{r_0}{1-r_0^2}\left[Tr[J_F(z_0)\overline{J_F(z_0)}'] + Tr[\overline{u}'\overline{J_F(z_0)}'J_F(z_0)u_0]\right] \\
+ \;& \frac{1}{1-r_0^2}Tr[J_F(z_0)\overline{J_F(z_0)}'D^2F(0)u_0].
\end{aligned} \tag{2.11}$$

Substituting (2.11) into (2.10), we get (2.8).

We now proceed with the proof of Theorem 2.1, beginning with the upper bound.

Proof of Theorem 2.1. The upper bound:Throughout the proof of the upper bound, F will be the extremal mapping for the maximum problem associated with a fixed point $z_0 \in \boldsymbol{B}^n$, i.e.,

$$Tr\left(\overline{J_F(z_0)}'J_F(z_0)\right) \geq Tr\left(\overline{J_f(z_0)}'J_f(z_0)\right)$$

holds for all $f \in \mathcal{S}$. We now assume that the upper bound in (2.3) is false. Hence there exists $r_0 \in (0,1)$, $z_0 \in \boldsymbol{B}^n, |z_0| = r_0$ such that

$$Tr\left(\overline{J_F(z_0)}'J_F(z_0)\right) > \frac{(1+r_0)^{2\alpha-2}}{(1-r_0)^{2\alpha+2}} + (n-1)\frac{(1+r_0)^{2\alpha-1}}{(1-r_0)^{2\alpha+1}}.$$

Since the function

$$\varphi(t,s)=\frac{(1+t)^{2s-2}}{(1-t)^{2s+2}}+(n-1)\frac{(1+t)^{2s-1}}{(1-t)^{2s+1}},\quad 0<t<1 \tag{2.12}$$

is an increasing function of s, there is a number $\beta_0>\alpha$, such that

$$Tr\left(\overline{J_F(z_0)}'J_F(z_0)\right)=\varphi(r_0,\beta_0). \tag{2.13}$$

Consequently, for small $\epsilon>0$ and $r_0-\epsilon<r<r_0+\epsilon$, there is a C^1 smooth function $\beta(r)$ such that $\beta(r_0)=\beta_0$ and

$$Tr\left(\overline{J_F(ru_0)}'J_F(ru_0)\right)=\frac{(1+r)^{2\beta(r)-2}}{(1-r)^{2\beta(r)+2}}+(n-1)\frac{(1+r)^{2\beta(r)-1}}{(1-r)^{2\beta(r)+1}}=\varphi(r,\beta(r)). \tag{2.14}$$

Thus

$$\frac{Tr\left(\overline{J_F(ru_0)}'J_F(ru_0)\right)-Tr\left(\overline{J_F(z_0)}'J_F(z_0)\right)}{r-r_0}=\frac{\varphi(r,\beta(r))-\varphi(r_0,\beta_0)}{r-r_0}.$$

By Lemma 2.5, we have

$$\begin{aligned}&\frac{\varphi(r,\beta(r))-\varphi(r_0,\beta_0)}{2(r-r_0)}\\ &=\frac{r_0}{1-r_0^2}\left[Tr\left(\overline{J_f(z_0)}'J_f(z_0)\right)+Tr[\overline{u_0}'\overline{J_F(z_0)}'J_F(z_0)u_0]\right]\\ &+\frac{1}{1-r_0^2}Tr\left(J_F(z_0)\overline{J_F(z_0)}'D^2F(0)u_0\right)+O(|r-r_0|).\end{aligned} \tag{2.15}$$

As $r\to r_0$, the limit of the left hand side of (2.15) is $\frac{1}{2}[\varphi_r+\beta'\varphi_\beta]_{r=r_0}$. We evaluate it.

Noting that

$$\begin{aligned}\varphi_r&=\frac{\partial}{\partial r}\left[\frac{(1+r)^{2\beta-2}}{(1-r)^{2\beta+2}}+(n-1)\frac{(1+r)^{2\beta-1}}{(1-r)^{2\beta+1}}\right]\\ &=\frac{(1+r)^{2\beta-2}}{(1-r)^{2\beta+2}}\frac{4\beta+4r}{1-r^2}+(n-1)\frac{(1+r)^{2\beta-1}}{(1-r)^{2\beta+1}}\frac{4\beta+2r}{1-r^2}\\ &=\frac{4\beta+4r}{1-r^2}\left[\frac{(1+r)^{2\beta-2}}{(1-r)^{2\beta+2}}+(n-1)\frac{(1+r)^{2\beta-1}}{(1-r)^{2\beta+1}}\right]-\frac{2r(n-1)}{1-r^2}\frac{(1+r)^{2\beta-1}}{(1-r)^{2\beta+1}},\end{aligned}$$

we have

$$\varphi_r\Big|_{r=r_0}=\frac{4\beta_0+4r_0}{1-r_0^2}Tr\left(\overline{J_F(z_0)}'J_F(z_0)\right)-\frac{2(n-1)r_0}{1-r_0^2}\frac{(1+r_0)^{2\beta_0-1}}{(1-r_0)^{2\beta_0+1}}$$

by (2.13).

Moreover,

$$\begin{aligned}\beta'\varphi_\beta &= \beta'\frac{\partial}{\partial\beta}\left[\frac{(1+r)^{2\beta-2}}{(1-r)^{2\beta+2}}+(n-1)\frac{(1+r)^{2\beta-1}}{(1-r)^{2\beta+1}}\right]\\ &= \frac{(1+r)^{2\beta-2}}{(1-r)^{2\beta+2}}2\beta' log\frac{1+r}{1-r}+\frac{(1+r)^{2\beta-1}}{(1-r)^{2\beta+1}}2\beta' log\frac{1+r}{1-r}\\ &= 2\beta' log\frac{1+r}{1-r}\left[\frac{(1+r)^{2\beta-2}}{(1-r)^{2\beta+2}}+(n-1)\frac{(1+r)^{2\beta-1}}{(1-r)^{2\beta+1}}\right],\end{aligned}$$

we have

$$\beta'\varphi_\beta\Big|_{r=r_0} = 2\beta'(r_0)log\frac{1+r_0}{1-r_0}Tr\left(\overline{J_F(z_0)}'J_F(z_0)\right)$$

by (2.13).

Thus

$$\begin{aligned}&\frac{1}{2}[\varphi_r+\beta'\varphi_\beta]_{r=r_0}\\ &= \left(\frac{2\beta_0+2r_0}{1-r_0^2}+\beta'(r_0)log\frac{1+r_0}{1-r_0}\right)Tr\left(\overline{J_F(z_0)}'J_F(z_0)\right)\\ &\quad -\frac{(n-1)r_0}{1-r_0^2}\frac{(1+r_0)^{2\beta_0-1}}{(1-r_0)^{2\beta_0+1}}.\end{aligned}\tag{2.16}$$

As $r\to r_0$, (2.15) becomes the following equality

$$\begin{aligned}&\left(\frac{2\beta_0+2r_0}{1-r_0^2}+\beta'(r_0)log\frac{1+r_0}{1-r_0}\right)Tr\left(\overline{J_F(z_0)}'J_F(z_0)\right)-\frac{(n-1)r_0}{1-r_0^2}\frac{(1+r_0)^{2\beta_0-1}}{(1-r_0)^{2\beta_0+1}}\\ &= \frac{r_0}{1-r_0^2}\left[Tr\left(\overline{J_F(z_0)}'J_F(z_0)\right)+Tr(\overline{u_0}'\overline{J_F(z_0)}'J_F(z_0)u_0)\right]\\ &\quad +\frac{1}{1-r_0^2}Tr(J_F(z_0)\overline{J_F(z_0)}'D^2F(0)u_0)\end{aligned}$$

by (2.16). After cancel the term $\frac{r_0}{1-r_0^2}Tr\left(\overline{J_F(z_0)}'J_F(z_0)\right)$ on both sides of the privious equality, we have

$$\begin{aligned}&\left(\frac{2\beta_0+r_0}{1-r_0^2}+\beta'(r_0)log\frac{1+r_0}{1-r_0}\right)Tr\left(\overline{J_F(z_0)}'J_F(z_0)\right)-\frac{(n-1)r_0}{1-r_0^2}\frac{(1+r_0)^{2\beta_0-1}}{(1-r_0)^{2\beta_0+1}}\\ &= \frac{r_0}{1-r_0^2}\overline{u_0}'\overline{J_F(z_0)}'J_F(z_0)u_0+\frac{1}{1-r_0^2}Tr(J_F(z_0)\overline{J_F(z_0)}'D^2F(0)u_0).\end{aligned}\tag{2.17}$$

By equality (2.17), we know that $Tr(J_F(z_0)\overline{J_F(z_0)}'D^2F(0)u_0)$ is real. $J_F(z_0)\overline{J_F(z_0)}'$ is a positive definite Hermitian matrix, we may decompose it as

$$J_F(z_0)\overline{J_F(z_0)}' = U\begin{pmatrix}\lambda_1^2 & & \\ & \ddots & \\ & & \lambda_n^2\end{pmatrix}\overline{U}',$$

where U is an unitary matrix. Thus

$$\begin{aligned}
&Tr(J_F(z_0)\overline{J_F(z_0)}'D^2F(0)u_0)\\
&= Tr\left(U\begin{pmatrix}\lambda_1^2 & & \\ & \ddots & \\ & & \lambda_n^2\end{pmatrix}\overline{U}'D^2F(0)u_0\right)\\
&= Tr\left(\left(\begin{pmatrix}\lambda_1^2 & & \\ & \ddots & \\ & & \lambda_n^2\end{pmatrix}\right)\overline{U}'D^2F(0)u_0U\right).
\end{aligned}$$

If

$$\overline{U}'D^2F(0)u_0U = \begin{pmatrix} a_{11} & \cdots & a_{1n} \\ \cdots & \cdots & \cdots \\ a_{n1} & \cdots & a_{nn}\end{pmatrix},$$

then

$$\left|Tr\left(\left(\begin{pmatrix}\lambda_1^2 & & \\ & \ddots & \\ & & \lambda_n^2\end{pmatrix}\right)\overline{U}'D^2F(0)u_0U\right)\right| = \left|\sum_{j=1}^{n}\lambda_j^2 a_{jj}\right| \le \sum_{j=1}^{n}\lambda_j^2|a_{jj}|.$$

But

$$|a_{jj}| \le \|\overline{U}'D^2F(0)u_0U\| = \|D^2F(0)u_0\| \le 2\alpha, j = 1,2,3,\cdots\cdots,n,$$

since the $\|ord\|S = \alpha$ and (1.20). Thus

$$|Tr(J_F(z_0)\overline{J_F(z_0)}'D^2F(0)u_0)| \le 2\alpha Tr\left(\overline{J_F(z_0)}'J_F(z_0)\right). \tag{2.18}$$

By (1.21)(Theorem 4.1 in [14]), and $\beta_0 > \alpha$, we have

$$\begin{aligned}
\overline{u_0}'\overline{J_F(z_0)}'J_F(z_0)u_0 \le\ & \text{The largest eigenvalue of } \overline{J_F(z_0)}'J_F(z_0)\\
\le\ & \frac{(1+r_0)^{2\alpha-2}}{(1-r_0)^{2\alpha+2}}\\
<\ & \frac{(1+r_0)^{2\beta_0-2}}{(1-r_0)^{2\beta_0+2}},
\end{aligned} \tag{2.19}$$

From (2.17), (2.18) and (2.19), we obtain the following inequality
(2.20)

$$\begin{aligned}
&\left(\frac{2\beta_0+r_0}{1-r_0^2}+\beta'(r_0)log\frac{1+r_0}{1-r_0}\right)Tr\left(\overline{J_F(z_0)}'J_F(z_0)\right)\\
\le\ & \frac{(1+r_0)^{2\beta_0-2}}{(1-r_0)^{2\beta_0+2}}\frac{r_0}{1-r_0^2}+\frac{(n-1)r_0}{1-r_0^2}\frac{(1+r_0)^{2\beta_0-1}}{(1-r_0)^{2\beta_0+1}}+\frac{2\alpha}{1-r_0^2}Tr\left(\overline{J_F(z_0)}'J_F(z_0)\right)\\
=\ & \frac{r_0}{1-r_0^2}Tr\left(\overline{J_F(z_0)}'J_F(z_0)\right)+\frac{2\alpha}{1-r_0^2}Tr\left(\overline{J_F(z_0)}'J_F(z_0)\right)
\end{aligned}$$

by (2.13). After cancel the term $\frac{r_0}{1-r_0^2}Tr\left(\overline{J_F(z_0)}'J_F(z_0)\right)$ on both sides of (2.20), we have

(2.21)

$$\left(\frac{2\beta_0}{1-r_0^2}+\beta'(r_0)log\frac{1+r_0}{1-r_0}\right)Tr\left(\overline{J_F(z_0)}'J_F(z_0)\right)\leq\frac{2\alpha}{1-r_0^2}Tr\left(\overline{J_F(z_0)}'J_F(z_0)\right).$$

From (2.21), we conclude $\beta'(r_0)<0$ since $\beta_0>\alpha$. Hence there is $r_1<r_0$ such that $\beta_1=\beta(r_1)>\beta(r_0)=\beta_0>\alpha$, and

$$\begin{aligned}Tr\left(\overline{J_F(r_1u_0)}'J_F(r_1u_0)\right) &= \frac{(1+r_1)^{2\beta_1-2}}{(1-r_1)^{2\beta_1+2}}+(n-1)\frac{(1+r_1)^{2\beta_1-1}}{(1-r_1)^{2\beta_1+1}}\\ &> \frac{(1+r_1)^{2\beta_0-2}}{(1-r_1)^{2\beta_0+2}}+(n-1)\frac{(1+r_1)^{2\beta_0-1}}{(1-r_1)^{2\beta_0+1}}\end{aligned}\tag{2.22}$$

by (2.14)

Let $u_1\in\partial\boldsymbol{B}^n$, such that

$$Tr\left(\overline{J_F(r_1u_1)}'J_F(r_1u_1)\right)=\max_{u\in\partial\boldsymbol{B}^n}Tr\left(\overline{J_F(r_1u)}'J_F(r_1u)\right).\tag{2.23}$$

Write $z_1=r_1u_1$, let F_1 be an extremal such that

$$Tr\left(\overline{J_{F_1}(z_1)}'J_{F_1}(z_1)\right)=\max_{f\in\mathcal{S}}Tr\left(\overline{J_f(z_1)}'J_f(z_1)\right).$$

By (2.22), (2.23), we have

$$\begin{aligned}Tr\left(\overline{J_{F_1}(z_1)}'J_{F_1}(z_1)\right) &\geq \frac{(1+r_1)^{2\beta_1-2}}{(1-r_1)^{2\beta_1+2}}+(n-1)\frac{(1+r_1)^{2\beta_1-1}}{(1-r_1)^{2\beta_1+1}}\\ &> \frac{(1+r_1)^{2\beta_0-2}}{(1-r_1)^{2\beta_0+2}}+(n-1)\frac{(1+r_1)^{2\beta_0-1}}{(1-r_1)^{2\beta_0+1}}.\end{aligned}$$

We can repeat the argument just given. Thus one proves that there are a sequence of numbers $\{r_i\}$ decreasing to zero and a sequence $\{F_j\}\subset\mathcal{S}$, such that

$$\begin{aligned}&Tr\left(\overline{J_{F_1}(z_1)}'J_{F_1}(z_1)\right)\\ > &\ \frac{(1+r_j)^{2\beta_0-2}}{(1-r_j)^{2\beta_0+2}}+(n-1)\frac{(1+r_j)^{2\beta_0-1}}{(1-r_j)^{2\beta_0+1}}\\ = &\ (1+(2\beta_0-2)r_j+\cdots)(1-(2\beta_0+2)r_j+\cdots)^{-1}\\ + &\ (n-1)(1+(2\beta_0-2)r_j+\cdots)(1-(2\beta_0+2)r_j+\cdots)^{-1}\\ = &\ n+4\beta_0nr_j+O(r_j^2),\end{aligned}\tag{2.24}$$

where $z_j=r_ju_j$, $u_j\in\partial\boldsymbol{B}^n$, $j=1,2,\cdots\cdots$.

Since $J_{F_j}(z_j)=I+D^2F_j(0)z_j+O(|z_j|^2)$, we have

$$\overline{J_{F_j}(z_j)}'J_{F_j}(z_j)=I+D^2F_j(0)z_j+\overline{D^2F_j(0)z_j}'+O(|z_j|^2),$$

$$D^2F_j(0)z_j + \overline{D^2F_j(0)z_j}' + O(|z_j|^2) = \overline{J_{F_j}(z_j)}' J_{F_j}(z_j) - I,$$

and hence
(2.25)
$$Tr\left(D^2F_j(0)z_j + \overline{D^2F_j(0)z_j}'\right) + O(|z_j|^2) = Tr\left(\overline{J_{F_j}(z_j)}' J_{F_j}(z_j)\right) - n > 4\beta_0 n r_j + O(|z_j|^2)$$

by (2.24)

Recall that $\mathcal{S}$ is a normal family when $\|ord\|\mathcal{S} = \alpha < \infty$, there is a subsequence of $\{F_j\}$ that converges locally uniformly to a mapping $f \in \mathcal{S}$ and u_j tend to a unit vector u. Thus dividing by r_j on both sides of (2.25) and letting $j \to \infty$ through the subsequence, we obtain

$$Tr\left(D^2f(0)u + \overline{D^2f(0)u}'\right) \geq 4\beta_0 n. \tag{2.26}$$

But

$$Tr\left(D^2f(0)u\right) + Tr\left(\overline{D^2f(0)u}'\right) \leq 2|TrD^2f(0)u| \leq 2n\|D^2f(0)u\|,$$

we have

$$2n\|D^2f(0)u\| \geq 4\beta_0 n > 4\alpha n,$$

i.e.,

$$\|D^2f(0)u\| > 2\alpha = 2\|ord\|\mathcal{S}.$$

by (2.26). This contradiction completes the proof of the upper bound inequality in Theorem 2.1.

The proof of the lower bound inequality in Theorem 2.1 is similar with the proof of upper bound inequality.

Throughout this argument, F will be an extremal mapping for the minimum problem associated with a fixed point $z_0 \in B^n$, i.e.,

$$Tr\left(\overline{J_F(z_0)}' J_F(z_0)\right) \leq Tr\left(\overline{J_f(z_0)}' J_f(z_0)\right) \tag{2.27}$$

holds for all $f \in \mathcal{S}$. We may prove the Lemma 2.2, 2.3, 2.4 and 2.5 under the assumption (2.27), and prove the lower bound of the inequality if we instead $\varphi(t,s)$ by

$$\eta(t,s) = \varphi(-t,s) = \frac{(1-t)^{2s-2}}{(1+t)^{2s+2}} + (n-1)\frac{(1-t)^{2s-1}}{(1+t)^{2s+1}}.$$

All the argument of the proof are similar with the proof of the upper bound. We omit the detail.

3. The estimations of the Jacobian of mappings in a B^n. We may use the method as we used in the previous section to estimate $det(J_f(z)\overline{J_f(z)}')$ of mapping f in a LIF on the unit ball B^n in $\mathbb{C}^n$. Actually, the following theorem about the estimations of $det(J_f(z)\overline{J_f(z)}')$ was known long time ago, which was proved by Barnard,FifzGerald and Gong [1], Liu [11], Gong and Zheng [8], and Pfaltzgraff [12]. In this section, the proof of Theorm 3.1 is different with the original proof. It also shows that using $C(\mathcal{S})$

which was defined by (1.5) and (1.19) to estimate $det(J_f(z)\overline{J_f(z)}')$ is the right choice. In this section, all the vectors are column vectors.

THEOREM 3.1. *If S is a LIF on the unit ball B^n in $\mathbb{C}^n$, then*

$$\frac{(1-|z|)^{2C(S)-(n+1)}}{(1+|z|)^{2C(S)+(n+1)}} \le det\left(\overline{J_f(z)}'J_f(z)\right) \le \frac{(1+|z|)^{2C(S)-(n+1)}}{(1-|z|)^{2C(S)+(n+1)}} \tag{3.1}$$

for all $f \in S$, and $z \in B^n$, where $C(S)$ is defined by (1.5) and (1.19).

First of all, we prove the following Lemmas.

LEMMA 3.2. *Fix $z_0 = r_0u_0 \in B^n$, $u_0 \in \partial B^n$. If F is the extremal mapping in S for the following maximum problem:*

$$det\left(\overline{J_F(z_0)}'J_F(z_0)\right) \ge det\left(\overline{J_f(z_0)}'J_f(z_0)\right)$$

holds for any $f \in S$, then

$$det(J_F(z_0)\overline{J_F(z_0)}') = max\left\{det(J_F(z)\overline{J_F(z)}')\ \Big|\ |z| = r_0\right\}. \tag{3.2}$$

Proof. If (3.2) were false, then there be a point $z_1 \ne z_0, |z_1| = r_0$ such that

$$det\left(\overline{J_F(z_1)}'J_F(z_1)\right) > det\left(\overline{J_F(z_0)}'J_F(z_0)\right).$$

There exists an unitary matrix U such that $Uz_0 = z_1$ where $z_0, z_1$are column vectors. Consider $F_U(z) = U^{-1}F(Uz) \in S$, then $J_{F_U}(z) = U^{-1}J_F(Uz)U$, and consequently

$$det\left(\overline{J_{F_U}(z)}'J_{F_U}(z)\right) = det\left(\overline{J_F(Uz)}'J_F(Uz)\right)$$

which yied the contradiction

$$det\left(\overline{J_{F_U}(z_0)}'J_{F_U}(z_0)\right) = det\left(\overline{J_F(z_1)}'J_F(z_1)\right) > det\left(\overline{J_F(z_0)}'J_F(z_0)\right).$$

LEMMA 3.3. *Fix $z_0 = r_0u_0 \in B^n$, $u_0 \in \partial B^n$. If F is the extremal mapping in S for the maximum problem in Lemma 3.2, then $Tr(D^2F(z_0)z_0(J_F(z_0))^{-1})$ is real, equivalently, $Tr(D^2F(z_0)u_0(J_F(z_0))^{-1})$ is real.*

Proof. Let t be a small real number, then by (2.5),

$$\begin{aligned}
&det\left(\overline{J_F(z_0e^{it})}'J_F(z_0e^{it})\right)\\
&= det\Big[\overline{J_F(z_0)}'\Big(I + itD^2F(z_0)z_0(J_F(z_0))^{-1}\\
&\qquad -it(\overline{J_F(z_0)}')^{-1}\overline{(D^2F(z_0)z_0)}' + O(t^2)\Big)J_F(z_0)\Big]\\
&= det\left(\overline{J_F(z_0)}'J_F(z_0)\right)\Big[1+tTr[itD^2F(z_0)z_0(J_F(z_0))^{-1} - i\overline{(J_F(z_0))^{-1}}'\overline{(D^2F(z_0)z_0)}'\Big]\\
&\quad +O(t^2)
\end{aligned}$$

since

$$det(I + tH) = 1 + tTrH + O(t^2)$$

when t is a small real number, H is a Hermitian matrix. By Lemma 3.2,

$$\begin{aligned} 0 &\geq det\left(\overline{J_F(z_0e^{it})}'J_F(z_0e^{it})\right) - det\left(\overline{J_F(z_0)}'J_F(z_0)\right) \\ &= det\left(\overline{J_F(z_0)}'J_F(z_0)\right) tTr\left[i\left(D^2F(z_0)z_0(J_F(z_0))^{-1}\right.\right. \\ &\quad \left.\left. - \left(\overline{J_F(z_0)}'\right)^{-1}\left(\overline{D^2F(z_0)z_0}\right)'\right)\right] + O(t^2). \end{aligned}$$

Dividing by $|t|$ and considering both $t \to 0^+$ and $t \to 0^-$, we have

$$Im\{Tr(D^2F(z_0)z_0(J_F(z_0))^{-1}\} = 0.$$

LEMMA 3.4. *Fix $z_0 = r_0u_0 \in \boldsymbol{B}^n$, $u_0 \in \partial\boldsymbol{B}^n$. If F is the extremal mapping in S for the maximum problem in Lemma 3.2, then*

$$\begin{aligned} Tr\left(D^2F(z_0)\nu(J_F(z_0))^{-1}\right) - \overline{z}'\nu Tr\left(D^2F(z_0)z_0(J_F(z_0))^{-1}\right) \\ -(n+1)\overline{z_0}'\nu - Tr(D^2F(0)\nu) = 0 \end{aligned} \tag{3.3}$$

holds for any $\nu \in \partial\boldsymbol{B}^n$, where z_0, ν are column vectors.

Proof. From (2.7), we have

$$\begin{aligned} & det\left(\overline{J_F(z;a)}'J_F(z;a)\right) \\ = \ & det\left(\overline{J_F(z)}'J_F(z)\right) det\left[I + 2\Re e\left((D^2F(z)(a-\overline{a}'zz) - J_F(z)z\overline{a}'\right.\right. \\ - \ & \left.\left.\overline{a}'zJ_F(z) - D^2F(0)aJ_F(z))(J_F(z))^{-1}\right) + O(|a|^2)\right] \\ = \ & det\left(\overline{J_F(z)}'J_F(z)\right)\left(1 + 2\Re e\left[Tr(D^2F(z)(a-\overline{a}'zz)(J_F(z))^{-1}\right.\right. \\ - \ & \left.\left.(n+1)\overline{a}'z - Tr(D^2F(0)a)\right] + O(|a|^2)\right) \end{aligned}$$

when $|a|$ is small. By the extremal property of F at z_0, we have

$$\begin{aligned} 0 \geq \ & det\left(\overline{J_F(z;a)}'J_F(z;a)\right) - det\left(\overline{J_F(z_0)}'J_F(z_0)\right) \\ = \ & det\left(\overline{J_F(z_0)}'J_F(z_0)\right)\cdot 2\Re e\left[Tr(D^2F(z_0)(a-\overline{a}'z_0z_0)(J_F(z_0))^{-1}\right) \\ - \ & (n+1)\overline{a}'z_0 - Tr(D^2F(0)a)\right] + O(|a|^2). \end{aligned}$$

By Lemma 3.3, we know that $Tr(D^2F(z_0)z_0(J_F(z_0))^{-1})$ is real, thus

$$\Re e[Tr(D^2F(z_0)\overline{a}'z_0z_0(J_F(z_0))^{-1})] = \Re e[Tr(D^2F(z_0)\overline{z_0}'az_0(J_F(z_0))^{-1})].$$

Substituting it into previous inequality, we have

$$\begin{aligned} 0 \geq \ & det\left(\overline{J_F(z_0)}'J_F(z_0)\right)\Re e\left[Tr(D^2F(z_0)(a-\overline{a}'z_0z_0)(J_F(z_0))^{-1}\right) \\ - \ & (n+1)\overline{a}'z_0 - Tr(D^2F(0)a)\right] + O(|a|^2). \end{aligned}$$

Dividing $|a|$ on both sides of the previous inequality, and denoting $\nu = \frac{a}{|a|}$, then let $|a| \to 0$, note that a can be replaced by $e^{it}a$ with arbitrary real t, we have proved (3.3).

LEMMA 3.5. *Fix $z_0 = r_0 u_0 \in \mathbf{B}^n$, $u_0 \in \partial \mathbf{B}^n$. If F is the extremal mapping in S for the maximum problem in Lemma 3.2, then*

$$\begin{aligned}(3.4)\qquad &\frac{\det\left(\overline{J_F(z)}' J_F(z)\right) - \det\left(\overline{J_F(z_0)}' J_F(z_0)\right)}{2(r-r_0)} \\ &= \det\left(\overline{J_F(z_0)}' J_F(z_0)\right) \frac{(n+1)r_0 + Tr(D^2F(0)u_o)}{1-r_0^2} + O(|r-r_0|)\end{aligned}$$

where $z = ru_0$, and r is a real number near r_0.

Proof. Expanding $J_F(z)$ at z_0, we have

$$\begin{aligned}&\det\left(\overline{J_F(z)}' J_F(z)\right) \\ &= \det\left(\overline{J_F(z_0)}' J_F(z_0)\right) (1 + 2\Re e[Tr(D^2F(z_0)(z-z_0)(J_F(z_0))^{-1})] + O(|z-z_0|^2)\end{aligned}$$

by (2.9). Dividing $2(r-r_0)$ on both sides, it is

$$\begin{aligned}&\frac{\det\left(\overline{J_F(z)}' J_F(z)\right) - \det\left(\overline{J_F(z_0)}' J_F(z_0)\right)}{2(r-r_0)} \\ &= \det\left(\overline{J_F(z_0)}' J_F(z_0)\right) \Re e[Tr(D^2F(z_0)u_0(J_F(z_0))^{-1})] + O(|r-r_0|).\end{aligned}$$

By Lemma 3.3, we know that

$$\Re e[Tr(D^2F(z_0)u_0(J_F(z_0))^{-1}] = Tr(D^2F(z_0)u_0(J_F(z_0))^{-1}).$$

Thus

$$\begin{aligned}(3.5)\qquad &\frac{\det\left(\overline{J_F(z)}' J_F(z)\right) - \det\left(\overline{J_F(z_0)}' J_F(z_0)\right)}{2(r-r_0)} \\ &= \det\left(\overline{J_F(z_0)}' J_F(z_0)\right) [Tr(D^2F(z_0)u_0(J_F(z_0))^{-1})] + O(|r-r_0|).\end{aligned}$$

Let $\nu = u_0$ in Lemma 3.4, (3.3) becomes

$$\begin{aligned}&Tr\left(D^2F(z_0)u_0(J_F(z_0))^{-1}\right) \\ &= \overline{z_0}' u_0 Tr\left(D^2F(z_0)z_0(J_F(z_0))^{-1}\right) + (n+1)\overline{z_0}' u_0 + Tr(D^2F(0)u_0) \\ &= r_0^2 Tr(D^2F(z_0)u_0(J_F(z_0))^{-1}) + (n+1)r_0 + Tr(D^2F(0)u_0).\end{aligned}$$

Hence

$$(3.6)\qquad Tr(D^2F(z_0)u_0(J_F(z_0))^{-1}) = \frac{(n+1)r_0 + Tr(D^2F(0)u_0)}{1-r_0^2}.$$

Substituting (3.6) into (3.5), we get (3.4).

Proof of Theorem 3.1. The upper bound: Let F be the extremal mapping for the maximum probrem in Lemma 3.2. We assume that the upper bound in (3.1) is false. Hence there exists $r_0 \in (0,1), z_0 \in \mathbf{B}^n, |z_0| = r_0$ such that

$$\det\left(\overline{J_F(z_0)}' J_F(z_0)\right) > \frac{(1+r_0)^{2C(S)-(n+1)}}{(1-r_0)^{2C(S)+(n+1)}}.$$

Since the function

$$\zeta(t,s)=\frac{(1+t)^{2s-(n+1)}}{(1-t)^{2s+(n+1)}},\qquad 0<t<1, \tag{3.7}$$

is an increasing function of s, there is a number $\beta_0>C(\mathcal{S})$ such that

$$det\left(\overline{J_F(z_0)}'J_F(z_0)\right)=\zeta(r_0,\beta_0). \tag{3.8}$$

Consequently, for small $\epsilon>0$ and $r_0-\epsilon<r<r_0+\epsilon$, there is a C^1 smooth function $\beta(r)$ such that $\beta(r_0)=\beta_0$ and

$$det\left(\overline{J_F(ru_0)}'J_F(ru_0)\right)=\frac{(1+r)^{2\beta(r)-(n+1)}}{(1-r)^{2\beta(r)+(n+1)}}=\zeta(r,\beta(r)). \tag{3.9}$$

Thus

$$\frac{det\left(\overline{J_F(ru_0)}'J_F(ru_0)\right)-det\left(\overline{J_F(z_0)}'J_F(z_0)\right)}{r-r_0}=\frac{\zeta(r,\beta(r))-\zeta(r_0,\beta_0)}{r-r_0}.$$

By Lemma 3.5, we have

(3.10)

$$\frac{\zeta(r,\beta(r))-\zeta(r_0,\beta_0)}{2(r-r_0)}=det\left(\overline{J_F(z_0)}'J_F(z_0)\right)\frac{(n+1)r_0+Tr(D^2F(0)u_0)}{1-r_0^2}+O(|r-r_0|).$$

As $r\to r_0$, the limit of the left hand side of (3.10) is $\frac{1}{2}[\zeta_r+\beta'\varphi_\beta]_{r=r_0}$. We evaluate it.

Noting that

$$\zeta_r=\frac{\partial}{\partial r}\frac{(1+r)^{2\beta-(n+1)}}{(1-r)^{2\beta+(n+1)}}=\frac{(1+r)^{2\beta-n-2}[4\beta+(2\beta+n+1)r]}{(1-r)^{2\beta+n+2}},$$

we have

$$\zeta_r\Big|_{r=r_0}=det\left(\overline{J_F(z_0)}'J_F(z_0)\right)\frac{4\beta_0+(2n+2)r_0}{1-r_0^2}$$

by (3.8). Moreover,

$$\beta'\varphi_\beta=2\beta'(r)\frac{(1+r)^{2\beta-(n+1)}}{(1-r)^{2\beta+(n+1)}}log\frac{1+r}{1-r},$$

we have

$$\beta'\varphi_\beta\Big|_{r=r_0}=2\beta(r_0)log\frac{1+r_0}{1-r_0}det\left(\overline{J_F(z_0)}'J_F(z_0)\right)$$

by (3.8). Thus

(3.11)

$$\frac{1}{2}[\zeta_r+\beta'\zeta_\beta]_{r=r_0}=\left(\frac{2\beta_0+(n+1)r_0}{1-r_0^2}+\beta'(r_0)log\frac{1+r_0}{1-r_0}\right)det\left(\overline{J_F(z_0)}'J_F(z_0)\right).$$

As $r \to r_0$, (3.10) becomes the following equality

$$\begin{aligned}&\left(\frac{2\beta_0+(n+1)r_0}{1-r_0^2}+\beta'(r_0)log\frac{1+r_0}{1-r_0}\right)det\left(\overline{J_F(z_0)}'J_F(z_0)\right)\\ =&\frac{(n+1)r_0+Tr(D^2F(0)u_0)}{1-r_0^2}det\left(\overline{J_F(z_0)}'J_F(z_0)\right)\end{aligned}$$

by (3.11). After cancel the term $\frac{(n+1)r_0}{1-r_0^2}det\left(\overline{J_F(z_0)}'J_F(z_0)\right)$ on both of the provious equality, we have

$$\begin{aligned}&\left(\frac{2\beta_0}{1-r_0^2}+\beta'(r_0)log\frac{1+r_0}{1-r_0}\right)det\left(\overline{J_F(z_0)}'J_F(z_0)\right)\\ =&\frac{Tr(D^2F(0)u_0)}{1-r_0^2}det\left(\overline{J_F(z_0)}'J_F(z_0)\right).\end{aligned}\tag{3.12}$$

By the definition of $C(\mathcal{S})$ (1.19), $|Tr(D^2F(0)u_0)|\le 2C(\mathcal{S})$. Hence

$$\begin{aligned}&\left(\frac{2\beta_0}{1-r_0^2}+\beta'(r_0)log\frac{1+r_0}{1-r_0}\right)det\left(\overline{J_F(z_0)}'J_F(z_0)\right)\\ \le&\frac{2C(\mathcal{S})}{1-r_0^2}det\left(\overline{J_F(z_0)}'J_F(z_0)\right)\end{aligned}$$

by (3.12). But $\beta_0 > C(\mathcal{S})$, it implies $\beta'(r_0)<0$. We may use the argument similar in the proof of Theorem 2.1. We prove that there are a sequence of number $\{r_j\}$ decreasing to zero and a sequence $\{F_j\}\subset\mathcal{S}$ such that

$$\begin{aligned}&det\left(\overline{J_{F_j}(z_j)}'J_{F_j}(z_j)\right)\\ >&\frac{(1+r_j)^{2\beta_0-(n+1)}}{(1-r_j)^{2\beta_0+(n+1)}}\\ =&(1+(2\beta-n-1)r_j+\cdots\cdots)(1-(2\beta+n+1)r_j+\cdots\cdots)^{-1}\\ =&1+4\beta_0r_j+O(r_j^2)\end{aligned}\tag{3.13}$$

where $z_j=r_ju_j, u_j\in\partial\boldsymbol{B}^n, j=1,2,\cdots\cdots$.

We know that

$$det\left(\overline{J_{F_j}(z_j)}'J_{F_j}(z_j)\right)=1+2Tr[\Re e(D^2F_j(0)z_j)]+O(r_j^2).$$

Thus

$$2Tr[\Re e(D^2F_j(0)z_j)]\ge 4\beta_0r_j+O(r_j^2),\qquad j=1,2,\cdots\cdots,\tag{3.14}$$

by (3.13). When $C(\mathcal{S})$ is finite, then $\|ord\|\mathcal{S}$ is finite, and $\mathcal{S}$ is a normal family. There is a subsequence of $\{F_j\}$ that converges locally uniformly to mapping $f\in\mathcal{S}$ and u_j tend to a unit vector u. Thus dividing by r_j on both sides of (3.14) and letting $j\to\infty$ through the subsequence, we obtain

$$|Tr(D^2F(0)u)|\ge 2\beta_0 > 2C(\mathcal{S}).$$

This contradiction complete the proof of the upper bound inequality in Theorem 3.1.

The proof of the lower bound inequality in Theorem 3.1 is similar with the proof of upper bound. We omit the detail of the proof.

4. Covariant derivative of biholomorphic convex mappings on unit ball. In Section 1, we already mentioned that (1.9) has many consequences, e.g., (1.11), (1.12), (1.13) and (1.14) are its consequences. In this section, we will give another consequence. It is about the covariant derivative of biholomorphic convex mappings on unit ball $\boldsymbol{B}^n$ in $\mathbb{C}^n$. In this section, all vectors are row vectiors.

THEOREM 4.1. *Let $f(z)$ be a normalized biholomorphic convex mapping on the unit ball $\boldsymbol{B}^n$ in $\mathbb{C}^n$, then for every $z \in \boldsymbol{B}^n$ and every vector $\eta(z) = (\eta_1(z), \cdots\cdots, \eta_n(z)) \in \mathbb{C}^n$, the inequality*

$$\left|\left(\eta^{[k]} \cdot (\frac{\delta}{\delta z})^{[k]}\right) f(z)\right| \leq k! \frac{1+|z|}{1-|z|} \left|\eta \frac{R}{\sqrt{1-z\overline{z}'}}\right|^k \tag{4.1}$$

holds, where $\frac{\delta}{\delta z} = (\frac{\delta}{\delta z_1}, \cdots\cdots, \frac{\delta}{\delta z_n})'$ is the covariant derivative operator with respect to the Bergman metric of $\boldsymbol{B}^n$, $\xi^{[k]}$ means $k-$th Kronecker product of vector $\xi \in \mathbb{C}^n$ and

$$R(z) = -I - \frac{\overline{z}' z(1-\sqrt{1-z\overline{z}'})}{z\overline{z}'\sqrt{1-z\overline{z}'}}, \tag{4.2}$$

I is the identity matrix.

In particular, if we let $\eta = e_j\sqrt{1-z\overline{z}'}$ in (4.1) where $e_j = (0, \cdots\cdots, 0, 1, 0, \cdots\cdots, 0)$ is the unit vector of $j-$ th coordinate, we have the estimation of $k-$th covariant derivative of f with respect to $z_j, j = 1, 2, \cdots\cdots, n; k = 1, 2, \cdots\cdots$.

COROLLARY 4.2. *Let $f(z)$ be a normalized biholomorphic convex mapping on the ball $\boldsymbol{B}^n$ in $\mathbb{C}^n$, then for every $z \in \boldsymbol{B}^n$, the inequality*

$$\left|\frac{\delta^k}{\delta z_j^k} f(z)\right| \leq \frac{k!(1-|z|^2+|z_j|^2)^{k/2}}{(1+|z|)^{k-1}(1-|z|)^{k+1}} \tag{4.3}$$

holds for $j = 1, 2, \cdots\cdots, n; k = 1, 2, \cdots\cdots$.

From (1.9), it is easy to verify

$$\frac{(1-|z|^2+|z_j|^2)^{\frac{1}{2}}}{(1+|z|)^2} \leq \left|\frac{\partial f}{\partial z_j}\right| \leq \frac{(1-|z|^2+|z_j|^2)^{\frac{1}{2}}}{(1-|z|)^2}, j = 1, 2, \cdots\cdots, n \tag{4.4}$$

holds.

When $k = 1$ in (4.3), it is the right hand side inequality of (4.4).

When $n = 1$, (4.3) is coincide with a result in [3].

If we let $\eta = z\sqrt{1-z\overline{z}'}$ in (4.1), we have the following

COROLLARY 4.3. *Let $f(z)$ be a normalized biholomorphic convex mapping on the unit ball $\boldsymbol{B}^n$ in $\mathbb{C}^n$, then for every $z \in \boldsymbol{B}^n$, the inequality*

$$\left|\left(z^{[k]} \cdot \left(\frac{\delta}{\delta z}\right)^{[k]}\right) f(z)\right| \leq \frac{k!|z|^k}{(1+|z|)^{k-1}(1-|z|)^{k+1}}$$

holds for $k = 1, 2, \cdots\cdots$.

Proof of Theorem 4.1. Gong and Yan [7] proved the following result.

Let $f(z)$ be a holomorphic mapping in $\boldsymbol{B}^n$, and $z = \varphi_a(w) \in Aut(\boldsymbol{B}^n)$ which maps $z = a$ to $w = 0$. We may express the Taylor expansion of $f(\varphi_a(w))$ at $w = 0$ as

$$F(w) = f(\varphi_a(w)) = \sum_{k=0}^{\infty} \frac{1}{k!}(1 - a\bar{a}')^{\frac{k}{2}}(wR^{-1})^{[k]} \cdot \left(\frac{\delta}{\delta a}\right)^{[k]} f(a),$$

where $R = R(a)$ is defined by (4.2) and $\left(\frac{\delta}{\delta a}\right)^{[k]} f(a)$ means the column vector $\left(\left(\frac{\delta}{\delta z}\right)^{[k]} f(z)\Big|_{z=a}\right)'$.

Normalizing the mapping $F(w)$, we have

$$\begin{aligned}(4.5)\qquad &\frac{F(w) - F(0)}{\sqrt{1 - a\bar{a}'}}(J_f(a))^{-1}R \\ &= w + \sum_{k=2}^{\infty} \frac{1}{k!}(\sqrt{1 - a\bar{a}'})^{k-1}(wR^{-1})^{[k]} \cdot \left(\frac{\delta}{\delta a}\right)^{[k]} f(a)(J_f(a))^{-1}R.\end{aligned}$$

It is a normalized biholomorphic convex mapping on $\boldsymbol{B}^n$ when $f(z)$ is a normalized biholomorphic convex mapping on $\boldsymbol{B}^n$.

We know that if $f(z)$ is a normalized biholomorphic convex mapping on $\boldsymbol{B}^n$, we may express it as

$$f(z) = z + \sum_{m=2}^{\infty} P_m(z),$$

where $P_m(z) = (P_m^{(1)}(z), \cdots\cdots, P_m^{(n)}(z))$, and $P_m^{(j)}(z), j = 1, 2, \cdots\cdots, n$ are homogeneous polynomial of the elements of z of degree m.

Gong and Liu [4] proved that

$$(4.6)\qquad |P_m(z)| \leq | z |^m, \qquad m = 2, 3, \cdots\cdots,$$

holds true for any $z \in \boldsymbol{B}^n$ if $f(z)$ is a normalized biholomorphic convex mapping on $\boldsymbol{B}^n$.

From (4.5) and (4.6), we have that for any $w \in \boldsymbol{B}^n$,

$$(4.7)\qquad \frac{(\sqrt{1 - a\bar{a}'})^{k-1}}{k!}\left|(wR^{-1})^{[k]} \cdot \left(\frac{\delta}{\delta a}\right)^{[k]} f(a)(J_f(a))^{-1}R\right| \leq | w |^k$$

holds for all $k = 2, 3, \cdots\cdots$.

It is easy to verify that

$$R\bar{R}' = \frac{\bar{a}'a + (1 - a\bar{a}')I}{1 - a\bar{a}'}.$$

Thus (4.7) is equivalent to

$$\begin{aligned}(4.8)\qquad &(wR^{-1})^{[k]} \cdot \left(\frac{\delta}{\delta a}\right)^{[k]} f(a)\,(J_f(a))^{-1}\,\frac{\bar{a}'a + (1 - a\bar{a}')I}{1 - a\bar{a}'} \\ &\cdot\overline{(J_f(a))^{-1}}'\,\overline{(wR^{-1})^{[k]} \cdot \left(\frac{\delta}{\delta a}\right)^{[k]} f(a)}' \leq \frac{(k!)^2 \mid w \mid^2 k}{(1 - a\bar{a}')^{k-1}}.\end{aligned}$$

From inequlities (1.9), we have

$$(4.9)\qquad (J_f(a))^{-1}(\bar{a}'a + (1 - a\bar{a}')I)\overline{J_f(a)}'^{-1} \geq (1 - | a |)^4 I.$$

Applying (4.9) to (4.8), we obtain

$$(wR^{-1})^{[k]}\cdot\left(\frac{\delta}{\delta a}\right)^{[k]}f(a)\overline{(wR^{-1})^{[k]}\cdot\left(\frac{\delta}{\delta a}\right)^{[k]}f(a)}\le\frac{(k!)^2\mid w\mid^{2k}}{(1-a\overline{a}')^{k-2}(1-\mid a\mid)^4}.$$

That is,

$$\left|(wR^{-1})^{[k]}\cdot\left(\frac{\delta}{\delta a}\right)^{[k]}f(a)\right|\le\frac{k!\mid w\mid^k}{(1-\mid z\mid)^2(\sqrt{1-a\overline{a}'})^{k-2}}. \tag{4.10}$$

Let $w=\eta R(a)$ in (4.10), we get

(4.11)

$$\left|\eta^{[k]}\cdot\left(\frac{\delta}{\delta a}\right)^{[k]}f(a)\right|\le\frac{k!\mid\eta R\mid^k}{(1-\mid a\mid)^2(\sqrt{1-a\overline{a}'})^{k-2}}=k!\frac{1+\mid a\mid}{1-\mid a\mid}\left|\frac{\eta R}{\sqrt{1-a\overline{a}'}}\right|^k$$

when $\eta R(a)\in\boldsymbol{B}^n$. Both sides of (4.11) are homogeneous polynomials of the elements of η of degree k, thus (4.11) holds true for any $\eta\in\mathbb{C}^n$.

Replace a by z at (4.11), we have proved Theorem 4.1.

REFERENCES

[1] R. W. BARNARD, C. H. FITZGERALD, AND S. GONG, *Distortion theorem for biholomorphic mappings in $\mathbb{C}^2$*, Tran. Amer. Math. Soc., 344 (1994), pp. 907–924.

[2] S. GONG, *Convex and Starlike Mappings in Several Complex Variables*, Science Press/Kluwer Academic Publisher, 1998.

[3] S. GONG, *The Bieberbach Conjecture*, AMS/International Press, 1999.

[4] S. GONG AND T. S. LIU, *The growth theorem of biholomorphic convex mappings on $\boldsymbol{B}_p$*, Chin. Quar. Jour. of Math., 6 (1991), pp. 78–82.

[5] S. GONG AND T. S. LIU, *Distortion theorems for biholomorphic convex mappings on bounded circular convex domains*, Chin. Ann. of Math., 20B (1999), pp. 297–304.

[6] S. GONG, S. K. WANG, AND Q. H. YU, *Biholomorphic convex mappings of ball in $\mathbb{C}^n$*, Pacific Jour. of Math., 161 (1993), pp. 287–306.

[7] S. GONG AND Z. M. YAN, *A note on Möbius transformations III*, Chin. Quar. Jour. of Math., 1 (1986), pp. 33–40.

[8] S. GONG AND X. A. ZHENG, *Distortion theorem for biholomorphic mappings in transitive domains I*, International Symposium in memory of L. K. Hua, Vol. II, Springer-Verlag, 1991, pp. 111–122.

[9] H. HAMADA AND G. KOHR, *Linear invariant families on the unit polydisc*, Preprint, 2000.

[10] L. LEMPERT, *La metrique Kobayashi et las representation des domains sur la boule*, Bull. Soc. Math. France, 109 (1981), pp. 427–474.

[11] TAISHUN LIU, *Doctor Thesis*, University of Sciece and Technology of China, 1989.

[12] J. A. PFALTZGRAFF, *Distortion of locally biholomorphic maps of the n-ball*, Complex Variables, 33 (1997), 239–253.

[13] J. A. PFALTZGRAFF AND T. S. SUFFRIDGE, *Linear invariant order and convex maps in $\mathbb{C}^n$*, Complex Variables, to appear.

[14] J. A. PFALTZGRAFF AND T. S. SUFFRIDGE, *Norm order and geometric properties of holomorphic mappings in $\mathbb{C}^n$*, Preprint, 1999.

ACADEMIC PRESS

Available online at www.sciencedirect.com
SCIENCE DIRECT·

J. Math. Anal. Appl. 274 (2002) 696–704

Journal of MATHEMATICAL ANALYSIS AND APPLICATIONS
www.academicpress.com

Criterion for the family of ε starlike mappings ☆

Sheng Gong * and Taishun Liu

Department of Mathematics, University of Science and Technology of China, Hefei, 230026, PR China

Received 8 May 2001

Submitted by S. Ruscheweyh

Abstract

We introduce the family of ε starlike mappings, in purpose to treat the family of convex mappings and the family of starlike mappings as one family, and to describe one family how to transit to another one. In this paper, we consider the criterion for the family of ε starlike mappings.

Keywords: ε starlikeness; Convexity; Starlikeness; Complex Banach space

1. Introduction

The family of convex mappings and the family of starlike mappings on the domains in complex plane, space of several complex variables $\mathbb{C}^n$, and complex Banach space are two important families in complex analysis. The family of convex mappings is a subfamily of the family of starlike mappings, but the properties of these two families are essentially different, and hence the methods to study these two families are different also. In this paper, we will introduce the family of ε starlike mappings, and will study the criterion for this family. This paper is one of a series of papers to study this family. Usually, the results which we obtain for this family cover the known results about the family of convex

☆ Project supported by the National Natural Science Foundation of China.
* Corresponding author.
E-mail addresses: gongs@ustc.edu.cn (S. Gong), lts@ustc.edu.cn (T. Liu).

0022-247X/02/$ – see front matter
PII: S0022-247X(02)00337-2

mappings and the family of starlike mappings, and can describe one family how to transit to another one.

In Section 2, we will give the criterion for the family of ε starlike mappings on the unit ball in the complex Banach space. This result implies the results of Suffridge [1,2], but the method of the proof of this result is different from his method.

In Section 3, we will give the criterion for the family of ε starlike mappings on bounded convex circular domains in $\mathbb{C}^n$. This result includes the results of Chen [3], Lin and Hong [4], and Liu [5] as special cases.

Definition. Let X, Y be two complex Banach spaces, and let $\Omega \subset X$ be a domain. Suppose that $f:\Omega \to Y$ be a locally biholomorphic mapping and $0 \in f(\Omega)$. Then f is said to be ε starlike, for an ε, $0 \leqslant \varepsilon \leqslant 1$, provided $f(\Omega)$ is starlike with respect to every point in $\varepsilon f(\Omega)$. All the ε starlike mappings on Ω form the family of ε starlike mappings on Ω.

When $\varepsilon = 0$, it is exactly the family of starlike mappings on Ω, and when $\varepsilon = 1$, it is the family of convex mappings on Ω.

Similarly, we may define the ε starlike mapping on a domain in $\mathbb{C}^n$.

Example 1. Let $D(0,1) = \{\zeta \in \mathbb{C}\colon |\zeta| < 1\}$, $0 < \varepsilon_1 < \varepsilon_2 \leqslant 1$. From Riemann mapping theorem, there exists a biholomorphic function $f(\zeta)$ on $D(0,1)$, $f(0) = 0$, such that $f(D(0,1)) = \{\zeta \in \mathbb{C}\colon |\operatorname{Im}\zeta| < 1 \text{ and } \operatorname{Re}\zeta < 1\} \cup \{\zeta \in \mathbb{C}\colon |\operatorname{Im}\zeta| < \varepsilon_1 \text{ and } \operatorname{Re}\zeta < \frac{1}{\varepsilon_1}\}$, so $\varepsilon_1 f(D(0,1)) = \{\zeta \in \mathbb{C}\colon |\operatorname{Im}\zeta| < \varepsilon_1 \text{ and } \operatorname{Re}\zeta < \varepsilon_1\} \cup \{\zeta \in \mathbb{C}\colon |\operatorname{Im}\zeta| < \varepsilon_1^2 \text{ and } \operatorname{Re}\zeta < 1\}$. It is clear that $f(\zeta)$ is an ε_1 starlike mapping on $D(0,1)$ but is not an ε_2 starlike mapping on $D(0,1)$.

Example 2. Let $\Omega_{n,p} = \{(z_1, z_2, \ldots, z_n) \in \mathbb{C}^n\colon |z_1|^2 + |z_2|^p + \cdots + |z_n|^p < 1\}$, $p \geqslant 1$, $0 < \varepsilon_1 < \varepsilon_2 \leqslant 1$, $f(z_1)$ is a normalized ε_1 starlike mapping on $D(0,1)$ but is not an ε_2 starlike mapping on $D(0,1)$ (Example 1). Then, $\Phi_{n,1/p}(f)(z) = (f(z_1), (f'(z_1))^{1/p} z_2, \ldots, (f'(z_1))^{1/p} z_n)$ is a normalized ε_1 starlike mapping on $\Omega_{n,p}$ but is not an ε_2 starlike mapping on $\Omega_{n,p}$ (cf. Gong and Liu [7]).

2. Criterion for ε starlikeness of mappings on the unit ball in complex Banach space

Let X be a complex Banach space, and let B be the open unit ball in X. When $x \neq 0$, we define

$$T(x) = \{T_x \in X^*\colon T_x(x) = \|x\| \text{ and } \|T_x\| = 1\} \tag{1}$$

where X^* is the dual space of X. $T(x)$ is not empty by functional analysis.

In 1973, Suffridge gave the criteria for starlikeness and convexity of mappings on the unit ball B in complex Banach space (cf. Suffridge [1,2]).

Theorem A. *Let X, Y be two complex Banach spaces, and let $B \subset X$ be the open unit ball. If $f : B \to Y$ is a locally biholomorphic mapping on B with $f(0) = 0$, then*

(1) *f is starlike if and only if*

$$\operatorname{Re}\{T_x[(Df(x))^{-1}f(x)]\} \geqslant 0 \tag{2}$$

holds for every $x \in B$;

(2) *f is convex if and only if*

$$\operatorname{Re}\{T_x[(Df(x))^{-1}(f(x) - f(y))]\} \geqslant 0 \tag{3}$$

holds for any $x \in B$ and $y \in B$ whenever $\|y\| \leqslant \|x\| < 1$, where $T_x \in T(x)$, $Df(x)$ is the Frechét derivative of f at x.

In this section, we generalize Theorem A as follows.

Theorem 1. *Let X, Y be two complex Banach spaces, and let $B \subset X$ be the open unit ball. If $f : B \to Y$ is a locally biholomorphic mapping on B with $f(0) = 0$, then f is ε starlike, $0 \leqslant \varepsilon \leqslant 1$, if and only if*

$$\operatorname{Re}\{T_x[(Df(x))^{-1}(f(x) - \varepsilon f(y))]\} \geqslant 0 \tag{4}$$

holds for every $x \in B$ and $y \in B$ wherever $\|y\| \leqslant \|x\| < 1$, where $T_x \in T(x)$, $Df(x)$ is the Frechét derivative of f at x.

In purpose to prove Theorem 1, we need the following two lemmas.

Lemma 1. *There exists a bounded linear functional T_x in $T(x)$ such that $T_{\zeta x} = (\bar{\zeta}/|\zeta|)T_x$, where ζ is any non-zero complex number and $x \in X$, $x \neq 0$.*

Proof. Let $x \in X, x \neq 0$ and $y \in X$, $y \neq 0$. We say x and y are equivalent if there exists a non-zero complex number ζ, such that $x = \zeta y$, we may classify all points in $X \setminus \{0\}$ by this equivalence relation, and use $[x]$ to denote the equivalence class whose representative element is x.

For x, there exists T_x in $T(x)$, then we let $T_y = (\bar{\zeta}/|\zeta|)T_x$ if $y = \zeta x$, where $\zeta \in \mathbb{C}$, $\zeta \neq 0$. Obviously, $\|T_y\| = \|T_x\| = 1$ and $T_y(y) = (\bar{\zeta}/|\zeta|)T_x(\zeta x) = |\zeta|\|x\| = \|y\|$. Hence $T_y \in T(x)$. Thus for every $y \in [x]$, T_y is well defined. We may define it for every point in every equivalence class. Hence it is well defined for every element $z \in X \setminus \{0\}$ and satisfies the condition in the lemma. □

Lemma 2. *Let B be the open unit ball in a complex Banach space X. If $h: B \times B \to B$ is a holomorphic mapping with $h(0,0) = 0$, then*

$$\|h(x, y)\| \leqslant \max(\|x\|, \|y\|) \tag{5}$$

holds for every $x \in B$ and $y \in B$.

Proof. Consider the holomorphic mapping $H = (h, h): B \times B \to B \times B$, then $H(0,0) = (0,0)$. The space $X \times X$ is a complex Banach space if we let $\|(x, y)\| = \max(\|x\|, \|y\|)$ as its norm, where $x \in X$ and $y \in X$, then $B \times B = \{(x, y) \in X \times X: \|(x, y)\| < 1\}$. By Schwarz' lemma (cf. Harris [8]), we have

$$\|H(x, y)\| \leqslant \|(x, y)\|.$$

It implies (5). □

Proof of Theorem 1. *Necessity.* If $f(x)$ is a ε starlike mappings on B, then for any $t \in [0, 1]$, $h(x, y) = f^{-1}[(1-t)f(x) + t\varepsilon f(y)]$ is a holomorphic mapping of $B \times B$ to B, and $h(0,0) = 0$, where $x \in B$ and $y \in B$. By Lemma 2, we have

$$\|h(x, y)\| \leqslant \max(\|x\|, \|y\|). \tag{6}$$

The expansion of $h(x, y)$ at $t = 0$ is

$$h(x, y) = x - t(Df(x))^{-1}(f(x) - \varepsilon f(y)) + O(t^2).$$

Operate T_x on both side of the previous equality, then take real part on both side, the right hand side becomes

$$\begin{aligned} &\operatorname{Re}\{T_x[x - t(Df(x))^{-1}(f(x) - \varepsilon f(y)) + O(t^2)]\} \\ &\quad = \|x\| - t\operatorname{Re}\{T_x[(Df(x))^{-1}(f(x) - \varepsilon f(y))]\} + O(t^2), \end{aligned} \tag{7}$$

and the left hand side becomes $\operatorname{Re}\{T_x[h(x, y)]\}$. By (6), we have

$$\operatorname{Re}\{T_x[h(x, y)]\} \leqslant |T_x[h(x, y)]| \leqslant \|h(x, y)\| \leqslant \max(\|x\|, \|y\|) = \|x\| \tag{8}$$

holds for every $x \in B$ and $y \in B$ whenever $\|y\| \leqslant \|x\| < 1$.

From (7) and (8), we have

$$\operatorname{Re}\{T_x[(Df(x))^{-1}(f(x) - \varepsilon f(y))]\} + O(t) \geqslant 0$$

holds for every $x \in B$ and $y \in B$ whenever $\|y\| \leqslant \|x\| < 1$.

We obtain (4) when we let $t \to 0$ in the previous inequality.

Sufficiency. Firstly, we prove that (4) implies the following inequality

$$\operatorname{Re}\{T_x[(Df(x))^{-1}(f(x) - \varepsilon f(y))]\} > 0 \tag{9}$$

when $x \in B$, $y \in B$ and $\|y\| < \|x\| < 1$.

Let $D(0, 1/\|x\|) = \{\zeta \in \mathbb{C}: |\zeta| < 1/\|x\|\}$, where $x \in B$, $x \neq 0$. By Lemma 1, there exists a bounded linear functional T_x in $T(x)$ such that $T_{\zeta x} = (\bar{\zeta}/|\zeta|)T_x$, where $\zeta \in \mathbb{C}$, $\zeta \neq 0$ and $x \in X$, $x \neq 0$. Using this T_x, we define a function of ζ,

$$g(\zeta) = \frac{T_x[(Df(\zeta x))^{-1}(f(\zeta x) - \varepsilon f(\zeta y))]}{\zeta} \tag{10}$$

on $D(0, 1/\|x\|)$, then $g(\zeta)$ is a holomorphic function of ζ on $D(0, 1/\|x\|)$. By the property of T_x, we have

$$g(\zeta) = \frac{T_{\zeta x}[(Df(\zeta x))^{-1}(f(\zeta x) - \varepsilon f(\zeta y))]}{|\zeta|},$$

and hence $\operatorname{Re} g(\zeta) \geqslant 0$ when $\zeta \in D(0, 1/\|x\|)$ by (4).

If (9) does not hold, then $\operatorname{Re} g(1) = 0$. By the extremum principle of harmonic function, $\operatorname{Re} g(\zeta) \equiv 0$ in the domain $D(0, 1/\|x\|)$ since $g(\zeta)$ is a holomorphic function on $D(0, 1/\|x\|)$. Hence $\operatorname{Re} g(0) = 0$. But

$$\operatorname{Re} g(0) = \operatorname{Re}\{T_x[x - \varepsilon y]\} = \operatorname{Re}\{\|x\| - \varepsilon T_x(y)\} \geqslant \|x\| - \varepsilon\|y\| > 0$$

by (10). We obtain contradiction. Thus (9) holds true.

Secondly, we prove that $f(rB)$ is starlike with respect to every point in $\varepsilon f(rB)$, where $r \in (0, 1)$.

If it is not true, then there exist points $p \in rB$ and $y \in rB$, such that a part of the line segment $\{tf(p) + (1-t)\varepsilon f(y) \in Y: 0 \leqslant t \leqslant 1\}$ is laid in $Y \setminus \overline{f(rB)}$. Thus there exists a point $x \in \partial(rB)$, and a number $\delta \in (0, 1)$, such that the open line segment

$$\{tf(x) + (1-t)\varepsilon f(y) \in Y: \delta < t < 1\} \subset f(B) \setminus \overline{f(rB)}.$$

Let

$$x(t) = f^{-1}[tf(x) + (1-t)\varepsilon f(y)], \quad \delta < t < 1, \tag{11}$$

then $x(t) \notin \overline{rB}$ and $y \in rB$. Hence

$$\|x(t)\| > r > \|y\|. \tag{12}$$

That means, $\|x(t)\|$ is not an increasing function in $\delta < t < 1$.

From (11), we have

$$f(x(t)) = tf(x) + (1-t)\varepsilon f(y),$$

thus

$$Df(x(t))\frac{dx(t)}{dt} = f(x) - \varepsilon f(y) \tag{13}$$

and

$$f(x(t)) - \varepsilon f(y) = t(f(x) - \varepsilon f(y)). \tag{14}$$

By (13) and (14), we have

$$\frac{dx(t)}{dt} = \frac{1}{t}\big(Df(x(t))\big)^{-1}\big(f(x(t)) - \varepsilon f(y)\big).$$

It implies

$$\liminf_{\substack{\Delta t\to 0\\ \Delta t>0}} \frac{\|x(t+\Delta t)\| - \|x(t)\|}{\Delta t} \geqslant \liminf_{\substack{\Delta t\to 0\\ \Delta t>0}} \frac{\operatorname{Re}\{T_{x(t)}(x(t+\Delta t)-x(t))\}}{\Delta t}$$
$$= \frac{1}{t}\operatorname{Re}\big\{T_{x(t)}\big[\big(Df(x(t))\big)^{-1}\big(f(x(t)) - \varepsilon f(y)\big)\big]\big\}. \tag{15}$$

By (9) and (12), the right-hand side of (15) is greater than zero when $\delta < t < 1$. That means, $\|x(t)\|$ is a strictly increasing function in $\delta < t < 1$. We obtain contradiction. Hence, $f(rB)$ is starlike with respect to every point in $\varepsilon f(rB)$, where $r \in (0, 1)$.

Finally, for any two points $x \in B$ and $y \in B$, there exists $r \in (0, 1)$ such that $x \in rB$ and $y \in rB$. Thus

$$tf(x) + (1-t)\varepsilon f(y) \in f(rB) \subset f(B)$$

when $t \in [0, 1]$. That means, $f(B)$ is starlike with respect to every point in $\varepsilon f(B)$. Let $y = 0$ in (4), and by Theorem A, we know that f is biholomorphic in B.

Theorem 1 have been proved.

3. Criterion for ε starlikeness of mappings on the bounded convex circular domains in $\mathbb{C}^n$

Let Ω be a bounded convex circular domain in $\mathbb{C}^n$ and $\rho(z)$ be the Minkowski functional on Ω. Then $\mathbb{C}^n$ is a complex Banach space with norm $\rho(z)$, and Ω is the open unit ball in this complex Banach space. In this setting, naturally, we expect a criterion for ε starlikeness formulated more concretely. But unfortunately, it can not be obtained as an immediate corollary of Theorem 1, since it is difficult to get the explicit expression of T_x in Theorem 1 unless Ω is a very special domain. We have to find a new approach.

Corresponding to Lemma 1, we have

Lemma 3. *For every bounded convex circular domain Ω in $\mathbb{C}^n$, there exists a mapping $\nu(z): \mathbb{C}^n \to \mathbb{C}^n$, satisfying the following conditions:*

(i) $\nu(0) = 0$;

(ii) *for every point $z \in \mathbb{C}^n$, $z \neq 0$, $\nu(z)$ is the outward normal vector of the convex domain*

$$\rho(z)\Omega = \{w \in \mathbb{C}^n\colon \rho(w) < \rho(z)\};$$

(iii) $\operatorname{Re}\{\overline{\nu(z)}'w\} \leqslant \rho(w)$ *holds for any* $z \in \mathbb{C}^n$ *and* $w \in \mathbb{C}^n$;
(iv) $\operatorname{Re}\{\overline{\nu(z)}'z\} = \rho(z)$ *holds for every* $z \in \mathbb{C}^n$;
(v) $\overline{\nu(\zeta z)}' = (\bar{\zeta}/|\zeta|)\overline{\nu(z)}'$ *holds for every* $z \in \mathbb{C}^n$ *and* $\zeta \in \mathbb{C}$, $\zeta \neq 0$.

Proof. We regard all points (except the origin) lying on a complex line through origin as an equivalent class. We may classify $\mathbb{C}^n \setminus \{0\}$ by this equivalent relation.

We take an outward normal vector $\nu(z)$ at the boundary point z of the convex domain $\rho(z)\Omega$, such that it satisfies the relation $\operatorname{Re}\{\overline{\nu(z)}'z\} = \rho(z)$. (it is possible if we justify the length of $\nu(z)$) and $\nu(0) = 0$. If w and z are two points in same class, i.e., $w = \zeta z$, $\zeta \in \mathbb{C}$, $\zeta \neq 0$, then we let $\nu(w) = (\zeta/|\zeta|)\nu(z)$. Hence $\operatorname{Re}\{\overline{\nu(w)}'w\} = \operatorname{Re}\{(\bar{\zeta}/|\zeta|)\overline{\nu(z)}'(\zeta z)\} = |\zeta|\rho(z) = \rho(w)$. That means, $\nu(w)$ satisfied (iv) again.

Since $\rho(z)\Omega$ is a circular domain, $e^{i\theta}\nu(z)$ is the outward normal vector at the boundary point $e^{i\theta}z$ of the convex domain $\rho(e^{i\theta}z)\Omega$ (it is exactly $\rho(z)\Omega$). If $\zeta = |\zeta|e^{i\theta}$, then $\nu(\zeta z) = (\zeta/|\zeta|)\nu(z)$ is the outward normal vector at the boundary point ζz of the convex domain $\rho(\zeta z)\Omega$. Thus $\nu(w)$ is well defined on all points which are equivalent to z, and hence $\nu(w)$ is well defined on every point $w \in \mathbb{C}^n \setminus \{0\}$. Obviously, $\nu(z)$ satisfies the conditions (i), (ii), (iv) and (v).

Now, we will verify $\nu(z)$ satisfies (iii).

If points $z, w \in \mathbb{C}^n$, $z \neq 0$ and $w \neq 0$, then $(\rho(z)/\rho(w))w \in \partial(\rho(z)\Omega)$, and

$$\operatorname{Re}\left\{\overline{\nu(z)}'\left(\frac{\rho(z)}{\rho(w)}w\right)\right\}$$

is equal to the projection of $(\rho(z)/\rho(w))w$ to the normal vector $\nu(z)$. It equals to the distance from 0 to the plane through the point $(\rho(z)/\rho(w))w$ with $\nu(z)$ as its normal vector.

Since $\rho(z)\Omega$ is a convex domain, we have

$$\operatorname{Re}\left\{\overline{\nu(z)}'\left(\frac{\rho(z)}{\rho(w)}w\right)\right\} \leqslant \operatorname{Re}\{\overline{\nu(z)}'z\},$$

then (iii) follows from (iv). □

Moreover, if $\rho(z)$ is differentiable everywhere in $\mathbb{C}^n$ except the origin, then we may get the concrete form of $\nu(z)$.

Lemma 4. *Let Ω be a bounded convex circular domain in $\mathbb{C}^n$, and let $\rho(z)$ be its Minkowski functional. If $\rho(z)$ is differentiable everywhere in $\mathbb{C}^n$ except the origin, then*

$$\nu(z) = 2\nabla\rho(z) \tag{16}$$

where $\nabla\rho(z)$ is the gradient of ρ.

Proof. At the boundary point z of the convex domain $\rho(z)\Omega$, the outward normal direction is exactly $\nabla\rho(z)$, the gradient of $\rho(z)$. We know that

$$\nabla\rho(z)=\left(\frac{\partial\rho}{\partial x_1}+i\frac{\partial\rho}{\partial y_1},\ldots,\frac{\partial\rho}{\partial x_n}+i\frac{\partial\rho}{\partial y_n}\right)'=2\left(\frac{\partial\rho}{\partial\bar{z}_1},\ldots,\frac{\partial\rho}{\partial\bar{z}_n}\right),$$

where $z_j=x_j+iy_j$, $j=1,\ldots,n$. We may find a unique positive number M, such that

$$v(z)=M\nabla\rho(z).$$

Hence

$$M=\frac{\operatorname{Re}\{\overline{v(z)}'z\}}{\operatorname{Re}\{\overline{\nabla\rho(z)}'z\}}. \tag{17}$$

Since $\rho(tz)=t\rho(z)$ when $t>0$, we have

$$\rho(z)=\frac{d\rho(tz)}{dt}\bigg|_{t=1}=\sum_{j=1}^{n}\frac{\partial\rho(z)}{\partial z_j}z_j+\sum_{j=1}^{n}\frac{\partial\rho(z)}{\partial\bar{z}_j}\bar{z}_j=2\operatorname{Re}\{\overline{\nabla\rho(z)}'z\}.$$

Combining this equality and Lemma 3(iv), and substituting these two results into (17), we have (16). □

$\mathbb{C}^n$ is a complex Banach space if we take the Minkowski functional $\rho(z)$ of Ω as its norm. By Lemma 2, we have

Lemma 5. *Let Ω be a bounded convex circular domain in $\mathbb{C}^n$. Let $h:\Omega\times\Omega\to\Omega$ be a holomorphic mapping with $h(0,0)=0$, then*

$$\rho(h(z,w))\leqslant\max(\rho(z),\rho(w)) \tag{18}$$

holds for every $z\in\Omega$ and $w\in\Omega$.

Using Lemmas 3, 4 and 5, we may prove the following results.

Theorem 2. *Let Ω be a bounded convex circular domain. Suppose that $f:\Omega\to\mathbb{C}^n$ is a normalized locally biholomorphic mapping, then f is a ε starlike mapping on Ω, $0\leqslant\varepsilon\leqslant1$, if and only if*

$$\operatorname{Re}\{\overline{v(z)}'[(J_f(z))^{-1}(f(z)-\varepsilon f(w))]\}\geqslant0 \tag{19}$$

holds for every $z\in\Omega$ and $w\in\Omega$ whenever $\rho(w)\leqslant\rho(z)<1$, where $J_f(z)$ is the Jacobian matrix of f at point z.

Moreover, if $\rho(z)$ is differentiable in $\mathbb{C}^n$ except the origin, then we may replace (19) by

$$\operatorname{Re}\left\{\frac{\partial\rho}{\partial z}(z)(J_f(z))^{-1}(f(z)-\varepsilon f(w))\right\}\geqslant0 \tag{20}$$

where $\frac{\partial\rho}{\partial z}(z)=(\frac{\partial\rho}{\partial z_1}(z),\ldots,\frac{\partial\rho}{\partial z_n}(z))$.

In the case $\rho(z)$ is differentiable in $\mathbb{C}^n \setminus \{0\}$, when $\varepsilon = 0$, i.e., the mapping is starlike, it is the results of Chen [3], Lin and Hong [4] (cf. [6, p. 21]); when $\varepsilon = 1$, i.e., the mapping is convex, it is the result of Liu [5] (cf. [6, p. 49]).

The proof of Theorem 2 is similar to the proof of Theorem 1, we omit the details.

References

[1] T.J. Suffridge, Starlike and convex maps in Banach space, Pacific J. Math. 46 (1973) 575–589.

[2] T.J. Suffridge, Starlikeness, convexity and other geometric properties of holomorphic maps in high dimensions, in: Lecture Notes in Math., Vol. 599, Springer, Berlin, 1977, pp. 146–159.

[3] Z. Chen, Criteria for starlikeness and univalency of maps on a class of bounded strictly balanced domains in $\mathbb{C}^n$, Chinese Ann. Math. 16A (1995) 230–237.

[4] Y.Y. Lin, Y. Hong, Some properties of holomorphic maps in Banach spaces, Acta Math. Sinica 38 (1995) 234–241.

[5] T.S. Liu, The growth theorems, covering theorems and distortion theorem for biholomorphic mappings in several complex variables, University of Science and Technology of China, Doctoral Thesis, 1989.

[6] S. Gong, Convex and Starlike Mappings in Several Complex Variables, Science Press/Kluwer Academic Publishers, 1998.

[7] S. Gong, T.S. Liu, On Roper–Suffridge extension operator, J. Anal. Math., to appear.

[8] L.A. Harris, Schwarz' lemma in normed linear spaces, Proc. Nat. Acad. Sci. USA 64 (4) (1969) 1014–1017.

ON THE ROPER–SUFFRIDGE EXTENSION OPERATOR

By

SHENG GONG AND TAISHUN LIU*

Abstract. In 1995, Roper and Suffridge defined an extension operator which maps a locally biholomorphic function on the unit disk D in $\mathbb{C}$ to a locally biholomorphic mapping on the unit ball B^n in $\mathbb{C}^n$. This extension operator preserves many important properties, e.g., convexity and starlikeness, etc. In this note, we introduce the family of ϵ starlike mappings, and prove that the Roper–Suffridge extension operator preserves the ϵ starlikeness on some Reinhardt domains. This result includes many known results and solves an open problem of Graham and Kohr.

1 The Roper–Suffridge extension operator

In 1995, Roper and Suffridge introduced an extension operator. This operator is defined for a normalized locally biholomorphic function f on the unit disk D in $\mathbb{C}$ by

$$\Phi_n(f)(z) = F(z) = \left(f(z_1),\ \sqrt{f'(z_1)}\, z_0\right), \tag{1}$$

where $z = (z_1, z_0)$ belongs to the unit ball B^n in $\mathbb{C}^n$, $z_1 \in D$, $z_0 = (z_2, \ldots, z_n) \in \mathbb{C}^{n-1}$, and we choose the branch of the square root such that $\sqrt{f'(0)} = 1$.

The Roper–Suffridge extension operator has remarkable properties: (i) If f is a normalized convex function on D, then F is a normalized convex mapping on B^n; (ii) if f is a normalized starlike function on D, then F is a normalized starlike mapping on B^n; (iii) if f is a normalized Bloch function on D, then F is a normalized Bloch mapping on B^n. (i) was proved by Roper and Suffridge themselves [10]; (ii) and (iii) were proved by Graham and Kohr [3]. Until now, we know only a few concrete examples of convex mappings, starlike mappings and Bloch mappings on B^n. Using the Roper–Suffridge extension operator, we may construct lots of concrete examples of these mappings on B^n. This is one reason why people are interested in this extension operator.

*Project supported by the National Science Foundation of China.

In [5], Graham, G. Kohr and M. Kohr generalized the operator (1) as

$$\Phi_{n,\alpha}(f)(z) = F_\alpha(z) = (f(z_1), (f'(z_1))^\alpha z_0), \tag{2}$$

where $\alpha \in [0, \frac{1}{2}]$, f, z_1, z_0, z are defined as above, and we choose the branch of the power function such that $(f'(0))^\alpha = 1$. They proved that this operator maps normalized starlike functions on D to normalized starlike mappings on B^n and maps normalized Bloch functions on D to normalized Bloch mappings on B^n, but it does not preserve convexity on B^n when $\alpha \in [0, \frac{1}{2})$.

In [4], Graham and Kohr introduced another extension operator for locally biholomorphic functions f on D by

$$\Psi_{n,\alpha}(f)(z) = G_\alpha(z) = \left(f(z_1), \left(\frac{f(z_1)}{z_1}\right)^\alpha z_0\right), \tag{3}$$

where $\alpha \in [0, 1]$; f, z_1, z_0, z are defined as above; $f(z_1) \neq 0$ when $z_1 \in D \setminus \{0\}$; and we choose the branch of the power function such that

$$\left(\frac{f(z_1)}{z_1}\right)^\alpha \bigg|_{z_1=0} = 1.$$

They proved that the operator (3) preserves starlikeness. When $\alpha = 1$, this was proved and discussed by Pfaltzgraff and Suffridge [8].

There are many interesting results for these operators; see, for example, [4], [9], etc.

In [10], Roper and Suffridge pointed out that if $\mathbb{C}^n$ is endowed with the p-norm

$$\|z\|_p = \begin{cases} \left(\sum_{j=1}^n |z_j|^p\right)^{1/p}, & 1 \le p < \infty, \\ \max\limits_{j=1,\ldots,n} \{|z_j|\}, & p = \infty, \end{cases} \tag{4}$$

then the Roper–Suffridge extension operator does not preserve convexity on the domain $\Omega_p = \{z \in \mathbb{C}^n : \|z\|_p < 1\}$ if $p \neq 2$. In [3], Graham and Kohr proved that the Roper–Suffridge extension operator does not preserve starlikeness in the bidisc $\mathbb{P}^2$ in $\mathbb{C}^2$. They proposed the following problem ([3], Open Problem 2.8). Consider the Reinhardt domain

$$\Omega_{2,p} = \{z \in \mathbb{C}^2 : |z_1|^2 + |z_2|^p < 1\},$$

where $p \ge 1$. Does the operator

$$\Phi_{2,1/p}(f)(z) = F_{1/p}(z) = \left(f(z_1), (f'(z_1))^{1/p} z_2\right)$$

extend convex functions on D to convex mappings on $\Omega_{2,p}$?

In this note, we introduce the family of ε-starlike mappings on the domain Ω in $\mathbb{C}^n$ and prove that the Roper–Suffridge extension operator preserves ε-starlikeness. This result solves the open problem of Graham and Kohr and has many known results as special cases.

2 The main result

Definition 1. Let Ω be a domain in $\mathbb{C}^n$, and let $f : \Omega \to \mathbb{C}^n$ be a locally biholomorphic mapping with $0 \in f(\Omega)$. We call f an ε-starlike mapping on Ω if there exists a positive number ε, $0 \le \varepsilon \le 1$, such that $f(\Omega)$ is starlike with respect to every point in $\varepsilon f(\Omega)$. All ε-starlike mappings on Ω form the family of ε-starlike mappings on Ω.

When $\varepsilon = 0$, this is the family of starlike mappings; and when $\varepsilon = 1$, it is the family of convex mappings.

Theorem 1. *Let $f(z_1)$ be a normalized biholomorphic εstarlike function on the unit disk $D = \{z_1 \in \mathbb{C} : |z_1| < 1\}$ in $\mathbb{C}$, $0 \le \varepsilon \le 1$. Then*

$$\Phi_{n,1/p}(f)(z) = F_{1/p}(z) = \left(f(z_1), (f'(z_1))^{1/p} z_0\right), \quad p \ge 1 \tag{5}$$

is a normalized εstarlike mapping on

$$\Omega_{n,p} = \{z \in \mathbb{C}^n : |z_1|^2 + \|z_0\|_p^p < 1\} \tag{6}$$

where $z_0 = (z_2, \ldots, z_n) \in \mathbb{C}^{n-1}$, $z = (z_1, z_0) \in \Omega_{n,p}$, and we choose the branch of the power function in (5) such that $(f'(z_1))^{1/p}|_{z_1=0} = 1$.

When $n = 2$, $\varepsilon = 1$, Theorem 1 solves the open problem of Graham and Kohr [3]. The answer is affirmative, and it holds true for any $n \ge 2$. When $p = 2$, $\varepsilon = 1$, Theorem 1 is the result of Roper and Suffridge [10]. When $p = 2$, $\varepsilon = 0$, Theorem 1 is the result of Graham and Kohr [3]. In the case $\varepsilon = 0$, Theorem 1 tells us that operator (5) not only preserves convexity on (6), but preserves starlikeness on (6) as well. Moreover, (6) is an important class of domains in several complex variables. When $p > 2$, (6) is a typical class of weak pseudoconvex domains. Theorem 1 tells us how to construct convex and starlike mappings on these domains.

The method of the proof of Theorem 1 is different from the proofs of the previous results mentioned above.

Proof of Theorem 1. For any $\lambda \in [0, 1]$, $z = (z_1, z_0) \in \Omega_{n,p}$, and $w = (w_1, w_0) \in \Omega_{n,p}$ where $z_1 \in D$, $w_1 \in D$ and $z_0 \in \mathbb{C}^{n-1}$, $w_0 \in \mathbb{C}^{n-1}$, if we can find

$(u_1, u_0) \in \Omega_{n,p}$ where $u_1 \in D, u_0 \in \mathbb{C}^{n-1}$, such that

$$(7) \qquad \left(f(u_1), (f'(u_1))^{1/p}u_0\right) = (1-\lambda)\left(f(z_1), (f'(z_1))^{1/p}z_0\right) + \varepsilon\lambda\left(f(w_1), (f'(w_1))^{1/p}w_0\right),$$

then Theorem 1 is proved.

Since f is a ε-starlike function on D, for any λ and z_1, w_1, there exists $u_1 \in D$, such that

$$(8) \qquad f(u_1) = (1-\lambda)f(z_1) + \varepsilon\lambda f(w_1).$$

Thus, the right-hand side of (7) is

$$\left(f(u_1), (1-\lambda)(f'(z_1))^{1/p}z_0 + \varepsilon\lambda(f'(w_1))^{1/p}w_0\right).$$

Let

$$(9) \qquad u_0 = \frac{(1-\lambda)(f'(z_1))^{1/p}z_0 + \varepsilon\lambda(f'(w_1))^{1/p}w_0}{(f'(u_1))^{1/p}}.$$

We need to prove the inequality

$$(10) \qquad \|u_0\|_p < (1-|u_1|^2)^{1/p}.$$

From (8), we have

$$(11) \qquad u_1(z_1, w_1) = f^{-1}[(1-\lambda)f(z_1) + \varepsilon\lambda f(w_1)].$$

Regarding (11) as a mapping from $D \times D$ to D, we prove the inequality

$$(12) \qquad \frac{\left|\frac{\partial u_1}{\partial z_1}\right||\xi_1| + \left|\frac{\partial u_1}{\partial w_1}\right||\xi_2|}{1-|u_1|^2} \le \max\left(\frac{|\xi_1|}{1-|z_1|^2}, \frac{|\xi_2|}{1-|w_1|^2}\right),$$

where ξ_1 and ξ_2 are any two arbitrary complex numbers.

Let F_C^D and $F_C^{D\times D}$ be the infinitesimal forms of the Carathéodory metric of D and $D \times D$, respectively. Then, by the contraction property of the Carathéodory metric, the inequality

$$(13) \qquad F_C^D(u_1, J_{u_1}(z_1, w_1)\xi) \le F_C^{D\times D}((z_1, w_1), \xi)$$

holds for any column vector $\xi = (\xi_1, \xi_2)' \in \mathbb{C}^2$, where $J_{u_1}(z_1, w_1)$ is the Jacobian matrix of u_1 with respect to (z_1, w_1). If $s \in D$ and $t \in D$, then

$$\varphi(s,t) = \left(\frac{z_1 - s}{1-\bar{z}_1 s}, \frac{w_1 - t}{1-\bar{w}_1 t}\right) \in \operatorname{Aut}(D \times D)$$

and $\varphi(z_1, w_1) = (0,0)$. Thus

$$F_C^{D\times D}((z_1,w_1),\xi) = F_C^{D\times D}((0,0), J_\varphi(z_1,w_1)\xi), \tag{14}$$

where $J_\varphi(z_1, w_1)$ is the Jacobian matrix of φ evaluated at (z_1, w_1). Since $D\times D$ is a bounded convex circular domain and $(0,0)\in D\times D$, we know that ([2])

$$F_C^{D\times D}((0,0), J_\varphi(z_1,w_1)\xi) = \rho(J_\varphi(z_1,w_1)\xi), \tag{15}$$

where $\rho(\cdot)$ is the Minkowski functional of $D\times D$.

Obviously,

$$J_{u_1}(z_1,w_1) = \left(\frac{\partial u_1}{\partial z_1}, \frac{\partial u_1}{\partial w_1}\right) \quad \text{and} \quad J_\varphi(z_1,w_1) = \begin{pmatrix} \dfrac{-1}{1-|z_1|^2} & 0 \\ 0 & \dfrac{-1}{1-|w_1|^2} \end{pmatrix}.$$

Since the left side of (13), F_C^D, is the Poincaré metric of the unit disk, we have

$$\frac{\left|\dfrac{\partial u_1}{\partial z_1}\xi_1 + \dfrac{\partial u_1}{\partial w_1}\xi_2\right|}{1-|u_1|^2} \le \max\left(\frac{|\xi_1|}{1-|z_1|^2}, \frac{|\xi_2|}{1-|w_1|^2}\right)$$

by (13), (14) and (15). This completes the proof of (12), as we may take any complex value of ξ_1 and ξ_2.

From (11), we know

$$\frac{\partial u_1}{\partial z_1} = \frac{1}{f'(u_1)}(1-\lambda)f'(z_1), \qquad \frac{\partial u_1}{\partial w_1} = \frac{1}{f'(u_1)}\varepsilon\lambda f'(w_1). \tag{16}$$

Substituting (16) into (9), we have

$$u_0 = (1-\lambda)^{1-1/p}\left(\frac{\partial u_1}{\partial z_1}\right)^{1/p} z_0 + (\varepsilon\lambda)^{1-1/p}\left(\frac{\partial u_1}{\partial w_1}\right)^{1/p} w_0.$$

Applying Hölders inequality to the previous equality yields

$$\begin{aligned} \|u_0\|_p &\le (1-\lambda)^{1/q}\left|\frac{\partial u_1}{\partial z_1}\right|^{1/p}\|z_0\|_p + (\varepsilon\lambda)^{1/q}\left|\frac{\partial u_1}{\partial w_1}\right|^{1/p}\|w_0\|_p \\ &\le (1-\lambda+\varepsilon\lambda)^{1/q}\left(\left|\frac{\partial u_1}{\partial z_1}\right|\|z_0\|_p^p + \left|\frac{\partial u_1}{\partial w_1}\right|\|w_0\|_p^p\right)^{1/p}, \end{aligned} \tag{17}$$

where $1/p+1/q=1$.

Let $\xi_1 = \|z_0\|_p^p$ and $\xi_2 = \|w_0\|_p^p$ in (12), and substitute it into (17). We obtain

$$\|u_0\|_p \le (1-|u_1|^2)^{1/p}\left\{\max\left(\frac{\|z_0\|_p^p}{1-|z_1|^2}, \frac{\|w_0\|_p^p}{1-|w_1|^2}\right)\right\}^{1/p} < (1-|u_1|^2)^{1/p}.$$

We have proved that (10) holds true, and hence Theorem 1 is proved.

3 Some remarks

I. In Section 1, we mentioned the following result. Roper and Suffridge [10] point out that the Roper–Suffridge extension operator does not preserve convexity on the domain $\Omega_p = \{z \in \mathbb{C}^n : \|z\|_p < 1\}$, where $1 \le p \le \infty$, $p \ne 2$ and $n \ge 2$, and $\|z\|_p$ is defined by (4). Actually, this is a direct consequence of a result of Liu and Zhang [7] when $p > 2$. They proved the following result.

Let $f(z)$ be a normalized biholomorphic convex mapping on $\Omega_p (p > 2)$, and let k be a positive integer such that $2 \le k < p \le k+1$. Then the homogeneous expansion of $f(z)$ at $z = 0$ is

$$f(z) = \begin{pmatrix} z_1 \\ z_2 \\ \vdots \\ z_n \end{pmatrix} + \begin{pmatrix} a_{12}z_1^2 \\ a_{22}z_2^2 \\ \vdots \\ a_{n2}z_n^2 \end{pmatrix} + \cdots + \begin{pmatrix} a_{1k}z_1^k \\ a_{2k}z_2^k \\ \vdots \\ a_{nk}z_n^k \end{pmatrix} + O(|z|^{k+1}), \tag{18}$$

where a_{ij}, $i = 1, \dots, n$, $j = 2, \dots, k$ are constant coefficients.

Obviously, the mapping (1) is different from (18); thus (1) is not a convex mapping on Ω_p when $p > 2$. Moreover, mapping (2) is different from (18) when $\alpha \in [0, \frac{1}{2}]$, $p > 2$; thus the operator $\Phi_{n,\alpha}$ does not preserve convexity on Ω_p when $\alpha \in [0, \frac{1}{2}]$, $p > 2$.

II. In Section 1, we mentioned another result. Graham, G. Kohr and M. Kohr [5] proved that the generalized Roper–Suffridge extension operator (2) $\Phi_{n,\alpha}$ preserves starlikeness on B^n, where $\alpha \in [0, \frac{1}{2}]$; they did not consider the case $\alpha \in (\frac{1}{2}, 1]$. Theorem 1 tells us that $\Phi_{n,\alpha}$ preserves starlikeness on $\Omega_{n,1/\alpha} = \{z \in \mathbb{C}^n : |z_1|^2 + \|z_0\|_{1/\alpha}^{1/\alpha} < 1\}$, which contains B^n when $\alpha \in (0, \frac{1}{2}]$. Of course, this result and the result of Graham, G. Kohr and M. Kohr are mutually independent. Moreover, by Theorem 1, $\Phi_{n,\alpha}$ preserves starlikeness on $\Omega_{n,1/\alpha}$ when $\alpha \in (\frac{1}{2}, 1]$. The domain $\Omega_{n,1/\alpha}$ is contained in B^n.

In [5], Graham, G. Kohr and M. Kohr show that $\Phi_{n,\alpha}(f)$ does not preserve convexity on B^n when $\alpha \in [0, \frac{1}{2})$. From Theorem 1, we know that $\Phi_{n,\alpha}(f)$ preserves convexity on $\Omega_{n,1/\alpha}$ when $\alpha \in [0, 1]$.

Thus, it seems more natural to consider the operator $\Phi_{n,\alpha}(f)$ on the domain $\Omega_{n,1/\alpha}$ than on the domain B^n.

III. Similar to the family of ε-starlike mappings, we may define the family of E-starlike mappings. These two families are closely related.

Definition 2. Let Ω be a starlike domain (with respect to the origin), $0 \in \Omega$, and let $f : \Omega \to \mathbb{C}^n$ be a locally biholomorphic mapping with $f(0) = 0$. We call

f a E-starlike mapping if there exists a positive number E, $0 \le E \le 1$, such that $f(\Omega)$ is starlike with respect to every point in $f(E\Omega)$.

When $E = 0$, we obtain the family of starlike mappings, and when $E = 1$ the family of convex mappings.

Theorem 2. *Let $f(z_1)$ be a normalized biholomorphic E-starlike function on D, $0 \le E \le 1$. Then (5) is a normalized E-starlike mapping on (6).*

Since the proof of Theorem 2 is similar to the proof of Theorem 1, we omit it.

IV. As an application of Theorems 1 and 2, we find the radius of convexity and radius of starlikeness of a subfamily of the family of the normalized biholomorphic mappings on $\Omega_{n,p}$.

Let S be the family of normalized biholomorphic functions on D. For any function in S, we obtain a normalized biholomorphic mapping on $\Omega_{n,p}$ by (5) when $p \ge 1$. Let

$$\mathfrak{S} = \{\Phi_{n,\frac{1}{p}}(f) : f \in S\}.$$

This is a subfamily of the family of normalized biholomorphic mappings on $\Omega_{n,p}$. Let

$$r^*(\mathfrak{S}) = \sup\{r : f \text{ is starlike on } \Omega^r_{n,p},\ f \in \mathfrak{S}\},$$
$$r_c(\mathfrak{S}) = \sup\{r : f \text{ is convex on } \Omega^r_{n,p},\ f \in \mathfrak{S}\},$$

where $\Omega^r_{n,p} = \{z \in \mathbb{C}^n : |z_1/r|^2 + \|z_0/r\|^p_p < 1\}$, and $0 < r < 1$, $p \ge 1$.

Theorem 3. *When $p \ge 1$,*

$$r_c(\mathfrak{S}) = 2 - \sqrt{3}, \tag{19}$$

$$r^*(\mathfrak{S}) = \tanh \tfrac{1}{4}\pi. \tag{20}$$

In [5], Graham, G. Kohr and M. Kohr proved that (19) holds when $p = 2$. Theorem 3 generalizes their result. In [5], they also proved that the radius of starlikeness of $\mathfrak{S}$ on B^n is $\tanh \frac{1}{4}\pi$ when $p \ge 2$. Neither their result nor (20) contains the other, but both results imply that the radius of starlikeness of $\mathfrak{S}$ on B^n is $\tanh \frac{1}{4}\pi$ when $p = 2$.

Proof of Theorem 3. The radius of convexity of S is $a = 2 - \sqrt{3}$ ([1]). For every $f \in S$, $g(z) = \frac{1}{a} f(az) \in S$. By Theorem 1 or 2,

$$\Phi_{n,1/p}(g)(z) = (g(z_1), (g'(z_1))^{1/p} z_0), \quad p \ge 1$$

is a convex mapping on $\Omega_{n,p}$. Thus $(f(az_1),(f'(az_1))^{1/p}(az_0))$ is a convex mapping on $\Omega_{n,p}$. Let $az_1 = w_1$, $az_0 = w_0$ and $w = (w_1, w_0)$; then $(f(w_1),(f'(w_1))^{1/p}w_0)$ is a convex mapping on $\{w \in \mathbb{C}^n : |w_1/a|^2 + \|w_0/a\|_p^p < 1\}$. We have proved (19).

The radius of starlikeness of S is $b = \tanh \frac{1}{4}\pi$. Using Theorem 1 or 2 again, we may prove (20) by the same method.

REFERENCES

[1] P. L. Duren, *Univalent Functions*, Springer-Verlag, Berlin, 1983.

[2] S. Gong, *Convex and starlike mappings in several complex variables*, Science Press/Kluwer Academic Publisher, Beijing/New York, 1998.

[3] I. Graham and G. Kohr, *Univalent mappings associated with the Roper–Suffridge extension operator*, J. Analyse Math. **81** (2000), 331–342.

[4] I. Graham and G. Kohr, *An extension theorem and subclass of univalent mappings in several complex variables*, Complex Variables **47** (2002), 59–72.

[5] I. Graham, G. Kohr and M. Kohr, *Loewner chains and Roper–Suffridge extension operator*, J. Math. Anal. Appl. **247** (2000), 448–465.

[6] S. Krantz, *Function Theory of Several Complex Variables*, 2nd edn., Wadsworth & Brooks/Cole, California, 1992.

[7] T. S. Liu and W. J. Zhang, *Homogeneous expansions of normalized biholomorphic convex mapping over* B^p, Science in China **40A** (1997), 799–806.

[8] J. Pfaltzgraff and T. J. Suffridge, *An extension theorem and linear invariant families generated by starlike maps*, Ann. Univ. Marie Curie-Skłodowska Sect. A **53** (1999), 193–207.

[9] J. Pfaltzgraff and T. J. Suffridge, *Linear invariance, order and convex maps in* $\mathbb{C}^n$, Complex Variables **40** (1999), 35–50.

[10] K. Roper and T. J. Suffridge, *Convex mappings on the unit ball in* $\mathbb{C}^n$, J. Analyse Math. **65** (1995), 333–347.

Sheng Gong and Taishun Liu
DEPARTMENT OF MATHEMATICS
UNIVERSITY OF SCIENCE AND TECHNOLOGY OF CHINA
HEFEI, ANHUI 230026, P. R. CHINA
email: gongs@ustc.edu.cn, lts@ustc.edu.cn

(Received May 14, 2001 and in revised form May 30, 2001)

ACADEMIC PRESS

J. Math. Anal. Appl. 284 (2003) 425–434

Journal of MATHEMATICAL ANALYSIS AND APPLICATIONS

www.elsevier.com/locate/jmaa

The generalized Roper–Suffridge extension operator ☆

Sheng Gong * and Taishun Liu

Department of Mathematics, University of Science and Technology of China, Hefei 230026, PR China

Received 10 August 2001

Submitted by S. Ruscheweyh

Abstract

Let $f(z)$ be a normalized convex (starlike) function on the unit disc D. Let $\Omega=\{z\in\mathbf{C}^n\colon |z_1|^2+|z_2|^{p_2}+\cdots+|z_n|^{p_n}<1\}$, where $z=(z_1,z_2,\ldots,z_n)$, $z_1\in D$, $(z_2,\ldots,z_n)\in\mathbf{C}^{n-1}$, $p_i\geqslant 1$, $i=2,\ldots,n$, are real numbers. In this note, we prove that $\Phi(f)(z)=(f(z_1),f'(z_1)^{1/p_2}z_2,\ldots,f'(z_1)^{1/p_n}z_n)$ is a normalized convex (starlike) mapping on Ω, where we choose the power function such that $(f'(z_1))^{1/p_i}|_{z_1=0}=1$, $i=2,\ldots,n$. Some other related results are proved.

Keywords: ε starlikeness; Roper–Suffridge extension operator; Carathéodory metric

1. Introduction

In 1995, Roper and Suffridge [1] introduced an extension operator. This operator is defined for normalized locally biholomorphic function f on the unit disc D in $\mathbf{C}$ by

$$\Phi_n(f)(z)=F(z)=\left(f(z_1),\sqrt{f'(z_1)}z_0\right), \tag{1}$$

where $z=(z_1,z_0)$ belongs to the unit ball B^n in $\mathbf{C}^n$, $z_1\in D$, $z_0=(z_2,\ldots,z_n)\in\mathbf{C}^{n-1}$, and we choose the branch of the square root such that $\sqrt{f'(0)}=1$.

Roper–Suffridge extension operator has remarkable properties:

☆ Project supported by the National Natural Science Foundation of China.

* Corresponding author.

E-mail addresses: gongs@ustc.edu.cn (S. Gong), lts@ustc.edu.cn (T. Liu).

0022-247X/$ – see front matter © 2003 Elsevier Inc. All rights reserved.
doi:10.1016/S0022-247X(02)00400-6

(i) If f is a normalized convex function on D, then F is a normalized convex mapping on B^n;
(ii) If f is a normalized starlike function on D, then F is a normalized starlike mapping on B^n;
(iii) If f is a normalized Bloch function on D, then F is a normalized Bloch mapping on B^n.

These results were proved by Roper and Suffridge [1], Graham and Kohr [2]. Until now, we only know a few concrete examples about the convex mappings, starlike mappings and Bloch mappings on B^n. By Roper–Suffridge extension operator, we may construct a lots of concrete examples about these mappings on B^n. This is one reason why people are interested in this extension operator.

After that there are many papers to discuss this operator (for example, [3–6], etc.). They generalized the Roper–Suffridge extension operators and discussed their properties.

In [3], Graham et al. generalized the operator (1) as

$$\Phi_{n,\alpha}(f)(z)=F_\alpha(z)=\big(f(z_1),\big(f'(z_1)\big)^{\alpha}z_0\big), \tag{2}$$

where $\alpha\in[0,1/2]$, f, z_1, z_0, z are defined as above, and we choose the branch of the power function such that $(f'(z_1))^{\alpha}|_{z_1=0}=1$. They proved that this operator maps the normalized starlike function on D to the normalized starlike mapping on B^n, and maps the normalized Bloch function on D to the normalized Bloch mapping on B^n, but it does not preserve convexity on B^n when $\alpha\in[0,1/2)$. In [2], Graham and Kohr proposed the following open problem: consider the Reinhardt domain

$$\Omega_{2,p}=\big\{z=(z_1,z_2)\in\mathbf{C}^2\colon |z_1|^2+|z_2|^p<1\big\},$$

where $p\geqslant 1$. Does the operator

$$\Phi_{2,1/p}(f)(z)=F_{1/p}(z)=\big(f(z_1),\big(f'(z_1)\big)^{1/p}z_2\big)$$

extend convex functions on D to the convex mappings on $\Omega_{2,p}$?

In [7], we defined the ε starlike mappings on a domain in $\mathbf{C}^n$.

Definition 1. Let Ω be a domain in $\mathbf{C}^n$, and let $f:\Omega\to\mathbf{C}^n$ be a locally biholomorphic mapping and $0\in f(\Omega)$. We say f is ε starlike mapping on Ω if there exists a positive number ε, $0\leqslant\varepsilon\leqslant 1$, such that $f(\Omega)$ is starlike with respect to every point in $\varepsilon f(\Omega)$. All ε starlike mappings on Ω form the family of ε starlike mappings on Ω.

When $\varepsilon=0$, it is exactly the family of starlike mappings, and when $\varepsilon=1$, it is exactly the family of convex mappings.

In [7], we proved the following result.

Theorem A. *Let $f(z_1)$ be a normalized biholomorphic ε starlike function on the unit disk $D=\{z_1\in\mathbf{C}\colon |z_1|<1\}$ in $\mathbf{C}$, $0\leqslant\varepsilon\leqslant 1$, then*

$$\Phi_{n,1/p}(f)(z)=F_{1/p}(z)=\big(f(z_1),\big(f'(z_1)\big)^{1/p}z_0\big),\quad p\geqslant 1, \tag{3}$$

is a normalized biholomorphic ε starlike mapping on

$$\Omega_{n,p} = \{z \in \mathbf{C}^n \colon |z_1|^2 + \|z_0\|_p^p < 1\}, \tag{4}$$

where $z_0 = (z_2, \dots, z_n) \in \mathbf{C}^{n-1}$, $z = (z_1, z_0) \in \Omega_{n,p}$, *we choose the branch of the power function in* (3) *such that* $(f'(z_1))^{1/p}|_{z_1=0} = 1$, *and*

$$\|z\|_p = \begin{cases} \left(\sum_{j=1}^n |z_j|^p\right)^{1/p}, & 1 \leqslant p < \infty; \\ \max_{j=1,\dots,n} |z_j|, & p = \infty. \end{cases}$$

When $n = 2$, $\varepsilon = 1$, Theorem A solved the open problem of Graham and Kohr [2]. The answer is affirmative, and it holds true for any $n \geqslant 2$. When $p = 2$, $\varepsilon = 1$, Theorem A is the result of Roper and Suffridge [1]. When $p = 2$, $\varepsilon = 0$, Theorem A is the result of Graham and Kohr [2].

Theorem A told how to construct concrete examples of convex mappings and starlike mappings on a class of Reinhardt domains (4). No doubt, it is an important class of Reinhardt domains in several complex variables, especially, it is a class of weak pseudoconvex domains when $p > 2$.

In Section 2, we will introduce some generalized Roper–Suffridge extension operator in purpose to construct the concrete convex mappings and starlike mappings on some class of more general Reinhardt domains. In Section 3, we will extend the Roper–Suffridge extension operator from on complex variable to several complex variables.

2. Generalized Roper–Suffridge operator on a class of Reinhardt domain

We have already known that for the class of Reinhardt domains (4), we may generalize the Roper–Suffridge extension operator as (3) such that we can use it to construct the convex mappings and the starlike mappings on (4). Now we consider the more general class of Reinhardt domains. Let

$$\Omega = \{z \in \mathbf{C}^n \colon |z_1|^{p_1} + \dots + |z_n|^{p_n} < 1\}, \tag{5}$$

where $p_i \geqslant 1$, $i = 1, 2, \dots, n$, $z = (z_1, \dots, z_n)$. How to generalize the Roper–Suffridge extension operator such that we can use it to construct the convex mappings and the starlike mappings on it? In general, we do not know how to do it. But we have the following result.

Theorem 1. *Let* $f(z_1)$ *be a normalized biholomorphic ε starlike function on the unit disk* D *in* $\mathbf{C}$, $0 \leqslant \varepsilon \leqslant 1$. *Then*

$$\begin{aligned} \Phi_{n,1/p_2,\dots,1/p_n}(f)(z) &= F_{1/p_2,\dots,1/p_n}(z) \\ &= \left(f(z_1), \left(f'(z_1)\right)^{1/p_2} z_2, \dots, \left(f'(z_1)\right)^{1/p_n} z_n\right) \end{aligned} \tag{6}$$

is a normalized biholomorphic ε starlike mapping on the Reinhardt domain

$$\Omega_{n,p_2,\dots,p_n} = \{z \in \mathbf{C}^n \colon |z_1|^2 + |z_2|^{p_2} + \dots + |z_n|^{p_n} < 1\}, \tag{7}$$

where $z = (z_1, \dots, z_n)$, $p_i \geqslant 1$, $i = 2, \dots, n$, *and we choose the branch of the power functions in* (6) *such that* $(f'(z_1))^{1/p_i}|_{z_1=0} = 1$, $i = 2, \dots, n$.

Theorem 1 is a special case of the following result.

Theorem 2. *Let $\|\cdot\|_i$ be the Banach norms of $\mathbf{C}^{n_i}$, $i = 1, 2, \ldots, k$, where n_i are positive integers. Let*

$$\Omega_N = \left\{(z_1, z, \ldots, w) \in \mathbf{C} \times \mathbf{C}^{n_1} \times \cdots \times \mathbf{C}^{n_k}\colon |z_1|^2 + \|z\|_1^{p_1} + \cdots + \|w\|_k^{p_k} < 1\right\}, \tag{8}$$

where $p_i \geqslant 1$, $i = 1, 2, \ldots, k$, $N = 1 + n_1 + \cdots + n_k$, $z_1 \in \mathbf{C}$, $z \in \mathbf{C}^{n_1}, \ldots, w \in \mathbf{C}^{n_k}$.

If $f(z_1)$ is a normalized biholomorphic ε starlike function on the unit disk D in $\mathbf{C}$ then

$$\begin{aligned}\Phi_{N,1/p_1,\ldots,1/p_k}(f)(z) &= F_{1/p_1,\ldots,1/p_k}(z)\\ &= \left(f(z_1), \left(f'(z_1)\right)^{1/p_1} z, \ldots, \left(f'(z_1)\right)^{1/p_k} w\right)\end{aligned} \tag{9}$$

is a normalized biholomorphic ε starlike mapping on Ω_N, where we choose the branch of the power function in (9) such that $(f'(z_1))^{1/p_i}|_{z_1=0} = 1$, $i = 1, \ldots, k$.

When $k = 1$, $\|\cdot\|_1$ is the p-norm, it is Theorem A. When $n_i = 1$, $\|\cdot\|_i$ is p-norm, $i = 1, \ldots, k$, it is Theorem 1.

Proof of Theorem 2. For any $\lambda \in [0, 1]$, $(z_1, z, \ldots, w) \in \Omega_N$ and $(a_1, a, \ldots, b) \in \Omega_N$, where $z_1 \in D$, $a_1 \in D$, $z \in \mathbf{C}^{n_1}$, $a \in \mathbf{C}^{n_1}$, $\ldots$, and $w \in \mathbf{C}^{n_k}$, $b \in \mathbf{C}^{n_k}$, if we can find $(u_1, u, \ldots, v) \in \Omega_N$, where $u_1 \in D$, $u \in \mathbf{C}^{n_1}, \ldots, v \in \mathbf{C}^{n_k}$, such that

$$\begin{aligned}&\left(f(u_1), \left(f'(u_1)\right)^{1/p_1} u, \ldots, \left(f'(u_1)\right)^{1/p_k} v\right)\\ &\quad = (1-\lambda)\left(f(z_1), \left(f'(z_1)\right)^{1/p_1} z, \ldots, \left(f'(z_1)\right)^{1/p_k} w\right)\\ &\qquad + \lambda\varepsilon\left(f(a_1), \left(f'(a_1)\right)^{1/p_1} a, \ldots, \left(f'(a_1)\right)^{1/p_k} b\right),\end{aligned} \tag{10}$$

then Theorem 2 has been proved.

Since f is a ε starlike function on D, for any λ, $\lambda \in [0, 1]$, and $z_1 \in D$, $a_1 \in D$, there exists $u_1 \in D$, such that

$$f(u_1) = (1-\lambda) f(z_1) + \lambda\varepsilon f(a_1). \tag{11}$$

Thus the right hand side of (10) is

$$\begin{aligned}&\Big(f(u_1), (1-\lambda)\left(f'(z_1)\right)^{1/p_1} z + \lambda\varepsilon\left(f'(a_1)\right)^{1/p_1} a, \ldots,\\ &\quad (1-\lambda)\left(f'(z_1)\right)^{1/p_k} w + \lambda\varepsilon\left(f'(a_1)\right)^{1/p_k} b\Big).\end{aligned}$$

Let

$$\begin{aligned}(u, \ldots, v) = \bigg(&\frac{(1-\lambda)(f'(z_1))^{1/p_1} z + \lambda\varepsilon(f'(a_1))^{1/p_1} a}{(f'(u_1))^{1/p_1}}, \ldots,\\ &\frac{(1-\lambda)(f'(z_1))^{1/p_k} w + \lambda\varepsilon(f'(a_1))^{1/p_k} b}{(f'(u_1))^{1/p_k}}\bigg).\end{aligned} \tag{12}$$

We need to prove the following inequality

$$\|u\|_1^{p_1} + \cdots + \|v\|_k^{p_k} < 1 - |u_1|^2 \tag{13}$$

holds. From (11), we have

$$u_1(z_1, a_1) = f^{-1}\big[(1-\lambda) f(z_1) + \lambda\varepsilon f(a_1)\big]. \tag{14}$$

Regarding (14) as a mapping from $D \times D$ to D, we have already proved [7] that

$$\frac{\left|\frac{\partial u_1}{\partial z_1}\right||\xi| + \left|\frac{\partial u_1}{\partial a_1}\right||\eta|}{1-|u_1|^2} \leqslant \max\left(\frac{|\xi|}{1-|z_1|^2}, \frac{|\eta|}{1-|a_1|^2}\right), \tag{15}$$

where ξ, η are any two arbitrary complex numbers. Moreover, by (14) we know that

$$\frac{\partial u_1}{\partial z_1} = \frac{1}{f'(u_1)}(1-\lambda) f'(z_1), \qquad \frac{\partial u_1}{\partial a_1} = \frac{1}{f'(u_1)}\lambda\varepsilon f'(a_1).$$

Substituting it into (12), we have

$$(u, \ldots, v) = \left((1-\lambda)^{1/q_1}\left(\frac{\partial u_1}{\partial z_1}\right)^{1/p_1} z + (\lambda\varepsilon)^{1/q_1}\left(\frac{\partial u_1}{\partial a_1}\right)^{1/p_1} a, \ldots,\right.$$
$$\left.(1-\lambda)^{1/q_k}\left(\frac{\partial u_1}{\partial z_1}\right)^{1/p_k} w + (\lambda\varepsilon)^{1/q_k}\left(\frac{\partial u_1}{\partial a_1}\right)^{1/p_k} b\right),$$

where $1/p_1 + 1/q_1 = 1, \ldots, 1/p_k + 1/q_k = 1$. By the triangle inequality of Banach norm and Hölder inequality, we have

$$\begin{aligned}\|u\|_1 &\leqslant (1-\lambda)^{1/q_1}\left|\frac{\partial u_1}{\partial z_1}\right|^{1/p_1}\|z\|_1 + (\lambda\varepsilon)^{1/q_1}\left|\frac{\partial u_1}{\partial a_1}\right|^{1/p_1}\|a\|_1 \\ &\leqslant (1-\lambda+\lambda\varepsilon)^{1/q_1}\left(\left|\frac{\partial u_1}{\partial z_1}\right|\|z\|_1^{p_1} + \left|\frac{\partial u_1}{\partial a_1}\right|\|a\|_1^{p_1}\right)^{1/p_1}.\end{aligned}$$

Thus

$$\|u\|_1^{p_1} \leqslant \left|\frac{\partial u_1}{\partial z_1}\right|\|z\|_1^{p_1} + \left|\frac{\partial u_1}{\partial a_1}\right|\|a\|_1^{p_1}.$$

Using the same process, we may obtain the estimations of the other terms, for example, the estimations of the last term is the following inequality:

$$\|v\|_k^{p_k} \leqslant \left|\frac{\partial u_1}{\partial z_1}\right|\|w\|_k^{p_k} + \left|\frac{\partial u_1}{\partial a_1}\right|\|b\|_k^{p_k}.$$

Thus,

$$\begin{aligned}\|u\|_1^{p_1} + \cdots + \|v\|_k^{p_k} \leqslant{} & \left|\frac{\partial u_1}{\partial z_1}\right|\left(\|z\|_1^{p_1} + \cdots + \|w\|_k^{p_k}\right) \\ & + \left|\frac{\partial u_1}{\partial a_1}\right|\left(\|a\|_1^{p_1} + \cdots + \|b\|_k^{p_k}\right).\end{aligned} \tag{16}$$

Let $\xi = \|z\|_1^{p_1} + \cdots + \|w\|_k^{p_k}$, $\eta = \|a\|_1^{p_1} + \cdots + \|b\|_k^{p_k}$ in (15); then

$$\begin{aligned}&\|u\|_1^{p_1}+\cdots+\|v\|_k^{p_k}\\&\leqslant(1-|u_1|^2)\max\left(\frac{\|z\|_1^{p_1}+\cdots+\|w\|_k^{p_k}}{1-|z_1|^2},\frac{\|a\|_1^{p_1}+\cdots+\|b\|_k^{p_k}}{1-|a_1|^2}\right)\\&<1-|u_1|^2\end{aligned}$$

by (16).

We have proved (13), and hence we have proved Theorem 2. □

3. Roper–Suffridge extension operator for several complex variables

The Roper–Suffridge extension operator and its generalizations which we mentioned above start from a locally biholomorphic function f of one complex variable on the unit disk in $\mathbf{C}$, by the Roper–Suffridge extension operator or its generalizations Φ, we get a locally biholomorphic mapping $\Phi(f)=F$ on some domain in $\mathbf{C}^n$, then we discussed the properties of F. Now we try to extend the Roper–Suffridge extension operator and its generalizations from one variable to several complex variables.

We start with a locally biholomorphic mapping $f:R\to\mathbf{C}^n$, where R is a domain in $\mathbf{C}^n$ and

$$f=\begin{pmatrix}f_1\\ \vdots\\ f_n\end{pmatrix},$$

then we construct a generalized Roper–Suffridge extension operator, using it we may get a locally biholomorphic mappings on some domain in $\mathbf{C}^m$ $(m>n)$. In this section, we give one example of such kind generalized Roper–Suffridge extension operator.

Let

$$D^n=\left\{z=\begin{pmatrix}z_1\\ \vdots\\ z_n\end{pmatrix}\in\mathbf{C}^n\colon |z_i|<1,\ i=1,\ldots,n\right\}$$

be the unit polydisk in $\mathbf{C}^n$. Let $f:D^n\to\mathbf{C}^n$ be a normalized biholomorphic convex mapping on D^n; then by Suffridge theorem [8],

$$f(z)=\begin{pmatrix}f_1(z_1)\\ \vdots\\ f_n(z_n)\end{pmatrix},$$

so

$$J_f(z)=\begin{pmatrix}f_1'(z_1)& &0\\ &\ddots& \\ 0& &f_n'(z_n)\end{pmatrix},$$

$J_f(z)$ is the Jacobi matrix of f at z, where $f_i(z_i)$, $i=1,2,\ldots,n$, are normalized biholomorphic convex functions on the unit disk. Let

$$\Omega_{2,p}^n=\left\{\begin{pmatrix}z\\ w\end{pmatrix}\in\mathbf{C}^{2n}\colon |z_i|^2+|w_i|^p<1,\ i=1,2,\ldots,n\right\},\quad p\geqslant 1,\tag{17}$$

where

$$w=\begin{pmatrix} w_1 \\ \vdots \\ w_n \end{pmatrix}\in \mathbf{C}^n.$$

As a consequence of Theorem A, we have the following result.

Corollary 1. *Let $f(z)$ be a normalized biholomorphic convex mapping on D^n, where $f(z): D^n \to \mathbf{C}^n$ is a column vector. Then*

$$\Phi_{2,1/p}(f)(z,w)=\begin{pmatrix} f(z) \\ (J_f(z))^{1/p}w \end{pmatrix} \tag{18}$$

is a normalized biholomorphic convex mapping on $\Omega^n_{2,p}$, where

$$w=\begin{pmatrix} w_1 \\ \vdots \\ w_n \end{pmatrix}\in \mathbf{C}^n, \qquad (J_f(z))^{1/p}=\begin{pmatrix} (f_1'(z_1))^{1/p} & & O \\ & \ddots & \\ O & & (f_n'(z_n))^{1/p} \end{pmatrix},$$

and we choose the branch of the power functions in (18) *such that $(f_i'(z_i))^{1/p}|_{z_i=0}=1$, $i=1,2,\dots,n$.*

Proof. From Theorem A,

$$\begin{pmatrix} f_i(z_i) \\ (f_i'(z_i))^{1/p}w_i \end{pmatrix} \quad (i=1,2,\dots,n)$$

is a normalized biholomorphic convex mapping on $\Omega_{2,p}$, so

$$\begin{pmatrix} f(z) \\ (J_f(z))^{1/p}w \end{pmatrix}=\begin{pmatrix} f_1(z_1) \\ \vdots \\ f_n(z_n) \\ (f_1'(z_1))^{1/p}w_1 \\ \vdots \\ (f_n'(z_n))^{1/p}w_n \end{pmatrix}$$

is a normalized biholomorphic convex mapping on $\Omega^n_{2,p}$. □

When $\varepsilon \neq 1$, we cannot define $(J_f(z))^{1/p}$ because $J_f(z)$ is not a diagonal matrix. But we have the following result.

Theorem 3. *Let $f(z): D^n \to \mathbf{C}^n$ be a normalized biholomorphic ε starlike mapping on D^n. Then $\Phi_{2,1}(f)(z,w)$ is a normalized biholomorphic ε starlike mapping on $\Omega^n_{2,1}$.*

In purpose to prove Theorem 3, we need the following lemma.

Lemma 1. *The infinitesimal form of the Carathéodory metric of D^n is*

$$F_C^{D^n}(z,\zeta)=\max\left(\frac{|\zeta_1|}{1-|z_1|^2},\dots,\frac{|\zeta_n|}{1-|z_n|^2}\right),$$

where

$$z=\begin{pmatrix} z_1 \\ \vdots \\ z_n \end{pmatrix}\in D^n, \qquad \zeta=\begin{pmatrix} \zeta_1 \\ \vdots \\ \zeta_n \end{pmatrix}\in \mathbf{C}^n.$$

Proof. Fix

$$z=\begin{pmatrix} z_1 \\ \vdots \\ z_n \end{pmatrix}\in D^n.$$

Let

$$s=\begin{pmatrix} s_1 \\ \vdots \\ s_n \end{pmatrix}\in D^n, \qquad \varphi(s)=\begin{pmatrix} \frac{z_1-s_1}{1-\bar{z}_1 s_1} \\ \vdots \\ \frac{z_n-s_n}{1-\bar{z}_n s_n} \end{pmatrix};$$

then $\varphi(s)\in \mathrm{Aut}(D^n)$, $\varphi(z)=0$,

$$J_\varphi(z)=\begin{pmatrix} \frac{-1}{1-|z_1|^2} & & O \\ & \ddots & \\ O & & \frac{-1}{1-|z_n|^2} \end{pmatrix} \quad \text{and} \quad J_\varphi(z)\zeta=\begin{pmatrix} \frac{-\zeta_1}{1-|z_1|^2} \\ \vdots \\ \frac{-\zeta_n}{1-|z_n|^2} \end{pmatrix}.$$

Let $F_C^{D^n}(\cdot,\cdot)$ be the infinitesimal form of Carathéodory metric, then $F_C^{D^n}(z,\zeta)=F_C^{D^n}(0,J_\varphi(z)\zeta)$. Since D^n is a bounded convex circular domain, $F_C^{D^n}(0,\zeta)=\rho(\zeta)$ [9, 10], where $\rho(\zeta)$ is the Minkowski functional of D^n. We already know that the Minkowski functional of D^n is $\max_{1\leqslant i\leqslant n}|z_i|$. Hence

$$F_C^{D^n}\big(0,J_\varphi(z)\zeta\big)=\max\left(\frac{|\zeta_1|}{1-|z_1|^2},\ldots,\frac{|\zeta_n|}{1-|z_n|^2}\right).$$

We have proved Lemma 1. □

By Lempert theorem [11], we know the infinitesimal form of Carathéodory metric of D^n and the infinitesimal form of Kobayashi metric of D^n are the same.

Proof of Theorem 3. For any $\lambda\in[0,1]$, $\binom{z}{w}\in\Omega^n_{2,1}$, $\binom{\xi}{\eta}\in\Omega^n_{2,1}$, if we can fine $\binom{u}{v}\in\Omega^n_{2,1}$ such that

$$\begin{pmatrix} f(u) \\ J_f(u)v \end{pmatrix}=(1-\lambda)\begin{pmatrix} f(z) \\ J_f(z)w \end{pmatrix}+\lambda\varepsilon\begin{pmatrix} f(\xi) \\ J_f(\xi)\eta \end{pmatrix},$$

then Theorem 3 have been proved.

Since f is a ε starlike mapping on D^n, for any $\lambda\in[0,1]$ and $z\in D^n$, $\xi\in D^n$, there exists $u\in D^n$, such that

$$f(u)=(1-\lambda)f(z)+\lambda\varepsilon f(\xi). \tag{19}$$

Let

$$v=(1-\lambda)J_f^{-1}(u)J_f(z)w+\lambda\varepsilon J_f^{-1}(u)J_f(\xi)\eta. \tag{20}$$

We need to prove $\binom{u}{v} \in \Omega^n_{2,1}$, i.e., the following inequality

$$\max\left(\frac{|v_1|}{1-|u_1|^2}, \ldots, \frac{|v_n|}{1-|u_n|^2}\right) < 1 \tag{21}$$

holds. From (19), we have

$$u(z,\xi) = f^{-1}\big[(1-\lambda)f(z) + \lambda\varepsilon f(\xi)\big]. \tag{22}$$

We regard it as a mapping from $D^n \times D^n$ to D^n, then by the contraction property of Carathéodory metric, the following inequality

$$F_C^{D^n}\left(u(z,\xi), J_u(z,\xi)\begin{pmatrix} w \\ \eta \end{pmatrix}\right) \leqslant F_C^{D^n \times D^n}\left(\begin{pmatrix} z \\ \xi \end{pmatrix}, \begin{pmatrix} w \\ \eta \end{pmatrix}\right) \tag{23}$$

holds for any column vector $\binom{w}{\eta} \in \mathbf{C}^{2n}$, where J_u is the Jacobi matrix of u. From (22),

$$J_u(z,\xi) = \big((1-\lambda)J_f^{-1}(u)J_f(z),\ \lambda\varepsilon J_f^{-1}(u)J_f(\xi)\big).$$

Hence

$$J_u(z,\xi)\begin{pmatrix} w \\ \eta \end{pmatrix} = (1-\lambda)J_f^{-1}(u)J_f(z)w + \lambda\varepsilon J_f^{-1}(u)J_f(\xi)\eta = v$$

by (20). Thus (23) becomes

$$F_C^{D^n}(u,v) \leqslant F_C^{D^n \times D^n}\left(\begin{pmatrix} z \\ \xi \end{pmatrix}, \begin{pmatrix} w \\ \eta \end{pmatrix}\right).$$

By Lemma 1, it is exactly the following inequality

$$\max\left(\frac{|v_1|}{1-|u_1|^2}, \ldots, \frac{|v_n|}{1-|u_n|^2}\right)$$
$$\leqslant \max\left(\frac{|w_1|}{1-|z_1|^2}, \ldots, \frac{|w_n|}{1-|z_n|^2}, \frac{|\eta_1|}{1-|\xi_1|^2}, \ldots, \frac{|\eta_n|}{1-|\xi_n|^2}\right) < 1.$$

Hence (21) holds true. We have proved the Theorem 3. □

References

[1] K. Roper, T.J. Suffridge, Convex mappings on the unit ball in $\mathbf{C}^n$, J. Anal. Math. 65 (1995) 333–347.

[2] I. Graham, G. Kohr, Univalent mappings associated with the Roper–Suffridge extension operator, J. Anal. Math. 81 (2000) 331–342.

[3] I. Graham, G. Kohr, M. Kohr, Loewner chains and Roper–Suffridge extension operator, J. Math. Anal. Appl. 247 (2000) 448–465.

[4] I. Graham, G. Kohr, An extension theorem and subclass of univalent mappings in several complex variables, Complex Variables, submitted for publication.

[5] J. Pfaltzgraff, T.J. Suffridge, An extension theorem and linear invariant families generated by starlike maps, Ann. Univ. Mariae Curie-Sklodowska Sect. A 53 (1999) 193–207.

[6] J. Pfaltzgraff, T.J. Suffridge, Linear invariance order and convex maps in $\mathbf{C}^n$, Complex Variables, submitted for publication.

[7] S. Gong, T.S. Liu, On Roper–Suffridge extension operator, J. Anal. Math, submitted for publication.

[8] T.J. Suffridge, The principle of subordination applied to function of several variables, Pacific. J. Math. 33 (1970) 241–248.
[9] S. Gong, T.S. Liu, Distortion theorems for biholomorphic convex mappings on bounded convex circular domains, Chinese Ann. Math. B 20 (1999) 297–304.
[10] S. Gong, Convex and Starlike Mappings in Several Complex Variables, Kluwer Academic, 1998.
[11] S. Krantz, Function Theory of Several Complex Variables, 2nd ed., Wadsworth & Brooks/Cole, 1992.

ELSEVIER

Available online at www.sciencedirect.com
SCIENCE DIRECT

J. Math. Anal. Appl. 299 (2004) 448–464

Journal of MATHEMATICAL ANALYSIS AND APPLICATIONS
www.elsevier.com/locate/jmaa

The decomposition theorem for the family of complete quasi-convex mappings ☆

Sheng Gong *, Taishun Liu

Department of Mathematics, University of Science and Technology of China, Hefei, 230026, PR China

Received 9 January 2003

Available online 23 September 2004

Submitted by T. Akahori

Abstract

In this paper, we will give a decomposition theorem for some type of holomorphic mappings. This result is the generalization of many known results.

Keywords: Bounded convex circular domain; Quasi-convex mapping; Complete quasi-convex mapping; Complete quasi-convex mapping of type $(n_1, n_2, \ldots, n_k)$; Decomposition theorem

1. Introduction

The family of biholomorphic convex mappings in several complex variables has many interesting properties. The decomposition theorem is one of them. In 1970, Suffridge [1] gave the first decomposition theorem.

Theorem A. *Let $f(z) = (f_1(z), \ldots, f_n(z))$ be a normalized locally biholomorphic mapping on the unit polydisk Δ^n in $\mathbb{C}^n$, $f : \Delta^n \to \mathbb{C}^n$, then f is a normalized biholomor-*

☆ Project supported by 973 plan, the National Natural Science Foundation of China and the Natural Science Foundation of Anhui Province.

* Corresponding author.

E-mail addresses: gongs@ustc.edu.cn (S. Gong), lts@ustc.edu.cn (T. Liu).

0022-247X/$ – see front matter © 2004 Elsevier Inc. All rights reserved.
doi:10.1016/j.jmaa.2004.03.075

phic convex mapping on Δ^n *if and only if there exist n biholomorphic convex functions* $g_1, \ldots, g_n$ *on the unit disk* Δ *in* $\mathbb{C}$ *such that*

$$f(z) = \big(g_1(z_1), \ldots, g_n(z_n)\big),$$

where $z = (z_1, \ldots, z_n) \in \Delta^n$.

Let $\Omega \subset \mathbb{C}^n$ be a domain with $0 \in \Omega$. We denote by $K(\Omega)$ the family of all normalized biholomorphic convex mappings on Ω, and denote by $D(\Omega)$ the family of all mappings which are in the form $(g_1(z_1), \ldots, g_n(z_n))$, where $g_i(z_i)$ is the normalized biholomorphic convex function on the domain $\Omega_i = \{z_i \in \mathbb{C}\colon$ there exists $(z_1, \ldots, z_{i-1}, z_{i+1}, \ldots, z_n)$ such that $(z_1, \ldots, z_{i-1}, z_i, z_{i+1}, \ldots, z_n) \in \Omega\}$, $i = 1, \ldots, n$. Then Theorem A tells us that $K(\Delta^n) = D(\Delta^n)$. In general, $K(\Delta^n) \neq D(\Delta^n)$ when $\Omega \neq \Delta^n$. For example, the mapping

$$\left(\frac{z_1}{1-z_1}, \frac{z_2}{1-z_2}, \ldots, \frac{z_n}{1-z_n}\right) \in D(B_p),$$

but $\notin K(B_p)$, where $B_p = \{(z_1, \ldots, z_n) \in \mathbb{C}^n\colon |z_1|^p + \cdots + |z_n|^p < 1\}$ $(p \geqslant 1)$ (cf. Gong [2]). For a domain Ω in $\mathbb{C}^n$, it is natural to search a subfamily of the family of biholomorphic mappings on Ω which is larger than $K(\Omega)$ and contains $D(\Omega)$. Recently, Taishun Liu and Wenjun Zhang [3] defined the family of complete quasi-convex mappings $CQ(\Omega)$ on the bounded convex complete circular domain Ω, $K(\Omega) \subset CQ(\Omega)$, and proved that $CQ(\Delta^n) = K(\Delta^n)(= D(\Delta^n))$. Thus they gave a more concise criterion of convexity for normalized locally biholomorphic mappings on Δ^n. In general, $CQ(\Omega) \neq K(\Omega)$, $K(\Omega) \neq D(\Omega)$ when $\Omega \neq \Delta^n$. In Section 2, we will prove that if Ω is a bounded convex complete circular domain, then $D(\Omega) \subset CQ(\Omega)$. Thus $CQ(\Omega)$ has the desired property that it has a relatively simple geometric definition and it contains both $K(\Omega)$ and $D(\Omega)$.

In 1998, Taishun Liu and Guangbin Ren [4] extended Theorem A from domain Δ^n to the domain which is the direct product of a finite number of bounded convex circular domains.

Theorem B. *Let* $\Omega_1 \subset \mathbb{C}^{n_1}, \ldots, \Omega_k \subset \mathbb{C}^{n_k}$ *be bounded convex circular domains and their Minkowski functionals* $p_1(z), \ldots, p_k(w)$ *are holomorphic functions of* $z, \bar{z}, \ldots, w, \bar{w}$ *on* $\Omega_1, \ldots, \Omega_k$*, respectively (except for a lower dimensional manifold), where* $z \in \Omega_1, \ldots, w \in \Omega_k$*. If*

$$f(z, \ldots, w) = \big(f_1(z, \ldots, w), \ldots, f_k(z, \ldots, w)\big)$$

is a normalized locally biholomorphic mappings on $\Omega = \Omega_1 \times \cdots \times \Omega_k$, $f : \Omega \to \mathbb{C}^{n_1} \times \cdots \times \mathbb{C}^{n_k}$*, then f is a normalized biholomorphic convex mapping on* Ω *if and only if there exist k normalized biholomorphic convex mappings* g_i *on* Ω_i, $i = 1, \ldots, k$*, such that*

$$f(z, \ldots, w) = \big(g_1(z), \ldots, g_k(w)\big).$$

Let $\Omega \subset \mathbb{C}^{n_1+\cdots+n_k}$ be a domain with $0 \in \Omega$. We denote by $D_k(\Omega)$ the family of all mappings which are in the form $(g_1(z), \ldots, g_k(w))$, where $g_1(z)$ is the normalized biholomorphic convex mapping on the domain $\Omega_1 = \{z \in \mathbb{C}^{n_1}\colon$ there exists $(\ldots, w)$ such that $(z, \ldots, w) \in \Omega\}, \ldots, g_k(w)$ is the normalized biholomorphic convex mapping on the domain $\Omega_k = \{w \in \mathbb{C}^{n_k}\colon$ there exists $(z,\ldots)$ such that $(z,\ldots,w) \in \Omega\}$. Then Theorem B tells us that $K(\Omega_1 \times \cdots \times \Omega_k) = D_k(\Omega_1 \times \cdots \times \Omega_k)$. In general, $K(\Omega) \neq D_k(\Omega)$, if Ω cannot be decomposed as the direct product of k bounded convex circular domains. For a domain $\Omega \subset \mathbb{C}^{n_1+\cdots+n_k}$, we also need to search a subfamily of the family of biholomorphic mappings on Ω which is larger than $K(\Omega)$ and contains $D_k(\Omega)$.

When the domain is the direct product of a finite number of bounded convex circular domains, we already know that the decomposition theorem is hold for the family of normalized biholomorphic convex mappings. It is natural to ask the following question. Does there exist more general family of mappings such that the decomposition theorem still holds? We will answer this question in this paper. The main result will be given in Section 3, and its proof will be given in Section 4.

2. Family of complete quasi-convex mappings

It is well known that: in one complex variable the Alexander theorem describes the relation between the family of convex functions and the family of starlike functions, but the analogous relationship in several complex variables is never true. Many families which are related with the family of convex mappings and family of starlike mappings were defined in several complex variables. The analytic condition for convexity and for starlikeness in several complex variables reduces to the usual conditions with positive real part when the dimension is 1.

Let $f(z)$ be a normalized locally biholomorphic function on Δ. It is known that the following statements are equivalent to each other:

(1) f is a normalized biholomorphic convex function on Δ;
(2) $\operatorname{Re}\{1 + \frac{zf''(z)}{f'(z)}\} \geqslant 0$ holds for all $z \in \Delta$;
(3) $\operatorname{Re}\{\frac{zf'(z)}{f(z)-f(\xi)}\} \geqslant 0$ holds for all $z, \xi \in \Delta$, where $|\xi| \leqslant |z|$, $z \neq \xi$;
(4) $\operatorname{Re}\{\frac{2zf'(z)}{f(z)-f(\xi)} - \frac{z+\xi}{z-\xi}\} \geqslant 0$ holds for all $z, \xi \in \Delta$.

Corresponding to (4), (2), (3) in the above, we have the following definition.

Definition 1. Let Ω be a bounded convex circular domain in $\mathbb{C}^n$, its Minkowski functional $p(z) \in C^1$ (except for a lower dimensional manifold), and $f(z)$ be a locally biholomorphic mapping on Ω, $f:\Omega \to \mathbb{C}^n$.

(1) If the inequality

$$\operatorname{Re}\left\{\frac{\alpha}{\frac{\partial p}{\partial z}(\mu)J_f^{-1}(\alpha\mu)(f(\alpha\mu) - f(\beta\mu))} - \frac{\alpha+\beta}{\alpha-\beta}\right\} \geqslant 0 \tag{2.1}$$

holds for any $\alpha, \beta \in \Delta$ and any $\mu \in \partial\Omega$, then f is said to be a quasi-convex mapping of type A on Ω.

(2) If the inequality

$$\mathrm{Re}\left\{\frac{\partial p(z)}{\partial z}J_f^{-1}(z)\left(\frac{d^2 f}{dz^2}(z,z)z^2+J_f(z)z\right)\right\}\geqslant 0 \tag{2.2}$$

holds for any $z\in\Omega$, then f is said to be a quasi-convex mapping of type B on Ω.

(3) If the inequality

$$\mathrm{Re}\left\{\frac{\partial p(z)}{\partial z}J_f^{-1}(z)\big(f(z)-f(\alpha z)\big)\right\}\geqslant 0 \tag{2.3}$$

holds for any $z\in\Omega$ and any $\alpha\in\bar{\Delta}$, then f is said to be a quasi-convex mapping of type C on Ω. where f, z are column vectors, $\frac{\partial p}{\partial z}=(\frac{\partial p}{\partial z_1},\ldots,\frac{\partial p}{\partial z_n})$, J_f is the Jacobi matrix of f, $\frac{d^2 f}{dz^2}$ is the Frechét derivative of order 2 of f. All quasi-convex mappings of type A, type B, and type C form the family of quasi-convex mappings $Q_A(\Omega)$, $Q_B(\Omega)$, and $Q_C(\Omega)$, respectively.

If the inequality (2.3) holds for any $z,w\in\Omega$ when $f(w)$ replaces $f(\alpha z)$, where $p(w)\leqslant p(z)<1$, then f is a convex mapping on Ω.

The relations

$$K(\Omega)\subset Q_A(\Omega)\subset S^*(\Omega),\qquad Q_A(\Omega)\subset Q_B(\Omega) \tag{2.4}$$

were proved, where $S^*(\Omega)$ is the family of normalized biholomorphic starlike mappings on Ω.

Roper and Suffridge [5] defined $Q_A(B^n)$ and $Q_B(B^n)$, and proved relations (2.4) when $\Omega=B^n$. They also conjectured $Q_A(B^n)=Q_B(B^n)$. Until now, it is an open problem. Hao Liu [6] extended these definitions and results from B^n to bounded convex circular domains. $Q_C(\Omega)$ was defined by Taishun Liu and Hao Liu [7], and proved that $K(\Omega)\subset Q_A(\Omega)\subset Q_C(\Omega)\subset S^*(\Omega)$, where Ω is a bounded convex circular domains. Taishun Liu and Wenjun Zhang [8] extended the definitions of $Q_A(\Omega)$ and $Q_C(\Omega)$ to the open unit ball B in complex Banach space, and proved that $K(B)\subset Q_A(B)=Q_C(B)\subset S^*(B)$, where $K(B)$ and $S^*(B)$ denote the family of normalized biholomorphic convex mappings on B and family of normalized biholomorphic starlike mappings on B. Since $Q_A=Q_C$, we can say it to be the family of quasi-convex mappings, and denote it by Q. Recently, Taishun Liu and Wenjun Zhang [3] defined the family of complete quasi-convex mappings on the bounded convex complete circular domains (cf. Krantz [9, p. 102]).

Definition 2. Let Ω be a bounded convex complete circular domain in $\mathbb{C}^n$, its Minkowski functional $p(z)\in C^1$ (except for a lower dimensional manifold), and $f(z)$ be a normalized locally biholomorphic mapping on Ω, $f:\Omega\to\mathbb{C}^n$. If the inequality

$$\mathrm{Re}\left\{\frac{\partial p(z)}{\partial z}J_f^{-1}(z)\big(f(z)-f(\xi_1 z_1,\ldots,\xi_n z_n)\big)\right\}\geqslant 0 \tag{2.5}$$

holds for any $z=(z_1,\ldots,z_n)\in\Omega$, and any $\xi=(\xi_1,\ldots,\xi_n)\in\bar{\Delta}^n$, then f is said to be a complete quasi-convex mapping on Ω, where f, z are column vectors, $\frac{\partial p(z)}{\partial z}=(\frac{\partial p(z)}{\partial z_1},\ldots,\frac{\partial p(z)}{\partial z_n})$. All complete quasi-convex mappings on Ω form the family of complete quasi-convex mappings $CQ(\Omega)$.

Comparing (2.3) and (2.5), we have $CQ(\Omega) \subset Q(\Omega)$. In [3], they gave the following examples.

Example 1. $(z_1 + \frac{1}{2}z_2^2, z_2) \in Q(\Delta^2)$, but $\notin CQ(\Delta^2)$.

Example 2. $(\frac{z_1}{1-z_1}, \frac{z_2}{1-z_2}, \dots, \frac{z_n}{1-z_n}) \in CQ(B_p)$, but $\notin K(B_p)$, where $B_p = \{(z_1, \dots, z_n) \in \mathbb{C}^n\colon |z_1|^p + \dots + |z_n|^p < 1\}$ $(p \geqslant 1)$. In [6] and [7], we already know that $(\frac{z_1}{(1-z_1)^2}, \dots, \frac{z_n}{(1-z_n)^2}) \in S^*(\Delta^n)$, but $\notin Q(\Delta^n)$. Thus, in general, we have the relations

$$K \subsetneqq CQ \subsetneqq Q \subsetneqq S^*.$$

In [3], they gave the following decomposition theorem for the family CQ.

Theorem C. *Let $f(z) = (f_1(z), \dots, f_n(z))$ be a normalized locally biholomorphic mapping on Δ^n, then $f \in CQ(\Delta^n)$ if and only if there exist n normalized biholomorphic convex functions $g_1, \dots, g_n$ on Δ, such that*

$$f(z) = \big(g_1(z_1), \dots, g_n(z_n)\big), \tag{2.6}$$

where $z = (z_1, \dots, z_n) \in \Delta^n$.

Theorem C tells us that $CQ(\Delta^n) = D(\Delta^n) = K(\Delta^n)$. From Example 1 in the above, we know that (2.6) does not hold for the family $Q(\Delta^n)$. Actually, for the family $Q(\Delta^n)$, it is impossible to express the mapping in $Q(\Delta^2)$ as $(z_1h_1(z), z_2h_2(z))$, where $h_1(z), h_2(z)$ are holomorphic functions of z.

The necessity part of Theorem C was proved in [3]. The proof of the sufficiency part of Theorem C is easy. We omit the detail.

In general, $CQ(\Omega) \neq K(\Omega)$, $D(\Omega) \neq K(\Omega)$. We will prove $D(\Omega) \subset CQ(\Omega)$ when Ω is a bounded convex complete circular domain. That means that $CQ(\Omega)$ is the family of mappings which we want to search.

Theorem 1. *Let Ω be a bounded convex complete circular domain in $\mathbb{C}^n$, and its Minkowski functional $p(z) \in C^1$ (except for a lower dimensional manifold), then*

$$K(\Omega) \cup D(\Omega) \subset CQ(\Omega).$$

Proof. Since Ω is a bounded convex complete circular domain, we have $p(z_1, \dots, z_{j-1}, z_je^{i\theta}, z_{j+1}, \dots, z_n) = p(z)$, where $z = (z_1, \dots, z_n) \in \Omega$, $\theta \in \mathbb{R}$. Obviously,

$$\begin{aligned}
0 &= \frac{d}{d\theta} p\big(z_1, \dots, z_{j-1}, z_je^{i\theta}, z_{j+1}, \dots, z_n\big)\big|_{\theta=0} \\
&= \frac{\partial p}{\partial z_j}\big(z_1, \dots, z_{j-1}, z_je^{i\theta}, z_{j+1}, \dots, z_n\big)e^{i\theta}z_ji \\
&\quad + \frac{\partial p}{\partial \bar{z}_j}\big(z_1, \dots, z_{j-1}, z_je^{i\theta}, z_{j+1}, \dots, z_n\big)e^{-i\theta}\bar{z}_j(-i)\big|_{\theta=0} \\
&= \frac{\partial p(z)}{\partial z_j}z_ji - \frac{\partial p(z)}{\partial \bar{z}_j}\bar{z}_ji.
\end{aligned}$$

Hence,

$$\frac{\partial p(z)}{\partial z_j} z_j = \frac{\partial p(z)}{\partial \bar{z}_j} \bar{z}_j,$$

i.e., $\frac{\partial p(z)}{\partial z_j} z_j$ is a real number. Now we will prove that $\frac{\partial p(z)}{\partial z_j} z_j$ is non-negative. For $z = (z_1, \dots, z_n) \in \Omega$, fixing j, there exists $\varepsilon > 0$, such that $(z_1, \dots, z_{j-1}, (1+\varepsilon) z_j, z_{j+1}, \dots, z_n) \in \Omega$. Thus $p(z_1, \dots, z_{j-1}, (1+\varepsilon) z_j, z_{j+1}, \dots, z_n) \geqslant p(z)$. Expanding the left-hand side of the above inequality, we have

$$p(z) + \left(\frac{\partial p(z)}{\partial z_j} z_j + \frac{\partial p(z)}{\partial \bar{z}_j} \bar{z}_j \right) \varepsilon + o(\varepsilon) \geqslant p(z),$$

since $p(z) \in C^1$. That is,

$$\frac{\partial p(z)}{\partial z_j} z_j + \frac{\partial p(z)}{\partial \bar{z}_j} \bar{z}_j \geqslant 0.$$

Thus $\frac{\partial p(z)}{\partial z_j} z_j$ is non-negative because $\frac{\partial p(z)}{\partial z_j} z_j$ is a real number, where $j = 1, \dots, n$.

If $f(z) = (f_1(z), \dots, f_n(z))' \in D(\Omega)$, it is easy to verify

$$\begin{aligned} & J_f^{-1}(z) \big(f(z) - f(\xi_1 z_1, \dots, \xi_n z_n) \big) \\ &= \left(\frac{f_1(z_1) - f_1(\xi_1 z_1)}{f_1'(z_1)}, \dots, \frac{f_n(z_n) - f_n(\xi_n z_n)}{f_n'(z_n)} \right)', \end{aligned}$$

where $z = (z_1, \dots, z_n) \in \Omega$, $\xi = (\xi_1, \dots, \xi_n) \in \bar{\Delta}^n$. We have

$$\begin{aligned} & \operatorname{Re} \left\{ \frac{\partial p(z)}{\partial z} J_f^{-1}(z) \big(f(z) - f(\xi_1 z_1, \dots, \xi_n z_n) \big) \right\} \\ &= \sum_{j=1}^n \operatorname{Re} \left\{ \frac{\partial p(z)}{\partial z_j} \cdot \frac{f_j(z_j) - f_j(\xi_j z_j)}{f_j'(z_j)} \right\} \\ &= \sum_{j=1}^n \operatorname{Re} \left\{ \frac{\partial p(z)}{\partial z_j} z_j \cdot \frac{f_j(z_j) - f_j(\xi_j z_j)}{z_j f_j'(z_j)} \right\}. \end{aligned}$$

Since $\frac{\partial p(z)}{\partial z_j} z_j$ $(j = 1, \dots, n)$ are real numbers, so

$$\sum_{j=1}^n \operatorname{Re} \left\{ \frac{\partial p(z)}{\partial z_j} z_j \cdot \frac{f_j(z_j) - f_j(\xi_j z_j)}{z_j f_j'(z_j)} \right\} = \sum_{j=1}^n \frac{\partial p(z)}{\partial z_j} z_j \operatorname{Re} \left\{ \frac{f_j(z_j) - f_j(\xi_j z_j)}{z_j f_j'(z_j)} \right\}.$$

Because Ω is a bounded convex complete circular domain, $\Omega_j = \{z_j \in \mathbb{C}\colon \exists (z_1, \dots, z_{j-1}, z_{j+1}, \dots, z_n)$ such that $(z_1, \dots, z_{j-1}, z_j, z_{j+1}, \dots, z_n) \in \Omega\}$ $(j = 1, \dots, n)$ are disks, $\Omega \subset \Omega_1 \times \dots \times \Omega_n$ and f_j $(j = 1, \dots, n)$ are convex functions on Ω_j, so we have

$$\operatorname{Re} \left\{ \frac{f_j(z_j) - f_j(\xi_j z_j)}{z_j f_j'(z_j)} \right\} \geqslant 0, \quad j = 1, \dots, n.$$

Because $\frac{\partial p(z)}{\partial z_j} z_j$ $(j = 1, \ldots, n)$ are non-negative real numbers, so

$$\sum_{j=1}^{n} \frac{\partial p(z)}{\partial z_j} z_j \operatorname{Re}\left\{ \frac{f_j(z_j) - f_j(\xi_j z_j)}{z_j f_j'(z_j)} \right\} \geqslant 0.$$

Hence (2.5) holds, that is, $D(\Omega) \subset CQ(\Omega)$. We have proved Theorem 1. □

3. The decomposition theorem for the family of complete multi-quasi-convex mappings

To answer the question at the end of Section 1, we need the following definition, and use it to establish a general decomposition theorem.

Definition 3. Let $\Omega_1 \subset \mathbb{C}^{n_1}, \ldots, \Omega_k \subset \mathbb{C}^{n_k}$ be bounded convex circular domains, $\Omega = \Omega_1 \times \cdots \times \Omega_k$, its Minkowski functional $p(\wedge) \in C^1$ (except for a lower dimensional manifold), where $\wedge = (z, \ldots, w)' \in \Omega$, $z = (z_1, \ldots, z_{n_1})' \in \Omega_1, \ldots, w = (w_1, \ldots, w_{n_k})' \in \Omega_k$, and $f(\wedge)$ be a normalized locally biholomorphic mapping on Ω, $f : \Omega \to \mathbb{C}^{n_1} \times \cdots \times \mathbb{C}^{n_k}$. If the inequality

$$\operatorname{Re}\left\{ \frac{\partial p(\wedge)}{\partial \wedge} J_f^{-1}(\wedge)\big(f(\wedge) - f(\xi_1 z. \ldots, \xi_k w)\big) \right\} \geqslant 0 \tag{3.1}$$

holds for any $\wedge = (z, \ldots, w)' \in \Omega$, $\xi = (\xi_1, \ldots, \xi_k) \in \bar{\Delta}^k$, then f is said to be a complete quasi-convex mapping of type $(n_1, \ldots, n_k)$, where f is a column vector,

$$\frac{\partial p(\wedge)}{\partial \wedge} = \left(\frac{\partial p(\wedge)}{\partial z_1}, \ldots, \frac{\partial p(\wedge)}{\partial z_{n_1}}, \ldots, \frac{\partial p(\wedge)}{\partial w_1}, \ldots, \frac{\partial p(\wedge)}{\partial z_{w_k}} \right).$$

All complete quasi-convex mappings of type $(n_1, \ldots, n_k)$ on Ω form the family of complete quasi-convex mappings of type $(n_1, \ldots, n_k)$ $MCQ(\Omega)$.

Obviously, $p(\wedge) = p(z, \ldots, w) = \max\{p(z), \ldots, p(w)\}$, $MCQ(\Omega) = CQ(\Omega)$ when $n_1 = \cdots = n_k = 1$, and $MCQ(\Omega) = Q(\Omega)$ when all n_i $(i = 1, \ldots, k)$ equal zero except one.

We will give a more general definition of the complete quasi-convex mappings of type $(n_1, \ldots, n_k)$.

Definition 4. Let $\Omega \subset \mathbb{C}^N$, $N = n_1 + \cdots + n_k$ is a decomposition of N, where n_i $(i = 1, \ldots, k)$ are positive integers. Any point $\wedge \in \Omega$ can be expressed as $(z, \ldots, w)$, where $z \in \mathbb{C}^{n_1}, \ldots, w \in \mathbb{C}^{n_k}$. If $(\xi_1 z, \ldots, \xi_k w) \in \Omega$ holds for any $\wedge = (z, \ldots, w) \in \Omega$ and any $\xi = (\xi_1, \ldots, \xi_k) \in \bar{\Delta}^k$, then we say Ω to be a complete circular domain of type $(n_1, \ldots, n_k)$.

Let $\Omega \subset \mathbb{C}^N$ be a bounded complete convex circular domain of type $(n_1, \ldots, n_k)$, its Minkowski functional $p(\wedge) \in C^1$ (except for a lower dimensional manifold), where $\wedge = (z, \ldots, w)' \in \Omega$, $z = (z_1, \ldots, z_{n_1})' \in \mathbb{C}^{n_1}, \ldots, w = (w_1, \ldots, w_{n_k})' \in \mathbb{C}^{n_k}$, and $f(\wedge)$ be a normalized locally biholomorphic mapping on Ω, $f : \Omega \to \mathbb{C}^N$. If for any $\wedge = (z, \ldots, w)' \in \Omega$, $\xi = (\xi_1, \ldots, \xi_k) \in \bar{\Delta}^k$, the inequality (3.1) holds, then we say f to be a complete quasi-convex mapping of type $(n_1, \ldots, n_k)$ on Ω. All of these mappings form the family of complete quasi-convex mappings of type $(n_1, \ldots, n_k)$ $MCQ(\Omega)$.

Obviously, $K(\Omega) \subset MCQ(\Omega)$. In general $D_k(\Omega) \not\equiv K(\Omega)$. But we can use the similar argument as we used in the proof of Theorem 1 to prove that $K(\Omega) \cup D_k(\Omega) \subset MCQ(\Omega)$. We omit the detail of the proof.

Now we go back to Definition 3.

The following two examples show the relations between $MCQ(\Omega)$ and other families.

Example 3. $f(z,w) = \left(\frac{z_1}{1-z_1}, \frac{z_2}{1-z_1}, \frac{w_1}{1-w_1}, \frac{w_2}{1-w_1}\right)' \in MCQ(\Delta^2 \times \Delta^2)$, but $f \notin CQ(\Delta^4)$, where $z = (z_1, z_2)'$, $w = (w_1, w_2)'$.

Proof. From Theorem C, we know $f(z,w) \notin CQ(\Delta^4)$. Now we are going to prove $f(z,w) \in MCQ(\Delta^2 \times \Delta^2)$.

It is easy to compute that

$$J_f^{-1}(z,w) = \begin{pmatrix} (1-z_1)^2 & 0 & 0 & 0 \\ -z_2(1-z_1) & (1-z_1) & 0 & 0 \\ 0 & 0 & (1-w_1)^2 & 0 \\ 0 & 0 & -w_2(1-w_1) & (1-w_1) \end{pmatrix}.$$

If $\xi = (\xi_1, \xi_2) \in \bar{\Delta}^2$, then

$$J_f^{-1}(z,w)\big(f(z,w) - f(\xi_1 z, \xi_2 w)\big) = \begin{pmatrix} \frac{z_1(1-z_1)(1-\xi_1)}{1-\xi_1 z_1} \\ \frac{z_2(1-z_1)(1-\xi_1)}{1-\xi_1 z_1} \\ \frac{w_1(1-w_1)(1-\xi_2)}{1-\xi_2 w_1} \\ \frac{w_2(1-w_1)(1-\xi_2)}{1-\xi_2 w_1} \end{pmatrix}.$$

If $\max\{|z_1|, |z_2|, |w_1|, |w_2|\} = |z_1| > 0$, then $p(\wedge) = |z_1|$, $\frac{\partial p(\wedge)}{\partial \wedge} = (\frac{1}{2}\frac{|z_1|}{z_1}, 0, 0, 0)$, where $\wedge = (z, w)$. (3.1) becomes

$$\operatorname{Re}\left\{\frac{(1-z_1)(1-\xi_1)}{1-\xi_1 z_1}\right\} \geqslant 0.$$

Of course it is true when $z_1 \in \Delta$, $\xi_1 \in \bar{\Delta}$. Thus (3.1) holds when $p(\wedge) = |z_1|$. By the same process, we can prove that (3.1) holds when $p(\wedge) = |z_2|, |w_1|$, or $|w_2|$ and > 0. Hence, $f(z,w) \in MCQ(\Delta^2 \times \Delta^2)$. □

Example 4. $g(z,w) = \left(\frac{z_1}{1-z_1}, \frac{z_2}{1-z_1}, \frac{w_1}{1-z_1}, \frac{w_2}{1-z_1}\right)' \in Q(\Delta^2 \times \Delta^2)$, but $g \notin MCQ(\Delta^2 \times \Delta^2)$, where $z = (z_1, z_2)'$, $w = (w_1, w_2)'$.

Proof. $g(z,w)$ is a normalized biholomorphic mapping on Δ^4. We prove $g(z,w) \notin MCQ(\Delta^2 \times \Delta^2)$ at first.

It is easy to compute that

$$J_g^{-1}(z,w) = \begin{pmatrix} (1-z_1)^2 & 0 & 0 & 0 \\ -z_2(1-z_1) & 1-z_1 & 0 & 0 \\ -w_1(1-z_1) & 0 & 1-z_1 & 0 \\ -w_2(1-z_1) & 0 & 0 & 1-z_1 \end{pmatrix}.$$

If $\xi = (\xi_1, \xi_2) \in \bar{\Delta}^2$, then

$$J_g^{-1}(z, w)\big(g(z, w) - g(\xi_1 z, \xi_2 w)\big) = \frac{1 - z_1}{1 - \xi_1 z_1} \begin{pmatrix} z_1(1-\xi_1) \\ z_2(1-\xi_1) \\ w_1(1-\xi_2) \\ w_2(1-\xi_2) \end{pmatrix}.$$

Taking $\wedge = (z, w)$ such that $\max\{|z_1|, |z_2|, |w_1|, |w_2|\} = |w_1| > 0$, then $p(\wedge) = |w_1|$, $\frac{\partial p(\wedge)}{\partial \wedge} = (0, 0, \frac{1}{2}\frac{|w_1|}{w_1}, 0)$. The left-hand side of (3.1) becomes

$$\frac{1}{2}|w_1| \operatorname{Re}\left\{\frac{(1-z_1)(1-\xi_2)}{1-\xi_1 z_1}\right\}.$$

Taking $(z_1, z_2, w_1, w_2) = (i(1-2\varepsilon), \star, 1-\varepsilon, \star)$ and $(\xi_1, \xi_2) = (-1, i)$, where $\star$ is any complex number, its absolute value is less than $1-\varepsilon$, $0 < \varepsilon < \frac{1}{2}$. Then

$$\operatorname{Re}\left\{\frac{(1-z_1)(1-\xi_2)}{1-\xi_1 z_1}\right\} \to -1$$

when $\varepsilon \to 0$. Obviously, the right-hand side is < 0 when ε is sufficiently small. Thus $g(z, w) \notin MCQ(\Delta^2 \times \Delta^2)$.

Next, we are going to prove $g(z, w) \in Q(\Delta^2 \times \Delta^2)$. For any $(z, w) \in \Delta^2 \times \Delta^2$, $\xi \in \bar{\Delta}$, we have

$$J_f^{-1}(z, w)\big(f(z, w) - f(\xi z, \xi w)\big) = \frac{(1-z_1)(1-\xi)}{1-\xi z_1} \begin{pmatrix} z_1 \\ z_2 \\ w_1 \\ w_2 \end{pmatrix}.$$

If $p(\wedge) = \max\{|z_1|, |z_2|, |w_1|, |w_2|\} = |z_1| > 0$, then (3.1) becomes

$$\frac{1}{2}|z_1| \operatorname{Re}\left\{\frac{(1-z_1)(1-\xi_1)}{1-\xi_1 z_1}\right\} \geqslant 0.$$

Of course, it is true. By the same process, we can prove (3.1) is true when $\max\{|z_1|, |z_2|, |w_1|, |w_2|\} = |z_2|, |w_1|$ or $|w_2|$ and > 0. Thus we have proved $g(z, w) \in Q(\Delta^2 \times \Delta^2)$.

From the above examples, we have the following relations:

$$K(\Omega) \subsetneqq CQ(\Omega) \subsetneqq MCQ(\Omega) \subsetneqq Q(\Omega) \subsetneqq S^*(\Omega). \qquad \square$$

For the family of complete quasi-convex mappings of type $(n_1, \ldots, n_k)$, we have the following decomposition theorem.

Theorem 2. *Let $\Omega_1 \subset \mathbb{C}^{n_1}, \ldots, \Omega_k \subset \mathbb{C}^{n_k}$ be bounded convex circular domains, $\Omega = \Omega_1 \times \cdots \times \Omega_k$, its Minkowski functional $p(\wedge)$ is a holomorphic function of $z, \bar{z}, \ldots, w, \bar{w}$ (except for a lower dimensional manifold), where $\wedge = (z, \ldots, w)' \in \Omega$, $z \in \Omega_1, \ldots, w \in \Omega_k$. If $f(z, \ldots, w) = (f_1(z, \ldots, w), \ldots, f_k(z, \ldots, w))$ be a normalized locally biholomorphic mapping on Ω, $f : \Omega \to \mathbb{C}^{n_1} \times \cdots \times \mathbb{C}^{n_k}$, then f is a normalized complete quasi-convex mapping of type $(n_1, \ldots, n_k)$ if and only if there exist k normalized biholomorphic quasi-convex mappings $g_1(z), \ldots, g_k(w)$ on $\Omega_1, \ldots, \Omega_k$ such that*

$$f(z, \ldots, w) = \big(g_1(z), \ldots, g_k(w)\big).$$

Theorem 2 tells us that: when the domain Ω is the direct product of a finite number of bounded convex circular domains, the decomposition theorem holds not only for the family of convex mappings $K(\Omega)$, but also for the family of complete quasi-convex mappings of type $(n_1, \ldots, n_k)$ $MCQ(\Omega)$. Thus Theorem 2 extends Theorem B. Notice that when the family is a family of convex mappings, it is decomposed as a direct product of families of convex mappings, and when the family is a family of complete quasi-convex mappings of type $(n_1, \ldots, n_k)$, it is decomposed as a direct product of families of quasi-convex mappings. This conclusion is reasonable if we observe that $K(\Omega) \subset MCQ(\Omega)$ and $K(\Omega_i) \subset Q(\Omega_i)$ $(i = 1, \ldots, k)$. Theorem 2 answers the question at the end of Section 1. Obviously, when $n_1 = \cdots = n_k = 1$, $\Omega_1 = \cdots = \Omega_k = \Delta$, Theorem 2 becomes Theorem C, i.e., $CQ(\Delta^n) = D(\Delta^n) = K(\Delta^n)$.

4. Proof of Theorem 2

Now we are going to prove Theorem 2. We only need to prove Theorem 2 in the case of $k=2$.

Proof of sufficiency part

If $f(z, w) = \binom{g_1(z)}{g_2(w)}$, where $g_1(z) \in Q(\Omega_1)$, $g_2(w) \in Q(\Omega_2)$. We need to prove $f(z, w) \in MCQ(\Omega_1 \times \Omega_2)$.

Let $\wedge = (z, w)$, then the Minkowski functional of Ω is $p(\wedge) = \max\{p_1(z), p_2(w)\}$, where $p_1(z)$, $p_2(w)$ are the Minkowski functional of Ω_1, Ω_2, respectively. If $\max\{p_1(z), p_2(w)\} = p_1(z) > 0$, then $\frac{\partial p(\wedge)}{\partial \wedge} = (\frac{\partial p(1z)}{\partial z}, 0)$. The left-hand side of (3.1) becomes

$$\begin{aligned} &\operatorname{Re}\left\{\frac{\partial p(\wedge)}{\partial \wedge} J_f^{-1}(\wedge)\big(f(\wedge) - f(\xi_1 z, \xi_2 w)\big)\right\} \\ &= \operatorname{Re}\left\{\frac{\partial p_1(z)}{\partial z} J_{g_1}^{-1}(z)\big(g_1(z) - g_1(\xi_1 z)\big)\right\}, \end{aligned} \tag{4.1}$$

where $\xi = (\xi_1, \xi_2) \in \bar{\Delta}^2$. By (2.3), the right-hand side of (4.1) is non-negative. Hence (3.1) holds when $\max\{p_1(z), p_2(w)\} = p_1(z) > 0$. By the same process, we can prove (3.1) when $\max\{p_1(z), p_2(w)\} = p_2(w) > 0$. Thus the sufficiency part of Theorem 2 has been proved.

Proof of necessity part

Let $f(z, w) = \binom{f_1(z,w)}{f_2(z,w)} \in MCQ(\Omega_1 \times \Omega_2)$. The necessity part of Theorem 2 has been proved if we can prove that there exist normalized holomorphic mappings $g_1(z)$ and $g_2(w)$ on Ω_1 and Ω_2, respectively such that $f_1(z, w) = g_1(z)$ and $f_2(z, w) = g_2(w)$. This assertion is true due to the following reason. If $f_1(z, w) = g_1(z)$, $f_2(z, w) = g_2(w)$ and $\max\{p_1(z), p_2(w)\} = p_1(z) > 0$, then (4.1) holds. Since $f(z, w) \in MCQ(\Omega_1 \times \Omega_2)$, the left-hand side of (4.1) is non-negative. Hence $g_1(z) \in Q(\Omega_1)$. By the same process, we can prove $g_2(w) \in Q(\Omega_2)$.

Now we are going to prove $f_1(z, w) = g_1(z)$ and $f_2(z, w) = g_2(w)$ when $f(z, w) \in MCQ(\Omega_1 \times \Omega_2)$. It needs a few steps.

(1) Denote

$$\sum_{j,k=1}^{n_1} \frac{\partial^2 f}{\partial z_j \partial z_k}(z,w) z_j z_k = \frac{\partial^2 f}{\partial z^2}(z,w) z^2, \tag{4.2}$$

$$\sum_{j=1}^{n_1}\sum_{k=1}^{n_2} \frac{\partial^2 f}{\partial z_j \partial w_k}(z,w) z_j w_k = \frac{\partial^2 f}{\partial z \partial w}(z,w) zw, \tag{4.3}$$

$$\sum_{j,k=1}^{n_2} \frac{\partial^2 f}{\partial w_j \partial w_k}(z,w) w_j w_k = \frac{\partial^2 f}{\partial w^2}(z,w) w^2, \tag{4.4}$$

where $z=(z_1,\ldots,z_{n_1})\in\Omega_1\subset\mathbb{C}^{n_1}$, $w=(w_1,\ldots,w_{n_2})\in\Omega_2\subset\mathbb{C}^{n_2}$. For any non-negative constant A_1, A_2, denote

$$\begin{pmatrix} P_{A_1,A_2}(z,w) \\ Q_{A_1,A_2}(z,w) \end{pmatrix} = \begin{pmatrix} A_1^2 z \\ A_2^2 w \end{pmatrix} + J_f^{-1}(z,w)\biggl(A_1^2 \frac{\partial^2 f}{\partial z^2}(z,w) z^2 + 2A_1A_2 \frac{\partial^2 f}{\partial z \partial w}(z,w) zw + A_2^2 \frac{\partial^2 f}{\partial w^2}(z,w) w^2\biggr). \tag{4.5}$$

We have

(i) The inequality

$$\mathrm{Re}\left\{\frac{\partial p_1}{\partial z}(z) P_{A_1,A_2}(z,w)\right\} \geqslant 0 \tag{4.6}$$

holds for any $z\in\Omega_1$, $w\in\Omega_2$ when $p_2(w)\leqslant p_1(z)<1$.

(ii) The inequality

$$\mathrm{Re}\left\{\frac{\partial p_2}{\partial w}(w) Q_{A_1,A_2}(z,w)\right\} \geqslant 0 \tag{4.7}$$

holds for any $z\in\Omega_1$, $w\in\Omega_2$ when $p_1(z)\leqslant p_2(w)<1$.

Proof. Let $\varphi(\theta)=J_f^{-1}(z,w)(f(z,w)-f(e^{iA_1\theta}z, e^{iA_2\theta}w))$, $\theta\in\mathbb{R}$, then

$$\varphi'(\theta) = -J_f^{-1}(z,w) J_f(e^{iA_1\theta}z, e^{iA_2\theta}w)\begin{pmatrix} iA_1 e^{iA_1\theta} z \\ iA_2 e^{iA_2\theta} w \end{pmatrix},$$

$$\begin{aligned}\varphi''(\theta) = &-J_f^{-1}(z,w) J_f(e^{iA_1\theta}z, e^{iA_2\theta}w)\begin{pmatrix} (iA_1)^2 e^{iA_1\theta} z \\ (iA_2)^2 e^{iA_2\theta} w \end{pmatrix} \\ &- J_f^{-1}(z,w)\biggl[\frac{\partial^2 f}{\partial z^2}(e^{iA_1\theta}z, e^{iA_2\theta}w)(iA_1 e^{iA_1\theta} z)^2 \\ &+ 2\frac{\partial^2 f}{\partial z\partial w}(e^{iA_1\theta}z, e^{iA_2\theta}w)(iA_1e^{iA_1\theta}z)(iA_2e^{iA_2\theta}w) \\ &+ \frac{\partial^2 f}{\partial w^2}(e^{iA_1\theta}z, e^{iA_2\theta}w)(iA_2e^{iA_2\theta}w)^2\biggr],\end{aligned}$$

by (4.2)–(4.4). Thus

$$\varphi(0)=\begin{pmatrix}0\\0\end{pmatrix},\qquad \varphi'(0)=-i\begin{pmatrix}A_1z\\A_2w\end{pmatrix},\qquad \varphi''(0)=\begin{pmatrix}P_{A_1A_2}(z,w)\\Q_{A_1A_2}(z,w)\end{pmatrix},$$

by (4.5). Let $\wedge=(z,w)$ and

$$\lambda(\theta)=\mathrm{Re}\left\{\frac{\partial p(\wedge)}{\partial\wedge}\varphi(\theta)\right\}=\mathrm{Re}\left\{\frac{\partial p(\wedge)}{\partial\wedge}J_f^{-1}(\wedge)\big(f(\wedge)-f\big(e^{iA_1\theta}z,e^{iA_2\theta}w\big)\big)\right\},\tag{4.8}$$

then $\lambda(\theta)\geqslant 0$ and $\lambda(\theta)\in C^\infty$ since $f(z,w)\in MCQ(\Omega_1\times\Omega_2)$ and (3.1).

When $p_2(w)\leqslant p_1(z)<1$, then $\frac{\partial p(\wedge)}{\partial\wedge}=\big(\frac{\partial p_1(z)}{\partial z},0\big)$. We can prove

$$\frac{\partial p_1(z)}{\partial z}z=\frac{1}{2}p_1(z).\tag{4.9}$$

By the definition of Minkowski functional and Ω_1 is a bounded convex circular domain, we have $p_1(e^{i\theta}z)=p_1(z)$ when $\theta\in\mathbb{R}$. Hence

$$0=\frac{d}{d\theta}p_1\big(e^{i\theta}z\big)\big|_{\theta=0}=\left(\frac{\partial p_1}{\partial z}\big(e^{i\theta}z\big)e^{i\theta}zi+\frac{\partial p_1}{\partial\bar z}\big(e^{i\theta}z\big)e^{-i\theta}\bar z(-i)\right)\bigg|_{\theta=0}$$
$$=\frac{\partial p_1(z)}{\partial z}zi-\frac{\partial p_1(z)}{\partial\bar z}\bar zi.$$

Thus $\frac{\partial p_1(z)}{\partial z}z=\frac{\partial p_1(z)}{\partial\bar z}\bar z$. By the definition of $p_1(z)$, we know $p_1(tz)=tp_1(z)$ when $t\in(0,+\infty)$, then

$$p_1(z)=\frac{d}{dt}p_1(tz)|_{t=1}=\left(\frac{\partial p_1}{\partial z}(tz)z+\frac{\partial p_1}{\partial\bar z}(tz)\bar z\right)\bigg|_{t=1}=\frac{\partial p_1(z)}{\partial z}z+\frac{\partial p_1(z)}{\partial\bar z}\bar z.$$

We have (4.9).

From (4.8), we have

$$\lambda(0)=\mathrm{Re}\left\{\left(\frac{\partial p_1(z)}{\partial z},0\right)\varphi(0)\right\}=0,$$

$$\lambda'(0)=\mathrm{Re}\left\{\left(\frac{\partial p_1(z)}{\partial z},0\right)\varphi'(0)\right\}=\mathrm{Re}\left\{\frac{\partial p_1(z)}{\partial z}(-iA_1z)\right\}=\mathrm{Re}\left\{-\frac{i}{2}A_1p_1(z)\right\}=0,$$

$$\lambda''(0)=\mathrm{Re}\left\{\frac{\partial p_1(z)}{\partial z}P_{A_1A_2}(z,w)\right\}.$$

Hence $\lambda''(0)\geqslant 0$ since $\lambda(\theta)\geqslant 0$ for any $\theta\in\mathbb{R}$. Similarly, we can prove (4.7). □

Before step (2), we need a lemma. Because the proof of the lemma is trivial, we only state the lemma and omit the detail of the proof.

Lemma. *Let Ω be a domain in $\mathbb{C}^n$, $z_0\in\Omega$. If $f(z)$ is a holomorphic function on Ω and satisfies the following two conditions:*

(i) $f(z_0)=0$;
(ii) *there exists a neighborhood U of z_0, $U\subset\Omega$, such that* $\mathrm{Re}\,f(z)\geqslant 0$ *when $z\in U$, then* $f(z)\equiv 0$ *on Ω.*

(2) The identifies

$$\frac{\partial p_1(z)}{\partial z} P_{0,1}(z,w) \equiv 0 \quad \text{and} \tag{4.10}$$

$$\frac{\partial p_2(w)}{\partial w} Q_{1,0}(z,w) \equiv 0 \tag{4.11}$$

hold for any $(z,w) \in \Omega$.

Proof. By (4.5), we have

$$\begin{pmatrix} P_{0,1}(z,w) \\ Q_{0,1}(z,w) \end{pmatrix} = \begin{pmatrix} 0 \\ w \end{pmatrix} + J_f^{-1}(z,w) \frac{\partial^2 f}{\partial w^2}(z,w) w^2.$$

So $P_{0,1}(z,0) = 0$ for any $z \in \Omega_1$. By (4.6), we know

$$\operatorname{Re}\left\{ \frac{\partial p_1(z)}{\partial z} P_{0,1}(z,w) \right\} \geqslant 0$$

for any $z \in \Omega_1$, $w \in \Omega_2$ when $p_2(w) \leqslant p_1(z) < 1$. Regarding $\frac{\partial p_1(z)}{\partial z} P_{0,1}(z,w)$ as a function of $w \in \Omega_2$, it satisfies the conditions in the lemma, thus we have (4.10). Similarly, we can prove (4.11). □

(3) The identities

$$P_{0,1}(z,w) \equiv 0 \quad \text{and} \tag{4.12}$$

$$Q_{1,0}(z,w) \equiv 0 \tag{4.13}$$

hold for any $(z,w) \in \Omega$.

Proof. The identity (4.10) means $P_{0,1}(z,w)$ is the tangent vector of convex domain $p_1(z)\Omega_1$ at the boundary point z. For any $t \in \mathbb{R}$, we have $p_1(z + tP_{0,1}(z,w)) \geqslant p_1(z)$. Expanding $p_1(z + tP_{0,1}(z,w))$ at $t = 0$,

$$\begin{aligned} p_1\big(z + tP_{0,1}(z,w)\big) = p_1(z) &+ \left[\frac{\partial p_1(z)}{\partial z} P_{0,1}(z,w) + \frac{\partial p_1(z)}{\partial \bar{z}} \overline{P_{0,1}(z,w)} \right] t \\ &+ \frac{1}{2!} H_2(z,w) t^2 + \frac{1}{3!} H_3(z,w) t^3 \\ &+ \frac{1}{4!} H_4(z,w) t^4 + \cdots, \end{aligned} \tag{4.14}$$

where

$$\begin{aligned} H_2(z,w) = {} & \frac{\partial^2 p_1(z)}{\partial z^2} P_{0,1}(z,w)^2 + 2 \frac{\partial^2 p_1(z)}{\partial z \partial \bar{z}} P_{0,1}(z,w) \overline{P_{0,1}(z,w)} \\ &+ \frac{\partial^2 p_1(z)}{\partial \bar{z}^2} \overline{P_{0,1}(z,w)}^2, \end{aligned} \tag{4.15}$$

$$H_3(z,w)=\frac{\partial^3 p_1(z)}{\partial z^3}P_{0,1}(z,w)^3+3\frac{\partial^3 p_1(z)}{\partial z^2\partial\bar z}P_{0,1}(z,w)^2\overline{P_{0,1}(z,w)}$$
$$+3\frac{\partial^3 p_1(z)}{\partial z\partial\bar z^2}P_{0,1}(z,w)\overline{P_{0,1}(z,w)}^2+\frac{\partial^3 p_1(z)}{\partial\bar z^3}\overline{P_{0,1}(z,w)}^3, \tag{4.16}$$

......,

the meaning of each term on the right-hand side of (4.15) is similar with (4.2)–(4.4). For example,

$$\frac{\partial^2 p_1(z)}{\partial z\partial\bar z}P_{0,1}(z,w)\overline{P_{0,1}(z,w)}=\sum_{j,k=1}^{n_1}\frac{\partial^2 p_1(z)}{\partial z_j\partial\bar z_k}P_{0,1}^{(j)}(z,w)\cdot\overline{P_{0,1}^{(k)}(z,w)},$$

where $P_{0,1}(z,w)=(P_{0,1}^{(1)}(z,w),\dots,P_{0,1}^{(n_1)}(z,w))$. Using similar notation, we can explain the meaning of each term on the right-hand side of (4.16).

By (4.10), (4.14), and $p_1(z+tP_{0,1}(z,w))\geqslant p_1(z)$, we get

$$\frac{1}{2!}H_2(z,w)t^2+\frac{1}{3!}H_3(z,w)t^3+\frac{1}{4!}H_4(z,w)t^4+\cdots\geqslant 0 \tag{4.17}$$

holds for any $(z,w)\in\Omega$. Hence $H_2(z,w)\geqslant 0$ holds for any $(z,w)\in\Omega$ since t is any real number. Differentiating both side of (4.10) by $\frac{\partial}{\partial\bar z}$, we have

$$\frac{\partial^2 p_1(z)}{\partial z\partial\bar z}P_{0,1}(z,w)\overline{P_{0,1}(z,w)}\equiv 0.$$

Thus

$$H_2(z,w)=2\,\mathrm{Re}\left\{\frac{\partial^2 p_1(z)}{\partial z^2}P_{0,1}(z,w)^2\right\}\geqslant 0.$$

Since $P_{0,1}(z,0)=0$, we have

$$\frac{\partial^2 p_1(z)}{\partial z^2}P_{0,1}(z,w)^2\equiv 0 \tag{4.18}$$

by the lemma, i.e., $H_2(z,w)\equiv 0$. From (4.17), we know $H_3(z,w)t^3+\cdots\geqslant 0$, hence $H_3(z,w)\geqslant 0$. Differentiating both side of (4.18) by $\frac{\partial}{\partial\bar z}$, we have

$$\frac{\partial^3 p_1(z)}{\partial z^2\partial\bar z}P_{0,1}(z,w)^2\overline{P_{0,1}(z,w)}\equiv 0.$$

Thus

$$H_3(z,w)=2\,\mathrm{Re}\left\{\frac{\partial^3 p_1(z)}{\partial z^3}P_{0,1}(z,w)^3\right\}\geqslant 0.$$

Since $P_{0,1}(z,0)=0$, we have

$$\frac{\partial^3 p_1(z)}{\partial z^3}P_{0,1}(z,w)^3\equiv 0$$

again by the lemma, i.e., $H_3(z,w)\equiv 0$. Using the same process, we can prove that all coefficients of t^m are identically equal to zero, $m=4,5,\dots$. Hence $p_1(z+tP_{0,1}(z,w))\equiv p_1(z)$ holds for any $t\in\mathbb{R}$. It implies $P_{0,1}(z,w)\equiv 0$ since Ω_1 is a bounded domain. Thus (4.12) has been proved. Similarly, we can prove (4.13). □

(4) Let

$$\begin{pmatrix} G_1(z,w) \\ G_2(z,w) \end{pmatrix} = J_f^{-1}(z,w)\frac{\partial^2 f}{\partial z \partial w}(z,w)zw, \tag{4.19}$$

then

(i) The inequality

$$\operatorname{Re}\left\{\frac{\partial p_1(z)}{\partial z}G_1(z,w)\right\} \geqslant 0 \tag{4.20}$$

holds for any $z \in \Omega_1$, $w \in \Omega_2$ when $p_2(w) \leqslant p_1(z) < 1$.

(ii) The inequality

$$\operatorname{Re}\left\{\frac{\partial p_2(w)}{\partial w}G_2(z,w)\right\} \geqslant 0 \tag{4.21}$$

holds for any $z \in \Omega_1$, $w \in \Omega_2$ when $p_1(z) \leqslant p_2(w) < 1$.

Proof. By (4.5) and (4.12), for any $\varepsilon > 0$, we have

$$\begin{aligned}
&\begin{pmatrix} P_{\varepsilon,1}(z,w) \\ Q_{\varepsilon,1}(z,w) \end{pmatrix} \\
&= \begin{pmatrix} \varepsilon^2 z \\ w \end{pmatrix} + J_f^{-1}(z,w)\left(\varepsilon^2\frac{\partial^2 f}{\partial z^2}(z,w)z^2 + 2\varepsilon\frac{\partial^2 f}{\partial z \partial w}(z,w)zw + \frac{\partial^2 f}{\partial w^2}(z,w)w^2\right) \\
&= \begin{pmatrix} 0 \\ w \end{pmatrix} + J_f^{-1}(z,w)\frac{\partial^2 f}{\partial w^2}(z,w)w^2 + 2\varepsilon J_f^{-1}(z,w)\frac{\partial^2 f}{\partial z \partial w}(z,w)zw + O(\varepsilon^2) \\
&= \begin{pmatrix} P_{0,1}(z,w) \\ Q_{0,1}(z,w) \end{pmatrix} + 2\varepsilon\begin{pmatrix} G_1(z,w) \\ G_2(z,w) \end{pmatrix} + O(\varepsilon^2) \\
&= \begin{pmatrix} 0 \\ Q_{0,1}(z,w) \end{pmatrix} + 2\varepsilon\begin{pmatrix} G_1(z,w) \\ G_2(z,w) \end{pmatrix} + O(\varepsilon^2).
\end{aligned}$$

By (4.6), the inequality

$$\operatorname{Re}\left\{\frac{\partial p_1(z)}{\partial z}(z)\left[2\varepsilon G_1(z,w) + O(\varepsilon^2)\right]\right\} \geqslant 0 \tag{4.22}$$

holds for any $z \in \Omega_1$, $w \in \Omega_2$ when $p_2(w) \leqslant p_1(z) < 1$.

Dividing ε on both side of (4.22), then let $\varepsilon \to 0$, we get (4.20). Similarly, we can prove (4.21). □

(5) By (4.20) and (4.21), using the lemma, the identities

$$\frac{\partial p_1(z)}{\partial z}(z)G_1(z,w) \equiv 0 \quad \text{and} \tag{4.23}$$

$$\frac{\partial p_2(w)}{\partial w}(w)G_2(z,w) \equiv 0 \tag{4.24}$$

holds for any $(z,w) \in \Omega$.

(6) Using the process as we used to get (4.12) and (4.13) from (4.10) and (4.11), we can get the following identities:

$$G_1(z, w) \equiv 0 \quad \text{and} \tag{4.25}$$

$$G_2(z, w) \equiv 0 \tag{4.26}$$

from (4.23) and (4.24).

(7) By (4.19), (4.25), and (4.26), we have

$$\frac{\partial^2 f}{\partial z \partial w}(z, w) z w \equiv 0. \tag{4.27}$$

If at the point $(z, w) = (0, 0)$, f has the expansion

$$f(z, w) = \begin{pmatrix} \sum_{\alpha \geqslant 0} \sum_{\beta \geqslant 0} A_{\alpha\beta}^{(1)} z^\alpha w^\beta \\ \sum_{\alpha \geqslant 0} \sum_{\beta \geqslant 0} A_{\alpha\beta}^{(2)} z^\alpha w^\beta \end{pmatrix}, \tag{4.28}$$

where $\alpha = (\alpha_1, \ldots, \alpha_{n_1}) \in Z^{n_1}$, $\beta = (\beta_1, \ldots, \beta_{n_2}) \in Z^{n_2}$, and the meaning of $A_{\alpha\beta}^{(i)} z^\alpha w^\beta$ $(i = 1, 2)$ are similar as (4.2)–(4.4). Then (4.27) becomes

$$\frac{\partial^2 f}{\partial z \partial w}(z, w) z w = \begin{pmatrix} \sum_{\alpha \geqslant 0} \sum_{\beta \geqslant 0} A_{\alpha\beta}^{(1)} |\alpha||\beta| z^\alpha w^\beta \\ \sum_{\alpha \geqslant 0} \sum_{\beta \geqslant 0} A_{\alpha\beta}^{(2)} |\alpha||\beta| z^\alpha w^\beta \end{pmatrix} \equiv 0, \tag{4.29}$$

where $|\alpha| = \alpha_1 + \cdots + \alpha_{n_1}$, $|\beta| = \beta_1 + \cdots + \beta_{n_1}$. (4.29) implies all $A_{\alpha\beta}^{(i)} = 0$ $(i = 1, 2)$ when $|\alpha||\beta| \neq 0$. Thus (4.28) becomes

$$f(z, w) = \begin{pmatrix} \sum_{\alpha \geqslant 0} A_{\alpha 0}^{(1)} z^\alpha + \sum_{\beta \geqslant 0} A_{0\beta}^{(1)} w^\beta \\ \sum_{\alpha \geqslant 0} A_{\alpha 0}^{(2)} z^\alpha + \sum_{\beta \geqslant 0} A_{0\beta}^{(2)} w^\beta \end{pmatrix} = \begin{pmatrix} g_1(z) + h_1(w) \\ g_2(w) + h_2(z) \end{pmatrix}, \tag{4.30}$$

where

$$g_1(z) = \sum_{\alpha \geqslant 0} A_{\alpha 0}^{(1)} z^\alpha, \qquad h_1(w) = \sum_{\beta \geqslant 0} A_{0\beta}^{(1)} w^\beta,$$

$$g_2(w) = \sum_{\beta \geqslant 0} A_{0\beta}^{(2)} w^\beta, \qquad h_2(z) = \sum_{\alpha \geqslant 0} A_{\alpha 0}^{(2)} z^\alpha.$$

We can prove $h_1(w) \equiv 0$ and $h_2(z) \equiv 0$ as follows. By (4.5), (4.12), and (4.30), we have

$$\begin{pmatrix} P_{0,1}(z, w) \\ Q_{0,1}(z, w) \end{pmatrix} = \begin{pmatrix} 0 \\ Q_{0,1}(z, w) \end{pmatrix} = \begin{pmatrix} 0 \\ w \end{pmatrix} + J_f^{-1}(0, w) \begin{pmatrix} \frac{\partial^2 h_1}{\partial w^2}(w) w^2 \\ \frac{\partial^2 g_2}{\partial w^2}(w) w^2 \end{pmatrix}. \tag{4.31}$$

Because f is a normalized locally biholomorphic mapping, it follows

$$J_f(0, w) = \begin{pmatrix} \frac{\partial g_1}{\partial z}(0) & \frac{\partial h_1}{\partial w}(w) \\ \frac{\partial h_2}{\partial z}(0) & \frac{\partial g_2}{\partial w}(w) \end{pmatrix} = \begin{pmatrix} I_{n_1} & \frac{\partial h_1}{\partial w}(w) \\ 0 & J_{g_2}(w) \end{pmatrix}.$$

Hence

$$J_f^{-1}(0, w) = \begin{pmatrix} I_{n_1} & -\frac{\partial h_1}{\partial w}(w) J_{g_2}^{-1}(w) \\ 0 & J_{g_2}^{-1}(w) \end{pmatrix}. \tag{4.32}$$

Substituting (4.32) into (4.31), we have

$$\frac{\partial h_1}{\partial w}(w)J_{g_2}^{-1}(w)\frac{\partial^2 g_2}{\partial w^2}(w)w^2 \equiv \frac{\partial^2 h_1}{\partial w^2}(w)w^2. \tag{4.33}$$

If $h_1(w) \not\equiv 0$, then the lowest degree of the homogenous polynomials in the expansion of $h_1(w)$ at $w=0$ is $m \geqslant 2$, because by the normalized condition of f, we know $h_1(0)=0$ and $\frac{\partial h_1}{\partial w}(0)=0$. The lowest degree of the homogenous polynomials in the expansion of the left-hand side of (4.33) is $\geqslant m+1$, and the lowest degree of the homogenous polynomials in the expansion of the right-hand side of (4.33) is m. We get a contradiction. Hence $h_1(w) \equiv 0$. Similarly, we can prove $h_2(z) \equiv 0$.

This complete the proof of the necessity part of Theorem 2.

References

[1] T.J. Suffridge, The principle of subordination applied to functions of several variables, Pacific J. Math. 33 (1970) 241–248.

[2] S. Gong, Convex and Starlike Mappings in Several Complex Variables, Kluwer Academic, Dordrecht, 1998.

[3] T. Liu, W. Zhang, Complete quasi-convex mappings in several complex variables, preprint, 2001.

[4] T. Liu, G. Ren, Decomposition theorem of normalized biholomorphic convex mappings, J. Reine Angew. Math. 496 (1998) 1–13.

[5] R.A. Roper, T.J. Suffridge, Convexity properties of holomorphic mappings in $\mathbb{C}^n$, Tran. Amer. Math. Soc. 351 (1999) 1803–1833.

[6] H. Liu, The family of starlike mappings in several complex variables, its extension and its subfamily, Doctor thesis, University of Science and Technology of China, 1999.

[7] T. Liu, H. Liu, The quasi-convex mappings on bounded convex circular domain, Acta Math. Sinica 44 (2001) 287–292.

[8] T. Liu, W. Zhang, The growth theorem and covering theorem of quasi-convex mappings on the unit ball in complex Banach space, Sci. China Ser. A 32 (2002) 1033–1041 (in Chinese).

[9] S.G. Krantz, Function Theory of Several Complex Variables, second ed., Wadsworth, 1992.

VII. 其　　他

关于Möbius变换的一点注记（三）*

龚　昇　严志敏

（中国科学技术大学）

一、引　言

在文献[1]中，讨论了单复变数的情形，指出：在单位园$|z|<1$中，解析函数经Möbius变换后展开的通项可以表为n阶协变导数，即若：$f(z)$在$|z|<1$中解析，则

$$f\left(\frac{\zeta+z}{1+\bar{z}\zeta}\right)=\sum_{n=0}^{\infty}\frac{(1-|z|^2)^n\nabla^n f(z)}{n!}\zeta^n \tag{1.1}$$

这里$|\zeta|<1$，∇是相对于Poincare度量的协变导数。或是可以更简单地表为

$$f\left(\frac{\zeta+z}{1+z\zeta}\right)=\sum_{n=0}^{\infty}\frac{1}{n!}\frac{\delta^n f(z)}{\delta\zeta^n}\zeta^n$$

这里$\frac{\delta}{\delta\zeta}$是相对于Poincarè度量的固有导数。

在文献[2]中，讨论了多复变数及多实变数的情形，指出：对于C^n中的复超球中的全纯函数，经Möbius变换后展开，其第二项系数可以表为二阶协变导数，即若$z=(z_1,\cdots z_n)\in C^n$，$f(z)$在超球B：$zz<1$中全纯，$z=\varphi_a(W)$为将$z=a$变为$W=0$的Möbius变换，这里$a=(a_1,\cdots a_n)\in C^n$，$W=(W_1,\cdots W_n)\in C^n$，则

$$f(z)=F(w)=f(a)+\sqrt{1-a\bar{a}'}\left(\frac{\partial f}{\partial a}R^{-1}{}'W'+\frac{1}{2}(1-a\bar{a}')WR^{-1}(\nabla^2 f(a))R^{-1}{}'W'+\cdots+\cdots\right. \tag{1.2}$$

这里

$$R=-\frac{\bar{a}'a(1-\sqrt{1-a\bar{a}'})}{aa\sqrt{1-aa}}-I,\quad\left(\frac{\partial f}{\partial a}\right)=\left(\frac{\partial f}{\partial a_1},\cdots,\frac{\partial f}{\partial a_n}\right) \tag{1.3}$$

$(\nabla^2 f(a))$为二阶协变导数矩阵在$Z=a$点之值。

同样对于R^n中的实超球$XX'<1$，$x=(x_1,\cdots x_n)\in R^n$也有同样的结果，不论其变换群为实射影群或是非欧运动群。当然，对于不同的运动群，其相应的不变微分度量也是不同的，而协变导数是分别相对于这不同的不变微分度量而言的。

自然要问，（1.2）的高阶项的系统是什么？是不是也可以象单复变数的结果（1.1）那样，可以用高阶的协变导数来表达之。本文就是要证明：在复超球的情形下，这是正确的。事实上，可以给出更一般的结果，在C^n中的有界域上，只要满足适当的条件，都可有

*1986年元月9日收到

这样的结果。典型域就是满足这些条件的例子。在R^n中，情形要复杂些，不能得到与C^n中相应的结论。

二 复超球的情形

若$D\subset C^n$为有界域，f为D上的全纯函数，其协变导数定义为

$$\frac{\delta f}{\delta z_\alpha}=\frac{\partial f}{\partial z_\alpha},\ \alpha=1,2,\cdots,n;$$

$$\frac{\delta}{\delta z_{\alpha_r}}\left(\frac{\delta^{r-1}f}{\delta z_{\alpha_{r-1}}\cdots\delta z_{\alpha_1}}\right)=\frac{\partial}{\partial z_{\alpha_r}}\left(\frac{\delta^{r-1}f}{\delta z_{\alpha_{r-1}}\cdots\delta z_{\alpha_1}}\right)-\sum_{j=1}^{r-1}\Gamma_{\alpha_j\alpha_r}^{\sigma}\frac{\delta^{r-1}f}{\delta z_{\alpha_1}\cdots\delta z_{\alpha_{j+1}}\cdot\delta z_\sigma\cdot\delta z_{\alpha_{j-1}}\cdots\delta z_{\alpha_1}}\tag{2.1}$$

这里$\alpha_1,\cdots,\alpha_r$取$1,\cdots,n$，$\Gamma_{\beta\gamma}^{\alpha}$为联络。这是一个权（0,0）的$\begin{pmatrix}0&0\\r&0\end{pmatrix}$型张量。

若$z=\varphi(w)$是D的解析自同胚，z，w都是D的座标，则有下列关系：

$$\frac{\delta^r f}{\delta w_{\alpha_r}\cdots\delta w_{\alpha_1}}=\frac{\delta^r f}{\delta z_{\beta_r}\cdots\delta z_{\beta_1}}\cdot\frac{\partial z_{\beta_r}}{\partial w_{\alpha_r}}\cdots\frac{\partial z_{\beta_1}}{\partial w_{\alpha_1}}\tag{2.2}$$

这里$\left(\frac{\partial z_\alpha}{\partial w_\beta}\right)$是坐标变换矩阵。

现在来讨论D是超球B的情形。若$z=\varphi_a(w)$为将$z=a$变为$w=0$的Möbius变换，由于B的Bergman度量

$$ds^2=(1-z\bar{z}')^{-1}dz(I-\bar{z}'z)^{-1}d\bar{z}'\tag{2.3}$$

在Möbius变换下不变，故在坐标w下，对应的联络仍为

$$\Gamma_{\alpha\gamma}^{\beta}=\frac{\delta_{\alpha\beta}\bar{w}_\gamma+\delta_{\beta\gamma}\bar{w}_\alpha}{1-w\bar{w}'}$$

显然当$s>0$

$$\frac{\partial^s\Gamma_{\alpha\gamma}^{\beta}}{\partial w_{l_1}\cdots\partial w_{l_s}}=\delta_{\alpha\beta}\bar{w}_\gamma\frac{\partial^s}{\partial w_{l_1}\cdots\partial w_{l_s}}\left(\frac{1}{1-w\bar{w}'}\right)+\delta_{\beta\gamma}\bar{w}_\alpha\frac{\partial^s}{\partial w_{l_1}\cdots\partial w_{l_s}}\left(\frac{1}{1-w\bar{w}'}\right)$$

所以当$w=0$时，对任意$s\geqslant1$，

$$\Gamma_{\alpha\gamma}^{\beta}=\frac{\partial^s\Gamma_{\alpha\gamma}^{\beta}}{\partial w_{l_1}\cdots\partial w_{l_s}}=0\tag{2.4}$$

由（2.1）可知

$$\frac{\delta^r f}{\partial w_{\alpha_r}\cdots\partial w_{\alpha_1}}=\frac{\partial^r f(\varphi_a(w))}{\partial w_{\alpha_r}\cdots\partial w_{\alpha_1}}+A$$

这里A为含有形如$\Gamma_{\alpha\gamma}^{\sigma}$或$\frac{\partial^s\Gamma_{\alpha\gamma}^{\sigma}}{\partial w_{l_1}\cdots\partial w_{l_s}}$的因子的和式。由（2.4）及（2.2）知

$$\left.\frac{\partial^r f(\varphi_a(w))}{\partial w_{\alpha_r}\cdots\partial w_{\alpha_1}}\right|_{w=0}=\left.\frac{\delta^r f}{\delta w_{\alpha_r}\cdots\delta w_{\alpha_1}}\right|_{w=0}=\left.\frac{\delta^r f}{\delta z_{\beta_r}\cdots\delta z_{\beta_1}}\frac{\partial z_{\beta_r}}{\partial w_{\alpha_r}}\cdots\frac{\partial z_{\beta_1}}{\partial w_{\alpha_1}}\right|_{z=\varphi_a(0)=a}$$

由[2]已知$z=\varphi_a(w)$在$w=0$点的变换矩阵为

$$\left(\frac{\partial z}{\partial w}\right)\bigg|_{w=0}=\sqrt{1-a\bar{a}}R^{-1} \tag{2.5}$$

这里R 由（1.3）所定义，于是$f(\varphi_a(w))$在$w=0$点展开的系数可以表达成各阶协变导数，即$f(\varphi_a(w))$等于

$$\sum_{\alpha\in N^n}\frac{1}{\alpha_1!\cdots\alpha_n!}\frac{\partial^{\alpha}f}{\delta z_{\beta_{1,1}}\cdots\delta z_{\beta_{1,\alpha_1}}\cdots\delta z_{\beta_{n,1}}\cdots\delta z_{\beta_{n,\alpha_n}}}\bigg|_{z=a}\cdot\frac{\partial z_{\beta_{1,1}}}{\partial w_1}\cdots\frac{\partial z_{\beta_{1,\alpha_1}}}{\partial w_1}\cdots\frac{\partial z_{\beta_{n,1}}}{\partial w_n}\cdots\frac{\partial z_{\beta_{n,\alpha_n}}}{\partial w_n}\bigg|_{w=0}\cdot w_1^{\alpha_1}\cdots w_n^{\alpha_n} \tag{2.6}$$

这里$\alpha=(\alpha_1,\cdots\alpha_n)$，$|\alpha|=\alpha_1+\cdots+\alpha_n$

若$w\in C^n$，$\frac{\delta}{\delta z}=\left(\frac{\delta}{\delta z_1},\cdots\frac{\delta}{\delta z_n}\right)$，记w的k 次Kronecker乘积为$w^{[k]}$，$\frac{\delta}{\delta z}$的k 次Kronecker 乘积为$\left(\frac{\delta}{\delta z}\right)^{[k]}$，则（2.6）式又可表为

$$f(\varphi_a(w))=\sum_{k=0}^{\infty}\frac{1}{k!}\left(\left(\frac{\delta}{\delta z}\right)^{[k]}f\right)\bigg|_{z=a}\left[\left(w\frac{\partial z}{\partial w}(0)\right)^{[k]}\right]'$$

这里$\left(\frac{\partial z}{\partial w}(0)\right)=\left(\frac{\partial z}{\partial w}\right)\bigg|_{w=0}$，由（2.5）确定。

于是总结起来得到

定理 1　若$B=\{z\in C^n|z\bar{z}'<1\}$是$C^n$中的单位球，$f(z)$ 在B 中全纯，$z=\varphi_a(w)$为将$z=a$映为$w=0$的Möbius变换，那么$F(w)=f(\varphi_a(w))$ 在$w=0$处的展开为

$$F(w)=f(\varphi_a(w))=\sum_{k=0}^{\infty}\frac{1}{k!}(\sqrt{1-a\bar{a}'})^{k}\left(\left(\frac{\delta}{\delta z}\right)^{[k]}f\right)\bigg|_{z=a}\cdot\left[(wR^{-1})^{[k]}\right]' \tag{2.7}$$

这里R 由（1.3）所定义，$\frac{\delta}{\delta z}=\left(\frac{\delta}{\delta z_1}\cdots,\frac{\delta}{\delta z_n}\right)$是相对于B 的Bergman度量的协变导数。

三　C^n中的有界域的情形

从定理1的证明中，可以看到这很容易地推广到一般的C^n中的有界域D。

定理 2　若$D\subset C^n$为有界域，$K(z,\bar{z})$为其Bergman核函数，$ds^2=h_{\alpha\bar{\beta}}dz_\alpha\overline{dz_\beta}$为相应的Bergman度量，其联络为$\Gamma_{\beta\gamma}^{\alpha}=h^{\bar{\sigma}\alpha}\frac{\partial h_{\beta\bar{\sigma}}}{\partial z_\gamma}$，这里$h_{\beta\bar{\sigma}}\cdot h^{\bar{\sigma}\alpha}=\delta_\beta^\alpha$。若$z=\varphi_{a,b}(w)$为将将D中的点$z=a$变到D中另一点$w=0$的一个全纯自同胚变换，如果$\Gamma_{\beta\gamma}^{\alpha}$及其对w的各阶微微商在$w=b$处均为零，若$f(z)$为D上的一个全纯函数，那末$F(w)=f(\varphi_{a,b}(w))$在$w=b$处的展开式为

$$\sum_{k=0}^{\infty}\frac{1}{k!}\left(\left(\frac{\delta}{\delta z}\right)^{[k]}f\right)\bigg|_{z=a}\left\{\left[(w-b)\frac{\partial z}{\partial w}(b)\right]^{[k]}\right\}' \tag{3.1}$$

这里$\left(\frac{\partial z}{\partial w}(b)\right)$是变换$z=\varphi_{a,b}(w)$在点$w=b$的变换矩阵，$\frac{\delta}{\delta z}$为相对于区域D的Bergman 度量的协变导数。

对于四类典型域，如果取$b=0$，则上述条件全满足，可验证如下：

在R_I：$I-Z\overline{Z}'>0$，$Z=Z^{(m,n)}$的情形，这时候Mobius变换为

$$W=Q(Z-Z_0)(I-\overline{Z}_0'Z)^{-1}R^{-1} \tag{3.2}$$

将$Z=Z_0\in R_I$映为$w=0$，这里Q，R为满足

$$\overline{Q}(I^{(m)}-\overline{Z}_0Z_0')Q'=I,\quad R(I^{(n)}-Z_0'\overline{Z}_0)R'=I$$

的方阵，记

$$z=(z_{11},\cdots z_{1n},z_{21},\cdots z_{2n},\cdots z_{m1},\cdots z_{mn}), \tag{3.3}$$
$$w=(w_{11},\cdots w_{1n},w_{21},\cdots w_{2n},\cdots w_{m1},\cdots w_{mn})。$$

R_I的Bergman核为$(V(R_I))^{-1}(\det(I-Z\overline{Z}'))^{-m-n}$，这里$V(R_I)$为$R_I$的体积，Bergman度量为

$$ds_I^2=(m+n)dz(I-\overline{Z}Z')^{-1}\cdot\times(I-\overline{Z}'Z)^{-1}d\overline{z}' \tag{3.4}$$

这里·×为Kronecker乘积。(3.2)的变换矩阵在$w=0$时为

$$\left(\frac{\partial z}{\partial w}\right)_{w=0}=(Q'\times\overline{R}')^{-1} \tag{3.5}$$

由于Bergman度量在Mobius变换下不变，故在此标w下，其Bergman度量仍为

$$ds_I^2=(m+n)dw((I-\overline{W}W')^{-1}\cdot\times(I-\overline{W}'W)^{-1}d\overline{w}'$$

记$(I-\overline{W}W')^{-1}=(a_{jk}),(I-\overline{W}'W)^{-1}=(b_{\alpha\beta})$，于是$(I-\overline{W}W')^{-1}\cdot\times(I-\overline{W}'W)^{-1}$中第$(j\alpha)$行第$(k\beta)$列的元素为

$$T^I_{(j\alpha)(\overline{k\beta})}=a_{jk}b_{\alpha\beta},$$

因之

$$\frac{\partial}{\partial w_{l\gamma}}T^I_{(j\alpha)(\overline{k\beta})}=\frac{\partial a_{jk}}{\partial w_{l\gamma}}b_{\alpha\beta}+a_{jk}\frac{\partial b_{\alpha\beta}}{\partial w_{l\gamma}}$$

然而

$$\frac{\partial a_{jk}}{\partial w_{l\gamma}}=\frac{\det(I-\overline{W}W')\dfrac{\partial\tilde{a}_{jk}}{\partial w_{l\gamma}}-\tilde{a}_{jk}\dfrac{\partial}{\partial w_{l\gamma}}\det(I-\overline{W}W')}{\det(I-\overline{W}W')^2}$$

这里$(\tilde{a}_{jk})$为$(I-\overline{W}W')$的伴随矩阵，显然$\det(I-\overline{W}W')$与$\tilde{a}_{jk}$都可以表为和式，各自和式中每项是形如$(1-\overline{w}_{ij}w_{kj})$或$(1-\overline{w}_{ij}w_{kj})$的乘积。故$\dfrac{\partial\tilde{a}_{jk}}{\partial w_{l\gamma}}$与$\dfrac{\partial}{\partial w_{l\gamma}}\det(I-\overline{W}W')$的各项都有形如$\overline{w}_{ij}$的因子。同样，$\dfrac{\partial b_{\alpha\beta}}{\partial w_{l\gamma}}$也有相同的结论。而$R_I$的联络为

$$\Gamma^{(j\alpha)}_{(k\beta)(l\gamma)}(w)=\sum_{h=1}^{m}\cdot\sum_{\lambda=1}^{n}T_I^{(\overline{h\lambda})(j\alpha)}\frac{\partial T^I_{(k\beta)(\overline{h\lambda})}}{\partial w_{l\gamma}}$$

故

$$\Gamma^{(j\alpha)}_{(k\beta)(l\gamma)}(w)\big|_{w=0}=0$$

同样$\Gamma^{(j\alpha)}_{(k\beta)(l\gamma)}(w)$对w的各阶导数在$w=0$的值也是零。因之，归纳起来可有

定理3 若$f(z)$在R_I全纯，经Mőbius变换(3.2)后，在$w=0$处展开，则有如下的展开式

$$\sum_{k=0}^{\infty}\frac{1}{k!}\left(\left(\frac{\delta}{\delta z}\right)^{(k)}f\right)\Bigg|_{z=z_0}[W\cdot(Q'\cdot\times R')^{-1}]^{[k]}$$

这里w由(3.3)所定义，$\dfrac{\delta}{\delta z}$为相对于R_I的Bergman度量的协变导数。

在R_I：$I-Z\overline{Z}>0, Z=Z', Z=Z^{(p)}$的情形，这时候Möbius 变换为

$$W=Q(Z-Z_0)(I-\overline{Z}_0Z)^{-1}\overline{Q}^{-1} \tag{3.6}$$

将$Z=Z_0\in R_I$映为$W=0$，这里Q 满足

$$\overline{Q}'Q=(I-Z_0\overline{Z}_0)^{-1}$$

记Z 为

$$\begin{pmatrix} z_{11}, & \frac{1}{\sqrt{2}}z_{12}, & \cdots\frac{1}{\sqrt{2}}z_{1p} \\ \frac{1}{\sqrt{2}}z_{21}, & z_{22}, & \cdots\frac{1}{\sqrt{2}}z_{2p} \\ \cdots & \cdots & \cdots \\ \frac{1}{\sqrt{2}}z_{p1}, & \frac{1}{\sqrt{2}}z_{p2}, \cdots & z_{pp} \end{pmatrix}$$

把$\frac{p(p+1)}{2}$个复变数$z_{\alpha\beta}=z_{\beta\alpha}(\alpha,\beta=1,\cdots,p)$排列成确定的单一次序，记为

$$z=(z_{11},\cdots z_{1p},\ z_{22}\cdots z_{2p},\ \cdots z_{pp})$$

同样记

$$w=(w_{11},\cdots w_{1p},\ w_{22}\cdots w_{2p},\ \cdots w_{pp}) \tag{3.7}$$

这时候R_I的Bergman　量为

$$ds^2_I=(p+1)dz[(I-\overline{Z}Z)^{-1}\cdot\times(I-\overline{Z}Z)^{-1}]_s\overline{dz}'$$

这里$[B\cdot\times B]_s$表示方阵B 的对称直乘积。（3.6）的变换矩阵在$w=0$时为

$$\left(\frac{\partial z}{\partial w}\right)_{w=0}=[Q'\cdot\times Q']_s^{-1}$$

在坐标w下，Bergman 度量仍为

$$ds^2_I=(p+1)dw[(I-\overline{W}W)^{-1}\cdot\times(I-\overline{W}W)^{-1}]_s\overline{dw}'$$

与前同理可证R_I的联络$\Gamma^{(\alpha\beta)}_{(\gamma\delta),(\epsilon\zeta)}(w)$以及对w的各阶导数在$w=0$处的值均为零。于是我们有

定理4　若$f(Z)$在R_I中全纯，经Möbius 变换（3.6）后，在$w=0$处展开，则有以下的展开式

$$\sum_{k=0}^{\infty}\frac{1}{k!}\left(\left(\frac{\delta}{\delta z}\right)^{(k)}f\right)\Bigg|_{z=z_0}((w\cdot[Q'\cdot\times Q']_s^{-1})^{(k)})'$$

w由（3.7）所定义，$\frac{\delta}{\delta z}$为相对于R_I的Bergman 度量的协变导数。

在R_{II}：$I+Z\overline{Z}>0$，$Z'=-Z$，$Z=Z^{(q)}$的情形，这时候Möbius变换为

$$W=Q(Z-Z_0)(I+\overline{Z}_0Z)^{-1}\overline{Q}^{-1} \tag{3.8}$$

将$Z=Z_0\in R_{II}$映为$w=0$，这里Q 满足$\overline{Q}'Q=(I+Z_0\overline{Z}_0)^{-1}$

把$\frac{q(q-1)}{2}$个复变数$z_{\alpha\beta}=-z_{\beta\alpha}$ $(\alpha,\beta=1,2,\cdots q)$排列成单一的次序，记为

$$z=(z_{12},\cdots z_{1q},\ z_{23},\cdots z_{2q},\ \cdots z_{q-1,q})$$

$$w=(w_{12},\cdots w_{1q},\ w_{23},\cdots w_{2q},\ \cdots w_{q-1,q}) \tag{3.9}$$

R_{II}的Bergman 度量为。

$$ds^2_{I}=2(q-1)dz[(I+\overline{Z}Z)^{-1}\cdot\times(I+\overline{Z}Z)^{-1}]_{A}dz'$$

这里$[A\cdot\times\cdot A]_{A}$表示方阵A的斜对称直乘积。(3.8)的变换矩阵在$w=0$时为

$$\left(\frac{\partial z}{\partial w}\right)\Big|_{w=0}=[Q'\cdot\times Q']_{A}^{-1}$$

在坐标w下，Bergman度量仍为

$$ds^2_{I}=2(q-1)dw[(I+\overline{W}W)^{-1}\cdot\times(I+\overline{W}W)^{-1}]_{A}\overline{dw}'$$

同样可以证明R_{I}的联络$\Gamma^{(\alpha\beta)}_{(\gamma\delta)(\varepsilon\eta)}(w)$以及对w的各阶导数在$w=0$均为零，于是我们有

定理5 若$f(z)$在R_{I}中全纯，经Möbius变换(3.8)后，在$w=0$处展开有如下的展开式

$$\sum_{k=0}^{\infty}\frac{1}{k!}\left(\left(\frac{\delta}{\delta z}\right)^{[k]}f\right)\Big|_{z=z_0}((W\cdot[Q'\cdot\times Q']_{A}^{-1})^{[k]})'$$

这里w由(3.9)定义，$\frac{\delta}{\delta z}$为相对于R_{I}的Bergman度量而言的协变导数

在R_{N}：$1+|zz'|-2z\overline{z}'>0$，$1-|zz'|>0$，$z\in C^n$的情形，这时候Möbius变换为

$$w=\{[(\frac{1+zz'}{2},\frac{1-zz'}{2})-zX_0']Q(\ )\}^{-1}\{z-(\frac{1+zz'}{2},\frac{1-zz'}{2i})X_0\}R \quad (3.10)$$

将$z=z_0$变为$w=0$，其中$z_0\in R_{N}$，

$$X_0=\frac{-1}{1-|z_0z_0'|^2}\begin{pmatrix}(\overline{z_0z_0'}-1)z_0+(z_0z_0'-1)\overline{z}_0\\ i(z_0z_0'+1)\overline{z}_0-i(\overline{z_0z_0'}+1)z_0\end{pmatrix}$$

Q和R都是实方阵，$\det Q>0$，且满足

$$QQ'=(I-X_0X_0')^{-1},\quad RR'=(I-X_0'X_0)^{-1}$$

R_{N}的Bergman度量为

$$ds^2_{N}=\frac{2N}{(1+|zz'|^2-2z\overline{z}')^2}[(1+|zz'|^2-2z\overline{z}')dz\overline{dz}'+4z\overline{z}'|zdz'|^2-2(\overline{zz}'zdz'\overline{zdz'}+zz'\overline{zdz'}\overline{z}dz')-2(|zdz'|^2-|z\overline{dz}'|^2)]$$

所以度量方阵为

$$T^N(z,\overline{z})=\frac{2N}{(1+|zz'|^2-2z\overline{z}')^2}[(1+|zz'|^2-2z\overline{z}')I^{(n)}+4z\overline{z}'z'\overline{z}-2(z\overline{z}'z'z'\overline{z}'z)-2(z'\overline{z}-\overline{z}'z)]$$

在坐标w下，度量方阵仍为

$$T^N(w,\overline{w})=\frac{2N}{(1+|ww'|^2-2w\overline{w}')^2}[(1+|ww'|^2-2w\overline{w}')I^{(n)}+4w\overline{w}'w'\overline{w}-2(\overline{ww}'w'w+ww'\overline{w}'\overline{w})-2(w'\overline{w}-\overline{w}'w)]$$

由联络$\Gamma^{\alpha}_{\beta\gamma}(w)$的定义，与前同理容易证明$\Gamma^{\alpha}_{\beta\gamma}(w)$以及对w的各阶导数在$w=0$点的值均为零。而(3.10)在$w=0$处的变换矩阵为

$$\left(\frac{\partial z}{\partial w}\right)\Big|_{w=0}=i(1+|z_0z_0'|^2-2z_0\overline{z}_0')R^{-1}\{I-2\frac{\overline{z}_0'z_0-z_0\overline{z}_0'z_0'z_0}{1-|z_0z_0'|^2}\}^{-1} \quad (3.11)$$

于是我们有

定理6 若$f(z)$在R_{N}中全纯，经Möbius变换(3.10)后，在$w=0$点展开，有如

如下展开式

$$\sum_{k=0}^{\infty}\frac{1}{k!}\left(\left(\frac{\delta}{\delta z}\right)^{(k)}f\right)\Bigg|_{z=z_0}\left(\left(i(1+|z_0z_0'|^2-2z_0z_0')WR^{-1}\{I-2\frac{z_0'z-z_0z_0'z_0'z_0}{1-|z_0z_0'|^2}\}^{-1}\right)^{(k)}\right)'$$

$\frac{\delta}{\delta z}$ 为相对于R_{I}的Bergman 度量的协变导数。

四　实超球的情形

在多实变数的情形，就与多复变数的情形不一样，这时相应联络的导数在一指定点的值往往不会全成为零。例如，在实超球的情形，不论取实射影变换群或非欧运动群，其相应的联络的导数在原点的值不是全为零，所以尽管在文献［2］中指出了：$\mathbf{R}^n$中实超球上解析函数在实射影变换和非欧运动变换后，其展开式的第二项系数可表示为二阶协变导数，但在以后的各项系数却不可能用相应的高阶协变导数来表达之。例如其展开式的第三项系数就较复杂。如讨论实射影变换群的情形，变换群是由把$y=a$ 变为$x=0$的解析自同构变换

$$x=\frac{\sqrt{1-aa'}(a-y)(I+\lambda a'a)}{1-ay'},\quad \lambda=\frac{1-\sqrt{1-aa'}}{aa'\sqrt{1-aa'}} \tag{4.1}$$

$$a=(a_1,\ \cdots a_n)\in B$$

以及使原点不变的变换

$$x=y\Gamma,\qquad \Gamma\Gamma'=I$$

所组成，这里B表示 $\mathbf{R}^n$中实超球$xx'<1,x=(x_1,\ \cdots x_n)\in \mathbf{R}^n$其不变微分度量为

$$ds^2=\frac{dy(I-y'y)^{-1}dy'}{1-yy'}=\frac{dx(I-x'x)^{-1}dx'}{1-xx'} \tag{4.2}$$

（4.2）的联络为

$$\Gamma_{ij}^k=\frac{\delta_{ik}x_j+\delta_{jk}x_i}{1-xx'}$$

经过复杂的计算，得到

若$f(x)$ 在 $\mathbf{R}^n$中的单位球$xx'<1$中解析，x经实射影群（4.1）变换后，$f(x)=F(y)$，于是$F(y)$ 可以在$y=0$ 处展开成

$$\begin{aligned}F(y)=f(x)&=f(a)+\sqrt{1-aa'}\left(\frac{\partial f}{\partial a}\right)R^{-1}y'+\frac{1}{2}(1-aa')y'R^{-1}(\nabla^2 f(a))R^{-1}y'\\&+\frac{1}{6}\sum_{j,k,l=1}^{n}\left[\frac{\delta^3 f(x)}{\delta x_p\delta x_q\delta x_r}\cdot\frac{\partial x_p}{\partial y_j}\cdot\frac{\partial x_q}{\partial y_k}\cdot\frac{\partial x_r}{\partial y_l}\Bigg|_{y=0}+\delta_{kl}\frac{\partial f(x)}{\partial x_p}\cdot\frac{\partial x_p}{\partial y_j}\Bigg|_{y=0}\right.\\&\left.+\delta_{jl}\frac{\partial f(x)}{\partial x_q}\frac{\partial x_q}{\partial y_k}\Bigg|_{y=0}\right]y_jy_ky_l+\cdots\end{aligned} \tag{4.5}$$

这里

$$\left(\frac{\partial x_i}{\partial y_j}\right)\Bigg|_{y=0}=R=-\frac{a'a(1-\sqrt{1-aa'})}{aa'\sqrt{1-aa'}}-1,\quad \nabla \text{及}\frac{\delta}{\delta x}$$

都是相对于度量（4.2）而言的协变导数

同样对于非欧运动群的情形

变换群是由把$y=a$ 变为$x=0$的变换

$$x=\frac{(y-a)(y-a)a'-(1-aa')(-a)}{1-2ay'+aa'yy'},\quad a=(a_1,\cdots,a_n)\in B \tag{4.6}$$

原点不变的变换

$$x=y\Gamma\ ,\quad \Gamma\Gamma'=I$$

所组成，其微分不变度量为

$$ds^2=\frac{dydy'}{(1-yy')^2}=\frac{dxdx'}{(1-xx')^2} \tag{4.7}$$

联络为

$$\Gamma^k_{ij}=2\frac{(\delta_{ki}x_j+\delta_{kj}x_i-\delta_{ij}x_k)}{1-xx'}$$

经过复杂计算得到：

若$f(x)$在R^n中单位球中解析，x经非欧运动群（4.6）变换后，$f(x)=F(y)$,$F(y)$可以在$y=0$处展开成

$$F(y)=f(x)=f(a)-(1-aa')(\frac{\partial f}{\partial a})y'+\frac{1}{2}(1-aa')^2y(\nabla^2f(a))y'+$$

$$\sum_{j,k,l=1}^{n}\frac{1}{6}[\frac{\delta^3f(a)}{\delta a_j\delta a_k\delta a_l}+2(\delta_{kl}\frac{\partial f(a)}{\partial a_j}+\delta_{jl}\frac{\partial f(a)}{\partial a_k}-\delta_{jk}\frac{\partial f(a)}{\partial a_l})]y_jy_ky_l+\cdots$$

这里∇及$\frac{\delta}{\delta a}$都是相对于度量（4.7）而言的协变导数。

参 考 文 献

[1] 龚 昇 关于Mobius变换的一点注记（一）,纯粹数学与应用数学，试刊号（1985）1—15

[2] 龚 昇 关于Mobius变换的一点注记（二）中国科技大学学报 14（1984）1—14

A REMARK ON MÖBIUS TRANSFORMATIONS (Ⅱ)

Gong Sheng Yan Zhimin

(University of Science and Technology of China)

In this note, We give the explicit expression of the expansion at the origin of a holomorphic fanction in the unit ball in C^n after Mobias transformations

This result can extend to the bounded domain in C^n satisfies certain conditions. For example, the classical domains satisfy these conditions. The explicit expression of the expansion at the origin of a holomorphic function in these classical domains after Mobius transformations are given also.

Vol. 32 No. 11　　KEXUE TONGBAO　　June 1987

COEFFICIENT INEQUALITIES AND GEOMETRICAL INEQUALITIES

GONG SHENG (SHENG KUNG 龚 昇)
(*University of Science and Technology of China, Hefei*)

Received January 6, 1986.

To give the geometrical meanings of some coefficient inequalities in classical function theory is the purpose of this note. It also offers some possibilities to extend these inequalities into a new point of view. We only consider the class S, for it can be done without any difficulties for the other function classes. The S is the class of functions $f(z) = z + \sum_{n=2}^{\infty} a_n z^n$ which is analytic and univalent in the unit disc $|z| < 1$. It is a great event in the recent years that Bieberbach Conjecture has been solved. The Bieberbach Conjecture states as follows: If $f(z) \in S$, then $|a_n| \leqslant n$ holds for all $n = 1, 2, \cdots$, equality holds iff $f(z)$ is the Koebe function $\frac{z}{(1-z)^2}$ and its rotations. This conjecture was proposed in 1916, and was solved 68 years later by de Branges[1] in 1984. What consequence can we get from de Branges Theorem? First of all, we get the precise estimate of the modulus of the derivatives of $f(z)$

$$|f^{(n)}(z)| \leqslant n! \frac{n+|z|}{(1-|z|)^{n+2}}, \quad n = 1, 2, \cdots \tag{1}$$

by Landau Theorem[2]. Of course, (1) is equivalent to the Bieberbach Conjecture. In [3], the author points out: If $f(z) = c_0 + \sum_{n=1}^{\infty} c_n z^n$ is analytic in $|z| < 1$ after the Möbius transformation of the independent variable, we can expand

$$f\left(\frac{z+\zeta}{1+\bar{z}\zeta}\right) = f(z) + \sum_{n=1}^{\infty} \frac{1}{n!} \frac{\delta^n f(z)}{\delta s^n} \zeta^n, \tag{2}$$

where $|\zeta| < 1$, $\frac{\delta}{\delta s}$ is the intrinsic derivative with respect to Poincaré Metric of the unit disc. As a consequence of (2), we know that Bieberbach Conjecture is equivalent to the following

Theorem 1. *If $f(z) \in S$, then*

$$\left|\frac{\delta^n f}{\delta s^n}\right| \leqslant n!\, n \left|\frac{\delta f}{\delta s}\right| \tag{3}$$

holds for all positive integers n.

Intrinsic derivatives are invariants. Thus (3) is the estimation of invariants. In this point of view, (3) is more interesting than (1).

From (3), we have

Theorem 2. *If* $f\in S$, $0\leqslant\alpha\leqslant\beta\leqslant 2\pi$, $0\leqslant r<1$, *then*

$$\left|\frac{\delta^n f(re^{i\alpha})}{\delta s^n}-\frac{\delta^n f(re^{i\beta})}{\delta s^n}\right|\leqslant(n+1)!(n+1)L(f;r;\alpha,\beta)\tag{4}$$

holds for all positive integers n, where $L(f;r;\alpha,\beta)$ *is the length of* $arc(re^{i\alpha},re^{i\beta})$ *after the mapping* $w=f(z)$.

Proof. According to the definition of covariant derivative, we have

$$\nabla^n f=\frac{(n-1)!}{A^{2n-2}}(A^{n-1}g_{n-1})',$$

where ∇ is the covariant derivative with respect to Poincaré Metric, $A=1-z\bar{z}$, ("$'$" is the derivative of z,)

$$g_n(z)=\sum_{j=0}^{n-1}\frac{(-1)^j\binom{n-1}{j}}{(n-j)!}\bar{z}^j(1-z\bar{z})^{n-j}f^{(n-j)}(z).$$

From (3), we obtain

$$|(A^{n-1}g_{n-1})'|\leqslant n^2A^{n-1}|f'(z)|.\tag{5}$$

Integrating (5) with respect to θ from α to β, we obtain

$$\int_\alpha^\beta|(A^{n-1}g_{n-1}(re^{i\theta}))'|d\theta\leqslant n^2\int_\alpha^\beta(1-r^2)^{n-1}|f'(re^{i\theta})|d\theta.$$

Thus we have

$$|g_{n-1}(re^{i\beta})-g_{n-1}(re^{i\alpha})|\leqslant r\cdot n^2\int_\alpha^\beta|f'(re^{i\theta})|d\theta.$$

This is (4).

Corollary 1. *If* $f\in S$, *then*

$$\left|\frac{\delta^n f(z)}{\delta s^n}-\frac{\delta^n f(-z)}{\delta s^n}\right|\leqslant\frac{(n+1)!(n+1)}{2}L(f,r)$$

holds for all positive integers n, where $L(r,f)$ *is the length of* $|z|=r$ *after mapping* $w=f(z)$.

Integrating (5) with respect to r from r_1 to r_2, we obtain

Theorem 3. *If* $f\in S$, $-1<r_1\leqslant r_2<1$, *then*

$$\left|(1-r_2^2)^n\frac{\delta^n f(r_2e^{i\theta})}{\delta s^n}-(1-r_1^2)^n\frac{\delta^n f(r_1e^{i\theta})}{\delta s^n}\right|$$

$$\leqslant(n+1)!(n+1)\int_{r_1}^{r_2}(1-t)^{n-3}(1+t)^{n+1}dt\tag{6}$$

holds for all positive integers n.

In particular, taking $r_1 = 0$, $r_2 = r$, we have

Corollary 2. *If* $f(z) = z + \sum_{n=2}^{\infty} a_n z^n \in S$, *then*

$$\left|(1-r^2)^n \frac{\delta^n f}{\delta s^n} - a_n\right| \leqslant (n+1)!(n+1)\int_0^r (1-t)^{n-3}(1+t)^{n+1}dt \tag{7}$$

holds for all positive integers n.

Moreover, if taking $r_2 = r$, $r_1 = -r$, we have

Corollary 3. *If* $f(z) \in S$, *then*

$$\left|\frac{\delta^n f(z)}{\delta s^n} - \frac{\delta^n f(-z)}{\delta s^n}\right| \leqslant \frac{(n+1)!(n+1)}{(1-r^2)^n}\int_0^r (1-t^2)^{n-3}(1+6t^2+t^4)dt$$

holds for all positive integers n.

If we regard $(f(z), 1)$ as the homogeneous coordinate of $\boldsymbol{CP}^1$, in [4] Flanders points out that Schwarz derivative of $f(z)$ is twice of the curvature K of the mapping of $|z| < 1$ by $w = f(z)$ at the point $(f(z), 1)$. If $f(z) = z + \sum_{n=2}^{\infty} a_n z^n \in S$, then $|a_3| \leqslant 3$ is equivalent to the following

Theorem 4. *If* $f(z) \in S$, *then*

$$\left|K + \frac{3}{4}\left(\frac{\nabla^2 f}{\nabla f}\right)^2\right| \leqslant 9\left|\frac{ds}{dz}\right|^2. \tag{8}$$

Proof. By Theorem 1, we get

$$\left|\frac{\delta^3 f}{\delta s^3}\right| \leqslant 18\left|\frac{\delta f}{\delta s}\right|.$$

We can prove that this inequality can be rewritten as

$$\left|\{f, z\} + \frac{3}{2}\left(\frac{\nabla^2 f}{\nabla f}\right)^2\right| \leqslant 18\left|\frac{ds}{dz}\right|^2,$$

where $\{f, z\}$ is Schwarz Derivative of f. We get (8) since $\{f, z\} = 2K$.

Let $f(z) \in S$, $\zeta = \frac{1}{z}$, then $F(\zeta) = \frac{1}{f(1/\zeta)} = \zeta + \alpha_0 + \sum_{n=1}^{\infty} \frac{\alpha_n}{\zeta^n}$ is analytic and univalent in $|\zeta| > 1$. It is well known that $|\alpha_1| \leqslant 1$ and $|\alpha_2| \leqslant \frac{2}{3}$. We can easily prove that $|\alpha_1| \leqslant 1$ is equivalent to

$$|K| \leqslant 3\left|\frac{ds}{dz}\right|^2. \tag{9}$$

$|a_3| \leqslant 3$, $|\alpha_1| \leqslant 1$ are classical coefficient inequalities, (8) and (9) are inequalities of differential geometry for the estimations of the curvature K. It is interesting that these two pairs of inequalities are equivalent to each other. The more general theorem containing the above theorems is the following Jenkins Theorem[5].

If $f(z)=z+\sum_{n=2}^{\infty} a_n z^n \in S$, then

$$|a_3-\mu a_2^2| \leqslant \begin{cases} 4\mu-3, & \text{when } \mu>1; \\ 1, & \text{when } \mu=1; \\ 1+2\exp\left(\dfrac{-2\mu}{1-\mu}\right), & \text{when } 0<\mu<1; \\ 3-4\mu, & \text{when } \mu \leqslant 0. \end{cases}$$

Using the same method, we can prove

Theorem 5. *If $f(z) \in S$, then*

$$\left|K+\alpha\left(\frac{\nabla^2 f}{\nabla f}\right)^2\right| \leqslant c_\alpha \left|\frac{ds}{dz}\right|^2, \tag{10}$$

where

$$C_\alpha = \begin{cases} 3-16\alpha & \text{when } \alpha<0, \\ 3 & \text{when } \alpha=0, \\ 3+6\exp\left(\dfrac{4\alpha-3}{2\alpha}\right) & \text{when } 0<\alpha<\dfrac{3}{4}, \\ 16\alpha-3 & \text{when } 3/4 \leqslant \alpha. \end{cases}$$

Moreover, the inequality $|a_2| \leqslant \frac{2}{3}$ is equivalent to the following

Theorem 6. *If $f \in S$, then*

$$\left|\frac{dk}{ds}-2\Gamma k\right| \leqslant 8\left|\frac{ds}{dz}\right|^3, \tag{11}$$

where Γ is the connection of Poincaré Metric.

Proof. Since $a_2=-a_4+2a_2a_3-a_2^3$, from (2) we have

$$|A_4(z)-2A_2(z)A_3(z)+A_2^3(z)| \leqslant \frac{2}{3},$$

where $A_n(z)=\frac{1}{n!}\frac{\delta^n f(z)}{\delta s^n}\Big/\frac{\delta f}{\delta z}$. Using the expressions A_2, A_3, A_4 and substituting them into the above inequality, we get (11).

We may rewrite (11) as

$$\left|\frac{d}{dz}[(1-z\bar{z})^4\{f,z\}]\right| \leqslant 16(1-z\bar{z}). \tag{12}$$

Integrating (12) with respect to θ from α to β, where $z=re^{i\theta}$, we have the following

Corollary 4. *If $f\in S$, $z=re^{i\theta}$, $\beta>\alpha$, then*

$$|K(re^{i\beta})-K(re^{i\alpha})|\leqslant\frac{8(\beta-\alpha)r}{(1-r^2)^5}.$$

In particular,

$$|K(z)-K(-z)|\leqslant\frac{8\pi|z|}{(1-|z|^2)^3}.$$

Integrating (12) with respect to r from r_1 to r_2, then we have the following

Corollary 5. *If $f\in S$, $-1<r_1\leqslant r_2<1$, then*

$$|(1-r_2^2)^4K(r_2e^{i\theta})-(1-r_1^2)^4K(r_1e^{i\theta})|\leqslant 8(r_2-r_1)\left(1-\frac{r_2^2+r_1r_2+r_1^2}{3}\right).$$

Moreover, taking $r_2=r$, $r_1=0$, then we have the following

Corollary 6. *If $f(z)\in S$, $\zeta=\frac{1}{z}$, $F(\zeta)=\frac{1}{f\left(\frac{1}{\zeta}\right)}=\zeta+\alpha_0+\sum_{n=1}^{\infty}\frac{\alpha_n}{\zeta^n}$, $|z|=r$,*

then

$$|(1-r^2)^4K(z)-3\alpha_1|\leqslant 8r\left(1-\frac{r^2}{3}\right).$$

Taking $r_1=-r$, $r_2=r$ in Corollary 5, we have

Corollary 7. *If $f(z)\in S$, $|z|=r$, then*

$$|K(z)-K(-z)|\leqslant 16r\left(1-\frac{r^2}{3}\right)(1-r^2)^{-4}.$$

References

[1] de Branges, *Acta Mathematica*, **154**(1985), 137—152.
[2] Landau, E., *Über Deutsch Math-Verein*, **34**(1925/26), 239—242.
[3] 龚　昇，纯粹数学与应用数学，试刊号，1985，1—19.
[4] Flanders, H., *J. of Differential Geometry*, **4**(1970), 515—519.
[5] Jenkins, J. A., *Analytic Functions*, 1960, 159—194.

Some open problems in geometric function theory in several complex variables

Sheng Gong

(received September 21, 2004)

The geometric function theory in several complex variables developed rapidly during last few decades. But we have a lots of open problems in this field which need to study and to solve. In this note, we give a few of them. These open problems seem simplelemantary and are interesting for us.

1. Let $\Omega \subset \mathbb{C}^n$ be a domain in $\mathbb{C}^n$ with $0 \in \Omega$. If $f : \Omega \to \mathbb{C}^n$ is a normalized (i.e. $f(0) = 0$, $J_f(0) = I$, where J_f is the Jacobian of f) holomorphic mapping on Ω. The Taylor epansion of f at $z = 0$ can be written as

$$f(z) = z + \sum_{j=1}^{\infty} \frac{1}{j!} D^j(0)(z, \cdots, z)$$

where $D^j f$ is the jth Fréchet derivative of f in $\mathbb{C}^n$, and $Df(z) = J_f(z)$.

All normalized biholomorphic convex mappings on Ω form a family, we denote it by $K(\Omega)$.

Theorem 1 (B-F-G, L) If $f \in K(B^n)$, where B^n is the unit ball in $\mathbb{C}^n$, then

$$\frac{(1-|z|)^{C(K)-\frac{n+1}{2}}}{(1+|z|)^{C(K)+\frac{n+1}{2}}} \leq |\det J_f(z)| \leq \frac{(1+|z|)^{C(K)-\frac{n+1}{2}}}{(1-|z|)^{C(K)+\frac{n+1}{2}}} \tag{1.1}$$

where

$$\begin{aligned} C(K) &= \sup_{f \in K(B^n)} \sup_{\lambda \in \partial B^n} \left| \frac{1}{2} tr(D^2 f(0)(\lambda, \cdot)) \right| \\ &= \sup_{f \in K(B^n)} \sup_{\lambda \in \partial B^n} \left| \frac{1}{2} tr \left(\sum_{k=1}^{n} \frac{\partial^2 f_i(0)}{\partial z_j \partial z_k} \lambda_k \right)_{i,j} \right| \end{aligned} \tag{1.2}$$

is called the trace order of the family, where $f = (f_1, \cdots, f_n)$, $\lambda = (\lambda_1, \cdots, \lambda_n)$. Moreover, $\frac{n+1}{2} \le C(K) \le 1 + \frac{\sqrt{2}(n-1)}{2}$.

This theorem was established in case $n = 2$ by R.W.Barnard, C.H.FitzGerald and S.Gong[1] (B-F-G), and generalized it to the case $n > 2$ by T.S.Liu[2]. B-F-G conjectured $C(K) = \frac{n+1}{2}$, and $\left(\frac{z_1}{1-z_1}, \frac{z_2}{1-z_1}, \cdots, \frac{z_n}{1-z_1}\right)$ is one of the extremal mapping, i.e. the following inequalitces

$$(1+|z|)^{-n-1} \le |\det J_f(z)| \le (1-|z|)^{-n-1}$$

hold.

In 1999, J.A.Pfaltzgraff and T.J.Suffridge[3] gave a countexample of this conjecture. The counterexample is

$$F(z) = \begin{pmatrix} \frac{z_1}{(1-xz_n)^2}\left(1 + z_n \dfrac{\frac{1}{\sqrt{2}} - x}{1 - \frac{x}{\sqrt{2}}}\right) \\ \vdots \\ \frac{z_{n-1}}{(1-xz_n)^2}\left(1 + z_n \dfrac{\frac{1}{\sqrt{2}} - x}{1 - \frac{x}{\sqrt{2}}}\right) \\ \frac{z_n}{1-xz_n} \end{pmatrix}$$

where $x = \sqrt{2} - \frac{\sqrt{2(n-1)}}{2\sqrt{n}}$. It is easy to check, for this mapping

$$\sup_{\lambda \in \partial B^n} \left|\frac{1}{2} tr(D^2F(0)(\lambda, 0))\right| = \frac{3n-1}{\sqrt{2}} - \sqrt{2n(n-1)} > \frac{n+1}{2}.$$

Thus, we have

$$\begin{aligned} \frac{n}{\sqrt{2}} + 1 - \frac{1}{\sqrt{2}} = 1 + \frac{\sqrt{2}(n-1)}{2} &\ge C(K) \ge \frac{3n-1}{\sqrt{2}} - \sqrt{2n(n-1)} \\ &= \frac{n}{\sqrt{2}} + \frac{1}{4\sqrt{2}n} + \frac{1}{8\sqrt{2}n^2} + \cdots \end{aligned}$$

Notice that the principal term of the upper bound and lower bound of $C(K)$ are same.

Problem 1 To find the exact value of $C(K)$.

2. In 1993, S.Gong, S.K.Wang and Q.H.Yu[4] (G-W-Y) established the following result.

Theorem 2 (G-W-Y) If $f \in K(B^n)$, then

$$\left(\frac{1-|z|}{1+|z|}\right)^2 G \le \overline{J_f(z)}' J_f(z) \le \left(\frac{1+|z|}{1-|z|}\right)^2 G \tag{2.1}$$

holds, where $z = (z_1, z_2, \cdots, z_n) \in B^n$, and

$$G = (g_{ij})_{1\le i,j\le n} = \left(\frac{1-|z|^2\delta_{ij} + \bar{z}_i z_j}{(1-|z|^2)^2}\right)_{1\le i,j\le n} \tag{2.2}$$

is the Bergman matrix of the Bergman metric of B^n. The estimate is sharp.

This theorem tells us: $\bar{J}_f' J_f$ is dominated by the Bergman metric of B^n.

In 1999, S.Gong and T.S.Liu[5] (G-L) extended it to the bounde convex circular domains in $\mathbb{C}^n$.

Theorem 3 (G-L) Let $\Omega \subset \mathbb{C}^n$ be a bounded convex circular domain. Its Minkowski functional $p(z) \in C^1(\Omega)$ except for a lower dimensional manifold. If $f \in K(\Omega)$, then for any $z \in \Omega$, $\xi \in \mathbb{C}^n$, the inequalities

$$\frac{1-p(z)}{1+p(z)} F[z,\xi] \le p(J_f(z)\xi) \le \frac{1+p(z)}{1-p(z)} F(z,\xi) \tag{2.3}$$

hold, where

$$F(z,\xi) = F_c(z,\xi) = F_k(z,\xi)$$

and $F_c(z,\xi]$ is the infinitesimal Carathéodory metric and $F_k(z,\xi)$ is the infinitesimal Kobayashi-Royden metric of Ω. (By a Theorem of L.Lempert(1981), $F_c = F_k$ when Ω is convex).

This theorem tells us: $p(J_f(z)\xi)$ is dominated by the infinitesimal Carathéodory metric and the infinitesimal Kobayashi-Royden metric of Ω.

When $\Omega = B^n, F_c, F_k$ are same as the Bergman metric, and it is easy to verify that (2.3) becomes (2.1). But for other bounded convex circular domains, (2.2) can not imply the estimates of $\overline{J_f(z)}' J_f(z)$.

Problem 2 If $f \in K(\Omega)$, and Ω is a bounded convex circular domain, can we give a precise estimates of $\overline{J_f(z)}' J_f(z)$ like (2.1) directly.

By the way, in 2002, H.Hamada and G.Kohr[6], T.S.Liu and W.J.Zhang[7] extended Theorem 3 to the unit ball in complex Banach space independently.

3. Let $\Omega \subset \mathbb{C}^n$ be a homogeneous domain with $O \in \Omega$. Let $Aut(\Omega)$ be the group of holomorphic automorphism of Ω. If S is a family of normalized holomorphic mappings on Ω, and for any $f \in S$ and any $\varphi \in Aut(\Omega)$, after normilized the composite mapping $f \circ \varphi$, it belongs S again, then we call S is a linear invariant family (L.I.F). For example, $K(\Omega)$ is L.I.F.

Let $M \subset \mathbb{C}^n$ be a bounded symmetric domain with $O \in M$. It is the Harish-Chandra canonical realization of Hermite symmetric space G/K, where $G = Aut(M)$ and K is the isotropy subgroup of G which fixes the point O. It is a maximal compact subgroup of G.

Let $\mathcal{G}$ be the Lie algebra of G (Of course, G is a semi-simple connected non-compact Lie group with finite center). Let $\mathcal{K}$ be the maximal compact subalgebra of $\mathcal{G}$ corresponding to K. Then $\mathcal{G}$ has the Cartan decomposition

$$\mathcal{G} = \mathcal{K} + \mathcal{P}.$$

Suppose $\mathcal{A}$ is a maximal Abelian subalgebra in $\mathcal{P}$ and A is the corresponding analytic subgroup in G. Then we have the Iwasawa decomposition $G = KAN$. For every $z \in M$, there exist $X \in \mathcal{A}$ and $k \in K$, such that

$$z = \xi(\exp Ad(k)X \cdot O)$$

where $O = eK$ is the identity coset in G/K and ξ is the holomorphic diffeomorphism from G/K onto M. We may choose a basis of $\mathcal{A}$ as $X_1, X_2, \cdots, X_q$ where $q = \dim \mathcal{A} =$ rank M. For every $X \in \mathcal{A}$, there exists an unique expression $X = x_1X_1 + x_2X_2 + \cdots + x_qX_q$.

In 1991, S.Gong and X.A.Zhang[8] (G-Z) extended the Theorem 1 to the following Theorem.

Theorem 4 (G-Z) Let S be a L.I.F. on a bounded symmetric domain M. If $f \in S$, then

$$\sqrt{\frac{K(z,\bar z)}{K(0,0)}}\left(\prod_{p=1}^{q}\frac{1-|\tan hx_p|}{1+|\tan hx_p|}\right)^{c(s)} \le |\det J_f(z)| \le \sqrt{\frac{K(z,\bar z)}{K(0,0)}}\left(\prod_{p=1}^{q}\frac{1+|\tan hx_p|}{1-|\tan hx_p|}\right)^{c(s)} \tag{3.1}$$

where $C(S)$ is defined by (1.2) (change $K(B^n)$ to S in (1.21)), and $K(z,\bar z)$ is the Bergman kernel function of M.

That means, the upper and lower bounds of $|\det J_f(z)|$ are expressed by the Lie algebra of $Aut(M)$ and the Bergman kernel function of M.

In 2000, J.A.Pfaltzgraff and T.J.Suffridge[9] introduced another order of L.I.F. on B^n, it called norm order. It was defined as

$$\|ord\|S = \sup_{f\in S}\left\{\frac{1}{2}\|D^2 f(0)\|\right\} \tag{3.2}$$

we have the following estimates about $\|J_f(z)\|$ and $Tr(\overline{J_f(z)}'J_f(z))$.

Theorem 5 (P-S, G-Y) Let S be a L.I.F. on B^n with $\|ord\|S=\alpha$. If $f\in S$, then

$$\frac{(1-|z|)^{\alpha-1}}{(1+|z|)^{\alpha+1}} \le \|J_f(z)\| \le \frac{(1+|z|)^{\alpha-1}}{(1-|z|)^{\alpha+1}} \tag{3.3}$$

and

$$\frac{(1-|z|)^{2\alpha-2}}{(1+|z|)^{2\alpha+2}}+(n-1)\frac{(1-|z|)^{2\alpha-1}}{(1+|z|)^{2\alpha+1}} \le tr(\overline{J_f(z)}'J_f|z|) \le \frac{(1+|z|)^{2\alpha-2}}{(1-|z|)^{2\alpha+2}}+(n-1)\frac{(1+|z|)^{2\alpha-1}}{(1-|z|)^{2\alpha+1}} \tag{3.4}$$

These estimates are sharp.

(3.3) was proved by J.A.Pfaltzgraff and T.J.Suffridge[9] (P-S) in 2000, and after that S.Gong and Q.H.Yu[10] (G-Y) proved (3.4) in 2000.

Problem 3 To generalized Theorem 5 to bounded symmetric domains, and to give the corresponding estimates by the Lie algebra of the group of holomorphic automorphism, Bergman kernel function of the bounded symmetric domains and the norm order of L.I.F.

4. In the case the L.I.F. is $K(B^n)$, then $\|ord\|K(B^n)=1$. (3.3) becomes

$$(1+|z|)^{-2} \le \|J_f(z)\| \le (1-|z|)^{-2} \tag{4.1}$$

and (3.4) becomes

$$(1+|z|)^{-4}+(n-1)\frac{1-|z|}{(1+|z|)^3} \le tr\left(\overline{J_f(z)}'J_f(z)\right) \le (1-|z|)^{-4}+(n-1)\frac{1+|z|}{(1-|z|)^3} \tag{4.2}$$

(4.1) and (4.2) are the consequences of (2.1). The upper bound of (4.1) is sharp and equality is achieved by $f(z)=(1-z_1)^{-1}z$, where $z=(z_1,\cdots,z_n)\in B^n$.

From the left inequality of (4.2), we have

$$n\|J_f(z)\|^2 \ge tr(\overline{J_f(z)}'J_f(z)) \ge (1+(z)^{-4}+(n-1))\frac{1-|z|}{(1+|z|)^3},$$

Thus:

$$\|J_f(z)\|^2 \geq \frac{1}{n(1+|z|)^4} + \frac{n-1}{n}\frac{1-|z|}{(1+|z|)^3} > \frac{1}{(1+|z|)^4},$$

when $n \geq 2$. That means the lower bound of (4.1) is not sharp.

In 2002, P.Liczberski and V.Starkov[11] made the following conjecture: $\|J_f(z)\| \geq \frac{1}{1+|z|}$.

Problem 4 (Liczberski-Starkov) Does the above conjecture true?

Moreover, we use the arithematic mean-geometric mean inequality, we have

$$\left|tr(\overline{J_f(z)}' J_f(z))\right| \geq n|\det J_f(z)|^{\frac{2}{n}} \geq \frac{n(1-|z|)^{\frac{2C(K)}{n}-1-\frac{1}{n}}}{(1+|z|)^{\frac{2C(K)}{n}+1+\frac{1}{n}}}$$

$$\geq \frac{n(1-|z|)^{\sqrt{2}-1+\frac{1-\sqrt{2}}{n}}}{(1+|z|)^{\sqrt{2}+1+\frac{3-\sqrt{2}}{n}}} > \frac{1}{(1+|z|)^4}$$

when $|z| < c_0$, where c_0 is a root of the equation $1+t-t^2+t^2-n^{\frac{-n}{n-1}}=0$, $n \geq 2$ in $(0,1)$.

Problem 5 To find the preceie estimates of $tr(\overline{J_f(z)}' J_f(z))$ when $f \in K(B^n)$.

Problem 1, 4, and 5 are problems about the estimates of the Jacobian of the convex mappings on the unit ball B^n in $\mathbb{C}^n$, $n \geq 2$. These are elementary and simple problems. As you know, convex mappings in several complex variables have a lots of very strong conclussion.

For example, in 1992, Ngaiming Mok and I-Hsum Tsai[12] (M-T) proved the following important result.

Theorem 6 (M-T) Let X_0 be an irreducible Hermitian symmetric manifold of non-compact type and of rank ≥ 2. Let $\tau: X_0 \to \Omega \subset \mathbb{C}^n$ be the Harish-Chandra embedding. Let D be a bounded convex domain in C^n, and let $f: X_0 \to D$ be a biholomorphism. Them f is the Harish-Chandra embedding up to automorphism of X_0 and affine linear transformation of $\mathbb{C}^n$. More precisely, f is of the form $T \circ \tau \circ \phi$, where T is an affice linear transformation of $\mathbb{C}^n$ and ϕ is an automorphism of X_0.

That means, there are no convex mapping on the irreducible bounded symmetric domain when rank ≥ 2 except the automorphism and affine linear transformation.

This theorem and some other theorems show that convex mapping has very strong conclusion. On the other hand, the unit ball is the only irreducible domain of rank $= 1$, it is the simplest domain in several complex variables. But some elemantary and simple problems about the convex mappings on the unit ball, like Problem 1, 4, and 5 are still open.

5. Even we have many open problems about distortion theorem for convex mappings, however, a lots of interesting and deep distortion theorems for convex mappings were established. Unfortunately, until now, we know nearly nothing about the distortion theorems for the family $S^*(\Omega)$ of all normalized biholomorphic starlike mappings on the domain Ω in $\mathbb{C}^n$, $n \geq 2$, even Ω is the unit ball B^n in C^n.

In 1999, J.A.Pfaltzgraff and T.J.Suffridge[13] proposed the following conjecture. If $f \in S^*(B^n)$, then

$$\frac{1-\|z\|}{(1+\|z\|)^{2n+1}} \leq |\det J_f(z)| \leq \frac{1+\|z\|}{(1-\|z\|)^{2n+1}} \tag{5.1}$$

The eotimates are sharp.

Problem 6 (Pfaltzgraff-Suffridge) Does the conjecture about the distortion theorem of the family $S^*(B^n)$ of all normalized biholomorphic starlike mppings on B^n true?

Let f_j be a normalized starlike function on the unit disk Δ in $\mathbb{C}$. In 1999, J.A.Pfaltzgraff and T.J.Suffridge[13] proved that the mapping

$$F(z) = z\prod_{j=1}^{n}\left(\frac{f_j(z_j)}{z_j}\right)^{\lambda_j}, \qquad z=(z_1,\cdots,z_n)$$

is a normalized starlike mapping on B^n, where $\lambda_j \geq 0$, $j=1,\cdots,n$ and $\sum_j \lambda_j = 1$.

All these mappings form a subfmily $S^*_\pi(B^n)$ of $S^*(B^n)$. They proved that (5.1) is true for $S^*_\pi(B^n)$.

Moreover, we may ask what kind conjectures about $\overline{J_f(z)}'J_f(z)$, $\|J_f(z)\|$ and $tr(\overline{J_f(z)}'J_f(z))$ we can propose when $f \in S^*(B^n)$?

We may also ask that: does the distortion theorem for $S^*(\Omega)$ exist or not when Ω is a bounded starlike domain? If it exists, what is the version of these distortion theorems?

6. In 2002, S.Gong and T.S.Liu[14] defined the fmily of ε-starlike mappings on domain in $\mathbb{C}^n$, in purpose to unify the theory about the family of convex mappings and the family of starlike mappings.

Let X be a complex Banach space. Let $\Omega \subset X$ be a domain in X with $0 \in \Omega$. If $f: \Omega \to Y$ is a biholomorphic mapping on Ω with $0 \in f(\Omega)$, where Y is another complex Banach space. If there exist a number $\varepsilon, 0 \leq \varepsilon \leq 1$, such that $f(\Omega)$ is starlike with respect to any point inside $\varepsilon f(\Omega)$, then we call f is a ε-starlike mapping on Ω. All these mappings on Ω form the family of ε-starlike mappings on Ω. When $\varepsilon = 0$, it is the family of starlike mappings with respect to the origin on Ω; when $\varepsilon = 1$, it is the fmily of convex mappings on Ω. Using this idea, we may unify and make more understanding some theories about the family of convex mappings and the family of starlike mappings. For example, we established the criterion of ε-starlike mappings on the unit ball B in a complex Banach space X. This result unified and made more understanding about the criterion of the family of starlike mappings and the criterion of the family of convex mappings. We also successful treat the Roper-Suffridge operator[15][16] by using ε-starlikeness. It unified a lots of results about this operator and solved an open problem.

However, it is just the beginning to study the theory of ε-starlike mappings. There are a lots of interesting problems which need us to study.

Problem 7. Try to establish the growth, distortion and covering theorems of ε-starlike mapping on the domains in $\mathbb{C}^n$, $n \geq 1$; and on the unit ball B in a complex Banach space.

7. C.H.FitzGerald[17] show that the Bieberbach conjecture does not true for the family of normalized biholomorphic mappings on any domain in $\mathbb{C}^n$, $n \geq 2$. In general, it is difficulty to study any coefficient problems for high dimension: But we still had some result. For example, in 1991, S.Gong and T.S.Liu[18] (G-L) proved the following result

Theorem 7 (G-L). Let $B_p = \{z = (z_1, \cdots, z_n) \in \mathbb{C}^n : \|z\|_p = \left(\sum\limits_{i=1}^{n} |z_i|^p\right)^{\frac{1}{p}} < 1\}$, $p \geq 1$. Let $f(z): B_p \to \mathbb{C}^n$ be a normalized biholomorphic convex mapping on B_p. The expansion of $f(z) = (f_1(z), \cdots, f_n(z))$ can be expressed as

$$f(z) = z + \sum_{k=2}^{\infty} \varphi^{(k)}(z)$$

where $\varphi^{(k)}(z) = (\varphi_1^{(k)}(z), \cdots, \varphi_n^{(k)}(z))$ and $\varphi_j^{(k)}(z)$ for $j = 1, 2, \cdots, n$ are homogeneous polynomials of the elements of $z = (z_1, \cdots, z_n)$ of degree k. Then

$$\|\varphi^{(k)}(z)\|_p \leq \|z\|_p^k$$

holds for all $k = 2, 3, 4, \cdots$, where

$$\|\varphi^{(k)}(z)\|_p = \left(\sum_{j=1}^{n} |\varphi_j^{(k)}(z)|^p\right)^{\frac{1}{p}}.$$

Obviously, this theorem is a generalization of the following classical result in one complex variable.

Let $f(z) = z + \sum a_n z^n$ be a convex univalent function on the unit disk $|z| < 1$, then $|a_n| \leq 1$ holds for all $n = 2, 3, \cdots$.

For the family $S^*(\Delta^n)$ of normalized starlike biholomorphic mappings on the unit polydisk Δ^n in $\mathbb{C}^n$. We proposed the following conjecture[19].

If $f(z) \in S^*(\Delta^n)$, then

$$\|\varphi^{(k)}(z)\| \leq k\|z\|^k$$

holds for $k = 2, 3, \cdots$, where $\| \quad \|$ is the norm of Δ^n.

Problem 8 Does the above Bieberbach conjecture for the family of normalized starlike bihomorphic mappings on Δ^n true?

When $k = 2$, M.Jahangiri, Y.P.Huang and T.S.Liu proved the conjecture is true independently. When $k = 3$, T.S.Liu proved the conjecture is true.

8. As early as in 1946, S.Bochner[20] already point out that the Bloch constant for the family of holomorphic mappings on B^n in $\mathbb{C}^n$, $n \geq 2$, does not exist. The Bloch constant for the high dimension exists only in the case that we need to ask more suitable assumptions on the family of mappings. The most important one is the family of quasiconformal mappings. Roughly speaking, the main requirement of it is that the largest eigen value of the Jacobian matrix of the mapping is dominated by the smallest eigen value of the Jacobian matrix of the mapping. Under this assumption, there are many interesting and important results which were given by S.S.Chern, H.Wu, L.A.Harris, R.T.Greene, K.T.Hahn, D.Minda, S.Krantz etc.

The another important one is to ask the mapping having some geometric property. Some results were obtained under this assumption. I.Graham and D.Varolin[21] proposed the following interesting problem in 1996.

Problem 9 (Graham-Varolin) If $f: B^n \to \mathbb{C}^n$, $n \geq 2$, is a normalized convex mapping, then does $f(B^n)$ contains a ball of radius $\pi/4$.

This is the Szegö Theorem when $n = 1$.

They proved that the conjecture is true when f is odd.

It is well known that the Bloch constant in one complex variable is still open. By Theorem 6, The solution of the Bloch constant for the family of convex mappings on bounded symmetric domains of rank ≥ 2 is trivial. Thus to ask the Bloch constant for the family of convex mappings on the unit ball B^n in $\mathbb{C}^n$, $n \geq 2$ is a very interesting problem and possible to solve it.

Of course, we may ask the exact value of the Bloch constant for the family of starlike mappings on the uni ball B^n in $\mathbb{C}^n$, $n \geq 2$.

References

[1] Barnard, R.W., FitzGerald, C.H. and Gong, S. Distortion therocm for biholomorphic mappinge in $\mathbb{C}^2$, Tran. Amer. Math. Soc. **344**(1994) 907–924.

[2] Liu, T.S., The distortion theorem for biholomorphic mappings in $\mathbb{C}^n$, 1989, preprint.

[3] Pfaltzgraff, J.A. and Suffridge, T.J., Linear invariance, order and convex maps in $\mathbb{C}^n$, Complex variables, **40**(1999) 35–50.

[4] Gong, S., Wang, S.K. and Yu, Q.H., Biholomorphic convex mappings of ball in $\mathbb{C}^n$, Pacif. J. Math., **161**(1993) 287-306.

[5] Gong, S. and Liu, T.S., Disfortion theorems for biholomorphic convex mappings on bounded convex circular domains, Proc. of the 5-th international Colloquium on complex analysis (1997) 73–80; Chin. Ann. Math. Ser. B **20**(1999), 297–304.

[6] Hamada, H. and Kohr, G., Φ-like and convex mappings in infinite dimensional spaces, Rev. Roum. Math. Pures Appl. to appear.

[7] Liu,T.S. and W.J.Zhang, On the distortion theorem for biholomorphic convex mappings on the unit ball in complex Banach space, to appear.

[8] Gong, S. and Zhang, X. A. Distortion theorem for biholomorphic mappings in transitive domains I. Inter. Symb. in memory of L.K. Hua, Vol.II. 111-121, Spring-Verlag, New York, 1991.

[9] Pfaltzgraff, J. A. and Suffridge, T. J., Norm order and geometric properties of holomorphic mappings in $\mathbb{C}^n$, J. Analyse Math., **82**(2000) 285-313.

[10] Gong, S. and Yu, Q.H., The distortion theorems of linear invariant family on the unit ball, Asian J. Math., 4(2000) 795-815.

[11] Liczberski, L. and Starkov, V.V., Distortion theorems for the biholomorphic convex mappings in $\mathbb{C}^n$, J. Math. Anal. Appl. **274**(2002) 495-504.

[12] Mok, N.M. and Tsai, I.H., Rigidity of convex realizations of irreducible bounded symmetric domains of rank ≥ 2, Jour. für die reine and angew. Math., **431**(1992) 91-122.

[13] Pfaltzgraff, J.A. and Suffridge, T.J., An extension theorem and linear invariant families generated by starlike maps, Ann. Univ. Mariae Curie Sklodovoska, Sect. A, **53**(1999), 193-207.

[14] Gong, S. and Liu, T.S., Criterion for the fmily of ε starlike mappings, J. Math. Anal and Appl. **274**(2002) 696-704.

[15] Graham, I. and Kohr, G., Geometric function theory in one and high dimensions, Marcel Dekker, Inc, New York, Basel, 2003.

[16] Roper, K. and Suffridge, T. J., Convex mappings on the unit ball of $\mathbb{C}^n$, J. Anal. Math., **65**(1995) 333-347.

[17] FitzGerald, C. H., Geometric function theory in one and several complex variables: parallels and problems, Compex analysis and its applications, Pitman Research Notes in Math. Series 305, 14-25, Longman Suentifi and Techincal, 1994.

[18] Gong, S. and Liu, T. S., The growth theorem of biholomorphic convex mappings on B_p, Chin. Quarterly Jour. of Math. 6(1991) 78-82 (in Chinese)

[19] Gong, S., The Bieberbach conjecture, AMS, International Press, 1999.

[20] Bochner, S., Bloch theorem for real variables, Bull. Amer. Math. Soc., **52**(1946) 715–718.

[21] Graham, I. and Varolin, D., Bloch constants in one and several variables, Pacif. J. Math. **174** (1996) 347–357.

Sheng Gong
Department of Mathematics
University of Science and Technology of China
Hefei, China

VIII. 附　　录

附录 1　著作目录 Publications(共 30 本)

I. 专著　Monographs

[M1] 龚昇. 多复变数的凸映照与星形映照. 2版. 纯粹数学与应用数学专著, 第 39 号. 北京：科学出版社, 2003 (M7 的第 2 版).

[M2] S. Gong. Convex and Starlike Mappings in Several Complex Variables. Kluwer Academic Publishers, 1998 (M7 的英文版).

[M3] S. Gong. Integrals of Cauchy Type on the Ball. International Press Co., 1993 (M9 的英文版).

[M4] S. Gong. Harmonic Analysis on Classical Groups. Springer-Verlag, 1991 (M8 的英文版).

[M5] S. Gong. The Bieberbach Conjecture. International Press Co., America Mathematical Society, 1999 (M10 的英文版).

[M6] 龚昇, 余其煌, 郑学安. 布洛赫常数与薛瓦兹导数. 现代数学丛书. 上海科学技术出版社, 1998.

[M7] 龚昇. 多复变数的凸映照与星形映照. 纯粹数学与应用数学专著, 第 34 号. 北京：科学出版社, 1995 (M2 的中文版) (M1 的第一版).

[M8] 龚昇. 典型群上的调和分析. 纯粹数学与应用数学专著, 第 12 号. 北京：科学出版社, 1983 (M4 的中文版).

[M9] 龚昇. 多复变数的奇异积分. 现代数学丛书. 上海科学技术出版社, 1982 (M3 的中文版).

[M10] 龚昇. 比贝尔巴赫猜想. 北京：科学出版社, 1989 (M5 的中文版).

II. **论文集** Proceedings

[P11] H. Halberstam 主编， S. Kung, Q. K. Look, Z. X. Wan and Y. Wang 参加. Loo-Keng Hua Selected Papers. Springer-Verlag, 1983.

[P12] S. Gong and C. C. Yang (Editors). Several Complex Variables in China. Comtemporary Mathematics Series, Vol.142, AMS, 1993.

[P13] S. Gong, Q. K. Lu, Y. Wang and L. Yang (Editors). International Symposium in Memory of Hua Loo Keng, Vol. I and II. Springer-Verlag, Science Press, 1991.

[P14] M. D. Cheng, D. G. Deng, S. Gong and C. C. Yang(Editors). Harmonic Analysis in China. Kluwer Academic Publishers, 1995.

[P15] C. H. FitzGerald and S. Gong. Proceeding of Geometrical Function Theory in Several Complex Variables. World Scientific, 2004.

III. **教科书** Textbooks

[T16] 龚昇. 简明复分析. 北京大学出版社，1995(1999, 2003 重印) (T18 的中文版).

[T17] 龚昇, 张声雷. 简明微积分. 北京：人民教育出版社，1977；2 版，合肥：中国科学技术大学出版社，1993；3 版，1997；4 版：普通高等教育“十五”国家级规划教材，北京：高等教育出版社，2006.

[T18] S. Gong. Concise Complex Analysis. World Scientific, 2001 (T16 的英文版).

[T19] S. Gong, Y. H. Gong. Concise Complex Analysis (Revised Edition). World Scientific, 2007 (T18 的修订版).

[T20] 龚昇. 微积分五讲. 北京：科学出版社，2004.

[T21] 龚昇. 线性代数五讲. 北京：科学出版社，2005.

IV. 科普书 Popular Science Books

[P22] 龚昇. 从刘徽割圆谈起. 数学小丛书第 8 号. 北京：人民教育出版社，1964；几十年来多次印刷，北京：科学出版社，2002.

[P23] 龚昇, 任福尧，何成奇. 复变函数概貌. 数学概貌丛书. 北京：科学技术文献出版社，1989.

[P24] 龚昇主编. 数学奥林匹克辅导讲座. 北京：科学出版社，1988.

[P25] 龚昇. 话说微积分. 合肥：中国科学技术大学出版社，1998.

[P26] 龚昇. 微积分杂谈. 北京：科学技术文献出版社，2002.

[P27] 龚昇, 林立军. 简明微积分发展史. 长沙：湖南教育出版社，2005.

V. 翻译 Translations

[T28] C. L. Siegel 著. 多复变数解析函数. 龚 , 译. 科学出版社，1960.

[T28] S. Bergman 著. 核函数和共形映照. 龚 , 陈希孺，译. 北京：科学出版社，1958.

[T28] C. Goffman 著. 多元微积分. 史济怀，彭家贵，何琛，李世雄，徐森林，龚昇, 译. 北京：人民教育出版社，1978.

附录 2 论文目录 (共 90 篇)

[1] S. Gong. Some open problems in geometric function theory in several complex variables. H. Kazama, M. Morimoto, C. C. Yang. Proceeding of the 12th international conference on finite and infinite dimensional complex analysis and applications. Kyushu Unversity Press, 2005: 85-96.

[2] S. Gong. The distortion theorems for convex mappings in several complex variables. C. H. FitzGerald, S. Gong. Geometric function theory in several complex variables. World Scientific, 2004: 143-150.

[3] 3.1 S. Gong, Taishun Liu. The decomposition theorem for the family of complete quasi-convex mappings. J. Math. Anal. Appl., 2004, 299: 448-464.

3.2 刘太顺，龚昇.完全准凸映照族的分解定理. 中国科学：A 辑，2004, 34: 243-256.

[4] 4.1 S. Gong, Taishun Liu. The generalized Roper-Suffridge extension operator. J. Math. Anal. Appl., 2003, 284: 425-434.

4.2 刘太顺，龚昇.ε 星形映照族 (II), Roper-Suffridge 算子. 数学年刊，2003, 24A: 625-638.

[5] S. Gong, Taishun Liu. On Roper-Suffridge extension operator. J. D'Analyse Math., 2002, 88: 397-404.

[6] 6.1 S. Gong, Taishun Liu. Criterion for the family of ε starlike mappings. J. Math. Anal. Appl., 2002, 274: 696-704.

6.2 刘太顺，龚昇. ε 星形映照族 (I), 数学年刊. 2002, 23A: 273-282.

[7] S. Gong, Xuean Zheng. The Bergman kernel function of some Reinhardt domains AMS/IP, Studies in Advanced Math., 2001, 20: 375-381.

[8] 殷承元，龚昇.多复变数的 Schwarz 导数 (V). 中国科学： A 辑，2001, 31: 481-488.

[9] 龚昇, 刘太顺. 复 Banach 空间中的 C-R 方程的全纯解. 数学学报, 2002, 45: 1-6.

[10] 10.1 S. Gong, Qihuang Yu. The distortion theorems on linear invariant family on the unit ball. The Asian Jour. of Math., 2000, 4: 795-816.

10.2 龚昇, 余其煌. 单位球上线性不变族的偏差定理. 数学年刊, 2001, 22A: 677-699.

[11] S. Gong, Xuean Zheng. The Bergman Kernel function of some Reinhardt domains II. Science in China: Serics A, 2000, 43: 458-469.

[12] 12.1 S. Gong, Taishun Liu. Distortion theorems for biholomorphic convex mappins on bounded convex circular domains, Chinese. Annals of Math., 1999, 20B: 297-304.

12.2 S. Gong, Taishun Liu. Distortion theorems for biholomorphic convex mappings on bounded convex circular domains. Proceeding of the 5th international colloquium on finite or infinite dimensional complex analysis, 1998: 73-80.

[13] 龚昇, 王世坤, 余其煌. 有界星形圆形域上全纯映照成为双全纯星形映照的判别准则. 数学学报, 1999, 42: 13-16.

[14] 龚昇, 郑学安, 余其煌. 多复变数的 Schwarz 导数 (V). 数学学报, 1999, 42: 215-226.

[15] 15.1 龚昇, 余其煌, 郑学安. 多复变数的 Schwarz 导数 (IV). 中国科学: A 辑, 1998, 28: 14-23.

15.2 S. Gong, Qihuang Yu, Xuean Zheng. The Schwarzian derivative in several complex variables (IV). Science in China: Series A, 1998, 41: 809-819.

[16] 16.1 龚昇, 郑学安, 余其煌. 多复变数的 Schwarz 导数 (III). 中国科学: A 辑, 1997, 27: 975-987.

16.2 S. Gong, Xuean Zheng, Qihang Yu. The Schwarzian derivative in several complex variables (III). Science in China: Series, 1998, A41: 158-171.

[17] S. Gong, Qihuang Yu, Xuean Zheng. The Schwarzian derivative in several complex variables (II). Chinese Annals of Math., 1998, 19B: 1-8.

[18] 18.1 S. Gong, Qihuang Yu, Jisheng Na. The Schwarzian derivatives in Kähler

manifolds (II). Science in China: Series A, 1997, 40: 30-36.

18.2 余其煌, 那吉生, 龚昇. Kähler 流形上的 Schwarz 导数 (II). 中国科学：A 辑，1996, 26: 793-798.

[19] Qihuang Yu, S. Gong, Xin Guo. Schwarzian derivatives of holomorphic mappings. Studies in Advanced Math., 1997, 5: 317-323. AMS/IP.

[20] 20.1 S. Gong, Qihuang Yu. The Schwarzian derivative in Kähler manifolds. Science in China: Series A, 1995, 38: 1033-1048.

20.2 余其煌，龚昇. Kähler 流形上的 Schwarz 导数. 中国科学：A 辑，1995, 25: 693-704.

[21] S. Gong. The Bloch constant of locally biholomorphic mappings on bounded symmetric domains. Chinese Annals of Math., 1996, 17B: 271-278.

[22] 22.1 S. Gong, Xuean Zheng. The Bergman Kernel function of some Reinhardt domains. Tran. Amer. Math. Soc., 1996, 348: 1771-1803.

22.2 郑学安，龚昇. Reinhardt 域的 Bergman 核函数 (一). 中国科学：A 辑，1995, 25: 580-589.

22.3 郑学安，龚昇. Reinhardt 域的 Bergman 核函数 (二). 中国科学：A 辑，1995, 25: 683-692.

22.4 S. Gong, Xuean Zheng. On Reinhardt domains (II). Chinese Science Bulletin, 1995, 40: 2025-2030.

22.5 郑学安，龚昇.关于 Reinhardt 域 (二). 科学通报，1995, 40: 1163-1167.

[23] 23.1 余其煌，龚昇. $\mathbb{C}^n$ 上的 Schwarz 导数. 数学年刊，1995, 16A: 454-460.

23.2 余其煌，龚 昇$\mathbb{C}^n$ 上的 Schwarz 导数. 科学通报，1995, 40: 966-967.

23.3 S. Gong, Qihuang Yu. The Schwarzian derivative in $\mathbb{C}^n$. Chinese Science Bulletin, 1965, 40: 1873-1874.

23.4 Qihuang Yu, S. Gong. The Schwarzian derivative in $\mathbb{C}^n$. Chinese Journal of Contemporary Math., 1995, 16: 301-314.

[24] Xuean Zheng, S. Gong. Distortion theorems for biholomorphic mappings in transitive domains (IV). Chinese Annals of Math, 1995, 16B: 203-212.

[25] C. H. FitzGerald, S. Gong. The locally biholomorphic Bloch constant and mar-

den constant of several complex variables. R. M. Ali, St. Ruscheweyh and E. B. Saff. Computational Methods and Function Theory 1994. Series in Approximations and decompositions, Vol.5. World Scientific Publishing Co., 1995:147-158.

[26] C. H. FitzGerald and S. Gong. Two-point distortion theorems for biholomorphic convex mappings on the ball. Preprint.

[27] 27.1 S. Gong. Biholomorphic mappings in several complex variables. AMS Contemporary Mathematics, 1993, 142: 15-48.

27.2 龚昇. 多复变数的双全纯映照. 数学进展，1994, 23: 115-141.

[28] 28.1 Zhimin Yan, S. Gong. Bloch constant of holomorphic mappings on bounded symmetric domains. Science in China: Series A, 1993, 36: 285-299.

28.2 严志敏，龚昇. 有界对称域上的全纯映照的 Bloch 常数. 中国科学：A 辑, 1992, 22: 1121-1134.

[29] 29.1 S. Gong, C. H. FitzGerald. The Schwarzian derivative in several complex variables, Science in China: Series A, 1993, 36: 513-523.

29.2 龚昇, C. H. FitzGerald. 多复变数的 Schwarz 导数. 科学通报，1989, 34: 391-392.

29.3 S. Gong, C. H. FitzGerald. The Schwarzian derivative in several complex variables (I). Chinese Science Bulletin, 1990, 35: 697-699.

[30] S. Gong, Shikun Wang, Qihuang Yu. Holomorphic convex mappings of ball in $\mathbb{C}^n$. Pacific Jour. of Math., 1993, 161: 287-306.

[31] 31.1 S. Gong, Qihuang Yu, Shikun Wang. On biholomorphic starlike mappings in several complex variables. Lecture Notes in Contemporary Math., 1991(2): 94-103, Instiute of Math. Science Press, Beijing, New York, 94-103.

31.2 龚昇, 王世坤，余其煌. 多复变数双全纯星形映照. 科学通报，1992, 37: 282.

31.3 S. Gong, Shikun Wang, Qihuang Yu. On holomorphic starlike mappings of several complex variables. Chinese Science Bulletin, 1992, 37: 1051-1052.

[32] 32.1 C. H. FitzGerald, S. Gong. Bloch constant in several complex variables, Journal of Geometric Analysis, 1994, 4: 35-38.

32.2 龚昇, C. H. FitzGerald. 多复变数的全纯映照的 Bloch 常数. 科学通报, 1992, 37: 485-487.

32.3 S. Gong, C. H. FitzGerald. The Bloch constant of holomorphic starlike mappings in several complex variables. Chinese Science Bulletin, 1992, 37: 1512-1515.

[33] 33.1 S. Gong, Qihuang Yu, Shikun Wang. The growth theorems in several complex variables. Chinese Annals of Math., 1993, 14B: 93-105.

33.2 龚昇, 王世坤, 余其煌. 多复变数中双全纯凸映照的增长型定理. 科学通报, 1990, 35: 1527-1528.

33.3 S. Gong, Shikun Wang, Qihuang Yu. The growth theorems in several complex variables. Chinese Science Bulletin, 1991, 36: 1148-1150.

[34] 34.1 S. Gong, Qihuang Yu, Shikun Wang. A necessary and sufficient condition that biholomorphic mappings be starlike in Reinhardt domains. Chinese Annals of Math., 1992, 13B: 95-104.

34.2 龚昇, 王世坤, 余其煌. 一类 Reinhardt 域上双全纯映照为星形的充要条件, 科学通报, 1990, 35: 1699-1700.

34.3 S. Gong, Shikun Wang, Qihuang Yu. A necessary and sufficient condition of biholomorphic mappings are starlike on a class of Reinhardt domians. Chinese Science Bulletin, 1991, 36: 1151-1153.

[35] 35.1 Xuean Zheng, S. Gong. Distortion theorems for biholomorphic mappings in transitive domains (III), Chinese Annals of Math. 1993, 14B: 367-386.

35.2 郑学安, 龚昇. 可递域上的双全纯映照的偏差定理 (III). 科学通报, 1990, 35: 1214-1217.

35.3 Xuean Zheng, S. Gong. Distortion theorems for biholomorphic mappings in transitive domains (III). Chinese Science Bulletin, 1991, 36: 542-546.

[36] 刘太顺, 龚昇. B^p 上双全纯凸映照的增长定理. 数学季刊, 1991, 6: 78-82.

[37] 37.1 Qihuang Yu, Shikun Wang, S. Gong. The growth and 1/4-theorem for starlike mappings in B^p. Chinese Annals of Math., 1990, 11B: 100-104.

37.2 余其煌, 王世坤, 龚昇. B^p 上星形映照的增长及 1/4 定理. 科学通报, 1989, 34: 15-19.

37.3 Qihuang Yu, Shikun Wang, S. Gong. The growth and 1/4-theorem for starlike mapping in B^p. Chinese Science Bulletin, 1990, 35: 2017-2018.

[38] 38.1 Xuean Zheng, S. Gong. Distortion theorems for biholomorphic mappings in transitive domains (II). Chinese Annals of Math., 1992, 13B: 471-484.

38.2 郑学安，龚昇. 可递域上的双全纯映照的偏差定理 (II). 科学通报，1990, 35: 330-332.

38.3 Xuean Zheng, S. Gong. Distortion theorems for biholomorphic mappings in transitive domains (II). Chinese Science Bulletin, 1991, 36: 538-541.

[39] 39.1 S. Gong, Xuean Zheng. Distortion theorems for biholomorphic mappings in transitive domains (I). S. Gong, Q. K. Lu, Y. Wang and L. Yang. International Symposium in memory of Hua Loo Keng: Vol.II. Springer-Verlag, Science Press, 1991: 111-122.

39.2 龚昇,郑学安. 可递域上的双全纯映照的偏差定理 (I). 科学通报，1989, 34: 1201-1202.

[40] 40.1 R. W. Barnard, C. H. FitzGerald, S. Gong. The growth and 1/4-theorem for starlike mappings in C^n. Pacific Journal of Math. 1991, 150: 13-22.

40.2 C. H. FitzGerald, 龚昇, R. W. Barnard. $\mathbb{C}^n$ 中星形映照的增长及 1/4 定理. 科学通报， 1989, 34: 161-162.

40.3 C. H. FitzGerald, S. Gong, R. W. Barnard. The growth and 1/4-theorems for starlike mappings in $\mathbb{C}^n$. Chinese Science Bulletin, 1990, 35: 357-359.

[41] 41.1 R. W. Barnard, C. H. FitzGerald, S. Gong. A distortion theorem for biholomorphic mappings in $\mathbb{C}^2$. Tran Amer. Math. Soc, 1994, 344: 907-924.

41.2 龚昇, C. H. FitzGerald, R. W. Barnard. $\mathbb{C}^2$ 中双全纯映照的偏差定理. 科学通报，1989, 34: 81-83.

41.3 S. Gong, C. H. FitzGerald, R. W. Barnard. A distortion theorem for biholomorphic mappings in $\mathbb{C}^2$. Chinese Science Bulletin, 1990, 35: 353-356.

[42] 42.1 S. Gong, Shixiong Li, Xuean Zheng. Harmonic analysis on classical groups. S. T. L. Choy, J. P. Jesudason, P. Y. Lee, Eds. Proceeding of the Analysis conference, Singapore 1986 Mathematics studies 150. North-Holland, 1988: 69-113.

42.2 S. Gong, Shixiong Li, Xuean Zheng. Harmonic analysis on classical groups. Chaohao Gu, Tyuan Wang. Advances in Science of China Mathematics: Vol.1. Science Press, John Wiley & Sons, 1986: 45-108.

[43] S. Gong, S. S. Miller. Partial differential subordinations inequalities defined on complete circular domains. Communications in Partial Diff. Equations, 1986, 11: 1243-1255.

[44] 44.1 龚昇. 系数不等式与几何不等式. 科学通报，1986, 31: 1209-1212.

44.2 S. Gong. Coefficient inequalities and geometric inequalities, Kexue Tongbao: Science Bulletin, 1987, 32: 732-736.

[45] 45.1 龚昇. 关于 Reinhardt 域 (一). 数学年刊，1987, 8A: 205-219.

45.2 S. Gong. On Reinhardt domains (I). Kexue Tongbao: Science Bulletin, 1986, 31: 433-438.

45.3 龚昇. 关于 Reinhardt 域 (一). 科学通报，1985, 30: 1365-1368.

[46] 46.1 龚昇, 严志敏. 关于 Möbius 变换的一点注记 (三). 数学季刊，1986,1: 33-40.

46.2 S. Gong, Zhimin Yan. A remark on Möbius transformations (III). Kexue Tongbao: Science Bulletin, 1987, 32: 1153-1156.

46.3 龚昇, 严志敏. 关于 Möbius 变换的一点注记 (三). 科学通报，1986, 31: 1601-1604.

[47] 47.1 龚昇. 关于 Möbius 变换的一点注记 (二). 中国科学技术大学学报，1984, 14: 1-14.

47.2 S. Gong. A remark on the Möbius transformations (II). Kexue Tongbao: Science Bulletin, 1984, 29: 293-297.

[48] 龚昇. 关于偏微分不等式的一点注记. 数学年刊，1984, 5A: 771-780.

[49] 龚昇. 关于 Möbius 变换的一点注记. 纯粹数学与应用数学，1985, 1: 1-15.

[50] Jihuai Shi, S. Gong. Singular integrals in several complex variables (IV), The derivative of integral of Cauchy type on ball. Chinese Annals of Math., 1984, 5B: 21-36.

[51] S. Gong, Jihuai Shi. Singular integrals in several complex variables (III), Integral

of Cauchy type on classical domains. Chinese Annals of Math., 1983, 4B: 467-484.

[52] S. Gong, J.H. Shi. Singular integrals in several complex variables (II), Hadamard principal value. Chinese Annals of Math., 1983, 4B: 307-318.

[53] S. Gong. On Aharonov pairs, Journal of Math. Analysis and Applications, 1982, 88: 8-24.

[54] 54.1 S. Gong, Jihuai Shi. Singular integrals in several complex variables (I), Integral of Henkin type on strictly pseudoconvex domains. Chinese Annals of Math., 1982, 3B: 483-502.

54.2 龚昇,史济怀. 多复变数的奇异积分 — 强拟凸域上的 Henkin 积分. 科学通报, 1982, 27: 769-770.

[55] S. Gong. A remark on Möbius transformations for high dimensions: Proc. of 1981 Shanghai-Hefei Symposium on Differential Geometry and Differential Equations, Science Press, 1984: 75-94.

[56] S. Gong. On FitzGerald inequalities. J. London Math. Soc., 1981, 24(2): 227-242.

[57] S. Gong. A remark on the integral of Cauchy type on strictly pseudoconvex domains: Proceeding of 1980 Beijing Symposium on Differential Geometry and Differential Equations. Science Press, Gordon and Breach, Science Publishers, 1982: 1183-1190.

[58] 龚昇. 旋转群上的富理埃级数的部分和. 中国科学技术大学学报, 1979, 9: 25-30.

[59] 59.1 龚昇. 关于 FitzGerald 不等式. 中国科学技术大学学报, 1979, 9: 1-24.

59.2 龚昇. FitzGerald 不等式的另一改进形式. 科学通报, 1980, 25: 673-675.

[60] 60.1 龚 . 第六项系数的 Bieberbach 猜想的一个简化证明. 中国科学, 1979: 1157-1170.

60.2 S. Gong. A simple proof on Bieberbach conjecture for sixth coefficient. Scintia Sinica, 1980, 23: 1-15.

[61] 龚昇. FitzGerald 不等式的改进. 中国科学, 1979: 237-246.

[62] 龚昇. 在第二系数限制下的 Bieberbach 猜想，中国科学：数学专辑 (I), 1979: 202-214.

[63] 63.1 龚昇, 孙继广. 复超球面上的奇异积分方程. 数学学报，1966, 16: 194-210.

63.2 龚昇, 孙继广. 复超球面上的奇异积分方程. 中国科学技术大学学报，1966, 2: 13-15.

63.3 Sheng Kung, Chi-Kuang Sun. Singular integral equations on hypersphere. Translated as Chinese Math. Acta., 1966, 8: 205-221.

[64] 龚昇, 史济怀. 关于多复变数可递域的 Poisson 公式的一点注记. 中国科学技术大学学报，1966, 2: 16-20.

[65] 65.1 龚昇, 孙继广. 多复变数的 Cauchy 型积分 (III) 矩阵双曲空间的 Cauchy 型积分. 数学学报，1965, 5: 800-811.

65.2 Sheng Kung, Chi-kuang Sun. The Cauchy type integrals of several complex variables III, Integrals of the Cauchy type on the hyperbolic space of matrices. Scientia Sinica, 1986, 14: 621-623.

65.3 Sheng Kung, Chi-kuang Sun. Integral of Cauchy type in several complex variables (III), Integrals of Cauchy type in hyperbolic spaces of matrices. Translated as Chinese Math. Acta., 1965, 7: 535-546.

[66] 66.1 龚昇, 孙继广. 多复变数的 Cauchy 型积分 (II) Lie 球双曲空间的 Cauchy 型积分. 数学学报，1965, 15: 775-799.

66.2 Sheng Kung, Chi-kuang Sun. Integrals of the Cauchy type of several complex variables II, Integrals of the Cauchy type on the hyperbolic space of a Lie sphere. Scientia Sinica, 1965, 14: 453-456.

66.3 龚昇, 孙继广. 多复变数典型域的 Cauchy 型积分 (II). 中国科学技术大学学报，1965, 1: 49-50.

66.4 Sheng Kung, Chi-kuang Sun. Integral of the Cauchy type in several complex variables (II), Integrals of Cauchy type in the hyperbolic spaces of Lie sphere. Translated as Chinese Math. Acta., 1965, 7: 509-534.

[67] 67.1 龚昇, 孙继广. 多复变数的 Cauchy 型积分 (I) 超球的 Cauchy 型积分. 数学学报，1965, 5: 431-441.

67.2 Sheng Kung, Chi-kuang Sun. The Cauchy type integral of several complex variables I, The Cauchy type integrals on a hypersphere. Scientia Sinica, 1965, 14: 300-301.

67.3 龚昇. 多复变数典型域的 Cauchy 型积分 (I): 中国科技大学建校五周年纪念科学论文集. 1963: 23-25.

67.4 Sheng Kung, Chi-kuang Sun. Integral of Cauchy type in several complex variables (I), Integrals of Cauchy type in ball. Translated as Chinese Math. Acta., 1965, 7: 138-151.

[68] 68.1 龚昇, 孙继广. 圆型域中解析函数的积分表示. 中国科学技术大学学报, 1965, 1: 219-226.

68.2 Sheng Kung, Chi-kuang Sun. Integral representations of analytic functions of several complex variables on circular domains. Scientia Sinica, 1965, 14: 302.

[69] 69.1 龚昇. 酉群上的富理埃分析 V, Fourier 级数的球求和及 Fourier 积分. 数学学报, 1965, 15: 305-325.

69.2 Sheng Kung. Fourier analysis on unitary group V. Spherical summability and Fourier integrals. Translated as Chinese Math., 1965, 7: 1-20.

[70] 70.1 龚昇. 酉群上的富里埃分析 IV, 关于 Peter-Weyl 定理. 数学学报, 1963, 13: 315-322.

70.2 Sheng Kung. Fourier analysis on unitary group IV. on Peter-Weyl theorem, Translated as Chinese Math., 1964, 4: 351-359.

[71] 71.1 龚昇. 酉群上的富里埃分析 III, 富里埃级数的收敛判别法. 数学学报, 1963, 13: 152-161.

71.2 Sun Kung. Fourier analysis on unitary group I. Partial sums and summability of Fourier series, Science Record: New Series, 1990, 4: 317-322.

71.3 龚昇. 酉群上的富里埃分析 I, 富里埃级数的部分和及求和. 科学记录: 新辑, 第 4 卷, 324-328.

71.4 Sheng Kung. Fourier analysis on unitary, group III, convergence criteria. Translated as Chinese Math., 1963, 169-178.

[72] 72.1 龚昇. 酉群上的富里埃分析 II, 富里埃级数的 Cesàro 求和. 数学学报, 1962, 12: 17-31.

72.2 Sun Kung. Fourier analysis on unitary group III. The Cesàro summability, Scientia Sinica, 1961, 10: 621-624.

[73] 73.1 龚昇. 酉群上的富里埃分析 I, 富里埃级数的 Abel 求和及 Dirichlet 核. 数学学报， 1960, 10: 239-261.

73.2 Sun Kung. Fourier analysis on unitary group II. Abel summability, Scientia Sinica, 1961, 10: 26-30.

73.3 Sheng Kung. Fourier analysis on unitary group I. Translated Chinese Math., 1962, 1: 249-272.

[74] Sheng Kung. Harmonic analysis on unitary group. Appendix III of the book "Harmoric analysis of functions of several complex variables in classical domains" by Hua Loo Keng. Transl. Amer. Math. Soc., Vol.6, 1963.

[75] 龚昇. 多连通区域的 Bloch 常数. 数学学报，1957, 7: 513-519.

[76] 龚昇. 多复变数的 Einstein 空间的 Schwarz 引理. 数学学报，1957, 7: 471-476.

[77] 龚昇. 单叶函数的系数 (III). 数学学报，1956, 6: 490-499.

[78] 龚昇. 单叶函数的系数 (II). 数学学报，1956, 6: 115-126.

[79] 龚昇. 富里埃级数的共轭级数的一个收敛条件. 复旦学报 :自然科学版, 1955, 1: 65-71.

[80] 龚昇. 富里埃级数的陈建功教授收敛判别法. 复旦学报 :自然科学版, 1955, 1: 58-64.

[81] Sun Kung. Contributions to the theory of Schlicht function II. The Coefficient Problem. Scientia Sinica, 1995, 4: 359-373.

[82] Sun Kung. Contributions to the theory of Schlicht function I. Distortion theorerns, Scientia Sinica, 1955, 4: 229-249.

[83] 龚昇. 关于平均 p 叶函数. 数学学报，1954, 4: 245-258.

[84] 龚昇. 单叶函数的开始多项式. 数学学报，1954, 4: 105-112.

[85] 龚昇. 单叶函数的系数 (I). 数学学报，1954, 4: 87-104.

[86] 龚昇. 对称单叶函数的二个定理. 数学学报，1953, 3: 251-260.

[87] 龚昇. 单叶函数之偏差定理及系数. 数学学报，1953, 3: 231-250.

[88] 龚昇. 关于戈鲁净一楼五纳微分方程. 数学学报，1953, 3: 225-230.

[89] Sun Kung. On the coefficient of odd schlicht functions. Science Record, 19
343-349.

[90] Sun Kung. The sections of odd schlicht functions. Science Record, 19
333-342.

附录 3　获奖项目

1．华罗庚奖，中国数学会，2002
2．中国科学院自然科学一等奖，2002
3．中国科学院自然科学一等奖，1996
4．国家自然科学三等奖，1989
5．中国科学院科技进步二等奖，1980

中国科学技术大学大师讲席 (华罗庚讲席) 教授

Hua Longgeng Chair Professorship USTC, 2004